Carbon Nanomaterials-Based Sensors

Carbon Nanomaterials-Based Sensors

Emerging Research Trends in Devices and Applications

Edited by

Jamballi G. Manjunatha
FMKMC College, Mangalore University, Madikeri, India

Chaudhery Mustansar Hussain
Department of Chemistry and Environmental Sciences, New Jersey Institute of Technology, Newark, NJ, United States

ELSEVIER

Elsevier
Radarweg 29, PO Box 211, 1000 AE Amsterdam, Netherlands
The Boulevard, Langford Lane, Kidlington, Oxford OX5 1GB, United Kingdom
50 Hampshire Street, 5th Floor, Cambridge, MA 02139, United States

Copyright © 2022 Elsevier Inc. All rights reserved.

No part of this publication may be reproduced or transmitted in any form or by any means, electronic or mechanical, including photocopying, recording, or any information storage and retrieval system, without permission in writing from the publisher. Details on how to seek permission, further information about the Publisher's permissions policies and our arrangements with organizations such as the Copyright Clearance Center and the Copyright Licensing Agency, can be found at our website: www.elsevier.com/permissions.

This book and the individual contributions contained in it are protected under copyright by the Publisher (other than as may be noted herein).

Notices

Knowledge and best practice in this field are constantly changing. As new research and experience broaden our understanding, changes in research methods, professional practices, or medical treatment may become necessary.

Practitioners and researchers must always rely on their own experience and knowledge in evaluating and using any information, methods, compounds, or experiments described herein. In using such information or methods they should be mindful of their own safety and the safety of others, including parties for whom they have a professional responsibility.

To the fullest extent of the law, neither the Publisher nor the authors, contributors, or editors, assume any liability for any injury and/or damage to persons or property as a matter of products liability, negligence or otherwise, or from any use or operation of any methods, products, instructions, or ideas contained in the material herein.

ISBN: 978-0-323-91174-0

For information on all Elsevier publications
visit our website at https://www.elsevier.com/books-and-journals

Publisher: Susan Dennis
Acquisitions Editor: Kathryn Eryilmaz
Editorial Project Manager: Kathrine Esten
Production Project Manager: Kumar Anbazhagan
Cover Designer: Greg Harris

Typeset by STRAIVE, India

Contents

Contributors xiii

Part I
Modern perspectives in sensor devices for sustainability

1. Sustainable development information management of carbon nanomaterial-based sensors

Kiran K. Somashekharappa and Shashanka Rajendrachari

Introduction	3
Types of carbon nanomaterials	3
Fullerenes	3
Carbon nanotubes (CNTs)	4
Graphene and related compounds	4
Carbon dots (CDs)	5
Carbon nanodiamonds	5
Carbon black	6
Carbon nanomaterial-based sensors for sustainable development	6
Environmental hazardous sensors	6
Electrochemical sensors	8
Biosensors	8
Gas sensors	8
Dopamine and other sensors	8
Conclusions	9
Conflicts of interest	9
References	9

Part II
Fabrication of carbon nanomaterials-based sensors platforms

2. Carbon nanomaterial-based sensor: Synthesis and characterization

Mohd Asyadi Azam and Muazzin Mupit

Introduction of carbon nanomaterial-based sensors	15
Carbon nanomaterial synthesis method for sensors	15
Top-down technique	16
Mechanical milling	16
Lithography	17
Electric arc deposition	18
Sputtering deposition	18
Quantum dots in analysis	18
Bottom-up technique	18
Sol-gel processes	18
Chemical vapor deposition	19
Micellization	19
Laser pyrolysis	20
Characterization of carbon nanomaterial-based sensors	20
X-ray diffraction spectroscopy (XRD)	20
Raman spectroscopy	21
Transmission electron microscopy (TEM)	22
Field emission scanning electron microscopy (FESEM)	22
Thermogravimetric analysis (TGA)	24
Brunauer-Emmet-Teller surface analysis (BET)	24
Electrical properties from I-V characterization	26
References	27

3. Novel trends for synthesis of carbon nanomaterial-based sensors

H.C. Ananda Murthy and H.P. Nagaswarupa

Introduction	29
Electrochemical sensors	30
Working principle and classification of electrochemical sensors	30
Potentiometric sensors	30
Conductometric sensors	31
Amperometric sensors	32

Carbon nanomaterial-based electrochemical sensors	33
Graphene	35
Carbon nanotube	35
Fullerene	36
Synthesis of CNTs	36
Arc discharge method	36
Laser ablation method	37
Chemical vapor deposition method	38
Green synthesis of CNTs	39
Synthesis of graphene-based nanomaterials	40
Conclusion	41
References	41

4. Development of new electroanalytical method based on graphene oxide-modified glassy carbon electrode for mephedrone illicit drug determination

*Maiyara Carolyne Prete,
Luana Rianne Rocha, and
César Ricardo Teixeira Tarley*

Introduction	43
Materials and methods	44
Reagents and solutions	44
Apparatus	44
Synthesis of graphene oxide	45
Preparation of GCE/GO-DHP sensor	45
Electrochemical measurements	46
Street sample simulation	46
Results and discussion	46
Characterization of GO	46
Electrochemical characterization of GCE/GO-DHP sensor	47
Electrochemical reduction of 4-MMC	49
Optimization of experimental parameters	50
Optimization of DPV technique and analytical features	50
Interference of adulterants in 4-MMC determination and application in a simulated sample	52
Conclusion	52
Acknowledgments	53
References	53

Part III
Carbon nanomaterials-based sensor applications (environment, biomedical, foods, point of care, etc.)

5. Carbon nanomaterials-based sensors for biomedical applications

*Amirreza Roshani,
Maryam Mousavizadegan, and
Morteza Hosseini*

Introduction	59
Biomarkers	59
Graphene	60
Carbon nanotubes	62
Fullerene	64
Nanodiamonds	67
Carbon quantum dots	69
Conclusion and outlook	70
References	71

6. Fabrication of disposable sensor strips for point-of-care testing of environmental pollutants

*Gnanesh Rao, Akhilesh Rao,
B.P. Nandeshwarappa, Raghu Ningegowda,
Kiran Kumar Mudnakudu-Nagaraju, and
Sandeep Chandrashekharappa*

Introduction	77
Substrate material for disposable sensor strips	77
CNT biosensor	78
Functionalization	79
Enzymatic biosensor	79
Affinity biosensors	79
Immunosensors	79
Pollutants	80
Fabrication	80
Standard materials for MEMS/NEMS	80
Methods used to define hydrophobic zones on the paper	81
Photolithography	81
Printing	82
Digital printing	83
Inkjet printing	84
Waxing method	84
Stamping	84

Chemical modification	85	**9. Carbon nanomaterials-based sensors for water treatment**	
Drawing	85		
Methods used for cutting the paper to obtain physical boundaries defined sensors	86	*Aniela Pop, Sorina Motoc, and Florica Manea*	
Conclusion	87	Introduction	126
References	87	Voltammetric and amperometric sensors for water treatment	127

7. Trends in carbon nanomaterial-based sensors in the food industry

Shridevi Doddamani, Vinusha Honnalagere Mariswamy, Vinay Karekura Boraiah, and Srikantamurthy Ningaiah

Introduction	95
Carbon nanomaterials in food technology	96
Carbon-based nanomaterials in the food industry	96
Carbon nanotubes in the food industry	97
Carbon dots in the food industry	98
Carbon nanosensors in the food industry	99
Conclusion	100
References	100

Voltammetric and amperometric techniques used for detection applications — 128

Configurations of the working electrode for voltammetric and amperometric sensors — 132

Nanostructured carbon-modified commercial carbon (NC-MCC) electrodes — 134

Nanostructured carbon paste (NC-P) electrodes — 137

Nanostructured carbon-epoxy composite (NC-EP) electrodes, including the electrodeposition of the silver/copper nanoparticles — 142

Conclusion and future perspectives — 145
Acknowledgments — 145
References — 145

8. Carbon nanomaterial-based sensors in air pollution remediation

Abdullah Al Mamun, Md Nafiujjaman, and A.J. Saleh Ahammad

Introduction	105
Nanomaterials in air pollution remediation	105
Brief history of carbon nanomaterials	106
Derivatives of carbon nanomaterials	108
Properties of carbon nanomaterials	108
Carbon nanotubes	108
Graphene	108
Fullerenes	109
Application of carbon nanomaterial-based sensors in air pollution remediation and relative mechanism	110
Process of pollutant gas sensing with carbon nanomaterials	110
Advantages of carbon nanomaterial-based sensors	111
Application and general mechanism	111
Conclusion	118
References	118

10. Carbon nanomaterial-based sensors: An efficient tool in the environmental sectors

Prashanth S. Adarakatti, K. Sureshkumar, and T. Ramakrishnappa

Introduction	149
Carbon-based nanomaterials used in sensor applications	150
Graphene	151
Graphene oxide (GO) and reduced graphene oxide (rGO)	151
Fullerenes	151
Carbon nanotube (CNTs)	152
Graphene quantum dots (GQDs)	152
Nanodiamonds (NDs)	153
Environmental sensor applications	154
Electrochemical sensor applications	155
Optical sensors	157
Colorimetric sensors	157
Fluorescence sensors	157
Conclusion	159
Acknowledgments	160
References	160

11. Graphene-based surface-enhanced Raman scattering as an efficient tool in the detection of toxic organic dyes in real industrial effluents

V. Poornima Parvathi, R. Parimaladevi, Vasant Sathe, and M. Umadevi

Introduction	167
Graphene-boosted silver nanocomposites for graphene-based surface-enhanced Raman scattering in the detection of toxic organic dyes in real industrial effluents	168
Optical studies	168
Vibrational spectroscopic studies	169
Morphological studies	169
Fluorescence suppression in dyes	171
SERS sensing of dye contaminants	171
Graphene-boosted silver nanocomposites for graphene-based surface-enhanced Raman scattering in the detection of toxic organic dyes in real industrial effluents	175
UV-visible spectral analysis	175
Vibrational spectral studies	175
Morphological studies	175
Spectral analysis of dyes and TE	176
Conclusion	185
References	185

Part IV
Role of carbon nanomaterials-based sensors in sustainability

12. Carbon nanomaterial-based sensors for the development of sensitive sensor platform

Hulya Silah, Ersin Demir, Sercan Yıldırım, and Bengi Uslu

Introduction	191
Overview of carbon nanomaterial chemistry	192
Carbon nanomaterial-based sensors	193
Electrochemical sensors based on carbon nanomaterials	193
Optical-based sensing systems using carbon nanomaterials	226
Conclusion and future prospects	232
References	232

13. Carbon nanomaterial-based sensors for wearable health and environmental monitoring

Maryam Rezaie and Morteza Hosseini

Introduction	247
Health monitoring	248
Metabolite sensing	248
Electrolytes sensing	251
Respiration monitoring	252
Pressure and strain wearable sensor	253
Temperature sensor	253
Pulse beat monitoring	254
Environmental monitoring	254
Heavy metal	254
Nitrogen monitoring	254
Air pollution monitoring	255
Current limitation	256
Conclusions and future perspectives	256
References	256

14. Advanced sensors based on carbon nanomaterials

Vinayak Adimule, Basappa C. Yallur, and Adarsha H.J. Gowda

Introduction	259
Materials and methods	260
Experimental	260
Synthesis of carbon nanotubes (CNTs)	260
Synthesis of graphene	260
Synthesis of graphitic carbon nitride	262
Results and discussion	262
Advancement in the electrochemical sensing performance of carbon nanotubes	262
Advancement in the electrochemical sensing performance of the graphene	264
Advancement in the electrochemical sensing performance of the graphitic carbon	264
Future aspects, limitations, and advantages of electrochemical sensors	264
Modification of the carbon electrodes, development and challenges	265
Conclusion	266
Conflict of interest	267
Authors' contributions	267
Acknowledgment	267
References	267

15. The role of carbon nanomaterial-based sensors in sustainability

*Doddahosuru M. Gurudatt,
S. Rajendra Prasad, Sneha S. Puttappa,
and Srikantamurthy Ningaiah*

Introduction	269
Sustainability of carbon nanomaterial-based sensors	270
Role of carbon nanomaterials as an electrochemical sensor	271
Dye-sensitized solar cells using carbon nanomaterials	271
Carbon-based material application in wastewater treatment/purifying	271
Carbon nanomaterials for biofuel cells and supercapacitors	272
Carbon-based materials in biodiesel production	272
Carbon materials in biochar-based functional materials	272
Conclusion	273
References	273

16. Microfluidic systems with amperometric and voltammetric detection and paper-based sensors and biosensors

Ebtesam Sobhanie, Amirreza Roshani, and Morteza Hosseini

Introduction	275
Basics of paper-based microfluidics	275
Formats of paper-based analytical devices	276
Fabrication methods for paper-based microfluidic devices	276
Detection techniques	277
Colorimetric detection	277
Luminescence detection	278
Electrochemical detection	278
Electrochemical detection in paper-based microfluidic devices and application	278
Voltammetric paper-based microfluidic sensors and biosensors and their application	280
Amperometric paper-based microfluidic sensors and biosensors and their application	282
Conclusion and prospects	284
References	284

Part V
Health, safety, and regulation issues of carbon nanomaterials-based sensors

17. Fundamentals of health, safety, and regulation issues of carbon nanomaterial-based sensors

Anila Hoskere Ashoka and Vadde Ramu

Introduction	291
Origin of toxicity	292
Physical characteristics	293
Metal impurities	294
Surface functionalization	294
Mode of administration	296
Toxicity mechanism of CNMs	296
Reactive oxygen species (ROS)	296
Inflammatory response	297
DNA damage	297
Biocompatibility and biodegradation of CNMs	297
Biodegradation by enzymes and microorganisms	298
Biocompatibility due to surface functionalization	299
Conclusion	299
Acknowledgments	299
References	300

18. Safety and ethics of carbon nanomaterial-based sensors

Monima Sarma and Shaik Mubeena

Introduction	303
Carbon nanomaterials as possible carcinogens?	308
An asthma angle	309
References	311

19. Carbon nanomaterial-based sensor safety in different fields

S. Pratibha and B. Chethan

Introduction	315
Nanostructure classification	315
Classification of nanoparticles based on the dimension	315
Classification of nanoparticles based on morphology and chemical properties	316

Synthesis, characterization, and applications of CNPs	319
Synthesis	319
Characterization of CNPs	319
Applications of CNPs	320
Carbon nanomaterial-based sensors	320
Physical sensors	320
Chemical sensors	320
CNPs in the environment: Discharge and destiny	322
Physical transformation	323
Chemical transformation	324
Biological transformation	324
Threats and endangerments in the context of the environment and econanotoxicology	324
Influences on terrestrial ecology	324
Influence on human health	325
Scope of future works	325
Conclusions	327
References	327

Part VI
Economics & comercialization of carbon nanomaterials-based sensors

20. The use of carbon nanotubes material in sensing applications for H_1-antihistamine drugs

Jessica Scremin, Bruna Coldibeli, Carlos Alberto Rossi Salamanca-Neto, Gabriel Rainer Pontes Manrique, Renan Silva Mariano, and Elen Romão Sartori

Introduction	335
Carbon nanotube-based sensors	335
H_1-antihistamine drugs: Definition and classes	337
Determination of H_1-antihistamine drugs	338
Working electrodes based on carbon nanotubes for the determination of H_1-antihistamines	338
Determination of first-generation H_1-antihistamines employing voltammetry	339
Determination of second-generation H_1-antihistamines employing voltammetry or amperometry	342
Conclusion	344
References	344

21. Carbon nanomaterial-based sensors: Emerging trends, markets, and concerns

Shalini Menon, Sonia Sam, K. Keerthi, and K. Girish Kumar

Introduction	347
Emerging trends in carbon nanomaterial-based optical sensors	348
Carbon nanomaterials as nanoprobes and sensing elements	349
In biosensors	350
In molecularly imprinted polymer-based sensors	352
Carbon nanomaterials as temperature sensors	352
Carbon nanomaterials as FL quenchers in biosensors	352
Carbon nanomaterials as catalysts	353
Carbon nanomaterials in bioimaging	353
Role of carbon nanomaterials in electrochemical sensing	353
Graphene and its derivatives	354
Carbon nanotubes	356
Fullerenes	357
Carbon nanodots	357
Carbon nanohorns	359
Nanodiamonds and carbon nano-onions	360
Market of carbon nanomaterials: Applications in sensors	360
Market overview	360
Market segmentation	361
Regional insights	362
Competitive market share	362
Impact of COVID-19 on the global market of carbon nanomaterials	362
Challenges/concerns of carbon nanomaterial-based sensors	363
Toxicity of carbon nanomaterials	363
Shortcomings in design and fabrication	364
Achieving stability, sensitivity, selectivity, and reproducibility	365
Concerns over biosensors	366
Lack of point-of-care devices	367
Conclusions	368
Conflict of interest	368
Acknowledgments	369
References	369

22. Challenges in commercialization of carbon nanomaterial-based sensors

Elif Esra Altuner, Merve Akin, Ramazan Bayat, Muhammed Bekmezci, Hakan Burhan, and Fatih Sen

Introduction	381
Sensors	381
The history of sensors	382
Types of sensors	383
Piezoelectric-based sensors	384
Carbon-based nanomaterials	384
Carbon nanotube chemical sensors	384
Ordered mesoporous and macroporous carbon-modified electrodes	385
Carbon quantum dots	385
Commercialization challenges based on carbon nanomaterials	385
Conclusion	387
References	389

Part VII
Future of sustainable sensors

23. Introduction and overview of carbon nanomaterial-based sensors for sustainable response

Tania Akter, Christopher Barile, and A.J. Saleh Ahammad

Introduction	395
Carbon-based nanomaterials	395
Carbon nanotubes for sensor applications	397
Carbon nanotubes in gas sensors	397
Carbon nanotubes in biosensors	397
Carbon nanotubes in drug delivery applications	399
Carbon nanotube sensors for food and environmental applications	399
Nanodiamonds for sensor applications	401
Nanodiamonds in gas sensors	401
Nanodiamonds in energy storage devices	401
Nanodiamonds in biological applications	402
Fullerenes for sensor applications	403
Fullerenes in fuel cells	403
Fullerene in biomedical applications	404
Fullerenes in biosensors	405
Graphenes for sensor applications	406
Graphenes in gas sensors	406
Graphene in biomedical applications	407
Carbon quantum dots in sensor applications	408
Conclusion	409
References	409

24. Prospects of carbon nanomaterial-based sensors for sustainable future

P. Karpagavinayagam, J. Antory Rajam, R. Baby Suneetha, and C. Vedhi

Introduction	417
Carbon and its transformation	417
Carbon transformation	418
Types of carbon nanomaterials (CNMs)	418
Sensors, carbon-based sensors	419
Types of sensors	419
Carbon-based sensors	419
Carbon nanotube sensors	419
Carbon nanofiber sensors	420
Sustainable carbon materials	420
Carbon quantum dots	421
Quantum dots in medicine	421
Quantum dots in photovoltaics	422
Graphene quantum dots	422
Perovskite quantum dots	422
Quantum dot TVs and displays	422
Applications of sensors	422
Nanosensor fabrication	423
Nanosensors based on nanowires, nanofibers, and carbon nanotubes	423
Nanosensors based on graphene	424
Optoelectronic and photonic applications	424
Nanoinks	424
Electrodes	424
Potential applications of CNMs	424
Environmental applications of carbon-based nanomaterials	425
Perspectives on sustainable graphitic structures	425
Sustainable discovery of CNMs	425
Conclusion	426
References	426
Further reading	428

25. Sustainable carbon nanomaterial-based sensors: Future vision for the next 20 years

S. Alwin David, R. Rajkumar, P. Karpagavinayagam, Jessica Fernando, and C. Vedhi

Introduction	429
Nanodiamond-based sensors	430
Carbon nanotube-based sensors	431
Buckypaper-based sensors	432
Graphene-based gas sensors	433
Graphene oxide-based sensors	434
Reduced graphene oxide-based sensors	434
Fullerene-based sensors	435
Mesoporous carbon-based sensors	436
Carbon quantum dot-based sensors	437
Sensors based on carbon hybrid structures	437
Vision for next 20 years	438
References	438

Index 445

Contributors

Numbers in parentheses indicate the pages on which the authors' contributions begin.

Prashanth S. Adarakatti (149), Department of Chemistry, SVM Arts, Science and Commerce College, Ilkal, Karnataka, India

Vinayak Adimule (259), Department of Chemistry, Angadi Institute of Technology and Management (AITM), Belagavi, Karnataka, India

A.J. Saleh Ahammad (105, 395), Department of Chemistry, Jagannath University, Dhaka, Bangladesh

Merve Akin (381), Sen Research Group, Department of Biochemistry, University of Dumlupinar, Kutahya, Turkey

Tania Akter (395), Department of Chemistry, University of Nevada, Reno, United States

Abdullah Al Mamun (105), Department of Chemistry, University of Houston, Houston, TX, United States

Elif Esra Altuner (381), Sen Research Group, Department of Biochemistry, University of Dumlupinar, Kutahya, Turkey

S. Alwin David (429), Department of Chemistry, V.O. Chidambaram College, Tuticorin, India

J. Antory Rajam (417), St. Mary's College, Tuticorin, India

Anila Hoskere Ashoka (291), Laboratory of Bioimaging and Pathologies, UMR 7021, University of Strasbourg, Illkirch, France

Mohd Asyadi Azam (15), Fakulti Kejuruteraan Pembuatan, Universiti Teknikal Malaysia Melaka, Melaka, Malaysia

R. Baby Suneetha (417), Department of Chemistry, V.O. Chidambaram College, Tuticorin, India

Christopher Barile (395), Department of Chemistry, University of Nevada, Reno, United States

Ramazan Bayat (381), Sen Research Group, Department of Biochemistry; Department of Materials Science & Engineering, Faculty of Engineering, University of Dumlupinar, Kutahya, Turkey

Muhammed Bekmezci (381), Sen Research Group, Department of Biochemistry; Department of Materials Science & Engineering, Faculty of Engineering, University of Dumlupinar, Kutahya, Turkey

Vinay Karekura Boraiah (95), Department of Mechanical Engineering, Vidyavardhaka College of Engineering, Visvesvaraya Technological University, Mysuru, Karnataka, India

Hakan Burhan (381), Sen Research Group, Department of Biochemistry, University of Dumlupinar, Kutahya, Turkey

Sandeep Chandrashekharappa (77), Department of Medicinal Chemistry, National Institute of Pharmaceutical Education and Research (NIPER) Raebareli, Lucknow, Uttar Pradesh; Institute for Stem Cell Science and Regenerative Medicine, NCBS, TIFR, GKVK-Campus, Bengaluru, Karnataka, India

B. Chethan (315), Department of Physics, IISC, Bangalore, Karnataka, India

Bruna Coldibeli (335), Laboratory of Electroanalytical and Sensors, Department of Chemistry, Exact Sciences Center, State University of Londrina, Londrina, Paraná, Brazil

Ersin Demir (191), Faculty of Pharmacy, Department of Analytical Chemistry, Afyonkarahisar Health Sciences University, Afyonkarahisar, Turkey

Shridevi Doddamani (95), Chemical Sciences and Technology Division, CSIR-NIIST, Thiruvananthapuram, Kerala, India

Jessica Fernando (429), Department of Chemistry, V.O. Chidambaram College, Tuticorin, India

K. Girish Kumar (347), Department of Applied Chemistry, Cochin University of Science and Technology, Kochi, Kerala, India

Adarsha H.J. Gowda (259), Centre for Research in Medical Devices, National University of Ireland, Galway, Ireland

Doddahosuru M. Gurudatt (269), Department of Studies in Organic Chemistry, University of Mysore, Mysuru, Karnataka, India

Morteza Hosseini (59, 247, 275), Department of Life Science Engineering, Faculty of New Sciences & Technologies, University of Tehran, Tehran, Iran

P. Karpagavinayagam (417, 429), Department of Chemistry, V.O. Chidambaram College, Tuticorin, India

K. Keerthi (347), Department of Applied Chemistry, Cochin University of Science and Technology, Kochi, Kerala, India

Florica Manea (125), Politehnica University of Timisoara, Timisoara, Romania

Gabriel Rainer Pontes Manrique (335), Laboratory of Electroanalytical and Sensors, Department of Chemistry, Exact Sciences Center, State University of Londrina, Londrina, Paraná, Brazil

Renan Silva Mariano (335), Laboratory of Electroanalytical and Sensors, Department of Chemistry, Exact Sciences Center, State University of Londrina, Londrina, Paraná, Brazil

Vinusha Honnalagere Mariswamy (95), PG Department of Chemistry, Sarada Vilas College, Mysuru, Karnataka, India

Shalini Menon (347), Department of Applied Chemistry, Cochin University of Science and Technology, Kochi, Kerala, India

Sorina Motoc (125), "Coriolan Dragulescu" Institute of Chemistry Timisoara of Romanian Academy, Timisoara, Romania

Maryam Mousavizadegan (59), Department of Life Science Engineering, Faculty of New Sciences & Technologies, University of Tehran, Tehran, Iran

Shaik Mubeena (303), Department of Chemistry, KL Deemed to be University (KLEF), Guntur, Andhra Pradesh, India

Kiran Kumar Mudnakudu-Nagaraju (77), Department of Biotechnology & Bioinformatics, Faculty of Life Sciences, JSS Academy of Higher Education & Research, Mysore, Karnataka, India

Muazzin Mupit (15), Polymer Department, Universiti Kuala Lumpur—Malaysian Institute of Chemical and Bio-engineering Technology, Melaka, Malaysia

H.C. Ananda Murthy (29), Department of Applied Chemistry, School of Natural Science, Adama Science and Technology University, Adama, Ethiopia

Md Nafiujjaman (105), Department of Biomedical Engineering, Michigan State University, East Lansing, MI, United States

H.P. Nagaswarupa (29), Department of Studies in Chemistry, Davangere University, Shivagangothri, Davangere, Karnataka, India

B.P. Nandeshwarappa (77), Department of Studies in Chemistry, Shivagangothri, Davangere University, Davangere, Karnataka, India

Srikantamurthy Ningaiah (95, 269), Department of Chemistry, Vidyavardhaka College of Engineering, Visvesvaraya Technological University, Mysuru, Karnataka, India

Raghu Ningegowda (77), Department of Chemistry, Jyoti Nivas College Autonomous, Bangalore, Karnataka, India

R. Parimaladevi (167), Department of Physics, Mother Teresa Women's University, Kodaikanal, Tamil Nadu, India

V. Poornima Parvathi (167), Department of Physics, Mother Teresa Women's University, Kodaikanal, Tamil Nadu, India

Aniela Pop (125), Politehnica University of Timisoara, Timisoara, Romania

S. Pratibha (315), Department of Physics, BMS Institute of Technology and Management, Bangalore, Karnataka, India

Maiyara Carolyne Prete (43), Chemistry Department, Londrina State University (UEL), Londrina, Paraná, Brazil

Sneha S. Puttappa (269), Department of Chemistry, Vidyavardhaka College of Engineering, Visvesvaraya Technological University, Mysuru, Karnataka, India

S. Rajendra Prasad (269), PG Department of Chemistry, Davangere University, Davangere, Karnataka, India

Shashanka Rajendrachari (3), Department of Metallurgical and Materials Engineering, Bartin University, Bartin, Turkey

R. Rajkumar (429), Department of Chemistry, V.O. Chidambaram College, Tuticorin, India

T. Ramakrishnappa (149), BMS Institute of Technology and Management, Bengaluru, Karnataka, India

Vadde Ramu (291), Laboratory of Biomolecules, University of Sorbonne, 24 RueLhmond, Paris, France

Akhilesh Rao (77), Department of Mechanical Engineering, Christ University, Bengaluru, Karnataka, India

Gnanesh Rao (77), Department of Biochemistry, Bangalore University, Bengaluru, Karnataka, India

Maryam Rezaie (247), Department of Life Science Engineering, Faculty of New Sciences & Technologies, University of Tehran, Tehran, Iran

Luana Rianne Rocha (43), Chemistry Department, Londrina State University (UEL), Londrina, Paraná, Brazil

Amirreza Roshani (59, 275), Department of Life Science Engineering, Faculty of New Sciences & Technologies, University of Tehran, Tehran, Iran

Carlos Alberto Rossi Salamanca-Neto (335), Laboratory of Electroanalytical and Sensors, Department of Chemistry, Exact Sciences Center, State University of Londrina, Londrina, Paraná, Brazil

Sonia Sam (347), Department of Applied Chemistry, Cochin University of Science and Technology, Kochi, Kerala, India

Monima Sarma (303), Department of Chemistry, KL Deemed to be University (KLEF), Guntur, Andhra Pradesh, India

Elen Romão Sartori (335), Laboratory of Electroanalytical and Sensors, Department of Chemistry, Exact Sciences Center, State University of Londrina, Londrina, Paraná, Brazil

Vasant Sathe (167), UGC-DAE Consortium for Scientific Research, Indore, Madhya Pradesh, India

Jessica Scremin (335), Laboratory of Electroanalytical and Sensors, Department of Chemistry, Exact Sciences Center, State University of Londrina, Londrina, Paraná, Brazil

Fatih Sen (381), Sen Research Group, Department of Biochemistry, University of Dumlupinar, Kutahya, Turkey

Hulya Silah (191), Faculty of Art & Science, Department of Chemistry, Bilecik Seyh Edebali University, Bilecik, Turkey

Ebtesam Sobhanie (275), Department of Life Science Engineering, Faculty of New Sciences & Technologies, University of Tehran, Tehran, Iran

Kiran K. Somashekharappa (3), Department of Chemistry, Kateel Ashok Pai Memorial College, Shivamogga, India

K. Sureshkumar (149), BMS Institute of Technology and Management, Bengaluru, Karnataka, India

César Ricardo Teixeira Tarley (43), Chemistry Department, Londrina State University (UEL), Londrina, Paraná; Bioanalytical National Institute of Science and Technology (INCTbio), Campinas State University (UNICAMP), Campinas, São Paulo, Brazil

M. Umadevi (167), Department of Physics, Mother Teresa Women's University, Kodaikanal, Tamil Nadu, India

Bengi Uslu (191), Faculty of Pharmacy, Department of Analytical Chemistry, Ankara University, Ankara, Turkey

C. Vedhi (417, 429), Department of Chemistry, V.O. Chidambaram College, Tuticorin, India

Basappa C. Yallur (259), Department of Chemistry, M. S. Ramaiah Institute of Technology, Bangalore, Karnataka, India

Sercan Yıldırım (191), Faculty of Pharmacy, Department of Analytical Chemistry, Karadeniz Techical University, Trabzon, Turkey

Part I

Modern perspectives in sensor devices for sustainability

Chapter 1

Sustainable development information management of carbon nanomaterial-based sensors

Kiran K. Somashekharappa[a] and Shashanka Rajendrachari[b]

[a]Department of Chemistry, Kateel Ashok Pai Memorial College, Shivamogga, India, [b]Department of Metallurgical and Materials Engineering, Bartin University, Bartin, Turkey

Introduction

In the last few decades, extensive research was done on miniaturized science and technology, especially on the carbon nanomaterials. The unique electrochemical characteristics have made carbon (Edwards, Marsh, & Menendez, 2013) to be used as an electrochemical-sensing material (Qureshi, Kang, Davidson, & Gurbuz, 2009). The recent breakthroughs in material design (Hung, Kuo, Ko, Tzeng, & Yan, 2009) and synthesis (Andrews, Jacques, Qian, & Rantell, 2002; Che, Lakshmi, Martin, Fisher, & Ruoff, 1998; Ebbesen & Ajayan, 1992; Li et al., 2011; Sahu, Behera, Maiti, & Mohapatra, 2012; Szabó et al., 2010; Wang, Lu, Tang, & Xu, 2017), particularly nanomaterials (Llobet, 2013; Zhang et al., 2019), have led to the development of reliable electrochemical-sensing devices with improved solutions and performances.

Due to the magnificent property of carbon nanomaterials (Hurt, Monthioux, & Kane, 2006; Llobet, 2013), many research communities have worked on these nanomaterials and found good scientific results in various fields. But the development of the carbon nanomaterial-based research should be sustainable, and those researches should address the environmental problems facing by the global communities such as climate change, environmental hazards etc. and the research motto is to have global peace by the help of scientific research and to find the solutions to those challenges.

The intensive scientific efforts have been made in understanding nanoscale phenomena and developing new nanofabrication processes; nanosensor technology has seen fast growth in recent years. The unusual chemical and physical characteristics of carbon-based nanomaterials, including fullerenes, graphene, nanodiamonds, carbon nanotubes, and carbon nanodots, have lately attracted the interest of the scientific community. Carbon-based nanomaterials have found their place in a wide range of applications as a result of extensive research.

In addition to their superior effectiveness in detecting gas molecules, heavy metal ions, food additives, antibodies, and hazardous pesticides, these carbon-based nanomaterials also serve as reporters for bioimaging, making them a unique kind of nanosensor.

Types of carbon nanomaterials

Fullerenes (0D), carbon nanotubes (1D), graphene (2D) and its derivatives, graphene oxide, nanodiamonds, and carbon-based quantum dots are all carbon-based nanomaterials. Many fields, including electrochemical sensing and biosensing, are interested in these materials due to their unique structural dimensions and exceptional thermal, chemical, optical, electrical, and mechanical characteristics.

Fullerenes

Carbon exists in a variety of forms, including diamond, graphite, and others. Fullerene is the third allotropic form of carbon. It's a term used to describe solely carbon-based molecules in various forms such as hollow spheres, tubes, and ellipsoid shapes. Fullerenes feature hexagonal and pentagonal rings of carbon atoms that are linked (Isaacson, Kleber, & Field, 2009;

Sano, Wang, Chhowalla, Alexandrou, & Amaratunga, 2001). Architect Buckminster Fuller created cage-like geodesic domes, and it was named Fullerene after him.

"Fullerenes" or C_{60} are a class of icosahedrally symmetrical molecules discovered in the 20th century. They are composed of 60 sp^2 hybridized carbon atoms and are sometimes dubbed "buckyballs." As catalysts, sensors, and photocatalysts, these fullerenes have been widely used in a wide range of disciplines, including energy generation and storage. Since fullerenes offer better conductivity and charge transfer characteristics, they have attracted a lot of attention, especially in the sensor field. For sensors, fullerene-to-sensor interactions are key. There are many different kinds of fullerenes, but C_{60} is one of the most widely used fullerenes.

As a result of their electron-accepting characteristics of fullerene-based sensors, much effort has been made subsequently to detect different analytes such as gases, volatile organic compounds, metal ions, anions, and biomolecules for diverse applications such as strain/gas sensors, electrochemical sensors, and optical sensors.

Fullerenes were discovered first in 1985 by HW Kroto, R. F. Curl, and RF Smalley that carbon may be modified in a way that makes it allotropic (fullerene) (Kroto, Heath, O'Brien, Curl, & Smalley, 1985). They are characterized by the development of C_n clusters of atoms ($n > 20$) of carbon on a sphere. As a result of fullerene, covalent bonds are formed between the carbon atoms by sp^2 hybridization.

Most often, they are found near the vertices of pentagons and hexagons on the surface of the sphere. A lot of research has been done on C_{60}, a fullerene where hexagons and pentagons are composed of 60 carbon atoms in highly symmetric spherical molecules or 60 carbon atoms in 12 five-member rings and 20 six-member rings (Paredes, Villar-Rodil, Martínez-Alonso, & Tascón, 2008). It has a diameter of 0.7 nm (Goel, Howard, & Sande, 2004).

Carbon nanotubes (CNTs)

In 1991, the Japanese scientist Ijima made the discovery of carbon nanotubes (CNTs) (Iijima, 1991). The sp^2 hybridization in CNTs allows each carbon atom with 3 electrons to establish trigonally co-ordinated s bonds with three other carbon atoms. As the name suggests, CNT is a single sheet of graphene that has been rolled into a hollow form. Carbon nanotubes are characterized by rolled graphene sheets piled in cylindrical/tubular formations with a diameter of several nanometers (Iijima, 1991; Kumar, Rani, Dilbaghi, Tankeshwar, & Kim, 2017; Yang et al., 2019).

There are several variables that might affect the CNTs, such as their length, diameter, and the number of layers (symmetry of the nulled graphite sheet). CNTs are mainly classified in the following paragraphs.

Single-walled carbon nanotubes (SWCNTs) (Iijima, 2002) and multi-walled carbon nanotubes (MWCNTs) (Hareesha, Manjunatha, Amrutha, Pushpanjali, et al., 2021; Hareesha, Manjunatha, Amrutha, Sreeharsha, et al., 2021; Iijima & Ichihashi, 1993)

SWCNTs have a diameter of 1–3 nm, whereas MWCNTs are 5–25 nm in diameter and are about 10 μm (Zhang et al., 2013). On the other hand, in the last several years, researchers have studied and reported on the syntheses of CNTs; when compared with other fibrous materials, CNTs offer superior physical characteristics such as stiffness, strength etc., and confirmed CNTs are better than other fibrous materials.

As a result of their unique characteristics, carbon nanotubes (CNTs) are among the most researched nanostructures. Particularly interesting in terms of electroanalysis is CNTs' capacity to enhance the electrochemical reactivity of significant biomolecules and to increase electron transfer processes in proteins. Adsorbate-sensitive CNTs can be used as very sensitive nanoscale sensors due to their extraordinary sensitivity to changes in surface conductivity.

Biomolecules such as nucleic acids can be held in place by CNT-modified electrodes, and surface fouling can be reduced. Scientists are therefore attracted to CNTs because they may be used to make a wide range of electrochemical sensors. Also, due to its excellent chemical inertness and biocompatibility, synthetic diamond's electrochemical characteristics make it desirable both as a (bio)chemical sensing medium and as an electrochemical interface for biological systems. Several electrochemical diamond-based biosensors have been developed recently to illustrate this point (Power, Gorey, Chandra, & Chapman, 2018).

Graphene and related compounds

The 2D allotropic carbon is made up of a single layer. There are 0.142 nm of space between the carbon hexagons in a hexagonal crystal lattice, linked by s and p bonds in sp^2 hybridization. This material was initially discovered by Canadian theoretical physicist P. R. Wallace in 1947. A Geim and K Novoselov studied graphene later (Geim, 2009; Novoselov,

2004; Novoselov et al., 2005). Graphene has been widely studied theoretically, but practically graphene materials are studied recently prior to few decades.

The study of graphene's properties is exceptionally dynamic. Despite its great mechanical rigidity and thermal stability, it is nevertheless important to note that the electrical characteristics of this carbon allotrope are fundamentally different from carbon nanotubes and 3D graphite may be made from graphene by wrapping, rolling, or stacking it (Geim & Novoselov, 2010). For example, fullerenes, CNTs, and graphite are structural elements of carbon.

Charge carrier mobility is more than 15,000 to 20,000 $cm^2\,V\,s^{-1}$ in graphene at normal temperature. A high degree of light transparency (97.7%) and outstanding chemical, physical, thermal, and mechanical, characteristics make it an ideal material. Therefore, it might be used in electrochemical sensors that require high sensitivity (Balandin et al., 2008; Chen, Zhang, Liu, & Li, 2013; Huang et al., 2011; Lee, Lee, Uhm, Lee, & Rhee, 2008). It is estimated that the mobility of electrons in graphene layers is 100 times greater than that of silicon (Bolotin et al., 2008). As a result, it is expected to replace silicon in the electronic manufacturing sector in the future. And also due to its enormous specific surface area and high charge carrier mobility, graphene is widely used in sensors (Fowler et al., 2009; Lu, Ocola, & Chen, 2009).

Graphene oxide

Graphene oxide (GO) has been synthesized at a very cheap cost by chemically oxidizing graphite. For the development of GO, the Hummers technique was used since it takes a relatively short time and does not dissolve hazardous compounds (Hummers & Offeman, 1958). Potassium permanganate and concentrated sulfuric acid were used as oxidation agents and for peeling off graphite to produce graphene oxide (GO) (Hummers & Offeman, 1958). Because of its hydrophilic nature, GO is acidic. The GO sheets are easily distributed in water and can be used (Paredes et al., 2008). GO may be chemically converted back to graphene by employing reduction techniques (Stankovich et al., 2007) and thermal reductions (Xu, Wang, & Zhu, 2008).

rGO

Graphene oxide reduced (rGO)—Graphene oxide has also been reduced using the electrochemical reduction technique (rGO). The 116 chemical, annealing, and laser radiation techniques have all been used to partly reduce GO to synthesize reduced graphene oxide (rGO) (Tung et al., 2009). Due to graphene's basal plane being harmed by strong chemicals used for oxidation, its characteristics deteriorate. Because of this, graphene from graphite is peeled off using appropriate solvents and surfactants to remove the graphite (Hernandez et al., 2008; Lotya et al., 2009). In some solvents, graphene tends to clump into graphite. Preparing single-layer graphene in solvents that is pure and evenly distributed is the tough task. For the mechanical peeling of the 2D graphite, adhesive tapes (Lim et al., 2015) with a lower flaw density are used.

Carbon dots (CDs)

They are made of tiny carbon atoms with a size of less than 10nm (0D). Because of their electrical and optical characteristics, quantum dots are a good example of these materials (Wang & Dai, 2015). For their use as electrochemical and biosensors, they have minimal toxicity, stability, and biocompatibility (Fan, Ma, Jayachandran, Li, & Luo, 2018; Liu et al., 2018; Song, Fan, Li, Gao, & Luo, 2018). The carbon atoms were laser-ablated to create the CDs (Zhou et al., 2014). Many other procedures have been utilized to create CDs, including pyrolysis (Zhu, Yin, Wang, & Chen, 2013), hydrothermal synthesis (Dong et al., 2018), electrochemical methods (Gao, Ding, Zhu, & Tian, 2014), and microwave synthesis (Liu et al., 2014). The soot from the candle flame may also be used to make them (Zhu et al., 2009).

Carbon nanodiamonds

In addition, carbon nanodiamonds (CNDs) are nanoparticles with the same crystal structure as diamonds and display the same outstanding characteristics as diamonds (Mochalin, Shenderova, Ho, & Gogotsi, 2012; Tang, Tsai, Gerberich, Kruckeberg, & Kania, 1995). An anion-like shell of graphite surrounds the diamond core of the CNDs. For chemical and biological adsorption purposes, they have a compact size, big surface area, and high adsorption capacity (Ansari et al., 2016). Their hardness, thermal conductivity, refractive index, and coefficients of friction are unmatched, and they display excellent insulating properties, as well as low toxicity (Schrand et al., 2007; Turcheniuk et al., 2015).

In order to synthesize CNDs, explosives are detonated (Danilenko, 2004). This is owing to a complicated defect (N-V) in diamond, which is composed of nitrogen (N) and vacancy (V). Fluorescent CNDs may be produced by doping N

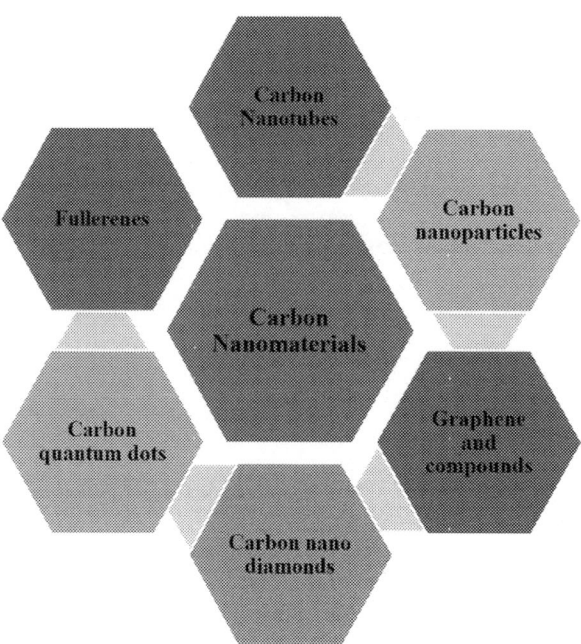

FIG. 1.1 Types of carbon nanomaterials. *(No permission required.)*

vacancies with electron beams and then annealing in the free space thereafter (Schirhagl, Chang, Loretz, & Degen, 2014). Fig. 1.1 depicts the different types of carbon nanomaterials.

Carbon black

Carbon black (CB) is a nanomaterial made from petroleum products that has been burned to create it. They are spherical nanoparticles that are firmly linked to one another to form aggregates, and they have a wide range of sizes of carbon black particles, ranging from 3.0 to 100 nm. Heating up to 7000°C can increase the conductivity of carbon black because more electrons in the hybridization state of the delocalized pi-bonds are accessible for current conduction when heated to this temperature. A significant number of oxygenated groups are generated at the margins of carbon black nanoparticles due to their enormous surface area.

It is the presence of sp^2 hybridized carbon atom edge planes and oxygenated groups over the carbon black nanomaterials (Bandosz, 2002) that make them capable to attach biomolecules on their surface to act as electrochemical and biosensors. They can be used for the detection of analytes for both electrochemical (Hareesha & Manjunatha, 2021; Hareesha, Manjunatha, Amrutha, Pushpanjali, et al., 2021; Hareesha, Manjunatha, Amrutha, Sreeharsha, et al., 2021) and biosensing (Pushpanjali, Manjunatha, Sreeharsha, Asdaq, & Anwer, 2021) applications.

Carbon nanomaterial-based sensors for sustainable development

Scientific research communities all over the world are striving hard to investigate/discover the material/substance which would helpful for the development, which should not harm the peaceful living nature of all living beings. Nanomaterials, particularly carbon nanomaterials, are widely recognized for having outstanding optical, catalytic, thermal, electrical, and mechanical characteristics, allowing for the development of carbon nanomaterial-based sensors and systems for monitoring environmental pollution in air, water, and soil, which are very essential for the sustainable development of not only human beings but also all living creators which are present in our ecosystem. Various carbon nanomaterials have been extensively investigated for their high sensitivity, selectivity, and simplicity in detecting and measuring toxic metal ions, toxic gases, pesticides, and hazardous industrial chemicals.

Environmental hazardous sensors

A large interfacial surface area, great electrical conductivity, high thermal and a high mechanical strength make CNTs and graphene perfect alternatives for environmental contamination monitoring, such as heavy metal ions like As^{3+}

(Kempahanumakkagari, Deep, Kim, Kumar Kailasa, & Yoon, 2017), Pb^{2+} (Ganjali, Motakef-Kazami, Faridbod, Khoee, & Norouzi, 2010), Hg^{2+} (Liu et al., 2013), and Ag^+ (Jiang et al., 2020) according to the researchers.

Yongming Guo et al., discovered that carbon quantum dot-based sensors were used to detect Fe^{3+}, Cu^{2+}, and Hg^{2+} in the range of nm to several pm by the fluorescence resonance energy transfer method (Guo et al., 2015). Bi- and Fe-modified carbon xerogel composites were assessed by modified electrodes for detecting biomarkers and Pb^{2+} (Fort et al., 2021), which is below the admittable limit in drinking water. Liu et al. (2021) investigated the sensing application of heavy metal ions and ethanol using targeted oriented synthesized carbon dots (ECDs-ethanol-soluble carbon dots and water-soluble carbon dots). Zero-dimensional, fluorescent carbon nanodot-based sensors were used to sense the heavy metal ions quenching mechanisms. Carbon nanodots were used as sensors because of their sensitive fluorescent sensor property even in the femtomolar concentrations (Sekar, Yadav, & Basavaraj, 2021). Fig. 1.2 depicts STEM microstructure and elemental mapping of C_xBiFe_x composite.

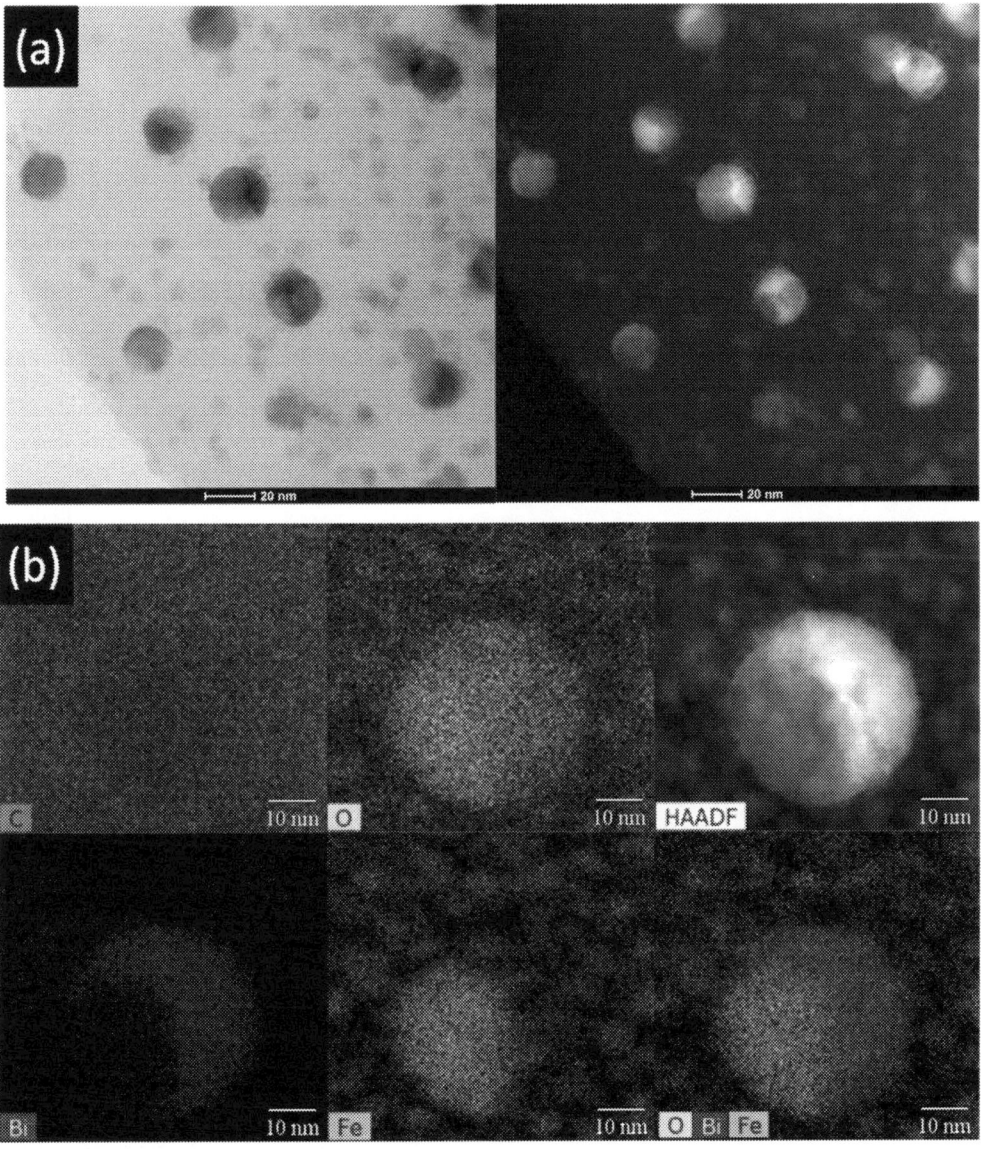

FIG. 1.2 (A) BF-STEM and HAADF-STEM images and (B) STEM-EDS elemental mapping of C_xBiFe_x composite. *(From Fort, C. I., Rusu, M. M., Cotet, L. C., Vulpoi, A., Florea, I., Tuseau-Nenez, S., Baia, M., Baibarac, M. & Baia, L. (2021). Carbon xerogel nanostructures with integrated Bi and Fe components for hydrogen peroxide and heavy metal detection. Molecules, 26 (1), 117. https://doi.org/10.3390/molecules26010117.)*

Electrochemical sensors

These sensors consist of a closed electrical circuit formed by at least two electrodes and a transducer where the charge transport (always electronic) takes place, whereas charge transport in the analyte sample might be either electronic or ionic.

Gao Shang et al. investigated the sensing of chloramphenicol, an antibiotic that is used to treat the eye infections; MoS_2-modified reduced graphene oxide electrodes were used to sense the chloramphenicol with a sensitivity of ~$3.4\,\mu A\,\mu M^{-1}\,cm^{-2}$ compared to other electrodes made with MWCNTs and carbon black (Gao et al., 2021).

In recent years, there has been a surge of interest in the electrochemical functionalization of carbon nanotubes with metallic nanoparticles and the use of the resultant metal decorated CNTs, notably in sensing and catalysis (Benadict Joseph et al., 2021; Charithra, Manjunatha, Sreeharsha, Asdaq, & Anwer, 2021; Shahsavar, Taghizadeh, & Kiadehi, 2021; Sonkar & Ganesan, 2021). Wang et al. (2015) developed a one-pot hot-solution synthesis technique for $Ni_{12}P_5$/CNTs hybrid nanostructures. When over-potentials of 65 and 129 mV were applied to hybrid structures, current densities of 2 and 10 mA cm^2 were achieved. In addition, when used as an anode material for lithium ion batteries, the hybrid structure exhibited improved electrochemical performance.

Biosensors

To manage and monitor health problems has received a lot of attention, and even more investment has been allocated to research and development of new technologies. The earlier changes in organism function are recognized, the more effective their control will be (Charithra & Manjunatha, 2021). When Palecek's discovery of DNA's electrochemical activity and the direct detection of DNA and its bases by electrochemical sensors (Paleček, 1960), genosensors/DNA-based sensors have been widely used in biomedical and environmental research for the detection of food and environmental pollutants, genetic diseases, and virus and bacteria identification (Robby et al., 2021; Yang, Denno, Pyakurel, & Venton, 2015). The combination of CNTs with DNA has earned a huge interest because it promotes the development of more sensitive electrochemical DNA detection techniques that are faster and less expensive. Furthermore, diagnosis methods with a low detection limit, which indicate greater sensitivity, enhance the chances of more effective medical intervention in medical treatment. In this approach, technologies for detecting and quantifying biomolecules are critical for improving life quality. Karami, Shamsipur, and Barati (2021) detected Cu^{2+} and aspartic acid in human serum by carbon dots sensors. Jiang et al. (2021) invented enzyme-free glucose sensors by flower-like carbon micro/nanostructures, which when integrated with CuO nanofilms showed the good sensitivity and stability.

Gas sensors

The population is the asset of every country only if they are healthy; concerning about the normal human health and country's development, there is a need to protect everyone from damage caused by air pollution and hazardous gases in our environment when the concentration of the hazardous gases increases. Carbon nanomaterials play an important role in sensing these hazardous gases due to their versatile structure, which makes them ideal for suitable gas sensors. The functionalization of these carbon nanomaterial-based sensors enhances the gas selectivity. These materials are suitable for sensing the gases and their quantification such as methane CH_4, ammonia NH_3, nitrogen oxide NO_X, carbon dioxide CO_2, carbon monoxide CO, and hydrogen sulfide H_2S. Ovsianytskyi et al. (2018) developed a gas sensor that includes graphene-decorated silver nanoparticles. This sensor was able to sense/detect less than 100 ppm of H_2S, CH_4, O_2, N_2, and CO_2 gases with immediate response. Ding et al. (2020) developed a sensor that includes 3-dimensional and porous graphene, which can sense gas and humidity as there is an increased active site compared to 2-dimensional structure. Ricciardella et al. (2021) investigated the effects of different graphene production methods on the morphology, crystalline structure, and, as a result, the performance of NO_2 sensors made from these materials. Mechanical exfoliation (ME) produced less-defective material with the largest value of signal variation than liquid-phase exfoliation (LPE) and CVD methods, indicating that a low defect level promotes a quicker interaction during the exposure period to the gas.

Dopamine and other sensors

Dopamine is a key neurotransmitter involved in the initiation of numerous behavioral reactions to diverse stimuli, as well as the central nervous, renal, hormonal systems, and cardiovascular, as well as emotional in vivo processes (Chandra, Miller, Bendavid, Martin, & Wong, 2014). Uric acid is a remnant of purine metabolism that may be found in biological fluids like blood or urine. Uric acid imbalances are signs of a variety of illnesses, including hyperuricemia and gout

(Manjunatha et al., 2014; Manjunatha, Deraman, Basri, & Talib, 2018). As a result, there has been a significant amount of study on the sensitive and selective detection of both the species in their physiological space. In this regard, nitrogen-incorporated ultrananocrystalline diamond electrodes have piqued curiosity. Shalini et al. (2013) evaluated them in the electrochemical detection of uric acid and dopamine. A glassy carbon electrode modified with poly(3-methylthiophene) and covered with a Nafion/single-walled carbon nanotubes sheet was produced and employed for extremely selective and sensitive dopamine detection by the differential pulse voltammetric (DPV) method with a detection limit of 5 nm (Wang, Li, Jia, & Xu, 2006). Li et al. (2021) investigated sensing of dopamine in human serum by using the modified boron-doped diamond BDD electrode limit with a detection limit of 54 nm for dopamine; this modified electrode showed good long-term functionality and stability.

Conclusions

Among the known elements, carbon appears as a magnificent element with unique characteristics, which is why it lies at the backbone of organic chemistry and, ultimately, life. Carbon's ability to generate a variety of structures with unique properties, allowing the development of new electrochemical/bio/gas sensors for the detection of a wide range of biological, organic, and inorganic analytes with desirable detection limits, levels of sensitivity, low and good stability, is one of these special properties. The ability to incorporate carbon nanostructures into other materials such as organic polymers or hydrogels, or to couple them to metal or oxide nanoparticles, might boost the sensors' detecting capabilities or improve their performance. Furthermore, carbon-based nanostructures may be chemically built into 2D and 3D structures, giving additional characteristics that increase sensing capabilities. Highly sensitive electrochemical/bio/gas sensors for the detection of pharmaceutically, environmentally, and physiologically significant substances were designed using composites of carbon nanomaterial-based compounds. For the modified sensors, carbon and its derivatives have good electrocatalytic features, such as increased high conductivity, electrocatalytic effects, detection sensitivity, and decreased contamination. These advancements in electrochemical/bio/gas sensors research pave the way for the creation of potential platforms for food quality, air pollution, control hazardous chemicals, and environmental monitoring. Thus, carbon nanomaterial-based sensors lead to the sustainable development.

Conflicts of interest

The authors declare no conflict of interest.

References

Andrews, R., Jacques, D., Qian, D., & Rantell, T. (2002). Multiwall carbon nanotubes: Synthesis and application. *Accounts of Chemical Research*, *35*(12), 1008–1017. https://doi.org/10.1021/ar010151m.

Ansari, S. A., Satar, R., Jafri, M. A., Rasool, M., Ahmad, W., & Zaidi, S. K. (2016). Role of nanodiamonds in drug delivery and stem cell therapy. *Iranian Journal of Biotechnology*, *14*(3), 70–81. https://doi.org/10.15171/ijb.1320.

Balandin, A. A., Ghosh, S., Bao, W., Calizo, I., Teweldebrhan, D., Miao, F., et al. (2008). Superior thermal conductivity of single-layer graphene. *Nano Letters*, *8*(3), 902–907. https://doi.org/10.1021/nl0731872.

Bandosz, T. J. (2002). On the adsorption/oxidation of hydrogen sulfide on activated carbons at ambient temperatures. *Journal of Colloid and Interface Science*, *246*(1), 1–20. https://doi.org/10.1006/jcis.2001.7952.

Benadict Joseph, X., Sriram, B., Wang, S. F., Baby, J. N., Hsu, Y. F., & George, M. (2021). Revealing the effect of multidimensional ZnO@CNTs/RGO composite for enhanced electrochemical detection of flufenamic acid. *Microchemical Journal*, *168*. https://doi.org/10.1016/j.microc.2021.106448.

Bolotin, K. I., Sikes, K. J., Jiang, Z., Klima, M., Fudenberg, G., Hone, J., et al. (2008). Ultrahigh electron mobility in suspended graphene. *Solid State Communications*, *146*(9–10), 351–355. https://doi.org/10.1016/j.ssc.2008.02.024.

Chandra, S., Miller, A. D., Bendavid, A., Martin, P. J., & Wong, D. K. Y. (2014). Minimizing fouling at hydrogenated conical-tip carbon electrodes during dopamine detection in vivo. *Analytical Chemistry*, *86*(5), 2443–2450. https://doi.org/10.1021/ac403283t.

Charithra, M. M., & Manjunatha, J. G. (2021). Electrochemical sensing of adrenaline using surface modified carbon nanotube paste electrode. *Materials Chemistry and Physics*, *262*, 124293. https://doi.org/10.1016/j.matchemphys.2021.124293.

Charithra, M. M., Manjunatha, J. G., Sreeharsha, N., Asdaq, S. M. B., & Anwer, M. K. (2021). Polymerized carbon nanotube paste electrode as a sensing material for the detection of adrenaline with folic acid. *Monatshefte für Chemie*, *152*(4), 411–420. https://doi.org/10.1007/s00706-021-02756-0.

Che, G., Lakshmi, B. B., Martin, C. R., Fisher, E. R., & Ruoff, R. S. (1998). Chemical vapor deposition based synthesis of carbon nanotubes and nanofibers using a template method. *Chemistry of Materials*, *10*(1), 260–267. https://doi.org/10.1021/cm970412f.

Chen, D., Zhang, H., Liu, Y., & Li, J. (2013). Graphene and its derivatives for the development of solar cells, photoelectrochemical, and photocatalytic applications. *Energy and Environmental Science*, *6*(6), 1362–1387. https://doi.org/10.1039/c3ee23586f.

Danilenko, V. V. (2004). On the history of the discovery of nanodiamond synthesis. *Physics of the Solid State, 46*(4), 595–599. https://doi.org/10.1134/1.1711431.

Ding, H., Wei, Y., Wu, Z., Tao, K., Ding, M., Xie, X., et al. (2020). Recent advances in gas and humidity sensors based on 3D structured and porous graphene and its derivatives. *ACS Materials Letters, 2*(11), 1381–1411. https://doi.org/10.1021/acsmaterialslett.0c00355.

Dong, M., Li, Q., Liu, H., Liu, C., Wujcik, E. K., Shao, Q., et al. (2018). Thermoplastic polyurethane-carbon black nanocomposite coating: Fabrication and solid particle erosion resistance. *Polymer, 158*, 381–390. https://doi.org/10.1016/j.polymer.2018.11.003.

Ebbesen, T. W., & Ajayan, P. M. (1992). Large-scale synthesis of carbon nanotubes. *Nature, 358*(6383), 220–222. https://doi.org/10.1038/358220a0.

Edwards, I. A. S., Marsh, H., & Menendez, R. (2013). *Introduction to carbon science*. Elsevier.

Fan, G. C., Ma, L., Jayachandran, S., Li, Z., & Luo, X. (2018). Separating photoanode from recognition events: Toward a general strategy for a self-powered photoelectrochemical immunoassay with both high sensitivity and anti-interference capabilities. *Chemical Communications, 54*(51), 7062–7065. https://doi.org/10.1039/c8cc02627k.

Fort, C. I., Rusu, M. M., Cotet, L. C., Vulpoi, A., Florea, I., Tuseau-Nenez, S., et al. (2021). Carbon xerogel nanostructures with integrated Bi and Fe components for hydrogen peroxide and heavy metal detection. *Molecules, 26*(1), 117. https://doi.org/10.3390/molecules26010117.

Fowler, J. D., Allen, M. J., Tung, V. C., Yang, Y., Kaner, R. B., & Weiller, B. H. (2009). Practical chemical sensors from chemically derived graphene. *ACS Nano, 3*(2), 301–306. https://doi.org/10.1021/nn800593m.

Ganjali, M. R., Motakef-Kazami, N., Faridbod, F., Khoee, S., & Norouzi, P. (2010). Determination of Pb^{2+} ions by a modified carbon paste electrode based on multi-walled carbon nanotubes (MWCNTs) and nanosilica. *Journal of Hazardous Materials, 173*(1–3), 415–419. https://doi.org/10.1016/j.jhazmat.2009.08.101.

Gao, X., Ding, C., Zhu, A., & Tian, Y. (2014). Carbon-dot-based ratiometric fluorescent probe for imaging and biosensing of superoxide anion in live cells. *Analytical Chemistry, 86*(14), 7071–7078. https://doi.org/10.1021/ac501499y.

Gao, S., Zhang, Y., Yang, Z., Fei, T., Liu, S., & Zhang, T. (2021). Electrochemical chloramphenicol sensors-based on trace MoS_2 modified carbon nanomaterials: Insight into carbon supports. *Journal of Alloys and Compounds, 872*, 159687. https://doi.org/10.1016/j.jallcom.2021.159687.

Geim, A. K. (2009). Graphene: Status and prospects. *Science, 324*(5934), 1530–1534. https://doi.org/10.1126/science.1158877.

Geim, A. K., & Novoselov, K. S. (2010). The rise of graphene. In P. Rodgers (Ed.), *Nanoscience and technology: A collection of reviews from nature journals* (pp. 11–19). Nature Publishing Group.

Goel, A., Howard, J. B., & Sande, J. B. V. (2004). Size analysis of single fullerene molecules by electron microscopy. *Carbon, 42*(10), 1907–1915. https://doi.org/10.1016/j.carbon.2004.03.022.

Guo, Y., Zhang, L., Zhang, S., Yang, Y., Chen, X., & Zhang, M. (2015). Fluorescent carbon nanoparticles for the fluorescent detection of metal ions. *Biosensors and Bioelectronics, 63*, 61–71. https://doi.org/10.1016/j.bios.2014.07.018.

Hareesha, N., & Manjunatha, J. G. (2021). Electro-oxidation of formoterol fumarate on the surface of novel poly(thiazole yellow-G) layered multi-walled carbon nanotube paste electrode. *Scientific Reports, 11*(1). https://doi.org/10.1038/s41598-021-92099-x.

Hareesha, N., Manjunatha, J. G., Amrutha, B. M., Pushpanjali, P. A., Charithra, M. M., & Prinith Subbaiah, N. (2021). Electrochemical analysis of indigo carmine in food and water samples using a poly(glutamic acid) layered multi-walled carbon nanotube paste electrode. *Journal of Electronic Materials, 50*(3), 1230–1238. https://doi.org/10.1007/s11664-020-08616-7.

Hareesha, N., Manjunatha, J. G., Amrutha, B. M., Sreeharsha, N., Basheeruddin Asdaq, S. M., & Anwer, M. K. (2021). A fast and selective electrochemical detection of vanillin in food samples on the surface of poly(glutamic acid) functionalized multiwalled carbon nanotubes and graphite composite paste sensor. *Colloids and Surfaces A: Physicochemical and Engineering Aspects, 626*, 127042. https://doi.org/10.1016/j.colsurfa.2021.127042.

Hernandez, Y., Nicolosi, V., Lotya, M., Blighe, F. M., Sun, Z., De, S., et al. (2008). High-yield production of graphene by liquid-phase exfoliation of graphite. *Nature Nanotechnology, 3*(9), 563–568. https://doi.org/10.1038/nnano.2008.215.

Huang, X., Li, S., Huang, Y., Wu, S., Zhou, X., Li, S., et al. (2011). Synthesis of hexagonal close-packed gold nanostructures. *Nature Communications, 2*(1). https://doi.org/10.1038/ncomms1291.

Hummers, W. S., & Offeman, R. E. (1958). Graphene oxide. *Journal of the American Chemical Society, 80*, 1339.

Hung, K. H., Kuo, W. S., Ko, T. H., Tzeng, S. S., & Yan, C. F. (2009). Processing and tensile characterization of composites composed of carbon nanotube-grown carbon fibers. *Composites Part A: Applied Science and Manufacturing, 40*(8), 1299–1304. https://doi.org/10.1016/j.compositesa.2009.06.002.

Hurt, R. H., Monthioux, M., & Kane, A. (2006). Toxicology of carbon nanomaterials: Status, trends, and perspectives on the special issue. *Carbon, 44*(6), 1028–1033. https://doi.org/10.1016/j.carbon.2005.12.023.

Iijima, S. (1991). Helical microtubules of graphitic carbon. *Nature, 354*(6348), 56–58. https://doi.org/10.1038/354056a0.

Iijima, S. (2002). Carbon nanotubes: Past, present, and future. *Physica B: Condensed Matter, 323*(1–4), 1–5. https://doi.org/10.1016/S0921-4526(02)00869-4.

Iijima, S., & Ichihashi, T. (1993). Single-shell carbon nanotubes of 1-nm diameter. *Nature, 363*(6430), 603–605. https://doi.org/10.1038/363603a0.

Isaacson, C. W., Kleber, M., & Field, J. A. (2009). Quantitative analysis of fullerene nanomaterials in environmental systems: A critical review. *Environmental Science and Technology, 43*(17), 6463–6474. https://doi.org/10.1021/es900692e.

Jiang, S., Chen, Q., Lin, J., Liao, G., Shi, T., & Qian, L. (2021). Thermal stress-induced fabrication of carbon micro/nanostructures and the application in high-performance enzyme-free glucose sensors. *Sensors and Actuators B: Chemical, 345*, 130364. https://doi.org/10.1016/j.snb.2021.130364.

Jiang, X., Huang, J., Chen, T., Zhao, Q., Xu, F., & Zhang, X. (2020). Synthesis of hemicellulose/deep eutectic solvent based carbon quantum dots for ultrasensitive detection of Ag^+ and L-cysteine with "off-on" pattern. *International Journal of Biological Macromolecules, 153*, 412–420. https://doi.org/10.1016/j.ijbiomac.2020.03.026.

Karami, S., Shamsipur, M., & Barati, A. (2021). Intrinsic dual-emissive carbon dots for efficient ratiometric detection of Cu2+ and aspartic acid. *Analytica Chimica Acta*, *1144*, 26–33. https://doi.org/10.1016/j.aca.2020.11.032.

Kempahanumakkagari, S., Deep, A., Kim, K. H., Kumar Kailasa, S., & Yoon, H. O. (2017). Nanomaterial-based electrochemical sensors for arsenic—A review. *Biosensors and Bioelectronics*, *95*, 106–116. https://doi.org/10.1016/j.bios.2017.04.013.

Kroto, H. W., Heath, J. R., O'Brien, S. C., Curl, R. F., & Smalley, R. E. (1985). C60: Buckminsterfullerene. *Nature*, 162–163. https://doi.org/10.1038/318162a0.

Kumar, S., Rani, R., Dilbaghi, N., Tankeshwar, K., & Kim, K. H. (2017). Carbon nanotubes: A novel material for multifaceted applications in human healthcare. *Chemical Society Reviews*, *46*(1), 158–196. https://doi.org/10.1039/c6cs00517a.

Lee, G. J., Lee, H. M., Uhm, Y. R., Lee, M. K., & Rhee, C. K. (2008). Square-wave voltammetric determination of thallium using surface modified thick-film graphite electrode with Bi nanopowder. *Electrochemistry Communications*, *10*(12), 1920–1923. https://doi.org/10.1016/j.elecom.2008.10.015.

Li, H., He, X., Liu, Y., Huang, H., Lian, S., Lee, S. T., et al. (2011). One-step ultrasonic synthesis of water-soluble carbon nanoparticles with excellent photoluminescent properties. *Carbon*, *49*(2), 605–609. https://doi.org/10.1016/j.carbon.2010.10.004.

Li, H., Zhou, K., Cao, J., Wei, Q., Lin, C. T., Pei, S. E., et al. (2021). A novel modification to boron-doped diamond electrode for enhanced, selective detection of dopamine in human serum. *Carbon*, *171*, 16–28. https://doi.org/10.1016/j.carbon.2020.08.019.

Lim, E. K., Kim, T., Paik, S., Haam, S., Huh, Y. M., & Lee, K. (2015). Nanomaterials for theranostics: Recent advances and future challenges. *Chemical Reviews*, *115*(1), 327–394. https://doi.org/10.1021/cr300213b.

Liu, R., Li, H., Kong, W., Liu, J., Liu, Y., Tong, C., et al. (2013). Ultra-sensitive and selective Hg2+ detection based on fluorescent carbon dots. *Materials Research Bulletin*, *48*(7), 2529–2534. https://doi.org/10.1016/j.materresbull.2013.03.015.

Liu, Z., Li, B., Shi, X., Li, L., Feng, Y., Jia, D., et al. (2021). Target-oriented synthesis of high synthetic yield carbon dots with tailored surface functional groups for bioimaging of zebrafish, flocculation of heavy metal ions and ethanol detection. *Applied Surface Science*, *538*, 148118. https://doi.org/10.1016/j.apsusc.2020.148118.

Liu, Y., Xiao, N., Gong, N., Wang, H., Shi, X., Gu, W., et al. (2014). One-step microwave-assisted polyol synthesis of green luminescent carbon dots as optical nanoprobes. *Carbon*, *68*, 258–264. https://doi.org/10.1016/j.carbon.2013.10.086.

Liu, Q., Zhang, N., Shi, H., Ji, W., Guo, X., Yuan, W., et al. (2018). One-step microwave synthesis of carbon dots for highly sensitive and selective detection of copper ions in aqueous solution. *New Journal of Chemistry*, *42*(4), 3097–3101. https://doi.org/10.1039/c7nj05000c.

Llobet, E. (2013). Gas sensors using carbon nanomaterials: A review. *Sensors and Actuators, B: Chemical*, *179*, 32–45. https://doi.org/10.1016/j.snb.2012.11.014.

Lotya, M., Hernandez, Y., King, P. J., Smith, R. J., Nicolosi, V., Karlsson, L. S., et al. (2009). Liquid phase production of graphene by exfoliation of graphite in surfactant/water solutions. *Journal of the American Chemical Society*, *131*(10), 3611–3620. https://doi.org/10.1021/ja807449u.

Lu, G., Ocola, L. E., & Chen, J. (2009). Reduced graphene oxide for room-temperature gas sensors. *Nanotechnology*, 445502. https://doi.org/10.1088/0957-4484/20/44/445502.

Manjunatha, J. G., Deraman, M., Basri, N. H., Nor, N. S. M., Talib, I. A., & Ataollahi, N. (2014). Sodium dodecyl sulfate modified carbon nanotubes paste electrode as a novel sensor for the simultaneous determination of dopamine, ascorbic acid, and uric acid. *Comptes Rendus Chimie*, *17*(5), 465–476. https://doi.org/10.1016/j.crci.2013.09.016.

Manjunatha, J. G., Deraman, M., Basri, N. H., & Talib, I. A. (2018). Fabrication of poly (Solid Red A) modified carbon nano tube paste electrode and its application for simultaneous determination of epinephrine, uric acid and ascorbic acid. *Arabian Journal of Chemistry*, *11*(2), 149–158. https://doi.org/10.1016/j.arabjc.2014.10.009.

Mochalin, V. N., Shenderova, O., Ho, D., & Gogotsi, Y. (2012). The properties and applications of nanodiamonds. *Nature Nanotechnology*, *7*(1), 11–23. https://doi.org/10.1038/nnano.2011.209.

Novoselov, K. S. (2004). Electric field effect in atomically thin carbon films. *Science*, 666–669. https://doi.org/10.1126/science.1102896.

Novoselov, K. S., Geim, A. K., Morozov, S. V., Jiang, D., Katsnelson, M. I., Grigorieva, I. V., et al. (2005). Two-dimensional gas of massless Dirac fermions in graphene. *Nature*, *438*(7065), 197–200. https://doi.org/10.1038/nature04233.

Ovsianytskyi, O., Nam, Y. S., Tsymbalenko, O., Lan, P. T., Moon, M. W., & Lee, K. B. (2018). Highly sensitive chemiresistive H2S gas sensor based on graphene decorated with Ag nanoparticles and charged impurities. *Sensors and Actuators, B: Chemical*, *257*, 278–285. https://doi.org/10.1016/j.snb.2017.10.128.

Paleček, E. (1960). Oscillographic polarography of highly polymerized deoxyribonucleic acid. *Nature*, *188*(4751), 656–657. https://doi.org/10.1038/188656a0.

Paredes, J. I., Villar-Rodil, S., Martínez-Alonso, A., & Tascón, J. M. D. (2008). Graphene oxide dispersions in organic solvents. *Langmuir*, *24*(19), 10560–10564. https://doi.org/10.1021/la801744a.

Power, A. C., Gorey, B., Chandra, S., & Chapman, J. (2018). Carbon nanomaterials and their application to electrochemical sensors: A review. *Nanotechnology Reviews*, *7*(1), 19–41. https://doi.org/10.1515/ntrev-2017-0160.

Pushpanjali, P. A., Manjunatha, J. G., Sreeharsha, N., Asdaq, S. M. B., & Anwer, M. K. (2021). A highly responsive voltammetric methodology for the sensing of antihistamine drug cetirizine on the surface of cetrimonium bromide immobilized multi-walled carbon nanotube electrode. *Journal of Materials Science: Materials in Electronics*. https://doi.org/10.1007/s10854-021-06751-3.

Qureshi, A., Kang, W. P., Davidson, J. L., & Gurbuz, Y. (2009). Review on carbon-derived, solid-state, micro and nano sensors for electrochemical sensing applications. *Diamond and Related Materials*, *18*(12), 1401–1420. https://doi.org/10.1016/j.diamond.2009.09.008.

Ricciardella, F., Vollebregt, S., Tilmann, R., Hartwig, O., Bartlam, C., Sarro, P. M., et al. (2021). Influence of defect density on the gas sensing properties of multi-layered graphene grown by chemical vapor deposition. *Carbon Trends*, *3*, 100024. https://doi.org/10.1016/j.cartre.2021.100024.

Robby, A. I., Kim, S. G., Lee, U. H., In, I., Lee, G., & Park, S. Y. (2021). Wireless electrochemical and luminescent detection of bacteria based on surface-coated CsWO3-immobilized fluorescent carbon dots with photothermal ablation of bacteria. *Chemical Engineering Journal, 403*. https://doi.org/10.1016/j.cej.2020.126351.

Sahu, S., Behera, B., Maiti, T. K., & Mohapatra, S. (2012). Simple one-step synthesis of highly luminescent carbon dots from orange juice: Application as excellent bio-imaging agents. *Chemical Communications, 48*(70), 8835–8837. https://doi.org/10.1039/c2cc33796g.

Sano, N., Wang, H., Chhowalla, M., Alexandrou, I., & Amaratunga, G. A. J. (2001). Synthesis of carbon "onions" in water. *Nature, 414*(6863), 506–507. https://doi.org/10.1038/35107141.

Schirhagl, R., Chang, K., Loretz, M., & Degen, C. L. (2014). Nitrogen-vacancy centers in diamond: Nanoscale sensors for physics and biology. *Annual Review of Physical Chemistry, 65*, 83–105. https://doi.org/10.1146/annurev-physchem-040513-103659.

Schrand, A. M., Huang, H., Carlson, C., Schlager, J. J., Ōsawa, E., Hussain, S. M., et al. (2007). Are diamond nanoparticles cytotoxic? *The Journal of Physical Chemistry. B, 111*, 2–7.

Sekar, A., Yadav, R., & Basavaraj, N. (2021). Fluorescence quenching mechanism and the application of green carbon nanodots in the detection of heavy metal ions: A review. *New Journal of Chemistry, 45*(5), 2326–2360. https://doi.org/10.1039/d0nj04878j.

Shahsavar, H., Taghizadeh, M., & Kiadehi, A. D. (2021). Effects of catalyst preparation route and promoters (Ce and Zr) on catalytic activity of CuZn/CNTs catalysts for hydrogen production from methanol steam reforming. *International Journal of Hydrogen Energy, 46*(13), 8906–8921. https://doi.org/10.1016/j.ijhydene.2021.01.010.

Shalini, J., Sankaran, K. J., Dong, C. L., Lee, C. Y., Tai, N. H., & Lin, I. N. (2013). In situ detection of dopamine using nitrogen incorporated diamond nanowire electrode. *Nanoscale, 5*(3), 1159–1167. https://doi.org/10.1039/c2nr32939e.

Song, Z., Fan, G. C., Li, Z., Gao, F., & Luo, X. (2018). Universal design of selectivity-enhanced photoelectrochemical enzyme sensor: Integrating photoanode with biocathode. *Analytical Chemistry, 90*(18), 10681–10687. https://doi.org/10.1021/acs.analchem.8b02651.

Sonkar, P. K., & Ganesan, V. (2021). *Metal oxide-carbon nanotubes nanocomposite-modified electrochemical sensors for toxic chemicals* (pp. 235–261). Elsevier BV. https://doi.org/10.1016/b978-0-12-820727-7.00006-9.

Stankovich, S., Dikin, D. A., Piner, R. D., Kohlhaas, K. A., Kleinhammes, A., Jia, Y., et al. (2007). Synthesis of graphene-based nanosheets via chemical reduction of exfoliated graphite oxide. *Carbon, 45*(7), 1558–1565. https://doi.org/10.1016/j.carbon.2007.02.034.

Szabó, A., Perri, C., Csató, A., Giordano, G., Vuono, D., & Nagy, J. B. (2010). Synthesis methods of carbon nanotubes and related materials. *Materials, 3*(5), 3092–3140. https://doi.org/10.3390/ma3053092.

Tang, L., Tsai, C., Gerberich, W. W., Kruckeberg, L., & Kania, D. R. (1995). Biocompatibility of chemical-vapour-deposited diamond. *Biomaterials, 16*(6), 483–488. https://doi.org/10.1016/0142-9612(95)98822-V.

Tung, V. C., Chen, L. M., Allen, M. J., Wassei, J. K., Nelson, K., Kaner, R. B., et al. (2009). Low-temperature solution processing of graphene-carbon nanotube hybrid materials for high-performance transparent conductors. *Nano Letters, 9*(5), 1949–1955. https://doi.org/10.1021/nl9001525.

Turcheniuk, V., Raks, V., Issa, R., Cooper, I. R., Cragg, P. J., Jijie, R., et al. (2015). Antimicrobial activity of menthol modified nanodiamond particles. *Diamond and Related Materials, 57*, 2–8. https://doi.org/10.1016/j.diamond.2014.12.002.

Wang, Z., & Dai, Z. (2015). Carbon nanomaterial-based electrochemical biosensors: An overview. *Nanoscale, 7*(15), 6420–6431. https://doi.org/10.1039/c5nr00585j.

Wang, C., Ding, T., Sun, Y., Zhou, X., Liu, Y., & Yang, Q. (2015). Ni12P5 nanoparticles decorated on carbon nanotubes with enhanced electrocatalytic and lithium storage properties. *Nanoscale, 7*(45), 19241–19249. https://doi.org/10.1039/c5nr05432j.

Wang, H. S., Li, T. H., Jia, W. L., & Xu, H. Y. (2006). Highly selective and sensitive determination of dopamine using a Nafion/carbon nanotubes coated poly(3-methylthiophene) modified electrode. *Biosensors and Bioelectronics, 22*(5), 664–669. https://doi.org/10.1016/j.bios.2006.02.007.

Wang, R., Lu, K. Q., Tang, Z. R., & Xu, Y. J. (2017). Recent progress in carbon quantum dots: Synthesis, properties and applications in photocatalysis. *Journal of Materials Chemistry A, 5*(8), 3717–3734. https://doi.org/10.1039/c6ta08660h.

Xu, C., Wang, X., & Zhu, J. (2008). Graphene—Metal particle nanocomposites. *Journal of Physical Chemistry C, 112*(50), 19841–19845. https://doi.org/10.1021/jp807989b.

Yang, C., Denno, M. E., Pyakurel, P., & Venton, B. J. (2015). Recent trends in carbon nanomaterial-based electrochemical sensors for biomolecules: A review. *Analytica Chimica Acta, 887*, 17–37. https://doi.org/10.1016/j.aca.2015.05.049.

Yang, N., Yu, S., MacPherson, J. V., Einaga, Y., Zhao, H., Zhao, G., et al. (2019). Conductive diamond: Synthesis, properties, and electrochemical applications. *Chemical Society Reviews, 48*(1), 157–204. https://doi.org/10.1039/c7cs00757d.

Zhang, D., Ye, K., Yao, Y., Liang, F., Qu, T., Ma, W., et al. (2019). Controllable synthesis of carbon nanomaterials by direct current arc discharge from the inner wall of the chamber. *Carbon, 142*, 278–284. https://doi.org/10.1016/j.carbon.2018.10.062.

Zhang, R., Zhang, Y., Zhang, Q., Xie, H., Qian, W., & Wei, F. (2013). Growth of half-meter long carbon nanotubes based on Schulz-Flory distribution. *ACS Nano, 7*(7), 6156–6161. https://doi.org/10.1021/nn401995z.

Zhou, J., Shan, X., Ma, J., Gu, Y., Qian, Z., Chen, J., et al. (2014). Facile synthesis of P-doped carbon quantum dots with highly efficient photoluminescence. *RSC Advances, 4*(11), 5465–5468. https://doi.org/10.1039/c3ra45294h.

Zhu, H., Wang, X., Li, Y., Wang, Z., Yang, F., & Yang, X. (2009). Microwave synthesis of fluorescent carbon nanoparticles with electrochemiluminescence properties. *Chemical Communications, 34*, 5118–5120. https://doi.org/10.1039/b907612c.

Zhu, L., Yin, Y., Wang, C. F., & Chen, S. (2013). Plant leaf-derived fluorescent carbon dots for sensing, patterning and coding. *Journal of Materials Chemistry C, 1*(32), 4925–4932. https://doi.org/10.1039/c3tc30701h.

Part II

Fabrication of carbon nanomaterials-based sensors platforms

Chapter 2

Carbon nanomaterial-based sensor: Synthesis and characterization

Mohd Asyadi Azam[a] and Muazzin Mupit[b]
[a]*Fakulti Kejuruteraan Pembuatan, Universiti Teknikal Malaysia Melaka, Melaka, Malaysia,* [b]*Polymer Department, Universiti Kuala Lumpur—Malaysian Institute of Chemical and Bio-engineering Technology, Melaka, Malaysia*

Introduction of carbon nanomaterial-based sensors

What is nano? Typically, "nano" means very small and tiny, something that cannot be seen with the naked eye. In ancient Greek, "nano" means "dwarf." Technically nano means a factor of 10^{-9} or a billionth, which is smaller than "micro." When we talk about nanotechnology, we are talking about engineering, science, and ultraprecise technology that involves a scale of 1–100nm. According to Dr. Noria Taniguchi, nanotechnology began in 1981 after the scanning tunneling microscope (STM) was developed to analyze the real-space surface structure in three-dimensional (3D) formation and unprecedented structure (Tersoff & Hamann, 1985).

After several decades, the word "nanotechnology" is now being printed and aired by the media regarding the existence of this technology that will change the world in every perspective. Currently, this technology exists around us in various ways such as nanomaterials in lotion or cosmetics which inviting the pros and cons issue (Fytianos, Rahdar, & Kyzas, 2020); dirt resistance on paint/glass by nanocoating, the so-called "lotus effect"; and the healthcare, military, and construction industries. These developments show that nanotechnology will bridge the gap between fantasy and reality (Sengupta, Kumar, & Editors, 2015).

Nanomaterials belong to an enormous segment of materials, and their existence has garnered great attention owing to their remarkable physicochemical properties such as electrical conductivity, mechanical stability, and thermal and catalytic properties. These offer great opportunities to fabricate nanomaterial-based sensors or monitoring devices to evaluate environmental contamination in the air, water, and soil. Various nanomaterial-based sensors such as carbon nanotubes (CNTs) (Tigari & Manjunatha, 2020), silicon nanowires, gold nanoparticles, and quantum dots have been explored as detection instruments to precisely and simply measure toxic metal ions, toxic gases, pesticides, and hazardous industrial chemical with high sensitivity (Su, Wu, Gao, Lu, & Fan, 2012).

Recently, nanostructured materials such as graphite (Charithra, Manjunatha, & Raril, 2020), graphene, quantum dots, carbon nanospheres, CNTs (Tigari, Manjunatha, Raril, & Hareesha, 2019), carbon black, and mesoporous carbon (Fig. 2.1) have been shown to be excellent potential carbon-based sensors due to their unique structures, large surface area, high chemical stability, excellent electrical and thermal conductivity, strong mechanical strength, excellent electrochemical properties, and high surface-to-volume ratio (S/V). What makes nanomaterials so interesting is that the greater the S/V, the more the molecules/atoms become exposed to the vicinity of the materials and the larger the number of dangling bonds at the surface, hence making the particles more chemically active.

Carbon nanomaterial synthesis method for sensors

Nowadays, carbon-based electrochemical sensors have aroused remarkable interest in electrochemical detection fields because of their outstanding in mechanical and thermal stability, large surface area, and great conductivity, which means they can accelerate the electron transfer reaction and synergistic response reaction on the electrode surface. The electrode as an electron communication so called medium conversion element to facilitate between electroactive species and data analyzer. Moreover, carbon-based nanomaterial electrodes can improve their electrochemical property functionalizing and consolidating among metal ions and organic metal skeletons (Hu & Zhang, 2020).

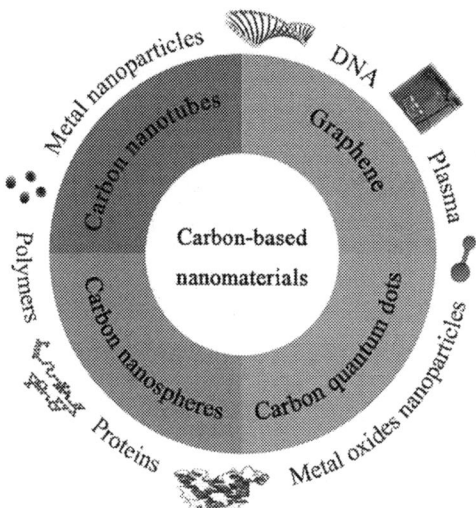

FIG. 2.1 Molecular modification of carbon-based nanomaterials on the detection ability of toxic substances (Xie et al., 2019).

When it comes to the synthesis of nanomaterials, there are two distinctive approaches: top-down or bottom-up techniques. The top-down approach is synthesis from bulk materials broken into smaller parts by exfoliation, either mechanical cleavage or the chemical approach. The bottom-up method is bridging the small collection of atoms that measure in nanometers, as shown in Fig. 2.2. The combination of both methods is expected to perform the best integration apparatus to obtain the elongation of nanoscale fabrication (Habiba, Makarov, & Weiner, 2014).

Top-down technique

This method often applies traditional microfabrication techniques, where the external controlled tools are used to mill, cut, and reshape into the desired formation. Starting the perimeter in the form of macrometer, millimeter or centimeter range size that broken down into smaller unit until the nanometer range that are expected. The top-down approach begins with the large scale of raw materials into the nanoscale dimension and thin film that are larger than 100 nm.

Mechanical milling

The ball mill process is a mechanical-chemical process that uses mechanical force to induce the chemical and structural change of certain materials (Amusat, Kebede, Dube, & Nindi, 2021). A mechanical mill is a type of grinder used to grind and blend a bulk material into extremely fine nanosize powder using different sizes of milling balls. Moreover, the ball milling process is able to reduce particle size and improve the specific surface area (SSA) as well as enhance the adsorption ability of the ball-milled materials (Wei, Wang, Gao, Zou, & Dong, 2020).

The rotating of the ball mill works with the impact and attrition principles between the balls and materials, as shown in Fig. 2.3. Attrition is the collision of metal, alumina, or tungsten carbide balls and materials. The amount of filled ball generally cannot exceed 30–35% from its mill volume to ensure the efficiency of the milling process. The ball mill consists of a

FIG. 2.2 Synthesis of carbon-based nanomaterials by top-down and bottom-up approaches (Soto et al., 2015).

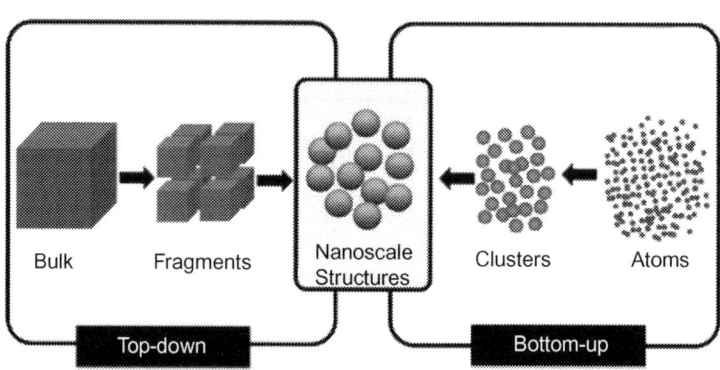

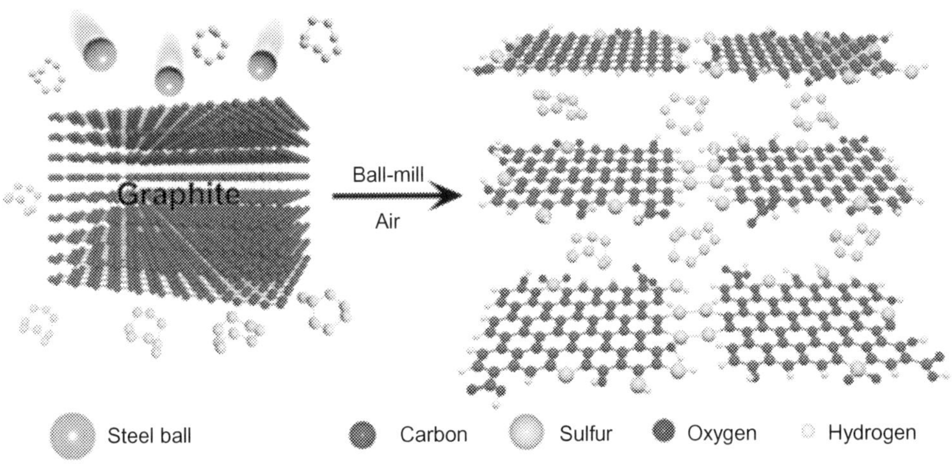

FIG. 2.3 The working principles of the ball mill process (Lithium et al., 2014).

hollow cylindrical shell rotated on the axis containing a stainless steel ball or rubber and the materials. It can be in the form of horizontal, attritor, planetary, or shaker. Some other processes also applied to the ball mill assist in the preparation of porous carbon to mix two or more different kinds of materials such as hexamethylenetetramine ($C_6H_{12}N_4$) and $FeCl_3 \cdot 6H_2O$ in an acetone solvent (Wang et al., 2021).

Lithography

Lithography comes from the Greek word "*lithos*," which means stone, and "*graphein*," which means make a pattern on the stone. In other words, this means layer by layer where technically utilize enhanced visual source in a short wavelength such as intense UV and X-ray for lithography printing to reach between 10 and 100 nm thickness (Sengupta et al., 2015). The main principles of lithography contain several steps. Initially, the photoresist material is spun onto a semiconductor wafer. Then, the photoresist is selectively exposed to ultraviolet (UV) rays through the patterned mask that contains the specific geometry, and the resist is developed. Finally, the specific geometry preserved after the etching process occurs, as shown in Fig. 2.4, where the chemical agent will remove the uppermost substrate layer not protected by the photoresist. Lithography is also considered a hybrid approach where the UV exposure on the patterned mask and the etching process occur by top-down techniques, whereas the growth of nanolayers is a bottom-up approach (Sumanth Kumar, Jai Kumar, & Mahesh, 2018).

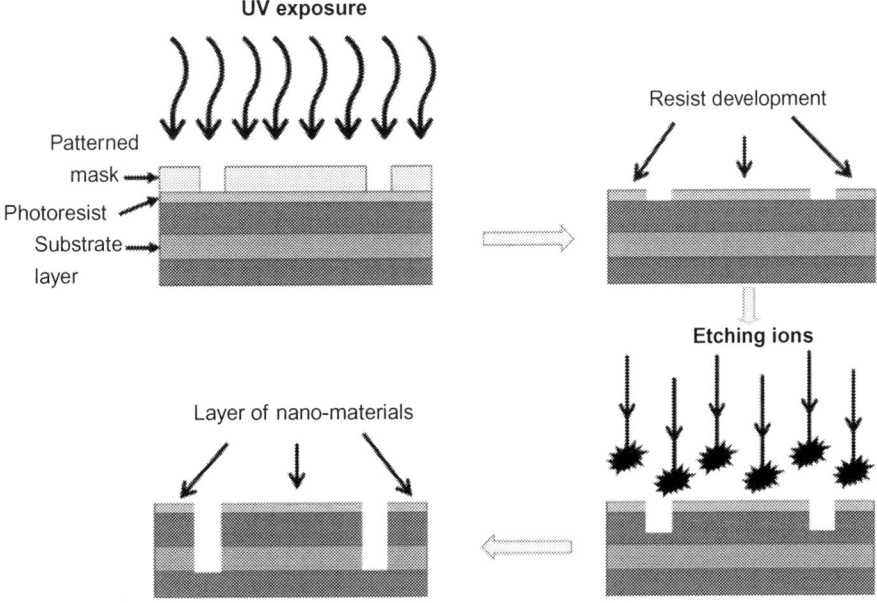

FIG. 2.4 The principle of lithography.

By this technique, the joint-circuit fabrication part is uniformly created and structured with no assemblage required. The lithography is divided into several subdivisions, which are electron-beam lithography (EBL), photolithography, X-ray lithography, soft lithography, nanoimprint lithography, colloidal lithography, scanning probe lithography, ion-beam lithography (IBL), atomic force microscope lithography, and others.

Photolithography was the first lithography technique. It has been applied for several decades to fabricate devices with a length from 15 μm to 180 nm. It was developed by establishing the etching resist directly in the projection of light or onto projected light containing the optical mask. Even though photolithography is good in high-resolution manufacturing, it has some drawbacks in term of cost effectiveness and time consumption. Furthermore, it requires a subtractive process in residual unwanted materials during developing and etching (Kanai & Itoh, 2014). Nowadays, with advancements in lithography, IBL, X-ray lithography, and EBL have been developed with better control and higher resolution (Sengupta et al., 2015).

Electric arc deposition

The cathodic arc deposition is a physical vapor deposition (PVD) technique in which the electric arc is used to vaporize the materials from a cathode target in a closed chamber filled with inert gas. The vaporized materials then condense on a substrate to establish a thin film.

Sputtering deposition

Sputtering deposition is the deposition of thin film by sputtering.

Quantum dots in analysis

In this world, we are surrounded by microscopic objects which some of it are existing in numerous phenomenon in this tiny, microscopic scale which very contradiction in our common-sense taboo and still become *enigma* in most of research entity. The main element to understand the principle of uniqueness of the microscopic world lies in *quantum physics*, as stated by Bohr (1922): "*If you are confused by quantum physics, then you are haven't really understood it*" (*The Nobel Price. The Nobel Prize in Physics 1922*, 2021).

Quantum dots (QDs) are nanoparticles in the range of 1–10 nm. The main element of quantum dots is the ability to manipulate their light absorbance in confined motion and emission frequencies (Haick, 2013).

Bottom-up technique

The bottom-up technique has become common in all aspects of physical and chemical research in the development of fabrication techniques for different types of nanostructures. It begins with single-molecule components that are reshaped into some useful configurations or designed as required from building blocks, either atom by atom or molecule by molecule, as shown in Fig. 2.2. This method utilized the self-assembly approach or molecular recognition approach. Various techniques have been established in the production of nanowires, QDs, thin film, and other derivative or hybrid structures such as nanoribbons and nanostructures that are less than 100 nm (Bezzon et al., 2019). This bottom-up technique can produce several types of structures by tuning the multiplicity of the defining parameters such as evaporation, sputtering, and chemical vapor deposition (CVD) (Azam, Isomura, Fujiwara, & Shimoda, 2012). It is easy to adapt into a variety of synthesis conditions and can be modified to suit many innovative synthesis protocols (Gregorczyk & Knez, 2016).

Sol-gel processes

The sol-gel process is a wet chemical technique that uses traditional methods such as observation to analyze materials (Azam, Zulkapli, Nawi, & Azren, 2015). It is called a wet chemical technique because most of the analysis is done in the liquid phase. This method allows the production of nanomaterials from metal oxide or colloidal-based solutions or tune the chemical compound in the solution in the form of pore volume, surface area and size of grain.

As shown in Fig. 2.5, the sol-gel process involves chemical solution deposition following the order of hydrolysis and condensation, gel formation, aging process, removing the solvent from materials by evaporation/calcination, and finally obtaining crystallization. However, this precursor method is costly with lower production and the possibility of dangerous conditions.

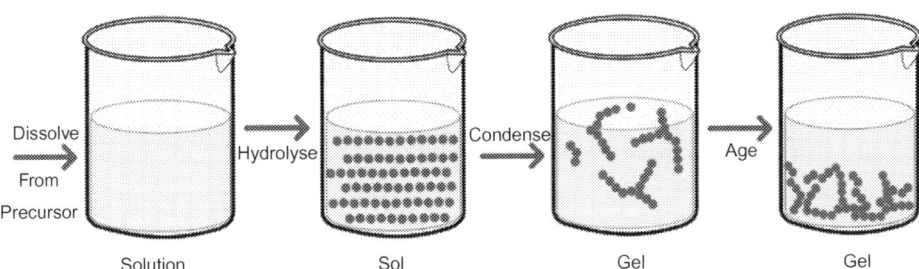

FIG. 2.5 The basic process of the sol-gel method.

Chemical vapor deposition

Chemical vapor deposition (CVD) is a multidirectional vapor deposition technique that applies the bottom-up method to synthesize industrial-scale graphene-based nanomaterials (Jacobberger et al., 2015). Fig. 2.6 shows that the chemical reaction from gases or the vapor phase precursor are pumped into the vacuum chamber and deposited onto the heated substrate to create a thin film in high-temperature conditions. Furthermore, by multidirectional deposition, the more conformal of thin layer deposition can be created even in the complex surfaces. In the electronics industry, the CVD approach is used to produce high performance and enhance the high quality of solid materials, which are mostly applied in semiconductors (Azam et al., 2012).

Micellization

The micellization process is an assembly surfactant molecule that is dispersible in water or liquid, which forms a colloidal suspension because the surfactant is able to lower the interfacial tension. The surfactant monomer consists of a surfactant head (hydrophilic) and tail (hydrophobic). When the surfactant head groups accumulate together, they become a micelle that is spherical in shape, as shown in Fig. 2.7. There are three types of micelles: water dispersed in oil (head-extended out), oil dispersed in water (tail-extended out), or bicontinuous. The micellization activity is done in an aqueous solution that disperses organic solvent that is controlled and stabilized by the surfactant (Fu et al., 2019).

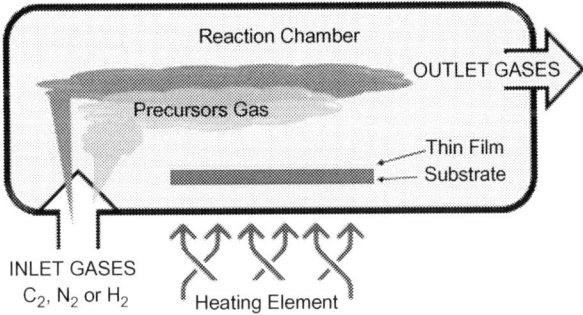

FIG. 2.6 The cross-sectional element in chemical vapor deposition.

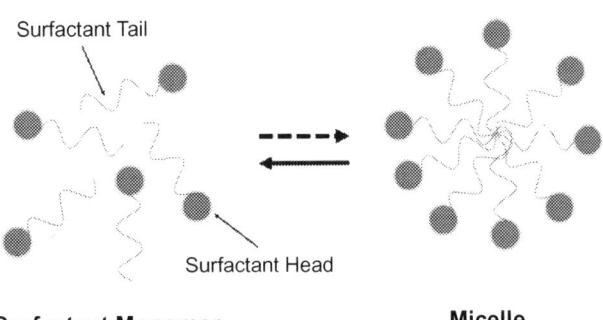

FIG. 2.7 The principle of micelle solubilization.

The microemulsion process (reverse micelle synthesis) uses the micelle method, where it is thermodynamically stable and only requires a simple mixing of components. This method delivers the homogeneous monodispersion of nanoparticles in a variety of metals, oxides, and chalcogenides (Ganguli, Ahmad, & Vaidya, 2008).

Laser pyrolysis

Pyrolysis is the thermal decomposition of substances in a nonreactive atmosphere at high temperature. Laser pyrolysis is the exchange process between the carbon dioxide laser and the movement of the reagent. The tremendous current energy beam between the two leads increases the temperature near the reactive zone by stimulating the degree of molecule vibration. It is a simple procedure that yields well-defined patterns. It has become the most promising technique due to its flexibility and the ability the ablate almost all kinds of material. However, it has more expensive processing costs than arc discharge and chemical vapor deposition.

Characterization of carbon nanomaterial-based sensors

Other than synthesis, in order to understand the nanoparticle morphology, properties, and material structure, the materials must be characterized by different methods. The material inspection can be started with simple visual inspection, physical inspection, or by mechanical properties. With advancements in science and technology, new techniques have been developed such as optical microscopes and the powerful techniques that can analyze and investigate nanomaterials. In this chapter, some of basic characterization techniques will be introduced such as X-ray diffraction, Raman spectroscopy, transmission electron microscopy (TEM), field emission scanning electron microscopy (FESEM), thermogravimetric analysis (TGA), and Brunauer-Emmet-Teller surface analysis (BET).

X-ray diffraction spectroscopy (XRD)

XRD is a method in material science to determine the crystalline structure or sample's composition of certain nanoscale materials (Azam et al., 2013). In a powder sample, the XRD characterization will provide information such as sample purity, phase identification, crystallite size, and, in some cases, the morphology structure (Cameron & Raymond, 2019). In principle, optic diffraction occurs when electromagnetic radiations by mean of incident X-ray irradiating the sample of materials. Then, the scattered X-ray intensity of that material will be measured in the sense of intensities and function of the scattering angle, as shown in Fig. 2.8. The material structure can also be determined from the analysis of the location in angle and the intensities of the scattered peak. The crystalline material pattern, the so-called fingerprint, has been characterized by the periodic arrangement of the atoms.

Because half the wavelength travels at the incident side and the other half at the scattered side, therefore the wavelength is divided into two and becomes:

$$d\,Sine\,\theta = \frac{\lambda}{2} \tag{2.1}$$

where:

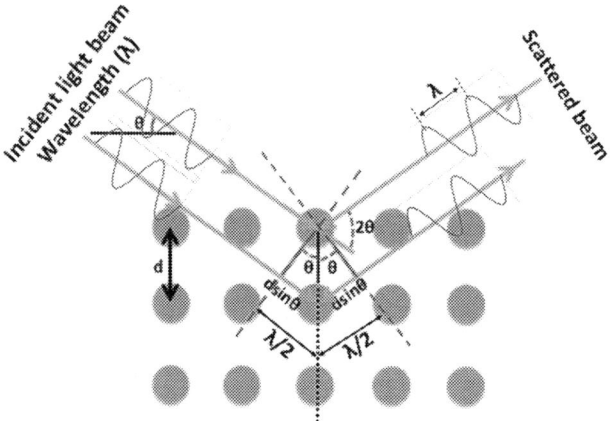

FIG. 2.8 Working principle of X-ray diffraction.

$$Sine\ominus = \frac{\lambda/2}{d} \qquad (2.2)$$

$$\therefore Sine\ominus = \frac{\lambda}{2d} \qquad (2.3)$$

There it become Bragg's law:

$$n\lambda = 2d\,Sine\ominus \qquad (2.4)$$

Where n is the number of wavelengths, λ (lambda) is the wavelength added, $\ominus$ is the angle of diffraction, and d is the distance between the atomic planes. $2\ominus$ is the angle between the incident and the reflected X-ray.

For instance, the XRD of graphite flakes and graphene oxide in Fig. 2.9 shows the intense peak of the (002) plane with a d-space of 3.4 Å at 26.23 degrees that represents the gradual weakening of graphite flakes, which increases the interplanar distance from 3.4 to 7.5 Å. Owing to the decreasing interplanar distance, the diffraction shifted from 26.23 degrees to 11.8 degrees. The enlargement of the interplanar distance shows the establishment of the oxygen functional group in graphene oxide. It can be confirmed by Bragg's equation of $n\lambda = 2d\,Sine\ominus$ where decreasing the interplanar distance is inversely proportional to the diffraction angle of $2\ominus$ (Saleem, Haneef, & Abbasi, 2018).

Raman spectroscopy

Raman spectroscopy is a nondestructive test to provide detailed information about chemical structure, crystallinity, and molecular interaction as well as the level of contamination and the impurities of carbon materials in terms of sp^2 and sp^3 bonds (Azam et al., 2012). Raman is a light-scattering technique from monochromatic radiation (photon emission) that is beamed on the samples and scattered through the materials after collision (absorbed) by the molecules inside. The *Rayleigh scatter* is the light that contains the same wavelength (λ) with the light source, which does not provide useful information.

However, the small amount of light at 0.0000001% scattered through the particles in different wavelengths (λ) of Raman scatter contains a bunch of useful information in the spectrum, as shown in Fig. 2.10. The spectrum comprised of multiple peaks dictates the intensity and wavelength of the Raman scatter. Each peak represents its molecular bond vibration caused by a distinctive scattered light wavelength compared with the source (*What Is Raman Spectroscopy?*, HORIBA, 2021). The common laser wavelength (light spectrum) for Raman is 785 nm near infrared, but recently a green laser with a wavelength of 532 nm has been progressively utilized in Raman spectroscopy (Pilkington, 2020).

Fig. 2.11 shows the Raman spectra of a carbon wood membrane (CWM) depicting two peaks at 1343 and 1582 cm^{-1} that represent the disorder D-band and the graphitic carbon of the G-band, respectively. Because CWM@PANI-70 electrodes show as almost identical for both bands, this means that there are no significant influences in the graphitization degree on the CWM matrix (Wang et al., 2021).

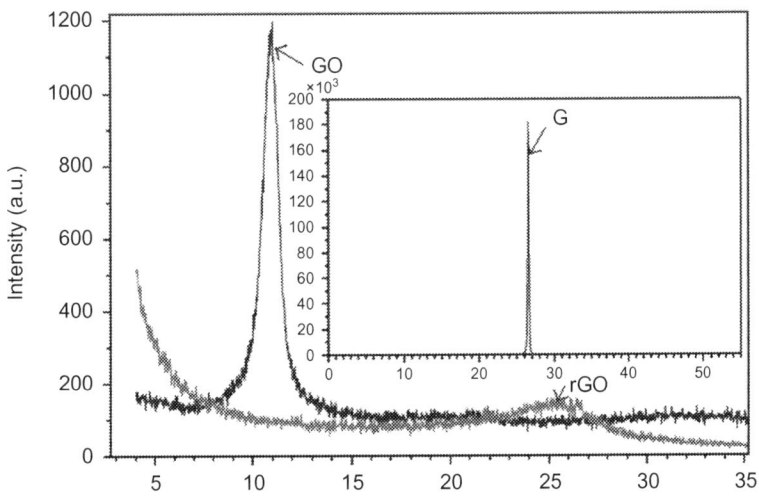

FIG. 2.9 The X-ray diffraction of graphite and graphene oxide (Strankowski, Damian, Piszczyk, & Strankowska, 2016).

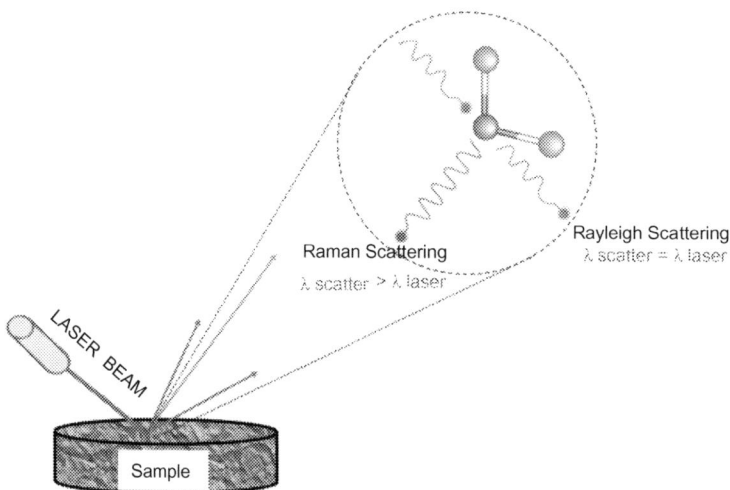

FIG. 2.10 The Raman scatter principle.

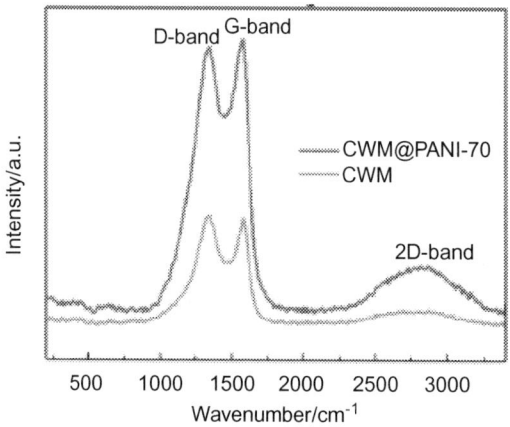

FIG. 2.11 Raman spectra of a carbon wood membrane (Cui, Wang, Zhang, & Min, 2020).

Transmission electron microscopy (TEM)

TEM is an electron microscope that consists of three main elements, as shown in Fig. 2.12. The first is an electron magnetic lens to project the electron beam and condenser lenses that focus the electron beam onto the specimen by controlling the illumination of the sample. Second, image producing consists of an objective lens, a specimen platform, an intermediate lens, and projector lenses below to form a high magnification image from 50 times to as high as 400,000 times. The specimen must be in a high vacuum container to avoid any air particles and gain a better image up to 0.2-nm resolution. The final part is the image-recording system on a fluorescent screen and the image captured on photographic film as permanent records. Fig. 2.13 shows the sample of porous carbon particles that contains several thin layers of nanosheets (a). In addition, high magnification shows several micropores less than 2 nm visualized on the nanosheet surface (Wang et al., 2021).

Field emission scanning electron microscopy (FESEM)

Field emission SEM is a type of electron microscope that visualizes the detailed topography of a material surface by a high-energy electron beam as small as 1 nm. In Fig. 2.14, the signal is derived from free electrons generated by an electron gun that produces an electron beam heated by current toward a tungsten wire. Then the electron beam accelerates the electron energies at 1–40 kV by an anode throughout the magnetic lenses, which focus on the sample in the raster scan pattern. When the beam touches the sample, it produces secondary electrons (SEs), back scattered electrons, and X-rays.

The secondary electron detector (SED) processes the sample interaction and unveils the sample information containing the surface morphology, crystalline structure, and chemical composition. The additional detector of BSE and X-ray will

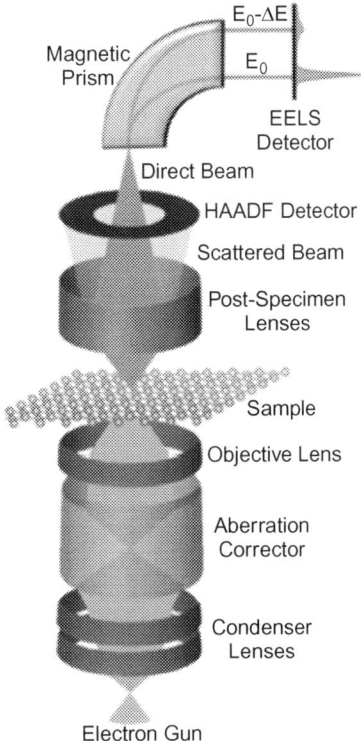

FIG. 2.12 The element in transmission electron microscopy (Hachtel, Lupini, & Idrobo, 2018).

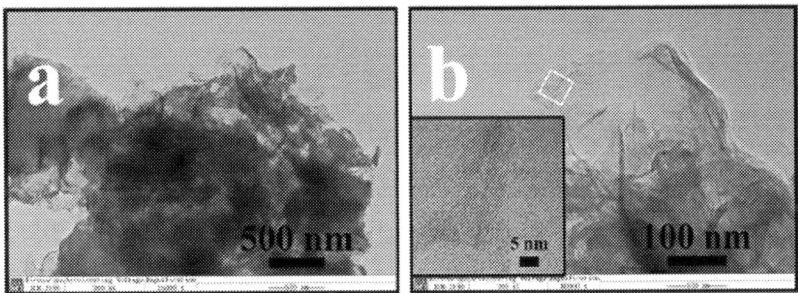

FIG. 2.13 Transmission electron microscopy images of porous carbon as a supercapacitor (Wang et al., 2021).

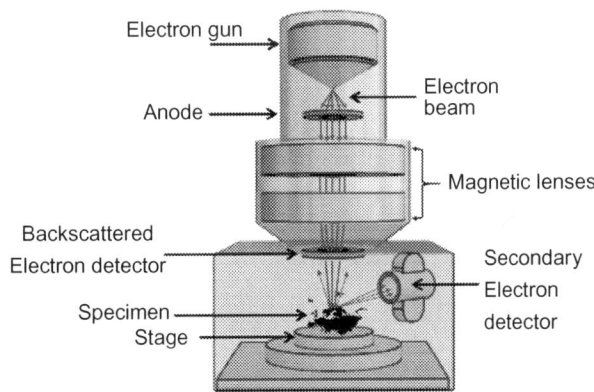

FIG. 2.14 The internal structure of field emission scanning electron microscopy.

produce the corresponding image. Fig. 2.15 shows the FESEM image of pure graphite and graphene oxide specimens, revealing the thin layer of a crumpled sheet-like structure after exposure to high temperature (>300°C). In pure graphite, it shows closely stacked carbon layers owing to the strong attraction of van der Waals forces (Sengupta et al., 2015).

Thermogravimetric analysis (TGA)

Thermogravimetric analysis (TGA) is a thermal analysis measured in time with respect to thermal changes. Fig. 2.16 illustrates the principle of TGA, where the sample is heated in specific gas environments such as N_2, air, CO_2, argon, or helium at a controlled flow rate. The balance operates on the null-balance principle where at zero position, the light shines on the two photodiodes. If an imbalance occurs due to weight changes, the uneven light is emitted toward the photodiodes. Then, the current applied meter return the balance back to null position. The amount of current applied is proportional with changes in the weight sample, either loss or gain due to decomposition, evaporation, desorption, reduction or oxidation, and adsorption/absorption.

An analysis of TGA, as shown in Fig. 2.17, involves graphene oxide (GO) and reduced graphene oxide (rGO). In the range of 30–177°C, there was a 30% mass reduction due to the absorbed moisture, hydroxyl, and carboxylic acid groups. The graph drastically falls at 177°C due to the rapid removal of functional groups, attributed to another 49% mass reduction. In the range of 200–700°C, another 9% mass reduction occurs due to the elimination of the remaining functional groups.

Brunauer-Emmet-Teller surface analysis (BET)

Surface area is a crucial parameter to study in optimizing porous materials. There is no specific method to determine the complex nature of *meso-* or micropore materials. To cater to the requirement above, Brunauer-Emmet-Teller (BET) is a method to calculate the specific surface area (m^2/g) and the pore size of sample powder under the high-vacuum condition of gas. Normally, nitrogen adsorption is used in cold to cryogenic temperature at 77 K to identify the behavior of certain nanomaterials.

The BET theory begins with exposing the sample with the adsorption of gas (nitrogen, argon, or krypton) onto a material's surface. The rise of gas pressure creates the monolayer of adsorbate that consist of ions, atoms, or molecules on the substance surface, which is used to calculate the BET surface area, as shown in Fig. 2.18. The filling of gas starts at the small

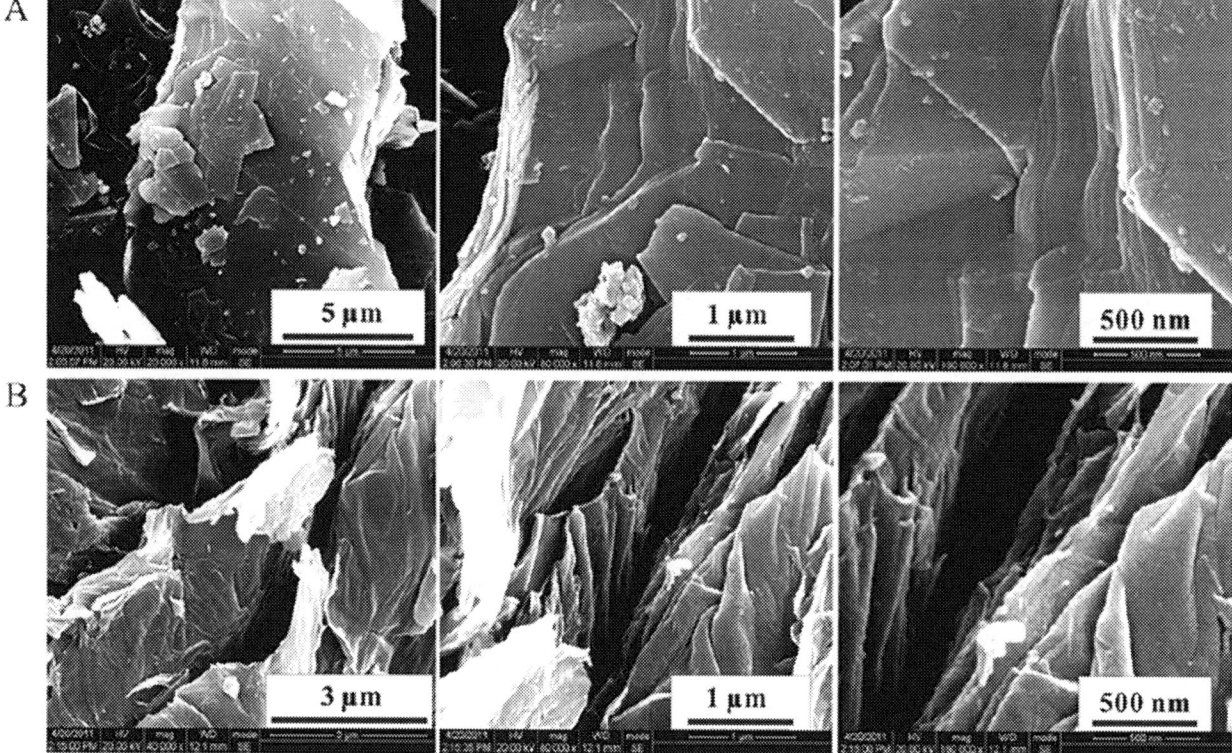

FIG. 2.15 Field emission scanning electron microscopy image of (A) pure graphite and (B) graphene oxide (Huang et al., 2012).

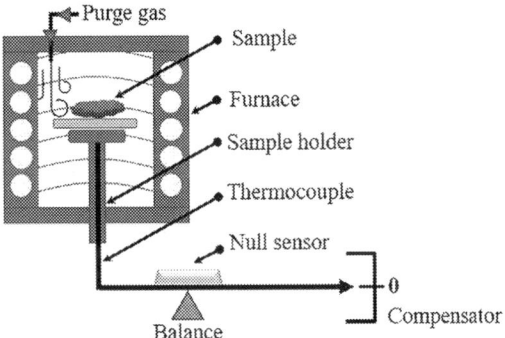

FIG. 2.16 The element of thermogravimetric analysis.

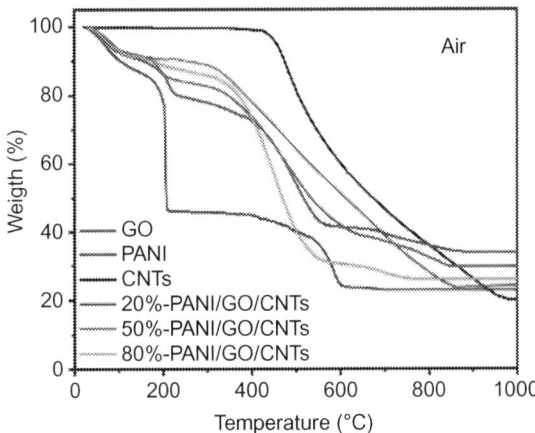

FIG. 2.17 Thermal reduction of polyaniline, GO, carbon nanotube, and the composites (Jiang et al., 2019).

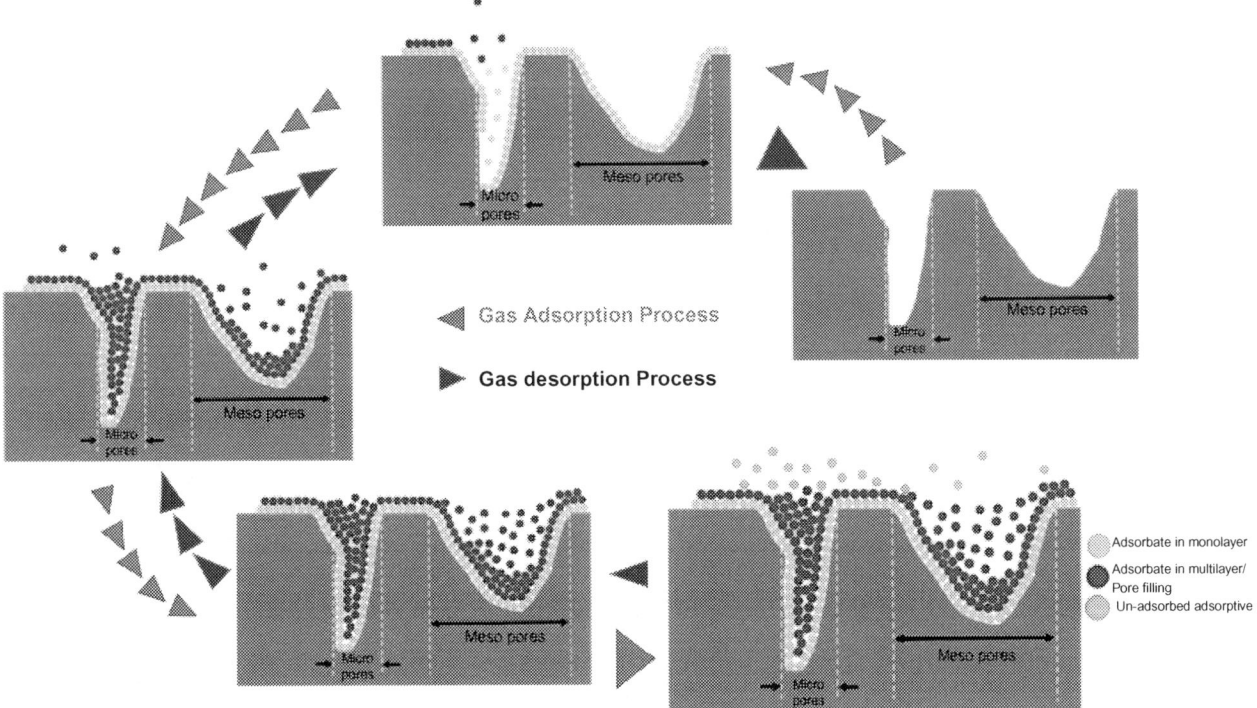

FIG. 2.18 The working principle of Brunauer-Emmet-Teller.

pore (micro) to the larger pore (meso) accordingly. Finally, at the gas desorption process, the real pore measurement is obtained. The amount of adsorbed gas onto adsorbent materials is associated with the surface area (Ambroz, Macdonald, Martis, & Parkin, 2018).

Electrical properties from *I-V* characterization

Cyclic voltammetry (CV) has become a popular and proven technique to study the initial performance of electrochemical capacitors and it is very useful in gaining information about fairly complicated electrode reactions (Bard, Faulkner, Swain, & Robey, 2001). CV is a powerful mechanism of electrochemical technique to measure and investigate the current and voltage by reduction and oxidation reactions as well as studying the qualitative aspect of the electrode process (Brownson & Banks, 2014). Furthermore, CV is also extremely useful in studying the transfer of electrons initiated by chemical reactions, which include catalysis (Carriedo, 1988; Rountree, Mccarthy, Rountree, Eisenhart, & Dempsey, 2017).

The typical cyclic voltammogram curve is shown in Fig. 2.19, where the *x*-axis represents the applied potential (V) while the *y*-axis represents the resulting current (i) passed. There are two conventions commonly used to report the CV data, which are the US convention and the International Union of Pure and Applied Chemistry (IUPAC) convention. Visual data report in both conventions are opposed 180 degrees among each other. The arrow in Fig. 2.19 shows the beginning of the sweep direction and the captions depict the experiment condition.

The CV is performed by cycling the potential of the working electrode and measuring the value of the current. The testing depicts the change of electric potential between the positive and negative electrodes for the two-electrode technique, or among the reference and working electrodes for the three-electrode configuration. By applying the CV, the instantaneous currents during the anodic and cathodic sweeps are recorded to characterize the reaction of the electrochemicals involved. The data are plotted as current (A) versus potential (V) or vice versa (Bard et al., 2001). Fig. 2.20 shows the typical cyclic voltammetry curve for several electrochemical processes.

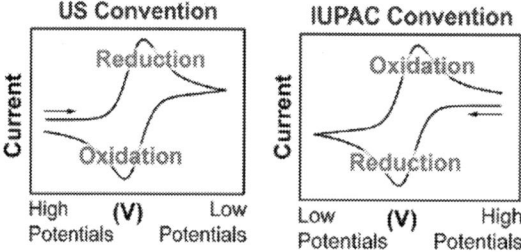

FIG. 2.19 The cyclic voltammogram of US and IUPAC convention (Rountree et al., 2017).

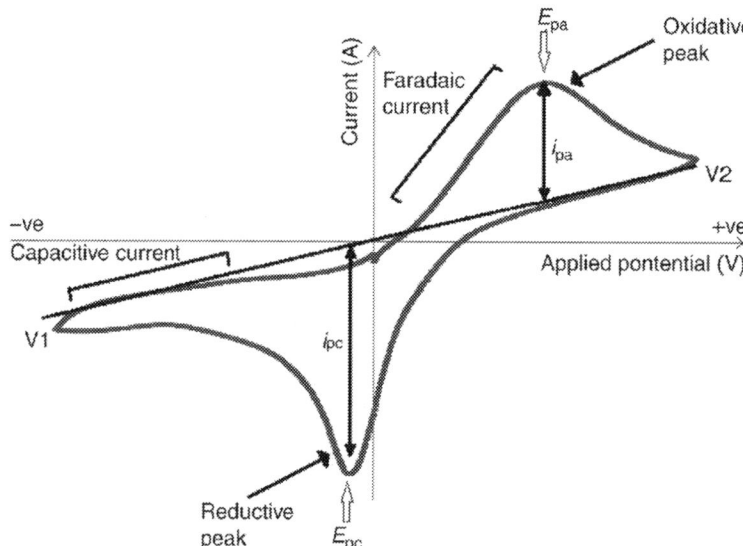

FIG. 2.20 Cyclic voltammogram depicting the peak potential E_P and peak current I_P. (Guy & Walker, 2016).

References

Ambroz, F., Macdonald, T. J., Martis, V., & Parkin, I. P. (2018). Evaluation of the BET theory for the characterization of meso and microporous MOFs. *Small Methods*, *2*(11). https://doi.org/10.1002/smtd.201800173.

Amusat, S. O., Kebede, T. G., Dube, S., & Nindi, M. M. (2021). Ball-milling synthesis of biochar and biochar-based nanocomposites and prospects for removal of emerging contaminants: A review. *Journal of Water Process Engineering*, *41*(2020), 101993. https://doi.org/10.1016/j.jwpe.2021.101993.

Azam, M. A., Rashid, A., Warikh, M., Isomura, K., Fujiwara, A., & Shimoda, T. (2013). X-ray and morphological characterization of Al-O thin films used for vertically aligned single-walled carbon nanotube growth. *Advanced Materials Research*, *620*, 213–218.

Azam, M. A., Isomura, K., Fujiwara, A., & Shimoda, T. (2012). Direct growth of vertically aligned single-walled carbon nanotubes on conducting substrate and its electrochemical performance in ionic liquids. *Physica Status Solidi (A)*, *209*(11), 2260–2266.

Azam, M. A., Zulkapli, N. N., Nawi, Z. M., & Azren, N. M. (2015). Systematic review of catalyst nanoparticles synthesized by solution process: Towards efficient carbon nanotube growth. *Journal of Sol-Gel Science and Technology*, *73*(2), 484–500.

Bard, A. J., Faulkner, L. R., Swain, E., & Robey, C. (2001). *Electrochemical methods: Fundamentals and applications*. John Wiley & Sons, Inc.

Bezzon, D. N., Montanheiro, L. A., De Menezes, B. R. C., Ribas, R. G., Righetti, V. A. N., Rodrigues, K. F., et al. (2019). Carbon nanostructure-based sensors: A brief review on recent advances. *Advances in Materials Science and Engineering*, *2019*, 2019.

Bohr, N. (1922). The Nobel Price. *The Nobel Price in Physic*, *1922*. https://www.nobelprize.org/prizes/physics/1922/bohr/biographical (Accessed 29 May 2021).

Brownson, D. A. C., & Banks, C. E. (2014). The handbook of graphene electrochemistry. In *The handbook of graphene electrochemistry*. https://doi.org/10.1007/978-1-4471-6428-9.

Cameron, F. H., & Raymond, E. S. (2019). Tutorial on powder x-ray diffraction for characterizing nanoscale materials. *ACS Nano*, *13*, 7359–7365. https://doi.org/10.1021/acsnano.9b05157.

Carriedo, G. A. (1988). The use of cyclic voltammetry in the study of the chemistry of metal-carbonyls: An introductory experiment. *Journal of Chemical Education*, *65*(11), 1020. https://doi.org/10.1021/ed065p1020.

Charithra, M. M., Manjunatha, J. G. G., & Raril, C. (2020). Surfactant modified graphite paste electrode as an electrochemical sensor for the enhanced voltammetric detection of estriol with dopamine and uric acid. *Advanced Pharmaceutical Bulletin*, *10*(2), 247.

Cui, M., Wang, F., Zhang, Z., & Min, S. (2020). Polyaniline-filled carbonized wood membrane as an advanced self-supported electrode for superior pseudocapacitive energy storage. *Electrochimica Acta*, *359*, 136961. https://doi.org/10.1016/j.electacta.2020.136961.

Fu, D., Gao, X., Huang, B., Wang, J., Sun, Y., & Zhang, W. (2019). Micellization, surface activities and thermodynamics study of pyridinium-based ionic liquid surfactants in aqueous solution. *Royal Society of Chemistry*, *9*, 28799–28807. https://doi.org/10.1039/c9ra04226a.

Fytianos, G., Rahdar, A., & Kyzas, G. Z. (2020). Nanomaterials in cosmetics: Recent updates. *Nanomaterials*, *10*(5), 1–16. https://doi.org/10.3390/nano10050979.

Ganguli, A. K., Ahmad, T., & Vaidya, S. (2008). Microemulsion route to the synthesis of nanoparticles *. *Pure and Applied Chemistry*, *80*(11), 2451–2477. https://doi.org/10.1351/pac200880112451.

Gregorczyk, K., & Knez, M. (2016). Hybrid nanomaterials through molecular and atomic layer deposition: Top down, bottom up, and in-between approaches to new materials. *Progress in Materials Science*, *75*, 1–37. https://doi.org/10.1016/j.pmatsci.2015.06.004.

Guy, O. J., & Walker, K. D. (2016). Graphene functionalization for biosensor applications. In *Silicon carbide biotechnology* (2nd ed.). Elsevier Inc. https://doi.org/10.1016/B978-0-12-802993-0/00004-6.

Habiba, K., Makarov, V. I., & Weiner, B. R. (2014). Fabrication of nanomaterials by pulsed laser synthesis. *Manufacturing Nanostructures*. https://doi.org/10.13140/RG.2.2.16446.28483.

Hachtel, J. A., Lupini, A. R., & Idrobo, J. C. (2018). Exploring the capabilities of monochromated electron energy loss spectroscopy in the infrared regime. *Scientific Reports*, 1–10. https://doi.org/10.1038/s41598-018-23805-5.

Haick, H. (2013). *Nanotechnology and nanosensors*. Israel Institute of Technology. Technion.

HORIBA. (2021). *What is Raman spectroscopy?*. https://www.horiba.com/en_en/raman-imaging-and-spectroscopy/. (Accessed 25 May 2021).

Hu, J., & Zhang, Z. (2020). Application of electrochemical sensors based on carbon nanomaterials for detection of flavonoids. *Nanomaterials*, *10*(10).

Huang, Y., Zeng, M., Ren, J., Wang, J., Fan, L., & Xu, Q. (2012). Preparation and swelling properties of graphene oxide/poly (acrylic acid-co- acrylamide) super-absorbent hydrogel nanocomposites. *Colloids and Surfaces A: Physicochemical and Engineering Aspects*, *401*(December 2017), 97–106. https://doi.org/10.1016/j.colsurfa.2012.03.031.

Jacobberger, R. M., Machhi, R., Wroblewski, J., Taylor, B., Gillian-daniel, A. L., & Arnold, M. S. (2015). Simple graphene synthesis via chemical vapor deposition. *Journal of Chemical Education*, *92*(11), 1902–1907. https://doi.org/10.1021/acs.jchemed.5b00126.

Jiang, Q., Shang, Y., Sun, Y., Yang, Y., Hou, S., Zhang, Y., et al. (2019). Flexible and multi-form solid-state supercapacitors based on polyaniline/graphene oxide/CNT composite films and fibers. *Diamond & Related Materials*, *92*, 198–207. https://doi.org/10.1016/j.diamond.2019.01.004.

Kanai, R., & Itoh, E. (2014). Fabrication of polymer-based transistors with source-drain electrodes made of carbon nanotubes and silver nanoparticles by soft lithography techniques. *Thin Solid Films*, *554*, 127–131. https://doi.org/10.1016/j.tsf.2013.08.107.

Lithium, P., Batteries, À. S., Xu, J., Shui, J., Wang, J., Wang, M., et al. (2014). Sulfur-graphene nanostructured cathodes via ball-milling for high-performance lithium-sulfur batteries. *ACS Nano*, *8*(10), 10920–10930.

Pilkington, B. B. (2020). *Green laser used for Raman spectroscopy*. https://www.azocleantech.com/article.aspx?ArticleID=1013.

Rountree, K. J., Mccarthy, B. D., Rountree, E. S., Eisenhart, T. T., & Dempsey, J. L. (2017). A practical beginner's guide to cyclic voltammetry. *Journal of Chemical Education*. https://doi.org/10.1021/acs.jchemed.7b00361.

Saleem, H., Haneef, M., & Abbasi, H. Y. (2018). Synthesis route of reduced graphene oxide via thermal reduction of chemically exfoliated graphene oxide. *Materials Chemistry and Physics*, *204*(October), 1–7. https://doi.org/10.1016/j.matchemphys.2017.10.020.

Sengupta, A., Kumar, C., & Editors, S. (2015). *Engineering materials introduction to nano*. http://www.springer.com/series/4288.

Soto, L., Li, K., Yang, M., Zhirkin, A. V., Alekseev, P. N., Batyaev, V. F., et al. (2015). Dense plasma focus—From alternative fusion source to versatile high energy density plasma source for plasma nanotechnology dense plasma focus—From alternative fusion source to versatile high energy density plasma source for plasma. *Journal of Physic*, *591*. https://doi.org/10.1088/1742-6596/591/1/012021, 012021.

Strankowski, M. B., Damian, W. B., Piszczyk, A., & Strankowska, J. (2016). Oxide, FTIR, Raman, and XRD studies. *Journal of Spectroscopy*, *2016*.

Su, S., Wu, W., Gao, J., Lu, J., & Fan, C. (2012). Nanomaterials-based sensors for applications in environmental monitoring. *Journal of Materials Chemistry*, *207890*.

Sumanth Kumar, D., Jai Kumar, B., & Mahesh, H. M. (2018). *Quantum nanostructures (QDs): An overview. In: Synthesis of inorganic nanomaterials*. Elsevier Ltd.

Tersoff, J., & Hamann, D. R. (1985). Theory of the scanning tunneling microscope. *Physical Review B*, *31*(2), 805–813. https://doi.org/10.1103/PhysRevB.31.805.

Tigari, G., & Manjunatha, J. G. (2020). A surfactant enhanced novel pencil graphite and carbon nanotube composite paste material as an effective electrochemical sensor for determination of riboflavin. *Journal of Science: Advanced Materials and Devices*, *5*(1), 56–64.

Tigari, G., Manjunatha, J. G., Raril, C., & Hareesha, N. (2019). Determination of riboflavin at carbon nanotube paste electrodes modified with an anionic surfactant. *ChemistrySelect*, *4*(7), 2168–2173.

Wang, Y., Bai, X., Wang, W., Lu, Y., Zhang, F., Zhai, B., et al. (2021). Ball milling-assisted synthesis and electrochemical performance of porous carbon with controlled morphology and graphitization degree for supercapacitors. *Journal of Energy Storage*, *38*(March). https://doi.org/10.1016/j.est.2021.102496, 102496.

Wei, X., Wang, X., Gao, B., Zou, W., & Dong, L. (2020). Facile ball-milling synthesis of CuO/biochar nanocomposites for efficient removal of reactive Red 120. *ACS Omega*, *5*(11), 5748–5755. https://doi.org/10.1021/acsomega.9b03787.

Xie, F., Yang, M., Jiang, M., Huang, X. J., Liu, W. Q., & Xie, P. H. (2019). Carbon-based nanomaterials – A promising electrochemical sensor toward persistent toxic substance. *TrAC—Trends in Analytical Chemistry*, *119*, 115624. https://doi.org/10.1016/j.trac.2019.11562.

Chapter 3

Novel trends for synthesis of carbon nanomaterial-based sensors

H.C. Ananda Murthy[a] and H.P. Nagaswarupa[b]
[a]Department of Applied Chemistry, School of Natural Science, Adama Science and Technology University, Adama, Ethiopia, [b]Department of Studies in Chemistry, Davangere University, Shivagangothri, Davangere, Karnataka, India

Introduction

Nowadays, with the advancement in industrial, mining, and agricultural activities, a large amount of pollutants were released into the environment causing serious environmental problems. Carbon-based nanomaterials have significantly enhanced the momentum of research work in the field of nanoscience and nanotechnology with beneficial applications in various sectors including catalysis, medicine, energy, and environment. Carbon nanotubes (CNTs) are versatile and have occupied significant position in the nanotechnology applications due to their special physicochemical properties. Carbon can form different types of open and closed cages of materials with a honeycomb structure. Graphene is such an open cage honeycomb carbon structure, and fullerene molecule is one of the closed carbon crates discovered for the first time. Theoretically, carbon nanotubes are obtained by rolling up graphene and are applicable in different research and technology areas. Graphene has received a lot of attention as a multifunctional material in recent years, which is possibly due to its extraordinary properties such as high current density, chemical stability, ballistic transport, optical property, high thermal conductivity, and superior hydrophobicity at the nanoscale. Sensor, in general, is a device that provides information about changes in its environment, whereas a chemical sensor converts chemical or physical properties into measurable signals. The sensors can be classified based on the property to be determined such as electrical, electrochemical, optical, mass-sensitive, magnetic, or thermal (thermometric). Owing to high sensitivity, long-term reliability, high accuracy, fast response, simplicity, low cost, and easy miniaturization, electrochemical sensors have received great attention for their applications in the analysis of species in biological, environmental, industrial, and pharmaceuticals. In the last few decades, various electrochemical sensors were developed from nanomaterials like metals, metal oxides, carbon-based nanomaterials, conductive polymers, and metal-organic frameworks. The surface modifications to these nanomaterials have been investigated to accommodate high loading, high selectivity, and specificity to the targets by recognition molecules such as enzymes, antibodies, and aptamers (Zou et al., 2017).

Nowadays, varieties of sensors have been developed for different applications; these use numerous detection approaches including colorimetry, surface-enhanced Raman scattering (SERS), chemiluminescence (CL), fluorescence and immune chromatographic assays (ICAs), and electrochemical methods(Majdinasab, Mitsubayashi, & Marty, 2019). Herein, we mainly discuss sensors based on carbon-based nanomaterials in which electrochemical methods are applied for different detections.

Nanomaterials are materials that possess nanodimensionality with unique internal structure. They exhibit tunable morphological, structural, morphology, surface chemistry, and bonding features. Therefore, their optoelectronic, physicochemical, and electrochemical properties can be adjusted as per the required application, such as electrochemical sensing of analytes. There are certain nanomaterials that exhibit distinct electrical properties, with superior analytical performance as electrochemical sensors. They also possess features such as biocompatibility and catalytic activity. Both carbon-based nanomaterials and metal-based nanomaterials have been widely utilized in the fabrication of electrodes for electrochemical sensor applications (Dervisevic, Dervisevic, & Şenel, 2019).

Electrochemical sensors

Electrochemical sensors are very much useful in chemistry and medicine to detect and determine the presence of chemical species. In the process of electrochemical sensing, the carbon-based nanomaterials can be used as electrodes or assisting matrices serving certain purposes such as improving the electrocatalytic property, enhancing electron transfer across the interface, and superior biocompatibility for signal enhancement (Cho, Kim, & Park, 2020).

In order to enhance the sensitivity and selectivity of electrochemical sensors, carbon-based nanomaterials can be fabricated for selective interplay with the target analytes and to reap a better signal. This will be achieved due to the increased active sites on the nanomaterial with increased active surface area, improved rate of mass transport, and accelerated rate of electron transfer (Hu & Zhang, 2020).

In addition, to determine the rate of electron transfer at the interface between the target analyte and the electrode and to eliminate the interferences of nontarget analytes, various interfacial materials like metal nanoparticles (noble metals), metal oxide nanoparticles, organic polymers, and carbon-based nanomaterials or their hybrids have been investigated and used for electrochemical sensor fabrications and modifications (Li et al., 2019).

So far, numerous analytical methodologies have been developed to identify and quantify different types of analytes in various sample matrices depending on their chemical characteristics. These analytical methods are believed to offer many advantages toward environmental, food, clinical, medical, and security utilities. At the same time, the advancement in the field of nanostructured materials facilitated the design and fabrication of integrated electrochemical sensors for the rapid and effective sensing of the target analyte with better accuracy and detection limits (Azzouz, Goud, et al., 2019).

This chapter mainly focuses on the detailed discussion on synthesis of advanced nanomaterials, especially carbon-based nanomaterials, which finds applications in different forms of electrochemical sensors to improve their performance in terms of sensitivity, response, and selectivity.

Working principle and classification of electrochemical sensors

Electrochemical sensors are a special type of sensors in which the electrode acts as transducer element and provides quantitative information about analyte chemical species. Electrochemical sensors are operated by reacting with the chemical solutions and by producing an electrical signal. It transforms electrochemical information into an applicable qualitative or quantitative signal. The reactions that occur at the interface of surface of an electrode between the recognition element and the target/binding analyte generate electrical double layer, and thus, this potential is measured after transforming these chemical reactions into this measurable electrochemical signal by a recognition element, and a transducer of the sensor. An electrochemical sensor mainly consists of two main components: One is chemical recognition system responsible for recognizing the analyte species, and the other is a physicochemical transducer, which converts chemical interactions into electrical signals that could be detected and displayed easily by modern electrical instruments (Azzouz, Kailasa, Kumar, Ballesteros, & Kim, 2019). The electrochemical sensors are mainly divided into three types. One of the recent trends in the development of electrochemical sensors is to use nanomaterials leading to substantial enhancement of their performance with respect to sensitivity, selectivity, and rate of detection. A large number of nanomaterials, including carbon nanostructures (e.g., graphite and CNTs), metal and metal oxide NPs, QDs, silica NPs, and other NPs, have been investigated for their applications as electrochemical sensors (Yadav, Chhillar, & Rana, 2020) (Fig. 5.1).

Electrochemical sensor working mechanism involves interaction of the target analyte material with the electrode surface and bringing a desired change as a consequence to a redox reaction, which generates an electrical signal that can be transformed to explore the nature of the analyte species (Fig. 5.2) (Cui, Wu, & Ju, 2015).

The nanomaterial electrode surface can be remarkably transformed by the process of functionalization, which can be executed by the attachment of biomolecules such as aptamers, peptides, and antigen/antibody that functions as the specific chemical recognition element.

Potentiometric sensors

In case of the potentiometric sensors, a potential difference across a pair of electrodes is determined for the analyte species of interest, when no current flows in the system. There are three main classes of potentiometric devices: ion-selective electrodes (ISEs), field-effect transistors (FETs), and coated wire electrodes (CWEs).

Fig. 5.3 (Chen & Chatterjee, 2013) shows the ion-selective electrode, the most common in environmental analysis and representative potentiometric sensor. ISEs are basically membrane-based devices, with a membrane that separates the

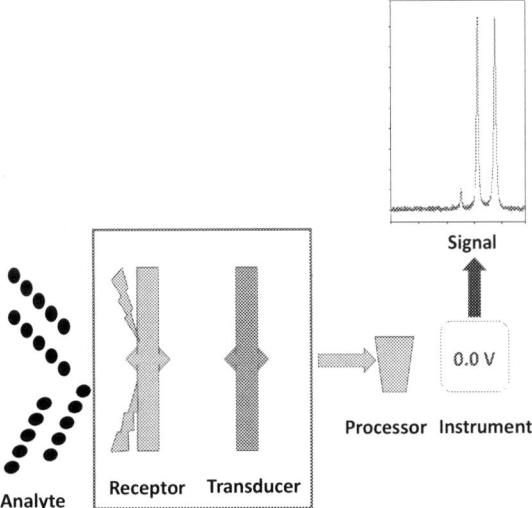

FIG. 5.1 Diagrammatic representation of working principle of electrochemical sensor. *(No Permission Required.)*

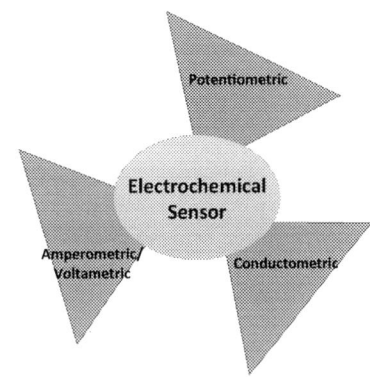

FIG. 5.2 Classification of electrochemical sensors. *(No Permission Required.)*

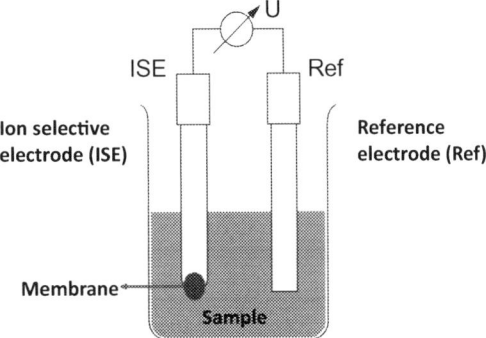

FIG. 5.3 Electrode in electrochemical cell used for potentiometry with ion-selective electrodes.

sample from the electrode. The ions of interest are measured based on the potential developed across the membrane between the analyte solution and the reference electrolyte.

Conductometric sensors

In the case of a conductometric sensor, changes in the electrical conductivity of the selective layer on the electrode are utilized to determine the concentration of the analyte species. The conductometric sensors are basically nonselective in nature.

Amperometric sensors

Amperometric sensor involves the application of small currents with a measurement at a fixed potential using a nonmercury working electrode. Thus, amperometric measurements were conducted by recording the current flow in the cell at a single applied potential. In this case, the analyte involves a faradic reaction at a desired polarity and magnitude of the potential applied. Moreover, faradic reaction is incomplete due to the availability of less surface area of the working electrode, and only a fraction of analyte reacts. A typical amperometric sensing system and its principle are shown in Fig. 5.4.

The detection of varieties of analytes from environment, industries, agricultural fields, pharmaceutical and medical sectors has been very much successful by the application of modern functionalized electrochemical sensors (Fig. 5.5).

In these applications, electrochemical sensors are used to transform electrochemical information into useful, understandable, and displayable signals.

Recently, nanomaterial-based electrochemical sensors having significant applications for different sensing purposes were investigated and reported. Due to their superior sensing phenomenon compared to other analytical approaches, electrochemical sensors based on various nanomaterials such as MOFs, graphene, CNTs, molecularly imprinted polymers (MIPs), metals, and metal oxides have been applied to quantitatively detect proteins, antibiotics, pesticides, heavy metal ions, bacteria, organic compounds, central nervous system (CNS), hormones, etc. (Azzouz, Kailasa, et al., 2019).

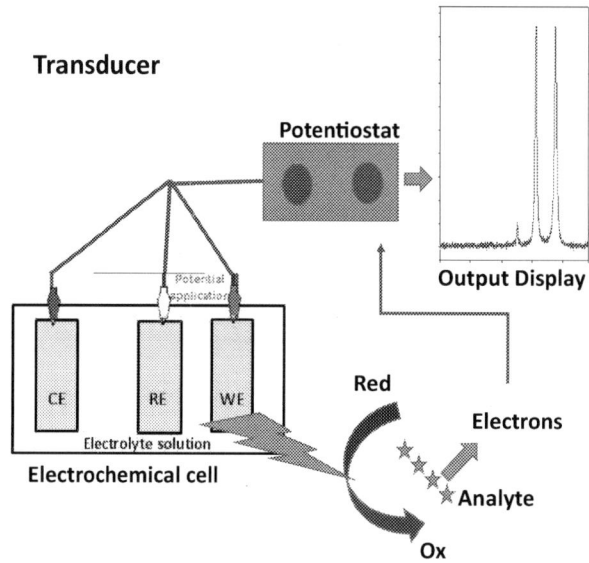

FIG. 5.4 Schematic representation of a typical amperometric sensing system.

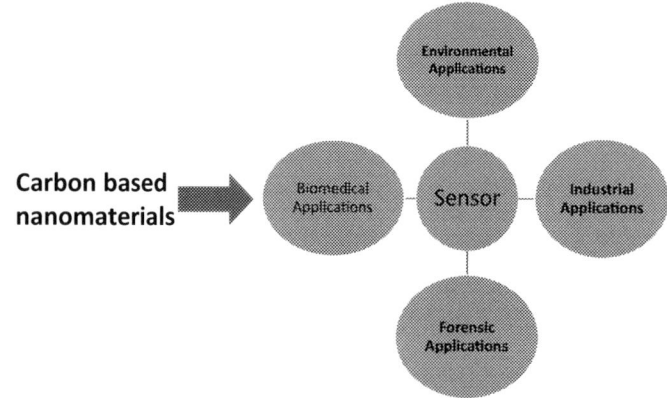

FIG. 5.5 Nanomaterial-based electrochemical sensing applications in different fields. *(No Permission Required.)*

Carbon nanomaterial-based electrochemical sensors

Carbon nanomaterials (CNMs) are commonly used in the fabrication of electrochemical sensors due to their unique properties, which include high levels of electroactive surface-to-volume ratio, electrical conductivity, chemical stability, wide potential window, biocompatibility, electrocatalytic activity for a variety of redox reactions, lower costs, and robust mechanical strength and flexibility. These CNM-based electrochemical sensors are found to exhibit greater sensitivity with low detection limits. The special morphological features of CNMs are believed to have created an extra critical factor that enables and enhances their functionality and stable operation in designing efficient electrochemical sensors. However, carbon-based nanomaterials were found to lack specific functional groups responsible for sensing target analytes. The absence of the functional groups makes them chemically inert. The presence of functional groups is very essential for immobilization of the recognition elements of the electrochemical sensors that are useful for the detection of the target analyte. Various techniques were implemented in the recent past to activate the carbon-based nanomaterials toward effective sensing of analyte species. The presence of functional groups at the surface of CNMs such as carboxyl, hydroxyl, amine, and thiol groups allows the covalent coupling of other molecules or biomolecules. This can be used to improve the analytical performance in electrochemistry, electron transfer efficiency, and sensitivity, response speed, anodic overpotential (lowering), and selectivity of the sensor (electrode) and its biocompatibility (Dang, Hu, Wang, & Hu, 2015). For instance, Fig. 5.6 (Bagheri, Hajian, Rezaei, & Shirzadmehr, 2017) indicates the fabrication of modified GCE with functionalized carbon nanomaterials that was applied for nitrite and nitrate detection and demonstrated in the concentration ranges from 0.1 to 75 M with low detection limits (Tajik et al., 2020).

In most cases, the fabrication of electrochemical sensors requires surface functionalization of the nanomaterials used in the fabrication process. The process of functionalizing nanomaterials involves covalent and noncovalent approaches as per the requirement of various intermolecular interactions. The covalent functionalization approach was found to significantly deteriorate the sp^2 hybrid structure of the carbon honeycomb lattice as a result of the covalent bonding and introduces defects on the structure of CNMs, consequently diminishing the electronic properties. On the other hand, the noncovalent functionalization approach involves the formation of pi bonds initiating pi-pi interaction, electron donor-acceptor

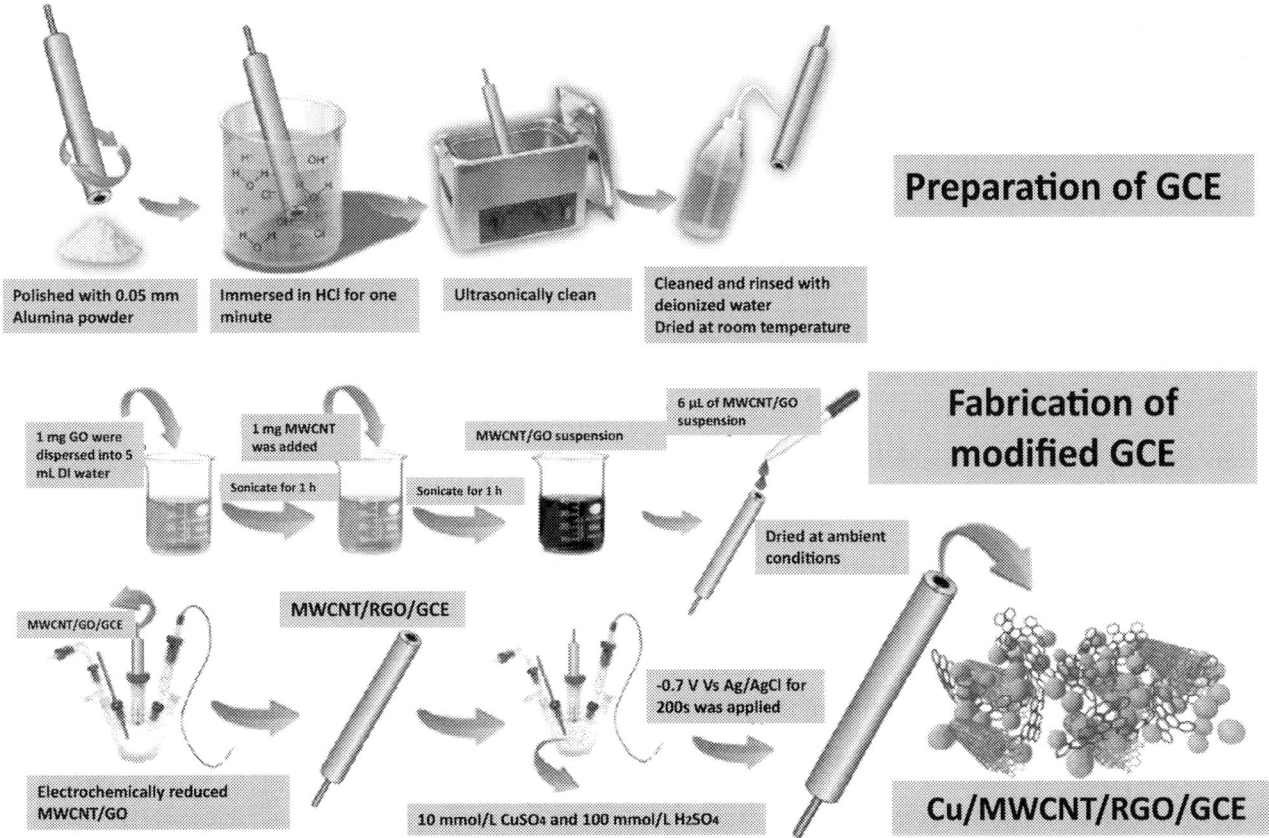

FIG. 5.6 Preparation of GCE and fabrication of modified GCE with Cu/MWCNT/rGO.

complexes, ionic interaction, and hydrogen bonding. This process of noncovalent functionalization does not affect the intrinsic structure, mechanical and electronic properties of CNMs. But, it accompanies the appearance of new functionalization groups at the surface of the material, resulting in the superior biocompatibility, improved dispensability, and sensing properties (Yang, Yang, Yang, & Yuan, 2018). Table 5.1 describes the current performance characteristics of electrochemical sensors based on CNMs.

TABLE 5.1 Recent performance characteristics of electrochemical sensors based on CNMs reported in the literature.

Modified electrode	Detection technique	Target analyte	LR	LOD
PEI-rGO-AuNCs	CV	β-Lactoglobulin	0.01–100 ng/mL	0.08 ng/mL
CTABMGPE	CV/DPV	VN	4×10^{-6}–1.5×10^{-5} and 2×10^{-5}–7×10^{-5} M	1.29 μM
PAMGPE	CV/DPV	CC	2×10^{-6}–8×10^{-6} and 1×10^{-5}–1.5×10^{-4} M	2.4×10^{-7}
OXL-9MGPE	CV/DPV	ERL	4×10^{-5}–1.2×10^{-4} M	1.4×10^{-6} M
PDLPAMCPS	DPV/CV	Riboflavin	6.0–30.0 μM	92 nM
OPEMCNTPE	CV	PL	10–90 μmol	0.71 μmol
PHLDMCPE	CV/DPV	RL	60–150 μM	4.02×10^{-8} M
SDS/CNTPE	CV/DPV	TY	2.0×10^{-6}–5×10^{-5} M	729 nM
SLSMCNTPE	CV	RF	2.0–200 μM	9.25×10^{-8} M
PGMCNTPE	CV	AC	4×10^{-6}–1×10^{-4} M	9.81×10^{-7} M
PEBSDSMCPE	CV	CFX	2×10^{-6}–4.5×10^{-5} M	1.83×10^{-7} M
TX-100 MCPE	CV	IG	1.5×10^{-6}–2×10^{-5} M	16×10^{-8} M
Cu/MWCNT/rGO	SWV	Nitrite	0.1–75 μM	30 nM
		Nitrate		20 nM
NG-PVP/AuNPs modified SPE	SWV	Hydrazine	2–300 μM	0.07 μM
AuNPs/3D-f-GE/GCE	DPV	Nitrite	0.125–20,375.98 μM	0.01 μM
PVP/Gr/GCE	CV	Dopamine	5×10^{-10}–1.13×10^{-3} mol/L	0.2 nM
CeOrGO/GCE	CV	Nitrite	0.7–385 μM	0.18 μM
CB/SPE	SWV	Catechol	1–50 μM	0.1 μM
		Gallic acid		0.8 μM
		Caffeic acid	10–100 μM	1 μM
		Tyrosol		2 μM
rGO/SPE			0.1–600 μM	0.1 ± 0.018 nM

3D CNTs, three-dimensional carbon nanotubes; *3Df-GE*, 3D flower-like structure graphene; *AC*, Alizarin Carmine; *ARS*, Alizarin Red S; *AuNCs*, gold nanoclusters; *AuNPs*, gold nanoparticles; *CB*, carbon black; *CC*, catechol; *CeO*, cerium dioxide; *CFX*, ciprofloxacin; *CTABMGPE*, CTAB-modified graphene paste electrode; *CuNPs*, Cu metal nanoparticles; *CV*, cyclic voltammetry; *DPV*, differential pulse voltammetry; *ERL*, estriol; *GCE*, glassy carbon electrode; *Gr*, graphene; *HQ*, hydroquinone; *CC*, catechol; *RC*, resorcinol; *IG*, indigotine; *MWCNT*, multiwalled carbon nanotubes; *NG*, nitrogen-doped graphene; *OPEMCNTPE*, octyl phenol ethoxylate-modified carbon nanotube paste electrode; *OXL-9MGPE*, octoxynol-9-modified graphite paste electrode; *PAMGPE*, poly-(adenine)-modified graphene paste electrode; *PDLPAMCPS*, poly-DL-phenylalanine-modified carbon paste sensor; *PEBSDSMCPE*, poly(Evans blue) sodium dodecyl sulfate doubly modified carbon paste electrode; *PEI*, polyethylenimine; *PGMCNTPE*, poly(glycine)-modified carbon nanotube paste electrode; *PHLDMCPE*, polymerized helianthin dye-modified carbon paste electrode; *PL*, phloroglucinol; *P-rGO*, porous reduced graphene oxide; *PVP*, polyvinylpyrrolidone; *PVP*, polyvinylpyrrolidone; *RF*, riboflavin; *rGO*, reduced graphene oxide; *rGONR*, reduced graphene oxide nanoribbon; *RL*, riboflavin; *SDS/CNTPE*, sodium dodecyl sulfate-modified carbon nanotube paste electrode; *SLSMCNTPE*, sodium lauryl sulfate-modified carbon nanotube paste electrode; *SPCE*, screen-printed carbon electrode; *SPE*, screen-printed electrode; *SWV*, square-wave voltammetry; *TX-100 MCNTPE*, TX-100-modified carbon nanotube paste electrode; *TX-100 MCPE*, TX-100-modified carbon paste electrode; *TY*, L-tyrosine; *VN*, vanillin.

Carbon nanomaterials can be classified into.

(i) Zero-dimensional (0D) nanoparticles such as quantum dots and fullerenes,
(ii) One-dimensional (1D) nanomaterials such as carbon nanotubes (CNTs) and carbon nanofibers,
(iii) Two-dimensional (2D) nanomaterials such as graphene (Gr), and
(iv) Three-dimensional (3D) nanomaterials such as Gr-CNT foams and hybrids (Kour et al., 2020).

Among these nanomaterials, the most commonly used carbon-based nanomaterials in the fabrications of electrochemical sensors were reviewed.

Graphene

Graphene is a 2D crystal of high quality obtained from a single atom with unique electronic properties. It is a building block for other forms of carbon. Due to its flatness and semiconductivity in addition to its high surface area, high mechanical rigidity, high thermal stability, superior thermal conductivity, electrical conductivity, good biocompatibility, and easy functionalization, graphene is the best candidate in making electrochemical sensors, being able to improve the electrochemical properties of electrodes to be highly sensitive and selective (Arduini, Cinti, Scognamiglio, & Moscone, 2019). These properties of graphene can be tuned and fit to the specifications of the designed electrochemical sensor (Muniandy et al., 2019).

Many types of graphene materials such as single- and multilayer graphene, graphene oxide (GO), reduced graphene oxide (rGO), and graphene quantum dots (GQDs) have been applied for the fabrication of varieties of electrochemical and biosensor applications. Every type of graphene material is found to exhibit different and unique tunable properties (physical, chemical, defect density, electrical, and mechanical), enabling it a futuristic hybrid material for applications as electrochemical sensors with tunable electrochemical properties (Hussain, 2020; Silva, Paschoalino, Damasceno, & Kubota, 2020).

The superior properties of modified graphene materials such as higher surface area and availability of functional groups at the surface compared to CNTs make them apply for enormous electrochemical sensor arrays. The availability of large surface area of graphene nanomaterials enables their ability to adsorb and conjugate with functional groups, or biomolecules, which leads to multifunctional sensor applications.

To improve the efficiency of graphene in making better electrochemical sensor in terms of functionality and applicability, surface modification with suitable functional groups, molecules, biomolecules, or nanomaterials, it is necessary for the fabrication of different electrochemical sensors (Kong et al., 2019). It was shown in some reports that the functionalization of graphene oxide can be accomplished via the formation of amide linkage between GO sheet edges and organic molecules, for example, p-amino thiophenol, polyethylenimine, or imidazole under mild conditions.

Manjunatha (2020a, 2020b) fabricated poly(tyrosine)-modified graphene paste electrode (PTMGPE) utilizing an electropolymerization technique and applied for voltammetric detection of catechol (CC) and its quantification in phosphate buffer solutions of pH 7.0 (PBS). According to his investigation, the detected cyclic voltammetric oxidation current of CC on PTMGPE was nearly 4 times higher with controlled overpotential when compared to bare graphene paste electrode (BGPE). The differential pulse voltammetry (DPV) showed two linear current responses in the concentration range of 2×10^{-6}–1×10^{-5} M and 1.5×10^{-5}–5×10^{-5} M with coefficients of correlation 0.9951 and 0.9976, respectively. The detection limit (DL) and quantification limit (QL) were found to be 3.04×10^{-7} and 10×10^{-7} mol/L, respectively. He also employed the as-prepared modified electrode for the detection of phloroglucinol (PG) and obtained a better result.

Carbon nanotube

Carbon nanotubes (CNTs), 1D carbon nanomaterials, are formed from graphite sheets rolled up into tubes consisting of hexagonal carbon ring units, having a diameter ranging between 1 and 10 nm, and a length ranging between 200 and 500 nm. Two categories of CNTs are mainly available: single-walled carbon nanotube (SWCNT) and multiwalled carbon nanotube (MWCNT). Like other types of carbon nanomaterials, CNTs are known for extraordinary features such as large surface areas, chemical stability, enhanced electronic properties, perfect absorption capability, high electrical conductivity, high aspect ratio, lower overvoltage, simplified functionalization, and rapid electrode kinetics. Owing to these unique properties, CNTs are exceptionally attractive for different types of electrochemical sensors (Baig, Sajid, & Saleh, 2019).

Even though CNTs have several attractive properties and advantages, they remain insoluble in water and other common solvents due to high hydrophobicity. Thus, functionalization is essential to improve the properties of CNTs based on the desired applications. Thus, CNTs' functional properties such as biocompatibility, dispensability, conductivity, mechanical strength, catalytic activity, and optical activity can be altered by fusing them with appropriate functional groups or biomolecules. The functional groups such as —OH, —COOH, and —NH$_2$ can be combined with nanomaterials such as metal nanoparticles, polymers, or supported nanocomposites to fabricate functionalized CNTs that subsequently exhibit enhanced solubility, catalytic activity, and surface area. The functionalized CNTs with few selected functional groups also pave a way for the effective adsorption of biological recognition of molecules for further sensing applications (Sireesha, Jagadeesh Babu, Kranthi Kiran, & Ramakrishna, 2018).

Amrutha, Manjunatha, Bhatt, Raril, and Pushpanjali (2019) fabricated TX-100 surfactant-modified carbon nanotube paste electrode (TX-100 MCNTPE) and with it studied the behavior of Alizarin Red S (ARS), an anthraquinone dye, using cyclic voltammetry (CV) that showed superior electrocatalytic activity toward ARS with two linear ranges observed in the array of 2×10^{-6}–10×10^{-6} M and 15×10^{-6}–35×10^{-6} M. The first linear range was assigned to a limit of detection (LOD) 1×10^{-6} M and limit of quantification (LOQ) 3×10^{-6} M. They also proposed that this sensor can be effectively employed for the simultaneous determination of ARS and Tartrazine (TZ).

The electrochemical oxidation technique is also employed to functionalize CNTs. It opens the closed tips of the pristine CNTs and simultaneously functionalizes them, and as a result, the electrochemical potentiality of carbon nanotubes could be improved (Mohajeri, Dolati, & Rezaie, 2019).

Fullerene

Like other carbon nanomaterials, another allotropic form of carbon, fullerene (C_{60}), has received remarkable importance recently for the development of various sensors. Fullerene (C_{60}) is a zero-dimensional nanomaterial, which consists of 5 and 6 membered rings of sp^2 hybrid carbon atoms with hexagons and pentagons, resulting in football-like structure. These are mostly used as electrode materials. The structural features of fullerenes are quite different from graphite, which consists of planar sheets of hexagons of sp^2 carbon atoms stacked on one other to produce 3D bulk architecture. The fullerenes exhibit superior electrochemical properties due to their tendency to accept and donate electrons under certain conditions. The fullerene-supported electrochemical sensors received significant popularity possibly due to their ability to reproduce catalytic responses, metallic free impurity, and high chemical stabilities. The electrochemical sensing ability of the fullerenes can be improved by functionalization through covalent or noncovalent interaction (Prasad, Kumar, & Singh, 2016).

Rather and De Wael (2013) reported the successful application of fullerene (C_{60}) as a sensor for the ultrasensitive detection of endocrine disruptor BPA. This study revealed excellent electrocatalytic activity by the fullerene sensor with the reduction in the anodic overpotential and leading to notable augmentation of the BPA anodic current in comparison with the electrochemical performance based on GCE. In addition, kinetic parameters such as the electron transfer number (n), electrode surface area (A), diffusion coefficient (D), and charge transfer coefficient (α) were also determined. Based on the results obtained after optimization, it was concluded that the oxidation peak current had a linear relationship in a concentration range between 74 nM and 0.23 μM, with a LOD equal to 3.7 nM. Thus, the authors claimed that fullerene sensor can be successfully applied to detect BPA molecule present in water in trace quantities. Table 5.1 describes the current performance characteristics of electrochemical sensors based on CNMs.

Synthesis of CNTs

CNTs have a proven record of significant applications in the research and development of modern technological applications in various fields. It is the most promising material in the application of nanoadsorbent, catalysis, and sensor for environmental protection. It exists in the form of single-walled nanotube and multiwalled nanotube (Ando, Zhao, Sugai, & Kumar, 2004). The three aforementioned synthesis techniques have been applied for many years for the fabrication of nanotubes. However, the industrial application of carbon nanotubes requires the development of techniques for large-scale and defect-free production where the vapor technique was considered as the promising fabrication technique for large-scale and high-purity production of the nanotube. In this following section, the three main production techniques such as electric arc discharge method, laser ablation method, and CVD method were described separately.

Arc discharge method

The preparation of new needle-like carbon nanotube was reported for the first time in 1991. The same way fullerene was synthesized, the carbon nanotube was also produced by arc discharge evaporation methods. The carbon nanotubes, having a diameter ranging from 3 to 29 nm and several micrometers long, were developed on the cathode end, and the graphite electrode provided for the direct current electric arc discharge evaporation of carbon in an inert gas-filled vessel at 110 Torr

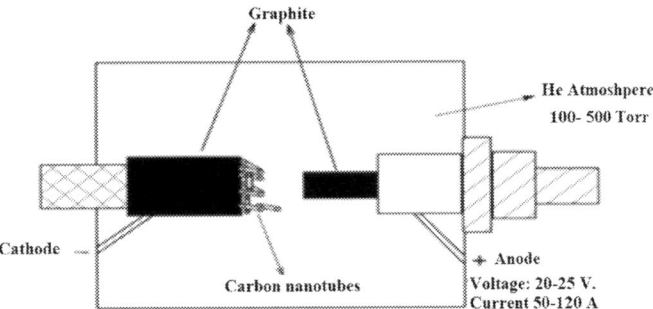

FIG. 5.7 A setup for an arc discharge technique.

(Ibrahim, 2013) as shown in Fig. 5.7. Carbon black was the first carbonaceous material, which was characterized by transmission electron microscopy (TEM), and the image shows that the synthesized nanotube consists of parallel aligned graphitic sheet numbered from 1 to 51 and it was later called multiwalled carbon nanotube. The arrangement of the carbon atom on each of the tubes is in a zigzag fashion about the center axis of the nanotube. The area of the nanotube formed was different from each other, and the edge was most of the time covered by pentagon-, heptagon-, and cone-shaped lids.

Therefore, the individual tube was seen to be grown in a helical fashion at the end of the tube and the TEM characterization result of the arc discharge-synthesized carbon microtubules showed that its growth morphology varied in shape typically around the tube tips. The most common tube observed by the construction of topological model was pentagons (positives) and heptagons (negatives) at the tube tips. When the carbon atoms are captured by dangling bonds, the growth happens layer by layer, and the nanotubes thickens gradually (Popov, 2004), as a result of an open-ended growth mechanism. The change in the growth direction was attained by the nucleation of pentagon and heptagon declinations on open tube, which resulted in giving different morphologies.

The above technique was used for large-scale production of multiwalled nanotubes. The synthesis of carbon nanotubes by using arc discharge method involves the filling of the bored anode electrode by transition metal catalysts such as Fe, Ni, and Co. The potential difference of 20 V and DC of 200 A were applied between two graphite electrodes anode and cathode under high vacuum in the presence of helium atmosphere to generate the arc discharge to synthesize the nanotubes. At a pressure of nearly 500 Torr, maximum yield of 75% was obtained (Das, Ali, Hamid, Ramakrishna, & Chowdhury, 2014). The three important raw materials for the synthesis of SWNTs are argon, iron, and methane. The spherical hat that fits to the cylindrically grown carbon nanotube was the fullerene molecule.

According to the electron microscopic analysis, the nanotube that was catalyzed by cobalt gave a single-walled layer of uniform diameter ranging from 1 to 0.05 nm. The deposited material observed by SEM was found to comprise a tremendous number of aggregated carbon nanotube ropes. The coherent interference of the wave scattered from all atoms in the sample resulted in a diffraction pattern, and the X-ray diffraction result (XRD) (Dore, Burian, & Tomita, 2000) revealed the alternative arrangement of the tube in the ropes. This characterization result indicates that the growth mechanisms of the nanotubes in arc discharge techniques do not depend on the detail of the experimental condition but depend more on the kinetics of carbon condensation under different rate of reaction condition.

Laser ablation method

Transition metals such as Ni and Co were used as catalysts to produce single-walled carbon nanotube by laser ablation vaporization method at 1200°C as presented in Fig. 5.8. The result showed that the nanotubes were highly uniform with

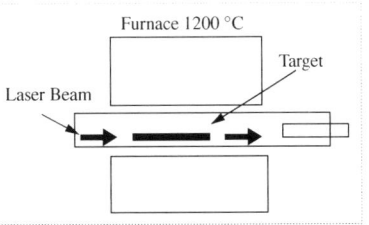

FIG. 5.8 A schematic of laser ablation method.

a diameter range of 3–18 nm and tens to hundreds micrometers long. The lattice constant of the two-dimensional triangular lattice rope formed was about 2 nm through a van der Waals bonding. It was supposed to be a particular tube of (10, 10) as a dominant component and metallic. The scooter mechanism was used to explain the growth of the nanotube where a transition metal atom Ni or Co was adsorbed chemically onto the active site of the nanotube. The formation of fullerene was inhibited by the metal atom having a sufficient electronegativity and catalyzes the reaction to favor the formation of the nanotube.

The growth of the tube was continued until a large amount of the catalyst atoms aggregated on it. Eventually, the catalysis was poisoned by the large particles when they were either detached or become overcoated with sufficient carbon. This way the termination of tube growth takes place with the formation of fullerene-like end or catalyst particles. The advantage of electric arc discharge and laser ablation techniques is the production of about 70% of SWNTs. There are also some disadvantages of these techniques including the use of extremely large temperature around 3000°C to evaporate a carbon atom from a solid target and the entanglement of the nanotube which is difficult to purify. The mixed carbon nanotubes formation has been a persisting problem for a long time. The separation of SWNTs from MWNTs by using a traditional technique has been a challenging activity since they have many common features. In laser ablation-synthesized nanotubes, it was observed that above 70% endless, tangled ropes of SWNTs and nanoscale impurities were present.

The purification method of the grown sample before the cutting of the nanotube was done according to the following procedure. This method consists of the refluxing in 2.6 M nitric acid and resuspension of the nanotube in pH 10 water with the surfactant, followed by a cross-flow filtration system. By using polytetrafluoroethylene as a filter, a freestanding mat of tangled SWNT ropes called buckypaper was formed from the resulting purified SWNT suspension. Of all the cutting techniques available, the prolonged sonification of the nitric acid-purified SWNT rope in a mixture of concentrated sulfuric acid at a temperature of 45oC was considered to be the most efficient one.

Chemical vapor deposition method

This technique is based on the thermal decomposition or chemical vapor deposition of hydrocarbon in the presence of the transition metal catalyst, which was used to produce carbon filament and fibers since the 1960s. This method was also applied to produce multiwalled nanotubes by using the chemical vapor deposition (CVD), which is easily improved and optimized. The hydrocarbon precursor is thermally decomposed under a temperature of about 600–1200°C, and hydrocarbon vapor was passed through a tubular reactor in the presence of the catalyst iron, nickel, and cobalt. Even if these metal catalysts were used in arc discharge and laser ablation techniques, the large-scale production of the nanotube in both cases was impossible. It has been reported that the interfacial force between the catalyst particles and the substrate influences the nanotube growth by tip or base growth mechanism from within the catalyst nanoparticles rooted in the perforation.

Application of transition metal, iron, as a catalyst in the chemical vapor deposition (CVD) technique resulted in large-scale aligned nanotube production. A mixture of 10% acetylene in nitrogen at a flow rate of 100 cm^3/min along with a substrate containing iron nanoparticles rooted in mesoporous silica was placed in the reaction chamber. Carbon nanotubes were formed on the substrate containing iron nanoparticles catalyst by the deposition of carbon atoms obtained from decomposition of acetylene at 850°C. About 40 shells of multiwalled nanotubes arrays with an outer diameter of 30 nm were formed with a consistent spacing of 100 nm between the pores on the substrate. The SEM result showed that the formation of thin film of nanotube grew continuously from the bottom to the top with a film length of 50 and 100 mm. The chemical vapor deposition techniques were used to grow consistent arrays of MWCNTs on a perforated silicon wafer substrate. The electrochemical etching of highly doped n-type Si wafer was used to obtain the porous silicon substrate. The substrate was decorated by Fe films by electron beam evaporation through shallow covered with squared openings having fixed side length and pitch distance. The nanotube with a diameter of 16 nm was formed on the top of the iron catalyst. The CVD process was performed in a tube reactor in the temperature range of 850–1000°C from the precursor ethylene under argon atmosphere. This CVD technique was also applied to produce high-quality SWCNTs on a silicon wafer substrate in the presence of transition metal catalyst under noble gas atmosphere. The most commonly used catalyst is Fe/Mo transition metal catalyst on alumina silica composite materials. The characterization of the sample with TEM revealed that a bundle and individual of SWCNTs were obtained. The SEM image of the sample also showed the high quality with a diameter distribution in the range between 0.7 and 5 nm with a peak at 1.7 nm. The increased metal support was the indicator of the large-scale production of the nanotube, but the weaker the metal support the larger the agglomeration formation (Kumar & Ando, 2010). The section of procedural activity and the schematic diagram of CVD experimental setup for the synthesis of carbon nanotube are shown in Fig. 5.9.

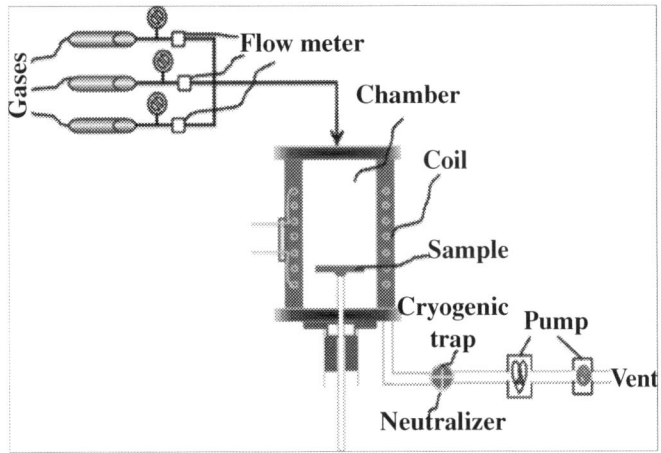

FIG. 5.9 Experimental CVD equipment setup.

Green synthesis of CNTs

Most of the physicochemical methods of synthesis of CNTs, discussed in the previous section, pose serious threat to the environment by introducing unwanted by-products by utilizing toxic chemicals with enhanced energy consumption. In an attempt to develop clean, cheap, biocompatible, and environmentally safe method, researchers ended with a novel process called green synthesis with an intention of exploring biological systems as source for possible conversion of materials into nanomaterials (Murthy, Desalegn, & Kassa, 2020). Green synthetic methods have gained greater significance in the last few decades due to their simple nature, nontoxic and environment-friendly attributes (Desalegn, Ravikumar, & Murthy, 2021).

Plant materials are available in plenty in nature, and hence, their usage for synthesis of CNTs has been explored by many researchers in the last few decades. Green materials provide better defined size and morphology for CNTs than conventional materials. The MWCNTs were recently synthesized by the application of green catalyst synthesized from garden grass (*Cynodon dactylon*), rose (*Rosa*), neem (*Azadirachta indica*), and walnut (*Juglans regia*) plant extracts (Tripathi, Pavelyev, & Islam, 2017). The microscopic images of MWCNTs grown by walnut extract revealed tube diameter of 8 to 15 nm with 3600 μm length. The FESEM images of CNTs demonstrated the formation of ultra-dense network as shown in Fig. 5.10A–D with different magnifications.

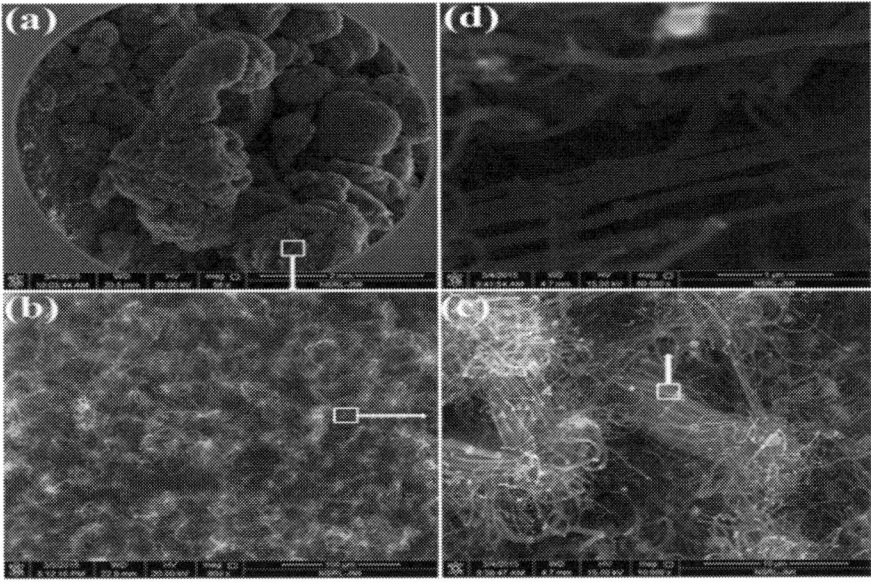

FIG. 5.10 FESEM images of CNTs obtained by green method using walnut extract.

In another study, CNT/PANI nanocomposites were prepared by in situ interfacial polymerization using two green solvents. This green route was successfully used to develop other PANI-based composites for multifunctional applications (Nguyen & Shim, 2015).

In order to overcome the disadvantages of physical methods, an environmentally friendly and convenient method called one-step water-assisted (quenching) green synthesis method was conducted from graphite flakes, obtained from coconut shell wastes to produce CNTs (Hakim, Yulizar, Nurcahyo, Surya, & Green, 2018). The TEM images of these CNTs revealed the presence of carbon tubes with irregular morphology and having average diameter of 123 nm. These CNTs obtained via coconut shell wastes were proved to be better adsorbents for Pb(II) ions. In the case of water treatment, functional CNTs modified with functional groups (chemical groups/polymers) were found to interact with undesirable metal ions. The modification also results in the solubility of the CNTs in respective solutions (Araga, Kali, & Sharma, 2019). Functionalized CNTs with silver nanoparticles were synthesized by the application of fresh holly leaf extract, where the phytochemicals of the extract serve the role of reducing agents. The average silver particle size in CNTs/Ag nanomaterials was found to vary from 7 to 11 nm, while the CNTs showed diameter in the range of 20–50 nm (Cao & He, 2021).

The use of environmentally friendly, renewable green methodologies remains at the core of sustainable development with respect to water, energy, and food security (WEF) issues. CNTs synthesized using green routes are believed to solve the problems associated with the WEF nexus without being harmful to the environment and life (Makgabutlane, Nthunya, Maubane-Nkadimeng, & Mhlanga, 2019). Hollow-structured carbon nanomaterials have attracted tremendous research activities due to the availability of interior void spaces in CNTs. Hierarchically porous CNTs (HPCNTs) were synthesized by green method following simple carbonization treatment without any assistance of soft/hard templates and activation procedures. These HPCNTs served as excellent electrode/host materials for high-performance supercapacitors and Li batteries (Wang et al., 2018).

Synthesis of graphene-based nanomaterials

Looking at the synthesis of graphene and its derivatives, it is important to note that in some simple top-down syntheses, the desired structure and properties are mainly contingent upon the size, shape, and functional groups attached to the surface of the material. Several synthetic routes have been reported through which metal-doped graphene (MDG) materials can be prepared for fabricating hybrid MDGs for electrocatalyst sensor applications. In the context of testing various methods, it is imperative to note the doping of transition metals by using the simple exfoliation of bulk graphite method. The doping of transition metals (Mn, Fe, Co, Ni) (Toh, Poh, Sofer, & Pumera, 2013), uranium (Ur) and thorium (Th) (Sofer et al., 2014), accumulation of platinum (Pt) nanoparticles on the surface of graphene nanosheet using a microwave-assisted method (Ullah et al., 2014) are among the methods in which the metal nanoparticles stuck/adhered to on the surface of graphene sheet. In some studies, transition metal-doped graphene hybrids with new properties were made by exfoliation of metal-doped graphite oxide precursors in either hydrogen or nitrogen atmospheres. Classifications of such materials suggest that the H_2 exfoliated materials had higher C/O ratios. With a similar notion, uranium graphene and thorium graphene hybrids were formed by the reaction of UO_2^{2+} and Th^{4+} ions with graphene oxide and subsequent exfoliation in hydrogen and nitrogen (H_2 and N_2) atmospheres. The characterization results revealed high C/O ratios, which confirm the formation of highly reduced graphene hybrids, predominantly in a hydrogen atmosphere. In a number of experiments, it was shown that graphene layers of very high-quality iron-doped thermally reduced graphene material were used as a magnetic modifier of a carbon-based electrode (Lim, Ambrosi, Sofer, & Pumera, 2014).

The findings of the amazing transport properties of graphene renewed interest in the examination of the metal/graphene interfaces. Here, some studies are ranged from the basic problems, such as the right description of the comparatively weak interaction between graphene and metal surfaces. This in most cases leads to the severe alteration of the electronic structure of graphene to the more useful issues like the organization of the ordered arrays of clusters on the graphene particularly on the surface of 4d and d5 metals. This can be used as an ideal system for the studying of, e.g., catalytic characteristics of MDG materials. In spite of several excellent studies focusing on the modification of MDG-based electrocatalysts and their applications in solving environmental issues, however, an extensive review by Hidalgo-Manrique et al. shows the versatile properties of copper-doped graphene composites synthesized by extensive mixing techniques like mechanical stirring, magnetic stirring, sonication and vortex mixing, electrochemical deposition, and chemical vapor deposition methods. According to the reviewers, in addition to the synthesizing method, the mechanical properties of copper-doped graphene mixtures are highly dependent on many factors. Some of these are processing conditions of graphene nanosheets, the graphene structural defects itself, effect of graphene modification when decorating graphene with nickel, ductility especially when adding high amount of graphene reduces the ductility of the composite. Moreover, microstructure properties of copper-doped graphene composites as well as the electronic and optical properties are affected by controlling the above external factors rather than changing synthetic routes (Hidalgo-Manrique et al., 2019).

Even though many efforts have been devoted on synthesizing MDG materials, considerable challenges are still existing to attain single sheets of graphene from bulk graphite. To solve this problem, many researchers prefer mechanical exfoliation technique. However, to have defect-free graphene, there have been even some difficulties like shape, size, edge, and layers of graphene due to random exfoliation.

Conclusion

Carbon-based nanomaterials have been widely utilized in the fabrication of electrodes for electrochemical sensor applications. The carbon nanotubes (CNTs) and graphene-based nanomaterials with remarkable physicochemical properties exhibit myriad applications, including supercapacitors, batteries, catalysts, and adsorption. Hence, further progress in this endeavor will need novel synthetic methods, which allow control over structure and properties of CNTs and graphene-based nanomaterials. More research could be undertaken to measure the effects of other exfoliation/reduction techniques, and the performances and size of other metals, such as base metals, held on graphene substances to produce at large scale for clinical diagnostic, food safety, pharmaceutical development, and environmental monitoring will progressively appear in the near future. This chapter summarizes the current research trends in the process of developing sensors based on carbon nanomaterials and presents modern synthetic methods for CNTs and graphene derivatives for sensing applications.

References

Amrutha, B. M., Manjunatha, J. G., Bhatt, A. S., Raril, C., & Pushpanjali, P. A. (2019). Electrochemical sensor for the determination of Alizarin Red-S at nonionic surfactant modified carbon nanotube paste electrode. *Physical Chemistry Research*, 7(3), 523–533. https://doi.org/10.22036/pcr.2019.185875.1636.

Ando, Y., Zhao, X., Sugai, T., & Kumar, M. (2004). Growing carbon nanotubes. *Materials Today*, 22–29. https://doi.org/10.1016/s1369-7021(04)00446-8.

Araga, R., Kali, S., & Sharma, C. S. (2019). Coconut-shell-derived carbon/carbon nanotube composite for fluoride adsorption from aqueous solution. *Clean: Soil, Air, Water*, 47(5). https://doi.org/10.1002/clen.201800286.

Arduini, F., Cinti, S., Scognamiglio, V., & Moscone, D. (2019). Nanomaterial-based sensors. In *Handbook of nanomaterials in analytical chemistry: modern trends in analysis* (pp. 329–359). Elsevier. https://doi.org/10.1016/B978-0-12-816699-4.00013-X.

Azzouz, A., Goud, K. Y., Raza, N., Ballesteros, E., Lee, S. E., Hong, J., et al. (2019). Nanomaterial-based electrochemical sensors for the detection of neurochemicals in biological matrices. *TrAC, Trends in Analytical Chemistry*, 110, 15–34. https://doi.org/10.1016/j.trac.2018.08.002.

Azzouz, A., Kailasa, S. K., Kumar, P., Ballesteros, E., & Kim, K. H. (2019). Advances in functional nanomaterial-based electrochemical techniques for screening of endocrine disrupting chemicals in various sample matrices. *TrAC, Trends in Analytical Chemistry*, 113, 256–279. https://doi.org/10.1016/j.trac.2019.02.017.

Bagheri, H., Hajian, A., Rezaei, M., & Shirzadmehr, A. (2017). Composite of Cu metal nanoparticles-multiwall carbon nanotubes-reduced graphene oxide as a novel and high performance platform of the electrochemical sensor for simultaneous determination of nitrite and nitrate. *Journal of Hazardous Materials*, 324, 762–772. https://doi.org/10.1016/j.jhazmat.2016.11.055.

Baig, N., Sajid, M., & Saleh, T. A. (2019). Recent trends in nanomaterial-modified electrodes for electroanalytical applications. *TrAC, Trends in Analytical Chemistry*, 111, 47–61. https://doi.org/10.1016/j.trac.2018.11.044.

Cao, D., & He, H.-Y. (2021). Eco-friendly synthesis and characterisations of single-wall carbon nanotubes/Ag nanoparticle heterostructures. *Materials Research Innovations*, 25(2), 76–82. https://doi.org/10.1080/14328917.2020.1740868.

Chen, A., & Chatterjee, S. (2013). Nanomaterials based electrochemical sensors for biomedical applications. *Chemical Society Reviews*, 42(12), 5425–5438. https://doi.org/10.1039/c3cs35518g.

Cho, I. H., Kim, D. H., & Park, S. (2020). Electrochemical biosensors: Perspective on functional nanomaterials for on-site analysis. *Biomaterials Research*, 24(1). https://doi.org/10.1186/s40824-019-0181-y.

Cui, L., Wu, J., & Ju, H. (2015). Electrochemical sensing of heavy metal ions with inorganic, organic and bio-materials. *Biosensors and Bioelectronics*, 63, 276–286. https://doi.org/10.1016/j.bios.2014.07.052.

Dang, X., Hu, H., Wang, S., & Hu, S. (2015). Nanomaterials-based electrochemical sensors for nitric oxide. *Microchimica Acta*, 182(3–4), 455–467. https://doi.org/10.1007/s00604-014-1325-3.

Das, R., Ali, M. E., Hamid, S. B. A., Ramakrishna, S., & Chowdhury, Z. Z. (2014). Carbon nanotube membranes for water purification: A bright future in water desalination. *Desalination*, 336(1), 97–109. https://doi.org/10.1016/j.desal.2013.12.026.

Dervisevic, M., Dervisevic, E., & Şenel, M. (2019). Recent progress in nanomaterial-based electrochemical and optical sensors for hypoxanthine and xanthine. A review. *Microchimica Acta*, 186(12). https://doi.org/10.1007/s00604-019-3842-6.

Desalegn, T., Ravikumar, C. R., & Murthy, H. C. A. (2021). Eco-friendly synthesis of silver nanostructures using medicinal plant *Vernonia amygdalina* Del. leaf extract for multifunctional applications. *Applied Nanoscience*, 11(2), 535–551. https://doi.org/10.1007/s13204-020-01620-7.

Dore, J., Burian, A., & Tomita, S. (2000). Structural studies of carbon nanotubes and related materials by neutron and x-ray diffraction. *Acta Physica Polonica A*, 98(5), 495–504. https://doi.org/10.12693/APhysPolA.98.495.

Hakim, Y., Yulizar, Y., Nurcahyo, A., Surya, M., & Green. (2018). Synthesis of carbon nanotubes from coconut shell waste for the adsorption of Pb(II) ions. *Acta Chimica Asiana*, 1.

Hidalgo-Manrique, P., Lei, X., Xu, R., Zhou, M., Kinloch, I. A., & Young, R. J. (2019). Copper/graphene composites: A review. *Journal of Materials Science*, 54(19), 12236–12289.

Hu, J., & Zhang, Z. (2020). Application of electrochemical sensors based on carbon nanomaterials for detection of flavonoids. *Nanomaterials, 10*.

Hussain, C. M. (2020). *Handbook of functionalized nanomaterials for industrial applications*. Elsevier.

Ibrahim, K. S. (2013). Carbon nanotubes-properties and applications: A review. *Carbon Letters*, 131–144. https://doi.org/10.5714/cl.2013.14.3.131.

Kong, F. Y., Li, R. F., Yao, L., Wang, Z. X., Li, H. Y., Wang, W. J., et al. (2019). A novel electrochemical sensor based on Au nanoparticles/8-aminoquinoline functionalized graphene oxide nanocomposite for Paraquat detection. *Nanotechnology, 30*(28).

Kour, R., Arya, S., Young, S.-J., Gupta, V., Bandhoria, P., & Khosla, A. (2020). Review—Recent advances in carbon nanomaterials as electrochemical biosensors. *Journal of the Electrochemical Society, 167*(037555).

Kumar, M., & Ando, Y. (2010). Chemical vapor deposition of carbon nanotubes: A review on growth mechanism and mass production. *Journal of Nanoscience and Nanotechnology, 10*(6), 3739–3758. https://doi.org/10.1166/jnn.2010.2939.

Li, X., Tian, A., Wang, Q., Huang, D., Fan, S., Wu, H., et al. (2019). An electrochemical sensor based on platinum nanoparticles and mesoporous carbon composites for selective analysis of dopamine. *International Journal of Electrochemical Science, 14*(1), 1082–1091. https://doi.org/10.20964/2019.01.112.

Lim, C. S., Ambrosi, A., Sofer, Z., & Pumera, M. (2014). Magnetic control of electrochemical processes at electrode surface using iron-rich graphene materials with dual functionality. *Nanoscale, 6*(13), 7391–7396. https://doi.org/10.1039/c4nr01985g.

Majdinasab, M., Mitsubayashi, K., & Marty, J. L. (2019). Optical and electrochemical sensors and biosensors for the detection of quinolones. *Trends in Biotechnology, 37*(8), 898–915. https://doi.org/10.1016/j.tibtech.2019.01.004.

Makgabutlane, B., Nthunya, L., Maubane-Nkadimeng, M., & Mhlanga, S. D. (2019). Green synthesis of carbon nanotubes to address the water-energy-food nexus: A critical review. *Journal of Environmental Chemical Engineering, 9*, 104736.

Manjunatha, J. G. (2020a). A promising enhanced polymer modified voltammetric sensor for the quantification of catechol and phloroglucinol. *Analytical and Bioanalytical Electrochemistry, 12*(7), 893–903. http://www.abechem.com/article_43499_14e9b1642faa3cfb0d7867d6452c363b.pdf.

Manjunatha, J. G. (2020b). Poly (adenine) modified graphene-based voltammetric sensor for the electrochemical determination of catechol, hydroquinone and resorcinol. *Open Chemical Engineering Journal, 14*(1), 52–62.

Mohajeri, S., Dolati, A., & Rezaie, S. S. (2019). Electrochemical sensors based on functionalized carbon nanotubes modified with platinum nanoparticles for the detection of sulfide ions in aqueous media. *Journal of Chemical Sciences, 131*(3).

Muniandy, S., Teh, S. J., Thong, K. L., Thiha, A., Dinshaw, I. J., Lai, C. W., et al. (2019). Carbon nanomaterial-based electrochemical biosensors for foodborne bacterial detection. *Critical Reviews in Analytical Chemistry, 49*(6), 510–533. https://doi.org/10.1080/10408347.2018.1561243.

Murthy, H., Desalegn, T., & Kassa, M. (2020). Synthesis of green copper nanoparticles using medicinal plant *Hagenia abyssinica* (Brace). JF. Gmel. Leaf extract: antimicrobial properties. *Journal of Nanomaterials*, 1–12.

Nguyen, V. H., & Shim, J. J. (2015). Green synthesis and characterization of carbon nanotubes/polyaniline nanocomposites. *Journal of Spectroscopy, 2015*. https://doi.org/10.1155/2015/297804.

Popov, V. N. (2004). Carbon nanotubes: Properties and application. *Materials Science & Engineering R: Reports, 43*(3), 61–102. https://doi.org/10.1016/j.mser.2003.10.001.

Prasad, B. B., Kumar, A., & Singh, R. (2016). Molecularly imprinted polymer-based electrochemical sensor using functionalized fullerene as a nanomediator for ultratrace analysis of primaquine. *Carbon, 109*, 196–207. https://doi.org/10.1016/j.carbon.2016.07.044.

Rather, J. A., & De Wael, K. (2013). Fullerene-C60 sensor for ultra-high sensitive detection of bisphenol-A and its treatment by green technology. *Sensors and Actuators, B: Chemical, 176*, 110–117. https://doi.org/10.1016/j.snb.2012.08.081.

Silva, A. D., Paschoalino, W. J., Damasceno, J. P. V., & Kubota, L. T. (2020). Structure, properties, and electrochemical sensing applications of graphene-based materials. *ChemElectroChem*, 4508–4525. https://doi.org/10.1002/celc.202001168.

Sireesha, M., Jagadeesh Babu, V., Kranthi Kiran, A. S., & Ramakrishna, S. (2018). A review on carbon nanotubes in biosensor devices and their applications in medicine. *Nanocomposites, 4*(2), 36–57. https://doi.org/10.1080/20550324.2018.1478765.

Sofer, Z., Jankovský, O., Šimek, P., Klímová, K., Macková, A., & Pumera, M. (2014). Uranium- and thorium-doped graphene for efficient oxygen and hydrogen peroxide reduction. *ACS Nano, 8*(7), 7106–7114. https://doi.org/10.1021/nn502026k.

Tajik, S., Beitollahi, H., Nejad, F. G., Zhang, K., Le, Q. V., Jang, H. W., et al. (2020). Recent advances in electrochemical sensors and biosensors for detecting bisphenol A. *Sensors, 20*(12), 3364.

Toh, R. J., Poh, H. L., Sofer, Z., & Pumera, M. (2013). Transition metal (Mn, Fe, Co, Ni)-doped graphene hybrids for electrocatalysis. *Chemistry—An Asian Journal, 8*(6), 1295–1300. https://doi.org/10.1002/asia.201300068.

Tripathi, N., Pavelyev, V., & Islam, S. S. (2017). Synthesis of carbon nanotubes using green plant extract as catalyst: Unconventional concept and its realization. *Applied Nanoscience (Switzerland), 7*(8), 557–566. https://doi.org/10.1007/s13204-017-0598-3.

Ullah, K., Ye, S., Zhu, L., Jo, S. B., Jang, W. K., Cho, K. Y., et al. (2014). Noble metal doped graphene nanocomposites and its study of photocatalytic hydrogen evolution. *Solid State Sciences, 31*, 91–98. https://doi.org/10.1016/j.solidstatesciences.2014.03.006.

Wang, J.-G., Liu, H., Zhang, X., Li, X., Liu, X., & Kang, F. (2018). Green synthesis of hierarchically porous carbon nanotubes as advanced materials for high-efficient energy storage. *Small, 14*(13), 1703950. https://doi.org/10.1002/smll.201703950.

Yadav, N., Chhillar, A. K., & Rana, J. S. (2020). Detection of pathogenic bacteria with special emphasis to biosensors integrated with AuNPs. *Sensors International, 1*, 100028.

Yang, Y., Yang, X., Yang, Y., & Yuan, Q. (2018). Aptamer-functionalized carbon nanomaterials electrochemical sensors for detecting cancer relevant biomolecules. *Carbon, 129*, 380–395. https://doi.org/10.1016/j.carbon.2017.12.013.

Zou, C., Yang, B., Bin, D., Wang, J., Li, S., Yang, P., et al. (2017). Electrochemical synthesis of gold nanoparticles decorated flower-like graphene for high sensitivity detection of nitrite. *Journal of Colloid and Interface Science, 488*, 135–141. https://doi.org/10.1016/j.jcis.2016.10.088.

Chapter 4

Development of new electroanalytical method based on graphene oxide-modified glassy carbon electrode for mephedrone illicit drug determination

Maiyara Carolyne Prete[a], Luana Rianne Rocha[a], and César Ricardo Teixeira Tarley[a,b]
[a]Chemistry Department, Londrina State University (UEL), Londrina, Paraná, Brazil, [b]Bioanalytical National Institute of Science and Technology (INCTbio), Campinas State University (UNICAMP), Campinas, São Paulo, Brazil

Introduction

The use and trafficking of illicit drugs are increasing constantly in the last decade, which consequently leads to serious harmful health effects and directly impacts society, economy, criminality, and environment (UNODC, 2019a; Zanfrognini, Pigani, & Zanardi, 2020). New psychoactive substances (NPS) are a complex and diverse group of synthetic substances created to mimic the actions and psychoactive effects of licensed medicines and other controlled substances; thereby, it is often called "legal highs." Nowadays, NPS play a big role in drug use and apprehensions and represent a challenge for researchers, forensic toxicologists, and drug control policy, being described as a growing worldwide epidemic (Shafi, Berry, Sumnall, Wood, & Tracy, 2020). Therefore, there is an increasing interest in developing rapid and accurate methods for the detection of illicit substances in biological fluids, wastewaters, and seized street samples.

Methods generally used to detect drugs of abuse are the ones based on chromatographic techniques, such as HPLC-MS and LC-MS (Bade et al., 2017; Couto et al., 2018; Olesti et al., 2016; Zuba & Adamowicz, 2018). However, these conventional methods involve expensive equipment, are time-consuming, usually require a tedious process for sample preparation, and require trained personnel for operation. Furthermore, these methods are more difficult to be automated and used for in situ analysis. For this reason, the development of methods based on electrochemical sensors is an interesting tool for the determination of drugs of abuse, since they usually are rapid, have high sensitivity, have a low cost, and can be automated (Radhakrishnan & Mathiyarasu, 2018; Shaw & Dennany, 2017; Zanfrognini et al., 2020).

In this sense, several carbon-based nanomaterials, such as carbon nanotubes, graphene, fullerenes, carbon black, and other carbon allotropes, are studied and applied as electrodic materials in electrochemical sensor development, due to their excellent electrical and mechanical properties, such as high surface area, superior electrical conductivity, and their excellent electrocatalytic activity for many electroactive species (Bezzon et al., 2019; Pasinszki, Krebsz, Tung, & Losic, 2017). However, the literature has reported that graphene and graphene-based materials such as graphene oxide (GO) and reduced graphene oxide (rGO) can exhibit better electroanalytical response when compared to other carbon allotropes (Li et al., 2012; Wang, Li, Tang, Lu, & Li, 2009; Yin et al., 2011).

Graphene oxide (GO) consists of pseudo-two-dimensional carbon layers decorated by oxygenated functional groups such as epoxy and hydroxyl on the basal plane of sheets and carboxyl groups at their edges. GO has performance similar to graphene owing to its large effective surface area, high adsorption capacity, strong mechanical strength, heat conductance, and high conductivity. Also, GO can be easily obtained by oxidation and exfoliation of graphite (Lakhe et al., 2020). Besides the presence of oxygenated functional groups generates considerable structural defects that can affect its electronic and chemical properties (Dideikin & Vul, 2019), they promote catalytic activities and make the material strongly hydrophilic. Thus, GO can be easily dispersed in solvents with long-term stability owing to hydrogen bonding between oxygen groups on their surface and solvent interface, which is of paramount importance for electrode modification by drop-casting method (Chen, Feng, & Li, 2012; Khan, Kausar, Ullah, Badshah, & Khan, 2016; Li et al., 2012). GO has been used beneficially in several areas within electrochemistry (Ambrosi, Chua, Bonanni, & Pumera, 2014; Lü, Feng, Zhang, Li, & Feng,

2010; Pumera, 2013), such as in the production of batteries (Li et al., 2012; Wang et al., 2009; Wang et al., 2010; Yin et al., 2011), capacitors (Pope, Punckt, & Aksay, 2011), solar cells (Kavan, Yum, & Grätzel, 2011), and electrochemical sensors (Li et al., 2018; Wang et al., 2009; Yin et al., 2011).

Recently, GO-based electrochemical sensors have been developed for sensitive determination of many species, such as heavy metal ions (Gong, Bi, Zhao, Liu, & Teoh, 2014; Nunes, Silva, & Cesarino, 2020; Vajedi & Dehghani, 2019), flavonoids (Hu & Zhang, 2020), environmental pollutants (Kumunda, Adekunle, Mamba, Hlongwa, & Nkambule, 2021; Li et al., 2012), endocrine disruptors (Gugoasa, Stefan-van Staden, van Staden, Coroş, & Pruneanu, 2019; Yu et al., 2017), pharmaceutical drugs (Mohamed, Atty, Salama, & Banks, 2017; Mohamed, El-Gendy, Ahmed, Banks, & Allam, 2018; Oghli & Soleymanpour, 2020; Qian, Thiruppathi, Elmahdy, van der Zalm, & Chen, 2020), cancer biomarkers (Akbari-Jonous et al., 2019; Crulhas, Basso, Parra, Castro, & Pedrosa, 2019; Zeng et al., 2018), and biosensors (Ahmadi & Ahour, 2020; Pasinszki et al., 2017). In a forensic context, GO-based electrochemical sensors have been developed for the detection of some psychotropic drugs such as cocaine (Hashemi, Bagheri, Afkhami, Ardakani, & Madrakian, 2017; Jiang, Wang, Chen, Xie, & Xiang, 2012), methcathinone (Zang, Yan, Ge, Ge, & Yu, 2013), and morphine (Atta, Hassan, & Galal, 2014; Maccaferri et al., 2019).

Mephedrone (4-methyl-methcathinone; 4-MMC) is an NPS derivative of cathinone naturally found in the *Catha edulis* plant, synthesized to be commercialized as a "legal" alternative to ecstasy (3,4-methylenedioxymethamphetamine, MDMA) for possessing amphetamine-like pharmacological properties, and provides a cheap alternative to psychoactive substances (EMCDDA, 2011; Gibbons & Zloh, 2010). 4-MMC is believed to have entered the market in mid-2007, when its consumption spread, especially in the United Kingdom, becoming the most common widely abused NPS in the recreational drug market (EMCDDA, 2011; Gibbons & Zloh, 2010; McNeill et al., 2021; UNODC, 2019b). Since 2010, the European Council has determined control measures for the 4-MMC across the European Union. Years later, in 2012, 4-MMC was classified as a "Schedule I Drug" in the US Controlled Substances Act (Sacco, 2016). In Brazil, psychotropic substances are described according to National Sanitary Vigilance Agency (ANVISA) (number 344/98), and 4-MMC was included in 2011 (ANVISA, 2020).

Despite the social and criminal problems associated with this drug (Anzillotti et al., 2020; Busardò, Kyriakou, Napoletano, Marinelli, & Zaami, 2015; Papaseit, Olesti, de la Torre, Torrens, & Farre, 2018; Peacock et al., 2019), few electroanalytical methods have been reported in the literature for 4-MMC determination (Krishnaiah, Reddy, Reddy, Reddy, & Rao, 2012; Razavipanah, Alipour, Deiminiat, & Rounaghi, 2018; Smith, Metters, Irving, Sutcliffe, & Banks, 2013; Smith, Metters, Khreit, Sutcliffe, & Banks, 2014; Tan, Smith, Sutcliffe, & Banks, 2015). Therefore, this study aims to develop a new, simple, low-cost, and rapid voltammetric method for the determination of 4-MMC based on GCE modified with GO and dihexadecyl hydrogen phosphate (DHP) with potential in situ detection on seized drugs, which is of great scientific law enforcement officials interest.

Materials and methods

Reagents and solutions

All reagents were of analytical grade, and the solutions were prepared using ultrapure water (18.2 MΩcm) from an ELGA PURELAB Maxima (Woodridge, IL) purification system. Graphite (GR) powder, sulfuric acid (H_2SO_4, 98%), potassium permanganate ($KMnO_4$ $\geq$99%), hydrogen peroxide (H_2O_2, 30%), hydrochloric acid (HCl $\geq$32%), dihexadecyl hydrogen phosphate (DHP), and isopropyl alcohol were purchased from Sigma-Aldrich. NaOH (99%) was purchased from Vetec. Britton-Robinson buffer (BRB) was prepared using boric acid (H_3BO_3, $\geq$99.5%, Vetec), acetic acid (CH_2COOH, $\geq$99.8%, Sigma-Aldrich), and phosphoric acid (H_3PO_4, $\geq$85.0%, Merck). Phosphate buffer solution (PBS) (NaH_2PO_4, 99.0%–102.0%, Merck) and KCl solutions (KCl, 99% Sigma-Aldrich) were prepared from their respective salts. Dimethyl sulfoxide (DMSO $\geq$99%) was purchased from Synth. Mephedrone (4-MMC) was obtained from Lipomed (Arlesheim, Switzerland), and its standard solution was prepared by dissolving its salt in methanol (MeOH, $\geq$99.9%, Merck). Benzocaine (BZC, 99%) and caffeine (CAF, 99%) were purchased from Sigma-Aldrich.

Apparatus

Functional groups of material were evaluated by Fourier-transform infrared spectrometry (FTIR) performed at 4000–400 cm^{-1} region using a spectrophotometer model 8300 Shimadzu (Tokyo, Japan), using KBr pellets conventional method. Raman analysis was measured using a DeltaNu Advantage1 laser Raman spectroscopy system in the range of 200–3400 cm^{-1}, and the wavelength of the excitation beam was fixed at 532 nm with an 8 cm^{-1} resolution. Voltammetric measurements were carried out with a potentiostat/galvanostat Metrohm Autolab BV Autolab PGSTAT101 (Eco Chemie BV, Utrecht, Netherlands), controlled by software Nova 2.1.4 using a conventional electrochemical cell containing three electrodes: a GCE modified with GO-DHP as the working electrode, Ag/AgCl (KCl $3.0\,\text{mol}\,\text{L}^{-1}$) as the reference electrode, and a platinum wire as the auxiliary electrode. The pH measurements of the working solutions were performed with a Metrohm pH 827 lab digital pH meter (Herisau, Switzerland). The suspensions were dispersed in a QUIMIS ultrasonic bath Model Q335D2 (50/60 Hz frequency with 2.4 L capacity) (Diadema, São Paulo, Brazil). Electrochemical impedance spectroscopy (EIS) measurements were performed using potentiostat/galvanostat Palm Instruments BV PalmSens4 (Houten, Netherlands), controlled by software Palm Instruments BV PSTrace 5.3 (Houten, Netherlands), with a $1.0\,\text{mmol}\,\text{L}^{-1}$ $[Fe(CN)_6]^{3-}/[Fe(CN)_6]^{4-}$ in $1.0\,\text{mol}\,\text{L}^{-1}$ KCl, in the frequency ranging from 0.1 to 100 kHz and an a.c. signal of 10.0 mV in amplitude as the perturbation. The fitting of the spectra to equivalent theoretical circuits was made using the EIS Spectrum Analyzer 1.0 software (Marques et al., 2021). Chromatographic measurements were performed on high-performance liquid chromatography (HPLC) with a diode array detector (DAD) (Shimadzu Prominence LC-20AD/T LPGE KIT, Tokyo, Japan) operating at 264 nm, equipped with a Gemini C_{18} chromatographic column ($5.0\,\mu m \times 250.0\,mm \times 4.6\,mm$) from Phenomenex (Torrance, California, EUA) using methanol and ammonium formate buffer (10.0 mM, pH 3.5) at 60:40 (% v/v) at a flow rate of $0.8\,\text{mL}\,\text{min}^{-1}$ as mobile phase with an injection volume of 20.0 μL. This procedure was modified from Smith et al. (2014).

Synthesis of graphene oxide

GO was synthesized directly from graphite by a modified Hummers' method (Oliveira, Braga, Tarley, & Pereira, 2018). Generally, 3.0 g of graphite (GR) was mixed with 3.0 g of $NaNO_3$. Then, 138.0 mL of H_2SO_4 (98%) was added to the mixture and it was stirred for 4 h into an ice bath. After this time, 18.0 g of $KMnO_4$ was added slowly into the solution under stirring for 1 h. The ice bath was removed, and the mixture was stirred at 35°C for 1 h. Then, 276.0 mL of H_2O was added to the mixture and kept at 90°C for 2 h. The mixture was stirred until it reached room temperature (25°C). Then, 600.0 mL of H_2O was added, and the mixture was stirred for 1 h. Finally, to eliminate the excess of $KMnO_4$, 300.0 mL of H_2O_2 30% (v/v) was added.

For impurities removal, the material was repeatedly washed with 5% (v/v) HCl and H_2O. For this task, centrifuge tubes with GO suspension were filled with 5% HCl solution, kept in an ultrasonic bath for 3 min, and then centrifuged for 6 min for four times. Then, the same procedure was repeated with H_2O until pH above 4.5 was reached. The schematic illustration of the GO synthesis is shown in Fig. 4.1.

Preparation of GCE/GO-DHP sensor

GO-DHP suspension was prepared by dispersing 2.0 mg of GO and 2.0 mg of DHP in 2 mL of DMSO under ultrasonication for 1 h. Before modification, a bare GCE was polished to form a mirror-like surface with 0.05 μm alumina slurry on microcloth pads and washed with isopropyl alcohol and water in an ultrasonic bath, and dried in air. Then, 8.0 μL of the prepared suspension was drop-cast onto the GCE surface and left to dry at 40°C in an oven. After dried, the electrode was rinsed with ultrapure water. The prepared electrode was denoted as GCE/GO-DHP. Before every measurement, the electrode was treated in the blank (supporting electrolyte) by cyclic voltammetry for 30 cycles in the range of −1.5 to 0.0 V with a scan rate of $50\,\text{mV}\,\text{s}^{-1}$.

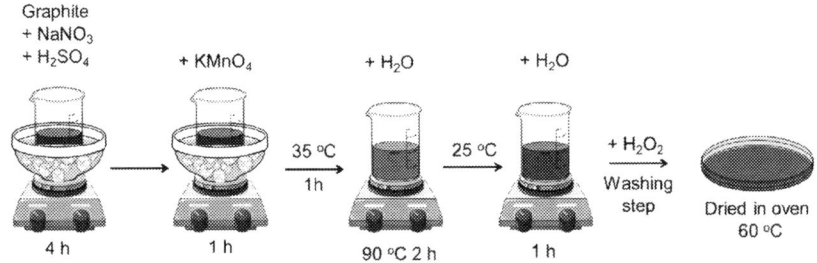

FIG. 4.1 Schematic illustration showing the steps undertaken for GO synthesis from graphite by a modified Hummers' method. *(No Permission Required.)*

Electrochemical measurements

Electrochemical characterization of GCE and GCE/GO-DHP was carried out by cyclic voltammetry (CV) and electrochemical impedance spectroscopy (EIS) using $1.0\,mmol\,L^{-1}$ of the redox probe $[Fe(CN)_6]^{3-}/[Fe(CN)_6]^{4-}$ in $1.0\,mol\,L^{-1}$ KCl. Electrochemical behavior of 4-MMC on GCE and GCE/GO-DHP, optimization of experimental parameters such as GO/DHP suspension proportion and concentration, preconcentration time and applied potential, pH, and supporting electrolyte type and concentration were evaluated by CV technique at a scan rate of $50.0\,mV\,s^{-1}$. Differential pulse voltammetry (DPV) was used for obtaining the analytical features at the following optimized conditions: pulse amplitude of $50.0\,mV$, pulse time of $50.0\,ms$, and scan rate of $50.0\,mV\,s^{-1}$.

Under optimized DPV parameters, an analytical curve was obtained, and the analytical performance of the method was assessed by determining the limit of detection (LOD), limit of quantification (LOQ), and interday/intraday precision. The LOD and LOQ were calculated according to IUPAC recommendation, described as *3std/m* and *10std/m*, respectively, where *m* is the slope of the analytical curve and *std* is the standard deviation of measurements of the blank (Long & Winefordner, 1983). The interday (two consecutive days, $n=2$) and intraday ($n=10$) precisions were calculated with solutions containing 5.0 and $10.0\,\mu mol\,L^{-1}$ of 4-MMC and assessed in terms of percentage of relative standard deviation (%RSD).

Street sample simulation

In order to verify the feasibility of the proposed method, street samples were simulated according to the samples analyzed by Smith et al. (2014). For this task, Sample A was prepared by dissolving 2.10 mg of 4-MMC and 27.9 mg of BZC in 10.0 mL of MeOH, and Sample B was made by dissolving 2.10 mg of 4-MMC and 17.9 mg of CAF in 10 mL of MeOH. Electrochemical measurements were carried out by transferring $50.0\,\mu L$ of each sample into the electrochemical cell containing 10.0 mL of BRB at pH 6.0 using the DPV technique at optimized conditions.

In order to check the accuracy of the proposed method, the simulated street sample was also analyzed by HPLC-DAD. Thus, $50.0\,\mu L$ of each sample was diluted in 10.0 mL of MeOH, filtered in $0.45\,\mu m$ nylon membrane, and injected directly into the chromatographic system.

Results and discussion

Characterization of GO

Structural changes after the oxidation process of graphite were evaluated by Raman and FT-IR measurements. The Raman spectra for GR and GO are presented in Fig. 4.2. The D band at $1358\,cm^{-1}$ is associated with the presence of sp^3 carbons and therefore with the defects or vacancies in the carbon structure. On the other hand, the G band at $1586\,cm^{-1}$ is related to the presence of sp^2 carbons, which are associated with a crystalline structure. The 2D band at $2700\,cm^{-1}$ corresponds to the overtone of D band and is related to the number of layers in the carbon structure. The I_D/I_G ratio is a measure of the disorder and defect sites in the graphite carbon (Amini, Kalaantari, Garay, Balandin, & Abbaschian, 2011; Ferreira Oliveira, César Pereira, Braga Bettio, & Teixeira Tarley, 2019).

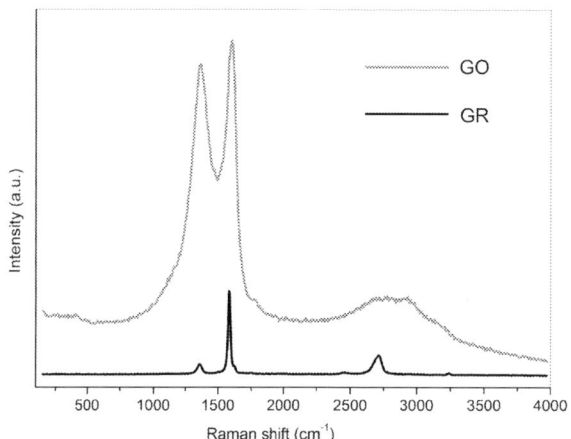

FIG. 4.2 Raman spectra for GR and GO materials. *(No Permission Required.)*

It is observed that GR spectrum exhibits an intense G band at 1586 cm^{-1} corresponding to the sp^2 hybridization of the graphene sheets in a two-dimensional network on the graphite structure. The D band can be observed at 1358 cm^{-1} with low intensity since the GR has low defects and vacancies in its structure. Regarding the 2D band, it is observed at 2700 cm^{-1} with medium intensity, which means that there are a high number of layers. The I_D/I_G ratio for GR is 0.67, indicating that the material structure is ordered.

After the oxidation process, it is observed an increase in the D band in the GO spectrum. During the oxidation, the introduction of functional groups transforms sp^2 bonds into sp^3, which creates defects and vacancies in the structure. The 2D band is inexistent due to the low number of layers in the GO structure. Also, the I_D/I_G ratio increases to 0.96, indicating that the GO structure becomes more disordered. These results confirm that the oxidation of GR occurred, and the GO was obtained.

FTIR spectra of GR and GO are presented in Fig. 4.3. In GR spectrum, it is observed a low intensity broad band at 3430 cm^{-1} assigned to the O—H stretching of adsorbed water (Oliveira et al., 2018). The bands at 1630 cm^{-1} and 1565 cm^{-1} are attributed to the stretching of the C=C and C=C—C bonds, respectively, from aromatic rings (Coates, 2006). The band at 1400 cm^{-1} may be attributed to the angular bending of O—H groups from carboxylic acids (Navarro-Pardo et al., 2013), while the band at 1050 cm^{-1} is attributed to the C—O bending from epoxy groups (Guo, Wang, Qian, Wang, & Xia, 2009). The presence of these two last bands at low intensity shows that few functional groups are presented in GR structure. At GO FTIR spectrum, it is observed an increase in band intensities at 3430 cm^{-1}, 1400 cm^{-1}, and 1050 cm^{-1} due to the insertion of oxygenated groups, such as alcohols, carboxylic acids, aldehydes, ketones, esters, and epoxies. The band at 1630 cm^{-1} can be attributed to the vibration of the nonoxidized graphitic chain and the stretching of the adsorbed water bonds, since the material becomes more hydrophilic (Navarro-Pardo et al., 2013). Also, it is possible to observe the appearance of an intense band at 1740 cm^{-1}, which refers to the stretching of the C=O bonds.

Electrochemical characterization of GCE/GO-DHP sensor

Upon GO obtention, a carbon glassy electrode (CGE) was modified with a GO/DHP (1:1 m/m) suspension at 1.0 mg L^{-1} concentration for electrochemical studies. Cyclic voltammetry (CV) and electrochemical impedance spectroscopy (EIS) in the presence of 1.0 mmol L^{-1} of the redox probe [Fe(CN)$_6$]$^{3-}$/[Fe(CN)$_6$]$^{4-}$ in 1.0 mol L^{-1} KCl were used to evaluate the behavior of GCE/GO-DHP and bare GCE toward redox process of [Fe(CN)$_6$]$^{3-}$/[Fe(CN)$_6$]$^{4-}$ and to evaluate the charge transfer resistance.

As observed in Fig. 4.4, there is a decrease in the redox peak when GCE was modified with GO-DHP. This can be explained due to the negative charge of oxygenated groups present on GO surface, which generate a repulsive force with the negative charge of the redox probe, as well as there may be the formation of an insulating layer formed by GO, which prevents the redox probe ions to migrate from the solution to the GCE surface (Azimzadeh, Rahaie, Nasirizadeh, Ashtari, & Naderi-Manesh, 2016; Golkarieh, Nasirizadeh, & Jahanmardi, 2021; Li et al., 2012; Qian et al., 2020).

Fig. 4.5 shows the EIS response for GCE/GO-DHP and bare GCE. The experimental data were fitted to the Randles equivalent circuit model (inset in Fig. 4.5), considering that the semicircle at the region of high frequencies is equivalent to the charge transfer resistance (R_{ct}), and the Warburg-type line simulates the diffusion process (Marques et al., 2021;

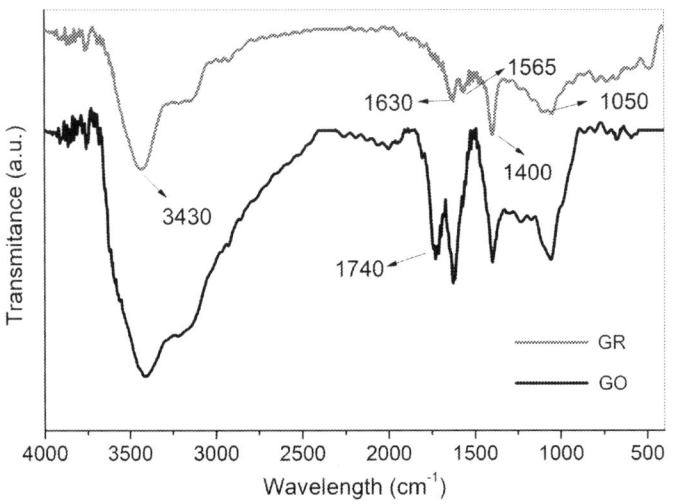

FIG. 4.3 FTIR spectra for GR and GO materials. *(No Permission Required.)*

FIG. 4.4 Cyclic voltammograms of 1.0 mmol L^{-1} of the redox probe [Fe(CN)$_6$]$^{3-}$/[Fe(CN)$_6$]$^{4-}$ in 1.0 mol L^{-1} KCl using CGE/DHP and CGE/GO-DHP. Scan rate of 20.0 mV s^{-1}. *(No Permission Required.)*

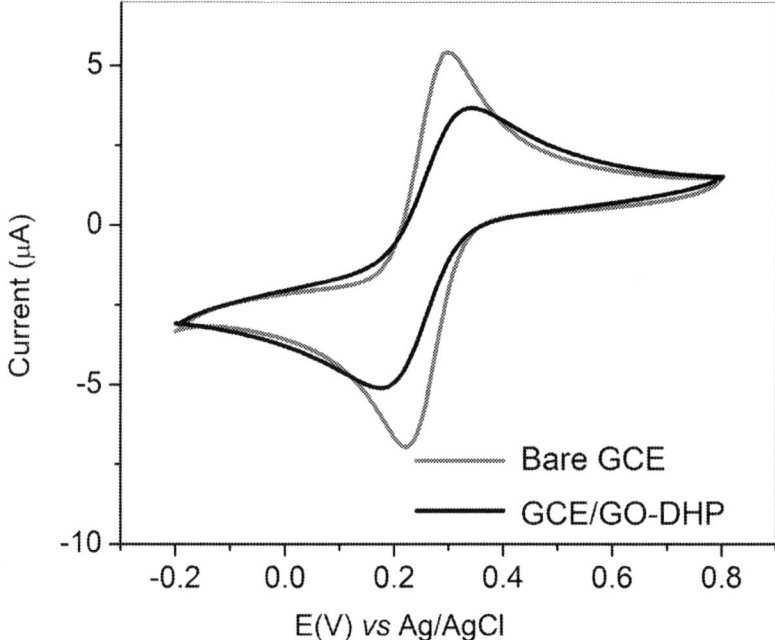

FIG. 4.5 Nyquist plots of bare GCE and GCE/GO-DHP in 1.0 mmol L^{-1} [Fe(CN)$_6$]$^{3-}$/[Fe(CN)$_6$]$^{4-}$ solution in 1.0 mol L^{-1} KCl. The frequency range of EIS was from 0.1 to 100.0 kHz. The GO/DHP suspension concentration was 1.0 mg L^{-1}. Inset: Randles equivalent circuit used to fit the experimental data. *(No Permission Required.)*

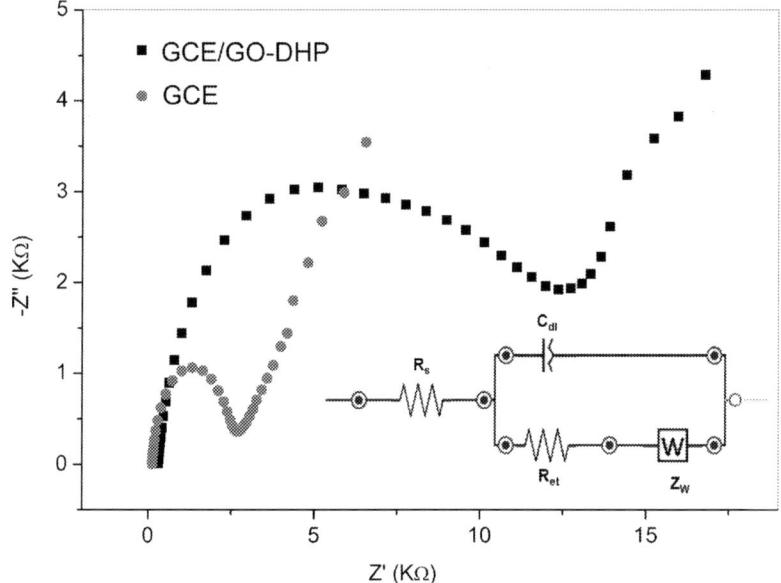

Mohamed et al., 2017, 2018). As can be seen, the bare GCE shows a small well-defined semicircle at high frequencies and a straight line at low frequencies with $R_{et} = 2429.0\,\Omega$. However, a larger semicircle than that at the bare GCE was observed for GCE/GO-DHP ($R_{et} = 10{,}000.0\,\Omega$), indicating a considerable increase in the resistance to electronic transfer. Such outcome can be attributed to the repulsive forces between [Fe(CN)$_6$]$^{3-}$/[Fe(CN)$_6$]$^{4-}$ probe and the excess of negatively charged oxygenated groups onto the surface of GO (Li et al., 2012; Marques et al., 2021; Mohamed et al., 2017, 2018). These results are in agreement with those presented previously by CV and also demonstrate that the GO was successfully immobilized on the GCE surface. Although a charge transfer resistance has been observed by using the GO compared to the bare GCE, as a result of repulsive forces with the probe at the interface solution/electrode, the GCE modified with GO presented better electrochemical performance toward 4-MMC, as will be further demonstrated.

Electrochemical reduction of 4-MMC

Electrochemical behavior of 5.0 µmol L^{-1} 4-MMC in the GCE and GCE/GO-DHP was evaluated by CV. Before the record of the cyclic voltammograms, it was applied −0.5 V during 180 s under magnetic stirring as pretreatment. Afterward, the cyclic voltammograms were recorded in the potential range of −1.5 to 0.0 V at scan rate of 50.0 mV s^{-1} (Fig. 4.6).

In spite of GCE/GO-DHP presenting lower conductivity than GCE, the presence of functional groups in GO surface is responsible for the preconcentration of 4-MMC molecules during the pretreatment. Therefore, in the potential range studied, an irreversible cathodic peak was observed at −1.26 V for GCE/GO-DHP electrode; however, no peak was observed for the GCE/DHP. The electrochemical reduction of 4-MMC on carbon electrodes near this potential has been already reported in the literature (Smith et al., 2014).

Despite some authors have recognized GO as a semiconductor (Brownson, Smith, & Banks, 2017; Dideikin & Vul, 2019; Eda, Mattevi, Yamaguchi, Kim, & Chhowalla, 2009; Qian et al., 2020), and GCE/GO-DHP presenting lower conductivity than GCE using the redox probe [Fe(CN)$_6$]$^{3-}$/[Fe(CN)$_6$]$^{4-}$, the electrocatalytic effects of GO, such as subtle electronic characteristics and strong adsorptive capability, are prevalent in favoring the reduction of 4-MMC (Golkarieh et al., 2021). Also, the presence of functional groups in GO surface helps the preconcentration of the 4-MMC molecules during the pretreatment, as a result of intermolecular interactions of hydrogen bonding between the oxygenated groups on the GO sheets with the NH group on 4-MMC molecule (Elbardisy et al., 2019; Li et al., 2012). In addition, there may be also electrostatic interaction between GO and 4-MMC, since there is a nitrogen atom positively charged in 4-MMC, as well as there may be π-π stacking between GO and 4-MMC due to the aromatic group present on 4-MMC molecule (Ambrosi et al., 2014; Khan et al., 2016; Li et al., 2012; Schifano et al., 2011).

Electroactive area of the bare GCE and GCE/GO-DHP was calculated by CV measurements of 1.0 mmol L^{-1} of the redox probe [Fe(CN)$_6$]$^{3-}$/[Fe(CN)$_6$]$^{4-}$ in 1.0 KCl with scan rates ranging from 10 to 100 mV s^{-1} and applying the Randles-Sevcik equation (Eq. 4.1):

$$I_p = 2.69 \times 10^5 \, n^{(3/2)} A_e D^{(1/2)} v^{(1/2)} C \tag{4.1}$$

where I_p is the anodic peak current (A), n is the number of electrons involved in the redox reaction (1), C is the concentration of the reduced species (mol cm^{-3}), D is the diffusion coefficient (in this condition 7.31×10^{-6}), $v^{1/2}$ is the scan rate square root (V s^{-1}), and A_e is the electroactive area of the electrode (cm^2), which can be calculated using the slope of the I_p versus $v^{1/2}$ graph (Bard & Faulkner, 1980; Konopka & McDuffie, 1970). The values of the electroactive areas were found to be 0.09 cm^2 and 0.04 cm^2 for the bare GCE and GCE/GO-DHP, respectively. From these results, it can be inferred that the better electrochemical behavior of GCE/GO-DHP toward 4-MMC is independent of the electroanalytical area, thereby confirming that the nature of functional groups on the GO surface plays a more important role in the performance of electrode than the electroactive area.

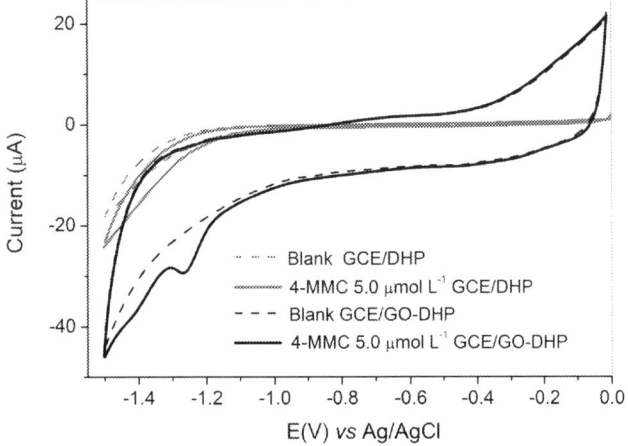

FIG. 4.6 Cyclic voltammograms in the absence *(dashed line)* and in the presence *(solid line)* of 5.0 mmol L^{-1} of 4-MMC using CGE/DHP and CGE/GO-DHP. Conditions: BR buffer 0.01 mol L^{-1} (pH 6.0) as support electrolyte, scan rate of 50.0 mV s^{-1}; pretreatment: −0.5 V for 180 s. *(No Permission Required.)*

Optimization of experimental parameters

Proportion of GO/DHP suspension and its concentration were investigated (data not shown). Proportions of 1:2, 1:1, and 2:1 of GO:DHP (m/m) at 1.0 mg L^{-1} in DMSO were evaluated, and the better response was obtained by using the 1:1 (m/m) GO:DHP suspension. It can be explained due to the excess of DHP in the suspension, which can make the electrode surface insulating; however, a low amount of DHP may cause instability to the film. Then, the concentration of the GO/DHP (1:1 m/m) suspension was evaluated from 0.5 to 2.0 mg mL^{-1} in DMSO. Higher cathodic peak current (Ipc) was obtained by using the 1.0 mg mL^{-1} concentration. A high concentration of GO may insulate the GCE surface, while a low concentration of GO is not enough to preconcentrate the 4-MMC molecule. Concerning the pretreatment step, the potential applied and the preconcentration time were evaluated (data not shown). Higher cathodic peak currents (Ipc) were obtained by applying negative potentials since the 4-MMC molecule is in its protonated form at pH 6.0. Thus, a better peak current value was obtained by applying −0.7 V for 360 s.

The effect of pH on the 4-MMC Ipc was investigated in the range of 4.0–8.0 by using BR buffer (0.01 mol L^{-1}) as support electrolyte, and the results obtained are shown in Fig. 4.7. Higher Ipc values were obtained at pHs above 6.0. The 4-MMC molecule presents pKa=8.81; thus, below this value, it starts to be protonated (Nowak, Woźniakiewicz, Mitoraj, Sagan, & Kościelniak, 2018). On the other hand, GO has two pKas (4.11 and 6.50), which refers to the carboxylic groups on its surface (Orth et al., 2016). Thus, above pH 5.0, the more acid OH groups are deprotonated, while the less acid groups remain partially deprotonated, which favors the electrostatic interaction with the protonated 4-MMC molecule. Besides, hydrophobic interactions such as π-π and van der Waals forces may also occur, which is pH-independent.

Fig. 4.7 also shows a linear dependence of the cathodic peak potential (Epc) with the pH (Epc(V) = −0.033pH −1.07; R^2 = 0.975), where the Epc shifts to less negative values as the pH decreases from 8.0 to 5.0, which indicates that the reaction is proton-dependent. Also, since the angular coefficient (0.033) is half of the theoretical value of 0.059 of the Nernstian system (Bard & Faulkner, 1980; Konopka & McDuffie, 1970), it can be assumed that, in that condition, the 4-MMC reduction process involves double the number of electrons over that of protons. This behavior has already been reported in the literature; however, the electrochemical mechanism is still unknown (Smith et al., 2014).

Type and concentration of supporting electrolyte were also studied. 0.01 mol L^{-1} of Britton-Robinson buffer (BRB), phosphate buffer solution (PBS), and KCl solution were evaluated. As can be seen in Fig. 4.8A, the use of KCl solution as supporting electrolyte provided a very low intensity peak for 4-MMC and shifted the Epc to higher negative values. The PBS provided a more defined 4-MMC peak with higher intensity and less capacitive current than BRB. Therefore, the PBS was chosen for further experiments. PBS concentrations of 0.01, 0.05, 0.10, and 0.50 were evaluated (Fig. 4.8B). It was observed that both 0.05 and 0.10 PBS concentrations resulted in similar Ipc values; however, 0.10 mol L^{-1} PBS concentration provided a more defined peak and less capacitive current. Therefore, the 0.10 mol L^{-1} PBS was chosen as a supporting electrolyte.

Optimization of DPV technique and analytical features

Differential pulse voltammetry (DPV) is a highly sensitive and low detection limit electrochemical technique and is more appropriate for irreversible systems, since the electron transfer kinetics in the electrode surface is slow (Bard & Faulkner,

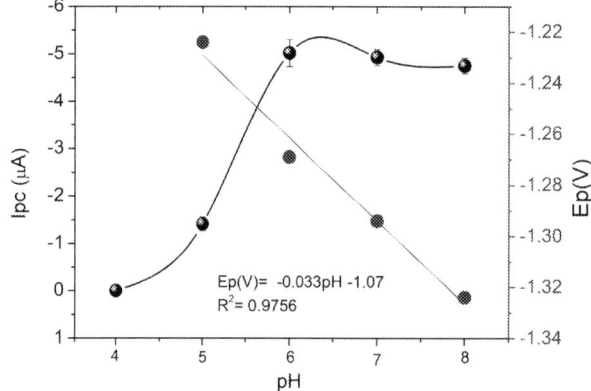

FIG. 4.7 Influence of pH on Ipc and Epc values for 5.0 μmol L^{-1} 4-MMC determination by GCE/GO-DHP. Conditions: BR buffer 0.01 mol L^{-1} as support electrolyte, scan rate of 50 mV s^{-1}. Pretreatment: −0.7 V for 360 s. *(No Permission Required.)*

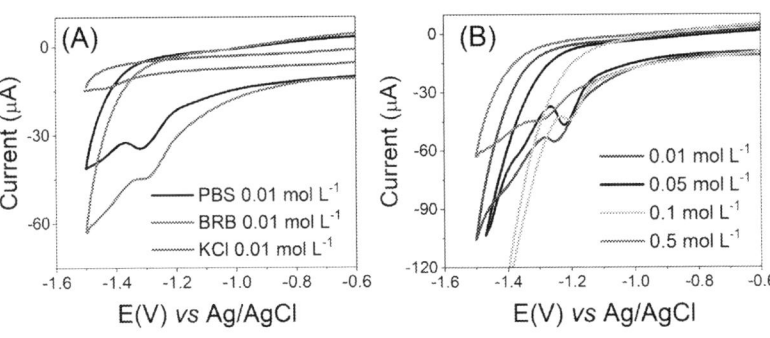

FIG. 4.8 Effect of supporting electrolyte type (phosphate solution buffer (PBS), Britton-Robinson buffer (BRB), and KCl solution) (A) and effect of PBS concentration (0.01–0.50 mol L^{-1}) (B) on 5.0 µmol L^{-1} 4-MMC determination by CV. *(No Permission Required.)*

1980; Konopka & McDuffie, 1970). Several methods using sensors based on GO for irreversible systems have used the DPV technique (Cheemalapati, Palanisamy, Mani, & Chen, 2013; Haldorai, Vilian, Rethinasabapathy, Huh, & Han, 2017; Sharma et al., 2016; Tajik & Beitollahi, 2019). Thus, DPV technique parameters of pulse amplitude (PA), scan rate (V), and pulse time (PT) were optimized using the Ipc of 4-MMC as analytical response. The range of each parameter studied and the better response obtained are summarized in Table 4.1.

After establishing the optimization of DPV parameters, the analytical curve for 4-MMC using the GCE/GO-DHP was obtained (Fig. 4.9). The linear regression equation was defined as Ipc = −1.01(±0.02)[4-MMC µmol L^{-1}] + 0.44(±0.30) with a regression coefficient $R^2 = 0.99$. The LOD and LOQ were found to be 1.23 and 4.10 µmol L^{-1}, respectively. The linearity of the analytical curve was validated by analysis of variance (ANOVA). The lack of fit and pure error ratio was found to be 1.77, which was smaller than the tabulated value ($F_{4,12}$) of 3.26, thus indicating the absence of lack of fit to the linear model (Bard & Faulkner, 1980; Konopka & McDuffie, 1970).

For comparison purposes, the analytical performance of other electrochemical methods reported in the literature for 4-MMC determination is shown in Table 4.2. Although some works present lower LOQ values, indirect determination of molecules may lead to interferences from the matrix. Another advantage that can be pointed out is that the GO-DHP sensor preparation is easy and has a low cost when compared to the f-MWCNT@AuNPs sensor, while dropping mercury electrode (DME) has environmental drawbacks.

TABLE 4.1 Optimization of operational parameters of DPV technique for 4-MMC determination by GCE/GO-DHP.

Parameter	Range studied	Better response
Pulse amplitude	10–300 (mV)	50 (mV)
Pulse time	5–100 (ms)	50 (ms)
Scan rate	10–100 (mV s^{-1})	50 (mV s^{-1})

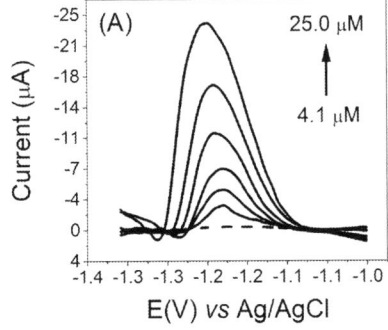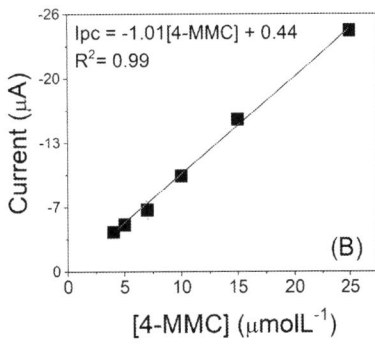

FIG. 4.9 (A) Differential pulse voltammetry (DPV) voltammograms for different 4-MMC concentrations and (B) Ipc values in function of 4-MMC concentration, at optimized conditions. DPV parameters: PA of 50 mV, PT of 50 ms, and V = 50 mV s^{-1}. Conditions: BRB buffer (pH 6.0) at 1.0 mol L^{-1}. Pretreatment: −0.7 V for 360 s. *(No Permission Required.)*

TABLE 4.2 Comparison of the proposed method with other electrochemical methods for determination of 4-MMC.

Sensor	Technique	Linear range ($\mu mol\,L^{-1}$)	LOQ ($\mu mol\,L^{-1}$)	Ref.
DME	DPP	0.0015–10	0.012	Krishnaiah et al. (2012)
GSPE	CV (oxidation)	91–1694 (pH 12)	74.58 (pH 12)	Smith et al. (2013)
		220–3770 (pH 2)	89.27 (pH 2)	
GSPE	CV (reduction)	0–1129	66.67	Smith et al. (2014)
Copper coin	CV (indirect determination)	0.0056–5.6	3.2	Tan et al. (2015)
f-MWCNT@AuNPs (MIP)	SWV (indirect determination)	0.001–0.010	0.0008	Razavipanah et al. (2018)
		0.010–0.100		
GCE/GO-DHP	DPV	4.1–25.0	4.1	This work

CV, cyclic voltammetry; *DME*, dropping mercury electrode; *DPP*, differential pulse polarography; *DPV*, differential pulse voltammetry; *GSPE*, graphite screen-printed electrode; *MIP*, molecularly imprinted polymer; *SWV*, square-wave voltammetry.

Concerning the methods using graphite screen-printed electrode (GSPE), it is very well-known that a new GSPE must be utilized for each measurement, which leads to a high-cost analysis and a large amount of residue. Also, the LOD obtained in this work is lower than that obtained in GSPE-based methods.

Intraday ($n=6$) and interday ($n=2$) precisions were calculated at two concentration levels of 4-MMC (5.0 and 10.0 $\mu mol\,L^{-1}$). The percentage of relative standard deviation (%RSD) for intraday precision was found to be 8.26% and 3.40%, respectively, and for interday precision, the %RSD was found to be 0.28% and 7.95%, respectively. The reproducibility of the electrode was attested by preparing it three times as described in "Apparatus" section. The %RSD obtained by measuring 5.0 $\mu mol\,L^{-1}$ of 4-MMC was 4.5%, which demonstrated a satisfactory reproducibility. The GCE/GO-DHP was utilized for up to 30 measurements without decreasing current, and the GO-DHP suspension remained stable for about 6 months under refrigeration.

Interference of adulterants in 4-MMC determination and application in a simulated sample

Street samples of 4-MMC typically present adulterants such as caffeine (CAF) and benzocaine (BZC), as reported by Smith et al. (2014). In this work, Smith and co-workers analyzed different "street" samples purchased from the Internet and found out that one 4-MMC sample was about 93% (m/m) benzocaine and the other one was about 89% (m/m) caffeine. Therefore, to evaluate if these adulterants interfere in 4-MMC determination using the method proposed, 5.0 $\mu mol\,L^{-1}$ of 4-MMC in the presence of 84.0 $\mu mol\,L^{-1}$ of BZC and in the presence of 46.0 $\mu mol\,L^{-1}$ of CAF was measured. As one can see from Fig. 4.10, the presence of these adulterants did not interfere in the cathodic peak of 4-MMC. Also, it is important to mention that no peak was observed when BZC and CAF were measured in the absence of 4-MMC.

Therefore, in order to evaluate the applicability of the method, it was applied in two simulated street samples. The accuracy was attested through addition and recovery tests as well as by using HPLC as the reference method. Sample A was composed of 7% of 4-MMC and 93% (m/m) of BZC, and Sample B was composed of 11% of 4-MMC and 89% (m/m) of CAF. The percentage of 4-MMC found in the simulated samples and the recoveries of the two standards additions obtained by the proposed method and HPLC (5.0 and 10.0 $\mu mol\,L^{-1}$) are arranged in Table 4.3.

The paired Student's *t*-test at a 95% confidence level was applied, and T_{calc} obtained ranged from 2.08 to 3.95, which is lower than $T_{tab}=4.30$. Thus, it can be attested that the proposed method is accurate for the determination of 4-MMC in seized street samples.

Conclusion

This work demonstrates the development of a simple and low-cost electrochemical sensor based on GCE modified with GO and DHP for the determination of the synthetic drug 4-MMC in seized samples. GO was easily obtained from graphite and dispersed in DMSO with DHP for GCE modification by the drop-casting method. Despite the insulating properties

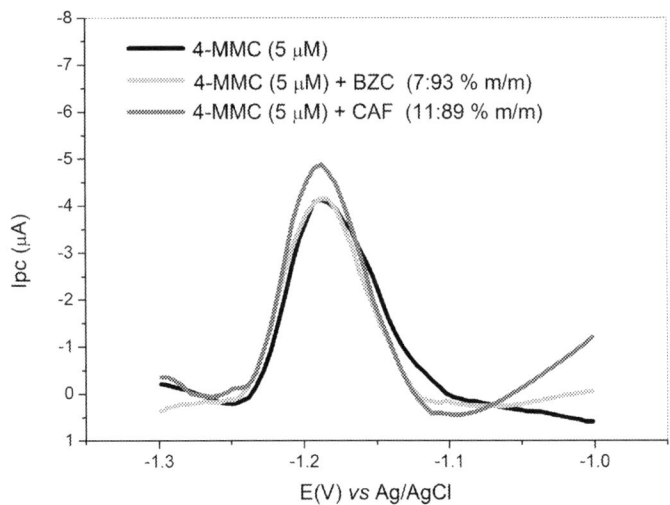

FIG. 4.10 DPV voltammogram of $5.0\,\mu mol\,L^{-1}$ 4-MMC in the absence and the presence of the adulterants BZC ($84.0\,\mu mol\,L^{-1}$) and CAF ($46.0\,\mu mol\,L^{-1}$). *(No Permission Required.)*

TABLE 4.3 Determination of 4-MMC and standard additions recoveries by the proposed method and HPLC as reference technique in simulated street samples.

		4-MMC found ($\mu mol\,L^{-1}$)		Recoveries (%)	
	4-MMC added ($\mu mol\,L^{-1}$)	GCE/GO-DHP	HPLC	GCE/GO-DHP	HPLC
Sample A	–	4.6 ± 0.3[a]	5.0 ± 0.1[b]	$91.9.8\pm4.0$	100.9 ± 0.1
	5.0	9.5 ± 0.4	11.2 ± 0.3	95.3 ± 3.8	111.51 ± 2.0
	10.0	13.9 ± 1.4	16.8 ± 0.1	92.7 ± 4.2	111.83 ± 0.1
Sample B	–	4.8 ± 0.2[c]	4.9 ± 0.3[d]	96.9 ± 2.8	97.8 ± 2.4
	5.0	10.2 ± 0.2	9.4 ± 0.3	102.5 ± 2.0	93.95 ± 1.7
	10.0	15.6 ± 0.3	16.4 ± 0.3	103.9 ± 3.0	109.03 ± 1.4

Equivalent concentration in solid samples: [a] $6.5\pm0.3\%$; [b] $7.2\pm0.1\%$; [c] $10.4\pm0.2\%$; [d] $10.5\pm0.3\%$ (m/m).

observed, the GCE/GO-DHP was able to reduce the 4-MMC molecule, while no response was obtained using the bare GCE. Also, the GCE/GO-DHP showed to be stable, reproducible, and accurate. The LOD obtained was sufficient for application in simulated street samples, and no interference was observed from common adulterants such as caffeine and benzocaine even at high proportions. Therefore, the method herein proposed can be used as an alternative for 4-MMC in situ determination in drug apprehensions by the drug control policy.

Acknowledgments

The authors acknowledge the financial support and fellowships of Coordenação de Aperfeiçoamento de Nível Superior (CAPES) (Project Pró-Forenses 3353/2014 Grant 23038.007082/2014-03), Conselho Nacional de Desenvolvimento Científico e Tecnológico (CNPq) (307432/2017-3), and Instituto Nacional de Ciência e Tecnologia de Bioanalítica (INCT) (FAPESP Grant No 2014/50867-3 and CNPq Grant No 465389/2014-7). This study was financed in part by the Coordenação de Aperfeiçoamento de Pessoal de Nível Superior—Brasil (CAPES)—Finance Code 001.

References

Ahmadi, M., & Ahour, F. (2020). An electrochemical biosensor based on a graphene oxide modified pencil graphite electrode for direct detection and discrimination of double-stranded DNA sequences. *Analytical Methods*, *12*(37), 4541–4550. https://doi.org/10.1039/d0ay01128b.

Akbari-Jonous, Z., Shayeh, J. S., Yazdian, F., Yadegari, A., Hashemi, M., & Omidi, M. (2019). An electrochemical biosensor for prostate cancer biomarker detection using graphene oxide–gold nanostructures. *Engineering in Life Sciences, 19*(3), 206–216. https://doi.org/10.1002/elsc.201800093.

Ambrosi, A., Chua, C. K., Bonanni, A., & Pumera, M. (2014). Electrochemistry of graphene and related materials. *Chemical Reviews, 114*(14), 7150–7188. https://doi.org/10.1021/cr500023c.

Amini, S., Kalaantari, H., Garay, J., Balandin, A. A., & Abbaschian, R. (2011). Growth of graphene and graphite nanocrystals from a molten phase. *Journal of Materials Science, 46*(19), 6255–6263. https://doi.org/10.1007/s10853-011-5432-9.

ANVISA. (2020). *Lista de substâncias sujeitas a controle especial no Brasil*. http://portal.anvisa.gov.br/lista-de-substancias-a-controle-especial.

Anzillotti, L., Calò, L., Banchini, A., Schirripa, M. L., Marezza, F., & Cecchi, R. (2020). Mephedrone and chemsex: A case report. *Legal Medicine, 42*. https://doi.org/10.1016/j.legalmed.2019.101640.

Atta, N. F., Hassan, H. K., & Galal, A. (2014). Rapid and simple electrochemical detection of morphine on graphene-palladium-hybrid-modified glassy carbon electrode. *Analytical and Bioanalytical Chemistry, 406*(27), 6933–6942. https://doi.org/10.1007/s00216-014-7999-x.

Azimzadeh, M., Rahaie, M., Nasirizadeh, N., Ashtari, K., & Naderi-Manesh, H. (2016). An electrochemical nanobiosensor for plasma miRNA-155, based on graphene oxide and gold nanorod, for early detection of breast cancer. *Biosensors and Bioelectronics, 77*, 99–106. https://doi.org/10.1016/j.bios.2015.09.020.

Bade, R., Bijlsma, L., Sancho, J. V., Baz-Lomba, J. A., Castiglioni, S., Castrignanò, E., … Hernández, F. (2017). Liquid chromatography-tandem mass spectrometry determination of synthetic cathinones and phenethylamines in influent wastewater of eight European cities. *Chemosphere, 168*, 1032–1041. https://doi.org/10.1016/j.chemosphere.2016.10.107.

Bard, A. J., & Faulkner, L. R. (1980). *Electrochemical methods: Fundamentals and Applications* (2nd ed.). John Wiley & Sons.

Bezzon, V. D. N., Montanheiro, T. L. A., De Menezes, B. R. C., Ribas, R. G., Righetti, V. A. N., Rodrigues, K. F., & Thim, G. P. (2019). Carbon nanostructure-based sensors: A brief review on recent advances. *Advances in Materials Science and Engineering, 2019*. https://doi.org/10.1155/2019/4293073.

Brownson, D. A. C., Smith, G. C., & Banks, C. E. (2017). Graphene oxide electrochemistry: The electrochemistry of graphene oxide modified electrodes reveals coverage dependent beneficial electrocatalysis. *Royal Society Open Science, 4*(11). https://doi.org/10.1098/rsos.171128.

Busardò, F. P., Kyriakou, C., Napoletano, S., Marinelli, E., & Zaami, S. (2015). Mephedrone related fatalities: A review. *European Review for Medical and Pharmacological Sciences, 19*(19), 3777–3790.

Cheemalapati, S., Palanisamy, S., Mani, V., & Chen, S. M. (2013). Simultaneous electrochemical determination of dopamine and paracetamol on multiwalled carbon nanotubes/graphene oxide nanocomposite-modified glassy carbon electrode. *Talanta, 117*, 297–304. https://doi.org/10.1016/j.talanta.2013.08.041.

Chen, D., Feng, H., & Li, J. (2012). Graphene oxide: Preparation, functionalization, and electrochemical applications. *Chemical Reviews, 112*(11), 6027–6053. https://doi.org/10.1021/cr300115g.

Coates, J. (2006). *Interpretation of infrared spectra, a practical approach*. John Wiley & Sons.

Couto, R. A. S., Gonçalves, L. M., Carvalho, F., Rodrigues, J. A., Rodrigues, C. M. P., & Quinaz, M. B. (2018). The analytical challenge in the determination of cathinones, key-players in the worldwide phenomenon of novel psychoactive substances. *Critical Reviews in Analytical Chemistry, 48*(5), 372–390. https://doi.org/10.1080/10408347.2018.1439724.

Crulhas, B. P., Basso, C. R., Parra, J. P. R. L. L., Castro, G. R., & Pedrosa, V. A. (2019). Reduced graphene oxide decorated with AuNPs as a new aptamer-based biosensor for the detection of androgen receptor from prostate cells. *Journal of Sensors, 2019*. https://doi.org/10.1155/2019/5805609.

Dideikin, A. T., & Vul, A. Y. (2019). Graphene oxide and derivatives: The place in graphene family. *Frontiers in Physics, 6*. https://doi.org/10.3389/fphy.2018.00149.

Eda, G., Mattevi, C., Yamaguchi, H., Kim, H., & Chhowalla, M. (2009). Insulator to semimetal transition in graphene oxide. *Journal of Physical Chemistry C, 113*(35), 15768–15771. https://doi.org/10.1021/jp9051402.

Elbardisy, H. M., García-Miranda Ferrari, A., Foster, C. W., Sutcliffe, O. B., Brownson, D. A. C., Belal, T. S., … Banks, C. E. (2019). Forensic electrochemistry: The electroanalytical sensing of mephedrone metabolites. *ACS Omega, 4*(1), 1947–1954. https://doi.org/10.1021/acsomega.8b02586.

EMCDDA. (2011). Report on the risk assessment of mephedrone in the framework of the Council Decision on new psychoactive substances. *Risk Assessments*. https://doi.org/10.2810/40800.

Ferreira Oliveira, A. E., César Pereira, A., Braga Bettio, G., & Teixeira Tarley, C. R. (2019). Synthesis, studies and structural characterization of thermal and hydrazine reduction of graphene oxide by Raman spectroscopy and infrared spectroscopy. *Revista Virtual de Química, 11*(3), 866–877. https://doi.org/10.21577/1984-6835.20190060.

Gibbons, S., & Zloh, M. (2010). An analysis of the "legal high" mephedrone. *Bioorganic and Medicinal Chemistry Letters, 20*(14), 4135–4139. https://doi.org/10.1016/j.bmcl.2010.05.065.

Golkarieh, A.-M., Nasirizadeh, N., & Jahanmardi, R. (2021). Fabrication of an electrochemical sensor with Au nanorods-graphene oxide hybrid nanocomposites for in situ measurement of cloxacillin. *Materials Science and Engineering: C, 118*, 111317. https://doi.org/10.1016/j.msec.2020.111317.

Gong, X., Bi, Y., Zhao, Y., Liu, G., & Teoh, W. Y. (2014). Graphene oxide-based electrochemical sensor: A platform for ultrasensitive detection of heavy metal ions. *RSC Advances, 4*(47), 24653–24657. https://doi.org/10.1039/C4RA02247E.

Gugoasa, L. A., Stefan-van Staden, R. I., van Staden, J. F., Coroș, M., & Pruneanu, S. (2019). Electrochemical determination of bisphenol A in saliva by a novel three-dimensional (3D) printed gold-reduced graphene oxide (rGO) composite paste electrode. *Analytical Letters, 52*(16), 2583–2606. https://doi.org/10.1080/00032719.2019.1620262.

Guo, H. L., Wang, X. F., Qian, Q. Y., Wang, F. B., & Xia, X. H. (2009). A green approach to the synthesis of graphene nanosheets. *ACS Nano, 3*(9), 2653–2659. https://doi.org/10.1021/nn900227d.

Haldorai, Y., Vilian, A. T. E., Rethinasabapathy, M., Huh, Y. S., & Han, Y. K. (2017). Electrochemical determination of dopamine using a glassy carbon electrode modified with TiN-reduced graphene oxide nanocomposite. *Sensors and Actuators, B: Chemical, 247*, 61–69. https://doi.org/10.1016/j.snb.2017.02.181.

Hashemi, P., Bagheri, H., Afkhami, A., Ardakani, Y. H., & Madrakian, T. (2017). Fabrication of a novel aptasensor based on three-dimensional reduced graphene oxide/polyaniline/gold nanoparticle composite as a novel platform for high sensitive and specific cocaine detection. *Analytica Chimica Acta*, *996*, 10–19. https://doi.org/10.1016/j.aca.2017.10.035.

Hu, J., & Zhang, Z. (2020). Application of electrochemical sensors based on carbon nanomaterials for detection of flavonoids. *Nanomaterials*, *10*(10), 1–14. https://doi.org/10.3390/nano10102020.

Jiang, B., Wang, M., Chen, Y., Xie, J., & Xiang, Y. (2012). Highly sensitive electrochemical detection of cocaine on graphene/AuNP modified electrode via catalytic redox-recycling amplification. *Biosensors and Bioelectronics*, *32*(1), 305–308. https://doi.org/10.1016/j.bios.2011.12.010.

Kavan, L., Yum, J. H., & Grätzel, M. (2011). Graphene nanoplatelets outperforming platinum as the electrocatalyst in co-bipyridine-mediated dye-sensitized solar cells. *Nano Letters*, *11*(12), 5501–5506. https://doi.org/10.1021/nl203329c.

Khan, Z. U., Kausar, A., Ullah, H., Badshah, A., & Khan, W. U. (2016). A review of graphene oxide, graphene buckypaper, and polymer/graphene composites: Properties and fabrication techniques. *Journal of Plastic Film & Sheeting*, *32*(4), 336–379. https://doi.org/10.1177/8756087915614612.

Konopka, S. J., & McDuffie, B. (1970). Diffusion coefficients of ferri- and ferrocyanide ions in aqueous media, using twin-electrode thin-layer electrochemistry. *Analytical Chemistry*, *42*(14), 1741–1746. https://doi.org/10.1021/ac50160a042.

Krishnaiah, V., Reddy, Y. V. R., Reddy, V. H., Reddy, M. T., & Rao, G. M. (2012). Electrochemical reduction behavior of mephedrone drug at a dropping mercury electrode and its pharmaceutical determination in spiked human urine samples. *International Journal of Scientific Research*, 14–17. https://doi.org/10.15373/22778179/SEP2012/5.

Kumunda, C, Adekunle, A. S., Mamba, B. B., Hlongwa, N. W., & Nkambule, T. T. I. (2021). Electrochemical detection of environmental pollutants based on graphene derivatives: A review. *Frontiers in Materials*, *7*. https://doi.org/10.3389/fmats.2020.616787.

Lakhe, P., Kulhanek, D. L., Zhao, X., Papadaki, M. I., Majumder, M., & Green, M. J. (2020). Graphene oxide synthesis: Reaction calorimetry and safety. *Industrial and Engineering Chemistry Research*, *59*(19), 9004–9014. https://doi.org/10.1021/acs.iecr.0c00644.

Li, J., Kuang, D., Feng, Y., Zhang, F., Xu, Z., & Liu, M. (2012). A graphene oxide-based electrochemical sensor for sensitive determination of 4-nitrophenol. *Journal of Hazardous Materials*, *201–202*, 250–259. https://doi.org/10.1016/j.jhazmat.2011.11.076.

Li, M.-F., Liu, Y.-G., Liu, S.-B., Zeng, G.-M., Hu, X.-J., Tan, X.-F., ... Liu, X.-H. (2018). Performance of magnetic graphene oxide/diethylenetriaminepentaacetic acid nanocomposite for the tetracycline and ciprofloxacin adsorption in single and binary systems. *Journal of Colloid and Interface Science*, *521*, 150–159. https://doi.org/10.1016/j.jcis.2018.03.003.

Long, G. L., & Winefordner, J. D. (1983). Limit of detection: A closer look at the IUPAC definition. *Analytical Chemistry*, *55*(7), 713–724. https://doi.org/10.1021/ac00258a724.

Lü, P., Feng, Y., Zhang, X., Li, Y., & Feng, W. (2010). Recent progresses in application of functionalized graphene sheets. *Science China Technological Sciences*, *53*(9), 2311–2319. https://doi.org/10.1007/s11431-010-4050-0.

Maccaferri, G., Terzi, F., Xia, Z., Vulcano, F., Liscio, A., Palermo, V., & Zanardi, C. (2019). Highly sensitive amperometric sensor for morphine detection based on electrochemically exfoliated graphene oxide. Application in screening tests of urine samples. *Sensors and Actuators, B: Chemical*, *281*, 739–745. https://doi.org/10.1016/j.snb.2018.10.163.

Marques, G. L., Rocha, L. R., Prete, M. C., Gorla, F. A., Moscardi dos Santos, D., Segatelli, M. G., & Tarley, C. R. T. (2021). Development of electrochemical platform based on molecularly imprinted poly(methacrylic acid) grafted on iniferter-modified carbon nanotubes for 17β-estradiol determination in water samples. *Electroanalysis*, *33*(3), 568–578. https://doi.org/10.1002/elan.202060270.

McNeill, L., Pearson, C., Megson, D., Norrey, J., Watson, D., Ashworth, D., ... Shaw, K. J. (2021). Origami chips: Development and validation of a paper-based Lab-on-a-Chip device for the rapid and cost-effective detection of 4-methylmethcathinone (mephedrone) and its metabolite 4-methylepherine in urine. *Forensic Chemistry*, *22*, 100293. https://doi.org/10.1016/j.forc.2020.100293.

Mohamed, M. A., Atty, S. A., Salama, N. N., & Banks, C. E. (2017). Highly selective sensing platform utilizing graphene oxide and multiwalled carbon nanotubes for the sensitive determination of tramadol in the presence of co-formulated drugs. *Electroanalysis*, *29*(4), 1038–1048. https://doi.org/10.1002/elan.201600668.

Mohamed, M. A., El-Gendy, D. M., Ahmed, N., Banks, C. E., & Allam, N. K. (2018). 3D spongy graphene-modified screen-printed sensors for the voltammetric determination of the narcotic drug codeine. *Biosensors and Bioelectronics*, *101*, 90–95. https://doi.org/10.1016/j.bios.2017.10.020.

Navarro-Pardo, F., Martínez-Barrera, G., Martínez-Hernández, A. L., Castaño, V. M., Rivera-Armenta, J. L., Medellín-Rodríguez, F., & Velasco-Santos, C. (2013). Effects on the thermo-mechanical and crystallinity properties of nylon 6,6 electrospun fibres reinforced with one dimensional (1D) and two dimensional (2D) carbon. *Materials*, *6*(8), 3494–3513. https://doi.org/10.3390/ma6083494.

Nowak, P. M., Woźniakiewicz, M., Mitoraj, M., Sagan, F., & Kościelniak, P. (2018). Thermodynamics of acid-base dissociation of several cathinones and 1-phenylethylamine, studied by an accurate capillary electrophoresis method free from the Joule heating impact. *Journal of Chromatography A*, *1539*, 78–86. https://doi.org/10.1016/j.chroma.2018.01.047.

Nunes, E. W., Silva, M. K. L., & Cesarino, I. (2020). Evaluation of a reduced graphene oxide-Sb nanoparticles electrochemical sensor for the detection of cadmium and lead in chamomile tea. *Chemosensors*, *8*(8). https://doi.org/10.3390/CHEMOSENSORS8030053.

Oghli, A. H., & Soleymanpour, A. (2020). Polyoxometalate/reduced graphene oxide modified pencil graphite sensor for the electrochemical trace determination of paroxetine in biological and pharmaceutical media. *Materials Science and Engineering: C*, *108*, 110407. https://doi.org/10.1016/j.msec.2019.110407.

Olesti, E., Pujadas, M., Papaseit, E., Pérez-Mañá, C., Pozo, Ó. J., Farré, M., & de la Torre, R. (2016). GC–MS quantification method for mephedrone in plasma and urine: Application to human pharmacokinetics. *Journal of Analytical Toxicology*. https://doi.org/10.1093/jat/bkw120.

Oliveira, A. E. F., Braga, G. B., Tarley, C. R. T., & Pereira, A. C. (2018). Thermally reduced graphene oxide: Synthesis, studies and characterization. *Journal of Materials Science*, *53*(17), 12005–12015. https://doi.org/10.1007/s10853-018-2473-3.

Orth, E. S., Ferreira, J. G. L., Fonsaca, J. E. S., Blaskievicz, S. F., Domingues, S. H., Dasgupta, A., ... Zarbin, A. J. G. (2016). PKa determination of graphene-like materials: Validating chemical functionalization. *Journal of Colloid and Interface Science*, *467*, 239–244. https://doi.org/10.1016/j.jcis.2016.01.013.

Papaseit, E., Olesti, E., de la Torre, R., Torrens, M., & Farre, M. (2018). Mephedrone concentrations in cases of clinical intoxication. *Current Pharmaceutical Design*, *23*(36). https://doi.org/10.2174/1381612823666170704130213.

Pasinszki, T., Krebsz, M., Tung, T. T., & Losic, D. (2017). Carbon nanomaterial based biosensors for non-invasive detection of cancer and disease biomarkers for clinical diagnosis. *Sensors (Switzerland)*, *17*(8). https://doi.org/10.3390/s17081919.

Peacock, A., Bruno, R., Gisev, N., Degenhardt, L., Hall, W., Sedefov, R., ... Griffiths, P. (2019). New psychoactive substances: Challenges for drug surveillance, control, and public health responses. *Lancet*, *394*, 32231–32237. https://doi.org/10.1016/S0140-6736(19.

Pope, M. A., Punckt, C., & Aksay, I. A. (2011). Intrinsic capacitance and redox activity of functionalized graphene sheets. *Journal of Physical Chemistry C*, *115*(41), 20326–20334. https://doi.org/10.1021/jp2068667.

Pumera, M. (2013). Electrochemistry of graphene, graphene oxide and other graphenoids: Review. *Electrochemistry Communications*, *36*, 14–18. https://doi.org/10.1016/j.elecom.2013.08.028.

Qian, L., Thiruppathi, A. R., Elmahdy, R., van der Zalm, J., & Chen, A. (2020). Graphene-oxide-based electrochemical sensors for the sensitive detection of pharmaceutical drug naproxen. *Sensors (Switzerland)*, *20*(5). https://doi.org/10.3390/s20051252.

Radhakrishnan, S., & Mathiyarasu, J. (2018). Graphene-carbon nanotubes modified electrochemical sensors. In *Graphene-based electrochemical sensors for biomolecules: A volume in micro and nano technologies* (pp. 187–205). Elsevier. https://doi.org/10.1016/B978-0-12-815394-9.00008-X.

Razavipanah, I., Alipour, E., Deiminiat, B., & Rounaghi, G. H. (2018). A novel electrochemical imprinted sensor for ultrasensitive detection of the new psychoactive substance "Mephedrone". *Biosensors and Bioelectronics*, *119*, 163–169. https://doi.org/10.1016/j.bios.2018.08.016.

Sacco, L. N. (2016). *Synthetic drugs: Overview and issues for congress.* Congressional Research Service Report.

Schifano, F., Albanese, A., Fergus, S., Stair, J. L., Deluca, P., Corazza, O., ... Ghodse, A. H. (2011). Mephedrone (4-methylmethcathinone; 'meow meow'): Chemical, pharmacological and clinical issues. *Psychopharmacology*, *214*(3), 593–602. https://doi.org/10.1007/s00213-010-2070-x.

Shafi, A., Berry, A. J., Sumnall, H., Wood, D. M., & Tracy, D. K. (2020). New psychoactive substances: A review and updates. *Therapeutic Advances in Psychopharmacology*, *10*. https://doi.org/10.1177/2045125320967197. 204512532096719.

Sharma, V., Hynek, D., Trnkova, L., Hemzal, D., Marik, M., Kizek, R., & Hubalek, J. (2016). Electrochemical determination of adenine using a glassy carbon electrode modified with graphene oxide and polyaniline. *Microchimica Acta*, *183*(4), 1299–1306. https://doi.org/10.1007/s00604-015-1740-0.

Shaw, L., & Dennany, L. (2017). Applications of electrochemical sensors: Forensic drug analysis. *Current Opinion in Electrochemistry*, *3*(1), 23–28. https://doi.org/10.1016/j.coelec.2017.05.001.

Smith, J. P., Metters, J. P., Irving, C., Sutcliffe, O. B., & Banks, C. E. (2013). Forensic electrochemistry: The electroanalytical sensing of synthetic cathinone-derivatives and their accompanying adulterants in "legal high" products. *Analyst*, *139*(2), 389–400. https://doi.org/10.1039/c3an01985c.

Smith, J. P., Metters, J. P., Khreit, O. I. G., Sutcliffe, O. B., & Banks, C. E. (2014). Forensic electrochemistry applied to the sensing of new psychoactive substances: Electroanalytical sensing of synthetic cathinones and analytical validation in the quantification of seized street samples. *Analytical Chemistry*, *86*(19), 9985–9992. https://doi.org/10.1021/ac502991g.

Tajik, S., & Beitollahi, H. (2019). A sensitive chlorpromazine voltammetric sensor based on graphene oxide modified glassy carbon electrode. *Analytical and Bioanalytical Chemistry Research*, *6*(1), 171–182. https://doi.org/10.22036/ABCR.2018.89229.1154.

Tan, F., Smith, J. P., Sutcliffe, O. B., & Banks, C. E. (2015). Regal electrochemistry: Sensing of the synthetic cathinone class of new psychoactive substances (NPSs). *Analytical Methods*, *7*(16), 6470–6474. https://doi.org/10.1039/c5ay01820j.

UNODC. (2019a). *Global overview of drug demand and supply.* World Drug Report https://wdr.unodc.org/wdr2019/en/drug-demand-and-supply.html.

UNODC. (2019b). *Stimulants.* World Drug Reports https://wdr.unodc.org/wdr2019/en/stimulants.html.

Vajedi, F., & Dehghani, H. (2019). The characterization of TiO2-reduced graphene oxide nanocomposites and their performance in electrochemical determination for removing heavy metals ions of cadmium(II), lead(II) and copper(II). *Materials Science & Engineering, B: Solid-State Materials for Advanced Technology*, *243*, 189–198. https://doi.org/10.1016/j.mseb.2019.04.009.

Wang, H., Cui, L-F., Yang, Y., Casalongue, H. S., Robinson, J. T., Liang, Y., ... Dai, H. (2010). Mn_3O_4-graphene hybrid as a high capacity anode material for lithium ion batteries. *Journal of the American Chemical Society*, *132*(40), 13978–13980. https://doi.org/10.1021/ja105296a.

Wang, Y., Li, Y., Tang, L., Lu, J., & Li, J. (2009). Application of graphene-modified electrode for selective detection of dopamine. *Electrochemistry Communications*, *11*(4), 889–892. https://doi.org/10.1016/j.elecom.2009.02.013.

Yin, H., Zhou, Y., Ma, Q., Liu, T., Ai, S., & Zhu, L. (2011). Electrochemical oxidation behavior of guanosine-5′-monophosphate on a glassy carbon electrode modified with a composite film of graphene and multi-walled carbon nanotubes, and its amperometric determination. *Microchimica Acta*, *172*(3–4), 343–349. https://doi.org/10.1007/s00604-010-0499-6.

Yu, H., Feng, X., Chen, X. X., Qiao, J. L., Gao, X. L., Xu, N., & Gao, L.-J. (2017). Electrochemical determination of bisphenol A on a glassy carbon electrode modified with gold nanoparticles loaded on reduced graphene oxide-multi walled carbon nanotubes composite. *Chinese Journal of Analytical Chemistry*, *45*(5), 713–720. https://doi.org/10.1016/S1872-2040(17)61014-4.

Zanfrognini, B., Pigani, L., & Zanardi, C. (2020). Recent advances in the direct electrochemical detection of drugs of abuse. *Journal of Solid State Electrochemistry*, *24*(11–12), 2603–2616. https://doi.org/10.1007/s10008-020-04686-z.

Zang, D., Yan, M., Ge, S., Ge, L., & Yu, J. (2013). A disposable simultaneous electrochemical sensor array based on a molecularly imprinted film at a NH2-graphene modified screen-printed electrode for determination of psychotropic drugs. *Analyst*, *138*(9), 2704–2711. https://doi.org/10.1039/c3an00109a.

Zeng, Y., Bao, J., Zhao, Y., Huo, D., Chen, M., Yang, M., ... Hou, C. (2018). A sensitive label-free electrochemical immunosensor for detection of cytokeratin 19 fragment antigen 21-1 based on 3D graphene with gold nanopaticle modified electrode. *Talanta*, *178*, 122–128. https://doi.org/10.1016/j.talanta.2017.09.020.

Zuba, D., & Adamowicz, P. (2018). Analytical methods used for identification and determination of synthetic cathinones and their metabolites. In *Synthetic cathinones* (pp. 41–69). Springer Nature. https://doi.org/10.1007/978-3-319-78707-7_4.

Part III

Carbon nanomaterials-based sensor applications (environment, biomedical, foods, point of care, etc.)

Chapter 5

Carbon nanomaterials-based sensors for biomedical applications

Amirreza Roshani, Maryam Mousavizadegan, and Morteza Hosseini
Department of Life Science Engineering, Faculty of New Sciences & Technologies, University of Tehran, Tehran, Iran

Introduction

Nowadays low cost, high speed, and high accuracy are among the necessities in diagnosing and treating diseases. With the increasing advancements in technology, especially in the field of medical diagnosis, we are observing the use of various biomedical sensors in everyday life, such as digital thermometers or wearable home-used blood glucose sensors. Biomedical sensors are widely used in medicine and biology to diagnose and measure a wide range of critical physiological variables. They are critical and precise medical diagnosis tools widely used in hospitals and biological laboratories (Mendelson, 2005). Some biomedical sensors also have a pivotal role in nonmedical applications like environmental monitoring or food analysis. The principle behind biomedical sensors is to turn the biological signal into a recognizable and visible optical or electrical signal. Biomedical sensors can be divided into three main groups: physical sensors, chemical sensors, and biological sensors (Manjunatha et al., 2018).

In recent years, carbon nanomaterials such as graphene, carbon nanotubes, and carbon quantum dots have drawn considerable attention due to their unique optical, structural, chemical, and electronic properties.

Biomarkers

In many authoritative articles, the term biomarker is defined as a characteristic that is measured as an indicator of normal biological processes, pathogenic processes, or responses to an exposure or intervention. Measuring biomarkers is essential for the diagnosis and prevention of diseases, evaluation of the healing process, and many other issues related to treatment and medicine. That is why rapid, accurate, and fast detection of biomarkers is vital in medicine; therefore, different types of sensors and biosensors have been developed to detect them in recent decades. In this part, we will briefly discuss four of the main biomarkers assessed in biomedicine (Adhikari, Govindhan, & Chen, 2015; Manjunatha et al., 2009).

Glucose

More than 85 percent of the global sensor and biosensors market is dedicated to sensing glucose. The increasing incidence of diabetes globally and that lack of on-time attention to diabetes which can lead to acute problems to the patient has made monitoring blood sugar level highly necessary for all people (Tiwari, Vij, Kemp, & Kim, 2015). According to WHO, by 2014, around 422 million people had diabetes worldwide, and it will most likely become the ninth fatal illness by 2030. Since the most critical issue in the diagnosis of blood sugar is the accuracy of measurement and due to the high electrical conductivity, high specific surface, and biocompatibility of carbon nanomaterials, they can be used to build appropriate sensors for sensing blood sugar levels (Ahammad, Islam, & Hasan, 2019).

Nucleic acids

Biomarkers can also be nucleic acids including micro-RNAs and DNAs. Micro-RNAs are short (19–25 nucleotides), noncoding (RNA) molecules that regulate gene expression by influence on mRNAs. More than 60% of human protein genes are under miRNA control. Furthermore, specific DNA sequences are another type of nucleic acid biomarker specifically used to detect pathogens in the genosensing process (Rivas et al., 2017).

Neurotransmitters

The human nervous system is responsible for the essential functions of the human body, such as perception, behavior, and movement. Neurotransmitters are essential components that are responsible for transferring signals between neurons and other cells. Detection and measurement of neurotransmitters is essential to understand brain functions and identify related diseases. The number of known neurotransmitters is more than 100 but the most important ones that have attracted the attention of scientists in the field of sensing and biosensing are dopamine, uric acid, and ascorbic acid which are small-sized biomolecules that play a pivotal role in sustaining biological functions at an appropriate level 2. Serotonin is a neurotransmitter controlling mood and sleep and is a primary target for depression pharmaceutical remedies. Epinephrine and norepinephrine are catecholamine neurotransmitters that play opposing roles (Manjunatha, Deraman, Basri, & Talib, 2018; Sanghavi, Wolfbeis, Hirsch, & Swami, 2015).

Proteins

Proteins are the main components of cells and play a pivotal role in many essential activities, and various biological pathways and processes in the human body are dependent on a variety of proteins. Many biomarkers are proteins, and the most important of which include carcinoembryonic antigen (CEA) that belongs to the family of cell-surface glycoproteins, prostate-specific antigen (PSA) that is considered as the best serum marker for prostate cancer detection, alpha-fetoprotein (AFP) and cytokines with essential functions like cell signaling and immunomodulating factors which are essential for early diagnosis and analysis of several diseases, and cardiac biomarkers that are a group of proteins that flow out of damaged myocardial cells into the bloodstream. Nevertheless, biomarkers are not limited to these proteins, other types of these macromolecules are involved in cell functions for the detection of many carbon-based sensors, and biosensors have been developed, the most important of which are peptides and amino acids, insulin, heme-containing proteins generally used for H_2O_2 detection, and antibodies (Baptista, 2015; Tiwari et al., 2015).

Graphene

Graphene is the most recently discovered part of the carbon nanomaterials family. Graphene is the first known two-dimensional material in the world that consists of a network of carbon atoms. This particular material with unique physical, electrical, and chemical properties comprises of a single-layer structure of hexagonal carbon atoms connected with an sp^2 bond (Anhar et al., 2017; Beigi, Mesgari, Hosseini, Aghazadeh, & Ganjali, 2019; Sobhanie, Faridbod, Hosseini, & Ganjali, 2020).

Due to its high potential and attractive properties, graphene and its derivatives are widely used in various fields such as sensors and biosensors, drug delivery, solar energy, medicine, biomedical imaging, ultrafast computing, and many more (Cernat, Tertiş, Fritea, & Cristea, 2016). The process from the discovery to the synthesis and manufacture of graphene are very long beginning in 1840. The history of this process has been summarized here (Salehnia, Hosseini, & Ganjali, 2018).

1840—The first reports of GO: a German scientist Schafhaeutl reported intercalation of graphite with sulfuric and nitric acid.
1948—G Ruess and F Vogt examined a dried drop of graphene oxide with TEM and witnessed folded thin nanometer layers.
1962—Following the studies of G Ruess and F Vogt, Ulrich Hofmann and coworkers searched the layers of graphene oxide and recognized some of their thin layers.
1968—Morgan and Somorjai investigated the adsorption of various organic molecules in the gas phase by low-energy electron diffraction (LEED) on the platinum's high-temperature surface.
1970—Blakely and coworkers prepared monolayer graphite by segregating carbon on the surface of Ni 100.
1975—van Pommel and coworkers produced a single layer of graphite using silicon sublimation of carbide monocrystals at ultrahigh temperature.
1986—H. P. Boehm and coworkers recommended the term (graphene) to describe a carbon monolayer that resembles graphite structure.
1997—The IUPAC system provided the following definition for the term graphene: "The term graphene should be used only when the reactions, structural relations, or other properties of individual layers are discussed."
1999—Rodney S. Ruoffb and coworkers used the micromechanical method to produce thin lamella that contained several layers of graphene by lithographic patterning of HOPG with oxygen plasma.
2004—By Geim, Novoselov, and coworkers, a thick layer of graphene was obtained by pressing the HOPG surface to the silicon wafer's surface.

Graphene is one of the most amazing substances all over the world with unique properties. It is the thinnest and at the same time the strongest material known in the world. Graphene is formed from a perfect two-dimensional carbon atomic network of sp^2 hybridization. It has amazing mechanical strength and is said to be up to 200 times stronger than steel. Its unique structure and characteristics have attracted the attention of many scientists in various fields in recent years. The exceptional physical characteristics of graphene have made it highly appealing for employment in electronics and other applications. Among all, the remarkably high carrier mobility of graphene has gained significant attention. Reports have shown its mobility is more than $100,000 \, cm^2/V \cdot s$ and saturation velocities are of about $5 \times 10^7 \, cm \, s^{-1}$ (Dreyer, Ruoff, & Bielawski, 2010; Mesgari, Beigi, Salehnia, Hosseini, & Ganjali, 2019).

Besides previous ones, it has a pretty high current carrying capacity (up to $10^9 \, A/cm^2$) and high thermal conductivity (up to $5000 \, Wm^{-1} \, K^{-1}$). Furthermore, the thinness, mechanical strength, and flexibility of graphene are exceptional (Avouris & Dimitrakopoulos, 2012); the fracture strength of graphene is about 130 GPa; the surface tension of graphene is about $54.8 \, mN \times m^{-1}$; and the flexural rigidity is $3.18 \, GPa \times nm^3$. In the form of graphite, the distance between the layers of graphene is 3.4 Å and it has $1.4 \times 10^{13} \, cm^{-2}$ charge carrier concentration and high Young's modulus (approximately 1100 GPa). Furthermore, graphene melting temperature is 350 °C, and it also has specific surface area which makes it appropriate for all kind of sensors. At room temperature, the electron mobility of graphene is $2.5 \times 10^5 \, cm^2 \, V^{-1} \, s^{-1}$, and it can absorb 2.3% of visible light (Bolotin et al., 2008).

Graphene is the mother of all graphitic forms, including zero-dimensional buckyballs, one-dimensional carbon nanotube, and three-dimensional graphites. Graphene can also be converted to other carbon materials with a similar structure like fullerene, carbon nanotubes, graphite, and carbon nanohorns through various chemical and physical methods. Since the discovery of graphene, several types of graphene and carbon-based nanomaterials have been made from it. Materials with structural properties of graphene are generally known as graphene derivatives (GDs). A brief description of GDs has been provided below (Salehnia, 2019; Seekaew, Arayawut, Timsorn, & Wongchoosuk, 2018).

Graphene oxide: Graphene oxide (GO) is an analog of graphene with various functional groups like hydroxyl and carbonyl that give it different properties compared to graphene. Since graphene oxide is the oxygenated form of graphene, it has a higher negative charge which gives it particular features. For example, it has lower electrical conductivity and higher electrical resistance in comparison with graphene. On the other hand, graphene oxide's negative charge makes it more soluble than graphene in water (Chen, Tang, & Li, 2010).

Reduced graphene oxide (RGO): It is in the form of graphene palates with areas containing oxidized chemical groups. It has similar properties as graphene because it owns a heterogeneous structure.

Graphene quantum dots (GQDs): Graphene quantum dots are nanometer-sized particles of graphene that exhibit unique properties with diameters below 20 nm. Graphene quantum dots have notable optical properties. Because of π-π interaction in their C=C bonds, they display outstanding photon trapping properties. As a result, a peak at 230 nm can be seen in the absorption spectra of GQDs. Photostability and nonblinking PL with high quantum yield (QY) are another outstanding feature of these nanomaterials (Tian, Tang, Teng, & Lau, 2018).

Graphene nanoribbons: These shapes are obtained by cutting graphene within a parallel axis with less than 100 nM width. Moreover, they can be attained by cutting graphene oxide or reduced graphene oxide (Ezawa, 2006).

In recent years, various methods have been established for the synthesis of graphene and its derivatives. Among all, the most common technics are mechanical and chemical exfoliation and CVD method. A brief explanation of the different conventional graphene synthesis methods has been presented below.

(1) Top-down methods:
   Chemical exfoliation: It is a two-stage process, in which, in the first step, the van der Waals forces between the graphite layers decrease by the rising of the distance between them, resulting in the formation of graphene intercalated compounds (GICs) (Rao, Maitra, & Matte, 2012).
   Mechanical exfoliation: It is the first identified process of graphene synthesis. Graphene sheets with different thicknesses can be obtained by peeling off layers from graphite-like substances like highly ordered pyrolytic graphite (HOPG), single crystals, or natural graphite. This process can be done by various operators like scotch tape, ultrasonication, and electric field (Cooper, 2012).
   Chemical synthesis: One of the methods of large-scale graphene production is the chemical reduction of graphite, which uses strong oxidants like sulfuric acid. Sonication and reduction of graphene oxide (GO) is another method for the production of graphene (Bhuyan, Uddin, Islam, Bipasha, & Hossain, 2016).
(2) Bottom-up methods:
   Epitaxial method: It is an expensive method in which silicon carbide (SiC) is subjected to very high temperatures and silicon sublimated. As a result, one or more layers of graphite remain on the surface (Kumar et al., 2019).

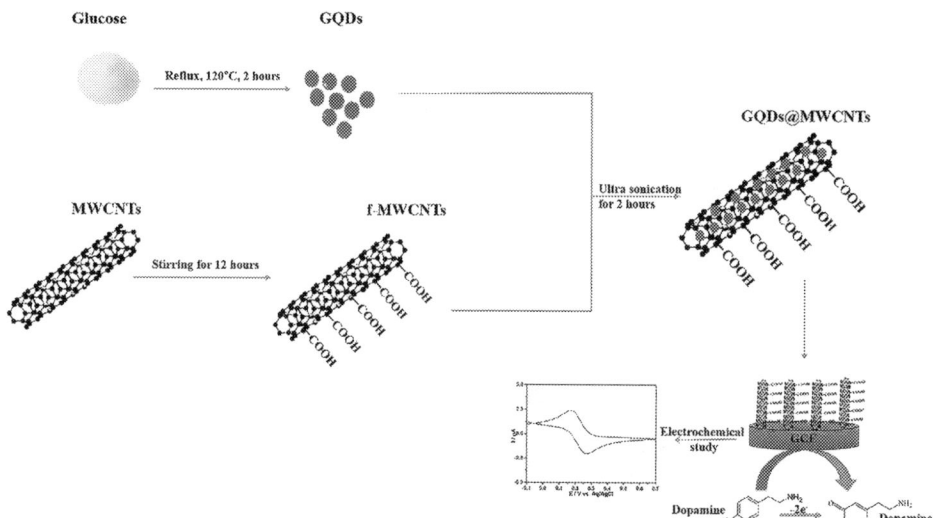

FIG. 5.1 GQD-based sensor. Schematic representation of the overall synthesis process of GQDs@MWCNTs and the electrochemical sensor used to detect dopamine. *From Arumugasamy, S. K., Govindaraju, S., & Yun, K. (2020). Electrochemical sensor for detecting dopamine using graphene quantum dots incorporated with multiwall carbon nanotubes. Applied Surface Science, 508. https://doi.org/10.1016/j.apsusc.2020.145294.*

CVD synthesis: Graphite evaporates at high temperatures and diffuses into the air during this process, thus forming graphene thin layers on the surface (Muñoz & Gómez-Aleixandre, 2013).

Graphene and its derivatives are more generally used in electrochemical sensors, although some, especially GQDs, have been utilized in the development of optical sensors. GQDs have many functional surface groups and attractive electrical properties, making them an exceptional candidate to be used in the development of electrochemical sensors and biosensors. Novel GQDs@MWCNTs biosensor was prepared for the rapid, simple, and sensitive detection of dopamine. Surface carboxyl groups of GQDs could significantly improve the redox response of dopamine. On the other hand, the nanotube's surface polar groups increase the adsorption affinity of the electron-donor or acceptor molecules and promote appropriate response; thus, GQDs@MWCNTs (graphene quantum dots incorporated multiwall carbon nanotubes) nanocomposite can be employed for the development of electrochemical dopamine sensors. The GQDs increase the electrode's electrochemically active surface area, and the MWCNTs enhance the conductivity of the electrode surface. In an article which is written in 2020 by Shiva Kumar and coworkers, a radiometric electrochemical method and GQDs@MWCNTs/GCE nanocomposite were employed to detect dopamine. Fig. 5.1 illustrates the overall synthesis process of GQDs@MWCNTs nanocomposite and the fabrication of the biosensor (Arumugasamy, Govindaraju, & Yun, 2020; Manjunatha & Deraman, 2017).

Upconversion nanoparticles (UCNPs) have outstanding chemical and photochemical stability, and low toxicity without blinking effects. UCNPs are lanthanide-doped nanoparticles, which can absorb low-energy photons and emit fluorescence at shorter wavelengths. UCNPs absorption is in the near-infrared region, and their emission is in the visible region. On the other hand, GO appears to be ideal due to its high solubility in water with high surface area and adequate quenching. In an article written in 2016 by Vilela and coworkers, graphene oxide and UCNPs were used in an optical platform to detect mRNA-related oligonucleotide markers in complex biological fluids as shown in Fig. 5.2 (Vilela et al., 2017).

Up to now, numerous sensors have been reported based on various forms of graphene and its derivatives. Table 5.1 lists some of the most common ones.

Carbon nanotubes

Among all the allotropes of carbon, the most encouraging and broadly studied allotrope is a carbon nanotube (CNT) (Prasek et al., 2011). In 1990, Sumio Iijima, a Japanese electron microscopist from the NEC laboratories, noticed significant graphical structure among fullerenes formed while examining electric discharge chamber walls (Sajid et al., 2016). In 1993, carbon nanotubes with monolayers were reported. CNTs are organized into two zones including tips and sidewalls, and also there are three ways that CNT molecules are rolled based on the crystallographic configurations: armchair, zigzag, and chiral. By taking a cross-section to the nanotube axis, these structures can be observed. Carbon nanotubes have an architecture of sp^2-bonded carbon (Gupta, Gupta, & Sharma, 2019).

To understand the construction of carbon nanotubes (CNTs), they can be considered as graphene sheets that are rolled into cylinders which can be either open or closed at their ends. Their length is about a few microns, and the width of a single-walled CNTs (SWCNTs) is about 0.4 to 2 nm, and that of multi-walled CNTs (MWCNTs) is about 2-100

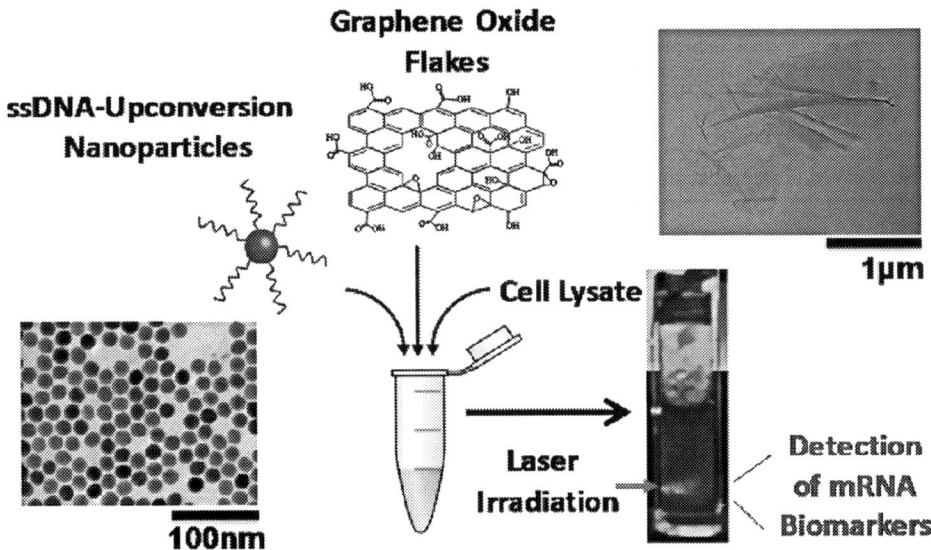

FIG. 5.2 Graphene oxide-based sensor. Principle and function of the sensing platform based on graphene oxide and upconversion nanoparticles (UCNPs) used for the detection of mRNA-related oligonucleotide markers in biological fluids. *From Vilela, P., El-Sagheer, A., Millar, T. M., Brown, T., Muskens, O. L., & Kanaras, A. G. (2017). Graphene oxide upconversion nanoparticle-based optical sensors for targeted detection of mRNA biomarkers present in Alzheimer's disease and prostate cancer. ACS Sensors, 2(1), 52–56. https://doi.org/10.1021/acssensors.6b00651.*

TABLE 5.1 Novel graphene-based sensor. List of graphene-based sensors and biosensors for the detection of biological markers gathered from recent studies.

Carbon-based nanomaterial	Detection method	Analyte	Limit of detection	Author and year
Platinum nanoparticles—carbon nanotubes and graphene oxide	Electrochemical amperometric sensor	Glucose	0.5 μM	Hrapovic, Liu, Male, and Luong, (2004)
Silver nanoparticles (AgNPs) decorated graphene	Field-effect transistor sensor	Glucose	0.0262 μM	Archana et al. (2020)
Graphene quantum dots	Photoluminescence optical sensor	Glucose	3.0 mM	Shehab, Ebrahim, and Soliman (2017)
Graphene quantum dots-wrapped square-plate-like MnO_2 nanocomposites	Optical fluorescent turn-on sensor	Glucose	48 nM	Wang et al. (2021)
Graphene nanosheets	Electrochemical biosensor	Serotonin	5.0×10^{-9} M	Kim, Kim, and Jeon (2012)
Graphene oxide	Electrochemical sensor	Uric acid	0.45 μM	Raj and John (2013)
Reduced graphene oxide/palladium composite	Electrochemical sensor	Epinephrine	30 nM	Renjini, Abraham, Kumar, Kumary, and Chithra (2019)
Graphene oxide-magnetite molecularly imprinted polymers	Chemiluminescence sensor	Epinephrine	1.09×10^{-9} M	Qiu et al. (2012)
Graphene-coated black phosphorus	Surface plasmon resonance sensor	DNA sensor	–	Pal, Verma, Raikwar, Prajapati, and Saini (2018)
Graphene oxide and fluorescent magnetic core-shell nanoparticles	Fluorescent sensor (FRET quenching mechanism)	DNA sensor	0.12 μM	Balaji, Yang, Wang, and Zhang (2019)
Reduced graphene oxide	Field-effect transistor biosensor	DNA sensor	1.0×10^{-7} μM	Cai et al. (2014)
Modified graphite electrode	Electrochemiluminescence biosensor	Glucose	0.04 μM	Hamtak, Hosseini, Fotouhi, and Aghazadeh (2018)

nanometer (Singh & Song, 2012). Singer-layer CNTs are made from rolling a single graphite sheet, while multilayers are formed from the rolling of several graphite sheets (Fakhri et al., 2019). MWCNTs are made up of two to several cylinders, each one made of a single layer of graphite and all surrounded by a central hollow core. The inner diameter of MWCNTs ranges from 1 to 3 nm. The bonds within CNT walls are of the covalent type, while the bonds between them, whether in the structure of one MWCNTs or between several SWCNTs, are of the van der Waals type (Sinnott & Andrews, 2001). CNTs bend and twist easily by mechanical pressure and then return to their normal position. The interesting fact is that the studies see no plastic deformation after returning to the tubular state, which indicates their high elasticity (Iijima, Brabec, Maiti, & Bernholc, 1996).

CNTs have attractive characteristics, including remarkably high tensile strength which is more than stainless steel (≈ 150 GPa), suitable chemical durability, high thermal conductivity (3000 W/m/K), and low density (1100–1300 kg/m^3), and also have many other unique features that have attracted many scientists. Unfortunately, carbon nanotubes are insoluble in water and many other solvents, which has limited their application in many biomedical areas. Convenient carbon nanotube production methods have been presented below (Singh & Song, 2012).

> **Arc discharge**: The first method for making carbon nanotubes was the arc discharge. A direct DC current between graphite electrodes occurs in an inert gas such as helium or argon in this method. In this method, metal catalysts are not needed to produce MWCNTs, while it requires metal catalysts to make SWCNTs (Ferreira et al., 2018).
> **Laser ablation**: During this process, a very high temperature is generated by a laser by which graphite is converted to metal-carbon and catalytic metal particles. Due to the laser beam's interaction with the target, the surface temperature rises to a critical point, and then the evaporation process takes place. With the continuation of laser radiation and increase of temperature, plasma mass is formed. As the plasma cools and the carbon atoms thicken, large carbon structures are formed (Govindaraj & Rao, 2006).
> **CVD method**: This is one of the most widely used methods for producing CNTs on an industrial scale. In this approach, with high temperature, carbon source molecules such as acetylene, CO, or CH_4 decompose and lead to the formation of CNTs (Shanmugam & Prasad, 2018).
> **Silane solution method**: In this method, as registered in the USA, there is a stainless-steel substrate. A carbon source such as ethylene then enters a solution containing cobalt and nickel as the catalysts. Heat is then generated through an electric current which causes the chemical reaction for the formation of CNTs.

Most CNTs production methods end in by-products such as graphite, fullerenes, amorphous carbons, and some impurities such as catalytic metal clusters. The post-production purification process is mainly performed to overcome this problem, which can be monitored by electron microscope examination (Chrzanowska et al., 2015).

CNT is an excellent material because of its excellent electrical conductivity, high surface area, appropriate corrosion resistance, and attractive chemical, physical, and mechanical features. As a result of their unique electro conductivity, CNTs are more commonly used to develop electrochemical sensors. Carbon nanomaterials such as multi-walled carbon nanotubes due to their properties have been embraced over the years as proper electrode materials for fabrication of electrochemical sensors. Rezaeinasab and coworkers presented a stable and selective nonenzymatic sensor by using a carbon paste electrode modified with multiwall carbon nanotubes and Ni(II)-SHP complex as a modifier in an alkaline solution for the determination of glucose in a real sample (Rezaeinasab et al., 2017).

Chitosan (Chit) can be used as a material for wrapping the CNT to make a unique CNT functionalization material. CNT-chitosan can be dispersed in water because they are hydrophilic, and it is also biocompatible and easily prepared. CNT-chitosan is a highly biocompatible composite that has been used for various biomedical analyses. Choi and coworkers in 2018 prepared a electrochemical glucose sensor, based on the chitosan-carbon nanotube hybrid as shown in Fig. 5.3 (Choi et al., 2019).

Table 5.2 provides a list of other novel CNT-based detection approaches for biological markers developed in the recent years.

Fullerene

Fullerene (C60) is a zero-dimensional carbon nanomaterial. It has attracted significant attention recently because of its excellent electrical conductivity, charge carrier, and photo-physical behavior. Among all carbon-based materials, buckminsterfullerene (C60) and its slightly larger homolog (C70) are the pioneers. Buckminsterfullerene is by far the most studied member of the carbon nanomaterial family (Wudl, 2002). C60 and C70 have only one stable isomer. However, two stable isomers are possible for C76 (Scott, 2004). Larger fullerenes have more stable isomers; for example, there are five isomers for C78, 24 isomers for C84, and 450 for C100. More than 1000 different types of fullerenes can be assembled from 100

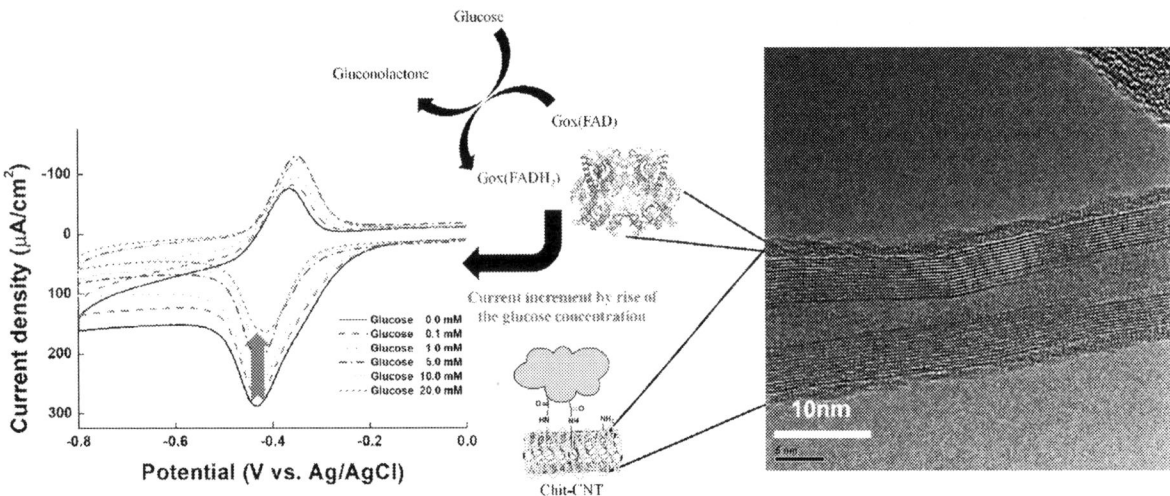

FIG. 5.3 Chitosan-carbon nanotube sensor. Representation of the performance of the electrochemical glucose sensor based on the chitosan-carbon nanotube hybrid. From Choi, Y. B., Kim, H. S., Jeon, W. Y., Lee, B. H., Shin, U. S., & Kim, H. H. (2019). The electrochemical glucose sensing based on the chitosan-carbon nanotube hybrid. Biochemical Engineering Journal, 144, 227–234. https://doi.org/10.1016/j.bej.2018.10.021.

TABLE 5.2 Carbon nanotube-based sensors. List of carbon nanotube-based sensors and biosensors developed for the detection of biological markers.

Carbon-based nanomaterial	Detection method	Analyte	Limit of detection	Author and year
SWCNT-modified carbon-ceramic electrode	Differential pulse voltammetry technique	Dopamine	48 nM	Habibi, Jahanbakhshi, and Pournaghi-Azar (2011)
Functionalized multiwalled carbon nanotubes composite	Electrochemical sensor	Epinephrine	10 nM	Yi, Zheng, Hu, and Hu (2008)
Co-Co_3O_4/carbon nanotube/carbon foam nanocomposites	Nonenzymatic electrochemical sensors	Glucose	0.4 µM	Han, Miao, and Song (2020)
Chitosan-carbon nanotube hybrid	Electrochemical sensor	Glucose	–	Choi et al. (2019)
Carbon nanotube yarn electrodes	Electrochemical voltammetric sensor	Neurotransmitter	0.021 µM	Schmidt, Wang, Zhu, and Sombers (2013)
Carbon nanotube/Prussian blue paste electrodes	Electrochemical amperometric sensor	Hydrogen peroxide	4.74×10^{-3} µM	Husmann, Nossol, and Zarbin (2014)
Monolithic and porous three-dimensional graphene	Enzymatic biosensor	Glucose	1.2 µM	Liu et al. (2014)

carbon atoms or fewer, and this amount increases with the number of carbons without limitation. Furthermore, each shape of fullerene has its unique properties (Zhilei, Zaijun, Xiulan, Yinjun, & Junkang, 2010).

A brief history of the significant points regarding the discovery and synthesis of fullerene has been provided (Mojica, Alonso, & Méndez, 2013; Wu et al., 2014).

1966—Deadalus predicted the construction of large carbon shelves, nowadays called giant fullerenes.
1985—There was a significant breakthrough in the experimental discovery of the fullerenes. At the time, Smalley and Kroto discovered a carbon allotrope consisting of sixty perfectly symmetrically carbon atoms.
1990—Kratschmer/Hoffman detected the first-gram quantities production of C60.
1992—Koruga/Hameroff research team captured the C60's first image.

Among the different forms of fullerene materials, C60 and C70 are more abundant. C60 fullerene is a shortened icosahedron and soccer ball shape molecule containing 60 carbon atoms with C5–C5 single bonds forming 12 pentagons and C5–C6 double bonds forming 20 hexagons (Wei & Shih, 2001). Solid C60 is a brown crystal powder that is insoluble in water and water-solvable polar solvents. However, it is somewhat soluble in hydrophobic solvents. C60 is the most ideally symmetrical structure in nature, with symmetry factors including a reversal center of symmetry, 30 twofold axes, 20 threefold axes, and 12 fivefold axes. It consists of 60 carbon atoms forming 12 pentagons and 20 hexagons, like a soccer ball. In its structure, carbon atoms are joined together by sp^2 bonds. Therefore, fullerene has thirty double bonds, which gives it an aromatic nature. The van der Waals diameter of the C60 molecule is 1 nm. Due to the existence of its 30 double bonds, C60 is the most productive free radical hunter among all identified compounds (Chibante, Thess, Alford, Diener, & Smalley, 1993). Therefore, it has good antioxidant properties. The C60 molecule absorbs light strongly in the ultraviolet and partially in the visible range. Furthermore, it should also be noted that C60 crystals are pretty soft compared to hard graphite crystals. Due to fullerene's hollow structure, it can be receptive to atoms and small molecules inside its structure. In this condition, molecules inside C60 were safe from the effects and influences of the outside environment (Lin & Shih, 2011).

Numerous experiments have shown that C60 and its derivatives can specifically cut DNA molecules due to their spherical shape, and they are also highly toxic when exposed to UV light. Another important feature of this particular molecule is that due to its size and hydrophobic properties, it can be located in the hydrophobic section of some specific enzymes and lead the inhibition of their activity (Moussa, 2018). Fullerene is incredibly active and has unique antioxidant properties because of its many free bonds that can react with numerous radicals. Fullerene can make holes in the cell membrane, modulate ion transportation, and pass the blood-brain barrier. The principal problem of fullerene in the field of biomedical application is the fact that fullerenes are insoluble in water and water-like solvents (Pochkaeva et al., 2020).

Different high-temperature methods that produce carbon atoms in the gas phase can be used to make carbon clusters to make a variety of types of fullerenes, and various approaches have been introduced to synthesize C60, but a few of them are efficient and valuable. Although the laser vaporization of graphite performs better than the arc method and archives a more significant amount of fullerene, the cost of using lasers is very high, so graphic laser vaporation is not a rational option for commercial production of fullerenes (Withers, Loutfy, & Löwe, 1997). Also, solar energy has been used to make fullerenes so that this energy is concentrated on carbon to vapor it. C60 and other types of fullerene fullerenes can be quickly provided by vaporation of carbon sources on an industrial scale. On the other hand, the chemical synthesis of fullerenes is considerable. The major challenge in the synthesis of fullerenes is the introduction of curvatures or pyramidalization in the carbon network (Wakabayashi & Achiba, 1992). The chemical synthesis method of fullerene, which takes place in 12 steps, has the advantage over graphite vaporization in that it happens in a controlled way. In this case, different types of fullerene can be directly obtained that did not produce in the previous one. Also, in fullerenes manufacture by targeted chemical method, no fullerene is formed as a by-product (Scott et al., 2002).

Fullerene materials have been extensively researched and studied in recent years because of their novel characteristics such as exceptional chemical stability, electrical conductivity, charge transport, photophysical behavior, and biocompatibility. These unique features have made them appropriate for using them in various sensing platforms, especially in electrochemical sensing. The first utilization of C60 fullerene was reported for chemosensing of adenosine-5-triphosphate (ATP).

Considering the suitable electrical properties of fullerene and the electrocatalytic properties of carbon nanotubes, fullerene (C60) and functionalized multi-walled carbon nanotubes can act together as active electrocatalytic electrodes. Taking advantage of these features, Mazloum-Ardakani and coworkers presented an electrochemical sensing approach to detect catecholamines based on fullerene-functionalized carbon nanotubes/ionic liquid in 2014 (Mazloum-Ardakani & Khoshroo, 2014).

In another work, Wei, Lei, Fu, and Yao (2012) produced an electrocatalytic sensor based on C60 hollow microspheres that were synthesized by a high-temperature reprecipitation method in alcohol bubbles for the detection of dopamine (DA) in the presence of ascorbic acid (AA), and uric acid (UA), and L-cysteine (RSH) as shown in Fig. 5.4 (Wei & Shih, 2001).

Many other sensors have been developed based on fullerene, the most recent of which have been listed in Table 5.3.

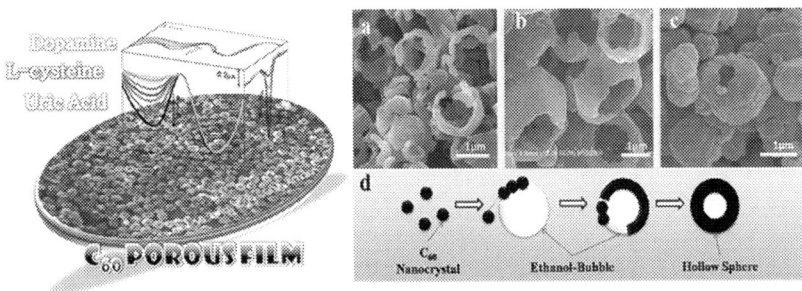

FIG. 5.4 C60 hollow microsphere-based sensor. Representation of an electrocatalytic sensor based on C60 hollow microspheres for the detection of dopamine. From Wei, L., Lei, Y., Fu, H., & Yao, J. (2012). Fullerene hollow microspheres prepared by bubble templates as sensitive and selective electrocatalytic sensor for biomolecules. ACS Applied Materials and Interfaces, 4(3), 1594–1600. https://doi.org/10.1021/am201769u.

TABLE 5.3 Fullerene-based sensors. List of fullerene-based sensors and biosensors for the detection of biological markers.

Carbon-based nanomaterial	Detection method	Analyte	Limit of detection	Author and year
Fullerene-C60-coated gold electrode	Electrochemical sensor	Dopamine	0.26×10^{-9} M	Goyal, Gupta, Bachheti, and Sharma (2008)
Fullerene-pyrrole-pyrrole-3-carboxylic acid nanocomposite	Electrochemical impedimetric sensor	Dopamine	8.77 ng/mL	Ertuğrul Uygun and Demir (2020)
Fullerene-C60	Electrochemical voltammetric sensor	Nandrolone	0.42 nM	Goyal, Gupta, and Bachheti (2007)
Fullerene glucose oxidase	Piezoelectric sensor	Glucose	3.9×10^{-5}	Chuang and Shih (2001)
Zinc porphyrin-fullerene (C60) derivative (ZnP-C60)	Nonenzymatic electrochemical sensor	Hydrogen peroxide	0.81 μM	Wu et al. (2014)
Fullerene C60-cryptand22 coated	Piezoelectric crystal urea sensor	Urea	10^{-4} M	Wei and Shih (2001)
C60 fullerene-glutathione self-assembled monolayer	Surface plasmon resonance sensor	L-histidine	10^{-10} M	Stefan-Van Staden (2014)
Fullerene C60	Enzyme-based electrochemical sensor	Glucose	1.6×10^{-6} M	Lin and Shih (2011)

Nanodiamonds

One of the forms of carbon nanomaterials is nanodiamond (ND) which is produced industrially on a large scale nowadays. ND is an allotropic nanocarbon with a length of 1–100 nm. Nanodiamonds are classified into three groups based on their initial particle size: nanocrystalline particles, ultrananocrystalline particles, and diamondoids, including diamondoids (1–2 nM), nanocrystalline particles (10–100 nM), and ultrananocrystalline (0–10 nM), classified by their size. The surface of nanodiamonds may contain many highly polar chemically reactive groups such as amides, amines, cyano, nitro, hydroxyl, carbonyl, and anhydride groups (Kausar, 2018a, 2018b). The average size of nanodiamond particles is about 4–5 nm. The structure of nanodiamonds consists of three parts: the inner core with sp^3 hybridized carbon, the middle layer of carbon around the core, and another carbon shells which contain sp^2 and sp^3 hybridized carbon with functional groups on its surface. Here a brief history of the discovery and key works of nanodiamonds has been presented (Danilenko & Shenderova, 2012; Danilenko, 2004; Kausar, 2018a; Kausar, 2018b).

1961—Diamond synthesis with preservation of shock-compressed graphite in a plane ampoule by P.J. De Carli and A.C. Jamisson was done.
1963—Diamond synthesis by compression of graphite and Me mixtures by V.V. Danilenko and V.I. Elin (VNIITP) and ultrananocrystalline particles synthesis from carbon of explosion products by K.V. Volkov, V.V. Danilenko, and V.I. Elin was done.
1976—Commercial production of diamond micropowder by compressing graphite and copper mixture was performed by Du Pont de Nemours.
1982—Independent work on ultrananocrystalline particles synthesis was carried out by detonation of carbon containing explosive by Staver and coworkers.
1983—Pilot-scale production of diamond micropowders by explosion of graphite or carbon black mixture with RDX was performed in a specific chamber by O.N. Breusov, V.N. Drobyshev, and G.A. Adadurov.
1994—Ampoule-free sintering of high-density ultrananocrystalline particle grains by explosion of a diamond was done by V.V. Danilenko and coworkers.
2008—Ultrananocrystalline particle synthesis in the presence of reducing agents was performed by V.Y. Dolmatov in the cooling media single crystals.

Different synthesis approaches for nanodiamonds can lead to variations in their size, shape, and surface functional groups. Therefore, the thermal stability of nanodiamonds is dependent on their synthesis methods (Wen & Tian, 2017). Nanodiamonds synthesis methods are basically classified into three general categories, although many new synthesis methods have been published in recent years (Shenderova, Zhirnov, & Brenner, 2002): low pressure and moderate temperature, high pressure and high temperature, and radiation or using high-energy particles. Here is a brief overview of conventional methods for the synthesis of carbon nanodiamonds (Gruen, Shenderova, & Vul, 2004; Ilie-Mihai, Gheorghe, Stefan-van Staden, & Bratei, 2021).

Shock wave compression: In this method, due to the explosion, a shock wave enters graphite, which results in a high temperature and pressure, and graphite is converted to diamond directly.
Detonation of carbon containing explosives: In this method, explosives are used instead of graphite, and as a result, due to high temperature and pressure, the soot from the explosion is turned into a diamond.
Chemical vapor deposition (CVD): Silicon wafer is used as a substrate, and a mixture of hydrogen and methane gases is placed in the reactor chamber. After reaching the optimal condition, 50–100 nm films of nanodiamonds will be created.
High-energy beam radiation: With high-energy beam radiation such as laser beams or high-energy electron beams, many carbon nanomaterials can be converted directly to nanodiamonds.
Reduction of carbides: Gogotsi and coworkers developed a reduction reaction in which β-SiC powders and Ar-Cl_2-H_2 gases are combined at high temperature and atmospheric pressure to form a layer of nanodiamond film on the surface.

Carbon nanodiamonds have numerous advantages including low cost and feasible large-scale synthesis. Due to the strong covalent bonds in their structure, they have excellent mechanical properties such as high hardness and low friction. Moreover, the lower toxicity of nanodiamonds compared with other carbon nanomaterials makes them a well-suited candidate for biomedical applications, such as biomedical imaging, drug delivery, and biosensing (Schrand, Hens, & Shenderova, 2009). Other features accounting for their suitability in biomedical applications include excellent Young's modulus, hardness, optical and fluorescence properties, chemical stability, thermal conductivity and stability, and biocompatibility (Simioni, Silva, Oliveira, & Fatibello-Filho, 2017). Electronic properties of nanodiamonds have been studied using X-ray absorption near-edge studies. Fluorescence properties of nanodiamond have also been found notable in magnetic sensing, resonance energy transfer, and biomedical imaging (Mochalin, Shenderova, Ho, & Gogotsi, 2012).

Due to the prominent electrical properties of carbon nanodiamonds, they have been used as suitable materials for the manufacturing of electrochemical sensors and electrodes. Recently, the surface-modified nanodiamond materials have shown significant potential for the creation of sensitive and selective biosensors. Huang and coworkers in 2018 developed a novel strategy for nonenzymatic glucose sensing using copper oxide (CuO)-deposited oxygen-doped nitrogen-incorporated nanodiamond (NOND)/Si pyramids (Pyr-Si). Because of the outstanding chemical stability of NONDs, coupled with the physicochemical properties of CuO, an outstanding nanosensor for the detection of glucose was developed as displayed in Fig. 5.5 (Huang et al., 2018).

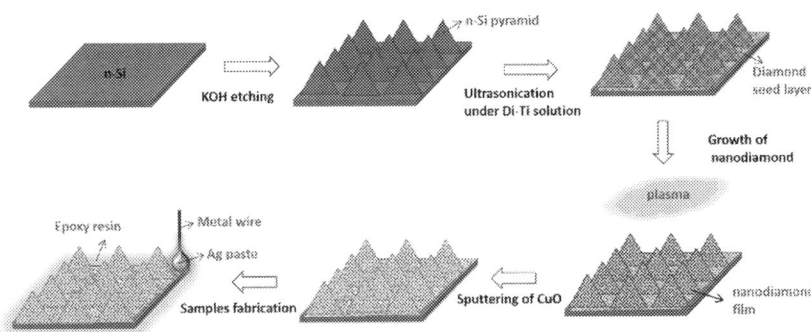

FIG. 5.5 Nanodiamond-based sensor. Representation of the fabrication of CuO/NOND/Pyr-Si electrode and the principle of oxygen-doped nanodiamond on CuO and micropyramidal silicon heterostructures nonenzymatic glucose sensor. *From Huang, B. R., Wang, M. J., Kathiravan, D., Kurniawan, A., Zhang, H. H., & Yang, W. L. (2018). Interfacial Effect of Oxygen-Doped Nanodiamond on CuO and Micropyramidal Silicon Heterostructures for Efficient Nonenzymatic Glucose Sensor. ACS Applied Bio Materials, 1(5), 1579–1586. https://doi.org/10.1021/acsabm.8b00454.*

Carbon quantum dots

Carbon quantum dots (CQDs) are described as a zero-dimensional class of core-shell materials containing a carbon core and surface with several functional groups, like hydroxyl, carboxyl, and amine. CQDs are a group of nanomaterials with comparatively better solubility and strong fluorescence (Zheng, Ananthanarayanan, Luo, & Chen, 2015). They are a class of carbon nanomaterials with spherical shape and sizes below 10 nm. There are two forms of CQDs: crystalline or amorphous. Graphene quantum dots have a crystalline structure and are built of single or a few graphene layers. The hybridization of carbon atoms in CQDs is sp^2, but sp^3 hybridization can also be found (Xu, Liu, Gao, & Wang, 2014). CQDs were first acquired through the purification of single-walled carbon nanotubes by preparative electrophoresis in 2004, and then again after a while in 2006 during laser ablation of graphite powder and cement. There are many carboxyl groups on the surface of carbon quantum dots, making them biocompatible and soluble in water. In fact, the toxicity of CQDs depends on the functional groups that cover its surface; for instance, PEGylated CQDs were found to be noncytotoxic (Wang & Hu, 2014). Moreover, their chemical surface plays a crucial role in their photoluminescence behavior. One of the most exciting characteristics of CQDs is their photoluminescence (PL), which has rendered them an ideal option for detection and real-time imaging as they show nonblinking PL abilities. CQDs typically show absorption in the UV region and partially in the visible range (Farshbaf et al., 2018). Carbon quantum dots maintain excellent properties, which has made them useful in many fields. The most important of these features are high solubility in water, easy surface modification, nontoxicity, biocompatibility, and appropriate photostability (Pebdeni, Hosseini, & Ganjali, 2020). The unique biological features of CQDs, such as biocompatibility and low toxicity, make them an appropriate nanomaterial for applications in biosensing, bioimaging, and drug delivery. Other important features of CQDs include efficient light harvesting, exceptional upconverted photoluminescence (UCPL), and excellent photo-induced electron transfer that has attracted significant attention in different photocatalytic applications (Wang & Hu, 2014). CQDs can be acceptors and donors of electrons and therefore be utilized as photosensitizers, electron mediators, and spectral converter in photocatalyst systems (Molaei, 2020a, 2020b).

Carbon quantum dot synthesis methods are divided into two general categories according to the size of performed materials: top-down methods and bottom-up methods. In the first approach, CQDs are synthesized from macroscopic carbon nanostructures such as graphite and its derivatives, while in the bottom-up methods, CQDs are made from molecular precursors. Here are some convenient methods for synthesizing carbon quantum dots (Molaei, 2020b).

(1) Top-down methods
   Laser ablation: In this method, using laser energy, the target surface is heated enough to create a plasma state. The resulting vapor then forms nanoparticles.
   Acidic oxidation: In this method, exfoliate, and decomposition of carbon atoms is performed, and as a result, carbon nanoparticles with hydrophobic surface groups are obtained.
   Arc discharge: In this method, in a reactor under specific conditions and by using gas plasma, decomposed carbon atoms from bulk precursors are rearranged.
(2) Bottom-up methods
   Microwave synthesis: In this method, the combination of polyethylene glycol and saccharides is exposed to microwave heat in an aqueous environment and leads to CQDs production. Design of carbon-based electrocatalysts is done this way.

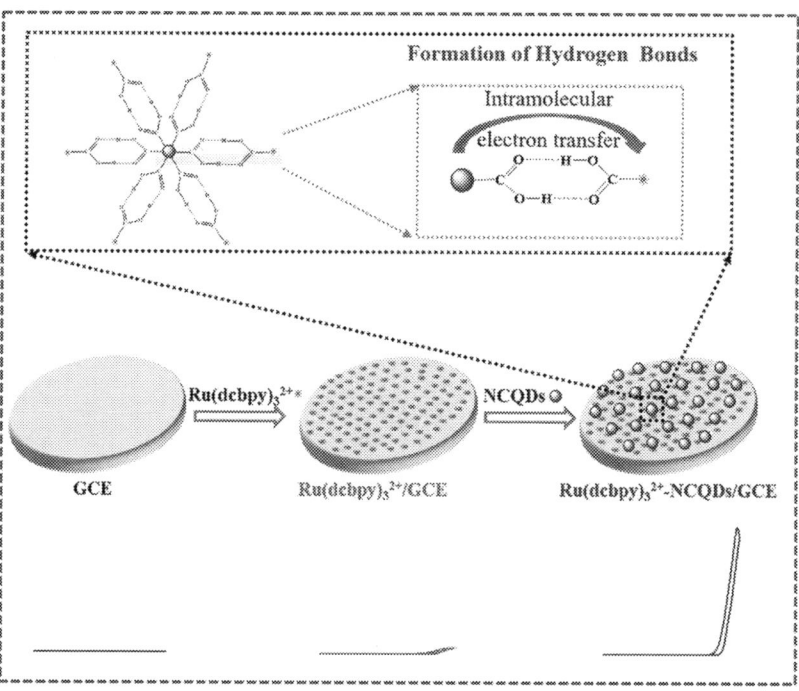

FIG. 5.6 Carbon quantum dot-based sensor. Representation of the principle of the electrochemiluminescence (ECL) sensor for 17b-estradiol detection using glassy carbon electrode and the formation of hydrogen bonds. *From Liu, X., Li, L., Luo, L., Bi, X., Yan, H., Li, X., & You, T. (2021). Induced self-enhanced electrochemiluminescence aptamer sensor for 17β-estradiol detection based on nitrogen-doped carbon quantum dots as Ru(dcbpy)32+ coreactant: What role of intermolecular hydrogen bonds play in the system? Journal of Colloid and Interface Science, 586, 103–109. https://doi.org/10.1016/j.jcis.2020.10.074.*

Hydrothermal treatment: Polymers and organic molecules are placed in an autoclave made of stainless steel and Teflon coating. These materials are then combined at high temperatures to form carbon seeding cores which will become CQDs.

Both carbon quantum dots and graphene quantum dots are carbon nanomaterials with unique properties. In addition to their numerous optical features such as nonblinking PL and remarkable photostability, they also have notable electrochemical and electrochemiluminescence (ECL) properties due to their surface functional groups (Zheng et al., 2015). Nitrogen-doped graphene quantum dots (NGQDs) and polyethylenimine (PEI)-functionalized NCQDs (NCQDs@PEI) have also been used as the practical coreactants of Ru(bpy)3 for developing ECL sensing platforms. They also have good biocompatibility and low toxicity, and appropriate luminescent functions. Xiaohong Liu and coworkers developed a novel solid-state electrochemiluminescence (ECL) sensor for 17b-estradiol detection based on an intermolecular hydrogen bonds-induced self-enhanced ECL composite (Ru(dcbpy)3 2+-NCQDs). Based on the carboxyl (–COOH) groups in Ru(dcbpy)3 2+ and oxygen, nitrogen-containing groups on NCQDs surface, an intermolecular hydrogen bonds-induced self-enhanced ECL composite was generated in the solid contact layer as shown in Fig. 5.6 (Liu et al., 2021).

Considering the excellent luminescence and great fluorescence stability of CQDs, a fluorescent sensing strategy was developed by Zhang and coworkers in 2020 using nitrogen-doped carbon quantum dots (N-CQDs) and efficient catalytic oxidation of tyrosinase (Tyr) for the detection of dopamine (DA) and alpha-lipoic acid (ALA) as depicted in Fig. 5.7 (Zhang & Fan, 2020).

Other recent CQD-based sensing approaches for biomedical purposes have been summarized in Table 5.4.

Conclusion and outlook

In recent decades, many sensors have been developed based on various carbon nanomaterials for the analysis and determination of biomedical compounds. Carbon nanomaterials possess unique electrical, mechanical, optical, and thermal properties such as high carrier mobility, optical properties, fluorescence, and thermal conductivity, which have rendered them extremely efficient in the design of biosensors. Carbon nanomaterial is particularly used in electrochemical sensors because of their electron transfer properties, resulting in higher sensitivity of the designed sensors. Another advantage of using carbon nanomaterials is that their nanometer-ranged thickness leads to the significant reduction of the size of sensors. For in vitro and in vivo analyses, especially with real samples, the flexibility of carbon-based nanomaterial sensors is essential and beneficial.

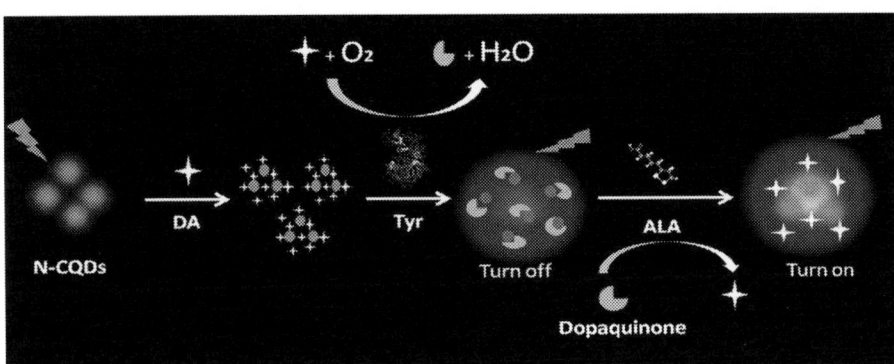

FIG. 5.7 N-doped carbon quantum dot-based sensor. Representation of the preparation and principle of the N-CQDs probe for DA and ALA sensing. *From Zhang, R., & Fan, Z. (2020). Nitrogen-doped carbon quantum dots as a "turn off-on" fluorescence sensor based on the redox reaction mechanism for the sensitive detection of dopamine and alpha lipoic acid. Journal of Photochemistry and Photobiology A: Chemistry, 392.* https://doi.org/10.1016/j.jphotochem.2020.112438.

TABLE 5.4 Carbon quantum dot-based sensors. List of carbon quantum dot-based sensors and biosensors for the detection of biological markers.

Carbon-based nanomaterial	Detection method	Analyte	Limit of detection	Author and year
Carbon quantum dots/octahedral Cu$_2$O nanocomposites	Nonenzymatic electrochemical amperometric sensor	Glucose	8.4 µM	Li, Zhong, Zhang, Weng, and Li (2015)
Nitrogen-doped carbon quantum dot	Fluorescence sensor	β-glucuronidase	0.3 U L^{-1}	Lu et al. (2016)
Surfaced passivation carbon quantum dot	Fluorescence sensor	Glutathione	0.943 µM	Pan et al. (2017)
Carbon quantum dots codoped with nitrogen, phosphorus, and chlorine (NPCl)	Ratiometric fluorescent sensors	Riboflavin	3.50 nM	Wang et al. (2021)
Nitrogen-doped carbon quantum dots	Electrochemiluminescence aptamer sensor	17b-estradiol	1.0×10^{-15} M	Liu et al. (2021)
Gold nanoparticles/carbon quantum dots composite	Colorimetric and fluorometric dual-signal sensor	Arginine	450 nM	Liu et al. (2017)

There are also many challenges to understand the properties and characteristics of carbon nanomaterials and their function in different environments. The toxicity and biocompatibility of these materials are still under investigation. Another critical challenge that researchers face is modifying and functionalizing the surface of carbon nanomaterials to control their properties and characteristics to improve their performance in different sensing approaches. Furthermore, the industrial production of carbon nanomaterials is one of the primary concerns due to the negative impact it can have on the environment and on human health.

In this chapter, we have provided an overview on the various carbon nanomaterials which are most commonly used in the fabrication of biomedical sensors, reflecting the positive characteristics of these nanomaterials which can result in highly sensitive and selective sensing platforms. With regard to their numerous unique attributes, it seems that carbon nanomaterials will be used in the future to build small, accurate, and biocompatible sensors to analyze biomarkers and biomolecules in humans in real time.

References

Adhikari, B. R., Govindhan, M., & Chen, A. (2015). Carbon nanomaterials based electrochemical sensors/biosensors for the sensitive detection of pharmaceutical and biological compounds. *Sensors (Switzerland)*, *15*(9), 22490–22508. https://doi.org/10.3390/s150922490.

Ahammad, A. J. S., Islam, T., & Hasan, M. M. (2019). Graphene-based electrochemical sensors for biomedical applications. In *Biomedical applications of graphene and 2D nanomaterials* (pp. 249–282). Elsevier. https://doi.org/10.1016/B978-0-12-815889-0.00012-X.

Anhar, N. A. M., Ramli, M. M., Halin, D. S. C., Isa, S. S. M., Hambali, N. A. M. A., & Abdullah, M. M. A. B. (2017). Reduced-graphene oxide in flexible substrate for wearable physiological sensor—A review. In *Vol. 1885. AIP conference proceedings* American Institute of Physics Inc. https://doi.org/10.1063/1.5002422.

Archana, R., Sreeja, B., Nagarajan, K., Radha, S., BalajiBhargav, P., Balaji, C., et al. (2020). Development of highly sensitive Ag NPs decorated graphene FET sensor for detection of glucose concentration. *Journal of Inorganic and Organometallic Polymers and Materials, 30*(9), 3818–3825. https://doi.org/10.1007/s10904-020-01541-6.

Arumugasamy, S. K., Govindaraju, S., & Yun, K. (2020). Electrochemical sensor for detecting dopamine using graphene quantum dots incorporated with multiwall carbon nanotubes. *Applied Surface Science, 508*. https://doi.org/10.1016/j.apsusc.2020.145294.

Avouris, P., & Dimitrakopoulos, C. (2012). Graphene: Synthesis and applications. *Materials Today, 15*(3), 86–97. https://doi.org/10.1016/S1369-7021(12)70044-5.

Balaji, A., Yang, S., Wang, J., & Zhang, J. (2019). Graphene oxide-based nanostructured DNA sensor. *Biosensors, 9*(2). https://doi.org/10.3390/bios9020074.

Baptista, F. R. (2015). Recent developments in carbon nanomaterial sensors. *Chemical Society Reviews, 44*.

Beigi, S. M., Mesgari, F., Hosseini, M., Aghazadeh, M., & Ganjali, M. R. (2019). An enhancement of luminol chemiluminescence by cobalt hydroxide decorated porous graphene and its application in glucose analysis. *Analytical Methods, 11*(10), 1346–1352. https://doi.org/10.1039/c8ay02727g.

Bhuyan, M. S. A., Uddin, M. N., Islam, M. M., Bipasha, F. A., & Hossain, S. S. (2016). Synthesis of graphene. *International Nano Letters*, 65–83. https://doi.org/10.1007/s40089-015-0176-1.

Bolotin, K. I., Sikes, K. J., Jiang, Z., Klima, M., Fudenberg, G., Hone, J., et al. (2008). Ultrahigh electron mobility in suspended graphene. *Solid State Communications, 146*(9–10), 351–355. https://doi.org/10.1016/j.ssc.2008.02.024.

Cai, B., Wang, S., Huang, L., Ning, Y., Zhang, Z., & Zhang, G. J. (2014). Ultrasensitive label-free detection of PNA-DNA hybridization by reduced graphene oxide field-effect transistor biosensor. *ACS Nano, 8*(3), 2632–2638. https://doi.org/10.1021/nn4063424.

Cernat, A., Tertiș, M., Fritea, L., & Cristea, C. (2016). Graphene in sensors design. In *Advanced 2D materials* (pp. 387–431). Wiley. https://doi.org/10.1002/9781119242635.ch10.

Chen, D., Tang, L., & Li, J. (2010). Graphene-based materials in electrochemistry. *Chemical Society Reviews, 39*(8), 3157–3180. https://doi.org/10.1039/b923596e.

Chibante, L. P. F., Thess, A., Alford, J. M., Diener, M. D., & Smalley, R. E. (1993). Solar generation of the fullerenes. *The Journal of Physical Chemistry, 97*(34), 8696–8700. https://doi.org/10.1021/j100136a007.

Choi, Y. B., Kim, H. S., Jeon, W. Y., Lee, B. H., Shin, U. S., & Kim, H. H. (2019). The electrochemical glucose sensing based on the chitosan-carbon nanotube hybrid. *Biochemical Engineering Journal, 144*, 227–234. https://doi.org/10.1016/j.bej.2018.10.021.

Chrzanowska, J., Hoffman, J., Małolepszy, A., Mazurkiewicz, M., Kowalewski, T. A., Szymanski, Z., et al. (2015). Synthesis of carbon nanotubes by the laser ablation method: Effect of laser wavelength. *Physica Status Solidi B, 252*(8), 1860–1867. https://doi.org/10.1002/pssb.201451614.

Chuang, C. W., & Shih, J. S. (2001). Preparation and application of immobilized C60-glucose oxidase enzyme in fullerene C60-coated piezoelectric quartz crystal glucose sensor. *Sensors and Actuators, B: Chemical, 81*(1), 1–8. https://doi.org/10.1016/S0925-4005(01)00914-5.

Cooper, D. R. (2012). Experimental review of graphene. *International Scholarly Research Notices, 2012*.

Danilenko, V., & Shenderova, O. A. (2012). *Advances in synthesis of nanodiamond particles. Ultrananocrystalline diamond: Synthesis, properties and applications* (2nd ed., pp. 133–164). Elsevier Inc. https://doi.org/10.1016/B978-1-4377-3465-2.00005-0.

Danilenko, V. V. (2004). On the history of the discovery of nanodiamond synthesis. *Physics of the Solid State, 46*(4), 595–599. https://doi.org/10.1134/1.1711431.

Dreyer, D. R., Ruoff, R. S., & Bielawski, C. W. (2010). From conception to realization: An historical account of graphene and some perspectives for its future. *Angewandte Chemie, International Edition, 49*(49), 9336–9344. https://doi.org/10.1002/anie.201003024.

Ertuğrul Uygun, H. D., & Demir, M. N. (2020). A novel fullerene-pyrrole-pyrrole-3-carboxylic acid nanocomposite modified molecularly imprinted impedimetric sensor for dopamine determination in urine. *Electroanalysis, 32*(9), 1971–1976. https://doi.org/10.1002/elan.202060023.

Ezawa, M. (2006). Peculiar width dependence of the electronic properties of carbon nanoribbons. *Physical Review B, 73*(4). https://doi.org/10.1103/physrevb.73.045432.

Fakhri, N., Salehnia, F., Mohammad Beigi, S., Aghabalazadeh, S., Hosseini, M., & Ganjali, M. R. (2019). Enhanced peroxidase-like activity of platinum nanoparticles decorated on nickel- and nitrogen-doped graphene nanotubes: Colorimetric detection of glucose. *Microchimica Acta, 186*(6). https://doi.org/10.1007/s00604-019-3489-3.

Farshbaf, M., Davaran, S., Rahimi, F., Annabi, N., Salehi, R., & Akbarzadeh, A. (2018). Carbon quantum dots: Recent progresses on synthesis, surface modification and applications. *Artificial Cells, Nanomedicine and Biotechnology, 46*(7), 1331–1348. https://doi.org/10.1080/21691401.2017.1377725.

Ferreira, F. V., Franceschi, W., Menezes, B. R. C., Biagioni, A. F., Coutinho, A. R., & Cividanes, L. S. (2018). Synthesis, characterization, and applications of carbon nanotubes. In *Carbon-based nanofillers and their rubber nanocomposites: Carbon nano-objects* (pp. 1–45). Elsevier. https://doi.org/10.1016/B978-0-12-813248-7.00001-8.

Govindaraj, A., & Rao, C. N. R. (2006). Synthesis, growth mechanism and processing of carbon nanotubes. In *Carbon nanotechnology* (pp. 15–51). Elsevier. https://doi.org/10.1016/B978-044451855-2/50005-X.

Goyal, R. N., Gupta, V. K., & Bachheti, N. (2007). Fullerene-C60-modified electrode as a sensitive voltammetric sensor for detection of nandrolone—An anabolic steroid used in doping. *Analytica Chimica Acta, 597*(1), 82–89. https://doi.org/10.1016/j.aca.2007.06.017.

Goyal, R. N., Gupta, V. K., Bachheti, N., & Sharma, R. A. (2008). Electrochemical sensor for the determination of dopamine in presence of high concentration of ascorbic acid using a fullerene-C60 coated gold electrode. *Electroanalysis*, *20*(7), 757–764. https://doi.org/10.1002/elan.200704073.

Gruen, D. M., Shenderova, O. A., & Vul, A. Y. (2004). Synthesis, properties and applications of ultrananocrystalline diamond. In *Vol. 192. Proceedings of the NATO ARW on synthesis, properties and applications of ultrananocrystalline diamond*.

Gupta, N., Gupta, S. M., & Sharma, S. K. (2019). Carbon nanotubes: Synthesis, properties and engineering applications. *Carbon Letters*, *29*(5), 419–447. https://doi.org/10.1007/s42823-019-00068-2.

Habibi, B., Jahanbakhshi, M., & Pournaghi-Azar, M. H. (2011). Simultaneous determination of acetaminophen and dopamine using SWCNT modified carbon-ceramic electrode by differential pulse voltammetry. *Electrochimica Acta*, *56*(7), 2888–2894. https://doi.org/10.1016/j.electacta.2010.12.079.

Hamtak, M., Hosseini, M., Fotouhi, L., & Aghazadeh, M. (2018). A new electrochemiluminescence biosensor for the detection of glucose based on polypyrrole/polyluminol/Ni(OH)$_2$-C$_3$N$_4$/glucose oxidase-modified graphite electrode. *Analytical Methods*, *10*(47), 5723–5730. https://doi.org/10.1039/c8ay01849a.

Han, J., Miao, L., & Song, Y. (2020). Preparation of co-Co3O4/carbon nanotube/carbon foam for glucose sensor. *Journal of Molecular Recognition*, *33*(3). https://doi.org/10.1002/jmr.2820.

Hrapovic, S., Liu, Y., Male, K. B., & Luong, J. H. T. (2004). Electrochemical biosensing platforms using platinum nanoparticles and carbon nanotubes. *Analytical Chemistry*, *76*(4), 1083–1088. https://doi.org/10.1021/ac035143t.

Huang, B. R., Wang, M. J., Kathiravan, D., Kurniawan, A., Zhang, H. H., & Yang, W. L. (2018). Interfacial effect of oxygen-doped nanodiamond on CuO and micropyramidal silicon heterostructures for efficient nonenzymatic glucose sensor. *ACS Applied Bio Materials*, *1*(5), 1579–1586. https://doi.org/10.1021/acsabm.8b00454.

Husmann, S., Nossol, E., & Zarbin, A. J. G. (2014). Carbon nanotube/Prussian blue paste electrodes: Characterization and study of key parameters for application as sensors for determination of low concentration of hydrogen peroxide. *Sensors and Actuators, B: Chemical*, *192*, 782–790. https://doi.org/10.1016/j.snb.2013.10.074.

Iijima, S., Brabec, C., Maiti, A., & Bernholc, J. (1996). Structural flexibility of carbon nanotubes. *Journal of Chemical Physics*, *104*(5), 2089–2092. https://doi.org/10.1063/1.470966.

Ilie-Mihai, R. M., Gheorghe, S. S., Stefan-van Staden, R. I., & Bratei, A. (2021). Electroanalysis of interleukins 1β, 6, and 12 in biological samples using a needle stochastic sensor based on nanodiamond paste. *Electroanalysis*, *33*(1), 6–10. https://doi.org/10.1002/elan.202060118.

Kausar, A. (2018a). Nanodiamond reinforcement in polyamide and polyimide matrices: Fundamentals and applications. *Journal of Plastic Film & Sheeting*, *34*(4), 439–458. https://doi.org/10.1177/8756087918773521.

Kausar, A. (2018b). Properties and applications of nanodiamond nanocomposite. *American Journal of Nanoscience and Nanotechnology Research*, *6*(1), 46–54.

Kim, S. K., Kim, D., & Jeon, S. (2012). Electrochemical determination of serotonin on glassy carbon electrode modified with various graphene nanomaterials. *Sensors and Actuators, B: Chemical*, *174*, 285–291. https://doi.org/10.1016/j.snb.2012.08.034.

Kumar, R., Sahoo, S., Joanni, E., Singh, R. K., Tan, W. K., Kar, K. K., et al. (2019). Recent progress in the synthesis of graphene and derived materials for next generation electrodes of high performance lithium ion batteries. *Progress in Energy and Combustion Science*, *100786*. https://doi.org/10.1016/j.pecs.2019.100786.

Li, Y., Zhong, Y., Zhang, Y., Weng, W., & Li, S. (2015). Carbon quantum dots/octahedral Cu2O nanocomposites for non-enzymatic glucose and hydrogen peroxide amperometric sensor. *Sensors and Actuators, B: Chemical*, *206*, 735–743. https://doi.org/10.1016/j.snb.2014.09.016.

Lin, L. H., & Shih, J. S. (2011). Immobilized fullerene C60-enzyme-based electrochemical glucose sensor. *Journal of the Chinese Chemical Society*, *58*(2), 228–235. https://doi.org/10.1002/jccs.201190081.

Liu, T., Li, N., Dong, J. X., Zhang, Y., Fan, Y. Z., Lin, S. M., et al. (2017). A colorimetric and fluorometric dual-signal sensor for arginine detection by inhibiting the growth of gold nanoparticles/carbon quantum dots composite. *Biosensors and Bioelectronics*, *87*, 772–778. https://doi.org/10.1016/j.bios.2016.08.098.

Liu, X., Li, L., Luo, L., Bi, X., Yan, H., Li, X., et al. (2021). Induced self-enhanced electrochemiluminescence aptamer sensor for 17β-estradiol detection based on nitrogen-doped carbon quantum dots as Ru(dcbpy)32 + coreactant: What role of intermolecular hydrogen bonds play in the system? *Journal of Colloid and Interface Science*, *586*, 103–109. https://doi.org/10.1016/j.jcis.2020.10.074.

Liu, J., Wang, X., Wang, T., Li, D., Xi, F., Wang, J., et al. (2014). Functionalization of monolithic and porous three-dimensional graphene by one-step chitosan electrodeposition for enzymatic biosensor. *ACS Applied Materials and Interfaces*, *6*(22), 19997–20002. https://doi.org/10.1021/am505547f.

Lu, S., Li, G., Lv, Z., Qiu, N., Kong, W., Gong, P., et al. (2016). Facile and ultrasensitive fluorescence sensor platform for tumor invasive biomarker β-glucuronidase detection and inhibitor evaluation with carbon quantum dots based on inner-filter effect. *Biosensors and Bioelectronics*, *85*, 358–362. https://doi.org/10.1016/j.bios.2016.05.021.

Manjunatha, J. G., & Deraman, M. (2017). Graphene paste electrode modified with sodium dodecyl sulfate surfactant for the determination of dopamine, ascorbic acid and uric acid. *Analytical and Bioanalytical Electrochemistry*, *9*(2), 198–213. https://www.scopus.com/inward/record.uri?eid=2-s2.0-85016763703&partnerID=40&md5=59de4d1c48cfc74c1e83b9b32c187eeb.

Manjunatha, J. G., Deraman, M., Basri, N. H., & Talib, I. A. (2018). Fabrication of poly (Solid Red A) modified carbon nano tube paste electrode and its application for simultaneous determination of epinephrine, uric acid and ascorbic acid. *Arabian Journal of Chemistry*, *11*(2), 149–158. https://doi.org/10.1016/j.arabjc.2014.10.009.

Manjunatha, J. G., Kumara Swamy, B. E., Mamatha, G. P., Raril, C., Nanjunda Swamy, L., Fattepur, S., et al. (2018). Carbon paste electrode modified with boric acid and TX-100 used for electrochemical determination of dopamine. *Materials Today Proceedings*, *5*(10), 22368–22375. https://doi.org/10.1016/j.matpr.2018.06.604.

Manjunatha, J. G., Swamy, B. E. K., Deepa, R., Krishna, V., Mamatha, G. P., Chandra, U., et al. (2009). Electrochemical studies of dopamine at eperisone and cetyl trimethyl ammonium bromide surfactant modified carbon paste electrode: A cyclic voltammetric study. *International Journal of Electrochemical Science*, *4*(5), 662–671. https://www.scopus.com/inward/record.uri?eid=2-s2.0-70149098025&partnerID=40&md5=11d77d94decb0609c463a9e01ec4d335.

Mazloum-Ardakani, M., & Khoshroo, A. (2014). High performance electrochemical sensor based on fullerene-functionalized carbon nanotubes/ionic liquid: Determination of some catecholamines. *Electrochemistry Communications*, *42*, 9–12. https://doi.org/10.1016/j.elecom.2014.01.026.

Mendelson, Y. (2005). Biomedical sensors. In *Introduction to biomedical engineering* (pp. 505–548). Elsevier Inc. https://doi.org/10.1016/B978-0-12-238662-6.50011-2.

Mesgari, F., Beigi, S. M., Salehnia, F., Hosseini, M., & Ganjali, M. R. (2019). Enhanced electrochemiluminescence of Ru(bpy)32 + by Sm2O3 nanoparticles decorated graphitic carbon nitride nano-sheets for pyridoxine analysis. *Inorganic Chemistry Communications*, *106*, 240–247. https://doi.org/10.1016/j.inoche.2019.05.023.

Mochalin, V. N., Shenderova, O., Ho, D., & Gogotsi, Y. (2012). The properties and applications of nanodiamonds. *Nature Nanotechnology*, *7*(1), 11–23. https://doi.org/10.1038/nnano.2011.209.

Mojica, M., Alonso, J. A., & Méndez, F. (2013). Synthesis of fullerenes. *Journal of Physical Organic Chemistry*, *26*(7), 526–539. https://doi.org/10.1002/poc.3121.

Molaei, M. J. (2020a). The optical properties and solar energy conversion applications of carbon quantum dots: A review. *Solar Energy*, *196*, 549–566. https://doi.org/10.1016/j.solener.2019.12.036.

Molaei, M. J. (2020b). Principles, mechanisms, and application of carbon quantum dots in sensors: A review. *Analytical Methods*, *12*(10), 1266–1287. https://doi.org/10.1039/c9ay02696g.

Moussa, F. (2018). [60]Fullerene and derivatives for biomedical applications. In *Nanobiomaterials: Nanostructured materials for biomedical applications* (pp. 113–136). Elsevier Inc. https://doi.org/10.1016/B978-0-08-100716-7.00005-2.

Muñoz, R., & Gómez-Aleixandre, C. (2013). Review of CVD synthesis of graphene. *Chemical Vapor Deposition*, *19*(10-11–12), 297–322. https://doi.org/10.1002/cvde.201300051.

Pal, S., Verma, A., Raikwar, S., Prajapati, Y. K., & Saini, J. P. (2018). Detection of DNA hybridization using graphene-coated black phosphorus surface plasmon resonance sensor. *Applied Physics A*, *124*(5). https://doi.org/10.1007/s00339-018-1804-1.

Pan, J., Zheng, Z., Yang, J., Wu, Y., Lu, F., Chen, Y., et al. (2017). A novel and sensitive fluorescence sensor for glutathione detection by controlling the surface passivation degree of carbon quantum dots. *Talanta*, *166*, 1–7. https://doi.org/10.1016/j.talanta.2017.01.033.

Pebdeni, A. B., Hosseini, M., & Ganjali, M. R. (2020). Fluorescent turn-on aptasensor of Staphylococcus aureus based on the FRET between green carbon quantum dot and gold nanoparticle. *Food Analytical Methods*, *13*(11), 2070–2079. https://doi.org/10.1007/s12161-020-01821-4.

Pochkaeva, E. I., Podolsky, N. E., Zakusilo, D. N., Petrov, A. V., Charykov, N. A., Vlasov, T. D., et al. (2020). Fullerene derivatives with amino acids, peptides and proteins: From synthesis to biomedical application. *Progress in Solid State Chemistry*, *57*. https://doi.org/10.1016/j.progsolidstchem.2019.100255.

Prasek, J., Drbohlavova, J., Chomoucka, J., Hubalek, J., Jasek, O., Adam, V., et al. (2011). Methods for carbon nanotubes synthesis—Review. *Journal of Materials Chemistry*, *21*(40), 15872–15884. https://doi.org/10.1039/c1jm12254a.

Qiu, H., Luo, C., Sun, M., Lu, F., Fan, L., & Li, X. (2012). A chemiluminescence sensor for determination of epinephrine using graphene oxide-magnetite-molecularly imprinted polymers. *Carbon*, *50*(11), 4052–4060. https://doi.org/10.1016/j.carbon.2012.04.052.

Raj, M. A., & John, S. A. (2013). Simultaneous determination of uric acid, xanthine, hypoxanthine and caffeine in human blood serum and urine samples using electrochemically reduced graphene oxide modified electrode. *Analytica Chimica Acta*, *771*, 14–20. https://doi.org/10.1016/j.aca.2013.02.017.

Rao, C. N. R., Maitra, U., & Matte, H. S. S. R. (2012). Synthesis, characterization, and selected properties of graphene. In *Graphene: Synthesis, properties, and phenomena* (pp. 1–47). Wiley-VCH. https://doi.org/10.1002/9783527651122.ch1.

Renjini, S., Abraham, P., Kumar, T. J., Kumary, V. A., & Chithra, P. G. (2019). Graphene oxide supported palladium nanoparticle as an electrochemical sensor for epinephrine. In *Vol. 2162. AIP conference proceedings* American Institute of Physics Inc. https://doi.org/10.1063/1.5130281.

Rezaeinasab, M., Benvidi, A., Tezerjani, M. D., Jahanbani, S., Kianfar, A. H., & Sedighipoor, M. (2017). An electrochemical sensor based on Ni(II) complex and multi wall carbon nano tubes platform for determination of glucose in real samples. *Electroanalysis*, *29*(2), 423–432. https://doi.org/10.1002/elan.201600162.

Rivas, G. A., Rodríguez, M. C., Rubianes, M. D., Gutierrez, F. A., Eguílaz, M., Dalmasso, P. R., et al. (2017). Carbon nanotubes-based electrochemical (bio)sensors for biomarkers. *Applied Materials Today*, *9*, 566–588. https://doi.org/10.1016/j.apmt.2017.10.005.

Sajid, M. I., Jamshaid, U., Jamshaid, T., Zafar, N., Fessi, H., & Elaissari, A. (2016). Carbon nanotubes from synthesis to in vivo biomedical applications. *International Journal of Pharmaceutics*, *501*(1–2), 278–299. https://doi.org/10.1016/j.ijpharm.2016.01.064.

Salehnia, F. (2019). Application of graphene materials in molecular diagnostics. In *Vol. 1. Handbook of graphene set* (pp. 535–560). Hoboken, NJ: Wiley.

Salehnia, F., Hosseini, M., & Ganjali, M. R. (2018). Enhanced electrochemiluminescence of luminol by an: In situ silver nanoparticle-decorated graphene dot for glucose analysis. *Analytical Methods*, *10*(5), 508–514. https://doi.org/10.1039/c7ay02375h.

Sanghavi, B. J., Wolfbeis, O. S., Hirsch, T., & Swami, N. S. (2015). Nanomaterial-based electrochemical sensing of neurological drugs and neurotransmitters. *Microchimica Acta*, *182*(1–2), 1–41. https://doi.org/10.1007/s00604-014-1308-4.

Schmidt, A. C., Wang, X., Zhu, Y., & Sombers, L. A. (2013). Carbon nanotube yarn electrodes for enhanced detection of neurotransmitter dynamics in live brain tissue. *ACS Nano*, *7*(9), 7864–7873. https://doi.org/10.1021/nn402857u.

Schrand, A. M., Hens, S. A. C., & Shenderova, O. A. (2009). Nanodiamond particles: Properties and perspectives for bioapplications. *Critical Reviews in Solid State and Materials Sciences*, *34*(1–2), 18–74. https://doi.org/10.1080/10408430902831987.

Scott, L. T. (2004). Methods for the chemical synthesis of fullerenes. *Angewandte Chemie, International Edition*, *43*(38), 4994–5007. https://doi.org/10.1002/anie.200400661.

Scott, L. T., Boorum, M. H., McMahon, B. J., Hagen, S., Mack, J., Blank, J., et al. (2002). A rational chemical synthesis of C60. *Science, 295*(5559), 1500–1503. https://doi.org/10.1126/science.1068427.

Seekaew, Y., Arayawut, O., Timsorn, K., & Wongchoosuk, C. (2018). Synthesis, characterization, and applications of graphene and derivatives. In *Carbon-based nanofillers and their rubber nanocomposites: carbon nano-objects* (pp. 259–283). Elsevier. https://doi.org/10.1016/B978-0-12-813248-7.00009-2.

Shanmugam, N. R., & Prasad, S. (2018). Carbon nanotubes: Synthesis and characterization. In *Nanopackaging: Nanotechnologies and electronics packaging* (2nd ed., pp. 575–596). Springer International Publishing. https://doi.org/10.1007/978-3-319-90362-0_17.

Shehab, M., Ebrahim, S., & Soliman, M. (2017). Graphene quantum dots prepared from glucose as optical sensor for glucose. *Journal of Luminescence, 184*, 110–116. https://doi.org/10.1016/j.jlumin.2016.12.006.

Shenderova, O. A., Zhirnov, V. V., & Brenner, D. W. (2002). Carbon nanostructures. *Critical Reviews in Solid State and Materials Sciences, 27*(3–4), 227–356. https://doi.org/10.1080/10408430208500497.

Simioni, N. B., Silva, T. A., Oliveira, G. G., & Fatibello-Filho, O. (2017). A nanodiamond-based electrochemical sensor for the determination of pyrazinamide antibiotic. *Sensors and Actuators, B: Chemical, 250*, 315–323. https://doi.org/10.1016/j.snb.2017.04.175.

Singh, C., & Song, W. (2012). Carbon nanotube structure, synthesis, and applications. In *Vol. 9781107008373. The toxicology of carbon nanotubes* (pp. 1–37). Cambridge University Press. https://doi.org/10.1017/CBO9780511919893.001.

Sinnott, S. B., & Andrews, R. (2001). Carbon nanotubes: Synthesis, properties, and applications. *Critical Reviews in Solid State and Materials Sciences, 26*(3). https://doi.org/10.1080/20014091104189.

Sobhanie, E., Faridbod, F., Hosseini, M., & Ganjali, M. R. (2020). An ultrasensitive ECL sensor based on conducting polymer/electrochemically reduced graphene oxide for non-enzymatic H2O2 detection in biological samples. *ChemistrySelect, 5*(17), 5330–5336. https://doi.org/10.1002/slct.202000233.

Stefan-Van Staden, R. I. (2014). Enantioselective surface plasmon resonance sensor based on C60 fullerene-glutathione self-assembled monolayer (SAM). *Chirality, 26*(3), 129–131. https://doi.org/10.1002/chir.22281.

Tian, P., Tang, L., Teng, K. S., & Lau, S. P. (2018). Graphene quantum dots from chemistry to applications. *Materials Today Chemistry, 10*, 221–258. https://doi.org/10.1016/j.mtchem.2018.09.007.

Tiwari, J. N., Vij, V., Kemp, K. C., & Kim, K. S. (2015). Engineered carbon-nanomaterial-based electrochemical sensors for biomolecules. *ACS Nano, 46*–80. https://doi.org/10.1021/acsnano.5b05690.

Vilela, P., El-Sagheer, A., Millar, T. M., Brown, T., Muskens, O. L., & Kanaras, A. G. (2017). Graphene oxide-upconversion nanoparticle based optical sensors for targeted detection of mRNA biomarkers present in Alzheimer's disease and prostate cancer. *ACS Sensors, 2*(1), 52–56. https://doi.org/10.1021/acssensors.6b00651.

Wakabayashi, T., & Achiba, Y. (1992). A model for the C60 and C70 growth mechanism. *Chemical Physics Letters, 190*(5), 465–468. https://doi.org/10.1016/0009-2614(92)85174-9.

Wang, Y., & Hu, A. (2014). Carbon quantum dots: Synthesis, properties and applications. *Journal of Materials Chemistry C, 2*(34), 6921–6939. https://doi.org/10.1039/c4tc00988f.

Wang, Z., Zhang, L., Hao, Y., Dong, W., Liu, Y., Song, S., et al. (2021). Ratiometric fluorescent sensors for sequential on-off-on determination of riboflavin, Ag+ and L-cysteine based on NPCl-doped carbon quantum dots. *Analytica Chimica Acta, 1144*, 1–13. https://doi.org/10.1016/j.aca.2020.11.054.

Wei, L., Lei, Y., Fu, H., & Yao, J. (2012). Fullerene hollow microspheres prepared by bubble-templates as sensitive and selective electrocatalytic sensor for biomolecules. *ACS Applied Materials and Interfaces, 4*(3), 1594–1600. https://doi.org/10.1021/am201769u.

Wei, L. F., & Shih, J. S. (2001). Fullerene-cryptand coated piezoelectric crystal urea sensor based on urease. *Analytica Chimica Acta, 437*(1), 77–85. https://doi.org/10.1016/S0003-2670(01)00941-2.

Wen, B., & Tian, Y. (2017). Synthesis, thermal properties and application of nanodiamond. In *Thermal transport in carbon-based nanomaterials* (pp. 85–112). Elsevier Inc. https://doi.org/10.1016/B978-0-32-346240-2.00004-2.

Withers, J. C., Loutfy, R. O., & Löwe, T. P. (1997). Fullerene commercial vision. *Fullerene Science and Technology, 5*(1), 1–31. https://doi.org/10.1080/15363839708011971.

Wu, H., Fan, S., Jin, X., Zhang, H., Chen, H., Dai, Z., et al. (2014). Construction of a zinc porphyrin-fullerene-derivative based nonenzymatic electrochemical sensor for sensitive sensing of hydrogen peroxide and nitrite. *Analytical Chemistry, 86*(13), 6285–6290. https://doi.org/10.1021/ac500245k.

Wudl, F. (2002). Fullerene materials. *Journal of Materials Chemistry, 12*(7), 1959–1963. https://doi.org/10.1039/b201196d.

Xu, Y., Liu, J., Gao, C., & Wang, E. (2014). Applications of carbon quantum dots in electrochemiluminescence: A mini review. *Electrochemistry Communications, 48*, 151–154. https://doi.org/10.1016/j.elecom.2014.08.032.

Yi, H., Zheng, D., Hu, C., & Hu, S. (2008). Functionalized multiwalled carbon nanotubes through in situ electropolymerization of brilliant cresyl blue for determination of epinephrine. *Electroanalysis, 20*(10), 1143–1146. https://doi.org/10.1002/elan.200704141.

Zhang, R., & Fan, Z. (2020). Nitrogen-doped carbon quantum dots as a "turn off-on" fluorescence sensor based on the redox reaction mechanism for the sensitive detection of dopamine and alpha lipoic acid. *Journal of Photochemistry and Photobiology A: Chemistry, 392*. https://doi.org/10.1016/j.jphotochem.2020.112438.

Zheng, X. T., Ananthanarayanan, A., Luo, K. Q., & Chen, P. (2015). Glowing graphene quantum dots and carbon dots: Properties, syntheses, and biological applications. *Small, 11*(14), 1620–1636. https://doi.org/10.1002/smll.201402648.

Zhilei, W., Zaijun, L., Xiulan, S., Yinjun, F., & Junkang, L. (2010). Synergistic contributions of fullerene, ferrocene, chitosan and ionic liquid towards improved performance for a glucose sensor. *Biosensors and Bioelectronics, 25*(6), 1434–1438. https://doi.org/10.1016/j.bios.2009.10.045.

Chapter 6

Fabrication of disposable sensor strips for point-of-care testing of environmental pollutants

Gnanesh Rao[a], Akhilesh Rao[b], B.P. Nandeshwarappa[c], Raghu Ningegowda[d], Kiran Kumar Mudnakudu-Nagaraju[e], and Sandeep Chandrashekharappa[f,g]

[a]*Department of Biochemistry, Bangalore University, Bengaluru, Karnataka, India,* [b]*Department of Mechanical Engineering, Christ University, Bengaluru, Karnataka, India,* [c]*Department of Studies in Chemistry, Shivagangothri, Davangere University, Davangere, Karnataka, India,* [d]*Department of Chemistry, Jyoti Nivas College Autonomous, Bangalore, Karnataka, India,* [e]*Department of Biotechnology & Bioinformatics, Faculty of Life Sciences, JSS Academy of Higher Education & Research, Mysore, Karnataka, India,* [f]*Department of Medicinal Chemistry, National Institute of Pharmaceutical Education and Research (NIPER) Raebareli, Lucknow, Uttar Pradesh, India,* [g]*Institute for Stem Cell Science and Regenerative Medicine, NCBS, TIFR, GKVK-Campus, Bengaluru, Karnataka, India*

Introduction

Rapid growth of industrialization and technology development has facilitated rising demand for basic requirements such as clothes, transport vehicles, and health care. However, these product-manufacturing industries destroy land, water, and air by discharging huge quantities of solid, gaseous, and wastewater resulting in environmental pollution (Dincer et al., 2019). This has led to not only increase in the rate of chemical pollution in the environment, but also enter the food chain, resulting in the biological magnification and serious health hazards.

For instance, dye-related pollution is a type of environmental pollution in which the chemical dye, which will be utilized to be imparted on to the fabrics, paper, plastic, and leather items (Khan & Malik, 2018; Sharma, Dangi, & Shukla, 2018) is being discharged into the flowing streams. As a result, they form a thick layer and block penetration of the sunlight into the water. This inhibits the aquatic life to resist photochemical and biological attacks (Pereira & Alves, 2012).

Further, pesticide pollution has become one of the major health concerns along with inorganic pollutants such as metals, inorganic phosphates, and nitrate (Verma & Bhardwaj, 2015). For instance, piezoelectric biosensors can detect organophosphate (Marrazza, 2014). Nowadays, there are many sensors in the market, which detect the presence of the various pollutants (Raril & Manjunatha, 2018; Raril et al., 2018; Raril, Manjunatha, Ravishankar, et al., 2020; Raril, Manjunatha, & Tigari, 2020). Also, disposable sensors are low-cost and easy-to-use sensing devices for short-term or rapid single-point measurements (Chauhan, Pandey, & Bhattacharya, 2019).

Analysis of natural resources like water and soil in the ecosystem for measuring pH sensitivity by litmus paper dates back to 1800 (Szabadvary & Oesper, 1964). Later, chemical analysis of metabolites from urine samples was started for estimating glucose and albumin concentrations in the 18th century. However, various chemical pollutants can be analyzed by most sensitive techniques today. Recent advances in sensor technology have resulted in the development of point-of-care testing, which is most convenient, fast, reliable, and disposable sensor strip (Benedict, 2002; Guthrie & Humphreys, 1988) (Fig. 6.1).

Substrate material for disposable sensor strips

Recent advances in the field of paper-based sensing technology have revolutionized environmental monitoring. Although paper-based materials are complex in its nature, the liquid flows through capillary action. Further, availability, low cost, and biocompatibility of the paper-based material add to the ease of generating simple and disposable sensor strips (Liana, Raguse, Gooding, & Chow, 2012).

FIG. 6.1 Graphical representation of the workflow for sensor-based analysis. *(No Permission Required.)*

Paper-based sensor was developed to monitor the glucose level in urine during early 1956 (Comer, 1956). Different types of paper are used for the fabrication of the disposable paper sensor. For example, filter papers and cellulose chromatography paper have been widely used as a substrate material for various purposes depending on fictionalization and include properties, such as thickness, particle retention, pore size, and flow rate (Irvine, Tan, & Stillman, 2017; Klug, Reynolds, & Yoon, 2018; Scala-Benuzzi, Raba, Soler-Illia, Schneider, & Messina, 2018).

Cellulose-derived nitrocellulose membrane, being one of the best materials as disposable sensors, is applicable for point-of-care testing (POCT) and used in commercially available lateral flow assays (LFAs) for home pregnancy or fertility testing (Posthuma-Trumpie, Korf, & Van Amerongen, 2009). These membranes have been demonstrated to be adequate for immobilization, which are applicable in the detection of proteins, enzymes, and DNA (Liana et al., 2012; Scida, Li, Ellington, & Crooks, 2013).

Cellophane, carbon nanotube or cellulose composite paper, cellulose-based woven textiles, and nanocellulose-based materials are also used as alternative materials due to its porosity, hydrophobicity, flexibility, and conductivity, which can be tuned depending on the percentage of the components and the chemical treatments. They serve as scalable fabrication of disposable sensors (Golmohammadi, Morales-Narváez, Naghdi, & Merkoçi, 2017; Morales-Narváez et al., 2015).

A hybrid material is made from two or more constituent materials with significantly different physical or chemical properties (Moschou & Tserepi, 2017) that, when combined, produce a material with characteristics different from the individual components. Also, in some cases, these are used to overcome the shortage of material. Comparing to the other materials, hybrid materials are having excellent performance in comparison with a single material and are low cost (Imani et al., 2016). The most common hybrid standard materials are known to be combined with polymers (Shiroma et al., 2016).

CNT biosensor

High electric conductivity of carbon and similar values in ionization potential and electron affinity contribute to its unique electric property to accept and donate electrons (Han, Kim, Li, & Meyyappan, 2014; Miyake, 2003).

CNTs are nanostructure materials, which are applied in the fields of agriculture, environmental monitoring, biofuel, chemical, pharmaceutical, cosmetics, electronics, pulping, enzyme production, etc. (Hussain, 2018; Panpatte & Jhala, 2019). Further, functionalized CNTs are suitable candidates for precise quantitative analysis (Im, Sterner, & Swager, 2016; Shim, Chen, Doty, Xu, & Kotov, 2008) and are used in the field of diagnosis and environmental monitoring as it provides the system with the increased surface area and the ability of the CNT to be covalently or noncovalently functionalized (Farma et al., 2013; Hareesha, Manjunatha, Raril, & Tigari, 2019a, 2019b; Tigari & Manjunatha, 2020a, 2020b; Tigari, Manjunatha, Raril, & Hareesha, 2019).

Moreover, functionalization of CNT by covalent or noncovalent modifications is used for the capture and quantification of a particular analyte (Hareesha & Manjunatha, 2020a, 2020b; Manjunatha, 2018; Manjunatha, Deraman, & Basri, 2015; Manjunatha et al., 2020; Prinith & Manjunatha, 2020; Pushpanjali, Manjunatha, Tigari, & Fattepur, 2020; Schroeder, Savagatrup, He, Lin, & Swager, 2019). The method of functionalization depends on the type of functionalization required. Functionalized CNTs are third-generation electrochemical biosensors capable of transferring the electron directly to the electrode (Ivnitski, Branch, Atanassov, & Apblett, 2006; Mehmeti et al., 2017). Linear sweep cyclic voltammetry is the most commonly applied technique for the determination of redox potential and electrochemical reaction rates of analyte solutions (Manjunatha, Deraman, Basri, & Talib, 2018; Manjunatha et al., 2020; Raril, Manjunatha, Ravishankar, et al., 2020; Wu et al., 2019). The transducer picks up the electric signals and forwards to the amplifier for amplification. Later, amplified signals are processed and then converted into interpretable value (Manjunatha, 2017a, 2017b; Manjunatha et al., 2014; Manjunatha et al., 2009).

Next, CNT-based optical biosensors can recognize and quantify specific analytes which can induce changes in the fluorescence, absorbance, optical resonance, and interferometric reflectance signal detected by optical sensors to quantify the analyte of interest (Hussain, 2020a, 2020b). This technique to be more precise requires accurate labeling with fluorophores (such as organic dyes, semiconductor nanocrystals, and quantum dots) and biological tags (enzymes with chromogens, green fluorescent proteins) (Singh, Truong, Reeves, & Hahm, 2017) (Fig. 6.2).

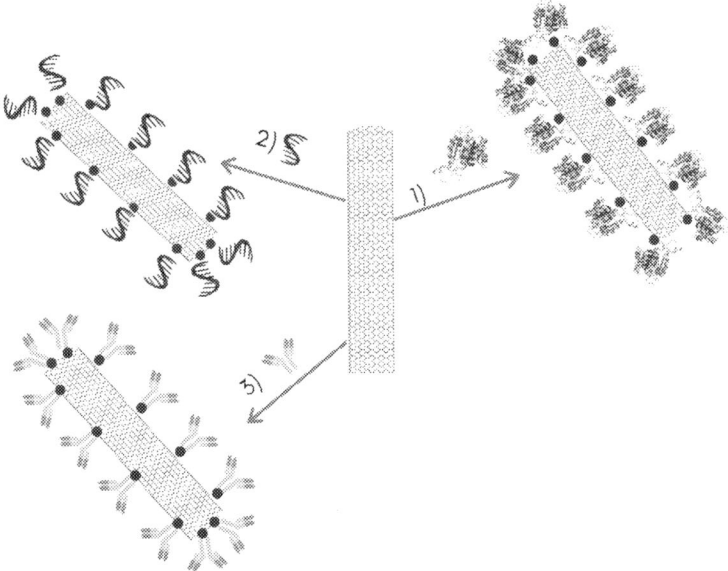

FIG. 6.2 Graphical representation of functionalization of CNT with (1) enzyme, (2) RNA, and (3) antibody. *(No Permission Required.)*

Functionalization

Enzymatic biosensor

Basic operation of enzyme-based electrochemical glucose biosensor is based on the fact that the enzyme glucose oxidase that catalyzes the oxidation of glucose to gluconolactone, which in turn reduces itself, has been demonstrated in 1962 by Leland C. Clark (Clark & Lyons, 1962).

Reduced state of CNT is responsible for the change in potentiometric, impedimetric, conductometric, and amperometric reading of sensors in the presence of the substrate (Charithra & Manjunatha, 2020; Rao, Suresh, Sharma, Gupta, & Vijayaraghavan, 2011). Colorimetric detection and quantification can be carried out by conjugating enzymes with peroxidase-like activity and substrate such as tetramethylbenzidine which upon oxidation leads to color formation (Chen et al., 2019; Hussain, 2019, 2020a, 2020d).

Affinity biosensors

The ability of a nucleotide strand to bind to the synthetic complementary strand to form a hybrid DNA is exploited in the quantification of target strands. The hybridization of the strand on the CNT surface brings about changes in the voltammetric and amperometric reading of sensors in the presence of the target strand (Amrutha, Manjunatha, Bhatt, & Hareesha, 2019; Amrutha, Manjunatha, Bhatt, Raril, & Pushpanjali, 2019; Ijeomah, Obite, & Rahman, 2016). The sample before loading, if tagged, can be optically detected and quantified.

Immunosensors

Antibodies against pathogens, proteins, and toxins are capable of binding specifically to form immunoprecipitation. The immunoprecipitation on CNT surface results in changed voltammetric and amperometric reading in the presence of a specific antigen (Chikkaveeraiah et al., 2009; Jacobs, Peairs, & Venton, 2010; Shim et al., 2008; Wang et al., 2009).

CNT functionalized with specific antibody immunoprecipitated with specific antigen. Further, fluorophore- or chromophore-conjugated secondary antibody produces fluorescence or develops color upon binding. Later, the amount of colorimetric change is proportional to the quantity of analyte, and the optical signal is converted into an electric signal, which will be processed, and converted into interpretable reading (Hussain, 2020c; Piao et al., 2011).

Many heteroatoms are used as substrate in pharmaceutical industries and as catalyst in many industrial processes. Many drugs are heterocyclic compounds (Sandeep, Padmashali, & Kulkarni, 2013; Sandeep et al., 2014, 2016, 2017; Venugopala, Al-Attraqchi, et al., 2019; Venugopala, Chandrashekharappa, et al., 2018, 2019; Venugopala, Khedr, et al., 2018; Venugopala, Sandeep, et al., 2019). Signal amplification might be necessary to detect particular analyte without the interference of heteroatoms, which are widely used as pharmaceutical drugs (Alwassil, Chandrashekarappa, Nayak, &

Venugopala, 2019; Mallikarjuna, Padmashali, & Sandeep, 2014; Mallikarjuna, Sandeep, & Padmashali, 2017; Nagesh et al., 2014; Padmashali et al., 2019; Rashmi et al., 2015; Siddesh, Padmashali, Thriveni, & Sandeep, 2014; Siddesh, Padmashali, Thriveni, Sandeep, & Goudarshivnnanava, 2013).

Pollutants

Inorganic chemicals such as metallic and nonmetallic elements, organic chemicals such as small molecules and pesticides, and biological contaminants like viruses, bacteria, fungi, etc., are the major water- and soil-borne environmental contaminants (Dincer et al., 2019).

Furthermore, harmful gases such as CO, CO_2, H_2, NO, NO_2, SO_2, H_2S, CH_4, C_2H_2, and O_3 are known to pollute ecosystem (Quang, Van Trinh, Lee, & Huh, 2006), followed by volatile organic compounds such as benzene, toluene, and xylenes (Katulski, Namiesnik, Sadowski, Stefanski, & Wardencki, 2011), which are produced by the industries and automobile. If these gases are not captured, it will pollute the atmosphere.

Pharmaceutical contaminants include ethinylestradiol and antibiotics (Peixoto, Machado, Oliveira, Bordalo, & Segundo, 2019), and synthetic dyes and heavy metals such as arsenic, mercury, lead, cadmium, iron, copper, chromium, and aluminum, when mixed with water used in the industrial processes such as textile coloring, tanning, steel making, and heat treatment (Chequer et al., 2013), are known sources of contaminants in the water bodies, soil, as well as groundwater.

Other than toxic elements produced by industries, the major contaminants of water and soil are persistent organic pollutants (POPs) such as dichlorodiphenyltrichloroethane (DDT), heptachlor, and chlordane, which are used as insecticide and pesticide (El-Shahawi, Hamza, Bashammakh, & Al-Saggaf, 2010). Along with these industrial chemicals, solvents like polychlorinated biphenyls (PCBs), dioxins, and polychlorinated dibenzofurans directly contaminate water and soil.

Contamination of drinking water by biological contaminants, microbes cause millions of infections worldwide, according to the World Health Organization (WHO). Microbial species belonging to the genus *Vibrio*, *Salmonella*, *Shigella*, and different variants of *Escherichia coli* are the leading case of infection. Biological contaminants also include microcystin, which is one of the most dangerous toxins and one of the leading causes of outbreaks of mass poisoning (Gupta, Pant, Vijayaraghavan, & Rao, 2003).

CNT sensors are reported for the detection and quantification of environmental pollutants such as nitrite, phosphate, pesticide, herbicide, phenolic compounds, dihydroxy benzene, mercury, cadmium, lead, arsenic, copper, ; nitrogen oxides, nitrogen dioxide, ammonia, sulfur dioxide, *E. coli*, microcystin, aflatoxin, etc. (Charithra, Manjunatha, & Raril, 2020; D'Souza et al., 2020; Pushpanjali, Manjunatha, & Shreenivas, 2019; Rengaraj, Cruz-Izquierdo, Scott, & Di Lorenzo, 2018; Tiwari, Vij, Kemp, & Kim, 2015) (Table 6.1).

Fabrication

Materials for disposable sensors: Most of the systems depend upon advance material science, but still it is not possible to produce a single one size fits material that meets all the requirements of the disposable sensors because of a wide range of applications (Dincer et al., 2019).

Standard materials for MEMS/NEMS

Advancement of the technology leads to the minimization for even smaller systems which is the emergence of nanoelectromechanical systems. Standard materials used in MEMS are silicon, glass, and ceramics, whereas carbon-based materials such as carbon nanotube (CNT), graphite, and diamond are commonly used in the NEMS fabrication (Madou, 2011).

Compared to other standard materials, NEMS/MEMS are having excellent mechanical, electrical, and thermal properties, but they are complex, expensive to process, costlier, and also produce hazardous chemical by-products. Drawback of the standard materials, MEMS/NEMS are limited to standards regarding their flexibility, stretchability, disposability, and biodegradability. To overcome this material based on cellulose, synthetic polymer and hybrid materials are used for fabrication (Kang et al., 2016).

Generally, the disposable electrochemical sensor consists of three main parts (Salvatore et al., 2017):

1. Sensing element
2. Interconnections
3. External wiring

TABLE 6.1 CNT-based sensors reported for monitoring environment pollutants.

Contaminants	Reference
Nitrite (NO_2)	Chang and Zen (2007), Plumeré et al. (2012), Quan et al. (2005)
Phosphate (PO_4)	Gilbert et al. (2011), Khaled et al. (2008)
Pesticide (organophosphate)	Alonso et al. (2013), Crew et al. (2011)
Herbicide	Baskeyfield et al. (2011), Ivanov et al. (2011)
Phenol and phenolic compounds	Nadifiyine et al. (2013), Su et al. (2011), Lu et al. (2010)
Dihydroxybenzene	Fotouhi et al. (2018)
Mercury (Hg)	Aragay et al. (2011), Kim et al. (2009)
Cadmium (Cd)	Wei et al. (2013), Xu et al. (2008)
Lead (Pb)	Fu et al. (2013), Injang et al. (2010)
Arsenic (As)	Sanllorente-Méndez et al. (2010)
Copper (Cu)	Andreuccetti et al. (2014), Bouden et al. (2014)
Nitrogen oxides (NO_x)	Ueda, Bhuiyan, et al. (2008), Ueda, Katsuki, et al. (2008)
Nitrogen dioxide (NO_2)	Kong et al. (2000), Valentini et al. (2003)
Ammonia (NH_3)	Nguyen et al. (2007)
Sulfur dioxide (SO_2)	Suehiro et al. (2005)
E. coli	So et al. (2008)
Microcystin	Chen et al. (2010), Yao et al. (2006)
Aflatoxin	Li et al. (2011), Singh et al. (2013)

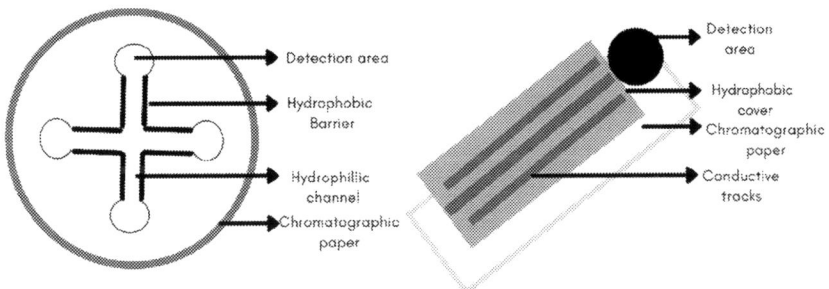

FIG. 6.3 Basic components of workflow electrochemical and colorimetric sensors. *(No Permission Required.)*

Colorimetric sensor consists of distinguished hydrophilic detection area and channel with characteristic hydrophobic barrier (Fig. 6.3).

Regardless of the detection method used, patterning of reaction zones and detection areas is required for fabrication of paper-based devices. Two main approaches are being followed which include: (1) defining hydrophobic zones on the paper, and (2) cutting the paper in such a way to delimit the channels and detection to define physical boundaries, followed by deposition of conductive electrode tracks on the paper surface for electrochemical detection.

Methods used to define hydrophobic zones on the paper

Photolithography

Patterning of paper into hydrophilic zones surrounded by hydrophobic polymeric barriers can be fabricated by photolithography, which is a standard method for printing circuit board and a convenient and quick technique suitable for large-scale industrial production of patterned paper (Martinez, Phillips, Butte, & Whitesides, 2007).

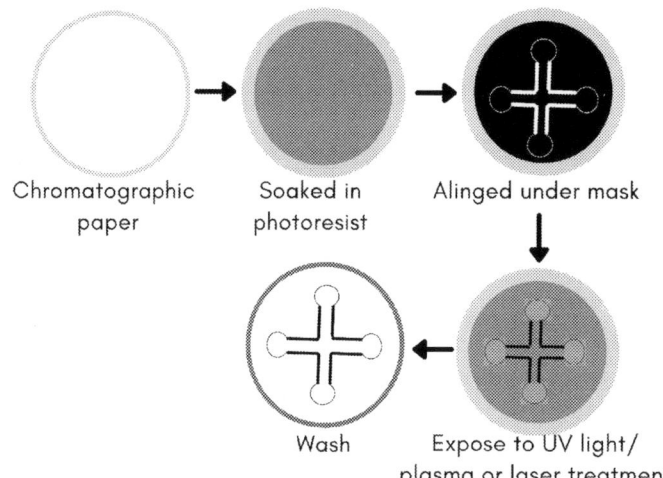

FIG. 6.4 Workflow of fabrication by photolithography. *(No Permission Required.)*

Photolithography patterning is based on complete hydrophobization of paper by soaking in a photoresist [octadecyltrichlorosilane (OTS), ultraviolet resin, SU8-2010 photoresist, SU8 2150 negative photoresist, TiO_2, or other materials] followed by selective dehydrophobization by exposing to UV light, plasma treatment, or laser treatment through a photomask to create a hydrophilic region. Further, photomasks are made of using a mixture of acrylic monomers, which can be custom-designed by 3D printer as per the need of application (Akyazi, Basabe-Desmonts, & Benito-Lopez, 2018).

Furthermore, photolithography provides high reproducibility and resolution in the fabrication of channels (Dungchai, Chailapakul, & Henry, 2011). However, the need for exposing the entire paper substrate including the converted to be hydrophilic channels to organic solvents and expensive photoresists is a drawback. Expensive photolithography equipment and the need for clean room facilities also make this method a little more complicated fabrication method (Yamada, Henares, Suzuki, & Citterio, 2015).

Fast lithographic activation of sheets (FLASH) is another rapid method for laboratory prototyping of microfluidic devices on paper which requires less than 30 min. Instead of sophisticated facilities, it requires a UV lamp and hot plate. Further, patterning can even be performed in sunlight instead of UV light (Martinez, Phillips, Wiley, Gupta, & Whitesides, 2008) (Fig. 6.4).

Printing

The printing process is one of the most commonly used techniques to minimize the consumption of expensive hydrophobic materials, making them suitable for large-scale production as they are cost-effective, reproducible, and simple to use (Akyazi et al., 2018; Jiang & Fan, 2016; Yamada et al., 2015).

Template-based methods (screen printing and flexographic printing) that use masks or stencils to delimit the regions where the hydrophobic material will be printed, and nontemplate methods (digital printing) are the printing techniques in use. Digital printing is the most popular method because of ease of transferring a computer image by digitally controlled deposition of ink onto paper substrate.

Screen printing method

Screen printing is a printing method that uses a blocking stencil to protect the area and mesh to transfer ink on the substrate. Adequate resolution can be obtained by simply adjusting the ink formulation to the desired viscosity.

Polydimethylsiloxane (PDMS) screen printing is a type of screen printing in which hydrophobic barriers and hydrophilic channels are developed on filter paper by dispensing PDMS dissolved in hexane solution, where the plotter continuously drew hydrophobic patterns on filter paper. The process involves developing mask using Adobe Illustrator software or alternative software that can be printed on screen stencil using a printer. Masks thus printed are placed on filter paper followed by pouring PDMS solution, through stencil rubbed using squeegee followed by gentle removal of stencil. The paper was kept in an oven for thermal curing followed by drying in an ambient temperature (Mohammadi et al., 2015).

Simple fabrication by applying polystyrene solution instead of PDMS through the patterned screen is also possible. As polystyrene solution penetrates the paper fibers to form the hydrophobic barrier and define a hydrophilic analysis zone, it is well suitable for colorimetric detection (Sameenoi, Nongkai, Nouanthavong, Henry, & Nacapricha, 2014).

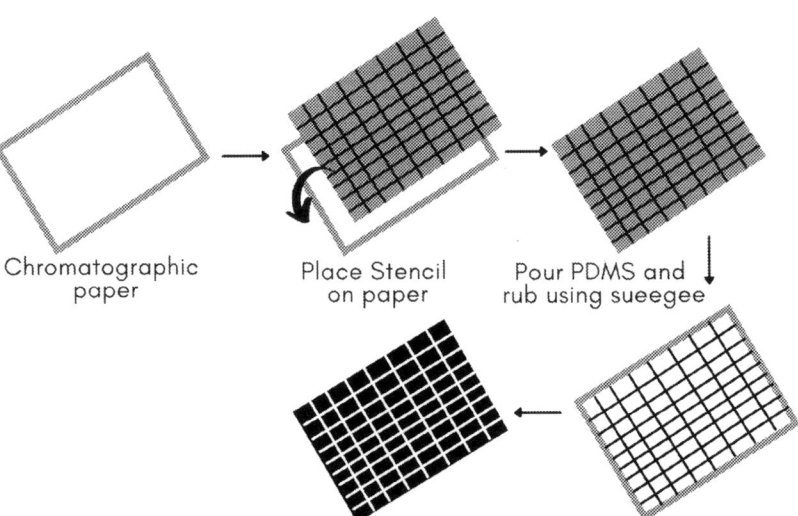

FIG. 6.5 General workflow of fabrication by screen printing technique. *(No Permission Required.)*

For electrochemical detection, the conductive track can be fabricated by screen printing. It is a three layer of carbon printing technology. The resistance in three layers will be in the reducing form. The electrodes are printed by squeegeeing carbon-based conductive ink on hydrophobic paper then dried at 60°C for around 30–60 min (Koo, Shanov, & Yun, 2016; Peixoto, Machado, Oliveira, Bordalo, & Segundo, 2019). To cross-link the cellulose fibers, the paper is submerged in the solution of ethylene glycol at a room temperature, and later thermal treated.

Screen printing is simple, which allows the incorporation of sensing materials. However, the resolution has been shown to be low and different blocking stencils are needed for different design and application to be made every time (Fig. 6.5).

Flexographic printing

High-throughput fabrication technique, allowing large-scale production of paper-based fluidic structures, is flexographic printing. This printing of polystyrene to fabricate microfluidics channel with hydrophobic boundaries on paper substrates is through roll-to-roll channel. This method can be easily scaled-up, versatile, and can be used to pattern both barriers and conductive tracks by just changing the stencils/mask layout and ink/cartridges. Additionally, these methods can be used to incorporate sensing materials (Olkkonen, Lehtinen, & Erho, 2010).

Disadvantage of printing is that one reagent at a time using different printing plates requires need of printing plates designed for specific design and application (Fig. 6.6).

Digital printing

Inkjet and wax printing are the most commonly used fabrication techniques, which are easy to use and commercially available.

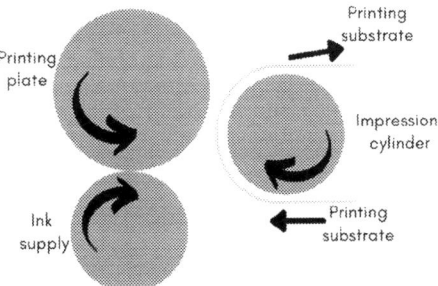

FIG. 6.6 Schematic representation of fabrication by flexographic printing technique. *(No Permission Required.)*

Inkjet printing

Inkjet printing overcomes the drawbacks of lithographic method. It is an inexpensive fabrication technique for direct patterning of hydrophilic channel and hydrophobic barrier on paper substrate. The technique has advantages like cost-effectiveness, high resolution, rapid fabrication, and large-scale production (Tenda et al., 2016).

Patterning of hydrophobic barriers using UV curable inks, polystyrene, silicone, alkyl ketene dimer (AKD), polyacrylate, and other materials directly on paper is possible without the need of specific mask and clean room facilities (Salentijn, Hamidon, & Verpoorte, 2016). However, several printed layers were needed to generate devices, which can interfere with the final print resolution. To achieve the required ink properties, it is often necessary to adjust the surface tension and viscosity using modifiers. Critical issue for printing reproducibility is ink composition which affects the resolution, speed of fabrication, and reliability of printing without clogging the nozzle. While using common inkjet printer, changing the cartridge materials is the requirement to resist the solvents and formulated inks. The pattern can be designed using personal computer and printed on filter paper by inkjet printer.

Papers fabricated by inkjet printing for hydrophobic barrier are suitable for colorimetric applications. Silver metallic conductive track can be fabricated by inkjet printing layers of silver ink and curing for electrochemical detection followed by subsequent deposition of CNTs for sensing (Kholghi Eshkalak et al., 2017; Lin et al., 2013).

Preparation of water-based CNT ink: Dried CNT powder dispersed in water is sonicated to obtain a uniform dispersion. To the final solution add glycerol in the weight ratio of 1:3 (glycerol:water) to meet the viscosity and surface tension required for inkjet printing (Lin et al., 2013). Temperature of printer plate is maintained at 60°C to assist evaporation of water and glycerol and the print head at 40°C, with suitable voltage for fabrication.

Inkjet etching

It is an indirect method of fabrication similar to photolithography, where inkjet printing used toluene to produce paper-based device to remove hydrophobic polystyrene previously patterned on the paper and create hydrophilic-desired designs (Abe, Suzuki, & Citterio, 2008).

Rapid fabrication method involves, using hexamethyldisilane (HMDS) and tetraethylorthosilicate (TEOS) to produce hydrophobic paper by three different methods such as heating, plasma treatment, and microwave irradiation. Inkjet printing is to develop hydrophilic channel on modified hydrophobic paper (Malekghasemi, Kahveci, & Duman, 2016).

Waxing method

In this fabrication technique, wax is used as a hydrophobic material that is patterned on paper and then thermal treatment or curing in oven is used to melt the wax and allow it to impregnate the cellulose substrate underneath to form hydrophobic barrier and reaction microzones or fluid reservoirs (Carrilho, Martinez, & Whitesides, 2009; Carrilho, Phillips, Vella, Martinez, & Whitesides, 2009). Wax-based deposition method is simple and low cost with advantages of wax being cheap, water insoluble, low viscosity needed for penetration into the filter paper after melting, and ease of disposable (Adkins et al., 2017; Kaewarsa, Laiwattanapaisal, Palasuwan, & Palasuwan, 2017).

Desired pattern can be developed by software like Auto CAD and Adobe Illustrator CC, and then pattern could be printed on filter paper by wax printer. However, the wax inevitably spread during thermal treatment and deformation/distortion of hydrophilic channel.

To deposit hydrophobic pattern barriers with reproducibility, controlling of temperature and heating time is imperative and also depends on density and porosity of the paper used.

Other patterning methods based on the use of wax include wax dipping and wax screen printing; however, most commonly mask-guided procedures like screen stencil as a mask to make a wax pattern on paper by rubbing the wax through stencil are extensively used (Dungchai et al., 2011; Songjaroen, Dungchai, Chailapakul, & Laiwattanapaisal, 2011) (Fig. 6.7).

Stamping

Stamping is a simple and low-cost method to fabricate hydrophobic barrier without expensive printing machines or laborious procedures with a moderate resolution. The process involves preheated handheld stainless steel stamp pressed onto paraffin paper to thermally transfer paraffin to other native paper, forming hydrophobic barriers in the latter (de Tarso Garcia, Garcia Cardoso, Garcia, Carrilho, & Tomazelli Coltro, 2014).

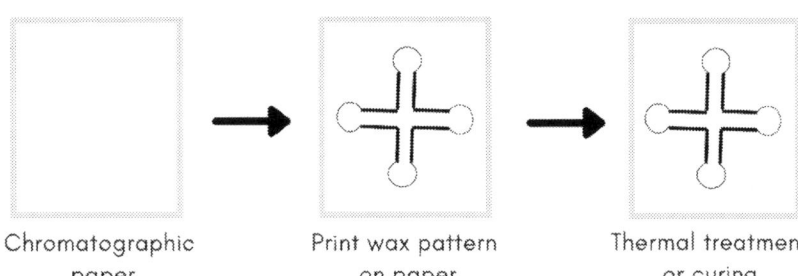

FIG. 6.7 General workflow of fabrication by waxing method. *(No Permission Required.)*

The other method is using custom-designed rubber stamps containing the desired pattern to contact print solution of PDMS and hexane onto chromatographic paper followed by curing to form hydrophobic barriers, after PDMS penetrating the paper (Dornelas, Dossi, & Piccin, 2015).

Chemical modification

This method is based on silanization procedures by chemical modifier like AKD or trimethoxyoctadecylsilane (TMOS) as patterning agent to define the hydrophobic boundary by direct use of masks 50 or wet etching using an appropriate etching reagents such as NaOH solution (containing 30% glycerol) and paper mask as template having the design (Cai et al., 2014).

Another strategy is based on hydrophobization of the entire paper followed by conversion of hydrophobic areas into hydrophilic ones by UV irradiation, plasma treatment, or laser treatment as done in case of photolithography (He, Ma, Hu, & Chen, 2013; Li, Tian, Nguyen, & Shen, 2008; Zhang, Liu, Wang, & Fan, 2019).

Drawing

In this method, the assay/test area will be directly drawn on to the paper with media such as correction pens, wax crayons, ballpoint pen containing custom ink/solvents, wax pen, and wax containing eyeliner pencils (Mani, Prabhu, Biswas, & Chakraborty, 2019; Zhong, Wang, & Huang, 2012).

Permanent marker pen

This method uses permanent markers (with black wax-based ink) as they are usually water resistant, but soluble in alcohol. Different designs can be drawn manually or digitally controlled plotter/cutter followed by drying or curing at room temperature to evaporate the solvent and the remaining resins to create hydrophobic barrier on filter paper due to the hydrophobic nature of wax (Dossi et al., 2017).

Patterns can be designed by software like Inkscape and printed on filter paper by electronic plotter/cutter using a permanent marker (Gallibu, Gallibu, Avoundjian, & Gomez, 2016; Nuchtavorn & Macka, 2016).

Eyeliner pencil

Rubbing with eyeliner pencil containing wax to develop patterns on the filter paper over white sticker or polyester screen mesh designed using Auto CAD or Corel Draw software, followed by curing at 150°C for few minutes to melt the wax forms hydrophobic and hydrophilic area on the paper (Abadian, Sepehri Manesh, & Jafarabadi Ashtiani, 2017).

As eyeliner pencils are inexpensive, easy to use, and eco-friendly, it can be used to develop sensors for colorimetric detection application (Ostad, Hajinia, & Heidari, 2017).

These sensors can be used for biosensing and colorimetric applications. This type of techniques is cheaper than other techniques and offers mass production, as the permanent marker ink and eyeliner pencil are very cheap. Normal pencil can be used to draw the electrodes, and electrical properties of graphite make this possible.

Pencil

Pencil-drawing technique has the simplicity to transfer carbon tracks onto paper surface with ready availability of all the raw materials and fast fabrication (Yu & Zhang, 2017).

Lower reproducibility is seen in this method comparatively due to the variation of pencil leads from batch to batch and the irregular characteristics of the electrode surface when drawn manually (Singhal, Prabhu, Giri Nandagopal, Dheivasigamani, & Mani, 2021).

Pencil-drawn electrodes can be improved for heterogeneous electron transfer by electrochemical treatment of electrode surface (Santhiago, Strauss, Pereira, Chagas, & Bufon, 2017). Drawing electrode using software-controlled cutting printer improves the reproducibility as the software also allows control of the writing speed and the applied pressure (Dossi et al., 2017).

Carbon nanotubes/cellulose composite paper

There are ways to fabricate carbon nanotubes or cellulose composite paper depending on the raw materials available and application. Cellulose suspension is by finely blending fibers in water, then, adding CNT and gelatin/sodium dodecyl sulfate (SDS) followed by sonication. Different concentrations of CNT depend on the requirement. Solution can be poured into a dish and dried at room temperature or vacuum-filtered to get a paper sheet or fiber mat (Maria & Mieno, 2017; Pang et al., 2015).

Methods used for cutting the paper to obtain physical boundaries defined sensors

Cutting the paper to define physical boundaries allows the fabrication of reagent-free devices, enabling greater patterning precision and ease of fabrication compared to other laborious procedures. In these approaches, the channels are generally obtained by cutting the paper either with a cutter plotter or with a CO_2 laser cutter. Also handheld blades, hole punches, craft knives, and other strategies can also be used (Fenton, Mascarenas, López, & Sibbett, 2009; Renault, Li, Fosdick, & Crooks, 2013).

In most cases, adhesive tape is used to cover the cut channels to impart certain microfluidic-like structure and higher strength. This method involves cutting a piece of paper according to the predesigned pattern, defining hydrophilic flowing path formed by the hollow microstructure. Main drawback of the cutting method for mass production is the need of specialized equipment (Akyazi et al., 2018).

Laser cutter

A microstructure-patterned paper is suitable as a sensor, and the required pattern is predesigned on the personal computer. By laser cutter, a hollow structure is created to avoid the chemicals. In most of the fabricating techniques, a hydrophobic wall is created within hydrophilic paper (Mehmeti et al., 2017). The CO_2 laser cutter is used, as the hollow microstructure-patterned paper acts as walls the fluid are distributed via capillary throughout the pattern design. This type of fabrication method depends on the resolution of the requirement. Also, they are suitable for the mass production of devices (Santhiago, Nery, Santos, & Kubota, 2014).

Cutter printer

Cutter printer is a low-cost technique to cut paper in desired pattern. One side of the paper will be covered with adhesive tape before cutting the design using cutter printer, then, heated on a hot plate at 120°C to melt the adhesive tape followed by sealing the back side of paper by heat press machine.

Laminated paper-based analytical devices (LPADs)

Initially, a pattern has to be created as per the required design and dimensions using CAD software then to keep the file in DXF file format for cutout and cover sheet pattern. It is then exported to the controller software of a craft cutter. A chromatography paper will be attached to the carrier section (Liu et al., 2014). As per the dimensions, the paper is cut using cutting plotter, unwanted curves and edges are removed, and paper strips are taken out. Using the design of the cover sheet pattern file, cover film is cut again using procedures as for the paper (Cassano & Fan, 2013). Later, paper strip at the center and bottom, and the cover sheet are aligned together. This alignment then let into the heated laminator of temperature of 220°F and rolling speed around. Due to the temperature, the polyester films will be heated and will cover the outline of the paper strip enhancing the flexibility and strength (Tenda et al., 2016).

Other techniques such as wax screen printing, oxygen plasma treatment, stamping, and permanent markers can also be used to establish hydrophilic channel and hydrophobic barrier (Jiang, He, Shen, Shi, & Liu, 2017; Kong, Wang, Zhang, Ge, & Yu, 2017; Wang, Chen, & Liao, 2017).

Thus, fabricated sensor using this type of technique can be used for biosensing and colorimetric applications, as technique is cheaper than other techniques and offers ease of mass production.

Conclusion

This is a brief review of CNT application in the field of environmental monitoring for contaminant analysis. CNT limited by its lower limit of detection and maximum precision can serve as sensors, to overcome the limitations and for easy, rapid detection of pollutants and periodic monitoring of air, soil, and water bodies near the industrial area and developing cities/towns. Improved sensitivity and specificity for particular analyte detection is possible by advances in the field of biosensors using nanomaterial. Paper-based sensors are cost-effective, portable, disposable, point-of-care testing, and lab-on-a-chip technology for medical, environmental, and industrial applications, which are accessible to the general population. Further in this chapter, the development/fabrication methods and analysis techniques of paper-based environmental sensing devices are covered. In reality, we are still far from scaling-up these concepts for the test through paper-based devices, which can be easily mass-produced and can integrate them into treatment plants for detecting and eliminating contaminants; however, disposable sensors will continue to play a central role in this regard. In conclusion, developments are required in the field of disposable sensing devices as it will be easier to repurpose for detection of new contaminants and multianalyte panel tests for complete analysis of contaminants.

References

Abadian, A., Sepehri Manesh, S., & Jafarabadi Ashtiani, S. (2017). Hybrid paper-based microfluidics: Combination of paper-based analytical device (μPAD) and digital microfluidics (DMF) on a single substrate. *Microfluidics and Nanofluidics*, 21(4), 65. https://doi.org/10.1007/s10404-017-1899-2.

Abe, K., Suzuki, K., & Citterio, D. (2008). Inkjet-printed microfluidic multianalyte chemical sensing paper. *Analytical Chemistry*, 80(18), 6928–6934. https://doi.org/10.1021/ac800604v.

Adkins, J. A., Boehle, K., Friend, C., Chamberlain, B., Bisha, B., & Henry, C. S. (2017). Colorimetric and electrochemical bacteria detection using printed paper- and transparency-based analytic devices. *Analytical Chemistry*, 89(6), 3613–3621. https://doi.org/10.1021/acs.analchem.6b05009.

Akyazi, T., Basabe-Desmonts, L., & Benito-Lopez, F. (2018). Review on microfluidic paper-based analytical devices towards commercialisation. *Analytica Chimica Acta*, 1001, 1–17. https://doi.org/10.1016/j.aca.2017.11.010.

Alonso, G. A., Muñoz, R., & Marty, J.-L. (2013). Automatic electronic tongue for on-line detection and quantification of organophosphorus and carbamate pesticides using enzymatic screen printed biosensors. *Analytical Letters*, 46(11), 1743–1757. https://doi.org/10.1080/00032719.2012.745087.

Alwassil, O. I., Chandrashekarappa, S., Nayak, S. K., & Venugopala, K. N. (2019). Design, synthesis and structural elucidation of novel NmeNANAS inhibitors for the treatment of meningococcal infection. *PLoS One*.

Amrutha, B. M., Manjunatha, J. G., Bhatt, S., & Hareesha, N. (2019). Electrochemical analysis of Evans blue by surfactant modified carbon nanotube paste electrode. *Journal of Materials and Environmental Science*, 10.

Amrutha, B. M., Manjunatha, J. G., Bhatt, A. S., Raril, C., & Pushpanjali, P. A. (2019). Electrochemical sensor for the determination of Alizarin Red-S at non-ionic surfactant modified carbon nanotube paste electrode. *Physical Chemistry Research*, 7(3), 523–533. https://doi.org/10.22036/pcr.2019.185875.1636.

Andreuccetti, C., Bettazzi, F., Giorgi, C., Laschi, S., Marrazza, G., Mascini, M., et al. (2014). *Macrocyclic polyamine modified screen-printed electrodes for copper(II) detection* (pp. 471–474). https://doi.org/10.1007/978-1-4614-3860-1_84.

Aragay, G., Pons, J., & Merkoçi, A. (2011). Enhanced electrochemical detection of heavy metals at heated graphite nanoparticle-based screen-printed electrodes. *Journal of Materials Chemistry*, 21(12), 4326. https://doi.org/10.1039/c0jm03751f.

Baskeyfield, D. E. H., Davis, F., Magan, N., & Tothill, I. E. (2011). A membrane-based immunosensor for the analysis of the herbicide isoproturon. *Analytica Chimica Acta*, 699(2), 223–231. https://doi.org/10.1016/j.aca.2011.05.036.

Benedict, S. R. (2002). A reagent for the detection of reducing sugars. 1908. *Journal of Biological Chemistry*, 277(16), e5.

Bouden, S., Bellakhal, N., Chaussé, A., & Vautrin-Ul, C. (2014). Performances of carbon-based screen-printed electrodes modified by diazonium salts with various carboxylic functions for trace metal sensors. *Electrochemistry Communications*, 41, 68–71. https://doi.org/10.1016/j.elecom.2014.01.028.

Cai, L., Xu, C., Lin, S., Luo, J., Wu, M., & Yang, F. (2014). A simple paper-based sensor fabricated by selective wet etching of silanized filter paper using a paper mask. *Biomicrofluidics*, 8(5), 056504. https://doi.org/10.1063/1.4898096.

Carrilho, E., Martinez, A. W., & Whitesides, G. M. (2009). Understanding wax printing: A simple micropatterning process for paper-based microfluidics. *Analytical Chemistry*, 81(16), 7091–7095. https://doi.org/10.1021/ac901071p.

Carrilho, E., Phillips, S. T., Vella, S. J., Martinez, A. W., & Whitesides, G. M. (2009). Paper microzone plates. *Analytical Chemistry*, 81(15), 5990–5998. https://doi.org/10.1021/ac900847g.

Cassano, C. L., & Fan, Z. H. (2013). Laminated paper-based analytical devices (LPAD): Fabrication, characterization, and assays. *Microfluidics and Nanofluidics*, 15(2), 173–181. https://doi.org/10.1007/s10404-013-1140-x.

Chang, J., & Zen, J. (2007). A poly(dimethylsiloxane)-based electrochemical cell coupled with disposable screen printed edge band ultramicroelectrodes for use in flow injection analysis. *Electrochemistry Communications*, 9(12), 2744–2750. https://doi.org/10.1016/j.elecom.2007.09.014.

Charithra, M. M., & Manjunatha, J. G. (2020). Enhanced voltammetric detection of paracetamol by using carbon nanotube modified electrode as an electrochemical sensor. *Journal of Electrochemical Science and Engineering, 10*(1), 29–40. https://doi.org/10.5599/jese.717.

Charithra, M. M., Manjunatha, J. G. G., & Raril, C. (2020). Surfactant modified graphite paste electrode as an electrochemical sensor for the enhanced voltammetric detection of estriol with dopamine and uric acid. *Advanced Pharmaceutical Bulletin, 10*(2), 247–253. https://doi.org/10.34172/apb.2020.029.

Chauhan, P. S., Pandey, M., & Bhattacharya, S. (2019). Paper based sensors for environmental monitoring. In *Paper microfluidics* (pp. 165–181). Springer Science and Business Media LLC. https://doi.org/10.1007/978-981-15-0489-1_10.

Chen, J., Liu, D., Li, S., & Yao, D. (2010). Development of an amperometric enzyme electrode biosensor for sterigmatocystin detection. *Enzyme and Microbial Technology, 47*(4), 119–126. https://doi.org/10.1016/j.enzmictec.2010.06.008.

Chen, Y., Zhong, Q., Wang, Y., Yuan, C., Qin, X., & Xu, Y. (2019). Colorimetric detection of hydrogen peroxide and glucose by exploiting the peroxidase-like activity of papain. *RSC Advances, 9*(29), 16566–16570. https://doi.org/10.1039/c9ra03111a.

Chequer, F. M., de Oliveira, G. A. R., Ferraz, E. R., Carvalho, J., Zanoni, M. V. B., & de Oliveir, D. P. (2013). Textile dyes: Dyeing process and environmental impact. In *Eco-friendly textile dyeing and finishing*. https://doi.org/10.5772/53659.

Chikkaveeraiah, B. V., Bhirde, A., Malhotra, R., Patel, V., Gutkind, J. S., & Rusling, J. F. (2009). Single-wall carbon nanotube forest arrays for immunoelectrochemical measurement of four protein biomarkers for prostate cancer. *Analytical Chemistry, 81*(21), 9129–9134. https://doi.org/10.1021/ac9018022.

Clark, L. C., & Lyons, C. (1962). Electrode systems for continuous monitoring in cardiovascular surgery. *Annals of the New York Academy of Sciences, 102*(1), 29–45. https://doi.org/10.1111/j.1749-6632.1962.tb13623.x.

Comer, J. P. (1956). Semiquantitative specific test paper for glucose in urine. *Analytical Chemistry, 28*(11), 1748–1750. https://doi.org/10.1021/ac60119a030.

Crew, A., Lonsdale, D., Byrd, N., Pittson, R., & Hart, J. P. (2011). A screen-printed, amperometric biosensor array incorporated into a novel automated system for the simultaneous determination of organophosphate pesticides. *Biosensors and Bioelectronics, 26*(6), 2847–2851. https://doi.org/10.1016/j.bios.2010.11.018.

de Tarso Garcia, P., Garcia Cardoso, T. M., Garcia, C. D., Carrilho, E., & Tomazelli Coltro, W. K. (2014). A handheld stamping process to fabricate microfluidic paper-based analytical devices with chemically modified surface for clinical assays. *RSC Advances, 4*(71), 37637–37644. https://doi.org/10.1039/C4RA07112C.

Dincer, C., Bruch, R., Costa-Rama, E., Fernández-Abedul, M. T., Merkoçi, A., Manz, A., et al. (2019). Disposable sensors in diagnostics, food, and environmental monitoring. *Advanced Materials, 31*(30), 1806739. https://doi.org/10.1002/adma.201806739.

Dornelas, K. L., Dossi, N., & Piccin, E. (2015). A simple method for patterning poly(dimethylsiloxane) barriers in paper using contact-printing with low-cost rubber stamps. *Analytica Chimica Acta, 858*, 82–90. https://doi.org/10.1016/j.aca.2014.11.025.

Dossi, N., Petrazzi, S., Toniolo, R., Tubaro, F., Terzi, F., Piccin, E., et al. (2017). Digitally controlled procedure for assembling fully drawn paper-based electroanalytical platforms. *Analytical Chemistry, 89*(19), 10454–10460. https://doi.org/10.1021/acs.analchem.7b02521.

D'Souza, E. S., Manjunatha, J. G., Raril, C., Tigari, G., Ravishankar, D. K., & Fattepur, S. (2020). Arginine electropolymerized carbon nanotube paste electrode as sensitive and selective sensor for electrochemical determination of vanillin. *Journal of Materials and Environmental Science, 11*, 512–521.

Dungchai, W., Chailapakul, O., & Henry, C. S. (2011). A low-cost, simple, and rapid fabrication method for paper-based microfluidics using wax screen-printing. *The Analyst, 136*(1), 77–82. https://doi.org/10.1039/C0AN00406E.

El-Shahawi, M. S., Hamza, A., Bashammakh, A. S., & Al-Saggaf, W. T. (2010). An overview on the accumulation, distribution, transformations, toxicity and analytical methods for the monitoring of persistent organic pollutants. *Talanta, 80*(5), 1587–1597. https://doi.org/10.1016/j.talanta.2009.09.055.

Farma, R., Deraman, M., Awitdrus, A., Talib, I. A., Taer, E., Basri, N. H., et al. (2013). Preparation of highly porous binderless activated carbon electrodes from fibres of oil palm empty fruit bunches for application in supercapacitors. *Bioresource Technology, 132*, 254–261.

Fenton, E. M., Mascarenas, M. R., López, G. P., & Sibbett, S. S. (2009). Multiplex lateral-flow test strips fabricated by two-dimensional shaping. *ACS Applied Materials & Interfaces, 1*(1). https://doi.org/10.1021/am800043z.

Fotouhi, L., Dorraji, P. S., Keshmiri, Y. S. S., & Hamtak, M. (2018). Electrochemical sensor based on nanocomposite of multi-walled carbon nanotubes/TiO2 nanoparticles in chitosan matrix for simultaneous and separate determination of dihydroxybenzene isomers. *Journal of the Electrochemical Society, 165*(5), B202–B211. https://doi.org/10.1149/2.0541805jes.

Fu, L., Li, X., Yu, J., & Ye, J. (2013). Facile and simultaneous stripping determination of zinc, cadmium and lead on disposable multiwalled carbon nanotubes modified screen-printed electrode. *Electroanalysis, 25*(2), 567–572. https://doi.org/10.1002/elan.201200248.

Gallibu, C., Gallibu, C., Avoundjian, A., & Gomez, F. (2016). Easily fabricated microfluidic devices using permanent marker inks for enzyme assays. *Micromachines, 7*(1), 6. https://doi.org/10.3390/mi7010006.

Gilbert, L., Jenkins, A. T. A., Browning, S., & Hart, J. P. (2011). Development of an amperometric, screen-printed, single-enzyme phosphate ion biosensor and its application to the analysis of biomedical and environmental samples. *Sensors and Actuators B: Chemical, 160*(1), 1322–1327. https://doi.org/10.1016/j.snb.2011.09.069.

Golmohammadi, H., Morales-Narváez, E., Naghdi, T., & Merkoçi, A. (2017). Nanocellulose in sensing and biosensing. *Chemistry of Materials, 29*(13), 5426–5446. https://doi.org/10.1021/acs.chemmater.7b01170.

Gupta, N., Pant, S. C., Vijayaraghavan, R., & Rao, P. V. L. (2003). Comparative toxicity evaluation of cyanobacterial cyclic peptide toxin microcystin variants (LR, RR, YR) in mice. *Toxicology, 188*(2–3), 285–296. https://doi.org/10.1016/S0300-483X(03)00112-4.

Guthrie, D. W., & Humphreys, S. (1988). Diabetes urine testing: An historical perspective. *The Diabetes Educator, 14*(6), 521–525. https://doi.org/10.1177/014572178801400615.

Han, J.-W., Kim, B., Li, J., & Meyyappan, M. (2014). A carbon nanotube based ammonia sensor on cellulose paper. *RSC Advances, 4*(2), 549–553. https://doi.org/10.1039/c3ra46347h.

Hareesha, N., & Manjunatha, J. G. (2020a). Fast and enhanced electrochemical sensing of dopamine at cost-effective poly(DL-phenylalanine) based graphite electrode. *Journal of Electroanalytical Chemistry, 878*, 114533. https://doi.org/10.1016/j.jelechem.2020.114533.

Hareesha, N., & Manjunatha, J. G. (2020b). Elevated and rapid voltammetric sensing of riboflavin at poly(helianthin dye) blended carbon paste electrode with heterogeneous rate constant elucidation. *Journal of the Iranian Chemical Society, 17*(6), 1507–1519. https://doi.org/10.1007/s13738-020-01876-4.

Hareesha, N., Manjunatha, J. G., Raril, C., & Tigari, G. (2019a). Design of novel surfactant modified carbon nanotube paste electrochemical sensor for the sensitive investigation of tyrosine as a pharmaceutical drug. *Advanced Pharmaceutical Bulletin, 9*.

Hareesha, N., Manjunatha, J. G., Raril, C., & Tigari, G. (2019b). Sensitive and selective electrochemical resolution of tyrosine with ascorbic acid through the development of electropolymerized alizarin sodium sulfonate modified carbon nanotube paste electrodes. *ChemistrySelect, 4*(15), 4559–4567. https://doi.org/10.1002/slct.201900794.

He, Q., Ma, C., Hu, X., & Chen, H. (2013). Method for fabrication of paper-based microfluidic devices by alkylsilane self-assembling and UV/O$_3$-patterning. *Analytical Chemistry, 85*(3), 1327–1331. https://doi.org/10.1021/ac303138x.

Hussain, C. M. (2018). *Handbook of nanomaterials for industrial applications* (pp. 1–1109). Elsevier. https://doi.org/10.1016/C2016-0-04427-3.

Hussain, C. M. (2019). *Handbook of environmental materials management*. Springer Nature Switzerland AG.

Hussain, C. M. (2020a). *Handbook of functionalized nanomaterials for industrial applications*. Elsevier.

Hussain, C. M. (Ed.). (2020b). *Handbook of manufacturing applications of nanomaterials*. Elsevier.

Hussain, C. M. (Ed.). (2020c). *Handbook of industrial applications of polymer nanocomposites* Elsevier.

Hussain, C. M. (2020d). *Handbook of nanomaterials for sensing applications*. Elsevier.

Ijeomah, G., Obite, F., & Rahman, O. (2016). Development of carbon nanotube-based biosensors. *International Journal of Nano and Biomaterials, 6*(2), 83–109. https://doi.org/10.1504/IJNBM.2016.079682.

Im, J., Sterner, E. S., & Swager, T. M. (2016). Integrated gas sensing system of SWCNT and cellulose polymer concentrator for benzene, toluene, and xylenes. *Sensors (Basel, Switzerland), 16*(2), 183. https://doi.org/10.3390/s16020183.

Imani, S., Bandodkar, A. J., Mohan, A. M. V., Kumar, R., Yu, S., Wang, J., et al. (2016). A wearable chemical-electrophysiological hybrid biosensing system for real-time health and fitness monitoring. *Nature Communications, 7*, 11650. https://doi.org/10.1038/ncomms11650.

Injang, U., Noyrod, P., Siangproh, W., Dungchai, W., Motomizu, S., & Chailapakul, O. (2010). Determination of trace heavy metals in herbs by sequential injection analysis-anodic stripping voltammetry using screen-printed carbon nanotubes electrodes. *Analytica Chimica Acta, 668*(1), 54–60. https://doi.org/10.1016/j.aca.2010.01.018.

Irvine, G. W., Tan, S. N., & Stillman, M. J. (2017). A simple metallothionein-based biosensor for enhanced detection of arsenic and mercury. *Biosensors, 7*(1). https://doi.org/10.3390/bios7010014.

Ivanov, A. N., Younusov, R. R., Evtugyn, G. A., Arduini, F., Moscone, D., & Palleschi, G. (2011). Acetylcholinesterase biosensor based on single-walled carbon nanotubes—Co phthalocyanine for organophosphorus pesticides detection. *Talanta, 85*(1), 216–221. https://doi.org/10.1016/j.talanta.2011.03.045.

Ivnitski, D., Branch, B., Atanassov, P., & Apblett, C. (2006). Glucose oxidase anode for biofuel cell based on direct electron transfer. *Electrochemistry Communications, 8*(8), 1204–1210. https://doi.org/10.1016/j.elecom.2006.05.024.

Jacobs, C. B., Peairs, M. J., & Venton, B. J. (2010). Review: Carbon nanotube based electrochemical sensors for biomolecules. *Analytica Chimica Acta, 662*(2), 105–127. https://doi.org/10.1016/j.aca.2010.01.009.

Jiang, X., & Fan, Z. H. (2016). Fabrication and operation of paper-based analytical devices. *Annual Review of Analytical Chemistry, 9*(1), 203–222. https://doi.org/10.1146/annurev-anchem-071015-041714.

Jiang, P., He, M., Shen, L., Shi, A., & Liu, Z. (2017). A paper-supported aptasensor for total IgE based on luminescence resonance energy transfer from upconversion nanoparticles to carbon nanoparticles. *Sensors and Actuators B: Chemical, 239*. https://doi.org/10.1016/j.snb.2016.08.005.

Kaewarsa, P., Laiwattanapaisal, W., Palasuwan, A., & Palasuwan, D. (2017). A new paper-based analytical device for detection of Glucose-6-phosphate dehydrogenase deficiency. *Talanta, 164*, 534–539. https://doi.org/10.1016/j.talanta.2016.12.026.

Kang, S. K., Murphy, R. K. J., Hwang, S. W., Lee, S. M., Harburg, D. V., Krueger, N. A., et al. (2016). Bioresorbable silicon electronic sensors for the brain. *Nature, 530*(7588), 71–76. https://doi.org/10.1038/nature16492.

Katulski, R. J., Namiesnik, J., Sadowski, J., Stefanski, J., & Wardencki, W. (2011). Monitoring of gaseous air pollution. In *The impact of air pollution on health, economy, environment and agricultural sources* IntechOpen.

Khaled, E., Hassan, H., Girgis, A., & Metelka, R. (2008). Construction of novel simple phosphate screen-printed and carbon paste ion-selective electrodes. *Talanta, 77*(2), 737–743. https://doi.org/10.1016/j.talanta.2008.07.018.

Khan, S., & Malik, A. (2018). Toxicity evaluation of textile effluents and role of native soil bacterium in biodegradation of a textile dye. *Environmental Science and Pollution Research, 25*(5), 4446–4458. https://doi.org/10.1007/s11356-017-0783-7.

Kholghi Eshkalak, S., Chinnappan, A., Jayathilaka, W. A. D. M., Khatibzadeh, M., Kowsari, E., & Ramakrishna, S. (2017). A review on inkjet printing of CNT composites for smart applications. *Applied Materials Today, 9*, 372–386. https://doi.org/10.1016/j.apmt.2017.09.003.

Kim, T. H., Lee, J., & Hong, S. (2009). Highly selective environmental nanosensors based on anomalous response of carbon nanotube conductance to mercury ions. *The Journal of Physical Chemistry C, 113*(45), 19393–19396. https://doi.org/10.1021/jp908902k.

Klug, K. E., Reynolds, K. A., & Yoon, J. Y. (2018). A capillary flow dynamics-based sensing modality for direct environmental pathogen monitoring. *Chemistry - A European Journal, 24*(23), 6025–6029. https://doi.org/10.1002/chem.201800085.

Kong, Q., Wang, Y., Zhang, L., Ge, S., & Yu, J. (2017). A novel microfluidic paper-based colorimetric sensor based on molecularly imprinted polymer membranes for highly selective and sensitive detection of bisphenol A. *Sensors and Actuators B: Chemical, 243*. https://doi.org/10.1016/j.snb.2016.11.146.

Kong, F., Zhou, C., Peng, C., & Dai. (2000). Nanotube molecular wires as chemical sensors. *Science (New York, N.Y.), 287*(5453), 622–625. https://doi.org/10.1126/science.287.5453.622.

Koo, Y., Shanov, V. N., & Yun, Y. (2016). Carbon nanotube paper-based electroanalytical devices. *Micromachines, 7*(4). https://doi.org/10.3390/mi7040072.

Li, S.c., Chen, J.h., Cao, H., Yao, D.s., & Liu, D.l. (2011). Amperometric biosensor for aflatoxin B1 based on aflatoxin-oxidase immobilized on multiwalled carbon nanotubes. *Food Control, 22*(1), 43–49. https://doi.org/10.1016/j.foodcont.2010.05.005.

Li, X., Tian, J., Nguyen, T., & Shen, W. (2008). Paper-based microfluidic devices by plasma treatment. *Analytical Chemistry, 80*(23), 9131–9134. https://doi.org/10.1021/ac801729t.

Liana, D. D., Raguse, B., Gooding, J. J., & Chow, E. (2012). Recent advances in paper-based sensors. *Sensors, 12*(9), 11505–11526. https://doi.org/10.3390/s120911505.

Lin, Z., Le, T., Song, X., Yao, Y., Li, Z., Moon, K. S., et al. (2013). Preparation of water-based carbon nanotube inks and application in the inkjet printing of carbon nanotube gas sensors. *Journal of Electronic Packaging, Transactions of the ASME, 135*(1). https://doi.org/10.1115/1.4023758.

Liu, W., Luo, J., Guo, Y., Kou, J., Li, B., & Zhang, Z. (2014). Nanoparticle coated paper-based chemiluminescence device for the determination of L-cysteine. *Talanta, 120*, 336–341. https://doi.org/10.1016/j.talanta.2013.12.033.

Lu, L., Zhang, L., Zhang, X., Huan, S., Shen, G., & Yu, R. (2010). A novel tyrosinase biosensor based on hydroxyapatite–chitosan nanocomposite for the detection of phenolic compounds. *Analytica Chimica Acta, 665*(2), 146–151. https://doi.org/10.1016/j.aca.2010.03.033.

Madou, M. J. (2011). *Manufacturing techniques for microfabrication and nanotechnology*. CRC Press.

Malekghasemi, S., Kahveci, E., & Duman, M. (2016). Rapid and alternative fabrication method for microfluidic paper based analytical devices. *Talanta, 159*, 401–411. https://doi.org/10.1016/j.talanta.2016.06.040.

Mallikarjuna, S. M., Padmashali, B., & Sandeep, C. (2014). Synthesis, anticancer and antituberculosis studies for [1-(4-chlorophenyl) cyclopropyl] (piperazine-yl) methanone derivatives. *International Journal of Pharmacy and Pharmaceutical Sciences, 6*(7), 423–427. http://innovareacademics.in/journals/index.php/ijpps/article/download/2230/9713.

Mallikarjuna, S. M., Sandeep, C., & Padmashali, B. (2017). Acid amine coupling of (1h-indole-6-yl) ppiperazin-1-yl) methanone with substituted acids using HATU coupling reagent and their antimicrobial and antioxidant activity. *International Journal of Pharmaceutical Sciences and Research, 8*, 2879–2885.

Mani, N. K., Prabhu, A., Biswas, S. K., & Chakraborty, S. (2019). Fabricating paper based devices using correction pens. *Scientific Reports, 9*(1), 1752. https://doi.org/10.1038/s41598-018-38308-6.

Manjunatha, J. G. (2017a). A new electrochemical sensor based on modified carbon nanotube-graphite mixture paste electrode for voltammetric determination of resorcinol. *Asian Journal of Pharmaceutical and Clinical Research, 10*.

Manjunatha, J. G. (2017b). Electroanalysis of estriol hormone using electrochemical sensor. *Sensing and Bio-Sensing Research, 16*, 79–84. https://doi.org/10.1016/j.sbsr.2017.11.006.

Manjunatha, J. G. G. (2018). A novel poly (glycine) biosensor towards the detection of indigo carmine: A voltammetric study. *Journal of Food and Drug Analysis, 26*(1), 292–299. https://doi.org/10.1016/j.jfda.2017.05.002.

Manjunatha, J. G., Deraman, M., & Basri, N. H. (2015). Electrocatalytic detection of dopamine and uric acid at poly (basic blue b) modified carbon nanotube paste electrode. *Asian Journal of Pharmaceutical and Clinical Research, 8*(5), 48–53. http://innovareacademics.in/journals/index.php/ajpcr/article/download/7586/2972.

Manjunatha, J. G., Deraman, M., Basri, N. H., Nor, N. S. M., Talib, I. A., & Ataollahi, N. (2014). Sodium dodecyl sulfate modified carbon nanotubes paste electrode as a novel sensor for the simultaneous determination of dopamine, ascorbic acid, and uric acid. *Comptes Rendus Chimie, 17*.

Manjunatha, J. G., Deraman, M., Basri, N. H., & Talib, I. A. (2018). Fabrication of poly (Solid Red A) modified carbon nano tube paste electrode and its application for simultaneous determination of epinephrine, uric acid and ascorbic acid. *Arabian Journal of Chemistry, 11*(2), 149–158. https://doi.org/10.1016/j.arabjc.2014.10.009.

Manjunatha, J. G., Raril, C., Hareesha, N., Charithra, M. M., Pushpanjali, P. A., Tigari, G., et al. (2020). Electrochemical fabrication of poly (niacin) modified graphite paste electrode and its application for the detection of riboflavin. *Open Chemical Engineering Journal, 14*, 90–98. https://doi.org/10.2174/1874123102014010090.

Manjunatha, J. G., Swamy, B. E., Mamatha, G. P., Gilbert, O., Shreenivas, T., & Sherigara, S. (2009). Electrocatalytic response of dopamine at mannitol and Triton X-100 modified carbon paste electrode: A cyclic voltammetric study. *International Journal of Electrochemical Science, 4*, 1706–1718.

Maria, K. H., & Mieno, T. (2017). Production and properties of carbon nanotube/cellulose composite paper. *Journal of Nanomaterials, 2017*. https://doi.org/10.1155/2017/6745029.

Marrazza, G. (2014). Piezoelectric biosensors for organophosphate and carbamate pesticides: A review. *Biosensors, 4*(3), 301–317. https://doi.org/10.3390/bios4030301.

Martinez, A. W., Phillips, S. T., Butte, M. J., & Whitesides, G. M. (2007). Patterned paper as a platform for inexpensive, low-volume, portable bioassays. *Angewandte Chemie International Edition, 46*(8), 1318–1320. https://doi.org/10.1002/anie.200603817.

Martinez, A. W., Phillips, S. T., Wiley, B. J., Gupta, M., & Whitesides, G. M. (2008). FLASH: A rapid method for prototyping paper-based microfluidic devices. *Lab on a Chip, 8*(12), 2146–2150. https://doi.org/10.1039/b811135a.

Mehmeti, E., Stanković, D. M., Chaiyo, S., Zavasnik, J., Žagar, K., & Kalcher, K. (2017). Wiring of glucose oxidase with graphene nanoribbons: An electrochemical third generation glucose biosensor. *Microchimica Acta, 184*(4), 1127–1134. https://doi.org/10.1007/s00604-017-2115-5.

Miyake, M. (2003). Electrochemical functions. In *Carbon alloys: Novel concepts to develop carbon science and technology* Elsevier.

Mohammadi, S., Maeki, M., Mohamadi, R. M., Ishida, A., Tani, H., & Tokeshi, M. (2015). An instrument-free, screen-printed paper microfluidic device that enables bio and chemical sensing. *The Analyst, 140*(19), 6493–6499. https://doi.org/10.1039/C5AN00909J.

Morales-Narváez, E., Golmohammadi, H., Naghdi, T., Yousefi, H., Kostiv, U., Horák, D., et al. (2015). Nanopaper as an optical sensing platform. *ACS Nano, 9*(7), 7296–7305. https://doi.org/10.1021/acsnano.5b03097.

Moschou, D., & Tserepi, A. (2017). The lab-on-PCB approach: Tackling the μTAS commercial upscaling bottleneck. *Lab on a Chip*, 1388–1405. https://doi.org/10.1039/C7LC00121E.

Nadifiyine, S., Haddam, M., Mandli, J., Chadel, S., Blanchard, C. C., Marty, J. L., et al. (2013). Amperometric biosensor based on tyrosinase immobilized on to a carbon black paste electrode for phenol determination in olive oil. *Analytical Letters, 46*(17), 2705–2726. https://doi.org/10.1080/00032719.2013.811679.

Nagesh, H. K., Padmashali, B., Sandeep, C., Yuvaraj, T. C. M., Siddesh, M. B., & Mallikarjuna, S. M. (2014). Synthesis and antimicrobial activity of benzothiophene substituted coumarins, pyrimidines and pyrazole as new scaffold. *International Journal of Pharmaceutical Sciences Review and Research, 28*(2), 6–10. http://globalresearchonline.net/journalcontents/v28-2/02.pdf.

Nguyen, L. H., Phi, T. V., Phan, P. Q., Vu, H. N., Nguyen-Duc, C., & Fossard, F. (2007). Synthesis of multi-walled carbon nanotubes for NH3 gas detection. *Physica E: Low-dimensional Systems and Nanostructures, 37*(1–2), 54–57. https://doi.org/10.1016/j.physe.2006.12.006.

Nuchtavorn, N., & Macka, M. (2016). A novel highly flexible, simple, rapid and low-cost fabrication tool for paper-based microfluidic devices (μPADs) using technical drawing pens and in-house formulated aqueous inks. *Analytica Chimica Acta, 919*, 70–77. https://doi.org/10.1016/j.aca.2016.03.018.

Olkkonen, J., Lehtinen, K., & Erho, T. (2010). Flexographically printed fluidic structures in paper. *Analytical Chemistry, 82*(24), 10246–10250. https://doi.org/10.1021/ac1027066.

Ostad, M. A., Hajinia, A., & Heidari, T. (2017). A novel direct and cost effective method for fabricating paper-based microfluidic device by commercial eye pencil and its application for determining simultaneous calcium and magnesium. *Microchemical Journal, 133*, 545–550. https://doi.org/10.1016/j.microc.2017.04.031.

Padmashali, B., Chidananda, B. N., Bhanuprakash, G., Basavaraj, S. M., Chandrashekharappa, S., & Venugopala, K. N. (2019). Synthesis and characterization of novel 1,6-dihydropyrimidine derivatives for their pharmacological properties. *Journal of Applied Pharmaceutical Science, 9*(05).

Pang, Z., Sun, X., Wu, X., Nie, Y., Liu, Z., & Yue, L. (2015). Fabrication and application of carbon nanotubes/cellulose composite paper. *Vacuum, 122*, 135–142. https://doi.org/10.1016/j.vacuum.2015.09.020.

Panpatte, D. G., & Jhala, Y. K. (2019). Nanotechnology for agriculture: Advances for sustainable agriculture. In *Nanotechnology for agriculture: Advances for sustainable agriculture* (pp. 1–305). Springer Singapore. https://doi.org/10.1007/978-981-32-9370-0.

Peixoto, P. S., Machado, A., Oliveira, H. P., Bordalo, A. A., & Segundo, M. A. (2019). *Paper-based biosensors for analysis of water*. IntechOpen. https://doi.org/10.5772/intechopen.84131.

Pereira, L., & Alves, M. (2012). Chapter 4: Dyes—Environmental impact and remediation. In A. Malik, & E. Grohmann (Eds.), *Environmental protection strategies for sustainable development, strategies for sustainability* (pp. 111–154). Dordrecht: Springer Science + Business Media B.V.

Piao, Y., Jin, Z., Lee, D., Lee, H. J., Na, H. B., Hyeon, T., et al. (2011). Sensitive and high-fidelity electrochemical immunoassay using carbon nanotubes coated with enzymes and magnetic nanoparticles. *Biosensors and Bioelectronics, 26*(7), 3192–3199. https://doi.org/10.1016/j.bios.2010.12.025.

Plumeré, N., Henig, J., & Campbell, W. H. (2012). Enzyme-catalyzed O2 removal system for electrochemical analysis under ambient air: Application in an amperometric nitrate biosensor. *Analytical Chemistry, 84*(5), 2141–2146. https://doi.org/10.1021/ac2020883.

Posthuma-Trumpie, G. A., Korf, J., & Van Amerongen, A. (2009). Lateral flow (immuno)assay: Its strengths, weaknesses, opportunities and threats. A literature survey. *Analytical and Bioanalytical Chemistry, 393*(2), 569–582. https://doi.org/10.1007/s00216-008-2287-2.

Prinith, N. S., & Manjunatha, J. G. (2020). Polymethionine modified carbon nanotube sensor for sensitive and selective determination of l-tryptophan. *Journal of Electrochemical Science and Engineering, 10*(4), 305–315. https://doi.org/10.5599/jese.774.

Pushpanjali, P. A., Manjunatha, J. G., & Shreenivas, M. T. (2019). The electrochemical resolution of ciprofloxacin, riboflavin and estriol using anionic surfactant and polymer-modified carbon paste electrode. *ChemistrySelect, 4*(46), 13427–13433. https://doi.org/10.1002/slct.201903897.

Pushpanjali, P. A., Manjunatha, J. G., Tigari, G., & Fattepur, S. (2020). Poly (niacin) based carbon nanotube sensor for the sensitive and selective voltammetric detection of vanillin with caffeine. *Analytical and Bioanalytical Electrochemistry, 12*(4), 553–568.

Quan, D., Shim, J. H., Kim, J. D., Park, H. S., Cha, G. S., & Nam, H. (2005). Electrochemical determination of nitrate with nitrate reductase-immobilized electrodes under ambient air. *Analytical Chemistry, 77*(14), 4467–4473. https://doi.org/10.1021/ac050198b.

Quang, N. H., Van Trinh, M., Lee, B. H., & Huh, J. S. (2006). Effect of NH3 gas on the electrical properties of single-walled carbon nanotube bundles. *Sensors and Actuators B: Chemical, 113*(1), 341–346. https://doi.org/10.1016/j.snb.2005.03.089.

Rao, V. K., Suresh, S., Sharma, M. K., Gupta, A., & Vijayaraghavan, R. (2011). Carbon nanotubes—A potential material for affinity biosensors. *Carbon nanotubes—Growth and applications*. IntechOpen.

Raril, C., & Manjunatha, J. G. (2018). Carbon nanotube paste electrode for the determination of some neurotransmitters: A cyclic voltammetric study. *Modern Chemistry & Applications*. https://doi.org/10.4172/2329-6798.1000263.

Raril, C., Manjunatha, J.G., Nanjundaswamy, L., Siddaraju, G., Ravishankar, D.K., Fattepur, S., & Niranjan, E. (2018). Surfactant immobilized electrochemical sensor for the detection of indigotine. Analytical and Bioanalytical Electrochemistry, 10(11), 1479–1490. http://www.abechem.com/No.%2011-2018/2018,%2010(11),%201479-1490.pdf.

Raril, C., Manjunatha, J. G., Ravishankar, D. K., Fattepur, S., Siddaraju, G., & Nanjundaswamy, L. (2020). Validated electrochemical method for simultaneous resolution of tyrosine, uric acid, and ascorbic acid at polymer modified nano-composite paste electrode. *Surface Engineering and Applied Electrochemistry*, *56*(4), 415–426. https://doi.org/10.3103/S1068375520040134.

Raril, C., Manjunatha, J. G., & Tigari, G. (2020). Low-cost voltammetric sensor based on an anionic surfactant modified carbon nanocomposite material for the rapid determination of curcumin in natural food supplement. *Instrumentation Science and Technology*, *48*(5), 561–582. https://doi.org/10.1080/10739149.2020.1756317.

Rashmi, S. K., Suresha Kumara, T. H., Sandeep, C., Nagendrappa, G., Sowmya, H. B. V., & Sujan Ganapathy, P. S. (2015). Synthesis, antimicrobial, antioxidant and docking study of (3-alkoxy-5-nitrobenzofuran-2-yl) (phenyl) methanone derivatives. *International Journal of Pharmacy and Pharmaceutical Sciences*, *7*(2), 493–497. http://innovareacademics.in/journals/index.php/ijpps/article/download/4062/pdf_653.

Renault, C., Li, X., Fosdick, S. E., & Crooks, R. M. (2013). Hollow-channel paper analytical devices. *Analytical Chemistry*, *85*(16). https://doi.org/10.1021/ac401786h.

Rengaraj, S., Cruz-Izquierdo, Á., Scott, J. L., & Di Lorenzo, M. (2018). Impedimetric paper-based biosensor for the detection of bacterial contamination in water. *Sensors and Actuators, B: Chemical*, *265*, 50–58. https://doi.org/10.1016/j.snb.2018.03.020.

Salentijn, G. I. J., Hamidon, N. N., & Verpoorte, E. (2016). Solvent-dependent on/off valving using selectively permeable barriers in paper microfluidics. *Lab on a Chip*, *16*(6), 1013–1021. https://doi.org/10.1039/C5LC01355K.

Salvatore, G. A., Sülzle, J., Dalla Valle, F., Cantarella, G., Robotti, F., Jokic, P., et al. (2017). Biodegradable and highly deformable temperature sensors for the internet of things. *Advanced Functional Materials*, *27*(35). https://doi.org/10.1002/adfm.201702390.

Sameenoi, Y., Nongkai, P. N., Nouanthavong, S., Henry, C. S., & Nacapricha, D. (2014). One-step polymer screen-printing for microfluidic paper-based analytical device (μPAD) fabrication. *The Analyst*, *139*(24), 6580–6588. https://doi.org/10.1039/C4AN01624F.

Sandeep, C., Padmashali, B., & Kulkarni, R. S. (2013). Synthesis of isomeric substituted 6-acetyl-3-benzoylindolizine-1-carboxylate and 8-acetyl-3-benzoylindolizine-1-carboxylate from substituteded 3-acetyl pyridinium bromides and their antimicrobial activity. *Journal of Applicable Chemistry*, 1049–1056.

Sandeep, C., Padmashali, B., Rashmi, S. K., Mallikarjuna, S. M., Siddesh, M. B., Nagesh, H. K., et al. (2014). Synthesis of substituted 5-acetyl-3-benzoylindolizine-1-carboxylates (2a-p) from subtitled 2-acetyl pyridinium bromides. *Heterocyclic Letters*, *4*, 371–376.

Sandeep, C., Padmashali, B., Venugopala, K. N., Kulkarni, R. S., Venugopala, R., & Odhav, B. (2016). Synthesis and characterization of ethyl 7-acetyl-2-substituted 3-(substituted benzoyl)indolizine-1-carboxylates for in vitro anticancer activity. *Asian Journal of Chemistry*, 1043–1048. https://doi.org/10.14233/ajchem.2016.19582.

Sandeep, C., Venugopala, K. N., Khedr, M. A., Padmashali, B., Kulkarni, R. S., Venugopala, R., et al. (2017). Design and synthesis of novel indolizine analogues as COX-2 inhibitors: Computational perspective and in vitro screening. *Indian Journal of Pharmaceutical Education and Research*, *51*(3), 452–460. https://doi.org/10.5530/ijper.51.3.73.

Sanllorente-Méndez, S., Domínguez-Renedo, O., & Arcos-Martínez, M. J. (2010). Immobilization of acetylcholinesterase on screen-printed electrodes. Application to the determination of arsenic(III). *Sensors*, *10*(3), 2119–2128. https://doi.org/10.3390/s100302119.

Santhiago, M., Nery, E. W., Santos, G. P., & Kubota, L. T. (2014). Microfluidic paper-based devices for bioanalytical applications. *Bioanalysis*, *6*(1), 89–106. https://doi.org/10.4155/bio.13.296.

Santhiago, M., Strauss, M., Pereira, M. P., Chagas, A. S., & Bufon, C. C. B. (2017). Direct drawing method of graphite onto paper for high-performance flexible electrochemical sensors. *ACS Applied Materials & Interfaces*, *9*(13), 11959–11966. https://doi.org/10.1021/acsami.6b15646.

Scala-Benuzzi, M. L., Raba, J., Soler-Illia, G. J. A. A., Schneider, R. J., & Messina, G. A. (2018). Novel electrochemical paper-based immunocapture assay for the quantitative determination of ethinylestradiol in water samples. *Analytical Chemistry*, *90*(6), 4104–4111. https://doi.org/10.1021/acs.analchem.8b00028.

Schroeder, V., Savagatrup, S., He, M., Lin, S., & Swager, T. M. (2019). Carbon nanotube chemical sensors. *Chemical Reviews*, *119*(1), 599–663. https://doi.org/10.1021/acs.chemrev.8b00340.

Scida, K., Li, B., Ellington, A. D., & Crooks, R. M. (2013). DNA detection using origami paper analytical devices. *Analytical Chemistry*, *85*(20), 9713–9720. https://doi.org/10.1021/ac402118a.

Sharma, B., Dangi, A. K., & Shukla, P. (2018). Contemporary enzyme based technologies for bioremediation: A review. *Journal of Environmental Management*, *210*, 10–22. https://doi.org/10.1016/j.jenvman.2017.12.075.

Shim, B. S., Chen, W., Doty, C., Xu, C., & Kotov, N. A. (2008). Smart electronic yarns and wearable fabrics for human biomonitoring made by carbon nanotube coating with polyelectrolytes. *Nano Letters*, *8*(12), 4151–4157. https://doi.org/10.1021/nl801495p.

Shiroma, L. S., Piazzetta, M. H. O., Duarte-Junior, G. F., Coltro, W. K. T., Carrilho, E., Gobbi, A. L., et al. (2016). Self-regenerating and hybrid irreversible/reversible PDMS microfluidic devices. *Scientific Reports*, *6*. https://doi.org/10.1038/srep26032.

Siddesh, M. B., Padmashali, B., Thriveni, K. S., & Sandeep, C. (2014). Synthesis of polynuclear pyrimidine derivatives and their pharmacological activities. *Heterocyclic Letters*, *4*, 503–514.

Siddesh, M. B., Padmashali, B., Thriveni, K. S., Sandeep, C., & Goudarshivnnanava, B. C. (2013). Synthesis and pharmacological evaluation of some novel pyrimidine derivatives. *Journal of Applied Chemistry*, 1281–1288.

Singh, C., Srivastava, S., Ali, M. A., Gupta, T. K., Sumana, G., Srivastava, A., et al. (2013). Carboxylated multiwalled carbon nanotubes based biosensor for aflatoxin detection. *Sensors and Actuators B: Chemical*, *185*, 258–264. https://doi.org/10.1016/j.snb.2013.04.040.

Singh, M., Truong, J., Reeves, W. B., & Hahm, J. I. (2017). Emerging cytokine biosensors with optical detection modalities and nanomaterial-enabled signal enhancement. *Sensors (Switzerland)*, *17*(2). https://doi.org/10.3390/s17020428.

Singhal, H. R., Prabhu, A., Giri Nandagopal, M. S., Dheivasigamani, T., & Mani, N. K. (2021). One-dollar microfluidic paper-based analytical devices: Do-It-Yourself approaches. *Microchemical Journal, 165*. https://doi.org/10.1016/j.microc.2021.106126.

So, H.-M., Park, D.-W., Jeon, E.-K., Kim, Y.-H., Kim, B. S., Lee, C.-K., et al. (2008). Detection and titer estimation of *Escherichia coli* using aptamer-functionalized single-walled carbon-nanotube field-effect transistors. *Small, 4*(2), 197–201. https://doi.org/10.1002/smll.200700664.

Songjaroen, T., Dungchai, W., Chailapakul, O., & Laiwattanapaisal, W. (2011). Novel, simple and low-cost alternative method for fabrication of paper-based microfluidics by wax dipping. *Talanta, 85*(5), 2587–2593. https://doi.org/10.1016/j.talanta.2011.08.024.

Su, W.-Y., Wang, S.-M., & Cheng, S.-H. (2011). Electrochemically pretreated screen-printed carbon electrodes for the simultaneous determination of aminophenol isomers. *Journal of Electroanalytical Chemistry, 651*(2), 166–172. https://doi.org/10.1016/j.jelechem.2010.11.028.

Suehiro, J., Zhou, G., & Hara, M. (2005). Detection of partial discharge in SF6 gas using a carbon nanotube-based gas sensor. *Sensors and Actuators B: Chemical, 105*(2), 164–169. https://doi.org/10.1016/S0925-4005(04)00415-0.

Szabadvary, F., & Oesper, R. E. (1964). Development of the pH concept: A historical survey. *Journal of Chemical Education, 41*(2), 105. https://doi.org/10.1021/ed041p105.

Tenda, K., Ota, R., Yamada, K., Henares, T., Suzuki, K., & Citterio, D. (2016). High-resolution microfluidic paper-based analytical devices for sub-microliter sample analysis. *Micromachines, 7*(5), 80. https://doi.org/10.3390/mi7050080.

Tigari, G., & Manjunatha, J. G. (2020a). A surfactant enhanced novel pencil graphite and carbon nanotube composite paste material as an effective electrochemical sensor for determination of riboflavin. *Journal of Science: Advanced Materials and Devices, 5*(1), 56–64. https://doi.org/10.1016/j.jsamd.2019.11.001.

Tigari, G., & Manjunatha, J. G. (2020b). Optimized voltammetric experiment for the determination of phloroglucinol at surfactant modified carbon nanotube paste electrode. *Instruments and Experimental Techniques, 63*(5), 750–757. https://doi.org/10.1134/S0020441220050139.

Tigari, G., Manjunatha, J. G., Raril, C., & Hareesha, N. (2019). Determination of riboflavin at carbon nanotube paste electrodes modified with an anionic surfactant. *ChemistrySelect, 4*(7), 2168–2173. https://doi.org/10.1002/slct.201803191.

Tiwari, J. N., Vij, V., Kemp, K. C., & Kim, K. S. (2015). Engineered carbon-nanomaterial-based electrochemical sensors for biomolecules. *ACS Nano, 10*(1), 46–80. https://doi.org/10.1021/acsnano.5b05690.

Ueda, T., Bhuiyan, M. M. H., Norimatsu, H., Katsuki, S., Ikegami, T., & Mitsugi, F. (2008). Development of carbon nanotube-based gas sensors for NOx gas detection working at low temperature. *Physica E: Low-dimensional Systems and Nanostructures, 40*(7), 2272–2277. https://doi.org/10.1016/j.physe.2007.12.006.

Ueda, T., Katsuki, S., Takahashi, K., Narges, H. A., Ikegami, T., & Mitsugi, F. (2008). Fabrication and characterization of carbon nanotube based high sensitive gas sensors operable at room temperature. *Diamond and Related Materials, 17*(7–10), 1586–1589. https://doi.org/10.1016/j.diamond.2008.03.009.

Valentini, L., Cantalini, C., Armentano, I., Kenny, J. M., Lozzi, L., & Santucci, S. (2003). Investigation of the NO[sub 2] sensitivity properties of multi-walled carbon nanotubes prepared by plasma enhanced chemical vapor deposition. *Journal of Vacuum Science & Technology B: Microelectronics and Nanometer Structures, 21*(5), 1996. https://doi.org/10.1116/1.1599858.

Venugopala, K. N., Al-Attraqchi, O., Tratrat, C., Nayak, S., Morsy, M., Aldhubiab, B., et al. (2019). Novel series of methyl 3-(substituted benzoyl)-7-substituted-2-phenylindolizine-1-carboxylates as promising anti-inflammatory agents: Molecular modeling studies. *Biomolecules, 11*.

Venugopala, K. N., Chandrashekharappa, S., Bhandary, S., Chopra, D., Khedr, M. A., Aldhubiab, B., et al..... (2018). Efficient synthesis and characterization of novel substituted 3-benzoylindolizine analogues via the cyclization of aromatic cycloimmoniumylides with electron-deficient alkenes. *Current Organic Synthesis, 15*, 400–407.

Venugopala, K. N., Chandrashekharappa, S., Pillay, M., Bhandary, S., Kandeel, M., Mahomoodally, F. M., et al. (2019). Synthesis and structural elucidation of novel benzothiazole derivatives as anti-tubercular agents: In-silico screening for possible target identification. *Medicinal Chemistry, 15*(3), 311–326. https://doi.org/10.2174/1573406414666180703121815.

Venugopala, K. N., Khedr, M. A., Pillay, M., Nayak, S. K., Chandrashekharappa, S., Aldhubiab, B. E., et al. (2018). Benzothiazole analogs as potential anti-TB agents: Computational input and molecular dynamics. *Journal of Biomolecular Structure and Dynamics*, 1–13. https://doi.org/10.1080/07391102.2018.1470035.

Venugopala, K. N., Chandrashekharappa, S., Pillay, M., Abdallah, H. H., Mahomoodally, F. M., Bhandary, S., et al. (2019). Computational, crystallographic studies, cytotoxicity and anti-tubercular activity of substituted 7-methoxy-indolizine analogues. *PLoS ONE, 14*(6). https://doi.org/10.1371/journal.pone.0217270, e0217270.

Verma, N., & Bhardwaj, A. (2015). Biosensor technology for pesticides—A review. *Applied Biochemistry and Biotechnology, 175*(6), 3093–3119. https://doi.org/10.1007/s12010-015-1489-2.

Wang, C.-M., Chen, C.-Y., & Liao, W.-S. (2017). Paper-polymer composite devices with minimal fluorescence background. *Analytica Chimica Acta, 963*. https://doi.org/10.1016/j.aca.2017.01.044.

Wang, L., Chen, W., Xu, D., Shim, B. S., Zhu, Y., Sun, F., et al. (2009). Simple, rapid, sensitive, and versatile SWNT-paper sensor for environmental toxin detection competitive with ELISA. *Nano Letters, 9*(12), 4147–4152. https://doi.org/10.1021/nl902368r.

Wei, Y., Yang, R., Liu, J.-H., & Huang, X.-J. (2013). Selective detection toward Hg(II) and Pb(II) using polypyrrole/carbonaceous nanospheres modified screen-printed electrode. *Electrochimica Acta, 105*, 218–223. https://doi.org/10.1016/j.electacta.2013.05.004.

Wu, H., Ma, Z., Lin, Z., Song, H., Yan, S., & Shi, Y. (2019). High-sensitive ammonia sensors based on tin monoxide nanoshells. *Nanomaterials (Basel, Switzerland), 9*(3). https://doi.org/10.3390/nano9030388.

Xu, H., Zeng, L., Xing, S., Xian, Y., Shi, G., & Jin, L. (2008). Ultrasensitive voltammetric detection of trace lead(II) and cadmium(II) using MWCNTs-nafion/bismuth composite electrodes. *Electroanalysis, 20*(24), 2655–2662. https://doi.org/10.1002/elan.200804367.

Yamada, K., Henares, T. G., Suzuki, K., & Citterio, D. (2015). Paper-based inkjet-printed microfluidic analytical devices. *Angewandte Chemie International Edition*, *54*(18), 5294–5310. https://doi.org/10.1002/anie.201411508.

Yao, D., Cao, H., Wen, S., Liu, D., Bai, Y., & Zheng, W. (2006). A novel biosensor for sterigmatocystin constructed by multi-walled carbon nanotubes (MWNT) modified with aflatoxin–detoxifizyme (ADTZ). *Bioelectrochemistry*, *68*(2), 126–133. https://doi.org/10.1016/j.bioelechem.2005.05.003.

Yu, Y., & Zhang, J. (2017). Pencil-drawing assembly to prepare graphite/MWNT hybrids for high performance integrated paper supercapacitors. *Journal of Materials Chemistry A*, *5*(9). https://doi.org/10.1039/C6TA10076G.

Zhang, Y., Liu, J., Wang, H., & Fan, Y. (2019). Laser-induced selective wax reflow for paper-based microfluidics. *RSC Advances*, *9*(20), 11460–11464. https://doi.org/10.1039/C9RA00610A.

Zhong, Z. W., Wang, Z. P., & Huang, G. X. D. (2012). Investigation of wax and paper materials for the fabrication of paper-based microfluidic devices. *Microsystem Technologies*, *18*(5), 649–659. https://doi.org/10.1007/s00542-012-1469-1.

Chapter 7

Trends in carbon nanomaterial-based sensors in the food industry

Shridevi Doddamani[a], Vinusha Honnalagere Mariswamy[b], Vinay Karekura Boraiah[c], and Srikantamurthy Ningaiah[d]

[a]Chemical Sciences and Technology Division, CSIR-NIIST, Thiruvananthapuram, Kerala, India, [b]PG Department of Chemistry, Sarada Vilas College, Mysuru, Karnataka, India, [c]Department of Mechanical Engineering, Vidyavardhaka College of Engineering, Visvesvaraya Technological University, Mysuru, Karnataka, India, [d]Department of Chemistry, Vidyavardhaka College of Engineering, Visvesvaraya Technological University, Mysuru, Karnataka, India

Introduction

In recent decades, sensors based on nanomaterials have gotten a lot of attention because they encourage researchers to experiment with new principles to enhance analytical results. To improve the efficiency of biosensing with improved precision, durability, and selectivity, functionalized nanocompounds are employed as catalytic tools and frameworks for immobilization. Nanotechnologies are thought to carry a lot of potential for producing new goods in almost every industry, and many implementations are already in the market. While the safety of nanomaterials is still an ongoing discussion, it should be geared toward developing a risk evaluation of nanomaterial's use in the food industry (de Otxoa, 2018). As a rapidly evolving area, nanotechnology offers the potential to solve some of the problems associated with food detection. Nanomaterial-based biosensors have more specific target identification, better selectivity and sensitivity, improved signal readout, and a shorter analysis time than standard chromatography approaches (Lv et al., 2018).

The expansion of novel sensors and biosensors with applications in the food sector is one of today's most crucial nanobiotechnology and nanomaterials research areas. The rapid evolution of nanotechnology has paved the way for the emergence of modern sensing and food processing solutions, solving long-standing problems in the food industry, for instance, extending the existence, reducing waste, determining food shelter, and improving food quality. Nanotechnology is utilized to develop the bioavailability, aroma, texture, and purity of food by altering particle size, cluster shape, and surface charge of food nanomaterials. Food packaging may benefit from nanomaterials to improve mechanical efficiency, improve gas barrier properties, increase water repellency, and have antimicrobial and scavenging action. Nanomaterials as food additives for food production, nanosensing, and nano-enabled packaging are the three types of nanotechnology applications in food. In food packaging, nanomaterials may be used as nutritional carriers, additives, or antimicrobial agents. The amount of food items designed to protect against microorganisms and impurity damage is on the rise.

Every year, millions of people fall infected, and hundreds of thousands die. Furthermore, foodborne illnesses caused by a lack of adequate food protection significantly impact developing countries' health and financial systems. As a result, food safety is more critical than ever before, and it has become a significant global concern (Manikandan, Adhikari, & Chen, 2018). Chemical additives are currently vital in the manufacture of food and beverages in the food industrialization process. Pesticides, fertilizers, and antibiotics are heavily used in modern agriculture around organic food crops. Cross-contamination, microbial development, and spoilage are constantly evolving due to the globalization of food production, which is linked to different treatments and fluctuating environmental factors.

Foodborne infections are transmitted by ingesting bacteria, viruses, or parasites infected foods or drinks. According to the Centers for Disease Control and Prevention's worldwide foodborne disease figures for 2011 to 2012, there were 1632 epidemics, 29,112 infected people, 1750 hospitalized, and 68 demises. Salmonella (31%), Listeria (28%), Campylobacter (5%), and *Escherichia coli* O157:H7 (3%) are a few of the bacterial pathogens that cause foodborne infections and fatality (Valdés, González, Calzón, & Díaz-García, 2009). Similarly, the Food and Drug Administration of Taiwan's Department of Health recorded 4284 epidemics and 82,342 cases between 1991 and 2010, with 285 outbreaks per year on average between 2001 and 2010 being significantly higher than the 143 outbreaks reported between 1991 and 2000.

Vibrio parahaemolyticus, *Staphylococcus aureus*, and *Bacillus cereus* are the three mainly prevalent foodborne pathogens accountable for these eruptions in Taiwan (Cheng et al., 2013a). Foodborne illness caused by pathogens and contaminants costs the food industry a lot of funds because a lot of time and money is spent researching and determining prevention steps for food safety (Leonard et al., 2003; Pérez-López & Merkoçi, 2011). As a result, developing a hurried, responsive, precise, and cost-effective analytical approach for detecting microbial pollutants is critical. Microscopy, nucleic acid, and immunoassay methods are also common techniques for detecting pathogens. To achieve these goals, super-precise, accurate, efficient, speedy, and cost-effective sensing systems, for instance, electrochemical sensors/biosensors, must be developed. Latest growths in the field of nanotechnology recommend numerous technological progressions for the recognition of foodborne pathogens and contaminants with ruination along with meat formulations (Ali et al., 2011; Ali et al., 2014; Ali, Hashim, Mustafa, Che Man, & Islam, 2012; Ali, Hashim, Mustafa, Che Man, & Yusop, 2011).

Carbon nanomaterials in food technology

Carbon nanowires, nanotubes, nanoparticles, metal nanocomposite, and nanostructured materials are used to develop food study sensors. In addition, these nanobiosystems can help with the development of new food detection strategies. Sensor systems' sensing capabilities are improved employing nanomaterials, including magnetic nanoparticles, carbon nanotubes, nanotubes, quantum dots, nanowires, nanochannels, and other nanomaterials. The invention of carbon nanotubes, also known as "buckyball fullerene," which is 1 nm in diameter, was a game-changing breakthrough that ushered in the era of nanoscience (Institute of Medicine, 2009). Carbon nanomaterials have tremendous promise in a diversity of areas, including the food sector.

It is raised to adjust electrode surfaces with new nanomaterials in the last few years to attain more rapid electron transfer of biomolecules and advanced specificity. This benefit has sparked studies into the use of nanomaterial-based biosensors in conjunction with biomolecules. Food safety, health research, and environmental monitoring are only a few of the applications for these biosensors (Pérez-López & Merkoçi, 2011). Integration of enzymes, antibodies, or DNA sequences with nanoparticles (e.g., gold nanoparticles (AuNPs), single-walled carbon nanotubes (SWCNTs), multiwalled carbon nanotubes (MWCNTs), cadmium telluride quantum dot nanoparticles (CdTe QDs), etc., results in new hybrid structures for food usages (Deo et al., 2005; Li et al., 2009; Vinayaka, Basheer, & Thakur, 2009; Viswanathan, Radecka, & Radecki, 2009; Yang, Huang, Meng, Shen, & Jiao, 2009).

In the last few years, the responsibility of food components has shifted from that of a nutrient supply to that of a contributor to consumer happiness. When immobilized in various customized carriers, some nutrients, for example, enzymes, may be resistant to proteases and additional denaturing compounds and also more stable to pH and temperature variations (Lopez-Rubio, Gavara, & Lagaron, 2006). Microencapsulated fish oil, for nutritional reasons, is added to bread (Chaudhry et al., 2008). This bread is commercially available, and the microencapsulation process masks the fish oil's unpleasant taste.

Gas chromatography (GC), high-performance liquid chromatography (HPLC), gas chromatography-mass spectrometry (GC-MS), liquid chromatography-mass spectrometry (LCMS), and enzyme-linked immunosorbent assay (ELISA) have all been employed to diagnose food protection (Malik, Blasco, & Picó, 2010; Patel, 2002; Rodríguez-Lázaro et al., 2007). The majority of these approaches have drawbacks, such as difficult sample pretreatment, lengthy recognition times, and costly equipment and well-trained technicians, limiting their use in countries with limited amenities and experts.

Nanotechnology, a rapidly evolving area, offers the potential to overcome these challenges in food safety detection. Nanotechnologies, in particular, have a considerable impact in the food industry, including industrialized, refining, protection, covering, transportation, storage, and distribution. In food safety, nanotechnology holds great promise for detecting food toxins, with nanomaterial-based biosensors playing a key position. In contrast to their bulk counterparts, nanomaterials have special physicochemical properties, making them useful for food safety detection.

Carbon-based nanomaterials in the food industry

Protection, tamper resistance, and unique physical, chemical, or biological parameters are all requirements for food packaging. It also displays the product's packaging, which has all nutritional facts about the food being eaten. The packaging is critical in preserving the product and ensuring its marketability. Packaging innovations have resulted in high-feature covering, a user-friendly approach to determining shelf life, biodegradable packaging, and many additional benefits. An exploit of nanotechnology in packaging is labeled according to the goal of the application.

Since their discovery, carbon-based nanomaterials have piqued the attention of scientists. Carbon-based nanomaterials have been employed to produce high-efficiency sensing modules for a food protection assessment to generate, recognize, and optimize sensing signals based on their spatial dimensions. In detail, analyses of modern carbon technologies, for example, graphene and carbon dots (CDs), have boosted the promise of carbon-based sensors and their application opportunities in the production of food shelter devices with high accuracy, high intervention immunity, and ease of use.

Although advanced food packaging technology such as vacuum packaging and pasteurization treatment enable food to be preserved for long periods, foodborne pathogens inevitably infect food, resulting in severe consumer diseases, such as acute emesis and acute abdominalgia. *E. coli*, Salmonella, and Listeria are the most common foodborne pathogens. In recent decades, outbreaks of foodborne pathogen contamination have occurred. It happened on occasion all over the world (Crim et al., 2006). These incidents have raised concerns that preventative measures, such as identifying pathogens in food, could be taken before consumption. Toxins created by pathogens were released into food in several cases of foodborne pathogen diseases.

Several techniques for detecting and quantifying pathogens and contaminants in food have been proposed. For now, a few difficulties remained, such as low precision, sensitivity, and detection time. Biosensors focused on nanomaterials provide an alternative approach for addressing these issues. Foodborne pathogens and toxins have been detected using biosensors made of nanoparticles, quantum dots (QDs), nanorods, and carbon nanotubes (CNTs) (Cho et al., 2014; Kaittanis, Santra, & Perez, 2010). Nanomaterial-based biosensors have successfully identified a variety of pathogens and toxins using a variety of biosensing techniques.

Nanoparticles have always been employed in food containers as antimicrobial agents and certain instances as food supplements. As nanoclay is used in the construction of a nanocomposite (bentonite) utilized in the manufacture of containers and additional food packaging products, the gas barrier properties strengthen, preventing oxygen and moisture entry, drink destabilization, and food spoilage (Sinha, Ma, & Yeow, 2006).

Clay nanoparticles embedded in an ethylene-vinyl alcohol copolymer in addition to a polylactic acid biopolymer have been found to strengthen the oxygen barrier properties of food products and increase their shelf life (Lagaron et al., 2005). As a result of their potential antimicrobial properties, carbon nanoparticles are processed tenfold the number of other nanomaterials. They are exercised in air purifiers, food preserving tanks, antiperspirants, bandages, toothpaste, paints, and further household implements.

The antimicrobial properties of biologically prepared metallic nanoparticles, besides regulating plant pathogens, also reduce contamination. Nanomaterials, such as carbon nanotubes, have antimicrobial properties. When carbon nanotube aggregates came into close contact with *E. coli*, they punctured the cell, causing cellular damage or death. Nanobiosensors have been extensively used in the identification of cancer-causing bacteria to prepare high-quality, infectivity-free food (Berekaa, 2015).

The addition of allyl isothiocyanate and carbon nanotubes to active packaging systems reduces microbial infections and color modifications, controls oxidation, and aids in the long-term preservation of shredded chicken meat that has been baked. Combining two or more nanoparticles in combination produces a synergistic impact with persuasive antimicrobial action, unlike the size of single nanoparticles. *E. coli* and *B. cereus* spores are successfully combated by silver nanoparticles mixed with titanium dioxide and carbon nanotubes (Rashidi & Khosravi-Darani, 2011).

Carbon nanotubes in the food industry

Food nanotechnologies have focused on zein, the main protein present in corn. Nanomaterials made of zein can form a microorganism-resistant tubular network. The employment of zein nanomaterials as a carrier for flavor composites and the nanoencapsulation of dietetic enhancements have been investigated (Ingale, 2018). A nanotube is a thin carbon-based wire-like structure. The cavity diameter of a-lactalbumin nanotubes is 8 nm, in which many food components like vitamins or enzymes can bind (Srinivas et al., 2010). These cavities may be utilized to encapsulate nutraceuticals or cover up unpleasant essences or aromas.

At the nanoscale, certain compounds are poisonous, but not at the macroscale. Single-walled carbon nanotubes prevented cell proliferation and had a detrimental effect on cell development and turnover of human embryonic kidney cells (Cui, Tian, Ozkan, Wang, & Gao, 2005). Although these nanomaterials are improbable to be used in the food sector, such contaminated consequences should be monitored at this early phase of technological advancement.

The World Health Organization (WHO) has indicated unequivocally that food protection is paramount. Food should be safe and non-toxic. Food protection is a multifaceted problem. Food manufacturing, planning, and storage are a subject of interest in learning more about reducing the risk of foodborne illness and learning that food safety regulations are also in place. BRC, FSSC 22000, IFS, and HACCP are examples of food quality principles that are now widely used in the global food industry (Aung & Chang, 2014; Chassy, Hlywka, Kleter, Kok, & Kuiper, 2004). However, global food safety actions continue to occur regularly, causing serious health, economic, and even social problems (Carvalho, 2006; Pena-Rosas, De-Regil, Rogers, Bopardikar, & Panisset, 2012).

The most popular use of nanotechnology in food technology is food processing. Nanoparticles have been shown to enhance the properties of formed substances and films, especially longevity, temperature tolerance, flame resistance, barrier properties, optical properties, and recycling. Nanopackaging might be intended to discharge vitamins, flavors, antimicrobials, antioxidants, and nutraceuticals to expand shelf life (Berekaa, 2015).

The use of nanoemulsions as transporters for lipophilic bioactive components, flavorings, additives, and medications is a promising field of nanotechnology throughout food technology (Silva, Cerqueira, & Vicente, 2012). Nanoemulsions have sparked interest in food, drinks, and pharmacy because they have certain possible advantages over traditional emulsions (Komaiko & Mcclements, 2016).

The use of carbon nanotubes allows for the removal of CO_2 or the absorption of unpleasant tastes. Moreover, nanoclay in nanocomposites (bentonite), which are utilized to make bottles and other food packaging products, greatly improves gas barrier properties, preventing oxygen, humidity diffusion, and imbibe destabilization (Huang, Mei, Chen, & Wang, 2018; Merkoçi, 2006). Carbon nanotubes, which are often used in food manufacturing, can be applied to measure hazards on human skin and lungs (Monteiro-Riviere, Nemanich, Inman, Wang, & Riviere, 2005; Warheit et al., 2004).

The most commonly used nanoparticles in producing antimicrobial active PNFP are metal nanoparticles, carbon nanotubes, and metal oxide nanomaterials. These particles work well when they are in close contact with each other, but they can also make gradual changes and respond to organics in food more favorably. Because of their antimicrobial properties, carbon nanotubes have also been used in food packaging (Silvestre, Duraccio, & Cimmino, 2011). It is worth noting that close interaction with food should be avoided because it can impair the system's efficiency or work and trigger migration issues. One of the key benefits of using these innovations is the potential to lower manufacturing costs by combining active packaging with less expensive conventional passive packaging. Furthermore, foods with active packaging systems are preferred or desired by consumers because they contain fewer additives, especially preservatives, and are of higher safety and quality.

For the detection of *E. coli* bacteria, nanobiosensors based on multiwalled carbon nanotubes immobilized galactosidase were used. The quantity of current generated was compared to the extent of hydrolysis of the substrate, which was used to measure *E. coli* infectivity. The identification limit for the *E. coli* inhabitants in the bacterial solution has been increased due to the electrochemical and mechanical nature of carbon nanomaterials and the enzyme's high selectivity and affinity. Nanobiosensors with 5 h showed a low limit of detection (10 colony-forming units/milliliter) (Cheng et al., 2013b).

The riboflavin estimation in the presence of dopamine was performed using a novel sensor made from anionic surfactant sodium lauryl sulfate-modified carbon nanotube and a pencil graphite composite paste electrode, which showed good recovery in B-complex pills and natural food supplements. With excellent recovery, the suggested sensing technique was used on pharmaceutical and food samples (Ho, Kung, Yang, & Wang, 2005; Tigari & Manjunatha, 2020). A new sensor electrode was prepared by electro-polymerization of L-methionine on carbon multiwalled nanotube paste surface (PMETCNTPS) and successfully applied to determine L-tryptophan (L-TPN). The projected sensor electrode disclosed a high percentage recovery, which may be applied in drug formulations and pharmacokinetics study (Prinith & Manjunatha, 2020). Using the cyclic voltammetric technique, a poly(adenine) film-modified carbon nanotube paste electrode (PAEMCNTPE) was created by electropolymerized adenine on a carbon nanotube paste electrode and used for the sensitive and selective analysis of dopamine (DA), uric acid (UA), and ascorbic acid (AA) (Prinith, Manjunatha, & Raril, 2019). By electrochemical polymerization of niacin on the surface of carbon nanotube paste electrode, a new sensor for the determination of vanillin (VN) was produced (Pushpanjali, Manjunatha, Tigari, & Fattepur, 2020). A highly sensitive and selective voltammetric technique was developed for the detection of natural curcumin in the food supplement (Raril, Manjunatha, & Tigari, 2020).

Carbon dots in the food industry

Furthermore, due to its low toxicity, high biocompatibility, easy preparation process, and low cost, CD-based sensing systems are being used in a variety of applications, especially in the field of food quality and safety. Sensing devices based on CDs have many uses in the food industry, including detecting nutrient compounds, contaminants, prohibited or minimal chemicals, pathogenic microorganisms, and microbial toxins. Bacterial identification and quantitative research is a critical aspect of food safety. Traditional approaches need elongated culture times, very qualified operators, or precise identification elements for every kind of bacteria. Using multiple cross-reactive receptors, sensor arrays provide a fast, cost-efficient, and easy solution. It was claimed that a fluorescence sensing array based on CDs functionalized during diverse receptors could be built quickly and easily for the identification of a variety of bacteria (Ho et al., 2005).

Owing to the physical and chemical structure of different bacterial surfaces, three varieties of receptors (boronic acid, polymyxin, and vancomycin) yielded CDs that can bind to a range of bacteria (Huang et al., 2018; Merkoçi, 2006). Carbon quantum dots (CQDs) are primarily used in the food industry to diagnose and identify bacteria, heavy metals, and additives. Its key benefits are low toxicity and high sensitivity (Ye et al., 2017).

In the food sector, pesticide residue is a major problem. The feasibility of CDs in pesticide identification in agriculture and food samples has been explored in several studies (Deo et al., 2005; Li et al., 2009; Vinayaka et al., 2009; Viswanathan et al., 2009; Yang et al., 2009). A sensitive and selective sensor for glyphosate determination with a limit of detection (LOD) of 8 ng/mL using a carbon dot is labeled with antibody, and the proposed procedure was extended to river water, tea, and soil samples with a satisfactory recovery ratio of 87.4% to 103.7% (Wang, Lin, Cao, Guo, & Yu, 2016).

For decades, pesticides, especially organophosphorus pesticides, have been widely used in agriculture, resulting in high pesticide residues in foods, particularly vegetables and fruits. Given the growing concern over pesticide residues' possible effects on human health, sensitive identification of these compounds is critical. Because of the sensitive variations in fluorescence in the presence of pesticides, researchers are interested in using CD-based fluorescence technologies to detect pesticides. An enzyme, metal ions, or nanoparticles are sometimes used in the sensitive identification of pesticides.

Veterinary drug residues, including drug prototypes and metabolites, accumulate in poultry tissue or remain in poultry products (eggs, milk, and meat) may be harmful to human health. Using CDs, it is possible to detect pet drugs (Deo et al., 2005; Li et al., 2009; Vinayaka et al., 2009; Viswanathan et al., 2009; Yang et al., 2009). Antibiotics are commonly used in the animal farming industry to treat pathogens due to their broad spectrum as antimicrobials and low cost. Antibiotic abuse in the farming sector, on the other hand, has resulted in elevated residue levels in animal-derived foods. In general, the quenching, recovery, or enhancement of CDs fluorescence is needed for responsive antibiotic detection. Antibiotics such as tetracycline (TC) and its derivatives are often used. The fluorescence quenching effect of TC against CDs was used to make a sensitive assessment of TC, with a detection limit of 7.5 nM. This method was employed to detect oxytetracycline (OTC), chlortetracycline (CTC), and doxycycline (DOXC). Similar assays for detecting DOXC and TC in raw milk and human urine (Feng, Zhong, Miao, & Yang, 2015; Song et al., 2017) yielded good results.

Banned chemicals, such as melamine, have received a lot of coverage in China in recent years. As a result, several CD-based nanosensors for sensitive melamine detection have been created. CDs are shown to be capable of quantifying melamine by restoring the fluorescence of quenched CD-Hg^{2+} and CD-AuNP complexes, which showed strong recoveries in raw milk and milk powder (Dai et al., 2014; Lei et al., 2016). The quantification of tetracycline (TC) in milk, honey, and fish samples was achieved using effective luminescence of carbon dots (Pan et al., 2019).

Carbon nanosensors in the food industry

Carbon nanostructure-based sensors are revolutionizing sensor science by introducing novel operating concepts. Nanosensors containing carbon black and polyaniline are capable of detecting and identifying three foodborne pathogens by creating a distinct pattern of response for each microbe (Arshak et al., 2007). The use of carbon nanotubes advances in protein crystallization and the construction of bioreactors and biosensors. The addition of silver to titanium dioxide nanoparticles improves its bactericidal properties against *E. coli*, whereas TiO_2 mixed with carbon nanotubes improves sterilizer properties besides sensing *B. cereus* spores substantially (Krishna et al., 2005).

In food technology, carbon nanostructure-based sensors are often used to identify various compounds associated with foodborne diseases. Pesticides, toxins (made by fungi), antibiotics, and bacteria products are the most familiar contaminants employed in food. Because of their inherent properties, carbon nanostructures can create compact biosensing systems for food quality control. High sensitivity, reliability, low power consumption, and integration capability are only a few advantages. Four pesticides were detected with the support of a nanobiosensor based on multiwalled carbon nanotube/chitosan immobilized acetylcholinesterase (Du, Huang, Cai, & Zhang, 2007). With glutaraldehyde as a cross-linker, the modified nanocomposite was used to immobilize enzymes covalently. The inhibition effectiveness of four pesticides, carbaryl, malathion, dimethoate, and monocrotophos, against acetylcholinesterase, was studied. The nanobiosensors sensitivity ranged from 0.96 to 4.28 M, depending on the application (Verma, 2017a, 2017b, 2017c). In the development of biosensors, carbon nanotube paste electrode (CNTPE) has been utilized owing to their good electrical and mechanical qualities, as well as low residual current and noise, quick surface renewal, and a wide range of anodic and cathodic potentials (Tigari, Manjunatha, Raril, & Hareesha, 2019).

Aflatoxin B1 was detected using a nanobiosensor mounted on multiwalled carbon nanotubes immobilized aflatoxin oxidase (Li, Chen, Cao, Yao, & Liu, 2011). The covalent immobilization technique used for enzyme binding to nanotubes was extremely reliable, preserving the enzyme's maximum function. With a detection limit of 1.6 nM, the nanobiosensor demonstrated high sensitivity. For the identification of fructose in honey, a nanobiosensor based on carbon nanotube immobilized fructose dehydrogenase was used (Antiochia, Lavagnini, & Magno, 2004).

Quinoxaline-2-carboxylic acid is a marker trace of carbadox, a food additive used in pork, chicken, and fish that has significant mutagenic and carcinogenic properties (Neethirajan, Jayas, & Sadistap, 2009). As a result, multiwalled carbon nanotubes are functionalized with chitosan fabricated on a glassy carbon electrode to detect the trace amount of quinoxaline-2-carboxylic acid.

Another method involves using a single-walled carbon nanotube to track D-fructose in fruit juices, honey, soft drinks, and energy drinks (Neethirajan et al., 2009). The electrons are shuttled between the immobilized fructose dehydrogenase enzyme and the single-walled carbon nanotube pasted on the working electrode by an immobilized fructose dehydrogenase enzyme-based biosensor built using osmium redox polymer as a redox mediator. Sudan I is also observed by a multiwalled carbon nanotube thin film-modified electrode, which is carcinogenic. Sudan I dye enhanced electrochemical oxidation on the electrode surface, as calculated by an electrochemical process with a detection limit of 5.0 g/L (Gan, Li, & Wu, 2008).

Nanosensors containing carbon black and polyaniline have been shown to detect three foodborne pathogens (*B. cereus*, *V. parahaemolyticus*, and Salmonella spp.) by creating a unique response pattern for each microbe (Arshak et al., 2007). Palytoxin, a marine poison, is commonly found in contaminated fish. Electrochemiluminescent sensors based on carbon nanotubes have been developed for the ultrasensitive identification of palytoxin in mussel meat (Li et al., 2016; Wang et al., 2016). Food dyes, sunset purple, and tartrazine have been detected using transformed carbon-ceramic electrodes made from multiwalled carbon nanotubes (MWCNT)–ionic liquid nanocomposites (Majidi, Fadakar Bajeh Baj, & Naseri, 2013).

Sudan I is a carcinogenic red dye used in chili powder as an adulterant. Sudan I adulteration in chili powder was detected using multiwalled carbon nanotubes (MWCNTs) (Yang, Zhu, & Jiang, 2010). Sudan I was also detected using nanoparticles in chili powder, egg yolk, ketchup, onion, chili, and strawberry sauce. Carbon dioxide (CO_2) sensors are in high demand in the bulk food storage industry, as they can be used to monitor incipient spoilage and measure CO_2 levels in modified atmosphere packages and storage systems (Zheng, Qi, & Zhang, 2019).

The detection of toxic cadmium ions has been reported using potentiometer sensors made of silica nanocomposites and multiwalled carbon nanotubes (Bagheri et al., 2013). The discovery of ascorbic acid and folic acid in different fruit juices, wheat flour, and milk samples has been identified using ionic liquid nanocomposites based on carbon nanotubes and nickel oxide nanoparticles (Karimi-Maleh et al., 2014; Wei, Zhao, Xu, & Zeng, 2006; Xiao et al., 2008). Improved voltammetric estimation of metanil yellow present in curcumin was developed based on calixarene and gold nanoparticles coated glassy carbon electrodes (Girigoswami, Ghosh, Pallavi, Ramesh, & Girigoswami, 2021).

Conclusion

In summary, carbon nanomaterials play a key role in food technology to ensure high-quality food through improved manufacturing, packaging, and long-term storage. Food consistency is being improved, resulting in increased growth in the food industry. Consumers benefit from nanomaterials and nanosensors because they know the condition of the food inside and its nutritional status and increased protection by pathogen identification. We have briefed the applications of various carbon nanomaterial-based sensors in food packing, food storage, and foodborne pathogens' identifications.

References

Ali, M. E., Hashim, U., Mustafa, S., Che Man, Y. B., Adam, T., & Humayun, Q. (2014). Nanobiosensor for the detection and quantification of pork adulteration in meatball formulation. *Journal of Experimental Nanoscience*, 9(2), 152–160. https://doi.org/10.1080/17458080.2011.640946.

Ali, M. E., Hashim, U., Mustafa, S., Che Man, Y. B., & Islam, K. N. (2012). Gold nanoparticle sensor for the visual detection of pork adulteration in meatball formulation. *Journal of Nanomaterials*, 2012. https://doi.org/10.1155/2012/103607.

Ali, M. E., Hashim, U., Mustafa, S., Che Man, Y. B., & Yusop, M. H. M. (2011). Nanobiosensor for the detection and quantification of specific DNA sequences in degraded biological samples. In *Vol. 35. IFMBE proceedings* (pp. 384–387). https://doi.org/10.1007/978-3-642-21729-6_99.

Ali, M. E., Hashim, U., Mustafa, S., Che Man, Y. B., Yusop, M. H. M., Bari, M. F., et al. (2011). Nanoparticle sensor for label free detection of swine DNA in mixed biological samples. *Nanotechnology*, 22(19). https://doi.org/10.1088/0957-4484/22/19/195503.

Antiochia, R., Lavagnini, I., & Magno, F. (2004). Amperometric mediated carbon nanotube paste biosensor for fructose determination. *Analytical Letters*, 37(8), 1657–1669. https://doi.org/10.1081/AL-120037594.

Arshak, K., Adley, C., Moore, E., Cunniffe, C., Campion, M., & Harris, J. (2007). Characterisation of polymer nanocomposite sensors for quantification of bacterial cultures. *Sensors and Actuators, B: Chemical*, 126(1), 226–231. https://doi.org/10.1016/j.snb.2006.12.006.

Aung, M. M., & Chang, Y. S. (2014). Traceability in a food supply chain: Safety and quality perspectives. *Food Control*, 39(1), 172–184. https://doi.org/10.1016/j.foodcont.2013.11.007.

Bagheri, H., Afkhami, A., Shirzadmehr, A., Khoshsafar, H., Khoshsafar, H., & Ghaedi, H. (2013). Novel potentiometric sensor for the determination of Cd2+ based on a new nano-composite. *International Journal of Environmental Analytical Chemistry*, 93(5), 578–591. https://doi.org/10.1080/03067319.2011.649741.

Berekaa, M. M. (2015). Nanotechnology in food industry; advances in food processing, packaging and food safety. *International Journal of Current Microbiology and Applied Sciences*, *4*(5), 345–357.

Carvalho, F. P. (2006). Agriculture, pesticides, food security and food safety. *Environmental Science and Policy*, *9*(7–8), 685–692. https://doi.org/10.1016/j.envsci.2006.08.002.

Chassy, B., Hlywka, J. J., Kleter, G. A., Kok, E. J., & Kuiper, H. A. (2004). Nutritional and safety assessments of foods and feeds nutritionally improved through biotechnology: An executive summary. *Comprehensive Reviews in Food Science and Food Safety*, *3*, 38–104.

Chaudhry, Q., Scotter, M., Blackburn, J., Ross, B., Boxall, A., Castle, L., et al. (2008). Applications and implications of nanotechnologies for the food sector. *Food Additives and Contaminants - Part A Chemistry, Analysis, Control, Exposure and Risk Assessment*, *25*(3), 241–258. https://doi.org/10.1080/02652030701744538.

Cheng, W. C., Kuo, C. W., Chi, T. Y., Lin, L. C., Lee, C. H., Feng, R. L., et al. (2013a). Investigation on the trend of food-borne disease outbreaks in Taiwan. *Journal of Food and Drug Analysis*, *21*(3), 261–267. https://doi.org/10.1016/j.jfda.2013.07.003.

Cheng, W.-C., Kuo, C.-W., Chi, T.-Y., Lin, L.-C., Lee, C.-H., Feng, R.-L., et al. (2013b). Investigation on the trend of food-borne disease outbreaks in Taiwan (1991–2010). *Journal of Food and Drug Analysis*, *21*(3), 261–267. https://doi.org/10.1016/j.jfda.2013.07.003.

Cho, I. H., Radadia, A. D., Farrokhzad, K., Ximenes, E., Bae, E., Singh, A. K., et al. (2014). Nano/micro and spectroscopic approaches to food pathogen detection. *Annual Review of Analytical Chemistry*, *7*, 65–88. https://doi.org/10.1146/annurev-anchem-071213-020249.

Crim, S. M., Iwamoto, M., Huang, J. Y., Griffin, P. M., Gilliss, D., Cronquist, A. B., et al. (2006). Incidence and trends of infection with pathogens transmitted commonly through food—Foodborne diseases active surveillance network. *MMWR.Morbidity and Mortality Weekly Report*, *63*(15).

Cui, D., Tian, F., Ozkan, C. S., Wang, M., & Gao, H. (2005). Effect of single wall carbon nanotubes on human HEK293 cells. *Toxicology Letters*, *155*(1), 73–85. https://doi.org/10.1016/j.toxlet.2004.08.015.

Dai, H., Shi, Y., Wang, Y., Sun, Y., Hu, J., Ni, P., et al. (2014). A carbon dot based biosensor for melamine detection by fluorescence resonance energy transfer. *Sensors and Actuators, B: Chemical*, *202*, 201–208. https://doi.org/10.1016/j.snb.2014.05.058.

de Otxoa, A. (2018). Nanotecnología y seguridad alimentaria. *Nutrición Hospitalaria*, *35*(spe4), 146–149.

Deo, R. P., Wang, J., Block, I., Mulchandani, A., Joshi, K. A., Trojanowicz, M., et al. (2005). Determination of organophosphate pesticides at a carbon nanotube/organophosphorus hydrolase electrochemical biosensor. *Analytica Chimica Acta*, *530*(2), 185–189. https://doi.org/10.1016/j.aca.2004.09.072.

Du, D., Huang, X., Cai, J., & Zhang, A. (2007). Comparison of pesticide sensitivity by electrochemical test based on acetylcholinesterase biosensor. *Biosensors and Bioelectronics*, *23*(2), 285–289. https://doi.org/10.1016/j.bios.2007.05.002.

Feng, Y., Zhong, D., Miao, H., & Yang, X. (2015). Carbon dots derived from rose flowers for tetracycline sensing. *Talanta*, *140*, 128–133. https://doi.org/10.1016/j.talanta.2015.03.038.

Gan, T., Li, K., & Wu, K. (2008). Multi-wall carbon nanotube-based electrochemical sensor for sensitive determination of Sudan I. *Sensors and Actuators, B: Chemical*, *132*(1), 134–139. https://doi.org/10.1016/j.snb.2008.01.013.

Girigoswami, A., Ghosh, M. M., Pallavi, P., Ramesh, S., & Girigoswami, K. (2021). Nanotechnology in detection of food toxins—Focus on the dairy products. *Biointerface Research in Applied Chemistry*, *11*(6), 14155–14172. https://doi.org/10.33263/BRIAC116.1415514172.

Ho, Y. P., Kung, M. C., Yang, S., & Wang, T. H. (2005). Multiplexed hybridization detection with multicolor colocalization of quantum dot nanoprobes. *Nano Letters*, *5*(9), 1693–1697. https://doi.org/10.1021/nl050888v.

Huang, Y., Mei, L., Chen, X., & Wang, Q. (2018). Recent developments in food packaging based on nanomaterials. *Nanomaterials*, *8*(10). https://doi.org/10.3390/nano8100830.

Ingale, A. G. (2018). Nanotechnology in the food industry. In *Nanotechnology, food security and water treatment* (pp. 87–128).

Kaittanis, C., Santra, S., & Perez, J. M. (2010). Emerging nanotechnology-based strategies for the identification of microbial pathogenesis. *Advanced Drug Delivery Reviews*, *62*(4–5), 408–423. https://doi.org/10.1016/j.addr.2009.11.013.

Karimi-Maleh, H., Moazampour, M., Yoosefian, M., Sanati, A. L., Tahernejad-Javazmi, F., & Mahani, M. (2014). An electrochemical nanosensor for simultaneous voltammetric determination of ascorbic acid and Sudan I in food samples. *Food Analytical Methods*, *7*(10), 2169–2176. https://doi.org/10.1007/s12161-014-9867-x.

Komaiko, J. S., & Mcclements, D. J. (2016). Formation of food-grade nanoemulsions using low-energy preparation methods: A review of available methods. *Comprehensive Reviews in Food Science and Food Safety*, *15*(2), 331–352. https://doi.org/10.1111/1541-4337.12189.

Krishna, V., Pumprueg, S., Lee, S. H., Zhao, J., Sigmund, W., Koopman, B., et al. (2005). Photocatalytic disinfection with titanium dioxide coated multi-wall carbon nanotubes. *Process Safety and Environmental Protection*, *83*(4 B), 393–397. https://doi.org/10.1205/psep.04387.

Lagaron, J. M., Cabedo, L., Cava, D., Feijoo, J. L., Gavara, R., & Gimenez, E. (2005). Improving packaged food quality and safety. Part 2: Nanocomposites. *Food Additives and Contaminants*, *22*(10), 994–998. https://doi.org/10.1080/02652030500239656.

Lei, C. H., Zhao, X. E., Jiao, S. L., He, L., Li, Y., Zhu, S. Y., et al. (2016). A turn-on fluorescent sensor for the detection of melamine based on the anti-quenching ability of Hg^{2+} to carbon nanodots. *Analytical Methods*, *8*(22), 4438–4444. https://doi.org/10.1039/c6ay01063f.

Leonard, P., Hearty, S., Brennan, J., Dunne, L., Quinn, J., Chakraborty, T., et al. (2003). Advances in biosensors for detection of pathogens in food and water. *Enzyme and Microbial Technology*, *32*(1), 3–13. https://doi.org/10.1016/S0141-0229(02)00232-6.

Li, S. C., Chen, J. H., Cao, H., Yao, D. S., & Liu, D. L. (2011). Amperometric biosensor for aflatoxin B1 based on aflatoxin-oxidase immobilized on multiwalled carbon nanotubes. *Food Control*, *22*(1), 43–49. https://doi.org/10.1016/j.foodcont.2010.05.005.

Li, H., Sun, C., Vijayaraghavan, R., Zhou, F., Zhang, X., & MacFarlane, D. R. (2016). Long lifetime photoluminescence in N, S co-doped carbon quantum dots from an ionic liquid and their applications in ultrasensitive detection of pesticides. *Carbon*, *104*, 33–39. https://doi.org/10.1016/j.carbon.2016.03.040.

Li, X., Zhou, Y., Zheng, Z., Yue, X., Dai, Z., Liu, S., et al. (2009). Glucose biosensor based on nanocomposite films of CdTe quantum dots and glucose oxidase. *Langmuir*, 25(11), 6580–6586. https://doi.org/10.1021/la900066z.

Lopez-Rubio, A., Gavara, R., & Lagaron, J. M. (2006). Bioactive packaging: Turning foods into healthier foods through biomaterials. *Trends in Food Science and Technology*, 17(10), 567–575. https://doi.org/10.1016/j.tifs.2006.04.012.

Lv, M., Liu, Y., Geng, J., Kou, X., Xin, Z., & Yang, D. (2018). Engineering nanomaterials-based biosensors for food safety detection. *Biosensors and Bioelectronics*, 106, 122–128. https://doi.org/10.1016/j.bios.2018.01.049.

Majidi, M. R., Fadakar Bajeh Baj, R., & Naseri, A. (2013). Carbon nanotube-ionic liquid (CNT-IL) nanocamposite modified sol-gel derived carbon-ceramic electrode for simultaneous determination of sunset yellow and Tartrazine in food samples. *Food Analytical Methods*, 6(5), 1388–1397. https://doi.org/10.1007/s12161-012-9556-6.

Malik, A. K., Blasco, C., & Picó, Y. (2010). Liquid chromatography-mass spectrometry in food safety. *Journal of Chromatography A*, 1217(25), 4018–4040. https://doi.org/10.1016/j.chroma.2010.03.015.

Manikandan, V. S., Adhikari, B. R., & Chen, A. (2018). Nanomaterial based electrochemical sensors for the safety and quality control of food and beverages. *Analyst*, 143(19), 4537–4554. https://doi.org/10.1039/c8an00497h.

Merkoçi, A. (2006). Carbon nanotubes: Exciting new materials for microanalysis and sensing. *Microchimica Acta*, 152(3–4), 155–156. https://doi.org/10.1007/s00604-005-0450-4.

Monteiro-Riviere, N. A., Nemanich, R. J., Inman, A. O., Wang, Y. Y., & Riviere, J. E. (2005). Multi-walled carbon nanotube interactions with human epidermal keratinocytes. *Toxicology Letters*, 155(3), 377–384. https://doi.org/10.1016/j.toxlet.2004.11.004.

Institute of Medicine. (2009). *Nanotechnology in food products: Workshop summary*.

Neethirajan, S., Jayas, D. S., & Sadistap, S. (2009). Carbon dioxide (CO_2) sensors for the agri-food industry-A review. *Food and Bioprocess Technology*, 2(2), 115–121. https://doi.org/10.1007/s11947-008-0154-y.

Pan, M., Yin, Z., Liu, K., Du, X., Liu, H., & Wang, S. (2019). Carbon-based nanomaterials in sensors for food safety. *Nanomaterials*, 9(9), 1330. https://doi.org/10.3390/nano9091330.

Patel, P. D. (2002). (Bio)sensors for measurement of analytes implicated in food safety: A review. *TrAC, Trends in Analytical Chemistry*, 21(2), 96–115. https://doi.org/10.1016/S0165-9936(01)00136-4.

Pena-Rosas, J. P., De-Regil, L. M., Rogers, L. M., Bopardikar, A., & Panisset, U. (2012). Translating research into action: WHO evidence-informed guidelines for safe and effective micronutrient interventions. *Journal of Nutrition*, 142(1). https://doi.org/10.3945/jn.111.138834.

Pérez-López, B., & Merkoçi, A. (2011). Nanomaterials based biosensors for food analysis applications. *Trends in Food Science and Technology*, 22(11), 625–639. https://doi.org/10.1016/j.tifs.2011.04.001.

Prinith, N. S., & Manjunatha, J. G. (2020). Polymethionine modified carbon nanotube sensor for sensitive and selective determination of l-tryptophan. *Journal of Electrochemical Science and Engineering*, 10(4), 305–315. https://doi.org/10.5599/jese.774.

Prinith, N. S., Manjunatha, J. G., & Raril, C. (2019). Electrocatalytic analysis of dopamine, uric acid and ascorbic acid at poly(adenine) modified carbon nanotube paste electrode: A cyclic voltammetric study. *Analytical and Bioanalytical Electrochemistry*, 11(6), 742–756. http://www.abechem.com/No.%206-2019/2019,%2011(6),%20742-756.pdf.

Pushpanjali, P. A., Manjunatha, J. G., Tigari, G., & Fattepur, S. (2020). Poly(niacin) based carbon nanotube sensor for the sensitive and selective voltammetric detection of vanillin with caffeine. *Analytical and Bioanalytical Electrochemistry*, 12(4), 553–568. http://www.abechem.com/article_39221_5b4cb9d77cdb1bfb6c48fc3dac8c7fc0.pdf.

Raril, C., Manjunatha, J. G., & Tigari, G. (2020). Low-cost voltammetric sensor based on an anionic surfactant modified carbon nanocomposite material for the rapid determination of curcumin in natural food supplement. *Instrumentation Science and Technology*, 48(5), 561–582. https://doi.org/10.1080/10739149.2020.1756317.

Rashidi, L., & Khosravi-Darani, K. (2011). The applications of nanotechnology in food industry. *Critical Reviews in Food Science and Nutrition*, 51(8), 723–730. https://doi.org/10.1080/10408391003785417.

Rodríguez-Lázaro, D., Lombard, B., Smith, H., Rzezutka, A., D'Agostino, M., Helmuth, R., et al. (2007). Trends in analytical methodology in food safety and quality: Monitoring microorganisms and genetically modified organisms. *Trends in Food Science and Technology*, 18(6), 306–319. https://doi.org/10.1016/j.tifs.2007.01.009.

Silva, H. D., Cerqueira, M. A., & Vicente, A. A. (2012). Nanoemulsions for food applications: Development and characterization. *Food and Bioprocess Technology*, 5(3), 854–867. https://doi.org/10.1007/s11947-011-0683-7.

Silvestre, C., Duraccio, D., & Cimmino, S. (2011). Food packaging based on polymer nanomaterials. *Progress in Polymer Science*, 36(12), 1766–1782. https://doi.org/10.1016/j.progpolymsci.2011.02.003.

Sinha, N., Ma, J., & Yeow, J. T. W. (2006). Carbon nanotube-based sensors. *Journal of Nanoscience and Nanotechnology*, 6(3), 573–590. https://doi.org/10.1166/jnn.2006.121.

Song, J., Li, J., Guo, Z., Liu, W., Ma, Q., Feng, F., et al. (2017). A novel fluorescent sensor based on sulfur and nitrogen co-doped carbon dots with excellent stability for selective detection of doxycycline in raw milk. *RSC Advances*, 7(21), 12827–12834. https://doi.org/10.1039/c7ra01074e.

Srinivas, P. R., Philbert, M., Vu, T. Q., Huang, Q., Kokini, J. L., Saos, E., et al. (2010). Nanotechnology research: Applications in nutritional sciences. *Journal of Nutrition*, 140(1), 119–124. https://doi.org/10.3945/jn.109.115048.

Tigari, G., & Manjunatha, J. G. (2020). A surfactant enhanced novel pencil graphite and carbon nanotube composite paste material as an effective electrochemical sensor for determination of riboflavin. *Journal of Science: Advanced Materials and Devices*, 5(1), 56–64. https://doi.org/10.1016/j.jsamd.2019.11.001.

Tigari, G., Manjunatha, J. G., Raril, C., & Hareesha, N. (2019). Determination of riboflavin at carbon nanotube paste electrodes modified with an anionic surfactant. *ChemistrySelect*, 4(7), 2168–2173. https://doi.org/10.1002/slct.201803191.

Valdés, M. G., González, A. C. V., Calzón, J. A. G., & Díaz-García, M. E. (2009). Analytical nanotechnology for food analysis. *Microchimica Acta*, *166*(1–2), 1–19. https://doi.org/10.1007/s00604-009-0165-z.

Verma, M. L. (2017a). Enzymatic nanobiosensors in the agricultural and food industry. *Nanoscience in Food and Agriculture*, *4*, 229–245. https://doi.org/2017.

Verma, M. L. (2017b). Nanobiotechnology advances in enzymatic biosensors for the agri-food industry. *Environmental Chemistry Letters*, *15*(4), 555–560. https://doi.org/10.1007/s10311-017-0640-4.

Verma, M. L. (2017c). *Springer Nature* (pp. 229–245). https://doi.org/10.1007/978-3-319-53112-0_7.

Vinayaka, A. C., Basheer, S., & Thakur, M. S. (2009). Bioconjugation of CdTe quantum dot for the detection of 2,4-dichlorophenoxyacetic acid by competitive fluoroimmunoassay based biosensor. *Biosensors and Bioelectronics*, *24*(6), 1615–1620. https://doi.org/10.1016/j.bios.2008.08.042.

Viswanathan, S., Radecka, H., & Radecki, J. (2009). Electrochemical biosensor for pesticides based on acetylcholinesterase immobilized on polyaniline deposited on vertically assembled carbon nanotubes wrapped with ssDNA. *Biosensors and Bioelectronics*, *24*(9), 2772–2777. https://doi.org/10.1016/j.bios.2009.01.044.

Wang, D., Lin, B., Cao, Y., Guo, M., & Yu, Y. (2016). A highly selective and sensitive fluorescence detection method of glyphosate based on an immune reaction strategy of carbon dot labeled antibody and antigen magnetic beads. *Journal of Agricultural and Food Chemistry*, *64*(30), 6042–6050. https://doi.org/10.1021/acs.jafc.6b01088.

Warheit, D. B., Laurence, B. R., Reed, K. L., Roach, D. H., Reynolds, G. A. M., & Webb, T. R. (2004). Comparative pulmonary toxicity assessment of single-wall carbon nanotubes in rats. *Toxicological Sciences*, *77*(1), 117–125. https://doi.org/10.1093/toxsci/kfg228.

Wei, S., Zhao, F., Xu, Z., & Zeng, B. (2006). Voltammetric determination of folic acid with a multi-walled carbon nanotube-modified gold electrode. *Microchimica Acta*, *152*(3–4), 285–290. https://doi.org/10.1007/s00604-005-0437-1.

Xiao, F., Ruan, C., Liu, L., Yan, R., Zhao, F., & Zeng, B. (2008). Single-walled carbon nanotube-ionic liquid paste electrode for the sensitive voltammetric determination of folic acid. *Sensors and Actuators, B: Chemical*, *134*(2), 895–901. https://doi.org/10.1016/j.snb.2008.06.037.

Yang, G. J., Huang, J. L., Meng, W. J., Shen, M., & Jiao, X. A. (2009). A reusable capacitive immunosensor for detection of Salmonella spp. based on grafted ethylene diamine and self-assembled gold nanoparticle monolayers. *Analytica Chimica Acta*, *647*(2), 159–166. https://doi.org/10.1016/j.aca.2009.06.008.

Yang, D., Zhu, L., & Jiang, X. (2010). Electrochemical reaction mechanism and determination of Sudan I at a multi wall carbon nanotubes modified glassy carbon electrode. *Journal of Electroanalytical Chemistry*, *640*(1–2), 17–22. https://doi.org/10.1016/j.jelechem.2009.12.022.

Ye, S.-L., Huang, J.-J., Luo, L., Fu, H.-J., Sun, Y.-M., Shen, Y.-D., et al. (2017). Preparation of carbon dots and their application in food analysis as signal probe. *Chinese Journal of Analytical Chemistry*, *45*(10), 1571–1581. https://doi.org/10.1016/s1872-2040(17)61045-4.

Zheng, L., Qi, P., & Zhang, D. (2019). Identification of bacteria by a fluorescence sensor array based on three kinds of receptors functionalized carbon dots. *Sensors and Actuators, B: Chemical*, *286*, 206–213. https://doi.org/10.1016/j.snb.2019.01.147.

Chapter 8

Carbon nanomaterial-based sensors in air pollution remediation

Abdullah Al Mamun[a], Md Nafiujjaman[b], and A.J. Saleh Ahammad[c]

[a]Department of Chemistry, University of Houston, Houston, TX, United States, [b]Department of Biomedical Engineering, Michigan State University, East Lansing, MI, United States, [c]Department of Chemistry, Jagannath University, Dhaka, Bangladesh

Introduction

In 1991, when Iijima was working in his laboratory to synthesize fullerenes using electric arc discharge technique, he noticed needle-shaped growth at the negative end of the electrode (Iijima, 1991). These needle-shaped materials, well known as carbon nanotubes, are one-sixth in weight but 100 times in strength compared to steel. As a matter of fact, these hollow needles have so many interesting properties that a new branch of knowledge "nanoscience" has emerged after their discovery. Human beings only then could properly realize the famous quote of Feynman "there is plenty of room at the bottom." It was the beginning of fabricating materials and devices at the atomic/molecular level that is the "bottom of a material." At that time, very few people imagined that it would be the "Next Industrial Revolution" called nanotechnology shortly, which would offer a new class of miniaturized instrumentation to manipulate, measure, and utilize these small "nano" structures (Roco, 2005).

At present, nanotechnology offers the best solution for various man-made and natural problems (Guerra, Attia, Whitehead, & Alexis, 2018; Hull & Bowman, 2018; Kumar et al., 2021). As an alarming combination of both types of problems, air pollution has become a global problem, and its scope is growing by the day. The rapid growth of industrialization and urbanization has already made it difficult to breathe clean air (Mehndiratta, Jain, Srivastava, & Gupta, 2013).

So, a lot of effort is put to remediate air pollutants. It is the first and crucial part of the pollution remediation process that we have enough information about the pollutants such as the presence and exact quantity of a particular pollutant in the environment. Only then proper action can be taken. Here come sensors that measure physical input from the surroundings and convert it into meaningful data that can be evaluated by a machine or human. As sensing materials, nanomaterials are highly investigated because they offer great responses, high sensitivity, selectivity, and low recovery time. Among all nanomaterials, carbon nanomaterials, for example, carbon nanotube (CNT), graphene (GPN), and fullerenes (FLN), have large surface-to-volume ratios and unique structures and electronic properties, which give them especially powerful pollutant gas adsorption and desorption capabilities. Thus, CNMs show their competencies to detect airborne gas pollutant gases such as sulfur-based oxides (SO_x), nitrogen-based oxides (NO_x), chlorofluorocarbons (CFCs), volatile organic compounds (VOCs), CO_2, CO, particulate matters (PM), etc. at ppb level. They have been in the high-interest point of the scientific research world for the last 20 years. As the title suggests, in this chapter we will explore the recent development of carbon nanomaterial-based sensors for the sensing part of the remediation of different inorganic and organic air pollutants and will give the underlying mechanism in this regard.

Nanomaterials in air pollution remediation

In general, there are two approaches to remediate air pollutants successfully. The first one is to solve the problem from the root, which focuses on controlling sources of pollution by reducing potential pollutant gas production and discharge. The second one is the end-of-pipe approach by remediating pollutants that are already in the air (Harrison, 2015). Remediation includes detection, collection, removal, or conversion of contaminant chemicals (Zhao, Li, Zhou, Wang, & Xu, 2019). There are several basic differences between traditional and nanomaterial-based remediation. Traditional methods do not decompose pollutants; rather, they separate pollutants by physical methods like filtration, adsorption, ventilation, etc. On the other hand, nanomaterials give the most practical treatment by both physical and chemical methods. For example, volatile organic compounds (VOC) can be removed by traditional methods, e.g., transferring from air to the solid phase by using adsorbents like activated carbon (González-García, 2018), fiber (Zhang, Gao, Creamer, Cao, & Li, 2017), biochar (Spokas et al., 2011), etc. In contrast, nano semiconductor catalysts can oxidize the same VOC to their benign constitutes, H_2O and CO_2. This approach by using nanomaterials is more efficient and cost-effective (Mo, Zhang, Xu, Lamson, & Zhao, 2009; Tompkins & Anderson, 2001; Yang et al., 2019). In general, the main advantages of nanomaterials are as follows: (i) they have a large surface-to-volume ratios, (ii) their distinct structures and electronic properties give them very effective gas adsorption and desorption capabilities, (iii) they are capable of adsorbing a wide variety of gaseous materials, (iv) they have fast reaction kinetics, (v) their electrical conductivity is high, (vi) they have a high rate of electron transfer and strong carrier mobility, and (vii) they have selectivity toward aromatic solute (Bosso & Enzweiler, 2002). Different morphologies of nanomaterials such as nanotubes, nanoparticles, nanowires, nanofibers, etc., and their composites are used as adsorbents and catalysts for detecting and removing air pollutants such as SO_x, NO_x, CO, and organic pollutants (Khin, Nair, Babu, Murugan, & Ramakrishna, 2012).

Nanomaterials increase surface area for gaseous reactions, which are used in different ways to purify the air from harmful particulates and gases. Nanocatalysts convert harmful gases to comparatively harmless gases. For example, Chen and Goodman studied that nanosized Au particles show exceptionally high catalytic activities to oxidize harmful CO to comparatively less harmful CO_2 (Chen & Goodman, 2006). They explained the unique catalytic properties of Au nanoparticles using quantum size effects. Another example is MnO nanoparticles that are used to remove VOC from industrial smokes (Das, Sen, & Debnath, 2015). Nanostructured membranes are another type of nanomaterials that have small pores and permit only specific molecules through them. These membranes can separate toxic gases like CO_2 and CH_4 from the original airspace. Nanostructured thin membranes made of SWNTs can be used to encapsulate super greenhouse gas tetrafluoromethane at 300 K temperature and 1 bar of external operating pressure (Kowalczyk & Holyst, 2008). The pore size of CNTs determines how much CF_4 is absorbed. The absorption rate by CNT is 100 times faster and larger in volume than that of the other traditional materials. Nanomaterial-based sensors are used to detect and quantify air pollutants, which is the first and crucial step of air pollution remediation (Fig. 8.1). Various nanomaterials and their applications in air pollution remediation are given in Table 8.1.

Brief history of carbon nanomaterials

The base of research on carbon nanomaterials developed in the 1950s and 1960s when interests in researches beyond germanium and silicon for electronic device application started to receive attention. The first carbon nanomaterial, graphene, was invented in 1962 by H.P. Boehm (Boehm, Clauss, Fischer, & Hofmann, 1962). During that time, it got little attention given focus was fixated on 3D graphene to facilitate various experimental studies and expand theoretical knowledge. The first carbon nanomaterial that won the warmth of immediate attention was fullerene after its discovery in 1984 by the Exxon research laboratory. Initially, it was observed as an anomaly in carbon cluster formation until Kroto, Smalley, and Curl came forward independently to explain the occurrence who was working in the relevant field since the 1970s (Kroto, 1997). They named the compound Buckminsterfullerene after the American architect R. Buckminster Fuller as the compound has topology similar to geodesic domes built by the architect. The discovery not only led to the winning of 1996's Nobel Prize but also cleared the path to the discovery of carbon nanotubes. Around 1990, the prospect of carbon fiber (multiwalled carbon nanotube) was being discussed in the form of extended fullerene. Although hollow carbon fibers may have originally been discovered by the Russians first, in the 1950s, the first HRTEM image of SWCNTs and MWCNTs is attributed to Morinobu Endo et al. from the middles of the 1970s (Oberlin, Endo, & Koyama, 1976). Mr. Endo was

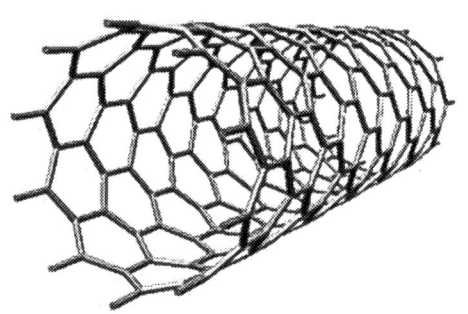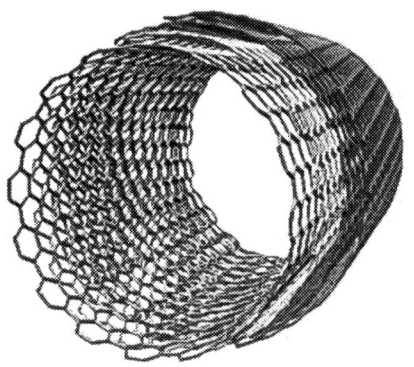

FIG. 8.1 The left one is the structure of a single-walled carbon nanotube (SWNT) and the right one is the structure of a multiwalled carbon nanotube (MWNT). *(From Gupta, S., Murthy, C. N., & Prabha, C. R. (2018). Recent advances in carbon nanotube based electrochemical biosensors. International Journal of Biological Macromolecules, 108, 687–703. https://doi.org/10.1016/j.ijbiomac.2017.12.038.)*

TABLE 8.1 Applications of different nanomaterials in air pollution remediation.

Type of nanomaterials	Application in air pollution remediation	Reference
CNTs	Sensor for H_2S, SO_2; absorption of tetrafluoromethane	Zhang, Janyasupab, et al. (2012)
CNTs	CO_2, CO, NO_2, NO, and SO_2 released as a result of fossil fuels burning	Mittal and Kumar (2014)
(CNTs-APTS), CNTs modified using 3-aminopropyltriethoxysilane	CO_2	Mittal and Kumar (2014)
CNTs deposited on the surface of quartz filters	VOCs	Amade, Hussain, Ocaña, and Bertran (2014)
CNT-10% SnO_2 nanoparticle composite	To detect methanol, ethanol, and H_2S to ppm levels	Mendoza et al. (2014)
SWNTs/NaClO	Isopropyl vapor	Hsu and Lu (2007)
SWCNTs	Encapsulation of tetrafluoromethane	Kowalczyk and Holyst (2008)
Boron-doped and Si-doped SWCNTs	CH_3OH and CO gases	Azam, Alias, Tack, Seman, and Taib (2017)
Cadmium arachidate/single-walled carbon nanotube (CdA/SWCNT) composites	Optical chemo-sensors to volatile organic compound (VOC) and gas exposure	Consales et al. (2009)
Cu NP-SWCNT	H_2S gas sensor	Asad and Sheikhi (2014)
Carbon nanoflake (CNFL)-SnO_2 composite	To detect NH_3	Lee, Chang, and Kim (2014)
$NiFe_2O_4$-MWCNT nanocomposite	To detect H_2S	Hajihashemi, Rashidi, Alaie, Mohammadzadeh, and Izadi (2014)
Graphene	Detection of CO_2	Yoon et al. (2011)
Molybdenum trioxide nanoparticle-decorated graphene oxide	For sensing H_2S	Malekalaie, Jahangiri, Rashidi, Haghighiasl, and Izadi (2015)
Fullerene B_{40}	CO_2	Dong et al. (2015)
Graphene oxide/nanocomposites	CO_2, SO_2, NH_3, H_2S, N_2	Seredych and Bandosz (2012)
Fullerene-like boron nitride nanocage	N_2O	Esrafili (2017)
Silica-titania nanocomposites	Mercury vapors removal from combustion sources	Pitoniak, Wu, Mazyck, Powers, and Sigmund (2005)
Materials based on metalorganic frameworks (MOFs)	Capture CO_2 efficiently and selectively	Yaghi, Li, and Li (1995)
TiO_2 nanofibers doped by Ag	Toluene and NO_x in the air	Srisitthiratkul, Pongsorrarith, and Intasanta (2011)
PAMAM dendrimer composite membrane with hyaluronic acid in a chitosan gutter layer	Separates CO_2 from a mixture of N_2 and CO_2 on porous substrates	Zhao and Vance (1998)
Chitosan and a dendrimer containing PAMAM dendrimer composite membrane	Separates CO_2 from fossil fuel emission	Arkas and Tsiourvas (2009)
Nano ZnO	To detect C_2H_2 and liquid petroleum gas	Dong, Cui, and Zhang (1997)
ZnO single crystals	for detecting CO, CH_4, and H_2 in air mixtures at 300–500°C	Bott, Jones, and Mann (1984)

No Permission Required.

analyzing the internal structure of carbon fibers developed from the pyrolysis of ferrocene and benzene and was trying to grow thin specimens for TEM studies while accidentally developed carbon nanotubes. The purported development of carbon nanotubes started in 1993 when two groups from Japan and IBM, United States, reported their success in bulk growth of SWNT. In 2004, Novoselov and Geim from the United Kingdom reported the simple exfoliation of graphene (Schedin et al., 2007) layer from graphite. This soon converted the attention back to graphene and led to the win of the 2006 Nobel Prize in Physics. Since then, a huge number of reports have been made on carbon nanomaterials particularly on carbon nanotubes and graphene. Researches are still ongoing to report the properties of these nanomaterials and their applications in all possible areas.

Derivatives of carbon nanomaterials

Carbon is one of the few elements that has been recognized in the prehistoric ages and since then has remained useful in its elemental as well as in various modified forms. The rapid developments in the field of nanomaterials have steered the attention of carbon studies to carbon-based nanomaterials due to their low price, availability, unique characteristics, and ubiquitous applications (Tan et al., 2017). The improved electrical, thermal, mechanical, and optical properties of these materials with chemical stability have inspired their reliable use in sensor applications (Yang et al., 2010). The development in research on carbon nanomaterials is presented by its three derivatives, e.g., 1. graphene, 2. fullerene, and 3. carbon nanotubes (CNT).

Properties of carbon nanomaterials

Carbon nanotubes

Cylindrical hollow structures with walls made of sp^2 hybridized carbon atoms are called carbon nanotubes. They are obtained when graphene sheets are rolled or fullerene is elongated with the two-side cut perpendicularly with the main hull. Depending upon the layer of carbon atoms in the wall, they are classified as single-walled carbon nanotube or in short SWNT and multiwalled carbon nanotube or in short MWNT (Fig. 8.1). The former has one layer of graphene structure, while the latter has more than one layer (Gupta, Murthy, & Prabha, 2018).

The diameter of single-walled carbon nanotubes varies from 0.5 to 50 nm and the length has no limit. Usually, it is within the micrometer range; however, reports have been found that it is as large as around 18 cm (Wang et al., 2009). The multiwalled carbon nanotubes have an inner diameter of 0.4 to few nanometers and the outer diameter varies between 2 and 20 nm. The SWCNT can form bundles to develop rope-like structures.

The CNTs have useful electrical properties. They can act as metal or semiconductors. The maximum current density is $1013 A/m^2$ and the resistivity is 10^{-4} Ohm cm. The thermal conductivity is 2000 W/m/K. They also have a very good elastic behavior. Young's modulus of MWCNT and SWCNT is 1.28 and 1 T, respectively. The maximum tensile strength is 100 GPa (Eatemadi et al., 2014). The CNTs can be modified to different functional groups and thus very useful in sensor-based applications. The CNTs are made following arc discharge, laser ablation, or catalytic growth method. The developed products are characterized following spectroscopic methods.

Graphene

Graphene is one of the most researched and applied nanomaterials. It is a single layer of graphite described as a 2D sheet of sp^2 hybridized carbon (Fig. 8.2) with a honeycomb type structure (Ahammad, Islam, & Hasan, 2019; Bonaccorso, Sun, Hasan, & Ferrari, 2010; Kim, Kim, Novoselov, & Hong, 2015). The brown-colored graphene forms the building block of many carbon materials and can be correlated with other carbon nanomaterials too. Graphite is nothing but stacked layers of graphene, carbon nanotubes are folded graphene, which when assumes a spherical shape called fullerene. Graphene has a zigzag, straight, and armchair geometry with the former as the dominated existence.

In graphene, the trigonal carbon atoms system is separated by sigma bonds with each other, while the remaining p orbital forms additional pi bonds. The circulation of pi electrons over the sheet gives it high electrical conductivity along with unique optoelectronic properties. The system has a binding energy of ~ 5.9 eV, and the electrical conductivity of graphene is $\sim 1 \times 10^8$ S/cm along with high thermal conductivity of 2000–4000 $W m^{-1} K^{-1}$.

The highest current density of graphene has been reported to be around $1.6 \times 10^9 A/cm^2$ and electron mobility is 200,000 $cm^2 V^{-1} s^{-1}$ at an electron density of around $2 \times 10^{11} cm^2$. These properties of graphene have made it suitable for various types of sensor applications, including electric and electrochemical sensors in various fields (Nag, Mitra, &

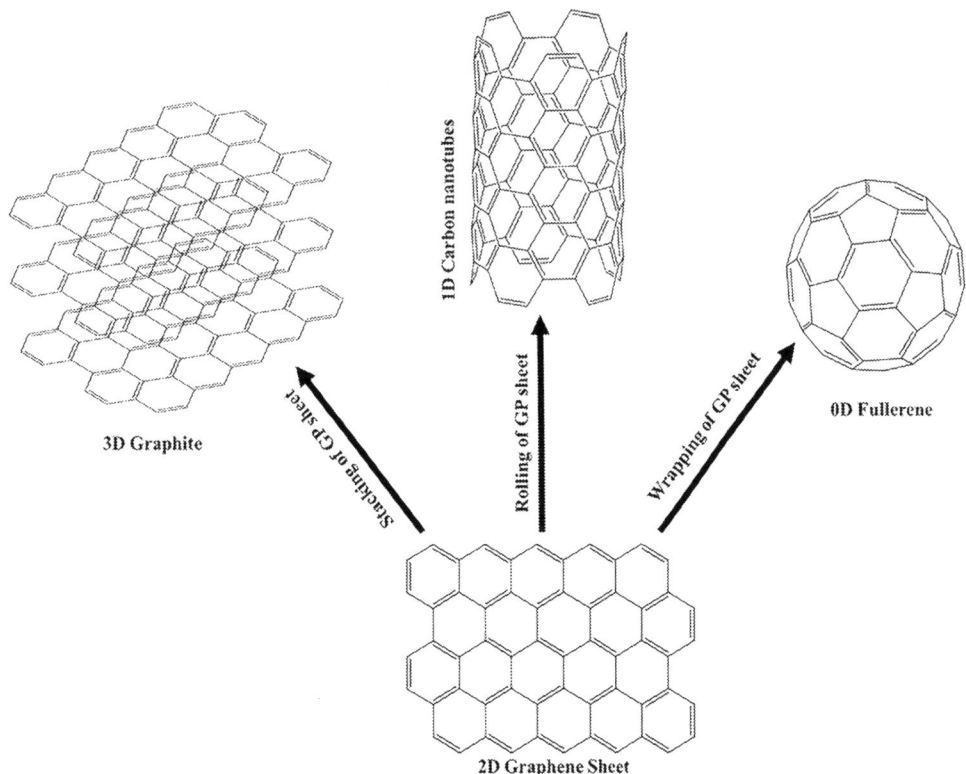

FIG. 8.2 Graphene and other carbon nanomaterials. *(From Ahammad, A. J. S., Islam, T., & Hasan, M. M. (2019). Graphene-based electrochemical sensors for biomedical applications. In Biomedical applications of graphene and 2D nanomaterials (pp. 249–282). Elsevier. https://doi.org/10.1016/B978-0-12-815889-0.00012-X.)*

Mukhopadhyay, 2018). Graphene is reported to be stronger than steel in terms of tensile strength (around 130 GPa). It is also unique based on elasticity as the report says; changes up to 20% in elasticity affect nominally in graphene properties.

Graphene has a very high theoretical surface area of 2600 m^2/g. Thus, it has a high potential in catalytic applications. The openness of graphene sheet from both sides makes it very reactive and suitable for various substitutions. In fact, this structure has allowed graphene oxide, the oxidized form of graphene, to bind with various metals and ligands forming a huge number of graphene-based composites with modified behavior. These modifications have led graphene to be applied in almost every known field.

Graphene is chemically produced using graphite powder. In this case, first graphene oxide is developed from graphite following the Hummer method or its modified forms. Later, it is reduced through chemical or thermal means. The synthesis is confirmed using techniques like FESEM, XRD, and Raman spectroscopy.

Fullerenes

The three-dimensional spherical structure (Fig. 8.3) with only carbon atoms is known as fullerene. The first observed fullerene was C_{60}. Although different classes of fullerene compounds are reported including buckyball clusters, nanotubes, mega tubes, polymer nano-onions, etc., the most common example is the C_{60} buckyball cluster. The cage-like structure has 12 pentagons and 20 hexagons (Kroto, 1997). This structure is called the most symmetric molecule. The carbon atoms in fullerene are sp^2 hybridized, and the compound is an alkene with electron deficiency. However, it is reported that it possesses an arrangement for the movement of electrons, leading to benzene-like additional stabilization from the structure and showing super-aromaticity (Malhotra & Ali, 2018; Nan'ya, Osawa, Obata, & Maekawa, 1970).

Physically, fullerene appears as dark needle-colored crystals with 1.65/cm^3 density. It is insoluble in water and sublimes at 600°C. Fullerene has useful electrochemical properties. It can absorb in the UV-Vis region and shows a photothermal effect. Due to its structural variation, it can accommodate multiple electrons as well as serve as an electron acceptor. Thus, it can play as both electrophile and nucleophile (Xu & Wang, 2019).

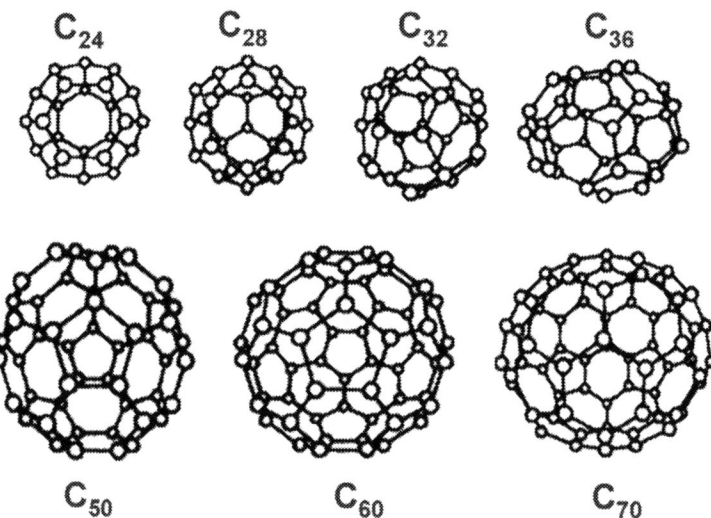

FIG. 8.3 Spherical structure of various types of fullerene. (From Malhotra, B. D., & Ali, Md. A. (2018). Functionalized carbon nanomaterials for biosensors (pp. 75–103). Elsevier BV. https://doi.org/10.1016/b978-0-323-44923-6.00002-9.)

Fullerene was first grown through laser irradiation of solid graphite in argon medium. Nowadays, it is done by setting a high electric current among graphite electrodes and collecting the residue from the water cool trap and then separated by HPLC.

Application of carbon nanomaterial-based sensors in air pollution remediation and relative mechanism

A sensor is a device that measures physical input from the surroundings and converts it into meaningful data that can be evaluated by a machine or human. A sensor's main component is an active surface that is physically modified when it comes into contact with a gas. Sensors are used to detect and quantify the amount of pollutant materials in the air. This is the first and crucial step of air pollution remediation. Carbon nanomaterials can efficiently detect most of the pollutants such as carbon monoxide (CO), carbon dioxide (CO_2), nitrogen oxides (NO_x), hydrogen sulfide (H_2S), methane (CH_4), and ammonia (NH_3) in ppb and ppm level as well as to quantify the amount of pollutant gas present in the surroundings. The main nanosized allotropes of carbon used for gas sensor materials are carbon nanotube (CNT), graphene nanoparticles (GPN), graphene oxide (GO), and fullerenes (FLN). The sensing process, application, and mechanism of carbon nanomaterial-based sensors will be discussed in the following sections.

Process of pollutant gas sensing with carbon nanomaterials

The actual mechanism of the sensing depends on the sensing material and the nature of the pollutant gas. However, in general, the process is pretty straightforward. The sensing materials interact with the sensing pollutant gas. The interaction may be physical adsorption, chemical adsorption, and/or van der Waals's force of attraction. Sometimes sensing materials alone do not give any significant interaction with the pollutant gas. In that case, the sensing material is modified with other materials. The interaction between the sensor surface and the gas converts chemical energy into electrical energy, which conserves energy (Moseley & Tofield, 1987). The interaction is measured in different quantities to detect the gas quantitatively. Sensors may be chemiresistive, chemicapacitive, and field-effect transistor (FET) based on the measurement parameters. Among them, CNM-based sensors are mainly chemiresistive where the principal mean parameter is the electrical resistance. When a certain gas is adsorbed on the sensing element from the surroundings, the device's output resistance changes. The output resistance of the sensor is provided by a transduction unit. Fig. 8.4 shows a schematic diagram of the whole process. When the gas molecules adsorbed on the sensor's surface are desorbed, the sensor returns to its original resistance.

Chemiresistive sensors are preferred because the resistance is simple to measure, they are easy to fabricate, and they have simple structures, which ensure sensing of gases with low power consumption. However, resistance is a second-order parameter, and sometimes it is considered as not so good indicator. To overcome this problem, scientists have tested numerous new materials and technologies. Among them, carbon nanomaterials such as CNT, graphene, and nanofibers are now the most promising and established materials for their unique, electrical, mechanical, and optical properties

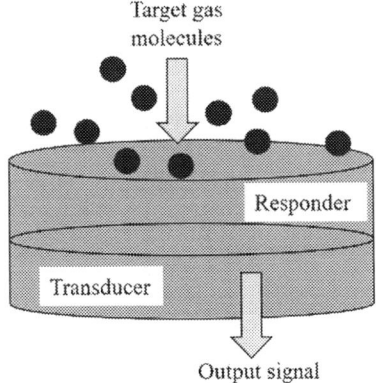

FIG. 8.4 Schematic diagram of a gas sensor working principle. *(Modified from Mittal, M., & Kumar, A. (2014). Carbon nanotube (CNT) gas sensors for emissions from fossil fuel burning. Sensors and Actuators, B: Chemical, 203, 349–362. htttps://doi.org/10.1016/j.snb.2014.05.080.)*

(Hanada, Okada, & Yase, 1999; Llobet, 2013). Finally, a few characteristics such as responsiveness, response time, sensitivity, selectivity, and recovery time characterize the sensor. These properties are determined by the sensing material's characteristics as well as the operating temperature.

Advantages of carbon nanomaterial-based sensors

To be an ideal sensor, a material should fulfill some requirements such as (i) high sensitivity, (ii) selectivity, (iii) fast recovery and response time, (iv) reversibility, and (v) stability. Carbon nanomaterial-based sensors possess all the requirements over traditional sensor materials. Additionally, they are versatile and may be utilized in a variety of sensing devices, including capacitance, ionization, sorption, resonance frequency shift, etc., their size is compact and portable, they have detection limits up to ppb or ppm ranges, their hollow structures increase bindings and signal response, and they can be modified easily by functionalization to increase pollutant gas sensitivity.

Application and general mechanism

CNT-based sensors in air pollution remediation

Both single-walled and multiwalled CNTs were tested and are being used to detect CO, CO_2, NO, NO_2, and SO_2. Besides, composite materials have shown better efficiency to detect different pollutant gases. For instance, a composite of carbon nanoflakes and SnO_2 can detect ammonia gas efficiently (Lee et al., 2014).

The sensing mechanism of CNT-based sensors

Carbon nanomaterials are used as the conducting channel between electrodes in the chemiresistive sensor. Carbon nanomaterials serve as the conducting channel among electrodes in the chemiresistive sensor. CNTs are deposited between two electrodes using chemical vapor deposition (CVD), drop-casting, printing, solid transfer, spraying, and other methods. To evaluate the sensory response, the conductance between two electrodes is measured. Because carbon nanotubes (CNTs) are virtually entirely made up of surface atoms, thus a slight alteration in the chemical environment has a big impact on the final response (Tang, Shi, Hou, & Wei, 2017). CNTs work efficiently because they can provide numerous binding sites to pollutant gas molecules. These binding sites come from the high surface-to-volume ratio, porous structure, and the defect sites such as impurities, Stone-Wales, pentagon, and heptagon. The most favored binding locations of gaseous molecules on carbon nanotubes are the axial sites, hollow sites, zig-zag, and the top surface, as shown in Fig. 8.5 (Durgun et al., 2003).

Fennell et al. have discussed extensively the CNT sensing mechanism (Fennell et al., 2016). In brief, the target pollutant gas is adsorbed chemically (chemisorption) or physically (physisorption) in the surface binding sites (Dekker, 1999; Zettl, 2000). For SWCNT, all atoms and for MWCNTs atoms from the outermost surface layer take part in the adsorption process (Bachtold et al., 1999). This results in the Schottky barrier at CNT-electrode junctions to be modulated, as well as charge transfer between the CNTs and pollutant gas molecules, which increases the distance between the CNT-CNT junctions. The type of the adsorbing gas determines electron transport from and to CNTs (Zhao, Buldum, Han, & Lu, 2002). Reducing gases, such as NH_3, which may donate free electrons, increase electrical resistance due mainly to free-electron recombination. Holes of CNTs also contribute to this (Chopra, McGuire, Gothard, Rao, & Pham, 2003; Villalpando-Páez et al.,

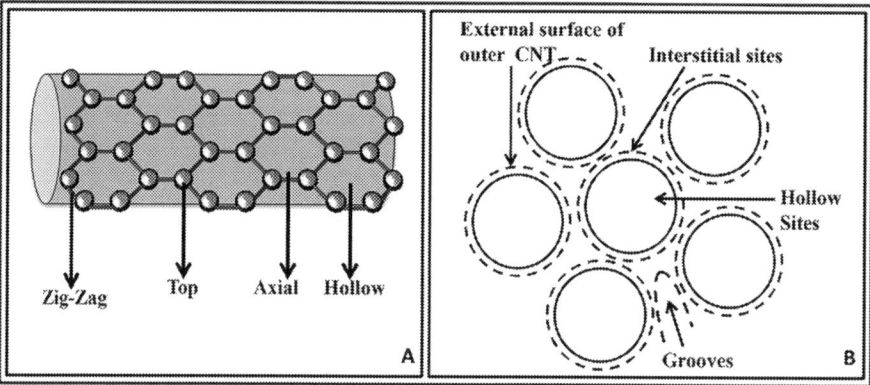

FIG. 8.5 Adsorption sites on (A) single CNT and (B) CNT bundles are shown graphically. *(From Mittal, M., & Kumar, A. (2014). Carbon nanotube (CNT) gas sensors for emissions from fossil fuel burning. Sensors and Actuators, B: Chemical, 203, 349–362. https://doi.org/10.1016/j.snb.2014.05.080.)*

2004). The opposite phenomenon occurs for the oxidizing gas like NO_2, and a noticeable decrease in the output resistance of the sensor occurs (Jung, Jung, Kim, & Suh, 2007; Kong et al., 2000; Zettl, 2000). From the change in the resistance, the adsorbed pollutant gas can be detected and quantified. Additionally, functionalization allows selectivity in the binding of pollutant gas molecules. Fig. 8.6 shows the mechanism of electron transfer between CNTs and adsorbed oxidizing and reducing gaseous molecules (Mittal & Kumar, 2014).

Due to weak interaction between CNTs and pollutant gas molecules, CNTs alone always can't show strong responses and good selectivity toward specific gas molecules. To solve this limitation, the functionalization of CNTs is used to improve both the selectivity and sensitivity of SWCNT sensors (Feng, Yi, Wu, Ozaki, & Yoshino, 2005; Hamwi, Alvergnat, Bonnamy, & Béguin, 1997; Hirsch & Vostrowsky, 2005; Holzinger et al., 2003; Hu, Zhang, Yuan, & Tang, 2017; Kim, Jeong, & Kim, 2014; Kong, Gao, & Yan, 2004; Ma, Siddiqui, Marom, & Kim, 2010; Mahouche Chergui et al., 2010; Paul, Lee, Choi, Kang, & Kim, 2010; Qin, Qin, Ford, Resasco, & Herrera, 2004; Shao et al., 2011; Sun, Fu, Lin, & Huang, 2002; Zhou, Zhang, Li, & Du, 2011). There are two types of functionalization to functionalize the CNTs: (i) noncovalent, which is based on supramolecular complexation via wrapping, π-stacking interactions, and Van der Waals forces, and (ii) covalent, which links metal nanoparticles, polymers, or organic molecules on the surface of CNTs (Balasubramanian & Burghard, 2008; Fennell et al., 2016; Ihsanullah et al., 2015; Liu, Ma, & Zhang, 2011). Common functionalization materials are oxygen-containing groups (such as OH, COOH), highly reactive species (for example, carbenes, diazonium ions, free radicals), conducting polymers (such as polyaniline, polypyrrole, polythiophene, and metallic

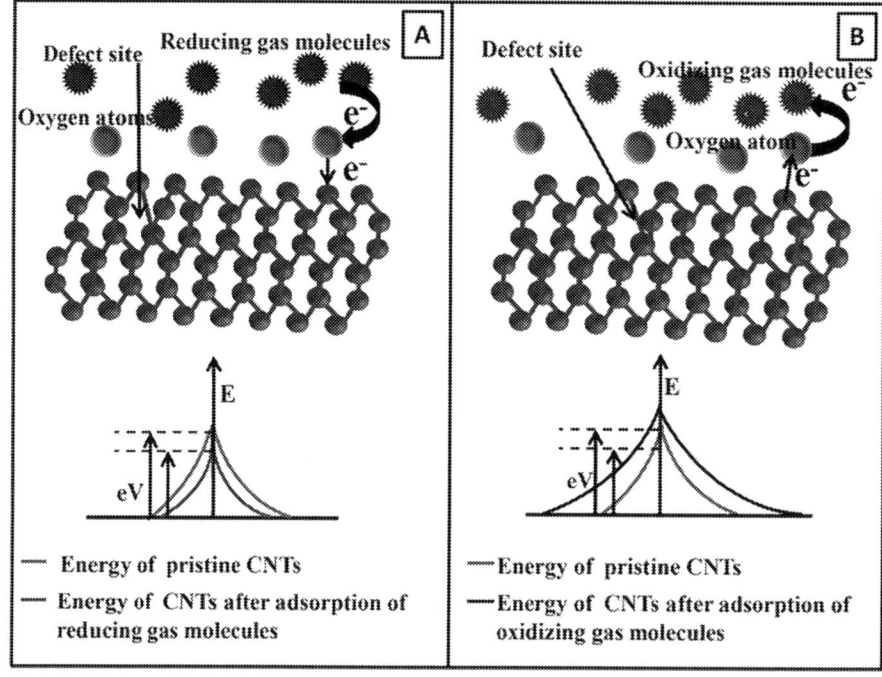

FIG. 8.6 Schematic diagram of electron transfer in and from pristine CNTs, (A) The surface of CNTs absorbs reducing gas molecules. As a result of chemical processes, electron transfer to CNT occurs through oxygen atoms and defects present on the surface, (B) CNT surface absorbs oxidizing gas molecules and donates electrons to the gas molecules (Mittal & Kumar, 2014). *(From Mittal, M., & Kumar, A. (2014). Carbon nanotube (CNT) gas sensors for emissions from fossil fuel burning. Sensors and Actuators, B: Chemical, 203, 349–362. https://doi.org/10.1016/j.snb.2014.05.080.)*

clusters) (Goldoni, Petaccia, Lizzit, & Larciprete, 2010; Srivastava et al., 2009; Yeung, Liu, & Wang, 2008; Zhao, Buongiorno Nardelli, Lu, & Bernholc, 2005). After functionalizing, CNTs show high sensitivity and selectivity toward some pollutant gas molecules. For example, boron- and nitrogen-functionalized CNT-based sensors show a significant change in the electrical response when exposed to various gases like O_2, NH_3, and NO_2 (Dresselhaus, 2001; Ouyang, Huang, Cheung, & Lieber, 2001).

Detection of carbon dioxide (CO_2) and carbon monoxide (CO)

Theoretical calculation and experimental results show that CNT alone is not very useful to detect CO due to the lack of charge transfer between CNTs and CO molecules (Peng & Cho, 2000). However, when CNTs are modified by the hydroxyl group, a significant interaction is found between CO and the modified sensor (Matranga & Bockrath, 2005). The —OH group acts as a bridge between the two molecules. Here is an illustration of the charge transfer between C-O molecule and CNTs.

Besides binding with the hydroxyl group, CNT can be modified with boron and nitrogen (Dong et al., 2013; Peng & Cho, 2003). In this case, chemical adsorption occurs for the first one, while the adsorption is physical for the second one. The modified CNTs show a significant decrease over time in the presence of CO, as shown in Fig. 8.7.

Unlike CO, CO_2 is adsorbed on the CNT surface directly via van der Waals force of attraction without any modification. After being adsorbed, CO_2 as a reductive gas transfers its electrons to the carbon nanotubes, which leads to electron and hole recombination inside the CNTs (Faizah, Fakhru'l-Razi, Sidek, & Abdullah, 2007). This results in a decrease in the hole current. Fig. 8.8 shows the response of CNT chemiresistor sensor at 150°C as a function of time for 10 ppm CO gas.

After each exposure cycle, the curves show incomplete CO_2 gas molecule desorption. Li et al. (2012) reported an ultrasensitive CO_2 sensor that has a detection limit of 500 ppt. As sensing materials, they used SWCNTs wrapped with poly(ionic liquid). The sensor had high selectivity for CO_2 and was resistive to relative humidity effects.

Detection of nitrogen dioxide (NO_2) and nitric oxide (NO)

NO_2 is an oxidizing gas. So, NO_2 can attach itself directly to the surface of CNTs and withdraws electrons from the surface. Theoretical calculations and experimental results (Kong et al., 2000) also suggest that there is a nonzero charge transfer value between NO_2 molecules and CNTs. After being adsorbed, NO_2 molecules decompose to NO and NO_3 as-

$$2NO_2 = NO + NO_3$$

NO_3 from this reaction bonds strongly with CNTs. This delays the desorption of NO_3 than that of NO. The adsorption of NO_2 molecules in CNTs resulted in the formation of new electronic states around the fermi-level, which altered their output resistance (Goldoni, Larciprete, Petaccia, & Lizzit, 2003; Santucci et al., 2003; Valentini et al., 2003). Detection of NO is pretty similar to NO_2 with just one extra step. Here, NO molecule at first reacts with O_2 molecule from the surrounding to make $NO + \frac{1}{2} O_2 \rightarrow NO_2$; then NO_2 adsorbs on the CNTs as described above.

Several CNT-based sensors have been designed to detect NO_2 (Dilonardo et al., 2016; Piloto et al., 2016). Valentini et al. (2003) demonstrated carbon nanotube (CNT)-based sensor produced on Si_3N_4/Si substrates using plasma-enhanced CVD for sensing NO_2 gas. They observed that the electrical resistance of CNTs decreases when they were exposed to NO_2. At an operating temperature of roughly 165°C, the highest variation in resistance to NO_2 was observed. The sensor designed was highly sensitive to NO_2 gas as low as 10 ppb concentrations. It also showed quick response time and excellent selectivity. They adjusted the resistance of the sensor film and optimized the sensing response to NO_2 by thermal treatment

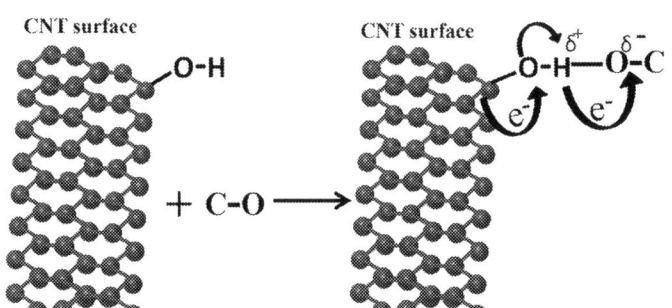

FIG. 8.7 CO gas molecule adsorption on CNTs functionalized with a hydroxyl group. The hydrogen atom of the hydroxyl group forms a hydrogen bond with the oxygen atom of carbon monoxide (Mittal & Kumar, 2014). *(From Mittal, M., & Kumar, A. (2014). Carbon nanotube (CNT) gas sensors for emissions from fossil fuel burning. Sensors and Actuators, B: Chemical, 203, 349–362. https://doi.org/10.1016/j.snb.2014.05.080.)*

FIG. 8.8 As a function of time, the response of the CNT chemiresistor sensor for 10 ppm CO gas is observed at 150°C (Choi, Jeong, Lee, Choi, & Choi, 2013). *(From Dong, K. Y., Choi, J., Lee, Y. D., Kang, B. H., Yu, Y. Y., Choi, H. H., et al. (2013). Detection of a CO and NH3 gas mixture using carboxylic acid-functionalized single-walled carbon nanotubes. Nanoscale Research Letters, 8(1), 1–6. https://doi.org/10.1186/1556-276X-8-12.)*

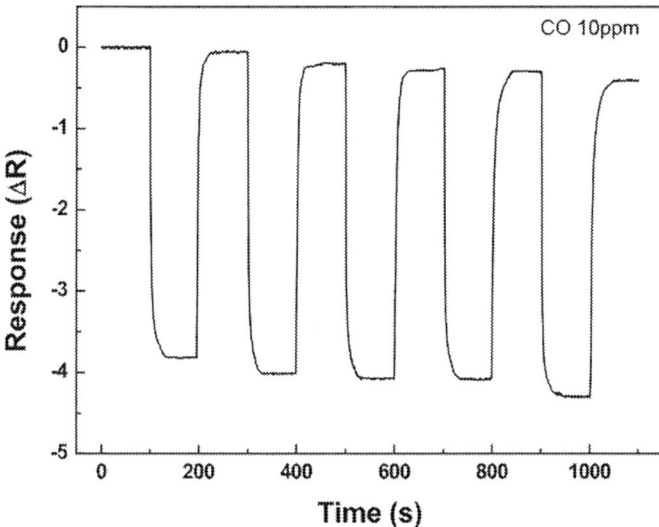

method based on repeated heating and cooling of the films. These sensors are more stable than metal oxide sensors because surface ions of metal oxide have grain and unreliable chemical properties. Additionally, driving voltage is comparatively less, making electronic circuitry more convenient and simple.

Gaikwad et al. (2015) developed a NO_2 chemiresistive sensor based on conducting polythiophene-SWCNTs. They used AC dielectrophoretic technique to align SWCNTs in the sensing film and charge-controlled potentiostatic deposition method for functionalizing the SWNT surfaces. The sensor they devised is cheap in cost and has an excellent response for a detection limit as low as 10 ppb.

$CuPcOC_8$/MWCNTs-COOH hybrid material was produced by Sharma, Kumar, Saini, Bedi, and Mahajan (2016) to detect numerous air pollutant gases such as NO_2, NH_3, and Cl_2 at 150°C. They functionalized $CuPcOC_8$ ($CuPcOC_8 = 2,3,9,10,16,17,23,24$-octakis (octyloxy)-29$H$, 31$H$) with MWCNTs-COOH and obtain $CuPcOC_8$/MWCNTs-COOH hybrid material. The π-π stacking interaction was crucial in the formation of the charge transfer conjugate.

Detection of ammonia (NH_3)

Several CNT-based sensors have been developed to detect NH_3 (Bannov et al., 2017; Datta, Ghosh, More, Shirsat, & Mulchandani, 2012; Guan et al., 2014; Piloto et al., 2016; Wang, Wang, et al., 2016; Yoon, Liu, & Swager, 2016). Mirica, Weis, Schnorr, Esser, and Swager (2012) developed sensors capable of detecting NH_3 gas at sub-ppm quantities that are created by mechanical abrasion of compressed single-walled carbon nanotubes (SWCNTs) on the surface of fibers of cellulose. This method of manufacture is simple, easy, and solvent-free, and it eliminates the problems that many SWCNT dispersions have due to their intrinsic instability. The detection limit is as low as 0.5 ppm.

Wang, Zhang, Yang, Sellin, and Zhong (2004) studied gas sensors based on MWNTs for detecting NH_3 molecules. Microwave plasma-enhanced CVD was used to make MWCNTs. The gas sensors were made using electron-beam lithography and lift-off processes, with the nanotubes laying beneath the electrodes. The conductivity of MWNTs is shown to decrease when the sensors are repeatedly exposed to NH_3 gas at room temperature. This shows that MWNTs would be a good candidate for detecting NH_3 at ambient temperature. Though the response time for this sensor was as low as ~180 s, it takes around 6 h to come back to the original state to reuse. Rigoni et al. (2013) designed a sensor by drop-casting SWCNTs. It has a detection limit of 3 ppb. He, Jia, Meng, Li, and Liu (2009) made a functionalized CNT sensor by polyaniline (PANI)-coated MWCNTs via in situ polymerization. At room temperature, response was fast and reproducibility was good. For NH_3, the linear response range was 0.2 ppm to 15 ppm. Zhang, Nix, Yoo, Deshusses, and Myung (2006) reported a PANI-SWNT network-based sensor with a sensitivity of 2.44% $\Delta R/R$ per ppm ammonia and a LOD of 50 ppb.

Detection of hydrogen sulfide (H_2S) and sulfur dioxide (SO_2)

Hydroxyl- and carboxyl-modified SWCNTs are used to detect SO_2. Due to the strong oxidizing capability of carboxyl-modified SWCNTs, they give a better response than that of hydroxyl-modified single-walled carbon nanotubes for both SO_2 and H_2S gases (Zhang, Yang, Dai, & Luo, 2012).

Detection of CH$_4$

SWCNTs have now direct sensing effect on CH$_4$ molecules. However, CH$_4$ can be detected by Li$^+$-doped CNT (Li$^+$/CNT) film. Chen et al. reported that this sensor has an excellent sensitivity of 14.5% at 500 ppm of CH$_4$ (Chen et al., 2018). Li et al. (2012) used SWNTs coated with Pd nanoparticles to detect methane levels from as low as 6–100 ppm at ambient temperature. Their Pd-SWNT sensors outperform traditional metal oxide and catalytic bead sensors for methane detection by a factor of 100 in terms of power consumption and size, and by a factor of 10 in terms of sensitivity. Sainato et al. (2016) showed the detection of methane using MWCNTs/ZnO composite-based chemiresistive sensor. They used the atomic layer deposition method to deposit ZnO as a functionalizing material to pretreat the surface of MWCNTs. The limit of detection is 10 ppm. For such a small LOD, the authors suggested few contributing factors such as a strong change in relative resistance of ZnO nanoparticles in response to low CH$_4$ concentrations, energetically advantageous electron transport at the MWCNTs/ZnO interface, and a substantial electrical current modulation potential due to ballistic transport of electrons through the MWCNTs.

Detection of VOC

Many methods were explored to analyze VOCs (Chiou, Wu, Huang, Chang, & Lin, 2017). Mostly functionalized CNTs are used for this purpose. For high-performance optical chemo-sensors, cadmium arachidate/single-walled carbon nanotube (CdA/SWCNT) composites were used. For the air environment, the estimated resolution was 30–250 ppb (Consales et al., 2009). Porphyrin is another good example of a functional molecule with CNTs (Rushi et al., 2016). Porphyrin's conductivity rises dramatically when it is used with CNTs as a hybrid material. Sarkar et al. created a combination of SWNTs and poly(tetraphenylporphyrin) to assess its capability as a low-power chemiresistor sensor to detect acetone vapor via an electrochemical method VOC (volatile organic compounds) model (Sarkar, Srinives, Sarkar, Haddon, & Mulchandani, 2014). The sensor had a large dynamic range of 50–230,000 ppm, with a limit of detection of 9 ppm. The results validated the device's good stability throughout a 180-day period. Liu, Moh, and Swager (2015) mixed SWCNTs with free-base and metalloporphyrins to detect and categorize five different kinds of VOCs, such as alkanes, aromatics, alcohols, amines, and ketones. Other sensors were also investigated to detect indoor air pollutant formaldehyde (Lu, Meyyappan, & Li, 2010; Shi et al., 2013; Wang, Zhang, Zhang, & Liu, 2006; Xie et al., 2012; Zhang, Zhang, Chen, Zhu, & Liu, 2014) and cyclohexane (Frazier & Swager, 2013). The DL was 0.2 ppm for formaldehyde (Zhang et al., 2014) and 5 ppm for cyclohexane (Frazier & Swager, 2013).

Graphene-based sensors for air pollution remediation

Sensing mechanism

Graphene has been deemed the "miracle material of the 21st century" due to its exceptional electrical, mechanical, chemical, optical, and thermal capabilities. The number of papers focused on graphene-based sensors is growing at an exponential rate, demonstrating graphene's growing interest in sensing applications (Varghese, Lonkar, Singh, Swaminathan, & Abdala, 2015). A large portion of these publications consists of pollutant gas sensing using graphene particles, graphene oxidase, and reduced graphene oxidase. The pollutant gas-sensing mechanism of graphene derivatives is almost similar to CNTs. At first, the pollutant gas molecules are adsorbed in the two-dimensional structure of graphene. This changes the carrier concentration and charge transfer occurs (Lu, Ocola, & Chen, 2009). Adsorbed gases might act as electron donors or acceptors according to their properties (Schedin et al., 2007). Further, graphene can be functionalized using different materials to increase electrical conductivity. In fact, graphene has high electrical conductivity and low intrinsic noise even in the absence of charge carriers. A few amounts of charge carriers created by the gas adsorbates result in noticeable changes in electrical conductivity. The conductivity is measured to get the final result.

Detection of NO$_2$

Graphene-based NO$_2$ demonstrated lower LOD and higher response, and they have more simple and practical preparation techniques. Seekaew, Phokharatkul, Wisitsoraat, and Wongchoosuk (2017) described a highly sensitive room-temperature gas sensor made of bilayer graphene. It was produced via interfacial transfer of CVD of graphene onto nickel-interdigitated electrodes. This bilayer sensor outperforms monolayer as well as multilayer graphene in terms of responsiveness, selectivity, and sensitivity to NO$_2$. The sensitivity found was 1.409 ppm^{-1} to NO$_2$ over a range of 1–25 ppm concentration. The direct charge transfer process caused by the adsorption of NO$_2$ molecules has been accounted to explain the NO$_2$ sensing mechanism of graphene-sensing film. The fabrication process of the graphene gas sensor is shown schematically in Fig. 8.9.

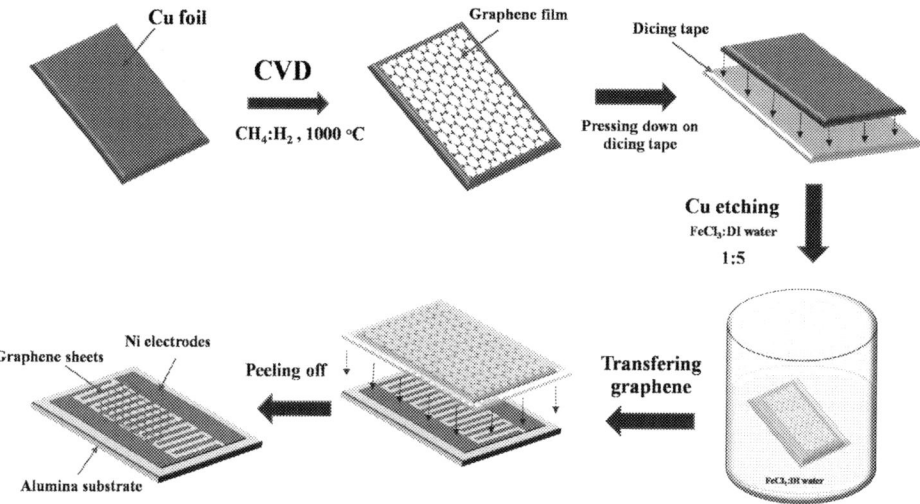

FIG. 8.9 Schematic diagram of fabrication process graphene gas sensor (Ricciardella et al., 2017). (From Seekaew, Y., Phokharatkul, D., Wisitsoraat, A., & Wongchoosuk, C. (2017). Highly sensitive and selective room-temperature NO2 gas sensor based on bilayer transferred chemical vapor deposited graphene. Applied Surface Science, 404, 357–363. https://doi.org/10.1016/j.apsusc.2017.01.286.)

The performance of GPN-based sensors depends on the synthesis route of GPN. Mechanical exfoliation (ME) gives a less-defective material and presents faster interaction between GPN and the gas (Ricciardella et al., 2017). Choi et al. (2013) presented a highly sensitive gas sensor to detect NO_2 based on multilayer graphene (MLG) films produced on a microheater-embedded flexible substrate using a chemical vapor deposition approach. Even at concentrations as low as 200 ppb, the MLG was able to detect NO_2. After exposing to 1 ppm NO_2 gas at room temperature for a minute, it showed 6% change in resistivity. Hoa, Tien, Luan, Chung, & Hur (2013) fabricated a low-temperature, low-cost but highly sensitive gas sensor using unique hybrid structures made of two-dimensional graphene and two-dimensional NiO nanosheets (NSs). On the surfaces of RGO, 2D NiO sheets with well-developed hierarchically porous architectures were produced. Even at 1 ppm, sensors made of hybrid structures have a sensitivity and responsivity two times higher than those made with NiO nanosheets alone. They also tested reducing gases like H_2, NH_3, and H_2S, where the sensing results are comparable, but NO_2 has the highest responsivity of all the gases examined.

Detection of NH_3

A tremendous deal of effort has recently resulted in a significant advancement in the development of graphene-based gas sensors for the detection of NH_3. Yavari, Castillo, Gullapalli, Ajayan, & Koratkar (2012) synthesized graphene by CVD to detect trace amounts of NO_2 and NH_3 in the air at room temperature and atmospheric pressure. The gas concentration was inversely related to the sensor response time. The chemisorbed molecules were ejected from the graphene surface when the film was heated, allowing for reversible functioning. The LOD was 500 ppb for NH_3 and 100 ppb for NO_2, which are far higher than those of commercially available NO_2 and NH_3 detectors. Gautam and his colleagues looked into the NH_3 gas-sensing characteristics of graphene made by CVD and found that the deposition of Au-NPs on the surface of graphene films improved sensitivity and recovery time (Gautam & Jayatissa, 2012a). Table 8.2 gives a summary of recent studies about graphene-based NH_3 detection (Wang, Wang, et al., 2016).

Detection of CO_2 and CO

A graphene-based high-performance CO_2 sensor was designed using mechanical cleavage by Yoon et al. (2011). They described that this sensor overcomes the limitations of other solid-state gas sensors because it can be used in ambient circumstances and at room temperature. Changes in device conductance are measured for various amounts of CO_2 gas adsorbed on the surface of graphene. The conductivity of the graphene gas sensor increases linearly as the concentration of CO_2 gas increases from 10 to 100 ppm. Low power consumption, quick response time, high sensitivity, and quick recovery time are all features of this sensor. Smith et al. (2017) presented a graphene-based sensor that can selectively detect CO_2 from other gases in the air such as argon (Ar), oxygen (O_2), and nitrogen (N_2). Liu, Jia, Ye, and Zeng (2013) examined the adsorption of a variety of gas molecules including common pollutants such as CO, CO_2, SO_2, and NH_3 on Li-decorated T graphene using DFT. They discovered that Li-decorated T graphene is more sensitive to CO.

TABLE 8.2 A summary of recent studies about NH3 detection based on graphene at room temperature.

Sensing material	Sensor structure	LOD	Reference
RGO/MnO$_2$+PANI	Chemiresistor	25%/5 ppm	Huang et al. (2012)
RGO/ANI	Chemiresistor	10.7%/5 ppm	Huang, Hu, Wang, and Zhang (2013)
RGO/ANI+PANI	Chemiresistor	20%/20 ppm	Huang, Hu, Zhang, et al. (2013)
RGO/Py	Chemiresistor	2.4%/1 ppb	Hu et al. (2014)
RGO/Py	Chemiresistor	4.2%/50 ppb	Wang et al. (2014)
GR+Au	Chemiresistor	1%/6 ppm	Gautam and Jayatissa (2012a)
GR	FET	0.49 V/ppm	Gautam and Jayatissa (2012b)
GR	Chemiresistor	3%/500 ppb	Yavari et al. (2012)
GR+PANI	Chemiresistor	0.7%/1 ppm	Wu et al. (2013a, 2013b)
GR supported by mica substrate	FET	4%/50 ppm	Ben Aziza, Zhang, and Baillargeat (2014)
GR gated by ionic liquid	FET	130 ppb	Inaba, Yoo, Takei, Matsumoto, and Shimoyama (2014)
Printed GR+PEDOT:PSS	Chemiresistor	25 ppm	Seekaew et al. (2014)
RGO+P3HT	Chemiresistor	7.15%/10 ppm	Ye, Jiang, Tai, and Yuan (2014)
RGO/tannic acid	Chemiresistor	9.3%/1310 ppm	Yoo et al. (2014)
RGO/Cu(OH)$_4^{2-}$+Cu$_2$O	Chemiresistor	80%/100 ppm	Meng, Yang, Ding, Feng, and Guan (2015)

ANI, aniline; *COP*, chemical oxidative polymerization; *DMMP*, dimethyl methyl phosphonate; *GR*, graphene; *PANI*, polyaniline; *PPD*, p-phenylenediamine; *Py*, pyrrole; *RGO*, reduced graphene oxide.
From Wang, T., Huang, D., Yang, Z., Xu, S., He, G., Li, X., et al. (2016). A review on graphene-based gas/vapor sensors with unique properties and potential applications. *Nano-Micro Letters*, 8(2), 95–119. https://doi.org/10.1007/s40820-015-0073-1.

TABLE 8.3 A summary of recent studies about gas sensors based on graphene for the detection of air pollutant gases.

Sensing material	Target gas	LOD	Reference
GR/PANI	CH$_4$	10 ppm	Wu et al. (2013b)
Edge-tailored GO	SO$_2$	5 ppm	Shen et al. (2013)
RGO+Cu$_2$O nanocrystal	H$_2$S	11%/5 ppb	Zhou et al. (2013)
GR/Ti or GR/Sn	SO$_2$/H$_2$S	–	Abdulkader Tawfik, Cui, Carter, Ringer, and Stampfl (2015)
RGO/Fe$_2$O$_3$	H$_2$S	15 ppm (1900C)	Jiang et al. (2014)
GR/porous WO$_3$ NFs	H$_2$S	3.9%/100 ppb (3000C)	Choi et al. (2014)

GO, graphene oxide; *GR*, graphene; *PANI*, polyaniline; *PSS*, poly 4-styrenesulfonic acid; *RGO*, reduced graphene oxide.
From Wang, T., Huang, D., Yang, Z., Xu, S., He, G., Li, X., et al. (2016). A review on graphene-based gas/vapor sensors with unique properties and potential applications. *Nano-Micro Letters*, 8(2), 95–119. https://doi.org/10.1007/s40820-015-0073-1.

Detection of other air pollutant gases

Other air pollutant gases such as SO$_2$, H$_2$S, VOC, and CH$_4$ are also widely investigated to sense by graphene-based sensors. A brief summary of recent studies is given in Table 8.3 (Wang, Huang, et al., 2016).

Fullerene (FLN)-based gas sensors

Fullerenes were also investigated as pollutant gas sensors. The sensing mechanism is also similar to the other nanomaterial-based sensors. Smazna et al. (2018) detected ethanol with C$_{60}$ FLN-decorated zinc oxide (ZnO) tetrapod compounds. Zhao,

Yang, Zou, Sun, and Ji (2017) investigated C_{20} molecular junction to detect NO and CO. Both of the authors found excellent responses from these sensors. Grynko et al. (2009) showed fullerene and fullerene-aluminum nanostructured films as sensitive layers for the detection of ethanol. Zhang and his co-workers created hollow fullerene nanostructures with custom openings and forms. They used DFT along with nonequilibrium Green's function formalism to investigate the transport properties and performance of two-node hollow-fullerene-based nanoelectronic devices to sense the greenhouse gas molecule, CF_4.

Conclusion

Various types of carbon nanomaterials and their applications in air pollution remediation by monitoring airborne pollutants were reviewed in this chapter. Three types of CNM, such as CNTs, fullerenes, and graphene, and their properties, as well as their sensing mechanism, were discussed. For their unique structures and electronic properties, CNM has especial powerful gas adsorption and desorption capabilities. Thus, CNM-based sensors show their competencies to detect and quantify airborne gas pollutants, which is the crucial step of air pollution remediation.

References

Abdulkader Tawfik, S., Cui, X. Y., Carter, D. J., Ringer, S. P., & Stampfl, C. (2015). Sensing sulfur-containing gases using titanium and tin decorated zigzag graphene nanoribbons from first-principles. *Physical Chemistry Chemical Physics, 17*(10), 6925–6932. https://doi.org/10.1039/c4cp05919k.

Ahammad, A. J. S., Islam, T., & Hasan, M. M. (2019). Graphene-based electrochemical sensors for biomedical applications. In *Biomedical applications of graphene and 2D nanomaterials* (pp. 249–282). Elsevier. https://doi.org/10.1016/B978-0-12-815889-0.00012-X.

Amade, R., Hussain, S., Ocaña, I. R., & Bertran, E. (2014). Growth and functionalization of carbon nanotubes on quartz filter for environmental applications. *Journal of Environmental Engineering and Ecological Science, 2*. https://doi.org/10.7243/2050-1323-3-2.

Arkas, M., & Tsiourvas, D. (2009). Organic/inorganic hybrid nanospheres based on hyperbranched poly (ethylene imine) encapsulated into silica for the sorption of toxic metal ions and polycyclic aromatic hydrocarbons from water. *Journal of Hazardous Materials, 170*(1), 35–42. https://doi.org/10.1016/j.jhazmat.2009.05.031.

Asad, M., & Sheikhi, M. H. (2014). Surface acoustic wave based H2S gas sensors incorporating sensitive layers of single wall carbon nanotubes decorated with Cu nanoparticles. *Sensors and Actuators, B: Chemical, 198*, 134–141. https://doi.org/10.1016/j.snb.2014.03.024.

Azam, M. A., Alias, F. M., Tack, L. W., Seman, R. N. A. R., & Taib, M. F. M. (2017). Electronic properties and gas adsorption behaviour of pristine, silicon-, and boron-doped (8, 0) single-walled carbon nanotube: A first principles study. *Journal of Molecular Graphics and Modelling, 75*, 85–93. https://doi.org/10.1016/j.jmgm.2017.05.003.

Bachtold, A., Strunk, C., Salvetat, J. P., Bonard, J. M., Forró, L., Nussbaumer, T., et al. (1999). Aharonov-Bohm oscillations in carbon nanotubes. *Nature, 397*(6721), 673–675. https://doi.org/10.1038/17755.

Balasubramanian, K., & Burghard, M. (2008). Electrochemically functionalized carbon nanotubes for device applications. *Journal of Materials Chemistry, 18*(26), 3071–3083. https://doi.org/10.1039/b718262g.

Bannov, A. G., Jašek, O., Manakhov, A., Márik, M., Nečas, D., & Zajíčková, L. (2017). High-performance ammonia gas sensors based on plasma treated carbon nanostructures. *IEEE Sensors Journal, 17*(7), 1964–1970. https://doi.org/10.1109/JSEN.2017.2656122.

Ben Aziza, Z., Zhang, Q., & Baillargeat, D. (2014). Graphene/mica based ammonia gas sensors. *Applied Physics Letters*, 254102. https://doi.org/10.1063/1.4905039.

Boehm, H. P., Clauss, A., Fischer, G. O., & Hofmann, U. (1962). Dünnste Kohlenstoff-Folien. *Zeitschrift Für Naturforschung B, 17*(3), 150–153. https://doi.org/10.1515/znb-1962-0302.

Bonaccorso, F., Sun, Z., Hasan, T., & Ferrari, A. C. (2010). Graphene photonics and optoelectronics. *Nature Photonics, 4*(9), 611–622. https://doi.org/10.1038/nphoton.2010.186.

Bosso, S. T., & Enzweiler, J. (2002). Evaluation of heavy metal removal from aqueous solution onto scolecite. *Water Research, 36*(19), 4795–4800. https://doi.org/10.1016/S0043-1354(02)00208-7.

Bott, B., Jones, T. A., & Mann, B. (1984). The detection and measurement of CO using ZnO single crystals. *Sensors and Actuators, 5*(1), 65–73. https://doi.org/10.1016/0250-6874(84)87007-9.

Chen, M., & Goodman, D. W. (2006). Catalytically active gold: From nanoparticles to ultrathin films. *Accounts of Chemical Research, 39*, 739–746. https://doi.org/10.1021/ar040309d.

Chen, X., Huang, Z., Li, J., Wu, C., Wang, Z., & Cui, Y. (2018). Methane gas sensing behavior of lithium ion doped carbon nanotubes sensor. *Vacuum, 154*, 120–128. https://doi.org/10.1016/j.vacuum.2018.04.053.

Chiou, J. C., Wu, C. C., Huang, Y. C., Chang, S. C., & Lin, T. M. (2017). Effects of operating temperature on droplet casting of flexible polymer/multi-walled carbon nanotube composite gas sensors. *Sensors (Switzerland), 17*(1). https://doi.org/10.3390/s17010004.

Choi, S. J., Jang, B. H., Lee, S. J., Min, B. K., Rothschild, A., & Kim, I. D. (2014). Selective detection of acetone and hydrogen sulfide for the diagnosis of diabetes and halitosis using SnO2 nanofibers functionalized with reduced graphene oxide nanosheets. *ACS Applied Materials and Interfaces, 6*(4), 2588–2597. https://doi.org/10.1021/am405088q.

Choi, H., Jeong, H. Y., Lee, D.-S., Choi, C.-G., & Choi, S.-Y. (2013). Flexible NO2 gas sensor using multilayer graphene films by chemical vapor deposition. *Carbon Letters*, 186–189. https://doi.org/10.5714/CL.2013.14.3.186.

Chopra, S., McGuire, K., Gothard, N., Rao, A. M., & Pham, A. (2003). Selective gas detection using a carbon nanotube sensor. *Applied Physics Letters, 83*(11), 2280–2282. https://doi.org/10.1063/1.1610251.

Consales, M., Crescitelli, A., Penza, M., Aversa, P., Veneri, P. D., Giordano, M., et al. (2009). SWCNT nano-composite optical sensors for VOC and gas trace detection. *Sensors and Actuators, B: Chemical, 138*(1), 351–361. https://doi.org/10.1016/j.snb.2009.02.041.

Das, S., Sen, B., & Debnath, N. (2015). Recent trends in nanomaterials applications in environmental monitoring and remediation. *Environmental Science and Pollution Research, 22*, 18333–18344. https://doi.org/10.1007/s11356-015-5491-6.

Datta, K., Ghosh, P., More, M. A., Shirsat, M. D., & Mulchandani, A. (2012). Controlled functionalization of single-walled carbon nanotubes for enhanced ammonia sensing: A comparative study. *Journal of Physics D: Applied Physics, 45*(35). https://doi.org/10.1088/0022-3727/45/35/355305.

Dekker, C. (1999). Carbon nanotubes as molecular quantum wires. *Physics Today, 52*, 22–30.

Dilonardo, E., Penza, M., Alvisi, M., Di Franco, C., Rossi, R., Palmisano, F., et al. (2016). Electrophoretic deposition of Au NPs on MWCNT-based gas sensor for tailored gas detection with enhanced sensing properties. *Sensors and Actuators, B: Chemical, 223*, 417–428. https://doi.org/10.1016/j.snb.2015.09.112.

Dong, K. Y., Choi, J., Lee, Y. D., Kang, B. H., Yu, Y. Y., Choi, H. H., et al. (2013). Detection of a CO and NH3 gas mixture using carboxylic acid-functionalized single-walled carbon nanotubes. *Nanoscale Research Letters, 8*(1), 1–6. https://doi.org/10.1186/1556-276X-8-12.

Dong, L. F., Cui, Z. L., & Zhang, Z. K. (1997). Gas sensing properties of nano-ZnO prepared by arc plasma method. *Nanostructured Materials, 8*(7), 815–823. https://doi.org/10.1016/S0965-9773(98)00005-1.

Dong, H., Lin, B., Gilmore, K., Hou, T., Lee, S. T., & Li, Y. (2015). B40 fullerene: An efficient material for CO2 capture, storage and separation. *Current Applied Physics, 15*(9), 1084–1089. https://doi.org/10.1016/j.cap.2015.06.008.

Dresselhaus, M. S. (2001). *Electronic properties of carbon nanotubes and applications* (pp. 11–49). Netherlands: Springer. https://doi.org/10.1007/978-94-010-0777-1_2.

Durgun, E., Dag, S., Bagci, V. M. K., Gülseren, O., Yildirim, T., & Ciraci, S. (2003). Systematic study of adsorption of single atoms on a carbon nanotube. *Physical Review B: Condensed Matter and Materials Physics, 67*(20). https://doi.org/10.1103/PhysRevB.67.201401.

Eatemadi, A., Daraee, H., Karimkhanloo, H., Kouhi, M., Zarghami, N., Akbarzadeh, A., et al. (2014). Carbon nanotubes: Properties, synthesis, purification, and medical applications. *Nanoscale Research Letters, 9*(1), 1–13. https://doi.org/10.1186/1556-276X-9-393.

Esrafili, M. D. (2017). N2O reduction over a fullerene-like boron nitride nanocage: A DFT study. *Physics Letters, Section A: General, Atomic and Solid State Physics, 381*(25–26), 2085–2091. https://doi.org/10.1016/j.physleta.2017.04.009.

Faizah, M. Y., Fakhru'l-Razi, A., Sidek, R. M., & Abdullah, A. G. L. (2007). Gas sensor application of carbon nanotubes. *International Journal of Engineering and Technology, 4*, 106–113.

Feng, W., Yi, W., Wu, H., Ozaki, M., & Yoshino, K. (2005). Enhancement of third-order optical nonlinearities by conjugated polymer-bonded carbon nanotubes. *Journal of Applied Physics, 98*(3). https://doi.org/10.1063/1.1954887.

Fennell, J. F., Liu, S. F., Azzarelli, J. M., Weis, J. G., Rochat, S., Mirica, K. A., et al. (2016). Nanowire chemical/biological sensors: status and a roadmap for the future. *Angewandte Chemie, International Edition, 55*(4), 1266–1281. https://doi.org/10.1002/anie.201505308.

Frazier, K. M., & Swager, T. M. (2013). Robust cyclohexanone selective chemiresistors based on single-walled carbon nanotubes. *Analytical Chemistry, 85*(15), 7154–7158. https://doi.org/10.1021/ac400808h.

Gaikwad, S., Bodkhe, G., Deshmukh, M., Patil, H., Rushi, A., Shirsat, M. D., et al. (2015). Chemiresistive sensor based on polythiophene-modified single-walled carbon nanotubes for detection of NO2. *Modern Physics Letters B, 29*, 6–7. World Scientific Publishing Co. Pte Ltd https://doi.org/10.1142/S0217984915400461.

Gautam, M., & Jayatissa, A. H. (2012a). Ammonia gas sensing behavior of graphene surface decorated with gold nanoparticles. In *Vol. 78. Solid-state electronics* (pp. 159–165). https://doi.org/10.1016/j.sse.2012.05.059.

Gautam, M., & Jayatissa, A. H. (2012b). Graphene based field effect transistor for the detection of ammonia. *Journal of Applied Physics, 112*(6). https://doi.org/10.1063/1.4752272.

Goldoni, A., Larciprete, R., Petaccia, L., & Lizzit, S. (2003). Single-wall carbon nanotube interaction with gases: Sample contaminants and environmental monitoring. *Journal of the American Chemical Society, 125*(37), 11329–11333. https://doi.org/10.1021/ja034898e.

Goldoni, A., Petaccia, L., Lizzit, S., & Larciprete, R. (2010). Sensing gases with carbon nanotubes: A review of the actual situation. *Journal of Physics. Condensed Matter, 22*(1). https://doi.org/10.1088/0953-8984/22/1/013001.

González-García, P. (2018). Activated carbon from lignocellulosics precursors: A review of the synthesis methods, characterization techniques and applications. *Renewable and Sustainable Energy Reviews, 82*, 1393–1414. https://doi.org/10.1016/j.rser.2017.04.117.

Grynko, D., Burlachenko, J., Kukla, O., Kruglenko, I., & Belyaev, O. (2009). Fullerene and fullerene-aluminum nanostructured films as sensitive layers for gas sensors. *Semiconductor Physics, Quantum Electronics and Optoelectronics*, 287–289. https://doi.org/10.15407/spqeo12.03.287.

Guan, L., Wang, S., Gu, W., Zhuang, J., Jin, H., Zhang, W., et al. (2014). Ultrasensitive room-temperature detection of NO2 with tellurium nanotube based chemiresistive sensor. *Sensors and Actuators, B: Chemical, 196*, 321–327. https://doi.org/10.1016/j.snb.2014.02.014.

Guerra, F. D., Attia, M. F., Whitehead, D. C., & Alexis, F. (2018). Nanotechnology for environmental remediation: Materials and applications. *Molecules, 23*(7). https://doi.org/10.3390/molecules23071760.

Gupta, S., Murthy, C. N., & Prabha, C. R. (2018). Recent advances in carbon nanotube based electrochemical biosensors. *International Journal of Biological Macromolecules, 108*, 687–703. https://doi.org/10.1016/j.ijbiomac.2017.12.038.

Hajihashemi, R., Rashidi, A. M., Alaie, M., Mohammadzadeh, R., & Izadi, N. (2014). The study of structural properties of carbon nanotubes decorated with NiFe2O4 nanoparticles and application of nano-composite thin film as H2S gas sensor. *Materials Science and Engineering C, 44*, 417–421. https://doi.org/10.1016/j.msec.2014.08.013.

Hamwi, A., Alvergnat, H., Bonnamy, S., & Béguin, F. (1997). Fluorination of carbon nanotubes. *Carbon, 35*(6), 723–728. https://doi.org/10.1016/S0008-6223(97)00013-4.

Hanada, T., Okada, Y., & Yase, K. (1999). Structure of multi-walled and single-walled carbon nanotubes. EELS study. In *The science and technology of carbon nanotubes* (pp. 29–39). Elsevier BV. https://doi.org/10.1016/b978-008042696-9/50004-5 (chapter 4).

Harrison, R. M. (2015). *Pollution: Causes, effects and control*. Royal Society of Chemistry.

He, L., Jia, Y., Meng, F., Li, M., & Liu, J. (2009). Gas sensors for ammonia detection based on polyaniline-coated multi-wall carbon nanotubes. *Materials Science and Engineering: B*, 76–81. https://doi.org/10.1016/j.mseb.2009.05.009.

Hirsch, A., & Vostrowsky, O. (2005). Functionalization of carbon nanotubes. *Topics in Current Chemistry*, 245, 193–237. https://doi.org/10.1007/b98169.

Hoa, L. T., Tien, H. N., Luan, V. H., Chung, J. S., & Hur, S. H. (2013). Fabrication of a novel 2D-graphene/2D-NiO nanosheet-based hybrid nanostructure and its use in highly sensitive NO2 sensors. *Sensors and Actuators, B: Chemical*, 185, 701–705. https://doi.org/10.1016/j.snb.2013.05.050.

Holzinger, M., Abraham, J., Whelan, P., Graupner, R., Ley, L., Hennrich, F., et al. (2003). Functionalization of single-walled carbon nanotubes with (R-) oxycarbonyl nitrenes. *Journal of the American Chemical Society*, 125(28), 8566–8580. https://doi.org/10.1021/ja029931w.

Hsu, S., & Lu, C. (2007). Modification of single-walled carbon nanotubes for enhancing isopropyl alcohol vapor adsorption from air streams. *Separation Science and Technology*, 42(12), 2751–2766. https://doi.org/10.1080/01496390701515060.

Hu, N., Yang, Z., Wang, Y., Zhang, L., Wang, Y., Huang, X., et al. (2014). Ultrafast and sensitive room temperature NH3 gas sensors based on chemically reduced graphene oxide. *Nanotechnology*, 25(2). https://doi.org/10.1088/0957-4484/25/2/025502.

Hu, H., Zhang, T., Yuan, S., & Tang, S. (2017). Functionalization of multi-walled carbon nanotubes with phenylenediamine for enhanced CO2 adsorption. *Adsorption*, 23(1), 73–85. https://doi.org/10.1007/s10450-016-9820-y.

Huang, X., Hu, N., Gao, R., Yu, Y., Wang, Y., Yang, Z., et al. (2012). Reduced graphene oxide-polyaniline hybrid: Preparation, characterization and its applications for ammonia gas sensing. *Journal of Materials Chemistry*, 22(42), 22488–22495. https://doi.org/10.1039/c2jm34340a.

Huang, X., Hu, N., Wang, Y., & Zhang, Y. (2013). Ammonia gas sensor based on aniline reduced graphene oxide. In *Vol. 669. Advanced materials research* (pp. 79–84). https://doi.org/10.4028/www.scientific.net/AMR.669.79.

Huang, X., Hu, N., Zhang, L., Wei, L., Wei, H., & Zhang, Y. (2013). The NH3 sensing properties of gas sensors based on aniline reduced graphene oxide. *Synthetic Metals*, 185–186, 25–30. https://doi.org/10.1016/j.synthmet.2013.09.034.

Hull, M., & Bowman, D. (2018). *Nanotechnology environmental health and safety: Risks, regulation, and management*. William Andrew.

Ihsanullah, Asmaly, H. A., Saleh, T. A., Laoui, T., Gupta, V. K., & Atieh, M. A. (2015). Enhanced adsorption of phenols from liquids by aluminum oxide/carbon nanotubes: Comprehensive study from synthesis to surface properties. *Journal of Molecular Liquids*, 206, 176–182. https://doi.org/10.1016/j.molliq.2015.02.028.

Iijima, S. (1991). Helical microtubules of graphitic carbon. *Nature*, 354(6348), 56–58. https://doi.org/10.1038/354056a0.

Inaba, A., Yoo, K., Takei, Y., Matsumoto, K., & Shimoyama, I. (2014). Ammonia gas sensing using a graphene field-effect transistor gated by ionic liquid. *Sensors and Actuators, B: Chemical*, 195, 15–21. https://doi.org/10.1016/j.snb.2013.12.118.

Jiang, Z., Li, J., Aslan, H., Li, Q., Li, Y., Chen, M., et al. (2014). A high efficiency H2S gas sensor material: Paper like Fe2O3/graphene nanosheets and structural alignment dependency of device efficiency. *Journal of Materials Chemistry A*, 2(19), 6714–6717. https://doi.org/10.1039/C3TA15180H.

Jung, H. Y., Jung, S. M., Kim, J., & Suh, J. S. (2007). Chemical sensors for sensing gas adsorbed on the inner surface of carbon nanotube channels. *Applied Physics Letters*, 90(15). https://doi.org/10.1063/1.2722196.

Khin, M. M., Nair, A. S., Babu, V. J., Murugan, R., & Ramakrishna, S. (2012). A review on nanomaterials for environmental remediation. *Energy and Environmental Science*, 5(8), 8075–8109. https://doi.org/10.1039/c2ee21818f.

Kim, E., Jeong, H. S., & Kim, B. M. (2014). Studies on the functionalization of MWNTs and their application as a recyclable catalyst for CC bond coupling reactions. *Catalysis Communications*, 46, 71–74. https://doi.org/10.1016/j.catcom.2013.11.028.

Kim, Y. J., Kim, Y., Novoselov, K., & Hong, B. H. (2015). Engineering electrical properties of graphene: Chemical approaches. *2D Materials*, 2(4). https://doi.org/10.1088/2053-1583/2/4/042001.

Kong, J., Franklin, N. R., Zhou, C., Chapline, M. G., Peng, S., Cho, K., et al. (2000). Nanotube molecular wires as chemical sensors. *Science*, 287(5453), 622–625. https://doi.org/10.1126/science.287.5453.622.

Kong, H., Gao, C., & Yan, D. (2004). Controlled functionalization of multiwalled carbon nanotubes by in situ atom transfer radical polymerization. *Journal of the American Chemical Society*, 126(2), 412–413. https://doi.org/10.1021/ja0380493.

Kowalczyk, P., & Holyst, R. (2008). Efficient adsorption of super greenhouse gas (tetrafluoromethane) in carbon nanotubes. *Environmental Science and Technology*, 42(8), 2931–2936. https://doi.org/10.1021/es071306+.

Kroto, H. (1997). Symmetry, space, stars, and C60 (Nobel lecture). *Angewandte Chemie International Edition in English*, 36(15), 1578–1593. https://doi.org/10.1002/anie.199715781.

Kumar, L., Paswan, S. K., Kumar, P., Singh, R. K., Kumar, R., & Shukla, S. K. (2021). *Nanotechnology-based filtration membranes for removal of pollutants from drinking water* (pp. 231–251). Elsevier BV. https://doi.org/10.1016/b978-0-12-823828-8.00011-6.

Lee, S. K., Chang, D., & Kim, S. W. (2014). Gas sensors based on carbon nanoflake/tin oxide composites for ammonia detection. *Journal of Hazardous Materials*, 268, 110–114. https://doi.org/10.1016/j.jhazmat.2013.12.049.

Li, Y., Li, G., Wang, X., Zhu, Z., Ma, H., Zhang, T., et al. (2012). Poly(ionic liquid)-wrapped single-walled carbon nanotubes for sub-ppb detection of CO2. *Chemical Communications*, 48(66), 8222–8224. https://doi.org/10.1039/c2cc33365a.

Liu, C. S., Jia, R., Ye, X. J., & Zeng, Z. (2013). Non-hexagonal symmetry-induced functional T graphene for the detection of carbon monoxide. *Journal of Chemical Physics*, 139(3). https://doi.org/10.1063/1.4813528.

Liu, L., Ma, W., & Zhang, Z. (2011). Macroscopic carbon nanotube assemblies: Preparation, properties, and potential applications. *Small*, 7(11), 1504–1520. https://doi.org/10.1002/smll.201002198.

Liu, S. F., Moh, L. C. H., & Swager, T. M. (2015). Single-walled carbon nanotube-metalloporphyrin chemiresistive gas sensor arrays for volatile organic compounds. *Chemistry of Materials*, 27(10), 3560–3563. https://doi.org/10.1021/acs.chemmater.5b00153.

Llobet, E. (2013). Gas sensors using carbon nanomaterials: A review. *Sensors and Actuators, B: Chemical, 179*, 32–45. https://doi.org/10.1016/j.snb.2012.11.014.

Lu, Y., Meyyappan, M., & Li, J. (2010). A carbon-nanotube-based sensor array for formaldehyde detection. *Nanotechnology, 22*.

Lu, G., Ocola, L. E., & Chen, J. (2009). Reduced graphene oxide for room-temperature gas sensors. *Nanotechnology*, 445502. https://doi.org/10.1088/0957-4484/20/44/445502.

Ma, P. C., Siddiqui, N. A., Marom, G., & Kim, J. K. (2010). Dispersion and functionalization of carbon nanotubes for polymer-based nanocomposites: A review. *Composites Part A: Applied Science and Manufacturing, 41*(10), 1345–1367. https://doi.org/10.1016/j.compositesa.2010.07.003.

Mahouche Chergui, S., Ledebt, A., Mammeri, F., Herbst, F., Carbonnier, B., Ben Romdhane, H., et al. (2010). Hairy carbon nanotube@nano-Pd heterostructures: Design, characterization, and application in Suzuki C−C coupling reaction. *Langmuir*, 16115–16121. https://doi.org/10.1021/la102801d.

Malekalaie, M., Jahangiri, M., Rashidi, A. M., Haghighiasl, A., & Izadi, N. (2015). Selective hydrogen sulfide (H2S) sensors based on molybdenum trioxide (MoO3) nanoparticle decorated reduced graphene oxide. *Materials Science in Semiconductor Processing, 38*, 93–100. https://doi.org/10.1016/j.mssp.2015.03.034.

Malhotra, B. D., & Ali, M. A. (2018). *Functionalized carbon nanomaterials for biosensors* (pp. 75–103). Elsevier BV. https://doi.org/10.1016/b978-0-323-44923-6.00002-9.

Matranga, C., & Bockrath, B. (2005). Hydrogen-bonded and physisorbed CO in single-walled carbon nanotube bundles. *Journal of Physical Chemistry B, 109*(11), 4853–4864. https://doi.org/10.1021/jp0464122.

Mehndiratta, P., Jain, A., Srivastava, S., & Gupta, N. (2013). Environmental pollution and nanotechnology. *Environmental Pollution*. https://doi.org/10.5539/ep.v2n2p49.

Mendoza, F., Hernández, D. M., Makarov, V., Febus, E., Weiner, B. R., & Morell, G. (2014). Room temperature gas sensor based on tin dioxide-carbon nanotubes composite films. *Sensors and Actuators, B: Chemical, 190*, 227–233. https://doi.org/10.1016/j.snb.2013.08.050.

Meng, H., Yang, W., Ding, K., Feng, L., & Guan, Y. (2015). Cu2O nanorods modified by reduced graphene oxide for NH3 sensing at room temperature. *Journal of Materials Chemistry A, 3*(3), 1174–1181. https://doi.org/10.1039/c4ta06024e.

Mirica, K. A., Weis, J. G., Schnorr, J. M., Esser, B., & Swager, T. M. (2012). Mechanical drawing of gas sensors on paper. *Angewandte Chemie*, 10898–10903. https://doi.org/10.1002/ange.201206069.

Mittal, M., & Kumar, A. (2014). Carbon nanotube (CNT) gas sensors for emissions from fossil fuel burning. *Sensors and Actuators, B: Chemical, 203*, 349–362. https://doi.org/10.1016/j.snb.2014.05.080.

Mo, J., Zhang, Y., Xu, Q., Lamson, J. J., & Zhao, R. (2009). Photocatalytic purification of volatile organic compounds in indoor air: A literature review. *Atmospheric Environment, 43*(14), 2229–2246. https://doi.org/10.1016/j.atmosenv.2009.01.034.

Moseley, P. T., & Tofield, B. C. (1987). *Solid state gas sensors* (pp. 12–31). Bristol/Philadelphia: Hilger.

Nag, A., Mitra, A., & Mukhopadhyay, S. C. (2018). Graphene and its sensor-based applications: A review. *Sensors and Actuators, A: Physical, 270*, 177–194. https://doi.org/10.1016/j.sna.2017.12.028.

Nan'ya, S., Osawa, H., Obata, K., & Maekawa, E. (1970). Synthesis of dibenzopyrromethene pigments from 1-phenylisoindole. *Nippon Kagaku Zasshi, 91*(2), 181–182. https://doi.org/10.1246/nikkashi1948.91.2_181.

Oberlin, A., Endo, M., & Koyama, T. (1976). Filamentous growth of carbon through benzene decomposition. *Journal of Crystal Growth, 32*(3), 335–349. https://doi.org/10.1016/0022-0248(76)90115-9.

Ouyang, M., Huang, J. L., Cheung, C. L., & Lieber, C. M. (2001). Energy gaps in "metallic" single-walled carbon nanotubes. *Science, 292*(5517), 702–705. https://doi.org/10.1126/science.1058853.

Paul, S., Lee, Y. S., Choi, J. A., Kang, Y. C., & Kim, D. W. (2010). Synthesis and electrochemical characterization of polypyrrole/multi-walled carbon nanotube composite electrodes for supercapacitor applications. *Bulletin of the Korean Chemical Society, 31*(5), 1228–1232. https://doi.org/10.5012/bkcs.2010.31.5.1228.

Peng, S., & Cho, K. (2000). Chemical control of nanotube electronics. *Nanotechnology, 11*(2), 57–60. https://doi.org/10.1088/0957-4484/11/2/303.

Peng, S., & Cho, K. (2003). Ab initio study of doped carbon nanotube sensors. *Nano Letters, 3*(4), 513–517. https://doi.org/10.1021/nl034064u.

Piloto, C., Mirri, F., Bengio, E. A., Notarianni, M., Gupta, B., Shafiei, M., et al. (2016). Room temperature gas sensing properties of ultrathin carbon nanotube films by surfactant-free dip coating. *Sensors and Actuators, B: Chemical, 227*, 128–134. https://doi.org/10.1016/j.snb.2015.12.051.

Pitoniak, E., Wu, C. Y., Mazyck, D. W., Powers, K. W., & Sigmund, W. (2005). Adsorption enhancement mechanisms of silica-titania nanocomposites for elemental mercury vapor removal. *Environmental Science and Technology, 39*(5), 1269–1274. https://doi.org/10.1021/es049202b.

Qin, S., Qin, D., Ford, W. T., Resasco, D. E., & Herrera, J. E. (2004). Functionalization of single-walled carbon nanotubes with polystyrene via grafting to and grafting from methods. *Macromolecules, 37*(3), 752–757. https://doi.org/10.1021/ma035214q.

Ricciardella, F., Vollebregt, S., Polichetti, T., Miscuglio, M., Alfano, B., Miglietta, M. L., et al. (2017). Effects of graphene defects on gas sensing properties towards NO2 detection. *Nanoscale, 9*(18), 6085–6093. https://doi.org/10.1039/c7nr01120b.

Rigoni, F., Tognolini, S., Borghetti, P., Drera, G., Pagliara, S., Goldoni, A., et al. (2013). Enhancing the sensitivity of chemiresistor gas sensors based on pristine carbon nanotubes to detect low-ppb ammonia concentrations in the environment. *Analyst, 138*(24), 7392–7399. https://doi.org/10.1039/c3an01209c.

Roco, M. C. (2005). The emergence and policy implications of converging new technologies integrated from the nanoscale. *Journal of Nanoparticle Research, 7*(2–3), 129–143. https://doi.org/10.1007/s11051-005-3733-0.

Rushi, A. D., Gaikwad, S., Deshmukh, M., Patil, H., Bodkhe, G., & Shirsat, M. D. (2016). Functionalized carbon nanotubes: Facile development of gas sensor platform. In *Vol. 1728. AIP conference proceedings*American Institute of Physics Inc. https://doi.org/10.1063/1.4946215.

Sainato, M., Humayun, M. T., Gundel, L., Solomon, P., Stan, L., Divan, R., et al. (2016). Parts per million CH4 chemoresistor sensors based on multi wall carbon nanotubes/metal-oxide nanoparticles. In *2016 IEEE Sensors* (pp. 1–3). https://doi.org/10.1109/ICSENS.2016.7808856.

Santucci, S., Picozzi, S., Di Gregorio, F., Lozzi, L., Cantalini, C., Valentini, L., et al. (2003). NO2 and CO gas adsorption on carbon nanotubes: Experiment and theory. *Journal of Chemical Physics*, *119*(20), 10904–10910. https://doi.org/10.1063/1.1619948.

Sarkar, T., Srinives, S., Sarkar, S., Haddon, R. C., & Mulchandani, A. (2014). Single-walled carbon nanotube-poly(porphyrin) hybrid for volatile organic compounds detection. *Journal of Physical Chemistry C*, *118*(3), 1602–1610. https://doi.org/10.1021/jp409851m.

Schedin, F., Geim, A. K., Morozov, S. V., Hill, E. W., Blake, P., Katsnelson, M. I., et al. (2007). Detection of individual gas molecules adsorbed on graphene. *Nature Materials*, *6*(9), 652–655. https://doi.org/10.1038/nmat1967.

Seekaew, Y., Lokavee, S., Phokharatkul, D., Wisitsoraat, A., Kerdcharoen, T., & Wongchoosuk, C. (2014). Low-cost and flexible printed graphene-PEDOT:PSS gas sensor for ammonia detection. *Organic Electronics*, *15*(11), 2971–2981. https://doi.org/10.1016/j.orgel.2014.08.044.

Seekaew, Y., Phokharatkul, D., Wisitsoraat, A., & Wongchoosuk, C. (2017). Highly sensitive and selective room-temperature NO2 gas sensor based on bilayer transferred chemical vapor deposited graphene. *Applied Surface Science*, *404*, 357–363. https://doi.org/10.1016/j.apsusc.2017.01.286.

Seredych, M., & Bandosz, T. J. (2012). Manganese oxide and graphite oxide/MnO2 composites as reactive adsorbents of ammonia at ambient conditions. *Microporous and Mesoporous Materials*, *150*(1), 55–63. https://doi.org/10.1016/j.micromeso.2011.09.010.

Shao, L., Mu, C., Du, H., Czech, Z., Du, H., & Bai, Y. (2011). Covalent marriage of multi-walled carbon nanotubes (MWNTs) and β-cyclodextrin (β-CD) by silicon coupling reagents. *Applied Surface Science*, *258*(5), 1682–1688. https://doi.org/10.1016/j.apsusc.2011.09.129.

Sharma, A. K., Kumar, P., Saini, R., Bedi, R. K., & Mahajan, A. (2016). Kinetic response study in chemiresistive gas sensor based on carbon nanotube surface functionalized with substituted phthalocyanines. In *Vol. 1728. AIP conference proceedings* American Institute of Physics Inc. https://doi.org/10.1063/1.4946544.

Shen, F., Wang, D., Liu, R., Pei, X., Zhang, T., & Jin, J. (2013). Edge-tailored graphene oxide nanosheet-based field effect transistors for fast and reversible electronic detection of sulfur dioxide. *Nanoscale*, *5*(2), 537–540. https://doi.org/10.1039/c2nr32752j.

Shi, D., Wei, L., Wang, J., Zhao, J., Chen, C., Xu, D., et al. (2013). Solid organic acid tetrafluorohydroquinone functionalized single-walled carbon nanotube chemiresistive sensors for highly sensitive and selective formaldehyde detection. *Sensors and Actuators, B: Chemical*, *177*, 370–375. https://doi.org/10.1016/j.snb.2012.11.022.

Smazna, D., Rodrigues, J., Shree, S., Postica, V., Neubüser, G., Martins, A. F., et al. (2018). Buckminsterfullerene hybridized zinc oxide tetrapods: Defects and charge transfer induced optical and electrical response. *Nanoscale*, *10*(21), 10050–10062. https://doi.org/10.1039/c8nr01504j.

Smith, A. D., Elgammal, K., Fan, X., Lemme, M. C., Delin, A., Råsander, M., et al. (2017). Graphene-based CO2 sensing and its cross-sensitivity with humidity. *RSC Advances*, 22329–22339. https://doi.org/10.1039/C7RA02821K.

Spokas, K. A., Novak, J. M., Stewart, C. E., Cantrell, K. B., Uchimiya, M., DuSaire, M. G., et al. (2011). Qualitative analysis of volatile organic compounds on biochar. *Chemosphere*, *85*(5), 869–882. https://doi.org/10.1016/j.chemosphere.2011.06.108.

Srisitthiratkul, C., Pongsorrarith, V., & Intasanta, N. (2011). The potential use of nanosilver-decorated titanium dioxide nanofibers for toxin decomposition with antimicrobial and self-cleaning properties. *Applied Surface Science*, *257*(21), 8850–8856. https://doi.org/10.1016/j.apsusc.2011.04.083.

Srivastava, S., Sharma, S. S., Kumar, S., Agrawal, S., Singh, M., & Vijay, Y. K. (2009). Characterization of gas sensing behavior of multi walled carbon nanotube polyaniline composite films. *International Journal of Hydrogen Energy*, *34*(19), 8444–8450. https://doi.org/10.1016/j.ijhydene.2009.08.017.

Sun, Y. P., Fu, K., Lin, Y., & Huang, W. (2002). Functionalized carbon nanotubes: Properties and applications. *Accounts of Chemical Research*, *35*(12), 1096–1104. https://doi.org/10.1021/ar010160v.

Tan, C., Cao, X., Wu, X. J., He, Q., Yang, J., Zhang, X., et al. (2017). Recent advances in ultrathin two-dimensional nanomaterials. *Chemical Reviews*, *117*(9), 6225–6331. https://doi.org/10.1021/acs.chemrev.6b00558.

Tang, R., Shi, Y., Hou, Z., & Wei, L. (2017). Carbon nanotube-based chemiresistive sensors. *Sensors*, *17*(4), 882. https://doi.org/10.3390/s17040882.

Tompkins, D. T., & Anderson, M. A. (2001). *Evaluation of photocatalytic air cleaning capability: A literature review & engineering analysis: Final report*. Atlanta, GA: ASHRAE. American Society of Heating, Refrigerating and Air-Conditioning Engineers.

Valentini, L., Armentano, I., Kenny, J. M., Cantalini, C., Lozzi, L., & Santucci, S. (2003). Sensors for sub-ppm NO2 gas detection based on carbon nanotube thin films. *Applied Physics Letters*, *82*(6), 961–963. https://doi.org/10.1063/1.1545166.

Varghese, S. S., Lonkar, S., Singh, K. K., Swaminathan, S., & Abdala, A. (2015). Recent advances in graphene based gas sensors. *Sensors and Actuators, B: Chemical*, *218*, 160–183. https://doi.org/10.1016/j.snb.2015.04.062.

Villalpando-Páez, F., Romero, A. H., Muñoz-Sandoval, E., Martínez, L. M., Terrones, H., & Terrones, M. (2004). Fabrication of vapor and gas sensors using films of aligned CNx nanotubes. *Chemical Physics Letters*, *386*(1–3), 137–143. https://doi.org/10.1016/j.cplett.2004.01.052.

Wang, T., Huang, D., Yang, Z., Xu, S., He, G., Li, X., et al. (2016). A review on graphene-based gas/vapor sensors with unique properties and potential applications. *Nano-Micro Letters*, *8*(2), 95–119. https://doi.org/10.1007/s40820-015-0073-1.

Wang, X., Li, Q., Xie, J., Jin, Z., Wang, J., Li, Y., et al. (2009). Fabrication of ultralong and electrically uniform single-walled carbon nanotubes on clean substrates. *Nano Letters*, *9*(9), 3137–3141. https://doi.org/10.1021/nl901260b.

Wang, F., Wang, Y. T., Yu, H., Chen, J. X., Gao, B. B., & Lang, J. P. (2016). One unique 1D silver(I)-bromide-thiol coordination polymer used for highly efficient chemiresistive sensing of ammonia and amines in water. *Inorganic Chemistry*, *55*(18), 9417–9423. https://doi.org/10.1021/acs.inorgchem.6b01688.

Wang, Y., Zhang, L., Hu, N., Wang, Y., Zhang, Y., Zhou, Z., et al. (2014). Ammonia gas sensors based on chemically reduced graphene oxide sheets self-assembled on Au electrodes. *Nanoscale Research Letters*, *9*(1), 1–12. https://doi.org/10.1186/1556-276X-9-251.

Wang, S. G., Zhang, Q., Yang, D. J., Sellin, P. J., & Zhong, G. F. (2004). Multi-walled carbon nanotube-based gas sensors for NH3 detection. *Diamond and Related Materials*, *13*(4–8), 1327–1332. https://doi.org/10.1016/j.diamond.2003.11.070.

Wang, R., Zhang, D., Zhang, Y., & Liu, C. (2006). Boron-doped carbon nanotubes serving as a novel chemical sensor for formaldehyde. *Journal of Physical Chemistry B*, *110*(37), 18267–18271. https://doi.org/10.1021/jp061766+.

Wu, Z., Chen, X., Zhu, S., Zhou, Z., Yao, Y., Quan, W., et al. (2013a). Enhanced sensitivity of ammonia sensor using graphene/polyaniline nanocomposite. *Sensors and Actuators, B: Chemical, 178*, 485–493. https://doi.org/10.1016/j.snb.2013.01.014.

Wu, Z., Chen, X., Zhu, S., Zhou, Z., Yao, Y., Quan, W., et al. (2013b). Room temperature methane sensor based on graphene nanosheets/polyaniline nanocomposite thin film. *IEEE Sensors Journal, 13*(2), 777–782. https://doi.org/10.1109/JSEN.2012.2227597.

Xie, H., Sheng, C., Chen, X., Wang, X., Li, Z., & Zhou, J. (2012). Multi-wall carbon nanotube gas sensors modified with amino-group to detect low concentration of formaldehyde. *Sensors and Actuators, B: Chemical, 168*, 34–38. https://doi.org/10.1016/j.snb.2011.12.112.

Xu, J., & Wang, L. (2019). Carbon nanomaterials. In *Vol. 29, Issue 4. Nano-inspired biosensors for protein assay with clinical applications* (pp. 3–38). Elsevier BV. https://doi.org/10.1016/b978-0-12-815053-5.00001-5.

Yaghi, O. M., Li, G., & Li, H. (1995). Selective binding and removal of guests in a microporous metal–organic framework. *Nature, 378*(6558), 703–706. https://doi.org/10.1038/378703a0.

Yang, C., Miao, G., Pi, Y., Xia, Q., Wu, J., Li, Z., et al. (2019). Abatement of various types of VOCs by adsorption/catalytic oxidation: A review. *Chemical Engineering Journal, 370*, 1128–1153. https://doi.org/10.1016/j.cej.2019.03.232.

Yang, W., Ratinac, K. R., Ringer, S. R., Thordarson, P., Gooding, J. J., & Braet, F. (2010). Carbon nanomaterials in biosensors: Should you use nanotubes or graphene. *Angewandte Chemie, International Edition, 49*(12), 2114–2138. https://doi.org/10.1002/anie.200903463.

Yavari, F., Castillo, E., Gullapalli, H., Ajayan, P. M., & Koratkar, N. (2012). High sensitivity detection of NO2 and NH3 in air using chemical vapor deposition grown graphene. *Applied Physics Letters, 100*(20). https://doi.org/10.1063/1.4720074.

Ye, Z., Jiang, Y., Tai, H., & Yuan, Z. (2014). The investigation of reduced graphene oxide/P3HT composite films for ammonia detection. *Integrated Ferroelectrics, 154*(1), 73–81. https://doi.org/10.1080/10584587.2014.904148.

Yeung, C. S., Liu, L. V., & Wang, Y. A. (2008). Adsorption of small gas molecules onto Pt-doped single-walled carbon nanotubes. *Journal of Physical Chemistry C, 112*(19), 7401–7411. https://doi.org/10.1021/jp0753981.

Yoo, S., Li, X., Wu, Y., Liu, W., Wang, X., & Yi, W. (2014). Ammonia gas detection by tannic acid functionalized and reduced graphene oxide at room temperature. *Journal of Nanomaterials, 2014*. https://doi.org/10.1155/2014/497384.

Yoon, H. J., Jun, D. H., Yang, J. H., Zhou, Z., Yang, S. S., & Cheng, M. M.-C. (2011). Carbon dioxide gas sensor using a graphene sheet. *Sensors and Actuators B: Chemical*, 310–313. https://doi.org/10.1016/j.snb.2011.03.035.

Yoon, B., Liu, S. F., & Swager, T. M. (2016). Surface-anchored poly(4-vinylpyridine)-single-walled carbon nanotube-metal composites for gas detection. *Chemistry of Materials, 28*(16), 5916–5924. https://doi.org/10.1021/acs.chemmater.6b02453.

Zettl, A. (2000). Extreme oxygen sensitivity of electronic properties of carbon nanotubes. *Science, 287*(5459), 1801–1804. https://doi.org/10.1126/science.287.5459.1801.

Zhang, X., Gao, B., Creamer, A. E., Cao, C., & Li, Y. (2017). Adsorption of VOCs onto engineered carbon materials: A review. *Journal of Hazardous Materials, 338*, 102–123. https://doi.org/10.1016/j.jhazmat.2017.05.013.

Zhang, Y., Janyasupab, Liu, C.-W., Lin, P.-Y., Wang, K.-W., Xu, et al. (2012). Improvement of amperometric biosensor performance for H2O2 detection based on bimetallic PtM (M = Ru, Au, and Ir) nanoparticles. *International Journal of Electrochemistry, 2012*. https://doi.org/10.1155/2012/410846, 410846.

Zhang, T., Nix, M. B., Yoo, B. Y., Deshusses, M. A., & Myung, N. V. (2006). Electrochemically functionalized single-walled carbon nanotube gas sensor. *Electroanalysis, 18*(12), 1153–1158. https://doi.org/10.1002/elan.200603527.

Zhang, X., Yang, B., Dai, Z., & Luo, C. (2012). Odpowiedź na obecność H2S lub SO2 czujników SWCNT modyfikowanych hydroxylem luv carboxylem. *Przeglad Elektrotechniczny, 88*(7 B), 311–314.

Zhang, Y. M., Zhang, J., Chen, J. L., Zhu, Z. Q., & Liu, Q. J. (2014). Improvement of response to formaldehyde at Ag-LaFeO3 based gas sensors through incorporation of SWCNTs. *Sensors and Actuators, B: Chemical, 195*, 509–514. https://doi.org/10.1016/j.snb.2014.01.031.

Zhao, J., Buldum, A., Han, J., & Lu, J. P. (2002). Gas molecule adsorption in carbon nanotubes and nanotube bundles. *Nanotechnology, 13*(2), 195–200. https://doi.org/10.1088/0957-4484/13/2/312.

Zhao, Q., Buongiorno Nardelli, M., Lu, W., & Bernholc, J. (2005). Carbon nanotube – metal cluster composites: A new road to chemical sensors? *Nano Letters, 5*(5), 847–851. https://doi.org/10.1021/nl050167w.

Zhao, Q., Li, X., Zhou, Q., Wang, D., & Xu, H. (2019). Nanomaterials developed for removing air pollutants. In *Advanced nanomaterials for pollutant sensing and environmental catalysis* (pp. 203–247). Elsevier. https://doi.org/10.1016/B978-0-12-814796-2.00006-X.

Zhao, H., & Vance, G. F. (1998). Sorption of trichloroethylene by organo-clays in the presence of humic substances. *Water Research, 32*(12), 3710–3716. https://doi.org/10.1016/S0043-1354(98)00172-9.

Zhao, W., Yang, C., Zou, D., Sun, Z., & Ji, G. (2017). Possibility of gas sensor based on C20 molecular devices. *Physics Letters, Section A: General, Atomic and Solid State Physics, 381*(21), 1825–1830. https://doi.org/10.1016/j.physleta.2017.03.038.

Zhou, L., Shen, F., Tian, X., Wang, D., Zhang, T., & Chen, W. (2013). Stable Cu2O nanocrystals grown on functionalized graphene sheets and room temperature H2S gas sensing with ultrahigh sensitivity. *Nanoscale, 5*(4), 1564–1569. https://doi.org/10.1039/c2nr33164k.

Zhou, H., Zhang, C., Li, H., & Du, Z. (2011). Fabrication of silica nanoparticles on the surface of functionalized multi-walled carbon nanotubes. *Carbon, 49*(1), 126–132. https://doi.org/10.1016/j.carbon.2010.08.051.

Chapter 9

Carbon nanomaterials-based sensors for water treatment

Aniela Pop[a], Sorina Motoc[b], and Florica Manea[a]
[a]*Politehnica University of Timisoara, Timisoara, Romania,* [b]*"Coriolan Dragulescu" Institute of Chemistry Timisoara of Romanian Academy, Timisoara, Romania*

List of Abbreviations

Ag-CNF-EP	silver-modified carbon nanofibers-epoxy electrode
Ag-CNT-EP	silver-modified carbon nanotubes-epoxy electrode
AgZ-CNF-EP	Ag-doped zeolite-modified carbon nanofibers-epoxy electrode
AgZ-CNT-EP	Ag-doped zeolite-modified carbon nanotubes-epoxy electrode
AOPs	advanced oxidation processes
ASSWV	anodic stripping square-wave voltammetry
ASV	anodic stripping voltammetry
AuNpCµMF	carbon fiber ultramicroelectrodes modified with gold nanoparticles
BDD	boron-doped diamond
BDD/Ag	silver-modified boron-doped diamond electrode
BDD/GR	graphene-modified boron-doped diamond electrode
BDD/GR/Ag	silver-/graphene-modified boron-doped diamond electrode
BIA	batch injection analysis
CA	chronoamperometry
CNF-EP	carbon nanofiber-epoxy
CNF	carbon nanofibers
CNT-EP	carbon nanotubes-epoxy
CNT-GP	carbon nanotubes-modified commercial graphite electrode
CNT-P	carbon nanotubes-paste electrode
CNT	carbon nanotubes
CV	cyclic voltammetry
DCF	diclofenac
DOX	doxorubicin
DPV	differential-pulsed voltammetry
FULL	fullerene
FULL-CNF-P	fullerenes-carbon nanofibers-paste
GCE	Glassy carbon electrode
GCE/AuNPs-rGO-MWCNTs	gold nanoparticles, reduced graphene oxide, and multiwalled carbon nanotubes-modified glassy carbon electrodes
GCE/CNTAu	carbon nanotube membrane decorated by gold nanoparticles-modified glassy carbon electrodes
GCE/CVD-CNT	chemical vapor deposition CNT-modified glassy carbon electrodes
GCE/GNs	graphene nanoribbons-modified glassy carbon electrodes
GCE/GO-COOH	carboxyl-functionalized graphene oxide-modified electrode
GCE/GO-IL	ionic-liquid-functionalized graphene oxide-modified electrode
4GCE/Gr–nPt	graphene-platinum nanocomposite sensor
GCE/MWCNT/CoPc	multiwalled carbon nanotube/cobalt phthalocyanine-modified glassy carbon electrode
GCE/MWCNTs-DHP	multiwalled carbon nanotubes dihexadecylhydrogenphosphate film-modified glassy carbon electrodes
GCE/NCQDs-GO	N-doped carbon quantum dots-graphene oxide hybrid-modified glassy carbon electrode
GCE/ZnSe-QDs-CNTs	ZnSe quantum dots decorated multiwall carbon nanotubes nanocomposite/glassy carbon electrode
GR	graphene

GR-CNT-P	graphene-carbon nanotubes-paste
GR-QD-CNT-P	graphene quantum dots-carbon nanotubes-paste
GR-QD-P	graphene quantum dots-paste
GR-QD	graphene quantum dots
HKUST-CNF-EP	HKUST-1 metal–organic framework-carbon nanofiber-epoxy-composite electrode
IBP	ibuprofen
LOD	the lowest limit of detection
MA	modulation amplitude
MWCNT/Cu/CPE	multiwalled carbon nanotube/copper nanoparticle composite paste electrodes
NC-EP	nanostructured carbon-epoxy composite electrodes
NC-MCC	nanostructured carbon-modified commercial carbon electrodes
NC-P	nanostructured carbon paste electrodes
NPX	naproxen
PA	pulsed amperometry
PBS	phosphate-buffered saline
PCP	pentachlorophenol
ppm	parts per million
SA	salicylic acid
SCBPEs	solid paste electrodes prepared using a nanostructured carbon black
SCE	saturated calomel electrode
SEM	scanning electron microscopy
SP	step potential
SWV	square-wave voltammetry
TC	tetracycline
ZXCPE	zeolite X-modified carbon paste electrode

Introduction

Water is the core element for life, and its quality is essential for the community and the public health. The rapid population growth, urbanization, and intensive agricultural and industrial activities have been consuming unsustainably water and/or negatively affected its quality generating the water pollution. Several contaminants or pollutants generated naturally or by the anthropic activity can be present in water bodies (e.g., rivers, lakes as surface waters, groundwaters, wastewaters), and their removal/destruction is required to assure the good quality of every water body, in accordance with Water Framework Directive (Directive 2000/60/EC of the European Parliament and of the Council Establishing a Framework for the Community Action in the Field of Water Policy, 2000).

In general, water treatment technologies are proposed in direct relation to the raw water quality and its specific use. Conventional and alternative advanced water treatment technologies are generally designed based on the wide spectrum of the unitary processes subjected to the contaminant/pollutant type and concentration, such as settling, coagulation/flocculation, filtering, ion-exchange, reverse osmosis, oxidation, adsorption, ultrafiltration, membrane, and advanced oxidation processes (AOPs) (Ahamad, Madnav, Singh, Kumar, & Singh, 2020; Nasrollahzadeh, Sajjadi, Iravani, & Varma, 2021).

Taking into account the ranges of the water contaminants/pollutants concentrations, the priority and emerging pollutants are more difficult to be quantitatively determined because their concentrations at trace levels necessitate the advanced analytical methods in comparison with the conventional ones, for which the concentrations are at parts per million (ppm) levels. However, the on-site or in-situ quantitative determination for all types of contaminants/pollutants or quality parameters for water including the assessment of the water treatment technology performance requires to develop alternative analytical methods.

The voltammetric and amperometric methods and sensors can be regarded as alternative analytical methods with great potential for the monitoring of the water quality in general, and for the real-time control of the water treatment processes and technology trains, taking into consideration the whole cycle of water use (Fig. 9.1).

In this chapter, presented are the comparative applications of the voltammetric and the amperometric sensors in detection of several priority pollutants (e.g., pentachlorophenol, carbaryl, and paraquat from pesticides class, and As (III) and Pb (II) from heavy metals class) and of a large spectrum of pharmaceuticals from emerging pollutants classes, besides some conventional water pollutants (S^{2-}, NO_2^-, NH_4^+).

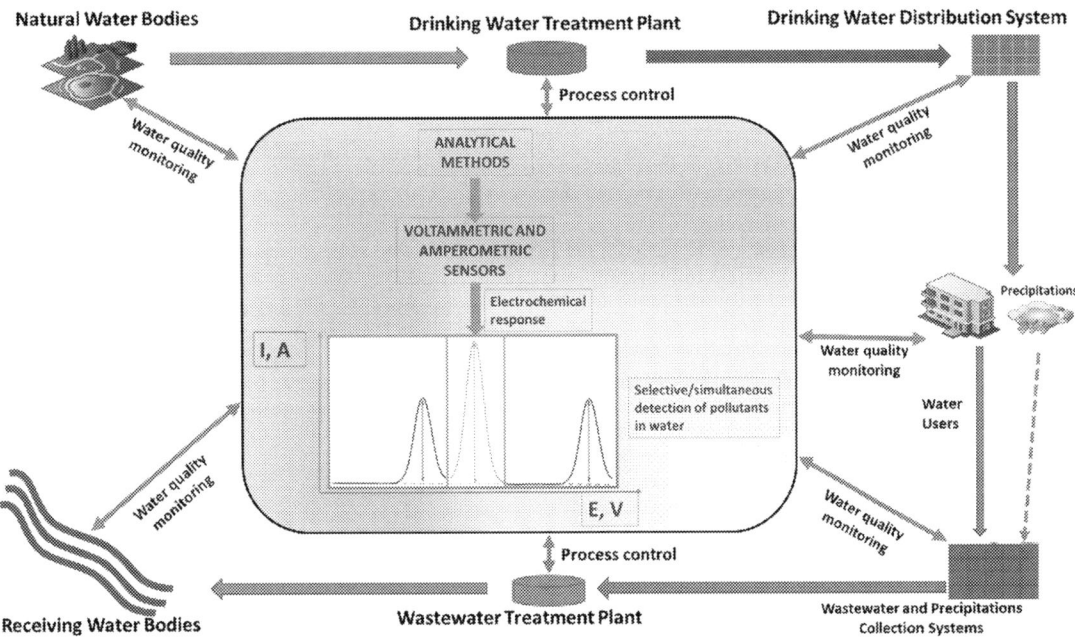

FIG. 9.1 Voltammetric and amperometric sensors in water use cycle.

Voltammetric and amperometric sensors for water treatment

Electrochemical sensors have been widely developed as an inexpensive and simple method to sensitively detect a large variety of analytes ranged from the environmental to food and medical fields. The amperometric and voltammetric sensors represent the important category of the electrochemical sensors used for the determination of a wide spectrum of the water pollutants: organics, heavy metals, and inorganics.

One of the most important characteristics responsible for the performance of the electrochemical sensors is given by the so-called working-electrode material. The working electrode should provide high signal-to-noise characteristics and stable and reproducible responses. However, its composition and selection depend on two factors: its electrochemical behavior in the presence of the water pollutant/contaminant considered the target analyte, and its background current within the potential range required for the measurement. Also, potential window, electrical conductivity, chemical and mechanical stability, surface reproducibility, and the lifetime represent the important characteristics of the working electrode.

Carbon-based electrodes are commonly used as working electrodes for the development of the electrochemical sensors because of their low cost, good electron transfer kinetics, good chemical stability, and biocompatibility. Traditional carbon-based sensors include glassy carbon electrodes, carbon fibers, and pyrrolytic graphite (Wang, 2006).

Recently, carbon nanomaterials have been incorporated into sensors development. The feature sizes of the nanomaterials are 1–100 nm, and they are advantageous because of their large surface-to-volume ratio and specific surface area. In addition, carbon nanomaterials have enhanced interfacial adsorption properties, better electrocatalytic activity, high biocompatibility, and faster electron transfer kinetics compared to many traditional electrochemical sensor materials (Yang, Denno, Pyakurel, & Venton, 2015).

Several designs of carbon nanomaterials, including carbon nanotubes (CNTs), carbon nanofibers (CNFs), fullerene (FULL), graphene (GR), and graphene quantum dots (GR-QDs), as different configurations of the electrode types (carbon nanomaterials-modified commercial carbon electrodes including silver nanoparticles, paste, epoxy-based composite) as sensing materials are discussed further for the development of very sensitive amperometric and voltammetric detection methods. The role of carbon nanomaterials in the enhancement of the electron transfer kinetics based on the electrocatalytic activity, and/or the improved mass transfer related to the morphostructural properties, is discussed subjected to the high performance of the electrochemical sensor in water quality monitoring, and the process control in water treatment technology.

Also, the importance of the electrochemical techniques used in detection, such as cyclic voltammetry (CV), differential-pulsed voltammetry (DPV), square-wave voltammetry (SWV), chronoamperometry (CA), and pulsed amperometry (PA), is presented, considering the operating parameters of each technique linked to the mechanistic aspects of the detection process. Their ability for selective and/or simultaneous detection of several pollutants in water is presented. In addition, the existing limitations and future challenges are discussed in the context of practical application in water quality monitoring and water treatment technology.

Voltammetric and amperometric techniques used for detection applications

Cyclic voltammetry (CV) technique represents the first electrochemical technique used for the study of each type of electrochemical experiment, including the detection applications subjected to the type of the electrode material. The electrochemical behavior of the electrode material must be studied before to propose its testing in the electrochemical detection application by the selection of appropriate supporting electrolyte in relation to the practical applications. The information about the oxidation or reduction process at the electrode-supporting electrolyte interface is provided through CV. The potential window, the capacitive component, and the electrochemical response toward the target analyte based on its oxidation or reduction process can be assessed, and these voltammetric parameters are very useful for further application of easy, conventional, and/or advanced electrochemical techniques in the voltammetric and amperometric detection. Some aspects about the mechanism elucidation related to the electrode process control and adsorption aspects, which are very useful in the detection application, are provided by this technique operated at different scan rates. In Fig. 1a-c, given is one example for electrochemical behavior of sulfide in simulated seawater on carbon nanofibers-epoxy (CNF-EP) composite electrode (Fig. 9.2). Two-step oxidation process of sulfide to sulfur involving the formation of Sx^2-polysulphide intermediates, in accordance with the reactions (9.1) and (9.2), was proposed for CNF-EP electrode (Ardelean, Manea, Vaszilcsin, & Pode, 2014).

$$xS^{2-} \to S_x^{2-} + (2x-2)e^- \tag{9.1}$$

$$S_x^{2-} \to xS + 2e^- \tag{9.2}$$

A good linearity for anodic peak current vs. square root of scan rates indicates a diffused-controlled oxidation process, which confirms that the CNF-EP electrode should voltammetrically and amperometrically detect the sulfide in water. The lack of the cathodic peak corresponding to the anodic one and random shifting of the peak potential value vs the logarithm of the scan rate showed the irreversibility character of the sulfide oxidation process.

The processes of the target analyte oxidation or reduction onto the electrode surface are often very complex, involving multisteps corresponding to different processes of sorption and/or further subproduct oxidation or reduction that should affect negatively or positively the electrode surface via its fouling or in-situ activation depending on the process type. The information about the interference in the detection application is very easily provided by the CV technique. The possibility of the selective or the simultaneous detection of multicomponents is checked by CV technique, as well.

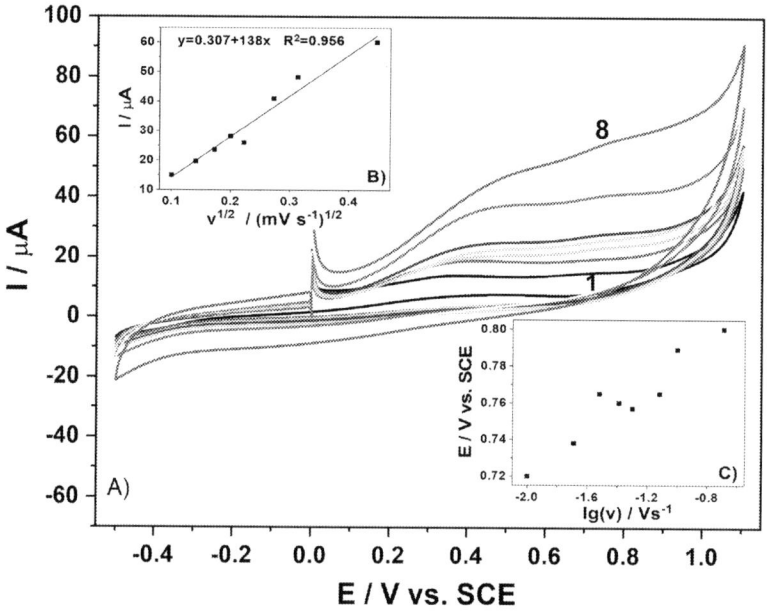

FIG. 9.2 (A) Cyclic voltammograms recorded on CNF-EP electrode in simulated seawater supporting electrolyte (curve 1) in the presence of 0.5 mM sulfide within a potential range from −0.5 to +1.1 V/SCE, at different scan rates: 0.01, 0.02, 0.03, 0.04, 0.05, 0.075, 0.1, 0.2 V s^{-1} (curves 1–8); insets: the calibration plots of the anodic peak current vs. square root of the scan rate (B); dependence of the peak potential E vs. log (v) (C).

Taking into account the possibility of nanostructured dispersion within insolated matrix based on a certain template or without specific template considering customized properties of the materials (e.g., self-organizing), CV should give information related to the electrochemical behaviors of the electrode in terms of ordered (array) or randomized micro/nanoelectrode ensembles characterized by the heterogeneous higher electroactivity by simple study of the scan rate influence (Ballarin et al., 2003; Cięciwa, Wüthrich, & Comninellis, 2006; Feeney & Kounaves, 2000; Manea et al., 2009; Ramírez-García, Alegret, Céspedes, & Forster, 2002; Simm et al., 2005; Štulík, Amatore, Holub, Mareček, & Kutner, 2000).

In the electrochemical detection applications, microelectrode arrays exhibit the advantage of individual microelectrodes to provide low charging current that favors the electroanalytical performance related to the sensitivity and the lowest limit of detection (LOD). Moreover, their use averts the disadvantages of single microelectrode subjected to the high susceptibility to the electrochemical noise due to low current output (Cięciwa et al., 2006; Feeney & Kounaves, 2000; Manea et al., 2009).

Sometimes, the voltammetric responses as ratio between Faradaic and background currents are more enhanced for carbon-based composite electrodes, which are quite similar to that reported for "edge effect" (Ramírez-García et al., 2002), manifested more evidently for carbon micro/nanoarrays and in certain situations even for random assembly. However, carbon-based composites obtained without a specific template are truly randomized ensembles, in which the different shapes and sizes of nano/microcarbon are separated and nonuniformly distributed over a wide range of insulating matrices. In specific situation, the random ensembles of micro/nanoelectrodes can exhibit array behavior, for which the electrode process is controlled by mass transport characterized by spherical diffusion in comparison with linear diffusion characteristics for macroelectrode. Another very important requirement is subjected to the insulating matrix and the minimum content of nanostructured carbon to prevent current leakage that resulted in the distortion of the cyclic voltammetry (Fig. 9.3A), and to assure the electrical conductivity and reduced ohmic drop (Fig. 9.3B).

CV technique is very easily to be applied in the voltammetric detection. Fig. 9.4 presents the results of the application of CV at the CNF-EP electrode for sulfide detection in simulated seawater. A linear dependence of the anodic peak currents vs the sulfide concentrations was determined for the concentrations ranged from 0.2mM to 1mM sulfide at the potential value of +0.8V/SCE, at which the second oxidation stage occurred. Also, the linear dependence of the anodic peak current at the potential value of +0.4V/SCE corresponding to the first stage of the sulfide oxidation at increasing concentration was achieved but with the lower sensitivity.

A voltammetric technique that exhibits the great utility in the voltammetric detection is differential-pulsed voltammetry (DPV), by which the background current given by the capacitive component is lowered and the useful signal is improved subjected to the operating conditions, that is, the modulation amplitude (MA) and the step potential (SP), which influenced the scan rate.

The height of the peak current from differential-pulsed voltammogram is directly proportional to the concentration of analyte (Wang, 2006).

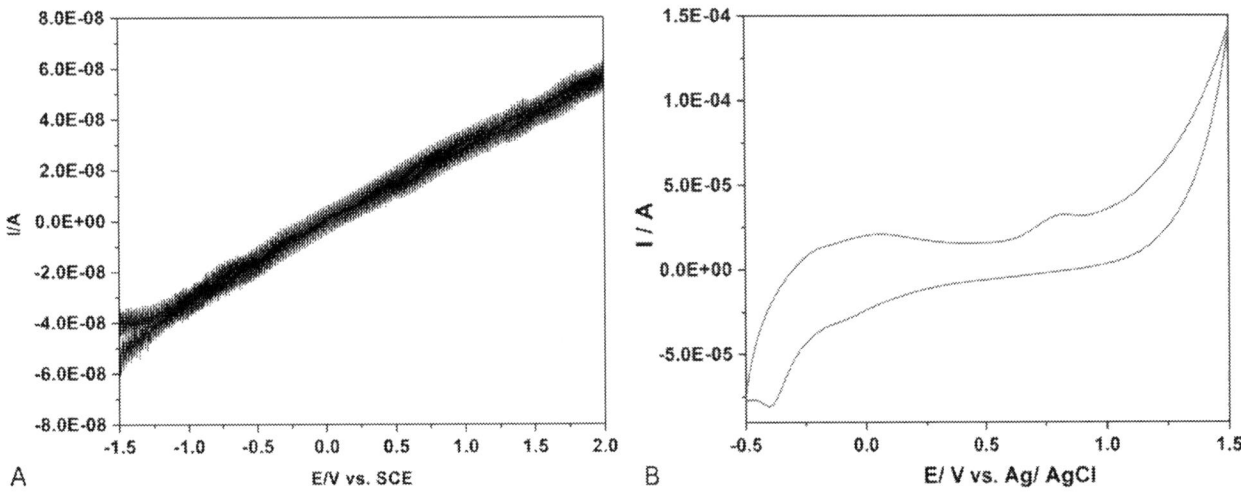

FIG. 9.3 Comparative CVs recorded at the scan rate of $+0.05\,\text{V}\,\text{s}^{-1}$ in 0.1M Na2SO4 supporting electrolyte at the paste electrodes: (A) graphene quantum dots paste (GR-QD-P) and (B) graphene quantum dots-carbon nanotubes paste (GR-QD-CNT-P).

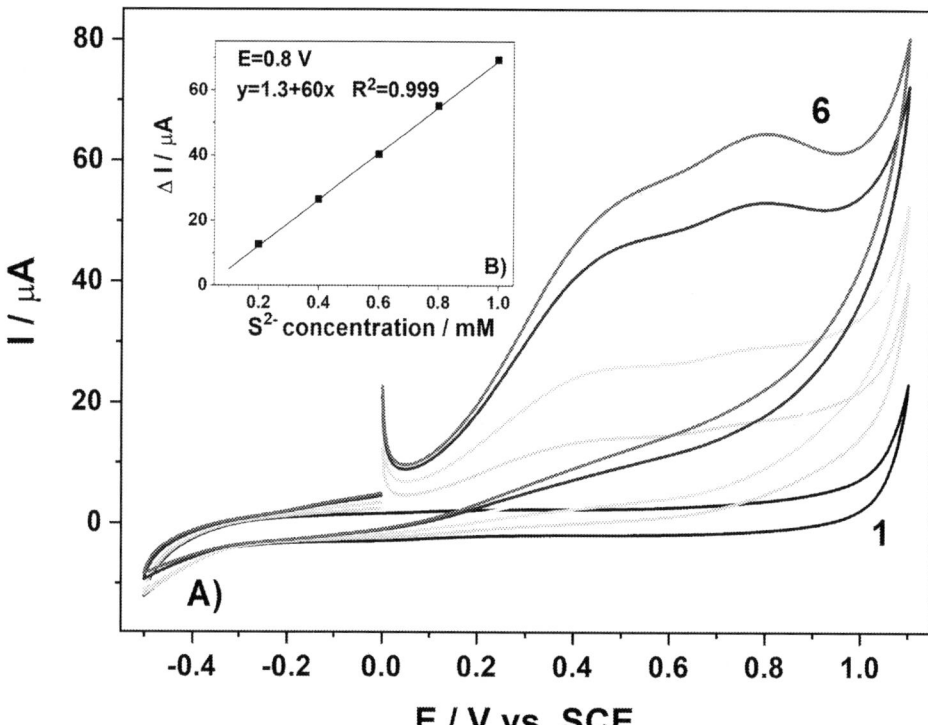

FIG. 9.4 (A) Cyclic voltammograms recorded on CNF-EP electrode in simulated seawater (curve 1) supporting electrolyte in the presence of 0.2–1 mM sulfide (curves 2–6) at a potential scan rate of $0.05\,V\,s^{-1}$ within a potential range from -0.5 to $+1.1$ V/SCE; inset: the calibration plots of the currents recorded at $E=+0.8$ V/SCE vs. sulfide concentration (B).

$$i_p = \frac{nFAD^{1/2}C}{\sqrt{\pi t_m}} \left(\frac{1-\delta}{1+\delta}\right) \tag{9.3}$$

where n is the number of electrons involved in the redox process, F is Faraday's constant, A is the active area of the working electrode (cm^2), D is the diffusion coefficient (cm^2 s^{-1}), C is the bulk concentration of the electroactive species (mol cm^{-3}), t_m modulation time (s), $\sigma = \exp[(nF/RT)(MA/2)]$, R is the universal gas constant, T is the temperature, and MA is the modulation amplitude.

The optimization of the operating conditions is required for DPV-based detection application. Fig. 9.5 shows comparatively the DPV response for sulfide detection in simulated seawater for the MA of 0.05, 0.1, and 0.2 V at the SP of 0.01 V, and it can be seen that increasing the MA enhanced the analytical signal for sulfide detection. Also, the first oxidation step is more evidenced at higher MA, while the second step decreased, which informed about a faster kinetics of the first oxidation step in comparison with the second one. For both oxidation stages and the detection values, the sensitivities are higher at increasing MA, whose value is limited by the detection response stability.

Another advanced pulsed technique, square-wave voltammetry (SWV), was tested in sulfide detection. Considering the above-presented optimized operating conditions established for DPV, the influence of the frequency as one of the main parameters of this technique should be considered in order to further enhance the electroanalytical performance. The SWV signals recorded at the frequency of 10, 25, and 50 Hz are presented comparatively in Fig. 9.6.

In comparison with DPV, better sensitivities for sulfide detection are achieved for SWV applying at each frequency.

Considering the amperometric detection, the current-time dependence is monitored in chronoamperometry (CA), which is considered as the easiest electrochemical detection technique and as the most suitable for the practical detection applications. This technique is working at the constant potential value for one and more potential levels, which are selected based on the existing well-established essential reference subjected to the oxidation/reduction processes, provided by the cyclic voltammograms. Taking into account the diffusion-controlled oxidation/reduction processes, which means that mass transport is controlled by diffusion, the current-time dependence reflects the change in the concentration gradient near to the electrode surface. For the mass transport characterized by linear diffusion, this technique can be used for diffusion coefficient calculation, and as a consequence, for the specific electroactive surface area determination using Cottrell equation (Wang, 2006).

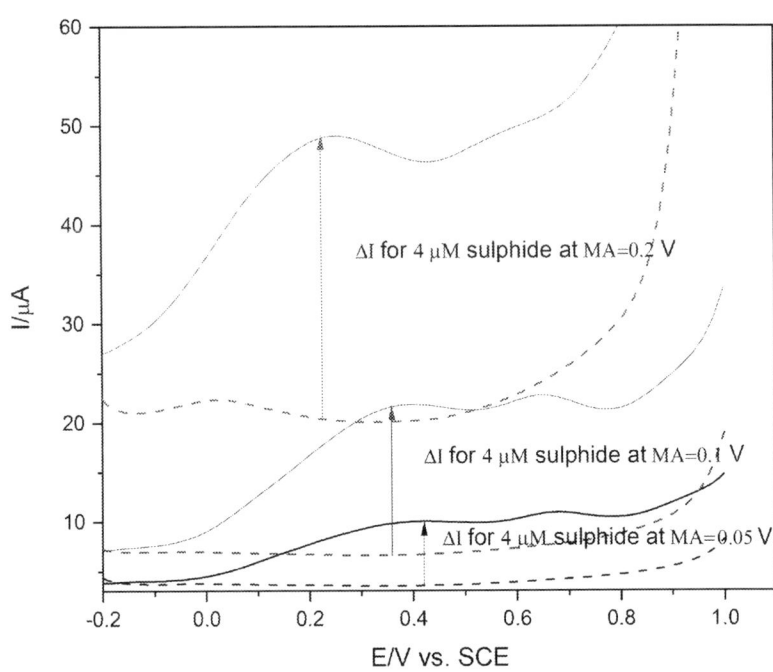

FIG. 9.5 DPVs recorded on CNF-EP composite electrode in simulated seawater (dash curves) and in the presence of 0.4 mM sulfide at the SP of 0.1 V and different MA (0.05 V—black curve, 0.1 V—blue curve (gray in print version); 0.2 V—green curve (light gray in print version)) with the scan rate of 0.05 V s^{-1}, within a potential range from −0.2 to +1 V/SCE.

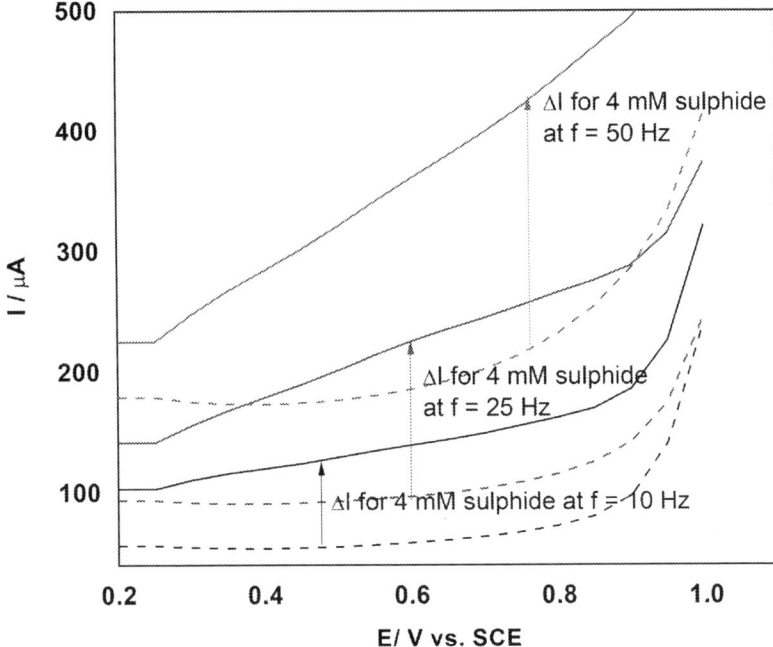

FIG. 9.6 SWVs recorded on CNF-EP composite electrode in simulated seawater (dash curves) and in the presence of 0.4 mM sulfide at the SP of 0.1 V, MA of 0.2 V, and various frequency (10 Hz—black curve; 25 Hz—blue curve (gray in print version); 50 Hz—green curve (light gray in print version)), and the scan rate of 0.05 V s^{-1}, in a potential range from +0.2 to +1 V/SCE.

$$i = \frac{nFACD^{1/2}}{\pi^{1/2}t^{1/2}} = k\,t^{1/2} \tag{9.4}$$

For the practical detection application, batch injection analysis (BIA) is the most common analytical method, where the amperometric response obtained for successive and continuous addition of a certain concentration of the analyte is recorded by chronoamperometry. Fig. 9.7 shows the amperometric response (BIA) recorded at CNF-EP electrode in 0.1 M Na$_2$SO$_4$ supporting electrolyte by adding 0.1 mM sulfide at one applied potential of +0.9 V vs. SCE. The amperometric response depends linearly on the S^{2-} concentration within the range of 0.1 to 0.6 mM (inset of Fig. 9.7).

FIG. 9.7 (A) Amperometric response-batch system analysis (BIA) recorded at applied potential of +0.9 V vs. SCE at the CNF-EP electrode for the successive and continuous addition of 0.1 mM S2-; inset: calibration plots of useful signal vs. S2- concentrations (B).

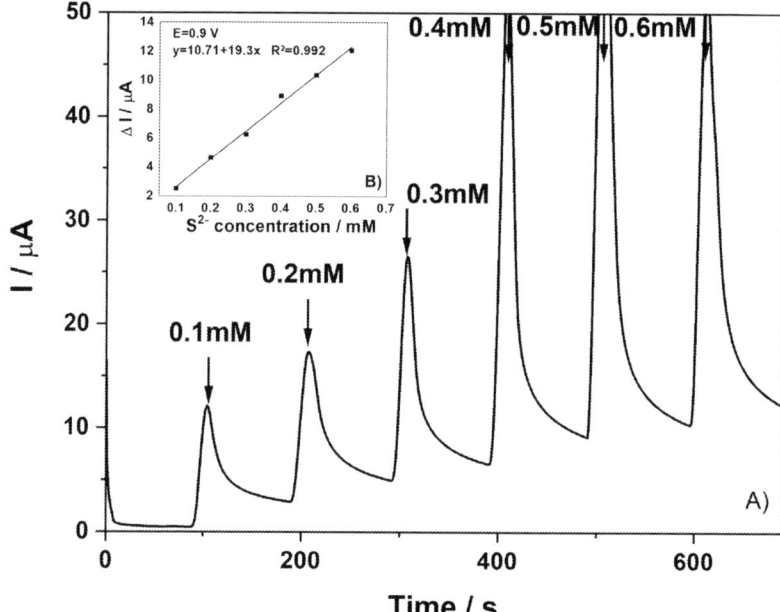

Pulsed amperometry (PA) is an advanced amperometric technique considered as an alternative for chronoamperometry to overcome the disadvantage of the electrode fouling during the detection application with the consequence of the amperometric response loss. The main principle of the PA detection consists of maintaining the measurement potential for a short time (measurement pulse) after the applications at different times of the electrode cleaning potential and conditioning potential pulses (Charoenraks, Chuanuwatanakul, Honda, Yamaguchi, & Chailapakul, 2005). Specially, this technique is more sensitive for the detection of organic molecules because the electrode fouling is avoided by in-situ cleaning and reactivating the electrode surface during the electroanalytical detection (Bebeselea, Manea, Burtica, Nagy, & Nagy, 2010).

In addition, PA detection can be used successfully in the simultaneous/selective amperometric detection of the analytes by selecting the appropriate number of the potential levels and the required time for each pulse. Fig. 9.7 gives one example of PA detection adapted for the simultaneous detection of sulfide and nitrite in 0.1 M Na_2SO_4 containing water with boron-doped diamond (BDD) electrode. The criteria for selecting the potential levels and values are different in comparison with the role of the PA detection subjected to the electrode surface avoiding, and it is based on the detection potentials of both sulfides and nitrites priorly established by CV technique besides the potential value for reactivation of the electrode surface (Baciu, Ardelean, Pop, Pode, & Manea, 2015). The pulses were applied continuously using the following scheme:

Pulse 1 operated at −0.5 V/SCE for a duration of 30 ms, where the electrode surface is activated by occurring the reduction process;
Pulse 2 operated at +0.85 V/SCE for a duration of 500 ms, for sulfide oxidation;
Pulse 3 operated at +1.25 V/SCE for a duration of 30 ms for nitrite oxidation.

The time for each pulse is optimized taking into account the electrochemical oxidation kinetics of each analyte to allow their simultaneous detection at two different potential values without each to other interference (Fig. 9.8).

Configurations of the working electrode for voltammetric and amperometric sensors

Three types of the nanostructured carbon-based electrodes are considered to develop voltammetric and amperometric sensors for water quality determination: carbon nanomaterials-modified commercial carbon electrodes, carbon paste, and carbon-based composites, including the electrodeposition of the silver/copper nanoparticles. Several carbon nanomaterials are presented considering their electrochemical and electrical characteristics, morphostructural and mechanical properties, which are derived from the carbon nanostructure type, structural conformation, and hybridization, subjected to the electroanalytical performance enhancement. A brief presentation of the main properties of the carbon nanostructures (Table 9.1) (Campuzano, Yáñez-Sedeño, & Pingarrón, 2019; Crevillen, Escarpa, & García, 2019; Eatemadi et al., 2014; Fan, Zhang, & Liu, 2017; Fan et al., 2020; Hiremath & Bhat, 2017; Kour et al., 2020; Krishnan, Singh, Singh, Meyyappan, & Nalwa, 2019; Mostofizadeh,

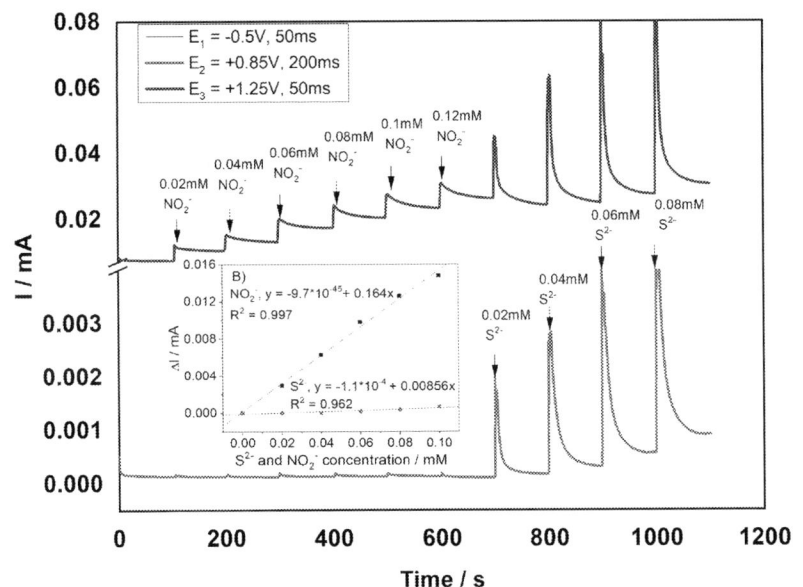

FIG. 9.8 PAs recorded at BDD electrode in 0.1 M Na2SO4 supporting electrolyte for the successive and continuous addition of 0.02 mM NO2- within the concentration range of 0.02 to 0.12 mM NO2-, followed by the successive and continuous addition of 0.02 mM S2- within the concentration range of 0.02 to 0.12 mM S2-, recorded at $E = +0.85$ V vs. SCE (red, light gray in print version), +1.25 vs. SCE (blue, gray in print version), and −0.1 V vs. SCE (not shown); inset: calibration plots of amperometric responses versus NO2- and S2-concentrations.

TABLE 9.1 Morphostructural, electrical, electrochemical, thermal, and mechanical properties of carbon nanostructures.

Carbon nanomaterials	Structural conformation/hybridization	Specific surface ($m^2 g^{-1}$)	Electrical properties ($S\ cm^{-1}$)	Electrochemical properties ($cm^2\ V^{-1}\ s^{-1}$)	Thermal conductivity ($W\ m^{-1}\ K^{-1}$)	Mechanical properties
Graphene	2D/sp^2	2630–2675 (Crevillen, Escarpa, & García, 2019; Kour et al., 2020; Raccichini et al., 2015)	2×10^3 (Nasir et al., 2018; Rauti et al., 2019)	10,000–15,000 (at room temperature) (Fan et al., 2017; Mostofizadeh et al., 2011)	4000–5000 (Eatemadi et al., 2014; Krishnan et al., 2019; Nasir et al., 2018; Rauti et al., 2019)	Flexible, elastic, very hard Young's modulus (GPa) of ≈ 1100
Carbon nanotubes	1D/sp^2	15–1375 (Crevillen, Escarpa, & García, 2019; Kour et al., 2020; Nasir et al., 2018; Osman et al., 2020)	10^3–10^7 (Rauti et al., 2019)	10,000 (Mostofizadeh et al., 2011)	2000–5000 (Crevillen, Escarpa, & García, 2019; Eatemadi et al., 2014; Nasir et al., 2018; Rauti et al., 2019)	Flexible, elastic, good strength Young's modulus (GPa) between 1000 and 1280
Fullerene	0D/sp^2	80–90 (Nasir et al., 2018; Raccichini et al., 2015)	10^{-5}–10^{-10} (Fan et al., 2020; Nasir et al., 2018; Raccichini et al., 2015; Rauti et al., 2019)	0.02 (at room temperature in vacuum) (Fan et al., 2020)	0.4 (Nasir et al., 2018; Notarianni et al., 2016; Raccichini et al., 2015)	Elastic good strength Young's modulus (GPa) between 0 and 14
Carbon nanofiber	1D/sp^2	300–900 (Wang et al., 2018)	1×10^4 (Hiremath & Bhat, 2017)	–	≈ 2000 (Hiremath & Bhat, 2017)	Good strength Young's modulus (GPa) of 240
Graphene quantum dots	0D/sp^2	Higher than graphene (Campuzano et al., 2019; Tajik et al., 2020; Tian et al., 2018)	Higher than graphene (Campuzano et al., 2019; Tajik et al., 2020; Tian et al., 2018)	Higher than graphene (Campuzano et al., 2019; Tajik et al., 2020; Tian et al., 2018)	Higher than graphene (Campuzano et al., 2019; Tajik et al., 2020; Tian et al., 2018)	–

Li, Song, & Huang, 2011; Nasir, Hussein, Zainal, & Yusof, 2018; Notarianni, Liu, Vernon, & Motta, 2016; Osman, Farrell, Al-Muhtaseb, Harrison, & Rooney, 2020; Raccichini, Varzi, Passerini, & Scrosati, 2015; Rauti, Musto, Bosi, Prato, & Ballerini, 2019; Tajik et al., 2020; Tian, Tang, Teng, & Lau, 2018; Wang, Zhang, Zhang, Liu, & Ma, 2018) allows to compare them for their selection in direct relation with the customized requirement, for example, electroactive surface area, specific surface area, electrocatalytic activity, higher electron transfer kinetics, and so on (Fig. 9.9).

Three types of the carbon-based electrodes with the surface geometry of the disc are considered for the development of the voltammetric and amperometric sensors for water quality determination: nanostructured carbon-modified commercial carbon (NC-MCC) electrodes, nanostructured carbon paste (NC-P) electrodes, and nanostructured carbon-epoxy composite (NC-EP) electrodes, including the electrodeposition of the silver/copper nanoparticles.

The morphostructural properties of the nanostructured carbon-based electrode surface depend on the preparation method subjected to the electrode type. The obtaining methods for the above-presented types of the electrodes are presented in Fig. 9.10, which presents the main stages of the preparation method and the SEM image of the electrode surface based on the carbon nanotubes.

The preparation method exhibits the advantages and disadvantages each to other subjected to the electrode surface morphology, more porous being achieved for paste. The commercial carbon electrode modified with the nanostructured carbon by simple immersion is the fastest and easiest way to test the influence of the nanostructured carbon and the longer lifetime, and the stability of the electrode surface is assured by the epoxy-based composite electrode.

Nanostructured carbon-modified commercial carbon (NC-MCC) electrodes

The simple immersion of the commercial carbon electrode into the nanostructured carbon suspension represents the fastest way of electrode modification, and it is used in general, as preliminary test considering the further development of modified sensor to improve its characteristics responsible for enhanced electroanalytical performance for the determination of the water quality. The modification of BDD electrode with the graphene (BDD/GR) allowed simultaneous detection of carbaryl and paraquat (Fig. 9.11), from pesticide class that belongs to the priority pollutants category (Pop, Manea, Flueras, & Schoonman, 2017). The further modification through the electrochemical deposition with silver nanoparticles led to improve the lowest limit of detection by about 65 times for carbaryl and by about 8 times for paraquat (Pop, Lung, Orha, & Manea, 2018).

Also, the nanostructured carbon can be used as the intermediate substrate for further modifying commercial substrate with the metallic nanoparticles, even if the commercial substrate modified only with nanostructured carbon did not improve the electroanalytical parameter for a certain pollutant. This behavior was found by Negrea et al. (2021) for the detection of

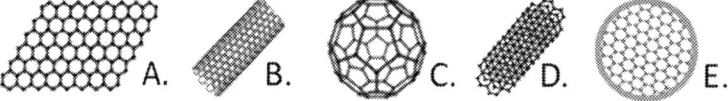

FIG. 9.9 Schematic structures of (A) graphene, (B) carbon nanotubes, (C) fullerene, (D) carbon nanofiber, and (E) graphene quantum dots.

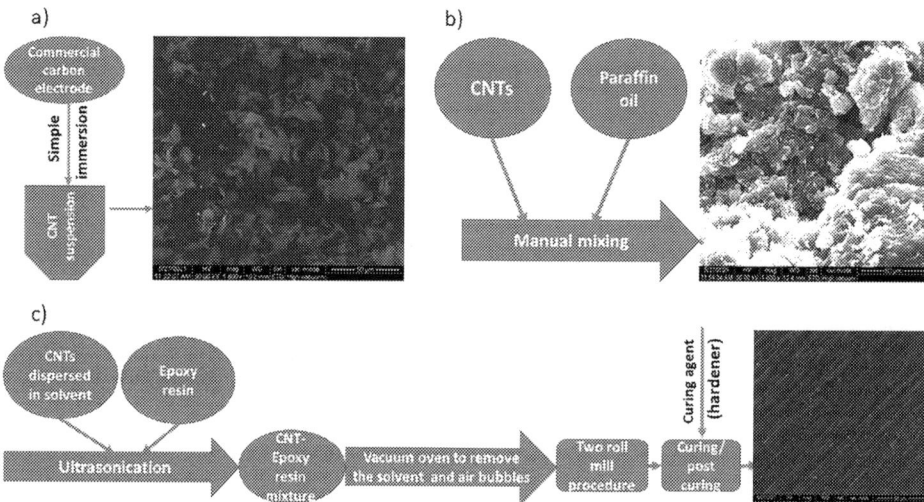

FIG. 9.10 Scanning electron microscopy (SEM) image-integrated scheme for electrode preparation: (A) carbon nanotubes-modified commercial graphite (CNT-GP) electrode; (B) carbon nanotubes-paste (CNT—P) electrode; (C) carbon nanotubes-epoxy (CNT-EP).

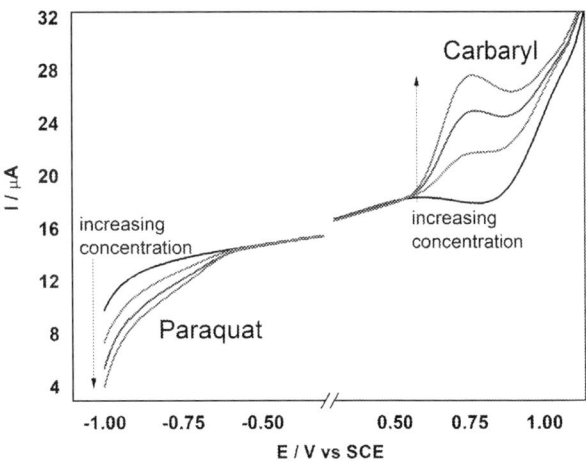

FIG. 9.11 DPVs recorded on BDD/GR electrode in acetic acid/sodium acetate buffer supporting electrolyte and in the presence of 2 μM CR, 4 μM CR, 6 μM CR, and 0.4 μM PQ, 0.8 μM PQ, 1.2 μM PQ, respectively; SP of 6 mV, MA of 800 mV; potential range: −1.0 to +1.75 V/SCE.

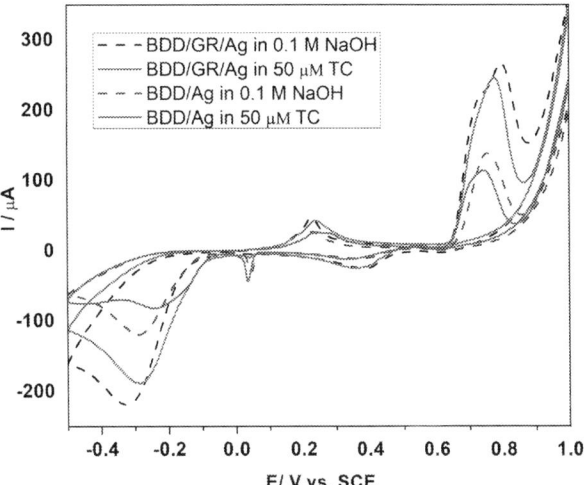

FIG. 9.12 CVs recorded at the scan rate of 0.05 V s^{-1} in 0.1 M NaOH supporting electrolyte and 50 μM tetracycline (TC) on GR-BDD and Ag-GR-BDD electrodes.

the tetracycline from water-emerging pollutants category. Using graphene for modification of commercial boron-doped diamond (BDD/GR) electrode did not improve the TC detection performance, but it influenced greatly the silver electrodeposition, resulting silver-modified graphene deposited onto BDD electrode characterized by enhanced electrocatalytic activity toward tetracycline oxidation and reduction, reflected in the enhanced detection performance.

The electrochemical deposition of graphene onto the BDD electrode surface influenced the background current through capacitive component. Also, the cathodic peak current characteristics to Ag_2O electroreduction increased with TC concentration for both Ag-based BDD electrodes but only the presence of graphene assured a linear increasing with TC concentrations, while for BDD//Ag electrode, a nonlinear increasing was noticed (Fig. 9.12). This should be explained by the surface-controlled processes that affected the Ag_2O-involved electroreduction process. The comparative sensitivities obtained for both electrodes are shown in Table 9.2, and it can be noticed that the graphene integration enhanced the sensitivity for TC detection.

A large spectrum of the contaminants/pollutants from different categories (priority micropollutants, emerging pollutants, and conventional pollutants/contaminants) have been reported to be detected in water using commercial carbon electrode modified with nanostructured carbon and metallic particles. The lowest limit of detections reported for several types of NC-MCC is presented in Table 9.3 (Buffa & Mandler, 2019; Fard, Alipour, & Ali Sabzi, 2016; Feng et al., 2015;

TABLE 9.2 Comparative sensitivity for TC detection at BDD/Ag and BDD/GR/Ag electrodes.

Electrode type	Potential branch	Detection potential (V/SCE)	Sensitivity ($\mu A \mu M^{-1} cm^{-2}$)
BDD/Ag	Anodic	+0.285	3.00
		+0.550	1.03
BDD/GR/Ag	Anodic	+0.296	3.17
		+0.550	2.86
	Cathodic	+0.039	2.16

TABLE 9.3 The lowest limits of detection reported for the detection of pollutants from water at NC-MCC electrodes.

Pollutant class	Pollutant	Electrode configuration	Detection method	LOD (ppb)	Detection matrix	Reference
Priority pollutants	PCP	GCE/ZnSe-QDs-CNTs	DPV	0.53	PBS (pH 4)	Feng et al. (2015)
	Carbaryl	GCE/CNT/CoPc	SWV	1.1	River water	Moraes et al. (2009)
	Carbaryl	GCE/GO-IL	SWV	4.0	Britton-Robinson (pH 5.0)	Liu et al. (2015)
	Carbaryl	BDD/GR	DPV	14.1	Acetic acid/sodium acetate buffer	Pop et al. (2017)
	Carbaryl	BDD/GR/Ag	DPV	0.24	pH 5.6 acetate buffer	Pop et al. (2018)
	Paraquat	GCE/MWCNTs-DHP	SWV	2.6	$NaNO_3$ (pH 6)	Garcia et al. (2013)
	Paraquat	BDD/GR	CV	2.6	Acetic acid/sodium acetate buffer	Pop et al. (2017)
	Paraquat	BDD/GR/Ag	DPV	0.32	Acetate buffer (pH 5.6)	Pop et al. (2018)
	Arsenic	GCE/CNTAu	LSV	0.75	0.1 M HCl	Buffa and Mandler (2019)
	Arsenic	GCE/Gr–nPt	SWV	0.001	1 M HCl	Kempegowda et al. (2014)
	Lead	GCE/NCQDs-GO	ASV	0.67	0.2 M acetate buffer (pH 5.0)	Li et al. (2018)
Emerging pollutants	Diclofenac	Pencil graphite electrode/CNT	DPV	0.017	Acetate buffer solution (pH 4)	Fard et al. (2016)
	Diclofenac	GCE/GO-COOH	LSV	26.7	PBS (pH 7)	Karuppiah et al. (2015)
	Sulfide	GCE/CVD-CNT	CA	1.6	0.05 M phosphate buffer	Lawrence et al. (2004)
Conventional water pollutants	Nitrite	GCE/GNs	CA	9.2	Britton-Robinson (pH 3)	Mehmeti et al. (2016)
	Nitrite	GCE/AuNPs-rGO-MWCNTs	CA	0.64	0.1 M phosphate buffer (pH 5.0)	Yu et al. (2019)

Garcia, Figueiredo-Filho, Oliveira, Fatibello-Filho, & Banks, 2013; Karuppiah, Cheemalapati, Chen, & Palanisamy, 2015; Kempegowda, Antony, & Malingappa, 2014; Lawrence, Deo, & Wang, 2004; Li, Liu, Shi, & You, 2018; Liu, Xiao, & Cui, 2015; Mehmeti, Stanković, Hajrizi, & Kalcher, 2016; Moraes, Mascaro, Machado, & Brett, 2009; Pop et al., 2018, 2017; Yu, Li, & Song, 2019).

Nanostructured carbon paste (NC-P) electrodes

The main characteristic of these electrode types refers to use of pasting liquids that are water-immiscible nonconducting organics, which should offer a mechanical and electrochemical stable composition of the electrode surface able to be used in sensing applications. The main advantage of the paste electrode is given by easily renewable and modified surface, low-cost and low background current contributions. Also, this composition is very easily to be further modified by simple mixing of carbon with the pasting liquid and modifying component without any template or matrix development. The most common pasting liquids used to develop electrochemical sensors include Nujol (mineral oil), paraffin oil, silicone grease, and bromonaphthalene (Wang, 2006). It is well known that the paste electrode performance depends on the pasting liquid content, which means decreasing the electron transfer rates, as well as the background current contributions with the increase in pasting liquid content. A disadvantage of carbon pastes is the possibility of the organic to be dissolved in solutions containing an appreciable fraction of organic solvent. However, for the aqueous solution, this problem does not appear, and the paste electrode should exhibit great potential for the development of the electrochemical sensors for the detection of different contaminants/pollutants in water. The paraffin oil is a common pasting liquid for carbon pastes in general and also for nanostructured carbon paste electrodes in special, and its consistency imposes the various contents of the nanostructured carbon related to its type. Examples of nanostructured carbon paste compositions are given in Table 9.4, and the weight ratios were established in direct relation with the mechanical stability and the electrochemical behavior of the electrode surface. For example, GR-QD-P electrode (GR-QD:paraffin oil of 1:1) without any nanostructured carbon integration did not show the electrochemical response in 0.1 M Na_2SO_4 supporting electrolyte (see Fig. 9.3A), probably due to very low electrical conductivity despite the high electrical conductivity reported as specific characteristic of GR-QD (Tajik et al., 2020).

The morphostructural, dimensionality, and hybridization type of nanostructured carbons are reflected in the electrochemical behaviors recorded with cyclic voltammetry (CV) (Fig. 9.13). CV is the first electrochemical technique used for electrode surface characterization, the mechanism elucidation for the electrode process, and also for the voltammetric detection, which are the important aspects that should be considered for the electrochemical sensors development. In comparison with carbon nanotubes, the higher specific surface of graphene modified the electrochemical behavior of the carbon paste electrode in terms of the higher background current due to the capacitive component and slight polarization effect in relation to the oxygen evolution (Fig. 9.13A). A significant change of the electrochemical behavior is given by the graphene quantum dots related to a great improvement of the background current and also, the depolarization effect that allows to shift the detection potential value to lower value and the interference mitigation. In addition, the anodic and cathodic peaks corresponding to the oxidation and reduction processes of the GR-QD are manifested within the potential range from -0.5 to $+1.5$ V vs Ag/AgCl, which should exhibit the electrocatalytic effect.

TABLE 9.4 Examples of nanostructured carbon paste electrodes.

Paste electrode type	Weight ratio					
	Paraffin oil (oil)	Carbon nanotubes (CNTs)	Carbon nanofibers (CNFs)	Fullerene (FULL)	Graphene (GR)	Graphene quantum dots (GR-QDs)
CNT-P	3	1	–	–	–	–
GR-CNT-P	3.5	1	–	–	1	–
GR-QD-CNT-P	2.5	1.75	–	–	–	0.75
FULL-CNF-P	1	–	2	1	–	–

–, No added.

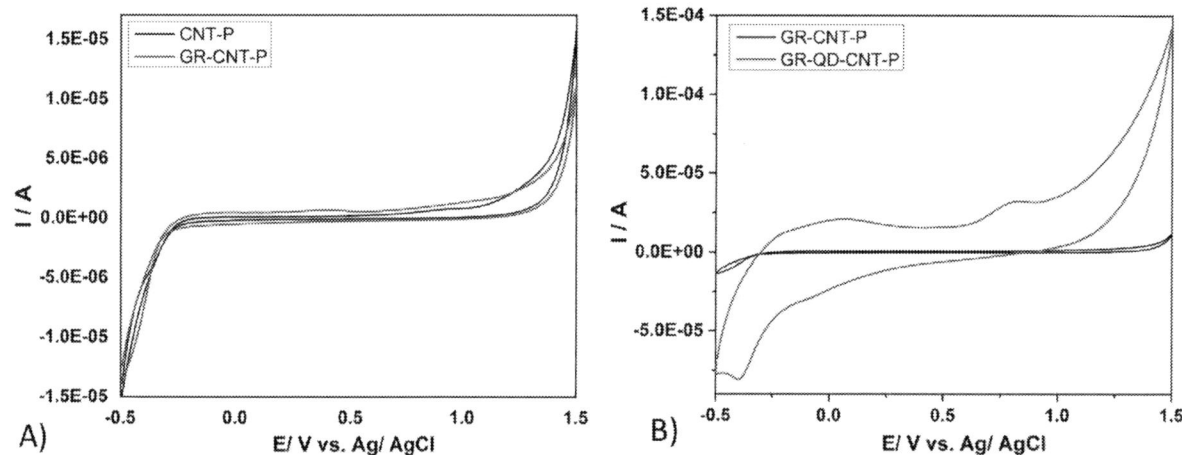

FIG. 9.13 Comparative CVs recorded at the scan rate of $0.05\,V\,s^{-1}$ in $0.1\,M\,Na_2SO_4$ supporting electrolyte at the paste electrodes: (A) CNT-P and GR-CNT-P; (B) GR-CNT-P and GR-QD-CNT-P.

Considering doxorubicin (DOX) from cytostatic class of pharmaceuticals from emerging pollutants of water, the comparative behaviors of CNT-P and GR-QD-CNT-P electrodes in the presence of 4 ppm DOX (Fig. 9.14) showed the major role of GR-QD into the enhancement of the electrochemical response to DOX presence related to the sensitivity and the detection potential value. The less positive potential value is desired taking into account the interference avoiding and the possibility to detect simultaneously more pharmaceutical pollutants without any electrode surface modification. To develop the voltammetric detection method using CV, the setup of the potential range is very important. Comparative anodic potential range with larger one involving the cathodic branch is shown in Fig. 9.15 for GR-QD-CNT-P. The electrode surface state is influenced by the potential range; thus, the cathodic branch allowed a pseudo-regeneration of the nanostructured carbon (GR-QD and CNT), while the oxidized state of the electrode surface is manifested within anodic potential range. The sorption process of DOX onto the electrode surface is favored by the cathodic branch, which enhanced the anodic peak of the nanostructured carbon oxidation.

The CVs recorded at various DOX concentrations within the anodic potential range allowed to develop a CV-based voltammetric detection of DOX with two detection potentials due to the reversibility of the DOX oxidation process (Fig. 9.16). The anodic peak corresponding to the DOX oxidation is shifted to more positive value within the anodic potential range but the oxidation process started earlier without no clear peak appearance. Also, for the backward scanning, the reduction process continued until 0 V vs Ag/AgCl, which means that the oxidation/reduction process of DOX occurred within 0 to +0.7 V vs Ag/AgCl. The linear calibrations of the anodic and respective cathodic peak currents vs DOX concentrations suggest the GR-QD-CNT-P electrode to be suitable for the voltammetric detection of DOX as anodic and

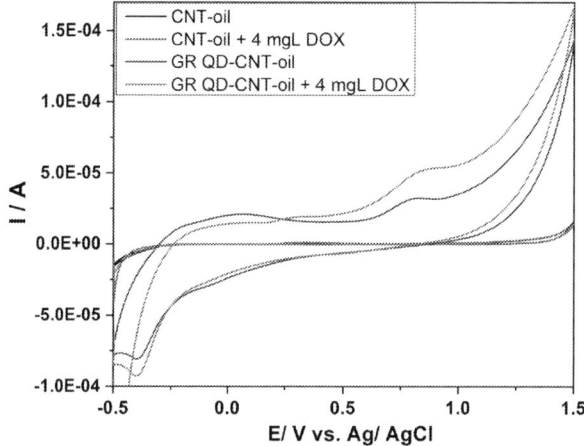

FIG. 9.14 Comparative CVs recorded with the scan rate of $0.05\,V\,s^{-1}$ in $0.1\,M\,Na_2SO_4$ supporting electrolyte and 4 ppm DOX on CNT-P and GR-CNT-P electrodes.

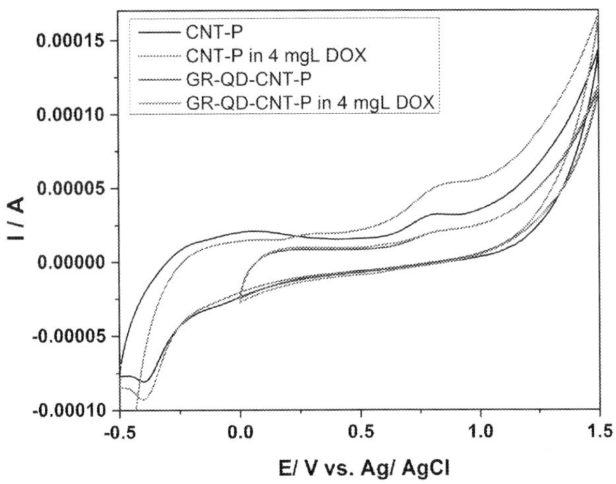

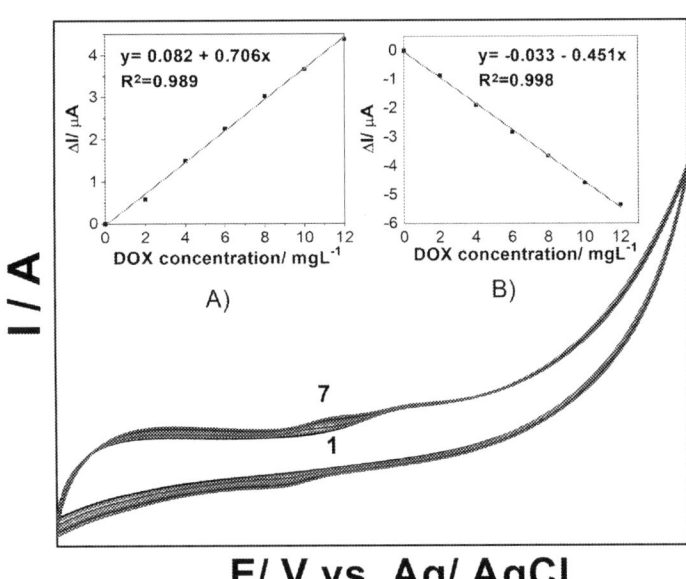

FIG. 9.15 Comparative CVs recorded with the scan rate of $0.05\,V\,s^{-1}$ in $0.1\,M$ Na_2SO_4 supporting electrolyte and 4 ppm DOX on GR-CNT-P paste electrodes within the potential range: $-0.5\,V$ to $+1.5\,V$ and 0 to $+1.5\,V$ vs Ag/AgCl.

FIG. 9.16 CVs recorded on GR-QD-P paste electrode in $0.1\,M$ Na_2SO_4 supporting electrolyte (curve 1) and in the presence of various DOX concentrations: $2-12\,mg\,L^{-1}$ (curves 2–7), potential scan rate: 0.05 V/s; potential range: 0 to $+1.5\,V$ vs Ag/AgCl. Insets: calibration plots of anodic current recorded at $+0.6\,V$ vs Ag/AgCl vs. DOX concentration (A) and of cathodic current recorded at $+0.53\,V$ vs Ag/AgCl vs. DOX concentration (B).

cathodic responses. The electroanalytical performance for cathodic detection is worse in comparison with anodic one, and sensitivity of 6.3 vs $10.0\,\mu A\,mg\,L^{-1}\,cm^{-2}$ and also the lowest limit of detection of 233 ppb in comparison with 65 ppb DOX were achieved.

Simultaneous detection of the contaminants/pollutants from water is a very interesting application of the voltammetric or amperometric sensors, which can be developed related to the electrode composition and besides the electrochemical techniques. Fullerene–carbon nanofiber–paraffin oil paste (FULL-CNF-P) electrode characterized by 25%, wt. CNF, 25% fullerene and 50%, wt. paraffin oil presents a great potential for the simultaneous detection of diclofenac (DCF), naproxen (NPX), ibuprofen (IBP), and anti-inflammatory pharmaceuticals from emerging pollutant class. CV application for the FULL-CNF-P electrode characterization in the presence of each pharmaceutical allowed their simultaneous detection (Figs. 9.17 and 9.18).

The electroanalytical performance of FULL-CNF-P paste electrode for simultaneous detection of DCF, NPX, and IBP using CV technique is presented in Table 9.5, and the best LOD was achieved for NPX that can be detected at two potential values, due to its oxidation occurred in two steps. Also, DCF that was oxidized easiest was detected at good limit of

140 PART | III Carbon nanomaterials-based sensor applications (environment, biomedical, foods, point of care, etc.)

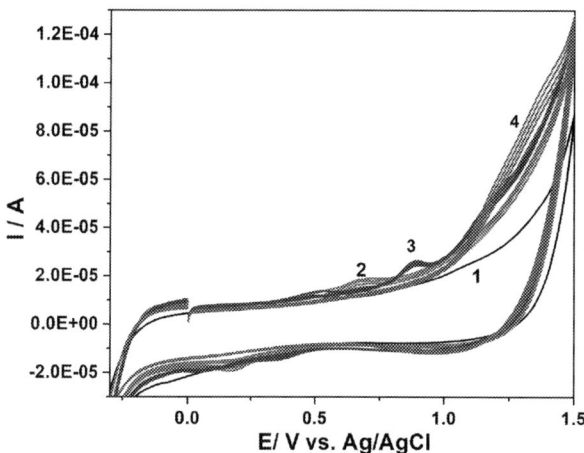

FIG. 9.17 Cyclic voltammogram recorded on FULL-CNF-P paste electrode in 0.1 M Na2SO4 supporting electrolyte (curve 1) and in the presence of: (1) DCF concentrations (3.4–17 µM), (2) NPX concentrations (6.8–17 µM), and (3) IBP concentrations (20.4–34 µM); potential scan rate: 0.05 V/s; potential range: −0.5 to +1.5 V vs Ag/AgCl.

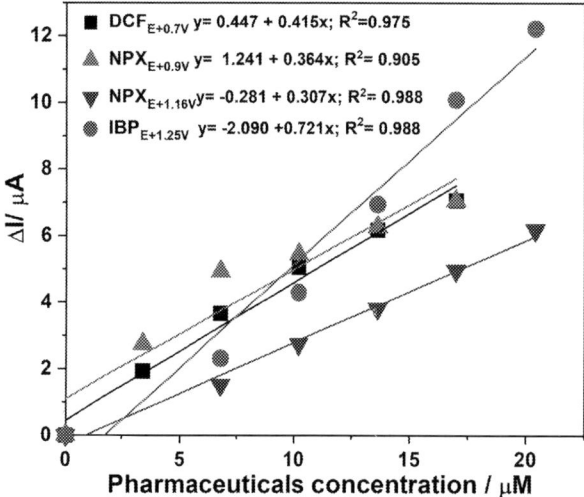

FIG. 9.18 Calibration plots of the current recorded at: +0.7 V vs. Ag/AgCl (for DCF); +0.9 V, +1.16 V vs. Ag/AgCl (for NPX), and +1.25 V vs. Ag/AgCl (for IBP) vs. pharmaceuticals concentration.

TABLE 9.5 Electrochemical performance for simultaneous detection of pharmaceuticals using FULL-CNF-P paste electrode.

Pharmaceutical	Detection potential (V vs Ag/AgCl)	Sensitivity ($\mu A \mu M^{-1} cm^{-2}$)	LOD (ppb)
DCF	+0.70	23.6	3.2
NPX	+0.90	20.7	1.8
	+1.16	17.4	3.4
IBP	+1.25	41.0	13.8

detection, while worse limit of detection was reached for IBP that was oxidized at more positive potential value. The best sensitivity was achieved for IBP detection, but because its detection occurred within oxygen evolution range, the relative standard deviation was affected negatively, was implicit, and has the lowest limit of detection.

In general, besides the electrode material, the electroanalytical performances are influenced by the electrochemical techniques, and one advanced electrochemical technique is square-wave voltammetry (SWV), which allowed improving the sensitivity and the lowest limit of detection. One example is given for GR-CNT-P applied for DCF detection using SWV operated at 50 mV SP, 200 mV MA, and 2 Hz frequency, which lowered the LOD to 5 ppb (Fig. 9.19).

Better improvement for DCF detection subjected to the LOD of 0.9 ppt has been reported by Motoc, Manea, Orha, and Pop (2019) with FULL-CNF-P electrode using SWV operated under at 2 mV SP, 10 mV MA, and 25 Hz frequency. It must be underlined that the optimized operating conditions for each advanced electrochemical technique are linked to the working electrode composition. A few examples of the detection results for different pollutants from water at nanostructured carbon paste electrodes (Malha, Mandli, Ourari, & Amine, 2013; Motoc et al., 2019; Salih et al., 2017; Valentini, Biagiotti, Lete, Palleschi, & Wang, 2007) are given in Table 9.6.

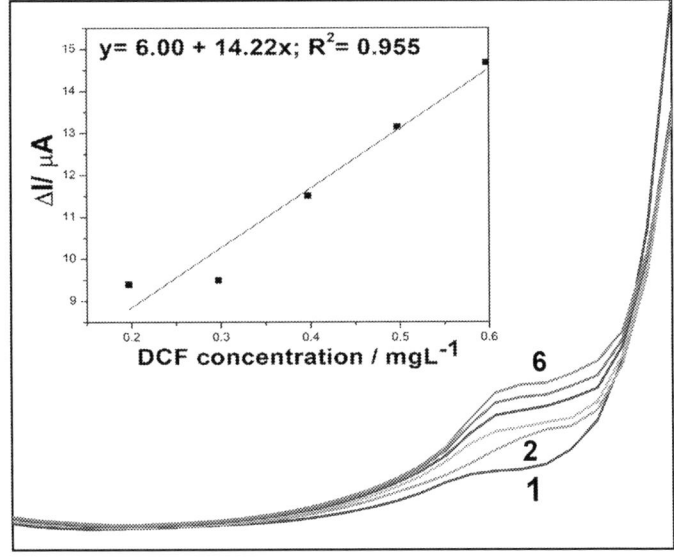

FIG. 9.19 SWVs recorded on GR-CNT-P paste electrode in 0.1 M Na2SO4 supporting electrolyte (curve 1) and in the presence of various DCF concentrations: curves 2–6: 0.197–0.597 mg L^{-1}; SP of 50 mV, MA of 200 mV; frequency of 2 Hz, and a scan rate of 100 mV s^{-1}, potential range: −0.5 to +1.5 V/Ag/AgCl.

TABLE 9.6 The lowest limits of detection (LODs) reported for the detection of pollutants from water at NC-P electrodes.

Pollutant class	Pollutant	Electrode configuration	Detection method	LOD (ppb)	Detection matrix	Reference
Priority pollutants	Carbaryl	ZXCPE	DPV	60.4	Acetate buffer (pH 4.3)	Salih et al. (2017)
Emerging pollutants	Diclofenac	FULL-CNF-P	SWV	0.0009	0.1 M Na$_2$SO$_4$	Motoc et al. (2019)
Conventional water pollutants	Ammonium	MWCNT/Cu/CPE	DPV	59.5	50 mM carbonate buffer (pH 10.0)	Valentini et al. (2007)
	Nitrite	SCBPEs	CA	0.23	0.1 M acetic acid	Malha et al. (2013)

Nanostructured carbon-epoxy composite (NC-EP) electrodes, including the electrodeposition of the silver/copper nanoparticles

Taking into account the main advantages of the composite electrodes related to the long lifetime and their behavior as microelectrode arrays or at least as random ensemble (Wang, 2006), the NC-EP electrodes belong to the new generation of the composite electrodes characterized by the enhanced detection characteristics due to their potential behavior as nanoelectrode arrays (Manea, 2014). NC-EP can be fabricated by mixing and pressing nanostructured carbon with epoxy insulator, assuring a good but nonhomogeneous dispersion and distribution of the nanostructured carbon-conductive zones within the epoxy insulating matrix. Their behavior as micro/nanoelectrode arrays was discussed in "Voltammetric and amperometric techniques used for detection applications" section, and their practical utility in water quality detection is highlighted. A very important characteristic of the microelectrodes array is given by their ability to be used directly in water without any supporting electrolyte adding (that is compulsory for the macroelectrode usage), which make them suitable for on-site and in-situ detection for water quality monitoring or during water treatment process/technology allowing the process control. An example of CNT-EP electrode tested in the detection of salicylic acid (SA) in real surface water (Bega river, Romania) and tap water without the supporting electrolyte is presented in Fig. 9.20A and B in comparison with the results obtained for Fig. 9.20C. Similar sensitivities were achieved for all tested situations, which confirmed CNT-EP utility for in-situ or on-site water pollution detection.

The composite electrode can be used also as a substrate to deposit electrochemically the metallic particles (e.g., Ag) that will be distributed onto carbon conductive zones to form the ensembles of the silver nanoelectrodes that exhibit the

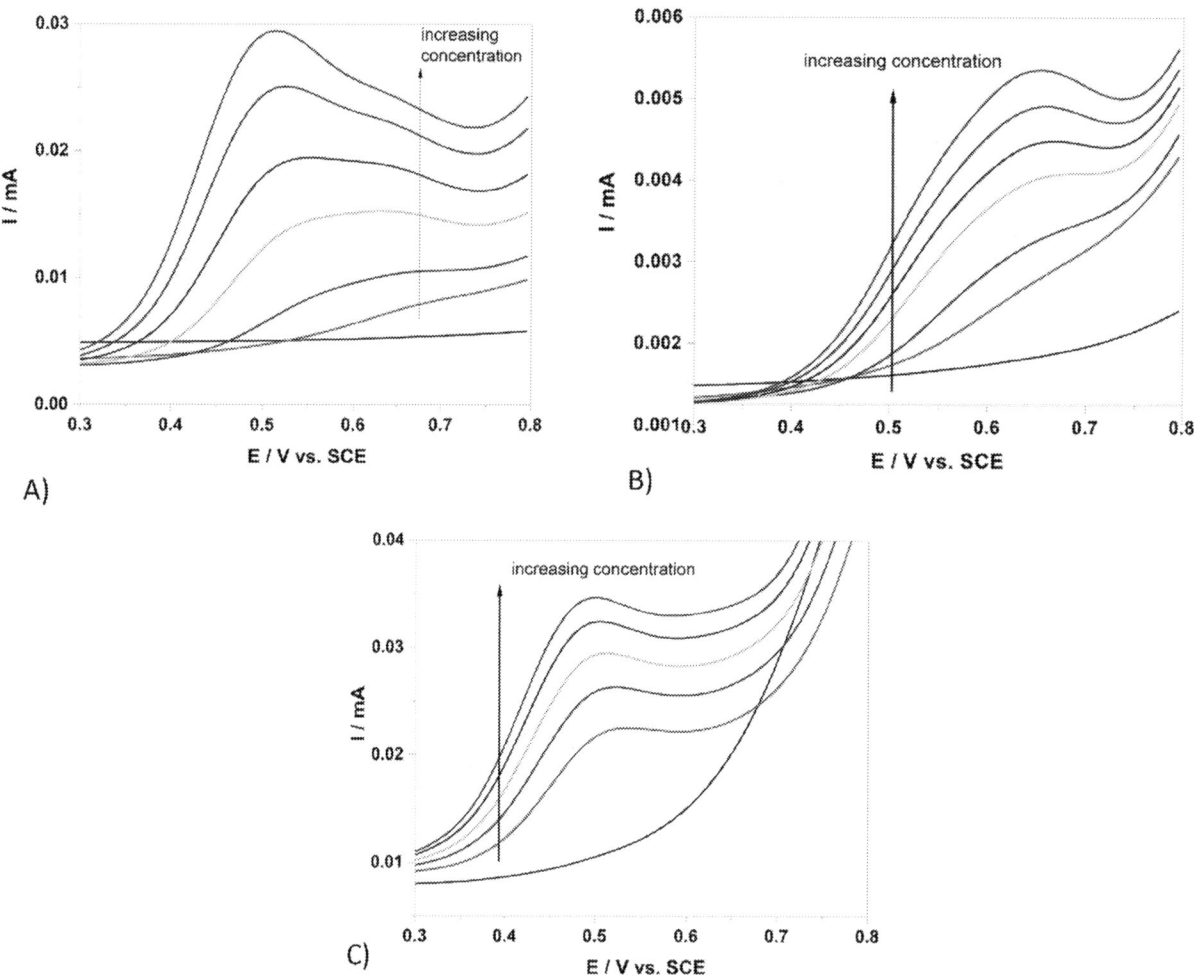

FIG. 9.20 DPVs recorded at 0.1 V MA and 0.01 V SP on CNT-EP electrode in: Bega River and SA concentrations ranged from 0.20 to 1.2 mM SA (A); tap water and SA concentrations ranged from 0.02 to 0.14 mM SA (B); 0.1 M Na2SO4 supporting electrolyte and SA concentrations ranged from 0.20 to 1.0 mM SA (C).

enhanced electrocatalytic activity and as a consequence the detection ability. The integration of Ag within the CNT-EP and CNF-EP electrodes allowed detecting IBP that cannot be detected onto simple nanostructured carbon-based electrodes (Table 9.7) (Motoc et al., 2016). Another interesting example of modified NC-EP electrode is CNF-EP modified with $[Cu_3(BTC)_2]$ metal–organic framework (BTC = 1,3,5 benzenetricarboxylate) that is known as HKUST, by simple mixing and pressing with CNF and epoxy insulating matrix, which allowed detecting simultaneously/selective DCF and IBP from the same anti-inflammatory pharmaceuticals class that belongs to the emerging pollutants of water. Fig. 9.21 presents the PAs recorded in the presence of DCF and IBP for their simultaneous detection using HKUST-CNF-EP electrode and the detection protocol subjected to set up the pulse level order and the pulse time (Motoc et al., 2016). Also, the detection protocols for the selective detection of IBP and for the selective detection of DCF from their mixture besides their simultaneous detection are presented in Table 9.7. The detection protocol consists of four levels of pulses setup based on CV behavior regarding the electrode processes occurring at each potential level. Thus, E_1: −0.20 V vs Ag/AgCl assure the electrode preconditioning/refreshing based on copper reduction process; E_2: +0.80 V vs Ag/AgCl fitting to DCF oxidation;

TABLE 9.7 Detection protocols for simultaneous detection of IBP and DCF, selective detection of DCF, and selective detection of IBP.

Detection type	Pulse potential order and value (V vs. Ag/AgCl)	Pulse time (s)
Simultaneous detection of DCF and IBP	$E_1 = -0.20$	0.50
	$E_2 = +0.80$	0.50
	$E_3 = +1.25$	0.10
	$E_4 = +0.02$	0.50
Selective detection of DCF	$E_1 = -0.20$	0.50
	$E_2 = +0.80$	0.10
	$E_3 = +1.25$	0.10
	$E_4 = +0.02$	0.50
Selective detection of IBP	$E_1 = -0.20$	0.50
	$E_2 = +1.25$	0.20
	$E_3 = +0.80$	0.20
	$E_4 = +0.02$	0.50

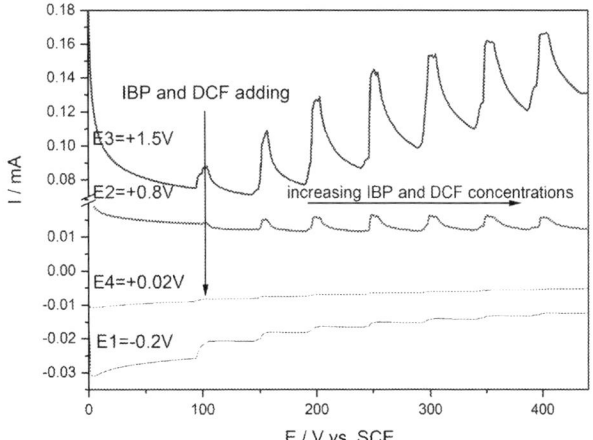

FIG. 9.21 Four-level PAs recorded on HKUST-CNF-EP electrode in 0.1 M Na2SO4 supporting electrolyte and adding continuously a mixture of 2 mg L^{-1} DCF and 2 mg L^{-1} IBP, recorded at $E_1 = -0.2$ V/SCE (0.5 s), $E_2 = +0.8$ V/SCE (0.5 s), $E_3 = +1.25$ V/SCE (0.1 s), and $E_4 = +0.02$ V/SCE (0.5 s).

E_3: +1.25 V vs Ag/AgCl fitting to IBP detection; E_4: +0.02 V vs Ag/AgCl corresponding to copper oxidation will be optimized to favor one or both pollutants oxidation processes responsible for their detection.

Several results of the NC-EP-based composite electrodes for the detection of different pollutants from priority class, emerging pollutants class, and conventional pollutants are presented in Table 9.8 (Baciu, Manea, Pop, Pode, & Schoonman, 2017; Baciu, Pop, Manea, & Schoonman, 2014; Carrera, Espinoza-Montero, Fernández, Romero, & Alvarado, 2017; Jakab

TABLE 9.8 The lowest limit of detection reported for NC-EP electrodes in the detection of water pollutants.

Pollutant class	Pollutant	Electrode configuration	Detection method	LOD (ppb)	Detection matrix	References
Priority pollutants	PCP	CNT-EP	CV	440	0.1 M Na_2SO_4	Jakab et al. (2013)
		CNT-EP	MPA	5.6	0.1 M Na_2SO_4	Remes et al. (2012)
	Arsenic	Ag-CNT-EP	ASSWV	3.3	0.09 M Na_2SO_4 0.01 M H_2SO_4	Baciu et al. (2014)
		Ag-CNF-EP	ASSWV	1.4	0.09 M Na_2SO_4 0.01 M H_2SO_4	Baciu et al. (2014)
		AuNpCµMF	DPV	0.90	0.1 M HCl	(Carrera et al., 2017)
	Lead	Ag-CNF-EP	ASSWV	61	0.09 M Na_2SO_4 0.01 M H_2SO_4	Baciu et al. (2014)
Emerging pollutants	Ibuprofen	AgZ-CNF-EP	DPV	8.0	0.1 M Na_2SO_4	Manea et al. (2012)
		Ag-CNF-EP	DPV	10	0.1 M Na_2SO_4	Manea et al. (2012)
		AgZ-CNT-EP	DPV	82	0.1 M Na_2SO_4	Motoc et al. (2013)
	Simultaneous diclofenac and ibuprofen	HKUST-CNF-EP	MPA	11.9 (DCF)	0.1 M Na_2SO_4	Motoc et al. (2016)
		HKUST-CNF-EP	MPA	3.20 (IBU)	0.1 M Na_2SO_4	(Motoc et al., 2016)
Conventional water pollutants	Simultaneous ammonium and nitrite	Ag-CNT-EP	DPV	17 (NH_4^+) 32 (NO_2^-)	0.1 M Na_2SO_4	(Baciu et al., 2017)

et al., 2013; Manea, Motoc, Pop, Remes, & Schoonman, 2012; Motoc et al., 2016; Motoc, Remes, Pop, Manea, & Schoonman, 2013; Remes et al., 2012).

In comparison with the voltammetric detection of the organic pollutants, an anodic stripping step is required prior to the voltammetric detection of the heavy metals, which leads to a preconcentration at the electrode surface of the heavy metal at zero oxidation state that is oxidized during voltammetric anodic scanning and generates a voltammetric signal. In general, advanced voltammetric techniques, for example, DPV and SWV, are applied, and the operational conditions are optimized to allow achieving good electroanalytical parameters. Also, the simultaneous or selective detection of heavy metals is possible through anodic stripping advanced voltammetric techniques. Simultaneous detection of As (III) and lead (II) through anodic stripping-square wave voltammetry using silver-electrodeposition onto CNF-EP electrode is reported (Baciu et al., 2014).

Conclusion and future perspectives

It is obviously that various nanostructured carbons should be considered excellent materials for the construction of the voltammetric and amperometric sensors, for the development of cheap and simple method for water quality monitoring, and for the control of water treatment, due to their outstanding performances. Generally, nanostructured carbon characterized by high specific surface area, excellent electrical and electrochemical properties, and good stability can accelerate the electron transfer reaction and the catalytic reaction of the electrode surface, which improve the sensing performance, avoiding the electrode fouling, as well. In addition, nanostructured carbon can be integrated within a common carbon-based electrode composition to enhance limited sensing properties or can be used as substrate for further modification with metal ions and organic metal skeletons, related to the practical specific needs. To obtain stable and reliable sensing data, it is compulsory to pay attention at similar degree for detection method optimization besides the electrode composition and design. The advanced electrochemical techniques have been developed, but their operating optimizations are in direct relation with the electrode composition.

The nanostructured carbon can contribute to the miniaturization of sensors, which make them suitable to serve the purpose and goal of in-situ and on-site water analysis, which is at early stage.

The presence of several pollutants in water at trace levels, for example micropollutants and emerging pollutants, represents a great challenge for the electrode sensitivity. In general, the working electrode characterized by high sensitivity does not exhibit the selectivity, and the interference aspect must be considered that requires further modification.

Also, it is important to highlight a clear limitation of electrochemical sensors related to the simultaneous analysis that is still restricted to a few number of pollutants that belong to same or different class of pollutants using the same electrode. Also, the selective detection of a pollutant from the same compound class (e.g., pharmaceutical) from emerging pollutants category is very challenging. Moreover, if considered the pharmaceutical analysis in wastewater characterized by very complex matrix, this becomes a serious issue considering many other pollutants interference potential and the possibility of more pollutants to be detected at the same potential value. A solution of this problem should be in the sample preparation, an aspect that has not been yet studied sufficiently because of the inherent complexity that embarrasses the practical utility of the electrochemical sensors. Another solution to eliminate the disadvantages should be the development of multiple different array electrodes customized to detect a variety of pollutants simultaneously and of micro or portable electrochemical devices to be able to in-situ and on-site detect selectively/simultaneously the specific pollutants from various categories of pollutants in different water bodies (from tap water to surface water, wastewaters, etc.). All aspects conclude that a continuous research in the sensing field is required to meet a mature technology ready to be applied in water quality monitoring in any stage of the cycle of water use.

Acknowledgments

This work was supported partially by a grant of the Romanian Ministry of Education and Research, CNCS—UEFISCDI, project code PN-III-P1-1.1-PD-2019-0676, project number PD 88/2020, within PNCDI III and partially by a grant of the Romanian Ministry of Research and Innovation, CCDI-UEFISCDI, project code PN-III-P2-2.1-PED-2019-4492, contract number 441PED / 2020 (3DSAPECYT), within PNCDI III.

References

Ahamad, A., Madnav, S., Singh, A. K., Kumar, A., & Singh, P. (2020). *Sensors in water pollutant monitoring: Role of material. Advanced functional materials and sensors.* Singapore: Springer. https://doi.org/10.1007/978-981-15-0671-0_1.

Ardelean, M., Manea, F., Vaszilcsin, N., & Pode, R. (2014). Electrochemical detection of sulphide in water/seawater using nanostructured carbon–epoxy composite electrodes. *Analytical Methods*, 6(13), 4775–4782. https://doi.org/10.1039/C4AY00732H.

Baciu, A., Ardelean, M., Pop, A., Pode, R., & Manea, F. (2015). Simultaneous voltammetric/amperometric determination of sulfide and nitrite in water at BDD electrode. *Sensors*, 15(6), 14526–14538. https://doi.org/10.3390/s150614526.

Baciu, A., Manea, F., Pop, A., Pode, R., & Schoonman, J. (2017). Simultaneous voltammetric detection of ammonium and nitrite from groundwater at silver-electrodecorated carbon nanotube electrode. *Process Safety and Environmental Protection*, 108, 18–25. https://doi.org/10.1016/j.psep.2016.05.006.

Baciu, A., Pop, A., Manea, F., & Schoonman, J. (2014). Simultaneous arsenic (III) and Lead (II) detection from aqueous solution by anodic stripping square-wave voltammetry. *Environmental Engineering and Management Journal*, 13(9), 2317–2323. https://doi.org/10.30638/eemj.2014.259.

Ballarin, B., Cordero-Rando, M. D. M., Blanco, E., Hidalgo-Hidalgo De Cisneros, J. L., Seeber, R., & Tonelli, D. (2003). New rigid conducting composites for electrochemical sensors. *Collection of Czechoslovak Chemical Communications*, 68(8), 1420–1436. https://doi.org/10.1135/cccc20031420.

Bebeselea, A., Manea, F., Burtica, G., Nagy, L., & Nagy, G. (2010). The electrochemical determination of phenolic derivatives using multiple pulsed amperometry with graphite based electrodes. *Talanta*, 80(3), 1068–1072. https://doi.org/10.1016/j.talanta.2009.07.036.

Buffa, A., & Mandler, D. (2019). Arsenic(III) detection in water by flow-through carbon nanotube membrane decorated by gold nanoparticles. *Electrochimica Acta*, 318, 496–503. https://doi.org/10.1016/j.electacta.2019.06.114.

Campuzano, S., Yáñez-Sedeño, P., & Pingarrón, J. M. (2019). Carbon dots and graphene quantum dots in electrochemical biosensing. *Nanomaterials*, 9(4). https://doi.org/10.3390/nano9040634.

Carrera, P., Espinoza-Montero, P. J., Fernández, L., Romero, H., & Alvarado, J. (2017). Electrochemical determination of arsenic in natural waters using carbon fiber ultra-microelectrodes modified with gold nanoparticles. *Talanta*, 166, 198–206. https://doi.org/10.1016/j.talanta.2017.01.056.

Charoenraks, T., Chuanuwatanakul, S., Honda, K., Yamaguchi, Y., & Chailapakul, O. (2005). Analysis of tetracycline antibiotics using HPLC with pulsed amperometric detection. *Analytical Sciences*, 21(3), 241–245. https://doi.org/10.2116/analsci.21.241.

Cięciwa, A., Wüthrich, R., & Comninellis, C. (2006). Electrochemical characterization of mechanically implanted boron-doped diamond electrodes. *Electrochemistry Communications*, 8(3), 375–382. https://doi.org/10.1016/j.elecom.2005.12.013.

Crevillen, A. G., Escarpa, A., & García, C. D. (2019). Chapter 1: Carbon-based nanomaterials in analytical chemistry. *RSC Detection Science*, 2019(12), 1–36. https://doi.org/10.1039/9781788012751-00001.

Directive 2000/60/EC of the European Parliament and of the Council establishing a framework for the Community action in the field of water policy. (2000). Off. J. Eur. Comm. L, 327(1), 1–72.

Eatemadi, A., Daraee, H., Karimkhanloo, H., Kouhi, M., Zarghami, N., Akbarzadeh, A., et al. (2014). Carbon nanotubes: Properties, synthesis, purification, and medical applications. *Nanoscale Research Letters*, 9(1), 1–13. https://doi.org/10.1186/1556-276X-9-393.

Fan, X., Soin, N., Li, H., Li, H., Xia, X., & Geng, J. (2020). Fullerene (C60) nanowires: The preparation, characterization, and potential applications. *Energy and Environmental Materials*, 3(4), 469–491. https://doi.org/10.1002/eem2.12071.

Fan, W., Zhang, L., & Liu, T. (2017). Graphene-carbon nanotube hybrids for energy and environmental applications, structures and properties of carbon nanomaterials. In *Springer Briefs in Molecular Science*.

Fard, G. P., Alipour, E., & Ali Sabzi, R. E. (2016). Modification of a disposable pencil graphite electrode with multiwalled carbon nanotubes: Application to electrochemical determination of diclofenac sodium in some pharmaceutical and biological samples. *Analytical Methods*, 8(19), 3966–3974. https://doi.org/10.1039/c6ay00441e.

Feeney, R., & Kounaves, S. P. (2000). Microfabricated ultramicroelectrode arrays: Developments, advances, and applications in environmental analysis. *Electroanalysis*, 12(9), 677–684. https://doi.org/10.1002/1521-4109(200005)12:9<677::AID-ELAN677>3.0.CO;2-4.

Feng, S., Yang, R., Ding, X., Li, J., Guo, C., & Qu, L. (2015). Sensitive electrochemical sensor for the determination of pentachlorophenol in fish meat based on ZnSe quantum dots decorated multiwall carbon nanotubes nanocomposite. *Ionics*, 21(12), 3257–3266. https://doi.org/10.1007/s11581-015-1512-1.

Garcia, L. L. C., Figueiredo-Filho, L. C. S., Oliveira, G. G., Fatibello-Filho, O., & Banks, C. E. (2013). Square-wave voltammetric determination of paraquat using a glassy carbon electrode modified with multiwalled carbon nanotubes within a dihexadecylhydrogenphosphate (DHP) film. *Sensors and Actuators, B: Chemical*, 181, 306–311. https://doi.org/10.1016/j.snb.2013.01.091.

Hiremath, N., & Bhat, G. (2017). High-performance carbon nanofibers and nanotubes. In *Structure and properties of high-performance fibers* (pp. 79–109). Elsevier Inc. https://doi.org/10.1016/B978-0-08-100550-7.00004-8.

Jakab, A., Manea, F., Bandas, C., Remes, A., Pop, A., Pode, R., et al. (2013). Unmodified/TiO2-modified carbon nanotubes composite electrodes for pentachlorophenol detection from water. *Environmental Engineering and Management Journal*, 12(5), 999–1005. https://doi.org/10.30638/eemj.2013.123.

Karuppiah, C., Cheemalapati, S., Chen, S. M., & Palanisamy, S. (2015). Carboxyl-functionalized graphene oxide-modified electrode for the electrochemical determination of nonsteroidal anti-inflammatory drug diclofenac. *Ionics*, 21(1), 231–238. https://doi.org/10.1007/s11581-014-1161-9.

Kempegowda, R., Antony, D., & Malingappa, P. (2014). Graphene-platinum nanocomposite as a sensitive and selective voltammetric sensor for trace level arsenic quantification. *International Journal of Smart and Nano Materials*, 5(1), 17–32. https://doi.org/10.1080/19475411.2014.898710.

Kour, R., Arya, S., Young, S.-J., Gupta, V., Bandhoria, P., & Khosla, A. (2020). Review—Recent advances in carbon nanomaterials as electrochemical biosensors. *Journal of the Electrochemical Society*, 167(037555), 1–23.

Krishnan, S. K., Singh, E., Singh, P., Meyyappan, M., & Nalwa, H. S. (2019). A review on graphene-based nanocomposites for electrochemical and fluorescent biosensors. *RSC Advances*, 9(16), 8778–8781. https://doi.org/10.1039/c8ra09577a.

Lawrence, N. S., Deo, R. P., & Wang, J. (2004). Electrochemical determination of hydrogen sulfide at carbon nanotube modified electrodes. *Analytica Chimica Acta, 517*(1–2), 131–137. https://doi.org/10.1016/j.aca.2004.03.101.

Li, L., Liu, D., Shi, A., & You, T. (2018). Simultaneous stripping determination of cadmium and lead ions based on the N-doped carbon quantum dots-graphene oxide hybrid. *Sensors and Actuators, B: Chemical, 255*, 1762–1770. https://doi.org/10.1016/j.snb.2017.08.190.

Liu, B., Xiao, B., & Cui, L. (2015). Electrochemical analysis of carbaryl in fruit samples on graphene oxide-ionic liquid composite modified electrode. *Journal of Food Composition and Analysis, 40*, 14–18. https://doi.org/10.1016/j.jfca.2014.12.010.

Malha, S. I. R., Mandli, J., Ourari, A., & Amine, A. (2013). Carbon black-modified electrodes as sensitive tools for the electrochemical detection of nitrite and nitrate. *Electroanalysis, 25*(10), 2289–2297. https://doi.org/10.1002/elan.201300257.

Manea, F. (2014). Electrochemical techniques for characterization and detection application of nanostructured carbon composite. In *Modern electrochemical methods in nano, surface and corrosion science*. Croatia: InTech.

Manea, F., Motoc, S., Pop, A., Remes, A., & Schoonman, J. (2012). Silver-functionalized carbon nanofiber composite electrodes for ibuprofen detection. *Nanoscale Research Letters, 7*. https://doi.org/10.1186/1556-276X-7-331.

Manea, F., Radovan, C., Pop, A., Corb, I., Burtica, G., Malchev, P., et al. (2009). Carbon composite electrodes applied for electrochemical sensors. In *NATO Science for Peace and Security Series C: Environmental Security* (pp. 179–189). Springer Science and Business Media LLC. https://doi.org/10.1007/978-1-4020-9009-7_11.

Mehmeti, E., Stanković, D. M., Hajrizi, A., & Kalcher, K. (2016). The use of graphene nanoribbons as efficient electrochemical sensing material for nitrite determination. *Talanta, 159*, 34–39. https://doi.org/10.1016/j.talanta.2016.05.079.

Moraes, F. C., Mascaro, L. H., Machado, S. A. S., & Brett, C. M. A. (2009). Direct electrochemical determination of carbaryl using a multi-walled carbon nanotube/cobalt phthalocyanine modified electrode. *Talanta, 79*(5), 1406–1411. https://doi.org/10.1016/j.talanta.2009.06.013.

Mostofizadeh, A., Li, Y., Song, B., & Huang, Y. (2011). Synthesis, properties, and applications of low-dimensional carbon-related nanomaterials. *Journal of Nanomaterials, 2011*, 1–21. https://doi.org/10.1155/2011/685081.

Motoc, S., Manea, F., Iacob, A., Martinez-Joaristi, A., Gascon, J., Pop, A., et al. (2016). Electrochemical selective and simultaneous detection of diclofenac and ibuprofen in aqueous solution using HKUST-1 metal-organic framework-carbon nanofiber composite electrode. *Sensors, 16*(10), 1719. https://doi.org/10.3390/s16101719.

Motoc, S., Manea, F., Orha, C., & Pop, A. (2019). Enhanced electrochemical response of diclofenac at a fullerene–carbon nanofiber paste electrode. *Sensors (Switzerland), 19*(6). https://doi.org/10.3390/s19061332.

Motoc, S., Remes, A., Pop, A., Manea, F., & Schoonman, J. (2013). Electrochemical detection and degradation of ibuprofen from water on multi-walled carbon nanotubes-epoxy composite electrode. *Journal of Environmental Sciences, 25*(4), 838–847. https://doi.org/10.1016/s1001-0742(12)60068-0.

Nasir, S., Hussein, M. Z., Zainal, Z., & Yusof, N. A. (2018). Carbon-based nanomaterials/allotropes: A glimpse of their synthesis, properties and some applications. *Materials, 11*(2). https://doi.org/10.3390/ma11020295.

Nasrollahzadeh, M., Sajjadi, M., Iravani, S., & Varma, R. S. (2021). Carbon-based sustainable nanomaterials for water treatment: State-of-art and future perspectives. *Chemosphere, 263*. https://doi.org/10.1016/j.chemosphere.2020.128005.

Negrea, S., Diaconu, L. A., Nicorescu, V., Motoc, S., Orha, C., & Manea, F. (2021). Graphene oxide electroreduced onto boron-doped diamond and electrodecorated with silver (Ag/GO/BDD) electrode for tetracycline detection in aqueous solution. *Nanomaterials, 11*(1566), 1–19. https://doi.org/10.3390/nano11061566.

Notarianni, M., Liu, J., Vernon, K., & Motta, N. (2016). Synthesis and applications of carbon nanomaterials for energy generation and storage. *Beilstein Journal of Nanotechnology, 7*(1), 149–196. https://doi.org/10.3762/bjnano.7.17.

Osman, A. I., Farrell, C., Al-Muhtaseb, A. H., Harrison, J., & Rooney, D. W. (2020). The production and application of carbon nanomaterials from high alkali silicate herbaceous biomass. *Scientific Reports, 10*(1). https://doi.org/10.1038/s41598-020-59481-7.

Pop, A., Lung, S., Orha, C., & Manea, F. (2018). Silver/graphene-modified boron doped diamond electrode for selective detection of carbaryl and paraquat from water. *International Journal of Electrochemical Science, 13*(3), 2651–2660. https://doi.org/10.20964/2018.03.02.

Pop, A., Manea, F., Flueras, A., & Schoonman, J. (2017). Simultaneous voltammetric detection of Carbaryl and Paraquat pesticides on graphene-modified boron-doped diamond electrode. *Sensors, 17*(9), 2033. https://doi.org/10.3390/s17092033.

Raccichini, R., Varzi, A., Passerini, S., & Scrosati, B. (2015). The role of graphene for electrochemical energy storage. *Nature Materials, 14*(3), 271–279. https://doi.org/10.1038/nmat4170.

Ramírez-García, S., Alegret, S., Céspedes, F., & Forster, R. J. (2002). Carbon composite electrodes: Surface and electrochemical properties. *Analyst, 127*(11), 1512–1519. https://doi.org/10.1039/b206201a.

Rauti, R., Musto, M., Bosi, S., Prato, M., & Ballerini, L. (2019). Properties and behavior of carbon nanomaterials when interfacing neuronal cells: How far have we come? *Carbon, 143*, 430–446. https://doi.org/10.1016/j.carbon.2018.11.026.

Remes, A., Pop, A., Manea, F., Baciu, A., Picken, S. J., & Schoonman, J. (2012). Electrochemical determination of pentachlorophenol in water on a multi-wall carbon nanotubes-epoxy composite electrode. *Sensors, 12*(6), 7033–7046. https://doi.org/10.3390/s120607033.

Salih, F. E., Achiou, B., Ouammou, M., Bennazha, J., Ouarzane, A., Younssi, S. A., et al. (2017). Electrochemical sensor based on low silica X zeolite modified carbon paste for carbaryl determination. *Journal of Advanced Research, 8*(6), 669–676. https://doi.org/10.1016/j.jare.2017.08.002.

Simm, A. O., Banks, C. E., Ward-Jones, S., Davies, T. J., Lawrence, N. S., Jones, T. G. J., et al. (2005). Boron-doped diamond microdisc arrays: Electrochemical characterisation and their use as a substrate for the production of microelectrode arrays of diverse metals (Ag, Au, Cu) via electrodeposition. *Analyst, 130*(9), 1303–1311. https://doi.org/10.1039/b506956d.

Štulík, K., Amatore, C., Holub, K., Mareček, V., & Kutner, W. (2000). Microelectrodes. Definitions, characterization, and applications (technical report). *Pure and Applied Chemistry, 72*(8), 1483–1492. https://doi.org/10.1351/pac200072081483.

Tajik, S., Dourandish, Z., Zhang, K., Beitollahi, H., Le, Q. V., Jang, H. W., et al. (2020). Carbon and graphene quantum dots: A review on syntheses, characterization, biological and sensing applications for neurotransmitter determination. *RSC Advances, 10*(26), 15406–15429. https://doi.org/10.1039/d0ra00799d.

Tian, P., Tang, L., Teng, K. S., & Lau, S. P. (2018). Graphene quantum dots from chemistry to applications. *Materials Today Chemistry, 10*, 221–258. https://doi.org/10.1016/j.mtchem.2018.09.007.

Valentini, F., Biagiotti, V., Lete, C., Palleschi, G., & Wang, J. (2007). The electrochemical detection of ammonia in drinking water based on multi-walled carbon nanotube/copper nanoparticle composite paste electrodes. *Sensors and Actuators, B: Chemical, 128*(1), 326–333. https://doi.org/10.1016/j.snb.2007.06.010.

Wang, J. (2006). *Analytical electrochemistry*. John Wiley & Sons, Inc.

Wang, J., Zhang, Z., Zhang, Q., Liu, J., & Ma, J. (2018). Preparation and adsorption application of carbon nanofibers with large specific surface area. *Journal of Materials Science, 53*(24), 16466–16475. https://doi.org/10.1007/s10853-018-2772-8.

Yang, C., Denno, M. E., Pyakurel, P., & Venton, B. J. (2015). Recent trends in carbon nanomaterial-based electrochemical sensors for biomolecules: A review. *Analytica Chimica Acta, 887*, 17–37. https://doi.org/10.1016/j.aca.2015.05.049.

Yu, H., Li, R., & Song, K. L. (2019). Amperometric determination of nitrite by using a nanocomposite prepared from gold nanoparticles, reduced graphene oxide and multi-walled carbon nanotubes. *Microchimica Acta, 186*(9). https://doi.org/10.1007/s00604-019-3735-8.

Chapter 10

Carbon nanomaterial-based sensors: An efficient tool in the environmental sectors

Prashanth S. Adarakatti[a], K. Sureshkumar[b], and T. Ramakrishnappa[b]
[a]Department of Chemistry, SVM Arts, Science and Commerce College, Ilkal, Karnataka, India, [b]BMS Institute of Technology and Management, Bengaluru, Karnataka, India

Introduction

The term "environmental pollution" causes sudden attention of every human being on this planet. This is because of fast deterioration of the air, water, and soil quality in the last two decades. The consequences of uncontrolled human activities, like deforestation, urbanization, industrial production, nuclear power plant emissions etc., depleted natural resources, resulting in an ecological imbalance. The major reason for disturbing environmental harmony is releasing the large fluxes of toxic materials into the environmental matrix due to the abovementioned human activities. The majority of the toxic substances released into the environment will come from industrial production, like inorganic heavy metal ions and hydrocarbons. The toxic metal ions that are found to be harmful are hexavalent Cr, pentavalent As, trivalent As, divalent Hg, Pb, Cd, Cu, and Ni ions (Nujić & Habuda-Stanić, 2019). The abovementioned heavy metal ions enter into nearby water bodies, percolate into the soil, and enter into the food chain through plants and seafood. The consumption of heavy metal-contaminated food induces the toxic effect that corresponds to heavy metals being found to be lethal to human beings. The Environmental Protection Agency (EPA) and World Health Organization (WHO) set the guidelines for threshold limit values (TLVs) for these toxic metal ions in drinking water. The TLVs of toxic metal ions according to EPA and WHO are As-10 µg/L, Cd-3 µg/L, Cr −50 µg/L, Cu-2000 µg/L, Pb-10 µg/L, Hg-610 µg/L, and Ni-7010 µg/L (Briffa, Sinagra, & Blundell, 2020). Similarly, the toxic hydrocarbon-based chemicals include hydrazine, nitrobenzene, phenols, aromatic aldehydes, insecticides, and pesticides (Cubo, 2010). Along with these other toxicants that pollute the environment are gases like NO_2, SO_2, and H_2S (Kailasa & Wu, 2012).

In order to combat environmental pollution and preserve its pace, it is very necessary to assess the degree of pollution of the particular environmental matrix by the pollutant. This requires highly sensitive and selective analytical devices for monitoring the pollutants in various environmental matrices with high accuracy. The registered methods for these toxic metal ions are AAS, AES, and ICP-AES. For gaseous and organic pollutants, it is HPLC, GC, and other sophisticated equipment, which demands the trained person and needs an expensive setup (Eckenrode, 2001; Kamala, Balaram, Dharmendra, Subramanyam, & Krishnaiah, 2014; West et al., 1984). From the last decade, the optical and electrochemical sensors have gained the attention of researchers due to their inherent properties like simplicity, portableness, low cost, and good sensitivities (Kamala et al., 2014). Numerous substrates have been reported as optical and electrochemical sensor platforms for monitoring environmentally important analytes (George et al., 2020; Pan, Cha, Chen, & Choi, 2013; Simões & Xavier, 2017). However, there are still some drawbacks in reported sensor systems, like expensiveness, lacking field portability, and cross-matrix when applied for sensing applications in complex environmental matrices.

In recent years, carbon-based nanostructures have been extensively used in almost all the sectors like human health diagnosis, energy production and conversion, environmental remediation etc. (Maiti, Tong, Mou, & Yang, 2019; Wang, Liu, Yao, & Hu, 2019; Wang, Pan, Chu, Vipin, & Sun, 2019; Wu et al., 2020). Furthermore, the properties of the carbon nanomaterials can be tuned, which enable to address and identify the various environmental challenges. Carbon nanomaterials can be used in a broad range of environmental applications. They can be used as depth filters, membranes, antimicrobial agents, and sorbents, in renewable energy technologies, sensors for toxic and harmful environmental substrates, and in pollution-controlling devices (Al-Jumaili, Alancherry, Bazaka, & Jacob, 2017; Sadik, Du, Yazgan, & Okello, 2014). The advent of nanotechnology makes researchers build every possible structure in the nanoscale and to compare the properties of their bulk counterparts. The uniqueness of carbon materials in the nanoscale is possible due to their special hybridization abilities, sensitivity of carbon structures for small perturbations, and their tailor-made structure tunability that is not easily possible with inorganic nanostructures. Even though the bulk and nanostructured carbon materials share similar bonding configurations, the properties and structures of nanocarbons are dominated by the resonance structures rather than their average crystalline forms.

The various possible nanoforms of carbon include carbon nanotubes (CNTs), graphene oxide (GO), reduced graphene oxide (rGO), graphene, graphene quantum dots (GQDs), fullerenes, and nanodiamonds (Perathoner & Centi, 2018). These nanomaterials can be classified into zero-dimensional (0-D) nanoparticles, 1-dimensional (1-D) nanotubes, and 2-dimensional (2-D) layered graphene materials (Mohamed, 2017). The properties of the carbon usefulness for environmental applications include their shape, size, surface area, molecular interactions, adsorption capacities, electronic, electrical, optical, and thermal properties (Rauti, Musto, Bosi, Prato, & Ballerini, 2019). Furthermore, carbon-based nanomaterials are ease to manipulate and exhibit robustness when fabricated into sensor devices. These features make carbon nanostructures their favorite substrates, like both optical and electrochemical sensor materials (Farrera, Torres Andón, & Feliu, 2017; Zhang & Du, 2020). The unique redox chemistry of fullerenes enables them to be used in electrochemical sensors (Rather et al., 2016). The carbon nano-onions were obtained by surrounding the fullerenes concentrically with graphitic carbon layers (Ugarte, 1992). These carbon nanostructures are used in electrochemical sensors due to their low density and high surface-to-volume ratios (Lettieri et al., 2017). The carbon nanodiamonds (CNDs) are optically transparent, chemically stable, along with all the properties of nanomaterials like small size, large surface areas, and high adsorption capabilities (Mochalin, Shenderova, Ho, & Gogotsi, 2012). All these features enable the researchers to use them in both electrochemical and optical sensors. The 0-D carbon dots possess their luminescent properties in the visible spectrum, and enable them to use as optical sensor materials (Baker & Baker, 2007). Largely, the reported carbon-based sensor materials are 1-D CNTs and 2-D graphene-based materials. Graphene is highly celebrated material in the material science and engineering field. The ability of this material to absorb light in the whole spectrum and its excellent electron mobility makes it an excellent material in optical and electrochemical sensing approaches (Geim & Novoselov, 2007). The analogs of graphene include photoluminescent GO obtained by oxidation of sp^2 carbon network of graphene, which was also reported widely as both an optical and electrochemical sensor material (Chien et al., 2012). The partial reduction of GO results in another grapheme analogue called rGO, which is also heavily used in sensor systems because of its higher conducting ability compared to its parent GO (Georgakilas et al., 2012). 1-D carbon materials like CNTs exist in two forms: single-walled carbon nanotubes (SWCNTs) and multiwalled carbon nanotubes (MWCNTs). Among the two forms, SWCNTs can act as metallic, semimetallic, and semiconductors, which are widely used in electrochemical sensor systems (Yang, Chang, et al., 2010; Yang, Ratinac, et al., 2010). The similar carbon nanostructures to SWCNTs are single-walled carbon nano horns (CNHs). These materials also possess good porosity along with the larger surface areas, which enables them to use as sensing materials (Zhu & Xu, 2010).

This chapter summarizes the recent advances in synthetic and sensor fabrication strategies of various carbon nanostructures for monitoring analytes of environmentally important areas, which will help in combating the present burning issues regarding environmental preservation.

Carbon-based nanomaterials used in sensor applications

Carbon is a nonmetallic element that is widely distributed throughout the world. Carbon's distinctiveness originates from its catenation feature, which refers to an element's ability to make bonds with like or unlike types of atoms. Carbon occurs primarily in two allotropic forms, crystalline and amorphous carbon, due to its catenation feature. As a result of the varied physical properties of each allotropic form of carbon, these substrate materials can be adapted to a specific application. Furthermore, carbon has a large potential window in both aqueous and nonaqueous media due to its high conductivity, facile surface modification, rich surface chemistry, high thermal stability, mechanical strength, and low background current. It outperforms metal electrodes due to its low cost and ability to adsorb analyte molecules from the bulk of the electrolytic solution to the surface, as it contains surface oxides that act as electron transfer sites, allowing electrocatalytic reactions to proceed.

Emphasis has been placed on the nanomaterials in terms of the synthesis of nanomaterials and the knowledge of their fundamental properties over the previous two decades. Carbon-based nanostructures, such as carbon nanotubes, graphene, and graphene oxide, reduced graphene oxide, and quantum dots, have been hot subjects in material science for quite some time. Because of their interesting physical, chemical, and electrical properties, carbon-based nanomaterials have been considered as powerful alternatives for the fabrication of the next miniaturized biosensors. Because of graphene's planar layout and carbon nanotubes' tubular form, nearly all surface atoms are exposed, enabling a substantial proportion of chemical species to bind to the transduction substrate. Other carbon-based nanomaterials, such as carbon nanofibers and nanocrystalline diamond, have increased electrochemical activity in addition to graphene and carbon nanotubes. However, there are few studies that have looked at the usage of these nanomaterials for electrochemical detection of environmental contaminants.

Graphene

Graphene is a two-dimensional carbon nanostructure that can be found in all types of carbon, including zero-dimensional fullerenes, one-dimensional nanotubes, and three-dimensional graphite (Cooper, 2012). Because of its massive theoretical surface area, thermal conductivity, remarkable electrical conductivity, outstanding biocompatibility, and high mechanical stiffness, graphene has sparked a lot of interest in numerous fields of research (Park & Ruoff, 2009; Rao, Sood, Voggu, & Subrahmanyam, 2010). It is made up of sp^2 hybridized carbon atoms that are arranged in sheets of limitless length. It is primarily defined by its basal and edge planes, which are involved in heterogeneous electron transfer (HET) activities. Further, it has been synthesized using various methods, because no single method can produce graphene, which is suitable for all possible applications. Chemical reduction of graphene oxide in the presence of reducing chemicals such as sodium borohydride, ethylene glycol, and hydrazine is an example of chemical process (Jeon et al., 2012; Li, Yan, Lu, & Su, 2018; Li, Yan, Qiao, Lu, & Su, 2018; Ren, Yan, Ji, Chen, & Li, 2011; Somani, Somani, & Umeno, 2006). The chemical vapor deposition (CVD), micromechanical cleavage (scotch tape method), and epitaxial graphene formation on SiC are examples of physical approaches. In addition to these approaches, a few other techniques have been used, including ball milling (Guo, Wang, Qian, Wang, & Xia, 2009; Zhao et al., 2010) and electrochemical reduction (Dreyer, Park, Bielawski, & Ruoff, 2010).

Graphene oxide (GO) and reduced graphene oxide (rGO)

Graphene oxide (GO) is a layered material made from the oxidation of natural graphite with the general chemical composition $C^+_x(OH)y(H_2O)_2$ (Dreyer et al., 2010). It can alternatively be thought of as an oxidized graphene having a large number of oxygen-containing functional groups in the basal and edge plane sites, such as epoxy, ether, carboxyl, and hydroxyl (Mkhoyan et al., 2009). Due to the presence of ionizable oxygen-containing functional moieties on the surface of GO, it is negatively charged over a pH range of 2 to 11, and this is also responsible for the creation of a stable aqueous colloidal solution that is known to be stable over several weeks (He et al., 2012). Further, GO occurs as a single layer in solution and can function as a cation exchange resin, allowing metal ions, polymers, biomolecules, and any other positively charged analyte species to be exchanged (Mishra & Ramaprabhu, 2011; Yang, Chang, et al., 2010; Yang, Ratinac, et al., 2010; Zuo et al., 2010). Graphite oxide is made using the Hummers process, which involves processing graphite with potassium permanganate and strong sulfuric acid for simultaneous oxidation and exfoliation (Hummers & Offeman, 1958). The acid treatment introduces polar oxygen functional groups such as epoxies, carbonyls, hydroxyls, and other polar oxygen functional groups, making GO hydrophilic (Ramesh, Bhagyalakshmi, & Sampath, 2004). The GO sheets can be easily dispersed in a variety of solvents, particularly water, using mild sonication (Paredes, Villar-Rodil, Martínez-Alonso, & Tascón, 2008). Following that, the colloidal graphene oxide solution can be reduced to graphene using a variety of methods, including chemical reduction with reducing agents such as hydrazine or thermal reduction at high temperatures (McAllister et al., 2007; Stankovich et al., 2007). The easiest and most straightforward way to create reduced graphene oxide is electrochemical reduction, which involves immobilizing the GO onto an electrode surface and performing consecutive reducing scans by sweeping the voltage from 0 to -1.5 V (Zhou et al., 2009). The resulting rGO flakes have typical diameters ranging from submicron to a few microns. Due to partial reduction, reduced graphene oxide still retains some defect sites in the form of sp^3 centers and oxygen functional groups. These functional groups have been shown to improve graphene's electrocatalytic characteristics and can be employed for direct covalent binding to biological receptors.

Fullerenes

The discovery of fullerenes, a novel class of carbon compounds, was done by Harry Kroto, Robert Curl, and Richard Smalley, who were awarded the Noble Prize in Chemistry in 1986 for their pioneering work (Kroto, Allaf, & Balm, 1991). This breakthrough has aided in the study of the structure–property link of this novel type of carbon allotropic form (Randić, Kroto, & Vukičević, 2007). The third allotropic type of carbon is represented by this family of spheroidal carbon cage molecules. It was first made by evaporating graphite with a strong laser, and then it was made in bulk by an electric arc discharge of graphite in a helium or argon atmosphere (Krätschmer, Lamb, Fostiropoulos, & Huffman, 1990). It has a spherical shape and solely comprises carbon atoms whose shape closely matches Richard Buckminster Fuller's highly symmetric architectonic geodesic domes, therefore the name "fullerene." And these are considered as the purest form of carbon as these contain no dangling bonds. It is made up of 60 carbon atoms organized in a spherical form, and the shapes are known as truncated icosahedrons, which look like soccer balls (Becker et al., 1994). Due to their spherical form, fullerenes are very reactive and can easily react with electrophiles and ligands. Fullerenes have empty lowest unoccupied molecular orbitals (LUMOs) on the electronic scale, allowing them to efficiently receive electrons from donors and perform reductive electrochemistry (Hirsch, 2008).

Carbon nanotube (CNTs)

Carbon nanotubes have become a major focus of interest since their discovery by Iijima in 1991, and they represent a new class of advanced materials in the creation of nanocomposites (Ajayan & Lijima, 1992). CNTs are distinguished by their size, shape, physical, and electrical properties. Depending on their structure, these materials can operate as either metallic or semiconductive (Agüí, Yáñez-Sedeño, & Pingarrón, 2008; Vairavapandian, Vichchulada, & Lay, 2008). Arc discharge, chemical vapor deposition (CVD), and laser ablation procedures were used to make them in general (Bethune et al., 1993; Dai et al., 1996; Guo, Nikolaev, Thess, Colbert, & Smalley, 1995). *Carbon-arc discharge* was used to create CNTs for the first time. This approach generates high yields of multiwalled carbon nanotubes (MWCNTs) or single-walled carbon nanotubes (SWCNTs), depending on the catalyst used, and allows more control over the diameters of produced tubes. This approach is suitable for large-scale commercial synthesis due to its excellent yield and reproducibility. *Chemical vapor deposition,* like arc discharge, is catalyst dependent and produces small-diameter carbon nanotubes. Despite the decreased output of CNTs, the resulting CNTs are cleaner, requiring less cleanup of soot and crude pollutants. Further, *laser ablation* produces the cleanest CNTs at a higher cost and lower yield with sizes of 1–2 nm (Ramnani, Saucedo, & Mulchandani, 2016). As nanotubes have a circular structure, the electron cloud will be deformed, resulting in a rich p electron conjugation on the nanotubes' surface, making them more electrically and thermally conductive. Furthermore, edges with a large surface area and probable topological flaws such as pentagons or heptagons can cause local perturbations in the electrical structure. Due to the existence of a cloud of electrons at the tube's walls, CNTs have a high electrical conductivity (Ju, Zhang, & Wang, 2011). As a result, electron transport from the electrode material to the electroactive moiety is facilitated. Because of these characteristics, CNTs have been used as substrate materials in a wide range of electrochemical applications. CNTs were divided into two components in terms of electrochemistry: a side wall that is similar to the basal plane (BPPG) and an end wall that is similar to the edge plane (EPPG) of highly ordered pyrolytic graphite. The Compton group evaluated the electrochemical response of the probe on a CNT-modified electrode using BPPG and EPPG electrodes (Banks, Moore, Davies, & Compton, 2004; Brownson et al., 2016; Ji, Kadara, Krussma, Chen, & Banks, 2010).

Graphene quantum dots (GQDs)

Due to the additional unique features derived from their nanoscale smallness, graphene quantum dots (GQDs) have numerous benefits as new carbon-based materials (Zhang, Zhang, Chen, & Qu, 2012). The application of GQDs in a variety of disciplines necessitates bulk manufacture. To present, a number of approaches for generating GQDs with good properties and a high yield have indeed been devised. Bottom-up and top-down approaches are the two major synthesis procedures. The bottom-up technique, which involves many chemical reactions and purification at each step, is appropriate for controlling the size of GQDs. However, breaking the carbon–carbon bonds of a big carbon source from the top-down technique is easy and straightforward, making it appropriate for mass manufacturing. Additional good qualities of GQDs, such as high transparency and high surface area, have recently been proposed for energy and display applications in recent studies (Bak, Kim, & Lee, 2016).

Electrodes made of GQDs are used in capacitors (Liu, Feng, Yan, Chen, & Xue, 2013) and batteries (Chao et al., 2015; Pan et al., 2013) because of their huge surface area, and GQDs have greater conductivity than graphene oxide, and also because of their excellent hole-transporting abilities, GQDs can be used as hole transport layers. Furthermore, GQDs are also well distributed in many organic solvents because of their nm size, allowing for a variety of organic reactions and solution processes. GQDs also have recently been chemically modified and employed for the first time in energy conversion, bioanalysis, and sensing applications. Synthesizing various quantum dots (QDs) can be done in a variety of ways. Top-down and bottom-up strategies are clearly defined among them (Bera, n.d.). In top-down processes, cheap, easily available bulk graphene-based materials, most often graphite, are decomposed and exfoliated in severe circumstances. Multiple processes involving concentrated acids, powerful oxidizing agents, and high temperatures are frequently required (Shin et al., 2014; Sun, Gao, et al., 2013; Sun, Wang, et al., 2013). However, the morphology and size distribution of the particles produced can't be precisely controlled using these approaches. Quantum dots (QDs) are synthesized from polycyclic aromatic compounds or other molecules with aromatic structures, such as fullerenes, in bottom-up processes (Liu, Wu, Feng, & Müllen, 2011; Yan, Cui, & Li, 2010). Despite their complexity, these procedures provide good control over the end product's properties, which are also directly connected to the band gap of QDs.

Pan et al. proposed a hydrothermal solution procedure for GQDs made from oxidized graphene sheets (GS) at 200 degrees Celsius (Pan, Zhang, Li, & Wu, 2010). They oxidized the graphene sheets to incorporate more oxygen-containing groups such as ketone, carboxyl, hydroxyl, and ether groups to boost the product yield. During the oxidation and reduction processes, the size of GS is reduced. After hydrothermal de-oxidation, GS is transformed into GQDs with an average diameter of 9.6 nm and a yield of about 5%. After hydrothermal treatment, the functional groups C=O (ketone/carboxyl) and epoxy are significantly reduced, according to FTIR and XPS. Furthermore, GQDs can be fine-tuned by utilizing solvothermal treatment using a variety of solvents. Due to the presence of oxygen and nitrogen atoms, GO ultrasonically cut in DMF can be cut into GQDs, which emit a green luminous light and are soluble in most polar solvents (Zhu et al., 2011). Furthermore, only mechanical force (ultrasonication method) can be used to reduce graphene derivatives to GQDs. Graphene sheets can be oxidized in this way with an ultrasonic treatment in a high acidic environment. GQDs with diameters of 3 to 5 nm can be created after several purifications. Ultrasonication can be used in conjunction with solvothermal and microwave treatments in most cases (Zhuo, Shao, & Lee, 2012).

Microwave methods make use of the heat released by polar solvent molecules having an electrical dipole moment as a result of molecular rotation caused by microwave energy (Piyasena, Dussault, Koutchma, Ramaswamy, & Awuah, 2003). This microwave method can be used to produce high yields of organic and inorganic compounds in a short amount of time (Kappe, 2004). Li et al. found that GO can be broken down into GQDs when exposed to strong acids (HNO_3 and H_2SO_4) using microwave-assisted reduction (Li et al., 2012). They also claimed that the reaction time was shorter than with a solvothermal technique. These GQDs have green fluorescence (gGQD), but following further reduction in sodium borohydrate, they shift to blue PL (bGQD) with greater intensity. Despite the fact that the PL characteristics have altered drastically, there is no apparent difference in size before and after $NaBH_4$ treatment.

Radical methods: Electrochemical methods can be used to create fluorescent carbon compounds from graphite electrodes (Lu et al., 2009). Li et al. used electrochemical techniques to make GQDs. GQDs are made by scanning reduced GO in phosphate buffer solution (PBS) using a cyclic voltammetry probe (Li et al., 2011). Zhou et al. used a photo-Fenton reaction to make GQDs. The Fenton reagent produces hydroxyl radicals that can cleave GOs into GQDs when exposed to UV (365 nm, 500 W). The reaction removes the majority of the oxygen groups in GOs, resulting in GQDs with blue photoluminescence (Zhou et al., 2012).

Nanodiamonds (NDs)

The first nanoscale diamond particles were created by detonation in the Soviet Union in the 1960s, but they were mostly unknown to the rest of the world until the late 1980s (Danilenko, 2004). Then, starting in the late 1990s, a series of significant advancements sparked a surge in interest in these nanodiamond-like particles (Greiner, Phillips, Johnson, & Volk, 1988). Nanodiamonds are diamond nanoparticles that exhibit many of the same characteristics as bulk diamonds but on a smaller scale. As a result, nanodiamonds (NDs) have sparked a lot of curiosity and, as an outcome, a great deal of research. High hardness is the most well-known property of the diamond phase, but other qualities include chemical stability, biocompatibility, a customizable surface structure, and tolerance to severe conditions. Individual diamond particles having a diameter of 4–5 nm ("single-digit" nanodiamonds) were first made available in colloidal solutions (Ozawa et al., 2007). Second, researchers began to employ fluorescent nanodiamonds in biomedical imaging as a nontoxic alternative to semiconductor quantum dots (Chang et al., 2008; Mochalin & Gogotsi, 2009). Finally, nanoscale magnetic sensors based on nanodiamonds have been constructed. Furthermore, researchers discovered that nanodiamond is less harmful than other carbon nanoparticles (Schrand, 2009; Schrand et al., 2007; Schrand, Hens, & Shenderova, 2009). As a consequence, it is being evaluated for use in diagnostic applications, drug development, and other areas of pharmaceuticals now (Mochalin et al., 2012).

At room temperature and pressure, graphite is the most stable allotrope of carbon. As shown in Fig. 10.1's carbon phase diagram, diamond formation necessitates extremely high temperatures and pressures. Natural diamonds, for example, are created in the Earth's mantle, 140–200 km below the surface (Shirey et al., 2013), at temperatures and pressures of 900–1400°C and 4.506 GPa, respectively (Davies, 1994). The high energy barrier for phase transitions prevents the transition back to graphite at ambient temperatures after the diamond phase has formed. Even if graphite is thermodynamically preferred (energy difference between diamond and graphite of 0.02 eV per atom), a 0.4 eV energy barrier must be crossed to transition from sp^2 to sp^3 chemical bonds (Mochalin et al., 2012). There are a variety of approaches for creating nanodiamonds artificially right now.

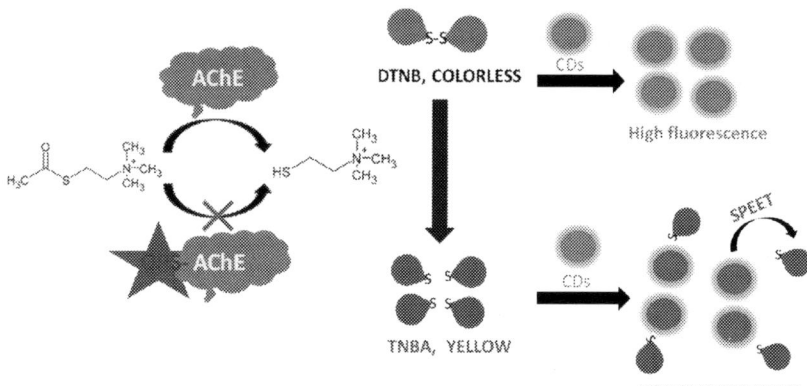

FIG. 10.1 The color and fluorescence change of carbon dots with pesticide. *(With permission from Li, H., Yan, X., Lu, G., & Su, X. (2018). Carbon dot-based bioplatform for dual colorimetric and fluorometric sensing of organophosphate pesticides. Sensors and Actuators, B: Chemical, 260, 563–570. https://doi.org/10.1016/j.snb.2017.12.170.)*

They have been made using the detonation technique (Fig. 10.1), laser ablation (Yang, Wang, & Liu, 1998), high-energy ball milling of high-pressure high-temperature (HPHT) diamond microcrystals (Boudou et al., 2009), plasma-assisted chemical vapor deposition (CVD) (Frenklach et al., 1991), autoclave synthesis from supercritical fluids (Gogotsi, 1996), chlorination of carbides (Welz, Gogotsi, & McNallan, 2003), ion irradiation of graphite (Daulton, Kirk, Lewis, & Rehn, 2001), electron irradiation of carbon "onions" (Banhart & Ajayan, 1996), and cavitation (Galimov et al., 2004). The first three of these technologies are commercially available.

Detonation, chemical vapor deposition (CVD), and milling of high-pressure, high-temperature (HPHT) micro-sized diamonds are the three primary commercially available processes covered in the following sections.

Detonation synthesis uses the energy of an explosion to cause the diamond phase to form. Explosives with a negative oxygen balance are detonated in a closed metallic chamber; most commonly, a mixture of 60% TNT ($C_6H_2(NO_2)CH_3$) and 40% hexogen ($C_3H_6N_6O_6$) is used. The explosive molecules or the precursor graphite deposited within the explosion chamber provides the carbon atoms that eventually form the NDs. When a gas (N_2, Ar, CO_2) or water (ice) is used as a coolant, the synthesis is referred to as "dry" or "wet." The carbon yield, which is typically around 10% of the explosive weight, is influenced by the cooling media (Dolmatov, 2018). The carbon atoms released during the breakup of the explosive molecule condense and crystallize into nanoclusters after an explosion. The detonation causes the carbon nanoclusters to crystallize into the diamond phase due to the pressure/temperature increase in the chamber. Finally, the generated NDs expand and agglomerate, forming NDs with a diameter of 4–5 nm (Ozawa et al., 2007).

The nanocrystalline diamond layer was created via chemical vapor deposition, which is one of the most used thin-film deposition processes. During the breakdown of a gas mixture with a carbon-containing component in excess of hydrogen, most often methane CH_4, carbon atoms are deposited. By utilizing a heated filament or a microwave plasma to breakdown the gas phase, radicals such as H˙ and CH_3 are formed, which are necessary for diamond formation. The ND film forms a continuous layer on a substrate, which is commonly a silicon wafer covered with a micrometer-sized diamond powder that functions as a seed for ND nucleation (Butler & Sumant, 2008). Depending on the relative concentrations of CH_4 and H_2, the grain size of the film varies from tens of microns to a few nanometers (May, Ashfold, & Mankelevich, 2007).

Milling of High-Pressure, High-Temperature (HPHT) Microdiamonds: The HPHT synthesis method is based on the natural process of diamond formation, which involves subjecting a carbon precursor, usually graphite, to tremendous pressure and heat. The temperature is raised to around 2000 degrees Celsius inside a chamber, and a set of anvils raises the pressure to several GPa (Akaishi, Kanda, & Yamaoka, 1990). This process produces bulk or microdiamonds, which must then be milled to produce NDs. Because the milling technique does not allow for precise control of nanoparticle size and form, more work must be done (Rehor & Cigler, 2014), for example, an acid treatment to eliminate milling-related impurities or centrifugation and filtering to isolate NDs with a tighter size distribution.

Environmental sensor applications

In this section, the different types of carbon nanomaterials which were prepared by various methods have been employed in the monitoring of the trace-level determination of various hazardous analytes by using modified electrodes.

Electrochemical sensor applications

The chemical reactions of the target analytes upon the electrodes will be converted into electrical signals by exhibiting conductivity, potential, and current. These changes can be measured with the aid of electrochemical sensor technology. Furthermore, this method offers high selectivity, sensitivity, and a very low detection limit and requires very few samples to analyze (Adarakatti & Kempahanumakkagari, 2019). Typically, this sensor setup consists of either two or three electrodes. A conventional three-electrode cell consists of a working, reference, and counter/auxiliary electrode, where the working electrode is made up of platinum, gold, or carbon. A reference electrode is usually a silver-silver chloride (Ag/AgCl) electrode, and a platinum wire will be used in the counter electrode. All the electrochemical reactions would occur in the electrode interface and electrolytic solution. This further exhibits current as a function of applied voltage to the working electrode (Adarakatti & Kumar, 2021).

To support electrochemical sensor applications by using carbon nanomaterial-based electrodes, researchers have developed many methods toward the trace-level determination of various target species, namely, ammonia, lead, cadmium, copper, mercury, and many more. For instance, glassy carbon spheres and anthraquinone moieties have been covalently modified via diazonium salt reduction toward the trace-level measurement of ammonia which was present in the natural samples, namely, soil and urine samples. In this study, the authors used a basal plane pyrolytic graphite electrode as a working electrode to measure the ammonia concentration level and the modified electrode showed a very low detection limit (Ramakrishnappa, Pandurangappa, & Nagaraju, 2011). In addition to this, nanoparticles of cerium dioxide were decorated with the graphene and subsequently utilized for the electrochemical determination of heavy metal ions such as lead, cadmium, copper, and mercury. In this method, the nanocomposite was prepared via hydrothermal treatment and a glassy carbon electrode was used to modify the electrode surface. Further, an electrochemical technique such as a differential pulse anodic stripping voltammetry method was employed to quantify the target species with high sensitivity (Xie et al., 2015). Similarly, thallium, lead, and mercury analytes were electrochemically determined by using graphene, 1-n-octylpyridinum hexafluorophosphate (OPFP), and [2,4- $Cl_2C_6H_3C(O)CHPPh_3$] (L), as a phosphorus ylide. The fabricated electrode showed good selectivity and stability and requires no separation of the three species from a complex matrix, as this modified electrode was successfully employed in the trace-level electrochemical determination of target analytes from water and soil samples with satisfactory results (Bagheri et al., 2015).

The simultaneous determination of lead and cadmium ions has been investigated by modifying the glassy carbon electrode surface with poly(amidoamine) dendrimer-functionalized magnetic graphene oxide (GO-Fe_3O_4-PAMAM). In this study, the quantification of metal ions was done by using the square wave anodic stripping voltammetry (SWASV) method. Furthermore, various experimental variables, namely, accumulation potential, preconcentration solution pH, concentration of the composite, and accumulation time, were investigated and optimized. The fabricated electrode showed the least interference from most of the common cations and anions and also exhibited good selectivity with high sensitivity toward target metal ions. The proposed electrode exhibited a wide dynamic working linear range and a low detection limit (Baghayeri et al., 2019). In addition to this, by using similar electrochemical techniques, reduced graphene oxide in combination with tin oxide, a modified glassy carbon electrode, was used for the simultaneous measurement of lead, cadmium, copper, and mercury in drinking water. The developed modified electrode showed a very low detection limit as well as a wide linear range (Wei et al., 2012).

Fullerene nanorods have also been used for the electrochemical detection of ethylparaben (EP). In this work, the authors synthesized the material by using a liquid–liquid interface method. Furthermore, the prepared material was covalently immobilized upon the glassy carbon electrode using diazonium salts as an interface. In this method, a nitrophenyl-modified electrode is prepared by electrografting the p-nitrophenyl diazonium salt at the electrode interface. Then, a sodium borohydride/gold–polyaniline solution was used to reduce the nitro group to produce phenylamine. The covalent functionalization of the fullerene nanorods occurred on the GCE via N—H addition reaction across the π-bond of fullerene. Then, the fabricated electrode was electrochemically reduced in potassium hydroxide solution to produce the highly conductive sensor toward the target analyte (Rather et al., 2016).

An antibacterial drug, namely, olaquindox, has been determined by using a simple and sensitive electrochemical sensor based on a super pure single-walled carbon nanotube-modified electrode. In this study, the gold electrodes were used to modify the surface of the electrode, and then it was subjected to cyclic voltammetry study. The developed electrode has shown superior electrocatalytic activity toward the reduction of olaquindox. Further, the bare electrode did not show its analytical response in the presence of the target analyte, whereas the modified electrode exhibited an improved analytical signal at a peak potential of -1.06 V and the reduction peak at -1.02 V (Wang, Liu, et al., 2019; Wang, Pan, et al., 2019). Finally, the modified electrode was successfully employed in the trace determination of olaquindox in pork samples.

Graphene quantum dots doped with nitrogen have gained a lot of attention within the electrochemical community as a sensor material. For instance, the indium tin oxide-conducting electrode was decorated with nitrogen-doped graphene quantum dots for the selective and sensitive determination of mercury ions. In this method, GQDs were synthesized with the aid of the infrared-assisted pyrolysis method, where citric acid and urea were used as fuels. The particle size of GQDs was found to be 4.5 nm. An electrochemical tool such as CV and electrochemical impedance spectroscopic techniques were used to investigate the electrochemical performance of the modified electrode toward mercury ions. The fabricated electrode displayed a very low detection limit with reduced accumulation time (Fu et al., 2020).

The use of nanodiamond (ND) in the development of an electrochemical-sensing platform of pyrazinamide (PZA), one of the most commonly used antibiotics for tuberculosis therapy, is described. A glassy carbon electrode (GCE) was modified with an aqueous ND dispersion to create the electrochemical sensor. In this study, scanning and transmission electron microscopy (SEM and TEM) techniques were used to characterize the produced ND film, and electrochemical characterization tests were carried by utilizing potassium ferricyanide as a redox probe for the bare GCE and modified ND-GCE. Upon this ND-GCE, this redox probe had a better response, a larger electroactive surface area, and a higher heterogeneous electron transfer rate constant. Having recovery percentage, the new voltammetric approach was successfully used in the analysis of PZA in biological systems (Simioni, Silva, Oliveira, & Fatibello-Filho, 2017). The comparison of various modified electrodes toward electrochemical sensing of different analytes is illustrated in Table 10.1.

TABLE 10.1 Comparison of various modified electrodes toward electrochemical sensing of different analytes.

Electrode	Technique	Analyte	Linear range (μg)	LOD (μg)	Real sample	References
AQ/GCE	CV	Ammonia	0.5–30	0.009	Urine and soil	Ramakrishnappa et al. (2011)
Graphene/CeO$_2$/GCE	DPASV	Cd^{2+}, Pb^{2+}, Cu^{2+}, and Hg^{2+}	0.2–2.5	0.0001944, 0.00010572, 0.0001636, and 0.0002771	NR	Xie et al. (2015)
IL/Gr-modified electrode	SWASV	Tl$^+$, Pb^{2+}, and Hg^{2+}	0.00125–0.2.00, 0.00125–0.2, and 0.00125–0.2	0.000357, 0.000450, and 0.000386	Tap water, River water and Soil	Bagheri et al. (2015)
GO-Fe$_3$O$_4$-PAMAM	DPASV	Pb^{2+} and Cd^{2+}	0.4–120 and 0.2–0.140	0.00130 and 0.007	Real water	Baghayeri et al. (2019)
SnO$_2$/rGO	SWASV	Cd^{2+}, Pb^{2+}, Cu^{2+}, and Hg^{2+}	0.3 1.2	0.001015, 0.001839, 0.002269, and 0.002789	NR	Wei et al. (2012)
ERC$_{60}$NRs–NH–Ph–GCE	SWASV	Ethylparaben	0.01–0.55	0.0038 M	OLAY cream	Rather et al. (2016)
spSWCNTs/GCE	DPASV	Olaquindox	0.0001–0.5	0.3	Pork	Wang, Liu, et al. (2019), Wang, Pan, et al. (2019)
N-doped GQDs/ITO	CV	Hg^{2+}	10–50 ppb	0.0010	NR	Fu et al. (2020)
ND-GCE	SWASV	Pyrazinamide	0.79–490	0.22	Urine and human serum	Simioni et al. (2017)

AQ = Anthraquinone; GCE = glassy carbon electrode; CeO$_2$ = cerium dioxide; IL/Gr = ionic liquid/graphene; GO-Fe$_3$O$_4$-PAMAM = poly(amidoamine) dendrimer-functionalized magnetic graphene oxide; SnO$_2$/rGO = tin oxide/reduced graphene oxide; ERC$_{60}$NRs–NH–Ph = electrochemically reduced fullerene phenylamine; spSWCNTs = super pure single-walled carbon nanotubes; N-doped GQDs/ITO = nitrogen-doped graphene quantum dots/indium tin oxide; ND = nanodiamond; CV = cyclic voltammetry; DPASV = differential pulse anodic stripping voltammetry; SWASV = square wave anodic stripping voltammetry; NR = not reported.

Optical sensors

Optical sensors are strong analytical tools used for monitoring analyzers remotely. Principally, they consist of two main units called molecular recognizers and signal transducers. The molecular recognizers will give information of an analysis like its quantitative, qualitative, and other physical properties by interacting with it. However, transducers transduce the signals obtained from the interaction of an analyte with molecular recognizers into measurable optical signals like absorbance, fluorescence, luminescence, and many more. The optical sensors have the ability to detect various parameters of light (UV and visible) like wavelength, frequency, and polarization of light and visualize to electrical signals due to the photoelectrical effect. The optical sensors are advantageous for multimodal and remote sensing compared to electronic and electrochemical sensors.

The generally used carbon materials as optical sensors are carbon nanotubes (CNTs), graphene, graphene oxide (GO), reduced graphene oxide (rGO), carbon dots, and graphene quantum dots (GQDs) (Anas et al., 2019; Farrera et al., 2017; Gao, Cheng, Jiang, Li, & Xing, 2021;Li, Yan, Lu, & Su, 2018; Li, Yan, Qiao, et al., 2018). Similarly, the CNTs emit light in the infrared region and also exhibit better fluorescence, making them good optical sensor materials (Farrera et al., 2017). The commonly employed optical sensing techniques using carbon substrates are fluorescence, colorimetric, surface plasmon resonance (SPR), and surface-enhanced Raman scattering (SERS). However, these sensor systems possess advantages along with some disadvantages when used for sensing of environmentally toxic metal ions. As fluorescence, sensors are sensitive, selective, and reproducible but they need long detection times and are restricted to small molecules. Similarly, colorimetric sensors are cost-effective, require fewer detection times, and are sensitive, but they are less reproducible, are less stable, and finally lack selectivity. The SPR sensors are found to be highly sensitive, cost-effective, label free, but their only drawback is their selectivity.

Colorimetric sensors

The colorimetric sensors have several advantages as mentioned above, and they can also be used as naked eye sensors for spot determination of analytes, especially in water samples. This dramatic color change of the solution indicates the presence of a particular analyte in the species, and measuring its absorbance using a simple colorimeter gives the concentration of analytes. However, there are not many reports regarding colorimetric sensors based on carbon substrates. In fact, some of the carbon substrates like GO, rGO, GQDs, and carbon dots are found to be excellent fluorescent sensors. Some of the carbon dots were reported as dual colorimetric as well as fluorescent sensors for environmentally important analytes like formaldehyde, pesticides, insecticides, and pH (Babazadeh, Moghaddam, Keshipour, & Mollazade, 2020; Li, Yan, Lu, & Su, 2018; Li, Yan, Qiao, et al., 2018; Qu et al., 2020; Wang, Li, et al., 2018; Wang, Liu, et al., 2018). The phosphorous- and nitrogen-doped carbon dots were synthesized using N-(phosphonomethyl)iminodiacetic acid (PMIDA) and branched polyethylenimine (BPEI) (Qu et al., 2020). The synthesized N,P-doped carbon dots exhibited naked eye color change from colorless to light yellow for formaldehyde (Qu et al., 2020). Similarly, carbon dots were used for monitoring the organophosphorus pesticide paraoxon for changes in intensity of yellow color. The sensing mechanism is given in Fig. 10.1. The color change is due to controlled blocking of acetylcholinesterase (Ache) activity by analyte paraoxon. The ACHe has proved to quench the fluorescence of carbon dots effectively. This was due to enzymatic reaction of ACHe with acetylthiocholine to produce thiocholine. The thiocholine specifically decomposes 5,5-dithiobis (2-nitrobenzoic acid) to yellow-colored 5-thio-2-nitrobenzoic acid (TNBA). The TNBA has the capacity to quench the strong fluorescence intensity of carbon dots. However, in the presence of organophosphorus pesticide paraoxon, the ACHe activity was blocked, resulting in an increase in fluorescence intensity and a decrease in yellow color according to the concentration of paraoxon (Li, Yan, Lu, & Su, 2018; Li, Yan, Qiao, et al., 2018). The carbon dot/Au(III)-based colorimetric sensor was reported for monitoring the insecticide imidacloprid in cucumber seeds (Babazadeh et al., 2020). This is based on the reduction reaction of the imidazole group of imidacloprid that results in converting Au(III) to Au(0) in the presence of carbon dots. The resulting Au(0) particle aggregation changes the color of Au(III) to gray or red (Babazadeh et al., 2020). The hydrothermally synthesized carbon dots were reported for pH monitoring (Wang, Li, et al., 2018; Wang, Liu, et al., 2018). In the solution phase, the color changed from red to yellow when the acidic solution was changed to alkaline. Similarly, when carbon dots were loaded on to test paper, the color of the paper turned purple, red to orange, and then yellow when the acidic solution was changed to alkaline (Fig. 10.2) (Wang, Li, et al., 2018; Wang, Liu, et al., 2018).

Fluorescence sensors

The first 2D material, graphene, emerged as my favorite optical sensor material because of its polarization-dependent effects, broadband light absorption, and fluorescence quenching properties. Graphene does not emit fluorescence because

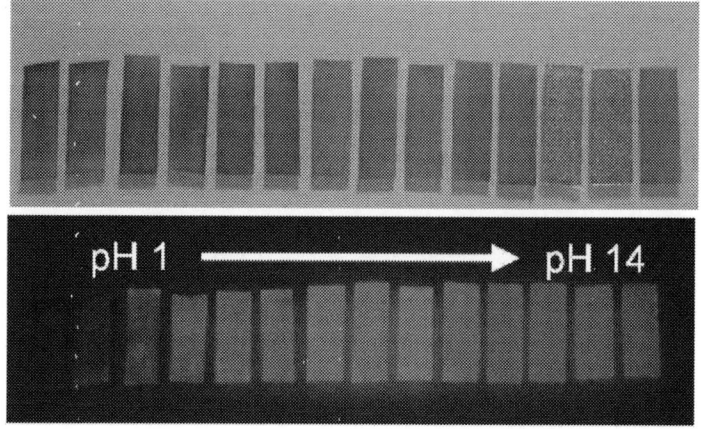

FIG. 10.2 The color change of carbon dots with pH change in solution and when loaded on to paper. *(Source: Wang, L., Li, M., Li, W., Han, Y., Liu, Y., Li, Z., Zhang, B., & Pan, D. (2018). Rationally designed efficient dual-mode colorimetric/fluorescence sensor based on carbon dots for detection of pH and Cu2 + ions. ACS Sustainable Chemistry & Engineering, 6(10), 12668–12674. https://doi.org/10.1021/acssuschemeng.8b01625.)*

of its zero band gap and because of its pure sp^2 hybridization (Zheng & Wu, 2017). However, its analogues, GO and rGO, are fluorescent due to their mixed sp^2 and sp^3 hybridization and nonzero band gaps (Huang, Li, She, & Wang, 2012). GO can absorb the analytes during fluorescence sensing through electrostatic, hydrogen bonding, and π- π interactions, and these interactions also enable them to interact between the analytes. The carbon/graphene QDs size ranges from 1 to 10 nm and proved to be versatile for fluorescence sensing due to their solubility in a wide range of solvents. The emission and energy of emitted light intensity of carbon QDs depend upon their size; the lesser the size, the higher will be the emission and energy of emitted light.

The GQDs synthesized by top-down (hydrothermal cutting, solvothermal cutting, microwave-assisted cutting, ultrasonic approach, and electrochemical/chemical oxidation etc.) and bottom-up approaches (hydrothermal treatment, pyrolysis, reduction of precursors, carbonization of organic precursors, and stepwise organic synthesis) were used for toxic metal ions sensing through fluorescence, colorimetric, and other optical sensing approaches (Anas et al., 2019). The GQD-based optical sensors are used for sensing the toxic metal ion analytes like Hg^{2+}, Pb^{2+}, Cu^{2+}, and Cd^{2+}, Ni^{2+}, and Co^{2+} ions (Anas et al., 2019). The type of GQDs used, synthetic approaches of GQDs, analyte sensed, optical sensor approach employed, and corresponding analytical features like linear range limit of detection (LODs) are outlined in Table 10.2. Other than toxic metal ions, reduced graphene quantum dots (rGQDs) were also reported for fluorescence sensing of the organopesticide diazinon and the sensor based on specific aptamer on fabricated rGQDs and MWCNTs (Talari, Bozorg, Faridbod, & Vossoughi, 2021).

Other than GQDS, various GO-based aptamer fluorescence sensors were also reported for monitoring environmentally toxic metal ions like Hg^{2+} ions (Khoshbin, Housaindokht, & Verdian, 2020; Li, Zhou, Ding, Guo, & Wu, 2013). The CNTs were reported as fluorescence signal enhancers for Cd^{2+} ion monitoring in tobacco (Carolina Talio, Alesso, Acosta, Olsina, & Fernández, 2013). A range of carbon nanostructures (mainly carbon dots) have been reported as fluorescent probes for hexavalent chromium ion detection from various matrixes (Bu et al., 2016; Chen et al., 2019; Fang et al., 2018; Gong et al., 2017; Liu et al., 2017; Liu et al., 2019; Sharma, Umar, Mehta, & Kansal, 2017; Singh et al., 2018; Wang et al., 2017; Wang, Li, et al., 2018; Wang, Liu, et al., 2018). These carbon dots were synthesized via microwave or hydrothermal routes to yield plain or N-doped carbon dots by using organic precursors like ionic liquids, ethylenediamine, citric acid, glycine, etc. (Jandera & Churáček, 1985; Liu et al., 2017; Liu et al., 2019; Wang, Li, et al., 2018; Wang, Liu, et al., 2018;Gong et al., 2017; Singh et al., 2018). The carbon dots doped with both phosphorus and nitrogen were also reported as fluorescent sensors for Cr^{6+} ions (Gong et al., 2017; Singh et al., 2018). A green hydrothermal approach was reported for the synthesis of carbon nanoglobules and their application for Cr^{6+} ion sensing. This was done by using pineapple juice as the carbon source, and one-pot hydrothermal process was employed to get the carbon nanostructures (Sharma et al., 2017). The organic molecule-appended carbon dots were also reported for Cr^{6+} ion sensing. The water-soluble fluorescent carbon dots were synthesized by the one-pot hydrothermal method using the cellulose as the carbon source. The obtained carbon dot materials were then linked with isophorone and β-cyclodextrin (Wang et al., 2017). The test papers based on fluorescent carbon dots were reported for spot testing the presence of Cr^{6+} ions in water samples through "naked eye color change". These test papers showed naked eye color changes in the presence of Cr^{6+} ions in water samples when illuminated under UV light (Bu et al., 2016). The green fluorescent carbon dots were synthesized by the one-pot hydrothermal technique using benzoxazine as the organic carbon precursor in basic conditions for Cr^{6+} ion sensing through fluorescence quenching (Fang et al., 2018).

TABLE 10.2 The type of GQDs, synthesis route, precursors used, analyte sensed, and their analytical features.

Sl. no.	Type of GQDs	Synthesis route	Precursors	Toxic metal ion	Optical method	LOD (μg)	Linear range (μg)	References
1	GQDs	Carbonization	Citric acid	Hg^{2+}	Fluorescence	3.36	NA	Chakraborti, Sinha, Ghosh, and Pal (2013)
2	Cys-GQDs	Carbonization	Citric acid/ cysteine	Hg^{2+}	Fluorescence	0.02	0–500	Tam, Hong, and Choi (2015)
3	GQDs	Pyrolysis	citric acid	Hg^{2+}	Dual fluorescence	0.00044	0.001–0.05	Li, Wang, Ni, and Kokot (2015)
4	DNA-GQDS	Hydrothermal cutting	Graphite powder	Hg^{2+}	Fluorescence	0.00025	0.001–10	Zhao et al. (2015)
5	NPS-GQDs	Electrochemical oxidation	Anthracite coal	Pb^{2+}	Fluorescence	0.75	0.001–0.02	Xu et al. (2018)
6	rGQDs	Oxidation reduction	Graphite powder	Pb^{2+}	Fluorescence	0.0006	0.0099–0.435	Qian, Shan, Chai, Chen, and Feng (2015)
7	GQDs	Hydrothermal	Reoxidized GO	Cu^{2+}	Fluorescence	0.226	0–15	Wang et al. (2014)
8	afGQDs	Microwave/ hydrothermal amination	GO	Cu^{2+}	Fluorescence	0.0069	0–100	Sun, Gao, et al. (2013), Sun, Wang, et al. (2013)
9	B-GQDs	Electrochemical exfoliation	Graphite rod	Al^{3+}	Fluorescence	3.64	0–100	Fan et al. (2014)
10	TMPyp/NGQDs	Hydrothermal oxidize	Nitrogen-doped graphene	Cd^{2+}	Fluorescence	0.088	0.5–9	Zhang, Peng, Liang, and Qiu (2015)

Cys=Cysteine; NPS=Nitrogen, phosphorous, and sulfur; af=amino-functionalized; B=Boron; TMPyp=5,10,15,20-tetrakis(1-methyl-4-pyridinio) porphyrin tetra(p-toluenesulfonate).

The fluorescent carbon dots were synthesized by green one-pot hydrothermal protocol using the waste tea extract as the carbon source and used for monitoring Cr^{6+} ions by fluorescence quenching (Chen et al., 2019). All the fluorescent carbon dots mentioned above were applied for monitoring the Cr^{6+} ions in real water samples like aquaculture water, biosystems, and soil samples. A similar carbon dot material was reported as a fluorescent probe for Cd^{2+} ions through a fluorescence quenching technique. The carbon nanodots were synthesized by carbonization of the presynthesized sago starch nanoparticles followed by their surface oxidation using nitric acid (Mohd Yazid, Chin, Pang, & Ng, 2013). The sago starch nanoparticles were synthesized by nanoprecipitation from aqueous solution suspensions of sago starch. The carbon nanodots were obtained from sago starch nanoparticles by dehydration in the presence of concentrated sulfuric acid (Mohd Yazid et al., 2013).

Conclusion

In conclusion, the present chapter describes the applications of various carbon-based nanomaterials like fullerene, CNTs, GO, rGO, graphene, and GQDs as electrochemical and optical sensing materials. Numerous environmental toxic substances like heavy metal ions, pesticides, and insecticides have been monitored in different environmental matrices by sensor systems developed using carbon nanomaterials. In the electrochemical systems, carbon nanomaterials like CNTS, graphene, and rGO, because of their excellent conducting properties, help in increasing the electron transfer rates, which are very necessary during sensing studies. The above-said carbon nanomaterials also served as electrocatalysts during electrochemical sensing studies, helping in decreasing the over-potentials, increasing the sensitivity and selectivity during electrochemical sensing applications. Similarly, carbon nanomaterials like graphene, GO, GQDs, and carbon dots were used as colorimetric and fluorescent sensor materials for environmental monitoring. The special optical properties of carbon nanomaterials like their wide spectrum of absorption and emission, as well as fluorescence quenching properties, enable them to be used as colorimetric and fluorescence sensor substrates. Furthermore, simple, naked eye sensor systems like paper-based devices were also developed for environmental monitoring. Finally, carbon-based nanomaterials are proved to be potential candidates in electrochemical and optical-based sensor systems, especially in environmentally monitoring applications.

Acknowledgments

Dr. Prashanth S A acknowledges SVM Arts, Science and Commerce College, ILKAL, for their support and encouragement. Dr. Sureshkumar K and Dr. Ramakrishnappa T acknowledge B.M.S. Institute of Technology and Management for their support and encouragement.

References

Adarakatti, P. S., & Kempahanumakkagari, S. K. (2019). Modified electrodes for sensing. *SPR Electrochemistry, 15*, 58–95. https://doi.org/10.1039/9781788013895-00058.

Adarakatti, P. S., & Kumar, K. S. (2021). Functionalized macromolecules-based disposable sensors. In *Disposable electrochemical sensors for healthcare monitoring* (pp. 381–409). Royal Society of Chemistry. https://doi.org/10.1039/9781839163364-00381.

Agüí, L., Yáñez-Sedeño, P., & Pingarrón, J. M. (2008). Role of carbon nanotubes in electroanalytical chemistry. A review. *Analytica Chimica Acta, 622*(1–2), 11–47. https://doi.org/10.1016/j.aca.2008.05.070.

Ajayan, P. M., & Lijima, S. (1992). Smallest carbon nanotube. *Nature*, 23. https://doi.org/10.1038/358023a0.

Akaishi, M., Kanda, H., & Yamaoka, S. (1990). Synthesis of diamond from graphite-carbonate system under very high temperature and pressure. *Journal of Crystal Growth, 104*(2), 578–581. https://doi.org/10.1016/0022-0248(90)90159-I.

Al-Jumaili, A., Alancherry, S., Bazaka, K., & Jacob, M. V. (2017). Review on the antimicrobial properties of carbon nanostructures. *Materials, 10*(9). https://doi.org/10.3390/ma10091066.

Anas, N. A. A., Fen, Y. W., Omar, N. A. S., Daniyal, W. M. E. M. M., Ramdzan, N. S. M., & Saleviter, S. (2019). Development of graphene quantum dots-based optical sensor for toxic metal ion detection. *Sensors, 19*(18), 3850. https://doi.org/10.3390/s19183850.

Babazadeh, S., Moghaddam, P. A., Keshipour, S., & Mollazade, K. (2020). Colorimetric sensing of imidacloprid in cucumber fruits using a graphene quantum dot/Au (III) chemosensor. *Scientific Reports, 10*(1). https://doi.org/10.1038/s41598-020-71349-4.

Baghayeri, M., Alinezhad, H., Fayazi, M., Tarahomi, M., Ghanei-Motlagh, R., & Maleki, B. (2019). A novel electrochemical sensor based on a glassy carbon electrode modified with dendrimer functionalized magnetic graphene oxide for simultaneous determination of trace Pb(II)and cd(II). *Electrochimica Acta, 312*, 80–88. https://doi.org/10.1016/j.electacta.2019.04.180.

Bagheri, H., Afkhami, A., Khoshsafar, H., Rezaei, M., Sabounchei, S. J., & Sarlakifar, M. (2015). Simultaneous electrochemical sensing of thallium, lead and mercury using a novel ionic liquid/graphene modified electrode. *Analytica Chimica Acta, 870*(1), 56–66. https://doi.org/10.1016/j.aca.2015.03.004.

Bak, S., Kim, D., & Lee, H. (2016). Graphene quantum dots and their possible energy applications: A review. *Current Applied Physics, 16*(9), 1192–1201. https://doi.org/10.1016/j.cap.2016.03.026.

Baker, S. N., & Baker, G. A. (2007). Luminescent carbon nanodots: Emergent nanolights. *Angewandte Chemie International Edition, 49*(38), 183–191.

Banhart, F., & Ajayan, P. M. (1996). Carbon onions as nanoscopic pressure cells for diamond formation. *Nature, 382*(6590), 433–435. https://doi.org/10.1038/382433a0.

Banks, C. E., Moore, M. R., Davies, T. J., & Compton, R. G. (2004). Investigation of modified basal plane pyrolytic graphite electrodes: Definitive evidence for the electrocatalytic properties of the ends of carbon nanotubes. *Chemical Communications, 16*, 1804–1805. https://doi.org/10.1039/b406174h.

Becker, L., Bada, J. L., Winans, R. E., Hunt, J. E., Bunch, T. E., & French, B. M. (1994). Fullerenes in the 1.85-billion-year-old sudbury impact structure. *Science, 265*(5172), 642–645. https://doi.org/10.1126/science.11536660.

Bera, D. (n.d.). Quantum dots and their multimodal applications: A review. Materials, 2010 (Vol. 3, pp. 2260–2345).

Bethune, D. S., Klang, C. H., De Vries, M. S., Gorman, G., Savoy, R., Vazquez, J., et al. (1993). Cobalt-catalysed growth of carbon nanotubes with single-atomic-layer walls. *Nature, 363*(6430), 605–607. https://doi.org/10.1038/363605a0.

Boudou, J. P., Curmi, P. A., Jelezko, F., Wrachtrup, J., Aubert, P., Sennour, M., et al. (2009). High yield fabrication of fluorescent nanodiamonds. *Nanotechnology, 20*(23). https://doi.org/10.1088/0957-4484/20/23/235602.

Briffa, J., Sinagra, E., & Blundell, R. (2020). Heavy metal pollution in the environment and their toxicological effects on humans. *Heliyon, 6*(9). https://doi.org/10.1016/j.heliyon.2020.e04691.

Brownson, D. A. C., Figueiredo-Filho, L. C. S., Riehl, B. L., Riehl, B. D., Gómez-Mingot, M., Iniesta, J., et al. (2016). High temperature low vacuum synthesis of a freestanding three-dimensional graphene nano-ribbon foam electrode. *Journal of Materials Chemistry A, 4*(7), 2617–2629. https://doi.org/10.1039/c5ta08561f.

Bu, L., Peng, J., Peng, H., Liu, S., Xiao, H., Liu, D., et al. (2016). Fluorescent carbon dots for the sensitive detection of Cr(VI) in aqueous media and their application in test papers. *RSC Advances, 6*(98), 95469–95475. https://doi.org/10.1039/c6ra19977a.

Butler, J. E., & Sumant, A. V. (2008). The CVD of nanodiamond materials. *Chemical Vapor Deposition, 14*(7–8), 145–160. https://doi.org/10.1002/cvde.200700037.

Carolina Talio, M., Alesso, M., Acosta, M., Olsina, R., & Fernández, L. P. (2013). Determination of cadmium in tobacco by solid surface fluorescence using nylon membranes coated with carbon nanotubes. *Talanta, 107*, 61–66. https://doi.org/10.1016/j.talanta.2012.12.048.

Chakraborti, H., Sinha, S., Ghosh, S., & Pal, S. K. (2013). Interfacing water soluble nanomaterials with fluorescence chemosensing: Graphene quantum dot to detect Hg2+ in 100% aqueous solution. *Materials Letters, 97*, 78–80. https://doi.org/10.1016/j.matlet.2013.01.094.

Chang, Y. R., Lee, H. Y., Chen, K., Chang, C. C., Tsai, D. S., Fu, C. C., et al. (2008). Mass production and dynamic imaging of fluorescent nanodiamonds. *Nature Nanotechnology, 3*(5), 284–288. https://doi.org/10.1038/nnano.2008.99.

Chao, D., Zhu, C., Xia, X., Liu, J., Zhang, X., Wang, J., et al. (2015). Graphene quantum dots coated VO2 arrays for highly durable electrodes for Li and Na ion batteries. *Nano Letters, 15*(1), 565–573. https://doi.org/10.1021/nl504038s.

Chen, K., Qing, W., Hu, W., Lu, M., Wang, Y., & Liu, X. (2019). On-off-on fluorescent carbon dots from waste tea: Their properties, antioxidant and selective detection of CrO 42−, Fe 3+, ascorbic acid and L-cysteine in real samples. *Spectrochimica Acta. Part A, Molecular and Biomolecular Spectroscopy, 213*, 228–234. https://doi.org/10.1016/j.saa.2019.01.066.

Chien, C. T., Li, S. S., Lai, W. J., Yeh, Y. C., Chen, H. A., Chen, I. S., et al. (2012). Tunable photoluminescence from graphene oxide. *Angewandte Chemie, International Edition, 51*(27), 6662–6666. https://doi.org/10.1002/anie.201200474.

Cooper, D. R. (2012). *Experimental review of graphene.* (International Scholarly Research Notices).

Cubo, E. (2010). Hydrocarbons. In *Encyclopedia of movement disorders* (pp. 49–51). Elsevier Inc. https://doi.org/10.1016/B978-0-12-374105-9.00035-6.

Dai, H., Rinzler, A. G., Nikolaev, P., Thess, A., Colbert, D. T., & Smalley, R. E. (1996). Single-wall nanotubes produced by metal-catalyzed disproportionation of carbon monoxide. *Chemical Physics Letters, 260*(3–4), 471–475. https://doi.org/10.1016/0009-2614(96)00862-7.

Danilenko, V. V. (2004). On the history of the discovery of nanodiamond synthesis. *Physics of the Solid State, 46*(4), 595–599. https://doi.org/10.1134/1.1711431.

Daulton, T. L., Kirk, M. A., Lewis, R. S., & Rehn, L. E. (2001). Production of nanodiamonds by high-energy ion irradiation of graphite at room temperature. In (Vols. 175–177). *Nuclear instruments and methods in physics research, section B: Beam interactions with materials and atoms* (pp. 12–20). https://doi.org/10.1016/S0168-583X(00)00603-0.

Davies, G. (1994). *Properties and growth of diamond.* London, UK: INSPEC, the Institution of Electrical Engineers.

Dolmatov, V. Y. (2018). The influence of detonation synthesis conditions on the yield of condensed carbon and detonation nanodiamond through the example of using TNT-RDX explosive mixture. *Journal of Superhard Materials, 40*(4), 290–294. https://doi.org/10.3103/S1063457618040093.

Dreyer, D. R., Park, S., Bielawski, C. W., & Ruoff, R. S. (2010). The chemistry of graphene oxide. *Chemical Society Reviews, 39*(1), 228–240. https://doi.org/10.1039/b917103g.

Eckenrode, B. A. (2001). Environmental and forensic applications of field-portable GC-MS: An overview. *Journal of the American Society for Mass Spectrometry, 12*(6), 683–693. https://doi.org/10.1016/S1044-0305(01)00251-3.

Fan, Z., Li, Y., Li, X., Fan, L., Zhou, S., Fang, D., et al. (2014). Surrounding media sensitive photoluminescence of boron-doped graphene quantum dots for highly fluorescent dyed crystals, chemical sensing and bioimaging. *Carbon, 70*, 149–156. https://doi.org/10.1016/j.carbon.2013.12.085.

Fang, B., Wang, P., Zhu, Y., Wang, C., Zhang, G., Zheng, X., et al. (2018). Basophilic green fluorescent carbon nanoparticles derived from benzoxazine for the detection of Cr(VI) in a strongly alkaline enVIronment. *RSC Advances, 8*(14), 7377–7382. https://doi.org/10.1039/c7ra10814a.

Farrera, C., Torres Andón, F., & Feliu, N. (2017). Carbon nanotubes as optical sensors in biomedicine. *ACS Nano, 11*(11), 10637–10643. https://doi.org/10.1021/acsnano.7b06701.

Frenklach, M., Howard, W., Huang, D., Yuan, J., Spear, K. E., & Koba, R. (1991). Induced nucleation of diamond powder. *Applied Physics Letters, 59*(5), 546–548. https://doi.org/10.1063/1.105434.

Fu, C. C., Hsieh, C. T., Juang, R. S., Gu, S., Ashraf Gandomi, Y., Kelly, R. E., et al. (2020). Electrochemical sensing of mercury ions in electrolyte solutions by nitrogen-doped graphene quantum dot electrodes at ultralow concentrations. *Journal of Molecular Liquids, 302*. https://doi.org/10.1016/j.molliq.2020.112593.

Galimov, E. M., Kudin, A. M., Skorobogatskiĭ, V. N., Plotnichenko, V. G., Bondarev, O. L., Zarubin, B. G., et al. (2004). Experimental corroboration of the synthesis of diamond in the cavitation process. *Doklady Physics, 49*(3), 150–153. https://doi.org/10.1134/1.1710678.

Gao, X. G., Cheng, L. X., Jiang, W. S., Li, X. K., & Xing, F. (2021). Graphene and its derivatives-based optical sensors. *Frontiers in Chemistry, 9*. https://doi.org/10.3389/fchem.2021.615164.

Geim, A. K., & Novoselov, K. S. (2007). The rise of graphene. *Nature Materials, 6*(3), 183–191. https://doi.org/10.1038/nmat1849.

Georgakilas, V., Otyepka, M., Bourlinos, A. B., Chandra, V., Kim, N., Kemp, K. C., et al. (2012). Functionalization of graphene: Covalent and non-covalent approaches, derivatives and applications. *Chemical Reviews, 112*(11), 6156–6214. https://doi.org/10.1021/cr3000412.

George, J., Halali, V. V., Sanjayan, C. G., Suvina, V., Sakar, M., & Balakrishna, R. G. (2020). Perovskite nanomaterials as optical and electrochemical sensors. *Inorganic Chemistry Frontiers, 7*(14), 2702–2725. https://doi.org/10.1039/d0qi00306a.

Gogotsi, Y. G. (1996). Structure of carbon produced by hydrothermal treatment of β-SiC powder. *Journal of Materials Chemistry, 6*(4), 595–604. https://doi.org/10.1039/JM9960600595.

Gong, X., Liu, Y., Yang, Z., Shuang, S., Zhang, Z., & Dong, C. (2017). An "on-off-on" fluorescent nanoprobe for recognition of chromium(VI) and ascorbic acid based on phosphorus/nitrogen dual-doped carbon quantum dot. *Analytica Chimica Acta, 968*, 85–96. https://doi.org/10.1016/j.aca.2017.02.038.

Greiner, N. R., Phillips, D. S., Johnson, J. D., & Volk, F. (1988). Diamonds in detonation soot. *Nature, 333*(6172), 440–442. https://doi.org/10.1038/333440a0.

Guo, T., Nikolaev, P., Thess, A., Colbert, D. T., & Smalley, R. E. (1995). Catalytic growth of single-walled manotubes by laser vaporization. *Chemical Physics Letters, 243*(1–2), 49–54. https://doi.org/10.1016/0009-2614(95)00825-O.

Guo, H. L., Wang, X. F., Qian, Q. Y., Wang, F. B., & Xia, X. H. (2009). A green approach to the synthesis of graphene nanosheets. *ACS Nano, 3*(9), 2653–2659. https://doi.org/10.1021/nn900227d.

He, T., Qi, X., Chen, R., Wei, J., Zhang, H., & Sun, H. (2012). Enhanced optical nonlinearity in noncovalently functionalized amphiphilic graphene composites. *ChemPlusChem, 77*(8), 688–693. https://doi.org/10.1002/cplu.201200113.

Hirsch, A. (2008). The chemistry of the fullerenes. In *The chemistry of the fullerenes* (pp. 1–203). Wiley. https://doi.org/10.1002/9783527619214.

Huang, H., Li, Z., She, J., & Wang, W. (2012). Oxygen density dependent band gap of reduced graphene oxide. *Journal of Applied Physics, 111*(5). https://doi.org/10.1063/1.3694665, 054317.

Hummers, W. S., & Offeman, R. E. (1958). Preparation of graphitic oxide. *Journal of the American Chemical Society, 80*(6), 1339. https://doi.org/10.1021/ja01539a017.

Jandera, P., & Churáček, J. (1985). Gradient elution in column liquid chromatography: Theory and practice. *Journal of Chromatography Library, 31*, 465–468.

Jeon, I.-Y., Shin, Y.-R., Sohn, G.-J., Choi, H.-J., Bae, S.-Y., Mahmood, J., et al. (2012). Edge-carboxylated graphene nanosheets via ball milling. *Proceedings of the National Academy of Sciences, 109*(15), 5588–5593. https://doi.org/10.1073/pnas.1116897109.

Ji, X., Kadara, R. O., Krussma, J., Chen, Q., & Banks, C. E. (2010). Understanding the physicoelectrochemical properties of carbon nanotubes: Current state of the art. *Electroanalysis, 22*(1), 7–19. https://doi.org/10.1002/elan.200900493.

Ju, H., Zhang, X., & Wang, J. (2011). *Springer science and business media LLC* (pp. 207–239). https://doi.org/10.1007/978-1-4419-9622-0_7.

Kailasa, S. K., & Wu, H. F. (2012). Inorganic contaminants: Sample preparation approaches. In *Vol. 3. Comprehensive sampling and sample preparation* (pp. 743–782). Elsevier Inc. https://doi.org/10.1016/B978-0-12-381373-2.00112-5.

Kamala, C. T., Balaram, V., Dharmendra, V., Subramanyam, K. S. V., & Krishnaiah, A. (2014). Application of microwave plasma atomic emission spectrometry (MP-AES) for environmental monitoring of industrially contaminated sites in Hyderabad City. *Environmental Monitoring and Assessment, 186*(11), 7097–7113. https://doi.org/10.1007/s10661-014-3913-4.

Kappe, C. O. (2004). Controlled microwave heating in modern organic synthesis. *Angewandte Chemie, International Edition, 43*(46), 6250–6284. https://doi.org/10.1002/anie.200400655.

Khoshbin, Z., Housaindokht, M. R., & Verdian, A. (2020). A low-cost paper-based aptasensor for simultaneous trace-level monitoring of mercury (II) and silver (I) ions. *Analytical Biochemistry, 597*. https://doi.org/10.1016/j.ab.2020.113689.

Krätschmer, W., Lamb, L. D., Fostiropoulos, K., & Huffman, D. R. (1990). Solid C60: A new form of carbon. *Nature, 347*(6291), 354–358. https://doi.org/10.1038/347354a0.

Kroto, H. W., Allaf, A. W., & Balm, S. P. (1991). C60: Buckminsterfullerene. *Chemical Reviews, 91*(6), 1213–1235. https://doi.org/10.1021/cr00006a005.

Lettieri, S., Camisasca, A., D'Amora, M., Diaspro, A., Uchida, T., Nakajima, Y., et al. (2017). Far-red fluorescent carbon nano-onions as a biocompatible platform for cellular imaging. *RSC Advances, 7*(72), 45676–45681. https://doi.org/10.1039/c7ra09442f.

Li, Y., Hu, Y., Zhao, Y., Shi, G., Deng, L., Hou, Y., et al. (2011). An electrochemical avenue to green-luminescent graphene quantum dots as potential electron-acceptors for photovoltaics. *Advanced Materials, 23*(6), 776–780. https://doi.org/10.1002/adma.201003819.

Li, L.-L., Ji, J., Fei, R., Wang, C.-Z., Lu, Q., Zhang, J.-R., et al. (2012). A facile microwave avenue to electrochemiluminescent two-color graphene quantum dots. *Advanced Functional Materials, 22*(14), 2971–2979. https://doi.org/10.1002/adfm.201200166.

Li, Z., Wang, Y., Ni, Y., & Kokot, S. (2015). A rapid and label-free dual detection of Hg (II) and cysteine with the use of fluorescence switching of graphene quantum dots. *Sensors and Actuators, B: Chemical, 207*, 490–497. https://doi.org/10.1016/j.snb.2014.10.071.

Li, H., Yan, X., Lu, G., & Su, X. (2018). Carbon dot-based bioplatform for dual colorimetric and fluorometric sensing of organophosphate pesticides. *Sensors and Actuators, B: Chemical, 260*, 563–570. https://doi.org/10.1016/j.snb.2017.12.170.

Li, H., Yan, X., Qiao, S., Lu, G., & Su, X. (2018). Yellow-emissive carbon dot-based optical sensing platforms: Cell imaging and analytical applications for biocatalytic reactions. *ACS Applied Materials and Interfaces, 10*(9), 7737–7744. https://doi.org/10.1021/acsami.7b17619.

Li, M., Zhou, X., Ding, W., Guo, S., & Wu, N. (2013). Fluorescent aptamer-functionalized graphene oxide biosensor for label-free detection of mercury(II). *Biosensors and Bioelectronics, 41*(1), 889–893. https://doi.org/10.1016/j.bios.2012.09.060.

Liu, S., Cui, J., Huang, J., Tian, B., Jia, F., & Wang, Z. (2019). Facile one-pot synthesis of highly fluorescent nitrogen-doped carbon dots by mild hydrothermal method and their applications in detection of Cr(VI) ions. *Spectrochimica Acta. Part A, Molecular and Biomolecular Spectroscopy, 206*, 65–71. https://doi.org/10.1016/j.saa.2018.07.082.

Liu, W. W., Feng, Y. Q., Yan, X. B., Chen, J. T., & Xue, Q. J. (2013). Superior micro-supercapacitors based on graphene quantum dots. *Advanced Functional Materials, 23*(33), 4111–4122. https://doi.org/10.1002/adfm.201203771.

Liu, X., Li, T., Wu, Q., Yan, X., Wu, C., Chen, X., et al. (2017). Carbon nanodots as a fluorescence sensor for rapid and sensitive detection of Cr(VI) and their multifunctional applications. *Talanta, 165*, 216–222. https://doi.org/10.1016/j.talanta.2016.12.037.

Liu, R., Wu, D., Feng, X., & Müllen, K. (2011). Bottom-up fabrication of photoluminescent graphene quantum dots with uniform morphology. *Journal of the American Chemical Society, 133*(39), 15221–15223. https://doi.org/10.1021/ja204953k.

Lu, J., Yang, J. X., Wang, J., Lim, A., Wang, S., & Loh, K. P. (2009). One-pot synthesis of fluorescent carbon nanoribbons, nanoparticles, and graphene by the exfoliation of graphite in ionic liquids. *ACS Nano, 3*(8), 2367–2375. https://doi.org/10.1021/nn900546b.

Maiti, D., Tong, X., Mou, X., & Yang, K. (2019). Carbon-based nanomaterials for biomedical applications: A recent study. *Frontiers in Pharmacology, 9*. https://doi.org/10.3389/fphar.2018.01401.

May, P. W., Ashfold, M. N. R., & Mankelevich, Y. A. (2007). Microcrystalline, nanocrystalline, and ultrananocrystalline diamond chemical vapor deposition: Experiment and modeling of the factors controlling growth rate, nucleation, and crystal size. *Journal of Applied Physics, 101*(5). https://doi.org/10.1063/1.2696363.

McAllister, M. J., Li, J. L., Adamson, D. H., Schniepp, H. C., Abdala, A. A., Liu, J., et al. (2007). Single sheet functionalized graphene by oxidation and thermal expansion of graphite. *Chemistry of Materials, 19*(18), 4396–4404. https://doi.org/10.1021/cm0630800.

Mishra, A. K., & Ramaprabhu, S. (2011). Removal of metals from aqueous solution and sea water by functionalized graphite nanoplatelets based electrodes. *Journal of Hazardous Materials, 185*(1), 322–328. https://doi.org/10.1016/j.jhazmat.2010.09.037.

Mkhoyan, K. A., Contryman, A. W., Silcox, J., Stewart, D. A., Eda, G., Mattevi, C., et al. (2009). Atomic and electronic structure of graphene-oxide. *Nano Letters, 9*(3), 1058–1063. https://doi.org/10.1021/nl8034256.

Mochalin, V. N., & Gogotsi, Y. (2009). Wet chemistry route to hydrophobic blue fluorescent nanodiamond. *Journal of the American Chemical Society, 131*(13), 4594–4595. https://doi.org/10.1021/ja9004514.

Mochalin, V. N., Shenderova, O., Ho, D., & Gogotsi, Y. (2012). The properties and applications of nanodiamonds. *Nature Nanotechnology, 7*(1), 11–23. https://doi.org/10.1038/nnano.2011.209.

Mohamed, E. F. (2017). Nanotechnology: Future of environmental air pollution control. *Environmental Management and Sustainable Development, 429*. https://doi.org/10.5296/emsd.v6i2.12047.

Mohd Yazid, S. N. A., Chin, S. F., Pang, S. C., & Ng, S. M. (2013). Detection of Sn(II) ions via quenching of the fluorescence of carbon nanodots. *Microchimica Acta, 180*(1–2), 137–143. https://doi.org/10.1007/s00604-012-0908-0.

Nujić, & Habuda-Stanić. (2019). Toxic metal ions in drinking water and effective removal using graphene oxide nanocomposite. In *A new generation material graphene: Applications in water technology* (pp. 373–395). Springer.

Ozawa, M., Inaguma, M., Takahashi, M., Kataoka, F., Krüger, A., & Osawa, E. (2007). Preparation and behavior of brownish, clear nanodiamond colloids. *Advanced Materials, 19*(9), 1201–1206. https://doi.org/10.1002/adma.200601452.

Pan, D., Zhang, J., Li, Z., & Wu, M. (2010). Hydrothermal route for cutting graphene sheets into blue-luminescent graphene quantum dots. *Advanced Materials, 22*(6), 734–738. https://doi.org/10.1002/adma.200902825.

Pan, J., Cha, T. G., Chen, H., & Choi, J. H. (2013). 10—Carbon nanotube-based optical platforms for biomolecular detection. In *Carbon nanotubes and graphene for photonic applications* (pp. 270–303). Elsevier.

Paredes, J. I., Villar-Rodil, S., Martínez-Alonso, A., & Tascón, J. M. D. (2008). Graphene oxide dispersions in organic solvents. *Langmuir, 24*(19), 10560–10564. https://doi.org/10.1021/la801744a.

Park, S., & Ruoff, R. S. (2009). Chemical methods for the production of graphenes. *Nature Nanotechnology, 4*(4), 217–224. https://doi.org/10.1038/nnano.2009.58.

Perathoner, S., & Centi, G. (2018). Advanced nanocarbon materials for future energy applications. In *Emerging materials for energy conversion and storage* (pp. 305–325). Elsevier. https://doi.org/10.1016/B978-0-12-813794-9.00009-0.

Piyasena, P., Dussault, C., Koutchma, T., Ramaswamy, H. S., & Awuah, G. B. (2003). Radio frequency heating of foods: Principles, applications and related properties—A review. *Critical Reviews in Food Science and Nutrition, 43*(6), 587–606. https://doi.org/10.1080/10408690390251129.

Qian, Z. S., Shan, X. Y., Chai, L. J., Chen, J. R., & Feng, H. (2015). A fluorescent nanosensor based on graphene quantum dots-aptamer probe and graphene oxide platform for detection of lead (II) ion. *Biosensors and Bioelectronics, 68*, 225–231. https://doi.org/10.1016/j.bios.2014.12.057.

Qu, J., Zhang, X., Liu, Y., Xie, Y., Cai, J., Zha, G., et al. (2020). N, P-co-doped carbon dots as a dual-mode colorimetric/ratiometric fluorescent sensor for formaldehyde and cell imaging via an aminal reaction-induced aggregation process. *Microchimica Acta, 187*(6). https://doi.org/10.1007/s00604-020-04337-0.

Ramakrishnappa, T., Pandurangappa, M., & Nagaraju, D. H. (2011). Anthraquinone functionalized carbon composite electrode: Application to ammonia sensing. *Sensors and Actuators, B: Chemical, 155*(2), 626–631. https://doi.org/10.1016/j.snb.2011.01.020.

Ramesh, P., Bhagyalakshmi, S., & Sampath, S. (2004). Preparation and physicochemical and electrochemical characterization of exfoliated graphite oxide. *Journal of Colloid and Interface Science, 274*(1), 95–102. https://doi.org/10.1016/j.jcis.2003.11.030.

Ramnani, P., Saucedo, N. M., & Mulchandani, A. (2016). Carbon nanomaterial-based electrochemical biosensors for label-free sensing of environmental pollutants. *Chemosphere, 143*, 85–98. https://doi.org/10.1016/j.chemosphere.2015.04.063.

Randić, M., Kroto, H. W., & Vukičević, D. (2007). Numerical Kekulé structures of fullerenes and partitioning of π-electrons to pentagonal and hexagonal rings. *Journal of Chemical Information and Modeling, 47*(3), 897–904. https://doi.org/10.1021/ci600484u.

Rao, C. N. R., Sood, A. K., Voggu, R., & Subrahmanyam, K. S. (2010). Some novel attributes of graphene. *Journal of Physical Chemistry Letters, 1*(2), 572–580. https://doi.org/10.1021/jz9004174.

Rather, J. A., Al Harthi, A. J., Khudaish, E. A., Qurashi, A., Munam, A., & Kannan, P. (2016). An electrochemical sensor based on fullerene nanorods for the detection of paraben, an endocrine disruptor. *Analytical Methods, 8*(28), 5690–5700. https://doi.org/10.1039/c6ay01489e.

Rauti, R., Musto, M., Bosi, S., Prato, M., & Ballerini, L. (2019). Properties and behavior of carbon nanomaterials when interfacing neuronal cells: How far have we come? *Carbon, 143*, 430–446. https://doi.org/10.1016/j.carbon.2018.11.026.

Rehor, I., & Cigler, P. (2014). Precise estimation of HPHT nanodiamond size distribution based on transmission electron microscopy image analysis. *Diamond and Related Materials, 46*, 21–24. https://doi.org/10.1016/j.diamond.2014.04.002.

Ren, P.-G., Yan, D.-X., Ji, X., Chen, T., & Li, Z.-M. (2011). Temperature dependence of graphene oxide reduced by hydrazine hydrate. *Nanotechnology*. https://doi.org/10.1088/0957-4484/22/5/055705, 055705.

Sadik, O. A., Du, N., Yazgan, I., & Okello, V. (2014). Nanostructured membranes for water purification. In *Nanotechnology applications for clean water: Solutions for improving water quality: Second edition* (pp. 95–108). Elsevier Inc. https://doi.org/10.1016/B978-1-4557-3116-9.00006-8.

Schrand, A. (2009). Safety of nanoparticles from manufacturing to medical applications. *Nanostructure Science and Technology*, *1*, 159–187.

Schrand, A. M., Hens, S. A. C., & Shenderova, O. A. (2009). Nanodiamond particles: Properties and perspectives for bioapplications. *Critical Reviews in Solid State and Materials Sciences*, *34*(1–2), 18–74. https://doi.org/10.1080/10408430902831987.

Schrand, A. M., Huang, H., Carlson, C., Schlager, J. J., Osawa, E., Hussain, S. M., et al. (2007). Are diamond nanoparticles cytotoxic? *Journal of Physical Chemistry B*, *111*(1), 2–7. https://doi.org/10.1021/jp066387v.

Sharma, S., Umar, A., Mehta, S. K., & Kansal, S. K. (2017). Fluorescent spongy carbon nanoglobules derived from pineapple juice: A potential sensing probe for specific and selective detection of chromium (VI) ions. *Ceramics International*, *43*(9), 7011–7019. https://doi.org/10.1016/j.ceramint.2017.02.127.

Shin, Y., Lee, J., Yang, J., Park, J., Lee, K., Kim, S., et al. (2014). Mass production of graphene quantum dots by one-pot synthesis directly from graphite in high yield. *Small*, *10*(5), 866–870. https://doi.org/10.1002/smll.201302286.

Shirey, S. B., Cartigny, P., Frost, D. J., Keshav, S., Nestola, F., Nimis, P., et al. (2013). Diamonds and the geology of mantle carbon. *Reviews in Mineralogy and Geochemistry*, *75*, 355–421. https://doi.org/10.2138/rmg.2013.75.12.

Simioni, N. B., Silva, T. A., Oliveira, G. G., & Fatibello-Filho, O. (2017). A nanodiamond-based electrochemical sensor for the determination of pyrazinamide antibiotic. *Sensors and Actuators, B: Chemical*, *250*, 315–323. https://doi.org/10.1016/j.snb.2017.04.175.

Simões, F. R., & Xavier, M. G. (2017). Electrochemical sensors. In *Nanoscience and its applications* (pp. 155–178). Elsevier Inc. https://doi.org/10.1016/B978-0-323-49780-0.00006-5.

Singh, V. K., Singh, V., Yadav, P. K., Chandra, S., Bano, D., Kumar, V., et al. (2018). Bright-blue-emission nitrogen and phosphorus-doped carbon quantum dots as a promising nanoprobe for detection of Cr(vi) and ascorbic acid in pure aqueous solution and in living cells. *New Journal of Chemistry*, *42*(15), 12990–12997. https://doi.org/10.1039/c8nj02126k.

Somani, P. R., Somani, S. P., & Umeno, M. (2006). Planer nano-graphenes from camphor by CVD. *Chemical Physics Letters*, *430*(1–3), 56–59. https://doi.org/10.1016/j.cplett.2006.06.081.

Stankovich, S., Dikin, D. A., Piner, R. D., Kohlhaas, K. A., Kleinhammes, A., Jia, Y., et al. (2007). Synthesis of graphene-based nanosheets via chemical reduction of exfoliated graphite oxide. *Carbon*, *45*(7), 1558–1565. https://doi.org/10.1016/j.carbon.2007.02.034.

Sun, H., Gao, N., Wu, L., Ren, J., Wei, W., & Qu, X. (2013). Highly photoluminescent amino-functionalized graphene quantum dots used for sensing copper ions. *Chemistry - A European Journal*, *19*(40), 13362–13368. https://doi.org/10.1002/chem.201302268.

Sun, Y., Wang, S., Li, C., Luo, P., Tao, L., Wei, Y., et al. (2013). Large scale preparation of graphene quantum dots from graphite with tunable fluorescence properties. *Physical Chemistry Chemical Physics*, *15*(24), 9907–9913. https://doi.org/10.1039/c3cp50691f.

Talari, F. F., Bozorg, A., Faridbod, F., & Vossoughi, M. (2021). A novel sensitive aptamer-based nanosensor using rGQDs and MWCNTs for rapid detection of diazinon pesticide. *Journal of Environmental Chemical Engineering*, *9*(1), 104878. https://doi.org/10.1016/j.jece.2020.104878.

Tam, T. V., Hong, S. H., & Choi, W. M. (2015). Facile synthesis of cysteine-functionalized graphene quantum dots for a fluorescence probe for mercury ions. *RSC Advances*, *5*(118), 97598–97603. https://doi.org/10.1039/c5ra18495a.

Ugarte, D. (1992). Curling and closure of graphitic networks under electron-beam irradiation. *Nature*, *359*(6397), 707–709. https://doi.org/10.1038/359707a0.

Vairavapandian, D., Vichchulada, P., & Lay, M. D. (2008). Preparation and modification of carbon nanotubes: Review of recent advances and applications in catalysis and sensing. *Analytica Chimica Acta*, *626*(2), 119–129. https://doi.org/10.1016/j.aca.2008.07.052.

Wang, F., Gu, Z., Lei, W., Wang, W., Xia, X., & Hao, Q. (2014). Graphene quantum dots as a fluorescent sensing platform for highly efficient detection of copper(II) ions. *Sensors and Actuators, B: Chemical*, *190*, 516–522. https://doi.org/10.1016/j.snb.2013.09.009.

Wang, L., Li, M., Li, W., Han, Y., Liu, Y., Li, Z., et al. (2018). Rationally designed efficient dual-mode colorimetric/fluorescence sensor based on carbon dots for detection of pH and Cu2+ ions. *ACS Sustainable Chemistry & Engineering*, *6*(10), 12668–12674. https://doi.org/10.1021/acssuschemeng.8b01625.

Wang, H., Liu, S., Xie, Y., Bi, J., Li, Y., Song, Y., et al. (2018). Facile one-step synthesis of highly luminescent N-doped carbon dots as an efficient fluorescent probe for chromium(VI) detection based on the inner filter effect. *New Journal of Chemistry*, *42*(5), 3729–3735. https://doi.org/10.1039/c8nj00216a.

Wang, H., Liu, Y., Yao, S., & Hu, G. (2019). Fabrication of super pure single — walled carbon nanotube electrochemical sensor and its application for picomole detection of olaquindox. *Analytica Chimica Acta*, *1049*, 82–90. https://doi.org/10.1016/j.aca.2018.10.024.

Wang, Y., Pan, C., Chu, W., Vipin, A. K., & Sun, L. (2019). Environmental remediation applications of carbon nanotubes and graphene oxide: Adsorption and catalysis. *Nanomaterials*, *9*(3). https://doi.org/10.3390/nano9030439.

Wang, J., Qiu, F., Wu, H., Li, X., Zhang, T., Niu, X., et al. (2017). Fabrication of fluorescent carbon dots-linked isophorone diisocyanate and β-cyclodextrin for detection of chromium ions. *Spectrochimica Acta. Part A, Molecular and Biomolecular Spectroscopy*, *179*, 163–170. https://doi.org/10.1016/j.saa.2017.02.031.

Wei, Y., Gao, C., Meng, F.-L., Li, H.-H., Wang, L., Liu, J.-H., et al. (2012). SnO2/reduced graphene oxide nanocomposite for the simultaneous electrochemical detection of cadmium(II), Lead(II), copper(II), and mercury(II): An interesting favorable mutual interference. *The Journal of Physical Chemistry C*, *116*(1), 1034–1041. https://doi.org/10.1021/jp209805c.

Welz, S., Gogotsi, Y., & McNallan, M. J. (2003). Nucleation, growth, and graphitization of diamond nanocrystals during chlorination of carbides. *Journal of Applied Physics*, *93*(7), 4207–4214. https://doi.org/10.1063/1.1558227.

West, N. G., Howe, A. M., Jackson, C. J., Smith, N. J., Fuller, C. W., & Clare, D. J. M. (1984). Applications of atomic spectroscopy to environmental analysis. *Analytical Proceedings, 21*(3), 102–115. https://doi.org/10.1039/AP9842100102.

Wu, M., Liao, J., Yu, L., Lv, R., Li, P., Sun, W., et al. (2020). 2020 roadmap on carbon materials for energy storage and conversion. *Chemistry - An Asian Journal, 15*(7), 995–1013. https://doi.org/10.1002/asia.201901802.

Xie, Y. L., Zhao, S. Q., Ye, H. L., Yuan, J., Song, P., & Hu, S. Q. (2015). Graphene/CeO2 hybrid materials for the simultaneous electrochemical detection of cadmium (II), lead(II), copper(II), and mercury(II). *Journal of Electroanalytical Chemistry, 757*, 235–242. https://doi.org/10.1016/j.jelechem.2015.09.043.

Xu, Y., Wang, S., Hou, X., Sun, Z., Jiang, Y., Dong, Z., et al. (2018). Coal-derived nitrogen, phosphorus and sulfur co-doped graphene quantum dots: A promising ion fluorescent probe. *Applied Surface Science, 445*, 519–526. https://doi.org/10.1016/j.apsusc.2018.03.156.

Yan, X., Cui, X., & Li, L. S. (2010). Synthesis of large, stable colloidal graphene quantum dots with tunable size. *Journal of the American Chemical Society, 132*(17), 5944–5945. https://doi.org/10.1021/ja1009376.

Yang, S. T., Chang, Y., Wang, H., Liu, G., Chen, S., Wang, Y., et al. (2010). Folding/aggregation of graphene oxide and its application in Cu2+ removal. *Journal of Colloid and Interface Science, 351*(1), 122–127. https://doi.org/10.1016/j.jcis.2010.07.042.

Yang, W., Ratinac, K. R., Ringer, S. R., Thordarson, P., Gooding, J. J., & Braet, F. (2010). Carbon nanomaterials in biosensors: Should you use nanotubes or graphene. *Angewandte Chemie, International Edition, 49*(12), 2114–2138. https://doi.org/10.1002/anie.200903463.

Yang, G. W., Wang, J. B., & Liu, Q. X. (1998). Preparation of nano-crystalline diamonds using pulsed laser induced reactive quenching. *Journal of Physics. Condensed Matter, 10*(35), 7923–7927. https://doi.org/10.1088/0953-8984/10/35/024.

Zhang, C., & Du, X. (2020). Electrochemical sensors based on carbon nanomaterial used in diagnosing metabolic disease. *Frontiers in Chemistry, 8*. https://doi.org/10.3389/fchem.2020.00651.

Zhang, L., Peng, D., Liang, R. P., & Qiu, J. D. (2015). Nitrogen-doped graphene quantum dots as a new catalyst accelerating the coordination reaction between cadmium(II) and 5,10,15,20-Tetrakis(1-methyl-4-pyridinio)porphyrin for cadmium(II) sensing. *Analytical Chemistry, 87*(21), 10894–10901. https://doi.org/10.1021/acs.analchem.5b02450.

Zhang, Z., Zhang, J., Chen, N., & Qu, L. (2012). Graphene quantum dots: An emerging material for energy-related applications and beyond. *Energy and Environmental Science, 5*(10), 8869–8890. https://doi.org/10.1039/c2ee22982j.

Zhao, W., Fang, M., Wu, F., Wu, H., Wang, L., & Chen, G. (2010). Preparation of graphene by exfoliation of graphite using wet ball milling. *Journal of Materials Chemistry, 20*(28), 5817–5819. https://doi.org/10.1039/c0jm01354d.

Zhao, X., Gao, J., He, X., Cong, L., Zhao, H., Li, X., et al. (2015). DNA-modified graphene quantum dots as a sensing platform for detection of Hg2+ in living cells. *RSC Advances, 5*(49), 39587–39591. https://doi.org/10.1039/c5ra06984j.

Zheng, P., & Wu, N. (2017). Fluorescence and sensing applications of graphene oxide and graphene quantum dots: A review. *Chemistry- -An Asian Journal, 12*(18), 2343–2353. https://doi.org/10.1002/asia.201700814.

Zhou, M., Wang, Y., Zhai, Y., Zhai, J., Ren, W., Wang, F., et al. (2009). Controlled synthesis of large-area and patterned electrochemically reduced graphene oxide films. *Chemistry - A European Journal, 15*(25), 6116–6120. https://doi.org/10.1002/chem.200900596.

Zhou, X., Zhang, Y., Wang, C., Wu, X., Yang, Y., Zheng, B., et al. (2012). Photo-Fenton reaction of graphene oxide: A new strategy to prepare graphene quantum dots for DNA cleavage. *ACS Nano, 6*(8), 6592–6599. https://doi.org/10.1021/nn301629v.

Zhu, S., & Xu, G. (2010). Single-walled carbon nanohorns and their applications. *Nanoscale, 2*(12), 2538–2549. https://doi.org/10.1039/c0nr00387e.

Zhu, S., Zhang, J., Qiao, C., Tang, S., Li, Y., Yuan, W., et al. (2011). Strongly green-photoluminescent graphene quantum dots for bioimaging applications. *Chemical Communications, 47*(24), 6858–6860. https://doi.org/10.1039/c1cc11122a.

Zhuo, S., Shao, M., & Lee, S. T. (2012). Upconversion and downconversion fluorescent graphene quantum dots: Ultrasonic preparation and photocatalysis. *ACS Nano, 6*(2), 1059–1064. https://doi.org/10.1021/nn2040395.

Zuo, X., He, S., Li, D., Peng, C., Huang, Q., Song, S., et al. (2010). Graphene oxide-facilitated electron transfer of metalloproteins at electrode surfaces. *Langmuir, 26*(3), 1936–1939. https://doi.org/10.1021/la902496u.

Chapter 11

Graphene-based surface-enhanced Raman scattering as an efficient tool in the detection of toxic organic dyes in real industrial effluents

V. Poornima Parvathi[a], R. Parimaladevi[a], Vasant Sathe[b], and M. Umadevi[a]
[a]*Department of Physics, Mother Teresa Women's University, Kodaikanal, Tamil Nadu, India,* [b]*UGC-DAE Consortium for Scientific Research, Indore, Madhya Pradesh, India*

Introduction

A sensor is an agent that responds to a chemical or physical stimulus, and responds by generating a signal that can be analyzed. Higher sensitivity, selectivity, reliability, reproducibility, stability, simple operation, and cost-effective sensors are the requirements for producing technologically viable sensors (Bezzon et al., 2019). Sensors and detectors are widely researched branches of nanotechnology as they find numerous applications (Manjunatha, 2017; Raril & Manjunatha, 2018; Raril et al., 2018; Tigari & Manjunatha, 2020b). Surface-enhanced Raman scattering (SERS) utilizing the amplified electric fields originating from the localized surface plasmon resonance (LSPR) of plasmonic nanoparticles (NPs), termed as "hotspots," has materialized as an efficient technique for selective and sensitive sensors. The sensitivity, selectivity, and efficiency of SERS technique are utilized in sensing of organic pollutants, dye molecules, explosives, carcinogens, pesticides, etc. (Arockia Jency, Umadevi, & Sathe, 2015; Prinith, Manjunatha, & Raril, 2019; Pushpanjali, Manjunatha, Tigari, & Fattepur, 2020; Raril, Manjunatha, & Tigari, 2020). Traditional metal NP substrates meet limitations such as heterogeneous adsorption, spectral fluctuation, possible chemical/photo reactions between analyte and substrate, continuum spectral background originating from the fluorescence of analytes, and the instability of metal NPs in atmospheric conditions (Gupta, Banaszak, Smith, & Dimakis, 2018). The sensitivity of plasmonic SERS materials can be boosted by the addition of carbon-based nanomaterials. Carbon-based materials such as carbon nanotubes (Charithra & Manjunatha, 2020; Manjunatha, 2017; Raril & Manjunatha, 2018; Raril et al., 2018, 2020; Tigari & Manjunatha, 2020a; Tigari, Manjunatha, Raril, & Hareesha, 2019), functionalized carbon nanotubes (Amrutha, Manjunatha, Bhatt, Raril, & Pushpanjali, 2019; Charithra, Manjunatha, & Raril, 2020; Hareesha, Manjunatha, Raril, & Tigari, 2019; Souza et al., 2020), graphene, graphene oxide, and reduced graphene oxide have proved to be the most promising materials owing to their unique structure. In graphene-boosted nanomaterials, the "hotspots" are converted to "hot surfaces" that aid the exceptional enhancement of signals giving rise to graphene-based surface-enhanced Raman scattering (G-SERS) (Poornima Parvathi, Parimaladevi, Sathe, & Umadevi, 2019; Poornima Parvathi, Parimaladevi, Sathe, & Mahalingam, 2019). Graphene oxide (GO) with its high specific surface area, abundant surface hydroxyl groups, fluorescence quenching ability, and biological compatibility can effectively overcome these limitations and mediate SERS enhancement (Li, Liu, Yu, Wu, & Su, 2017; Liang et al., 2017; Sharma, Prakash, & Mehta, 2017). The importance of detection and removal of organic dyes in effluents from textile, leather, paper, rubber, plastic, food, and cosmetic industries arises due to the pollutant's mutagenic and carcinogenic effects (Jafari et al., 2016). Additionally, dye-based industrial effluents decrease the penetration of light in water resources and therefore photosynthetic activity, causing oxygen shortage that is a serious hazard to aquatic organisms (Bel Hadjltaief, Ben Ameur, Da Costa, Ben Zina, & Elena Galvez, 2018). This toxic nature of organic dyes exhibits itself even in trace-level quantities, thereby necessitating sensitive detection techniques. Various techniques such as absorption, fluorescence, and infrared spectroscopy, gas chromatography, mass spectrometry, and electrochemical sensing have been researched as dye sensors (Pinheiro, Touraud, & Thomas, 2004). However, the application of G-SERS on industrial effluents that contain dye toxins can yield selective and enhanced results. This chapter discusses briefly the synthesis and characterization of graphene-boosted silver and gold nanoparticles. Further commercially available dyes and real textile industrial effluent have been analyzed using graphene-based surface-enhanced Raman scattering technique.

Graphene-boosted silver nanocomposites for graphene-based surface-enhanced Raman scattering in the detection of toxic organic dyes in real industrial effluents

Untreated textile industry effluent, emptied into water resources, remains a source of carcinogenic toxins. Trace-level detection of these toxins can be attained through G-SERS. The combination between nanoparticles (NPs) and functionalized graphene oxide sheets could not only integrate the merits of NPs and graphene, but also provides additional novel properties and increased stabilities (Xiao, Liu, Lin, Lin, & Fang, 2017). SERS has been widely employed to study the structural and bonding natures of various dye molecules such as rhodamine 6G (R6G), methyl orange, and Janus green B (Bonancêa, do Nascimento, de Souza, Temperini, & Corio, 2008; Chen, Li, Zhao, Chang, & Qi, 2015; Sharma et al., 2017; Si, Kang, & Zhang, 2009). Silver colloids were used successfully to identify colorants in actual artwork (Brosseau, Casadio, & Van Duyne, 2011). GO was prepared by the modified Hummers method (Li et al., 2017; Liang et al., 2017; Sharma et al., 2017). To a glass beaker, 0.5 g of graphite flakes, 0.5 g of sodium nitrate, and 23 mL of 98% sulfuric acid were added and mixed under stirring in an ice bath. Then, 3 g of potassium permanganate was added to the mixture. 40 mL of water was added to the formed paste and stirred at 90°C. Finally, 100 mL of water was added, followed by the slow addition of 3 mL of hydrogen peroxide. The suspension was washed and centrifuged. The sediment was dried and labeled as GO. The obtained powder was re-dispersed in water under sonication. The obtained solution was added to 0.1 mM silver nitrate and stirred. 0.1 mM sodium borohydride solution was added dropwise till a brown color was obtained. The solution was washed and dried to obtain Ag-rGO. 10 mL of substrate dispersed in distilled water was added to 25 mL of samples and stirred. The resultant mixture was used for fluorescence analysis. For SERS analysis, 0.1 mL of R6G, CR, and MG of varying concentrations (100, 10, and 1 μM) was added to 0.5 mL Ag NPs and subjected to Raman analysis. Similarly, 0.1 mL of R6G, CR, and MG of concentrations 100, 50, 10, 5, and 1 μM was added to 0.5 mL Ag-rGO NPs and subjected to Raman analysis. 2 μL, 4 μL, and 6 μL of textile effluent (TE) were used with 0.5 mL of Ag NPs and Ag-rGO.

Optical studies

Fig. 11.1 shows the UV-visible spectra of Ag NPs, GO, and Ag-rGO, respectively. The absorbance peak observed at 392 nm is characteristic of the LSPR bands of Ag NPs. The peak appearing at 231 nm in the synthesized GO sheets is due to the π-π* transitions of aromatic C=C bonds. Ag-rGO exhibits two peaks; the maximum at 249 nm corresponds to mildly deoxygenated sp^2 structures of rGO and at 400 nm represents Ag NPs, which were slightly red-shifted. The red shift (light gray in print version) observed can be due to the deposition of Ag NPs on the surface of GO sheets (Vinoth, Wu, Asiri, & Anandan, 2017).

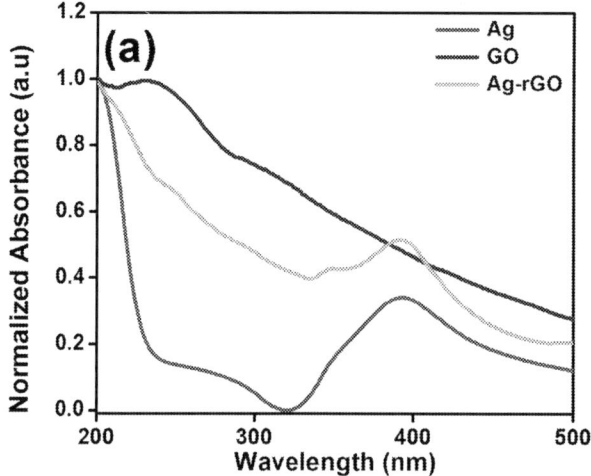

FIG. 11.1 UV-visible spectra of Ag NPs, GO, and Ag-rGO. *(From Poornima Parvathi, V., Parimaladevi, R., Sathe, V., & Mahalingam, U. (2019). Graphene boosted silver nanoparticles as surface enhanced Raman spectroscopic sensors and photocatalysts for removal of standard and industrial dye contaminants. Sensors and Actuators B: Chemical, 281, 679–688. https://doi.org/10.1016/j.snb.2018.11.007.)*

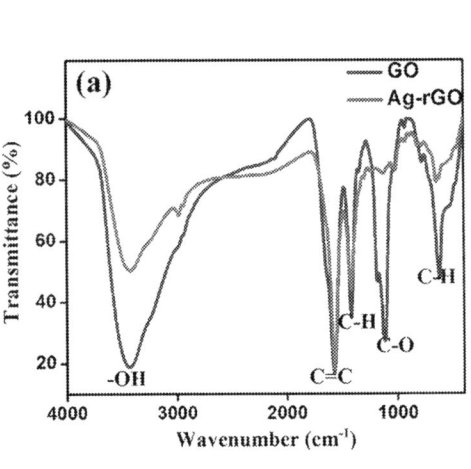
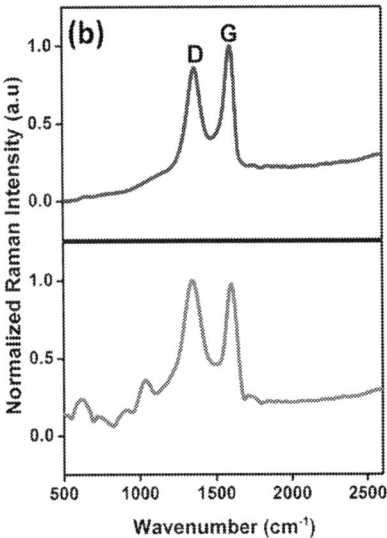

FIG. 11.2 (A) FTIR spectra (B) Raman spectra of GO and Ag-rGO. *(From Poornima Parvathi, V., Parimaladevi, R., Sathe, V., & Mahalingam, U. (2019). Graphene boosted silver nanoparticles as surface enhanced Raman spectroscopic sensors and photocatalysts for removal of standard and industrial dye contaminants. Sensors and Actuators B: Chemical, 281, 679–688. https://doi.org/10.1016/j.snb.2018.11.007.)*

Vibrational spectroscopic studies

Fig. 11.2A shows the FTIR spectra of GO and Ag-rGO, respectively. C=C skeletal vibration of GO was observed at 1639 cm^{-1} in both samples, conforming the presence of GO. This, along with the intensity-reduced C—O bands at 1050 cm^{-1} in Ag-rGO, illustrates that NaBH$_4$ used in the reduction of Ag NPs has reduced GO to rGO through deoxygenation. These results indicate the presence of strong interactions between Ag NPs and the remaining surface hydroxyl oxygen atoms. Stretching vibration of —OH group, observed at 3400 cm^{-1}, is very strong oxidizing and highly reactive agent that can degrade organic pollutant into harmless compounds during photocatalysis (Acar Bozkurt, 2017; Saleh, Al-Shalalfeh, & Al-Saadi, 2018). Fig. 11.2B shows the normalized Raman spectra of GO and Ag-rGO, respectively. Two prominent bands located near 1330 cm^{-1} and 1590 cm^{-1} corresponding to the D and G bands of GO were observed in both spectra. The G band and D band arise due to the first-order scattering of E$_{2g}$ phonons and from the breathing mode of k-point phonons of A$_{1g}$ symmetry, respectively. The intensity ratio for the D and G bands, I$_D$ and I$_G$, respectively, was calculated as 0.98 for GO and 1.04 for Ag-rGO. This increase in the I$_D$/I$_G$ value is a GO reduction signature that illustrates the formation of newly isolated smaller graphitic domain (Vinoth et al., 2017).

Morphological studies

Fig. 11.3A–C shows the TEM images, (D) EDAX elemental mapping (inset particle size distribution curve), and (E) SAED pattern of Ag NPs, respectively. Spherical Ag NPs were observed with slight aggregation. The average size was calculated from the particle size distribution curve as 15.01 nm. The SAED pattern depicts good crystallinity that can be indexed to the face-centered cubic structure characteristic of Ag NPs. The observed d values of 0.304, 0.246, 0.123, and 0.118 nm correspond to the (111), (200), (220), and (311) lattice planes of face-centered cubic Ag NPs (Vinoth et al., 2017). Fig. 11.3F–H shows the TEM images, (I) EDAX elemental mapping (inset particle size distribution curve), and (J) SAED pattern of Ag-rGO, respectively. On modification by rGO sheets, the average size of Ag NPs reduced further to 8.4 nm. TEM images show the presence of transparent wrinkled sheets with folded regions that are characteristic of GO layers, with spherical Ag particles dispersed on its surface. The images signify that the GO sheets have prevented the aggregation of Ag NPs, thereby increasing their stability. The diffuse SAED pattern observed can be attributed to the amorphous sixfold structure typical of GO, while the clear spots arise from the crystalline Ag NPs dispersed along the layers of GO. The EDAX elemental mapping confirms the presence of elemental Ag in Ag NPs and elemental Ag, C, and O in Ag-rGO. From TEM images and the SAED pattern, it could be expected that an intimate interfacial contact exists in Ag-rGO, which may favor the easy transfer of charge carriers, thus providing a suitable environment for both SERS and photocatalytic performance (Marcano et al., 2010; Sharma et al., 2017).

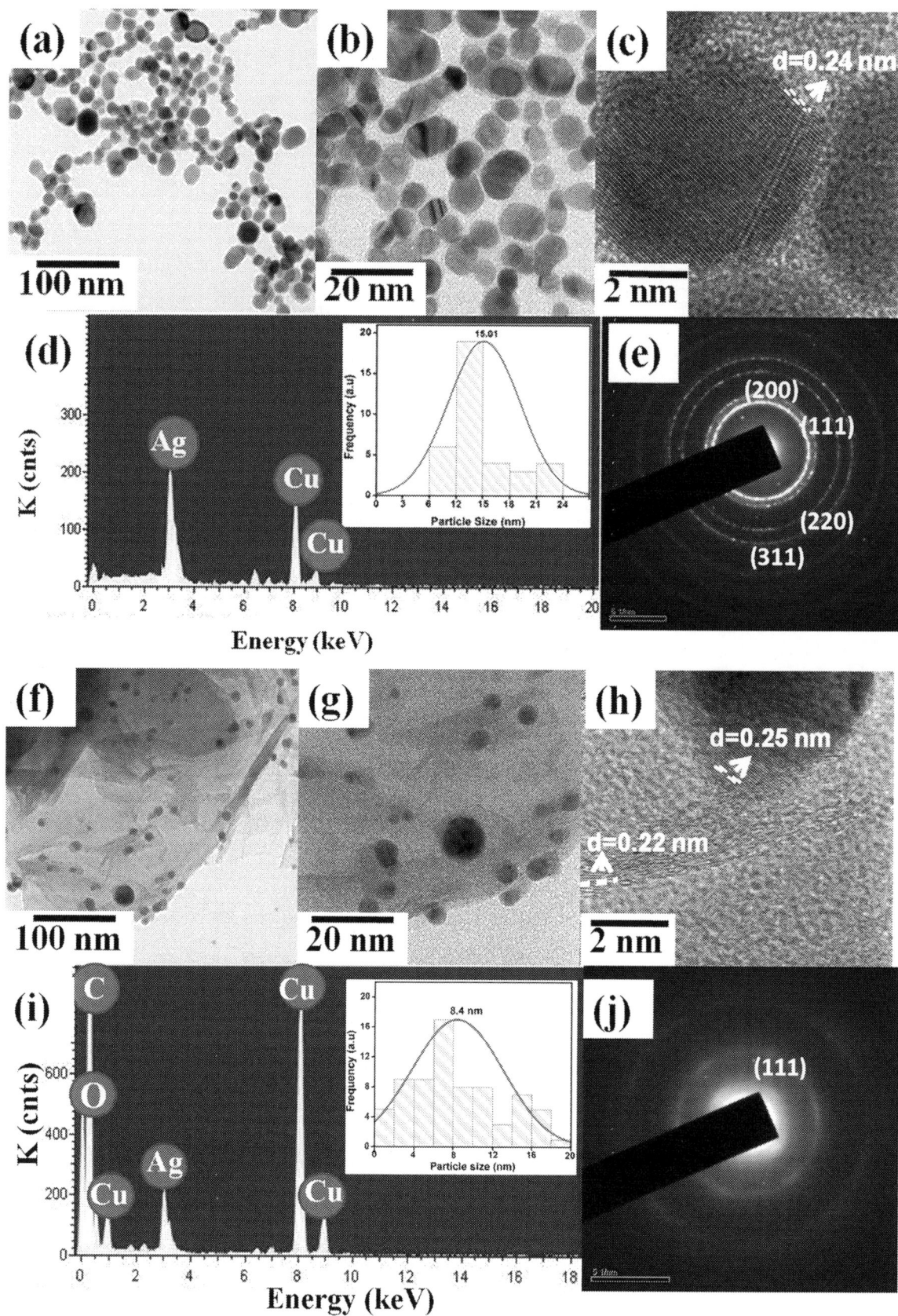

FIG. 11.3 (A–C) TEM images, (D) EDAX elemental mapping (inset particle size distribution curve), (E) SAED pattern of Ag NPs, respectively. (F–H) TEM images, (I) EDAX elemental mapping (inset particle size distribution curve), and (J) SAED pattern of Ag-rGO, respectively. *(From Poornima Parvathi, V., Parimaladevi, R., Sathe, V., & Mahalingam, U. (2019). Graphene boosted silver nanoparticles as surface enhanced Raman spectroscopic sensors and photocatalysts for removal of standard and industrial dye contaminants.* Sensors and Actuators B: Chemical, *281, 679–688. https://doi.org/10.1016/j.snb.2018.11.007.)*

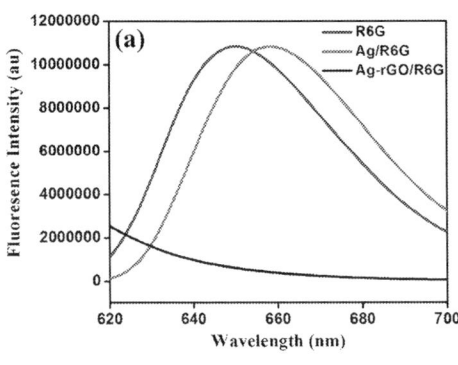
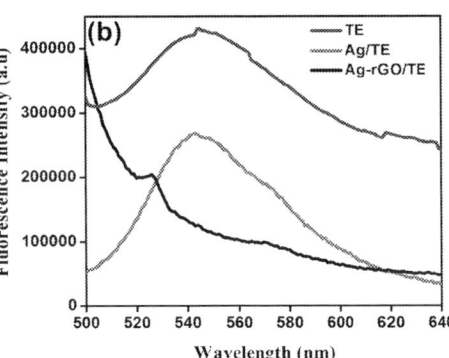

FIG. 11.4 (A) Fluorescence spectra of R6G-, Ag NP-, and Ag-rGO-treated R6G, respectively; (B) fluorescence spectra of TE-, Ag NP-, and Ag-rGO-treated TE, respectively. *(From Poornima Parvathi, V., Parimaladevi, R., Sathe, V., & Mahalingam, U. (2019). Graphene boosted silver nanoparticles as surface enhanced Raman spectroscopic sensors and photocatalysts for removal of standard and industrial dye contaminants. Sensors and Actuators B: Chemical, 281, 679–688. https://doi.org/10.1016/j.snb.2018.11.007.)*

Fluorescence suppression in dyes

R6G, a xanthene dye, is a widely used fluorescent molecule, and hence, this dye was utilized to probe the fluorescence suppression ability of Ag-rGO. Fig. 11.4A shows fluorescence spectra of R6G-, Ag NP-, and Ag-rGO-treated R6G, respectively. R6G showed strong fluorescence emission when excited at 550 nm. On introduction of Ag-rGO, a drastic decrease in the fluorescence intensity of R6G was observed. Fig. 11.4B shows fluorescence spectra of TE-, Ag NP-, and Ag-rGO-treated TE, respectively. Similar results with quenched fluorescence were observed for TE. Changes in the fluorescence spectra can be observed when the dye molecules come in close proximity to plasmonic Ag NPs due to resonant energy transfer. A decrease in the chromophoric nature of dyes occurs when the dye molecules are adsorbed on the surface of substrates and form aggregates, evidencing the significance of the adsorptive nature of rGO. This advances SERS signal enhancement by quenching the redundant fluorescence background. Furthermore, the quenched fluorescence intensity depicts the retardation of the electron-hole recombination rate due to the interfacial charge transfer at the surface of the substrate. The retarded electron-hole recombination rate aids the intended charge separation during photocatalysis (Liu, Liu, & Liu, 2011).

SERS sensing of dye contaminants

The SERS activities of the synthesized substrates were verified using the well-established model dyes from different dye families, namely, R6G, a xanthene dye; CR a diazo dye; and MG, a triarylmethane dye.

R6G

Fig. 11.5A shows the normal Raman (nR) spectra of 100 μM R6G and SERS spectra of R6G on Ag NPs with different concentrations and (B) shows the SERS spectra of R6G on Ag-rGO with different concentrations. Comparatively, the Raman spectrum of R6G does not provide any vital spectral information regarding the dye molecule. Adsorption on SERS substrates showed an improved enhancement with the appearance of well-established vibrations of R6G at 616 cm^{-1}, 773 cm^{-1}, 1185 cm^{-1}, 1311 cm^{-1}, 1371 cm^{-1}, 1519 cm^{-1}, 1584 cm^{-1}, and 1647 cm^{-1}. These bands are assigned to the xanthene moiety stretch, ethylamine group wag, carbon–oxygen stretch, and aromatic carbon–carbon stretch of R6G (Lee et al., 2007; Si et al., 2009). More importantly, two characteristic bands at 1185 cm^{-1} and 1647 cm^{-1} were observed only after adsorption on Ag-rGO, thus presenting information-rich spectra than traditional noble metal substrates. Fig. 11.5C shows the linear plot of concentration vs SERS peak intensities for R6G at 1317 cm^{-1} on Ag-rGO substrate. Conducive linear regression coefficient of 0.9953 was observed. The stability of SERS signals of R6G on Ag-rGO is illustrated in Fig. 11.5D. The consistent peak intensities and Raman bands even over a time period of 3 months reiterate the stability of the substrate.

CR

Fig. 11.6A shows the normal Raman (nR) spectra of 100 μM CR and SERS spectra of CR on Ag NPs with different concentrations and (B) shows the SERS spectra of CR on Ag-rGO with different concentrations. Similar results with enhanced efficiency on Ag-rGO were observed. Raman bands at 805 cm^{-1}, 917 cm^{-1}, and 1621 cm^{-1} corresponding to C—H out-

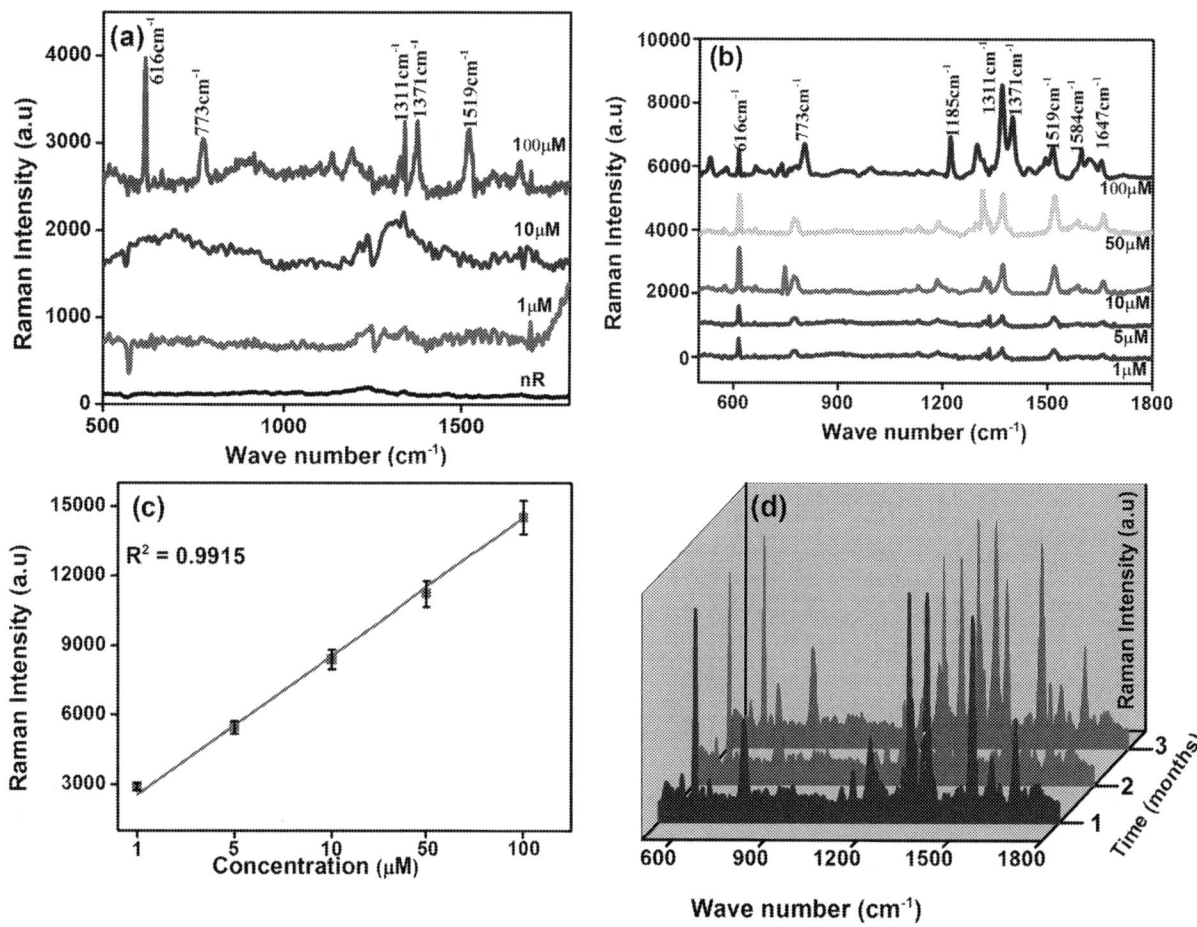

FIG. 11.5 (A) Normal Raman (nR) spectrum of 100 μM R6G and SERS spectra of R6G on Ag NPs with different concentrations, (B) SERS spectra of R6G on Ag-rGO with different concentrations, (C) linear plot of concentration vs SERS peak intensities for R6G on Ag-rGO substrate, and (D) stability of SERS signals of R6G on Ag-rGO. *(From Poornima Parvathi, V., Parimaladevi, R., Sathe, V., & Mahalingam, U. (2019). Graphene boosted silver nanoparticles as surface enhanced Raman spectroscopic sensors and photocatalysts for removal of standard and industrial dye contaminants.* Sensors and Actuators B: Chemical, *281, 679–688. https://doi.org/10.1016/j.snb.2018.11.007.)*

of-plane bending, ring skeletal stretching, and C—C stretching were observed. The di-azo nature of CR was established in the C—N and N=N bands observed at 1174 cm^{-1} and 1386 cm^{-1}, respectively. Regression coefficient of 0.9915 was calculated from Fig. 11.6C (the linear plot for concentration of CR vs SERS band intensity at 1174 cm^{-1} in the presence of Ag-rGO) (Bonancêa et al., 2008). Ag-rGO showed an excellent stability for CR also (Fig. 11.6D).

MG

Fig. 11.7A shows the normal Raman (nR) spectra of 100 μM MG and SERS spectra of MG on Ag NPs with different concentrations and (B) shows the SERS spectra of MG on Ag-rGO with different concentrations. The C—N vibration was observed around 1367 cm^{-1}. Typical C—C stretching was observed at 1618 cm^{-1}. The presence of Raman band at 1170 cm^{-1} corresponds to the characteristic C—H in-plane bending (Cesaratto, Centeno, Lombardi, Shibayama, & Leona, 2017). This, along with the C—H out-of-plane bending observed at 802 cm^{-1} and in 917 cm^{-1} MG, originates from the triarylmethane backbone. Fig. 11.7C is the linear plot of the concentrations vs SERS intensities of MG at 1618 cm^{-1} on Ag-rGO substrate. A consistently good regression coefficient value of 0.9795 was observed for MG (Fan et al., 2014). Good stability in Ag-rGO toward MG is observed in Fig. 11.7D.

TE

Supportive results demonstrated in the SERS detection of commercial dyes by Ag-rGO inspired the detection of dye contaminants in TE. Fig. 11.8A and B shows the normal Raman spectrum of 2 μL TE (nR) and SERS spectra of TE on Ag NPs and Ag-rGO, respectively, with different concentrations. As expected from the linearity plots, the surface-enhanced

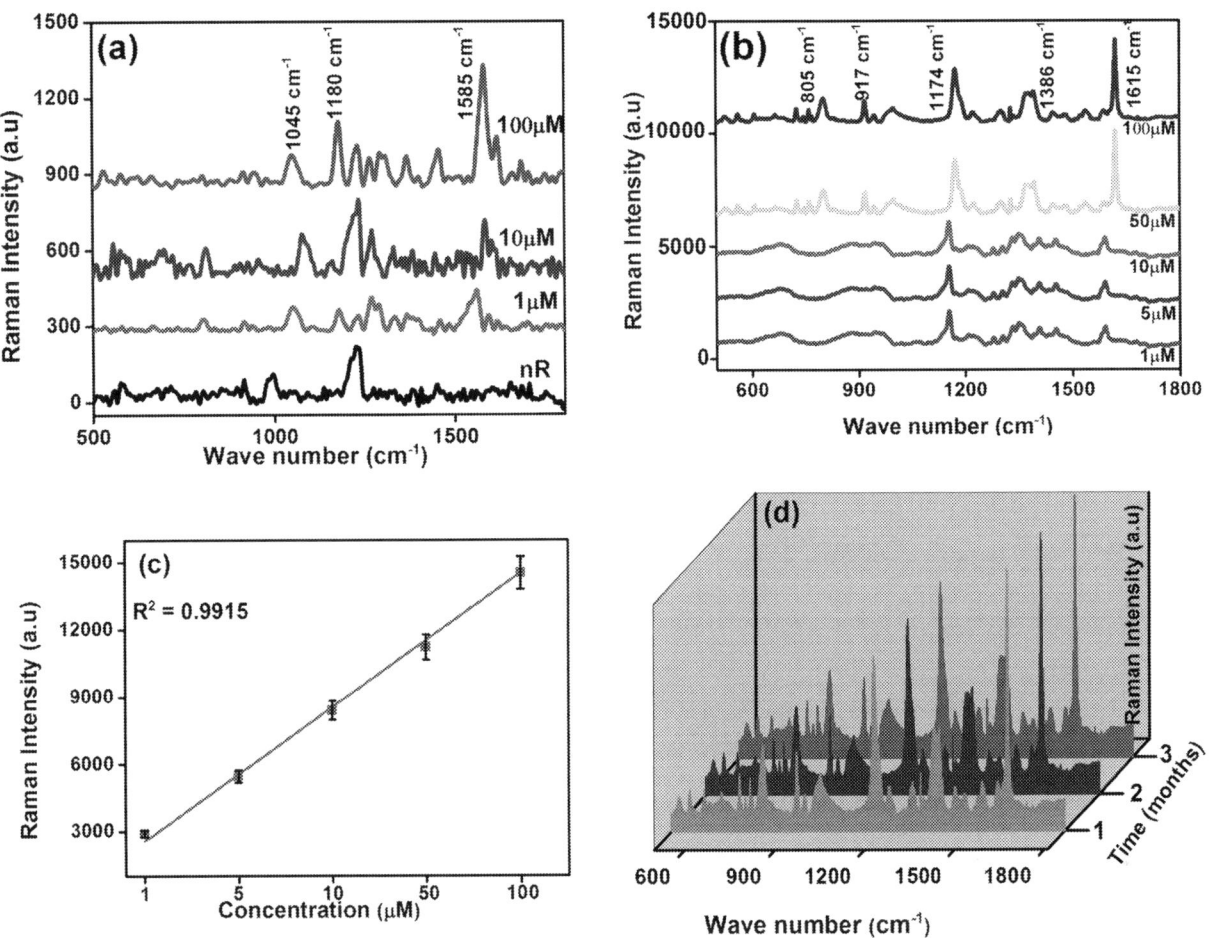

FIG. 11.6 (A) Normal Raman spectrum (nR) of 100 μM CR and SERS spectra of CR on Ag NPs with different concentrations, (B) SERS spectra of CR on Ag-rGO with different concentrations, (C) linear plot of concentration vs SERS peak intensities for CR on Ag-rGO substrate, and (D) stability of SERS signals of CR on Ag-rGO. *(From Poornima Parvathi, V., Parimaladevi, R., Sathe, V., & Mahalingam, U. (2019). Graphene boosted silver nanoparticles as surface enhanced Raman spectroscopic sensors and photocatalysts for removal of standard and industrial dye contaminants.* Sensors and Actuators B: Chemical, *281, 679–688. https://doi.org/10.1016/j.snb.2018.11.007.)*

spectral activity was predictably highest when 6 μL of TE was used. Typical bands as observed in the model dyes R6G, CR, and MG were readily visible in the SERS spectra of TE. The characteristic C—C stretching of aromatic molecules was observed around 1580–1630 cm^{-1} and 1353 cm^{-1}. The bands occurring in the region 1380–1450 cm^{-1} are due to the N=N stretching vibrations of azo compounds as observed in the SERS spectra of CR. In TE, such bands arise at 1420–1425 cm^{-1} corresponding to the signature N=N stretch of azo dyes. Similar C—N stretching and bending modes were observed around 1190–1185 cm^{-1}, 1155–1160 cm^{-1}, and 1130–1136 cm^{-1}, respectively. The distinctive band around 1290 cm^{-1} is due to the C—C vibrations. This is typical for diarylide azo pigments arising from the C—C bridge between two phenyl molecules. The phenyl OH modes in the region 1220–1240 cm^{-1} are characteristically observed in various acid red dyes such as acid red 14, 18, 27, and 73. In this study, the phenyl OH modes were observed around 1223 cm^{-1} (Bonancêa et al., 2008; Cesaratto et al., 2017; Vandenabeele, Moens, Edwards, & Dams, 2000). The resulting Raman bands correlating with a wide range of commercial dyes and pigments provide suitable grounds to establish the presence of dye contaminants in TE. Evidently, Ag-rGO-based SERS is a viable source for the sensing of dye contaminants in the textile waste effluents. Two main mechanisms generally contribute to the Raman scattering enhancement observed in SERS: (i) electromagnetic mechanism (EM) and (ii) chemical enhancement mechanism (CM). The former is based on the enhancement of the local electromagnetic field, whereas the latter is based on charge transfer between absorbed molecules and metal surface. In the plasmonic Ag NPs, "hotspots," i.e., region of enhanced local electromagnetic field, are formed in the nanosized junctions between the NPs. This contributes to the EM of SERS and thus elicits an enhanced Raman signal when the probe molecules are in these "hot regions." However, GO plays a crucial role in contributing to both EM and CM. The hot spots of Ag NPs are transformed into "hot surfaces" when the enhanced electromagnetic fields of plasmonic nanoparticles pass through the graphene layers. GO with its high adsorption ability enables an easy route for charge transfer by

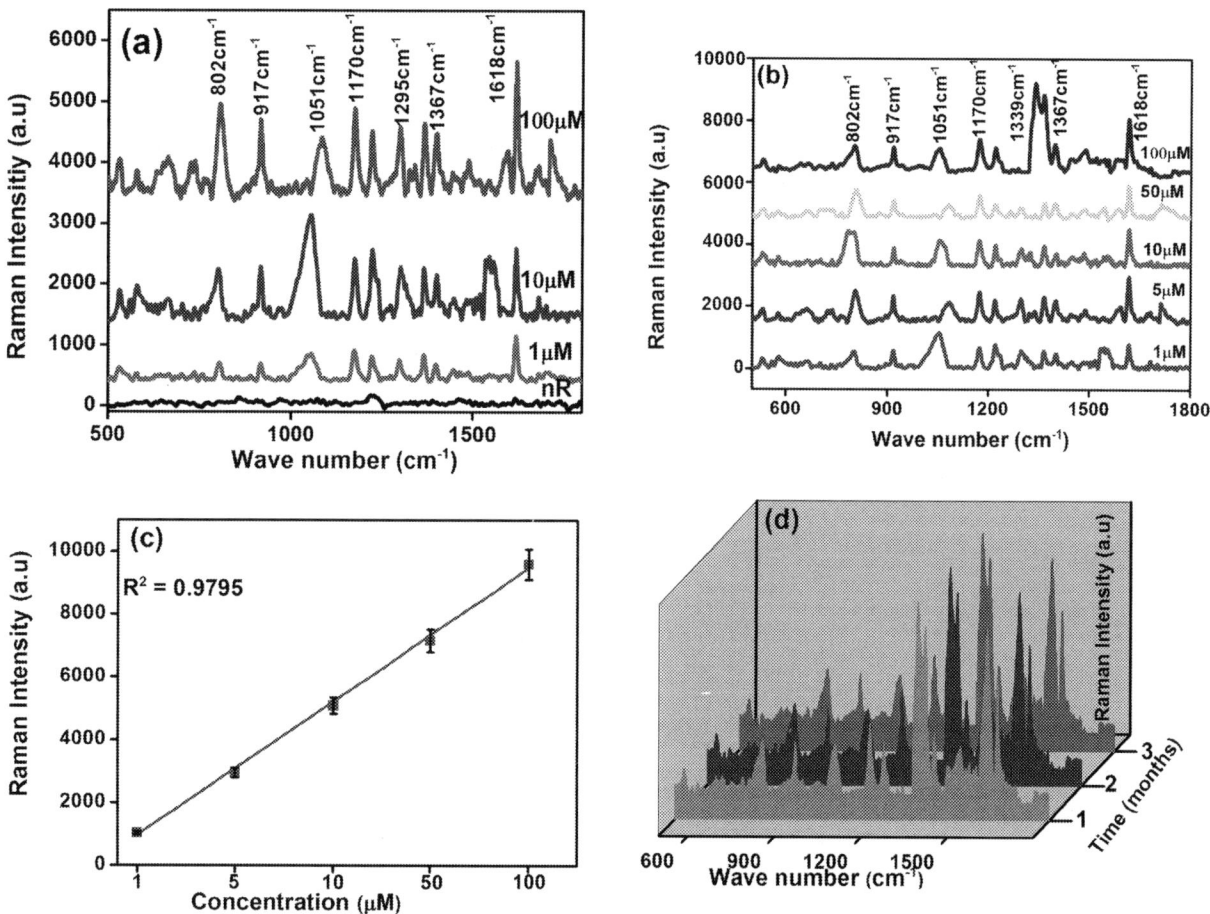

FIG. 11.7 (A) Normal Raman spectrum (nR) of 100 μM MG and SERS spectra of MG on Ag NPs with different concentrations, (B) SERS spectra of MG on Ag-rGO with different concentrations, (C) linear plot of concentration vs SERS peak intensities for MG on Ag-rGO substrate, and (D) stability of SERS signals of MG on Ag-rGO. *(From Poornima Parvathi, V., Parimaladevi, R., Sathe, V., & Mahalingam, U. (2019). Graphene boosted silver nanoparticles as surface enhanced Raman spectroscopic sensors and photocatalysts for removal of standard and industrial dye contaminants.* Sensors and Actuators B: Chemical, *281, 679–688. https://doi.org/10.1016/j.snb.2018.11.007.)*

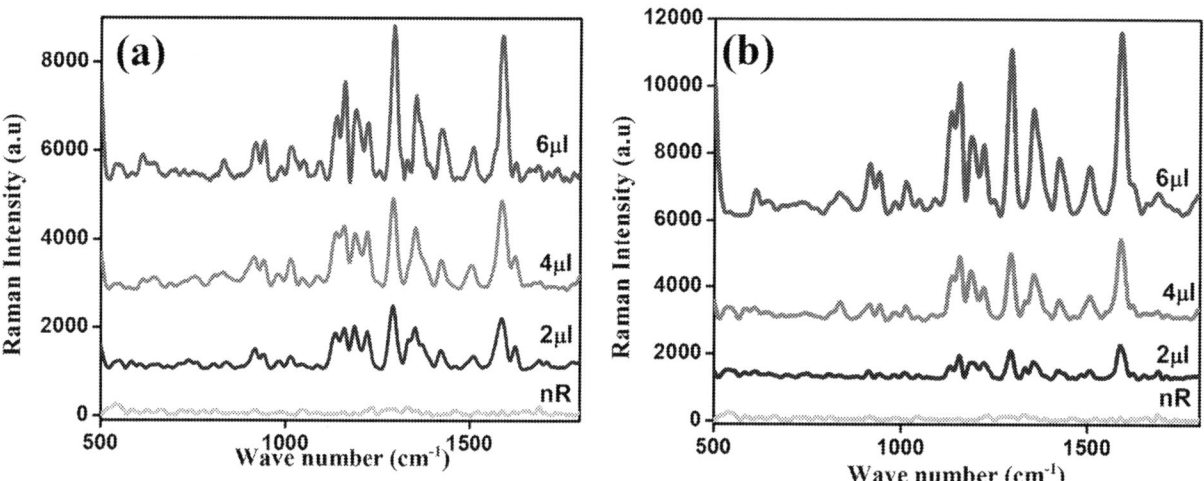

FIG. 11.8 (A) Normal Raman spectrum (nR) of 2 μL TE and SERS spectra of TE on Ag NPs with different concentrations and (B) normal Raman (nR) spectrum of 2 μL TE and SERS spectra of TE on Ag-rGO with different concentrations. *(From Poornima Parvathi, V., Parimaladevi, R., Sathe, V., & Mahalingam, U. (2019). Graphene boosted silver nanoparticles as surface enhanced Raman spectroscopic sensors and photocatalysts for removal of standard and industrial dye contaminants.* Sensors and Actuators B: Chemical, *281, 679–688. https://doi.org/10.1016/j.snb.2018.11.007.)*

the vibrational coupling arising from the π-π stacked nature of GO and aromatic nature of dyes, thus contributing to the enhancement through CM. The suppression of fluorescence (as discussed in the section "Fluorescence suppression in dyes") eliminates the unnecessary background signals. Reduced aggregation of Ag NPs in Ag-rGO provides greater dye exposure to the crystal facets of Ag NPs. Synergistically, the electromagnetic enhancement from the hot spots and hot surfaces arising from the chemical enhancement offered by rGO result in well-defined, enhanced SERS signals (Fan et al., 2014; Yang et al., 2011).

Graphene-boosted silver nanocomposites for graphene-based surface-enhanced Raman scattering in the detection of toxic organic dyes in real industrial effluents

Widely used azo dyes and their metabolites, aromatic amines, are mutagenic and are listed as category 3 carcinogens by the International Agency for Research on Cancer (IARC) (Jahn et al., 2015). Various techniques such as gas chromatography mass spectroscopy, near-infrared spectroscopy, fluorescence spectroscopy, and absorption spectroscopy are used in the detection of dyes. Au/GO NPs and Au/GO NPs showed an enhancement in the Raman spectra of p-amino-thiophenol in comparison with Au metal NPs (Huang et al., 2010). Au NPs/graphene nanohybrid showed high-sensitivity molecule detection toward rhodamine 6G (R6G) owing to its large gap coverage area and small gap width between Au NPs (Lu et al., 2015). SERS signals obtained for crystal violet (CV) on GO-Au nanorod substrates suggest that the observed enhancement arises from both the chemical enhancement offered by GO and the electromagnetic enhancement arising from the Au nanorods (Hu et al., 2013). Further, Au/GO composites find applications in bioimaging, catalysis, and biomedical therapy (Turcheniuk, Boukherroub, & Szunerits, 2015). GO was prepared by the modified Hummers method (Li et al., 2014). 0.01 M of hydrogen tetrachloroaurate and 0.01 M of sodium citrate in water were added to calculate quantities of GO dispersed in water under sonication. 0.1 M sodium borohydride was added to the solution dropwise until a slight red color corresponding to the reduction of Au NPs was observed. The resultant solution was washed, dried, and labeled as Au/GO NCs. 4 μL dye solutions at 10^{-6} M concentration was added to 0.5 mL substrates and subjected to fluorescence and Raman analyses. Similarly, for TE, 4 μL of sample was used.

UV-visible spectral analysis

Fig. 11.9A shows the UV-visible spectrum of the prepared Au/GO NCs. The absorption band arising at 261 nm indicates the extended conjugation in basal plane carbon network resulting from the reduction of GO. The surface plasmon (a quantum of plasma oscillation) absorption of Au NPs is characterized by the distinctive absorbance peak at 531 nm (Lu et al., 2015). Similar absorbance peak was observed at 538 nm in contrast to the red-shifted plasmon bands arising from electronic couplings between particles in metal NCs by Bramhaiah et al., suggesting that the plasmon band positions in rGO-metal NC films belong to isolated particles. Thus, the NCs are formed by the incorporation of Au NPs into rGO layers (Bramhaiah & John, 2013).

Vibrational spectral studies

Fig. 11.9B shows the FT-IR spectrum of Au/GO NCs. The in situ reduction of GO was evidenced by the absence of functional groups. The FT-IR bands at 3450 cm^{-1}, 1636 cm^{-1}, and 1106 cm^{-1} correspond to the O—H stretching, C=C skeletal vibrations of graphene and C—O stretching vibrations (Poornima Parvathi, Umadevi, & Bhaviya Raj, 2015). Fig. 11.9C shows the Raman spectrum of Au/GO NCs. Characteristic D and G bands of GO arise at 1337 cm^{-1} and 1600 cm^{-1}, respectively. The D band originates from the edge states or symmetry breaking of the hexagonal graphite and is considered as an indication of the disorder in graphene system (Bramhaiah & John, 2013). The G band is due to the sp_2 carbons. The G band of GO has shifted to a higher wave number in comparison with the typical values around $1590-1600 \text{ cm}^{-1}$ (Liu et al., 2011). A positional shift or a splitting of G band denotes the substitution of the basal plane carbon or electron or hole doping by other atoms. This shift is consistent with several other studies where GO was chemically reduced (Poornima Parvathi et al., 2015). The intensity ratio D/G of the Raman bands calculated as 1.080 further supports that surface modification extent is high in Au/GO NCs.

Morphological studies

Fig. 11.10A–D shows the TEM images of Au/GO NCs at various magnifications. The TEM images showed an intense distribution of Au NPs on the surface of rGO sheets justifying the LSPR band observed at 531 nm. The Au NPs were

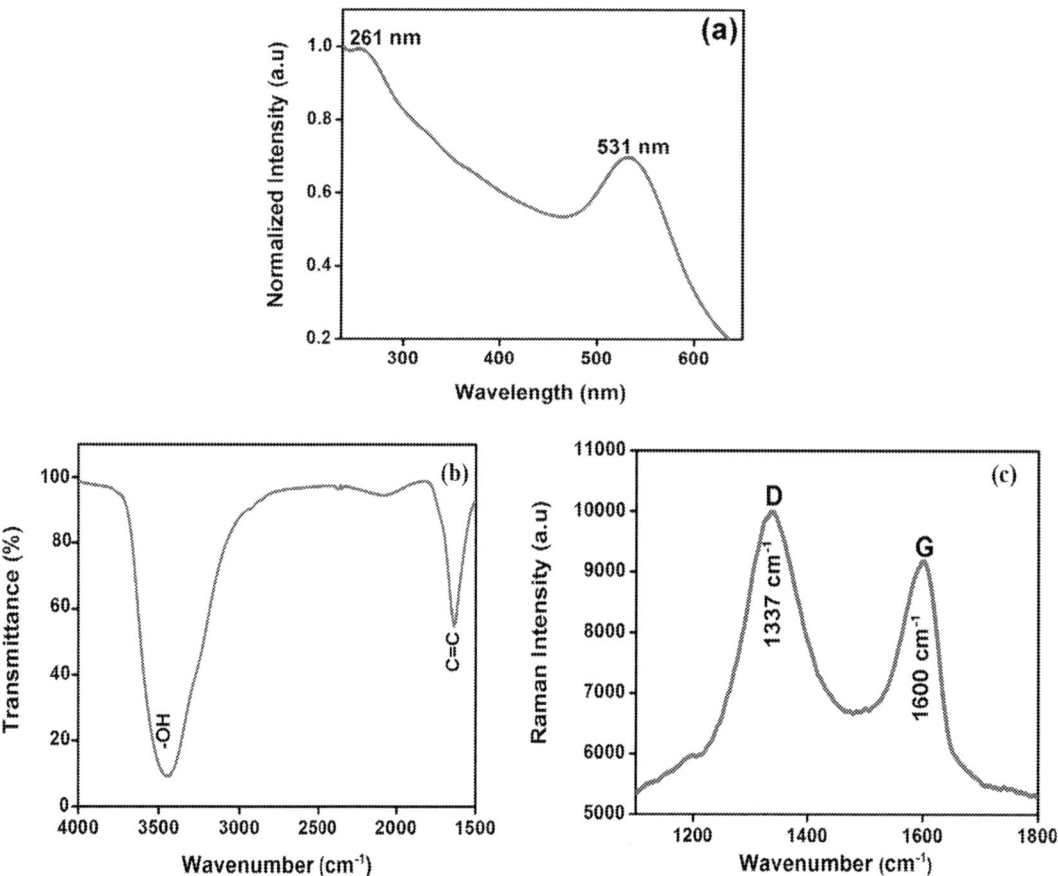

FIG. 11.9 (A) UV-visible spectrum, (B) FTIR spectrum, and (C) Raman spectrum of Au/GO NCs. *(From Poornima Parvathi, V., Parimaladevi, R., Sathe, V., & Umadevi, M. (2019). Application of G-SERS for the efficient detection of toxic dye contaminants in textile effluents using gold/graphene oxide substrates.* Journal of Molecular Liquids, 273, 203–214. https://doi.org/10.1016/j.molliq.2018.10.027.)

incorporated between the multiple layers of rGO. Slight aggregation of Au NPs on multiple-layered rGO sheets increases the local electromagnetic fields. This, along with the strong interaction along the Au NPs and GO layers, renders this substrate conducive for SERS measurements. The diffuse SAED pattern (Fig. 11.10E) observed can be attributed to the amorphous sixfold structure typical of GO, while the clear spots arise from the crystalline Au NPs dispersed along the layers of GO. The average particle size was estimated to be 19.1 nm from the particle size distribution curve (Fig. 11.10F).

Spectral analysis of dyes and TE

Raman spectral analysis of dyes and TE on Au/GO was studied. Fluorescence quenching was quantitatively analyzed by calculating the relative fluorescence quantum yield (Q). Rhodamine 6G was used as the fluorescence reference ($Q_R = 0.94$). Q of the sample in terms of the reference sample (Q_R) is: $Q = Q_R$, where OD and OD_R are the optical densities, n and n_R are the refractive indices, and I and I_R are the integrated fluorescence intensities of the sample and Rhodamine 6G, respectively (Kavitha, Umadevi, Janani, Balakrishnan, & Ramanibai, 2014).

Mono azo dyes

Mono azo dyes are a commercially important family of dyes with a single C—N=N—C linkage. Their bright colors and easy binding to textiles ensure their wide usage. Certain azo dyes are banned in many world countries owing to their carcinogenic and mutagenic nature. Here, MO and MY were chosen to represent mono azo dyes. Fig. 11.11(a) shows the absorption spectrum of MO, (b), (c) fluorescence spectrum of MO before and after the addition of Au/GO, (d) Raman spectrum, and (e) G-SERS spectrum of MO, respectively. Similarly, Fig. 11.12(a) shows the absorption spectrum of MY, (b), (c) fluorescence spectrum of MY before and after the addition of Au/GO, (d) Raman spectrum, and (e) G-SERS spectrum of MY, respectively. Table 11.1 gives the Q values of dyes and TE before and after the addition of Au/GO. Strong

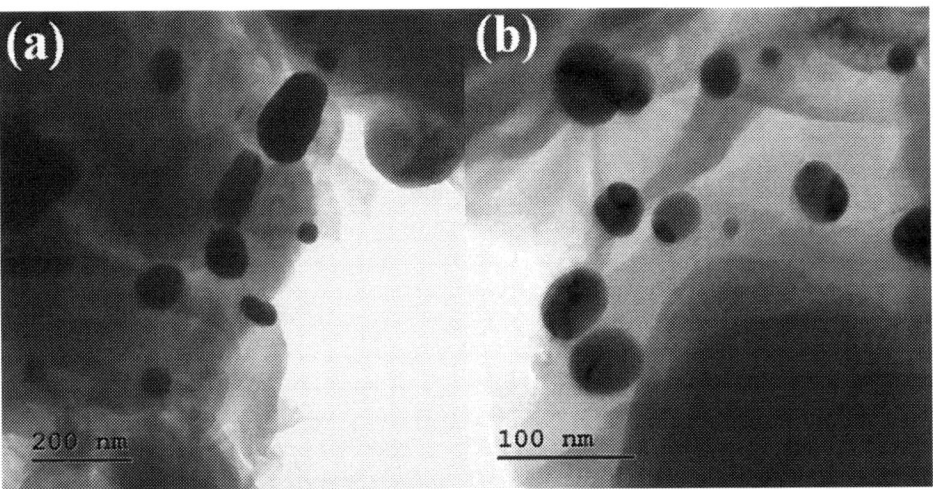
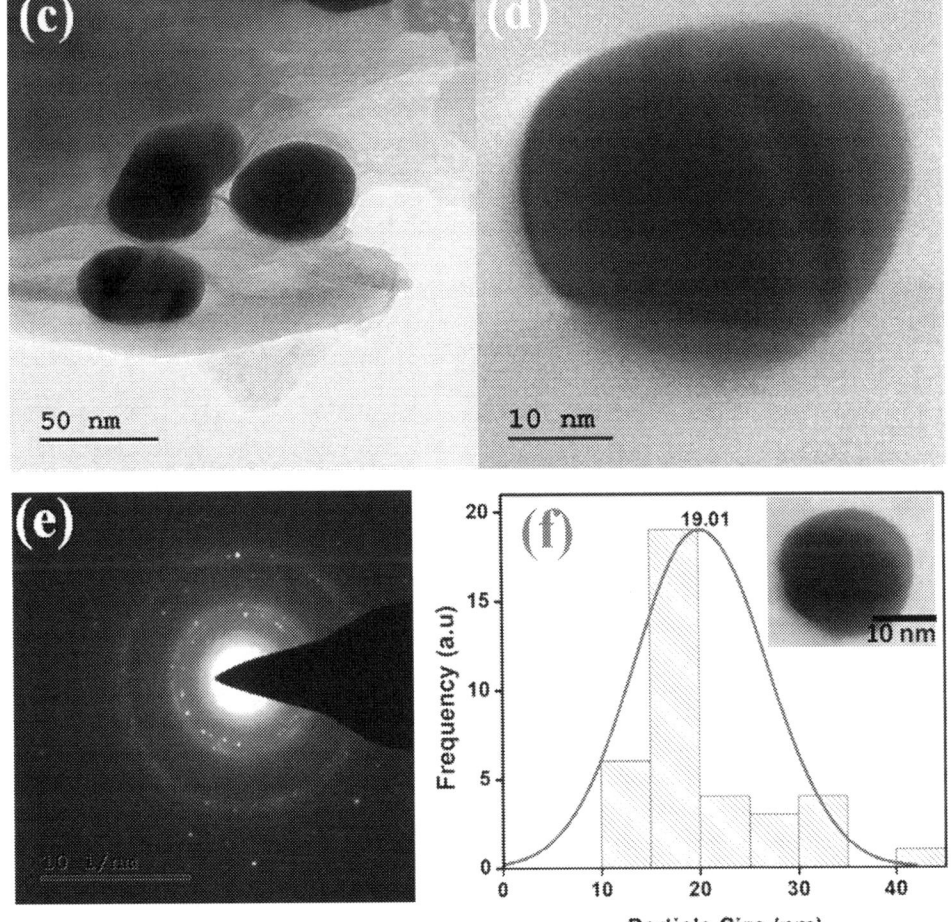

FIG. 11.10 (A–D) Typical TEM images, (E) SAED pattern of Au/GO NCs, and (F) particle size distribution curve (inset: TEM image at 10 nm) of Au/GO NCs. *(From Poornima Parvathi, V., Parimaladevi, R., Sathe, V., & Umadevi, M. (2019). Application of G-SERS for the efficient detection of toxic dye contaminants in textile effluents using gold/graphene oxide substrates. Journal of Molecular Liquids, 273, 203–214. https://doi.org/10.1016/j.molliq.2018.10.027.)*

absorption was observed at 464 and 450 nm for MO and MY, respectively. For MO and MY, quenched fluorescence illustrated by a 10 times decrease in Q was observed. The adsorption of fluorophore in the dye molecules onto the surface of Au/GO quenches the fluorescence. The Raman spectrum of MO provided no discernible information, while few Raman bands were observed for MY. In G-SERS spectra, the azo nature of both dyes is evident from the N=N stretching vibrations observed at 1368 cm^{-1}, and the C—N stretching vibrations arise at 1167 cm^{-1} in MO and at 1177 cm^{-1} in MY. Strong C—C stretching band was observed in MO and MY at 1623 cm^{-1} and 1620 cm^{-1} (Si et al., 2009).

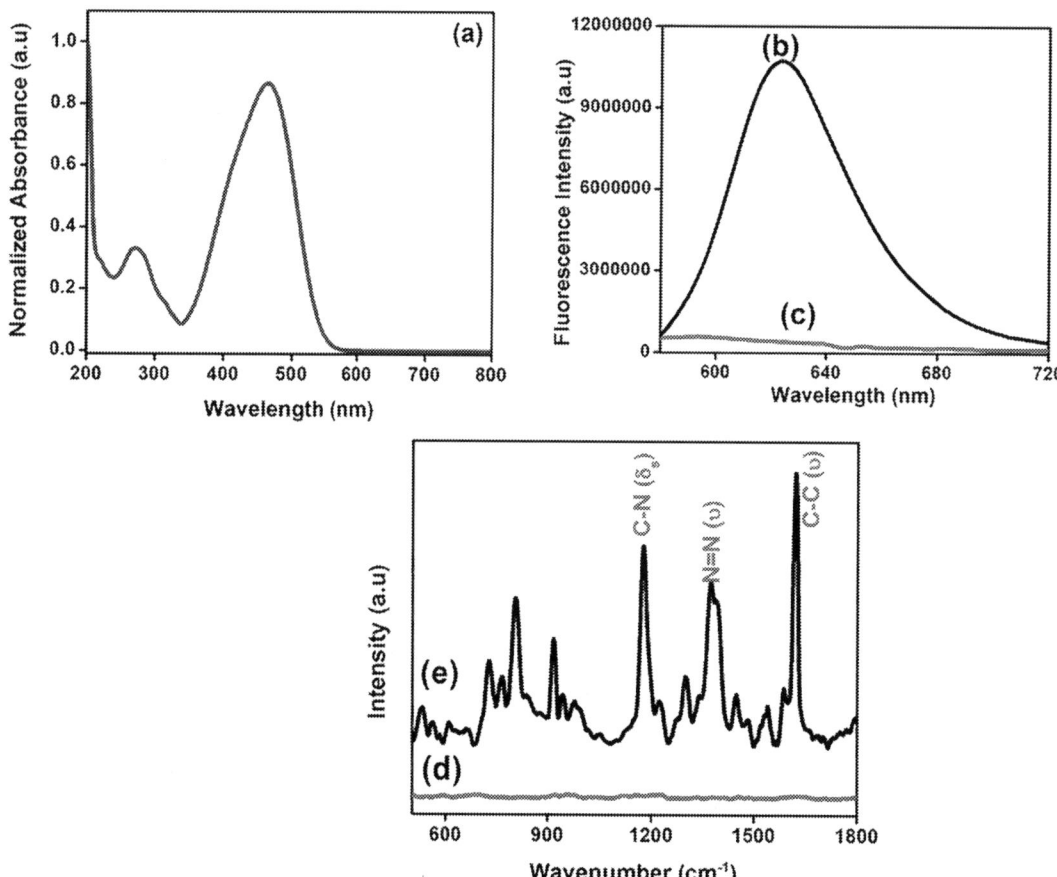

FIG. 11.11 (a) Absorption spectrum of MO, (b), (c) fluorescence spectrum of MO before and after addition of Au/GO, (d) Raman spectrum, and (e) G-SERS spectrum of MO, respectively. *(From Poornima Parvathi, V., Parimaladevi, R., Sathe, V., & Umadevi, M. (2019). Application of G-SERS for the efficient detection of toxic dye contaminants in textile effluents using gold/graphene oxide substrates.* Journal of Molecular Liquids, *273, 203–214. https://doi.org/10.1016/j.molliq.2018.10.027.)*

Di-azo dyes

The presence of double C—N=N—C linkage in di-azo dyes as their chromophore results in an enhanced permanence and hence is employed in printing inks, paper, textile, and paints. The two diazo dyes used in this study are BB and CR. Fig. 11.13(a) shows the absorption spectrum of BB, (b), (c) fluorescence spectrum of BB before and after the addition of Au/GO, (d) Raman spectrum, and (e) G-SERS spectrum of BB, respectively. Fig. 11.14(a) shows the absorption spectrum of CR, (b), (c) fluorescence spectrum of CR before and after the addition of Au/GO, (d) Raman spectrum, and (e) G-SERS spectrum of CR, respectively. Strong absorption was observed at 464 and 498 nm for BB and CR, respectively, arising from the azo nature. In both diazo dyes, a similar loss in fluorescence to almost 10 times was observed. The di-azo nature of the dyes was characterized by the Raman bands at 1382 cm^{-1}, 1479 cm^{-1} in BB and at 1387 cm^{-1}, and 1480 cm^{-1} in CR. C—C stretching band in both dyes was observed at 1619 cm^{-1}. The C—N symmetric bending modes and C—N stretching modes were distinctly exhibited in the 1160 cm^{-1}-1200 cm^{-1} region of both di-azo dyes (Bonancêa et al., 2008).

Triarylmethane dye

Triarylmethane dyes are synthetic organic dyes containing triphenylmethane backbones. These dyes are predominantly used in textile dyeing, including wool, silk, cotton, and leather. Here, CV and MG were analyzed. Fig. 11.15(a) shows the absorption spectrum of CV, (b), (c) fluorescence spectrum of CV before and after the addition of Au/GO, (d) Raman spectrum, and (e) G-SERS spectrum of CV, respectively. Fig. 11.16(a) shows the absorption spectrum of MG, (b), (c) fluorescence spectrum of MG before and after the addition of Au/GO, (d) Raman spectrum, and (e) G-SERS spectrum of MG, respectively. Strong absorption was observed at 589 and 616 nm for MO and MY, respectively. On observing

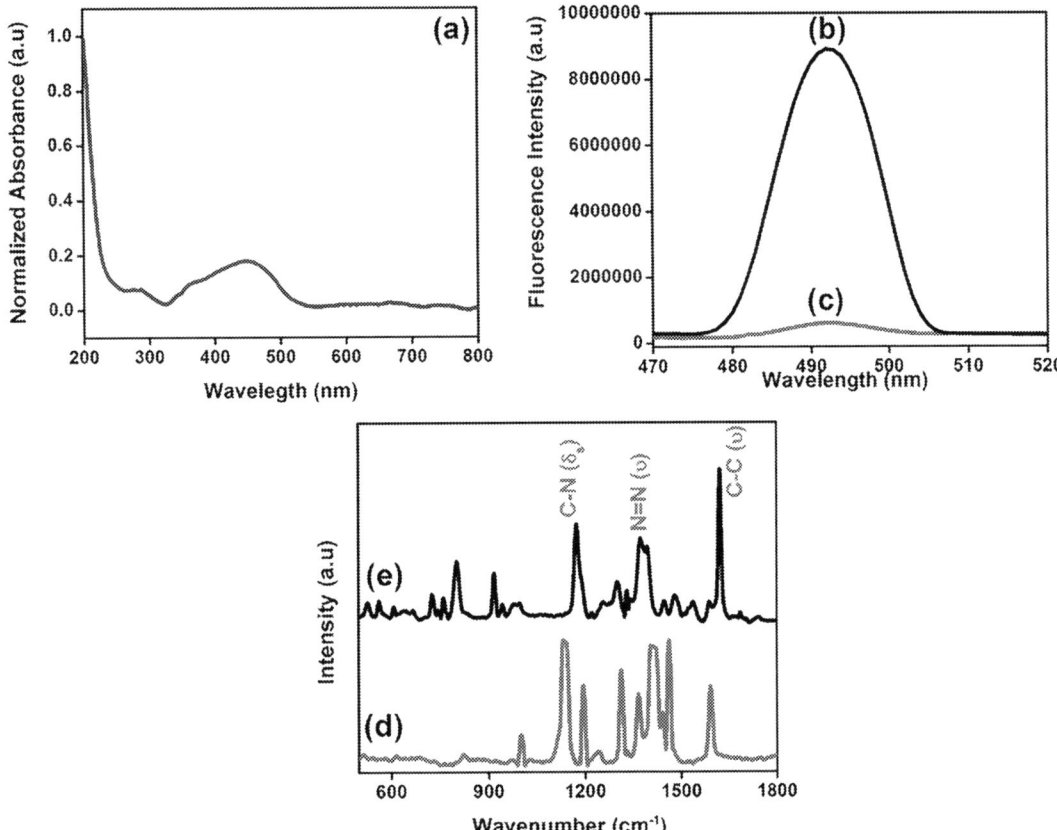

FIG. 11.12 (a) Absorption spectrum of MY, (b), (c) fluorescence spectrum of MY before and after the addition of Au/GO, (d) Raman spectrum, and (e) G-SERS spectrum of MY, respectively. *(From Poornima Parvathi, V., Parimaladevi, R., Sathe, V., & Umadevi, M. (2019). Application of G-SERS for the efficient detection of toxic dye contaminants in textile effluents using gold/graphene oxide substrates.* Journal of Molecular Liquids, 273, *203–214. https://doi.org/10.1016/j.molliq.2018.10.027.)*

TABLE 11.1 Relative fluorescence quantum yield of dyes and TE before and after the addition of Au/GO NCs.

DYE	Q	
	Before the addition of Au/GO NCs	After the addition of Au/GO NCs
MO	0.047	0.002
MY	0.230	0.047
BB	0.041	0.006
CR	0.129	0.060
CV	0.044	0.019
MG	0.062	0.009
MB	0.951	0.463
Rh	0.954	0.582
TE	0.417	0.118

From Poornima Parvathi, V., Parimaladevi, R., Sathe, V., & Umadevi, M. (2019). Application of G-SERS for the efficient detection of toxic dye contaminants in textile effluents using gold/graphene oxide substrates. *Journal of Molecular Liquids, 273,* 203–214. https://doi.org/10.1016/j.molliq.2018.10.027.

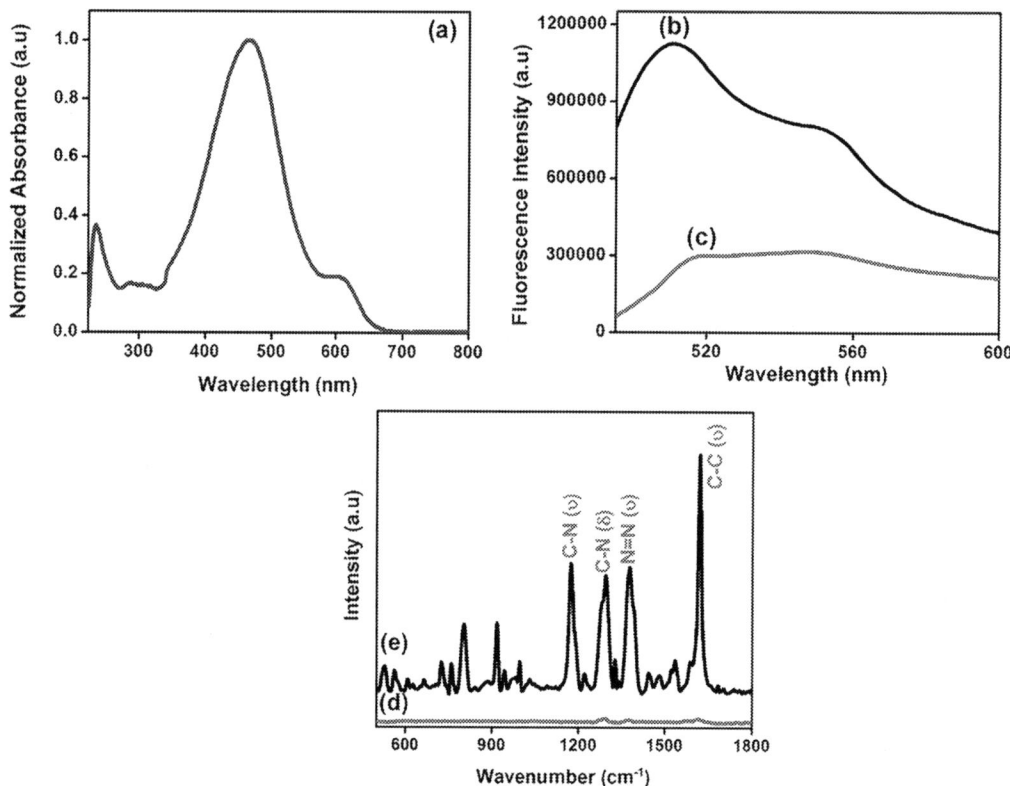

FIG. 11.13 (a) Absorption spectrum of BB, (b), (c) fluorescence spectrum of BB before and after the addition of Au/GO, (d) Raman spectrum, and (e) G-SERS spectrum of BB, respectively. *(From Poornima Parvathi, V., Parimaladevi, R., Sathe, V., & Umadevi, M. (2019). Application of G-SERS for the efficient detection of toxic dye contaminants in textile effluents using gold/graphene oxide substrates.* Journal of Molecular Liquids, 273, 203–214. https://doi.org/10.1016/j.molliq.2018.10.027.)

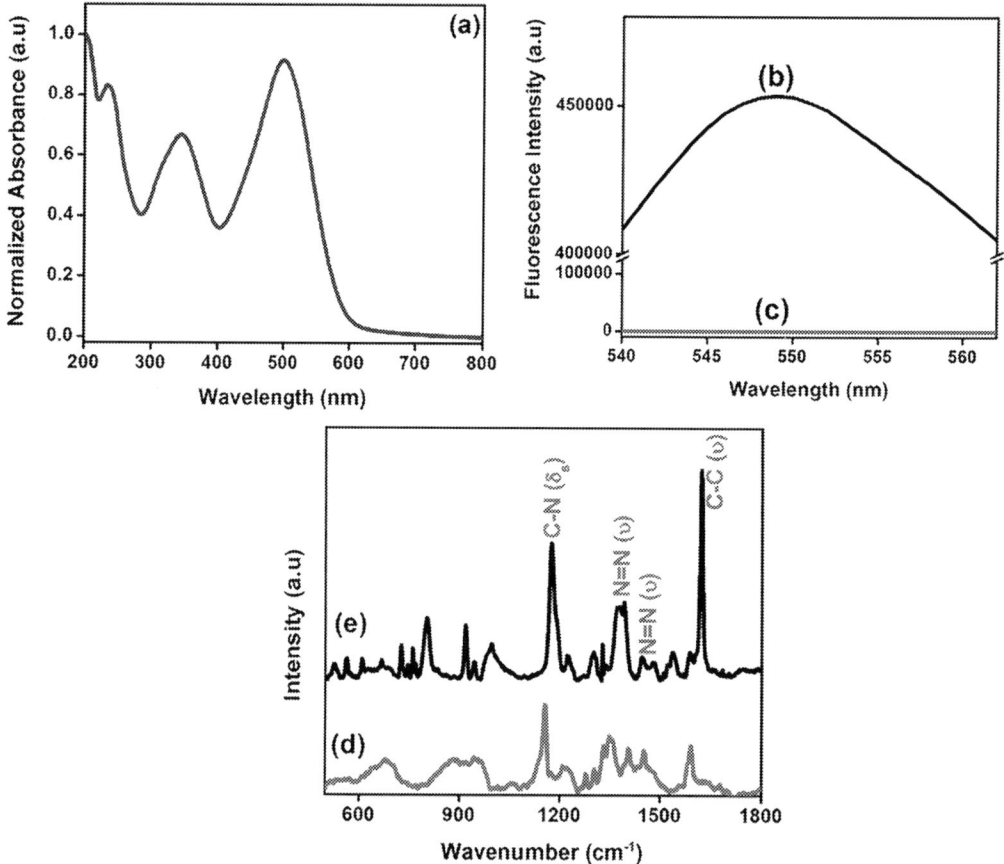

FIG. 11.14 (a) Absorption spectrum of CR, (b), (c) fluorescence spectrum of CR before and after the addition of Au/GO, (d) Raman spectrum, and (e) G-SERS spectrum of CR, respectively. *(From Poornima Parvathi, V., Parimaladevi, R., Sathe, V., & Umadevi, M. (2019). Application of G-SERS for the efficient detection of toxic dye contaminants in textile effluents using gold/graphene oxide substrates.* Journal of Molecular Liquids, 273, 203–214. https://doi.org/10.1016/j.molliq.2018.10.027.)

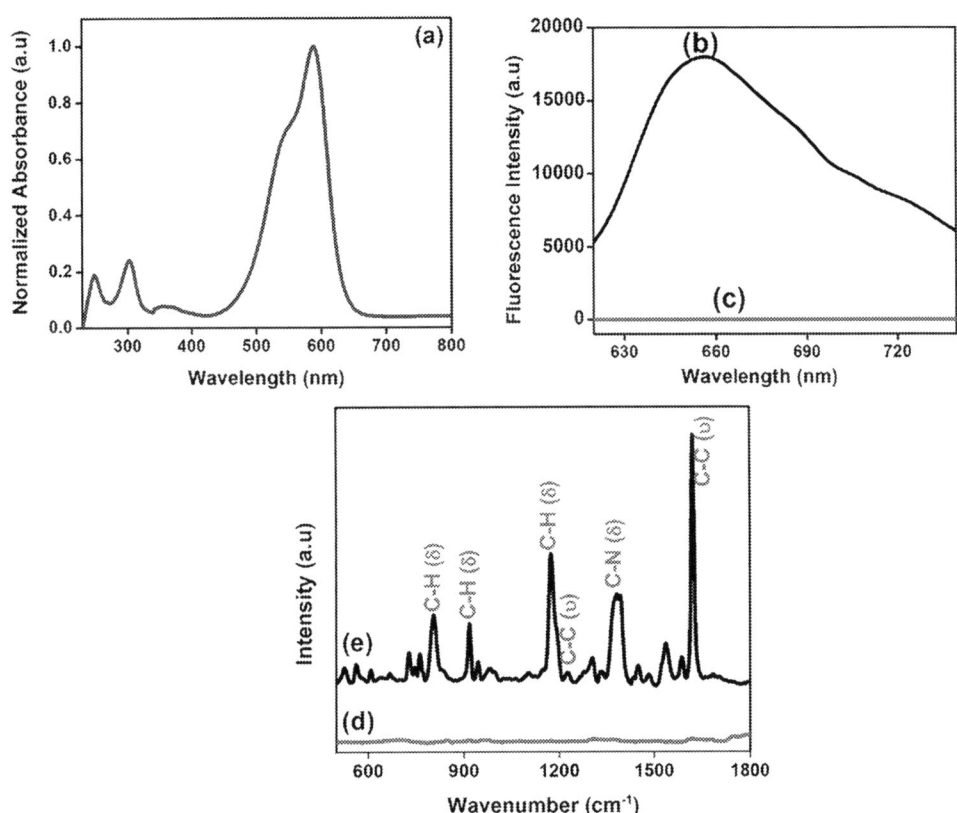

FIG. 11.15 (a) Absorption spectrum of CV, (b), (c) fluorescence spectrum of CV before and after the addition of Au/GO, (d) Raman spectrum, and (e) G-SERS spectrum of CV, respectively. *(From Poornima Parvathi, V., Parimaladevi, R., Sathe, V., & Umadevi, M. (2019). Application of G-SERS for the efficient detection of toxic dye contaminants in textile effluents using gold/graphene oxide substrates. Journal of Molecular Liquids, 273, 203–214. https://doi.org/10.1016/j.molliq.2018.10.027.)*

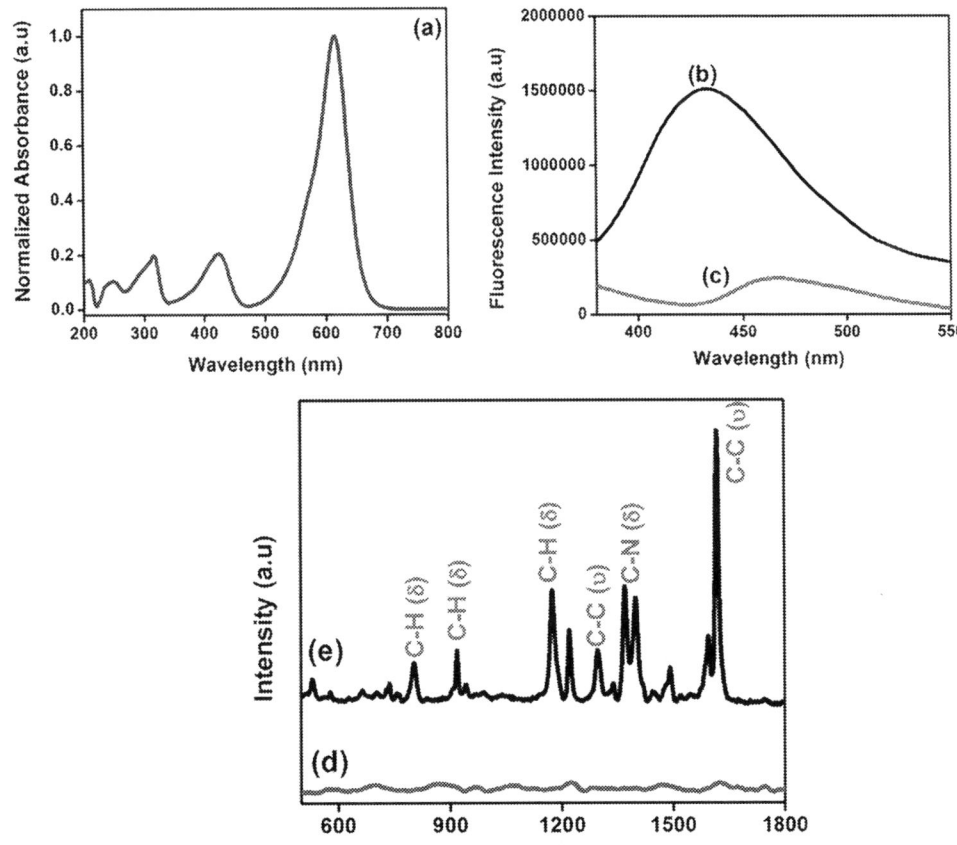

FIG. 11.16 (a) Absorption spectrum of MG, (b), (c) fluorescence spectrum of MG before and after the addition of Au/GO, (d) Raman spectrum, and (e) G-SERS spectrum of MG, respectively. *(From Poornima Parvathi, V., Parimaladevi, R., Sathe, V., & Umadevi, M. (2019). Application of G-SERS for the efficient detection of toxic dye contaminants in textile effluents using gold/graphene oxide substrates. Journal of Molecular Liquids, 273, 203–214. https://doi.org/10.1016/j.molliq.2018.10.027.)*

the Q values of CV and MG, it can be seen that a better fluorescence quenching was observed for MG. This may arise from the structural differences in CV and MG, namely, the dimethylamino and monomethylamino groups, respectively. The C—N vibrations were observed around 1370 cm^{-1}. Typical C—C stretching was observed at 1619 cm^{-1}. Bands at 1298 cm^{-1} and 1290^{-1} in CV and MG are due to the C—C vibrations. This is typical for aryl dyes arising from the C—C bridge between two phenyl molecules (Fan et al., 2014). The presence of Raman bands with medium intensities at 1175 cm^{-1} and 1171 cm^{-1} in CV and MG corresponds to the characteristic C—H in-plane bending. This, along with the C—H out-of-plane bending observed at 727 cm^{-1}, 765 cm^{-1}, 807 cm^{-1} in CV and 735 cm^{-1}, 802 cm^{-1} in MG, originates from the of triarylmethane backbone (Smitha, Gopchandran, Smijesh, & Philip, 2013).

Thiazine dye

Heterocyclic series with the presence of nitrogen and sulfur characterize thiazine dyes. MB, the thiazine dye studied here, is a medication dye with a wide range of side effects. Fig. 11.17(a) shows the absorption spectrum of MB, (b), (c) fluorescence spectrum of MB before and after the addition of Au/GO, (d) Raman spectrum, and (e) G-SERS spectrum of MB, respectively. In the UV-visible spectrum, the band at 599 nm corresponding to MB was observed. The Q value of MB was reduced to 0.436 from 0.951 on the addition of Au/GO. In MB, strong Raman bands were observed at 804 cm^{-1}, 920 cm^{-1}, 1178 cm^{-1}, 1218 cm^{-1}, 1286 cm^{-1}, 1366 cm^{-1}, 1403 cm^{-1}, 1591 cm^{-1}, and 1616 cm^{-1} corresponding to the C—H out-of-plane bending, ring skeletal stretching, C—H in-plane bending, C—N stretching and bending modes, and C—C stretching vibrations. The C-S-C skeletal deformations were observed around 623 cm^{-1}.

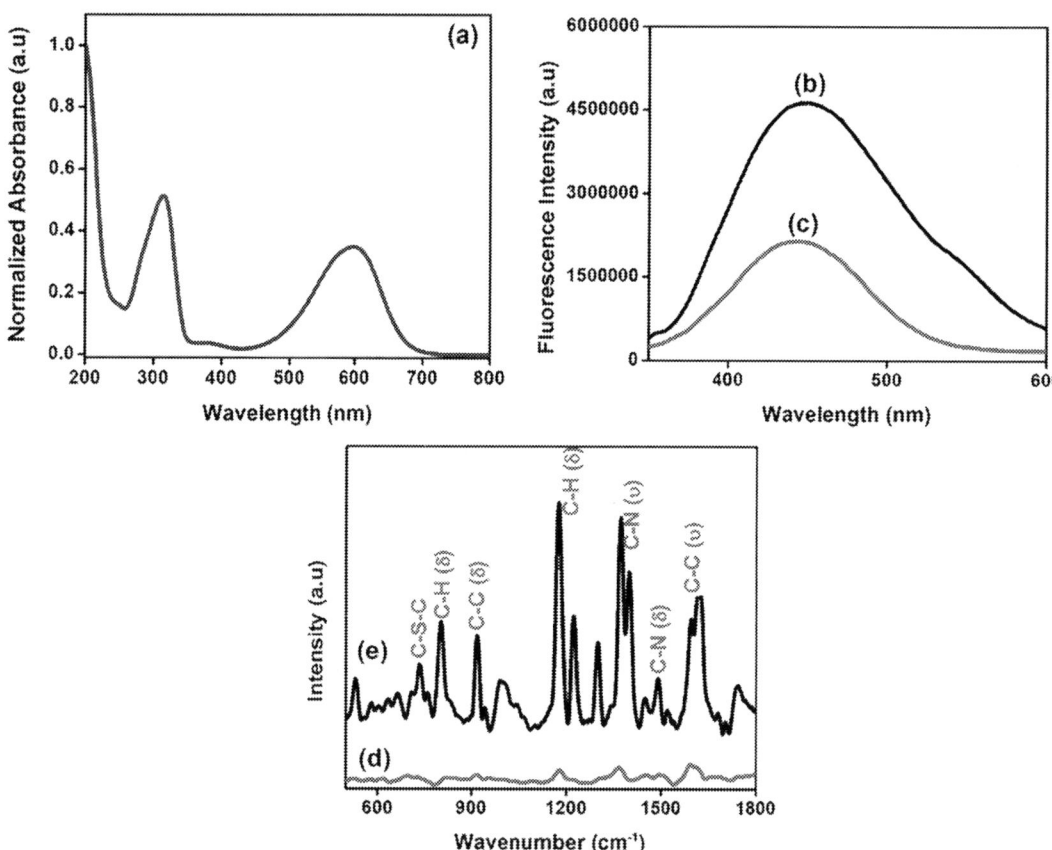

FIG. 11.17 (a) Absorption spectrum of MB, (b), (c) fluorescence spectrum of MB before and after the addition of Au/GO, (d) Raman spectrum, and (e) G-SERS spectrum of MB, respectively. *(From Poornima Parvathi, V., Parimaladevi, R., Sathe, V., & Umadevi, M. (2019). Application of G-SERS for the efficient detection of toxic dye contaminants in textile effluents using gold/graphene oxide substrates.* Journal of Molecular Liquids, 273, 203–214. *https://doi.org/10.1016/j.molliq.2018.10.027.)*

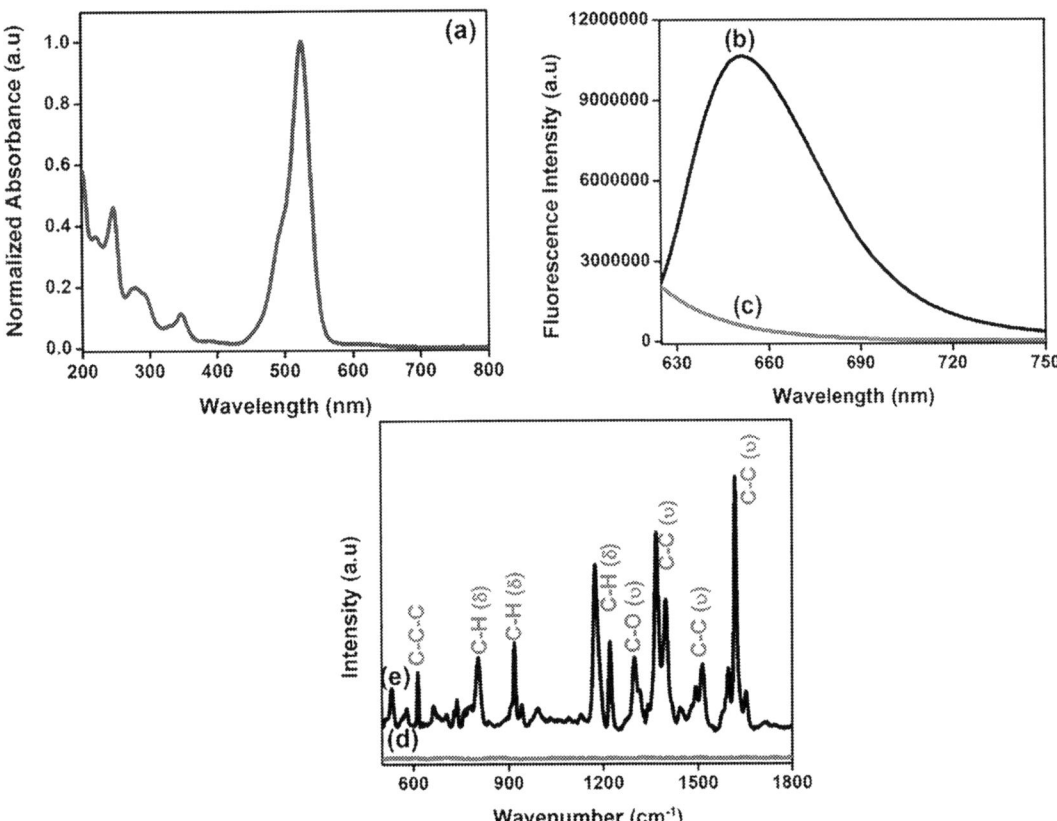

FIG. 11.18 (a) Absorption spectrum of R6G, (b), (c) fluorescence spectrum of R6G before and after the addition of Au/GO, (d) Raman spectrum, and (e) G-SERS spectrum of R6G, respectively. *(From Poornima Parvathi, V., Parimaladevi, R., Sathe, V., & Umadevi, M. (2019). Application of G-SERS for the efficient detection of toxic dye contaminants in textile effluents using gold/graphene oxide substrates.* Journal of Molecular Liquids, 273, 203–214. https://doi.org/10.1016/j.molliq.2018.10.027.)

Xanthene dye

R6G, containing the xanthene core ($CH_2(C_6H_4)_2O$), was studied here due to its strong fluorescence property. Fig. 11.18(a) shows the absorption spectrum of R6G, (b), (c) fluorescence spectrum of R6G before and after the addition of Au/GO, (d) Raman spectrum, and (e) G-SERS spectrum of R6G, respectively. For R6G, the absorption maximum was observed at 525 nm. Evidencing the fluorescence quenching ability of Au/GO, the Q value of R6G decreased to 0.582 from 0.954. The well-established vibrations of R6G at 616 cm^{-1}, 773 cm^{-1}, 1171 cm^{-1}, 1311 cm^{-1}, 1371 cm^{-1}, 1519 cm^{-1}, and 1619 cm^{-1} were observed. These bands are assigned to the xanthene moiety stretch, ethylamine group wag, carbon—oxygen stretch, and aromatic carbon—carbon stretch of R6G (Chen et al., 2015).

Textile effluent

Among the multifarious utilities of dyes, textile industries are the major contributors toward environmental pollution. Hence, real effluent samples from textile dyeing industry at Erode, India, were studied in this work. Fig. 11.19(a) shows the absorption spectrum of TE, (b), (c) fluorescence spectrum of TE before and after the addition of Au/GO, (d) Raman spectrum, and (e) G-SERS spectrum of TE, respectively. Here, a decrease in the Q from 0.417 to 0.118 was observed. No significant information was gained from the Raman spectra of TW. As expected from the G-SERS spectra of dyes, TE also presented enhanced, information-rich spectrum on adsorption to Au/GO NCs. The aromatic nature of the real sample was evidenced by the Raman bands observed around 1610 cm^{-1} corresponding to C—C stretching modes. This mode was observed in all tested dyes. As illustrated in the mono and di-azo dyes, the N=N stretching vibrations of azo compounds 1380–1425 cm^{-1} were witnessed in TW along with the C—N stretching and bending modes observed around 1190–1185 cm^{-1}, 1155–1160 cm^{-1}, and 1130–1136 cm^{-1}, respectively. The distinctive band around 1290 cm^{-1} evidences the utilization of aryl dyes. The xanthenes stretch was also observed at 614 cm^{-1} and 773 cm^{-1} in TW. The absence of any typical C-S-C skeletal deformations may suggest the absence of thiazine type of dyes. Thus, it is possible to conclude that

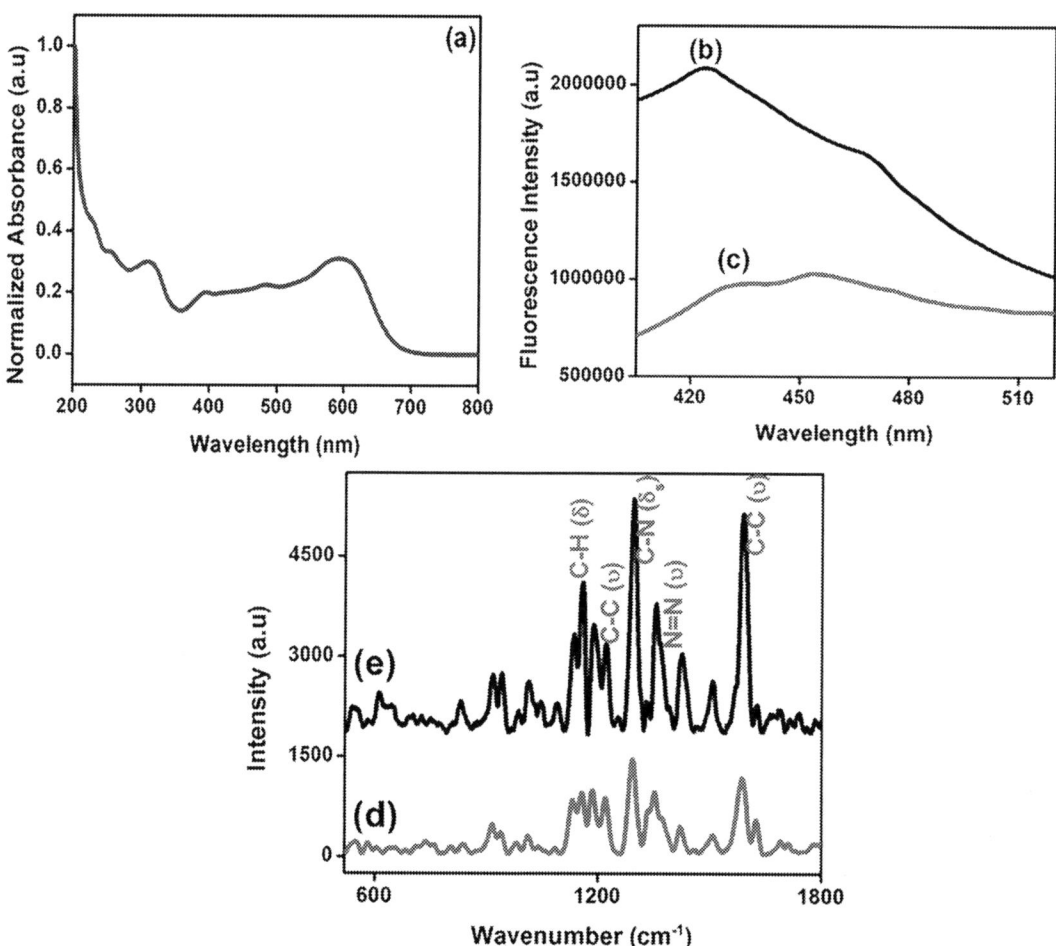

FIG. 11.19 (a) Absorption spectrum of TE, (b), (c) fluorescence spectrum of TE before and after the addition of Au/GO, (d) Raman spectrum, and (e) G-SERS spectrum of TE, respectively. *(From Poornima Parvathi, V., Parimaladevi, R., Sathe, V., & Umadevi, M. (2019). Application of G-SERS for the efficient detection of toxic dye contaminants in textile effluents using gold/graphene oxide substrates. Journal of Molecular Liquids, 273, 203–214. https://doi.org/10.1016/j.molliq.2018.10.027.)*

the real waste effluent may be a mixture of azo, aryl, and xanthenes dyes. The identification of hazardous C—C, N=N and C—N vibrations also supports the feasibility of Au/GO in sensing of carcinogenic dye contaminants from waste effluents. There are two main mechanisms at work for SERS: the chemical mechanism (CM) and the electromagnetic mechanism (EM). EM mechanism is owing to the enhanced local electric field that results in a significant increase in the cross section of the Raman scattering. This enormous EM enhancement of SERS is derived from the "hotspots" on the metal surfaces (Yi et al., 2016). "Hotspots" are highly localized regions of intense local field enhancement caused by LSPR in metal NPs. It has been demonstrated that in ZnO@Ag nanorod arrays the existence of nanogaps has aided SERS not only from the strong local electromagnetic field strength of adjacent Ag NPs but also by providing probe molecules the opportunity to directly adsorb onto ZnO-nanorods (Li et al., 2017; Liang et al., 2017; Sharma et al., 2017). These critically important hotspots formed within the interstitial crevices or "nanogaps" present in metallic NPs define the efficiency of any SERS substrates. The gap regions observed in the TEM images of Au/GO NCs form hotspots that induce additional strong field enhancement. In G-SERS, these hotspots pass through the graphene oxide layers, resulting in atomically flat "hot surface" for Raman signal enhancement. Aromatic dye molecules studied here are adsorbed on the rGO layers parallel to the π-π stacked surface of rGO, thereby reducing the distance between them. This allows for an easy charge transfer, which can induce chemical enhancement. SERS activity, being a surface phenomenon, also depends on the size of the nanoparticles. Additively, the similarity of the chemical structure of dye molecules used and rGO, the vibrational coupling between them, may be another contributor. Synergistically, the electromagnetic enhancement from the hotspots and hot surfaces and the chemical enhancement offered by rGO result in well-defined, enhanced G-SERS signals.

Conclusion

This chapter has demonstrated the efficiency of graphene-based nanoparticles as effective sensors for the detection of dyes in textile industrial waste effluents. TEM images show the possibilities in the formation of critically important "hotspots" in the nanogaps along plasmonic NPs and the possible "hot surface" due to the overlapping rGO layers. This efficient detection of four different classes of dyes describes the efficiency of G-SERS as a platform to replace traditional SERS. Despite the unsimulated nature of TE, favorable results for detection were observed in the presence of graphene-boosted nanoparticles. The toxicity of the studied dyes further necessitates the development of commercially viable G-SERS platform as an effective pollutant sensor in water processing and treatment plants for industrial effluents.

References

Acar Bozkurt, P. (2017). Sonochemical green synthesis of Ag/graphene nanocomposite. *Ultrasonics Sonochemistry, 35*, 397–404. https://doi.org/10.1016/j.ultsonch.2016.10.018.

Amrutha, B. M., Manjunatha, J. G., Bhatt, A. S., Raril, C., & Pushpanjali, P. A. (2019). Electrochemical sensor for the determination of Alizarin Red-S at non-ionic surfactant modified carbon nanotube paste electrode. *Physical Chemistry Research, 7*(3), 523–533. https://doi.org/10.22036/pcr.2019.185875.1636.

Arockia Jency, D., Umadevi, M., & Sathe, G. V. (2015). SERS detection of polychlorinated biphenyls using β-cyclodextrin functionalized gold nanoparticles on agriculture land soil. *Journal of Raman Spectroscopy, 46*(4), 377–383. https://doi.org/10.1002/jrs.4654.

Bel Hadjltaief, H., Ben Ameur, S., Da Costa, P., Ben Zina, M., & Elena Galvez, M. (2018). Photocatalytic decolorization of cationic and anionic dyes over ZnO nanoparticle immobilized on natural Tunisian clay. *Applied Clay Science, 152*, 148–157. https://doi.org/10.1016/j.clay.2017.11.008.

Bezzon, V. D. N., Montanheiro, T. L. A., de Menezes, B. R. C., Ribas, R. G., Righetti, V. A. N., Rodrigues, K. F., et al. (2019). Carbon nanostructure-based sensors: A brief review on recent advances. *Advances in Materials Science and Engineering*, 1–21.

Bonancêa, C. E., do Nascimento, G. M., de Souza, M. L., Temperini, M. L. A., & Corio, P. (2008). Surface-enhanced Raman study of electrochemical and photocatalytic degradation of the azo dye Janus Green B. *Applied Catalysis B: Environmental, 77*(3–4), 339–345. https://doi.org/10.1016/j.apcatb.2007.07.026.

Bramhaiah, K., & John, N. S. (2013). Hybrid films of reduced graphene oxide with noble metal nanoparticles generated at a liquid/liquid interface for applications in catalysis. *RSC Advances, 3*(21), 7765–7773. https://doi.org/10.1039/c3ra23324c.

Brosseau, C. L., Casadio, F., & Van Duyne, R. P. (2011). Revealing the invisible: Using surface-enhanced Raman spectroscopy to identify minute remnants of color in Winslow Homer's colorless skies. *Journal of Raman Spectroscopy, 42*(6), 1305–1310. https://doi.org/10.1002/jrs.2877.

Cesaratto, A., Centeno, S. A., Lombardi, J. R., Shibayama, N., & Leona, M. (2017). A complete Raman study of common acid red dyes: Application to the identification of artistic materials in polychrome prints. *Journal of Raman Spectroscopy, 48*(4), 601–609. https://doi.org/10.1002/jrs.5082.

Charithra, M. M., & Manjunatha, J. G. (2020). Enhanced voltammetric detection of paracetamol by using carbon nanotube modified electrode as an electrochemical sensor. *Journal of Electrochemical Science and Engineering, 10*(1), 29–40. https://doi.org/10.5599/jese.717.

Charithra, M. M., Manjunatha, J. G. G., & Raril, C. (2020). Surfactant modified graphite paste electrode as an electrochemical sensor for the enhanced voltammetric detection of estriol with dopamine and uric acid. *Advanced Pharmaceutical Bulletin, 10*(2), 247–253. https://doi.org/10.34172/apb.2020.029.

Chen, S., Li, X., Zhao, Y., Chang, L., & Qi, J. (2015). Graphene oxide shell-isolated Ag nanoparticles for surface-enhanced Raman scattering. *Carbon, 81*(1), 767–772. https://doi.org/10.1016/j.carbon.2014.10.021.

Fan, W., Lee, Y. H., Pedireddy, S., Zhang, Q., Liu, T., & Ling, X. Y. (2014). Graphene oxide and shape-controlled silver nanoparticle hybrids for ultrasensitive single-particle surface-enhanced Raman scattering (SERS) sensing. *Nanoscale, 6*(9), 4843–4851. https://doi.org/10.1039/c3nr06316j.

Gupta, S., Banaszak, A., Smith, T., & Dimakis, N. (2018). Molecular sensitivity of metal nanoparticles decorated graphene-family nanomaterials as surface-enhanced Raman scattering (SERS) platforms. *Journal of Raman Spectroscopy, 49*(3), 438–451. https://doi.org/10.1002/jrs.5318.

Hareesha, N., Manjunatha, J. G., Raril, C., & Tigari, G. (2019). Sensitive and selective electrochemical resolution of tyrosine with ascorbic acid through the development of electropolymerized alizarin sodium sulfonate modified carbon nanotube paste electrodes. *ChemistrySelect, 4*(15), 4559–4567. https://doi.org/10.1002/slct.201900794.

Hu, C., Rong, J., Cui, J., Yang, Y., Yang, L., Wang, Y., et al. (2013). Fabrication of a graphene oxide-gold nanorod hybrid material by electrostatic self-assembly for surface-enhanced Raman scattering. *Carbon, 51*(1), 255–264. https://doi.org/10.1016/j.carbon.2012.08.051.

Huang, J., Zhang, L., Chen, B., Ji, N., Chen, F., Zhang, Y., et al. (2010). Nanocomposites of size-controlled gold nanoparticles and graphene oxide: Formation and applications in SERS and catalysis. *Nanoscale, 2*(12), 2733–2738. https://doi.org/10.1039/c0nr00473a.

Jafari, Z., Mokhtarian, N., Hosseinzadeh, G., Farhadian, M., Faghihi, A., & Shojaie, F. (2016). Ag/TiO2/freeze-dried graphene nanocomposite as a high performance photocatalyst under visible light irradiation. *Journal of Energy Chemistry, 25*(3), 393–402. https://doi.org/10.1016/j.jechem.2016.01.013.

Jahn, M., Patze, S., Bocklitz, T., Weber, K., Cialla-May, D., & Popp, J. (2015). Towards SERS based applications in food analytics: Lipophilic sensor layers for the detection of Sudan III in food matrices. *Analytica Chimica Acta, 860*, 43–50. https://doi.org/10.1016/j.aca.2015.01.005.

Kavitha, S. R., Umadevi, M., Janani, S. R., Balakrishnan, T., & Ramanibai, R. (2014). Fluorescence quenching and photocatalytic degradation of textile dyeing waste water by silver nanoparticles. *Spectrochimica Acta Part A: Molecular and Biomolecular Spectroscopy, 127*, 115–121. https://doi.org/10.1016/j.saa.2014.02.076.

Lee, S., Choi, J., Chen, L., Park, B., Kyong, J. B., Seong, G. H., et al. (2007). Fast and sensitive trace analysis of malachite green using a surface-enhanced Raman microfluidic sensor. *Analytica Chimica Acta, 590*(2), 139–144. https://doi.org/10.1016/j.aca.2007.03.049.

Li, X., Li, J., Zhou, X., Ma, Y., Zheng, Z., Duan, X., et al. (2014). Silver nanoparticles protected by monolayer graphene as a stabilized substrate for surface enhanced Raman spectroscopy. *Carbon, 66,* 713–719. https://doi.org/10.1016/j.carbon.2013.09.076.

Li, D., Liu, T., Yu, X., Wu, D., & Su, Z. (2017). Fabrication of graphene-biomacromolecule hybrid materials for tissue engineering application. *Polymer Chemistry, 8*(30), 4309–4321. https://doi.org/10.1039/c7py00935f.

Liang, A., Li, C., Wang, X., Luo, Y., Wen, G., & Jiang, Z. (2017). Immunocontrolling graphene oxide catalytic nanogold reaction and its application to SERS quantitative analysis. *ACS Omega, 2*(10), 7349–7358. https://doi.org/10.1021/acsomega.7b01335.

Liu, Y., Liu, C. Y., & Liu, Y. (2011). Investigation on fluorescence quenching of dyes by graphite oxide and graphene. *Applied Surface Science, 257*(13), 5513–5518. https://doi.org/10.1016/j.apsusc.2010.12.136.

Lu, R., Konzelmann, A., Xu, F., Gong, Y., Liu, J., Liu, Q., et al. (2015). High sensitivity surface enhanced Raman spectroscopy of R6G on in situ fabricated Au nanoparticle/graphene plasmonic substrates. *Carbon, 86,* 78–85. https://doi.org/10.1016/j.carbon.2015.01.028.

Manjunatha, J. G. (2017). A new electrochemical sensor based on modified carbon nanotube-graphite mixture paste electrode for voltammetric determination of resorcinol. *Asian Journal of Pharmaceutical and Clinical Research, 10*(12), 295–300. https://doi.org/10.22159/ajpcr.2017.v10i12.21028.

Marcano, D. C., Kosynkin, D. V., Berlin, J. M., Sinitskii, A., Sun, Z., Slesarev, A., et al. (2010). Improved synthesis of graphene oxide. *ACS Nano, 4*(8), 4806–4814. https://doi.org/10.1021/nn1006368.

Pinheiro, H. M., Touraud, E., & Thomas, O. (2004). Aromatic amines from azo dye reduction: Status review with emphasis on direct UV spectrophotometric detection in textile industry wastewaters. *Dyes and Pigments, 61*(2), 121–139. https://doi.org/10.1016/j.dyepig.2003.10.009.

Poornima Parvathi, V., Parimaladevi, R., Sathe, V., & Mahalingam, U. (2019). Graphene boosted silver nanoparticles as surface enhanced Raman spectroscopic sensors and photocatalysts for removal of standard and industrial dye contaminants. *Sensors and Actuators B: Chemical, 281,* 679–688. https://doi.org/10.1016/j.snb.2018.11.007.

Poornima Parvathi, V., Parimaladevi, R., Sathe, V., & Umadevi, M. (2019). Application of G-SERS for the efficient detection of toxic dye contaminants in textile effluents using gold/graphene oxide substrates. *Journal of Molecular Liquids, 273,* 203–214. https://doi.org/10.1016/j.molliq.2018.10.027.

Poornima Parvathi, V., Umadevi, M., & Bhaviya Raj, R. (2015). Improved waste water treatment by bio-synthesized graphene sand composite. *Journal of Environmental Management, 162,* 299–305. https://doi.org/10.1016/j.jenvman.2015.07.055.

Prinith, N. S., Manjunatha, J. G., & Raril, C. (2019). Electrocatalytic analysis of dopamine, uric acid and ascorbic acid at poly(adenine) modified carbon nanotube paste electrode: A cyclic voltammetric study. *Analytical and Bioanalytical Electrochemistry, 11*(6), 742–756. http://www.abechem.com/No.%206-2019/2019,%2011(6),%20742-756.pdf.

Pushpanjali, P. A., Manjunatha, J. G., Tigari, G., & Fattepur, S. (2020). Poly(niacin) based carbon nanotube sensor for the sensitive and selective voltammetric detection of vanillin with caffeine. *Analytical and Bioanalytical Electrochemistry, 12*(4), 553–568. http://www.abechem.com/article_39221_5b4cb9d77cdb1bfb6c48fc3dac8c7fc0.pdf.

Raril, C., & Manjunatha, J. G. (2018). Carbon nanotube paste electrode for the determination of some neurotransmitters: A cyclic voltammetric study. *Modern Chemistry & Applications.* https://doi.org/10.4172/2329-6798.1000263.

Raril, C., Manjunatha, J. G., Nanjundaswamy, L., Siddaraju, G., Ravishankar, D. K., Fattepur, S., et al. (2018). Surfactant immobilized electrochemical sensor for the detection of indigotine. *Analytical and Bioanalytical Electrochemistry, 10*(11), 1479–1490. http://www.abechem.com/No.%2011-2018/2018,%2010(11),%201479-1490.pdf.

Raril, C., Manjunatha, J. G., Ravishankar, D. K., Fattepur, S., Siddaraju, G., & Nanjundaswamy, L. (2020). Validated electrochemical method for simultaneous resolution of tyrosine, uric acid, and ascorbic acid at polymer modified nano-composite paste electrode. *Surface Engineering and Applied Electrochemistry, 56*(4), 415–426. https://doi.org/10.3103/S1068375520040134.

Raril, C., Manjunatha, J. G., & Tigari, G. (2020). Low-cost voltammetric sensor based on an anionic surfactant modified carbon nanocomposite material for the rapid determination of curcumin in natural food supplement. *Instrumentation Science and Technology, 48*(5), 561–582. https://doi.org/10.1080/10739149.2020.1756317.

Saleh, T. A., Al-Shalalfeh, M. M., & Al-Saadi, A. A. (2018). Silver loaded graphene as a substrate for sensing 2-thiouracil using surface-enhanced Raman scattering. *Sensors and Actuators, B: Chemical, 254,* 1110–1117. https://doi.org/10.1016/j.snb.2017.07.179.

Sharma, S., Prakash, V., & Mehta, S. K. (2017). Graphene/silver nanocomposites-potential electron mediators for proliferation in electrochemical sensing and SERS activity. *TrAC, Trends in Analytical Chemistry, 86,* 155–171. https://doi.org/10.1016/j.trac.2016.10.004.

Si, M. Z., Kang, Y. P., & Zhang, Z. G. (2009). Surface-enhanced Raman scattering (SERS) spectra of Methyl Orange in Ag colloids prepared by electrolysis method. *Applied Surface Science, 255*(11), 6007–6010. https://doi.org/10.1016/j.apsusc.2009.01.055.

Smitha, S. L., Gopchandran, K. G., Smijesh, N., & Philip, R. (2013). Size-dependent optical properties of Au nanorods. *Progress in Natural Science: Materials International, 23*(1), 36–43. https://doi.org/10.1016/j.pnsc.2013.01.005.

Souza, E. S., Manjunatha, J. G., Raril, C., Tigari, G., Ravishankar, D. K., & Fattepur, S. (2020). Arginine electropolymerized carbon nanotube paste electrode as sensitive and selective sensor for electrochemical determination of vanillin. *Journal of Materials and Environmental Science, 11,* 512–521.

Tigari, G., & Manjunatha, J. G. (2020a). A surfactant enhanced novel pencil graphite and carbon nanotube composite paste material as an effective electrochemical sensor for determination of riboflavin. *Journal of Science: Advanced Materials and Devices, 5*(1), 56–64. https://doi.org/10.1016/j.jsamd.2019.11.001.

Tigari, G., & Manjunatha, J. G. (2020b). Optimized voltammetric experiment for the determination of phloroglucinol at surfactant modified carbon nanotube paste electrode. *Instruments and Experimental Techniques, 63*(5), 750–757. https://doi.org/10.1134/S0020441220050139.

Tigari, G., Manjunatha, J. G., Raril, C., & Hareesha, N. (2019). Determination of riboflavin at carbon nanotube paste electrodes modified with an anionic surfactant. *ChemistrySelect*, *4*(7), 2168–2173. https://doi.org/10.1002/slct.201803191.

Turcheniuk, K., Boukherroub, R., & Szunerits, S. (2015). Gold-graphene nanocomposites for sensing and biomedical applications. *Journal of Materials Chemistry B*, *3*(21), 4301–4324. https://doi.org/10.1039/c5tb00511f.

Vandenabeele, P., Moens, L., Edwards, H. G. M., & Dams, R. (2000). Raman spectroscopic database of azo pigments and application to modern art studies. *Journal of Raman Spectroscopy*, *31*(6), 509–517. https://doi.org/10.1002/1097-4555(200006)31:6<509::AID-JRS566>3.3.CO;2-S.

Vinoth, V., Wu, J. J., Asiri, A. M., & Anandan, S. (2017). Sonochemical synthesis of silver nanoparticles anchored reduced graphene oxide nanosheets for selective and sensitive detection of glutathione. *Ultrasonics Sonochemistry*, *39*, 363–373. https://doi.org/10.1016/j.ultsonch.2017.04.035.

Xiao, Y., Liu, J., Lin, Y., Lin, W., & Fang, Y. (2017). Novel graphene oxide-silver nanorod composites with enhanced photocatalytic performance under visible light irradiation. *Journal of Alloys and Compounds*, *698*, 170–177. https://doi.org/10.1016/j.jallcom.2016.12.160.

Yang, Y. K., He, C. E., He, W. J., Yu, L. J., Peng, R. G., Xie, X. L., et al. (2011). Reduction of silver nanoparticles onto graphene oxide nanosheets with N, N-dimethylformamide and SERS activities of GO/Ag composites. *Journal of Nanoparticle Research*, *13*(10), 5571–5581. https://doi.org/10.1007/s11051-011-0550-5.

Yi, Z., Niu, G., Luo, J., Kang, X., Yao, W., Zhang, W., et al. (2016). Ordered array of Ag semishells on different diameter monolayer polystyrene colloidal crystals: An ultrasensitive and reproducible SERS substrate. *Scientific Reports*, *6*. https://doi.org/10.1038/srep32314.

Part IV

Role of carbon nanomaterials-based sensors in sustainability

Chapter 12

Carbon nanomaterial-based sensors for the development of sensitive sensor platform

Hulya Silah[a], Ersin Demir[b], Sercan Yıldırım[c], and Bengi Uslu[d]

[a]Faculty of Art & Science, Department of Chemistry, Bilecik Seyh Edebali University, Bilecik, Turkey, [b]Faculty of Pharmacy, Department of Analytical Chemistry, Afyonkarahisar Health Sciences University, Afyonkarahisar, Turkey, [c]Faculty of Pharmacy, Department of Analytical Chemistry, Karadeniz Techical University, Trabzon, Turkey, [d]Faculty of Pharmacy, Department of Analytical Chemistry, Ankara University, Ankara, Turkey

Introduction

The development of new techniques for carrying out fast in situ determinations can be contemplated as a major challenge because it requires high accuracy and sensitivity for the detection of different species with various properties in real samples. Sensors and sensor systems are important for our awareness of surroundings and provide security, safety, and surveillance, as well as enable monitoring of our health and environment (Hunter, Stetter, Hesketh, & Liu, 2020). The use of sensors is necessary since many of their functions affect the health and life quality of the population. For example, early diagnosis of various diseases such as cancers, chemical contamination determination, and control of agriculture production and foodstuffs are among the important practices. The significance of the sensors in clinical applications and medicine is demonstrated by the effective progress in point-of-care diagnostic devices and lab-on-chip platforms that combine the steps of sampling, sample treatment, separation, and analysis (Evtugyn, Porfireva, Shamagsumova, & Hianik, 2020). Sensors as instrumental devices in analytical chemistry are employed for the conversion of chemical information into an electronic signal, using two basic components, such as a transducer and a receptor connected in series (Ganesan & Paramathevar, 2020). The most significant requirements of sensors are sensitivity, diversity, selectivity, accuracy of data extracted, short response time, and stability (Karim, Reda, & Fattah, 2020). Sensors are among the most popular devices presently available and have been applied in many important areas, such as clinical, industrial, agricultural, military, environmental, and aerospace analyses.

The commonly employed analytical methods for the detection of organic and inorganic species are atomic absorption spectroscopy, ultraviolet spectroscopy, capillary electrophoresis, high-performance liquid chromatography, gas chromatography, fluorimetry, chemiluminescence, mass spectrometry, infrared spectroscopy, voltammetry, amperometry, etc. In comparison with electrochemical techniques, the other analytical methods are characterized as less sensitive and time-consuming procedures. Besides, they require very expensive equipment with complicated sample preparation processes, professional testing personnels, systematized laboratories, etc. (Antherjanam, Saraswathyamma, Krishnan, & Gopakumar, 2021; Hu & Zhang, 2020) In this context, electrochemical techniques emerge as advantageous alternatives.

Several interfacial electrochemical techniques are employed in the sensing procedures depending upon the required information, such as amperometry, potentiometry, voltammetry, and impedimetry. The basic factor underlying an electrochemical sensor selectivity and sensitivity is the electron-driving performance of the materials used at an electrode or probe surface. Generally, materials with high electrical conductivity, such as carbon, might easily promote electron transfer on electrochemical devices (Morais, Rijo, Hernan, & Nicolai, 2020).

Nowadays, nanomaterials have received global attention because of their unique size, shape, and special characteristics (Gopal & Reddy, 2018). Nanomaterials are being commonly used to improve products and production processes with better characteristics or new functionalities in different areas, such as medicine, cosmetics, drug designing, electronics, and treatment of wastewater. The nanosized character of nanomaterials and their particular electronic, optical, mechanical, and magnetic properties improve many applications through the use of sensors with extreme sensitivity (Ehtesabi, 2020). Carbon-based nanomaterials have been frequently investigated and used in nanotechnology due to their extraordinary properties. Carbonaceous structures present myriad superiorities compared to other usually studied materials.

Carbon materials usually display advantageous properties because of their characteristics, such as affordable cost, wide potential range, wide linear concentration range, chemical and physical stability, etc. Electrocatalytic characteristics of carbonaceous materials contribute to electrochemical analysis for determining and identifying organic and inorganic compounds, such as metals and biological and environmental materials. Additionally, these nanomaterials have been frequently preferred for the construction of optical sensors because of their outstanding optical properties, such as high light absorptivity, surface plasmon resonance, and photoluminescence. Single-walled carbon nanotube (SWCNT), double-walled carbon nanotube (DWCNT), multiwalled carbon nanotube (MWCNT), graphene, fullerene, carbon dots (CDs), graphene dots (GDs), boron-doped diamond (BDD), carbon black, carbon nano-onions, and carbon fiber are examples of novel carbon materials (Yence, Cetinkaya, Ozcelikay, Kaya, & Ozkan, 2021).

In this chapter, selected recent trends in carbon nanomaterial-based sensors over the last 5 years have been presented. Signal enhancement strategies together with a comprehensive literature review of sensor applications were provided. Moreover, highlights on the application of carbon nanomaterial-based sensors for nucleic acids, environmental pollutants, vitamins, hormones, drugs, enzymes, metal ions, pesticides, gases, and cancer markers have been given here. Current challenges and future directions on the development of carbon nanomaterial-based sensors were also discussed.

Overview of carbon nanomaterial chemistry

As an abundant element, carbon is widely used in scientific and technological fields. Carbon is an indispensable element not only for living things but also for all compounds. Most of the objects we see around contain carbon or carbon derivatives. The discovery of the first carbon nanomaterial dates back to the mid-18th century. Graphite oxide (GO) known as "Graphon" was first synthesized by Brodie in 1859 (Brodie, 1859). In the following years, the structural and electronic properties of GO were discovered by Bernal and Wallace, respectively (Wallace, 1947). Ruess and Vogt successfully imaged several layers of graphite through transmission electron microscopy (TEM) in the 1950s (Ruess & Vogt, 1948). After the 1950s, intensive studies began to be carried out for carbonaceous materials, especially graphite (DiVincenzo & Mele, 1984; Hummers & Offeman, 1958; Kroto, Heath, O'Brien, Curl, & Smalley, 1985). First, Hummers and Offeman developed a new method of producing GO that is more reliable and highly efficient than Brodie's method in 1957 (Hummers & Offeman, 1958). The discovery of another carbonaceous material was made by Smally in 1985 (Kroto et al., 1985). It entered the literature with the name fullerene, which represents the name eureka in the scientific world. This is the first carbon allotropy to be groundbreaking in the scientific world. Intensive studies on carbonaceous materials continued to yield incredible results. In 1991, Iijima discovered carbon allotropy, carbon nanotubes (CNTs) with extraordinary properties (Iijima, 1991). In 2004, Geim and Novoslov together produced a new carbonaceous material called graphene in their surprising work, and afterward, they won the Nobel Prize in 2010 (Novoselov et al., 2004). Today, carbon nanomaterials are available in many different forms, such as SWCNT, DWCNT, MWCNT, fullerene, graphene, GO, quantum dots such as graphene quantum dots (GQDs), carbon quantum dots (CQDs), and polymer dots, carbon black, carbon nano-onions (CNOs), etc. (Fig. 12.1).

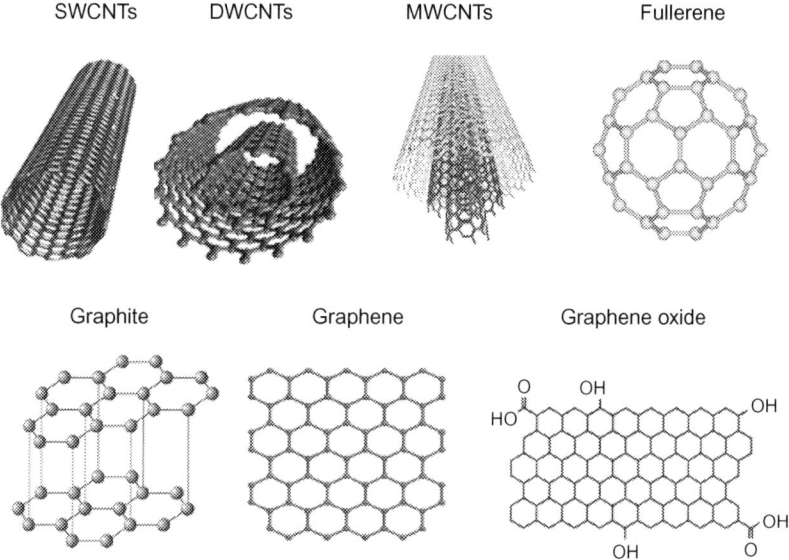

FIG. 12.1 Different structures of carbon nanomaterials. *Reprinted with permission from Lalabadi MA, Hashemi H, Feng J, Jafari SM. Carbon nanomaterials against pathogens; the antimicrobial activity of carbon nanotubes, graphene/graphene oxide, fullerenes, and their nanocomposites. Advances in Colloid and Interface 2020;284:102250.*

Extensive various uses of carbon nanomaterials are related to the unrivaled chemical and physical properties of carbon as well as the variety of possible carbon nanostructures. The variety of carbon nanostructures, as well as their chemical, physical, and structural properties, depends on variations in sp, sp^2, and sp^3 hybridizations. Typical hexagonal lattices of carbon nanomaterials have occurred as a result of sp^2 hybridization where carbon atoms bind together with two single bonds and a double bond in a trigonal planar fashion to form the classic two-dimensional hexagonal lattice of many carbon nanomaterials. Classic cubic crystal structures of diamonds are formed by sp^3 hybridization, where carbon atoms have four single bonds in a tetrahedral arrangement. The ratio of $sp/sp^2/sp^3$ hybridizations in carbon nanomaterials defines formation of flat two-dimensional (2D) nanomaterials, one-dimensional (1D) nanomaterials, and zero-dimensional (0D) nanomaterials (Loh et al., 2018). These three primary classes of carbon nanostructures are widely utilized in sensors. For example, GQDs and CQDs, represented as zero-dimensional nanomaterials, have acquired increasing attention in the past years. Their particular fluorescent, electronic, chemiluminescent, photoluminescent, and electrochemiluminescent properties make them very attractive alternatives in the development of sensor designs. Other advantages of CDs are cost-effective synthesis, simple synthetic paths, cheapness, low toxicity, high water solubility, physical/chemical stability, and straightforward functionalization. CD-based chemosensors and biosensors function by different modes of action, namely static quenching, fluorescence quenching, dynamic quenching, inner filter effect, energy transfer, photo-induced electron transfer, and fluorescence resonance energy transfer (Ehtesabi, 2020). Nanotubes consist of one-dimensional carbon nanostructures. Graphene and its derivatives are two-dimensional carbon nanostructures with impressive specifications (Ehtesabi, 2020). Additionally, the ratio of $sp/sp^2/sp^3$ hybridizations determines other features of carbon nanomaterials, including structural strength, chemical, electrical, and magnetic properties that all contribute to the unmatched advantages of various carbon nanomaterials toward different applications (Loh et al., 2018).

Carbon nanomaterial-based sensors

In this chapter, sensors based on carbon nanomaterials were categorized into two groups according to their transduction mechanism: (i) electrochemical sensors employing nanocarbon material-modified electrodes and (ii) optical sensors based on colorimetry, fluorescence quenching including chemiluminescence resonance energy transfer (CRET), Förster resonance energy transfer (FRET), and surface plasmon resonance (SPR) (Panwar et al., 2019).

Electrochemical sensors based on carbon nanomaterials

Carbon nanomaterials are favorable for electrochemical sensor systems because they increase the electroactive surface area, improve electron transfer between the analytes and transducers, decrease overpotentials, and support the adsorption of analyte molecules. Electrochemical sensors are constituted of an electrochemical cell which includes a minimum of two electrodes to form a closed electrical circuit and a transducer where the charge transportation (which is always electronic) occurs, whereas the charge transport in the analyte sample can be either electronic, ionic, or mixed (Power, Gorey, Chandra, & Chapman, 2018). Electrochemical sensor systems present significant knowledge about the analytes in real environments, such as blood, urine, saliva, water, etc., without using complex sample preparation techniques. Also, multianalytes can be determined simultaneously. Simple measurement of biological systems or living organisms can be achieved, thanks to the miniaturization of sensor systems (Huang, Xu, Liu, Wang, & Chen, 2017).

Electrochemical sensor systems can be divided into three categories based on the employed method: conductometric, potentiometric, and voltammetric. Voltammetry and potentiometry are interfacial techniques in which controlled potential is implemented. On the other hand, conductometry is a bulk electrochemical method (Huang, Xu, et al., 2017). The voltammetric techniques typically consist of a three-electrode system including a counter, reference, and working electrode. The working electrode plays a crucial role in measuring the current flow in the cell (Gopal & Reddy, 2018).

To improve the active surface area of electrodes, the application of nano and three-dimensional porous materials has been remarkably increased (Kamyabi & Moharramnezhad, 2021). Nanostructured materials used in electrode surfaces have displayed prospects in increasing the sensitivity and selectivity of electrochemical sensors because of their increased surface area to volume ratio, advanced access of target molecules to bio-recognition elements, the incremented density of bio-recognition elements (active sites), and higher catalytic activity (Morais et al., 2020). The common nanomaterials employed in electrochemical biosensors are metal nanomaterials (Au, Ag, Pt, Pd, etc.), metal oxide nanomaterials (CeO_2, CuO, NiO, Fe_2O_3, Co_3O_4, ZnO, SnO_2, TiO_2, etc.), transition metal dichalcogenides (MoS_2, etc.), polymer nanomaterials (polyaniline, polypyrrole, poly[3,4-ethylenedioxythiophene], etc.), and carbon-based nanomaterials (CNTs, carbon black, graphene, carbon fiber, fullerenes, nanodiamonds, etc.) (Zhang et al., 2021).

Since the discovery of carbon nanomaterials, their structural properties and applications have been studied in detail. They have unlimited usage areas due to their number of features such as large surface area, high conductivity, thermal and chemical stability, durability, optical properties, unique structural dimensions, temperature resistance, low cost, low residual current, biocompatibility, component diversity, and easy modification. Additionally, carbonaceous materials can be accepted as alternatives to currently employed high-cost electronic materials and being considered environment-friendly materials.

Although it is not possible to make a complete classification of carbon nanomaterials, we can divide carbon nanomaterial-based sensors into main groups as follows: fullerenes, SWCNT, MWCNT, CQDs, nanodiamonds, and graphene.

Sensor-based fullerene

Fullerene (C_{60}) is a member of the nanocarbon material class, which has a one-dimensional structure and consists of only sp^2 hybridized carbon atoms (Toghan, Abd-Elsabour, & Abo-Bakr, 2021). The simplest definition of fullerene is closed lattice forms of carbon molecules containing pentagonal and hexagonal rings arranged to have the general formula C_{20+m}. Although it was named "Buckminster fullerene" in honor of the famous architect Buckminster Fuller, who designed the geodesic domes, it was discovered in 1985 by Koroto et al. (Kroto et al., 1985). Herein, fullerene as a nanostructured carbon-based material is obtained by the evaporation of graphite exposed to electric arc in a helium atmosphere. Fullerenes with a wide variety of isomer structures such as C_{240}, C_{540}, C_{720}, C_{60}, and C_{70} are available today. C_{60} materials, the large π bond conjugated system composed of 60 p-orbitals makes it have a perfect electron acceptance and donation capabilities; in this way, the dope of metal nanoparticles is promising in improving the sensitivity of C_{60} composites to analytes (Toghan et al., 2021). The reasons why this carbon nanomaterial is of great interest are its use in medical diagnostics, anticancer drugs, nuclear magnetic resonance (NMR) systems, cosmetics, fuel cells, solar cells, transistors, electronics, and optics. Due to their extraordinary properties such as very high thermal and electrical conductivity, hardness, and stability, electrochemical sensor studies are intense (Table 12.1). Fullerenes have a wide potential range, easy preparation, very low cost, and high stability and reproducibility, making them a great difference to carbon allotropes in electrochemical sensor platform (Toghan et al., 2021). The usages of fullerene based on sensors have been surveyed in Table 12.1.

Taouri et al. designed a new selective, sensitive, economic, and simple electrochemical sensor for the determination of vanillin in commercial vanilla sugar samples using an electrode consisting of a carbon paste electrode (CPE) nanostructured with fullerene and functionalized MWCNTs (f-MWCNTs) (Fig. 12.2). The developed electroanalytical method was implemented using cyclic voltammetry (CV) after the investigation of experimental conditions, such as effects of supporting electrolyte type, pH of solution, and accumulation time. After optimization, the proposed nanostructured sensor demonstrated wide linear concentration responses between 5×10^{-8} to $9-10^{-6}$ M for vanillin trace levels and from 10^{-5} to 10^{-4} M for higher vanillin concentrations, with a low limit of detection (LOD) of 3.4×10^{-8} M. The designed electrode has several advantages over other analytical methods used for vanillin detection, including the facility of sensor preparation, low cost, rapidity with very good analytical characteristics, such as accuracy, selectivity, and sensitivity. The most significant advantage is the applicability to the determination of real vanillin concentrations in commercial vanilla sugar samples as a real sample matrix, without any pretreatment by following a simple procedure. The proposed electrochemical method was validated in comparison to the ultra-performance liquid chromatography method with a satisfactory recovery and good accuracy. The CPE nanostructured with fullerene and f-MWCNTs showed good stability, repeatability, and reproducibility (less than 7%) (Taouri et al., 2021).

Sensor-based single- and multiwalled carbon nanotube electrodes

CNTs can find implementation in numerous fields, such as sensors, nanoprobes, supercapacitors, hydrogen storage, lithium battery anodes, energy conversion, additives to polymers and catalysts, composite materials (fillers or coatings), flat displays, absorption, screening of electromagnetic waves, gas discharge tubes in telecommunication networks, etc. (Ehtesabi, 2020). In fact, MWCNTs were first obtained by Ijima using the arc discharge method. Later, the discovery of SWCNT was carried out by Ijima and Ichihashi and Bethune et al. CNTs are contemplated as graphite sheets that are rolled up into cylindrical forms, which have taken enormous attention because of their extraordinary optical, physical, mechanical, electronic, and thermal properties. Especially, the electrocatalytic properties of CNTs have driven researchers to investigate these nanomaterials in electrochemical sensor systems (Taouri et al., 2021). Up to date, it has been possible to produce these carbon materials by three techniques: (i) carbon arc discharge, (ii) laser ablation, and (iii) chemical vapor deposition. SWCNTs only have a single graphene cylinder layer, while MWCNTs have 2 or more layers (around 50).

TABLE 12.1 Some reported fullerene-based electrochemical sensors for the detection of different analytes.

Analyte	Nanosensor type	Method	Linear range	Limit of detection	Limit of quantification	Application type	Ref.
Amitriptyline	C_{60}-rGO-H_2O-NH_2/CoNPs	Immunosensor	1 nM–100 μM	0.5 nM	—	Urine	(Medyantseva et al., 2020)
Catechol	Pt/C_{60}/PGE	DPV	50.0–1500.0 μM	2.97 μM	—	Local river sanitary wastewater	(Zhu et al., 2020)
Cd(II)	C_{60}-chitosan-modified GCE	DPASV	0.5–9.0 μM	21 nM	—	Foodstuff	(Han, Meng, Zhang, & Zheng, 2018)
Cu(II)	C_{60}-chitosan-modified GCE	DPASV	0.05–6.0 μM	14 nM	—	Foodstuff	(Han et al., 2018)
Chlorambucil	C_{60}-monoadduct-micellar MIP/IL-CCE	DPASV	1.47–247.20 μg/L	0.36 μg/L	—	Blood plasma, urine, and pharmaceutics	(Prasad, Singh, & Kumar, 2017)
Ciprofloxacin	Nafion and fullerenes	DPV	1.0–18 μM	1.0 μM	3.0 μM	Beef	(Hernández, Aguilar-Lira, Islas, & Rodriguez, 2021)
Diazepam	C_{60}–CNT/IL/GCE	DPV	0.3–700.0 μM	87 nM	—	Tablet, urine, and serum	(Rahimi-Nasrabadi, Khoshroo, & Mazloum-Ardakani, 2017)
DNA	[Ru(dcbpy)2dppz]$^{2+}$/rose bengal dyes/fullerene/GCE	Photoelectrochemical	0.1 fM–1 nM	37 aM	—	Human blood serum and hela cell	(Wang et al., 2018)
DNA	Methylene blue/Au/C_{60}/GCE	Photoelectrochemical	10 fM–100 nM	3.3 fM	—	—	(Long et al., 2019)
Dopamine	SPCE-fullerene-Py-Py-3-COOH	EIS	25–250 μg/L	8.77 μg/L	26.6 μg/L	Urine	(Ertuğrul Uygun & Demir, 2020)
Dopamine	C_{70}-COOH/Au electrode	DPV	1.0–13.0 μM	10 nM	—	Dopamine hydrochloride injection	(Han et al., 2016)
Doxorubicin	N-CNOs/GCE	DPV	0.2 nM–10.0 μM	60.0 pM	—	Human blood serum	(Ghanbari & Norouzi, 2020)
Heat shock protein 70 (HSP70)	Fullerene/GCE	EIS	0.8–12.8 ng/L	0.273 ng/L	0.909 ng/L	Human blood serum	(Demirbakan & Sezgintürk, 2016)

Continued

TABLE 12.1 Some reported fullerene-based electrochemical sensors for the detection of different analytes—cont'd

Analyte	Nanosensor type	Method	Linear range	Limit of detection	Limit of quantification	Application type	Ref.
Hg(II)	C_{60}-chitosan-modified GCE	DPASV	0.01–6.0 μM	3 nM	—	Foodstuff	(Han et al., 2018)
Homovanilic acid	C_{60}–GO/nitrophenyl/GCE	SWV	0.1–7.2 μM	0.03 μM	—	Urine samples	(Rather, Khudaish, Munam, Qurashi, & Kannan, 2016)
Hydroquinone	Pt/C_{60}/PGE	DPV	50.0–1100.0 μM	2.19 μM	—	Local river sanitary waste water	(Zhu, Huai, et al., 2020)
Levedopa	C_{60}–CNT/GCE	DPV	0.5–2000 μM	0.035 μM	—	Synthetic solutions	(Mazloum-Ardakani, Ahmadi, Safaei Mahmoudabadi, & Khoshroo, 2016)
Mangiferin	AuPt NPs/Chitosan-IL/C_{60}/MWCNTs-IL/GCE	Amperometry	2–40 nM	0.61 nM	—	Human plasma samples	(Jalalvand, 2020)
Mycobacterium tuberculosis MPT64	AuNPs-C_{60}-PANI	DPV	0.02–1000 ng/L	20 ng/L	—	Human serum samples	(Bai et al., 2017)
M. tuberculosis IS6110	Au-nano-C_{60} NGS/GCE	DPV	10 fM–10 nM	3 fM	—	Nontuberculosis patients and tuberculosis patient	(Bai et al., 2019)
Paracetamol	CuNPs/C_{60}/MWCNT/CPE	SWV	4.0–400.0 nM	73 pM	2.4 nM	Serum, plasma, and urine	(Brahman, Suresh, Lokesh, & Nizamuddin, 2016)
Prostate-specific antigen	C_{60}/methylene blue/Au electrode	Amperometry	0.015–8 μg/L	1.7 ng/L	—	—	(Hao et al., 2017)
Prostate-specific antigen	Au NPs/C_{60}-chitosan-IL/MWCNTs/SPCE	DPV	2.5–90 μg/L	1.5 ng/L	—	Serum samples	(Jalalvand, 2019)
Prostate-specific antigen	Au NPs/C_{60}-chitosan-IL/MWCNTs/SPCE	EIS	1–200 ng/L	0.5 ng/L	—	Serum samples	(Jalalvand, 2019)

Analyte	Electrode	Method	Linear range	LOD	Sample	Reference
Prostate-specific antigen	Pd NP@PANI-C_{60}/GCE	Immunosensor	0.16 ng/L–38 µg/L	19.5 pg/L	Serum and urine samples	(Suresh, Bondili, & Brahman, 2020)
Pb(II)	C_{60}-chitosan-modified GCE	DPASV	0.005–6.0 µM	1 nM	—	Foodstuff	(Han et al., 2018)
Primaquine	Fullerene C_{60}-MIP/PGE	DPASV	2.7–848.5 nM	0.80 nM	Human blood plasma, urine, and pharmaceutics	(Prasad, Kumar, & Singh, 2016)
Ractopamine	$PMEO_2MA/C_{60}$-rGO/GCE	DPV	0.1–3.1 µM	82 nM	—	Spiked pork samples	(Chen, Zhang, Li, Xie, & Fei, 2018)
Rifaximin	MnO_2NPs/C_{60}/CPE	DPV	0.8–31.5 mg/L	0.76 mg/L	2.31 mg/L	Pharmaceutical samples	(Abdellatef, Khaled, Hendawy, & Hassan, 2021)

CCE: Carbon ceramic electrode, DPASV: Differential pulse adsorptive stripping voltammetry, DPV: Differential pulse voltammetry, EIS: Electrochemical impedance spectroscopy, GCE: Glassy carbon electrode, IL: Ionic liquid, MIP: Moleculer imprinted polymer, N-CNOs: Nitrogen-doped carbon nano-onions, NGS: Nitrogen-doped graphene nanosheet, NPs: Nanoparticles, PANI: Polyaniline, PGE: Pencil graphite electrode, $PMEO_2MA$: poly(2-(2-methoxyethoxy)ethyl methacrylate), Py-Py-3-COOH: pyrrole-pyrrole-3-carboxylic acid, [Ru(dcbpy)$_2$dppz]$^{2+}$: Bis(4,4'-dicarboxyl-2,2'-bibpyridyl) (4,5,9,14-tetraaza-benzo[b]-triphenylene) ruthenium(II), SPCE: Screen-printed carbon electrode, SWV: Square wave voltammetry.

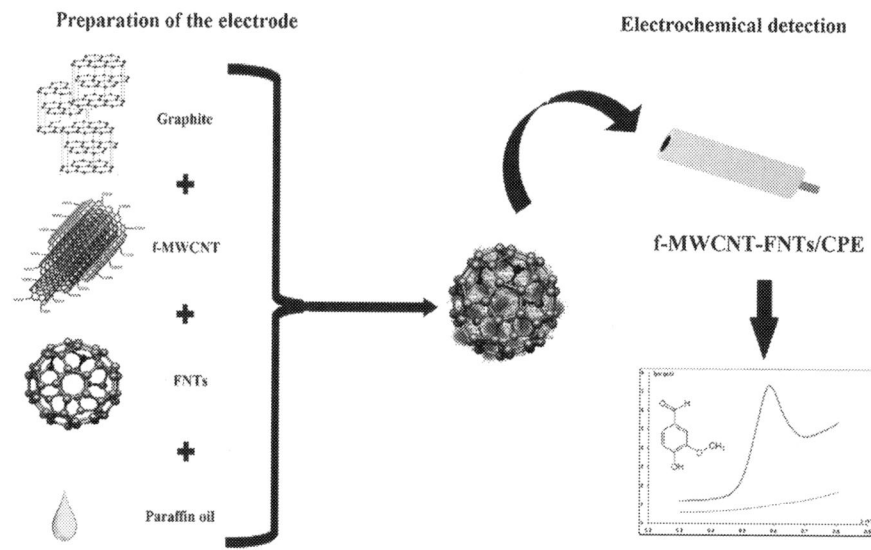

FIG. 12.2 Graphical presentation of *f*-MWCNTs-FNTs/CPE sensor fabrication and scanning electron microscopy (SEM) images for (A) graphite (×10.000), (B) *f*-MWCNTs-FNTs/CPE (×10.000), (C) FNTs (×10.000), (D) FNTs (×50.000), (E) MWCNT (×30.000), and (F) *f*-MWCNT (×30.000). *Reprinted with permission from Taouri L, Bourouina M, Bacha SB, Hauchard D. Fullerene-MWCNT nanostructured-based electrochemical sensor for the detection of vanillin as food additive. Journal of Food Composition and Analysis 2021;100:103811.*

They are among the most important nanomaterials due to their unique physical, electrical, and mechanical properties, and miniature size. Using these features, they enable the production and development of microsensors, obtaining very fast, sensitive, and reliable responses in complex matrix environments for the analysis of electroactive compounds. Moreover, MWCNT with functional groups, such as —COOH and —NH$_2$, has given new dimensions to analyte-sensor interactions. Weak interactions occur between analyte and functional end groups, causing them to migrate faster on the electrode surface. This allows lower LODs of the analyte to be achieved on the sensor. In addition, new sensors based on carbon nanomaterials are obtained by direct chemical reactions between the end groups and the modified material. Thus, it is possible to develop more robust, stable, selective, and sensitive sensors using these carbonaceous sensor materials.

SWCNTs and MWCNTs are commonly used in sensor designs for the determination of substances including drugs, heavy metals, vitamins, antioxidants, explosives, nucleic acids, and environmental pollutants, such as pesticides, polycyclic aromatic hydrocarbons, organochlorine compounds, etc. CNT materials and electrochemical parameters of the selected electrochemical sensors used for sensing of various analytes are listed in Table 12.2.

DNA and RNA are genetic molecules, which are made up of bases and nucleotides, and they have an important role in the lifecycle process (Chetankumar, Swamy, & Naik, 2021). Since DNA is electrochemically active, DNA and its bases can be determined by electrochemical biosensors directly. In this manner, DNA-based sensors or genosensors have been widely utilized in biomedical and environmental applications for the determination of food and environmental pollutants and genetic diseases and the identification of viruses and bacteria. The combination of CNTs with DNA is an important field as it enables the design of faster and more cost-effective electrochemical DNA determination methods with improved sensitivity (Power et al., 2018).

Chetankumar and coworkers prepared a new sensor using magnesium oxide and MWCNT-modified CPE (MgO-MWCNTs-MCPE) for specific and simultaneous detection of guanine, adenine, and epinephrine (Fig. 12.3). In the study, MgO nanoparticles were prepared by a mechanochemical method. The electrochemical properties of guanine, adenine, and epinephrine were studied by different voltammetric techniques including CV, linear sweep voltammetry (LSV), and differential pulse voltammetry (DPV) at MgO-MWCNTs-MCPE. The effects of particular experimental parameters such as pH of the solution, speed rate, and analyte concentration were investigated and optimized. The electrode process was characterized to be adsorption-controlled and the LOD for guanine, adenine, and epinephrine was calculated as 0.92, 1.49, and 0.83 μM, respectively. The developed electrode also displayed very good stability and reproducibility (Chetankumar et al., 2021).

A novel, selective, and sensitive electrochemical sensor for the determination of dopamine using zinc-layered hydroxide-sodium dodecyl sulphate-isoprocarb/MWCNTs was developed by Ahmad et al. The surface morphology and electrochemistry features of prepared nanocomposite material were investigated using CV, SWV, TEM, SEM, and electrochemical impedance spectroscopy (EIS). Under optimized experimental conditions, the linear analytical curve for the detection of dopamine was between 1.0 μM and 0.3 mM with the LOD of 0.43 μM. In addition, the developed sensor demonstrated good selectivity, sensitivity, and stability and was found to be applicable for dopamine determination in the commercial dopamine hydrochloride injection samples (Ahmad et al., 2019).

TABLE 12.2 Some reported SWCNT- and MWCNT-based electrochemical sensors for the detection of different analytes.

Analyte	Nanosensor type	Method	Linear range	Limit of detection	Application type	Ref.
Acetaminophen	MWCNT/ZnO NPs/Au NPs/GCE	SWV	0.05–20 μM	9 nM	Tablets	(Kenarkob & Pourghobadi, 2019)
Acetaminophen	SWCNTs-Chitosan-RTIL/GCE	DPV	2–200 μM	0.11 μM	Human serum and urine	(Afrasiabi, Kianipour, Babaei, Nasimi, & Shabanian, 2016)
Acid red 52	MWCNT/CPE	DPV	1.0–45.0 μM	30.0 nM	Water and hair	(Golestaneh, 2020)
Adefovir dipivoxil	MIP/chitosan-MWCNTs +C_3N_4/GCE	DPV	0.1–9.9 μM	0.05 μM	Drug formulations, human serum, and urine	(Dorraji, Noori, & Fotouhi, 2019)
Adenine	DomP/SWNTs/GCE	DPV	0.5–360 μM	0.5 μM	Denatured DNA sample	(Ji et al., 2019)
Alfuzosin hydrochloride	MWCNTP/MCPE	Potentiometry	10 μM – 10 mM	9.5 μM	Xatral tablet, human serum, and urine	(El Badry, Frag, & Gamal Eldin, 2020)
Alloxan	SWCNT/GCE	Derivative voltammetry	0.3 μM–3 mM	0.05 μM	—	(Murthy, Duraimurugan, Sridhar, & Madhavan, 2019)
Amodiaquine	MWCNT/PMO/GCE	DPV	0.1–3.5 μM	89 nM	Pharmaceutical formulations and human urine	(Chiwunze et al., 2019)
Artemisinin	FeGd-MOF/hemin/MWCNTs/GCE	DPV	0.3–350 μM	0.17 μM	Dried leaves of *Artemisia apiacea*	(Jiang, Sun, Wei, & Li, 2021)
Ascorbic acid	PDY/MWCNTs/GCE	DPV	5–1110 μM	9.9 nM	Human blood serum	(Hatefi-Mehrjardi, Karimi, Soleymanzadeh, & Barani, 2020)
Ascorbic acid	TOAB/YD/MWCNT/GCE	Amperometry	18.72 μM–1.85 mM	0.18 μM	Vitamin C tablets	(Huang et al., 2019)
Ascorbic acid	PAMAM/AgNPs-MWCNTs/PNR	DPV	0.16–2500 μM	0.053 μM	Vitamin C tablets and lemon and orange fruit juices	(Devi & Narayanan, 2019)
Ascorbic acid	MWCNT/GO/AuNR/GCE	DPV	1 nM–8 mM	0.85 nM	Spiked serum	(Zhao et al., 2019)
Ascorbic acid	SWCNTs-chitosan-RTIL/GCE	DPV	20–950 μM	4.3 μM	Human serum and urine	(Afrasiabi et al., 2016)

Continued

TABLE 12.2 Some reported SWCNT- and MWCNT-based electrochemical sensors for the detection of different analytes—cont'd

Analyte	Nanosensor type	Method	Linear range	Limit of detection	Application type	Ref.
Asulam	CuNPs/MWCNTs-IL-chitosan/GCE	Amperometry	1–200 μM	0.33 nM	Water	(Roushani, Mohammadi, & Valipour, 2020)
Bilastine	MWCNTs-SPCE	LSV	0.1–2.0 mg/L	0.13 μM	Tablets and spiked urine	(Teixeira & Oliveira, 2021)
Bisphenol	MWCNTs/CuFe$_2$O$_4$/GCE	DPV	0.01–120 μM	3.2 nM	Water and baby bottles	(Baghayeri, Amiri, Fayazi, Nodehi, & Esmaeelnia, 2021)
Bisphenol A	Au NPs/NH$_2$-Fe$_3$O$_4$/MWCNT/GCE	DPV	10^{-19}–10^{-14} M	0.08 aM	Human serum and lake water	(Zhao, Zheng, et al., 2020)
Bisphenol A	GO-IL-f-COOH-SWCNT-AuNPs/GCE	DPV	5–100 nM	1.5 nM	Plastic water bottle and milk carton	(Tian et al., 2017)
Bisphenol A	Carbon black/f-MWCNTs/GCE	Amperometry	0.1–130 μM	0.08 μM	Plastic drinking water bottles	(Thamilselvan, Rajagopal, & Suryanarayanan, 2019)
Bisphenol A	Zn/Al-LDH quinclorac/MWCNT/CPE	SWV	30 nM–0.3 mM	4.4 nM	Baby bottle and mineral water	(Zainul et al., 2019)
Bisphenol B	Prussian blue/MWCNTs-COOH/GCE	CV	0.5–175 μM	0.5 nM	Real samples	(Xing et al., 2019)
Bisphenol S	MWCNTs/SPCE	Potentiometry	0.25 μM–1 mM	0.02 mg/L	Baby bottles	(Abd-Rabboh & Kamel, 2020)
Caffeic acid	MWCNT/SPE	DPV	2.0–50 μM	0.2 μM	Mate, fennel, and white tea	(Araújo et al., 2020)
Capsaicin	CeO$_2$-CPB/SWNT-COOH/GCE	DPV	0.10–500 μM	28 nM	Red hot pepper and tinctures of *Capsicum annuum* L.	(Ziyatdinova, Ziganshina, Shamsevalieva, & Budnikov, 2020)
Cardiac troponin T	MIP/PMB/f-MWCNTs/SPCE	DPV	0.10–8.0 ng/L	0.040 ng/L	Human plasma	(Phonklam, Wannapob, Sriwimol, Thavarungkul, & Phairatana, 2020)
Carvacrol	MWCNTs/GCE	DPV	0.1–150 μM	0.075 μM	Oregano spices	(Ziyatdinova & Budnikov, 2021)

Analyte	Electrode	Method	Linear range	LOD	Sample	Reference
Catechol	Cellulase/MWCNTs/GCE	Amperometry	10–160 μM	4 nM	Green tea	(Wang, Wang, Li, & Yang, 2021)
Catechol	SiSG-TYR/Fe$_3$O$_4$-MWCNTs/GCE	DPV	1.5–30 μM	0.055 μM	Spiked local tap water	(Shaikshavali et al., 2021)
Catechol	Co$_3$O$_4$/MWCNTs/GCE	DPV	10–700 μM	8.5 μM	Seawater	(Song et al., 2019)
Catechol	MWCNTs/eosin Y/GCE	DPV	0.4–50 μM	15 nM	Tap water	(Zheng, 2019)
Catechol	HKUST-1/SWCNT-Nafion/GCE	DPV	0.4–600 μM	0.13 μM	Rainwater, tap water, sewage	(Ai et al., 2020)
Cd(II)	MWCNTs–COOH/UiO-66-NH$_2$/MWCNTs-COOH/GCE	DPASV	1–511 ppb	0.09 ppb	Seawater, rainwater, and air conditioning	(Lu et al., 2021)
Cd(II)	Sb$_2$O$_3$/MWCNTs/CPE	SWASV	10–100 ppb	1.7 ppb	Tap water	(Majidian et al., 2020)
Cd(II)	Bi/PPy/MWCNT/CPE	SWASV	0.16–120 μg/L	0.157 μg/L	Tap water	(Oularbi, Turmine, & El Rhazi, 2019)
Cd(II)	BiNP/MWCNT/Nafion/CPE	ASV	5–1000 ppb	1.06 ppb	Herbal food supplement	(Palisoc, Vitto, & Natividad, 2019)
Cd(II)	MWCNTs/Nafion/Hg	DPASV	0.2–100 μg/L	0.02 μg/L	Tap and lake waters	(Lai, Lin, Huang, Wang, & Chen, 2021)
Ceftizoxime	MWCNT/P-Cys@MIP/GCE	DPV	2–100 nM	0.1 nM	Serum and urine	(Ali et al., 2021)
Chloramphenicol	MoS$_2$-MWCNTs/GCE	DPV	1–35 μM	0.4 μM	—	(Gao et al., 2021)
Chlorogenic acid	Polysunset yellow/MWCNTs/GCE	DPV	0.10–4.0 μM	0.076 μM	Espresso and instant coffee	(Ziyatdinova, Guss, Morozova, & Budnikov, 2021)
Cocaine	Poly (MAA)/MWCNT/β-CD/GCE	DPV	12.2–200.0 μM	3.6 μM	Seized street	(Capelari, De Cássia, Da Rocha, et al., 2021)
Codeine phosphate	MWCNT paste electrode/SPE	Potentiometry	4.88 μM–10 mM	3.45 μM	Human plasma and tablet	(Essam, Bassuoni, Elzanfaly, Zaazaa, & Kelani, 2020)
Cr(III)	MWCNTs–COOH/CPE	Potentiometry	0.1 nM–10 mM	30 pM	EDTA titration	(Fakhrabad, Khoshnood, Ebrahimi, & Abedi, 2020)
Cr(VI)	PANI@ZrO$_2$-SO$_4{}^{2-}$@MWCNT/GCE	DPV	0.55–39.5 μM	64.3 nM	Wastewater	(Motaghedifard, Pourmortazavi, & Mirsadeghi, 2021)
Creatinine	Folic acid/MWCNTP/AgNPs/CPE	DPV	0.01–200 μM	8 nM	Urine	(Fekry, Abdel-Gawad, Tammam, & Zayed, 2020)

Continued

TABLE 12.2 Some reported SWCNT- and MWCNT-based electrochemical sensors for the detection of different analytes—cont'd

Analyte	Nanosensor type	Method	Linear range	Limit of detection	Application type	Ref.
Cs(I)	MWCNTs@Cs-IIP/GCE	Potentiometry	0.1–100 µM	0.04 µM	River; industrial wastewater and brine	(Wang, Wang, Zhou, & Sun, 2021)
Cu(II)	SSA/MoS$_2$/o-MWCNTs/GCE	DPV	0.1–11 µM	0.057 µM	Lake Water and milk	(Wang, Li, et al., 2019)
Cu(II)	f-MWCNT-Calix/CPE	DPASV	0.1–1.6 µg/L	0.096 µg/L	Wastewater, plant leaves, and soft drinks	(Kucukkolbasi, Sayin, & Yilmaz, 2019)
Cu(II)	CPE/MWCNTs	Potentiometry	10 µM–1 M	3.3 µM	Real water samples and urine and serum	(Frag, Mohamed, Ali, & Mohamed, 2020)
Cu(II)	3DOM/chitosan/PB/SWCNTs/Au electrode	Amperometry	10^{-18}–10^{-5} M	10^{-19} M	River water	(Tian, Chen, Liu, & Yao, 2016)
Curcumin	MWCNT paste/GCE	SWASV	0.010–5.00 µM	5 nM	Turmeric extract	(Cittan, Altuntaş, & Çelik, 2020)
Cyprodinil	MWCNT paste electrode	SWSV	0.25–4.0 mg/L	0.076 mg/L	Commercial formulation, apple juice, and tap water	(Ayhan & Inam, 2020)
Daclatasvir	Ni-NPs/MWCNTs/GCE	SWV	0.024–300 µM	15.82 nM	Tablet and serum	(Joghani, Rafati, Ghodsi, Assari, & Feizollahi, 2020)
Diclofenac	Au-PtNPs/f-MWCNTs/Au electrode	DPV	0.5 µM–1000 µM	0.3 µM	Tablet and urine	(Eteya, Rounaghi, & Deiminiat, 2019)
Dopamine	MWCNT-COOH/SWCNT-OH/GCE	DPV	2–150 µM	0.37 µM	Bovine serum	(Guan, Zou, Liu, Jiang, & Yu, 2020)
Dopamine	MIP-MWCNTs/GCE	DPV	0.04–70 µM	0.015 µM	Rat plasma	(Liu et al., 2020)
Dopamine	PDY/MWCNTPs/GCE	DPV	0.0007–2.5 µM	0.7 nM	Human blood serum	(Hatefi-Mehrjardi et al., 2020)
Dopamine	ZnSe/GO@MWNTs/Ru(bpy)$_3^{2+}$/GCE	Electrochemiluminescence	1 nM–100 µM	0.6 nM	Serum	(Tian et al., 2019)
Dopamine	SWCNTs-PGE	DPV	0.5–1000 mg/L	50 µM	—	(Muti & Cantopcu, 2018)
Doxycycline hyclate	MIP/MWCNTs/GCE	DPV	0.05–0.5 µM	13 nM	Human serum	(Xu, Jiang, Liu, & Yang, 2020)

Analyte	Electrode	Technique	Linear range	LOD	Sample	Reference
Empagliflozin	Co$_3$O$_4$NPs/MWCNTs/CPE	SWV	7.94–107 μM	1.71 μM	Tablet dosage, Jardiance tablet, and human urine	(Rizk, Attia, Mohamed, & Elshahed, 2020)
Entacapone	NH$_2$-f-MWCNT/GCE	DPV	0.5–50 nM	31.3 pM	Tablet, human serum, and urine	(Aftab et al. 2019)
Ephedrine hydrochloride	MIP/Nafion-MWCNTs/GCE	DPV	0.18–75 μM	72 nM	Spiked saliva	(Jia et al., 2021)
Epinephrine	MIP-MWCNTs/GCE	DPV	0.04–70 μM	0.023 μM	Rat plasma	(Liu et al., 2020)
Epinephrine	Pt-NiO NPs/MWCNTs/GCE	DPV	0.5–300 μM	0.035 μM	Human serum, urine, and injection samples	(Dehdashti & Babaei, 2020)
Eriochrome black T	MWCNTs/GCE	LSASV	0.1–10 μM	0.04 μM	Well water	(Cittan & Çelik, 2019)
Ertapenem	ZnONPs/MWCNT/CPE	SWASV	0.4–3.2 μM	78.0 nM	Pharmaceutical and plasma	(Fayed, Youssif, Salama, Elzanfaly, & Hendawy, 2021)
Estradiol	AuNP/MWCNT/GCE	LSV	1–20 μM	0.7 μM	Tap and wastewater	(Masikini, Ghica, Baker, Iwuoha, & Brett, 2019)
Fentanyl	MWCNTs/GCE	DPASV	0.1–100 μM	0.1 μM	Injection, serum, and urine	(Najafi, Sohouli, & Mousavi, 2020)
Ferulic acid	Polysunset yellow/MWCNTs/GCE	DPV	0.5–4.0 μM	0.098 μM	Espresso and instant coffee	(Ziyatdinova et al., 2021)
Fludarabine	NH$_2$-MWCNTs/GCE	DPASV	0.2–4.0 μM	29 nM	Pharmaceutical formulations	(Dogan Topal, Bakirhan, Tok, & Ozkan, 2020)
Flufenamic acid	Ru-TiO$_2$ NPs/MWCNTs-CPE	SWV	0.01–0.9 μM	0.68 nM	Human urine	(Shetti et al., 2019)
Fluoroquinolone	ZnO NPs-MWCNT/MCPE	DPV	0.1–2.4 μM	45.2 nM	Avalox tablet and human urine	(Siddegowda et al., 2020)
Folic acid (vitamin B9)	Ni(OH)$_2$)/f-MWCNTs/GCE	DPV	0.5–26 μM	0.095 μM	Dietary supplement, fortifier compound, and wheat flour	(Winiarski, Rampanelli, Bassani, Mezalira, & Jost, 2020)
Furazolidone	MIP-CPE-MWCNTs	DPV	0.01 μM–1 μM	0.03 μM	River and tap water	(Rebelo et al., 2021)
Guanine	DomP/SWNTs/GCE	DPV	0.5–440 μM	0.25 μM	Denatured DNA sample	(Ji et al., 2019)
Guanosine	PAPBA-MWCNTs-PGE	DPV	1–120 μM	0.14 μM	Urine	(Kuralay & Gürsoy, 2020)
Heregulin-A	Mn(TPP)Cl/MWCNT paste electrode	ELISA	102 pg/L–5 mg/L	102 pg/L	Blood and tumor brain tissues	(Stefan-Van Staden, Negut, Gheorghe, & Cioriță, 2021)

Continued

TABLE 12.2 Some reported SWCNT- and MWCNT-based electrochemical sensors for the detection of different analytes—cont'd

Analyte	Nanosensor type	Method	Linear range	Limit of detection	Application type	Ref.
Homovanillic acid	MWCNTs-PtNPs/GCE	DPV	0.2–80 μM	0.08 μM	Human urine	(Fu et al., 2020)
Hydrogen peroxide	PPY/MWCNTs/PB/GCE	EIS	5–503 μM	1.7 μM	Milk	(Yang, Wang, Lü, & Hui, 2021)
Hydrogen peroxide	2D Cu-TCPP/MWCNT/GCE	Amperometry	0.001–8.159 mM	0.70 μM	Serum and beer	(Wang, Huang, et al., 2019)
Hydrogen peroxide	H2-MIL53(Fe)/HRP/MWCNTs/GCE	Amperometry	0.1–600 μM	0.028 μM	Cell culture solutions	(Jiang, Sun, Wei, & Li, 2020)
Hydroquinone	MWCNTs/Eosin Y/GCE	DPV	0.4–50 μM	10 nM	Tap water	(Zheng, 2019)
Hydroquinone	Co_3O_4/MWCNTs/GCE	DPV	10–800 μM	5.6 μM	Seawater	(Song et al., 2019)
Hydroquinone	SiSG-TYR/Fe_3O_4-MWCNTs/GCE	DPV	1.5–40 μM	0.057 μM	Spiked local tap water	(Shaikshavali et al., 2021)
Hydroquinone	HKUST-1/SWCNT-Nafion/GCE	DPV	0.1–300 μM	0.033 μM	Rainwater, tap water, and sewage	(Ai et al., 2020)
Hyperin	Three-dimensional graphene-MWCNTs/GCE	DPV	5 nM–1.5 μM	1 nM	Actual sample (*Abelmoschus manihot*)	(Wang, 2019)
Insulin	CoNPs/chitosan-MWCNTs/SPCE	CV	0.05 μM–5 μM	25 nM	—	(Šišoláková, Hovancová, Oriňaková, et al., 2020)
Insulin	ZnNPs/chitosan-MWCNTs/SPCE	CV	0.5–5 μM	0.28 μM	Blood serum	(Šišoláková, Hovancová, Chovancová, et al., 2021)
Isoniazid	GCE/MWCNT/CuOx	LSV	0.2–50 μM	10 nM	Commercial tablets, artificial human serum, and urine	(Özdokur, 2020)
Ivabradine hydrochloride	t-Butyl calixarene/Fe_2O_3@MWCNTs/CPE	Potentiometry	0.1 μM–1 mM	36 nM	Pharmaceutical formulations	(Abdel-Haleem et al., 2020)
La(III)	Chitosan/MWCNTs/CPE	Potentiometry	1 μM–10 mM	1 μM	Zoshk river and tap water	(Rajabi, Masrournia, & Abedi, 2020)
Lactic acid	Nano-MIP/MWCNTs/CRE	Potentiometry	1 μM–1 M	0.73 μM	Milk and yogurt	(Alizadeh, Nayeri, & Mirzaee, 2019)

Analyte	Electrode	Method	Linear range	LOD	Sample	Reference
Levodopa	GCE/MWCNTs/Cu (TPA) MOF	SWV	0.9–35 μM	2 nM	Human serum	(Asadpour Joghani, Abbas Rafati, Ghodsi, Assari, & Feizollahi, 2020)
Linagliptin	Co$_3$O$_4$NPs/MWCNTs/CPE	SWV	39.8 μM–1.53 mM	11.3 μM	Tablet dosage, Jardiance tablet, and human urine	(Rizk et al., 2020)
Mefenamic acid	Ru-TiO$_2$/MWCNTs-CPE	SWV	0.01–0.9 μM	0.45 nM	Human urine	(Shetti et al., 2019)
Mercury	PANI/MWCNTs/AuNPs/ITO	DPV	0.01–10.00 ppm	0.08 ppm	Cosmetic products	(Bohari, Siddiquee, Saallah, Misson, & Arshad, 2021)
m-Tolyl hydrazine	CdO/MWCNTs/Nafion/GCE	Amperometry	0.01 nM–0.1 mM	4.0 pM	Industrial effluent, sea water, and PVC-food packaging bag	(Rahman, Alam, & Alamry, 2019)
Methadone	CPE/MWCNTs/Fe$_3$O$_4$@SiO$_2$ NPs	SWV	10 nM–8 μM	3 nM	Urine	(Yousefi, Irandoust, & Haghighi, 2020)
Methamphetamine	GCE (GCE/MWCNT/Au-NPs-SH-(CH$_2$)$_3$-Si-SiO$_2$@Fe$_3$O$_4$)	SWV	0.5 nM–50 μM	16 nM	Human urine	(Haghighi, Shahlaei, Irandoust, & Hassanpour, 2020)
Methamphetamine	MIP/MWCNT-CPE	Fast Fourier transform-SWV	0.1–100 μM	0.83 nM	Urine and serum	(Akhoundian, Alizadeh, Ganjali, & Norouzi, 2019)
Methotrexate	f-MWCNT paste electrode	DPV	0.4–5.5 μM	100 nM	Pharmaceutical drug, human serum, and urine	(Kummari, Kumar, Satyanarayana, & Gobi, 2019)
Methotrexate	f-MWCNT paste electrode	SWV	0.01–1.5 μM	2.9 nM	–	(Kummari et al., 2019)
Methotrexate	f-MWCNT paste electrode	Amperometry	10 nM–1.5 μM	10 nM	–	(Kummari et al., 2019)
Methyl parathion	Halloysite-MWCNTs/GCE	DPV	0.5–11 μM	0.034 μM	Romaine and kiwi fruit	(Zhao et al., 2021)
microRNA	PPY/MWCNTs/PB/GCE	EIS	0.1 pM–1 nM	33.4 fM	Human serum	(Yang et al., 2021)
Mitomycin C	PoPD-MWCNT/PGE	DPV	0.5–25 mg/L	0.012 mg/L	Urine	(Kuralay & Bayramlı, 2021)
Morphine	CPE/MWCNTs/Fe$_3$O$_4$@SiO$_2$ NPs	SWV	5.0 nM–1.8 μM	1.6 nM	Urine	(Yousefi et al., 2020)

Continued

TABLE 12.2 Some reported SWCNT- and MWCNT-based electrochemical sensors for the detection of different analytes—cont'd

Analyte	Nanosensor type	Method	Linear range	Limit of detection	Application type	Ref.
Morphine	La^{3+}-CuO/MWCNTs/CPE	DPV	0.05–325.0 µM	8.0 nM	Ampules and urine	(Rajaei, Foroughi, Jahani, Shahidi Zandi, & Hassani, 2019)
N-acetylcysteine	GCE/MWCNT-Nafion	DPV	20–500 µM	0.43 µM	Urine and serum	(Kuyumcu, 2019)
Neotame	Chitosan/NiNPs/MWCNTs/GCE	SWV	2 µM–50 µM	0.84 µM	Fruit juice	(Incebay & Saylakci, 2021)
Nitrate	GCE/MWCNTs/NO_3^--ISE	Potentiometry	0.08 µM–10 mM	0.028 µM	Real wastewater	(Hassan et al., 2019)
Nitrite	Cu-Co PBA/MWCNTs/GCE	DPV	10–2100 µM	0.5 µM	Ham sausage and mustard samples	(Liu, Wen, et al., 2019)
Nitrite	AuNP-rGO-MWCNTs/GCE	Amperometry	50 nM–2.2 mM	14 nM	Local river water	(Yu, Li, & Song, 2019)
Nitrite	Chitosan-AgNPs/MWCNT paste electrode	CV	100 nM–50 µM	30 nM	River water	(Bibi et al., 2019)
Nitrocellulose	MWCNTs-MIP/GCE	DPV	0–70 g/L	0.345 ng/L	Propellant sample	(Meng, Xiao, & Scott, 2019)
Nitrofurantoin	ErGO/MWCNTs/GCE	Amperometry	0.005–2.81 µM	1.87 nM	Human serum	(Kummari, Sunil Kumar, & Vengatajalabathy, 2020)
Norepinephrine	MWCNTs-Fe_3O_4@MCM-48-SO_3H/GCE	DPV	0.4–600 µM	0.19 µM	Human blood serum and human urine	(Yousefi & Babaei, 2019)
Noscapine	MWCNTs/TiO_2NPs/CPE	DPV	4–600 µM	3.5 µM	Urine	(Sharifi, Zarei, & Asghari, 2021)
Olanzapine	ISS-NH_2-TiO_2/MWCNTs/GCE	SWV	0.05–10 µM	8 nM	Tablet and blood serum	(Arvand & Pourhabib, 2019)
Oxymetazoline	TiO_2NPs/f-MWCNTs/GCE	DPASV	0.12–1.5 µM	4.40 nM	Nasal spray	(Munir, Bozal-Palabiyik, Khan, Shah, & Uslu, 2019)
Oxytetracycline	AuNPs/MWCNTs/cDNA@thionine/TFGE	DPV	0.1 pg/L–10 µg/L	0.031 pg/L	Chicken sample	(He et al., 2019)
p-nitrophenol	$La(OH)_3$-OxMWCNTs/GCE	LSV	1.0–30.0 µM	0.27 µM	Industrial wastewater and Xiangjiang river	(Yuan, Zou, Guan, Huang, & Yu, 2020)

Analyte	Electrode	Method	Linear range	LOD	Sample	Reference
Papaverine	MWCNTs/TiO$_2$NPs/CPE	DPV	5–400 μM	4.6 μM	Urine	(Sharifi et al., 2021)
Pb(II)	Bi/PPy/MWCNT/CPE	SWASV	0.11–120 μg/L	0.099 μg/L	Tap water	(Oularbi et al., 2019)
Pb(II)	f-MWCNT-Calix/CPE	DPASV	0.1–1.6 μg/L	0.061 μg/L	Wastewater, plant leaves, and soft drinks	(Kucukkolbasi et al., 2019)
Pb(II)	BiNP/MWCNT/Nafion/CPE	ASV	5–1000 ppb	0.72 ppb	Herbal food supplement	(Palisoc, Vitto, & Natividad, 2019)
Pb(II)	Sb$_2$O$_3$/MWCNTs/CPE	SWASV	10–100 ppb	1.2 ppb	Tap water	(Majidian et al., 2020)
Pb(II)	TCPP-MWCNTs@Fe$_3$O$_4$/MGCE	FSV	0.2–2.0 ng/L	0.067 ng/L	Seawater	(Zhang et al., 2020)
Pb(II)	Co-Nx-C@/MWCNTs/GCE	DPASV	2–60 μg/L	0.7 μg/L	Tap and river water	(Zhao, Wu, et al., 2020)
Pb(II)	MWCNTs/Nafion/Hg	DPASV	0.2–100 μg/L	0.17 μg/L	Tap and lake waters	(Lai et al., 2021)
Pb(II)	MWCNTs-COOH/UiO-66-NH$_2$/MWCNTs-COOH/GCE	DPASV	1–271 ppb	0.071 ppb	Seawater, rainwater, and air conditioning	(Lu et al., 2021)
Pb(II)	AuNP/SWCNT/SPCE	DPASV	2–60 nM	0.321 nM	Sea water	(Molinero-Abad, Izquierdo, Pérez, Escudero, & Arcos-Martínez, 2018)
Pimozide	NH$_2$f-MWCNT/ZnONPs/GQDs/GCE	DPV	62.5 pM–120 nM	10.2 pM	Drug and serum and urine	(Aftab, Kurbanoglu, Ozcelikay, Shah, & Ozkan, 2019)
Primaquine	MWCNTs/GCE	SWV	0.1–5.0 μM	28 nM	Urine	(Pedrozo-Peñafiel et al., 2019)
Propafenone	NH$_2$f-MWCNTs/GCE	DPV	0.1–10 μM	0.03 μM	Tablet dosage	(Farag, Bakirhan, Švancara, & Ozkan, 2019)
Propranolol	TiO$_2$/MWCNTs/PGE	DPV	0.085–6.5 μM	21 nM	Tablet and blood serum and urine	(Dehnavi & Soleymanpour, 2021)
Prostate-specific antigen (PSA)	GCE/COOH-MWCNTs/PANI/AuNPs	DPV	1.66 ag/mL–1.3 ng/mL	0.5 pg/mL	Human serum	(Assari, Rafati, Feizollahi, & Joghani, 2020)
Rapamycin	Pyrolle/MWCNT/CPE	DPV	0.1–20 μM	0.06 μM	Pharmaceutical formulations	(Zrinski et al., 2021)
Resorcinol	HKUST-1/SWCNT-Nafion/GCE	DPV	0.5–200 μM	0.17 μM	Rainwater, tap water, and sewage	(Ai et al., 2020)

Continued

TABLE 12.2 Some reported SWCNT- and MWCNT-based electrochemical sensors for the detection of different analytes—cont'd

Analyte	Nanosensor type	Method	Linear range	Limit of detection	Application type	Ref.
Rhodamine B	MWCNT/CPE	DPV	0.1–15.0 μM	20.0 nM	Water and hair color	(Golestaneh & Ghoreishi, 2020)
Rifampicin	MWCNTs/GCE	DPV	0.04–10 μM	7.51 nM	Capsule	(Kul, 2020)
Rifampicin	MWCNTs/GCE	SWV	0.04–10 μM	11.3 nM	Capsule	(Kul, 2020)
Salicylic acid	Silica gel/MWCNTs/CPE	SWV	3.0–70 nM.	0.9 nM	Tomato	(Rabie, Assaf, Abo Elmaaref, & Khodari, 2019)
Sarcosine	MWCNT/Nafion®/Ni(OH)$_2$/SPE	FIA	3.2–25.0 μM	0.96 μM	Synthetic urine	(De Cássia, Da Rocha, Capelari, et al., 2020)
Serotonin	AA-MWCNT/PEDOT/PGE	DPV	0.1–100 μM	0.092 μM	Human blood serum	(Gorduk, 2020)
Staphylococcus aureus	AuPdPt/MWCNTs/GCE	Amperometry	1.1×10^2–1.1×10^7 CFU/mL	39 CFU/mL	Yogurt, pure milk, and infant milk powder	(Han, Zhang, Cai, & Zhang, 2021)
Sulfamethazine	Aptamer/MCH/cDNA/AuNP/MWCNT@GONR/Au electrode	DPV	0.01–50 μg/L	5.2 ng/L	Spiked chicken	(He, Li, & Li, 2020)
Sulfentrazone	MWCNT/SPE	SWV	1.0–25 μM	0.8 μM	Soy milk and groundwater	(Silva, De, Da Silva, Fiorucci, & Ferreira, 2019)
Tadalafil	MWCNT	LSV	0.85–8.9 μM	78 nM	Pharmaceutical drugs	(Zambianco, Da Silva, Orzari, et al., 2020)
Tamoxifen	CeO$_2$-MWCNTs/CCE	DPV	0.2–40 nM	0.132 nM	Human serum	(Shafaei, Hosseinzadeh, Kanberoglu, Khalilzadeh, & Mohammad-Rezaei, 2021)
Tramadol	Pt-NiO/MWCNTs/GCE	DPV	1.0–240 μM	0.084 μM	Human serum and urine and injection samples	(Dehdashti & Babaei, 2020)
Tetracycline	AuNP/MWCNTs/GCE	DPV	0.2–0.6 ppm	42 ppb	Chicken meat and eggs	(Palisoc, De Leon, Alzona, Racines, & Natividad, 2019)

Trolox	CPE/MWCNTs/tyrosinase/Nafion®	Amperometry	—	—	Berries	(Frangu, Ashrafi, Sýs, et al., 2020)
Trospium hydrochloride	MWCNT paste/SPE	Potentiometry	1 µM–10 mM	0.8 µM	Pharmaceutical formulations and biological fluids	(El Attar, Hassan, & Khaled, 2020)
Tyrosine	MWCNT/PSF/GCE	CV	1.96–394 µM	0.3 nM	—	(Phelane et al., 2020)
Norfloxacin	fMWCNTs-MIP/GCE	DPV	3 nM–3.125 µM	1.58 nM	Pharmaceutical formulations and rat plasma	(Liu, Jin, Lu, et al., 2019)
Tyrosine	MWCNT-COOH/ABPE	SDLSV	0.04–600 µM	0.02 µM	Milk, yogurt, beer, and cheese	(Feng, Deng, Xiao, et al., 2021)
Uranium (VI)	IL/MWCNTP/SPCE	Potentiometry	0.47 µM–0.1 M	0.47 µM	Spiked waters	(Ali & Akl, 2021)
Uric acid	g-C_3N_4/MWCNT/GCE	DPV	0.2–20 µM	0.139 µM	Human serum	(Lv, Li, Feng, et al., 2019)
Uric acid	MWCNT-COOH/SWCNT-OH/GCE	DPV	2–150 µM	0.61 µM	Bovine serum	(Guan et al., 2020)
Uric acid	PDY/MWCNT/GCE	DPV	0.1–7.5 µM	8.9 nM	Human blood serum	(Hatefi-Mehrjardi et al., 2020)
Uric acid	SWCNTs-Chitosan-RTIL/GCE	DPV	3–320 µM	0.27 µM	Human serum and urine	(Afrasiabi et al., 2016)
Vanillylmandelic acid	MWCNTs-PtNPs/GCE	DPV	0.5–80 µM	0.173 µM	Human urine	(Fu et al., 2020)
Vardenafil	NH_2-MWCNT/ZnO/GPE	SWSV	0.02–1.0 mg/L	13.6 µg/L	Tablets and synthetic serum	(Demir, Bozal-Palabıyık, Uslu, & Inam, 2019)
Vardenafil	NH_2-MWCNT/ZnO/GPE	DPSV	0.01–0.5 mg/L	4.38 µg/L	Tablets and synthetic serum	(Demir et al., 2019)
Vitamin B2	f-MWCNTs/GCE	DPV	0.004–1.0 µM	1.0 nM	Ampule and tablet	(Zeybek & Üğe, 2021)
Vitamin B6	f-MWCNTs/GCE	DPV	8.0–1000 µM	2.8 nM	Ampule and tablet	(Zeybek & Üğe, 2021)
Fluometuron	FePc/MWCNTPE	DPV	69.8–233.0 µg/L	101 µg/L	Tap water and commercial pesticide formulation	(Demir, Goktug, Inam, & Doyduk, 2021)
Xanthine	Au NPs/MWCNT/SPE	Amperometry	0.1–10 pM	1.14 nM	Fish meat freshness	(Sharma, Monika, et al., 2021)
2-Aminophenol	MWCNT-CoTSAPc/GCE	CV	0.1–30 µM	0.03 µM	Tap and mineral water	(Sanna Jilani, Mruthyunjayachari, Malathesh, Mounesh, & Reddy, 2019)

Continued

TABLE 12.2 Some reported SWCNT- and MWCNT-based electrochemical sensors for the detection of different analytes—cont'd

Analyte	Nanosensor type	Method	Linear range	Limit of detection	Application type	Ref.
2-Aminophenol	MWCNT-CoTSAPc/GCE	DPV	0.1–38 μM	0.03 μM	–	(Sanna Jilani et al., 2019)
2-Aminophenol	MWCNT-CoTSAPc/GCE	Chronomperometry	0.05–20 μM	0.016 μM	–	(Sanna Jilani et al., 2019)
2,4-Dichlorophenol	Polyeosin Y/MWCNTs-OH/GCE	DPV	0.005–40.0 μM	1.5 nM	Yitong River	(Zhu et al., 2021)
β-agonists	2D-hBN/f-MWCNTs/GCE	DPV	1pM–10nM	0.1 pM	Urine	(Yola & Atar, 2019a)

AA-MWCNTs: Acid-activated multiwalled carbon nanotubes, ABPE: Acetylene black paste electrode, ASV: Anodic stripping voltammetry, AuNRs: Gold nanorods, BPY: bibpyridyl, CCE: Ceramic carbon electrode, CoTSAPc: Cobalt tetra-substituted sorbaamide phthalocyanine, CPE: Carbon paste electrode, CPB: Cationic cetylpyridinium bromide, CRE: Carbon rod electrode, DomP: 1-Docosyloxylmethylpyrene, DPASV: Differential pulse adsorptive stripping voltammetry, DPV: Differential pulse voltammetry, ErGO: Electrochemically reduced graphene oxide, f-COOH: Functionalized-COOH, FIA: Flow injection analysis, f-MWCNT: Functionalized-MWCNT, FSV: Fast scan voltammetry, GONR: Graphene oxide nanoribbons, GQDs: Graphene quantum dots, hBN: Two-dimensional hexagonal boron nitride, HKUST-1: Hong Kong University of Science and Technology-1, HRP: Horseradish peroxidase, IIP: Ion-imprinted polymer composite, ISE: Ion selective electrode, ISS: In situ surfactant, ITO: Indium tin oxide, LDH: Lactate dehydrogenase, LSASV: Linear sweep adsorptive stripping voltammetry, MCH: 6-mercapto-1-hexanol, MCPE: Modified carbon paste electrode, MGCE: Magnetic glass carbon electrode, MIP: Molecular-imprinted polymer, MMA: Methacrylic Acid, Mn(TPP)Cl: 2,3,7,8,12,13,17,18- octaethyl-21H,23H-porphine manganese (III) chloride, MOF: Metal organic framework, Ox-MWCNTS: Oxidized multiwalled carbon nanotubes, PAMAM: Poly(amido amine) dendrimer, PANI: Polyaniline, PAPBA: Poly(3-amino phenyl boronic acid), PB: Prussian blue, PBA: Prussian blue analog, P-Cys: Poly-cysteine, PDY: Polymer dianix yellow, PEDOT: Poly (3,4-ethylenedioxythiophene), PGE: Pencil graphite electrode, PMB: Polymethylene blue, PMO: Polymerized methyl orange, PNR: Polyneutral red, PoPD: Poly(o-phenylenediamine), PPY: Polypyrrole film, PSF: Polysulfone, RTIL: Room temperature ionic liquid, SDLSV: Second-order derivative linear sweep voltammetry, SiSG: Silica sol-gel, SPCE: Screen-printed carbon electrode, SPE: Screen-printed electrode, SSA: 5-Sulfosalicylic Acid, SWASV: Square wave adsorptive stripping voltammetry, SWSV: Square wave stripping voltammetry, TCPP: Tetra(4-carboxyphenyl) porphine, TFGE: Thin film gold electrode, TOAB: Tetraoctylammonium bromide, TYR: Tyrosinase, TPA: Terephthalic acid, YD: Porphyrin dye, β-CD: β-Cyclodextrin, 3DOM: Three-dimensional ordered macroporous.

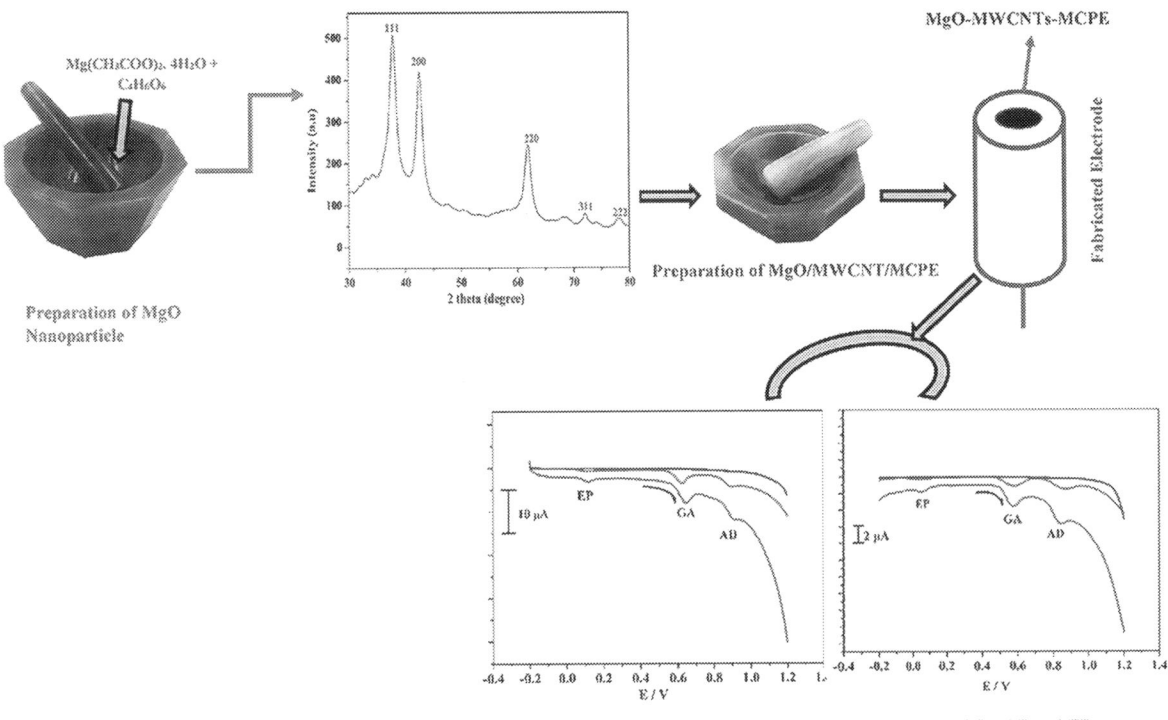

FIG. 12.3 Preparation of MgO-MWCNTs-MCPE electrode, linear sweep, and differential pulse voltammograms for simultaneous determination of guanine (GA), adenine (AD), and epinephrine (EP). *Reprinted with permission from Chetankumar K, Swamy BE, Naik HSB. MgO and MWCNTs amplified electrochemical sensor for guanine, adenine and epinephrine. Materials Chemistry and Physics 2021;267:124610.*

Sensor-based graphene and graphene quantum dots

Graphene is the thinnest two-dimensional (2D) carbon material with excellent electronic, mechanical, optical, and thermal properties. Graphene exhibits a zero-spacing semimetal property with little overlap between its valence and conduction bands. This unique and Nobel Prize-winning carbon material is produced using three main techniques: (i) chemical vapor deposition, (ii) mechanical exfoliation, and (iii) Hummers' method. The most striking feature of this carbon nanomaterial is its use in many industrial areas, such as light-emitting diodes, touch screens, transistors, solar cells, supercapacitors, and batteries. Also, graphene has been utilized in medical applications, including drug and gene delivery, cancer remedy, bioimaging and biosensing, and tissue/cell culture applications (Lalabadi et al., 2020). Graphene's outstanding conductivity, high charge carrier mobilities of $\approx 10,000\,cm^2/V\,s$, and high surface area make it a very promising nanomaterial in the development of nanosensors. Up till now, graphene electrodes as carbon nanomaterial have been used in the sensor platform for the determination of a wide range of organic materials. Also, graphene and its derivatives have a great potential to modify the main electrodes based on carbon nanomaterials used in electrochemical detection. Graphene and its derivatives play an important role in the development of metal oxide-based sensors, such as MoO_2, MnO_2, Fe_3O_4, and Cu_2O. As a result, the development of graphene-based materials has opened up more opportunities for its wider applications in sensor designs.

Lately, among diverse carbon nanomaterials, graphene oxide with a 2D nanostructure has been extensively used as a supporting material for sensor systems owing to its active functional groups, high surface area, and conductivity. This nanomaterial is of low cost and easily dispersible in water. Further, GO develops electrocatalytic activity and avoids accumulation of the deposited nanoparticles on its surface (Kamyabi & Moharramnezhad, 2021). Selected applications of graphene-based electrochemical sensors are given in Table 12.3.

The measurement of renal function by significant clinical parameters such as creatinine clearance and glomerular filtration rate often goes wrong vis-à-vis the creatinine level in the human body. Therefore, improving an accurate determination system over a wide concentration range of creatinine in both urine and blood is especially essential. Ponnaiah and Periakaruppan developed a new nonenzymatic electrochemical sensor based on CD-doped tungstic anhydride embedded on GO nanopanels (CDs/WO_3@GO) for picomolar-level creatinine determination in blood and urine with a wide linear

TABLE 12.3 Some reported graphene-based electrochemical sensors for the detection of different analytes.

Analyte	Nanosensor type	Method	Linear range	Limit of detection	Application type	Ref.
Acetaminophen	Cu-Ni/rGO/GCE	SWV	0.10–50.00 µM	2.13 nM	Pharmaceutical samples	(Dou & Qu, 2020)
Acetamiprid	Chitosan/rGO/GCE	SWV	0.1 pM–0.1 µM	71.2 fM	Tea samples	(Yi et al., 2020)
Adenine	p-GLY/GO/GCE	DPV	0.090–103 µM	0.030 µM	Urine and fish sperm DNA samples	(He et al., 2018)
Amoxicillin	MIP-GO/GCE	DPV	5.0×10^{-10} – 9.1×10^{-7} M	2.94×10^{-10} M	Pharmaceutical antibiotic tablet	(Güney, Arslan, Yanık, & Güney, 2021)
Amoxicillin	MIP/rGO/overoxidized polypyrrole/AuNPs/GCE	CV	1×10^{-6} – 1×10^{-3} M	1.22×10^{-6} M	Human serum and milk samples	(Essousi, Barhoumi, Karastogianni, & Girousi, 2020)
Arsenite	SB@SiO$_2$@GO/ITO	SWV	0.150–0.930 nM	156 pM	Soil and link samples	(Kaur, Rana, Singh, Kaur, & Narula, 2019)
Ascorbic acid	g-C$_3$N$_4$/NC@GC/h-ATS/ITO	DPV	0.1–200 µM	0.02 µM	Human urine samples	(Krishnan, Tong, Liu, & Xing, 2020)
Azathioprine	Mn$_2$O$_3$-rGO/SPCE	DPV	9 nM–0.57 mM	4 nM	Human blood serum and urine	(Selvi, Nataraj, Chen, & Prasannan, 2021)
C-reactive protein	BSA/rGO/antibody/GO/SPCE	SWV	1–100,000 ng/mL	0.38 ng/mL	Human serum sampes	(Boonkaew et al., 2021)
Carbofuran	GO/AuNPs	DPV	0.2–50 nM	67 pM	Cabbage, chili, lettuce, tomato, apple, banana, tangerine, and watermelon samples	(Li, Li, Luo, Xu, & Ma, 2018)
Cardiac troponin I	BSA/rGO/antibody/GO/SPCE	SWV	0.001–250 ng/mL	0.16 pg/mL	Human serum sampes	(Boonkaew et al., 2021)
Cd(II)	PrGO/AuNPs/Sal-Cys/GCE	SWASV	1–10 nM	0.06 nM	Lake, sewage, tap and groundwaters	(Priya, Dhanalakshmi, Thennarasu, Karthikeyan, & Thinakaran, 2019)
Cd(II)	Bi$_2$O$_3$/Fe$_2$O$_3$@GO	SWV	6.2–1160.2 ng/L	1.85 ng/L	Fruit, water, soil, and biological samples	(Das & Sharma, 2019)

Analyte	Electrode	Method	Linear range	LOD	Real sample	Reference
Cd(II)	GO-chitosan/poly-L-lysine/GCE	DPASV	0.05–10.0 μg/L	0.01 μg/L	Tap water samples	(Guo et al., 2017)
Cholesterol	Ag/GO/Au NPs/SPE	LSV	0.01–5000 μg/mL	0.001 μg/mL		(Huang et al., 2018)
Chloramphenicol	Chlorine-doped rGO/GCE	DPV	2–35 μM	1.0 μM	Milk, calf plasma, water, and pharmaceutical samples	(Wang, Zhang, Zhang, & Shen, 2019)
Chlorpyrifos	BNQDs/GO/GCE	DPV	1.0×10^{-12}–1.0×10^{-8} M	3.3×10^{-14} M	Water and apple juice samples	(Yola, 2019)
Chlorpyrifos	RuNBs/AgNPs/GO/GCE	Electrochemiluminescence	5.0×10^{-15}–4.2×10^{-9} M	6.5×10^{-16} M	Water and fruit samples	(Kamyabi & Moharramnezhad, 2021)
Ciprofloxacin	rGO/polyphenol red/GCE	DPV	0.002–0.05 μM 0.05–400 μM	2 nM	Animal serum samples	(Chauhan, Gill, Nate, & Karpoormath, 2020)
Clozapine	GO/Fe$_3$O$_4$/SiO$_2$/SPE	DPV	0.1–700.0 μM	0.03 μM	Pharmaceutical drug and urine samples	(Beitollahi et al., 2020)
Codeine	Modified graphene with Zn$_2$SnO$_4$/CPE	DPV	0.020–15 μM	0.0090 μM	Human body fluids and pharmaceutical samples	(Bagheri, Khoshsafar, Afkhami, & Amidi, 2016)
Cu(II)	GO-chitosan/poly-L-lysine/GCE	DPASV	0.05–10.0 μg/L	0.02 μg/L	Tap water samples	(Guo et al., 2017)
Cu(II)	GO/ZnS/GCE	Electrochemiluminescence	1×10^{-10}–1×10^{-7} M	2.2×10^{-11} M	Tap and lake water samples	(Ye, Wang, Liu, & Li, 2020)
Daidzein	Poly(sodium 4-styrenesulfonate)-rGO/GCE	DPV	0.1–20.0 nM	0.5 nM	Human serum samples	(Liang, Qu, Yang, Qu, & Li, 2017)
DNA	Graphene/GCE	EIS	2.0×10^{-18}–1.0×10^{-12} M	1.0×10^{-18} M		(Benvidi et al., 2018)
DNA bases	GCE/GONRs-chitosan	DPV	0.05–342 μM	0.002 μM	Single nucleotides and dsDNA	(Zhou, Li, Noroozifar, & Kerman, 2020)
Diazinon	BNQDs/GO/GCE	DPV	1.0×10^{-12}–1.0×10^{-8} M	6.7×10^{-14} M	Water and apple juice samples	(Yola, 2019)
Dopamine	p-GLY/GO/GCE	DPV	20–62 μM	0.011 μM	Urine and fish sperm DNA samples	(He et al., 2018)

TABLE 12.3 Some reported graphene-based electrochemical sensors for the detection of different analytes—cont'd

Analyte	Nanosensor type	Method	Linear range	Limit of detection	Application type	Ref.
Dopamine	Graphene foam	Amperometry	0.01–10 µM	2.0 nM	Drug injection	(Huang et al., 2014)
Dopamine	NiO-rGO/ITO	DPV	1–60 µM	1 µM	Human urine samples	(Yue, Zhang, et al., 2019)
Dopamine	rGO-NiO/ITO	SWV	0.5–50 µM	0.495 µM	Serum samples	(Roychoudhury, Prateek, Basu, & Jha, 2018)
Dopamine	MoS$_2$ NFs-rGO/ITO	DPV	5–60 µM	0.12 µM	Human serum samples	(Guo, Yue, Song, et al., 2020)
Dopamine	3D HGBs/ITO	DPV	1–60 µM	1 µM	Urine samples	(Yue et al., 2018)
Dopamine	GCE/rGO-Zn(I/)TPEBiPc	CV	0.2–160 µM	60 nM	Pharmaceutical drug samples	(Pari, Reddy, Fasiulla, & Chandrakala, 2020)
Dopamine	GCE/rGO-Zn(I/)TPEBiPc	DPV	0.05–08 µM	16 nM	Pharmaceutical drug samples	(Pari et al., 2020)
Dopamine	GCE/rGO-Zn(I/)TPEBiPc	Chronoamperometry	0.02–1.0 µM	6 nM	Pharmaceutical drug samples	(Pari et al., 2020)
Dopamine	ZnO-rGO	DPV	5–70 µM	0.167 µM	Human serum samples	(Cao et al., 2020)
Dopamine	ErGO/MWCNTs/PPy	Amperometry	25–1000 nM	2.3 nM	—	(Kathiresan et al., 2021)
Dopamine	AuNPs@3-dimensional graphene/ITO	DPV	0.1–5 µM 5–60 µM	0.1 µM	Serum samples	(Wang, Yue, Huang, et al., 2019)
Epinephrine	rGO-NiO/ITO	SWV	0.5–50 µM	0.423 µM	Serum samples	(Roychoudhury et al., 2018)
Epithelial cell adhesion molecules; EpCAM tumor biomarker	rGO@TiO$_2$/ITO	DPV	0.01–60 ng/mL	0.0065 ng/mL	Spiked human serum samples	(Jalil, Pandey, & Kumar, 2020)
Free chlorine	rGO/MWCNT/PB/ITO	Amperometry	50–600 µM	2.1 µM	Tap water	(Silva, Cardoso, Richter, Munoz, & Nossol, 2020)

Analyte	Electrode	Technique	Linear range	LOD	Sample	Reference
Flutamide	rGO/CuO/GCE	Amperometry	0.005–71.32 μM	0.001 μM	Human serum and urine samples	(Sakthinathan et al., 2019)
Gallic acid	nano-GO-SiO$_2$NP/GCE	DPV	6.25–1000 μM	2.09 μM	Red wine, white wine, and orange juice	(Chikere, Faisal, Lin, & Fernandez, 2019)
Gallic acid	AuCMs/SF-graphene/GCE	DPV	0.05–8.0 μM	0.0107 μM	Black tea, Cortex Moutan, and urine samples	(Liang et al., 2016)
Gatifloxacin	GCE/rGO-CuS/g-C$_3$N$_4$	Electrochemiluminescence	10nM–100 μM	3.5 nM	Mouse plasma	(Jiang, Mo, et al., 2019)
Gatifloxacin	β-cyclodextrin/rGO/GCE	DPV	0.05–150 μM	21.0 nM	Pharmaceutical and human urine samples	(Jiang, Li, & Zhang, 2016)
Glucose	GOx/rGO/GCE	SWV	0.5–3.0 mM	0.51 mM	2G ethanol production	(Donini, Silva, Bronzato, Leao, & Cesarino, 2020)
Glucose	GOx/SPG-AuNPs-chitosan/ITO	Amperometry	0.5–22.2 μM	510 μM	Human serum samples	(Singh et al., 2013)
Glucose	MPC-chitosan-GOx/SPCE	Amperometry	0.25–3 mM	4.1 μM	Saliva, human serum, and urine samples	(Barathi, Thirumalraj, Chen, & Angaiah, 2019)
Glucose	N-doped graphene-CNTs-gold NPs	Amperometry	2 μM–19.6 mM	500nM	—	(Jeong et al., 2018)
Guanine	p-GLY/GO/GCE	DPV	0.15–48 μM	0.026 μM	Urine and fish sperm DNA samples	(He et al., 2018)
Heavy metals	FGP/AuNC/GCE	SWASV	4–7000 μg/L	0.01 μg/L	Peanut, rape bolt, and tea	(Tan, Wu, Feng, Wu, & Zhang, 2020)
Honokiol	GO-COO-4-hydroxy-1-methyl-[1-(3-pyrrolyl-propyl)]-piperidinium bromide	DPV	1.0×10^{-8}–1.0×10^{-5} M	1.53×10^{-9} M	Pharmaceutical samples	(Zhang et al., 2015)
Hydrazine	CuO/CNTs-rGO/GCE	Amperometry	1.2–430 μM	0.20 μM	—	(Zhao, Wang, et al., 2020)
Hydrogen peroxide	rGO/MWCNT/PB/ITO	Amperometry	50–1000 μM	8.7 μM	Tap water	(Silva et al., 2020)
Hydrogen peroxide	Ag-Fe$_2$O$_3$/POM/rGO/GCE	Amperometry	300–3000 μM	0.2 μM	—	(Ross & Nqakala, 2020)
L-Cysteine	CoHCF@(PEI/GO)$_5$/ITO	DPV	50–2000 μM	0.46 μM	Pharmaceutical cosmetic sample	(Hao, Li, Chen, Xu, & Feng, 2019)

Continued

TABLE 12.3 Some reported graphene-based electrochemical sensors for the detection of different analytes—cont'd

Analyte	Nanosensor type	Method	Linear range	Limit of detection	Application type	Ref.
Levodopa	In_2S_3 NSPs/3D graphene/ITO	DPV	0–60 µM	87 nM	Human serum samples	(Guo, Yue, Huang, et al., 2020)
Levodopa	ZnO NFs-rGO nanosheets/ITO	DPV	1–60 µM	1 µM	Human urine samples	(Yue, Wu, et al., 2019)
Levodopa	3D GF/ITO	DPV	1–60 µM	1 µM	Human urine samples	(Wang, Yue, Yu, et al., 2019)
Levodopa	3D HGBs/ITO	DPV	1–60 µM	1 µM	Human plasma samples	(Gao et al., 2018)
Levodopa	MoS_2-graphene/ITO	DPV	5–60 µM	0.3 µM	Human plasma samples	(Yue et al., 2017)
Levofloxacin	rGO/GCE	DPV	0.2–3000 µM	60 nM	Blood serum	(Ghanbari, Nasrabadi, & Sobati, 2019)
Maltose	SPCE/CuO/GOx/maltase/SiO_2	CV	10–100 µM	5 µM	Beer	(Kornii, Saska, Lisnyak, & Tananaiko, 2020)
Magnolol	GO-COO-4-hydroxy-1-methyl-[1-(3-pyrrolyl-propyl)]-piperidinium bromide	DPV	$7 \times 10^{-8} – 1 \times 10^{-5}$ M	8.27×10^{-9} M	Pharmaceutical samples	(Zhang et al., 2015)
Methyl parathion	BNQDs/GO/GCE	DPV	$1.0 \times 10^{-12} – 1.0 \times 10^{-8}$ M	3.1×10^{-13} M	Water and apple juice samples	(Yola, 2019)
Morphine	Modified graphene with Zn_2SnO_4/CPE	DPV	0.020–15 µM	0.011 µM	Human body fluids and pharmaceutical samples	(Bagheri et al., 2016)
Nitrite	AuNPs-WNPs@graphene-chitosan/PGE	CV	10–250 µM	0.12 µM	Water, milk, and fruit juices	(Lavanya, Kumari, Prasad, & Brahman, 2021)
Norepinephrine	Poly(L-arginine)/ErGO/GCE	SWV	$9.00 \times 10^{-6} – 8.00 \times 10^{-7} – 7.00 \times 10^{-5} – 2.00 \times 10^{-5}$ M	4.22×10^{-8} M	Blood serum and urine samples	(Anand, Mathew, Radecki, Radecka, & Kumar, 2020)
Paracetamol	Hematite(α-Fe_2O_3)/graphene/GCE	DPV	0.3–1800 µM	95 nM	Tablet formulations	(Haridas, Yaakob, Renuka, Sugunan, & Narayanan, 2021)

Analyte	Electrode	Method	Linear range	LOD	Real sample	Reference
Pb (II)	GO-chitosan/poly-L-lysine/GCE	DPASV	0.05–10.0 µg/L	0.02 µg/L	Tap water samples	(Guo et al., 2017)
Procalcitonin	BSA/rGO/antibody/GO/SPCE	SWV	0.0005–250 ng/mL	0.27 pg/mL	Human serum sampes	(Boonkaew et al., 2021)
Procalcitonin	rGO/AuNPs/GCE	Amperometry	0.05–100 ng/mL	0.0001 ng/mL	Human serum sampes	(Liu, Li, et al., 2019)
Progesterone	AuNPs/5-amino-2-mercaptobenzimidazole/rGO/SPCE	SWV	0.9×10^{-9}–27×10^{-6} M	0.28×10^{-9} M	Calf and milk serum samples	(Zhao, Zheng, Yan, Cao, & Zhang, 2021)
Puerarin	Poly(sodium 4-styrenesulfonate)-graphene/WO$_3$/GCE	LSV	0.06–0.6 µM 0.6–60 µM	0.04 µM	Pharmaceutical and human plasma samples	(Jing, Zheng, Zhao, Qu, & Yu, 2017)
Rizatriptan benzoate	Graphene/αFe$_2$O$_3$/CPE	DPV	2–50 µM	0.42 µM	Human urine and pharmaceutical	(Nouri, Rahimnejad, Najafpour, & Moghadamnia, 2019)
Rutin	Au-Ag nanothorns/N-doped-graphene	DPV	0.1–420 µM	0.015 µM	Pharmaceutical samples	(Yang, Bin, Zhang, Du, & Majima, 2018)
Serotonin	NDs-AuNPs-graphene-CSN/GCE	DPV	0.3–3.0 µM	0.1 µM	Synthetic urine samples	(Ramos, Carvalho, Oliveira, & Janegitz, 2020)
Terbutaline	MWCNTs/graphene/GCE	DPV		6.527×10^{-7} M	Cough syrups, injections, and tablets	(Gopal & Reddy, 2018)
Thrombin	Graphene@Fe$_2$O$_3$/GCE	DPV	1.0×10^{-11}–4.0×10^{-9} M	1.0×10^{-12} M	—	(Zhang, Zhang, Chen, & Qin, 2017)
Thrombin	Graphene oxide/PGE	DPV	0.1–10 nM	0.07 nM	Human serum samples	(Ahour & Ahsani, 2016)
Tyramine	Poly(L-arginine)/ErGO/GCE	SWV	1.00×10^{-5}–6.00×10^{-7} M 7.00×10^{-5}–2.00×10^{-5} M	1.45×10^{-7} M	Blood serum and urine samples	(Anand et al., 2020)
Tryptophan	PSS-graphene/GCE	LSV	0.04–10.0 µM	0.02 µM	Human serum samples	(Wang, Yang, Li, Qu, & Harrington, 2019)
4-Aminophenol	Cu-Ni/rGO/GCE	SWV	0.10–30.00 µM	2.19 nM	Pharmaceutical samples	(Dou & Qu, 2020)

Continued

TABLE 12.3 Some reported graphene-based electrochemical sensors for the detection of different analytes—cont'd

Analyte	Nanosensor type	Method	Linear range	Limit of detection	Application type	Ref.
4-Nitrophenol	AgInS$_2$/rGO/GCE	DPV	1.43–2696.14 µM	0.078 µM	Tap and river waters	(Li, Shi, et al., 2019)
Urapidil	Poly(sodium 4-styrenesulfonate)-Gr/GCE	LSV	2.0 nM–200 µM	0.8 nM	Tablet formulations	(Gao, Li, Wang, Gao, & Ye, 2019)
Uric acid	Uricase/magnetic beads/rGO/NiO	DPV	0.2–50 µM	0.12 µM	Urine samples	(Chou et al., 2021)
Uric acid	MoS$_2$ NFs-rGO/ITO	DPV	5–60 µM	0.14 µM	Human serum samples	(Guo, Yue, Song, et al. 2020)
Uric acid	3D HGBs/ITO	DPV	1–60 µM	1 µM	Urine samples	(Yue et al., 2018)
Uric acid	p-GLY/GO/GCE	DPV	0.10–105 µM	0.061 µM	Urine and fish sperm DNA samples	(He et al., 2018)
Uric acid	UOx/Au-rGO/ITO	DPV	50–800 µM	7.32 µM	Human serum samples	(Verma, Choudhary, Singh, Chandra, & Singh, 2019)
Uric acid	Au NPs@3D graphene/ITO	DPV	0.1–5 µM, 5–60 µM	0.1 µM	Serum samples	(Wang, Yue, Huang, et al., 2019)
Uric acid	MoS$_2$-graphene/ITO	DPV	5–60 µM	0.4 µM	Human plasma samples	(Yue et al., 2017)

AuCMs: Gold microclusters, BNQD: Boron nitrite quantum dots, BSA: Bovine serum albumin, COHCF: Cobalt hexacyanoferrate, CPE: Carbon paste electrode, CSN: Casein, DPASV: Differential pulse adsorptive stripping voltammetry, EIS: Electrochemical impedance spectroscopy, ErGO: Electrochemically reduced graphene oxide, FGP/AuNC: Fluorinated graphene/gold nanocage, GONRs: Graphene oxide nanoribbons, GOx: Glucose oxidase, h-ATS: Ag–(TiO$_2$/SnO$_2$), HGB: Hollow graphene ball, ITO: Indium tin oxide, MIP: Molecularly imprinted polymer, MoS$_2$ NFs: MoS$_2$ nanoflowers, MPC: Mesoporous carbon, ND: Nanodiamond, rGO: Reduced graphene oxide, PB: Prussian blue, PEI: Polyethyleneimine, p-GLY: Poly(glycine), POM: Polyoxometalate, PPy: Polypyrrole, PSS: Polysodium 4-styrenesulfonate) functionalized, PrGO: Porous reduced graphene oxide, RuNBs: Ruthenium nanobeads, Sal-Cys: Salicylaldehyde- L-cysteine, SB: Schiff base, SF-graphene: Sulfonate-functional graphene, SPCE: Screen-printed carbon electrode, SPE: Screen-printed electrode, SPG: Sulfonated graphene, SWASV: Square wave adsorptive stripping voltammetry, UOx: Uricase, ZnO NFs: ZnO nanoflowers, Zn(II)TPEBiPc: Zinc tetra [4-[2-[(E)-2-phenylethenyl]-1H-benzimidazol-1-yl]]phthalocyanine.

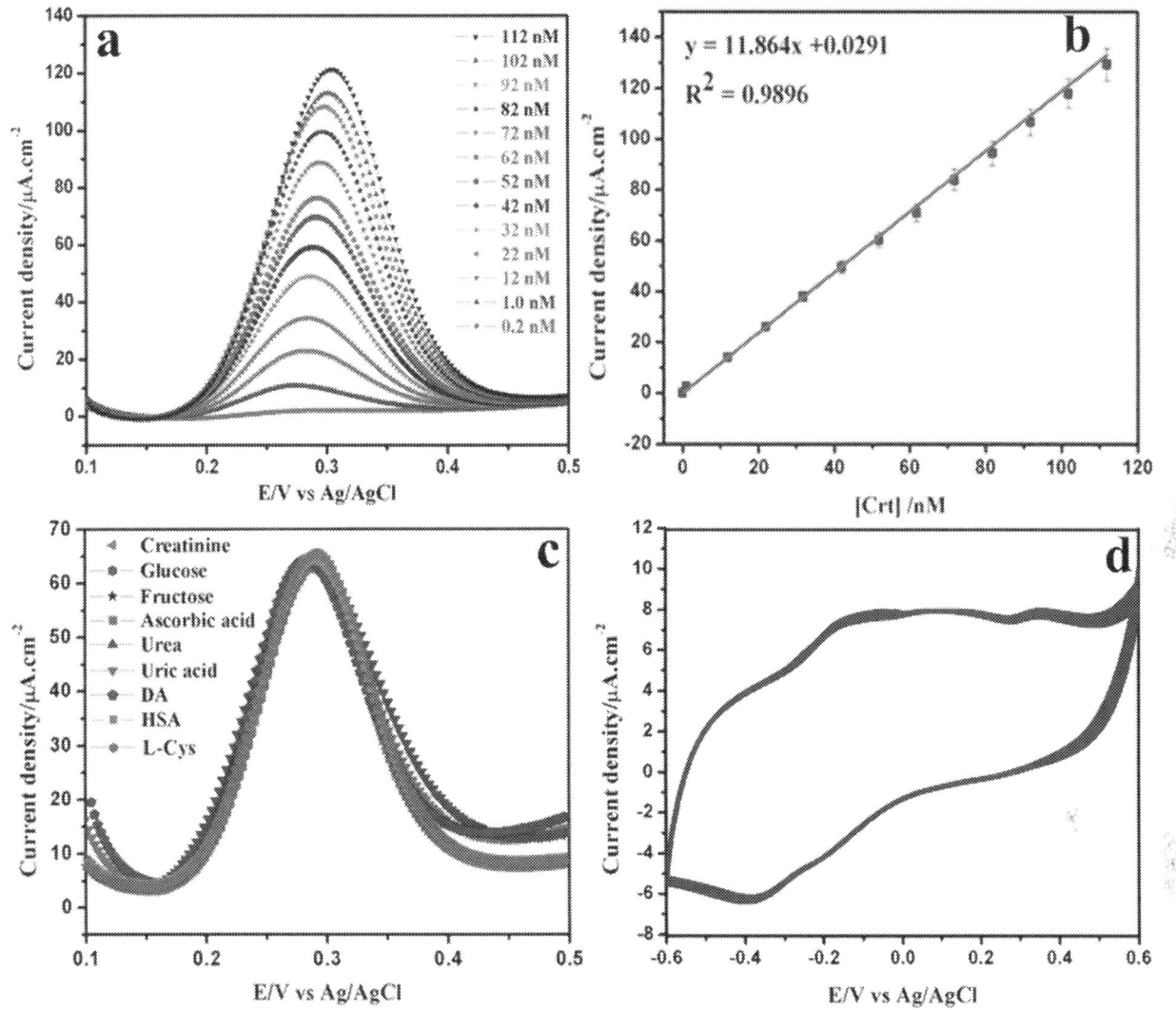

FIG. 12.4 (A) DPV response of CDs/WO$_3$@GO-modified GCE in 0.05 M PB (pH 7.0), [Crt]: 0.2–112.0 nM, (B) plot of oxidation peak current density response vs. creatinine concentration, (C) DPV response of CDs/WO$_3$@GO/GCE to successive addition of common interferences in the presence of 50 nM creatinine, and (D) CV cycles (30 nos.) with the addition of 50 μM creatinine. *Reprinted with permission from Ponnaiah SK, Periakaruppan P. Carbon dots doped tungstic anhydride on graphene oxide nanopanels: A new picomolar-range creatinine selective enzymeless electrochemical sensor. Materials Science and Engineering 2020;113:111010.*

concentration range from 0.2 to 112.0 nM (Fig. 12.4). The proposed sensor is of low cost, reproducible, stable, and interference-free. Properties of CDs/WO$_3$@GO were investigated using different analytical techniques including Fourier transform infrared spectroscopy, ultraviolet visible spectroscopy, field emission scanning electron microscopy, high-resolution transmission electron microscopy, field emission scanning electron microscopy, elemental mapping analysis, X-ray diffraction, and thermogravimetric analysis. Experimental results showed that the developed electrochemical creatinine sensor can be used as a sustainable alternative for diagnostic use in medical applications with a wide detection range and lower detection limit (Ponnaiah & Periakaruppan, 2020). DPV is more sensitive and has a lower background current density compared to some techniques, such as cyclic voltammetry. Also, the reduction or oxidation peak currents can be determined even at lower analyte concentrations. Therefore, DPV was utilized to determine creatinine under optimized experimental conditions using the developed sensor system (CDs/WO$_3$@GO/GCE). The response of oxidation current values was obtained for creatinine concentrations between 0.2 nM and 112.0 nM in 0.05 M phosphate buffer, pH 7.0, and potential range from 0 to 0.6 V. With the increase of creatinine concentration, the peak current density at 0.32 V increases linearly as shown in Figs. 12.4A and B. This result suggests that the developed sensor system with CDs/WO$_3$@GO can be used as a prominent sensor material for creatinine determination. LOD was found as 220 pM, with a correlation coefficient of 0.9896, and the sensitivity of the CDs/WO$_3$@GO/GCE is calculated to be 1.5325 µA µM^{-1} cm^{-2} (Ponnaiah & Periakaruppan, 2020). Two significant elements that identify the selectivity of a sensor toward the analyte under study are functional groups of the related molecules and their size. In analytical methods, molecules with similar molecular sizes and functional groups are usually more likely to interfere with the recognition of target analyte molecules. Therefore, the sensor system for creatinine was investigated by widespread interferences of glucose, fructose, ascorbic acid, urea, uric acid, dopamine, human serum albumin, and L-cysteine. The interference results are as shown in Fig. 12.4C, which demonstrates that the sensor system is interferents-free and very selective to creatinine determination. The stability of the developed electrode was investigated by protecting the CDs/WO$_3$@GO/GCE electrode in a refrigerator at 4°C for 4 weeks. The decrease in response was 2.3% compared to the initial current (Fig. 12.4D) (Ponnaiah & Periakaruppan, 2020).

A new electrochemical sensor approach based on monodisperse boron nitride quantum dots (BNQDs) on GO was developed for simultaneous determination of some organophosphate pesticides, such as methyl parathion, diazinon, and chlorpyrifos in water samples by Yola. Transmission electron microscope, scanning electron microscope, X-ray diffraction method, X-ray photoelectron spectroscopy, cyclic voltammetry, and electrochemical impedance spectroscopy methods were employed for the characterization of employed nanomaterials. Boron nitride quantum dots/graphene oxide composite-modified glassy carbon electrode was prepared. The sensor was found to be linear in the range of 1.0×10^{-12}–1.0×10^{-8} M. The detection limits for methyl parathion, diazinon, and chlorpyrifos were calculated to be 3.1×10^{-13} M, 6.7×10^{-14} M, and 3.3×10^{-14} M, respectively. Finally, the prepared electrochemical sensor was applied for water and apple juice sample analysis (Yola, 2019).

GQDs have been demonstrated to be derivations of graphite/graphene and different graphitic three-dimensional materials through top-down synthetic studies. GQDs have physicochemical properties as the same as properties of graphene. The comprehensive features of GQDs can be used for various applications in different areas such as biological imaging, drug/gene delivery, antioxidant and antibacterial activity, and sensor applications in photoluminescence sensors, electrochemical sensors, and electrochemiluminescence sensors (Table 12.4). According to the literature, GQDs are one of the interesting topics in sensor studies because of their high speed electron transfers and better conductivity, strong chemical inertness, great biocompatibility, and high photostability (Ngo, Jana, Chung, & Hur, 2020; Tajik et al., 2020).

Sensor-based carbon and carbon quantum dots

CDs are novel species of carbon nanomaterials that have taken much interest because of their perfect optical properties and potential applications in the areas of sensing, catalysis, nanomedicine, and bioimaging (Li et al., 2021). CQDs are generally defined as remarkable nanomaterial made up of carbons with dimensions of about 10 nm. Furthermore, CQDs are known as luminescent nanoparticles with photoluminescent and optoelectronic properties. CQDs with these unique properties are produced using techniques such as laser ablation, hydrothermal electrochemical, microwave irradiation, and ultrasonication. Compared with traditional fluorescence probes, such as semiconductor quantum dots and organic dyes, CDs have advantages with the inclusion of low toxicity, facilitation of synthesis and functionalization, low aqueous solubility, photochemical stability (resistance to photobleaching), and good biocompatibility (Li et al., 2021). CQDs appear to be used in many application areas, such as photocatalysis, biological imaging, heavy metal determination, supercapacitor, and pollution removal in water due to their prominent properties, such as high stability, excellent conductivity, good biocompatibility, low toxicity, cheapness, and excellent photostability. Moreover, due to their incredible physical and chemical properties, carbon dots are potentially a common nanomaterial in sensor platforms. They are included in the literature not only as direct indicator electrodes but also as very good modifiers. Since CQDs are semiconductor nanoparticles made up entirely of carbonaceous materials, they have a wide potential to work in the electroanalytical field. Some reported carbon quantum dot-based electrochemical sensors for the detection of different analytes are given in Table 12.5.

Sensor-based nanodiamond, carbon nano-onion, and carbon black electrodes

Nanodiamonds were initially synthesized in the 1960s and have been used in different areas such as in the production of cosmetics, delivery of drugs, and development of electrochemical probes and sensors (Baccarin, Neale, Cavalheiro, Smith, & Banks, 2019). Nanodiamonds consist of crystalized sp^3 diamond carbon doped with boron, making it a conductive material, known as a boron-doped diamond (BDD) (Jiang, Santiago, & Foord, 2021). They have several functional groups, such as —OH, —C=O, —COOH, —C—O—C, and —CN, which can be responsible for increasing the electric conductivity of the nanodiamonds (Simioni, Oliviera, et al., 2017). Carbon nano-onions (CNOs) are one of the newest carbon allotropes which are considered the zero-dimensional carbon nanostructure. CNOs as a member of the fullerene family consist of graphitic layers with quasi-spherical- and polygonal-shape close to one another. These layers compose of many defects and holes which can be filled by various heptagonal and pentagonal carbon rings to create amorphous or crystalline

TABLE 12.4 Some reported graphene quantum dot-based electrochemical sensors for the detection of different analytes.

Analyte	Nanosensor type	Method	Linear range	Limit of detection	Application type	Ref.
Aflatoxin B1	Au nanorods/GQDs/polyindole carboxylic acid/F-Au	Electrochemiluminescence	0.01–100 ng/mL	3.75 pg/mL	Maize, peanut, and wheat samples	(Lu et al., 2020)
Ascorbic acid	NSGQDs-PEI-luminol-Pt	Electrochemiluminescence	0.01–0.36 μM	0.0033 μM	Human serum samples	(Liu et al., 2021)
Bisphenol A	AuNPs/N,S-doped-GQDs/aptamer probe	DPV	0.1–10.0 μM	0.03 μM	Tap water samples	(Yao, Li, Xie, Yang, & Liu, 2020)
Caffeic acid	CSPE-MoS$_2$-GQDs-TvL	Chronoamperometry	0.38–10.00 μM 10.00–100.00 μM	0.32 μM	Red wine samples	(Vasilescu et al., 2016)
Cardiac troponin I	ABEI@GQDs/FTO	Electrochemiluminescence	1.0 fg/mL–5.0 pg/mL	0.35 fg/mL	Human serum samples	(Guo et al., 2021)
Cardiac troponin I	AuNPs@GQDs-modified screen-printed gold electrode	SWV	1–1000 pg/mL	0.1 pg/mL	Human serum samples	(Mansuriya & Altıntas, 2021)
Cd(II)	GCE/Gqds-Nafion	SWASV	20–200 μg/L	11.30 μg/L	Certified reference material	(Pizarro et al., 2020)
Chlorogenic acid	CSPE-MoS$_2$-GQDs-TvL	Chronoamperometry	0.38–8.26 μM 8.26–100.00 μM	0.19 μ	Red wine samples	(Vasilescu et al., 2016)
Creatinine	Au/Nafion-GQDs-Cu electrode	SWV	0.11–50.9 mg/L	–	Human urine samples	(Pedrozo-Peñafiel et al., 2020)
Diazinon	NMO/GQDs/chitosan/GCEox	DPV	0.1–330 μM	27 nM	Cucumber and tomato samples	(Ghiasi, Ahmadi, Ahmadi, Olyai, & Khodadadi, 2021)
Diethylstilbestrol	SiO$_2$-GOQDs/CPE	DPV	0.15–0.5 μM	0.18 μM	–	(Mikhraliieva et al., 2020)
Dopamine	GQDs/GCE	DPV	04.100 μM	50 μM	Pharmaceutical samples	(Zheng et al., 2018)
Dopamine	Chitosan/N-doped-GQDs@SPCE	DPV	1–100 μM 100–200 μM	0.145 μM	Human urine samples	(Aoun, 2017)
Dopamine	GQD-N[3-(trimethoxysilyl)propyl]ethylenediamine-Au nanocrystal	Amperometry	0.005–2.1 μM	5 nM	Pharmaceutical and urine samples	(Vinoth, Natarajan, Mangalaraja, Valdes, & Anandan, 2019)

Continued

TABLE 12.4 Some reported graphene quantum dot-based electrochemical sensors for the detection of different analytes—cont'd

Analyte	Nanosensor type	Method	Linear range	Limit of detection	Application type	Ref.
Epicatechin	CSPE-MoS$_2$-GQDs-TvL	Chronoamperometry	2.86–100.00 μM	2.04 μM	Red wine samples	(Vasilescu et al., 2016)
Epinephrine	GQD-N[3 (trimethoxysilyl)propyl] ethylenediamine-Au nanocrystal	Amperometry	0.01–4.0 μM	10 nM	Pharmaceutical and urine samples	(Vinoth et al., 2019)
Estriol	SiO$_2$-GOQDs/CPE	DPV	0.01–0.6 μM	0.009 μM	—	(Mikhraliieva et al., 2020)
Fumarate	Ni-CoS/GQDs/GCE	DPV	5–18 μM	12.1 pM	Pharmaceutical formulation	(Chihava et al., 2020)
Glucose	NiCo$_2$O$_4$/GQDs/CC	Amperometry	$1-159 \times 10^{-6}$ M $159-949 \times 10^{-6}$ M	0.27×10^{-6} M	Blood serum samples	(Wu et al., 2020)
Glucose	Nafion/GOx/ErGO-PLL/GCE	SWV	5–9000 μM	2.0 μM	—	(Zhang et al., 2019)
Human IgG	MoS$_2$@N-GQDs-IL MIP/GCE	DPV	0.1–50 ng/mL	0.02 ng/mL	Serum samples	(Liang, Hou, Tang, Sun, & Luo, 2021)
Hydrogen peroxide	GQDs/AuNPs-PDA-GQD/ITO	Amperometry	0.1–40 μM 40–20.000 μM	5.8 nM	—	(Zhu et al., 2017)
Hydrogen peroxide	GQDs-chitosan/MB/GCE	Amperometry	$1.0 \times 10^{-6}-2.9 \times 10^{-3}$ M 2.9–11.78 mM	0.7 μM	Food and water samples	(Mollarasouli, Asadpour-Zeynali, Campuzano, Yanez-Sedeno, & Pingarron, 2017)
Hydrogen peroxide	Gold-graphene	Amperometry	—	—	—	(Terzi et al., 2013)
Levofloxacin	Poly(o-aminophenol)/GQDs/GCE	LSV	0.05–100 μM	10 nM	Milk samples	(Huang, Bao, Hu, Wen, & Zhang, 2017)
Oxytetracycline	Bi$_4$VO$_8$Cl/N doped GQDs/ITO	Photoelectrochemical	0.1–150 nM	0.03 nM	Tomato samples	(You et al., 2020)
Paracetamol	GQDs/deep eutectic solvent/MWCNTs-COOH	DPV	0.030–110 μM	0.010 μM	Human fluid samples	(Arap, Fotouhi, Salis, & Dorraji, 2020)
Pb(II)	GCE/Gqds-Nafion	SWASV	20–200 μg/L	8.49 μg/L	Certified reference material	(Pizarro et al., 2020)

Analyte	Electrode	Technique	Linear range	LOD	Real samples	Reference
Patulin	MIP/Au@Cu-MOF/N-GQDs/GCE	DPV	0.001–70.0 ng/mL	0.0007 ng/mL	Apple juice samples	(Hatamluyi, Rezayi, Beheshti, & Boroushaki, 2020)
Serotonin	MIP/GQDs/2D-hBN/GCE	DPV	1.0×10^{-12}–1.0×10^{-8} M	2.0×10^{-13} M	Human urine samples	(Yola & Atar, 2018)
Sotalol	MIP/AuNPs/GQD-SH/SPCE	DPV	0.1–250 µM	0.035 µM	Blood serum samples and tablet formulations	(Roushani, Jalilian, & Nezhadali, 2019)
Sulfamethoxazole	SiO$_2$-GOQDs/CPE	DPV	4–20 µM	0.46 µM	–	(Mikhraliieva et al., 2020)
Thiomersal	GCE/GQDs	SWV	3–32 µM	0.9 µM	Influenza vaccine samples	(Pedrozo-Penafel, Miranda-Andrades, Gutierrez-Beleno, Larrude, & Aucelio, 2020)
Trichloroacetic acid	Nafion/hemoglobin/B-GQDs/carbon ionic liquid electrode	CV	0.1–300 mM	0.053 mM	Food and drug samples	(Chen et al., 2018)
Triclosan	AgNPs/C$_3$N$_4$NTs@GQDs/ionic liquid/GCE	DPV	1.0×10^{-12}–1.0×10^{-8} M	2.0×10^{-12} M	Waste water samples	(Akyıldırım, 2020)
Trifluralin	Ag-citrate/GQDs nanoink/leaf	SWV	0.01–1 mM	10 µM	Apple skin samples	(Saadati, Hassanpour, & Hasanzadeh, 2020)
Trimethoprim	SiO$_2$-GOQDs/CPE	DPV	0.7–3.5 µM	0.191 µM	–	(Mikhraliieva et al., 2020)
17-α-ethynylestradiol	(mag@NIP)-GQDs-FG-NF/SPE	SWV	10 nmol/L–2.5 µmol/L	2.6 nmol/L	River water and serum and urine samples	(Santos et al., 2021)
2-nitroaniline	β-cyclodextrins-GQDs/GCE	DPV	0.06–200 µg/mL	0.0066 µg/mL	Pharmaceutical samples	(Dong, Bi, Qiao, Shao, & Lu, 2020)
3-nitroaniline	β-cyclodextrins-GQDs/GCE	DPV	0.06–200 µg/mL	0.1005 µg/mL	Pharmaceutical samples	(Dong et al., 2020)
4-nitroaniline	β-cyclodextrins-GQDs/GCE	DPV	0.06–200 µg/mL	0.0019 µg/mL	Pharmaceutical samples	(Dong et al., 2020)
4-aminophenol	GQDs/deep eutectic solvent/MWCNTs-COOH	DPV	0.050–100 µM	0.017 µM	Human fluid samples	(Arap et al., 2020)
24-HIV protein	GQD-SPE/aptamer	CV	0.93 ng/mL–93 mg/mL	51.7 pg/mL	Spiked human serum samples	(Gogola et al., 2021)

ABEI: Nanoluminophore N-(4-aminobutyl)-N-ethylisoluminol, CC: Carbon cloth, CPE: Carbon paste electrode, CSPE: Carbon screen-printed electrode, DPASV: Differential pulse adsorptive stripping voltammetry, F-Au: Flower-gold nanocomposite, FG: Functionalized graphene, FTO: Fluorine-doped tin oxide, GCEox: Activated GCE, GOOQDs: Graphene oxide quantum dots, GQD-SH: Thiol graphene quantum dots, GOx: Glucose oxidase, hBN: Two-dimensional hexagonal boron nitride, IL: Ionic liquid, ITO: Indium tin oxide, mag@MIP: Magnetic nanoparticles coated with molecularly imprinted polymers, MB: Methylene blue, MOF: Metal organic framework, MIP: Molecularly imprinted polymer, NMO: Nickel molybdate nanocomposite, NSGQD: Nitrogen sulfur-doped graphene quantum dots, PDA: Polydopamine, PEI: polyetherimide, PLL: Poly-L-lysine, SPCE: Screen-printed carbon electrode, SPE: Screen-printed electrode, SWASV: Square wave adsorptive stripping voltammetry, TvL: Trametes versicolor laccase.

TABLE 12.5 Some reported carbon quantum dot-based electrochemical sensors for the detection of different analytes.

Analyte	Nanosensor type	Method	Linear range	Limit of detection	Application type	Ref.
Adrenaline	CQDs/CPE	Chronoamperometry	0.02–0.8 μM 0.8–20 μM	0.006 μM	Pharmaceutical sample	(Shankar et al., 2019)
Ascorbic acid	CQDs-rGO/GCE	DPV	60–1000 μM 2000–7000 μM	3.33 μM	Fetal bovine serum samples	(Wei et al., 2020)
Ascorbic acid	NH$_2$-CQDs/GCE	DPV	0–1000 μM	2.7 μM	Vitamin C tablet samples	(Zhou et al., 2020)
Bisphenol A	Anti-bisphenol A/AuNPs/N,S,P-doped CDs/GCE	DPV	0.01–120 μM	0.53 nM	Plastic spoons, plastic lunch boxes, tap water, milk, and mineral water samples	(Yao, Liu, & Yang, 2019)
Cardiac troponin I	SnO$_2$/N-doped CQDs/BiOI/ITO	Photoelectrochemical	0.001–100 ng/mL	0.3 pg/mL	Human serum samples	(Fan et al., 2020)
Chlorpromazine	N-doped-CDs/Cu$_2$O/GCE	DPV	0.001–230 μM	25 nM	Human urine and pharmaceutical formulation samples	(Palakollu, Karpoormath, Wang, Tang, & Liu, 2020)
Cu(II)	Phosphorus-doped CQDs/GCE	Electrochemiluminescence	1–1000 nM	0.27 nM	Tap and drinking water samples	(Raju, Kalaiyarasan, Paramasivam, Joseph, & Kumar, 2020)
Dopamine	CQDs-rGO/GCE	DPV	1–10 μM 27–80 μM	0.0167 μM	Fetal bovine serum samples	(Wei et al., 2020)
Dopamine	GCE/CD	DPV	0.05×10^{-6}–2.0×10^{-6} M	4.6 nM	Human urine samples	(Canevari, Nakamura, Cincotto, Melo, & Toma, 2016)
Dopamine	CD/ionic liquid-graphene/GCE	DPV	0.1–600 μM	30 nM	Fetal bovine serum samples	(Zhuang, Wang, He, & Chen, 2016)
Dopamine	B-cyclodextrins/CQDs/GCE	Amperometry	4–220 μM	0.45 μM	Urine samples	(Chen et al., 2017)
Dopamine	rGO/CD/carbon fiber	DPV	0.1–100 μM	0.02 μM	—	(Fang, Xie, Wallace, & Xungai, 2017)
Epinephrine	Graphitic carbon nitride/N-doped CD composite	DPV	1.0×10^{-12}–1.0×10^{-9} M	3.0×10^{-13} M	Human urine samples	(Yola & Atar, 2019b)

Analyte	Electrode	Method	Linear range	LOD	Sample	Reference
Epinephrine	GCE/CD	DPV	0.05×10^{-6}–2.0×10^{-6} M	6.1 nM	Human urine samples	(Canevari et al., 2016)
Glucose	PDDA/CuO-C-dot on SPCE	Amperometry	0.5–2 mM 2–5 mM	0.2 mM	Human blood serum samples	(Sridara et al., 2020)
Glucose	GO/CQDs/gold nanoparticles/gold microdisk array electrode	Chronoamperometry	0.16–4.32 mM	17 µM	Commercial sweet wine samples	(Buk & Pemble, 2019)
Glucose	CQDs-AuNPs/GOx-Au electrode	Chronoamperometry	0.05–2.85 mM	17 µM	Human serum samples	(Buk, Pemble, & Twomey, 2019)
Hydrogen peroxide	Nitrogen-doped carbon/GCE	Amperometry	0.05 µM–2.25 mM	41 nM	Disinfectant samples	(Fu et al., 2018)
Hydrogen peroxide	AgNPs/CDs/GCE	Amperometry	0.2–27.0 µM	80 nM	Fetal bovine serum samples	(Jahanbakhshi & Habibi, 2016)
Hydrogen peroxide	CDs/MWCNTs/GCE	Amperometry	3.5 µM–30 mM	0.25 µM	Human serum samples	(Bai, Sun, & Jiang, 2016)
Hydrazine	Polypyrrole/CDs/prussian blue/single gold nanopore electrode	Amperometry	0.5–80 µM	0.18 µM	Human urine samples	(Chen et al., 2019)
Lysozyme	rGO/MWCNT/chitosan/CQD	DPV	20 fM–10 nM	3.7 fM	Spiked human and urine samples	(Rezai, Jamei, & Ensafi, 2018)
Mesalazine	Chitosan/CDs-hexadecyltrimethyl ammonium bromide/GCE	Amperometry	0.1–10.0 µM	0.05 µM	Human blood serum samples	(Jalali, Hassanvand, & Baratti, 2020)
Paracetamol	Nitrogen-doped carbon/GCE	DPV	0.5–600 µM	157 nM	Tablet formulations	(Fu et al., 2018)
Progesterone	CDs/GO	Photoelectrochemical	0.5–180 nM	0.17 nM	Human serum samples	(Zhu, Xu, Gao, Ji, & Zhang, 2020)
Tryptophan	B-cyclodextrins/CQDs/GCE	Amperometry	5–270 µM	0.50 µM	Urine samples	(Chen et al., 2017)
Uric acid	CQDs-rGO	DPV	20–90 µM 100–500 µM	1.33 µM	Fetal bovine serum samples	(Wei et al., 2020)
Uric acid	B-cyclodextrins/CQDs/GCE	Amperometry	0.3–200 µM	0.04 µM	Urine samples	(Chen et al., 2017)
Vitamin D2	CQD@Fe$_3$O$_4$-chitosan/ITO	DPV	10–100 ng/mL	–	–	(Sarkar, Dhiman, Sajwan, Sri, & Solanki, 2020)

GOx: Glucose oxidase, PDDA: Poly-(dimethyldiallylammonium chloride), SPCE: Screen-printed carbon electrode.

quasi-spherical onions. The studies state that the conductivity of CNOs with such specific structures is close to the metallic behavior and their catalytic activity generally depends on the high surface area. It is noteworthy that other advantages of CNOs are namely including high corrosion resistance, high surface to volume ratio, good thermal stability, excellent controlled size distribution, and cost effectiveness. Accordingly, they are used in various applications including catalysis, electronics, energy conversion and storage, sensors, and environmental applications (Sohouli et al., 2020).

Jiang et al. used nanodiamonds to accomplish the determination of bisphenol A with a low detection limit at 5 nM. The method was linear in the concentration range between 0.1 and 80 mM with an LOD down to a few nanomolars. Carbon-based sensors showed in this work are capable of determining bisphenol A at concentrations lower than the daily dose limits set by recent medical advice. The main feature of these three sensors is the evasion of electrode surface fouling, which is the main difficulty for electrochemical bisphenol A sensors. Jiang's work demonstrated different methods that are suitable for diverse bisphenol A determination situations, either developing a highly sensitive, selective, and reliable continuous bisphenol A monitoring system or an easy-to-use, low-cost bisphenol A sensor. Also, based on the nanodiamond electrodes, they can be applied for portable on-site continuous bisphenol monitoring. Additionally, nanocarbon electrodes for single-use rapid tests and home use can improve further strategies (Jiang, Santiago, & Foord, 2021) (Table 12.6).

A fentanyl electrochemical sensor based on a GCE modified with carbon nano-onions was developed by Sohouli et al. (Fig. 12.5). Electrochemical studies showed that the sensor is capable of voltammetric determination of trace concentrations of fentanyl at a working potential of 850 mV (vs. Ag/AgCl). DPV currents related to fentanyl determination were linear with an increase in fentanyl concentrations in a linear range between 1 µM and 60 µM with an LOD of 300 nM. Obtained experimental results indicated that using CNOs in the electrode design is very appealing due to their advantages including high surface area, good electrocatalytic activity, and excellent electrical conductivity (Sohouli et al., 2020).

Optical-based sensing systems using carbon nanomaterials

Although sensors based on electrochemical transduction mechanisms make up a large part of the literature, carbon nanomaterials are also frequently employed in the development of optical sensors because of their outstanding physicochemical properties. Optical sensors are usually divided into 5 main groups, namely colorimetry, fluorescence, surface plasmon resonance (SPR), chemiluminescence (CL), and surface-enhanced Raman scattering (SERS) (Liu et al., 2018). Analytical parameters of some selected carbon nanomaterial-based optical sensors are presented in Table 12.7 and discussed in this section. They were applied to determine various analytes, including drugs, ions, gases, pesticides, proteins, and enzymes. Among the studies conducted, fluorescence sensors are the most frequently employed. Graphene, graphene oxide (GO), CNTs, QDs obtained from different precursors, and fullerene were among the carbon materials utilized to establish optical sensing strategies.

Graphene is a kind of two-dimensional material composed of sp^2-hybridized carbon atoms. After its introduction in 2004 by Andre Geim and Konstantin Novoselov, who received the 2010 Nobel Prize, graphene has taken a great deal of attention. It has outstanding physicochemical properties, such as high mechanical stability, thermal conductivity, light absorptivity, surface plasmon resonance, and photoluminescence. Therefore, graphene and its derivatives, namely GO and reduced GO, have been frequently used in the optical sensor design (Deng, Tang, & Jiang, 2014; Zhao, Li, Zhou, & Zhang, 2016). Wang et al. developed a sensitive SPR biosensor for the sensitive determination of IgG, which was based on graphene oxide/silver-coated polymer cladding silica fiber (Wang & Wang, 2018). In the development of optical fiber SPR sensors, the fiber surface is coated with a gold film. On the other hand, it has some disadvantages, including inadequate biomolecule immobilization efficiency and the need for expensive systems for production. In their work, the authors used a cost-effective chemical method to produce a silver film on the surface of the fiber as an alternative. Compared to fiber coated with only silver film, approximately 15% improvement in sensitivity was achieved with graphene oxide/silver film-coated fiber due to excellent properties of graphene oxide, including large surface area, high carrier mobility, and superior surface plasmon resonance. Furthermore, since GO can interact with analytes through ionic, covalent, and non-covalent bonds, the extraction efficiency per unit area of IgG was improved. Under optimum conditions, LOD of the sensor was calculated to be 0.04 µg/mL and the obtained results showed applicability of the platform.

Some polymeric materials such as polyimides and polyesters are used for humidity sensing. It has been reported that the humidity-sensing properties of pure polymers can be enhanced by fabricating composites with other materials. In this manner, Hernaez et al. developed a fiber-optic lossy mode resonance sensor based on a thin film consisting of polyethylenimine and GO (Hernaez et al., 2019). The sensing region was obtained by depositing five layers of thin film on a SnO_2-sputtered fiber core with a layer-by-layer assembly. The developed sensor allowed humidity determination in the range of 10%–90% with a response time of 160 ms, proving the appropriateness of the proposed composite for observing quick humidity variations.

TABLE 12.6 Some reported other carbon structure-based electrochemical sensors for the detection of different analytes.

Analyte	Nanosensor type	Method	Linear range	Limit of detection	Application type	Ref.
Acetaminophen	GO-CB-poly(3,4 ethylenedioxythiophene)-poly (styrenesulfonate/GCE	SWV	0.9–7.0 μM	0.23 μM	Urine and human serum samples	(Wong, Santos, Silva, & Filho, 2018)
Caffeine	GO-CB-poly(3,4 ethylenedioxythiophene)-poly (styrenesulfonate/GCE	SWV	11–64 μM	3.4 μM	Urine and human serum samples	(Wong et al., 2018)
Casein	Graphene/CNF/gelatin methacryloyl	DPV	1×10^{-7}–1×10^{-6} g/mL	3.2×10^{-8} g/mL	—	(Jiang, Ge, et al., 2019)
Cd(II)	3D-printed polylactic acid material containing CB nanoparticles	SWASV	30–270 μg/L	2.9 μg/L	Urine and saliva samples	(Rocha, Squissato, Silva, Richter, & Munoz, 2020)
Cetirizine	CB-GCE	SWASV	4.97×10^{-7}–1.08×10^{-5} M	4.00×10^{-7} M	Spiked artificial cerebrospinal fluid and urine samples	(Lourencao, Silva, Santos, Ferreira, & Filho, 2017)
Codeine	ND-dihexadecyl phosphate/GCE	SWV	0.299–10.8 μM	54.5 μM	Commercial formulations and urine and human serum samples	(Simioni, Oliviera, et al., 2017)
Dopamine	CB-GCE	SWV	0.1–20.0 μM	0.06 μM	—	(Jiang, Nelson, Abda, & Foord, 2016)
Dopamine	ND-SPE	DPV	—	5.7×10^{-7} M	—	(Baccarin et al., 2019)
Dopamine	g-C$_3$N$_4$/nitrogen-doped carbon@GC/h-ATS/ITO	DPV	25–100 μM	0.01 μM	Human urine samples	(Krishnan et al., 2020)
D-Tryptophan	GCE modified with CB	DPV	1–1400 μM	0.33 μM	Tryptophan isomer mixture solution sample	(Kingsford, Zhang, Ma, Wu, & Zhu, 2019)
Fentanyl	GCE modified with CNOs	DPV	1–60 μM	300 nM	Human blood serum, ampule, and urine samples	(Sohouli et al., 2020)
Folic acid	GO-CB-poly(3,4 ethylenedioxythiophene)-poly (styrenesulfonate/GCE	SWV	5.0–31 μM	1.0 μM	Urine and human serum samples	(Wong et al., 2018)
Glucose	Co-Co$_3$O$_4$/CNT/CF	Amperometry	1.2–2290 μM	0.4 μM	Human blood serum samples	(Han, Miao, & Song, 2020)
Hydrogen peroxide	Myoglobin/mesoporous CF/GCE	Amperometry	1.0–80 μM	180 nM	Spiked fetal bovine serum samples	(Jahanbakhshi, 2018)
Hydroxyzine	CB-GCE	SWASV	2.99×10^{-7}–9.81×10^{-6} M	1.00×10^{-7} M	Spiked artificial cerebrospinal fluid and urine samples	(Lourencao et al., 2017)

Continued

TABLE 12.6 Some reported other carbon structure-based electrochemical sensors for the detection of different analytes—cont'd

Analyte	Nanosensor type	Method	Linear range	Limit of detection	Application type	Ref.
Isoproterenol	GO-CB-poly(3,4 ethylenedioxythiophene)-poly (styrenesulfonate)/GCE	SWV	8.0–50 μM	1.9 μM	Urine and human serum samples	(Wong et al., 2018)
L-Tryptophan	GCE modified with CB	DPV	1–1400 μM	0.45 μM	Tryptophan isomer mixture solution sample	(Kingsford et al., 2019)
Lercanidipine	CB: dihexadecylphosphate-GCE	DPV	—	0.058 μM	Spiked Milli-Q water and tap water samples	(Fernandes, Paquim, & Brett, 2020)
Pb(II)	3D-printed polylactic acid material containing CB nanoparticles	SWASV	30–270 μg/L	2.6 μg/L	Urine and saliva samples	(Rocha et al., 2020)
Propranolol	GO-CB-poly(3,4 ethylenedioxythiophene)-poly (styrenesulfonate)/GCE	SWV	0.5–2.9 μM	0.18 μM	Urine and human serum samples	(Wong et al., 2018)
Pyrazinamide	ND-GCE	SWV	7.9×10^{-7}–4.9×10^{-5} M	2.2×10^{-7} M	Pharmaceutical drug samples	(Simioni, Silva, & Oliveira, 2017)
Uric acid	ND-SPE	DPV		8.9×10^{-7} M		(Baccarin et al., 2019)
Uric acid	g-C_3N_4/nitrogen-doped carbon@GC/h-ATS/ITO	DPV	2.5–500 μM	0.06 μM	Human urine samples	(Krishnan et al., 2020)

CB: Carbon black, CF: Carbon foam, CNF: Carbon nanofiber, CNO: Carbon nano-onion, GF: Graphene foam, h-ATS: Ag-(TiO_2/SnO_2), ITO: Indium tin oxide, ND: Nanodiamond, SPE: Screen-printed electrode.

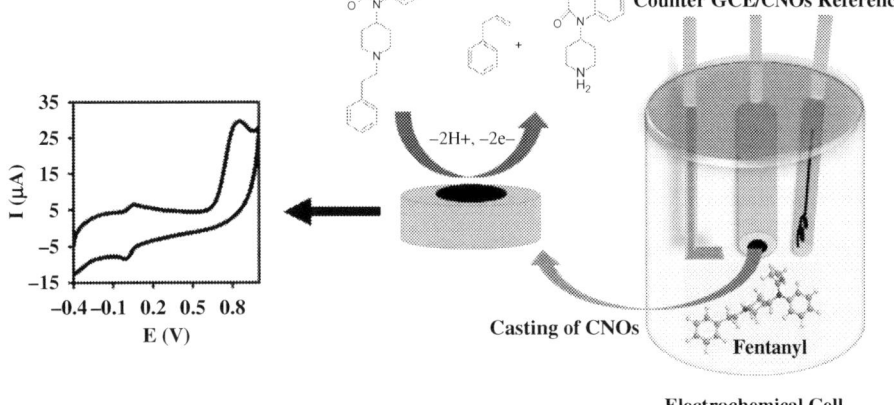

FIG. 12.5 The fabrication process of (A) CNOs from NDs and (B) electrochemical sensor for the fentanyl detection. *Reprinted with permission from Sohouli E, Keihan AH, Shahdost-Fard S, Naghian E, Plonska-Brzezinska M, Rahmi-Nasrabi M, Ahmadi F. A glassy carbon electrode modified with carbon nanoonions for electrochemical determination of fentanyl. Materials Science and Engineering 2020;110:110684.*

As the zero-dimensional carbon-based nanomaterials with diameters less than 10 nm, carbon dots play an important role in the optical sensor design due to their low toxicity, superior biocompatibility, and high quantum efficiency. CDs are generally classified into three groups in terms of the nature of the carbon precursor, core structure, and quantum effect, which are GQDs, CNDs, and CQDs (Ji, Zhou, Leblanc, & Peng, 2020; Li & Zhang, 2021; Lin et al., 2021). One of the most valuable properties of CDs is that they show fluorescence in the near IR and visible region. Additionally, their fluorescence properties can be tuned by changing the initial components and experimental conditions, which make them extremely suitable nanoprobes for biosensing. It has been reported that functionalization of QD surfaces with nitrogen can improve quantum yield, optical properties, reactivity, and catalytic activity. Consequently, there has been an increased interest in the synthesis and utilization of heteroatom-doped QDs. Jiang et al. employed chitin as cheap, abundant, and green biomass to produce nitrogen-doped quantum dots (N-CQDs) with a high quantum yield (Jiang, Jing, et al., 2020). Prepared QDs displayed a blue fluorescence in the visible region (433) when they were excited at 360 nm. Fluorescence intensity was higher in the acidic medium. Among the examined various metal ions and anions, synthesized QDs showed an excellent selectivity against ClO^-. A linear decrease in the fluorescence intensity of QDs was observed with the addition ClO^-. LOD of the developed method was found to be 1.47 μM, which was sufficient to monitor the ClO^- concentration in the water. Nair et al. synthesized sulfur-doped graphene quantum dots (S-GQDs) for the fluorescence sensing of carbamate pesticides, namely carbofuran and thiram (Nair et al., 2020). They used H_2SO_4 and $KMnO_4$ as the sulfur precursor and oxidizing agents, respectively. Sulfur doping led to achieving a relatively high quantum yield (27.8%). Compared with bare GQDs, emission wavelengths shifted from blue to green with S-GQDs, indicating the success of heteroatom doping in modifying the fluorescence features of QDs. The authors also produced poly(vinyl alcohol)/S-GQD composite films to construct a simple solid-state fluorescence sensor. Linear decreases in the response were observed for both sensors with the LODs in the range of 0.45–210 ppb, demonstrating the sensitive and ultrasensitive determination of analytes. The quenching mechanism for both sensors was proposed to be due to complexation followed by the photo-induced electron transfer.

TABLE 12.7 Some reported carbon nanomaterial-based optic sensors for the detection of different analytes.

Sensor type	Analyte	Matrix	Sensing materials	Linear range	LOD	Ref.
Fiber optic-based biosensor	C-reactive protein	Serum	Long period grating in double cladding fiber coated with GO	1 ng/mL–100 µg/mL	0.15 ng/mL	(Esposito et al., 2021)
Fiber optic Fabry-Perot sensor	Humidity	Air	Graphene QDs	11%–85%	—	(Wang et al., 2021)
Fiber optic-LMR sensor	Humidity	Air	Optical fiber coated with polyethylenimine and GO	20%–90%	—	(Hernaez, Acevedo, Mayes, & Melendi-Espina, 2019)
Fiber optic-SPR sensor	Nitrate	Standard solutions	Optical fiber coated with silver and nanocomposite of CNT/Cu-nanoparticles	1 µM–5 mM	4 nM	(Parveen, Pathak, & Gupta, 2017)
Fiber optic-SPR sensor	Dopamine	Synthetic cerebrospinal fluid	Optical fiber coated with permselective Nafion membrane and surface imprinted MWCNTs-PPy matrix	10^{-9}–10^{-5} M	18.9 pM	(Pathak & Gupta, 2019)
Fiber optic-SPR sensor	Human IgG	Standard solutions	GO/silver-coated polymer	5–30 µg/mL	0.04 µg/mL	(Wang & Wang, 2018)
Fiber optic-fluorescence sensor	Glucose	Standard solutions	Carbon QDs-glucose oxidase/cellulose acetate complex	10–200 µM	6.43 µM	(Yu, Ding, Lin, Wu, & Huang, 2019)
Fluorescence sensor	Fe^{3+}	Water	Nitrogen and chlorine co-doped CDs	0.3–50 µM	95.6 nM	(Li, Tang, et al., 2019)
Fluorescence sensor	Ag^{+}	River and tap water samples	Sulfur and nitrogen co-doped photoluminescence CDs	0.1–25 µM	37 nM	(Lu, Li, Wang, Wang, & Xu, 2019)
Fluorescence sensor	Fe^{3+} Pb^{2+} Hg^{2+}	Water	Carboxylate group-terminated CDs	0.4–100 µM for Fe^{3+} 100–600 µM for Pb^{2+} 20–200 µM for Hg^{2+}	2.8 µM for Fe^{3+} 7.2 µM for Pb^{2+} 5.7 µM for Hg^{2+}	(Li et al., 2017)
Fluorescence sensor	Sialic acid	Human serum	Boronic-acid-functionalized CDs	80–4000 µM	54 µM	(Xu et al., 2019)
Fluorescence sensor	ClO^{-}	Water	Carbon QDs	0–50 µM	1.47 µM	(Jiang, Jing, Ni, Gao, & Zhou, 2020)

Sensor type	Analyte	Sample	Sensing material	Linear range	LOD	Reference
Fluorescence sensor	Isotretinoin	Blood serum and pharmaceutical samples	Amino-functionalized carbon QDs	0.08–70.0 μM	0.03 μM	(Madrakian, Maleki, Gilak, & Afkhami, 2017)
Fluorescence sensor	Glutathione	Grape and cucumber	Dual-colored CD-encapsulated zeolitic imidazolate framework	3–25 nM	0.90 nM	(Jalili, Khataee, Rashidi, & Luque, 2017)
Fluorescence sensor	Nitrite	Water	Polythienothiophene-fullerene thin film	1.1–28 μM	<0.55 μM	(Pires, Dong, & Yang, 2019)
Fluorescence sensor	Urokinase plasminogen activator uPA	Blood	SWCNTs functionalized with ssDNA and anti-uPA antibody	–	50 nM	(Williams, Lee, & Heller, 2018)
Fluorescence sensor	HIV	Serum	DNA-CNT complexes	–	–	(Harvey, Baker, Ortiz, Kentsis, & Heller, 2019)
Fluorescence sensor	Carbamate Pesticides	Apple	Sulfur-doped graphene QDs	–	0.45 ppb for carbofuran and 1.6 ppb for thiram	(Nair, Thomas, Mohamed, & Pillai, 2020)
Fluorescence sensor	Glutathione	Cell lines	Graphene QDs wrapped square-plate-like MnO_2 nanocomposite	0.07–70 μM	48 nM	(Wang, Li, Wang, Dong, & Shuang, 2020)
NIR optical biosensor	2,4,6-trinitrotoluene	Standard solutions	Peptide-functionalized SWCNT hybrids	–	0.01 ppm	(Esposito et al., 2021)
NIR optical biosensor	Protease	Urine	Peptide-encapsulated SWCNTs	1–20 μg/mL	1 μg/mL	(Shumeiko, Paltiel, Bisker, Hayouka, & Shoseyov, 2020)
Optical emission spectrometry sensor	H_2	Air	Pt/CNT nanocomposite-decorated FTO electrode	0.17%–4.0% (v/v)	–	(Shen, Chen, Yan, & Zhang, 2015)
QCM sensor	Humidity	Air	Fullerene/GO nanocomposites	11%–97%	–	(Ding, Chen, Chen, Zhao, & Li, 2018)
SERS sensor	Cardiac troponin I	Serum	GO/gold nanoparticles	0.01–1000 ng/mL	5 pg/mL	(Fu et al., 2019)
SPR sensor	Dengue virus E-protein	Standard solutions	Cadmium sulfide QDs composited with amine-functionalized GO	Up to 0.01 nM	1 pM	(Omar et al., 2019)
TCF sensor	Humidity	Air	TCF coated with a CNT/PVA film	–	–	(Ma et al., 2018)

CDs: Carbon dots, FTO: Fluorine-doped tin oxide, GO: Graphene oxide, HIV: Human immunodeficiency virus, LMR: Lossy mode resonance, NIR: Near infrared, PVA: Polyvinyl alcohol, QCM: Quartz crystal microbalance, QDs: Quantum dots, SERS: Surface-enhanced Raman scattering, SPR: Surface plasmon resonance, TCF: Thin core fiber, uPA: Urokinase plasminogen activator.

As a different type of nanomaterials, CNTs are often preferred in SPR and near-infrared sensor design (Table 12.7). They have several outstanding features such as high surface area, electrical conductivity, rapid response, and adequate chemical stability. CNTs produce fast reactions following the interaction with target analytes by charge transformation. Furthermore, improved selectivity and sensitivity can be achieved by surface modification with other substances, e.g., polymers, metallic nanoparticles, and biological substances. Parveen et al. proposed an SPR sensor employing a fiber optic probe coated with silver and nanocomposite of CNT/Cu-nanoparticles for the sensitive determination of nitrate (Parveen et al., 2017). The measurements were based on the reduction of nitrate and formation of ammonium ions following the contact with Cu-nanoparticles. The ammonium ions formed were adsorbed on the probe surface and caused a shift in the resonance wavelength. The total analysis time was 30 s, indicating the high throughput of the method. LOD was found to be 4 nM, which was lower than the most reported in the literature. Wang(Wang, 2018) proposed a NIR optical biosensor for the determination of 2,4,6-trinitrotoluene (TNT). The sensing mechanism was based on TNT-binding peptide-functionalized single-walled CNTs. Following the interaction between TNT and the TNT recognition peptide-based SWCNT hybrids via π-electron-mediated effects and hydrogen bond, the shift in the NIR spectra was examined. The developed biosensor was found to be highly selective and sensitive for the determination of TNT.

Conclusion and future prospects

Carbon nanomaterial-based sensors enable a selective, sensitive, repeatable, and reliable analysis of many analytes at a very sensitive level on wide sensor platforms using bare or modified carbon nanomaterials with unique properties together with the developing technology. Hereby, we conducted a comprehensive chapter on nanocarbon-based sensors which includes single and multiwalled carbon nanotubes, graphene, pyrolytic graphite, fullerenes, carbon dots, graphene dots, boron-doped diamond, carbon black, carbon nano-onions, and carbon fibers. With the development of nanotechnology in recent years, incredible sensors based on carbonaceous materials with extraordinary properties, such as selective, sensitive, low cost, environmentally friendly, miniature, and portable, have been manufactured. We focused on the application of these nanocarbon materials in electrochemical and optic sensors with the aim of detecting different organic and inorganic compounds such as drugs, heavy metals, vitamins, antioxidants, explosives, nucleic acids, pesticides, polycyclic aromatic hydrocarbons, organochlorine compounds, etc. Based on the reported literature, especially bioassay applications of nanocarbons have introduced a major way to meet these increasing demands.

Using nanoengineering, nanocarbon molecules can be transformed into various forms and these materials demonstrate high chemical and physical stability, excellent phosphor luminescence characteristics, and low cytotoxicity, making them a perfect nominee for chemical and biochemical sensing and imaging implementation. Also, to improve the electrochemical features of nanocarbon electrodes, impressive developments were made in the application such as doping nanocarbon structures with heteroatoms or modifying with other nanomaterials such as Au, Ag, Pt, Pd, etc., metal oxide nanomaterials (CeO_2, CuO, NiO, Fe_2O_3, Co_3O_4, ZnO, SnO_2, TiO_2, etc.), transition metal dichalcogenides (MoS_2, etc.), polymer nanomaterials (polyaniline, polypyrrole, poly(3,4-ethylenedioxythiophene, etc.)) (Zhang et al., 2021). Additionally, carbonaceous materials can be accepted as an alternative to the currently high-cost electronic materials, being considered as environment-friendly. For this reason, carbon structures have been researched to be used in powerful sensor devices since they present excellent chemical and physical characteristics yielding high-quality sensing properties.

References

Abd-Rabboh, H. S. M., & Kamel, A. H. (2020). Novel potentiometric screen-printed carbon electrodes for bisphenol S detection in commercial plastic samples. *Analytical Sciences, 36*(11), 1359–1364. https://doi.org/10.2116/Analsci.20P143.

Abdel-Haleem, F. M., Gamal, E., Rizk, M. S., El Nashar, R. M., Anis, B., Elnabawy, H. M., et al. (2020). T-butyl calixarene/Fe_2O_3@MWCNTS composite-based potentiometric sensor for determination of ivabradine hydrochloride in pharmaceutical formulations. *Materials Science and Engineering: C, 116*, 111110. https://doi.org/10.1016/j.msec.2020.111110.

Abdellatef, R., Khaled, E., Hendawy, H. A., & Hassan, R. Y. A. (2021). Manganese dioxide (MnO_2)/fullerene-C_{60}-modified electrodes for the voltammetric determination of rifaximin. *Journal of Analysis and Testing.* https://doi.org/10.1007/s41664-020-00151-y.

Afrasiabi, M., Kianipour, S., Babaei, A., Nasimi, A. A., & Shabanian, M. (2016). A new sensor based on glassy carbon electrode modified with nanocomposite for simultaneous determination of acetaminophen, ascorbic acid and uric acid. *Journal of Saudi Chemical Society, 20*, S480–S487. https://doi.org/10.1016/J.Jscs.2013.02.002.

Aftab, S., Ozcelikay, G., Kurbanoglu, S., Shah, A., Iftikhar, F. J., & Ozkan, S. A. (2019). A novel electrochemical nanosensor based on NH_2-functionalized multi walled carbon nanotubes for the determination of catechol-orto-methyltransferase inhibitor entacapone. *Journal of Pharmaceutical and Biomedical Analysis, 165*, 73–81. https://doi.org/10.1016/J.Jpba.2018.11.050.

Aftab, S., Kurbanoglu, S., Ozcelikay, G., Shah, A., & Ozkan, S. A. (2019). NH$_2$-functionalized multi walled carbon nanotubes decorated with ZnO nanoparticles and graphene quantum dots for sensitive assay of pimozide. *Electroanalysis, 31*(6), 1083–1094. https://doi.org/10.1002/elan.201800788.

Ahmad, M. S., Isa, I. M., Hashim, N., Saidin, M. I., Si, S. M., Zainul, R., et al. (2019). Zinc layered hydroxide-sodium dodecyl sulphate-isoprocarb modified multiwalled carbon nanotubes as sensor for electrochemical determination of dopamine in alkaline medium. *International Journal of Electrochemical Science, 14*, 9080–9091.

Ahour, F., & Ahsani, M. K. (2016). An electrochemical label-free and sensitive thrombin aptasensor based on graphene oxide modified pencil graphite electrode. *Biosensors & Bioelectronics, 86*, 764–769.

Ai, Y., Gao, N., Wang, Q., Gao, F., Hibbert, D. B., & Zhao, C. (2020). Electrosynthesis of HKUST-1 on a carbon-nanotube-modified electrode and its application for detection of dihydroxybenzene isomers. *Journal of Electroanalytical Chemistry, 872*. https://doi.org/10.1016/J.Jelechem.2020.114161, 114161.

Akhoundian, M., Alizadeh, T., Ganjali, M. R., & Norouzi, P. (2019). Ultra-trace detection of methamphetamine in biological samples using FFT-square wave voltammetry and nano-sized imprinted polymer/MWCNTS-modified electrode. *Talanta, 200*, 115–123. https://doi.org/10.1016/j.talanta.2019.02.027.

Akyıldırım, O. (2020). A sensitive voltammetric sensor based on silver nanoparticles/carbon nitride nanotubes@graphene quantum dots/a novel organic liquid: Determination of triclosan in wastewater. *Bulletin of Materials Science, 43*, 195.

Ali, T. A., & Akl, Z. F. (2021). Ionic liquid-multi-walled carbon nanotubes modified screen-printed electrodes for sensitive electrochemical sensing of uranium. *Journal of Radioanalytical and Nuclear Chemistry, 328*(1), 267–276. https://doi.org/10.1007/S10967-020-07573-Z.

Ali, M. R., Bacchu, M. S., Al-Mamun, M. R., Rahman, M. M., Ahommed, M. S., Aly, M. A. S., et al. (2021). Sensitive MWCNT/P-Cys@MIP sensor for selective electrochemical detection of ceftizoxime. *Journal of Materials Science, 56*(22), 12803–12813. https://doi.org/10.1007/S10853-021-06115-6.

Alizadeh, T., Nayeri, S., & Mirzaee, S. (2019). A high performance potentiometric sensor for lactic acid determination based on molecularly imprinted polymer/MWCNTS/PVC nanocomposite film covered carbon rod electrode. *Talanta, 192*, 103–111. https://doi.org/10.1016/j.talanta.2018.08.027.

Anand, S. K., Mathew, M. R., Radecki, J., Radecka, H., & Kumar, K. G. (2020). Individual and simultaneous voltammetric sensing of norepinephrine and tyramine based on poly(L-arginine)/reduced graphene oxide composite film modified glassy carbon electrode. *Journal of Electroanalytical Chemistry, 878*, 114531.

Antherjanam, S., Saraswathyamma, B., Krishnan, R. G., & Gopakumar, G. M. (2021). Electrochemical sensors as a versatile tool for the quantitative analysis of vitamin B$_{12}$. *Chemical Papers, 75*, 2981–2995.

Aoun, S. B. (2017). Nanostructured carbon electrode modified with N-doped graphene quantum dots-chitosan nanocomposite: A sensitive electrochemical dopamine sensor. *Royal Society Open Science, 4*, 171199.

Arap, N., Fotouhi, L., Salis, A., & Dorraji, P. S. (2020). An amplified electrochemical sensor employing a polymeric film and graphene quantum dots/multiwall carbon nanotubes in a deep eutectic solvent for sensitive analysis of paracematol and 4-aminophenol. *New Journal of Chemistry, 44*, 15742–15751.

Araújo, D. A. G., Camargo, J. R., La, P.-F., Lima, A. P., RAA, M., Takeuchi, R. M., et al. (2020). A lab-made screen-printed electrode as a platform to study the effect of the size and functionalization of carbon nanotubes on the voltammetric determination of caffeic acid. *Microchemical Journal, 158*, 105297. https://doi.org/10.1016/J.Microc.2020.105297.

Arvand, M., & Pourhabib, A. (2019). Surfactant-assisted voltammetric determination of olanzapine at amine functionalized TiO$_2$/multi-walled carbon nanotubes nanocomposite. *Journal of Analytical Chemistry, 74*(11), 1096–1103. https://doi.org/10.1134/s1061934819110030.

Asadpour Joghani, R., Abbas Rafati, A., Ghodsi, J., Assari, P., & Feizollahi, A. (2020). First report for levodopa electrocatalytic oxidation based on copper metal-organic framework (MOF): Application in a voltammetric sensor development for levodopa in real samples. *ChemistrySelect, 5*(28), 8532–8539. https://doi.org/10.1002/slct.202001781.

Assari, P., Rafati, A. A., Feizollahi, A., & Joghani, R. A. (2020). Fabrication of a sensitive label free electrochemical immunosensor for detection of prostate specific antigen using functionalized multi-walled carbon nanotubes/polyaniline/AuNps. *Materials Science and Engineering: C, 115*. https://doi.org/10.1016/j.msec.2020.111066, 111066.

Ayhan, E. A., & İnam, R. (2020). Square wave stripping voltammetric determination of cyprodinil fungicide in food samples by nanostructured multi walled carbon nanotube paste electrode. *Journal of Food Measurement and Characterization, 14*(3), 1333–1343. https://doi.org/10.1007/S11694-020-00381-9.

Baccarin, M., Neale, S. J. R., Cavalheiro, E. T. G., Smith, G. C., & Banks, C. E. (2019). Nanodiamond based surface modified scree-printed electrodes for the simultaneous voltammetric determination of dopamine and uric acid. *Microchimica Acta, 186*, 200.

Baghayeri, M., Amiri, A., Fayazi, M., Nodehi, M., & Esmaeelnia, A. (2021). Electrochemical detection of bisphenol a on a MWCNTs/CuFe$_2$O$_4$ nanocomposite modified glassy carbon electrode. *Materials Chemistry and Physics, 261*. https://doi.org/10.1016/j.matchemphys.2021.124247, 124247.

Bagheri, H., Khoshsafar, H., Afkhami, A., & Amidi, S. (2016). Sensitive and simple simultaneous determination of morphine and codeine using a Zn$_2$SnO$_4$ nanoparticle/graphene composite modified electrochemical sensor. *New Journal of Chemistry, 40*, 7102–7112.

Bai, L., Chen, Y., Bai, Y., Chen, Y., Zhou, J., & Huang, A. (2017). Fullerene-doped polyaniline as new redox nanoprobe and catalyst in electrochemical aptasensor for ultrasensitive detection of mycobacterium tuberculosis MPT64 antigen in human serum. *Biomaterials, 133*, 11–19. https://doi.org/10.1016/j.biomaterials.2017.04.010.

Bai, L., Chen, Y., Liu, X., Zhou, J., Cao, J., Hou, L., et al. (2019). Ultrasensitive electrochemical detection of mycobacterium tuberculosis IS6110 fragment using gold nanoparticles decorated fullerene nanoparticles/nitrogen-doped graphene nanosheet as signal tags. *Analytica Chimica Acta, 1080*, 75–83. https://doi.org/10.1016/j.aca.2019.06.043.

Bai, J., Sun, C., & Jiang, X. (2016). Carbon dots-decorated multiwalled carbon nanotubes nanocomposites as a high-performance electrochemical sensor for detection of H_2O_2 in living cells. *Analytical and Bioanalytical Chemistry, 408*, 4705–4714.

Barathi, P., Thirumalraj, B., Chen, S. M., & Angaiah, S. (2019). A simple and flexible enzymatic glucose biosensor using chitosan entrapped mesoporous carbon nanocomposite. *Microchemical Journal, 147*, 848–856.

Beitollahi, H., Tajik, S., Aflatoonian, M. R., Nejad, F. G., Zhang, K., Asl, M. S., et al. (2020). A novel screen-printed electrode modified by graphene nanocomposite for detecting clozapine. *International Journal of Electrochemical Science, 15*, 9271–9281.

Benvidi, A., Saucedo, N. M., Ramnani, P., Villarreal, C., Mulchandani, A., Tezerjani, M. D., et al. (2018). Electro-oxidized monolayer CVD graphene film transducer for ultrasensitive impedimetric DNA biosensor. *Electroanalysis, 30*, 1791–1800.

Bibi, S., Zaman, M. I., Niaz, A., Rahim, A., Nawaz, M., & Bilal, A. M. (2019). Voltammetric determination of nitrite by using a multiwalled carbon nanotube paste electrode modified with chitosan-functionalized silver nanoparticles. *Microchimica Acta, 186*(9), 595. https://doi.org/10.1007/s00604-019-3699-8.

Bohari, N. A., Siddiquee, S., Saallah, S., Misson, M., & Arshad, S. E. (2021). Electrochemical behaviour of real-time sensor for determination mercury in cosmetic products based on PANI/MWCNTS/AuNPs/ITO. *Cosmetics, 8*(1), 17. https://doi.org/10.3390/cosmetics8010017.

Boonkaew, S., Jang, I., Noviana, E., Siangproh, W., Chailapakul, O., & Henry, C. S. (2021). Electrochemical paper-based analytical device for multiplexed, point-of-care detection of cardiovacular disease biomarkers. *Sensors and Actuators B: Chemical, 330*, 129336.

Brahman, P. K., Suresh, L., Lokesh, V., & Nizamuddin, S. (2016). Fabrication of highly sensitive and selective nanocomposite film based on CuNPs/fullerene-C_{60}/MWCNTs: An electrochemical nanosensor for trace recognition of paracetamol. *Analytica Chimica Acta, 917*, 107–116. https://doi.org/10.1016/j.aca.2016.02.044.

Brodie, B. C. (1859). On the atomic weight of graphite. *Philosophical Transactions of the Royal Society, 149*, 249.

Buk, V., & Pemble, M. E. (2019). A highly sensitive glucose biosensor based on a micro disk array electrode desing modified with carbon quantum dots and gold nanoparticles. *Electrochimica Acta, 298*, 97–105.

Buk, V., Pemble, M. E., & Twomey, K. (2019). Fabrication and evaluation of a carbon quantum dot/gold nanoparticle nanohybrid material intagrated onto planar micro gold electrodes for potential bioelectrochemical sensing applications. *Electrochimica Acta, 293*, 307–317.

Canevari, T. C., Nakamura, M., Cincotto, F. H., Melo, F. M., & Toma, H. E. (2016). High performance electrochemical sensor for dopamine and epinephrine using naoncrystalline carbon quantum dots obtanied under controlled chronoamperometric conditions. *Electrochimica Acta, 209*, 464–470.

Cao, M., Zheng, L., Gu, Y., Wang, Y., Zhang, H., & Xu, X. (2020). Electrostatic self-assembly to fabricate ZnO nanoparticles/reduced graphene oxide composites for hypersensitivity detection of dopamine. *Microchemical Journal, 159*, 105465.

Capelari, T. B., De Cássia, M. J., Da Rocha, L. R., et al. (2021). Synthesis of novel poly(methacrylic acid)/β-cyclodextrin dual grafted MWCNT-based nanocomposite and its use as electrochemical sensing platform for highly selective determination of cocaine. *Journal of Electroanalytical Chemistry, 880*. https://doi.org/10.1016/J.Jelechem.2020.114791, 114791.

Chauhan, R., Gill, A. A. S., Nate, Z., & Karpoormath, R. (2020). Highly selective electrochemical detection of ciprofloxacin using reduced graphene oxide/poly(phenol red) modified glassy carbon electrode. *Journal of Electroanalytical Chemistry, 871*, 114254.

Chen, J., He, P., Bai, H., He, S., Zhang, T., Zhang, X., et al. (2017). Poly(β-cylodextrin)/carbon quantum dots modified glassy carbon electrode: Preparation, characterization and simultaneous electrochemical determination of dopamine, uric acid and tryptophan. *Sensors and Actuators B: Chemical, 252*, 9–16.

Chen, W., Weng, W., Niu, X., Li, X., Men, Y., Sun, W., et al. (2018). Boron-doped graphene quantum dots modified electrode for electrochemistry and electrocatalysis of hemoglobin. *Journal of Electroanalytical Chemistry, 823*, 137–145.

Chen, W., Wang, H., Tang, H., Yang, C., Guan, X., & Li, Y. (2019). Amperometric sensing of hydrazine by using single gold nanopore electrodes filled with prusian blue and coated with polypyrrole and carbon dots. *Microchimica Acta, 186*, 350.

Chen, C., Zhang, M., Li, C., Xie, Y., & Fei, J. (2018). Switched voltammetric determination of ractopamine by using a temperature-responsive sensing film. *Microchimica Acta, 185*(2), 155. https://doi.org/10.1007/s00604-018-2680-2.

Chetankumar, K., Swamy, B. E., & Naik, H. S. B. (2021). MgO and MWCNTs amplified electrochemical sensor for guanine, adenine and epinephrine. *Materials Chemistry and Physics, 267*, 124610.

Chihava, R., Apath, D., Moyo, M., Shumba, M., Chista, V., & Tshuma, P. (2020). One-pot synthesized nickel-cobalt sulfide-decorated graphene quantum dot composite for simultaneous electrochemical determination of antiretroviral drugs: Lamivudine and tenofovir disoproxil fumarate. *Journal of Sensors*, 3124102.

Chikere, C. O., Faisal, N. H., Lin, P. K. T., & Fernandez, C. (2019). The synergistic effect between graphene oxide nanocolloids and silicon dioxide nanoparticles for gallic acid sensing. *Journal of Solid State Electrochemistry, 23*, 1795–1809.

Chiwunze, T. E., Palakollu, V. N., Gill, A. A. S., Kayamba, F., Thapliyal, N. B., & Karpoormath, R. (2019). A highly dispersed multi-walled carbon nanotubes and poly(methyl orange) based electrochemical sensor for the determination of an anti-malarial drug: Amodiaquine. *Materials Science and Engineering: C, 97*, 285–292. https://doi.org/10.1016/J.Msec.2018.12.018.

Chou, J. C., Lai, T. Y., Lin, S. H., Kuo, P. Y., Lai, C. H., Nien, Y. H., et al. (2021). Characteristics and stability of a flexible arrayed uric acid biosensor based on NiO film modified by graphene and magnetic beads. *IEEE Sensors Journal, 21*, 7218–7225.

Cittan, M., & Çelik, A. (2019). An electrochemical sensing platform for trace analysis of eriochrome black T using multi-walled carbon nanotube modified glassy carbon electrode by adsorptive stripping linear sweep voltammetry. *International Journal of Environmental Analytical Chemistry, 99*(15), 1540–1552. https://doi.org/10.1080/03067319.2019.1625342.

Cittan, M., Altuntaş, E., & Çelik, A. (2020). Multi-walled carbon nanotube modified glassy carbon electrode as curcumin sensor. *Monatshefte für Chemie - Chemical Monthly, 151*(6), 881–888. https://doi.org/10.1007/S00706-020-02615-4.

Das, T. R., & Sharma, P. K. (2019). Sensitive and selective electrochemical detection of Cd^{2+} by using bimetal oxide decorated graphene oxide (Bi_2O_3/Fe_2O_3@GO) electrode. *Microchemical Journal, 147,* 1203–1214.

De Cássia, M. J., Da Rocha, L. R., Capelari, T. B., et al. (2020). Design and performance of novel molecularly imprinted biomimetic adsorbent for pre-concentration of prostate cancer biomarker coupled to electrochemical determination by using multi-walled carbon nanotubes/nafion®/Ni$(OH)_2$-modified screen-printed electrode. *Journal of Electroanalytical Chemistry, 878.* https://doi.org/10.1016/j.jelechem.2020.114582, 114582.

Dehdashti, A., & Babaei, A. (2020). Designing and characterization of a novel sensing platform based on Pt doped NiO/MWCNTs nanocomposite for enhanced electrochemical determination of epinephrine and tramadol simultaneously. *Journal of Electroanalytical Chemistry, 862.* https://doi.org/10.1016/J.Jelechem.2020.113949, 113949.

Dehnavi, A., & Soleymanpour, A. (2021). Titanium dioxide/multi-walled carbon nanotubes composite modified pencil graphite sensor for sensitive voltammetric determination of propranolol in real samples. *Electroanalysis, 33*(2), 355–364. https://doi.org/10.1002/elan.202060132.

Demir, E., Bozal-Palabıyık, B., Uslu, B., & Inam, R. (2019). Voltammetric determination of Vardenafil on modified electrodes constructed by graphite, metal oxides and functionalized multi-walled carbon nanotubes. *Revue Roumaine de Chimie, 64*(1), 45–54. https://doi.org/10.33224/rrch.2019.64.1.04.

Demir, E., Goktug, Z., İnam, R., & Doyduk, D. (2021). Development and characterization of iron (III) phthalocyanine modified carbon nanotube paste electrodes and application for determination of fluometuron herbicide as an electrochemical sensor. *Journal of Electroanalytical Chemistry, 895.* https://doi.org/10.1016/j.jelechem.2021.115389, 115389.

Demirbakan, B., & Sezgintürk, M. K. (2016). A novel immunosensor based on fullerene C_{60} for electrochemical analysis of heat shock protein 70. *Journal of Electroanalytical Chemistry, 783,* 201–207. https://doi.org/10.1016/j.jelechem.2016.11.020.

Deng, X., Tang, H., & Jiang, J. (2014). Recent progress in graphene-material-based optical sensors. *Analytical and Bioanalytical Chemistry, 406,* 6903–6916. https://doi.org/10.1007/s00216-014-7895-4.

Devi, C. L., & Narayanan, S. S. (2019). Poly(amido amine) dendrimer and silver nanoparticle-multi-walled carbon nanotubes composite with poly(neutral red)-modified electrode for the determination of ascorbic acid. *Bulletin of Materials Science, 42*(2), 73. https://doi.org/10.1007/S12034-019-1775-7.

Ding, X., Chen, X., Chen, X., Zhao, X., & Li, N. (2018). A QCM humidity sensor based on fullerene/graphene oxide nanocomposites with high quality factor. *Sensors and Actuators B: Chemical, 266,* 534–552. https://doi.org/10.1016/j.snb.2018.03.143.

DiVincenzo, D. P., & Mele, E. (1984). Self-consistent effective-mass theory for intralayer screening in graphite intercalation compounds. *Physical Review B, 29,* 1685–1694.

Dogan Topal, B., Bakirhan, N. K., Tok, T. T., & Ozkan, S. A. (2020). Electrochemical determination and in silico studies of fludarabine on NH_2 functionalized multiwalled carbon nanotube modified glassy carbon electrode. *Electroanalysis, 32*(1), 37–49. https://doi.org/10.1002/Elan.201900347.

Dong, S., Bi, Q., Qiao, J., Shao, S., & Lu, X. (2020). Simultaneous determination of three nitroaniline isomers by β-cyclodextrins (β-CDs) and graphene quantum dots (GQDs) composite modified glassy carbon electrodes. *International Journal of Electrochemical Science, 15,* 8552–8562.

Donini, C. A., Silva, M. K. L., Bronzato, G. R., Leao, A. L., & Cesarino, I. (2020). Evaluation of a biosensor based on reduced graphene oxide and glucose oxidase enzyme on the monitoring of second-generation ethanol production. *Journal of Solid State Electrochemistry, 24,* 2011–2018.

Dorraji, P. S., Noori, M., & Fotouhi, L. (2019). Voltammetric determination of adefovir dipivoxil by using a nanocomposite prepared from molecularly imprinted poly(o-phenylenediamine), multi-walled carbon nanotubes and carbon nitride. *Microchimica Acta, 186*(7), 427. https://doi.org/10.1007/S00604-019-3538-Y.

Dou, N., & Qu, J. (2020). Fast synthesis of copper-nickel bimetal and reduced graphene hybrid nanomaterial for sensitive sensing of 4-aminophenol and acetaminophen simultaneously. *Journal of the Electrochemical Society, 167*(13), 136513.

Ehtesabi, H. (2020). Application of carbon nanomaterials in human virus detection. *Journal of Science: Advanced Materials and Devices, 5,* 436–450.

El Attar, R., Hassan, H., & Khaled, E. (2020). Novel electroanalytical technique for determination of trospium hydrochloride. *Egyptian Journal of Chemistry, 63*(10), 3915–3923. https://doi.org/10.21608/ejchem.2020.23769.2428.

El Badry, M. M., Frag, E. Y., & Gamal Eldin, Y. M. (2020). A validated potentiometric method for determination of alfuzosin hydrochloride in pharmaceutical and biological fluid samples. *Journal of the Iranian Chemical Society, 17*(9), 2257–2265. https://doi.org/10.1007/S13738-020-01922-1.

Ertuğrul Uygun, H. D., & Demir, M. N. (2020). A novel fullerene-pyrrole-pyrrole-3-carboxylic acid nanocomposite modified molecularly imprinted impedimetric sensor for dopamine determination in urine. *Electroanalysis, 32*(9), 1971–1976. https://doi.org/10.1002/elan.202060023.

Esposito, F., Sansone, L., Srivastava, A., Baldini, F., Campopiano, S., Chiavaioli, F., et al. (2021). Long period grating in double cladding fiber coated with graphene oxide as high-performance optical platform for biosensing. *Biosensors & Bioelectronics, 172.* https://doi.org/10.1016/j.bios.2020.112747, 112747.

Essam, H. M., Bassuoni, Y. F., Elzanfaly, E. S., Zaazaa, H. E. S., & Kelani, K. M. (2020). Potentiometric sensing platform for selective determination and monitoring of codeine phosphate in presence of ibuprofen in pharmaceutical and biological matrices. *Microchemical Journal, 159.* https://doi.org/10.1016/J.Microc.2020.105286, 105286.

Essousi, H., Barhoumi, H., Karastogianni, S., & Girousi, S. T. (2020). An electrochemical sensor based on reduced graphene oxide, gold nanoparticles and molecular imprinted over oxidized polypyrrole for amoxicillin determination. *Electroanalysis, 32,* 1546–1558.

Eteya, M. M., Rounaghi, G. H., & Deiminiat, B. (2019). Fabrication of a new electrochemical sensor based on Au Pt bimetallic nanoparticles decorated multi-walled carbon nanotubes for determination of diclofenac. *Microchemical Journal, 144,* 254–260. https://doi.org/10.1016/J.Microc.2018.09.009.

Evtugyn, G., Porfireva, A., Shamagsumova, R., & Hianik, T. (2020). Advanced in electrochemical aptasensors based on carbon nanomaterials. *Chemosensors, 8*(4), 96.

Fakhrabad, A. H., Khoshnood, R. S., Ebrahimi, M., & Abedi, M. R. (2020). Fabrication of carbon paste electrodes modified with multi-walled carbon nanotubes for the potentiometric determination of chromium(III). *Journal of Analytical Chemistry, 75*(7), 951–957. https://doi.org/10.1134/S1061934820070072.

Fan, D., Liu, X., Shao, X., Zhang, Y., Zhang, N., Wang, X., et al. (2020). A cardiac troponin I photoelectrochemical immunosensor: Nitrogen-doped carbon quantum dots-bismuth oxyiodide-flower-like SnO_2. *Microchimica Acta, 187,* 187–332.

Fang, J., Xie, Z., Wallace, G., & Xungai, W. (2017). Co-deposition of carbon dots and reduced graphene oxide nanosheets on carbon-fiber microelectrode surface for selective detection of dopamine. *Applied Surface Science, 412,* 131–137.

Farag, A. S., Bakirhan, N. K., Švancara, I., & Ozkan, S. A. (2019). A new sensing platform based on NH_2-f-MWCNTs for the determination of antiarrhythmic drug propafenone in pharmaceutical dosage forms. *Journal of Pharmaceutical and Biomedical Analysis, 174,* 534–540. https://doi.org/10.1016/j.jpba.2019.06.026.

Fayed, A. S., Youssif, R. M., Salama, N. N., Elzanfaly, E. S., & Hendawy, H. A. M. (2021). Ultra-sensitive stripping SWV for determination of ertapenem via ZnONPs/MWCNT/CP sensor: Greenness assessment. *Microchemical Journal, 162.* https://doi.org/10.1016/J.Microc.2020.105752, 105752.

Fekry, A. M., Abdel-Gawad, S. A., Tammam, R. H., & Zayed, M. A. (2020). An electrochemical sensor for creatinine based on carbon nanotubes/folic acid/silver nanoparticles modified electrode. *Measurement, 163.* https://doi.org/10.1016/J.Measurement.2020.107958, 107958.

Feng, J., Deng, P., Xiao, J., et al. (2021). New voltammetric method for determination of tyrosine in foodstuffs using an oxygen-functionalized multiwalled carbon nanotubes modified acetylene black paste electrode. *Journal of Food Composition and Analysis, 96.* https://doi.org/10.1016/j.jfca.2020.103708, 103708.

Fernandes, I. G., Paquim, A. M. C., & Brett, A. M. O. (2020). Calcium channel blocker lercanidipine electrochemistry using a carbon black-modified glassy carbon electrode. *Analytical and Bioanalytical Chemistry, 412,* 6381–6389.

Frag, E., Mohamed, M., Ali, A., & Mohamed, G. (2020). Potentiometric sensors selective for Cu(II) determination in real water samples and biological fluids based on graphene and multi-walled carbon nanotubes modified graphite electrodes. *Indian Journal of Chemistry -Section A, 59*(02), 162–173.

Frangu, A., Ashrafi, A. M., Sýs, M., et al. (2020). Determination of trolox equivalent antioxidant capacity in berries using amperometric tyrosinase biosensor based on multi-walled carbon nanotubes. *Applied Sciences, 10*(7), 2497. https://doi.org/10.3390/app10072497.

Fu, L., Wang, A., Lai, G., Te-Lin, C., Yu, J., Yu, A., et al. (2018). A glassy carbon electrode modified with N-doped carbon dots for improved detection of hydrogen peroxide and paracetamol. *Microchimica Acta, 185,* 87.

Fu, X., Wang, Y., Liu, Y., Liu, H., Fu, L., Wen, J., et al. (2019). A graphene oxide/gold nanoparticle-based amplification method for SERS immunoassay of cardiac troponin I. *Analyst, 144,* 1582–1589. https://doi.org/10.1039/c8an02022a.

Fu, B., Chen, H., Yan, Z., Zhang, Z., Chen, J., Liu, T., et al. (2020). A simple ultrasensitive electrochemical sensor for simultaneous determination of homovanillic acid and vanillyl mandelic acid in human urine based on MWCNTS-Pt nanoparticles as peroxidase mimics. *Journal of Electroanalytical Chemistry, 866.* https://doi.org/10.1016/j.jelechem.2020.114165, 114165.

Ganesan, M., & Paramathevar, N. (2020). Quantum dots as nanosensors for detection of toxics: A literature review. *Analytical Methods, 12*(35), 4254–4275.

Gao, X., Yue, H., Song, S., Huang, S., Li, B., Lin, X., et al. (2018). 3-dimensional hollow graphene balls for voltammetric sensing of levodopa in the presence of uric acid. *Microchimica Acta, 185,* 91.

Gao, S., Zhang, Y., Yang, Z., Fei, T., Liu, S., & Zhang, T. (2021). Electrochemical chloramphenicol sensors-based on trace MoS_2 modified carbon nanomaterials: Insight into carbon supports. *Journal of Alloys and Compounds, 872.* https://doi.org/10.1016/J.Jallcom.2021.159687, 159687.

Gao, Y., Li, H., Wang, L., Gao, Y., & Ye, B. (2019). A simple method for determination of urapidil at a glassy carbon electrode modified with poly (sodium4-styrenesulfonate) functionalized graphene. *International Journal of Environmental Analytical Chemistry, 99,* 1471–1483.

Ghanbari, M. H., & Norouzi, Z. (2020). A new nanostructure consisting of nitrogen-doped carbon nanoonions for an electrochemical sensor to the determination of doxorubicin. *Microchemical Journal, 157.* https://doi.org/10.1016/j.microc.2020.105098, 105098.

Ghanbari, M. H., Nasrabadi, M. R., & Sobati, H. (2019). Modifying a glassy carbon electrode with reduced graphene oxide for the determination of levofloxacin with a glassy. *Analytical and Bioanalytical Electrochemistry, 11,* 189–200.

Ghiasi, T., Ahmadi, S., Ahmadi, E., Olyai, M. R. T. B., & Khodadadi, Z. (2021). Novel electrochemical sensor based on modified glassy carbon electrode with graphene quantum dots, chitosan and nickel molybdate nanocomposites for diazinon and optimal desing by the Taguchi method. *Microchemical Journal, 160,* 105628.

Gogola, J. L., Martins, G., Gevaerd, A., Blanes, L., Cardoso, J., Marchini, F. K., et al. (2021). Label-free aptasensor for p24-HIV protein detection based on graphene quantum dots as an electrochemical signal amplifier. *Analytica Chimica Acta, 1166,* 338548.

Golestaneh, M. (2020). A simple and fast electrochemical nano-structure approach for the determination of acid red 52 in real samples. *Microchemical Journal, 158.* https://doi.org/10.1016/J.Microc.2020.105281, 105281.

Golestaneh, M., & Ghoreishi, S. M. (2020). Sensitive determination of rhodamine B in real samples at the surface of a multi-walled carbon nanotubes paste electrode. *Analytical and Bioanalytical Electrochemistry, 12,* 81–92. https://doi.org/10.1080/00032719108052988.

Gopal, P., & Reddy, T. M. (2018). Fabrication of carbon-based nanomaterial composite electrochemical sensor for the monitoring of terbutaline in pharmaceutical formulations. *Colloid Surfaces A, 538,* 600–609.

Gorduk, O. (2020). Differential pulse voltammetric determination of serotonin using an acid-activated multiwalled carbon nanotube-over-oxidized poly (3,4-ethylenedioxythiophene) modified pencil graphite electrode. *Analytical Letters, 53*(7), 1034–1052. https://doi.org/10.1080/00032719.2019.1693583.

Guan, J. F., Zou, J., Liu, Y. P., Jiang, X. Y., & Yu, J. G. (2020). Hybrid carbon nanotubes modified glassy carbon electrode for selective, sensitive and simultaneous detection of dopamine and uric acid. *Ecotoxicology and Environmental Safety, 201.* https://doi.org/10.1016/J.Ecoenv.2020.110872, 110872.

Güney, S., Arslan, T., Yanık, S., & Güney, O. (2021). An electrochemical sensing platform based on graphene oxide and molecular imprinted polymer modified electrode of selective detection of amoxicillin. *Electroanalysis, 33,* 46–56.

Guo, Z., Li, D., Luo, X. K., Li, Y. H., Zhao, Q. N., Li, M. M., et al. (2017). Simultaneous determination of trace cd(II), Pb(II) and cu(II) by differential pulse anodic stripping voltammetry using a reduced graphene oxide-chitosan/poly-L-lysine nanocomposite modified glassy carbon electrode. *Journal of Colloid and Interface Science, 490*, 11–12.

Guo, X., Yue, H., Huang, S., Gao, X., Chen, H., Wu, P., et al. (2020). Electrochemical method for determination of levodopa in the presence of uric acid using In_2S_3 nanspheres on 3D graphene-modified ITO glass electrode. *Journal of Materials Science: Materials in Electronics, 31*, 13680–13687.

Guo, X., Yue, H., Song, S., Huang, S., Gao, X., Chen, H., et al. (2020). Simultaneous electrochemical determination of dopamine and uric acid based on MoS_2 nanoflowers-graphene/ITO electrode. *Microchemical Journal, 154*, 104527.

Guo, M., Shu, J., Du, D., Haghighatbin, M. A., Yang, D., Bian, Z., et al. (2021). A label-free three potential ratiometric electrochemiluminscence immunosensor for cardiac troponin I based on N-(4-aminobutyl)-N-ethylisoluminol functionalized graphene quantum dots. *Sensors and Actuators B: Chemical, 334*, 129628.

Haghighi, M., Shahlaei, M., Irandoust, M., & Hassanpour, A. (2020). New and sensitive sensor for voltammetry determination of methamphetamine in biological samples. *Journal of Materials Science: Materials in Electronics, 31*(14), 10989–11000. https://doi.org/10.1007/s10854-020-03647-6.

Han, D. W., Zhang, X. L., Yu, B., Cong, H. L., Mu, C. B., & Yuan, H. (2016). Electrochemical behavior of dopamine on La@C_{82}-COOH/C_{60}-COOH/C_{70}-COOH modified electrodes. *Integrated Ferroelectrics, 171*(1), 131–139. https://doi.org/10.1080/10584587.2016.1174563.

Han, X., Meng, Z., Zhang, H., & Zheng, J. (2018). Fullerene-based anodic stripping voltammetry for simultaneous determination of Hg(II), cu(II), Pb(II) and Cd(II) in foodstuff. *Microchimica Acta, 185*(5), 274. https://doi.org/10.1007/s00604-018-2803-9.

Han, J., Miao, L., & Song, Y. (2020). Preparation of co-Co_3O_4/carbon nanotube/carbon foam for glucose sensor. *Journal of Molecular Recognition, 33*, e2820.

Han, E., Zhang, Y., Cai, J., & Zhang, X. (2021). Development of highly sensitive immunosensor for detection of Staphylococcus aureus based on AuPdPt Trimetallic nanoparticles functionalized nanocomposite. *Micromachines, 12*(4), 446. https://doi.org/10.3390/mi12040446.

Hao, Y., Yan, P., Zhang, X., Shen, H., Gu, C., Zhang, H., et al. (2017). Ultrasensitive amperometric determination of PSA based on a signal amplification strategy using nanoflowers composed of single-strand DNA modified fullerene and methylene blue, and an improved surface-initiated enzymatic polymerization. *Microchimica Acta, 184*(11), 4341–4349. https://doi.org/10.1007/s00604-017-2476-9.

Hao, Y., Li, Z., Chen, C., Xu, Z., & Feng, S. (2019). Polyethyleneimine/graphene multilayer film supported ferric cobalt modified electrode for high-performance sensing of L-cysteine. *Journal of the Electrochemical Society, 166*, B1408.

Haridas, V., Yaakob, Z., Renuka, N. K., Sugunan, S., & Narayanan, B. N. (2021). Selective electrochemical determination of paracetamol using hematite/graphene nanocomposite modified electrode prepared in green chemical route. *Materials Chemistry and Physics, 263*, 124379.

Harvey, J. D., Baker, H. A., Ortiz, M. V., Kentsis, A., & Heller, D. A. (2019). HIV detection via a carbon nanotube RNA sensor. *ACS Sensors, 4*, 1236–1244. https://doi.org/10.1021/acssensors.9b00025.

Hassan, S. S. M., Eldin, A. G., Amr, A. E., Al-Omar, M. A., Kamel, A. H., & Khalifa, N. M. (2019). Improved solid-contact nitrate ion selective electrodes based on multi-walled carbon nanotubes (MWCNTS) as an ion-to-electron transducer. *Sensors, 19*(18), 3891. https://doi.org/10.3390/s19183891.

Hatamluyi, B., Rezayi, M., Beheshti, H. R., & Boroushaki, M. T. (2020). Ultra-sensitive molecularly imprinted electrochemical sensor for patulin detection based on a novel assembling strategy using au@cu-MOF/N-GQDs. *Sensors and Actuators B: Chemical, 318*, 128219.

Hatefi-Mehrjardi, A., Karimi, M. A., Soleymanzadeh, M., & Barani, A. (2020). Highly sensitive detection of dopamine, ascorbic and uric acid with a nanostructure of dianix yellow/multi-walled carbon nanotubes modified electrode. *Measurement, 163*. https://doi.org/10.1016/J.Measurement.2020.107893, 107893.

He, S., He, P., Zhang, X., Zhang, X., Liu, K., Jia, L., et al. (2018). Poly(glycine)/graphene oxide modified glassy carbon electrode: Preparation, characterization and simultaneous electrochemical determination of dopamine, uric acid, guanine and adenine. *Analytica Chimica Acta, 1031*, 75–82.

He, B., Wang, L., Dong, X., Yan, X., Li, M., Yan, S., et al. (2019). Aptamer-based thin film gold electrode modified with gold nanoparticles and carboxylated multi-walled carbon nanotubes for detecting oxytetracycline in chicken samples. *Food Chemistry, 300*. https://doi.org/10.1016/j.foodchem.2019.12517, 125179.

He, B., Li, M., & Li, M. (2020). Electrochemical determination of sulfamethazine using a gold electrode modified with multi-walled carbon nanotubes, graphene oxide nanoribbons and branched aptamers. *Microchimica Acta, 187*(5), 274. https://doi.org/10.1007/s00604-020-04244-4.

Hernaez, M., Acevedo, B., Mayes, A. G., & Melendi-Espina, S. (2019). High-performance optical fiber humidity sensor based on lossy mode resonance using a nanostructured polyethylenimine and graphene oxide coating. *Sensors and Actuators B: Chemical, 286*, 408–414. https://doi.org/10.1016/j.snb.2019.01.145.

Hernández, P., Aguilar-Lira, G. Y., Islas, G., & Rodriguez, J. A. (2021). Development of a new voltammetric methodology for the determination of ciprofloxacin in beef samples using a carbon paste electrode modified with nafion and fullerenes. *Electroanalysis*, 1–9. https://doi.org/10.1002/elan.202060525 (Published online).

Hu, J., & Zhang, Z. (2020). Application of electrochemical sensors based on carbon nanomaterials for detection of flavonoids. *Nanomaterials, 10*(10), 2020.

Huang, S., Yue, H., Zhou, J., Zhang, J., Zhang, C., Gao, X., et al. (2014). Higly selective and sensitive determination of dopamine in the presence of ascorbic acid using a 3D graphene foam electrode. *Electroanalysis, 26*, 184–190.

Huang, Y., Tan, J., Cui, L., Zhou, Z., Zhou, S., Zhang, Z., et al. (2018). Graphene and Au NPs co-mediated enzymatic silver deposition for the ultrasensitive electrochemical detection of cholesterol. *Biosensors & Bioelectronics, 102*, 560–567.

Huang, D., Li, X., Chen, M., Chen, F., Wan, Z., Rui, R., et al. (2019). An electrochemical sensor based on a porphyrin dye-functionalized multi-walled carbon nanotubes hybrid for the sensitive determination of ascorbic acid. *Journal of Electroanalytical Chemistry, 841*, 101–106. https://doi.org/10.1016/J.Jelechem.2019.04.041.

Huang, J. Y., Bao, T., Hu, T. X., Wen, W., & Zhang, X. H. (2017). Voltammetric determination of levofloxacin using a glassy carbon electrode modified with poly(o-aminophenol) and graphene quantum dots. *Microchimica Acta, 184*, 127–135.

Huang, Y., Xu, J., Liu, J., Wang, X., & Chen, B. (2017). Disease related detection with electrochemical biosensors: A review. *Sensors, 17*, 2375.

Hummers, W. S., & Offeman, R. E. (1958). Preparation of graphitic oxide. *Journal of the American Chemical Society, 80*, 1339.

Hunter, G. W., Stetter, J. R., Hesketh, P. J., & Liu, C. C. (2020). Smart sensor system. *Electrochemical Society Interface*, 29–34.

Iijima, S. (1991). Helical microtubules of graphitic carbon. *Nature, 354*, 56–58.

Incebay, H., & Saylakci, R. (2021). Voltammetric determination of neotame by using chitosan/nickelnanoparticles/multi walled carbon nanotubes biocomposite as a modifier. *Electroanalysis, 33*(6), 1451–1460. https://doi.org/10.1002/elan.202100021.

Jahanbakhshi, M. (2018). Myoglobin immobilized of mesoporous carbon foam in a hydrogel (selep) dispersant for voltammetric sensing of hydrogen peroxide. *Microchimica Acta*. https://doi.org/10.1007/s00604-017-2654-9.

Jahanbakhshi, M., & Habibi, B. (2016). A novel and facile of carbon quantum dots via salep hydrothermal treatment as the silver nanoparticles support: Application to electroanalytical determination of H_2O_2 in fetal bovine serum. *Biosensors & Bioelectronics, 81*, 143–150.

Jalali, F., Hassanvand, Z., & Baratti, A. (2020). Electrochemical sensor based on a nanocomposite of carbon dots, hexadecyltrimethylammonium bromide and chitosan for mesalazine determination. *Journal of Analytical Chemistry, 75*, 544–552.

Jalalvand, A. R. (2019). Fabrication of a novel and ultrasensitive label-free electrochemical aptasensor for detection of biomarker prostate specific antigen. *International Journal of Biological Macromolecules, 126*, 1065–1073. https://doi.org/10.1016/j.ijbiomac.2019.01.012.

Jalalvand, A. R. (2020). Fabrication of a novel amperometric sensing platform for determination of mangiferin. *Sensing and Bio-Sensing Research, 29*. https://doi.org/10.1016/j.sbsr.2020.100352, 100352.

Jalil, O., Pandey, C. M., & Kumar, D. (2020). Electrochemical biosensor for the epithelial cancer biomarker EpCAM based on reduced graphene oxide modified with nanostructured titanium dioxide. *Microchimica Acta, 187*, 275.

Jalili, R., Khataee, A., Rashidi, M. R., & Luque, R. (2017). Dual-colored carbon dot encapsulated metal-organic framework for ratiometric detection of glutathione. *Sensors and Actuators B: Chemical, 297*. https://doi.org/10.1016/j.snb.2019.126775, 126775.

Jeong, H., Nguyen, D. M., Lee, M. S., Kim, H. G., Ko, S. C., & Kwac, L. K. (2018). N-doped graphene-carbon nanotube hybrid networks attaching with gold nanoparticles for glucose non enzymatic sensor. *Materials Science and Engineering: C, 90*, 38–45.

Ji, L., Yu, S., Zhou, X., Bao, Y., Yang, F., Kang, W., et al. (2019). Modification of electron structure on the semiconducting single-walled carbon nanotubes for effectively electrosensing guanine and adenine. *Analytica Chimica Acta, 1079*, 86–93. https://doi.org/10.1016/j.aca.2019.06.027.

Ji, C., Zhou, Y., Leblanc, R. M., & Peng, Z. (2020). Recent developments of carbon dots in biosensing: A review. *ACS Sensors, 5*, 2724–2741. https://doi.org/10.1021/acssensors.0c01556.

Jia, L., Mao, Y., Zhang, S., Li, H., Qian, M., Liu, D., et al. (2021). Electrochemical switch sensor toward ephedrine hydrochloride determination based on molecularly imprinted polymer/nafion-MWCNTS modified electrode. *Microchemical Journal, 164*. https://doi.org/10.1016/J.Microc.2021.105981, 105981.

Jiang, D., Ge, P., Wang, L., Jiang, H., Yang, M., Yuan, L., et al. (2019). A novel electrochemical mast cell-based paper biosensor for the rapid detection of milk allergen casein. *Biosensors & Bioelectronics, 130*, 299–306.

Jiang, L., Mo, G., Yu, C., Ya, D., He, X., Mo, W., et al. (2019). Based on reduced graphene oxide-copper sulfide-carbon nitride nanosheets composite electrochemiluminescence sensor for determination of gatifloxacin in mouse plasma. *Colloids and Surfaces. B, Biointerfaces, 173*, 378–385.

Jiang, Q., Jing, Y., Ni, Y., Gao, R., & Zhou, P. (2020). Potentiality of carbon quantum dots derived from chitin as a fluorescent sensor for detection of ClO. *Microchemical Journal, 157*. https://doi.org/10.1016/j.microc.2020.105111, 105111.

Jiang, Z., Li, G., & Zhang, M. (2016). Electrochemical sensor based on electro-polymerization of β-cylodextrin and reduced-graphene oxide on glassy carbon electrode for determination of gatifloxacin. *Sensors and Actuators B: Chemical, 228*, 59–65.

Jiang, L., Nelson, G. W., Abda, N. J., & Foord, J. S. (2016). Novel modifications to carbon-based electrodes to improve the electrochemical detection of dopamine. *Applied Materials & Interfaces, 8*, 28338–28348.

Jiang, L., Santiago, I., & Foord, J. (2021). A comparative study of fouling-free nanodiamond and nanocarbon electrochemical sensors for sensitive bisphenol A detection. *Carbon, 174*, 390–395.

Jiang, T., Sun, X., Wei, L., & Li, M. (2020). Determination of hydrogen peroxide released from cancer cells by a Fe-organic framework/horseradish peroxidase-modified electrode. *Analytica Chimica Acta, 1135*, 132–141. https://doi.org/10.1016/j.aca.2020.09.040.

Jiang, T., Sun, X., Wei, L., & Li, M. (2021). Electrochemical determination of artemisinin based on signal inhibition for the reduction of hemin. *Analytical and Bioanalytical Chemistry, 413*(2), 565–576. https://doi.org/10.1007/S00216-020-03028-2.

Jing, S., Zheng, H., Zhao, L., Qu, L., & Yu, L. (2017). A novel electrochemical sensor based on WO_3 nanorods-decorated poly(sodium 4-styrenesulfonate) functionalized graphene nanocomposite modified electrode for detecting of puerarin. *Talanta, 174*, 477–485.

Joghani, R. A., Rafati, A. A., Ghodsi, J., Assari, P., & Feizollahi, A. (2020). A sensitive voltammetric sensor based on carbon nanotube/nickel nanoparticle for determination of daclatasvir (an anti-hepatitis c drug) in real samples. *Journal of Applied Electrochemistry, 50*(11), 1199–1208. https://doi.org/10.1007/S10800-020-01478-1.

Kamyabi, M. A., & Moharramnezhad, M. (2021). An enzyme-free electrochemiluminescence sensing probe based on ternary nanocomposite for ultrasensitive determination of chlorpyrifos. *Food Chemistry, 351*, 129252.

Karim, R. A., Reda, Y., & Fattah, A. A. (2020). Review-nanostructured materials-based nanosensors. *Journal of the Electrochemical Society, 167*, 037554.

Kathiresan, V., Thirumalai, D., Rajarathinam, T., Yeom, M., Lee, J., Kim, S., et al. (2021). A simple one-step electrochemical deposition of bioinspired nanocomposite for the non-enzymatic detection of dopamine. *Journal of Analytical Science and Technology, 12*(5).

Kaur, R., Rana, S., Singh, R., Kaur, V., & Narula, P. (2019). A schiff base modified graphene oxide film for anodic stripping voltammetric determination of arsenite. *Microchimica Acta, 186*, 741.

Kenarkob, M., & Pourghobadi, Z. (2019). Electrochemical sensor for acetaminophen based on a glassy carbon electrode modified with ZnO/Au nanoparticles on functionalized multi-walled carbon nano-tubes. *Microchemical Journal, 146*, 1019–1025. https://doi.org/10.1016/J.Microc.2019.02.038.

Kingsford, O. J., Zhang, D., Ma, Y., Wu, Y., & Zhu, G. (2019). Electrochemically recognizing tryptophan enantiomers based on carbon black/poly-L-cysteine modified electrode. *Journal of the Electrochemical Society, 166*, 13.

Kornii, A., Saska, V., Lisnyak, V. V., & Tananaiko, O. (2020). Carbon nanostructured screen-printed electrodes modified with CuO/glucose oxidase/maltase/SiO$_2$ composite film for maltose determination. *Electroanalysis, 32*, 1468–1479.

Krishnan, S., Tong, L., Liu, S., & Xing, R. (2020). A mesoporous silver-doped TiO$_2$-SnO$_2$ nanocomposite on g-C$_3$N$_4$ nanosheets and decorated with a hierarchical core-shell metal-organic framework for simultaneous voltammetric determination of ascorbic acid, dopamine and uric acid. *Microchimica Acta, 187*, 82.

Kroto, H. W., Heath, J. R., O'Brien, S. C., Curl, R. F., & Smalley, R. E. (1985). C60: Buckminsterfullerene. *Nature, 318*, 162–163.

Kucukkolbasi, S., Sayin, S., & Yılmaz, M. (2019). Fabrication and application of a new modified electrochemical sensor using newly synthesized calixarene-grafted MWCNTS for simultaneous determination of Cu(II) and Pb(II). *Acta Chimica Slovenica, 66*(4), 839–849. https://doi.org/10.17344/Acsi.2018.4922.

Kul, D. (2020). Electrochemical determination of rifampicin based on its oxidation using multi-walled carbon nanotube-modified glassy carbon electrodes. *Turkish Journal Of Pharmaceutical Sciences, 17*(4), 398–407. https://doi.org/10.4274/tjps.galenos.2019.33600.

Kummari, S., Kumar, V. S., Satyanarayana, M., & Gobi, K. V. (2019). Direct electrochemical determination of methotrexate using functionalized carbon nanotube paste electrode as biosensor for in-vitro analysis of urine and dilute serum samples. *Microchemical Journal, 148*, 626–633. https://doi.org/10.1016/j.microc.2019.05.054.

Kummari, S., Sunil Kumar, V., & Vengatajalabathy, G. K. (2020). Facile electrochemically reduced graphene oxide-multi-walled carbon nanotube nanocomposite as sensitive probe for in-vitro determination of nitrofurantoin in biological fluids. *Electroanalysis, 32*(11), 2452–2462. https://doi.org/10.1002/elan.202060157.

Kuralay, F., & Bayramlı, Y. (2021). Electrochemical determination of mitomycin C and its interaction with double-stranded DNA using a poly(o-phenylenediamine)-multi-walled carbon nanotube modified pencil graphite electrode. *Analytical Letters, 54*(8), 1295–1308. https://doi.org/10.1080/00032719.2020.1801710.

Kuralay, F., & Gürsoy, T. (2020). Direct electrochemistry and sensitive detection of guanosine on nanopolymeric surfaces bearing boronic acid groups. *ChemistrySelect, 5*(29), 9134–9142. https://doi.org/10.1002/slct.202001812.

Kuyumcu, S. E. (2019). Electrochemical determination of N-acetyl cysteine in the presence of acetaminophen at multi-walled carbon nanotubes and nafion modified sensor. *Sensors and Actuators B: Chemical, 282*, 500–506. https://doi.org/10.1016/j.snb.2018.11.092.

Lai, Z., Lin, F., Huang, Y., Wang, Y., & Chen, X. (2021). Automated determination of Cd^{2+} and Pb^{2+} in natural waters with sequential injection analysis device using differential pulse anodic stripping voltammetry. *Journal of Analysis and Testing, 5*(1), 60–68. https://doi.org/10.1007/S41664-021-00165-0.

Lalabadi, M. A., Hashemi, H., Feng, J., & Jafari, S. M. (2020). Carbon nanomaterials against pathogens; the antimicrobial activity of carbon nanotubes, graphene/graphene oxide, fullerenes, and their nanocomposites. *Advances in Colloid and Interface, 284*, 102250.

Lavanya, A. L., Kumari, G. B., Prasad, K. R. S., & Brahman, P. K. (2021). Development of pen-type portable electrochemical sensor based on Au-W bimetallic nanoparticles decorated graphene-chitosan nanocomposite film for the detection of nitrite in water, milk and fruit juices. *Electroanalysis, 33*, 1096–1106.

Li, S., & Zhang, Z. (2021). Recent advances in the construction and analytical applications of carbon dots-based optical nanoassembly. *Talanta, 223*. https://doi.org/10.1016/j.talanta.2020.121691, 121691.

Li, C., Liu, W., Ren, Y., Sun, X., Pan, W., & Wang, J. (2017). The selectivity of the carboxylate groups terminated carbon dots switched by buffer solutions for the detection of multi-metal ions. *Sensors and Actuators B: Chemical, 240*, 941–948. https://doi.org/10.1016/j.snb.2016.09.068.

Li, J., Tang, K., Yu, J., Wang, H., Tu, M., & Wang, X. (2019). Nitrogen and chlorine co-doped carbon dots as probe for sensing and imaging in biological samples. *Royal Society Open Science, 6*, 1–10. https://doi.org/10.1098/rsos.181557.

Li, X., Shi, Y., Chen, P., Bai, Y., Li, G., Shu, H., et al. (2019). Multifunctional electrochemical application of a novel 3D AgInS$_2$/rGO nanohybrid for electrochemical detection and HER. *Journal of Chemical Technology and Biotechnology, 94*, 3713–3724.

Li, B., Suo, T., Xie, S., Xia, A., Ma, Y., Huang, H., et al. (2021). Rational desing, synthesis and applications of carbon dots@metal-organic frameworks (CD@MOF) based sensors. *TrAC Trends in Analytical Chemistry, 135*, 116163.

Li, S., Li, J., Luo, J., Xu, Z., & Ma, X. (2018). A microfluidic chip containing a molecularly imprinted polymer and a DNA aptamer for voltammetric determination of carbofuran. *Microchimica Acta, 185*, 295.

Liang, Z., Zhai, H., Chen, Z., Wang, H., Wang, S., Zhou, Q., et al. (2016). A simple, ultrasensitive sensor for gallic acid and uric acid based on gold microclusters/sulfonate functionalized graphene modified glassy carbon electrode. *Sensors and Actuators B: Chemical, 224*, 915–925.

Liang, A. A., Hou, B. H., Tang, C. S., Sun, D. L., & Luo, E. A. (2021). An advanced molecularly imprinted electrochemical sensor for the highly sentive and selective detection and determination of human IgG. *Bioelectrochemistry, 137*, 107671.

Liang, Y., Qu, C., Yang, R., Qu, L., & Li, J. (2017). Molecularly imprinted electrochemical sensor for daidzein recognition and detection based on poly(sodium 4-styrenesulfonate) functionalized graphene. *Sensors and Actuators B: Chemical, 251*, 542–550.

Lin, X., Xiong, M., Zhang, J., He, C., Ma, X., Zhang, H., et al. (2021). Carbon dots based on natural resources: Synthesis and applications in sensors. *Microchemical Journal, 160*. https://doi.org/10.1016/j.microc.2020.105604, 105604.

Liu, X., Huang, D., Lai, C., Zeng, G., Qin, L., Zhang, C., et al. (2018). Recent advances in sensors for tetracycline antibiotics and their applications. *TrAC, Trends in Analytical Chemistry, 109*, 260–274. https://doi.org/10.1016/j.trac.2018.10.011.

Liu, H. Y., Wen, J. J., Huang, Z. H., Ma, H., Xu, H., Qiu, Y., et al. (2019). Prussian blue analogue of copper-cobalt decorated with multi-walled carbon nanotubes based electrochemical sensor for sensitive determination of nitrite in food samples. *Chinese Journal of Analytical Chemistry, 47*(6), E19066–E19072. https://doi.org/10.1016/s1872-2040(19)61168-0.

Liu, P., Li, C., Zhang, R., Tang, Q., Wei, J., Lu, Y., et al. (2019). An ultrasensitive electrochemical immunosensor for procalcitonin detection based on the gold nanoparticles-enhanced tyramide signal amplification strategy. *Biosensors & Bioelectronics, 126*, 543–550.

Liu, W., Cui, F., Li, H., Wang, S., Zhuo, B., & Wang, S. (2020). Three-dimensional hybrid networks of molecularly imprinted poly(9-carbazoleacetic acid) and MWCNTS for simultaneous voltammetric determination of dopamine and epinephrine in plasma sample. *Sensors and Actuators B: Chemical, 323*. https://doi.org/10.1016/J.Snb.2020.128669, 128669.

Liu, P., Meng, H., Han, Q., Zhang, G., Wang, C., Song, L., et al. (2021). Determination of ascorbic acid using electrochemiluminescence sensor based on nitrogen and sulfur doping graphene quantum dots with luminol as internal standard. *Microchimica Acta, 188*, 120.

Liu, Z., Jin, M., Lu, H., et al. (2019). Molecularly imprinted polymer decorated 3D-framework of functionalized multi-walled carbon nanotubes for ultrasensitive electrochemical sensing of norfloxacin in pharmaceutical formulations and rat plasma. *Sensors and Actuators B: Chemical, 288*, 363–372. https://doi.org/10.1016/j.snb.2019.02.097.

Loh, K. P., Ho, D., Chiu, G. N. C., Leong, D. T., Pastorin, G., & Chow, E. K. H. (2018). Clinical applications of carbon nanomaterials in diagnostics and therapy. *Advanced Materials, 30*, 1802368.

Long, D., Li, M., Wang, H., Wang, H., Chai, Y., & Yuan, R. (2019). A photoelectrochemical biosensor based on fullerene with methylene blue as a sensitizer for ultrasensitive DNA detection. *Biosensors & Bioelectronics, 142*. https://doi.org/10.1016/j.bios.2019.111579, 111579.

Lourencao, B. C., Silva, T. A., Santos, M. S., Ferreira, A. G., & Filho, O. F. (2017). Sensitive voltammetric determination of hydroxyzine and its main metabolite cetirizine and identification of oxidation products by nuclear magnetic resonance spectroscopy. *Journal of Electroanalytical Chemistry, 807*, 187–195.

Lu, Y., Zhao, X., Tian, Y., Guo, Q., Li, C., & Nie, G. (2020). An electrochemiluminescence aptasensor for the ultrasensitive detection of aflotoxin B1 based on gold nanorods/graphene quantum dots-modified poly (indole-6-carboxylic acid)/flower-gold nanocomposite. *Microchemical Journal, 157*, 104959.

Lu, Z., Zhao, W., Wu, L., He, J., Dai, W., Zhou, C., et al. (2021). Tunable electrochemical of electrosynthesized layer-by-layer multilayer films based on multi-walled carbon nanotubes and metal-organic framework as high-performance electrochemical sensor for simultaneous determination cadmium and lead. *Sensors and Actuators B: Chemical, 326*. https://doi.org/10.1016/J.Snb.2020.128957, 128957.

Lu, H., Li, C., Wang, H., Wang, X., & Xu, S. (2019). Biomass-derived sulfur, nitrogen co-doped carbon dots for colorimetric and fluorescent dual mode detection of silver (I) and cell imaging. *ACS Omega, 4*, 21500–21508. https://doi.org/10.1021/acsomega.9b03198.

Lv, J., Li, C., Feng, S., et al. (2019). A novel electrochemical sensor for uric acid detection based on PCN/MWCNT. *Ionics, 25*(9), 4437–4445. https://doi.org/10.1007/s11581-019-03010-8.

Ma, Q. F., Tou, Z. Q., Ni, K., Lim, Y. Y., Lin, Y. F., Wang, Y. R., et al. (2018). Carbon-nanotube/polyvinyl alcohol coated thin core fiber sensor for humidity measurement. *Sensors and Actuators B: Chemical, 257*, 800–806. https://doi.org/10.1016/j.snb.2017.10.121.

Madrakian, T., Maleki, S., Gilak, S., & Afkhami, A. (2017). Turn-off fluorescence of amino-functionalized carbon quantum dots as effective fluorescent probes for determination of isotretinoin. *Sensors and Actuators B: Chemical, 247*, 428–435. https://doi.org/10.1016/j.snb.2017.03.071.

Majidian, M., Raoof, J. B., Hosseini, S. R., Ojani, R., Barek, J., & Fischer, J. (2020). Novel type of carbon nanotube paste electrode modified by Sb_2O_3 for square wave anodic stripping voltammetric determination of Cd^{2+} and Pb^{2+}. *Electroanalysis, 32*(10), 2260–2265. https://doi.org/10.1002/Elan.20206013.

Mansuriya, B. D., & Altıntas, Z. (2021). Enzyme-free electrochemical nano-immunosensor based on graphene quantum dots and gold nanoparticles for cardiac biomarker determination. *Nanomaterials, 11*, 578.

Masikini, M., Ghica, M. E., Baker, P. G. L., Iwuoha, E. I., & Brett, C. M. A. (2019). Electrochemical sensor based on multi-walled carbon nanotube/gold nanoparticle modified glassy carbon electrode for detection of estradiol in environmental samples. *Electroanalysis, 31*(10), 1925–1933. https://doi.org/10.1002/Elan.201900190.

Mazloum-Ardakani, M., Ahmadi, S. H., Safaei Mahmoudabadi, Z., & Khoshroo, A. (2016). Nano composite system based on fullerene-functionalized carbon nanotubes for simultaneous determination of levodopa and acetaminophen. *Measurement, 91*, 162–167. https://doi.org/10.1016/j.measurement.2016.05.035.

Medyantseva, E. P., Brusnitsyn, D. V., Gazizullina, E. R., Varlamova, R. M., Konovalova, O. A., & Budnikov, H. C. (2020). Hybrid nanocomposites as electrode modifiers in amperometric immunosensors for the determination of amitriptyline. *Journal of Analytical Chemistry, 75*(4), 536–543.

Meng, X., Xiao, Z., & Scott, S. K. (2019). Preparation and application of electrochemical sensor based on molecularly imprinted polymer coated multiwalled carbon nanotubes for nitrocellulose detection. *Propellants, Explosives, Pyrotechnics, 44*(10), 1337–1346. https://doi.org/10.1002/prep.201900055.

Mikhraliieva, A., Zaitsev, V., Tkachenko, O., Nazarkovsky, M., Xing, Y., & Benvenutti, E. V. (2020). Graphene oxide quantum dots immobilized on mesoporus silica: Preparation, characterization and electroanalytical application. *RSC Advances, 10*, 31305–31315.

Molinero-Abad, B., Izquierdo, D., Pérez, L., Escudero, I., & Arcos-Martínez, M. J. (2018). Comparison of backing materials of screen printed electrochemical sensors for direct determination of the sub-nanomolar concentration of lead in seawater. *Talanta, 182*, 549–557. https://doi.org/10.1016/j.talanta.2018.02.005.

Mollarasouli, F., Asadpour-Zeynali, K., Campuzano, S., Yanez-Sedeno, P., & Pingarron, J. M. (2017). Non-enzymatic hydrogen peroxide sensor based on graphene quantum dots-chitosan/methylene blue hybrid nanostructures. *Electrochimica Acta, 246*, 303–314.

Morais, A. L., Rijo, P., Hernan, M. B. B., & Nicolai, M. (2020). Biomolecules and electrochemical tools in chronic non-communicable disease surveillance: A systematic review. *Biosensors, 10*, 221.

Motaghedifard, M. H., Pourmortazavi, S. M., & Mirsadeghi, S. (2021). Selective and sensitive detection of Cr(VI) pollution in waste water via polyaniline/sulfated zirconium dioxide/multi walled carbon nanotubes nanocomposite based electrochemical sensor. *Sensors and Actuators B: Chemical, 327*. https://doi.org/10.1016/J.Snb.2020.128882, 128882.

Munir, A., Bozal-Palabiyik, B., Khan, A., Shah, A., & Uslu, B. (2019). A novel electrochemical method for the detection of oxymetazoline drug based on MWCNTS and TiO_2 nanoparticles. *Journal of Electroanalytical Chemistry, 844*, 58–65. https://doi.org/10.1016/j.jelechem.2019.05.017.

Murthy, A. P., Duraimurugan, K., Sridhar, J., & Madhavan, J. (2019). Application of derivative voltammetry in the quantitative determination of alloxan at single-walled carbon nanotubes modified electrode. *Electrochimica Acta, 317*, 182–190. https://doi.org/10.1016/J.Electacta.2019.05.163.

Muti, M., & Cantopcu, M. (2018). Nanosensing platform for the electrochemical determination of dopamine. *Journal of Analytical Chemistry, 73*(8), 809–816. https://doi.org/10.1134/S1061934818080075.

Nair, R. V., Thomas, R. T., Mohamed, A. P., & Pillai, S. (2020). Fluorescent turn-off sensor based on Sulphur-doped graphene quantum dots in colloidal and film forms for the ultrasensitive detection of carbamate pesticides. *Microchemical Journal, 157*. https://doi.org/10.1016/j.microc.2020.104971, 104971.

Najafi, M., Sohouli, E., & Mousavi, F. (2020). An electrochemical sensor for fentanyl detection based on multi-walled carbon nanotubes as electrocatalyst and the electrooxidation mechanism. *Journal of Analytical Chemistry, 75*(9), 1209–1217. https://doi.org/10.1134/S1061934820090130.

Ngo, Y. L. T., Jana, J., Chung, J. S., & Hur, S. H. (2020). Electrochemical biosensors based on nanocomposites of carbon-based dots. *Korean Chemical Engineering Research, 54*(4), 499–513.

Nouri, M., Rahimnejad, M., Najafpour, G., & Moghadamnia, A. A. (2019). A gr/αFe$_2$O$_3$/carbon paste electrode developed as an electrochemical sensor for determination of rizatriptan benzoate: An antimigraine drug. *ChemistrySelect, 4*, 13421–13426.

Novoselov, K. S., Geim, A. K., Morozov, S. V., Jiang, D., Zhang, Y., Dubonos, S. V., et al. (2004). Electric field effect in atomically thin carbon films. *Science, 306*, 666–669.

Omar, N. A. S., Fen, Y. W., Abdullah, J., Zaid, M. H. M., Daniyal, W. M. E. M. M., & Mahdi, M. A. (2019). Sensitive surface plasmon resonance performance of cadmium sulfide quantum dots-amine functionalized graphene oxide based thin film towards dengue virus E-protein. *Optics and Laser Technology, 114*, 204–208. https://doi.org/10.1016/j.optlastec.2019.01.038.

Oularbi, L., Turmine, M., & El Rhazi, M. (2019). Preparation of novel nanocomposite consisting of bismuth particles, polypyrrole and multi-walled carbon nanotubes for simultaneous voltammetric determination of cadmium(II) and lead(II). *Synthetic Metals, 253*, 1–8. https://doi.org/10.1016/J.Synthmet.2019.04.011.

Özdokur, K. V. (2020). Voltammetric determination of isoniazid drug in various matrix by using CuO$_x$ decorated MW-CNT modified glassy carbon electrode. *Electroanalysis, 32*(3), 489–495. https://doi.org/10.1002/elan.201900307.

Palakollu, V. N., Karpoormath, R., Wang, L., Tang, J. N., & Liu, C. (2020). A versatile and ultrasensitive electrochemical sensing platform for detection of chlorpromazine based on nitrogen-doped carbon dots/cuprous oxide composite. *Nanomaterials, 10*, 1513.

Palisoc, S., De Leon, P. G., Alzona, A., Racines, L., & Natividad, M. (2019). Highly sensitive determination of tetracycline in chicken meat and eggs using AuNP/MWCNT-modified glassy carbon electrodes. *Heliyon, 5*(7), E02147. https://doi.org/10.1016/j.heliyon.2019.E02147.

Palisoc, S., Vitto, R. I. M., & Natividad, M. (2019). Determination of heavy metals in herbal food supplements using bismuth/multi-walled carbon nanotubes/nafion modified graphite electrodes sourced from waste batteries. *Scientific Reports, 9*(1), 18491. https://doi.org/10.1038/S41598-019-54589-X.

Panwar, N., Soehartono, A. M., Chan, K. K., Zeng, S., Xu, G., Qu, J., et al. (2019). Nanocarbons for biology and medicine: Sensing, imaging, and drug delivery. *Chemical Reviews, 119*(16), 9559–9656.

Pari, M., Reddy, K. R. V., Fasiulla, & Chandrakala, K. B. (2020). Amperometric determination of dopamine based on an interface platform comprising tetra-substituted Zn^{2+} phthalocyanine film layer with embedment of reduced graphene oxide. *Sensors and Actuators, A: Physical, 316*, 112377.

Parveen, S., Pathak, A., & Gupta, B. D. (2017). Fiber optic SPR nanosensor based on synergistic effects of CNT/cu-nanoparticles composite for ultratrace sensing of nitrate. *Sensors and Actuators B: Chemical, 246*, 910–919. https://doi.org/10.1016/j.snb.2017.02.170.

Pathak, A., & Gupta, B. D. (2019). Ultra-selective fiber optic SPR platform for the sensing of dopamine in synthetic cerebrospinal fluid incorporating permselective nafion membrane and surface imprinted MWCNTs-PPy matrix. *Biosensors & Bioelectronics, 133*, 205–214. https://doi.org/10.1016/j.bios.2019.03.023.

Pedrozo-Penafel, M. J., Miranda-Andrades, J. R., Gutierrez-Beleno, L. M., Larrude, D. G., & Aucelio, R. Q. (2020). Indirect voltammetric determination of thiomersal in influenza vaccine using photo-degradation and graphene quantum dots modified glassy carbon electrode. *Talanta, 215*, 120938.

Pedrozo-Peñafiel, M. J., Almeida, J. M. S., Toloza, C. A. T., Larrudé, D. G., Pacheco, W. F., & Aucelio, R. Q. (2019). Square-wave voltammetric determination of primaquine in urine using a multi-walled carbon nanotube modified electrode. *Microchemical Journal, 150*. https://doi.org/10.1016/Jj.microc.2019.104201, 104201.

Pedrozo-Penafiel, M. J., Lopes, T., Gutierrez-Beleno, L. M., MEH, M. D. C., Larrude, D. G., & Aucelio, R. Q. (2020). Voltammetric determination of creatinine using a gold electrode modified with nafion mixed with graphene quantum dots-copper. *Journal of Electroanalytical Chemistry, 878*, 114561.

Phelane, L., Gouveia-Caridade, C., Barsan, M. M., Baker, P. G. L., Brett, C. M. A., & Iwuoha, E. I. (2020). Electrochemical determination of tyrosine using a novel tyrosinase multi-walled carbon nanotube (MWCNT) polysulfone modified glassy carbon electrode (GCE). *Analytical Letters, 53*(2), 308–321. https://doi.org/10.1080/00032719.2019.1649417.

Phonklam, K., Wannapob, R., Sriwimol, W., Thavarungkul, P., & Phairatana, T. (2020). A novel molecularly imprinted polymer PMB/MWCNTs sensor for highly-sensitive cardiac troponin T detection. *Sensors and Actuators B: Chemical, 308*. https://doi.org/10.1016/J.Snb.2019.127630, 127630.

Pires, N. M. M., Dong, T., & Yang, Z. (2019). A fluorimetric nitrite biosensor with polythienothiophene-fullerene thin film detectors for on-site water monitoring. *Analyst, 144*, 4342–4350. https://doi.org/10.1039/c8an02441c.

Pizarro, J., Segura, R., Tapia, D., Navarro, F., Fuenzalida, F., & Aguirre, M. J. (2020). Inexpensive and green electrochemical sensor for the determination of Cd(II) and Pb(II) by square wave anodic stripping voltammetry in bivalve mollusks. *Food Chemistry, 321*, 126682.

Ponnaiah, S. K., & Periakaruppan, P. (2020). Carbon dots doped tungstic anhydride on graphene oxide nanopanels: A new picomolar-range creatinine selective enzymeless electrochemical sensor. *Materials Science and Engineering, 113*, 111010.

Power, A. C., Gorey, B., Chandra, S., & Chapman, J. (2018). Carbon nanomaterials and their application to electrochemical sensors: A review. *Nanotechnology Reviews, 7*(1), 19–41.

Prasad, B. B., Kumar, A., & Singh, R. (2016). Molecularly imprinted polymer-based electrochemical sensor using functionalized fullerene as a nanomediator for ultratrace analysis of primaquine. *Carbon NY, 109*, 196–207. https://doi.org/10.1016/j.carbon.2016.07.044.

Prasad, B. B., Singh, R., & Kumar, A. (2017). Synthesis of fullerene (C_{60}-monoadduct)-based water-compatible imprinted micelles for electrochemical determination of chlorambucil. *Biosensors & Bioelectronics, 94*, 115–123. https://doi.org/10.1016/j.bios.2017.02.040.

Priya, T., Dhanalakshmi, N., Thennarasu, S., Karthikeyan, V., & Thinakaran, N. (2019). Ultra sensitive electrochemical detection of Cd^{2+} an Pb^{2+} using penetrable nature of graphene/gold nanoparticles/modified L-cysteine nanocomposite. *Chemical Physics Letters, 731*, 136621.

Rabie, E., Assaf, H., Abo Elmaaref, A., & Khodari, M. (2019). Fabrication of a new electrochemical sensor based on carbon paste electrode modified by silica gel/MWCNTs for the voltammetric determination of salicylic acid in tomato. *Egyptian Journal of Chemistry, 62*, 4–5. https://doi.org/10.21608/ejchem.2019.17087.2048.

Rahimi-Nasrabadi, M., Khoshroo, A., & Mazloum-Ardakani, M. (2017). Electrochemical determination of diazepam in real samples based on fullerene-functionalized carbon nanotubes/ionic liquid nanocomposite. *Sensors and Actuators B: Chemical, 240*, 125–131. https://doi.org/10.1016/j.snb.2016.08.144.

Rahman, M. M., Alam, M. M., & Alamry, K. A. (2019). Sensitive and selective *m*-tolyl hydrazine chemical sensor development based on CdO nanomaterial decorated multi-walled carbon nanotubes. *Journal of Industrial and Engineering Chemistry, 77*, 309–316. https://doi.org/10.1016/j.jiec.2019.04.053.

Rajabi, N., Masrournia, M., & Abedi, M. (2020). Potentiometric determination of La(III) using chitosan modified carbon paste electrode with an experimental design. *Chemical Methodologies, 4*, 660–670. https://doi.org/10.22034/chemm.2020.109975.

Rajaei, M., Foroughi, M. M., Jahani, S., Shahidi Zandi, M., & Hassani, N. H. (2019). Sensitive detection of morphine in the presence of dopamine with La^{3+} doped fern-like CuO nanoleaves/MWCNTS modified carbon paste electrode. *Journal of Molecular Liquids, 284*, 462–472. https://doi.org/10.1016/j.molliq.2019.03.135.

Raju, C. V., Kalaiyarasan, G., Paramasivam, S., Joseph, J., & Kumar, S. S. (2020). Phosphorous doped carbon quantum dots as an efficient solid state electrochemiluminescence platform for highly sensitive turn-on detection of Cu^{2+} ions. *Electrochimica Acta, 331*, 135391.

Ramos, M. M. V., Carvalho, J. H. S., Oliveira, P. R., & Janegitz, B. C. (2020). Determination of serotonin by using a thin film containing graphite, nanodiamonds and gold nanoparticles anchored in casein. *Measurement, 149*, 106979.

Rather, J. A., Khudaish, E. A., Munam, A., Qurashi, A., & Kannan, P. (2016). Electrochemically reduced fullerene-graphene oxide interface for swift detection of Parkinsons disease biomarkers. *Sensors and Actuators B: Chemical, 237*, 672–684. https://doi.org/10.1016/j.snb.2016.06.137.

Rebelo, P., Pacheco, J. G., Voroshylova, I. V., Melo, A., Cordeiro, M. N. D. S., & Delerue-Matos, C. (2021). Rational development of molecular imprinted carbon paste electrode for furazolidone detection: Theoretical and experimental approach. *Sensors and Actuators B: Chemical, 329*. https://doi.org/10.1016/j.snb.2020.129112, 129112.

Rezai, B., Jamei, H. R., & Ensafi, A. A. (2018). An ultrasensitive and selective electrochemical aptasensor based on rGO-MWCNTs/chitosan/carbon quantum dot for the detection of lysozyme. *Biosensors & Bioelectronics, 115*, 37–44.

Rizk, M., Attia, A. K., Mohamed, H. Y., & Elshahed, M. S. (2020). Validated voltammetric method for the simultaneous determination of anti-diabetic drugs, linagliptin and empagliflozin in bulk, pharmaceutical dosage forms and biological fluids. *Electroanalysis, 32*(8), 1737–1753. https://doi.org/10.1002/Elan.202000007.

Rocha, D. P., Squissato, A. L., Silva, S. M., Richter, E. M., & Munoz, R. A. A. (2020). Improved electrochemical detection of metals in biological samples using 3D-printed electrode: Chemical/electrochemical treatment exposes carbon-black conductive sites. *Electrochimica Acta, 335*, 135688.

Ross, N., & Nqakala, N. C. (2020). Electrochemical determination of hydrogen peroxide by a nonenzymatic catalytically enhanced silver-iron(III) oxide/polyoxometalate reduced graphene oxide modified glassy carbon electrode. *Analytical Letters, 53*, 2445–2464.

Roushani, M., Jalilian, Z., & Nezhadali, A. (2019). Screen printed carbon electrode sensor with thiol graphene quantum dots and gold nanoparticles for voltammetric determination of solatol. *Heliyon, 5*, e1984.

Roushani, M., Mohammadi, F., & Valipour, A. (2020). Electroanalytical sensing of asulam based on nanocomposite modified glassy carbon electrode. *Journal of Nanostructures, 10*(1), 128–139. https://doi.org/10.22052/JNS.2020.01.014.

Roychoudhury, A., Prateek, A., Basu, S., & Jha, S. K. (2018). Preparation and characterization of reduced graphene oxide supported nickel oxide nanoparticle-based platform for sensor applications. *Journal of Nanoparticle Research, 20*, 70.

Ruess, G., & Vogt, F. (1948). Hochstlamellarer kohlenstoff aus graphitoxyhydroxyd. *Monatshefte für Chemie, 78*, 222–242.

Saadati, A., Hassanpour, S., & Hasanzadeh, M. (2020). Lab-on-fruit skin and lab-on leaf towards recognition of trifluralin using Ag-citrate/GQDs nanocomposite stabilized on the flexible substrate: A new platform for the electroanalysis of herbicides using direct writing of nano-inks and pen-on paper technology. *Heliyon, 6*, e05779.

Sakthinathan, S., Kokulnathan, T., Chen, S. M., Karthik, R., Tamizhdurai, P., Chiu, T. W., et al. (2019). Simple sonochemical synthesis of cupric oxide sphere decorated reduced graphene oxide composite for the electrochemical detection of flutamide drug in biological samples. *Journal of the Electrochemical Society, 166*, B68–B75.

Sanna Jilani, B., Mruthyunjayachari, C. D., Malathesh, P., Mounesh, S. T. M., & Reddy, K. R. V. (2019). Electrochemical sensing based MWCNT-cobalt tetra substituted sorbaamide phthalocyanine onto the glassy carbon electrode towards the determination of 2-amino phenol: A voltammetric study. *Sensors and Actuators B: Chemical, 301*, 127078. https://doi.org/10.1016/J.Snb.2019.127078.

Santos, A. M., Wong, A., Prado, T. M., Fava, E. L., Fatibelo-Filho, O., Sotomayor, M. D. P. T., et al. (2021). Voltammetric determination of ethinylestradiol using screen-printed electrode modified with functionalized graphene, graphene quantum dots and magnetic nanoparticles coated with molecularly imprinted polymers. *Talanta, 224*, 121804.

Sarkar, T., Dhiman, T. K., Sajwan, R. K., Sri, S., & Solanki, P. R. (2020). Studies on carbon-quantum-dot-embedded iron oxide nanoparticles and their electrochemical response. *Nanotechnology*, *31*, 355502.

Selvi, S. V., Nataraj, N., Chen, S. M., & Prasannan, A. (2021). Electrochemical platform for the selective detection of azathioprine utilizing a screen-printed carbon electrode modified with manganese oxide/reduced graphene oxide. *New Journal of Chemistry*, *45*, 3640–3651.

Shafaei, S., Hosseinzadeh, E., Kanberoglu, G. S., Khalilzadeh, B., & Mohammad-Rezaei, R. (2021). Preparation of cerium oxide-MWCNTs nanocomposite bulk modified carbon ceramic electrode: A sensitive sensor for tamoxifen determination in human serum samples. *Journal of Materials Science: Materials in Electronics*. https://doi.org/10.1007/S10854-021-06019-W. Published Online.

Shaikshavali, P., Madhusudana Reddy, T., Venu Gopal, T., Venkataprasad, G., Narasimha, G., Narayana, A. L., et al. (2021). Development of carbon-based nanocomposite biosensor platform for the simultaneous detection of catechol and hydroquinone in local tap water. *Journal of Materials Science: Materials in Electronics*, *32*(4), 5243–5258. https://doi.org/10.1007/S10854-021-05256-3.

Shankar, S. S., Shereema, R. M., Ramachandran, V., Sruthi, T. V., Kumar, V. B. S., & Rakhi, R. B. (2019). Carbon quantum dot-modified carbon paste electrode-based sensor for selective and sensitive determination of adrenaline ACS. *Omega*, *4*, 7903–7910.

Sharifi, S., Zarei, E., & Asghari, A. (2021). Surfactant assisted electrochemical determination of noscapine and papaverine by TiO_2 nanoparticles/multi-walled carbon nanotubes modified carbon paste electrode. *Russian Journal of Electrochemistry*, *57*(2), 183–196. https://doi.org/10.1134/s1023193521020129.

Sharma, N. K., Monika, K. A., et al. (2021). Nanohybrid electrochemical enzyme sensor for xanthine determination in fish samples. *3. Biotech*, *11*(5), 212. https://doi.org/10.1007/S13205-021-02735-6.

Shen, L., Chen, P., Yan, B., & Zhang, C. (2015). A miniaturized optical emission spectrometry based on a needle-plate electrode discharge as a light source for room temperature hydrogen sensing with Pt/CNT nanocomposites. *Sensors and Actuators B: Chemical*, *215*, 9–14. https://doi.org/10.1016/j.snb.2015.03.044.

Shetti, N. P., Nayak, D. S., Malode, S. J., Kakarla, R. R., Shukla, S. S., & Aminabhavi, T. M. (2019). Sensors based on ruthenium-doped TiO_2 nanoparticles loaded into multi-walled carbon nanotubes for the detection of flufenamic acid and mefenamic acid. *Analytica Chimica Acta*, *1051*, 58–72. https://doi.org/10.1016/J.Aca.2018.11.041.

Shumeiko, V., Paltiel, Y., Bisker, G., Hayouka, Z., & Shoseyov, O. (2020). A paper-based near-infrared optical biosensor for quantitative detection of protease activity using peptide-encapsulated SWCNTs. *Sensors*, *20*, 1–14. https://doi.org/10.3390/s20185247.

Siddegowda, K. S., Mahesh, B., Chamaraja, N. A., Roopashree, B., Kumara Swamy, N., & Nanjundaswamy, G. S. (2020). Zinc oxide nanoparticles supported on multi-walled carbon nanotube modified electrode for electrochemical sensing of a fluoroquinolone drug. *Electroanalysis*, *32*(10), 2183–2192. https://doi.org/10.1002/elan.202000010.

Silva, S. C., Cardoso, R. M., Richter, E. M., Munoz, R. A. A., & Nossol, E. (2020). Reduced graphene oxide/multi-walled carbon nanotubes/prussian blue nanocomposites for amperometric detection of strong oxidants. *Materials Chemistry and Physics*, *250*, 123011.

Silva, R., De, O., Da Silva, É. A., Fiorucci, A. R., & Ferreira, V. S. (2019). Electrochemically activated multi-walled carbon nanotubes modified screen-printed electrode for voltammetric determination of sulfentrazone. *Journal of Electroanalytical Chemistry*, *835*, 220–226. https://doi.org/10.1016/j.jelechem.2019.01.018.

Simioni, N. B., Oliviera, G. G., Vicentini, F. C., Lanza, M. R. V., Janegitz, B. C., & Filho, O. F. (2017). Nanodiamonds stabilized in dihexadecyl phosphate film for electrochemical study and quantification of codeine in biological and pharmaceutical samples. *Diamond and Related Materials*, *74*, 191–196.

Simioni, N. B., Silva, T. A., & Oliveira, G. G. (2017). A nonodiamond-based electrochemical sensor for the determination of pyrazinamine antibiotic. *Sensors and Actuators B: Chemical*, *250*, 315–323.

Singh, J., Khanra, P., Kuila, T., Srivastava, M., Das, A. K., Kim, N. H., et al. (2013). Preparation of sulfonated poly(ether-ether-ketone) functionalized ternary graphene/AuNPs/chitosan nanocomposite for efficient glucose biosensor. *Process Biochemistry*, *48*, 1724–1735.

Šišoláková, I., Hovancová, J., Chovancová, F., et al. (2021). Zn nanoparticles modified screen printed carbon electrode as a promising sensor for insulin determination. *Electroanalysis*, *33*(3), 627–634. https://doi.org/10.1002/elan.202060417.

Šišoláková, I., Hovancová, J., Oriňaková, R., et al. (2020). Electrochemical determination of insulin at CuNPs/chitosan-MWCNTS and CoNPs/chitosan-MWCNTS modified screen printed carbon electrodes. *Journal of Electroanalytical Chemistry*, *860*. https://doi.org/10.1016/j.jelechem.2020.113881, 113881.

Sohouli, E., Keihan, A. H., Shahdost-Fard, S., Naghian, E., Plonska-Brzezinska, M., Rahmi-Nasrabi, M., et al. (2020). A glassy carbon electrode modified with carbon nanoonions for electrochemical determination of fentanyl. *Materials Science and Engineering*, *110*, 110684.

Song, Y., Zhao, M., Wang, X., Qu, H., Liu, Y., & Chen, S. (2019). Simultaneous electrochemical determination of catechol and hydroquinone in seawater using Co_3O_4/MWCNTs/GCE. *Materials Chemistry and Physics*, *234*, 217–223. https://doi.org/10.1016/J.Matchemphys.2019.05.071.

Sridara, T., Upan, J., Saianand, G., Tuantranont, A., Karuwan, C., & Jakmune, J. (2020). Non-enzymatic amperometric glucose sensor based on carbon nanodots and copper oxide nanocomposites electrode. *Sensors*, *20*, 808.

Stefan-Van Staden, R. I., Negut, C. C., Gheorghe, S. S., & Ciorîță, A. (2021). 3D stochastic microsensors for molecular recognition and determination of Heregulin-A in biological samples. *Analytical and Bioanalytical Chemistry*, *413*(13), 3487–3492. https://doi.org/10.1007/s00216-021-03295-7.

Suresh, L., Bondili, J. S., & Brahman, P. K. (2020). Fabrication of immunosensor based on polyaniline, fullerene-C_{60} and palladium nanoparticles nanocomposite: An electrochemical detection tool for prostate cancer. *Electroanalysis*, *32*(7), 1439–1448. https://doi.org/10.1002/elan.201900659.

Tajik, S., Dourandish, Z., Zhang, K., Beitollahi, H., Le, Q. V., Jang, H. W., et al. (2020). Carbon and graphene quantum dots: A review on syntheses, characterization, biological, and sensing applications for neurotransmitter determination. *RCS Advances*, *10*, 15406–15429.

Tan, Z., Wu, W., Feng, C., Wu, H., & Zhang, Z. (2020). Simultaneous determination of heavy metals by an electrochemical method based on a nanocomposite consisting of fluorinated graphene and gold nanocage. *Microchimica Acta*, *187*, 414.

Taouri, L., Bourouina, M., Bacha, S. B., & Hauchard, D. (2021). Fullerene-MWCNT nanostructured-based electrochemical sensor for the detection of vanillin as food additive. *Journal of Food Composition and Analysis*, *100*, 103811.

Teixeira, J. G., & Oliveira, J. (2021). Voltammetric study of the antihistamine drug bilastine: Anodic characterization and quantification using a reusable MWCNTs modified screen printed carbon electrode. *Electroanalysis*, *33*(4), 891–899. https://doi.org/10.1002/elan.202060494.

Terzi, F., Pelliciari, J., Zanardi, C., Pigani, L., Viinikanoja, A., Lukkari, J., et al. (2013). Graphene-modified electrode. Determination of hydrogen peroxide at high concentrations. *Analytical and Bioanalytical Chemistry*, *405*, 3579–3586.

Thamilselvan, A., Rajagopal, V., & Suryanarayanan, V. (2019). Highly sensitive and selective amperometric determination of BPA on carbon black/f-MWCNT composite modified GCE. *Journal of Alloys and Compounds*, *786*, 698–706. https://doi.org/10.1016/J.Jallcom.2019.02.020.

Tian, Y., Li, J. B., Wang, Y. H., Ding, C. F., Sun, Y. L., Sun, W. Y., et al. (2017). Gold nanoparticle-dotted, ionic liquid-functionalised, carbon hybrid material for ultra-sensitive detection of bisphenol a. *Environment and Chemistry*, *14*(6), 385–393. https://doi.org/10.1071/EN17081.

Tian, L., Wang, X., Wu, K., Hu, Y., Wang, Y., & Lu, J. (2019). Ultrasensitive electrochemiluminescence biosensor for dopamine based on ZnSe, graphene oxide@multi walled carbon nanotube and $Ru(Bpy)_3^{2+}$. *Sensors and Actuators B: Chemical*, *286*, 266–271. https://doi.org/10.1016/J.Snb.2019.01.161.

Tian, R., Chen, X., Liu, D., & Yao, C. (2016). A sensitive biosensor for determination of Cu^{2+} by one-step electrodeposition. *Electroanalysis*, *28*(7), 1617–1624. https://doi.org/10.1002/Elan.201501070.

Toghan, A., Abd-Elsabour, M., & Abo-Bakr, A. M. (2021). A novel electrochemical sensor based on EDTA-NQS/GC for simultaneous determination of heavy metals. *Sensors and Actuators, A: Physical*, *322*, 112603.

Vasilescu, I., Eremia, S. A. V., Kusko, M., Radoi, A., Vasile, E., & Radu, G. L. (2016). Molybdenum disulphide and graphene quantum dots as electrode modifiers for laccase biosensor. *Biosensors & Bioelectronics*, *75*, 232–237.

Verma, S., Choudhary, J., Singh, K. P., Chandra, P., & Singh, S. P. (2019). Uricase grafted nanoconducting matrix based electrochemical biosensor for ultrafast uric acid detection in human serum samples. *International Journal of Biological Macromolecules*, *130*, 333–341.

Vinoth, V., Natarajan, L. N., Mangalaraja, R. V., Valdes, H., & Anandan, S. (2019). Simultaneous electrochemical determination of dopamine and epinephrine using gold nanocrystals capped with graphene quantum dots in a silica network. *Microchimica Acta*, *186*, 681.

Wallace, P. R. (1947). The band theory of graphite. *Physics Review*, *71*, 622–634.

Wang, J. (2018). Near infrared optical biosensor based on peptide functionalized single-walled carbon nanotubes hybrids for 2,4,6-trinitrotoluene (TNT) explosive detection. *Analytical Biochemistry*, *550*, 49–53. https://doi.org/10.1016/j.ab.2018.04.011.

Wang, L. (2019). Direct fabrication of 3D graphene-multi walled carbon nanotubes network and its application for sensitive electrochemical determination of hyperin. *International Journal of Electrochemical Science*, *14*(1), 481–493. https://doi.org/10.20964/2019.01.42.

Wang, Q., & Wang, B. T. (2018). Surface plasmon resonance biosensor based on graphene oxide/silver coated polymer cladding silica fiber. *Sensors and Actuators B: Chemical*, *275*, 332–338. https://doi.org/10.1016/j.snb.2018.08.065.

Wang, H., Li, M., Zheng, Y., Hu, T., Chai, Y., & Yuan, R. (2018). An ultrasensitive photoelectrochemical biosensor based on $[Ru(dcbpy)_2dppz]^{2+}$/Rose Bengal dyes co-sensitized fullerene for DNA detection. *Biosensors & Bioelectronics*, *120*, 71–76. https://doi.org/10.1016/j.bios.2018.08.035.

Wang, C., Huang, S., Luo, L., Zhou, Y., Lu, X., Zhang, G., et al. (2019). Ultra thin two-dimension metal-organic framework nanosheets/multi-walled carbon nanotube composite films for the electrochemical detection of H_2O_2. *Journal of Electroanalytical Chemistry*, *835*, 178–185. https://doi.org/10.1016/j.jelechem.2019.01.030.

Wang, S., Li, J., Qiu, Y., Zhuang, X., Wu, X., & Jiang, J. (2019). Facile synthesis of oxidized multi-walled carbon nanotubes functionalized with 5-sulfosalicylic acid/MoS_2 nanosheets nanocomposites for electrochemical detection of copper ions. *Applied Surface Science*, *487*, 766–772. https://doi.org/10.1016/J.Apsusc.2019.05.161.

Wang, Z., Yue, H. Y., Huang, S., Yu, Z. M., Gao, X., Chen, H. T., et al. (2019). Gold nanoparticles anchored onto three-dimensional graphene: Simultaneous voltammetric determination of dopamine and uric acid. *Microchimica Acta*, *186*, 573.

Wang, Z., Yue, H. Y., Yu, Z. M., Huang, S., Gao, X., Wang, B., et al. (2019). A novel 3D porous graphene foam prepared by chemical vapor deposition using nickel nanoparticles: Electrochemical determination of levodopa in the presence of uric acid. *Microchemical Journal*, *147*, 163–169.

Wang, N., Tian, W., Zhang, H., Yu, X., Yin, X., Du, Y., et al. (2021). An easily fabricated high performance fabry-perot optical fiber humidity sensor filled with graphene quantum dots. *Sensors*, *21*, 1–11. https://doi.org/10.3390/s21030806.

Wang, Q., Li, L., Wang, X., Dong, C., & Shuang, S. (2020). Graphene quantum dots wrapped square-plate-like MnO_2 nanocomposite as a fluorescent turn-on sensor for glutathione. *Talanta*, *219*. https://doi.org/10.1016/j.talanta.2020.121180, 121180.

Wang, J., Wang, J., Li, W., & Yang, C. (2021). Study on enzymatic and electrochemical properties of cellulase immobilized with multi-walled carbon nanotubes as sensor for catechol. *International Journal of Electrochemical Science*, *16*, 1–12. https://doi.org/10.20964/2021.04.62.

Wang, Z., Wang, L., Zhou, C., & Sun, C. (2021). Determination of cesium ions in environmental water samples with a magnetic multi-walled carbon nanotube imprinted potentiometric sensor. *RSC Advances*, *11*(17), 10075–10082. https://doi.org/10.1039/D0RA09659H.

Wang, L., Yang, R., Li, J., Qu, L., & Harrington, P. B. (2019). A highly selective and sensitive electrochemical sensor for tryptophan based on the excellent surface adsorption and electrochemical properties of PSS functionalized graphene. *Talanta*, *196*, 309–316.

Wang, K. P., Zhang, Y. C., Zhang, X., & Shen, L. (2019). Green preparation of chlorine-doped graphene and its application in electrochemical sensor for chloramphenicol detection. *SN Applied Sciences*, *1*, 157.

Wei, Y., Xu, Z., Wang, S., Liu, Y., Zhang, D., & Fang, Y. (2020). One-step preparation of carbon quantum dots-reduced graphene oxide nanocomposite-modified glass carbon electrode for the simultaneous detection of ascorbic acid, dopamine, and uric acid. *Ionics*, *26*, 5817–5828.

Williams, R. M., Lee, C., & Heller, D. A. (2018). A fluorescent carbon nanotube sensor detects the metastatic prostate cancer biomarker uPA. *ACS Sensors*, *3*, 1838–1845. https://doi.org/10.1021/acssensors.8b00631.

Winiarski, J. P., Rampanelli, R., Bassani, J. C., Mezalira, D. Z., & Jost, C. L. (2020). Multi-walled carbon nanotubes/nickel hydroxide composite applied as electrochemical sensor for folic acid (vitamin B_9) in food samples. *Journal of Food Composition and Analysis*, *92*. https://doi.org/10.1016/j.jfca.2020.103511, 103511.

Wong, A., Santos, A. M., Silva, T. A., & Filho, O. F. (2018). Simultaneous determination of isoproterenol, acetaminophen, folic acid, prophanolol and caffeine using a sensor platform based on carbon black, graphene oxide, copper nanoparticles and PEDOT:PSS. *Talanta*, *83*, 329–338.

Wu, M., Zhu, J., Ren, Y., Yang, N., Hong, Y., Wang, W., et al. (2020). NH_2-GQDs-doped nickel-cobalt oxide deposited on carbon cloth for nonenzymatic detection of glucose. *Advanced Materials Interfaces, 7*, 1901578.

Xing, Y., Wu, G., Ma, Y., Yu, Y., Yuan, X., & Zhu, X. (2019). Electrochemical detection of bisphenol B based on poly(prussian blue)/carboxylated multi-walled carbon nanotubes composite modified electrode. *Measurement, 148*. https://doi.org/10.1016/J.Measurement.2019.106940, 106940.

Xu, S., Che, S., Ma, P., Zhang, F., Xu, L., Liu, X., et al. (2019). One-step fabrication of boronic-acid-functionalized carbon dots for the detection of sialic acid. *Talanta, 197*, 548–552. https://doi.org/10.1016/j.talanta.2019.01.074.

Xu, Z., Jiang, X., Liu, S., & Yang, M. (2020). Sensitive and selective molecularly imprinted electrochemical sensor based on multi-walled carbon nanotubes for doxycycline hyclate determination. *Chinese Chemical Letters, 31*(1), 185–188. https://doi.org/10.1016/J.Cclet.2019.04.026.

Yang, B., Bin, D., Zhang, K., Du, Y., & Majima, T. (2018). A seed-mediated method to desing N-doped graphene supported gold-silver nanothorns sensor for rutin dedection. *Journal of Colloid and Interface Science, 512*, 446–454.

Yang, L., Wang, J., Lü, H., & Hui, N. (2021). Electrochemical sensor based on prussian blue/multi-walled carbon nanotubes functionalized polypyrrole nanowire arrays for hydrogen peroxide and microRna detection. *Microchimica Acta, 188*(1), 25. https://doi.org/10.1007/s00604-020-04673-1.

Yao, J., Li, Y., Xie, M., Yang, Q., & Liu, T. (2020). The electrochemical behaviors and kinetics of AuNPs/N,S-GQDs composite electrode: A novel label-free amplified BPA aptasensor with extreme sensitivity and selectivity. *Journal of Molecular Liquids, 320*, 114384.

Yao, J., Liu, C., & Yang, M. (2019). An ultrasensitive and highly selective electrochemical aptasensor for environmental endocrine disrupter bisphenol A determination using gold nanoparticles/nitrogen, sulfur, and phosphorus co-doped carbon dots as signal enhancer and its electrochemical kinetic research. *Journal of the Electrochemical Society, 166*(13), B1161–B1170.

Ye, Y., Wang, L., Liu, K., & Li, J. (2020). A label-free and sensitive electrochemiluminescence sensor based on a simple one-step electrodeposition of GO/ZnS modified electrode for trace copper ions detection. *Microchemical Journal, 155*, 104749.

Yence, M., Cetinkaya, A., Ozcelikay, G., Kaya, S. I., & Ozkan, S. A. (2021). Boron-doped diamond electrodes: Recent developments and advances in view of electrochemical drug sensors. *Critical Reviews in Analytical Chemistry*. https://doi.org/10.1080/10408347.2020.1863769.

Yi, J., Liu, Z., Liu, J., Liu, H., Xia, F., Tian, D., et al. (2020). A label-free electrochemical aptasensor based on 3D porous CS/rGO/GCE for acetamiprid residue detection. *Biosensors & Bioelectronics, 148*, 111827.

Yola, M. L. (2019). Electrochemical activity enhancement of monodisperse boron nitride quantum dots on graphene oxide: Its application for simultaneous detection of organophosphate pesticides in real samples. *Journal of Molecular Liquids, 277*, 50–57.

Yola, M. L., & Atar, N. (2018). A novel detection approach for serotonin by graphene quantum dots/two dimensional (2D) hexagonal boron nitride nanosheets with molecularly imprinted polymer. *Applied Surface Science, 458*, 648–655.

Yola, M. L., & Atar, N. (2019a). Simultaneous determination of β-agonists on hexagonal boron nitride nanosheets/multi-walled carbon nanotubes nanocomposite modified glassy carbon electrode. *Materials Science and Engineering: C, 96*, 669–676. https://doi.org/10.1016/j.msec.2018.12.004.

Yola, M. L., & Atar, N. (2019b). Development of molecular imprinted sensor including graphitic carbon nitride/N-doped carbon dots composite for novel recognition of epinephrine. *Composites Part B Engineering, 175*, 107113.

You, F., Wei, J., Cheng, Y., Wen, Z., Ding, C., Guo, Y., et al. (2020). A sensitive and stable visible-light-driven photoelectrochemical aptasensor for determination of oxytetracycline in tomato samples. *Journal of Hazardous Materials, 398*, 122944.

Yousefi, A., & Babaei, A. (2019). A new sensor based on glassy carbon electrode modified with Fe_3O_4@MCM-48-SO_3H/multi-wall carbon nanotubes composite for simultaneous determination of norepinephrine and tyrosine in the presence of ascorbic acid. *Ionics, 25*(6), 2845–2856. https://doi.org/10.1007/s11581-018-2815-9.

Yousefi, N., Irandoust, M., & Haghighi, M. (2020). New and sensitive magnetic carbon paste electrode for voltammetry determination of morphine and methadone. *Journal of the Iranian Chemical Society, 17*(11), 2909–2922. https://doi.org/10.1007/s13738-020-01962-7.

Yu, S., Ding, L., Lin, H., Wu, W., & Huang, J. (2019). A novel optical fiber glucose biosensor based on carbon quantum dots-glucose oxidase/cellulose acetate complex sensitive film. *Biosensors & Bioelectronics, 146*. https://doi.org/10.1016/j.bios.2019.111760, 111760.

Yu, H., Li, R., & Song, K. (2019). Amperometric determination of nitrite by using a nanocomposite prepared from gold nanoparticles, reduced graphene oxide and multi-walled carbon nanotubes. *Microchimica Acta, 186*(9), 624. https://doi.org/10.1007/s00604-019-3735-8.

Yuan, M. M., Zou, J., Guan, J. F., Huang, Z. N., & Yu, J. G. (2020). Highly sensitive and selective determination of p-nitrophenol at an interpenetrating networks structure of self-assembled rod-like lanthanum hydroxide-oxidized multi-walled carbon nanotubes nanocomposite. *Ecotoxicology and Environmental Safety, 201*. https://doi.org/10.1016/j.ecoenv.2020.110862, 110862.

Yue, H. Y., Song, S. S., Huang, S., Zhang, H., Gao, X. P. A., Gao, X., et al. (2017). Preparation of MoS_2-graphene hybrid nanosheets and simultaneously electrochemical determination of levodopa and uric acid. *Electroanalysis, 29*, 2565–2571.

Yue, H. Y., Song, S. S., Huang, S., Gao, X., Wang, B., Guan, E. H., et al. (2018). Preparation of three-dimensional hollow graphene balls and simultaneous electrochemical determination of dopamine and uric acid. *Journal of Materials Science: Materials in Electronics, 29*, 12330–12339.

Yue, H. Y., Wu, P. F., Huang, S., Gao, X., Wang, Z., Wang, W. Q., et al. (2019). Electrochemical determination of levodopa in the presence of uric acid using ZnO nanoflowers-reduced graphene oxide. *Journal of Materials Science: Materials in Electronics, 30*, 3984–3993.

Yue, H. Y., Zhang, H. J., Huang, S., Gao, X., Song, S. S., Zhao, W., et al. (2019). A novel non-enzymatic dopamine sensors based on NiO-reduced graphene oxide hybrid nanosheets. *Journal of Materials Science: Materials in Electronics, 30*, 5000–5007.

Zainul, R., Azis, N. A., Md Isa, I., Hashim, N., Ahmad, M. S., Saidin, M. I., et al. (2019). Zinc/aluminium-quinclorac layered nanocomposite modified multi-walled carbon nanotube paste electrode for electrochemical determination of bisphenol A. *Sensors, 19*(4), 941. https://doi.org/10.3390/S19040941.

Zambianco, N. A., Da Silva, V. A. O. P., Orzari, L. O., et al. (2020). Determination of tadalafil in pharmaceutical samples by vertically oriented multi-walled carbon nanotube electrochemical sensing device. *Journal of Electroanalytical Chemistry, 877*. https://doi.org/10.1016/j.jelechem.2020.114501, 114501.

Zeybek, B., & Üğe, A. (2021). Development of carbon-based sensors for electrochemical quantification of vitamins B_2 and B_6 at nanomolar levels. *Chemical Papers, 75*(4), 1323–1339. https://doi.org/10.1007/S11696-020-01387-9.

Zhang, S., Chen, X., Liu, G., Hou, X., Huang, Y., Chen, J., et al. (2015). A novel sensing platform based on ionic liquid integrated carboxylic-functionalized graphene oxide nanosheets for honokiol determination. *Electrochimica Acta, 155*, 45–53.

Zhang, D., Chen, X., Ma, W., Yang, T., Li, D., Dai, B., et al. (2019). Direct electrochemistry of glucose oxidase based on one step electrodeposition of reduced graphene oxide incorporating polymerized L-lysine and its application in glucose sensing. *Materials Science and Engineering, 104*, 109880.

Zhang, C., Hao, T., Lin, H., Qi, W., Wu, Y., Kang, K., et al. (2020). One-step electrochemical sensor based on an integrated probe toward sub-ppt level Pb^{2+} detection by fast scan voltammetry. *Analytica Chimica Acta, 1128*, 174–183. https://doi.org/10.1016/j.aca.2020.07.007.

Zhang, W., Xiao, G., Chen, J., Wang, L., Hu, Q., Wu, J., et al. (2021). Electrochemical biosensors for measurement of colorectal cancer biomarkers. *Analytical and Bioanalytical Chemistry, 413*, 2407–2428.

Zhang, H., Zhang, B., Chen, A., & Qin, Y. (2017). Controllable n-Fe_2O_3@graphene nanomaterials by ALD applied in an aptasensor with enhanced electrochemical performance for thrombin detection. *Dalton Transactions, 46*, 7434–7440.

Zhao, Y., Qin, J., Xu, H., Gao, S., Jiang, T., Zhang, S., et al. (2019). Gold nanorods decorated with graphene oxide and multi-walled carbon nanotubes for trace level voltammetric determination of ascorbic acid. *Microchimica Acta, 186*(1), 17. https://doi.org/10.1007/S00604-018-3138-2.

Zhao, R., Wu, X., Gao, Y., Liu, Y., Gao, J., Chen, Y., et al. (2020). A unique bimetallic MOF derived carbon-MWCNTS hybrid structure for selective electrochemical determination of lead ion in aqueous solution. *Microchemical Journal, 158*. https://doi.org/10.1016/j.microc.2020.105271, 105271.

Zhao, Z., Wang, W., Tang, W., Xie, Y., Li, Y., Song, J., et al. (2020). Synthesis and electrochemistry performance of CuO-functionalized CNTs-rGO nanocomposites for highly sensitive hydrazine detection. *Ionics, 26*, 2599–2609.

Zhao, Z., Zheng, J., Nguyen, E. P., Tao, D., Cheng, J., Pan, H., et al. (2020). A novel SWCNT-amplified "signal-on" electrochemical aptasensor for the determination of trace level of bisphenol A in human serum and lake water. *Microchimica Acta, 187*(9), 500. https://doi.org/10.1007/S00604-020-04475-5.

Zhao, H., Ma, H., Li, X., Liu, B., Liu, R., & Komarneni, S. (2021). Nanocomposite of halloysite nanotubes/multi-walled carbon nanotubes for methyl parathion electrochemical sensor application. *Applied Clay Science, 200*. https://doi.org/10.1016/j.clay.2020.105907, 105907.

Zhao, Y., Li, X. G., Zhou, X., & Zhang, Y. N. (2016). Review on the graphene based optical fiber chemical and biological sensors. *Sensors and Actuators B: Chemical, 231*, 324–340. https://doi.org/10.1016/j.snb.2016.03.026.

Zhao, X., Zheng, L., Yan, Y., Cao, R., & Zhang, J. (2021). An electrocatalytic active AuNPs/5-Amino-2-mercaptobenzimidazole/rGO/SPCE composite electrode for ultrasensitive detection of progesterone. *Journal of Electroanalytical Chemistry, 882*, 115023.

Zheng, S. (2019). Electrochemical determination of hydroquinone and catechol using multi-walled carbon nanotubes/eosin Y modified glassy carbon electrode. *International Journal of Electrochemical Science, 14*(7), 6234–6246. https://doi.org/10.20964/2019.07.71.

Zheng, S., Huang, R., Ma, X., Tang, J., Li, Z., Wang, X., et al. (2018). A highly sensitive dopamine sensor based on graphene quantum dots modified glassy carbon electrode. *International Journal of Electrochemical Science, 13*, 5723–5735.

Zhou, X., Qu, Q., Wang, L., Li, L., Li, S., & Xia, K. (2020). Nitrogen dozen carbon quantum dots as one dual function sensing platform for electrochemical and fluorescent detecting ascorbic acid. *Journal of Nanoparticle Research, 22*, 20.

Zhou, J., Li, S., Noroozifar, M., & Kerman, K. (2020). Graphene oxide nanoribbons in chitosan for simultaneous electrochemical detection of guanine, adenine, thymine and cytosine. *Biosensors, 10*, 30.

Zhu, Y., Lu, S., Manohari, A. G., Dong, X., Chen, F., Xu, W., et al. (2017). Polydopamine interconnected graphene quantum dots and gold nanoparticles for enzymeless H_2O_2 detection. *Journal of Electroanalytical Chemistry, 796*, 75–81.

Zhu, Y., Huai, S., Jiao, J., Xu, Q., Wu, H., & Zhang, H. (2020). Fullerene and platinum composite-based electrochemical sensor for the selective determination of catechol and hydroquinone. *Journal of Electroanalytical Chemistry, 878*. https://doi.org/10.1016/j.jelechem.2020.114726, 114726.

Zhu, X., Zhang, K., Wang, C., Guan, J., Yuan, X., & Li, B. (2021). Quantitative determination and toxicity evaluation of 2,4-dichlorophenol using poly(eosin Y)/hydroxylated multi-walled carbon nanotubes modified electrode. *Scientific Reports, 11*(1), 5864. https://doi.org/10.1038/S41598-021-84930-2.

Zhu, Y., Xu, Z., Gao, J., Ji, W., & Zhang, J. (2020). An antibody-aptamer sandwich cathodic photoelectrochemical biosensor for the detection of progesterone. *Biosensors & Bioelectronics, 160*, 112210.

Zhuang, X., Wang, H., He, T., & Chen, L. (2016). Enhanced voltammetric determination of dopamine using a glassy carbon electrode modified with ionic liquid-functionalized graphene and carbon dots. *Microchimica Acta, 183*, 3177–3182.

Ziyatdinova, G., & Budnikov, H. (2021). MWNT-based electrode for the voltammetric quantification of carvacrol. *Food Analytical Methods, 14*(2), 401–410. https://doi.org/10.1007/S12161-020-01895-0.

Ziyatdinova, G. K., Guss, E. V., Morozova, E. V., & Budnikov, H. C. (2021). An electrode based on electropolymerized sunset yellow for the simultaneous voltammetric determination of chlorogenic and ferulic acids. *Journal of Analytical Chemistry, 76*(3), 371–380. https://doi.org/10.1134/S1061934821030163.

Ziyatdinova, G., Ziganshina, E., Shamsevalieva, A., & Budnikov, H. (2020). Voltammetric determination of capsaicin using CeO_2-surfactant/SWNT-modified electrode. *Arabian Journal of Chemistry, 13*(1), 1624–1632. https://doi.org/10.1016/J.Arabjc.2017.12.019.

Zrinski, I., Martinez, S., Ortner, A., Samphao, A., Zavasnik, J., Kalcher, K., et al. (2021). A novel sensor based on carbon paste electrode modified with polypyrrole/multi-walled carbon nanotubes for the electrochemical detection of cytostatic drug rapamycin. *Electroanalysis, 33*(5), 1325–1332. https://doi.org/10.1002/elan.202060527.

Chapter 13

Carbon nanomaterial-based sensors for wearable health and environmental monitoring

Maryam Rezaie and Morteza Hosseini

Department of Life Science Engineering, Faculty of New Sciences & Technologies, University of Tehran, Tehran, Iran

Introduction

With increasing life expectancy and the continuing development of state-of-the-art technologies, humankind's living standards also rise, creating a growing demand for personalized health and environmental monitoring handy equipment. In the modern decades, scientists have shown a great interest to wearable tools due to their tremendous promise for a plethora of health and environmental applications. This type of wireless sensing device has opened up a new exciting door in the healthcare and environmental industry. So, a large number of wearable sensors and biosensors have been fabricated to monitor in real time a variety of biological analytes that exist in sweat, tears, or saliva as indicators. However, current wireless health-monitoring devices have technical limitations in fabrication and design (Heo, Hossain, & Kim, 2020).

Considerable progress in sensing technology has led to the development of commercial hand-held sensors, such as iSTAT by Abbot, ACCU-CHEK by Roche Diagnostics, and Lactate Scout by Sports Resource Group to measure metabolites and electrolytes (Bandodkar & Wang, 2014). Many glucometers also are in the market that people are using them, such as CONTOURNEXT EZ meter, which can monitor glucose level for 3 months, and one of its types is a tiny fluorescent sensor placed just under the skin, which is a small incision closed with Steri-Strips. Manufacturers of the AUVON DS-W test strips use automatic carbon printing techniques to be much more stable and accurate. OneTouch Ultra 2 m is a fast and simple tool to see the effect of food on blood sugar results. Prodigy is an innovative product to meet the evolving needs of consumers and with a mission to set standards of excellence for providing blood glucose monitoring systems and other diabetes products to distributors for their patients. Hence, all of these devices have paid the way for designing the new generation of wearable sensors for a dozen of biological analytes, increasing the standard of life.

The up-to-date wearable sensors are light and low-power enough to be embedded in not only smartphones but electronic watches, glasses, and bands that can be easily carried. To reach this purpose in practical wearable health-monitoring sensors, it is essential to utilize advanced functional materials and up-to-date technologies for manufacturing. Among the various materials for these handy tools, carbon-based nanomaterials are considered a promising candidate to improve the devices' electrical and mechanical performance. These new types of sensors have significantly improved portability, sensitivity, selectivity, and signal-to-noise ratio. Carbon nanotubes (CNTs) (Jian et al., 2017a, 2017b), graphene-based materials (Pur et al., 2016) (An et al., 2016), carbon black (CB) (Hu et al., 2020), single-walled carbon nanotubes (SWCNTs) (Tahernejad-Javazmi, Shabani-Nooshabadi, & Karimi-Maleh, 2018), and other two-dimensional (2-D) carbon allotropes (Mesgari et al., 2020) have been extensively used in wearable sensors because of their low cost and potentials for mass production, high chemical and thermal stabilities, inherent flexibility, and ease of chemical functionalization. Such real-time monitoring can provide information on wellness and health, enhance the management of chronic diseases, and alert the user or medical professionals of abnormal or unforeseen situations. Wearable biosensors can prevent painful and risky blood sampling procedures and can be readily blended with a wearer's daily routine (Fig. 13.1).

It should be mentioned that, among various advantages of 2-D structures, the remarkable characteristics of graphene are high electron mobility of $\sim 200,000\,cm^2/V\,s$, thermal conductivity of $5300\,W\,m^{-1}\,1\,K^{-1}$ (Neto et al., 2009), surface area-to-volume ratio of $2630\,m^2/g$, Young's modulus of $0.5\,TPa$, and resistivity of $10^{-6}\,W\,cm$ (Qian, Ismail, & Stein, 2014). It has now been used as a sensor and sometimes as a signal transducer in different prototypes for wearable and mobile health devices (Peña-Bahamonde et al., 2018).

248 PART | IV Role of carbon nanomaterials-based sensors in sustainability

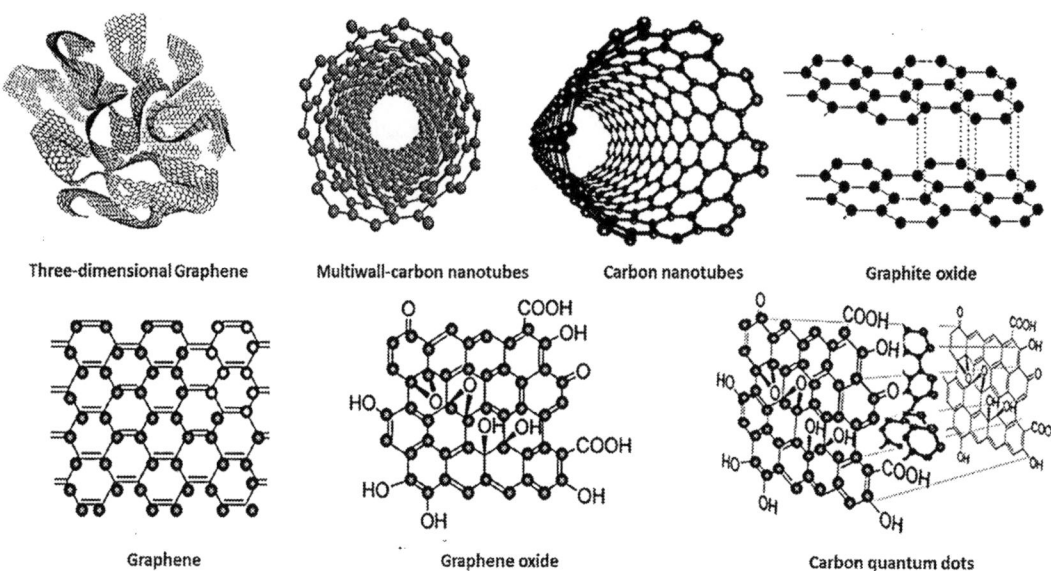

FIG. 13.1 Various types of carbon-based materials. *(No permission required.)*

This chapter classifies applications of wearable sensors into the two main groups in the health and environment sector. In each part, several types of sensors and biosensors have been accounted for, and the role and the effect of carbon-based material were revealed. In the end, the current limitation hindering the commercialization of wireless sensors and biosensors will be briefly discussed.

Health monitoring

The human body is a biological system that includes different sensors such as the nose, the fingers, and the mouth. Thousands of physiological signals are being created every second in the molecular and biological paths, describing the body's condition (Yang & Yang, 2006). The quality and the quantity of the physiological signals are markedly noteworthy for early diagnoses and therapies' operation (Bao, Zhang, & Shen, 2006). In new decades, wearable health-monitoring sensors have become a part of our daily life. The real-time monitoring of physiological data using wireless sensors provides a great deal of essential and vital information regarding people's bodies. These sensors commonly interact with an external or host computer to provide information about the human's physical and physiological conditions such as activities, stress, sleep, and blood pressure (Stoppa & Chiolerio, 2014). To improve the data quality, it is important to detect these signals in real time. Therefore, utilizing a material with excellent properties that meet these needs is a pressing issue.

Metabolite sensing

The metabolites concentration in biofluid is one of the most critical factors for assessing the clinical risks, diagnosis, and monitoring the treatment process. Moreover, circulating nutrients and metabolites in the biological system provide valuable indicators of biological processes. The abnormal level of their nutrition concentrations is associated with health conditions such as metabolic syndrome and cardiovascular disease (Cool et al., 2006).

What is more, metabolites are intermediates or products of biological metabolic reactions in the various molecular pathways. Any imbalance of these metabolic pathways, either owing to abrupt fluctuations or underlying severe chronic disease, can significantly pass the way to identify irregularities in metabolic patterns. The major metabolites that act as the probe in the body include glucose, lactate, ions, and electrolytes (sodium, potassium, chloride, and hydrogen/pH) (Wiorek et al., 2020; Yokus et al., 2020). Simultaneous monitoring of several metabolic analytes provides a comprehensive and accurate metabolic profile, describing personal control and timely medical intervention. Therefore, because of healthcare costs and the increasing world's aging population, there is a pressing need to personalize wireless devices for monitoring the health status of patients while they are out of the hospital. Metabolite wearable sensors markedly help people control or monitor their health conditions in every situation. These personalized home-based tools offer users several fantastic options: high sensitivity, fast response to cost-effectiveness, and user-friendly (Bhide et al., 2021).

Glucose

Diabetes mellitus, commonly known as diabetes, is a metabolic disease that is one of the most widely spread diseases in the world (Salek-Maghsoudi et al., 2018). This disease affects thousands of people's lives, and unfortunately, it ranks among the leading causes of death globally. Diabetic patients usually check their blood glucose levels both before and after administrating routine insulin shots. Monitoring blood glucose levels is crucial for patients because maintaining glucose levels within the physiological range helps them understand diabetes progression toward optimal disease management (Kim, Campbell, & Wang, 2018). Such continuous monitoring provides a wealth of information regarding the direction, duration, and fluctuations of glucose level, improving the treatment quality of people with diabetes.

Considerable research from the 1980s has paid close attention to introduce self-testing blood glucometers (Clarke & Foster, 2012). Since then, various types of glucose sensors have developed, and some of them are commercialized in the world. A new generation of these portable, wearable, and even implantable sensing devices is being designed and fabricated by emerging a myriad of carbon structures. Carbon-based networks have a dramatic effect on the sensitivity of the device so that it can detect glucose concentration in low amounts due to high electron mobility, large surface area, and high conductivity.

Tear is one of the biofluids that the level of glucose in it is detectable. A study developed a wearable sensor to monitor glucose within tears by a contact lens, and according to the in vivo and in vitro tests on a live rabbit, the results were reliable. The graphene-based hybrid metal nanowire was a prominent part of this sensor, which this structure offers sufficient transparency (>91%) and stretchability (~25%) for a reliable and comfortable vision for users (Kim et al., 2017). A wearable sweat patch is developed in the other study based on screen-printing technology on a silicon-based platform (polydimethylsiloxane, PDMS). Glucose oxidase (GOx) and ion-selective electrodes (ISE) were immobilized on the electrode surface to simultaneously detect glucose, sodium, potassium, and pH value. The reproducibility results were impressive for Na, K, and pH, respectively, RSD of 1.8%, 2.3%, and 1.5%. One of the crucial points in this research is that temperature had a negligible influence on the ISE performance, so on-body tests reflected the individual sweat physiological characteristics (Xu et al., 2019). Different wearable sensors based on carbon structures have been fabricated to monitor glucose, summarized in Table 13.1.

Uric acid (UA)

Uric acid is the end product of purine metabolic pathways. The existence of uric acid in human pathological findings in serum and urine samples included severe clinical implications for patients. UA is an important risk factor for diagnosing cardiovascular disease, diabetes especially type 2, and renal disease, and above all of them, it is practical in healthcare clinical to manage gout that is affecting millions of people (Baker et al., 2005). Hence, a wearable sensor can pave this way for doctors and themselves patients and affect the quality of their life directly.

TABLE 13.1 Several types of wireless sensors for glucose monitoring.

Electrode	Technique	Analyte	Linear range	LOD	Ref.
rGO/Au/Pt	Chronoamperometry	sweat	0.005–0.3 mM	5 μm	Xuan, Yoon, and Park (2018)
Pt NPs-rGO	Chronoamperometric	Interstitial fluid	0.006–0.7 mM	2.8 μM	Lipani et al. (2018)
Single-layer graphene/AuNPs	Amperometric	Sweat	0–162 mg/dL	1.44 mg/dL	Pu et al. (2016)
SWCN/Nafion	Amperometric	Sweat	50–1 mM	50 μM	Kang, Park, and Ha (2019)
Gold-doped CVD graphene	CV	Sweat	10 μM–0.7 mM	–	Lee et al. (2016)
Graphene-AgNWs hybrid	Amperometric	Tear fluid	1 μM–10 mM	0.4 μM	Kim et al. (2017)

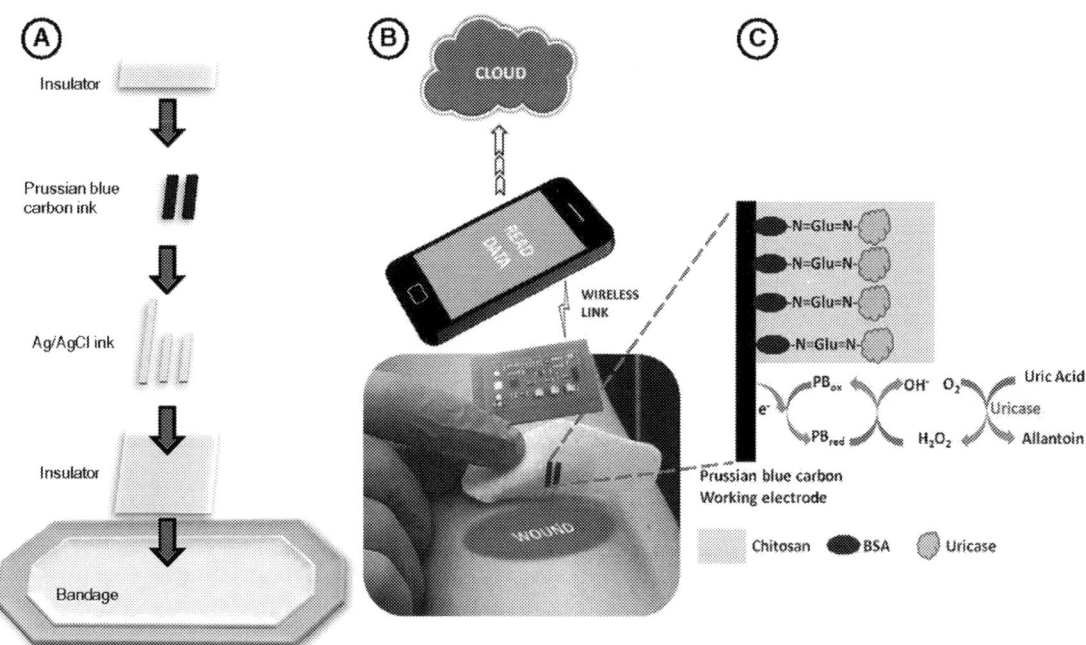

FIG. 13.2 Schematic of the wireless bandage biosensor to monitor uric acid. *(No permission required.)*

A sensing bandage for uric acid with a wireless connection feature was constructed by Kassal et al. The commercial bandage was decorated with Prussian blue (PB), and also a carbon electrode was screen-printed on it. The enzyme uricase was immobilized on the electrode surface to act specifically. Oxidation of UA produces H_2O_2, and PB-carbon electrode reduces this mediator. The working potential for the detection of UA was achieved at a low negative potential. The fabricated bandage was connected to a potentiostat device to measure and store the biosensor's current output and transfer data wirelessly (Kassal et al., 2015).

What is more, the level of UA in wound fluid is correlated with wound severity and its healing, and in consequence, it has turned into a promising biomarker to assess wound status and also monitoring of wound-healing progress (RoyChoudhury et al., 2018). For example, a UA sensor was fabricated by highly conductive materials, a polyester substrate, carbon ink, and uricase enzyme. The designed platform has a steady and constant output in laboratory-simulated conditions of wound fluid (Liu & Lillehoj, 2017). This biosensor showed high selectivity, excellent linear range, from 0 to 800 μM, and good reproducibility. Repeatability tests of it also were performed, revealing accurate measurements for up to 7h (Fig. 13.2).

Urea

Urea is a vital nitrogen metabolite in body fluids, including serum, urine, and sweat. This metabolic product is also a marker to examine renal functions and metabolic heart disorders in clinical tests (Matsumoto et al., 2019). Higher levels of toxic urea were found in the sweat of people with hepatic or renal failure or uremia than those of ordinary people.

Although a wealth of knowledge was achieved for fabricating a wireless sensor to detect and measure biological analytes, no device had been witnessed to measure urea levels in body fluids until the last few years. Therefore, developing a wearable device for urea analysis is a great achievement for personalized healthcare monitoring for humankind (Senel, Dervisevic, & Voelcker, 2019). A wearable sweat urea sensor based on a new technology of surface molecularly imprinted electrodes (MIP) was designed. The electrode was printed by the polymerized network of carbon nanotubes (CNTs), gold nanotubes (Au NTs), and urea-PEDOT. After templating, a low detection limit of 0.1 mM was observed. Also, according to the detected range of the sensor during on-body testing, it has the potential to screen for levels of sweat urea (Liu & Lillehoj, 2017).

Lactate monitoring

Clinically, blood lactate levels are frequently used to monitor tissue hypoxia. The utilization of pyruvate depends on oxygen presence, so that decreases in cellular oxygen delivery should increase lactate production and blood lactate levels (Mason et al., 2017). In 1927, the relationship between increased blood lactate levels and oxygen debt in patients with circulatory shock was introduced, and detection of lactate levels has been used in patients' treatment (Bakker, Nijsten, & Jansen, 2013). However, lactate is a typical metabolism product so the relationship between raised lactate levels in patients and shock could be a severe warning (Tur-García et al., 2017). Therefore, it is no surprise that clinical signs of critical illness have not met the reflection of this complex metabolic disturbance. The mentioned points highlight the importance of measuring lactate levels rather than estimating them from other variables.

Because of the labor-intensive aspects of the early and fast detection techniques, utilizing them has been roughly limited in the world. So, a great deal of time needs to make a decision about therapeutic. Fabricating a decorated wearable electrode provides a situation for doctors to measure lactate levels within approximately 2 min in a minimal amount (130 μL) of biofluid, while the average concentration of lactate in the blood is 1.3 mmol/L (Karpova et al., 2020). Using such a device can simultaneously give a wealth of information regarding hemoglobin levels, oxygen saturation, and blood lactate levels. It should be mentioned that the technology used to the development of lactate biosensors is very similar to glucose sensors, paving the way for detecting lactate levels quickly. Reduction and oxidation of H_2O_2 are related to L-lactate concentration. The saliva-based lactate has modified by ZnO NPs, MWCNT, and LOx on a GCE surface and then indicated good repeatability and quick detection time (20s), and versatility (Alam et al., 2018).

Ascorbic acid

Vitamin C, or ascorbic acid, plays a significant role in several functions in the biological system, such as the production process of collagen, wound healing process, and skincare, the absorption of iron, the treatment of cold (DePhillipo et al., 2018). This vitamin plays a part in helping the immune system protect the body against viral infections and neurological disorders, cardiac, gum, and even cancer diseases. Vitamin C is also involved in gene expression as well as various cellular metabolic processes in a nonantioxidant role (Kim et al., 2018).

On the other hand, it is a water-soluble vitamin that is naturally present in some foods and is added to foodstuffs as a conservative as well as being available as a dietary supplement. Unlike most animals, humans themselves are unable to synthesize vitamin C in biological cycles, so they must obtain it from their diet (Collie et al., 2020).

The ability to monitor the intake of food and vitamins continuously could greatly affect people's lifestyles. Wearable sensors for nutrient monitoring can help prevent and manage nutritional imbalance and provide personalized guidance toward maintaining nutrient levels. To be more precise, this information could be urgently needed for maintaining balanced vitamin levels in the body. However, conventional vitamin monitoring techniques are usually time-consuming, rely on expensive facilities, and are invasive methods. In addition, the properties of this vitamin change significantly in environmental conditions, such as temperature, pH, and pressure; stabilizing the samples right after sampling is crucial in the laboratory. Wearable sensors introduce a new generation of clinical tools to monitor ascorbic acid levels and detect it in a few time, without extra high-tech instruments.

Electrolytes sensing

Analyzing electrolytes such as sodium, potassium, calcium, chloride, and nitrite in bodily fluids has a significant diagnostic value early and fast detecting a dozen biological conditions, from kidney disorder and urinary stone disease to urinary tract infection and cystic fibrosis. These days, sodium and potassium intake is commonly analyzed through urine samples. Chloride ion is one factor for diagnosing several disorders such as cystic fibrosis and diabetic acidosis, while the measurement of nitrite is also performed to estimate the high probability of a urinary tract infection (UTI) as nitrite, which is a metabolic output of pathogens of the urinary tract (Ghaderinezhad et al., 2020). Various analytical techniques are practical to determine the cation and anion concentrations in the urine sample. Flame emission spectrophotometry, ion-selective electrodes, and ion chromatography are prevalent clinical methods. Yet, all of these tests need a long time to perform and analyze, and also people have to spend good money on them. Portable sensors provide a condition that everybody is able to monitor these ions regularly via their handy device such as cellphones, controlling the consumption of these ingredients. These biosensors are sensitive and simple to use so that people, without considering their place, even in far-flung corners of the world, and time, can use them. Wearable sensors based on carbon structure nanomaterials have met all these properties, developing quickly.

Pirovano's group has fabricated a platform with the name of "SwEatch," a biomedical device that can draw sweat from the person's skin and measure Na^+ ions concentration continuously through designed electrodes. They have reported an improved platform model, which opens a new window to the simultaneous detection and measurement of the multiple

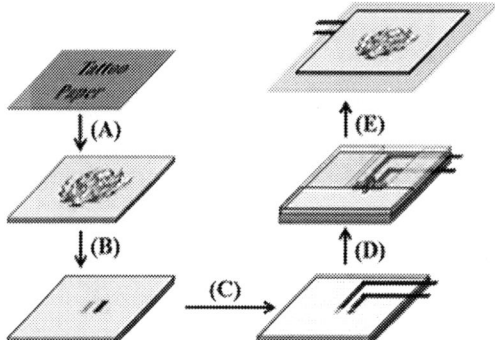

FIG. 13.3 Schematic revealing the fabrication steps for the Na-tattoo wireless sensor. The first step involves (A) screen printing an insulator coating on the tattoo platform. (B) Printed two electrodes and a layer of ink. (C) Printed another layer of insulator. (D) Modification of two electrodes. (E) The Na-tattoo sensor is applied to the PET. *(No permission required.)*

analytes concentration such as Na^+, K^+, and Cl^-. In this research, carbon ink that acts as a conductive layer and the dielectric layer were screen-printed on PET substrate to fabricate a working modified electrode.

Sodium is the primary cation located within the human body's extracellular fluid space and accounts for 80%–90% of plasma osmolality. The lack of electrolytes in sweat has been related to several homeostatic physiological imbalances (Brian et al., 2018), for instance, hyponatremia, cognitive impairment, or fatigue resulting from the loss of these ions. Moreover, sodium loss is highly variable owing to individual factors such as age, gender, diet, genetics, and environmental conditions (Khaw & Barrett-Connor, 1988).

Utilizing commercial carbon fibers (CCF) to build the wireless potentiometric sensor's electrodes to the real-time monitoring of sodium concentration in sweat during exercise is presented by Parrilla. CCF is made from carbon fibers with micron-sized carbon structures extruded with a polymeric matrix made of the epoxide. CCF is one of the ideal networks for the development of wearable biosensors because it exhibits a high electrical conductivity (approximately $50\,\Omega\,cm^{-1}$), good stiffness and tensile strength, low weight, both chemical and mechanical resistance, and the most important property is its manufacturing cost that is relatively low (Parrilla et al., 2016).

The other work developed an epidermal tattoo-based sensor in which a miniaturized wearable wireless transceiver is embedded in the structure to monitor sodium in human sweat (Bandodkar et al., 2014). During physical activities, sodium, which is a biological marker for electrolyte imbalance, is secreted. Its level provides a piece of valuable information about an individual's physical and mental well-being. Carbon ink was pasted on a flexible polyethylene terephthalate (PET) substrate and the Na-tattoo platform connect to the electrochemical analyzer or the wireless transceiver (Fig. 13.3).

Respiration monitoring

Respiration signals are vital signs of people's health, giving a piece of valuable detailed information from health assessment and treatment for a myriad of diseases such as asthma, chronic bronchitis, and emphysema (Al-Khalidi et al., 2011). Temperature, humidity, airflow, and oxygen gas are physiological outputs of respiration. Therefore, there is a great demand for a wearable sensing device to monitor respiration, providing the patterns and respiration rates relevant to people's health (Ayala et al., 2014).

Prevalent methods to assess these signals show the limitations of portability for chronic dyspepsia patients. Various sensors based on wireless sensing mechanisms have recently been used for breathing monitoring (Xue et al., 2017; Kano, Dobashi, & Fujii, 2017). Due to the exceptional mechanical strength, large surface area, and high conductivity properties, carbon structures such as graphene bilayer (Chen, Hsu, & Hsueh, 2014; Smith et al., 2015), reduced graphene oxide (rGO), three-dimensional graphene (Mirzaie, Rezaie, & Nekahi, 2019), and their hybrid composites (Ho et al., 2016) offer a fantastic opportunity to fabricate wearable sensing tool. Pang et al. (2018) synthesized a porous graphene network via CVD technology that graphene sheets were grown on the nickel foam. The thickness of the as-prepared porous graphene network is roughly 1.5 μm. In addition, the GO, PEDOT/PSS, and AC have been used to modify the porous graphene structure. The sensor's humidity results pointed out that the porous graphene/PEDOT/PSS reveals the fast response time and recovery time of 31 s and 72 s, respectively. Also, the porous network exhibited good sensitivity and a wide linear range. This humidity sensor was successfully applied to detect human respiration, from fast, normal, and deep breathing to mouse and nose breathing.

Pressure and strain wearable sensor

Strain sensors are smart electronic devices that can translate an external applied tensile force into an observable electrical signal. Two-dimensional (2D) conductive carbon materials such as graphene and carbon nanotube can also be embedded in the polymeric matrix to develop an accurate biosensor. The degree of contact between the conductive fillers is modulated by strain exerted to the matrix material, and the connection between the conductors is off/on when the material is stretched and released. The one-atom carbon network, graphene, which was coated on the electrode surface of sensors, has attracted attention for several decades. Much research has been carried out to examine remedy effects on mechanical substrate's properties to give a convenient feeling for wearers (Lee et al., 2018).

Lv and co-workers developed a stable and extremely sensitive strain sensor decorated with hydrogel graphene in which glycerol was used as a co-solvent during two steps. This hydrogel-based sensor effectively keeps water in the polymer structure because a strong hydrogen bonding was created between glycerol and water. Consequently, it prevents water releasing from the polymer composite and guarantees the long-term stability of the sensor. The addition of glycerol not only enhances the stability of the hydrogel over a wider temperature range but also increases the stretchability of the hydrogel by 1200%. The improved sensitivity can be referred to the graphene structure as these sheets have a straight effect on optimizing the contact surface under different strains. The particular design enables this device to be practical in stretching and curving modes into the polymer network (Lv et al., 2019). Zhang et al. reported a nanocomposite of silver nanoparticle-coated carbon nanowires (Ag@CNTs) for high stretchability and good linearity of the strain sensor to develop an electronic wearable skin sensor (Zhang et al., 2017). A stretchable and tunable strain sensor was designed using an environmentally friendly water-based solution process. This composite has consisted mainly of SWCNT and conductive elastomeric polymers, polyurethane (PU), and poly(3,4-ethylene dioxythiophene) polystyrene sulfonate (PEDOT:PSS) to detect minor strains on human skin. The fabricated sensor results were ultrasensitive, allowing users to detect subtle changes such as eye movement and emotional expression (Roh et al., 2015).

A wireless pressure sensor is a transducer that transforms mechanical forces into electrical signals. This highly useful tool is an operative part for detection force and pressure in health and environmental monitoring. A large number of wearable pressure sensors based on carbon networks have been developed. The first type of fabricated pressure sensor was introduced in 2011. It was made via the sandwich technique to place a carbon black fabric between two tooth-structured PDMS layers. The pressure sensors were stable even after 100,000 compressive cycles. It was able to measure pressure from 0 to 2 MPa with a sensitivity of $2.98\,MPa^{-1}$, covering the required monitoring range of human foot pressure (Wang, n.d.).

SWCNT and PMDS have been used to fabricate a pressure sensor with the sensitivity of $1.8\,kPa^{-1}$ via a silk-molded micropattern method. Speech recognition and wrist pulse detection are quickly and accurately visible (Wang et al., 2014). SWCTN, Au, and polyimide (PI) modified tissue paper for the detection of pressure sensor-based assembling technique (Zhan et al., 2017). The sensor showed a wide range of detection, from 2.5 to 35 kPa and a sensitivity of $2.2\,kPa^{-1}$, offering a high potential device to monitor human physiological signals. In several studies, due to the weakness and low stability of CNTs, bending with graphene sheets is one effective treatment, resulting in a large surface area and excellent strength. The CNT/graphene hybrid composite (ACNT/Gr) was placed on PDMS (m-PDMS) substrate, and a high-performance wearable pressure sensor was fabricated. This hybrid helped the sensor to detect bending, torsion, and acoustic signals and also revealed an ultrasensitivity of $19.8\,kPa^{-1}$ (<0.3 kPa) and high stability (>35,000 cycles).

Temperature sensor

Temperature is a fundamental physical parameter that affects both physical and mental health directly. Therefore, sensing it can be helpful to reflect health conditions (Chen et al., 2014; Smith et al., 2015). At present, real-time temperature sensors are commonly used in daily life. However, these sensors might be large or not convenient to use in every situation and time, and also some types of them need a specialist to analyze data. Emerging wireless sensors address these challenges with a practical and feasible solution, and it is earning extraordinary interest in various applications. The precise, soft, and lightweight wearable temperature sensor could attach to the user's body and monitor the temperature via an intelligent way. Different types of such sensors are fabricated with diverse nanomaterials, but most of them generated a very low thermal response, and the other has not well biocompatible to attach. Hence, to reach an accurate response, a sensitive electronic circuit is required, making the sensor ultrasensitive to temperature changes (Kuo, Yu, & Meng, 2012). Graphene has excellent electrical features due to outstanding properties such as high thermal conductivity. Based on those properties, graphene is a potential candidate for accurate temperature sensors.

Graphene nanowalls (GNWs) were synthesized by the CVD technique and were placed on polydimethylsiloxane (PDMS) via polymer-assisted transfer method, a facile, biocompatible, and cost-effective method, to build a temperature sensor. The results showed that the resistance rose gradually with increasing the temperature from 25 to 120°C, which was caused by the bending of heat-sensitive and stretchable GNWs and also the thermal expansion effect of the PDMS networks (Yang & Yang, 2006). The positive temperature coefficient of resistivity (TCR) is the other impressive result, as large as $0.214°C^{-1}$, which is three times higher than conventional counterparts. Moreover, the sensor is capable of monitoring body temperature in real time and presents relatively fast response/recovery speed as well as long-term stability. Such wearable sensing devices could take a significant step toward emerging personalized healthcare and human-machine interface systems.

Pulse beat monitoring

The pulse is the heart rate or the number of times the heart beats in one minute. Pulse rates vary from person to person, but in general, the pulse is lower at rest and increases with exertion (Chellappa, Lasauskaite, & Cajochen, 2017). How the heart works during various activities can tell a lot about a person's health; monitoring it significantly affects people's performance. For the first time, Xie has employed polyoxometalate as a highly conductive material along with graphene sheets to develop a supercapacitor for the detection of pulse beat rate (Xie et al., 2019). Dispersing the POM into the 2D GNs can significantly enhance the cycling stability of the composite. This study not only introduced a novel conductive network, POM/2D GNs nanocomposites, but also provided new insights into how to improve the electrochemical performance, paving up a pathway to design a wearable sensor based on carbon-based materials. The structure of polyoxometalate/2D graphene is also used for pulse-beat applications (Shi et al., 2019). The capacitance of the composite after 6000 cycles was reached 2 times as high as that of the 2D GNs after 2000 cycles, and this sensor could linearly detect a bending radius from 10 to 50 mm. Hence, the combination of these materials has a marked effect on the cycling stability of the sensor.

Environmental monitoring

Heavy metal

Because of globalization and industrialization and human activities, releasing heavy metals have a marked effect on the ecosystem. Lead, zinc, cadmium, mercury, copper, and other types of metals may cause severe health difficulties, such as cancers, brain damages, cardiovascular diseases, and kidney damage (Ha et al., 2017). People are usually exposed to the mentioned heavy metals via breathing, drinking, or eating. These days, a piece of research has been concentrated on wearable sensing devices in healthcare applications, while monitoring harmful chemicals and toxic pollutants in water, food, and soil is also essential, preventing people from health problems.

A large number of scientific works have been proved that most toxic forms of metals in the environment, environmental waters, and even in food chains may vary in concentration, and existing of them in air and water can have adverse effects on health (Engwa et al., 2019). Environmental sampling techniques often involve large laboratory setups with expert operators and are highly priced to analyze samples utilizing equipment including GC-MS and HPLC. Sensors can play a prominent role in monitoring security hazards and effectively turn away humankind from the epidermis.

From a chemistry perspective, due to the environmental conditions such as insufficient long-term stability of the sensing interface, drift over time, and difficulties in repetitive calibration, there are substantial demands for the development of wearable sensors to monitor the local environment, recognizing health and safety hazards.

Copper is a basic ingredient of antifouling paints used for marine applications. Releasing the copper from these paints pollutes the water and soil, affecting greatly ecology. Thus, a wearable sensor for trace-metal monitoring would be a high-priority need for detecting copper in an aquatic environment. Tracing copper was performed using a square-wave stripping technique that screen-printed carbon-based sensor with a gold film to address this issue. It is the first example of utilizing a wireless sensor for monitoring metal voltammetric measurements (Malzahn et al., 2011).

Nitrogen monitoring

Ammonia (NH_3) has been widely utilized in several industries, from chemistry, food, and chemical fertilizer to pharmaceutical and synthetic fiber. This material has an irritating odor that can rapidly stimulate the eyes and skin at even a low concentration, causing lung damage or even death with a great deal of inhalation (Engwa et al., 2019). Developed sensors manufactured based on different metals usually suffer from a heavy price and an extreme working temperature, and the most important defect has a short service life. Hence, it is a pressing need to develop an ammonia sensor with ultra-sensitive performance that can overcome on these drawbacks. That overcomes these disadvantages.

A large number of studies have been carried out on wireless carbon nanotube (CNT)-based gas sensors that most of the methods that relied on the shift of the resonant frequency are the most effective. However, the frequency shift does not vary significantly as the amplitude changes for long-distance wireless detection. Nanocomposites of phosphate-functionalized rGO/polyaniline were synthesized and immobilized on microstrip resonators for ammonia detection. Polyaniline-rGO was incorporated in PANI and generated a positive synergistic effect on sensing performance. One of the toxic and harmful gases from the combustion of fossil fuels is nitrogen dioxide gas. Short-term exposure to people at less than 1 ppm doses does not cause serious long-term health problems, but long-term exposure may pose a severe threat for humans, such as impairing breathe, especially in individuals with asthma or allergies. Above 10 ppm concentration of nitrogen dioxide can cause prominent nose and throat irritation, and above 100 ppm will result in asphyxiation (Larkin et al., 2017).

Ammu et al. fabricated a novel sensor based on inkjet-printed single-walled CNT films on PET and cellulosic substrates to simultaneously detect nitrogen dioxide and chlorine. The penetration of the gas molecules into the CNT networks raises the distance between the conducting pathways of CNTs; as a consequence, it causes the sensor's resistance changes. The films exhibited a decline in resistance for NO_2 and Cl_2 as compared to an increase in resistance for other common gas vapors (Ammu et al., 2012).

Air pollution monitoring

Nowadays, air pollution has been considered one of the major concerns in the 21st century because it causes widespread damage to humans' environment and adverse health effects. There are a dozen of air pollutants sources in both indoor (homes and workplaces) and outdoor environments. In addition to various contaminates with different particle sizes, several toxic gases such as sulfur dioxide (SO_2) and carbon monoxide (CO) have been found in the air that can highly decrease air quality (Singh, Meyyappan, & Nalwa, 2017). Progressing in developing wireless sensing devices can make a revolution in improving air quality and because it continuously monitors different types of air pollutants using portable sensors, providing a fast warn when their concentrations reached levels beyond safety limitations.

Among other toxic air pollutants, the detection of CO remains challenging due to its odorless and colorless characteristics. This gas has typically been produced by incomplete burning of fossil fuels, vehicle emissions, and industrial combustions (Nandy, Coutu, & Ababei, 2018). CO concentration and the exposure time in this gas can determine health difficulties. Generally, moderate exposure to CO leads to breathing problems, headache, nausea, dizziness, and tiredness. Vomiting, reduced muscular coordination, loss of consciousness, and even death are the results of being exposed to a higher concentration of CO (Safarianzengir et al., 2020).

It is helpful to monitor the concentration of CO on time and fast to decrease its problems. Hence, CO monitoring has been considered a promising practical way that is combining a wireless sensor platform with the Internet of Things. This up-to-date network provides real-time tracking of multiple nodes and then estimates environmental conditions in the area (Yang et al., 2015a, 2015b).

Various metal oxides have been widely employed as sensing materials due to their excellent electrical, optical, and gas-sensing properties. Among all used metal oxides, zinc oxide (ZnO) has been considered the most promising sensing material due to its low cost, high sensitivity, good chemical and thermal stabilities, and high electron transfer. Nevertheless, the poor selectivity and high operating temperature of this metal often limit the gas-sensing performance. Other novel materials with designed architectures have been synthesized to address these challenges. A wearable resistive gas sensor was fabricated on a flexible cotton fabric based on a nanocomposite of graphene sheets and zinc oxide (ZnO), which was employed to monitor odorless CO gas. The ZnO nanorod was grown on the ZnO-coated fabrics with graphene structures by the chemical bath deposition (CBD) method. Electrical characterization results exhibit that the hybrid graphene/ZnO composite has a lower surface resistance at room temperature than the surface resistance of the ZnO layer. This is because of electron transfer from graphene to ZnO. The presence of graphene sheets underneath the ZnO has enhanced the sensing performance of ZnO-based sensors.

Work (2011) presented an electrochemical biosensor on textiles based on printable enzyme-containing inks. Such a new type of design for wearable sensors integrated with an encapsulated potentiostat gives the wearer a visual indication regarding the level of a specific target hazard (e.g., phenols). The user would be alerted about their harmful situation to relocate or clean environment, and some information about water quality, warning health difficulties.

Current limitation

The limited availability of both wearable chemical and physical sensors has hampered further progress in field of health and environment. This is because of several key issues that have yet to be addressed, such as immobilization of enzyme on the electrode surface, repeatability and biocompatibility of the sensors, and being cost-effectiveness. Attention should be given to the development of related storage and stability issues, enzyme-free calibration, and user-independent wearable sensing devices. Designing a new generation of embedded healthcare sensor over teeth, skin, and eyes will face several challenges because of biofluid that may hinder the performance of wearable sensors. According to the studied research, the electrode surface coverage with protective or antimicrobial coating can minimize these difficulties. Efforts should also be devoted to improving the sensitivity and selectivity of the devices. This is often considered via carbon-based nanomaterial signal amplification, but the nanomaterials' toxicity should be taken into account before their on-body application.

Conclusions and future perspectives

Wearable sensors have brought many opportunities for humankind to monitor a broad range of biomedical and environmental analytes continuously. With the entry of the Internet of Things, tremendous progress in wearable, noninvasive, and wearer sensors has been rapidly created, and a wide range of new innovative wireless devices have been entered the market. Integrating Bluetooth/smart devices with wearable chemical and physical sensors will enable users to easily gain a great deal of information via wirelessly transmit data to their cellphones in a user-friendly fashion. These types of real-time monitoring devices offer users a more comprehensive assessment of their healthcare status. The commercialization of wearable biosensors based on biological analytes is substantially more challenging than physical monitoring types due to the immobilization of biological items. Such devices must monitor continuous on-body biochemical responses and give a reliable and accurate measurement of a biorecognition element. A designed robust sensor should overcome such challenges, from the limited stability of several bioreceptors and exact calibration curve to being simple to use without a specialist and analyzing the outputs. The tremendous interest in wireless sensors reflects growing shifts from clinical-based patient care to home-based personal care to decline healthcare costs and greatly alleviate mental for patients. Carbon-based structures with outstanding chemical and physical properties pave the way for the new generation of the sensor to face a dozen of challenges. Researchers have developed commercialized carbon-based pressure, temperature, strain, and integrated sensor in recent years. Carbon-based wearable sensors with various signal sensing applications are expected to become the leading promising in this path.

References

Alam, F., et al. (2018). Lactate biosensing: The emerging point-of-care and personal health monitoring. *Biosensors and Bioelectronics, 117*, 818–829.
Al-Khalidi, F. Q., et al. (2011). Respiration rate monitoring methods: A review. *Pediatric Pulmonology, 46*(6), 523–529.
Ammu, S., et al. (2012). Flexible, all-organic chemiresistor for detecting chemically aggressive vapors. *Journal of the American Chemical Society, 134*(10), 4553–4556.
An, B., et al. (2016). Three-dimensional multi-recognition flexible wearable sensor via graphene aerogel printing. *Chemical Communications, 52*(73), 10948–10951.
Ayala, G. X., et al. (2014). A prospective examination of asthma symptom monitoring: Provider, caregiver and pediatric patient influences on peak flow meter use. *Journal of Asthma, 51*(1), 84–90.
Baker, J. F., et al. (2005). Serum uric acid and cardiovascular disease: Recent developments, and where do they leave us? *The American Journal of Medicine, 118*(8), 816–826.
Bakker, J., Nijsten, M. W., & Jansen, T. C. (2013). Clinical use of lactate monitoring in critically ill patients. *Annals of Intensive Care, 3*(1), 1–8.
Bandodkar, A. J., & Wang, J. (2014). Non-invasive wearable electrochemical sensors: A review. *Trends in Biotechnology, 32*(7), 363–371.
Bandodkar, A. J., et al. (2014). Epidermal tattoo potentiometric sodium sensors with wireless signal transduction for continuous non-invasive sweat monitoring. *Biosensors and Bioelectronics, 54*, 603–609.
Bao, S.-D., Zhang, Y.-T., & Shen, L.-F. (2006). Physiological signal based entity authentication for body area sensor networks and mobile healthcare systems. In *2005 IEEE engineering in medicine and biology 27th annual conference* IEEE.
Bhide, A., et al. (2021). Next-generation continuous metabolite sensing toward emerging sensor needs. *ACS Omega, 6*(9), 6031–6040.
Brian, M. S., et al. (2018). The influence of acute elevations in plasma osmolality and serum sodium on sympathetic outflow and blood pressure responses to exercise. *Journal of Neurophysiology, 119*(4), 1257–1265.
Chellappa, S. L., Lasauskaite, R., & Cajochen, C. (2017). In a heartbeat: Light and cardiovascular physiology. *Frontiers in Neurology, 8*, 541.
Chen, M.-C., Hsu, C.-L., & Hsueh, T.-J. (2014). Fabrication of humidity sensor based on bilayer graphene. *IEEE Electron Device Letters, 35*(5), 590–592.
Clarke, S., & Foster, J. (2012). A history of blood glucose meters and their role in self-monitoring of diabetes mellitus. *British Journal of Biomedical Science, 69*(2), 83–93.

Collie, J. T. B., et al. (2020). Vitamin C measurement in critical illness: Challenges, methodologies and quality improvements. *Clinical Chemistry and Laboratory Medicine, 58*(4), 460–470.

Cool, B., et al. (2006). Identification and characterization of a small molecule AMPK activator that treats key components of type 2 diabetes and the metabolic syndrome. *Cell Metabolism, 3*(6), 403–416.

DePhillipo, N. N., et al. (2018). Efficacy of vitamin C supplementation on collagen synthesis and oxidative stress after musculoskeletal injuries: A systematic review. *Orthopaedic Journal of Sports Medicine, 6*(10), 2325967118804544.

Engwa, G. A., et al. (2019). Mechanism and health effects of heavy metal toxicity in humans. In *Vol. 10. Poisoning in the modern world-new tricks for an old dog* Medical Toxicology.

Ghaderinezhad, F., et al. (2020). Sensing of electrolytes in urine using a miniaturized paper-based device. *Scientific Reports, 10*(1), 1–9.

Ha, E., et al. (2017). Current progress on understanding the impact of mercury on human health. *Environmental Research, 152*, 419–433.

Heo, J. S., Hossain, M. F., & Kim, I. (2020). Challenges in design and fabrication of flexible/stretchable carbon- and textile-based wearable sensors for health monitoring: A critical review. *Sensors, 20*(14), 3927.

Ho, D. H., et al. (2016). Stretchable and multimodal all graphene electronic skin. *Advanced Materials, 28*(13), 2601–2608.

Hu, J., et al. (2020). Nano carbon black-based high performance wearable pressure sensors. *Nanomaterials, 10*(4), 664.

Jian, M., et al. (2017a). Advanced carbon materials for flexible and wearable sensors. *Science China Materials, 60*(11), 1026–1062.

Jian, M., et al. (2017b). Flexible and highly sensitive pressure sensors based on bionic hierarchical structures. *Advanced Functional Materials, 27*(9), 1606066.

Kang, B.-C., Park, B.-S., & Ha, T.-J. (2019). Highly sensitive wearable glucose sensor systems based on functionalized single-wall carbon nanotubes with glucose oxidase-nafion composites. *Applied Surface Science, 470*, 13–18.

Kano, S., Dobashi, Y., & Fujii, M. (2017). Silica nanoparticle-based portable respiration sensor for analysis of respiration rate, pattern, and phase during exercise. *IEEE Sensors Letters, 2*(1), 1–4.

Karpova, E. V., et al. (2020). Relationship between sweat and blood lactate levels during exhaustive physical exercise. *ChemElectroChem, 7*(1), 191–194.

Kassal, P., et al. (2015). Smart bandage with wireless connectivity for uric acid biosensing as an indicator of wound status. *Electrochemistry Communications, 56*, 6–10.

Khaw, K., & Barrett-Connor, E. (1988). The association between blood pressure, age, and dietary sodium and potassium: A population study. *Circulation, 77*(1), 53–61.

Kim, J., Campbell, A. S., & Wang, J. (2018). Wearable non-invasive epidermal glucose sensors: A review. *Talanta, 177*, 163–170.

Kim, J., et al. (2017). Wearable smart sensor systems integrated on soft contact lenses for wireless ocular diagnostics. *Nature Communications, 8*, 14997.

Kuo, J. T., Yu, L., & Meng, E. (2012). Micromachined thermal flow sensors—A review. *Micromachines, 3*(3), 550–573.

Larkin, A., et al. (2017). Global land use regression model for nitrogen dioxide air pollution. *Environmental Science & Technology, 51*(12), 6957–6964.

Lee, H., et al. (2016). A graphene-based electrochemical device with thermoresponsive microneedles for diabetes monitoring and therapy. *Nature Nanotechnology, 11*(6), 566–572.

Lee, H., et al. (2018). Preparation of fabric strain sensor based on graphene for human motion monitoring. *Journal of Materials Science, 53*(12), 9026–9033.

Lipani, L., et al. (2018). Non-invasive, transdermal, path-selective and specific glucose monitoring via a graphene-based platform. *Nature Nanotechnology, 13*(6), 504–511.

Liu, X., & Lillehoj, P. B. (2017). Embroidered electrochemical sensors on gauze for rapid quantification of wound biomarkers. *Biosensors and Bioelectronics, 98*, 189–194.

Lv, J., et al. (2019). Wearable, stable, highly sensitive hydrogel–graphene strain sensors. *Beilstein Journal of Nanotechnology, 10*(1), 475–480.

Malzahn, K., et al. (2011). Wearable electrochemical sensors for in situ analysis in marine environments. *Analyst, 136*(14), 2912–2917.

Mason, A., et al. (2017). Noninvasive in-situ measurement of blood lactate using microwave sensors. *IEEE Transactions on Biomedical Engineering, 65*(3), 698–705.

Matsumoto, S., et al. (2019). Urea cycle disorders—Update. *Journal of Human Genetics, 64*(9), 833–847.

Mesgari, F., et al. (2020). Paper-based chemiluminescence and colorimetric detection of cytochrome c by cobalt hydroxide decorated mesoporous carbon. *Microchemical Journal, 157*, 104991.

Mirzaie, M. A. M., Rezaie, M., & Nekahi, A. (2019). Non-enzymatic reusable graphene-based electrodes by a simple scalable preparation method and superior electrochemical property. *Materials Research Express, 6*(11), 1150b9.

Nandy, T., Coutu, R. A., & Ababei, C. (2018). Carbon monoxide sensing technologies for next-generation cyber-physical systems. *Sensors, 18*(10), 3443.

Neto, A. C., et al. (2009). The electronic properties of graphene. *Reviews of Modern Physics, 81*(1), 109.

Pang, Y., et al. (2018). Wearable humidity sensor based on porous graphene network for respiration monitoring. *Biosensors and Bioelectronics, 116*, 123–129.

Parrilla, M., et al. (2016). Wearable potentiometric sensors based on commercial carbon fibres for monitoring sodium in sweat. *Electroanalysis, 28*(6), 1267–1275.

Peña-Bahamonde, J., et al. (2018). Recent advances in graphene-based biosensor technology with applications in life sciences. *Journal of Nanobiotechnology, 16*(1), 1–17.

Pu, Z., et al. (2016). A continuous glucose monitoring device by graphene modified electrochemical sensor in microfluidic system. *Biomicrofluidics, 10*(1), 011910.

Pur, M. R. K., et al. (2016). A novel solid-state electrochemiluminescence sensor for detection of cytochrome c based on ceria nanoparticles decorated with reduced graphene oxide nanocomposite. *Analytical and Bioanalytical Chemistry, 408*(25), 7193–7202.

Qian, Y., Ismail, I. M., & Stein, A. (2014). Ultralight, high-surface-area, multifunctional graphene-based aerogels from self-assembly of graphene oxide and resol. *Carbon, 68*, 221–231.

Roh, E., et al. (2015). Stretchable, transparent, ultrasensitive, and patchable strain sensor for human–machine interfaces comprising a nanohybrid of carbon nanotubes and conductive elastomers. *ACS Nano, 9*(6), 6252–6261.

RoyChoudhury, S., et al. (2018). Continuous monitoring of wound healing using a wearable enzymatic uric acid biosensor. *Journal of the Electrochemical Society, 165*(8), B3168.

Safarianzengir, V., et al. (2020). Monitoring, analysis and spatial and temporal zoning of air pollution (carbon monoxide) using Sentinel-5 satellite data for health management in Iran, located in the Middle East. *Air Quality, Atmosphere and Health, 13*, 709–719.

Salek-Maghsoudi, A., et al. (2018). Recent advances in biosensor technology in assessment of early diabetes biomarkers. *Biosensors and Bioelectronics, 99*, 122–135.

Senel, M., Dervisevic, M., & Voelcker, N. H. (2019). Gold microneedles fabricated by casting of gold ink used for urea sensing. *Materials Letters, 243*, 50–53.

Shi, Y., et al. (2019). Fabrication and characterization of polyoxometalate/2D graphene-based flexible supercapacitors for wearable electronic pulse-beat application. *Journal of Materials Science: Materials in Electronics, 30*(4), 3692–3700.

Singh, E., Meyyappan, M., & Nalwa, H. S. (2017). Flexible graphene-based wearable gas and chemical sensors. *ACS Applied Materials & Interfaces, 9*(40), 34544–34586.

Smith, A. D., et al. (2015). Resistive graphene humidity sensors with rapid and direct electrical readout. *Nanoscale, 7*(45), 19099–19109.

Stoppa, M., & Chiolerio, A. (2014). Wearable electronics and smart textiles: A critical review. *Sensors, 14*(7), 11957–11992.

Tahernejad-Javazmi, F., Shabani-Nooshabadi, M., & Karimi-Maleh, H. (2018). Analysis of glutathione in the presence of acetaminophen and tyrosine via an amplified electrode with MgO/SWCNTs as a sensor in the hemolyzed erythrocyte. *Talanta, 176*, 208–213.

Tur-García, E. L., et al. (2017). Novel flexible enzyme laminate-based sensor for analysis of lactate in sweat. *Sensors and Actuators B: Chemical, 242*, 502–510.

Wang, X., et al. (2014). Silk-molded flexible, ultrasensitive, and highly stable electronic skin for monitoring human physiological signals. *Advanced Materials, 26*(9), 1336–1342.

Wiorek, A., et al. (2020). Epidermal patch with glucose biosensor: pH and temperature correction toward more accurate sweat analysis during sport practice. *Analytical Chemistry, 92*(14), 10153–10161.

Xie, T., et al. (2019). Graphene-based supercapacitors as flexible wearable sensor for monitoring pulse-beat. *Ceramics International, 45*(2, Part A), 2516–2520.

Xu, G., et al. (2019). Battery-free and wireless epidermal electrochemical system with all-printed stretchable electrode array for multiplexed in situ sweat analysis. *Advanced Materials Technologies, 4*(7), 1800658.

Xuan, X., Yoon, H. S., & Park, J. Y. (2018). A wearable electrochemical glucose sensor based on simple and low-cost fabrication supported micro-patterned reduced graphene oxide nanocomposite electrode on flexible substrate. *Biosensors and Bioelectronics, 109*, 75–82.

Xue, H., et al. (2017). A wearable pyroelectric nanogenerator and self-powered breathing sensor. *Nano Energy, 38*, 147–154.

Yang, G.-Z., & Yang, G. (2006). *Body sensor networks*. Vol. 1. Springer.

Yang, J., et al. (2015a). Wearable temperature sensor based on graphene nanowalls. *RSC Advances, 5*(32), 25609–25615.

Yang, J., et al. (2015b). A real-time monitoring system of industry carbon monoxide based on wireless sensor networks. *Sensors, 15*(11), 29535–29546.

Yokus, M. A., et al. (2020). Wearable multiplexed biosensor system toward continuous monitoring of metabolites. *Biosensors and Bioelectronics, 153*, 112038.

Zhan, Z., et al. (2017). Paper/carbon nanotube-based wearable pressure sensor for physiological signal acquisition and soft robotic skin. *ACS Applied Materials & Interfaces, 9*(43), 37921–37928.

Zhang, Q., et al. (2017). Highly sensitive and stretchable strain sensor based on Ag@ CNTs. *Nanomaterials, 7*(12), 424.

Chapter 14

Advanced sensors based on carbon nanomaterials

Vinayak Adimule[a], Basappa C. Yallur[b], and Adarsha H.J. Gowda[c]

[a]Department of Chemistry, Angadi Institute of Technology and Management (AITM), Belagavi, Karnataka, India, [b]Department of Chemistry, M. S. Ramaiah Institute of Technology, Bangalore, Karnataka, India, [c]Centre for Research in Medical Devices, National University of Ireland, Galway, Ireland

Introduction

Qualitative and quantitative determination of the electroactive components present in the solution can be determined by low-cost, simple electrochemical methods. Apart from the electrochemical methods, other methods such as fluorescence, luminescence, and chromatography are used to detect the species in low concentration; they are more accurate and user-friendly. Many techniques are available for the researchers to carry out the investigation of the electroactive species; among them, some of the analytical methods, such as cyclic voltammetry (CV), linear sweep voltammetry, stripping voltammetry, find effective since they have better response and selective in electrochemistry. The response and selectivity can be increased by several factors including size of the species, its concentration, type of the electrodes, and choice of the electrolytes. Among the various other factors, morphology and size play very important roles in voltammetric response of the system (Campbell & Compton, 2010). The ease of detection depends on the parameters such as physical and chemical properties of the electrodes, applied potential, coatings on electrode surface, etc. Carbon nanomaterials in different forms are widely used in electrochemistry for the detection of electroactive species since they possess inertness and operate with low applied potential and current and good lifetime compared to other electrode systems (Ndamanisha & Guo, 2012). The carbon-based nanomaterials operate under critical environments and possess stability with dynamics of temperature. Literature varieties of materials used in the electroanalytical methods are Pt, Au, and various other doped forms of carbon nanomaterials for electrochemical detection (Amor-Gutiérrez, Costa Rama, Costa-García, & Fernández-Abedul, 2017). In recent years potential implications of carbon derived materials such as graphene, diamond etc. which are mainly used for the electrode material in electrochemical investigations. The polymorphous materials are excellent in their electrochemical sensing performance due to a large surface-to-volume ratio, high conductivity, and electron mobility. The carbon nanomaterial-based electrochemical sensors are also used in the detection of DNA, proteins, gases, immune sensors, etc. (Kroto, Heath, O'Brien, Curl, and Smalley (1985) discovered C60 nanomaterial with carbon crystal structure along with the attached graphene molecule led to the development of CNTs by Iijima. CNTs attracted much attention due to their applications in the field of physics (De Volder, Tawfick, Baughman, & Hart, 2013), chemistry (Pan et al., 2007), and materials sciences (Dastjerdi & Montazer, 2010). The CNTs possess distinguishable mechanical, electrical, and chemical properties and distinguishable crystal structures. CNTs possess hexagonal honeycomb-like lattices fabricated with sp^2 hybridized carbon atoms, and the topology with the diameter of the carbon nanotubes consists of two defined crystal groups such as single-walled carbon nanotubes (SWCNTs) and multiwalled carbon nanotubes (MWCNTs) (Iijima, 2002). Single-walled carbon nanotubes consist of a closed single sheet of graphene rolled and made in the form of cylinders with a diameter of 1–2 nm, whereas multiwalled carbon nanotubes consist of many sheets of graphene rolled and made in the form of a cylinder with a diameter ranging from 2 to 30 nm and the gap between the tubes is similar to interlayer spacing in graphite of approximately 0.34 nm. Depending upon the diameter, CNTs act similarly as semiconductors and exhibit electrical behavior as bulk metals (Arnold, Green, Hulvat, Stupp, & Hersam, 2006). CNTs in such cases which consist of NPs with a high surface-to-volume ratio, spatial confinements, and completely altered physical and chemical properties have resulted in wide-ranging applications. The carbon-based nanomaterials find applications in the field of chemical sensors (Akbari et al., 2014), catalyst scaffolds (Chung, Won, & Zelenay, 1922), energy storage and conversion (Wang & Su, 2014), and electronic devices. CNTs also show greater electron transfer capabilities when incorporated in electrochemical applications. They show significant potential in the biosensor applications and immobilization of proteins

(Marega et al., 2013). CNTs behave differently in aqueous solutions and involve in electron transfer to the electrode surface and develop an interface between the electrode surface and electroactive species in the solution (Zhao et al., 2016). In the literature reports, electrochemical behavior of CNTs showed a remarkable performance than other electrodes. The CNTs display high conductivity and chemical stability. Electrochemical transducers that consist of CNTs exhibited improved electrochemical performance with respect to amperometric enzyme electrodes, immunosensors, and nucleic acid sensitivity explained in detail.

Materials and methods

Chemicals and reagents used without any further purification. Hitachi S-4800 was used to analyze the size and morphology of the grains using scanning electron microscopy (SEM). data recovery service (DRS), Shimadzu, and JASCO FP-8500 were employed for the detection of optical absorptivity and bandgap of the nanomaterials. BJH method used a NOVA 4200 surface area and pore size analyzer; by using it, the specific area was measured. TG-DTA analysis was carried out by using Rigaku TG8120 with a heating rate of 10°C/min. Yanaco MT-6 instrument was used to analyze the elemental analysis. Cyclic voltammetry (CV) experiments were carried out by using Autolab PGSTAT204 FRA32M 10 Hz to 32 MHz using NOVA software for Bode, electrochemical impedance spectroscopy (EIS), and CV analyses, 0.0002–0.35 times current range in galvanostatic mode, 0.2 mV to 0.35 V rms in potentiostatic mode. Current gain, cyclic stability, and EIS were calculated.

Experimental

Synthesis of carbon nanotubes (CNTs)

Many methods are incorporated in the synthesis of CNTs, such as carbon arc discharge, chemical vapor deposition (CVD), and laser ablation (Yu, Shearer, & Shapter, 2016). In general, carbon arch method was used to synthesize high-yield, uniform carbon NPs, especially SWCNTs and MWCNTs (Deng et al., 2009). CVD method gives the NPs with low yield and smaller sizes. Fine-sized carbon NPs can be obtained by laser ablation method (Guo et al., 1995). Conducting and metallic nature of the carbon NPs can be synthesized by electrical heating method (Collins & Avouris, 2000). Electrical heating method selectively synthesizes the carbon NPs. CVD makes use of transition metal catalysts in order to produce high-quality and smaller grain-sized NPs (Li et al., 1996). Synthesis by CVD method involves simple equipment, mild temperature, and mild pressure. Vertical aligned array of carbon NPs can be synthesized by using quasicrystalline substrate (Talapatra et al., 2006). Pyrolysis of metal carbonyl compounds in the presence of hydrocarbons is also reported for the synthesis of CNTs. NPs such as Ni, Co, and Fe have been widely used as catalysts in CVD method to synthesize MWCNTs of the diameter from 0.5 nm to 5 nm. In the CVD method, inert gas methane was used for the production of SWCNTs. Thermal decomposition of hydrocarbons in the presence of catalysts forms fullerene caps that in turn act as basic units for the synthesis of CNTs. The obtained CNTs need to be purified to remove amorphous carbon materials. The potential drawback of the synthesis is to remove the residual C-atom that needs a high temperature of the order 3000°C which limits the applications of the CNTs (C. Deng et al., 2009). The potential advantage of the CNTs is having a large external surface area and acts as a carrier for the different drugs to the tissues (Liu, Tabakman, Welsher, & Dai, 2009). Schematic representation of the various allotropic forms of the C-nanomaterials is given in Fig. 14.1. Table 14.1 summarizes the significant electronic, mechanical, and physical properties of CNTs. Fig. 14.2 represents the SEM images of the C-nanomaterials and their sensing performance characteristics.

Synthesis of graphene

The graphene in layered form was first synthesized by Novaselov and Geim in 2004 with the help of the peeling off method with a Scotch tape (Novoselov, 2004). The method produces high-quality graphene without any defects. Moreover,

FIG. 14.1 Different allotropes of carbon nanomaterials. *(From Carbon nanomaterials and their application to electrochemical sensors: a review. (2018). Nanotechnology Reviews, 7 (1), 19–44. https://doi.org/10.1515/ntrev-2017-0160.)*

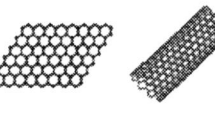   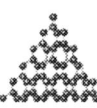 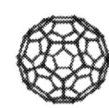

Graphene　　CNTs　　Carbon Dots　　MWCNTs　　C-Diamonds　　Fullerenes

TABLE 14.1 Significant physical, electronic, and mechanical characteristic features of CNTs.

Serial number	Physical parameters	Obtained values
1	Specific surface area	200–900 m^2/J
2	Specific gravity	0.8–2/g/cm^2
3	Electrical conductivity	$2 \times 10^{-2} - 0.25$ S/cm
4	Thermal conductivity	6600 W/m/K
5	Elastic modulus	>1 TPA
6	Tensile strength	>100 GPa

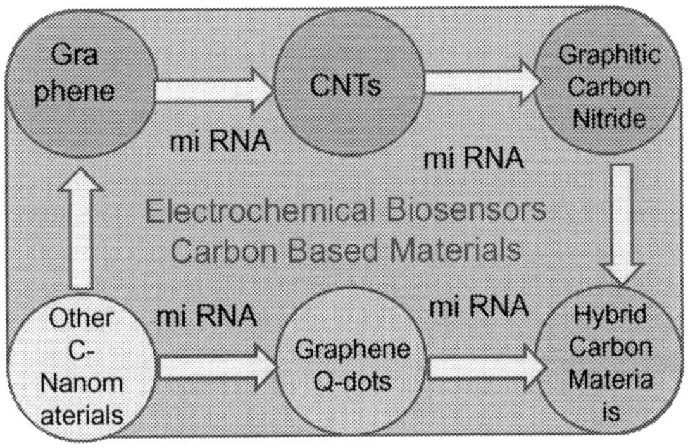

FIG. 14.2 Selectivity and sensitivity of the different allotropic forms of C-nanomaterials toward miRNA molecules. *(From Carbon nanomaterials and their application to electrochemical sensors: a review. (2018). Nanotechnology Reviews, 7 (1), 19–44. https://doi.org/10.1515/ntrev-2017-0160.)*

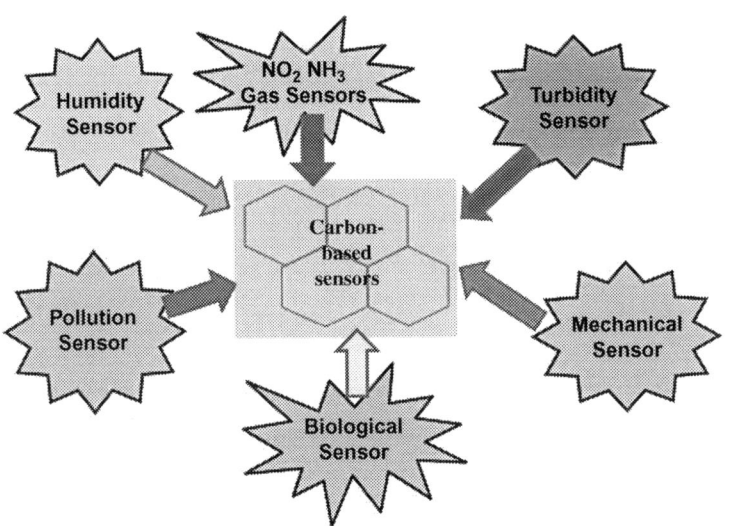

FIG. 14.3 Illustration of different sensing performances of C-nanomaterials. *(No permission required.)*

formation of graphite in solvents like *N*-methyl pyrrolidone (Hernandez et al., 2008) and surfactant sodium dodecyl benzene sulfonate has been reported (Lotya et al., 2009). Layered graphene can be fabricated with less cost and less number of steps by CVD method using graphite. The process can be archived by epitaxial method (Sutter, Flege, & Sutter, 2008), microwave-assisted exfoliation (Zhu & Xu, 2010), solvothermal production, etc. Defected graphene can also be synthesized from the graphene oxide. In electrochemical sensor performance, graphene-modified electrodes are synthesized by the CVD method and fabricated using Ni, Po, Pt, Cu, etc. The epitaxial graphene can be synthesized by graphitization of doped SiC at high temperatures (Berger et al., 2004). Fig. 14.3 represents the sensitivity and selectivity of the

TABLE 14.2 Comparison between some of the biomolecule sensor systems described in this review and those already reported.

Sensor compounds	Sensitivity/linear range	Response	Response for other systems	Response from others
GPN/Au-SPR	7.01 nm	Glucose	Sensor coated with Au	2.98 nm
QD-C60	100 fM	DNA	MB-based DNA detection	4.4–8.5 nM
MCNTFET	1 pM	Influenza A	–	–
Graduated neutral density (GND)/ polyaniline (PANI)	381.5 µA	Urea	Zinc oxide-chitosan	0.13 µA·m/M/cm^{-2}

C-nanomaterials toward various sensing devices. Table 14.2 represents the comparison between some of the biomolecule sensor systems described in this review and those already reported (Choucair, Thordarson, & Stride, 2009).

Synthesis of graphitic carbon nitride

Melamine, dicyandiamide, and guanidine carbonate were used as a starting material for the synthesis of graphitic carbon nitride (g-C$_3$N$_4$). The synthesis was carried out by thermal polymerization reaction. The polymerization reaction involves the starting materials such as g-C$_3$N$_4$ placed in a crucible and heated at 350°C. The materials were heated at the rate of 50°C/min. After 2 h of heating the reaction mixture at 350°C, the temperature was raised to 500°C and kept at that temperature for 30 min. The products were separated by fractional thermal evaporation and then purified. The pure powdered form of graphitic carbon nitride (g-C$_3$N$_4$) was used for the fabrication.

Results and discussion

Advancement in the electrochemical sensing performance of carbon nanotubes

In the last few years, CNTs attracted many researchers because of stress resistance and electrical properties (Dang et al., 2016). Diversified properties of CNTs (large surface-to-volume ratio, few nanometers in diameter, high conductivity, and mechanical deformations) make them to use in miniaturized sensors (Llobet, 2013). Electronic sensitivity of CNTs is due to the formation of graphene-like sheets incorporated in electrochemical sensors that consisted of two electrodes and a charge transporter in the form of transducers. Electrons present in the C-atom are distributed uniformly in graphene sheets making electron cloud distortion, and hence, CNTs become electrochemically active (He et al., 2013). Increase or decrease in the conductivity in CNTs can be achieved by doping electron realizing or withdrawing groups such as NO$_2$, NH$_3$, and O$_2$. Metal-decorated CNTs used in high-efficient electrochemical sensors resulted in an increase in interest due to high sensitivity and catalysis properties (Zhang et al., 2016). Wang et al. designed a one-pot hot-solution synthesis method for Ni$_2$P$_5$/CNTs hybrid nanostrctures (NS). These hybrid NSs demonstrated an enhanced electrochemical sensitivity and are used as anode materials in Li-ion batteries. Due to the internal conductivity of CNTs, they are used to detect the DNA/RNA and their base pairs. NPs or enzymes promote the electron transfer rate during electrochemical sensing of DNA/RNA, thus enhancing the sensitivity of the CNTs. Li et al. (1996) demonstrated the improved DNA detection by MW CNTs, and amplification depends upon electron/hole transport rate and improves the fabrication time and applications in miniaturization of biosensors. The technique employed was immobilization of DNA with CNT for the effective detection of oligonucleotide and its complementary DNA sequence. Guo et al. outlined the fabrication of a simple 8-hydroxy-2′-deoxyguanosine (8-OHDG) as a biomarker selectively sensitive in electrochemical response for the oxidation. Liu and coworkers demonstrated a highly sensitive electrochemical sensor for the detection of pathogenic bacteria and the detection depends on the nanophase of the materials. They also reported Au-dispersed MWCNTs used for the rapid detection and diagnosis of tetanus making its applications in the ultraselective bioassay determination (Liu et al., 2009). Fig. 14.4 demonstrates the AuNP-functionalized MWCNTs for ultraselectivity, ultrasensitivity, and their morphology. Li et al. developed an electrochemical sensor (label free) for the effective detection of miRNA-24 by monitoring the oxidation signal of

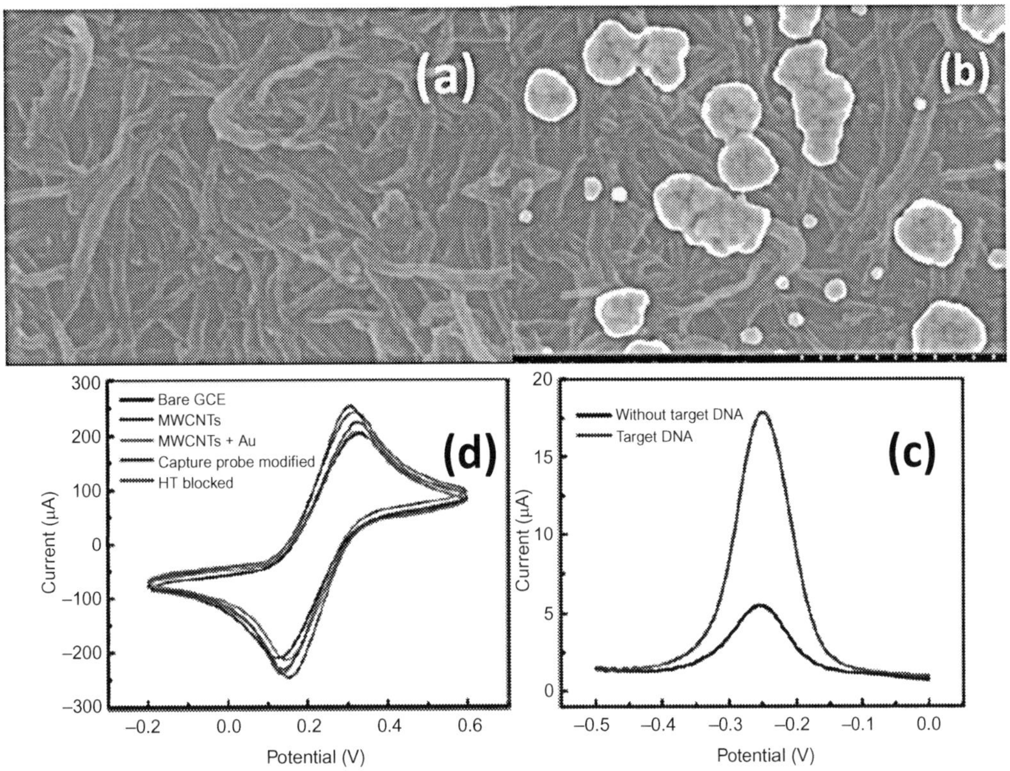

FIG. 14.4 (A) SEM images of MWCNTs, (B) SEM images of MWCNTs doped with Au, (C) galvanostatic charge discharge (GCD) curve CV, (D) sensitivity of the DNA (1.0 × 10–12 m). *(From Carbon nanomaterials and their application to electrochemical sensors: a review. (2018). Nanotechnology Reviews, 7 (1), 19–44. https://doi.org/10.1515/ntrev-2017-0160.)*

guanine. The detection can be improved by doping other carbon nanomaterials over CNTs. The performance like cyclic stability and sensitivity showed by miRNA-21 led to the development of secondary antibody detection reported by Tran et al. using horseradish peroxidase (HRP) to set up an electrochemical enzyme-linked immunosorbent assay (ELISA)-like amplification system. During the detection of miRNA, remarkable conductivity of the CNTs with improved signal amplification resulted in increased stability. Table 14.3 depicts the comparison between some of the molecules and chemical elements sensor systems described in this review and different systems found in the literature.

TABLE 14.3 Comparison between some of the molecules and chemical element sensor systems described in this review and different systems found in the literature.

Sensor compounds	Sensitivity	Great response	Other reported systems	Responses from other systems
GPN polyimide	21.69 ppm	NO_3-N	Laboratory method	21.5 ppm
rGO	1.1 μg/L	NO_3-N phenol disulfonic acid method	Phenol disulfonic acid method	0.02–2 μg/L
CNTs	5×10^{-8} M	Cr (III)	Chromium carbon paste sensor	0.16–1.0 μM
Boron-dope NDs	0.12 μM	1,2-Dihydroxybenzene	Graphene-polyoxometalate	0.05 μM
GPN film	0.5 to 5.0 μg/L	Cd^{2+}, Pb^{2+}	Screen-printed CNTs	0.7 μg/L
GPN/Nafion	0.021 μmol/L	Caffeine	Reduction of graphene oxide (GO)	0.02–0.43 μmol/L

Advancement in the electrochemical sensing performance of the graphene

Graphene was successfully prepared in the year 2004 (Azimzadeh, Nasirizadeh, Rahaie, & Naderi-Manesh, 2017) and it consists of a single layer of C-atoms. In the field of electrochemical biosensors, graphene and its derivatives exhibit excellent biocompatibility due to their large specific surface area and electrical conductivity. Moreover, for the early detection of tumor, a significant progress was achieved in the electrochemical biosensor properties of graphene and effective sensing of miRNA structure. Kilic et al. reported effective detection of miRNA using pencil graphite electrode (GME) and compared the results with modified electrodes during electrochemical biosensing (Kilic et al., 2015). Moreover, such electrodes detected breast cancer cell lines effectively with a lower detection limit of the order of 2.09 mg/mL.

Abdullah Salimi et al. reported amino graphene (AGr)-based sensing electrode for the effective detection of miRNA-155 and analyzed its complementary target substrates. By making use of nanocompounds (NCs) of graphene and their metal oxides, the biosensors can able to detect the sensitivity in the NG (nanogram) scale. Graphene and WO_3 attracted much interest due to their low cost, nontoxic, and chemical stability (Geng et al., 2017). Shuai's group designed an ultrasensitive electrochemical sensor using WO_3-graphene composite to detect miRNA (Shuai, Huang, Xing, & Chen, 2016). Graphene also contains π-π stacking interactions and excellent covalent compatibility, and the derivative of graphene exhibits superior electrochemical sensing properties, especially toward the detection of miRNA. The derivatives of graphene fabrication are cheaper, free from metallic impurities, and cause domination of electrochemical signals.

Advancement in the electrochemical sensing performance of the graphitic carbon

One of the important structural forms of carbon nanomaterials is graphitic carbon nitride (g-C_3N_4). They possess a medium bandgap of the order of 2.68 eV and form 2D-layered structure with controllable morphology and sp^2 hybridization in C and N results from electron-coupled system. They also have good chemical and thermal stability, are nontoxic, and can be used for the effective detection of miRNA. The graphitic carbon shows the photoelectric properties and can be used in electrochemiluminescence (ECL) biosensors and photoelectric chemicals (PECs) (Bettazzi et al., 2018). Additionally, g-C_3N_4 is a new ECL emitter with high quantum yield attracted much interest to further explore the possible luminescent properties and expand the scope of its electrochemical applications by combining with metals and their oxides. Shao et al. reported a novel cage-type ECL biosensor, and in this biosensor, DNA1 (HP1) and hairpin DNA2 (HP2) were immobilized against the Fe_3O_4/Au NCs and Ru-based organic framework used as signal units with the detection limit of 0.3 FM (Liu et al., 2017). In order to improve the detection performance of the g-C_3N_4, AuNPs coined with organic framework are also employed for electrochemical sensing performance of miRNA. Wang's group reported a novel PEC biosensor for miRNA detection based on a ternary MoS_2/g-C_3N_4/black TiO_2 heterojunction as the photoactive material, and the AuNPs carrying Histostar antibodies (Wang & Su, 2014). The detection of miRNA-396a at low concentration can be done in the presence of HRP and H_2O_2 accelerated oxidation of 4-chloro-1-naphthol. Similarly, Yin et al. proposed a more simple detection method by using the antibody to detect miRNA. The synergistic effect of g-C_3N_4 combined with other NPs such as Au, MOS_2 can improve the light absorption capacity, facile transfer of electrons, and improved signal performance. The 2D structure and high specific surface area of g-C_3N_4 NPs resulted in improved performance of CDs and GQDs in ECL or PEC sensing (Inagaki et al., 2019). Fig. 14.5 represents schematically the diversified applications of C-nanomaterials.

Future aspects, limitations, and advantages of electrochemical sensors

Nature and composition of the electrodes play vital roles in electrochemical performance. The research in the area has opened up many applications, especially in the analysis of the analyte. The composition of the material, selection of the dopant material, structure, and morphology of the membrane play important role in the detection of ions. In recent years, miniaturization and microelectronics gained much attention due to their versatile properties in the detection of ions selectively and sensitively. For quantitative detection of biological species in the presence of biospecific reagents immobilized over the electrode converts quantitative process of the response as recorded. The development of screen-printed electrodes fulfills the need of the analytical determinations that are potentially portable devices. The highly sensitive biosensors developed as a result of the new methods incorporated in the fabrication of the devices. The newer developed electrode systems such as microelectrodes, microelectrode arrays, and modified chemical electrodes. The industrial and clinical usage of the electrochemical sensors, especially biosensors attracted much interest owing to diversified applications. Examples of the applications are in the development of improved electrocatalytic systems with a high rate of chemical stability and sensitivity. Coating is the electrode process over the semiconducting electrodes, photosensitization, and corrosion resistance properties. Such devices find robust applications in the field of electrochromic displays,

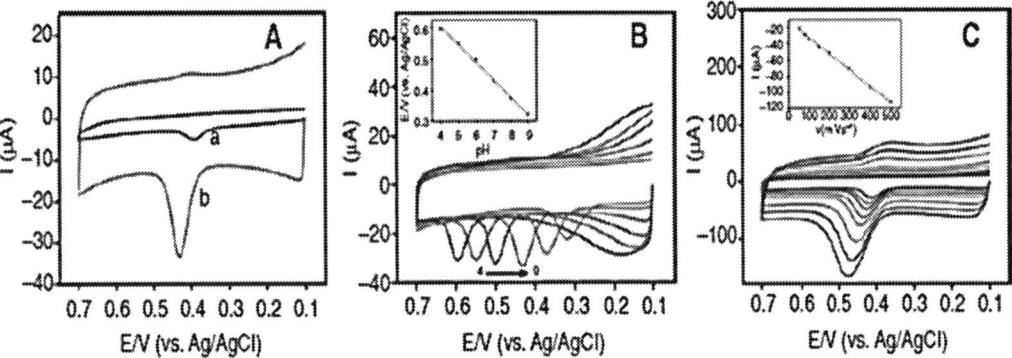

FIG. 14.5 (A) CV of bare GCE (a), MWCNTs/GCE and (b) at 100 mV s^{-1}. (B) CV curve at different pH 4, 5, 6, 7, 8, and 9 at 100 mV s^{-1}, EPA Vs pH (inset). (C) CV curve at different scan rates (50–500 mV s^{-1}) EPA Vs v (inset). *(No permission required.)*

microelectrochemical devices, etc. The analytical application involves the acceleration of the electron transfer reactions, involving preferential accumulation of the ions, selective membrane permeation, and interfering exclusion. Such newer applications impart high selectivity and sensitivity incorporating the modified surface area of the thin films. The construction of biosensors involves modification in the surface area of the electrodes, improvising the solubility, adsorption, etc. Different types of inorganic films, such as metal oxide, clay, zeolite, and hybrid nanomaterials, will be used as surface modifiers. The modified thin films consist of increased morphology, well-defined structures, thermally and chemically stable, inexpensive, and readily available. Table 14.4 summarizes the advantages and disadvantages of electrochemical techniques over widely used conventional methods.

Modification of the carbon electrodes, development and challenges

Critical challenge involved in the development of bioelectrochemical (BEC) systems is to improve the interaction of microbial system in electrolysis or electrochemical analysis. The development of new varieties of electroanalytical systems wherein electrode and microorganisms is involved in the effective transfer of exchangeable electrons between the microbial solution or surface and the electrode. Carbon nanomaterials are well-known biocatalysts because of their structural features, shapes, size, porosity, and thermally, electrically conducting surfaces (Tigari & Manjunatha, 2020; Tigari, Manjunatha, Raril, & Hareesha, 2019). The nanostructures derived from the carbon nanomaterials are low cost and chemically stable; however, the performance in the case of BECs retarded due to low current gain and decrease in the mass transportation of the ions. The major challenges involved are decreasing the overpotential and increasing the selectivity and sensitivity of the systems (Charithra, Manjunatha, & Raril, 2020; Hareesha & Manjunatha, 2020a). Recent electrodes developed by the researchers involve the detection of identification mechanism by the microbes in aqueous solutions of low ionic strengths. In this respect, current research focused on improving the surface chemistry of the carbon-based nanostructures during the sensing performance in BECs. The newer design, new capabilities, synthesis, and preparations in material chemistry question BECs can improve its detection range and limit. Furthermore, development of biofilms and biofilms developed

TABLE 14.4 Advantages and disadvantages of electrochemical techniques over widely used conventional methods.

Methods	Advantages	Disadvantages
Cell culture systems	Highly sensitive over rapid antigen tests	Antiviral susceptibility testing
epidemiologic studies	Highly selective	Technical expertise required to read
Immunofluorescence assay	Usually exhibits good sensitivity and excellent specificity	Requires trained experienced hands in reading results poor
Molecular approach	Excellent specificity and sensitivity of the Food and Drug Administration (FDA)-cleared kits	Approved protocols not commonly available for most viruses
Electrochemical biosensors	Conversion to a sensor device significantly reduces	Sensitive to sample matrix effects

on the electrode surfaces plays a vital role in the modifications of the electrode system in BECs. By the modifications in the electrode surface, electronic and electrical conductivity, adhesion, the sensing performance can be improved (Charithra et al., 2020; Hareesha & Manjunatha, 2020a). The modifications are categorized under the following headings: (1) increase in the crystal porosity for the most effective utilization of the surface area; (2) various functional groups present on the surfaces responsible for the selectivity and sensitivity of the electrochemical systems, such as O and N; (3) increase in the surface conductivity and biocompatibility. Bacterial adhesion and formation of biofilm and electron transport nature significantly affect the detection limit and performance of the electrodes. Surface treatments also improve sensitivity and low-level detection. The surface treatments to the electrode commonly done are ammonia treatment, modification in the polymeric structure, surface oxidation, etc. Recent reports have demonstrated an increase in the current density by the presence of oxygen and nitrogen functional groups at the electrode surface of carbon nanostructures. These surface modifications are likely to improve biofilm-electrode interactions. Further development of the functional porous solids having porosity, active surface area, and tailored pore structures and pore topologies improves the substantial current gain in the carbon nanostructures, especially hierarchical porous materials which consist of mesopores (porosity of the size 5–50 nm), and graphitic degrees will improve the surface sensitivity of the materials. Among the many carbon-based nanostructures, improving its advance in the microbial detection demonstrated increases the current density of up to $6.8\,mA\,cm^{-2}$ and fatty volatile acids production such as formic acids and acetate. Very recently, porous carbon electrodes containing carbon nanotubes (CNTs) have brought considerable attention due to the combination of high surface area, enhancing the biofilm growth and the current density generated, and the high conductivity of the CNTs. To provide novel properties, the design of new electrodes combining surface chemistry and porosity remains crucial for new generation BECs Hareesha & Manjunatha, 2020b; Manjunatha, 2020a). The electrode material is a key component that determines a BEC's performance. However, for that, the mechanistic understanding of the interaction between microbes and such modified electrodes is still needed.

In particular, among various forms of carbon nanomaterials, activated carbon fibers (ACFs) are mainly used for conductivity. ACFs possess outstanding physicochemical properties as it contains microporous materials, large surface area, and high conductivity. Especially when the ACFs used in the wastewater treatment process wherein even with the contaminated water, it is able to sense the materials and species electrochemically (Raril & Manjunatha, 2020). Increased sensitivity and selectivity of the ACFs are due to the presence of porous texture and large surface-to-volume ratio of the materials. In the case of ACFs, the specific capacitance of each of the carbons is found to increase with the increase in the oxidation of the carbon nanomaterials. Additionally, Surface complex analysis using temperature-programmed desorption showed that the double-layer capacitance was enhanced due to the presence of CO-desorbing complexes while CO_2-desorbing complexes exhibited a negative effect. The micropore resistance for ion migration was low for these carbons. Even carbon cloths can be directly employed as electrodes, its specific area is $1000-000\,m^2/g$. A capacitance increase of more than 40% has been achieved (Hareesha & Manjunatha, 2020b; Manjunatha, 2020a, 2020b). The use of an insulating binder will increase the internal resistance of the electrode, and it will affect the rate of performance of the electrochemical double layers (EDLCs). The use of binders block anables the porous sites, deteriate and thus diminishing the double-layer performance in the case of ACFs electrodes, due the presence of fibrous surface electrode exhibit the higher capacitance and better electrochemical performance. Carbon fiber surface modifications can be done with the help of electrooxidation and reduction. The method is done by making carbon fiber anodic and phenolic –OH group that was introduced and made in the form of cathode in the presence of various functional groups such as alkyl halides and alkyl tosylates. These electrochemical surface modification methods are applicable for fixing various functional groups, namely alkyl amide groups, haloalkyl amides, and tosylates. This kind of carbon fiber is used in the determination of some species in electroanalytical techniques.

Conclusion

Carbon nanostructures are versatile due to their physicochemical characteristics. The limitation in the electrochemical applications of carbon-based nanomaterials is due to the presence of defects, structural properties, and agglomerations of the NPs. However, limitations are related to the sensor systems and their responses. Different carbon allotropic forms can be used in several sensor fields with good response among them, and graphene, graphitic carbon, and hybrid nanostructures are robustly used due to their thermal and mechanical stability and electrical and electronic properties. Carbon nanostructures have a high degree of selectivity, especially when functionalized with other nanomaterials because of their cage-type structure, high electrochemical reactivity, biocompatibility, etc. Many remained challenges were regarded on the carbon-based manufacture to phase out contaminants as well as to improve the functionalization process. Improving these aspects will increase the sensors' selectivity and control defect densification, which is strictly related to physical and chemical properties that must be overcome.

Conflict of interest

All authors declare that they do not have any conflict of interest.

Authors' contributions

Dr. Vinayak Adimule performed the manuscript writing and synthesis and characterization of the samples; Dr. Basappa C Yallur and Mr. Adarsha HJG contributed for the grammar correction, creation of figures and tables, and revision of the manuscript.

Acknowledgment

All the authors are thankful to the Centre for Advanced Material Technology, MSRIT, Bangalore, Karnataka, India.

References

Akbari, E., Buntat, Z., Ahmad, M. H., Enzevaee, A., Yousof, R., Iqbal, S. M. Z., et al. (2014). Analytical calculation of sensing parameters on carbon nanotube based gas sensors. *Sensors (Switzerland)*, *14*(3), 5502–5515. https://doi.org/10.3390/s140305502.

Amor-Gutiérrez, O., Costa Rama, E., Costa-García, A., & Fernández-Abedul, M. T. (2017). Paper-based maskless enzymatic sensor for glucose determination combining ink and wire electrodes. *Biosensors and Bioelectronics*, *93*, 40–45. https://doi.org/10.1016/j.bios.2016.11.008.

Arnold, M. S., Green, A. A., Hulvat, J. F., Stupp, S. I., & Hersam, M. C. (2006). Sorting carbon nanotubes by electronic structure using density differentiation. *Nature Nanotechnology*, *1*(1), 60–65. https://doi.org/10.1038/nnano.2006.52.

Azimzadeh, M., Nasirizadeh, N., Rahaie, M., & Naderi-Manesh, H. (2017). Early detection of Alzheimer's disease using a biosensor based on electrochemically-reduced graphene oxide and gold nanowires for the quantification of serum microRNA-137. *RSC Advances*, *7*(88), 55709–55719. https://doi.org/10.1039/c7ra09767k.

Berger, C., Song, Z., Li, T., Li, X., Ogbazghi, A. Y., Feng, R., et al. (2004). Ultrathin epitaxial graphite: 2D electron gas properties and a route toward graphene-based nanoelectronics. *Journal of Physical Chemistry B*, *108*(52), 19912–19916. https://doi.org/10.1021/jp040650f.

Bettazzi, F., Palchetti, I. Y., Lin, L., Zou, Y. T., Huang, Jiang, J., et al. (2018). The combination of ternary electrochemiluminescence system of g-C3N4 nanosheet/TEA/Cu@Cu2O and G-quadruplex-driven regeneration strategy for ultrasensitive bioanalysis. *Current Opinion in Electrochemistry*, *12*.

Campbell, F. W., & Compton, R. G. (2010). The use of nanoparticles in electroanalysis: An updated review. *Analytical and Bioanalytical Chemistry*, *396*(1), 241–259. https://doi.org/10.1007/s00216-009-3063-7.

Charithra, M. M., Manjunatha, J. G. G., & Raril, C. (2020). Surfactant modified graphite paste electrode as an electrochemical sensor for the enhanced voltammetric detection of estriol with dopamine and uric acid. *Advanced Pharmaceutical Bulletin*, *10*(2), 247–253. https://doi.org/10.34172/apb.2020.029.

Choucair, M., Thordarson, P., & Stride, J. A. (2009). Gram-scale production of graphene based on solvothermal synthesis and sonication. *Nature Nanotechnology*, *4*(1), 30–33. https://doi.org/10.1038/nnano.2008.365.

Chung, H., Won, J., & Zelenay, P. (1922). Active and stable carbon nanotube/nanoparticle composite electrocatalyst for oxygen reduction. *Nature Communications*, *4*.

Collins, P. G., & Avouris, P. (2000). Nanotubes for electronics. *Scientific American*, *283*(6), 62–69. https://doi.org/10.1038/scientificamerican1200-62.

Dang, V. T., Nguyen, D. D., Cao, T. T., Le, P. H., Tran, D. L., Phan, N. M., et al. (2016). Recent trends in preparation and application of carbon nanotube–graphene hybrid thin films. *Advances in Natural Sciences: Nanoscience and Nanotechnology*, *7*(3). https://doi.org/10.1088/2043-6262/7/3/033002, 033002.

Dastjerdi, R., & Montazer, M. (2010). A review on the application of inorganic nano-structured materials in the modification of textiles: Focus on antimicrobial properties. *Colloids and Surfaces B: Biointerfaces*, *79*(1), 5–18. https://doi.org/10.1016/j.colsurfb.2010.03.029.

De Volder, M. F. L., Tawfick, S. H., Baughman, R. H., & Hart, A. J. (2013). Carbon nanotubes: Present and future commercial applications. *Science*, *339*(6119), 535–539. https://doi.org/10.1126/science.1222453.

Deng, C., Chen, J., Wang, M., Xiao, C., Nie, Z., & Yao, S. (2009). A novel and simple strategy for selective and sensitive determination of dopamine based on the boron-doped carbon nanotubes modified electrode. *Biosensors and Bioelectronics*, *24*(7), 2091–2094. https://doi.org/10.1016/j.bios.2008.10.022.

Geng, X., You, J., Wang, J., Zhang, C., Malefane, Ntsendwana, B., et al. (2017). Situ synthesis of tetraphenylporphyrin/tungsten (VI) oxide/reduced graphene oxide (TPP/WO3/RGO) nanocomposite for visible light photocatalytic degradation of acid blue 25. *Materials Chemistry and Physics*, *191*, 8379–8389.

Guo, T., Nikolaev, P., Rinzler, A. G., Tomanek, D., Colbert, D. T., & Smalley, R. E. (1995). Self-assembly of tubular fullerenes. *Journal of Physical Chemistry*, *99*(27), 10694–10697. https://doi.org/10.1021/j100027a002.

Hareesha, N., & Manjunatha, J. G. (2020a). A simple and low-cost poly (DL-phenylalanine) modified carbon sensor for the improved electrochemical analysis of Riboflavin. *Journal of Science: Advanced Materials and Devices*, *5*(4), 502–511. https://doi.org/10.1016/j.jsamd.2020.08.005.

Hareesha, N., & Manjunatha, J. G. (2020b). Surfactant and polymer layered carbon composite electrochemical sensor for the analysis of estriol with ciprofloxacin. *Materials Research Innovations*, *24*(6), 349–362. https://doi.org/10.1080/14328917.2019.1684657.

He, Y., Chen, W., Gao, C., Zhou, J., Li, X., & Xie, E. (2013). An overview of carbon materials for flexible electrochemical capacitors. *Nanoscale, 5*(19), 8799–8820. https://doi.org/10.1039/c3nr02157b.

Hernandez, Y., Nicolosi, V., Lotya, M., Blighe, F. M., Sun, Z., De, S., et al. (2008). High-yield production of graphene by liquid-phase exfoliation of graphite. *Nature Nanotechnology, 3*(9), 563–568. https://doi.org/10.1038/nnano.2008.215.

Iijima, S. (2002). Carbon nanotubes: Past, present, and future. *Physica B: Condensed Matter, 323*(1–4), 1–5. https://doi.org/10.1016/S0921-4526(02)00869-4.

Inagaki, M., Tsumura, T., Kinumoto, T., Toyoda, Li, Y., Bu, Y., et al. (2019). Fabrication of ultra-sensitive photoelectrochemical aptamer biosensor: Based on semiconductor/DNA interfacial multifunctional reconciliation via 2D-C3N4. *Biosensors and Bioelectronics, 141*, 111903.

Kilic, T., Erdem, A., Erac, Y., Seydibeyoglu, M. O., Okur, S., & Ozsoz, M. (2015). Electrochemical detection of a cancer biomarker mir-21 in cell lysates using graphene modified sensors. *Electroanalysis*, 317–326. https://doi.org/10.1002/elan.201400518.

Kroto, H. W., Heath, J. R., O'Brien, S. C., Curl, R. F., & Smalley, R. E. (1985). C60: Buckminsterfullerene. *Nature*, 162–163. https://doi.org/10.1038/318162a0.

Li, W. Z., Xie, S. S., Qian, L. X., Chang, B. H., Zou, B. S., Zhou, W. Y., et al. (1996). Large-scale synthesis of aligned carbon nanotubes. *Science, 274*(5293), 1701–1703. https://doi.org/10.1126/science.274.5293.1701.

Liu, Q., Peng, Y. J., Xu, J. C., Ma, C., Li, L., Mao, C. J., et al. (2017). Label-free electrochemiluminescence aptasensor for highly sensitive detection of acetylcholinesterase based on Au-nanoparticle-functionalized g-C3N4 nanohybrid. *ChemElectroChem, 4*(7), 1768–1774. https://doi.org/10.1002/celc.201700035.

Liu, Z., Tabakman, S., Welsher, K., & Dai, H. (2009). Carbon nanotubes in biology and medicine: In vitro and in vivo detection, imaging and drug delivery. *Nano Research, 2*(2), 85–120. https://doi.org/10.1007/s12274-009-9009-8.

Llobet, E. (2013). Gas sensors using carbon nanomaterials: A review. *Sensors and Actuators, B: Chemical, 179*, 32–45. https://doi.org/10.1016/j.snb.2012.11.014.

Lotya, M., Hernandez, Y., King, P. J., Smith, R. J., Nicolosi, V., Karlsson, L. S., et al. (2009). Liquid phase production of graphene by exfoliation of graphite in surfactant/water solutions. *Journal of the American Chemical Society, 131*(10), 3611–3620. https://doi.org/10.1021/ja807449u.

Manjunatha, J. G. (2020a). Poly (Adenine) modified graphene-based voltammetric sensor for the electrochemical determination of catechol, hydroquinone and resorcinol. *Open Chemical Engineering Journal, 14*, 52–62. https://doi.org/10.2174/1874123102014010052.

Manjunatha, J. G. (2020b). Fabrication of efficient and selective modified graphene paste sensor for the determination of catechol and hydroquinone. *Surfaces*, 473–483. https://doi.org/10.3390/surfaces3030034.

Marega, R., De Leo, F., Pineux, F., Sgrignani, J., Magistrato, A., Naik, A. D., et al. (2013). Functionalized Fe-filled multiwalled carbon nanotubes as multifunctional scaffolds for magnetization of cancer cells. *Advanced Functional Materials, 23*(25), 3173–3184. https://doi.org/10.1002/adfm.201202898.

Ndamanisha, J. C., & Guo, L.p. (2012). Ordered mesoporous carbon for electrochemical sensing: A review. *Analytica Chimica Acta, 747*, 19–28. https://doi.org/10.1016/j.aca.2012.08.032.

Novoselov, K. S. (2004). Electric field effect in atomically thin carbon films. *Science*, 666–669. https://doi.org/10.1126/science.1102896.

Pan, X., Fan, Z., Chen, W., Ding, Y., Luo, H., & Bao, X. (2007). Enhanced ethanol production inside carbon-nanotube reactors containing catalytic particles. *Nature Materials, 6*(7), 507–511. https://doi.org/10.1038/nmat1916.

Raril, C., & Manjunatha, J. G. (2020). A simple approach for the electrochemical determination of vanillin at ionic surfactant modified graphene paste electrode. *Microchemical Journal, 154*. https://doi.org/10.1016/j.microc.2019.104575.

Shuai, H. L., Huang, K. J., Xing, L. L., & Chen, Y. X. (2016). Ultrasensitive electrochemical sensing platform for microRNA based on tungsten oxide-graphene composites coupling with catalyzed hairpin assembly target recycling and enzyme signal amplification. *Biosensors and Bioelectronics, 86*, 337–345. https://doi.org/10.1016/j.bios.2016.06.057.

Sutter, P. W., Flege, J. I., & Sutter, E. A. (2008). Epitaxial graphene on ruthenium. *Nature Materials, 7*(5), 406–411. https://doi.org/10.1038/nmat2166.

Talapatra, S., Kar, S., Pal, S. K., Vajtai, R., Ci, L., Victor, P., et al. (2006). Direct growth of aligned carbon nanotubes on bulk metals. *Nature Nanotechnology*, 112–116. https://doi.org/10.1038/nnano.2006.56.

Tigari, G., & Manjunatha, J. G. (2020). Optimized voltammetric experiment for the determination of phloroglucinol at surfactant modified carbon nanotube paste electrode. *Instruments and Experimental Techniques, 63*(5), 750–757. https://doi.org/10.1134/S0020441220050139.

Tigari, G., Manjunatha, J. G., Raril, C., & Hareesha, N. (2019). Determination of riboflavin at carbon nanotube paste electrodes modified with an anionic surfactant. *ChemistrySelect, 4*(7), 2168–2173. https://doi.org/10.1002/slct.201803191.

Wang, D. W., & Su, D. (2014). Heterogeneous nanocarbon materials for oxygen reduction reaction. *Energy and Environmental Science, 7*(2), 576–591. https://doi.org/10.1039/c3ee43463j.

Yu, L., Shearer, C., & Shapter, J. (2016). Recent development of carbon nanotube transparent conductive films. *Chemical Reviews, 116*(22), 13413–13453. https://doi.org/10.1021/acs.chemrev.6b00179.

Zhang, L., Wang, B., Ding, Y., Wen, G., Hamid, S. B. A., & Su, D. (2016). Disintegrative activation of Pd nanoparticles on carbon nanotubes for catalytic phenol hydrogenation. *Catalysis Science and Technology, 6*(4), 1003–1006. https://doi.org/10.1039/c5cy02165k.

Zhao, Y., Fan, L., Hong, B., Ren, J., Zhang, M., Que, Q., et al. (2016). Nonenzymatic detection of glucose using three-dimensional PtNi nanoclusters electrodeposited on the multiwalled carbon nanotubes. *Sensors and Actuators, B: Chemical, 231*, 800–810. https://doi.org/10.1016/j.snb.2016.03.115.

Zhu, S., & Xu, G. (2010). Single-walled carbon nanohorns and their applications. *Nanoscale, 2*(12), 2538–2549. https://doi.org/10.1039/c0nr00387e.

Chapter 15

The role of carbon nanomaterial-based sensors in sustainability

Doddahosuru M. Gurudatt[a], S. Rajendra Prasad[b], Sneha S. Puttappa[c], and Srikantamurthy Ningaiah[c]

[a]Department of Studies in Organic Chemistry, University of Mysore, Mysuru, Karnataka, India, [b]PG Department of Chemistry, Davangere University, Davangere, Karnataka, India, [c]Department of Chemistry, Vidyavardhaka College of Engineering, Visvesvaraya Technological University, Mysuru, Karnataka, India

Introduction

Environmental, technological, socioethical, and economic concerns can all be classified as components of sustainability, and they are all interlinked. Sustainability is defined as the ability to remain indefinitely and meet human requirements without compromising future generations' well-being, i.e., extending material lifetimes and repurposing materials to preserve "modern life comfort." Technologies that clean water and air and reduce unnecessary overdose and agricultural delivery methods all contribute to sustainability. Products are meant to last and be optimized in the circular economy so that, in theory, waste does not exist. To improve resource productivity, the product's life cycle should be extended by reusing, repairing, remanufacturing, and recycling. This kind of approach is not only sustainable but also covers economic growth (Wintzheimer et al., 2021). The major concern while using nanomaterial will be its long-term stability (Ferrag & Kerman, 2020).

In scientific and technological development, nanotechnology has a great impact. Because of their emanating and unique properties, nanomaterials are the most studied materials and have attracted many users in various fields. In a recent development, the subject of ongoing research is carbon nanomaterial-based sensors and their exposures. Once the new nanomaterial is developed, one should investigate its effects and sustainability on health and the environment. What amount of health and safety consequences on the environment is acceptable? During the life cycle of a nanomaterial product, exposure sources should be evaluated to assure health and safety.

Carbon nanomaterials are the most renowned products of nanotechnology. Carbon nanostructures are made up mostly of carbon atoms with a large surface area, distinct chemical and physical characteristics, and a low density (Marinho, Ghislandi, Tkalya, Koning, & de With, 2012). The modern sensor devices must have few attributes: quick reaction times, quick recovery times, low cost, small size, in situ analysis, simple operation, high sensitivity, and dependability.

One of the most studied sensors is based on carbon nanomaterials due to their remarkable properties. Because of their excellent performance and superior physical and chemical properties, these carbon nanomaterial-based sensors are used as powerful sensor devices for currently expensive electronic compounds, yielding a high-quality sensing property, and considered as an environment-friendly material (Bezzon et al., 2019).

Carbon nanotubes (CNTs) are one of nanotechnology's most appealing nanomaterials. Because of their excellent features, CNTs have drawn the most interest in various research domains, including agriculture and biosensors. Polymer-modified electrodes have gotten a lot of attention from researchers in biosensors because the polymer layer is very stable and reproducible. To measure the sensing performance, the modified electrode's stability and reproducibility are tested (Raril & Manjunatha, 2018).

The simple definition for sustainability signifies the ability to continue to live indefinitely for the benefit of future generations (Kuhlman & Farrington, 2010). This chapter aims to explain how these materials contribute to the sustainable development of society and the technical systems to fulfill human needs in modern life.

Sustainability of carbon nanomaterial-based sensors

Naturally occurring carbon-based nanomaterials exist in small quantities, and most of them are artificially synthesized (Zaytseva & Neumann, 2016). For meeting the needs of modern life, from health to housing, nanotechnology plays a vital role. As the requirement of carbon nanomaterial-based sensors is increasing in materials and manufacturing processes, it is better to understand their robustness (Lynch & Hazebrouck, 2015). Once its safety measures are known, it is easy to define its sustainability. Safety measures to be taken depend on few factors like probability, the extent of exposure, and ingestion. Safety measures that can recommend are the systematic methods for the production and handling of the product, reproducibility, and release of the amount of these particles. As all these processes initially begin at the working environment, an overview data of the exposure to these particles should be estimated (Borm et al., 2006).

The major strategies to be involved in the research of these materials include various nanomaterial types and variants, identifying the consequences of these materials on the environment and quantifying the hazards and exposures of these materials. Once the sources are identified, their exposure needs to be assessed. The exposure routes may differ, and the interactions between organism, population, and ecosystem responses are complicated. Providing input to nanomaterial makers will improve data quality and consistency, allowing for a better understanding of sustainability. The changes in protocol settings, instrumentation, environment, media, and models result in roughness and robustness. This necessitates the use of established validation methods that include extensive precision, accuracy, dependability, and reproducibility testing (A Research Strategy for Environmental, Health, and Safety Aspects of Engineered Nanomaterials, 2012).

Carcinogenicity, specific target organ toxicity, germ cell mutagenicity, respiratory or skin sensitization, and significant eye damage are the most common dangers following repeated exposures. By evaluating the root of the problem, some control measures should be taken to avoid exposure (World Health Organization, 2017). Because some of these particles can cause explosions and fires, their usage as a chemical catalyst can cause unanticipated reactions; hence, they should maintain exposure limits (Groso, Petri-Fink, Magrez, Riediker, & Meyer, 2010).

Although these carbon nanomaterial-based sensors appear to be functional in their current condition, their development has been hampered by a lack of assurance. For long-term expansion, further nanomaterial environmental health and safety research is required. Electronics, energy, and biological applications are three of the most common uses for these materials. And, in the current global economic climate, these three will account for roughly 70% of the worldwide economy's nanomarket. It is vital to secure the long-term development of these materials to be used in various industries around the world (Liu & Xia, 2020).

The consolidated approaches responsible for these materials' sustainable production guarantee their satisfactory, safe utilization, and clearance. There is the biggest challenge in their advancement of effectiveness. There is an adverse impact on the surroundings because of their increased use (Saleem & Zaidi, 2020). It is vital to understand their toxicological effects on human health and the environment, as well as their interactions at the interface, to forecast their implications on environmental sustainability (He, Aker, Fu, & Hwang, 2015).

For the sustainable energy of these materials, some of the measures to be taken are reducing costs, emissions, and improving the security of supply (Yi et al., 2018). Because of their affordability and impressive energy conversion, this material has an impact on a variety of fields; thus, their long-term sustainability on health and the environment is a problem due to their renewable energy source technologies to satisfy their long-term demand (Ong, Ahmad, Zein, & Tan, 2010). For safe and sustainable use of these materials in future research, the main focus is multifunctional nanomaterials, cost-effective, environmentally friendly, and appropriate regulatory framework. So, selecting the most appropriate materials for future use is important (Shafique & Luo, 2019). With the increase in the amount of research on these materials, many different areas will question sustainability. Sustainability refers to the impact of energy policy on the economy, society, and environment. Product creation and the economic operation of these materials with pollution reduction are the integrated components of sustainability. It is critical to have a better understanding of the robust risk management concept. The probability of occurrence and magnitude of damage and incertitude, ubiquity, persistency, and reversibility are all part of the risk concept. For this, a complete assessment of the life cycle flow of material production and consumption is to be carried out to avoid future environmental issues (Helland & Kastenholz, 2008).

Coupling technological growth with sociological, environmental, and economic implications can benefit society and ensure the long-term sustainability of these materials. To examine and develop their sustainability implications, these materials are analyzed to map the most important focus areas. Sustainability is assessed in terms of performance in the economy, environmental ramifications, assessment and management of environmental risks, assessment and control of human health risks, societal ramifications, and technical capability (Cinelli, Coles, Sadik, Karn, & Kirwan, 2016).

The sustainability of nanomaterials depends on applying these materials in various fields like electronics, health care, and clean energy, and their feature development is aimed at commercial objectives.

Role of carbon nanomaterials as an electrochemical sensor

Electrochemical sensors are one of the important applications of carbon nanomaterials. Electrochemical science and engineering aim to provide an overview of current technologies for incorporating various forms of carbon nanomaterials, such as carbon nanotubes, graphene derivatives, carbon nanodots, and active screen-printed carbon electrodes, to design new electrochemical sensors. Carbon-based sensing elements are used as a multicomponent or hybrid electrode combination for medical analysis, food safety, soil quality, drug detection, environmental monitoring, and other applications. Researchers and analytical chemists have virtually endless options for directing sensor development toward more accurate and nondisposable devices useful in critical domains of interest such as human health and environmental protection, thanks to various sensor design methodologies (Bobacka, 2020).

This material has recently generated robust electrochemical sensing devices with high analytical performance; the capacity of these materials to enhance the electrochemical reactivity of biomolecules and stimulate electron transfer processes of proteins is of particular interest. The presence of adsorbates causes changes in surface conductivity, which makes them sensitive. This electrochemical analysis efficiently determines the levels of electroactive species in a solution quantitatively and qualitatively because it is simple, low cost, easy to use, accurate, and reliable. As a result, the long-term viability of these carbon nanostructures is owing to their outstanding electrochemical properties, which have led to their widespread utilization. Because of their structure and mechanical deformations, these materials have an electronic property that is unique to the carbon family, making them valuable in the construction of tiny sensors that are sensitive to chemical, mechanical, and physical environments (Power, Gorey, Chandra, & Chapman, 2018).

Dye-sensitized solar cells using carbon nanomaterials

The sharp increase in the cost of fossil fuels and the growing threat of carbon dioxide emissions cause a greenhouse effect. Alternative energy resources are in high demand; thus, sunlight is the most plentiful carbon-free energy source. Solar (or photovoltaic-cell-based) technologies are currently available and vital because they can harness the sun's energy. Compared to traditional photovoltaic technologies, dye-sensitized solar cells (DSSCs) are of great interest due to several advantages. The future development of carbon nanomaterial composites as prospective materials for DSSCs is expected to play a significant role in the future solar cell market. Solar photovoltaics will provide a solution to dwindling energy sources as well as climatic change concerns. In this DSSC technology, the absorption of a photon by a dye molecule allows for the generation of an electrical current. Carbon nanomaterials can replace both the transparent electrode and the nontransparent counter electrode, and they are a good alternative to expensive platinum catalysts. Carbon nanomaterials have also shown great promise as an alternate material for counter electrodes, allowing for more durable and affordable electrodes in the future (Brennan, Byrne, Bari, & Gun'ko, 2011).

Carbon-based material application in wastewater treatment/purifying

The carbon nanomaterial has an application in water purification. As a response to human society's water scarcity for a clean environment, hazardous organic and inorganic pollutants from water should be removed. Adsorption-based water technologies are widely employed due to their high efficiency, low cost, and lack of need for expensive infrastructure. This activation technique produces nanocomposites with inorganic and organic components with increased adsorptive characteristics and separation efficiency. These findings may help to shape future studies and practical applications of carbon-based nanoadsorbents in the field of water security. The use of pesticides in agriculture and industrial discharge pollutes water and harms human health. Adsorption, photoelectrocatalysis, photocatalysis, reverse osmosis, flocculation, and membrane separation are some of the processes used to treat water. Adsorption is utilized to remove dangerous organic and inorganic contaminants from wastewater because it is cost-effective, rapid, and easy. The adsorbent can be reused numerous times without losing its efficiency. Various materials have been investigated for water contamination remediation, with carbon-based nanomaterials being one of the most commonly used nanomaterials for contaminant adsorption. Nanomaterials used for water treatment are regenerable and minimize the formation of contaminants, thanks to their size and shape-dependent features, friendly to the environment, availability, and ease of handling. A range of technical analyses should be used to determine whether this material can operate as a viable alternative for present technologies. For the safety and sustainability of these materials on the environment and human health, it is critical to evaluate the synthetic routes and the effects of postsynthesis processing, purification, and separation on purity during their manufacture. They can be utilized as an adsorbent to remove contaminants from wastewater because of their high regeneration capability. Nanomaterials based on graphene and graphene oxide, as well as nanomaterials from carbon and graphene quantum dots, have shown tremendous promise in the treatment and purification of water and wastewater, notably for industrial and pharmaceutical wastes (Gusain, Kumar, & Ray, 2020).

The advantage of nanotechnology-enabled water treatment is that it can replace existing water treatment procedures, which have sustainability issues. Nanomaterials offer a clear potential because their costs are lower than the cost of conventional water treatment. Nanotechnology promises effective, safe, and sustainable technology for the treatment of worldwide water and wastewater (Falinski et al., 2020).

Carbon nanomaterials for biofuel cells and supercapacitors

These materials will also find application for biofuel cells (BFCs) and supercapacitors attributed to carbon materials. Because of their new fascinating capabilities, they find applications in various sources such as biodevices, wearable, and other implantable devices in biomedical, fitness, academic, and industrial fields. These materials find applications in various fields, boosting energy conversion and producing different material devices, all of which contribute to developing sustainable and suitable energy alternatives for biodevices in the human body. Personalized energy-harvesting technologies such as tiny, portable, wearable, or implantable applications are particularly appealing due to their remarkable electrochemical characteristics, electrical conductivity, and adaptability. These energy devices maximize electron transfer processes in BFCs, which involve an electrochemical link between enzymes and carbon nanomaterials, between the enzymatic site and the electrode. They are durable and long-lasting when in contact with biofluids and human tissues. Wearable devices such as smartwatches, armbands, glasses, artificial skin, and soft robotics allow some biosensors to interact and communicate with their surroundings. The device's performance in real-world scenarios, such as biocompatibility, compact size, lightweight, and stretchability, is checked to ensure the long-term survivability of the biosensors. Sweat, tears, interstitial fluid, and blood all contain useful metabolites like glucose and lactate that can be used as fuel for personalized BFCs to create electricity (Jeerapan & Ma, 2019).

Carbon-based materials in biodiesel production

Carbon nanomaterials also find applications in biodiesel production, and for these, catalyst plays an important role. With their inexpensive cost, large surface area, and thermal stability, carbon nanostructures are regarded as good catalysts. Because activated carbon is exceptionally porous and has a large surface area available for adsorption or chemical reactions, it acts as a catalyst. Thermal decomposition in a furnace utilizing a regulated environment and heat converts carbon-based products to activated carbon. The starting material was glucose with sucrose, glutinous rice starch, maize starch, and amylase, which was then pyrolyzed at 400°C for 60, 75, and 90 min, resulting in the formation of polycyclic aromatic carbon rings. By sulfonating the sample combination with H_2SO_4 at 150°C for 5 h, these rings are integrated with the SO_3H group. This catalyst, which has a high free fatty acid (FFA) of about 55.2 wt percent, was utilized to make biodiesel from waste cottonseed oil. Biodiesel synthesis utilizing carbon-based materials that act as catalysts reduces the cost of producing ecologically friendly biodiesel while also eliminating the issues that come with the traditional technique (Konwar, Boro, & Deka, 2014).

Carbon materials in biochar-based functional materials

Energy sources that are both economically viable and environmentally benign will aid in the creation of a sustainable future. The most important part is to provide a material that is an ecologically friendly energy source that will aid in the creation of a sustainable future and resolve the numerous issues that are current and in the future. Biomass may be transformed into biofuels, which can be used as a source of renewable energy. Carbon materials can be used to solve real challenges such as pollution and global warming. Biochar produced by biomass pyrolysis is gaining attraction and recognition as a versatile and efficient platform, signaling a substantial shift in research focus on biochar's potential applications in catalysis, energy storage, and environmental protection (Liu & Xia, 2020).

The focus is on developing highly efficient and environmentally friendly sustainable nanomaterials with unique qualities such as high efficiency and selectivity, earth abundance, recyclability, low-cost manufacturing procedures, and stability at an affordable cost for a sustainable future (Nasrollahzadeh, Sajjadi, Iravani, & Varma, 2021). As a result, they are used in a variety of fields, including the development of a new generation of environmental sensing systems, chemical and biological sensors, fiber-optic systems, and magneto-optic and optical devices, as well as solar cells for producing clean energy, in coatings for building exterior surfaces, and sonochemical decolorization of dyes by nanocomposite. From surface chemistry to large-scale production, the science and applications of carbon nanomaterial-based sensors will contribute to nanotechnology, which will supply the idea of product sustainability. They are also employed in medicine, electronic devices, sunscreens, military applications, paints, and catalysts. Some of these will have a little environmental impact, while others will have a significant one (Ali, 2020).

Conclusion

In summary, though carbon nanomaterials contribute directly or indirectly to the sustainable development of society, we can see a very few reports on the safety and long-term attainment as for health and the environment is concerned. In the present scenario, it is critical to be aware of the potential dangers associated with exposure to these nanomaterials to protect researchers, laboratory personnel, and those who work in the industries. Depending on the exposure, carbon nanomaterial-based particles can enter the body through the skin, lungs, or stomach. Hence, it is essential to examine the various physicochemical features of these carbon nanomaterial-based sensors before commercialization.

This chapter also necessitates enhanced quantitative characterization methods to be implemented in production processes, as well as health and safety regulations to restrict nanomaterial exposures and prevent detrimental effects from impacting sustainability. Also, more focused research needs to be considered to assure the long-term market success of these nanomaterials. As the use of these materials expands, it is critical to comprehend their toxicity to assess their long-term viability. This, in turn, increases energy demands and sustainability concerns, some of which were briefly discussed in this chapter.

References

A Research Strategy for Environmental, Health, and Safety Aspects of Engineered Nanomaterials. (2012).

Ali, A. S. (2020). Application of nanomaterials in environmental improvement. In *Nanotechnology and the environment* IntechOpen.

Bezzon, V. D. N., Montanheiro, T. L. A., De Menezes, B. R. C., Ribas, R. G., Righetti, V. A. N., Rodrigues, K. F., et al. (2019). Carbon nanostructure-based sensors: A brief review on recent advances. *Advances in Materials Science and Engineering, 2019*. https://doi.org/10.1155/2019/4293073.

Bobacka, J. (2020). Electrochemical sensors for real-world applications. *Journal of Solid State Electrochemistry, 24*(9), 2039–2040. https://doi.org/10.1007/s10008-020-04700-4.

Borm, P. J. A., Robbins, D., Haubold, S., Kuhlbusch, T., Fissan, H., Donaldson, K., et al. (2006). The potential risks of nanomaterials: A review carried out for ECETOC. *Particle and Fibre Toxicology, 3*. https://doi.org/10.1186/1743-8977-3-11.

Brennan, L. J., Byrne, M. T., Bari, M., & Gun'ko, Y. K. (2011). Carbon nanomaterials for dye-sensitized solar cell applications: A bright future. *Advanced Energy Materials, 1*(4), 472–485. https://doi.org/10.1002/aenm.201100136.

Cinelli, M., Coles, S. R., Sadik, O., Karn, B., & Kirwan, K. (2016). A framework of criteria for the sustainability assessment of nanoproducts. *Journal of Cleaner Production, 126*, 277–287. https://doi.org/10.1016/j.jclepro.2016.02.118.

Falinski, M. M., Turley, R. S., Kidd, J., Lounsbury, A. W., Lanzarini-Lopes, M., Backhaus, A., et al. (2020). Doing nano-enabled water treatment right: Sustainability considerations from design and research through development and implementation. *Environmental Science: Nano, 7*(11), 3255–3278. https://doi.org/10.1039/d0en00584c.

Ferrag, C., & Kerman, K. (2020). Grand challenges in nanomaterial-based electrochemical sensors. *Frontiers in Sensors*. https://doi.org/10.3389/fsens.2020.583822.

Groso, A., Petri-Fink, A., Magrez, A., Riediker, M., & Meyer, T. (2010). Management of nanomaterials safety in research environment. *Particle and Fibre Toxicology, 7*. https://doi.org/10.1186/1743-8977-7-40.

Gusain, R., Kumar, N., & Ray, S. S. (2020). Recent advances in carbon nanomaterial-based adsorbents for water purification. *Coordination Chemistry Reviews, 405*. https://doi.org/10.1016/j.ccr.2019.213111.

He, X., Aker, W. G., Fu, P. P., & Hwang, H. M. (2015). Toxicity of engineered metal oxide nanomaterials mediated by nano-bio-eco-interactions: A review and perspective. *Environmental Science: Nano, 2*(6), 564–582. https://doi.org/10.1039/c5en00094g.

Helland, A., & Kastenholz, H. (2008). Development of nanotechnology in light of sustainability. *Journal of Cleaner Production, 16*(8–9), 885–888. https://doi.org/10.1016/j.jclepro.2007.04.006.

Jeerapan, I., & Ma, N. (2019). Challenges and opportunities of carbon nanomaterials for biofuel cells and supercapacitors: Personalized energy for futuristic self-sustainable devices. *C—Journal of Carbon Research*, 62. https://doi.org/10.3390/c5040062.

Konwar, L. J., Boro, J., & Deka, D. (2014). Review on latest developments in biodiesel production using carbon-based catalysts. *Renewable and Sustainable Energy Reviews, 29*, 546–564. https://doi.org/10.1016/j.rser.2013.09.003.

Kuhlman, T., & Farrington, J. (2010). What is sustainability? *Sustainability, 2*(11), 3436–3448. https://doi.org/10.3390/su2113436.

Liu, S., & Xia, T. (2020). Continued efforts on nanomaterial-environmental health and safety is critical to maintain sustainable growth of nanoindustry. *Small, 16*(21). https://doi.org/10.1002/smll.202000603.

Lynch, I., & Hazebrouck, B. (2015). *Engineered nanomaterial mechanisms of interactions with living systems and the environment: A universal framework for safe nanotechnology* (pp. 1–4). NanoMILE, European Union's Seventh Framework Program for Research, Technological Development and Demonstration. www.nanomile.eu.

Marinho, B., Ghislandi, M., Tkalya, E., Koning, C. E., & de With, G. (2012). Electrical conductivity of compacts of graphene, multi-wall carbon nanotubes, carbon black, and graphite powder. *Powder Technology, 221*, 351–358. https://doi.org/10.1016/j.powtec.2012.01.024.

Nasrollahzadeh, M., Sajjadi, M., Iravani, S., & Varma, R. S. (2021). Carbon-based sustainable nanomaterials for water treatment: State-of-art and future perspectives. *Chemosphere, 263*. https://doi.org/10.1016/j.chemosphere.2020.128005.

Ong, Y. T., Ahmad, A. L., Zein, S. H. S., & Tan, S. H. (2010). A review on carbon nanotubes in an environmental protection and green engineering perspective. *Brazilian Journal of Chemical Engineering, 27*(2), 227–242. https://doi.org/10.1590/S0104-66322010000200002.

Power, A. C., Gorey, B., Chandra, S., & Chapman, J. (2018). Carbon nanomaterials and their application to electrochemical sensors: A review. *Nanotechnology Reviews, 7*(1), 19–41. https://doi.org/10.1515/ntrev-2017-0160.

Raril, C., & Manjunatha, J. G. (2018). Carbon nanotube paste electrode for the determination of some neurotransmitters: A cyclic voltammetric study. *Modern Chemistry & Applications.* https://doi.org/10.4172/2329-6798.1000263.

Saleem, H., & Zaidi, S. J. (2020). Recent developments in the application of nanomaterials in agroecosystems. *Nanomaterials, 10*(12), 1–34. https://doi.org/10.3390/nano10122411.

Shafique, M., & Luo, X. (2019). Nanotechnology in transportation vehicles: An overview of its applications, environmental, health and safety concerns. *Materials, 12*(15). https://doi.org/10.3390/ma12152493.

Wintzheimer, S., Reichstein, J., Groppe, P., Wolf, A., Fett, B., Zhou, H., et al. (2021). Supraparticles for sustainability. *Advanced Functional Materials, 31*(11), 2011089. https://doi.org/10.1002/adfm.202011089.

World Health Organization. (2017). *WHO guidelines on protecting workers from potential risks of manufactured nanomaterials.*

Yi, H., Huang, D., Qin, L., Zeng, G., Lai, C., Cheng, M., et al. (2018). Selective prepared carbon nanomaterials for advanced photocatalytic application in environmental pollutant treatment and hydrogen production. *Applied Catalysis B: Environmental, 239*, 408–424. https://doi.org/10.1016/j.apcatb.2018.07.068.

Zaytseva, O., & Neumann, G. (2016). Carbon nanomaterials: Production, impact on plant development, agricultural and environmental applications. *Chemical and Biological Technologies in Agriculture, 3*(1). https://doi.org/10.1186/s40538-016-0070-8.

Chapter 16

Microfluidic systems with amperometric and voltammetric detection and paper-based sensors and biosensors

Ebtesam Sobhanie, Amirreza Roshani, and Morteza Hosseini

Department of Life Science Engineering, Faculty of New Sciences & Technologies, University of Tehran, Tehran, Iran

Introduction

Microfluidics is a powerful technology that links different scientific disciplines, such as chemistry, mechanics, biology, control systems, microscale physics and thermal/fluidic transport, simulation of microflows, and materials science with engineering. It has developed as a reliable and expanding technology for creating inexpensive and field-deployable sensors for various applications in the past two decades. With the development of microfluidic devices with the advantages of valuable microscale fluid physics, a new era of sensing has begun in which microfluidic devices reduce the sample, reagent use, size, and cost of tests, and also improve overall automation (Fakhri et al., 2020; Hasanzadeh & Hashemzadeh, 2021). However, microfluidics has faced some challenges, such as costly equipment, complex manufacturing techniques, complicated flow control methods, and nondegradable substrates. Due to these limitations, microfluidic paper-based analytical devices (μPADs) were introduced in 2007 (Martinez, Phillips, Butte, & Whitesides, 2007). Paper is a suitable substrate for microfluidic devices because it is low-cost, flexible, abundant, thin, lightweight, and good biological compatibility for applications in clinical diagnosis. Furthermore, the use of capillary action to drive fluid flow within the paper is one of the significant advantages of paper-based microfluidic devices that avoid the use of external pumps, which increase the device's cost (Jafry, Lim, & Lee, 2020). Accordingly, μPADs are being developed for various applications, including determination of air, water, and food quality and the respective application for biochemical, immunological, and molecular detection.

As we mentioned, microfluidics permits reducing the size of devices. Therefore, minimizing the device's size suggested the miniaturization of the detection system to be simply integrable (Palit & Hussain, 2020). Electrochemical methods can provide high selectivity and sensitivity (Alavi-Tabari, Khalilzadeh, & Karimi-Maleh, 2018; Karimi-Maleh et al., 2020, 2021; Raril & Manjunatha, 2018). They are also easily miniaturized without losing the analytical performance, so coupling microfluidics and electrochemical sensors is of great importance because they act as the achievement of point-of-care (POC) and laboratory-on-a-chip (LOC) systems (Noviana, McCord, Clark, Jang, & Henry, 2020). This chapter first provides an overview of the basics of paper-based microfluidics and its recent advances in miniaturized electrochemical sensors for biosensing applications. At last, the main limitations of coupling microfluidics and electrochemical techniques are summarized. An outlook on the current technology status and prospects is given.

Basics of paper-based microfluidics

The construction and use of analytical techniques based on paper dates back many years ago and has a long history. In the early 1800s, Gay-Lussac describes the litmus paper test for acids; in 1859, Hugo Schiff reported the first spot test for uric acid, using silver carbonate. Chromatographic papers used to separate small molecules by capillary force were introduced in the 19th century. In 1988, the first commercial survey was launched for home pregnancy test kits. In the year 2007 and the year 2008, the first international conference on bioactive paper was held in Finland, and paper-based microfluidic devices were introduced. Paper is now considered an attractive and efficient material for microfluidic devices for its magnetic properties and mechanical characteristics such as flexibility, lightness, and low thickness. The paper has unique features for developing microanalytical devices; here are some of the most important ones: Papers are very light and thin

(0.07–1 mm), making them very suitable for fabricating lightweight and portable devices. Another essential point about papers is that they are highly adaptable with different functional groups to bind to biomolecules. Papers are mainly made of random network fibers of cellulose during the pressure drying of the aqueous cellulose suspension; therefore, by considering cellulose as a biocompatible compound, it can be used for various biomedical sensing approaches such as health diagnosis, environmental monitoring, immunoassays, food safety, and many other areas. Moreover, these structural fibers and pores are involved in the filtration of separated suspended solids inside the sample (Martinez, 2011; Schilling, Lepore, Kurian, & Martinez, 2012).

Formats of paper-based analytical devices

Recently, paper is known as a user-friendly and universal substrate for the development of microfluidic devices. There are several techniques for designing and manufacturing paper-based sensors. Here is a brief description of the main paper-based methods.

Paper devices based on dipsticks

In this platform, which is the simplest type, reagents are deposited on the paper to be used for qualitative evaluation. They are inadequate for multistep detection procedures because they have limited capability to handle fluids and are usually known for simple detection approaches.

Lateral flow assays (LFAs)

Lateral flow assays (LFAs) are low-cost, simple, rapid, and portable detection devices typically composed of nitrocellulose membranes popular in various diagnostic fields. The sample containing the desired analyte passes through a polymer strip consisting of four zones by capillary in this platform: (1) Sample pad is located at the beginning of the strip, and the sample containing the analyte is absorbed. (2) The conjugate release pad is attached to the sample pad and contains the dried reagents specific to the target analyte. (3) Detection zone: The sample and reagents attached to the analyte move to this region, a porous membrane with specific biological elements. In this region, the reagents are immobilized on nitrocellulose for signal development. (4) Absorbent pad: This pad has loose nitrocellulose that provides the capillary force needed to move the liquid containing the analyte (Bahadır & Sezgintürk, 2016; Koczula & Gallotta, 2016).

Paper devices based on the μ-PAD

The first microfluidic device was developed in 1975. Microfluidic paper-based analytical devices (μPADs) were introduced in 2007 (Martinez et al., 2007). Significant advances in microfluidic technology have occurred in recent decades. Microfluidic devices are a potentially efficacious analytical platform because they can provide the facilities and performances of a diagnostic laboratory in the micro- or nanoscale. That is why, they are called laboratory-on-chip (LOC). They are composed of hydrophilic and hydrophobic network channels. The μPADs, fabricated by paper, are inexpensive, user-friendly, and do not require external apparatuses and complicated fabrication methods. This technology provides control and manipulation of liquids on a microscopic scale, which is a valid option in fabricating accurate sensors with high sensitivity and selectivity, and so the loss of the sample is minimized. Due to the fantastic properties of paper materials mentioned in the previous section, μPADs are attractive, convenient platforms for analyzing and determining organic and inorganic compounds with a broad application of various fields. These low-cost, portable, simple, and light devices give considerable diagnostic facilities that could change medicine and the pharmaceutical industry (He, Wu, Fu, & Wu, 2015; Lisowski & Zarzycki, 2013; Xia, Si, & Li, 2016).

Fabrication methods for paper-based microfluidic devices

Paper has been in the world of analytical chemistry for ages. Inactive transfer of liquid, which is the most critical feature of papers, has made them alternative technology for manufacturing simple, low-cost, and disposable analytical tools for several applications. Paper patterning can be discovered back to the early 19th century. In those years, paraffin was used to draw hydrophobic patterns on paper. Nowadays, various examples of paper-based sensors are being performed. A paper device is made up of hydrophilic and hydrophobic regions to limit fluid containing the analyte and its controlled conductivity. Therefore, various hydrophobic materials such as PDMS, paraffin, polystyrene, alkenyl ketene dimer, silicones, methylcellulose, and hydrophobic gels are used to block paper pores physically and create fluid passage channels. Here is a summary of the most common technologies and their advantages and limitations for the fabrication of paper-based

TABLE 16.1 Fabrication methods for paper-based microfluidic devices (Akyazi, Basabe-Desmonts, & Benito-Lopez, 2018; Almeida, Jayawardane, Kolev, & McKelvie, 2018; Lim, Jafry, & Lee, 2019).

Method of fabrication	Advantages and disadvantages
Laser cutting: In the laser cutting method, by using CO_2 gas, the desired cuts and patterns are created on the paper. The hydrophilic patterns are then reconstructed to perform a hydrophobic coating.	Large-scale production. High resolution of channel-expensive instruments. Fragile while bending
Laser printing: In this method, which is commonly used to compose paper-based microfluidic devices, the toner is laser-placed on a transparent paper surface.	Requires special equipment. Simple and inexpensive
Wax printing: In this method, which is mainly used for making colorimetric or electrochemical papers, wax is placed on the filter or chromatography paper by a printer or specific pen to form desired patterns. The flow and rate of the wax are controllable.	Fast and simple. Environmentally friendly heating step after wax deposition required. Inadequate for prototyping
Screen printing: The liquid material is manually or automatically placed on paper's surface by screen and eventually dried by heat.	Low price with a simple process. Low resolution. Different printing screens for different patterns
Wax dipping: In this method, hydrophobic barriers are created by melting wax on paper. A temporary iron mold is also used; when the paper is attached to the iron, it is impregnated with molten wax	Fast, cheap, and simple. Not suitable for large-scale production
Photolithography: In this method, ultraviolet light is irradiated into a light-resistant polymer. As a result, hydrophobic areas are created due to cross-linking of the polymer.	Sharp barriers. Large-scale production. Expensive instruments
Inkjet printing: A noncontact approach that the desired material is sprayed on the paper's surface through a narrow nozzle.	Large-scale production fast. Requires a customized inkjet printer extra heating step for fixation
PDMS plotting: Polydimethylsiloxane, also known as dimethylpolysiloxane, is a silicon-based organic polymer. It is a hydrophobic component that can be placed on the surface of the paper and create hydrophobic channels	Low cost. Low resolution. Required modification
Plasma treatment: In this method, the paper is first impregnated with alkyl ketene dimer (AKD)-heptane and then placed in the hood to evaporate the heptane. Then, paper becomes hydrophobic using the temperature of the oven between two metal masks with the desired patterns and submitted to plasma treatment to create the hydrophilic regions	Inexpensive. Each pattern requires a specific photomask.

devices (Table 16.1). Most of these methods make hydrophobic barriers in the hydrophilic paper, except for some that use the opposite approach (Nery & Kubota, 2013; Xia et al., 2016).

Detection techniques

The fabrication method for paper-based microfluidic and design patterns created on paper is highly influenced by the detection method. The read-out techniques are chosen according to the analyte type and concentration to be detected and desired accuracy. Therefore, detection of reactions in paper-based microfluidic devices has been developed in recent years. Several detection methods have been used in paper-based microfluidic devices, such as colorimetric detection, luminescence detection, and electrochemical detection (EC).

Colorimetric detection

Colorimetric detection is defined by the reactions that lead to a change of color of the substrate due to chemical or biochemical interaction between the analyte and reagents. The results and color change analysis can be determined by the naked eye, cameras, or smartphones (Naderi, Hosseini, & Ganjali, 2018). So colorimetric is an ideal method for simple, semiquantitative explanation or when a yes or no answer is needed. Furthermore, colorimetric detection is one of the most common microfluidic paper-based analysis methods due to its low cost, ease of use, simple fabrication, and fast detection efficiency (Bagheri pebdeni & Hosseini, 2020; Jafari et al., 2021). Nevertheless, colorimetric sensors are naturally affected by the background of the paper or the sample. Also, because of the inhomogeneity of the color distribution on the paper, sometimes the decision about the final color with the naked eye is difficult.

Luminescence detection

Luminescence detection, which includes fluorescence, chemiluminescence (CL), and electrogenerated chemiluminescence (ECL), has been coupled with paper-based microfluidic devices. Fluorescence measurement is based on evaluating the intensity of light emitted by a substance that has absorbed light or other electromagnetic radiation from another source. Fluorescence sensing provides new capabilities to µPADs (Zhang et al., 2020); however, this method often faces some challenges for µPADs, since fluorescence sensing's efficiency will depend on cost reduction and miniaturization of fluorescence readers (Akyazi et al., 2018).

CL methods use the emission of light from excited species to quantify the target's concentration in paper-based platforms. CL is widely used as a detection technique for µPADs because of its wide linear range, inexpensive reagents, and high sensitivity (Mesgari et al., 2020). As a result, various chemiluminescence-based µPAD methods for the fabrication of different sensors and biosensors have been proposed.

ECL is one kind of luminescence that active species generate at or near the electrode's surface due to electrochemical reactions and luminescence reactions (Karimi, Husain, Hosseini, Azar, & Ganjali, 2020; Pur, Hosseini, Faridbod, Dezfuli, & Ganjali, 2016; Sobhanie, Faridbod, Hosseini, & Ganjali, 2020). Due to the integration of advantages of these two techniques, such as small sample volume needed, high sensitivity, low cost, and rapid detection rate, microfluidic paper-based ECL sensor and biosensors have been developed rapidly in recent years (Gross, Durant, Hipp, & Lai, 2017).

Electrochemical detection

Electrochemical detection is a redox-based method. The electric potential applied between the working and the reference electrode drives an electrochemical reaction at the working electrode's surface. The electric current generated from oxidative or reductive reactions at the working electrode appears as a peak on the recording device. Electrochemical detection is described by simplicity, low cost, portability, good selectivity and sensitivity, low electrical power consumption, and minimal instrumentation (Alavi-Tabari et al., 2018; Karimi-Maleh et al., 2020, 2021). This method, against the colorimetric method, can determine the exact quantification of the analyte's concentration and, compared with the luminescence techniques, has higher sensitivity and less background interference. Therefore, the employment of µPADs for electrochemical sensing has become one of the most popular techniques that provide more stable and quantifiable signals.

Electrochemical detection in paper-based microfluidic devices and application

Among all the analytical detection methods, electrochemical detection is the simplest to couple with microfluidics devices, making it ideal for point-of-care applications (Prinith, Manjunatha, & Raril, 2019; Raril & Manjunatha, 2018). Therefore in this section, we discuss the coupling of electrochemical detection with microfluidics. The incorporation of paper microfluidics with electrochemistry is attractive as it combines two low-cost and sensitive technologies (Manjunatha, 2017; Shen, Zhang, & Etzold, 2020). A typical electrochemical cell is a three-electrode system that consists of a working electrode (WE), a counter electrode, and a reference electrode. The intended electrochemical reaction occurs on the working electrode's surface; therefore, the material of WE is very significant in the design and improvement of the performance of the electrochemical system (Tigari & Manjunatha, 2020b). In recent years, the use of paper as a platform for electrochemical detection gained more attention, and several good reviews with helpful information about microfluidic electrochemical paper-based analytical devices were published.

The first microfluidic paper-based electrochemical device (µPED) was developed by Dungchai et al. in 2009 to measure the level of glucose and lactate in biological samples (Dungchai, Chailapakul, & Henry, 2009). The electrochemical performance of the system was demonstrated by using cyclic voltammetry methods and screen-printed electrodes on paper. For the first time, this work showed that the concept of fabrication of electrochemical analytical devices using paper microfluidic is possible. Hence, since 2009, the integration of microfluidics and electrochemical devices has been widely studied to develop the next generation of POC testing systems.

Researchers have investigated different types of electrochemical techniques (Raril, Manjunatha, Ravishankar, et al., 2020; Raril, Manjunatha, & Tigari, 2020; Tigari & Manjunatha, 2020a), including voltammetry, amperometry, and impedance spectroscopy (Pradela-Filho, Araújo, Takeuchi, Santos, & Henry, 2021; Xing et al., 2021). In Table 16.2,

TABLE 16.2 Summary of some typical μPADs with electrochemical sensing.

References	LOD	Analyte	Electrochemical analysis method	Paper substrate	Fabrication method	Working electrode
Punjiya, Moon, Matharu, Rezaei Nejad, and Sonkusale (2018)	5–17.5 mM	Dopamine, glucose, and pH	CV, amperometry, and potentiometry	Whatman grade 1	Wax printing, screen printing	Carbon
Xing et al. (2021)	$0.38\,ng\,mL^{-1}$–$0.27\,pg\,mL^{-1}$	CRP, cTnI, and PCT	SWV	Whatman grade 1	Wax printing	Carbon
Jemmeli, Marcoccio, Moscone, Dridi, and Arduini (2020)	$0.03\,\mu mol\,L^{-1}$	Bisphenol A	SWV	Labor filter paper	Wax printing	Carbon black
Cinti, Minotti, Moscone, Palleschi, and Arduini (2017)	$3\,\mu g\,L^{-1}$	Nerve agents (paraoxon)	Chronoamperometric	Filter paper	Wax printing, screen printing	Carbon ink
Zhang, Bao, Luo, and Ding (2018)	$0.03\,pg\,mL^{-1}$	CEA	ECL	Whatman chromatography paper	Wax printing, screen printing	PG-Fe_3O_4/GCE
Nie et al. (2010)	0.22 mM	Glucose	Chronoamperometry	Whatman grade 1	Photolithography, wax printing	Carbon enzyme
Medina-Sánchez, Cadevall, Ros, and Merkoçi (2015)	7–11 ppb	Cd^+, Pb^{2+}	SWV	Whatman grade 1	Wax printing, screen printing	Graphite
Kaur, Tomar, and Gupta (2017)	0.12–10.23 mM	Cholesterol	Amperometry	Whatman grade 1	Photolithography	Graphite
Ding, He, Lisak, Qin, and Bobacka (2017)	10 μM	Cl^-	Potentiometric	Whatman grade 1, 3		GCE
Fava et al. (2019)	0.3 μM	Glucose	CV, SWV, DPV, and chronoamperometry	Whatman grade 1	Paper cutting, screen printing	Carbon
Ding et al. (2020)	$10^{-3.3}$–$10^{-4.1}$ $mol\,L^{-1}$	K^+, Na^+, Cl^-	Potentiometric	Filter paper Whatman	Wax printing	GCE
Cao, Han, Xiao, Chen, and Fang (2020)	$25\,\mu mol\,L^{-1}$	Glucose	Chronoamperometry	Whatman grade 1	Photolithography and screen printing	Carbon
Zhang, Yang, Liu, and Liang (2019)	$0.8\,mg\,kg^{-1}$	Biotoxicities	CV	Whatman grade 1	Screen printing	Carbon
Fava et al. (2020)	0.012–0.120 mM	Glucose, creatinine, and uric acid	CV, DPV, and chronoamperometry	Whatman grade 1	Screen printing	Carbon

we summarized some typical paper-based microfluidic electrochemical sensing devices for various applications. The table illustrated that amperometric and voltammetric methods are mainly used as a detection technique for µPADs. Also, wax printing techniques are typically used to pattern the various areas on the paper, and electrodes were fabricated mainly onto paper substrates by screen-printing processes.

Voltammetric paper-based microfluidic sensors and biosensors and their application

The voltammetric technique is a category of electrochemical methods in which details about an analyte are achieved by measuring the current while the potential is changed (Charithra, Manjunatha, & Raril, 2020; Pushpanjali, Manjunatha, Tigari, & Fattepur, 2020). Due to the high efficiency, sensitivity, and selectivity of the voltammetric method, various µPAD sensors and biosensors have recently been developed, using several voltammetry types, such as square wave voltammetry (SWV) (Cao et al., 2017a; Jemmeli et al., 2020; Shi et al., 2012), cyclic voltammetry (CV) (Nechaeva, Shishov, Ermakov, & Bulatov, 2018), stripping voltammetry (Kokkinos, Economou, & Giokas, 2018; Pokpas, Jahed, & Iwuoha, 2019), and differential pulse voltammetry (DPV) (Cao et al., 2017b; Jemmeli et al., 2020; Saadati, Hassanpour, Bahavarnia, & Hasanzadeh, 2020; Shi et al., 2012). The application of µPADs based on voltammetric detection can be categorized into two groups, as we described below.

Clinical analysis

Research in the clinical study focuses on paper-based microfluidic devices due to their critical importance and high commercial outlook to produce low-cost and usable biosensing devices for numerous applications (Abarghoei, Fakhri, Borghei, Hosseini, & Ganjali, 2019; Pradela-Filho et al., 2021; Tomei et al., 2019). Many µPADs based on voltammetric detection have been suggested in this field. For instance, Cincotto, Fava, Moraes, Fatibello-Filho, and Faria (2019) developed a µPAD for creatinine and uric acid's simultaneous determination. The µPAD was used to determine those clinical biomarkers in human urine samples using one working electrode divided into two spots. They fabricated the device using a cutter printer from cutting the filter paper to form the microfluidic channel through the electrode by making the screen-printed electrodes over the polyester film. Thus, this sensitive electrochemical µPAD was recognized to determine two biological biomarkers by SWV with a low detection limit of $8.4\,nmol\,L^{-1}$ to uric acid and $3.7\,nmol\,L^{-1}$ to creatinine.

Wang et al. (2019) and colleagues have constructed multiparameter µPAD for simultaneous detection of two lung cancer biomarkers (carcinoembryonic antigen (CEA) and neuron-specific enolase (NSE)) in a clinical sample. The microfluidic channels of the paper-based aptasensor were fabricated through wax printing and screen printing. The aptasensor can simultaneously detect two analytes in one sample as two working electrodes were combined onto one device. A schematic diagram of the aptasensor is shown in Fig. 16.1. Under optimal conditions, the responses of the multiplexed detection aptasensor for CEA and NSE were studied by measuring the change in peak currents of DPV responses. Suggested microfluidic paper-based aptasensor could provide a low-cost and easy-to-carry diagnostics platform for cancer biomarkers.

Yakoh, Chaiyo, Siangproh, and Chailapakul (2019) reported a 3D µPAD for electrochemical detection of biological species. The device eliminates the undesirable procedure of multiple-step reagent manipulation in a complex assay due to storing and delivering the sequence of analytes to the detection channel without the need for agents' manipulation by the users. A wax-printing technique was used to pattern Whatman grade 1 chromatography paper. The device was employed for the amperometric determination of ascorbic acid, differential-pulse voltammetric determination of serotonin, and an impedimetric label-free α-fetoprotein immunosensor.

Kokkinos, Economou, and Giokas (2018) reported a µPAD for the voltammetric determination of DNA. The microfluidic channels were wax-printed on the paper substrate. Then, the sputtering process was performed on the reverse side of the paper, and sputtering Sn was deposited on the working electrode. For the biosensor preparation, capture DNA was immobilized in the circular assay zone of the µPAD, then hybridized with biotinylated target oligonucleotide, and labeled with streptavidin-conjugated CdSe/ZnS quantum dots. Eventually, after the acidic dissolution of the QDs, the released Cd(II) is detected with ASV to quantify the target DNA.

A recent report by Xing et al. (2021) has also utilized µPAD for label-free electrochemical detection of programmed death-ligand 1 (PD-L1) in body fluids. PD-L1 aptamer was used as a biorecognition molecule, and the DPV method was adopted for electrochemical measurement. Nanocomposites of amine-functionalized single-walled carbon nanotube, methylene blue, and Ag nanoparticles were employed to modify the working electrode and help aptamer binding. The aptasensor was used for the detection of serum samples, and the LOD was $10\,pg\,mL^{-1}$.

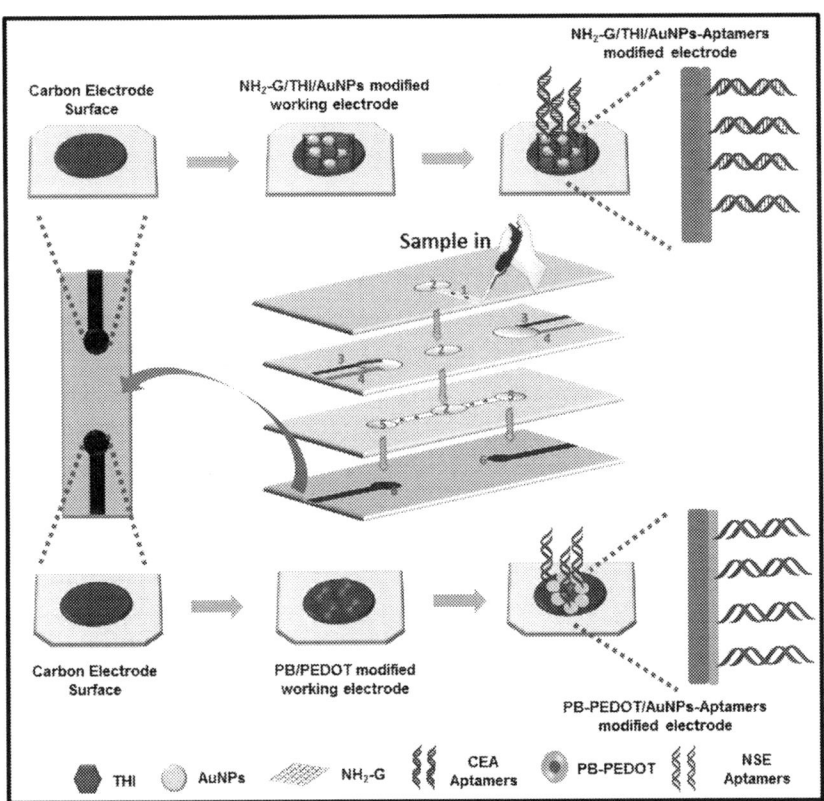

FIG. 16.1 Schematic diagrams of the voltammetric μPAD, configuration and modification of electrochemical aptasensor: (1) Sample inlet hole, (2) filter hole, (3) screen-printed counter electrode, (4) screen-printed reference electrode, (5) detection zone, and (6) screen-printed working electrode (Wang et al., 2019). *(No Permission Required.)*

Environmental analysis

Environmental pollution is a significant problem that has an effect on the ecosystem and human health worldwide (Hussain, 2018, 2020; Hussain, Hoque, & Balachandran, 2019). Compounds such as toxic and heavy metal ions are the primary source of environmental pollution and an important cause of environmental concern. μPAD is an attractive device for rapid and sensitive sensing of environmental pollution. Therefore, several μPADs have been developed for ecological analysis—for example, Jemmeli et al. (2020) reported an electrochemical microfluidic paper-based device to detect bisphenol A (BPA) in drinking water. The working electrode was fabricated using a wax printer, and then the three-electrode system was screen-printed on filter paper. The sample was added on the backside of the printed sensor and then detected via SWV as an electrochemical technique (Fig. 16.2). The proposed sensor achieved a detection limit of 0.03 μM for sensitive and inexpensive BPA determination.

Christos Kokkinos et al. also reported μPAD for the ASV detection of heavy metals such as Cd(II) and Zn(II) at trace levels. Microfluidic channel was fabricated by wax-printing method and, on the reverse side of the paper, sputtered thin

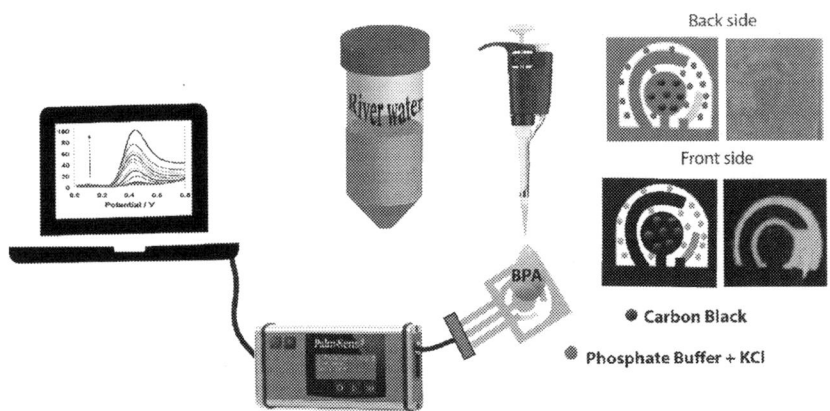

FIG. 16.2 Fabrication of μPAD and measurement methods for reagent-free detection of bisphenol A (Jemmeli et al., 2020). *(No Permission Required.)*

films of Sn, was used as the working electrode. This ready-to-use μPAD presents significant advantages over regular electrochemical sensors in terms of cost, the scope for mass production, and simplicity of operation in the determination of heavy metal (Kokkinos, Giokas, Economou, Petrou, & Kakabakos, 2018; Pokpas et al., 2019).

Amperometric paper-based microfluidic sensors and biosensors and their application

In the amperometric technique, the current response in constant potential is measured to detect the concentration of an analyte. A constant potential is applied on the working electrode whenever the current response is related to the analyte concentration due to the oxidation or reduction of the target. The amperometric technique was employed as a detection method in μPADs owing to its potential for miniaturization and portability, high sensitivity, and low fabrication cost. Various μPADs based on amperometric methods have been reported for detection in clinical, environmental, and biological applications (Julika et al., 2020; Lee, Mohd-Naim, Tamiya, & Ahmed, 2018).

Clinical analysis

Several biosensors used amperometric methods in clinical applications due to the dynamic characteristics and the current measurement under the constant optimal potential (Ataide, Mendes, Gama, De Araujo, & Paixaõ, 2020). Cao et al. (2019) developed a 3D μPAD for the wearable electrochemical detection of glucose in human sweat metabolites. The device was created by wax screen-printing patterns on paper and then a 3D-flow channel formed by four times folding a patterned paper. That included five equal layers as sweat collector, vertical channel, transverse channel, an electrode layer, and sweat evaporator. The electrodes were screen-printed using Prussian blue/graphite ink to create the counter and working electrode and silver/silver chloride ink to fabricate the reference electrode. The device was designed using amperometric measurement and achieved high sensitivity and low detection limit ($5\,\mu mol\,L^{-1}$). The results showed that the proposed device is low cost and easy to use, and also, the problem of sweat accumulation under the device was solved by the microfluidic properties of the paper.

Fava et al. (2019) proposed a disposable electrochemical μPAD for multiplexed glucose determination in human urine samples, based on 16 independent microfluidic channels coupled to the same number of individual electrochemical cells. The patterning process of Whatman filter paper using a craft cutter printer was carried out for device fabrication (Fig. 16.3). The significant advantage of this work was the nonnecessary use of wax printing to produce the 16-channel paper component, which reduces the cost of operation. The electrochemical performance of the device was examined by employing

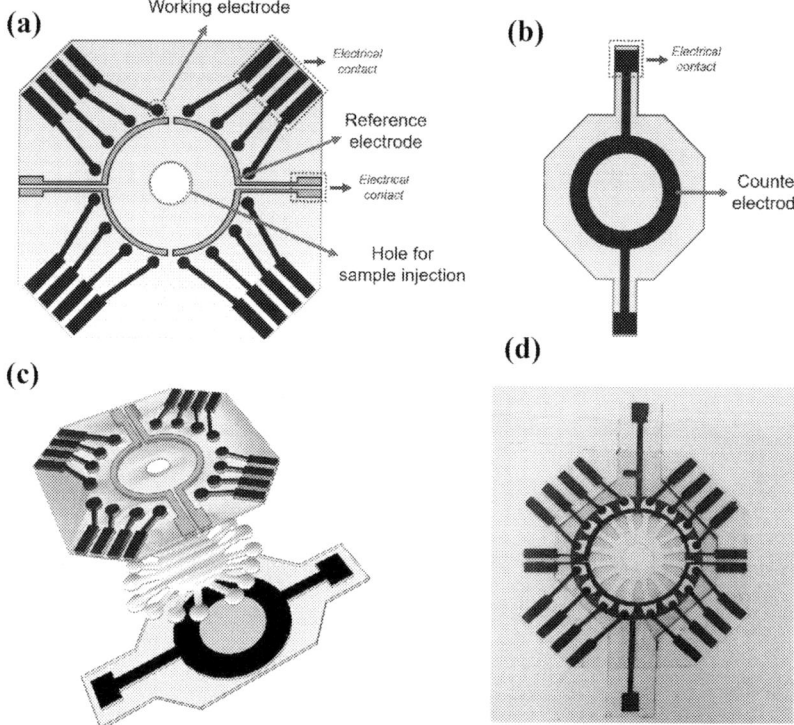

FIG. 16.3 (A) Design of the WE/RE and (B) CE layers (C) Steps for device layer assembly. (D) Photography of assembled Mpad (Fava et al., 2019). *(No Permission Required.)*

CV, SWV, DPV, and chronoamperometry. The detection limit of $0.3\,\mu mol\,L^{-1}$ for glucose determination was obtained using chronoamperometry. The same group also reported a similar disposable electrochemical μPAD for the simultaneous determination of uric acid and creatinine in human urine samples (Cincotto et al., 2019). A more specific device was also constructed to determine glucose, creatinine, and uric acid in urine samples (Fava et al., 2020).

Environmental analysis

Arduini et al. (2019) reported a 3D origami μPAD to detect various types of pesticide (paraoxon, 2,4-dichlorophenoxyacetic acid, and atrazine) in river water by integrating different enzyme inhibition biosensors. They fabricated the device by combining two different office paper-based electrodes printed on the front and the backside of the origami system and multiple filter paper-based pads to load enzymes and enzymatic substrates (Fig. 16.4). The presence of pesticide was detected using a portable potentiostat and chronoamperometry technique via monitoring the enzymatic activity in the absence and presence of pesticides. The LOD of the proposed device was found to be 2 ppb and demonstrated cost-effective, rapid, and accurate pesticide detection in surface water.

Dabhade, Jayaraman, and Paramasivan (2021) recently developed a paper-based electrochemical biosensor to determine hexavalent chromium as a toxic heavy metal in potable water. They immobilized glucose oxidase enzyme on filter paper using chitosan as an entrapping agent and integrated it with a screen-printed carbon electrode for amperometric. They obtained the detection limit of 0.05 ppm for Cr(VI), and the fabricated biosensor showed good reproducibility.

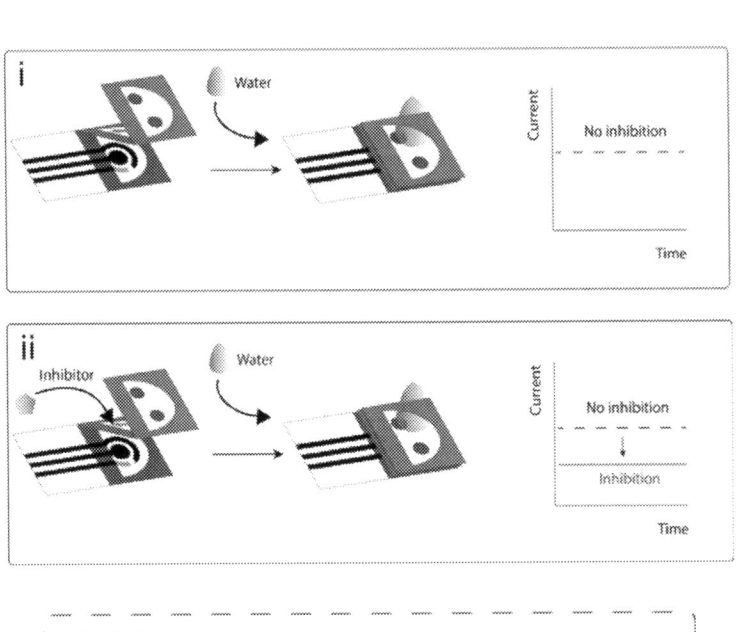

FIG. 16.4 Configuration and quantification of origami paper-based electrochemical biosensor (Arduini et al., 2019). *(No Permission Required.)*

Conclusion and prospects

Herein, we described some properties that make electrochemical analysis a suitable detection method for microfluidic paper-based devices. The number of different applications indicates how important and essential the development of electrochemical µPADs is. Many factors are investigated in µPAD fabrication, including lower cost, ease of fabrication, efficiency, and reproducible for POC applications. The development of electrochemical paper-based microfluidic remains a point of attention and is commercialized in applications in the near future. The electrochemical µPADs still face some difficulty to become commercialized. For example, microfluidics within paper mostly depends on the paper properties and configuration. Hence, investigating the paper properties, influencing the behavior of the paper properties on the electrochemical performance, and developing inexpensive and environmentally friendly electrode materials are necessary. However, in µPAD, the electrode substance or its modifier might be toxic or costly. The analysis processes are also done with a potentiostat, which is not regularly available and familiar to general people. So for portable applications, the electrochemical sensing system should be developed.

References

Abarghoei, S., Fakhri, N., Borghei, Y. S., Hosseini, M., & Ganjali, M. R. (2019). A colorimetric paper sensor for citrate as biomarker for early stage detection of prostate cancer based on peroxidase-like activity of cysteine-capped gold nanoclusters. *Spectrochimica Acta Part A: Molecular and Biomolecular Spectroscopy, 210*, 251–259. https://doi.org/10.1016/j.saa.2018.11.026.

Akyazi, T., Basabe-Desmonts, L., & Benito-Lopez, F. (2018). Review on microfluidic paper-based analytical devices towards commercialisation. *Analytica Chimica Acta, 1001*, 1–17. https://doi.org/10.1016/j.aca.2017.11.010.

Alavi-Tabari, S. A. R., Khalilzadeh, M. A., & Karimi-Maleh, H. (2018). Simultaneous determination of doxorubicin and dasatinib as two breast anticancer drugs uses an amplified sensor with ionic liquid and ZnO nanoparticle. *Journal of Electroanalytical Chemistry, 811*, 84–88. https://doi.org/10.1016/j.jelechem.2018.01.034.

Almeida, M. I. G. S., Jayawardane, B. M., Kolev, S. D., & McKelvie, I. D. (2018). Developments of microfluidic paper-based analytical devices (µPADs) for water analysis: A review. *Talanta, 177*, 176–190. https://doi.org/10.1016/j.talanta.2017.08.072.

Arduini, F., Cinti, S., Caratelli, V., Amendola, L., Palleschi, G., & Moscone, D. (2019). Origami multiple paper-based electrochemical biosensors for pesticide detection. *Biosensors and Bioelectronics, 126*, 346–354. https://doi.org/10.1016/j.bios.2018.10.014.

Ataide, V. N., Mendes, L. F., Gama, L. I. L. M., De Araujo, W. R., & Paixaõ, T. R. L. C. (2020). Electrochemical paper-based analytical devices: Ten years of development. *Analytical Methods, 12*(8), 1030–1054. https://doi.org/10.1039/c9ay02350j.

Bagheri pebdeni, A., & Hosseini, M. (2020). Fast and selective whole cell detection of *Staphylococcus aureus* bacteria in food samples by paper based colorimetric nanobiosensor using peroxidase-like catalytic activity of DNA-Au/Pt bimetallic nanoclusters. *Microchemical Journal, 159*, 105475. https://doi.org/10.1016/j.microc.2020.105475.

Bahadır, E. B., & Sezgintürk, M. K. (2016). Lateral flow assays: Principles, designs and labels. *TrAC - Trends in Analytical Chemistry, 82*, 286–306. https://doi.org/10.1016/j.trac.2016.06.006.

Cao, L., Fang, C., Zeng, R., Zhao, X., Jiang, Y., & Chen, Z. (2017a). Paper-based microfluidic devices for electrochemical immunofiltration analysis of human chorionic gonadotropin. *Biosensors and Bioelectronics, 92*, 87–94. https://doi.org/10.1016/j.bios.2017.02.002.

Cao, L., Fang, C., Zeng, R., Zhao, X., Zhao, F., Jiang, Y., et al. (2017b). A disposable paper-based microfluidic immunosensor based on reduced graphene oxide-tetraethylene pentamine/Au nanocomposite decorated carbon screen-printed electrodes. *Sensors and Actuators, B: Chemical, 252*, 44–54. https://doi.org/10.1016/j.snb.2017.05.148.

Cao, L., Han, G. C., Xiao, H., Chen, Z., & Fang, C. (2020). A novel 3D paper-based microfluidic electrochemical glucose biosensor based on rGO-TEPA/PB sensitive film. *Analytica Chimica Acta, 1096*, 34–43. https://doi.org/10.1016/j.aca.2019.10.049.

Cao, Q., Liang, B., Tu, T., Wei, J., Fang, L., & Ye, X. (2019). Three-dimensional paper-based microfluidic electrochemical integrated devices (3D-PMED) for wearable electrochemical glucose detection. *RSC Advances, 9*(10), 5674–5681. https://doi.org/10.1039/c8ra09157a.

Charithra, M. M., Manjunatha, J. G. G., & Raril, C. (2020). Surfactant modified graphite paste electrode as an electrochemical sensor for the enhanced voltammetric detection of estriol with dopamine and uric acid. *Advanced Pharmaceutical Bulletin, 10*(2), 247–253. https://doi.org/10.34172/apb.2020.029.

Cincotto, F. H., Fava, E. L., Moraes, F. C., Fatibello-Filho, O., & Faria, R. C. (2019). A new disposable microfluidic electrochemical paper-based device for the simultaneous determination of clinical biomarkers. *Talanta, 195*, 62–68. https://doi.org/10.1016/j.talanta.2018.11.022.

Cinti, S., Minotti, C., Moscone, D., Palleschi, G., & Arduini, F. (2017). Fully integrated ready-to-use paper-based electrochemical biosensor to detect nerve agents. *Biosensors and Bioelectronics, 93*, 46–51. https://doi.org/10.1016/j.bios.2016.10.091.

Dabhade, A., Jayaraman, S., & Paramasivan, B. (2021). Development of glucose oxidase-chitosan immobilized paper biosensor using screen-printed electrode for amperometric detection of Cr(VI) in water. *3 Biotech, 11*(4). https://doi.org/10.1007/s13205-021-02736-5.

Ding, R., Fiedoruk-Pogrebniak, M., Pokrzywnicka, M., Koncki, R., Bobacka, J., & Lisak, G. (2020). Solid reference electrode integrated with paper-based microfluidics for potentiometric ion sensing. *Sensors and Actuators B: Chemical, 323*, 128680. https://doi.org/10.1016/j.snb.2020.128680.

Ding, J., He, N., Lisak, G., Qin, W., & Bobacka, J. (2017). Paper-based microfluidic sampling and separation of analytes for potentiometric ion sensing. *Sensors and Actuators, B: Chemical, 243*, 346–352. https://doi.org/10.1016/j.snb.2016.11.128.

Dungchai, W., Chailapakul, O., & Henry, C. S. (2009). Electrochemical detection for paper-based microfluidics. *Analytical Chemistry, 81*(14), 5821–5826. https://doi.org/10.1021/ac9007573.

Fakhri, N., Abarghoei, S., Dadmehr, M., Hosseini, M., Sabahi, H., & Ganjali, M. R. (2020). Paper based colorimetric detection of miRNA-21 using Ag/Pt nanoclusters. *Spectrochimica Acta Part A: Molecular and Biomolecular Spectroscopy, 227*, 117529. https://doi.org/10.1016/j.saa.2019.117529.

Fava, E. L., Martimiano do Prado, T., Almeida Silva, T., Cruz de Moraes, F., Censi Faria, R., & Fatibello-Filho, O. (2020). New disposable electrochemical paper-based microfluidic device with multiplexed electrodes for biomarkers determination in urine sample. *Electroanalysis, 32*(5), 1075–1083. https://doi.org/10.1002/elan.201900641.

Fava, E. L., Silva, T. A., Prado, T. M.d., Moraes, F. C.d., Faria, R. C., & Fatibello-Filho, O. (2019). Electrochemical paper-based microfluidic device for high throughput multiplexed analysis. *Talanta, 203*, 280–286. https://doi.org/10.1016/j.talanta.2019.05.081.

Gross, E. M., Durant, H. E., Hipp, K. N., & Lai, R. Y. (2017). Electrochemiluminescence detection in paper-based and other inexpensive microfluidic devices. *ChemElectroChem, 4*(7), 1594–1603. https://doi.org/10.1002/celc.201700426.

Hasanzadeh, A., & Hashemzadeh, I. (2021). Microfluidic paper-based devices. In M. R. Hamblin, & M. Karimi (Eds.), *Biomedical applications of microfluidic devices* (pp. 257–274). (chapter 11).

He, Y., Wu, Y., Fu, J. Z., & Wu, W. B. (2015). Fabrication of paper-based microfluidic analysis devices: A review. *RSC Advances, 5*(95), 78109–78127. https://doi.org/10.1039/c5ra09188h.

Hussain, C. M. (2018). Handbook of nanomaterials for industrial applications. In *Handbook of nanomaterials for industrial applications* (pp. 1–1109). Elsevier. https://doi.org/10.1016/C2016-0-04427-3.

Hussain, C. M. (2020). *Handbook of polymer nanocomposites for industrial applications*. Elsevier.

Hussain, C. M., Hoque, R., & Balachandran, S. (2019). *Handbook of environmental materials management*. Springer.

Jafari, P., Beigi, S. M., Yousefi, F., Aghabalazadeh, S., Mousavizadegan, M., Hosseini, M., et al. (2021). Colorimetric biosensor for phenylalanine detection based on a paper using gold nanoparticles for phenylketonuria diagnosis. *Microchemical Journal, 163*, 105909. https://doi.org/10.1016/j.microc.2020.105909.

Jafry, A. T., Lim, H., & Lee, J. (2020). *Basic paper-based microfluidics/electronics theory* (pp. 7–39). Springer Science and Business Media LLC. https://doi.org/10.1007/978-981-15-8723-8_2.

Jemmeli, D., Marcoccio, E., Moscone, D., Dridi, C., & Arduini, F. (2020). Highly sensitive paper-based electrochemical sensor for reagent free detection of bisphenol A. *Talanta, 216*, 120924. https://doi.org/10.1016/j.talanta.2020.120924.

Julika, W. N., Ajit, A., Syazlin, N., Ariff, A., Naila, A., & Sulaiman, A. Z. (2020). Development of paper based amperometric biosensor for glucose content measurement in Malaysian Stingless Bee Honey. *Journal of Physics: Conference Series, 1529*(5), 052035. https://doi.org/10.1088/1742-6596/1529/5/052035.

Karimi, A., Husain, S. W., Hosseini, M., Azar, P. A., & Ganjali, M. R. (2020). A sensitive signal-on electrochemiluminescence sensor based on a nanocomposite of polypyrrole-Gd2O3 for the determination of L-cysteine in biological fluids. *Microchimica Acta, 187*(7). https://doi.org/10.1007/s00604-020-04372-x.

Karimi-Maleh, H., Karimi, F., Malekmohammadi, S., Zakariae, N., Esmaeili, R., Rostamnia, S., et al. (2020). An amplified voltammetric sensor based on platinum nanoparticle/polyoxometalate/two-dimensional hexagonal boron nitride nanosheets composite and ionic liquid for determination of N-hydroxysuccinimide in water samples. *Journal of Molecular Liquids, 310*, 113185. https://doi.org/10.1016/j.molliq.2020.113185.

Karimi-Maleh, H., Karimi, F., Orooji, Y., Mansouri, G., Razmjou, A., Aygun, A., et al. (2020). A new nickel-based co-crystal complex electrocatalyst amplified by NiO dope Pt nanostructure hybrid; A highly sensitive approach for determination of cysteamine in the presence of serotonin. *Scientific Reports, 10*(1). https://doi.org/10.1038/s41598-020-68663-2.

Karimi-Maleh, H., Yola, M. L., Atar, N., Orooji, Y., Karimi, F., Senthil Kumar, P., et al. (2021). A novel detection method for organophosphorus insecticide fenamiphos: Molecularly imprinted electrochemical sensor based on core-shell Co3O4@MOF-74 nanocomposite. *Journal of Colloid and Interface Science, 592*, 174–185. https://doi.org/10.1016/j.jcis.2021.02.066.

Kaur, G., Tomar, M., & Gupta, V. (2017). A simple paper based microfluidic electrochemical biosensor for point-of-care cholesterol diagnostics. *Physica Status Solidi (A) Applications and Materials Science, 214*(12). https://doi.org/10.1002/pssa.201700468.

Koczula, K. M., & Gallotta, A. (2016). Lateral flow assays. *Essays in Biochemistry, 60*(1), 111–120. https://doi.org/10.1042/EBC20150012.

Kokkinos, C., Economou, A., & Giokas, D. (2018). Paper-based device with a sputtered tin-film electrode for the voltammetric determination of Cd(II) and Zn(II). *Sensors and Actuators, B: Chemical, 260*, 223–226. https://doi.org/10.1016/j.snb.2017.12.182.

Kokkinos, C. T., Giokas, D. L., Economou, A. S., Petrou, P. S., & Kakabakos, S. E. (2018). Paper-based microfluidic device with integrated sputtered electrodes for stripping voltammetric determination of DNA via quantum dot labeling. *Analytical Chemistry, 90*(2), 1092–1097. https://doi.org/10.1021/acs.analchem.7b04274.

Lee, V. B. C., Mohd-Naim, N. F., Tamiya, E., & Ahmed, M. U. (2018). Trends in paper-based electrochemical biosensors: From design to application. *Analytical Sciences, 34*(1), 7–18. https://doi.org/10.2116/analsci.34.7.

Lim, H., Jafry, A. T., & Lee, J. (2019). Fabrication, flow control, and applications of microfluidic paper-based analytical devices. *Molecules, 24* (16). https://doi.org/10.3390/molecules24162869.

Lisowski, P., & Zarzycki, P. K. (2013). Microfluidic paper-based analytical devices (μPADs) and micro total analysis systems (μTAS): Development, applications and future trends. *Chromatographia, 76*(19–20), 1201–1214. https://doi.org/10.1007/s10337-013-2413-y.

Manjunatha, J. G. (2017). A new electrochemical sensor based on modified carbon nanotube-graphite mixture paste electrode for voltammetric determination of resorcinol. *Asian Journal of Pharmaceutical and Clinical Research, 10*(12), 295–300.

Martinez, A. W. (2011). Microfluidic paper-based analytical devices: From POCKET to paper-based ELISA. *Bioanalysis, 3*(23), 2589–2592. https://doi.org/10.4155/bio.11.258.

Martinez, A. W., Phillips, S. T., Butte, M. J., & Whitesides, G. M. (2007). Patterned paper as a platform for inexpensive, low-volume, portable bioassays. *Angewandte Chemie*, 1340–1342. https://doi.org/10.1002/ange.200603817.

Medina-Sánchez, M., Cadevall, M., Ros, J., & Merkoçi, A. (2015). Eco-friendly electrochemical lab-on-paper for heavy metal detection. *Analytical and Bioanalytical Chemistry*, 407(28), 8445–8449. https://doi.org/10.1007/s00216-015-9022-6.

Mesgari, F., Beigi, S. M., Fakhri, N., Hosseini, M., Aghazadeh, M., & Ganjali, M. R. (2020). Paper-based chemiluminescence and colorimetric detection of cytochrome c by cobalt hydroxide decorated mesoporous carbon. *Microchemical Journal*, 157.

Naderi, M., Hosseini, M., & Ganjali, M. R. (2018). Naked-eye detection of potassium ions in a novel gold nanoparticle aggregation-based aptasensor. *Spectrochimica Acta Part A: Molecular and Biomolecular Spectroscopy*, 195, 75–83. https://doi.org/10.1016/j.saa.2018.01.051.

Nechaeva, D., Shishov, A., Ermakov, S., & Bulatov, A. (2018). A paper-based analytical device for the determination of hydrogen sulfide in fuel oils based on headspace liquid-phase microextraction and cyclic voltammetry. *Talanta*, 183, 290–296. https://doi.org/10.1016/j.talanta.2018.02.074.

Nery, E. W., & Kubota, L. T. (2013). Sensing approaches on paper-based devices: A review. *Analytical and Bioanalytical Chemistry*, 405(24), 7573–7595. https://doi.org/10.1007/s00216-013-6911-4.

Nie, Z., Nijhuis, C. A., Gong, J., Chen, X., Kumachev, A., Martinez, A. W., et al. (2010). Electrochemical sensing in paper-based microfluidic devices. *Lab on a Chip*, 10(4), 477–483. https://doi.org/10.1039/b917150a.

Noviana, E., McCord, C. P., Clark, K. M., Jang, I., & Henry, C. S. (2020). Electrochemical paper-based devices: Sensing approaches and progress toward practical applications. *Lab on a Chip*, 20(1), 9–34. https://doi.org/10.1039/c9lc00903e.

Palit, S., & Hussain, C. M. (2020). *Modern manufacturing and nanomaterial perspective* (pp. 3–20). Elsevier BV. https://doi.org/10.1016/b978-0-12-821381-0.00001-6.

Pokpas, K., Jahed, N., & Iwuoha, E. (2019). Tuneable, pre-stored paper-based electrochemical cells (μPECs): An adsorptive stripping voltammetric approach to metal analysis. *Electrocatalysis*, 352–364. https://doi.org/10.1007/s12678-019-00516-7.

Pradela-Filho, L. A., Araújo, D. A. G., Takeuchi, R. M., Santos, A. L., & Henry, C. S. (2021). Thermoplastic electrodes as a new electrochemical platform coupled to microfluidic devices for tryptamine determination. *Analytica Chimica Acta*, 1147, 116–123. https://doi.org/10.1016/j.aca.2020.12.059.

Prinith, N. S., Manjunatha, J. G., & Raril, C. (2019). Electrocatalytic analysis of dopamine, uric acid and ascorbic acid at poly(adenine) modified carbon nanotube paste electrode: A cyclic voltammetric study. *Analytical and Bioanalytical Electrochemistry*, 11(6), 742–756. http://www.abechem.com/No.%206-2019/2019,%2011(6),%20742-756.pdf.

Punjiya, M., Moon, C. H., Matharu, Z., Rezaei Nejad, H., & Sonkusale, S. (2018). A three-dimensional electrochemical paper-based analytical device for low-cost diagnostics. *Analyst*, 143(5), 1059–1064. https://doi.org/10.1039/c7an01837a.

Pur, M. R. K., Hosseini, M., Faridbod, F., Dezfuli, A. S., & Ganjali, M. R. (2016). A novel solid-state electrochemiluminescence sensor for detection of cytochrome c based on ceria nanoparticles decorated with reduced graphene oxide nanocomposite. *Analytical and Bioanalytical Chemistry*, 408(25), 7193–7202. https://doi.org/10.1007/s00216-016-9856-6.

Pushpanjali, P. A., Manjunatha, J. G., Tigari, G., & Fattepur, S. (2020). Poly(niacin) based carbon nanotube sensor for the sensitive and selective voltammetric detection of vanillin with caffeine. *Analytical and Bioanalytical Electrochemistry*, 12(4), 553–568. http://www.abechem.com/article_39221_5b4cb9d77cdb1bfb6c48fc3dac8c7fc0.pdf.

Raril, C., & Manjunatha, J. G. (2018). Carbon nanotube paste electrode for the determination of some neurotransmitters: A cyclic voltammetric study. *Modern Chemistry & Applications*. https://doi.org/10.4172/2329-6798.1000263.

Raril, C., Manjunatha, J. G., Ravishankar, D. K., Fattepur, S., Siddaraju, G., & Nanjundaswamy, L. (2020). Validated electrochemical method for simultaneous resolution of tyrosine, uric acid, and ascorbic acid at polymer modified nano-composite paste electrode. *Surface Engineering and Applied Electrochemistry*, 56(4), 415–426. https://doi.org/10.3103/S1068375520040134.

Raril, C., Manjunatha, J. G., & Tigari, G. (2020). Low-cost voltammetric sensor based on an anionic surfactant modified carbon nanocomposite material for the rapid determination of curcumin in natural food supplement. *Instrumentation Science and Technology*, 48(5), 561–582. https://doi.org/10.1080/10739149.2020.1756317.

Saadati, A., Hassanpour, S., Bahavarnia, F., & Hasanzadeh, M. (2020). A novel biosensor for the monitoring of ovarian cancer tumor protein CA 125 in untreated human plasma samples using a novel nano-ink: A new platform for efficient diagnosis of cancer using paper based microfluidic technology. *Analytical Methods*, 12(12), 1639–1649. https://doi.org/10.1039/d0ay00299b.

Schilling, K. M., Lepore, A. L., Kurian, J. A., & Martinez, A. W. (2012). Fully enclosed microfluidic paper-based analytical devices. *Analytical Chemistry*, 84(3), 1579–1585. https://doi.org/10.1021/ac202837s.

Shen, L. L., Zhang, G. R., & Etzold, B. J. M. (2020). Paper-based microfluidics for electrochemical applications. *ChemElectroChem*, 7(1), 10–30. https://doi.org/10.1002/celc.201901495.

Shi, J., Tang, F., Xing, H., Zheng, H., Bi, L., & Wang, W. (2012). Electrochemical detection of Pb and Cd in paper-based microfluidic devices. *Journal of the Brazilian Chemical Society*, 23(6), 1124–1130. https://doi.org/10.1590/S0103-50532012000600018.

Sobhanie, E., Faridbod, F., Hosseini, M., & Ganjali, M. R. (2020). An ultrasensitive ECL sensor based on conducting polymer/electrochemically reduced graphene oxide for non-enzymatic H2O2 detection in biological samples. *ChemistrySelect*, 5(17), 5330–5336. https://doi.org/10.1002/slct.202000233.

Tigari, G., & Manjunatha, J. G. (2020a). Optimized voltammetric experiment for the determination of phloroglucinol at surfactant modified carbon nanotube paste electrode. *Instruments and Experimental Techniques*, 63(5), 750–757. https://doi.org/10.1134/S0020441220050139.

Tigari, G., & Manjunatha, J. G. (2020b). A surfactant enhanced novel pencil graphite and carbon nanotube composite paste material as an effective electrochemical sensor for determination of riboflavin. *Journal of Science: Advanced Materials and Devices*, 5(1), 56–64. https://doi.org/10.1016/j.jsamd.2019.11.001.

Tomei, M. R., Cinti, S., Interino, N., Manovella, V., Moscone, D., & Arduini, F. (2019). Paper-based electroanalytical strip for user-friendly blood glutathione detection. *Sensors and Actuators, B: Chemical*, 294, 291–297. https://doi.org/10.1016/j.snb.2019.02.082.

Wang, Y., Luo, J., Liu, J., Sun, S., Xiong, Y., Ma, Y., et al. (2019). Label-free microfluidic paper-based electrochemical aptasensor for ultrasensitive and simultaneous multiplexed detection of cancer biomarkers. *Biosensors and Bioelectronics*, *136*, 84–90. https://doi.org/10.1016/j.bios.2019.04.032.

Xia, Y., Si, J., & Li, Z. (2016). Fabrication techniques for microfluidic paper-based analytical devices and their applications for biological testing: A review. *Biosensors and Bioelectronics*, *77*, 774–789. https://doi.org/10.1016/j.bios.2015.10.032.

Xing, Y., Liu, J., Sun, S., Ming, T., Wang, Y., Luo, J., et al. (2021). New electrochemical method for programmed death-ligand 1 detection based on a paper-based microfluidic aptasensor. *Bioelectrochemistry*, *140*, 107789. https://doi.org/10.1016/j.bioelechem.2021.107789.

Yakoh, A., Chaiyo, S., Siangproh, W., & Chailapakul, O. (2019). 3D capillary-driven paper-based sequential microfluidic device for electrochemical sensing applications. *ACS Sensors*, *4*(5), 1211–1221. https://doi.org/10.1021/acssensors.8b01574.

Zhang, X., Bao, N., Luo, X., & Ding, S. N. (2018). Patchy gold coated Fe_3O_4 nanospheres with enhanced catalytic activity applied for paper-based bipolar electrode-electrochemiluminescence aptasensors. *Biosensors and Bioelectronics*, *114*, 44–51. https://doi.org/10.1016/j.bios.2018.05.016.

Zhang, Z., Ma, X., Li, B., Zhao, J., Qi, J., Hao, G., et al. (2020). Fluorescence detection of 2,4-dichlorophenoxyacetic acid by ratiometric fluorescence imaging on paper-based microfluidic chips. *Analyst*, *145*(3), 963–974. https://doi.org/10.1039/c9an01798d.

Zhang, J., Yang, Z., Liu, Q., & Liang, H. (2019). Electrochemical biotoxicity detection on a microfluidic paper-based analytical device via cellular respiratory inhibition. *Talanta*, *202*, 384–391. https://doi.org/10.1016/j.talanta.2019.05.031.

Part V

Health, safety, and regulation issues of carbon nanomaterials-based sensors

Chapter 17

Fundamentals of health, safety, and regulation issues of carbon nanomaterial-based sensors

Anila Hoskere Ashoka[a] and Vadde Ramu[b]
[a]Laboratory of Bioimaging and Pathologies, UMR 7021, University of Strasbourg, Illkirch, France, [b]Laboratory of Biomolecules, University of Sorbonne, 24 RueLhmond, Paris, France

Introduction

Carbon nanomaterials (CNMs) such as graphene (GRA), carbon nanotubes (CNTs), and fullerenes are allotropes of carbon. CNMs possess exceptional mechanical, thermal, chemical, and optical properties (Kotchey et al., 2012). GRA is made up of a single layer of sp^2-bonded carbon atoms arranged in a honeycomb-like lattice. GRA has excellent electrical conductivity and mechanical strength (Peng et al., 2020). GRA and its derivatives such as graphene oxide (GO), reduced graphene oxide (RGO), GRA nanoribbons are widely used in various electronic devices (Han, Kim, Kwon, & Lee, 2017), batteries (Fu, Xu, Li, Sun, & Peng, 2018), and sensors (Choi et al., 2020). Fullerenes possess a three-dimensional closed mesh-like structure. Fullerenes are used in electronic devices (Colbert & Smalley, 1999), photocatalysis (Vorobiev et al., 2017), and biomedical applications (Castro, Garcia, Zavala, & Echegoyen, 2017). Among the various CNMs, carbon nanotubes (CNTs) are unique and are widely used in various applications ranging from architecture to medicine. CNTs are typically graphene sheets rolled up into cylinders. CNTs are divided into two types: CNTs with single layers are called single-walled carbon nanotubes (SWCNTs) and CNTs with two are more layers are called multiwalled carbon nanotubes (MWCNTs) (Saito et al., 2014). CNTs exhibit unique electronic and optical properties. There is a great deal of interest to use CNTs in sensing and biomedical fields due to their small size that can be comparable to cellular organelles. Due to their extremely small size, CNTs can be used inside the cells without much perturbation to vital cellular components (Saito et al., 2014). CNTs emit highly photostable near-infrared (NIR) photoluminescence that is sensitive to local environment. Due to their unique optical properties, CNTs are widely used in biosensing and medical imaging applications (Jena et al., 2017). Even though biosensing and imaging research is dominated by a wide variety of fluorescent dyes (Ashoka et al., 2020; Ashoka, Ashokkumar, Kovtun, & Klymchenko, 2019; Ramu et al., 2015; Walker, Ramu, Meijer, Das, & Thomas, 2017), fluorescent materials (Ashoka et al., 2020), quantum dots (Pinaud et al., 2006), and conjugated polymer nanoparticles (Wu, Bull, Christensen, & McNeill, 2009), extensive research efforts are underway to develop CNM-based biosensors and imaging reagents. As a result, a wide variety of sensors for various analytes and imaging agents to target various cellular organelles and processes have already been developed based on CNMs in general and CNTs in particular (Pan, Li, & Choi, 2017). For example, CNTs such as SWCNTs are successfully used in DNA sensing through surface functionalization with complementary oligonucleotides (Heller et al., 2006). CNTs functionalized with short DNA oligonucleotide were used as lipid reporters to probe the lipid flux in lysosomes (Jena et al., 2017). Functionalized CNTs are also used for targeted imaging and therapeutic applications (Grodzinski, Silver, & Molnar, 2006). With the recent developments in fluorescence imaging in the second near-infrared window (NIR II), CNTs have become a hot research topic to develop various NIR II imaging agents for potential biomedical applications (Welsher, Sherlock, & Dai, 2011). On the contrary, CNTs are extensively used to develop electrochemical sensors for various analytes having biological and medical significance (Amrutha, Manjunatha, Bhatt, Raril, & Pushpanjali, 2019; Jg, 2017; Prinith, Manjunatha, & Raril, 2019; Pushpanjali, Manjunatha, Tigari, & Fattepur, 2020; Raril et al., 2018; Raril et al., 2020; Raril, Manjunatha, & Tigari, 2020; Tigari & Manjunatha, 2020a, 2020b; Tigari, Manjunatha, Raril, & Hareesha, 2019).

Due to the widespread use of CNMs in various applications, their commercial production is also rapidly increasing worldwide. Despite the extensive potential, the commercial translation of carbon-based nanomaterials, in general, is not fully explored due to the underlying debate about their potential toxicity and biocompatibility issues. If the CNMs are used in various products, then they most likely come in contact with humans and animals and accumulate in the environments. This will definitely increase the risk of potential exposure to such materials. CNMs can enter the living organisms through various routes such as the respiratory tract through inhalation or respiration and the digestive tract through ingestion. CNMs can also enter the body through the skin (dermal exposure), eyes, and also blood circulation by injection. In occupational settings, a high concentration of CNMs can enter the respiratory airways of workers and accumulate in the lungs. If the CNMs are used in cosmetic products, they can come directly in contact with the skin. Dermal exposure to CNMs may lead to skin irritation and dermal toxicity (Murray et al., 2009). If the CNMs are used as careers in drug/vaccine delivery systems, they can directly enter the blood circulation and accumulate in various vital organs and cells. Accumulation of CNMs in the lungs causes inflammation, fibrosis, and granulomas (Poland et al., 2008). Various transition metal catalysts such as Fe, Ni, and Co are used in the production of CNMs. Metal impurities can increase the toxicity of CNMs through the generation of reactive oxygen species (ROS), which can induce DNA damage and mutation (Kagan et al., 2006). Various functionalization techniques are used to improve the biocompatibility and solubility of CNMs. Such functionalization could also elevate the toxicity (Magrez et al., 2006). To date, only a little is known about the effect of nanosized materials on human health and the environments. In contemporary literature, there are many conflicting pieces of evidence regarding the toxicity of CNMs. Some reports suggest acute toxicity, whereas some suggest no toxicity at all. Therefore, systematic toxicological studies are still needed to understand the toxicity profile of CNMs. In this chapter, we discuss some fundamental health, safety, and regulation issues of CNMs. We discuss various sources of CNM toxicity and toxicity mechanisms. We also discuss some recent developments on the biodegradation, biocompatibility of CNMs, and potential future directions to improve the biocompatibility of CNMs or CNM-based sensors.

Origin of toxicity

An overview of this chapter is depicted in Scheme 1 (Fig. 17.1) CNMs are widely used in various applications due to their unique chemical, physical, and optical properties. Their widespread usage also increased toxicity concerns. A number of studies have been conducted to assess the toxicity of CNMs. Various physicochemical factors such as size, length, diameter, and aggregation of CNMs can elevate the toxicity of CNMs. Various metal catalysts are used in the production of CNMs. The presence of metal impurities can also induce toxicity.

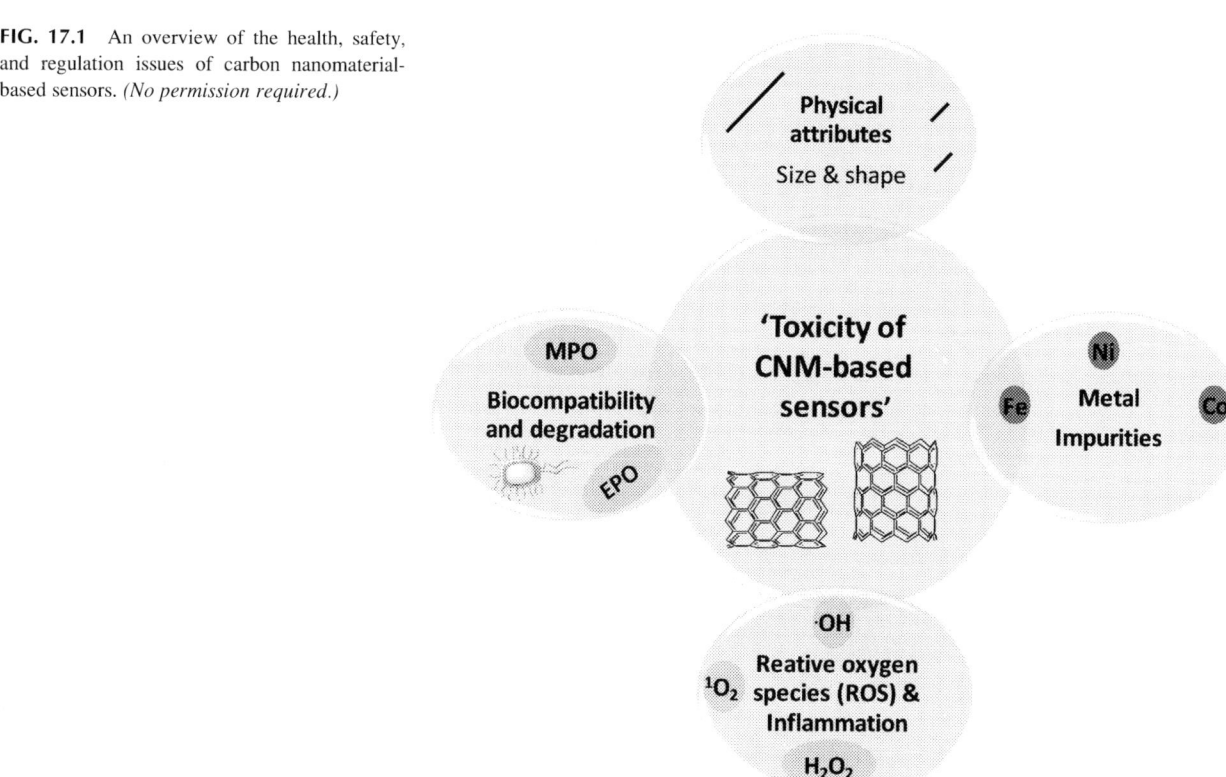

FIG. 17.1 An overview of the health, safety, and regulation issues of carbon nanomaterial-based sensors. *(No permission required.)*

CNMs are functionalized with various substituents to increase their biocompatibility and solubility. Analyte-specific functional groups are also introduced on CNM surfaces in order to use them for specific sensing applications. Surface modification of CNMs can also trigger toxicity. CNMs can enter the body through various routes such as inhalation, intravenous injection, and oral and dermal routes. Rapid clearance of CNMs from the body is very important to reduce unwanted side effects. However, CNMs accumulate in certain organelles, such as lungs, liver, and spleen, and in cells and induce toxicity. Among the various CNMs, the toxicity of CNTs is widely studied by the scientific community. Some existing pieces of evidence for CNMs toxicity are discussed in this section.

Physical characteristics

Size and shape

CNMs of various sizes are produced in the manufacturing process. Among the various CNMs, CNTs with large size and long fiber-like morphology pose potential health risks. Several studies suggest that exposure to long, straight, fiber-like CNTs can induce pulmonary toxicity similar to asbestos. For example, Poland et al. observed that the exposure of the lungs in mice to long MWCNTs resulted in acute inflammation and granuloma formation (Fig. 17.1) similar to the one caused by asbestos fibers (Poland et al., 2008) (Fig. 17.2).

The pathogenicity and toxicity of CNTs are due to length-dependent retention in the lungs. For instance, small and short CNTs that enter the lungs are easily cleared by macrophages through phagocytosis but long, fiber-like CNTs remain in the lungs and cause inflammation and fibrosis (Fig. 17.2). It can also lead to mesothelioma cancer in the inner lining of the lungs (Murphy et al., 2011) (Fig. 17.3).

Agglomeration

CNMs such as CNTs and graphene have the tendency to agglomerate due to Van der Waals forces between the carbon atoms. CNTs exhibit pronounced toxicity in the agglomerated state compared to well-dispersed CNTs (Wick et al.,

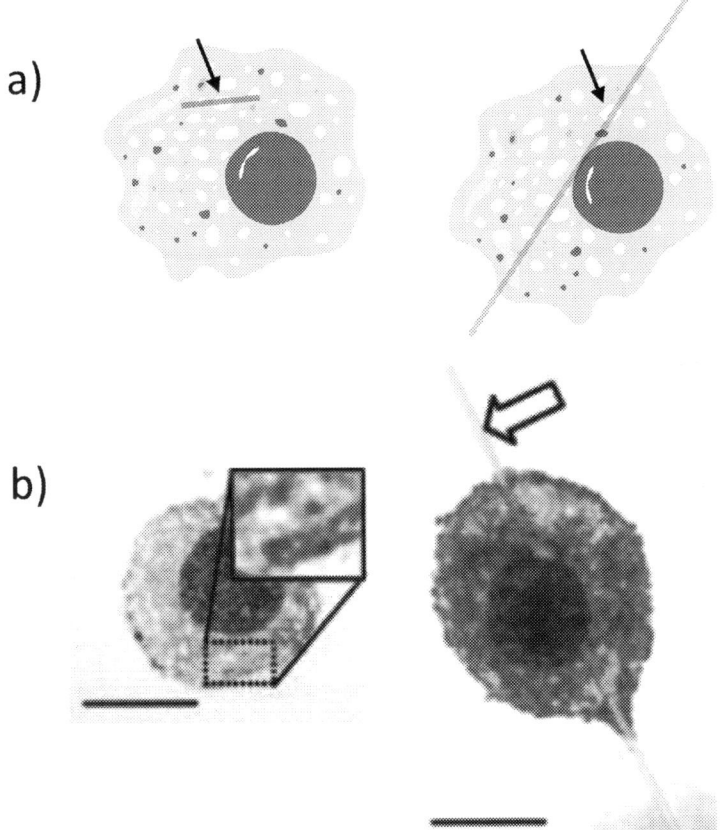

FIG. 17.2 Size-dependent internalization of CNTs. (A) Schematic representation of the effect of CNT size on the internalization. (B) Size-dependent phagocytosis of CNTs by macrophages. Small CNTs are phagocytosed by cells, whereas the long, fiber-like CNTs cannot be phagocytosed by macrophages, thus leading to frustrated phagocytosis. *Reprinted with permission from Springer Nature, Copyright 2008. From Poland, C.A., Duffin, R., Kinloch, I., Maynard, A., Wallace, W.A.H., & Seaton, A., et al. (2008). Carbon nanotubes introduced into the abdominal cavity of mice show asbestos-like pathogenicity in a pilot study. Nature Nanotechnology, 3(7), 423–428. https://doi.org/10.1038/nnano.2008.111.*

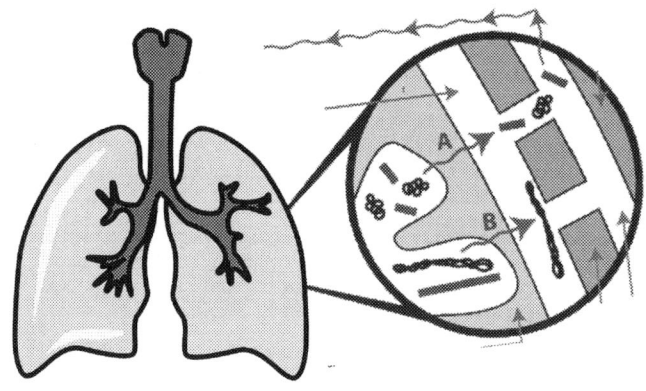

FIG. 17.3 Length-dependent clearance of CNTs from lungs. Small and tangled CNTs cleared from the pleural space of lungs, whereas the long, fiber-like CNTs retain in lungs. *Reprinted with permission from Elsevier, Copyright 2011. From Murphy, F.A., Poland, C.A., Duffin, R., Al-Jamal, K.T., Ali-Boucetta, H., & Nunes, A., et al. (2011). Length-dependent retention of carbon nanotubes in the pleural space of mice initiates sustained inflammation and progressive fibrosis on the parietal pleura. American Journal of Pathology, 178(6), 2587–2600. https://doi.org/10.1016/j.ajpath.2011.02.040.*

2007). When compared to single CNTs, agglomerates are tough to clear off and stimulate the formation of granulomas (Dong & Ma, 2015). The interaction of cells with agglomerated CNTs is different from well-dispersed CNTs. A study conducted on the effect of the degree of agglomeration of SWCNTs on cells suggests that agglomerated SWCNTs significantly decreased the overall DNA content (Van Daele & Cassell, 2009).

Metal impurities

Metal catalysts such as Fe, Ni, Mo, and Co are routinely used in the manufacture of CNMs. Metal residues that remain after the purification of CNMs can enhance toxicity. Metal contaminants can damage cells and DNA through the generation of reactive oxygen species (ROS). Kagan and colleagues studied the effect of Fe impurity in SWCNTs on RAW 264.7 macrophages (Kagan et al., 2006). They utilized nonpurified SWCNTs containing 26% Fe impurity and purified SWCNTs containing 0.23% Fe impurity and demonstrated that nonpurified SWCNTs effectively generate ROS and induce oxidative stress in RAW 264.7 macrophages. Shvedova and coworkers studied the exposure of human epidermal keratinocytes (HacaT) to SWCNTs containing Fe impurity and observed the accelerated oxidative stress and cytotoxicity (Shvedova et al., 2003). They also observed the dramatic reduction in the cytotoxicity in the presence of the iron-chelating agent, deferoxamine, suggesting the toxic effect of metal impurities in SWCNTs.

Respiratory exposure of SWCNTs with metal impurities could cause lung inflammation and damage (Fig. 17.3). Cuicui Ge et al. studied the respiratory exposure of SWCNTs containing different metal contents (Ge et al., 2011). They observed a significant increase in inflammation and oxidative stress biomarkers after 24-h exposure. These studies clearly suggest that the presence of metal impurities increases the toxicity of CNMs. Since metal catalysts are essential for the large-scale production of CNMs, the presence of metal impurities is inevitable. Toxic metal impurities could reduce the biocompatibility of CNMs. So, it is very important to remove metal impurities from CNMs postproduction without affecting the other properties of CNMs. Different purification processes such as acid treatment and heat treatment are developed to reduce the metal impurities in CNMs. Moreover, there is a pressing need to follow a standard protocol to analyze metal impurities in CNMs postproduction. Manufacturers and commercial suppliers should take utmost care to specify the amount of metal impurities present in their samples (Fig. 17.4)

Surface functionalization

CNMs are hydrophobic in nature. CNMs are extensively modified to increase their biocompatibility and biodistribution. Functional groups such as hydroxyl (-OH) and carboxyl (-COOH) have been introduced to improve the hydrophilicity of the CNMs. Moreover, these functional groups are also used to attach other analyte-specific sensor moieties, antibodies, drugs, and other biomolecules to CNMs in order to use them for sensing, imaging, and therapeutic purposes. Polyethylene glycol (PEG) coatings on CNMs are also used to increase hydrophilicity and to reduce the nonspecific interaction of CNMs with cells and biomolecules (Heister et al., 2010). On the one hand, surface modification is a powerful method to make CNMs functional for various applications. On the other hand, such modifications can alter the toxicity of CNMs (increase or decrease). The toxicity of surface-modified CNMs depends on the functional groups that are introduced during the course of modification. Magrez and coworkers studied the toxicity effect of surface-modified carbon fibers, carbon nanoparticles, and carbon nanotubes (Magrez et al., 2006). They found the toxicity of carbon nanotubes increased significantly after the introduction of carbonyl (—C=O), carboxyl (—COOH), and hydroxyl (—OH) groups through acid treatment. Lung tumor

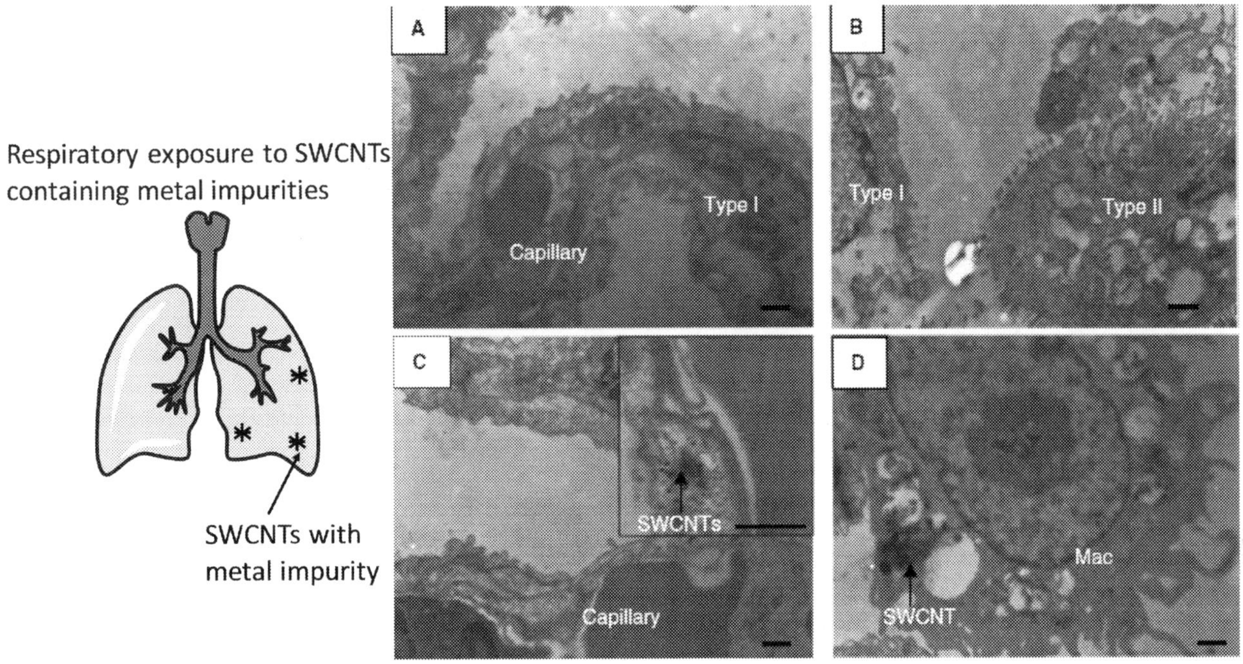

FIG. 17.4 Respiratory exposure to SWCNTs containing metal impurities. Ultrastructure observation of lung tissues of rats exposed to SWCNTs. A) Control, B) lung tissue exposed to SWCNTs containing less metal impurity, and C and D) lung tissue exposed to metal-rich SWCNTs causing inflammation and lung damage. Type I and Type II represent epithelium cells. Mac represents macrophages. *Reprinted with permission from Taylor & Francis, Copyright 2012. From Dong, J., & Ma, Q. (2015). Advances in mechanisms and signaling pathways of carbon nanotube toxicity. Nanotoxicology, 9(5), 658–676. https://doi.org/10.3109/17435390.2015.1009187.*

cells treated with oxidized CNTs are found to lose cell-to-cell contact in retracted cytoplasm and condensed nucleus (Fig. 17.4). Bottini and coworkers compared the toxicity of pristine (unmodified) and oxidized MWCNTs (Bottini et al., 2006). Oxidized MWCNTs showed significantly higher cell death than the pristine MWCNTs (Fig. 17.5).

On the contrary, some surface modifications can improve the biocompatibility of CNMs. Carboxylated SWCNTs undergo biodegradation in the presence of oxidative enzymes (discussed in detail in the later section—The Enzyme-Catalyzed Degradation of Carbon Nanomaterials) (Kotchey et al., 2012). Singh and coworkers functionalized SWCNTs and MWCNTs with a chelating agent diethylenetriaminepentaacetic dianhydride (DTPA-dianhydride) and labeled them

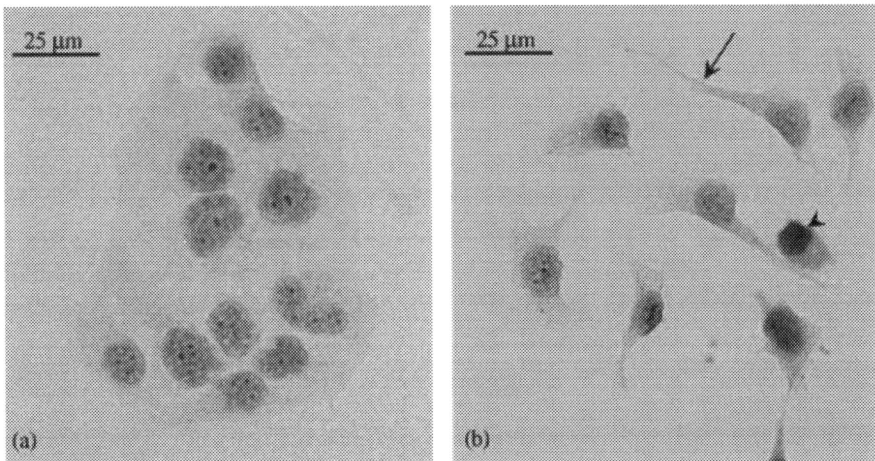

FIG. 17.5 Effect of functionalization of MWCNTs on lung tumor cells. (A) Control cells showing prominent cell-to-cell attachment and regular morphology for nucleus and cytoplasm. (B) Cells treated with oxidized MWCNTs after 24-h incubation showing no cell-to-cell contact, retracted cytoplasm, and condensed nucleus. *Reprinted with permission from American Chemical Society, Copyright 2006. From Magrez, A., Kasas, S., Salicio, V., Pasquier, N., Seo, J.W., & Celio, M., et al. (2006). Cellular toxicity of carbon-based nanomaterials. Nano Letters, 6(6), 1121–1125. https://doi.org/10.1021/nl060162e.*

with a radiotracer, ^{111}In (Singh et al. (2006). Following intravenous administration in BALB/c mice, they monitored the radioactivity of functionalized SWCNTs (f-SWCNTs). They observed a rapid clearance of f-SWCNTs from blood circulation through the renal clearance route and there was no accumulation in the liver or spleen of mice. They also observed the presence of entire CNTs in urine samples, further suggesting renal clearance. Moreover, functionalized CNTs showed a significantly improved toxicity profile compared to nonfunctionalized CNTs. Doping with other elements is also found to alter the toxicity profile of CNMs. MWCNTs doped with nitrogen showed better tolerance and low toxicity in mice compared to pure MWCNTs (Sayes et al., 2004).

So, concerning the toxicity of CNMs due to surface modification, there is no clear-cut conclusion. Some literature reports suggest a clear elevation in the toxicity after surface modification, whereas some other reports suggest a completely opposite phenomenon. An important aspect to consider here is the body/cell's response to modified and unmodified CNMs. Nontoxic CNMs can be toxic after surface functionalization. So, the nature of substituents also contributes to the overall toxicity of surface-modified CNMs. Since CNMs are surface modified with different functionalities for specific applications, their biodistribution/target sites could be different. So, there is no general rule of thumb, the toxicity of CNMs should be analyzed on a case-by-case basis by keeping in mind their intended application and biodistribution profiles.

Mode of administration

CNMs can enter the body by various routes such as inhalation, intravenous injection, and dermal exposure. After administration, their biopersistence determines the toxicity. On the one hand, CNMs can be cleared from the body through various mechanisms such as renal clearance, phagocytosis, and enzymatic degradation, and on the other hand, CNMs can accumulate in different organelles such as the liver, lungs, kidneys, and cells and induce toxicity. For example, the inhalation of CNMs such as CNTs, GRA, and C_{60} causes lung inflammation, granuloma, and fibrosis formation (Myojo & Ono-Ogasawara, 2018). Dermal exposure to CNTs can lead to skin irritation and CNTs containing metal impurities showed dermal toxicity (Murray et al., 2009). MWCNTs upon intraperitoneal injection cause inflammation and granuloma formation in mice (Takagi et al., 2008).

Toxicity mechanism of CNMs

CNMs enter the body by respiration/inhalation, intravenous injection, and dermal and oral routes. CNMs can cause oxidative stress through the generation of ROS, inflammation in lungs and tissues, DNA damage, carcinogenesis, etc. The toxicity mechanism of CNMs is explained in this section.

Reactive oxygen species (ROS)

Reactive oxygen species (ROS) such as $^1O_2$, H_2O_2, OH, are generally produced in cells. ROS is responsible for normal cell functioning and homeostasis. ROS also prevents cells from invading microbes. ROS levels inside the cells are maintained by an antioxidant defense system driven by intracellular thiols, especially glutathione. Oxidative stress is a condition, wherein the cells produce excess ROS due to external stimuli. When the cells produce excess ROS, the intracellular antioxidant defense system fails to neutralize them. Overproduction of ROS can damage cellular macromolecules such as proteins, lipids, and DNA and ultimately leads to cell death. Oxidative stress is also related to several adverse health problems such as aging, cancer, and neural disorders. CNMs may multiply such health complications by inducing oxidative stress through the overproduction of ROS.

It has been observed by several researchers that the cells that are exposed to CNMs, especially SWCNTs and MWCNTs, can cause oxidative stress and provoke the cells to produce excess ROS. One of the main reasons for the overproduction of ROS in the presence of transition metal impurities in the CNM samples. As described in the earlier section, transition metal catalysts are extensively used in the production of CNMs. Metal impurities can stimulate the cells to overproduce ROS. Morphology and dimensions of CNMs can also influence the cells to produce excess ROS. For example, MWCNTs that are larger in size with long fiber-like morphology cannot be phagocytosed by macrophages (a type of cells that engulf and eliminate foreign particles and diseased cells) (Brown et al., 2007). This so-called "frustrated" phagocytosis stimulates the cells to produce excess ROS. Some in vivo studies also evident the size-dependent retention of CNMs (Murphy et al., 2011). Lung tissues of mice exposed to CNMs such as carbon black particles, small CNT tangles, and short and long CNTs showed size-dependent retention in lungs. Only small particles and short fibers were cleared from the lungs postexposure, whereas long CNTs are retained in the lung tissue. Long CNTs retained in the lungs cause acute inflammation, suggesting the size-dependent retention and pathogenicity of CNTs.

Inflammatory response

The respiratory route is one of the major pathways for the entry of CNMs into the body. Due to their small size and low density, CNMs can easily circulate in the air and enter the respiratory system through inhalation. Upon entry into the lungs, their physical and chemical properties such as size, surface area, and composition influence their distribution in different parts of the lungs. People who are working in CNM manufacturing settings are highly susceptible to CNM exposure. CNMs having a smaller size can be cleared from the lungs by macrophages through phagocytosis, whereas those with larger sizes accumulate in the lungs and cause adverse immune responses. CNM exposure to lung cells stimulates the release of proinflammatory cytokines such as tumor necrosis factor-α (TNF-α). TNF-α can cause lung inflammation. CNTs have a high aspect ratio. According to the World Health Organization, CNTs with a minimum aspect ratio of 3:1 can enter the lungs as particles through inhalation (Oberdörster, Castranova, Asgharian, & Sayre, 2015). The toxicity of CNTs is due to their fiber-like structure that resembles asbestos fibers. Exposure to asbestos fibers causes many health complications such as inflammation, mesothelioma, asbestosis, and pleural fibrosis (Lam, James, McCluskey, Arepalli, & Hunter, 2006). Several literature reports suggest that the exposure of lung tissue lining to long, multiwalled CNTs results in asbestos-like pathogenic behavior. Exposure to large MWCNTs causes lung inflammation and granulomas similar to asbestos fibers (Poland et al., 2008). Moreover, metal impurities present in CNMs also cause inflammatory responses (Ge et al., 2011). Exposure to SWCNTs containing metal impurities can induce an inflammatory response, oxidative stress, and cardiovascular lesions (Ge et al., 2011). So, extreme care should be taken in occupational settings to minimize exposure to CNTs.

DNA damage

One of the principal reasons for the DNA damage by CNMs is the overproduction of ROS. CNMs stimulate cells in several ways to produce excess ROS. Excessive ROS can cause DNA mutation and DNA strand break and can ultimately lead to cell death. CNMs, especially CNTs, stimulate the cells to produce excess ROS. Toxicity of CNTs often leads to elevated levels of oxidative stress markers and depletion of antioxidants such as glutathione at the cellular level (Kagan et al., 2006). Moreover, the metal impurities such as transition metals (Fe, Ni, and Co) that are often used in the production of CNMs stimulate the cells to produce excess ROS. Transition metal impurities can produce free radicals that can cause DNA damage, DNA mutation, and lipid peroxidation. The surface functionalization of CNMs can also influence their toxic potential. Surface functionalization with —COOH/—OH groups increases their water solubility and dispersion (Jiang et al., 2020). Higher solubility and dispersion can boost CNM interaction with the biomolecules, proteins, and DNA and thus damage them either by direct interaction or by the production of ROS (Patlolla, Patra, Flountan, & Tchounwou, 2016). CNMs such as SWCNTs functionalized with carboxylate and hydroxyl groups showed a significant DNA damage compared to pristine SWCNTs (Jiang et al., 2020). MWCNTs' accumulation in cells leads to overexpression DNA double-strand break repair protein Rad 51, activation of tumor suppressor proteins, p53, and apoptosis (Zhu, Chang, Dai, & Hong, 2007). MWCNTs increase mutation frequency by twofold when compared to normal mutation frequency in mouse embryonic stem cells (Zhu et al., 2007).

Biocompatibility and biodegradation of CNMs

Biocompatibility is one of the fundamental criteria for any material to be used in real-time biological and clinical applications. In order to consider a material biocompatible, the material should have the following characteristics: (a) material should have better solubility and dispersion in the aqueous environment, (b) material should not induce any adverse inflammatory response upon interaction with biomolecules, (c) material and its substituents including degradation by-products should not induce any toxic side effects at cellular and organelle levels, and (d) material should possess rapid clearance from the body without any accumulation in the vital organelles and cells. Moreover, from the clinical standpoint, any approval for the clinical use of CNMs depends on the demonstration of their biodegradation and clearance from the body. Due to their higher stability and resistance to chemicals, CNMs can accumulate in tissues and organelles like a nonbiodegradable material and cause severe side effects.

Most of the as-prepared CNMs are insoluble in the aqueous medium; several approaches such as noncovalent and covalent functionalization, encapsulation in biopolymers, wrapping with surfactants, and amphiphilic molecules have been explored to improve the solubility and, hence, the biocompatibility of CNMs (Tasis, Tagmatarchis, Bianco, & Prato, 2006). As discussed in the earlier section, surface functionalization of CNMs can be a double-edged sword; such modifications can either improve the biocompatibility or increase the toxicity. So, the biocompatibility of the material should purely be dependent on the application of modified materials, type of study, and the organs/cells exposed. Recently, researchers discovered the biodegradation of CNMs by oxidative enzymes such as peroxidases. Biodegradation by naturally occurring enzymes could enable the safe use of CNMs in biological and clinical settings. In this section, various strategies that have been explored to improve the solubility, biocompatibility, and biodegradation of CNMs are discussed.

Biodegradation by enzymes and microorganisms

Biodegradation by enzymes

Over the past years, several studies explored the biodegradation of CNMs by enzymes. Enzymatic degradation of CNMs is mainly catalyzed by oxidative enzymes such as horseradish peroxidase (HRP), myeloperoxidase (MPO), and eosinophil peroxidase (EPO). Enzymes such as HRP can biodegrade SWCNTs in the presence of a small amount of hydrogen peroxide after 12-week incubation (Fig. 17.5) (Allen et al., 2008) (Fig. 17.6).

Reactive intermediates and hypochlorous acid (HOCl) generated by myeloperoxidase (MPO) enzymes can biodegrade CNTs. For example, SWCNTs incubated with MPO along with H_2O_2 showed degradation of CNTs after 24-h incubation (Kagan et al., 2010). Interestingly, inhalation of biodegraded nanotubes did not show any inflammatory response in mice lungs. This important observation suggests that enzymatic degradation of CNTs made them noninflammatory or nontoxic postdegradation.

EPO is an oxidative enzyme released from eosinophils. EPO is mainly released during inflammation by foreign bodies and they are responsible for killing invading parasites. EPOs in the presence of H_2O_2 are also shown to biodegrade carboxylated SWCNTs (Fig. 17.6) (Andõn et al., 2013). EPO also catalyzes the oxidation of halides into hypohalous acids. EPO preferentially oxidizes bromine into hypobromous acid (HOBr), and along with EPO, such oxidation products also contribute to the degradation of CNTs (Andõn et al., 2013). Most of the enzymatic biodegradation studies were carried out on carboxylated CNTs. Carboxylation or oxidation of CNTs creates more defect sites and thus improves the interaction between the enzyme and substrate to facilitate biodegradation.

Biodegradation by microbes

Researchers over the years have discovered several species of bacteria and fungi that are capable of degrading CNMs. For example, a naphthalene-degrading bacteria species isolated from a graphite mine is shown to degrade various CNMs such as graphite, graphene oxide (GO), and reduced graphene oxide (RGO) through oxidation. Interestingly, RGO with a higher number of defect sites was shown to undergo a higher degree of oxidation compared to graphite (Liu et al., 2015).

A combination of three bacterial species, *Burkholderia kururiensis*, *Delftia acidovorans*, and *Stenotrophomonas maltophilia*, has been shown to biodegrade MWCNTs in the presence of carbon dioxide. Even though individual species of these bacteria are not capable to biodegrade the CNTs, the cooperation of several species could biodegrade MWCNTs into CO_2 and several degradation products such as cinnamaldehyde, isophthalic acid, naphthol (Zhang, Petersen, Habteselassie, Mao, & Huang, 2013).

Along with bacteria, some fungal species are also known to biodegrade CNMs. For example, lignin peroxidase (LiP), secreted by edible mushroom *Sparassis latifolia*, can biodegrade both raw-grade and thermally treated SWCNTs. The radicals produced by LiP played a major role in the degradation of SWCNTs (Zhang et al., 2013).

Manganese peroxidase (MnP) released from white-rot fungus, and *Phanerochaete chrysosporium* has been shown to biodegrade pristine SWCNTs. Interestingly, MnP could only degrade pristine SWCNTs but not carboxylated SWCNTs; this is due to the binding of surface carboxylates in oxidized SWCNTs could to Mn^{2+}, which inhibits the enzymatic activity (Zhang et al., 2013).

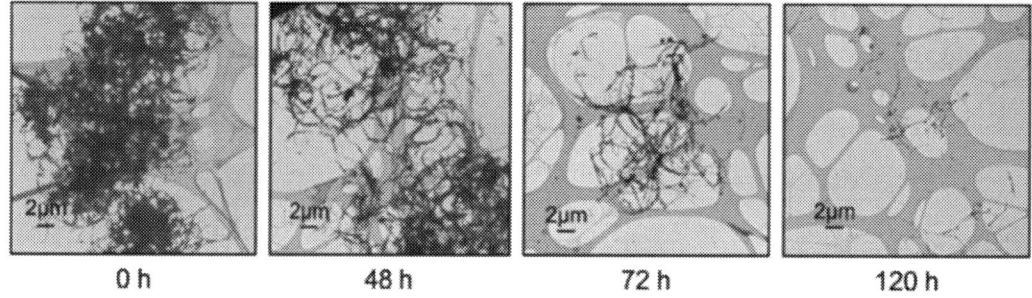

FIG. 17.6 TEM images of the biodegradation of SWCNTs over time in the presence of EPO and H_2O_2. *Reprinted with permission from Wiley-VCH, Copyright 2013. From Andõn, F.T., Kapralov, A.A., Yanamala, N., Feng, W., Baygan, A., & Chambers, B.J., et al. (2013). Biodegradation of single-walled carbon nanotubes by eosinophil peroxidase. Small, 9(16), 2721–2729. https://doi.org/10.1002/smll.201202508.*

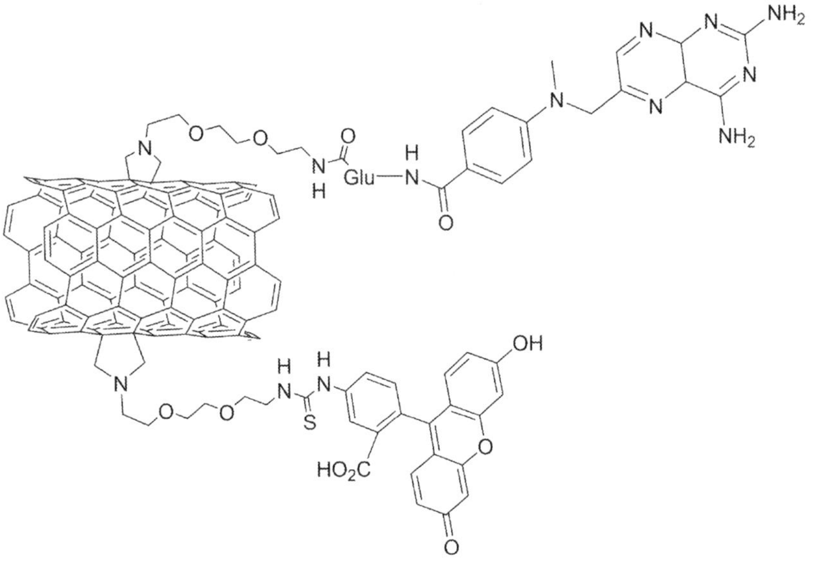

FIG. 17.7 MWCNT dual functionalized with a drug and a dye as theranostics application.

Biocompatibility due to surface functionalization

It has been proved by several research groups that the chemical functionalization of CNMs improves their biocompatibility (Bianco, Kostarelos, & Prato, 2011). Chemically modified CNMs are now widely used in the biomedical field for therapeutic and imaging applications. For example, a theranostic agent composed of MWCNTs covalently functionalized with anticancer drug methotrexate (MTX) and a fluorescent probe fluorescein isothiocyanate (FITC) via 1,3-dipolar cycloaddition of azomethine ylides was developed to achieve therapeutic and imaging applications (Fig. 17.7) (Pastorin et al., 2006).

Surface functionalization also influences the clearance of administered CNMs from the body. For example, MWCNTs surface functionalized with diethylenetriaminepentaacetic dianhydride (DTPA-MWCNT) and radiolabeled with Indium-111 (^{111}In) upon intravenous administration in rats showed urinary excretion within 24 h postinjection (Lacerda et al., 2008) (Fig. 17.7).

These studies suggest that surface functionalization is one of the promising ways forward to improve the biocompatibility and thus the clinical potential of CNMs.

Conclusion

There is no doubt that CNMs have immense potential to be used in various applications ranging from electronics to biomedicine. However, their potential health impact, biopersistence, and environmental accumulation could hinder their commercial translation. Literature reports on the toxicity of CNMs are often conflicting. There is a clear inconsistency between the studies. Moreover, there is a pressing need to have some standard materials to compare the toxicity. It is difficult to draw a conclusion on the toxicity of CNMs based on the conflicting evidence. There should be a uniform and well-defined limit on the level of impurities present in commercial CNM samples. It is recommended that the commercial suppliers should specify different impurities and their levels present in the commercial CNM samples. Standard test protocols are needed to uniformly assess the health risks associated with various CNMs. Appropriate safety protocols must be followed in occupational settings to protect the employees from potential exposure to CNMs. Recent advances in the functionalization of CNMs to improve biocompatibility is a promising way forward but more in vivo studies are necessary for the risk assessment. Recent discoveries on the biodegradation of CNMs by microbes are significant but such studies are mostly limited to a particular class of CNTs. More research on biodegradation of a wide variety of CNMs is still needed to determine the overall efficacy. If the toxicity issues are satisfactorily addressed, then CNMs can become an invaluable tool in biosensing and biomedicine.

Acknowledgments

AHA acknowledges the Laboratory of Bioimaging and Pathologies, UMR 7021 CNRS, the Université de Strasbourg, for all the resources and support. VR acknowledges the Department of Chemistry, Sorbonne University, for all the support.

References

Allen, B. L., Kichambare, P. D., Gou, P., Vlasova, I. I., Kapralov, A. A., Konduru, N., et al. (2008). Biodegradation of single-walled carbon nanotubes through enzymatic catalysis. *Nano Letters*, *8*(11), 3899–3903. https://doi.org/10.1021/nl802315h.

Amrutha, B. M., Manjunatha, J. G., Bhatt, A. S., Raril, C., & Pushpanjali, P. A. (2019). Electrochemical sensor for the determination of Alizarin Red-S at non-ionic surfactant modified carbon nanotube paste electrode. *Physical Chemistry Research*, *7*(3), 523–533. https://doi.org/10.22036/pcr.2019.185875.1636.

Andõn, F. T., Kapralov, A. A., Yanamala, N., Feng, W., Baygan, A., Chambers, B. J., et al. (2013). Biodegradation of single-walled carbon nanotubes by eosinophil peroxidase. *Small*, *9*(16), 2721–2729. https://doi.org/10.1002/smll.201202508.

Ashoka, A. H., Ashokkumar, P., Kovtun, Y. P., & Klymchenko, A. S. (2019). Solvatochromic near-infrared probe for polarity mapping of biomembranes and lipid droplets in cells under stress. *Journal of Physical Chemistry Letters*, *10*(10), 2414–2421. https://doi.org/10.1021/acs.jpclett.9b00668.

Ashoka, A. H., Kong, S. H., Seeliger, B., Andreiuk, B., Soares, R. V., Barberio, M., et al. (2020). Near-infrared fluorescent coatings of medical devices for image-guided surgery. *Biomaterials*, 261. https://doi.org/10.1016/j.biomaterials.2020.120306.

Bianco, A., Kostarelos, K., & Prato, M. (2011). Making carbon nanotubes biocompatible and biodegradable. *Chemical Communications*, *47*(37), 10182–10188. https://doi.org/10.1039/c1cc13011k.

Bottini, M., Bruckner, S., Nika, K., Bottini, N., Bellucci, S., Magrini, A., et al. (2006). Multi-walled carbon nanotubes induce T lymphocyte apoptosis. *Toxicology Letters*, *160*(2), 121–126. https://doi.org/10.1016/j.toxlet.2005.06.020.

Brown, D. M., Kinloch, I. A., Bangert, U., Windle, A. H., Walter, D. M., Walker, G. S., et al. (2007). An in vitro study of the potential of carbon nanotubes and nanofibres to induce inflammatory mediators and frustrated phagocytosis. *Carbon*, *45*(9), 1743–1756. https://doi.org/10.1016/j.carbon.2007.05.011.

Castro, E., Garcia, A. H., Zavala, G., & Echegoyen, L. (2017). Fullerenes in biology and medicine. *Journal of Materials Chemistry B*, *5*(32), 6523–6535. https://doi.org/10.1039/c7tb00855d.

Choi, J. H., Lee, J., Byeon, M., Hong, T. E., Park, H., & Lee, C. Y. (2020). Graphene-based gas sensors with high sensitivity and minimal sensor-to-sensor variation. *ACS Applied Nano Materials*, 2257–2265. https://doi.org/10.1021/acsanm.9b02378.

Colbert, D. T., & Smalley, R. E. (1999). Fullerene nanotubes for molecular electronics. *Trends in Biotechnology*, *17*(2), 46–50. https://doi.org/10.1016/S0167-7799(98)01256-6.

Dong, J., & Ma, Q. (2015). Advances in mechanisms and signaling pathways of carbon nanotube toxicity. *Nanotoxicology*, *9*(5), 658–676. https://doi.org/10.3109/17435390.2015.1009187.

Fu, X., Xu, L., Li, J., Sun, X., & Peng, H. (2018). Flexible solar cells based on carbon nanomaterials. *Carbon*, *139*, 1063–1073. https://doi.org/10.1016/j.carbon.2018.08.017.

Ge, C., Meng, L., Xu, L., Bai, R., Du, J., Zhang, L., et al. (2011). Acute pulmonary and moderate cardiovascular responses of spontaneously hypertensive rats after exposure to single-wall carbon nanotubes. *Nanotoxicology*, *6*(5), 526–542. https://doi.org/10.3109/17435390.2011.587905.

Grodzinski, P., Silver, M., & Molnar, L. K. (2006). Nanotechnology for cancer diagnostics: Promises and challenges. *Expert Review of Molecular Diagnostics*, *6*(3), 307–318. https://doi.org/10.1586/14737159.6.3.307.

Han, T. H., Kim, H., Kwon, S. J., & Lee, T. W. (2017). Graphene-based flexible electronic devices. *Materials Science & Engineering R: Reports*, *118*, 1–43. https://doi.org/10.1016/j.mser.2017.05.001.

Heister, E., Lamprecht, C., Neves, V., Tîlmaciu, C., Datas, L., Flahaut, E., et al. (2010). Higher dispersion efficacy of functionalized carbon nanotubes in chemical and biological environments. *ACS Nano*, *4*(5), 2615–2626. https://doi.org/10.1021/nn100069k.

Heller, D. A., Jeng, E. S., Yeung, T. K., Martinez, B. M., Moll, A. E., Gastala, J. B., et al. (2006). Optical detection of DNA conformational polymorphism on single-walled carbon nanotubes. *Science*, *311*(5760), 508–511. https://doi.org/10.1126/science.1120792.

Jena, P. V., Roxbury, D., Galassi, T. V., Akkari, L., Horoszko, C. P., Iaea, D. B., et al. (2017). A carbon nanotube optical reporter maps endolysosomal lipid flux. *ACS Nano*, *11*(11), 10689–10703. https://doi.org/10.1021/acsnano.7b04743.

Jg, M. (2017). A new electrochemical sensor based on modified carbon nanotube-graphite mixture paste electrode for voltammetric determination of resorcinol. *Asian Journal of Pharmaceutical and Clinical Research*, *10*(12), 295–300.

Jiang, T., Amadei, C. A., Gou, N., Lin, Y., Lan, J., Vecitis, C. D., et al. (2020). Toxicity of single-walled carbon nanotubes (SWCNTs): Effect of lengths, functional groups and electronic structures revealed by a quantitative toxicogenomics assay. *Environmental Science: Nano*, *7*(5), 1348–1364. https://doi.org/10.1039/d0en00230e.

Kagan, V. E., Konduru, N. V., Feng, W., Allen, B. L., Conroy, J., Volkov, Y., et al. (2010). Carbon nanotubes degraded by neutrophil myeloperoxidase induce less pulmonary inflammation. *Nature Nanotechnology*, *5*(5), 354–359. https://doi.org/10.1038/nnano.2010.44.

Kagan, V. E., Tyurina, Y. Y., Tyurin, V. A., Konduru, N. V., Potapovich, A. I., Osipov, A. N., et al. (2006). Direct and indirect effects of single walled carbon nanotubes on RAW 264.7 macrophages: Role of iron. *Toxicology Letters*, *165*(1), 88–100. https://doi.org/10.1016/j.toxlet.2006.02.001.

Kotchey, G. P., Hasan, S. A., Kapralov, A. A., Ha, S. H., Kim, K., Shvedova, A. A., et al. (2012). A natural vanishing act: The enzyme-catalyzed degradation of carbon nanomaterials. *Accounts of Chemical Research*, *45*(10), 1770–1781. https://doi.org/10.1021/ar300106h.

Lacerda, L., Soundararajan, A., Singh, R., Pastorin, G., Al-Jamal, K. T., Turton, J., et al. (2008). Dynamic imaging of functionalized multi-walled carbon nanotube systemic circulation and urinary excretion. *Advanced Materials*, *20*(2), 225–230. https://doi.org/10.1002/adma.200702334.

Lam, C. W., James, J. T., McCluskey, R., Arepalli, S., & Hunter, R. L. (2006). A review of carbon nanotube toxicity and assessment of potential occupational and environmental health risks. *Critical Reviews in Toxicology*, *36*(3), 189–217. https://doi.org/10.1080/10408440600570233.

Liu, L., Zhu, C., Fan, M., Chen, C., Huang, Y., Hao, Q., et al. (2015). Oxidation and degradation of graphitic materials by naphthalene-degrading bacteria. *Nanoscale*, *7*(32), 13619–13628. https://doi.org/10.1039/c5nr02502h.

Magrez, A., Kasas, S., Salicio, V., Pasquier, N., Seo, J. W., Celio, M., et al. (2006). Cellular toxicity of carbon-based nanomaterials. *Nano Letters*, 6(6), 1121–1125. https://doi.org/10.1021/nl060162e.

Murphy, F. A., Poland, C. A., Duffin, R., Al-Jamal, K. T., Ali-Boucetta, H., Nunes, A., et al. (2011). Length-dependent retention of carbon nanotubes in the pleural space of mice initiates sustained inflammation and progressive fibrosis on the parietal pleura. *American Journal of Pathology*, 178(6), 2587–2600. https://doi.org/10.1016/j.ajpath.2011.02.040.

Murray, A. R., Kisin, E., Leonard, S. S., Young, S. H., Kommineni, C., Kagan, V. E., et al. (2009). Oxidative stress and inflammatory response in dermal toxicity of single-walled carbon nanotubes. *Toxicology*, 257(3), 161–171. https://doi.org/10.1016/j.tox.2008.12.023.

Myojo, T., & Ono-Ogasawara, M. (2018). Review; risk assessment of aerosolized SWCNTs, MWCNTs, fullerenes and carbon black. *Kona Powder and Particle Journal*, 2018(35), 80–88. https://doi.org/10.14356/kona.2018013.

Oberdörster, G., Castranova, V., Asgharian, B., & Sayre, P. (2015). Inhalation exposure to carbon nanotubes (CNT) and carbon nanofibers (CNF): Methodology and dosimetry. *Journal of Toxicology and Environmental Health, Part B*, 18(3–4), 121–212. https://doi.org/10.1080/10937404.2015.1051611.

Pan, J., Li, F., & Choi, J. H. (2017). Single-walled carbon nanotubes as optical probes for bio-sensing and imaging. *Journal of Materials Chemistry B*, 5(32), 6511–6522. https://doi.org/10.1039/c7tb00748e.

Pastorin, G., Wu, W., Wieckowski, S., Briand, J. P., Kostarelos, K., Prato, M., et al. (2006). Double functionalisation of carbon nanotubes for multimodal drug delivery. *Chemical Communications*, 11, 1182–1184. https://doi.org/10.1039/b516309a.

Patlolla, A. K., Patra, P. K., Flountan, M., & Tchounwou, P. B. (2016). Cytogenetic evaluation of functionalized single-walled carbon nanotube in mice bone marrow cells. *Environmental Toxicology*, 31(9), 1091–1102. https://doi.org/10.1002/tox.22118.

Peng, Z., Liu, X., Zhang, W., Zeng, Z., Liu, Z., Zhang, C., et al. (2020). Advances in the application, toxicity and degradation of carbon nanomaterials in environment: A review. *Environment International*, 134. https://doi.org/10.1016/j.envint.2019.105298.

Pinaud, F., Michalet, X., Bentolila, L. A., Tsay, J. M., Doose, S., Li, J. J., et al. (2006). Advances in fluorescence imaging with quantum dot bio-probes. *Biomaterials*, 27(9), 1679–1687. https://doi.org/10.1016/j.biomaterials.2005.11.018.

Poland, C. A., Duffin, R., Kinloch, I., Maynard, A., Wallace, W. A. H., Seaton, A., et al. (2008). Carbon nanotubes introduced into the abdominal cavity of mice show asbestos-like pathogenicity in a pilot study. *Nature Nanotechnology*, 3(7), 423–428. https://doi.org/10.1038/nnano.2008.111.

Prinith, N. S., Manjunatha, J. G., & Raril, C. (2019). Electrocatalytic analysis of dopamine, uric acid and ascorbic acid at poly(adenine) modified carbon nanotube paste electrode: A cyclic voltammetric study. *Analytical and Bioanalytical Electrochemistry*, 11(6), 742–756. http://www.abechem.com/No.%206-2019/2019,%2011(6),%2020742-756.pdf.

Pushpanjali, P. A., Manjunatha, J. G., Tigari, G., & Fattepur, S. (2020). Poly(Niacin) based carbon nanotube sensor for the sensitive and selective voltammetric detection of vanillin with caffeine. *Analytical and Bioanalytical Electrochemistry*, 12(4), 553–568. http://www.abechem.com/article_39221_5b4cb9d77cdb1bfb6c48fc3dac8c7fc0.pdf.

Ramu, V., Ali, F., Taye, N., Garai, B., Alam, A., Chattopadhyay, S., et al. (2015). New imaging reagents for lipid dense regions in live cells and the nucleus in fixed MCF-7 cells. *Journal of Materials Chemistry B*, 3(36), 7177–7185. https://doi.org/10.1039/c5tb01309g.

Raril, C., Manjunatha, J. G., Nanjundaswamy, L., Siddaraju, G., Ravishankar, D. K., Fattepur, S., et al. (2018). Surfactant immobilized electrochemical sensor for the detection of indigotine. *Analytical and Bioanalytical Electrochemistry*, 10(11), 1479–1490. http://www.abechem.com/No.%2011-2018/2018,%2010(11),%201479-1490.pdf.

Raril, C., Manjunatha, J. G., Ravishankar, D. K., Fattepur, S., Siddaraju, G., & Nanjundaswamy, L. (2020). Validated electrochemical method for simultaneous resolution of tyrosine, uric acid, and ascorbic acid at polymer modified nano-composite paste electrode. *Surface Engineering and Applied Electrochemistry*, 56(4), 415–426. https://doi.org/10.3103/S1068375520040134.

Raril, C., Manjunatha, J. G., & Tigari, G. (2020). Low-cost voltammetric sensor based on an anionic surfactant modified carbon nanocomposite material for the rapid determination of curcumin in natural food supplement. *Instrumentation Science and Technology*, 48(5), 561–582. https://doi.org/10.1080/10739149.2020.1756317.

Saito, N., Haniu, H., Usui, Y., Aoki, K., Hara, K., Takanashi, S., et al. (2014). Safe clinical use of carbon nanotubes as innovative biomaterials. *Chemical Reviews*, 114(11), 6040–6079. https://doi.org/10.1021/cr400341h.

Sayes, C. M., Fortner, J. D., Guo, W., Lyon, D., Boyd, A. M., Ausman, K. D., et al. (2004). The differential cytotoxicity of water-soluble fullerenes. *Nano Letters*, 4(10), 1881–1887. https://doi.org/10.1021/nl0489586.

Shvedova, A., Castranova, V., Kisin, E., Schwegler-Berry, D., Murray, A., Gandelsman, V., et al. (2003). Exposure to carbon nanotube material: Assessment of nanotube cytotoxicity using human keratinocyte cells. *Journal of Toxicology and Environmental Health-Part A*, 66(20), 1909–1926. https://doi.org/10.1080/713853956.

Singh, R., Pantarotto, D., Lacerda, L., Pastorin, G., Klumpp, C., Prato, M., et al. (2006). Tissue biodistribution and blood clearance rates of intravenously administered carbon nanotube radiotracers. *Proceedings of the National Academy of Sciences of the United States of America*, 103(9), 3357–3362. https://doi.org/10.1073/pnas.0509009103.

Takagi, A., Hirose, A., Nishimura, T., Fukumori, N., Ogata, A., Ohashi, N., et al. (2008). Induction of mesothelioma in p53+/− mouse by intraperitoneal application of multi-wall carbon nanotube. *Journal of Toxicological Sciences*, 33(1), 105–116. https://doi.org/10.2131/jts.33.105.

Tasis, D., Tagmatarchis, N., Bianco, A., & Prato, M. (2006). Chemistry of carbon nanotubes. *Chemical Reviews*, 106(3), 1105–1136. https://doi.org/10.1021/cr050569o.

Tigari, G., & Manjunatha, J. G. (2020a). A surfactant enhanced novel pencil graphite and carbon nanotube composite paste material as an effective electrochemical sensor for determination of riboflavin. *Journal of Science: Advanced Materials and Devices*, 5(1), 56–64. https://doi.org/10.1016/j.jsamd.2019.11.001.

Tigari, G., & Manjunatha, J. G. (2020b). Optimized voltammetric experiment for the determination of phloroglucinol at surfactant modified carbon nanotube paste electrode. *Instruments and Experimental Techniques, 63*(5), 750–757. https://doi.org/10.1134/S0020441220050139.

Tigari, G., Manjunatha, J. G., Raril, C., & Hareesha, N. (2019). Determination of riboflavin at carbon nanotube paste electrodes modified with an anionic surfactant. *ChemistrySelect, 4*(7), 2168–2173. https://doi.org/10.1002/slct.201803191.

Van Daele, D. J., & Cassell, M. D. (2009). Multiple forebrain systems converge on motor neurons innervating the thyroarytenoid muscle. *Neuroscience, 162*(2), 501–524. https://doi.org/10.1016/j.neuroscience.2009.05.005.

Vorobiev, A. K., Gazizov, R. R., Borschevskii, A. Y., Markov, V. Y., Ioutsi, V. A., Brotsman, V. A., et al. (2017). Fullerene as photocatalyst: visible-light induced reaction of perfluorinated α,ω-diiodoalkanes with C60. *Journal of Physical Chemistry A, 121*(1), 113–121. https://doi.org/10.1021/acs.jpca.6b10718.

Walker, M. G., Ramu, V., Meijer, A. J. H. M., Das, A., & Thomas, J. A. (2017). A ratiometric sensor for DNA based on a dual emission Ru(dppz) light-switch complex. *Dalton Transactions, 46*(18), 6079–6086. https://doi.org/10.1039/c7dt00801e.

Welsher, K., Sherlock, S. P., & Dai, H. (2011). Deep-tissue anatomical imaging of mice using carbon nanotube fluorophores in the second near-infrared window. *Proceedings of the National Academy of Sciences, 108*(22), 8943–8948. https://doi.org/10.1073/pnas.1014501108.

Wick, P., Manser, P., Limbach, L. K., Dettlaff-Weglikowska, U., Krumeich, F., Roth, S., et al. (2007). The degree and kind of agglomeration affect carbon nanotube cytotoxicity. *Toxicology Letters, 168*(2), 121–131. https://doi.org/10.1016/j.toxlet.2006.08.019.

Wu, C., Bull, B., Christensen, K., & McNeill, J. (2009). Ratiometric single-nanoparticle oxygen sensors for biological imaging. *Angewandte Chemie, International Edition, 48*(15), 2741–2745. https://doi.org/10.1002/anie.200805894.

Zhang, L., Petersen, E. J., Habteselassie, M. Y., Mao, L., & Huang, Q. (2013). Degradation of multiwall carbon nanotubes by bacteria. *Environmental Pollution, 181*, 335–339. https://doi.org/10.1016/j.envpol.2013.05.058.

Zhu, L., Chang, D. W., Dai, L., & Hong, Y. (2007). DNA damage induced by multiwalled carbon nanotubes in mouse embryonic stem cells. *Nano Letters, 7*(12), 3592–3597. https://doi.org/10.1021/nl071303v.

Chapter 18

Safety and ethics of carbon nanomaterial-based sensors

Monima Sarma and Shaik Mubeena
Department of Chemistry, KL Deemed to be University (KLEF), Guntur, Andhra Pradesh, India

Introduction

There are over 100 elements in the periodic table. Of all these elements, carbon has the most diversifying chemistry. This single element is the center of a whole branch of chemistry—organic chemistry. However, carbon science is not limited to only organic molecules, rather a broad spectrum of scientists cheer on this element. The comprehensiveness of carbon is due to its flexibility toward adopting different hybridizations (sp, sp^2, and sp^3) because of the low energy difference between 2s and 2p orbitals. Hybridization variability is the reason for the existence of various allotropes of carbon, each with distinctive properties and applications. The allotropes have different crystal structures that determine their physicochemical properties. For example, diamond has a three-dimensional tetrahedral network of sp^3-hybridized carbon atoms, whereas graphite is a stacked layer of sheets consisting of a hexagonal network of sp^2-hybridized carbon atoms. Breaking down the macroscopic structures of carbon into the nanoscopic structures explores the most fascinating world of it. Ironically, this nanoworld of carbon is vast as the universe and has been continuously captivating scientists since its discovery. Exfoliation of one layer of carbon sheet from graphite creates a new material with exciting properties—graphene—which is the basis of all carbon nanomaterials. Graphene can be rolled into carbon nanotubes (CNTs) or wrapped into fullerenes—two other carbon nanomaterials with distinct properties and applications. Novoselov and Geim were jointly awarded the physics Nobel Prize in 2010 for their exfoliation of single-layer graphene sheets. Other popular carbon-based nanomaterials are carbon dots and carbon nano-horns. These carbon nanomaterials have unique physical, chemical, electronic, and mechanical properties and are the subject of intensive modern research. For instance, CNTs have huge mechanical strength, can withstand extreme strain, and is one of the most durable materials known. Also, CNTs have exciting electrical properties, such as enormous electrical conductivity. Graphene is the subject of intensive research due to its unique electronic, mechanical, and physicochemical properties. It has higher specific surface area of all materials which means that it has a substantial fraction of exposed surface atoms. The electron mobility of graphene is as high as $2 \times 10^5 \, \text{cm}^2 \, \text{V}^{-1} \, \text{s}^{-1}$ even at room temperature (Geim & Novoselov, 2007).

Sensors, on the other hand, are mechanical devices that can "feel" impulses, such as light, temperature, magnetism, pressure, electric field, movement, pH, chemical reactions, and so on. Besides biological sensors, human life has been heavily dependent on artificial sensors lately, especially after the discovery of the Internet of Things (IoT). Any sensor has two major components—the receptor and the transducer. Chemosensors are types of sensors that provide information about the chemical composition of their surroundings. The receptor fragment recognizes the analyte by supramolecular interactions or chemical transformations which modify the electronic environment of the receptor. The transducer identifies this chemical change and produces an output signal. A converter converts this analog signal into a digital one which is then amplified by an amplifier. This electric impulse is analyzed by the integrated computer and a report is generated. The most important factors that determine the applicability of chemosensors are selectivity, sensitivity, detection limit, response time, and the size of the device. Ideally, the sensor should have excellent selectivity and sensitivity, fast response time, and portable packaging size. In the past few decades, the world has witnessed the evolution of miniature microelectronic devices with high selectivity and sensitivity. The operating mechanisms are different among various categories of sensors and they have target-specific operating principles, such as absorption or emission of light, electrochemical responses, absorption, release of heat, etc. Sensors have widespread applications in detecting the level of carbon monoxide, analyzing water samples for heavy metal pollutants, automobile exhaust analysis, environmental monitoring, food and health monitoring, and medical diagnostics, to name a few (Vetelino & Reghu, 2017).

The application of carbon nanomaterials in sensors unveiled a new domain of research—carbon nanomaterial-based sensors. The development of new devices often requires hybrid materials. The properties of hybrid materials or nanocomposites consisting of organic and inorganic counterparts are often better than the individual materials. Interestingly, carbon nanostructures have some reactive positions that can be chemically modified. Application of these nanomaterials is expanded by surface modification including functionalization, capping, coating, etc., through appropriate chemical reactions leading to new materials with different physicochemical properties. One such example is the attachment of a target-specific binding group to increase the selectivity of the relevant sensors. Similarly, enhanced solubility in specific solvents is obtained by functionalizing with an appropriate group. The carbon nanomaterial-based sensors rarely comprise pristine materials. Instead, the devices use their chemically modified analogs, such as nanocomposites of graphene oxide, CNTs, fullerenes, etc. The uniqueness of each of these carbon nanocomposites makes them suitable for specific sensing mechanisms. A typical example is sensors based on nano-diamonds containing nitrogen-vacancy defect centers. This type of defect is common in the diamond structure and is responsible for the fluorescence of the material. The sensors based on nano-diamonds exhibit high sensitivity and are useful in biological environments (Ermakova et al., 2013). Graphene derivatives are popular materials in sensors for quantifying gases due to their exceptional properties, such as high thermal and electrical conductivity, large surface area, fast electron transfer rate, high mechanical strength, etc. (Choi et al., 2020). Fullerene nanocomposites have a high surface-to-volume ratio, fast electron transferability, and biocompatibility which make them useful in biosensors to detect a range of biomolecules, such as drugs, glucose, DNA, ATP, and so on. The confinement of π-electrons in zero dimension imparts incredible electronic properties in fullerenes. They have a strong electron affinity and are chemically reactive, although they are physically stable (Uygun & Uygun, 2020). CNTs are another carbon-based nanomaterials with distinct properties, such as hollow tubular structure, high surface-to-volume ratio, etc. They are brilliant materials in chemo- and biosensors owing to their ability to adsorb different types of molecules on their surface (Manjunatha, 2016, 2017; Manjunatha, 2018a; Manjunatha, 2018b; Manjunatha, 2019; Manjunatha & Deraman, 2017; Manjunatha, Deraman, & Basri, 2015; Manjunatha, Deraman, Basri, & Talib, 2014; Manjunatha, Deraman, Basri, & Talib, 2018; Raril & Manjunatha, 2020; Zaporotskova, Boroznina, Parkhomenko, & Kozhitov, 2016).

The evolution of nanomaterials and nanotechnology led to many benefits and opportunities. Having said that, the rapid growth of engineered nanomaterials raises concerns about safety issues. During the production, handling, and application stages, these nanomaterials could slip into the environment making them exposed to humans and other organisms. Earlier, hazards and risks were assumed to be independent of the particle size of the nanomaterials which now is proved incorrect. Unfortunately, there is a lack of engineered nanomaterial safety-related data as this area has remained relatively less explored. The available data are mostly centered in pristine materials whereas most of the devices use functionalized nanomaterials. Also, these studies mostly concern inhalation exposure of the airborne nanoparticles to human health and there is an uncertainty in the risk estimation. The workers in laboratories are primarily exposed to these nanoparticles which could complicate their health through ingestion, besides inhalation. The nanomaterials exhibit both intrinsic and extrinsic properties. The intrinsic properties do not change much during their measurement and are more predictable. However, a potential problem is the irreversible transformation of the nanoparticles into aggregates and agglomerates between the airborne nanoparticles and other particles through chemical interactions. These secondary particles often display a different level of toxicity than the original particles. On balance, humans are not exposed to a single type of nanoparticle, rather are exposed to a variety of them having different shapes and sizes, surface coating, etc. Cellular damages by the nanomaterials might be induced by various mechanisms, such as membrane impairment including cationic phagolysosome damage, production of reactive oxygen species, photocatalytic and redox activities, oxidative stress, activation of inflammasome and production of cytokine and chemokine, cytotoxicity of toxic ions, DNA damage, and so on. Apart from human toxicity and cytotoxicity, there might be other types of hazards such as ecotoxicity associated with the application of nanomaterials (Reijnders, 2014).

Graphene and CNTs are the most used variants of carbon nanomaterials in nanotechnology. Therefore, it is important to consider the safety-related issues of these two distinct nanomaterials. They have completely different structures and dimensions. Graphene has a planar, two-dimensional structure, whereas CNTs are tubular one-dimensional. Their dispersion in common solvents is quite different too, graphene having an edge, as CNTs require a surfactant to ease their dispersion. Graphene sheets tend to form a multilayer stack while CNTs tend to form bundles. Graphene materials are usually obtained in a purer form than CNTs which are often contaminated with metal nanoparticles sourced from the metal catalysts. The various physicochemical properties of these two materials are summarized in Table 18.1 (Bussy, Ali-Boucetta, & Kostarelos, 2013).

This chapter systematically reviews the health issues of engineered nanomaterials. Of the various carbon nanomaterials, microelectronic devices like sensors mostly use graphene and CNTs. Graphenes are relative newcomers in the carbon nanomaterial family, and their safety considerations are often overlooked among the exuberance over their exciting properties.

TABLE 18.1 Physicochemical properties of graphene and carbon nanotubes.

Physicochemical property	Graphene	Carbon nanotube	Similarity
Shape	Planar—monolayer or multilayer	Cylindrical (single-walled and multiwalled CNTs)	Different structures
Dimensions	Lateral size: 0.3–10 µm Thickness: 0.34–100 nm	Length: 10 nm–1 cm (SWCNTs), 10 nm–few microns (MWCNTs) Diameter: 0.4–3 nm (SWCNTs), 2–200 nm (MWCNTs)	Comparable dimensions
Surface area	Up to 2675 m^2/g, decreases with layers	>1000 m^2/g (SWCNTs), 100–500 m^2/g (MWCNTs)	Different surface areas
Surface charge	Depends on functionalization or coating	Depends on functionalization or coating	Comparable
Elasticity	Young modulus of 1100 GPa, flexible; addition of layers increases stiffness	Young modulus 1–5 TPa (SWCNT), 0.2–0.95 TPa (MWCNTs), flexible	Comparable
Colloidal stability	More dispersibility, stacked aggregate formation	Less dispersibility, bundled aggregate formation	Different dispersibility
Durability	Enzymatic degradation by defects in the plane	Enzymatic degradation by unzipping	Different durability
Impurities	Mainly graphite and chemical residues	Metal and carbon nanoparticles, amorphous C	Different impurities

Adapted with permission from Bussy, C., Ali-Boucetta, H., & Kostarelos, K. (2013). Safety considerations for graphene: Lessons learnt from carbon nanotubes. Accounts of Chemical Research, 46(3), 692–701. Copyright © 2013, American Chemical Society.

There are some reports of other carbon nanomaterial-based sensors, such as carbon dots, but their safety considerations were rarely examined. Many countries started examining nanomaterial-containing products and populations exposed to nanomaterials by bringing in information-gathering legislation. Therefore, a thorough analysis and understanding of all the physicochemical properties of any material are paramount to assessing their safety issues. The carbon nanomaterial-based sensors have brilliant applications which are nicely described in the other chapters of this book and are beyond the scope of the present chapter. The objective of the present chapter is to summarize the various safety issues of carbon nanoparticle-based sensors and their ethical use.

Impact on biodiversity: Breaking down the bulk materials into the nanosized particles unravels exciting properties completely different than the bulk. However, sometimes this poses a problem. The nanoparticles being more penetrating pass through the cell membrane more easily than the bulk particles and can interact with the proteins and other biomolecules in the cell. The explosive growth of nanotechnology in the past few decades perhaps took a toll on the environment. Research in nanotechnology has been historically focused on the bright side of nanomaterials. Environmental impact assessment of this emerging technology has nonetheless gained only a little attention, and it is about time we began to estimate the threat to biodiversity. Apart from being released into the air, the nanomaterials could contaminate the water and soil resources through their disposal and affect the growth of plants. The materials could be subsequently accumulated in the plants and develop phytotoxicity. The composition of microbes in the soil such as the biomass of gram-positive and gram-negative bacteria, fungi, etc., is depleting with the rapid utilization of engineered nanomaterials. Exposure to SWCNTs causes irreversible damage to the soil fungal and bacterial community affecting the nutrient cycling in soil. An interesting study conducted by Chinese scientists compared the toxicity of fullerenes, reduced graphene oxide, and MWCNTs on rice grown in loamy potting soil. Just 30 days of exposure to the CNTs significantly affected the shoot height and root length of rice by shrinking the root cortical cells. High exposure dose increased concentrations of phytohormones in the roots along with enhanced activities of antioxidant enzymes, such as peroxidase (POD) and superoxide dismutase (SOD) (Hao et al., 2018).

Research suggests that graphene and MWCNTs impede the growth of red spinach, tomato, and cabbage. There are some reports about the negative impact of carbon nanomaterials on the animal kingdom too. For example, SWCNTs are found to cause embryo death, angiogenesis inhibition, and growth retardation in chicken; MWCNTs and fullerenes are found to be fatal for zebrafish and mice, respectively. Some reports suggest carbon nanomaterials hinder the reproduction and growth

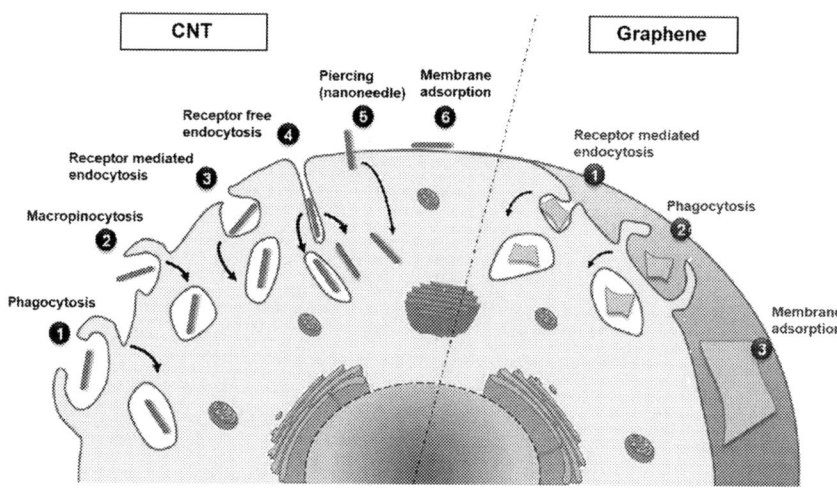

FIG. 18.1 Cellular uptake mechanism of carbon nanotubes and graphene. *Adapted with permission from Bussy, C., Ali-Boucetta, H., & Kostarelos, K. (2013). Safety considerations for graphene: Lessons learnt from carbon nanotubes. Accounts of Chemical Research, 46(3), 692–701. Copyright © 2013, American Chemical Society.*

of invertebrates, such as earthworms. As the whole ecosystem is intricately interconnected, damage to some classes in the living kingdom might pose a serious threat to biodiversity, putting the future generation at risk (Chen, Qin, & Zeng, 2017).

Interaction of carbon nanomaterials with cells: To get into the cells, the carbon nanomaterials must get through the cell membranes first. Simulation studies proposed that carbon nanotubes could behave like needles to pierce the lipid bilayers. Various proposed mechanisms consider the amphiphilicity of the chemically modified CNTs and the surface charge as the crucial factors behind the interaction between the CNTs and the cell membrane. A positively charged carbon nanotube surface facilitates the interaction due to electrostatic attraction. The ammonium-functionalized CNTs can directly pierce the cell membrane and translocate freely into the cytoplasm. Cell internalization can follow different types of mechanisms (Fig. 18.1). It was proved that the cellular uptake of carbon nanomaterials was controlled by their size and individualization. For example, the insertion of CNTs could be blocked if they are coated with large molecules, such as block copolymers, proteins, polyethylene glycol, etc. The size and individuality of the carbon nanotubes could be chemically modified. Size could be altered by carboxylation reactions and individuality could be induced by different types of chemical functionalization. This brings up an interesting point that the toxicity of carbon nanomaterial-based sensors perhaps could be minimized if they are coated with macromolecules.

Once the nanomaterials penetrate the cell membrane and are in the cytoplasm, they can further enter the intracytoplasmic compartments, such as the endoplasmic reticulum, lysosome, mitochondria, Golgi, and nucleus, and accumulate there. Fortunately, CNTs were not observed inside the mitochondria and nucleus, excluding the possibility of genomic alteration or respiratory blueprint of the cell. On balance, the intracellular activity of the carbon nanomaterials is mostly dependent on their size, surface properties, functionalization, and ability to interact with the membranes. For example, short ammonium-functionalized MWCNTs can easily escape the endosomes. The cytotoxicity of the materials is collectively controlled by the above-mentioned processes, i.e., membrane insertion, cellular uptake, and the intracellular activity of the materials. The toxicity mainly stems from oxidative stress through the generation of reactive oxygen species which could lead to genotoxic, inflammatory, and cytotoxic damages. The physicochemical properties, such as dimension, colloidal properties, surface properties, metal impurities, etc., are found to play an important role in the biological responses of carbon nanomaterials. Monitoring intracellular activities of graphene oxide is relatively daunting due to the challenge of imaging and visualizing graphene sheets intracellularly (Bussy et al., 2013).

The surface properties of carbon nanomaterials on their cytotoxicity nonetheless caught special attention. In general, the cationic nanomaterials stimulate membrane leakage and cytotoxicity by binding to the negatively charged cell membranes. Interestingly, the impact of negatively charged nanomaterials on membranes has remained elusive. The functionalized carbon nanotubes used in various applications normally have a negatively charged surface due to the presence of the functionalities, such as —COOH, —OH, —C—O—C—, etc. Understanding their interaction with cell membranes is still unclear because of mechanistic uncertainty. Based on computer simulations, it was suggested that CNTs bring about a change in the cell structure and their mode of interaction with the lipid bilayer (electrostatic attractive or repulsive) remaining debatable though. Surface defects of CNTs are found to play an important role in their biological toxicity. The nature of defects includes incomplete bonds, functional groups, sp^3 hybridized carbon atoms, distortion from the hexagonal network shapes, and so on. A recent study attempted to clear up the vagueness of the interaction of MWCNTs with positively and negatively charged cell membranes by using giant unilamellar vesicles (GUVs) and supported lipid bilayers

(SLBs) as the model cell membranes. MWCNTs were stabilized using bovine serum albumin (BSA) to mimic the biological environment. GUVs were subsequently exposed to the BSA-MWCNTs and the morphologies of the positively and negatively charged GUVs were imaged under bright field observations. The electrophoretic mobilities of both the untreated and BSA-MWCNTs indicated a net negative charge on the carbon nanotube surface thereby pointing to electrostatically mediated adhesion of the MWCNTs on the membrane. The negatively charged CNTs with a high degree of defect stuck to the positively charged lipid membrane disrupting its structure and inducing cytotoxicity. On the other hand, the carbon nanotubes extracted phospholipids from the negatively charged membranes. CNTs with a diminished degree of defect were less harmful to the membranes (Jiang et al., 2017).

Protein-carbon nanomaterial interaction: Proteins are the fundamental structural and functional units of organisms. The carbon nanomaterials can interact with the proteins after they are in the cells, and this might be a lethal encounter. Therefore, systematic investigations of the interactions between proteins and carbon nanomaterials (CNMs) are crucial to estimate their toxicity. Though carbon nanomaterials interact with the proteins in separate ways depending on their shape and size, their interactions are mostly driven by van der Walls, π-π stacking, and electrostatic and hydrophobic interaction forces. Sometimes, the carbon nanomaterials can trigger a conformational change of the proteins and the extent of conformational change is reported to be nonuniform for various CNMs. Fullerenes are believed to be less severely biotoxic among the common carbon nanomaterials. The toxicity of fullerenes might stem from impurities instead. CNTs were found to be tougher on certain enzymes than the hydration of C_{60}. Also, SWCNTs are found to be relatively less toxic than MWCNTs. The level of toxicity among the various functionalized carbon nanomaterials varies with the functional groups. Another mechanism of interaction between the proteins and carbon nanomaterials is the adsorption of the former on the latter. The degree of adsorption is found to depend on the surface curvature of the nanomaterials. For example, graphene with a flat surface was found to adsorb bovine serum proteins more powerfully than carbon nanotubes. Let us explore the protein-CNM interactions in more detail (Wang et al., 2019).

Studies showed that proteins in an aqueous solution tend to form a complex with carbon nanotubes and adsorb on their surface. This complex formation changes the conductivity of CNTs perhaps due to the charge transfer between CNTs and the electron-rich or deficient moieties in the proteins (Yu, Chattopadhyay, Galeska, Papadimitrakopoulos, & Rusling, 2003). The protein-CNT interaction could be heavily influenced by the charge transfer which in turn could stimulate the protein conformational change. Molecular dynamics simulation studies pointed out the role of the functional density of SWCNTs on the orientation of the proteins and active site conformation. If the CNTs are carboxyl functionalized, then the binding interaction might depend on the oxygen density of the nanomaterials (Geldert, Liu, Loh, & Teck Lim, 2017). CNTs can elevate or impede biodegradation by adjusting the toxic effects with alteration in bioavailability originating from adsorption or desorption. The presence of SWCNTs is found to trigger a conformational transformation in maleylpyruvate isomerase and lignin peroxidase enzymes with subsequent alteration of protein cavity dimension. Also, the CNTs are found to change the behavior of water molecules near manganese peroxidase and its substrates. There was an attempt to measure out the enzyme activity, adsorption, thermostability, and conformational changes based on the interaction of carboxyl-functionalized MWCNTs and porcine trypsin. The study revealed that the carboxylic functionalized MWCNTs had only a little effect on the conformational change of the protein (Noordadi, Mehrnejad, Sajedi, Jafari, & Ranjbar, 2018). The cytotoxicity of carbon nanotubes at variable degrees of agglomeration was studied that discovered less harmfulness of dispersed CNTs than agglomerates. Another molecular dynamic simulation of the interaction between dopamine-related proteins and SWCNTs suggested adsorption of the protein on the nanotubes where the interaction was more serious in the case of aggregated SWCNTs. In a nutshell, the fate of protein-CNT interaction is affected by both the polymerization and size of the nanomaterials (Yu et al., 2017).

Both graphene and graphene oxides are heavily used in carbon nanomaterial-based sensors. Therefore, it is important to examine the toxicity of both of these materials which differ in their physicochemical properties. Graphene is hydrophobic and has higher conductivity than graphene oxide while the latter is soluble in the aqueous medium and has a lower electrical conductivity than the former. Like carbon nanotubes, the toxicity of graphene varies with factors, such as structure, surface properties, size, exposure time, etc. A molecular dynamics simulation was set up to understand the nature of toxicity of graphene through the hydrophobic protein-protein interactions. This study observed disruption of the protein-protein interactions by graphene nanosheets which could be deadly to the cells (Luan, Huynh, Zhao, & Zhou, 2015). Another molecular dynamics simulation hypothesized containment of toxic changes of amyloid beta-peptide by graphene oxide and reduced graphene oxide where the latter was found to be more efficient (Baweja, Balamurugan, Subramanian, & Dhawan, 2015). Fluorescence and circular dichroism spectroscopic studies of the affinity, thermal stability, conformational structure, and nonenzymatic glycosylation of graphene oxide-hemoglobin binding discovered that hemoglobin became more prone to thermal denaturation in the presence of graphene oxide. Additionally, graphene oxide prevented the secondary structure of hemoglobin by forming aggregates with hemoglobin (Wang, 2016).

Fullerenes are phenomenal carbon nanomaterials. Modification of the fullerene skeleton by tethering with biomolecules induces the properties of the biomolecules in fullerenes. The nature of the interaction between fullerenes and proteins or enzymes is controlled by the type of functional groups attached to the fullerenes. Fullerenes are considerably less toxic than the other carbon nanomaterials, such as graphene or carbon nanotubes.

Interaction of carbon nanomaterials with tissue: The carbon nanomaterials that make their way into the human body could either be eliminated by excretion or they could be accumulated in tissues. In the latter case, it is important to identify the location of the deposition, residence time of accumulation, and any subsequent cell damage which could take a toll on the local histopathological reactions. The prothrombic activity of the nanomaterials should be assessed once they are in the bloodstream. The blood compatibility of the SWCNTs is believed to improve if they are coated with human serum proteins. Hemolytic activity of small graphene oxide sheets on human red blood cells was detected. Interestingly, this is almost eliminated by enhancing their dispersion using chitosan. The hemocompatibility and hemotoxicity profile of both carbon nanotubes and graphene are reported to improve on amidation. Intravenous injection of pristine MWCNTs coated with serum proteins leads to their accumulation in the organs, such as lungs, liver, and spleen. On the contrary, individualization of the nanotubes in vivo is observed in the case of covalently functionalized MWCNTs with enhanced surface functionalization. The tissue accumulation could be lowered by increasing the extent of amino functionalization. On a positive note, no histological or physiological damage to the major tissue the CNTs passed through or accumulated in is observed. The graphene materials, on the other hand, do not have a common distribution profile where the crucial factors are the lateral size, surface modification, and individualization of the graphene sheets. The inadequately individualized materials normally have poor dispersibility and tend to form aggregates in vivo which affects their blood circulation and their ability to interact with biological barriers (Bussy et al., 2013).

Effect of carbon nanomaterials on the human body: Now let us put things together and discuss the effect of carbon nanomaterials on the human body. The severity of graphene on the human body incorporates two major aspects, i.e., immunoreactivity and inflammogenicity. Carbon nanomaterials are often marketed with a vague safety assessment although they are known to produce asbestos-like lung cancer, mesothelioma, asthma, etc., when chronically inhaled.

Carbon nanomaterials as possible carcinogens?

Due to their small sizes, CNTs could be easily inhaled to the lung alveolar region upon their exposure to the human body. The common sources of exposure are research laboratories, production facilities, recycling facilities, waste disposal, consumer goods containing carbon nanomaterials such as carbon nanosensors, and so on. Once they are in the lungs, they could be retained there for as long as a year, inducing adverse health effects such as cancer. A carcinogen is a type of substance that directly or indirectly promotes one or all the stages of carcinogenesis (initiation, promotion, and progression). The mechanism of action could be either damage to the genome or modify cellular metabolic processes. Genotoxic substances are a type of mutagens that disrupt the genetic materials, such as DNA, RNA, chromosomes which ultimately develop cancer. There is another type of substance that is not carcinogenic but could stimulate carcinogens, much like the co-catalysts do to the catalysts. They are called co-carcinogens (Luanpitpong, Wang, Davidson, Riedel, & Rojanasakul, 2016).

The inflammatory effects of carbon nanotubes are correlated with the pathogenic fibers. Long, thin, and bio-persistent fibers tend to induce inflammation if they persist in the lungs and pleural cavity. The pathogenic effect of long asbestos fibers on the lungs and the mesothelium and the similarity in shape between asbestos fibers and carbon nanotubes prompted scientists to investigate the possible carcinogenic effect of CNTs. Various in vitro and in vivo studies indicate possible genotoxicity and carcinogenicity of carbon nanotubes. The nonfunctionalized CNTs are relatively longer than the short entangled CNTs and the former was found to stimulate macrophage phagocytosis causing local inflammation with subsequent formation of mesothelial granuloma (Poland et al., 2008). Therefore, chemically modified MWCNTs that are shorter in length are found to be relatively safer than their longer pristine counterparts. The pathogenicity of the long fibers might be due to length-dependent retention of the materials on the parietal pleura which could ultimately cause inflammation, granuloma, and fibrosis. A series of cationic, anionic, and neutral functionalized CNTs were prepared from pristine MWNNTs to evaluate the effect of the surface charge of functionalized carbon nanotubes on fibrosis. Their pro-inflammatory and pro-fibrogenic activities were examined in vitro and assessed in vivo. The relevant study revealed significant lung fibrosis induced by strong cationic MWCNTs whereas weak cationic to neutral species were relatively safer (Li et al., 2013). Another experiment on long-term (~6 months) in vitro exposure to SWCNTs exhibited malignant transformation of human lung epithelial cells. A variety of genes participating in apoptosis, cell cycle control, and oncogenic development were found to be influenced by the SWCNT treatment. The immune response represented the most affected biological process. The oncogenic phenotype alluded to pAkt signaling, destabilized p53, and enhanced expression of Ras family genes, together with downregulation of the anti-apoptotic genes (Chen, 2015).

However, the majority of these data are based on experiments carried out on mice, which though could be extrapolated to the human body and are not sufficient to classify CNTs as carcinogens with certainty. The risk assessment of the carbon nanomaterials is nevertheless based on models and there is a paucity of specific guidelines to do so. The carcinogenic studies of nanomaterials come with some obvious challenges though. Carcinogenesis is a multistep long-term process that stems from long-term exposure. In a typical laboratory testing facility, the dose amount is relatively low and requires even longer exposure which is not often achievable for general research purposes. Another problem is the interspecies differences in inhalation and respiration between humans and rodents. For example, CNTs with a size smaller than 10 μm are inhalable to humans but not necessarily to mice. Thus, long CNTs and agglomerates present in the workplace atmosphere are more likely to be inhaled by humans and less likely by mice, complicating the comparison. Therefore, setting up new methodologies for reliable prediction of the health hazards of carbon or any other nanomaterials to humans concerning dosing, exposure time, route of exposure, etc., is extremely difficult. Each of these nanomaterials has diverse properties such as dimension, functionalization, surface properties, etc., which make the scenario even more complicated. A probable solution might be high-throughput screening assays for assessing cancer and mesothelioma pathogenicity (Luanpitpong et al., 2016).

Carbon nanofibers (CNFs) are relatively new entrants to the carbon nanomaterial family and they are being tried out as an alternative to carbon nanotubes. They have unique physicochemical properties and have superior applications in materials science including sensors. Task-specific carbon nanofibers could be synthesized through surface engineering which caught up with researchers' attention to employing them in a variety of gas sensors, pressure sensors, strain sensors, small and macromolecule sensors, etc. The advantages of using CNFs as the sensory material include enhanced surface area of electrode materials, easily manipulable properties that boost their sensory activity. CNF-based sensors offer high stability and high selectivity because of their high mechanical strength and chemical inertness and their capability to mitigate the oxidation overpotential (Wang, Zhu, et al., 2019; Wang, Wu, Wang, Yu, & Wei, 2019). With the emergence of carbon nanofibers as a potential material in various applications, concerns about their possible toxicity run in parallel. A recent study aimed to understand the molecular mechanism of the interaction of graphene carbon nanofibers (GCNF) with in vitro human lung adenocarcinoma cells using a pulmonary-like cell system. Exposure to GCNF was found to disrupt autophagic flux and reduce the ATP level. Besides, the autophagy flux was observed to be terminated by the reduced activity of lysosomes and cytoskeleton breakdown leading to apoptotic cell death. The whole process was reported to be controlled by an enhanced level of reactive oxygen species (Mittal et al., 2017).

An asthma angle

Asthma is an inflammatory disorder in the lungs characterized by frequent airway obstruction, chronic bronchial inflammation, hyperreactivity to bronchoconstrictor, etc. The disease is stimulated by recurrent hypersensitivity and allergic reactions to the inhaled allergens and is often fatal. A major source of the disease is exposure to fumes, gases, or dust in the workplace. An existing asthma condition could be worsened by stimuli, such as allergen-specific immunoglobulins, cold, certain drugs, or sometimes by exercise. The hypertrophy and hyperreactivity of bronchial smooth muscle cells are prompted by cytokines and mediators secreted by lymphocytes. On the other hand, constriction of the airway smooth muscles is generally caused by eosinophil, basophil, and mast cell secretions. There is a vague understanding of the molecular mechanisms regarding allergic and nonallergic asthma which must be cleared up to evaluate the environmental factors and conditions. Industrial high aspect ratio and biopersistent nanomaterials, such as carbon nanotubes, carbon black, etc., which could get into the human body through inhalation, are suspected to cause asthma. Various in vitro and in vivo models were studied for a thorough comprehension of the effect of carbon nanomaterials on asthma. Unfortunately, these models do not imitate the complex biological pathways leading to asthma, and thus, no concrete conclusion was established (Dobrovolskaia, Shurin, Kagan, & Shvedova, 2017).

Enzymatic degradation of carbon nanomaterials: The full commercial potential of CNMs remains unclear unless their toxicological issues are subdued, and their biopersistence is completely understood. Tissue accumulation and clearance of CNMs are two processes closely related to biodegradation. If the CNMs were nonbiodegradable, their toxic effect on the human body would be lessened only if they could be readily excreted. However, if they were biodegradable, they would be relatively risk-free provided any degradation by-product was nontoxic. There are some reports of elimination of carbon nanomaterials from the body after exposure to them. On a positive note, some researchers discovered the degradation of CNMs catalyzed by some naturally occurring enzymes in plants and the human body. This is nonetheless a silver lining in mitigating the perilous effect of the CNMs on the environment. Strong oxidizing agents, such as sulfuric acid, nitric acid, etc., oxidize CNMs by introducing oxygen atoms in the subsequent lattice. This process stimulates structural and morphological changes of the materials, such as shortening of the length of CNTs, disruption of graphene network and

TABLE 18.2 Summary of ex vitro enzymatic degradation of carbon nanomaterials.

Carbon nanomaterial	Functionalization	Enzyme	Oxidant	Degradation
SWCNT	Pristine carboxylated carboxylated carboxylated /phosphatidylserine	HRP HRP MPO MPO	$H_2O_2 H_2O_2 H_2O_2$, NaCl H_2O_2, NaCl	No Yes Yes Yes
MWCNT	Pristine carboxylated 1,3-dipolar cycloaddition $CONH(CH_2CH_2O)_2CH_2CH_2NH_3^+$ Nitrogen-doped $CONH(CH_2CH_2O)_2CH_2CH_2NH_3^+$	HRP HRP HRP HRP HRP MPO	$H_2O_2 H_2O_2 H_2O_2 H_2O_2 H_2O_2 H_2O_2$, NaCl	No Yes Yes Yes Yes Yes
Graphene	Reduced graphene oxide (RGO) Graphene oxide (GO)	HRP HRP	$H_2O_2 H_2O_2$	No Yes

HRP, horseradish peroxidase; *MPO*, myeloperoxidase; *MWCNT*, multi-walled carbon nanotube; *SWCNT*, single walled carbon nanotube.
Adapted with permission from Kotchey, G.P., Hasan, S.A., Kapralov, A.A., Ha, S.H., Kim, K., & Shvedova, A.A. (2012). A natural vanishing act: The enzyme-catalyzed degradation of carbon nanomaterials. Accounts of Chemical Research, 45(10), 1770–1781. Copyright © 2012, American Chemical Society.

appearance of holes in it, etc. Prolonged oxidation leads to the loss of material from the carbon networks in the form of carbon dioxide, carbon monoxide, and hydrocarbons. Though the strength of biological oxidants is nowhere close to the above-mentioned inorganic oxidants, the idea of degradation of CNMs by oxidation was nevertheless extrapolated to test their biopersistence by considering the possibility of enzymatic degradation. Table 18.2 summarizes the enzymatic degradation of carbon nanomaterials using two enzymes—horseradish peroxidase (HRP) and myeloperoxidase (MPO). As scientists are progressively delving into the biodegradation studies of carbon nanomaterials, hope arises about labeling carbon nanomaterial-based sensors as "green devices" soon (Kotchey et al., 2012).

Safety management of engineered carbon nanomaterials: The risk of any chemical substance is a combined expression of the consequences of an event and the likelihood of occurrence. The major workplace concern about the engineered nanomaterials is their risk evaluation and moderation. Management of any organization is responsible to assess the risks their employees might face in their workplace and take necessary actions to maintain a relatively hazard-free work environment. The risk evaluation is a meticulous process that needs the compilation of accurate detailed information about possible health complications upon exposure to chemical substances. The process is complicated by the fact that engineered nanomaterials mostly exhibit properties that are significantly different than their bulk counterparts. The development of a handy tool for rapid screening of nanomaterials in the workplace that also provides protective and preventive measures is an important research topic these days. However, safety in the laboratory is not only about the knowledge of materials and equipment but also involves human behavior and ethics that could be improved by proper chemical education (Groso, Petri-Fink, Rothen-Rutishauser, Hofmann, & Meyer, 2016).

The risk assessment studies conducted so far are mostly based on in vitro and in vivo animal studies and there is a serious limitation of data regarding possible health hazards of carbon nanomaterials on humans. Solubility is an important parameter here as the particles to be biotoxic must get into the biofluids. The solubility of spherical nanomaterials in water or any other solvent depends on various factors, such as their shape and size, surface area, crystallinity, etc. (Misra, Dybowska, Berhanu, Luoma, & Valsami-Jones, 2012). The influence of particle size on solubility could be monitored using the Ostwald-Freundlich equation:

$$S = S_0 \exp\left(\frac{2\gamma V}{RTr}\right) \tag{18.1}$$

where S is the solubility, r is the radius, S_0 is the solubility of the bulk material, V is the molecular volume, γ is the surface tension, R is the gas constant, and T is temperature (Baweja et al., 2015). The solubility of nanomaterials above 10 nm particle size is not expected to have drastic values as compared to their bulk analog. In this regard, soluble fullerenes are categorized as potentially hazardous. Nevertheless, there is a lack of data related to the long-term health hazard of fullerenes on humans. Though the pristine fullerenes are known to be less toxic than their functionalized analogs, their toxicity would be challenging to correlate.

The toxicity of the carbon nanofibres follows a different mechanism than spherical particles like fullerenes due to the very different sizes of the materials. Nanofibers have one dimension substantially longer than two others and their toxicity is often correlated with that of the asbestos fibers. The common types of fiber toxicity are fibrosis, pleural plaques, mesothelioma, etc., where the toxicity is ascribed to fiber length, thickness, and biopersistence. Spherical particles such as fullerenes could be incorporated into the cells but macrophages usually do not surround the fibers and undergo frustrated phagocytosis worsening their biopersistence scenario. This might cause retention of the fibers in the lungs despite the physiological clearance mechanism (Brown et al., 2007).

Outlook and Conclusions: Carbon nanomaterials are wonderful materials that have diverse applications. The unique physicochemical properties including interesting surface, electronic properties, etc., make them a suitable material for electrode modification and subsequent development of cheaper electrochemical sensing technology. The pristine carbon nanotubes graphene, fullerenes are rarely used as electrode modifiers, but their functionalized analogs to achieve desired electrochemical properties. Interestingly, these materials offer synthetic flexibility to attach various functional groups to them via appropriate chemical reactions. The popularity of electrochemical sensing of heavy metal ion-based toxic pollutants is mostly driven by the low cost of the technique where carbon nanomaterials play an important role. Apart from the detection of heavy metal environmental contaminants, carbon nanomaterials are also known for sensing pesticides. These materials have great optical properties that make them good candidates for optical sensing of toxic organic molecules such as pesticides. The fluorescent carbon nanomaterials are reported to be useful as indicators to improve the performance of such sensors. Another important application of carbon nanomaterials is in the food safety analysis where fluorescent carbon dots are found to be a promising alternative to other fluorescent materials such as heavy metal quantum dots (CdSe, CdTe), conventional fluorescent molecules (rhodamines, fluorescence, etc.) either due to their toxicity or other issues.

In a nutshell, carbon nanomaterials are fabulous materials for sensing applications. However, be that as it may, they come with their own set of cons. These materials are being increasingly engineered to develop task-specific materials for appropriate goals and that raises many concerns. The safety, handling, and storage of these engineered materials are often overlooked. Turned out, the engineered carbon nanoparticles could have toxic effects if inhaled and ingested, which appeals to careful investigation for any hazardous health effects before practical applications. The workers are primarily exposed to airborne carbon nanoparticles. Uncontrolled use of hazardous materials also passes to the consumers who are passively exposed. Another point of concern is the biopersistence of carbon nanomaterials which might add to the toxicity of these nanomaterials on the human body. The discovery of enzymatic degradation of carbon nanomaterials is a silver lining in this area. As every coin has two sides, some negative aspects of carbon nanomaterials come as no surprise. However, a proper risk assessment will reduce the hazardous possibilities. It is hoped that better performing and more biocompatible carbon nanomaterials will be realized soon that could be comfortably used in various sensors.

References

Baweja, L., Balamurugan, K., Subramanian, V., & Dhawan, A. (2015). Effect of graphene oxide on the conformational transitions of amyloid beta peptide: A molecular dynamics simulation study. *Journal of Molecular Graphics and Modelling, 61*, 175–185. https://doi.org/10.1016/j.jmgm.2015.07.007.

Brown, D. M., Kinloch, I. A., Bangert, U., Windle, A. H., Walter, D. M., Walker, G. S., et al. (2007). An in vitro study of the potential of carbon nanotubes and nanofibres to induce inflammatory mediators and frustrated phagocytosis. *Carbon, 45*(9), 1743–1756. https://doi.org/10.1016/j.carbon.2007.05.011.

Bussy, C., Ali-Boucetta, H., & Kostarelos, K. (2013). Safety considerations for graphene: Lessons learnt from carbon nanotubes. *Accounts of Chemical Research, 46*(3), 692–701. https://doi.org/10.1021/ar300199e.

Chen, D., et al. (2015). Gene expression profile of human lung epithelial cells chronically exposed to single-walled carbon nanotubes. *Nanoscale Research Letters, 10*, 12. https://doi.org/10.1186/s11671-014-0707-0.

Chen, M., Qin, X., & Zeng, G. (2017). Biodiversity change behind wide applications of nanomaterials? *Nano Today, 17*, 11–13. https://doi.org/10.1016/j.nantod.2017.09.001.

Choi, J. H., Lee, J., Byeon, M., Hong, T. E., Park, H., & Lee, C. Y. (2020). Graphene-based gas sensors with high sensitivity and minimal sensor-to-sensor variation. *ACS Applied Nano Materials*, 2257–2265. https://doi.org/10.1021/acsanm.9b02378.

Dobrovolskaia, M. A., Shurin, M. R., Kagan, V. E., & Shvedova, A. A. (2017). Ins and outs in environmental and occupational safety studies of asthma and engineered nanomaterials. *ACS Nano, 11*(8), 7565–7571. https://doi.org/10.1021/acsnano.7b04916.

Ermakova, A., Pramanik, G., Cai, J. M., Algara-Siller, G., Kaiser, U., Weil, T., et al. (2013). Detection of a few metallo-protein molecules using color centers in nanodiamonds. *Nano Letters, 13*(7), 3305–3309. https://doi.org/10.1021/nl4015233.

Geim, A. K., & Novoselov, K. S. (2007). The rise of graphene. *Nature Materials, 6*(3), 183–191. https://doi.org/10.1038/nmat1849.

Geldert, A., Liu, Y., Loh, K. P., & Teck Lim, C. (2017). Nano-bio interactions between carbon nanomaterials and blood plasma proteins: Why oxygen functionality matters. *NPG Asia Materials*, e422. https://doi.org/10.1038/am.2017.129.

Groso, A., Petri-Fink, A., Rothen-Rutishauser, B., Hofmann, H., & Meyer, T. (2016). Engineered nanomaterials: Toward effective safety management in research laboratories. *Journal of Nanobiotechnology*, *14*(1). https://doi.org/10.1186/s12951-016-0169-x.

Hao, Y., Ma, C., Zhang, Z., Song, Y., Cao, W., Guo, J., et al. (2018). Carbon nanomaterials alter plant physiology and soil bacterial community composition in a rice-soil-bacterial ecosystem. *Environmental Pollution*, *232*, 123–136. https://doi.org/10.1016/j.envpol.2017.09.024.

Jiang, W., Wang, Q., Qu, X., Wang, L., Wei, X., Zhu, D., et al. (2017). Effects of charge and surface defects of multi-walled carbon nanotubes on the disruption of model cell membranes. *Science of the Total Environment*, *574*, 771–780. https://doi.org/10.1016/j.scitotenv.2016.09.150.

Kotchey, G. P., Hasan, S. A., Kapralov, A. A., Ha, S. H., Kim, K., Shvedova, A. A., et al. (2012). A natural vanishing act: The enzyme-catalyzed degradation of carbon nanomaterials. *Accounts of Chemical Research*, *45*(10), 1770–1781. https://doi.org/10.1021/ar300106h.

Li, R., Wang, X., Ji, Z., Sun, B., Zhang, H., Chang, C. H., et al. (2013). Surface charge and cellular processing of covalently functionalized multiwall carbon nanotubes determine pulmonary toxicity. *ACS Nano*, *7*(3), 2352–2368. https://doi.org/10.1021/nn305567s.

Luan, B., Huynh, T., Zhao, L., & Zhou, R. (2015). Potential toxicity of graphene to cell functions via disrupting protein-protein interactions. *ACS Nano*, *9*(1), 663–669. https://doi.org/10.1021/nn506011j.

Luanpitpong, S., Wang, L., Davidson, D. C., Riedel, H., & Rojanasakul, Y. (2016). Carcinogenic potential of high aspect ratio carbon nanomaterials. *Environmental Science: Nano*, *3*(3), 483–493. https://doi.org/10.1039/c5en00238a.

Manjunatha, J. G. (2016). Poly (nigrosine) modified electrochemical sensor for the determination of dopamine and uric acid: A cyclic voltammetric study. *International Journal of ChemTech Research*, *9*(2), 136–146. http://sphinxsai.com/2016/ch_vol9_no2/1/(136-146)V9N2CT.pdf.

Manjunatha, J. G. (2017). A new electrochemical sensor based on modified carbon nanotube-graphite mixture paste electrode for voltammetric determination of resorcinol. *Asian Journal of Pharmaceutical and Clinical Research*, *10*(12), 295–300. https://doi.org/10.22159/ajpcr.2017.v10i12.21028.

Manjunatha, J. G. (2018a). A novel voltammetric method for the enhanced detection of the food additive tartrazine using an electrochemical sensor. *Heliyon*, *4*(11). https://doi.org/10.1016/j.heliyon.2018.e00986, e00986.

Manjunatha, J. G. G. (2018b). A novel poly (glycine) biosensor towards the detection of indigo carmine: A voltammetric study. *Journal of Food and Drug Analysis*, *26*(1), 292–299. https://doi.org/10.1016/j.jfda.2017.05.002.

Manjunatha, J. G. (2019). Electrochemical polymerised graphene paste electrode and application to catechol sensing. *Open Chemical Engineering Journal*, *13*(1), 81–87. https://doi.org/10.2174/1874123101913010081.

Manjunatha, J. G., & Deraman, M. (2017). Graphene paste electrode modified with sodium dodecyl sulfate surfactant for the determination of dopamine, ascorbic acid and uric acid. *Analytical and Bioanalytical Electrochemistry*, *9*(2), 198–213. http://www.abechem.com/No.%202-2017/2017,9(2),198-213.pdf.

Manjunatha, J. G., Deraman, M., & Basri, N. H. (2015). Electrocatalytic detection of dopamine and uric acid at poly (basic blue b) modified carbon nanotube paste electrode. *Asian Journal of Pharmaceutical and Clinical Research*, *8*(5), 48–53. http://innovareacademics.in/journals/index.php/ajpcr/article/download/7586/2972.

Manjunatha, J. G., Deraman, M., Basri, N. H., & Talib, I. A. (2014). Selective detection of dopamine in the presence of uric acid using polymerized phthalo blue film modified carbon paste electrode. *Advanced Materials Research*, *895*, 447–451. https://doi.org/10.4028/www.scientific.net/AMR.895.447.

Manjunatha, J. G., Deraman, M., Basri, N. H., & Talib, I. A. (2018). Fabrication of poly (solid red A) modified carbon nano tube paste electrode and its application for simultaneous determination of epinephrine, uric acid and ascorbic acid. *Arabian Journal of Chemistry*, *11*(2), 149–158. https://doi.org/10.1016/j.arabjc.2014.10.009.

Misra, S. K., Dybowska, A., Berhanu, D., Luoma, S. N., & Valsami-Jones, E. (2012). The complexity of nanoparticle dissolution and its importance in nanotoxicological studies. *Science of the Total Environment*, *438*, 225–232. https://doi.org/10.1016/j.scitotenv.2012.08.066.

Mittal, S., Sharma, P. K., Tiwari, R., Rayavarapu, R. G., Shankar, J., Chauhan, L. K. S., et al. (2017). Impaired lysosomal activity mediated autophagic flux disruption by graphite carbon nanofibers induce apoptosis in human lung epithelial cells through oxidative stress and energetic impairment. *Particle and Fibre Toxicology*, *14*(1). https://doi.org/10.1186/s12989-017-0194-4.

Noordadi, M., Mehrnejad, F., Sajedi, R. H., Jafari, M., & Ranjbar, B. (2018). The potential impact of carboxylic-functionalized multi-walled carbon nanotubes on trypsin: A comprehensive spectroscopic and molecular dynamics simulation study. *PLoS One*, *13*(6). https://doi.org/10.1371/journal.pone.0198519.

Poland, C. A., Duffin, R., Kinloch, I., Maynard, A., Wallace, W. A. H., Seaton, A., et al. (2008). Carbon nanotubes introduced into the abdominal cavity of mice show asbestos-like pathogenicity in a pilot study. *Nature Nanotechnology*, *3*(7), 423–428. https://doi.org/10.1038/nnano.2008.111.

Raril, C., & Manjunatha, J. G. (2020). A simple approach for the electrochemical determination of vanillin at ionic surfactant modified graphene paste electrode. *Microchemical Journal*, *154*. https://doi.org/10.1016/j.microc.2019.104575, 104575.

Reijnders, L. (2014). The safety of emerging inorganic and carbon nanomaterials. In *Vol. 9781119977926. Aerosol science: Technology and applications* (pp. 327–344). Wiley Blackwell. https://doi.org/10.1002/9781118682555.ch13.

Uygun, H. D. E., & Uygun, Z. O. (2020). Fullerene based sensor and biosensor technologies. In T. Shabatina, & V. Bochenkov (Eds.), *Smart nanosystems for biomedicine, optoelectronics and catalysis* Intech Open. https://doi.org/10.5772/intechopen.93316.

Vetelino, J. F., & Reghu, A. (2017). Introduction to sensors. In *Introduction to sensors* (pp. 1–180). CRC Press. https://doi.org/10.1201/9781315218274.

Wang, X., Zhu, Y., Chen, M., Yan, M., Zeng, G., & Huang, D. (2019). How do proteins 'response' to common carbon nanomaterials? *Advances in Colloid and Interface Science*, *270*, 101–107. https://doi.org/10.1016/j.cis.2019.06.002.

Wang, Y., et al. (2016). Investigation on the conformational structure of hemoglobin on graphene oxide. *Materials Chemistry and Physics*, *182*, 272–279. https://doi.org/10.1016/j.matchemphys.2016.07.032.

Wang, Z., Wu, S., Wang, J., Yu, A., & Wei, G. (2019). Carbon nanofiber-based functional nanomaterials for sensor applications. *Nanomaterials*, *9*(7), 1045. https://doi.org/10.3390/nano9071045.

Yu, X., Chattopadhyay, D., Galeska, I., Papadimitrakopoulos, F., & Rusling, J. F. (2003). Peroxidase activity of enzymes bound to the ends of single-wall carbon nanotube forest electrodes. *Electrochemistry Communications*, *5*(5), 408–411. https://doi.org/10.1016/S1388-2481(03)00076-6.

Yu, Y., Sun, H., Gilmore, K., Hou, T., Wang, S., & Li, Y. (2017). Aggregated single-walled carbon nanotubes absorb and deform dopamine-related proteins based on molecular dynamics simulations. *ACS Applied Materials & Interfaces*, *9*(38), 32452–32462. https://doi.org/10.1021/acsami.7b05478.

Zaporotskova, I. V., Boroznina, N. P., Parkhomenko, Y. N., & Kozhitov, L. V. (2016). Carbon nanotubes: Sensor properties. A review. *Modern Electronic Materials*, 95–105. https://doi.org/10.1016/j.moem.2017.02.002.

Chapter 19

Carbon nanomaterial-based sensor safety in different fields

S. Pratibha[a] and B. Chethan[b]

[a]*Department of Physics, BMS Institute of Technology and Management, Bangalore, Karnataka, India,* [b]*Department of Physics, IISC, Bangalore, Karnataka, India*

Introduction

At the beginning of the last decay, the word "nano" flourished, in every discipline of knowledge and found its widespread applications. Nanoscience, nanotechnology, nanoengineering, nanomaterials, nanorobotics, and nanophysics are the terminologies used worldwide in these years in research articles, books, reports, and papers this made nano so familiar to the common man. The word nano came from the ancient Latin word "nanus" which means "dwarf." In the S.I. system, nano is taken as a reduction factor of 10^9 times. Nanoparticles are measured in nanometres which means 1 nm corresponding to 10^{-9} m. Nano view of the world includes the size above molecular dimension and below macroscopic. The science that deals with small and very small things is called nanotechnology. When the material is in the nano form, its physical, chemical, and mechanical properties get changed and give a variety of astonishing and advantageous applications. Today, nanotechnology and nanoscience have emerged as the dominant branches of science because they provide an opportunity for the fabrication of nanomaterials that are helpful for medical applications (Kumar et al., 2018; Mondal, Balasubramaniam, Gupta, Lahcen, & Kwiatkowski, 2019). Carbon-based nanoparticles (CNPs) (molar mass: 12.01 g/mol; melting and boiling point above 3500°C) appear as black powder and are spherical. Due to the more advantageous properties of these CNPs, they find widespread applications in a variety of fields.

In this era, the increased population in the world wants to lead a sophisticated life with the use of technical devices, and then the demand for nano carbon-based sensors gained the attention of the world (D'Souza et al., 2021; Hareesha et al., 2021; Manjunatha et al., 2014; Manjunatha, Deraman, Basri, & Talib, 2018; Manjunatha, Pushpanjali, & Hareesha, 2021; Manjunatha et al., 2021). The world wants the fastest sensing device with more accuracy, reliability, and stability. This motivates the researchers to run behind the fabrication of various sensing devices using carbon materials with attractive physical and chemical properties, along with the use of the best preparation methodology to enhance sensitivity and quick timing behavior. Recently, these gas, electrochemical, and moisture sensors find widespread applications in many industrial and agricultural applications. This chapter explores the overview of the carbon-based nanomaterials as various sensors with their fundamental application in many disciplines of life. A recent picture of the classification and synthesis of carbon-based nanomaterials is discussed. The later sections of this chapter discuss the various types of carbon-based sensors and a detailed description of the hazards is dealt with carbon-based sensors. At the end of the chapter, the remedial measures taken to prevent hazards are discussed.

Nanostructure classification

Classification of nanoparticles based on the dimension

There are various approaches for the classification of nanoparticles. Among them, based on the dimension, NPs are classification as (Maynard, 2006).

One-dimension nanoparticles

Thin film and monolayer are one-dimension nanoparticles and these classes of materials are used in chemistry, biology, electronics, and engineering fields for many centuries. This type of monolayer nanoparticles is usually used in catalysis and solar cell applications. This type of nanoparticle finds widespread applications in storage devices, chemical sensors, and magneto-optic devices (Pal, Jana, Manna, Mohanta, & Manavalan, 2011).

Two-dimension nanoparticles

Carbon nanotubes (CNTs) are two-dimensional nanoparticles that are in the dimension of 1 nm in diameter and 100 nm in length. CNT nanoparticles are of two types: single-walled carbon nanotubes (SWCNTs) and multiwalled carbon nanotubes (MWCNTs) and these nanoparticles find application in various fields (Kohler & Fritzsche, 2008).

Three-dimension nanoparticles

Dendrimers, quantum dots (QDs), and fullerene (carbon-60) are three-dimensional nanoparticles. These nanoparticles find application in nanoelectronics, the medical field, anticancer drug industries, and tissue engineering.

Classification of nanoparticles based on morphology and chemical properties

Nanoparticles are classified into inorganic, organic, and carbon nanoparticles.

Inorganic NPs

Inorganic NPs consist of metal, metal oxide, ceramic, and semiconductor NPs.

Metal-based NPs: These types of NPs are prepared by metal precursor solutions. Cu, Ag, and Au are metal-based NPs. The shape- and size-controlled synthesis of these NPs find importance in recent cutting-edge materials (Dreaden, Alkilany, Huang, Murphy, & El-Sayed, 2012). NPs like Cu, Ag, and Au are having broad absorption bands in the visible zone and so offer unique optoelectronic properties (Salavati-Niasari, Davar, & Mir, 2008). Table 19.1 depicts the advantageous properties and applications of metal-based NPs (Bogutska et al., 2013; Harshiny et al., 2015; Hulteen et al., 1999; Ryu et al., 2016; Syed et al., 2016).

Ceramic-based NPs: Ceramics are inorganic solids that are usually nonmetallic and synthesized by combustion and cooling methods. These NPs find wide applications because they are dense, porous, amorphous, and polycrystalline solids (Sigmund et al., 2006). These are used as catalysis, photocatalysis, and imaging applications (Thomas, Harshita, Mishra, & Talegaonkar, 2015).

Metal oxides based: To modify the properties of the metal NPs, metal oxide NPs are synthesized. To enhance the reactivity and efficiency, metal NPs are mainly converted into metal oxide NPs. Iron NPs are less reactive but when they are oxidized as iron oxide, they are more reactive to the environment (Tai, Tai, Chang, & Liu, 2007). The most widely synthesized and used NPs are silicon dioxide, tantalum oxide, titanium oxide, cerium oxide, zinc oxide, magnesium oxide, aluminum oxide, niobium oxide, perovskites (Pratibha, Chethan, Ravikiran, Dhananjaya, & Jagadeesh Angadi, 2020; Pratibha, Dhananjaya, & Lokesh, 2020; Pratibha, Dhananjaya, Manjunatha, & Narayana, 2020; Pratibha, Dhananjaya, Manohara, & Yadav, 2019; Pratibha, Dhananjaya, Pasha, & Khasim, 2020; Pratibha, Dhananjaya, Shabaaz Begum, & Halappa, 2020). Because of their more reactivity, they are excellently used instead of metal NPs. Table 19.2 (Chen,

TABLE 19.1 Physical and chemical properties of metal-based NPs.

Sl. no.	Material	Properties	Ref.
1.	Gold NPs	Shining property, interactive to visible light	Syed, Prasad, and Satisha (2016)
2.	Silver NPs	Antibacterial activity scatters light, stable	Hulteen et al. (1999)
3.	Iron NPs	Aggressively reactive, sensitive to air and water, unstable	Harshiny, Iswarya, and Matheswaran (2015)
4.	Copper NPs	Extensive thermal and electrical conductivity, high ductile, extensively flammable solid	Ryu, Joo, and Kim (2016)
5.	Zinc NPs	Antifungal, anticorrosive, antibacterial	Bogutska, Sklyarov, and Prylutskyy (2013)

No permission required.

TABLE 19.2 Physical and chemical properties of metal oxide-based NPs.

Sl. no.	Material	Properties	Ref.
1.	Tantalum oxide NPs	High thermal stability and chemical resistance, wide bandgap, good mechanical strength, high melting point	Chen (2011)
2.	Yttrium oxide NPs	High temperature resistor and has good thermal stability up to 2200°C	Curtis (1957)
3.	Titanium oxide NPs	Stops bacterial growth, high surface, and magnetic property	Kaynar, Şabikoğlu, Kaynar, and Eral (2016)
4.	Silicon dioxide NPs	Less toxic, stable	Laad and Jatti (2018)
5.	Aluminum oxide NPs	Large surface to volume ratio. Sensitive to humidity, sunlight and heat	Munuswamy, Madhavan, and Mohan (2005)

No permission required.

2011; Curtis, 1957; Kaynar et al., 2016; Laad & Jatti, 2018; Munuswamy et al., 2005) provides the properties and applications of metal oxide-based NPs.

Semiconductor-based NPs: These are the NPs that found applications in many areas because their properties lie between the nonmetals and metals (Ali et al., 2016). They possess a tunable bandgap, hence finding applications in electronic devices and photocatalytic activities (Sun, Murray, Weller, Folks, & Moser, 2000).

Organic NPs

These are the NPs that contain polymer, dendrimers, micelles, liposomes, and ferritin NPs.

Polymer-based NPs: These are organic NPs whose mechanical and electrical properties can be tailored by many routes by many ways of making their combinations and composites by various techniques of preparation (Megha et al., 2018). These NPs are mostly nanospheres and nano capsular shaped (Mansha, Khan, Ullah, & Qurashi, 2017). Polymer nanoparticles having functionality find bundles of applications in many fields.

Dendrimer-based NPs: These are a new class of controlled structure materials with nano dimensions of the order of 10–80nm. Since these NPs are biodegradable, they are nontoxic having special characteristics and multiple functional groups on the surface, hence used in drug delivery systems (Tiwari, Behari, & Sen, 2008). The efficient drug-carrying capacity, its stability, and delivery functioning made it widely applicable in the biomedical field.

Carbon-based NPs

Carbon nanoparticles are useful in many aspects and in different fields. Many researchers have reported the use of different forms of carbon nanoparticles for sensing applications (Charithra & Manjunatha, 2021a, 2021b; Manjunatha, Pushpanjali, & Hareesha, 2021; Manjunatha, Raril, et al., 2021; Prinith, Manjunatha, & Hareesha, 2021; Pushpanjali et al., 2021). Carbon-based nanoparticles are synthesized naturally as well as they can be designed, hence on the basis of the synthesis procedure they are classified as natural CNPs and engineered CNPs.

Natural sources of CNPs

Wildfire charcoal: Charcoal is one of the major sources of carbon. Charcoal is formed when the water is removed from animal and vegetable substrates. In the absence of oxygen, charcoal is formed (Forbes, Raison, & Skjemstad, 2006). Based on the temperature of the fire, the physical and chemical properties of the charcoal depend. Approximately 15% of the total CNPs are derived from wildfire charcoal (Uhelski & Miesel, 2017).

Fuel combustion: The fossil fuel combustion in the existence of oxygen to generate CO_2 and H_2O with so many byproducts. Along with this, many more byproducts are formed namely CO and lead.

Fossil coal: Most countries depend on fossil coal for energy production. There is very little amount of CNPs that are formed due to the burning of fossil coal. CNPs are released into the atmosphere while transporting the fossil coal (Sigmund, Jiang, Hofmann, & Chen, 2018).

Biochar: Due to the pyrolysis of the biomass, biochar was formed. The heterogeneous nature of the biochar is due to the pyrolysis temperature and due to the duration of pyrolysis. When pyrolysis is slow, then the biochar formation is fast and it yields about 35% but it only yields 12% when pyrolysis is fast. Biochar is broadly used in the agriculture field. It is being mixed with other organic fertilizers, heavy metals, and polyaromatic hydrocarbons (PAHs). Biochar contaminates the soil and poses a maximum environmental risk (Alvarez-Puebla, Goulet, & Garrido, 2005).

Soot: Due to the burning of natural gas, soot is produced. Carbon soot and CNPs obtained from Castrol oil soot were altered and used in many electrochemical applications (Liu et al., 2018).

Engineered CNPs

These are the CNPs that are engineered/designed by humans using conventional methods like arc discharge, laser ablation, vapor deposition, etc., that can withstand comparatively high temperatures and energies. The important classes of CNPs are graphene, carbon nanotube (CNT), fullerene (C60), carbon nanofibers, and carbon black as shown in Fig. 19.1 and their physical and chemical properties are represented in Table 19.3 (De Volder et al., 2013; Fawole et al., 2016; Mondal & Jana, 2014; Tenne, 2002; Zhang, Aboagye, et al., 2014; Zhang, Chen, & Alvarez, 2014).

Graphene-based NPs: Graphene is an allotropic form of carbon and it is a hexagonal linkage of honeycomb-structured two-dimensional material. Graphene has evolved as an exotic material of the 21st century. It has gained the attention of the world due to its attractive thermal, optical, mechanical, physical, and charge transport properties. These sheets are around 1 nm in thickness (Ankamwar & Surti, 2012; Zhang et al., 2018).

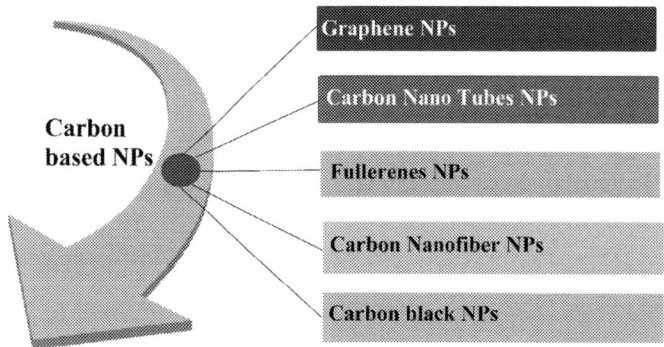

FIG. 19.1 A pictorial representation of the classification of the engineered CNPs. *(No permission required.)*

TABLE 19.3 Chemical and physical properties of carbon-based NPs.

Sl. no.	Material	Dimension	Chemical and physical properties	Ref.
1.	Graphene NPs	2D	Thermally stable, electrically conductive, light adsorption capacity, high strength	Mondal and Jana (2014)
2.	Carbon nanofibers	2D	High mechanical strength, more electrical and thermal shielding	Zhang, Aboagye, Kelkar, Lai, and Fong (2014) and Zhang, Chen, and Alvarez (2014)
3.	Carbon nanotubes	1D	Elastic, flexible, high tensile strength, extremely high thermally and electrically conductive	De Volder, Tawfick, Baughman, and Hart (2013)
4	Fullerene NPs	0D	Transmits light based on intensity, conductor, superconductor	Tenne (2002)
5.	Carbon black NPs	0D	Resistive to UV degradation, more electrical conductivity, high strength	Fawole, Cai, and Mackenzie (2016)

No permission required.

Carbon nanotubes (CNTs): These are a class of materials that are one of the allotropy forms of carbon which are made up of a hexagonal two-dimensional lattice of C-atom which joined together to form a hollow cylinder (Ibrahim, 2013). These nanotubes are of the diameter of 0.7 nm for single-walled carbon nanotubes (SWCNTs) and about 100 nm for multiwalled carbon nanotubes (MWCNTs). The length of these nanotubes varies from micrometers to millimeters (Aqel, El-Nour, Ammar, & Al-Warthan, 2012).

Fullerenes: These are the SP^2-hybridized carbon atoms that are held together in a spherical shape. The single layers are 8.3 nm in diameter whereas the multilayered fullerenes are about the diameter of 35 nm. Around 20–1500 carbon atoms are found in each spherical structure of fullerene (Ealias & Saravanakumar, 2017).

Carbon nanofibers: These are similar structures as CNTs but these are not in the form of regular cylindrical tubes; these are wounded into one another to form cone- or cup-shaped tubes. Graphene nano foils are used to produce these nanofibers (Khan, Saeed, & Khan, 2017).

Carbon black: These are amorphous materials made up of carbon atoms. These are spherical with a diameter of about 10–60 nm.

Synthesis, characterization, and applications of CNPs

Synthesis

CNP synthesis can be broadly classified into two types viz., bottom-up and top-down methods as shown in Fig. 19.2.

The bottom-up approach is also called a constructive approach or built-up approach. In this approach, NPs are formed from atoms—cluster—NPs, i.e., from very simpler substances. This approach includes the most important methods, such as spinning, sol-gel, pyrolysis, biosynthesis, and chemical vapor deposition (CVD). But the CNPs are synthesized mainly by sol-gel, CVD, and pyrolysis.

The top-down approach is called a destructive or reduction approach. In this approach, bulk to simple scale production of NPs is used. Important methods of this approach are nanolithography, mechanical milling, sputtering, thermal decomposition, and laser ablation.

Characterization of CNPs

CNPs are characterized based on different phases of their availability. Based on the size, surface area, composition, surface morphology, surface charge, crystallography, and concentration of the nanoparticles, they are characterized using different techniques as tabulated in Table 19.4 (Manjunatha, 2017, 2020a, 2020b, 2020c; Tigari & Manjunatha, 2020).

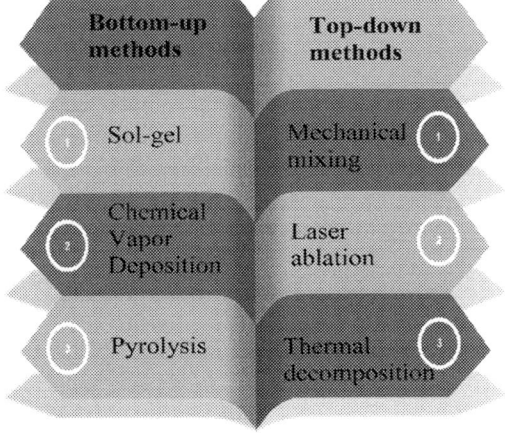

FIG. 19.2 Various methods of carbon-based nanoparticle synthesis. *(No permission required.)*

TABLE 19.4 Various characterization techniques used for CNPs.

Sl. no.	The phase of availability of CNPs	Characterization techniques		
		Solid CNPs	Liquid CNPs	Gas CNPs
1.	Size	Transmission electron microscope and laser diffraction	Photon correlation spectroscopy	Scanning mobility, particle size, and optical particle counter
2.	Surface morphology	Scanning electron microscope and atomic force microscopy	Deposition onto a surface for electron microscopy	Filtration for imaging using electron microscopy
3.	Composition	X-ray photoelectron spectroscopy and EDAX	Mass spectrometry, chromatography	Wet chemical techniques
4.	Surface area	Brunauer-Emmett-Teller studies	Nuclear magnetic resonance spectroscopy	Differential mobility analyzer
5.	Surface charge	Zeta potential	Zeta potential	Differential mobility analyzer
6.	Crystallography	Powder X-ray or neutron diffraction	–	–
7.	Concentration	–	–	Condensation particle counter

No permission required.

Applications of CNPs

Due to the fascinating electrical, mechanical, physical, and chemical properties, CNPs find widespread applications in many fields of life. The special properties like high surface-to-volume ratio and carbonyl functional groups make them a good candidate for device fabrication applications. The applications of CNPs are as revealed in Fig. 19.3.

Carbon nanomaterial-based sensors

The sensor is a small device that converts a physical impulse into an electrical signal. The variation in impulse can be detected by the computer interfaced with the sensor. It is easy to measure electrical impulses than physical ones. Once the data has been obtained, it is analyzed and information is gathered. Sensors can detect various physical parameters namely pressure-temperature, displacement, and chemical parameters, such as ions, molecules, concentration, and interaction between molecules. Nowadays, scientists are trying to develop a cost-effective, small, and very efficient sensor to cope with today's needs.

The sensors are broadly classified into two types as shown in Fig. 19.4.

Physical sensors

These are the sensors used for monitoring the environmental atmosphere (indoor and outdoor) and detecting the surrounding moments. They detect real-time data about the environment (Chethan, Prakash, Ravikiran, Vijayakumari, & Thomas, 2019; Chethan et al., 2020, 2019; Chethan, Raj Prakash, Vijayakumari, & Ravikiran, 2018; Manjunatha, Chethan, Ravikiran, & Machappa, 2018; Manjunatha, Deraman, et al., 2018; Pratibha, Chethan, Ravikiran, et al., 2020; Pratibha, Dhananjaya, & Lokesh, 2020; Pratibha, Dhananjaya, Manjunatha, & Narayana, 2020; Pratibha, Dhananjaya, Pasha, & Khasim, 2020; Pratibha, Dhananjaya, Shabaaz Begum, & Halappa, 2020; Sunilkumar, Manjunatha, Chethan, Ravikiran, & Machappa, 2019).

Examples: Temperature sensors, humidity sensors, and pressure sensors.

Chemical sensors

These are the devices that convert the chemical state into electrical signals.

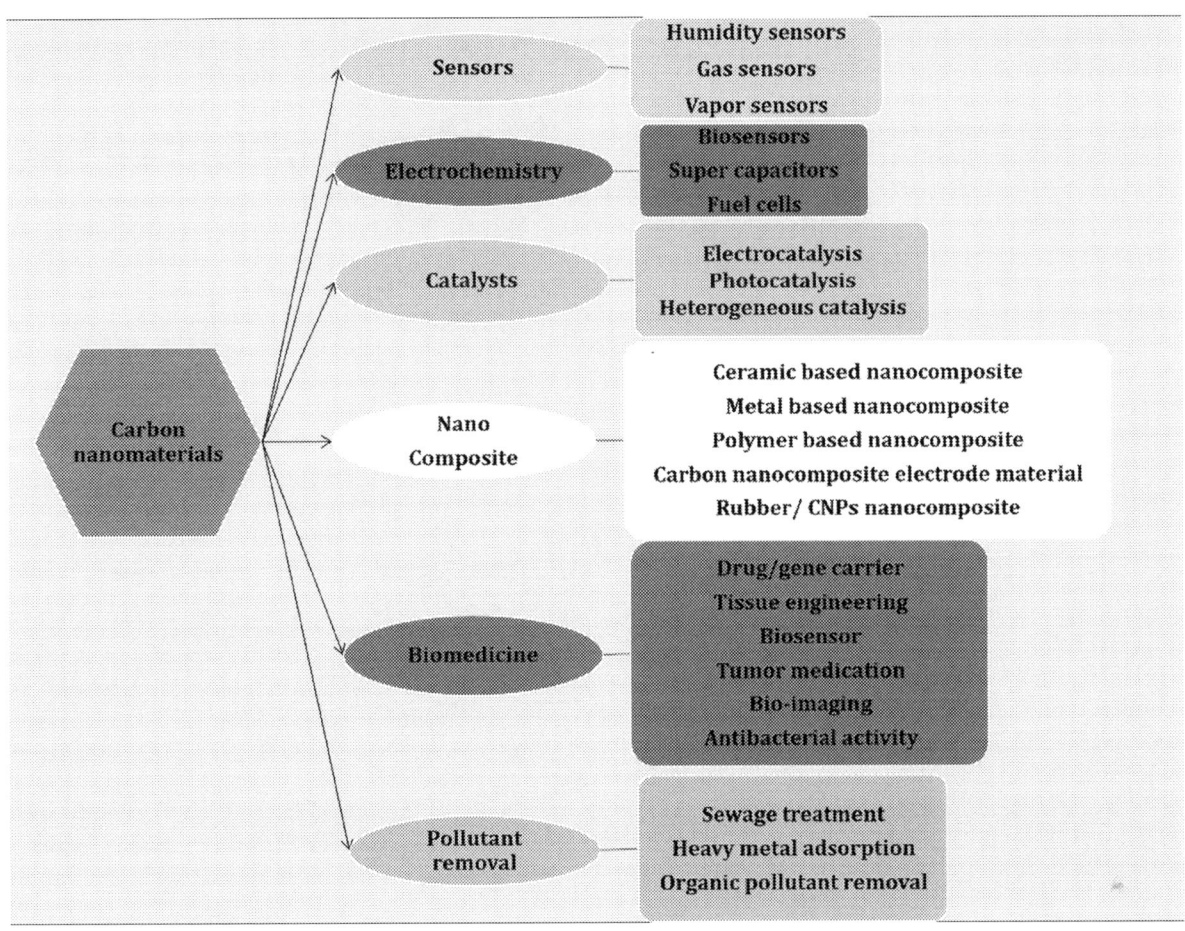

FIG. 19.3 Some widespread applications of the CNPs. *(No permission required.)*

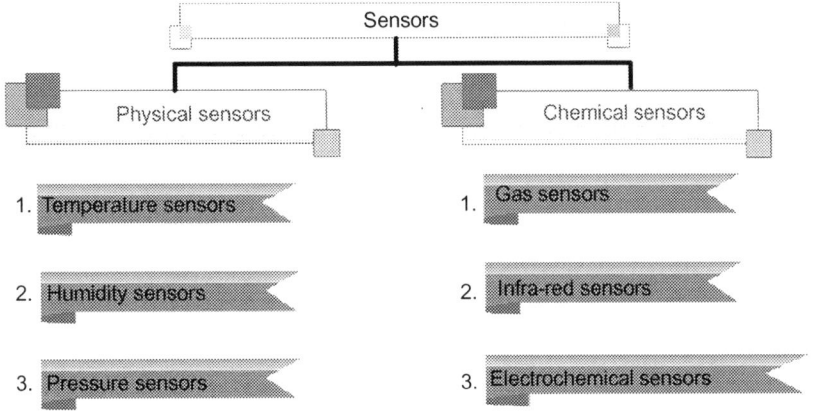

FIG. 19.4 Classification of the different sensors. *(No permission required.)*

Examples: Gas sensors, electrochemical sensors, infrared sensors

To judge the viability and reliability of any sensor, the important parameters that one must consider are:

1. Accuracy of the sensor
2. Calibration (materials and method of preparation)
3. Size
4. Cost of the sensor

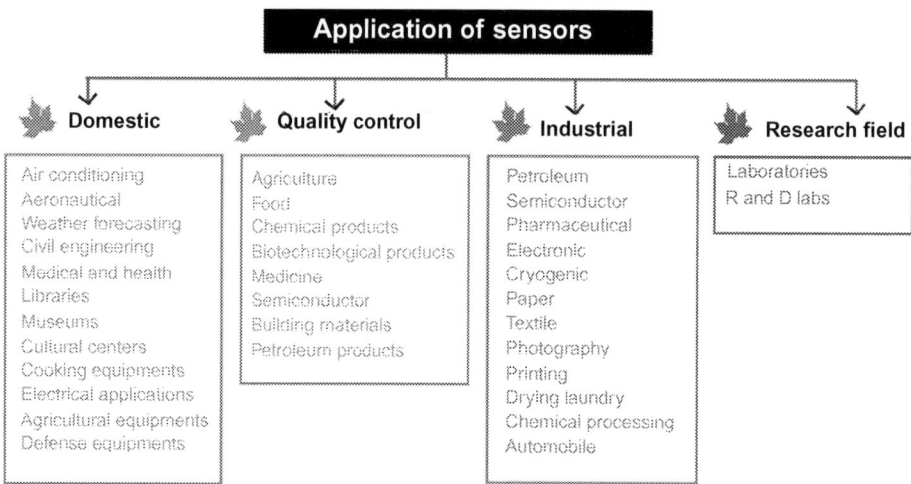

FIG. 19.5 Application of physical and chemical sensors in various fields. *(No permission required.)*

5. Repeatability
6. Resistance for adulterants
7. Validity

The physical and chemical sensors find widespread applications in many areas as shown in Fig. 19.5.

Industrial fabrication of physical and chemical sensors using other than carbon material has many disadvantages like:

Not room temperature operable.
When these sensors are exposed for a long duration of time, degradation of sensitivity occurs.
These sensors are the least sensitive.
The distance between the sensor and the signal circuit is limited.

To overcome these drawbacks, carbon-based nanoparticle sensors were fabricated which are more advantageous due to their:

Simple fabrication methods
Room temperature operability
Easy processability
Hygroscopicity

The advantageous physical and chemical properties of carbon-based nanoparticles that are used to fabricate different types of sensors are tabulated as follows in Table 19.5 (Alahi et al., 2018; Chethan, Prakash, et al., 2019; Chethan, Raj Prakash, et al., 2019; Li et al., 2017; Majumdar et al., 2014; Rahimi-Nasrabadi et al., 2017; Wei et al., 2018).

Based on the several advantages of CNPs and their easy fabrication techniques, there are multiple sensors that are designed in various fields. But the CNPs get disintegrated over time and get discharged to the environment and interact with soil, water, and air, and might be toxic for the ecosystem and may cause hazardous health issues in humans and other species.

CNPs in the environment: Discharge and destiny

Natural and engineered CNPs are discharged in the environment in varying proportions from various resources, as detailed in previous sections. Further, despite having a few chemical similarities like strong aromaticity, electron acceptors, and donor capacity, these particles differ in structural, surface, and porosity features. CNPs' fate and modification after treatment and disposal may be determined by their molecular similarities and structural differences because there is presently hardly any information on the environmental discharge and disposition of CNPs and shortage of research interest on discharge and destiny of CNPs. Also, only some limited techniques for detecting and quantifying CNPs have been planned and implemented, like laser fluorescence, microscopy, and quantification assessing concentrations of distinctive catalyst impurities, though none of these techniques are effective in distinguishing between engineered and natural CNPs

TABLE 19.5 Types of sensors and their sensitivity of few CNP-based sensors.

Sl. no.	Nanostructure material	Sensitivity	Type of sensor	Ref.
1.	Polyaniline/water-soluble graphene oxide	99.99% in the RH range of 11%–97%	Humidity sensor	Chethan, Prakash, et al. (2019) and Chethan, Raj Prakash, et al. (2019)
2.	CNT/SnO$_2$	208%–359% in the range 20–1000 ppm (100°C)	Gas sensor	Majumdar, Nag, and Devi (2014)
3.	GPN onto polyimide substrate	21.69 ppm (LOD)	Chemical element sensor	Alahi, Nag, Mukhopadhyay, and Burkitt (2018)
4.	MWCNT/PDMS substrate	−9.95 kPa−1	Biosensor	Li et al. (2017)
5.	CNT-C$_{60}$-IL	0.087 μM (LOD)	Drug sensor	Wei et al. (2018)
6.	GPN/Au-SPR	7.01 nm/(mg/mL)	Biomolecule sensor	Rahimi-Nasrabadi, Khoshroo, and Mazloum-Ardakani (2017)

No permission required.

(Chowdhury et al., 2015; Lowry, Gregory, Apte, & Lead, 2012; Petersen et al., 2016; Sigmund, Jiang, Hofmann, & Chen, 2018). As a result, technologies for identifying and quantifying both engineered and natural CNPs in the environment are urgently needed. CNPs operate differently in the environment than bulk carbon materials because of their differences in chemical and physical features. CNPs can change and transition into other forms after being released into the environment, changing their chemical and physical behaviors through photochemical responses, numerous ecological conditions, organic alterations, and genetic interfaces. These possible transformations are explained in detail.

Physical transformation

Physical transformation relates to changes in porosity and particle size, along with interfaces with new particles, as a consequence of pH changes, coagulation, abrasion, ionic strength alters, condensation, and intrinsic composition. Physical exfoliation, physical disintegration, and interaction with CNPs to diverse ecological elements like water and soil are all involved in the physical alteration and dissolution of CNPs. Physical disintegration transforms natural CNPs derived from biochar, wildfires, or fossil coal, while chemical flaking is intermediate by the soil's alkaline states (Braadbaart, Poole, & van Brussel, 2009; Li et al., 2016; Liu et al., 2018; Mitrano, Motellier, Clavaguera, & Nowack, 2015; Pulleman, Bouma, Van Essen, & Meijles, 2000; Zhao, Wang, White, & Xing, 2014). Furthermore, sandy soils disintegrate at a faster rate than other soil types. Furthermore, contact with water and dirt accelerates CNP dissolution and decomposition. Carbonization and aromaticity are inversely proportional to the degree of degeneration and generation of natural CNPs. Comparatively, CNPs made up of cellulose-rich plants (e.g., grass) degenerate to bigger particles than those formed from lignin-rich feedstock (e.g., wood) (Mitrano et al., 2015; Spokas et al., 2014). CNPs interact themselves and when discharged into the environment they interact with atmospheric humidity, sea salt, nitrates, sulfate, fly ash, dust particles, and forms dissolved, colloidal, and granular elements. This can cause condensation, coagulation, and packing of uncovered voids, which can help stabilize them and prevent further disintegration (Kerminen, Lehtinen, Anttila, & Kulmala, 2004; Mitrano et al., 2015; Tiwari & Marr, 2010; Wang et al., 2017). In the atmosphere, when the quantity and polydispersibility of CNPs are increased, they coagulate into spherical structures initially and subsequently form chain-like clusters (D'Anna, 2009; Kerminen et al., 2004; Kulmala et al., 2004; Ni et al., 2014). Condensation can also happen when CNPs directly come in contact with the environment as in the case of volcanic eruptions, neighboring highways, or industrial pollution. Nucleation of chemicals such as polyaromatic hydrocarbons, water, and acetylene from the gas phase results in natural CNPs originating from soot. Due to CNP accumulation and spreading, natural CNPs may develop beyond the nanoscale as a result of coagulation and condensation (Kerminen et al., 2004; Lighty, Veranth, & Sarofim, 2000). Interactions with other gas-phase elements, variations in temperature, humidity, and physical stress all alter the CNPs' surface charge, particle size, shape, and aggregation throughout their generation and movement into the atmosphere (Czimczik & Masiello, 2007; Ni et al., 2014; Ruiz-Dueñas & Martínez, 2009; Spokas et al., 2014; Zhang, Chen, & Alvarez, 2014; Zhang et al., 2008; Zhang, Yu, Colvin, & Monteiro-Riviere, 2008).

Chemical transformation

Despite the fact that there are a few data regarding the chemical transformation of CNPs, the existing sources imply that there is a change in surface features including a charge on the surface leading to change in surface functioning (Chowdhury et al., 2015; Mitrano et al., 2015). Redox reactions and photolysis are the two major mechanisms that mediate the chemical change of CNPs. Environmental elements, such as reductants and oxidants, as well as organic material, have been discovered to influence the chemical change of CNPs in the environment (Braadbaart et al., 2009; Li et al., 2016; Liu et al., 2018; Pulleman et al., 2000; Zhao et al., 2014). External factors such as soil chemical composition, pH, and type are reported to affect CNP chemical transition, although the process is unknown. For a better understanding of the mechanism, further studies in the future should focus on the chemical modification of CNPs in habitats like aquatic ecosystems, soil, sediment, and air. The discharge and destiny of CNPs produced from various sources and their mechanism of interaction along with the impacts on ecology should also be considered for the investigations.

Biological transformation

Biological transformations are changes in the physicochemical properties of CNPs caused by interactions with microbes, enzymes, and specific cells (Chowdhury et al., 2015). In microorganisms, both extracellular and intracellular biological transformations can occur. Several studies have been published on the biotransformation of lignin and other highly aromatic compounds by fungi, bacteria, enzymes, and actinomycetes derived from microorganisms (Han, Sun, Jin, & Xing, 2016) using extracellular and intracellular biotransformation enzymes, such as Mn peroxidase, laccase, lignin peroxidase, and others (Hilscher, Heister, Siewert, & Knicker, 2009; Saquing, Yu, & Chiu, 2016). Even though they have not been identified yet, these play a vital part in the biological alteration of CNPs. All of these enzymes are oxidoreductases, which are thought to be responsible for CNP reduction and oxidation. Also, they alter the surface characteristics of CNP. Until now, a study on this topic has been restricted to carbonaceous constituents (Allen et al., 2009; Flores-Cervantes, Maes, Schäffer, Hollender, & Kohler, 2014) who revealed a decrease in the aromaticity for few carbonaceous materials (Allen et al., 2009) and quinone groups in biochar's carbonized phase being able to operate both as electron acceptors and donors for microbiota (Flores-Cervantes et al., 2014). CNP enzymatic conversions have been little studied in the laboratory and are hence not symbolic of enzymatic transformations in natural surroundings. Horseradish peroxidase and Mn peroxidase were used in a small study to better understand the key enzymatic transformation pathways (Kotchey et al., 2011; Navarro et al., 2008; Spokas, 2010; Zhang, Khalizov, et al., 2008; Zhang, Yu, et al., 2008). However, because our understanding of the biological transition of natural CNPs and engineered CNPs is limited, a thorough examination is required. In spite of this, widespread recognition has been given for the importance of fungal hyphae in the physical degradation of biochar particles and massive fossil coal, as well as the role of fungal extracellular enzymes in chemical decomposition (Hildebrandt, Regvar, & Bothe, 2007). The physicochemical properties of CNPs, their interactions and aggregation, composition and microbial diversity, the consequence of various parameters, microbial activity, nutrient availability, and seasonal disparities on microbial activity and nutritional accretion are all factors that must be considered in CNP biological transformation research. The transfer and possible cellular uptake of CNPs by microbial cells, as well as the mechanisms that drive the associated gene expression of CNP-degrading enzymes, deserve more research. A greater perception of the mechanisms influencing biological transformation methods, besides the fate and potential hazards connected with CNPs, will be aided by such systematically conducted investigations.

Threats and endangerments in the context of the environment and econanotoxicology

Carbon nanoparticles (CNPs) have a wide range of applications in the medical, industrial, environmental, and agricultural sectors, with a growing customer market. If managed recklessly or irrationally, this could pose a higher danger to human health and the environment. Econanotoxicology is a nascent nanotechnology science that gives perception into NP manufacturing processes, lifespan, and possible consequences and risks when discharged into the environment. The ecotoxicological characteristics of CNPs when interacted with various environmental sources, as well as their later environmental consequences, will be discussed in this section.

Influences on terrestrial ecology

Animals, plants, fungi, and bacteria are all part of the terrestrial ecosystem at both the micro and macro levels, and they all play important roles in the environment. The toxicity of CNPs concerning several constituents of the terrestrial ecosystem is determined by the dose-response relationship (frequency concentration, at which CNPs are discharged into the surroundings). There has been almost minimal research on the effects on the soil by CNPs to date (Guzman, Taylor, & Banfield, 2000). Bacteria are common in ecosystems and play an important role in the nutrient cycle. However, ecotoxicity studies on plants, fungi, and bacteria have found no indication of detrimental impacts on relationship with endangering nitrogen-fixing, free-living bacteria and synergetic relationships including lichens, rhizobia, mycorrhiza, legumes, and other organisms. Decreased plant nutrient accessibility and oxidative emphasis would occur, while important ecological processes like organic material decomposition, biomass production, purification of groundwater, and soil characteristics would be hampered (Hardman, 2006). Because of their prolonged residence time in the cell, some CNPs, like quantum dots, may be collected by bacteria and transported through the food web. Because of its high partitioning into membranes, this can result in considerable CNP bioaccumulation. Plants are particularly important in econanotoxicology because they are exposed to CNPs both in the atmosphere and on the ground. Roots will interact with watery, soil material-associated CNPs, whereas airborne CNPs will adhere to leaves and other aerial plant parts (Imahori, Mori, & Matano, 2003; Köstner, 2001). Airborne CNPs have a larger interception potential in plant communities with higher leaf area indices, boosting their entry into trophic webs (Maurer-Jones, Gunsolus, Murphy, & Haynes, 2013). The harmful effects of CNPs on organisms are shown by changes in chemical composition and surface reactivity. The photo-induced electron transfer capacity of some CNPs, such as fullerene, suggests that CNPs penetrating the cell wall and membrane, then reaching the cytoplasm, will have an impact on the photosynthetic or respiratory processes. By measuring physiological parameters on seedlings of tomato, cabbage, lettuce, and red spinach, the effects of aqueous suspensions of graphene oxide (500–2000 mg/L) were observed. Plant growth was found to be hampered by nanoparticle exposure at the earliest stages of plant development, except for lettuce. Furthermore, in soil ecology research, earthworms are regarded as model detrivores. One of the few nanotoxicological investigations that used 14C-labeled CNPs found DNA damages in earthworms as a result of greater CNP bioaccumulation levels in earthworms (Colvin, 2003; Petersen et al., 2011; Price, Haberstroh, & Webster, 2004).

Influence on human health

CNPs are increasingly being used in sensors and pharmacological, industrial, electronic, and biological applications, also for medication and gene delivery because of their unique electrical, cytocompatibility, and mechanical features (Helland, Wick, Koehler, Schmid, & Som, 2007). CNPs have been recommended as a promising new dental/orthopedic implantation material. Nanomaterials may pose a significant concern to public health, given the growing scope of engineered CNPs and the common route of exposure. Chronic exposure to particulate matter of a nanosize causes various disorders linked with pulmonary toxicity, putting both producers and users of CNPs at risk. Through skin or inhalation or oral ingestion, workers of SWCNT-producing industries get exposed to CNPs (Nogueira, Nakabayashi, & Zucolotto, 2015). Hence, skin penetration is the main path for the CNP contact (Allen et al., 2009; Saquing et al., 2016). Individual forms of CNPs, on the other hand, can have varying impacts on organisms depending on their coating, functionalization, length, life cycle, exposure dose, aggregation qualities, and many other features (Chen et al., 2019; Chung, Son, Yoon, Kim, & Kim, 2011; Deryabin et al., 2015; Jin et al., 2014; Liu, Cui, Li, & Jin, 2014; Myojo & Ono-Ogasawara, 2018; Su, Ku, Kulkarni, & Cheng, 2016; Yang et al., 2019). CNTs may cause inflammation, oxidative stress, and cell malfunction and damage, and also if they are orally ingested, they might cause fibrosis, granulomas, and cell wall thickening in humans and other species. Different toxicity issues dealt with by humans and other species are listed in Table 19.6 (Chng & Pumera, 2013; da Rocha et al., 2019; Efremova et al., 2015; Giełdoń et al., 2017; Jang & Hwang, 2018; Kalman et al., 2019; Kayat et al., 2011; Keelan, 2011; Martínez-Paz et al., 2019; Nogueira et al., 2015; Panchapakesan et al., 2005; Qinglin et al., 2015; Shvedova et al., 2008; Song et al., 2018; Tabei et al., 2019; Wang et al., 2016).

Scope of future works

The use of various forms of CNPs in the environment is growing by the day. The unintended release of these organisms and their subsequent exposure to plants, bacteria, algae, fungi, animals, and humans are becoming a reality. The development of a more complicated system is needed to examine the impacts, transport, and fate of CNPs in soil and aquatic systems (oceans, rivers, lakes, etc.). Toxicity and nanoecotoxicity of CNPs to critical ecosystem organisms are difficult to understand and measure due to the lack of data on these key characteristics. Transformations and aging are important factors to consider when assessing the environmental implications of CNPs. Normalized tests, which are a type of ecotoxicological study that focuses on a single species, may help researchers better understand toxicity mechanisms. Critical criteria, such as bioaccumulation, biomagnification, biotransformation, and the impacts of abiotic variables, on the other hand, should be

TABLE 19.6 Toxic influence of CNPs.

Materials	Results	Ref.
Cytotoxicity		
C60	Unchanged C60 molecules can interchange electrons with nucleotides	Nogueira et al. (2015)
C60	Cellular and electrostatic structures initiating damage	Giełdoń, Witt, Gajewicz, and Puzyn (2017)
GR	The cell wall of algae cell can be cut by the sharp edge of the graphene	Song et al. (2018)
GR	Metabolic activity of fish cells is affected	Kalman, Merino, Fernández-Cruz, and Navas (2019)
GR	Oxidative stress reaction causes cytotoxicity	Chng and Pumera (2013)
GO	Noteworthy mechanical injury to cells	Efremova, Vasilchenko, Rakov, and Deryabin (2015)
MWCNTs	Affects DNA repair mechanism	Martínez-Paz et al. (2019)
MWCNTs	Cytotoxicity and phagocytic activity against HL-60 cells	Tabei et al. (2019)
Toxicity to humans and other species		
SWCNTs	Excessive influence on the growth of Artemia salina (Brine shrimp) in seawater	Jang and Hwang (2018)
SWCNTs	Initiates the hormone control influencing the brain of zebrafish	da Rocha et al. (2019)
CNTs	Increases the toxicity of Cu and Cd	Shvedova et al. (2008) and Wang et al. (2016)
CNTs	Gets into the body across the respiratory tract and accumulates in the lungs, initiating fibrosis, granuloma, or inflammation in the lungs	Kayat, Gajbhiye, Tekade, and Jain (2011)
GR	Terminating the physiological features and ultrastructure of the lungs, playing a toxic role	Panchapakesan et al. (2005)
GR	Neurotoxic for zebrafish embryo growth	Qinglin, Bao, Xinghua, and Guoqing (2015)
GR	Passes to fetal blood through placenta and influences fetal development	Keelan (2011)
C60	Creates swelling in the tissue of rat lung	Yang et al. (2019)
Nano C60	Nonderivative C60 suspension toxic to fish by oxidative stress	Su et al. (2016)
Nano C60	Creates chronic and acute toxicity to large cockroaches	Myojo and Ono-Ogasawara (2018)
GR	Chronic toxicity to large leeches	Liu et al. (2014)
CNPs	Affecting microbial communities by changing biomass and community structure	Deryabin et al. (2015)
MWCNMs	Constrains the development of some fungi and bacteria	Chen et al. (2019)
MWCNTs	High concentration of MWCNTs constrains the biomass and activity of microorganisms in the soil	Chung et al. (2011) and Jin et al. (2014)

Notations: C60, fullerene; GR, graphene; GO, graphene oxide; CNPs, carbon nanoparticles; MWCNTs, multiwalled carbon nanotubes; SWCNTs, single-walled carbon nanotubes.
No permission required.

considered into account. Because the possible influence of CNPs may be more than anticipated, it is past time for the traditional toxicological method to be replaced with a more appropriate ecoevaluation of CNP impact.

Some major recommendations for the safety measures to be focused on while using carbon nanomaterial-based sensors are as follows:

For safe nanotechnology applications, incorporating nanotoxicology with life cycle research should be regarded as a requirement.

CNP functionalization has an impact on toxicity along with its mobility; so, it is an important factor of consideration. To address critical difficulties, a multidisciplinary approach involving analytical researchers, toxicologists, ecotoxicologists, biophysicists, chemists, and biologists should be adopted.

Creating frameworks that allow in vitro results to be extrapolated to natural systems established on prior information. A more practical solution would be to standardize CNP toxicity testing methodologies.

Creating an integrative approach for evaluating possible dangers associated with CNPs.

Conclusions

In conclusion, this chapter critically examines the structure, properties, behavior, applications, fate, and toxicity of several forms of CNPs. While the scientific community has made various efforts to focus on the environmental consequences of CNPs, there are still loos in the presently available literature. It is critical to (i) handle the difficulties of CNPs and their alterations; (ii) assimilate quantitative and qualitative data; (iii) make modeling tools easier to use to fill in data gaps and give both qualitative and quantitative data; and (iv) reduce reliance on expert judgment. To adopt nanotechnology sustainably that has no negative impact on the environment, it is critical that experts from a wide range of disciplines collaborate and contribute to a better understanding of nano-bio interactions, which will help drive this emergent field in the correct path.

References

Alahi, M. E. E., Nag, A., Mukhopadhyay, S. C., & Burkitt, L. (2018). A temperature-compensated graphene sensor for nitrate monitoring in real-time application. *Sensors and Actuators, A: Physical, 269*, 79–90. https://doi.org/10.1016/j.sna.2017.11.022.

Ali, A., Zafar, H., Zia, M., ul Haq, I., Phull, A. R., Ali, J. S., et al. (2016). Synthesis, characterization, applications, and challenges of iron oxide nanoparticles. *Nanotechnology, Science and Applications, 9*, 49–67. https://doi.org/10.2147/NSA.S99986.

Allen, B. L., Kotchey, G. P., Chen, Y., Yanamala, N. V. K., Klein-Seetharaman, J., Kagan, V. E., et al. (2009). Mechanistic investigations of horseradish peroxidase-catalyzed degradation of single-walled carbon nanotubes. *Journal of the American Chemical Society, 131*(47), 17194–17205. https://doi.org/10.1021/ja9083623.

Alvarez-Puebla, R. A., Goulet, P. J. G., & Garrido, J. J. (2005). Characterization of the porous structure of different humic fractions. *Colloids and Surfaces A: Physicochemical and Engineering Aspects, 256*(2–3), 129–135. https://doi.org/10.1016/j.colsurfa.2004.12.062.

Ankamwar, B., & Surti, F. (2012). Water soluble graphene synthesis. *Chemical Science Transactions*, 500–507. https://doi.org/10.7598/cst2012.155.

Aqel, A., El-Nour, K. M. M. A., Ammar, R. A. A., & Al-Warthan, A. (2012). Carbon nanotubes, science and technology part (I) structure, synthesis and characterisation. *Arabian Journal of Chemistry, 5*(1), 1–23. https://doi.org/10.1016/j.arabjc.2010.08.022.

Bogutska, K. I., Sklyarov, Y. P., & Prylutskyy, Y. I. (2013). Zinc and zinc nanoparticles: Biological role and application in biomedicine. *Ukrainica Bioorganica Acta, 1*, 9–16.

Braadbaart, F., Poole, I., & van Brussel, A. A. (2009). Preservation potential of charcoal in alkaline environments: An experimental approach and implications for the archaeological record. *Journal of Archaeological Science, 36*(8), 1672–1679. https://doi.org/10.1016/j.jas.2009.03.006.

Charithra, M. M., & Manjunatha, J. G. (2021a). Electroanalytical determination of acetaminophen using polymerized carbon nanocomposite based sensor. *Chemical Data Collections, 33*. https://doi.org/10.1016/j.cdc.2021.100718, 100718.

Charithra, M. M., & Manjunatha, J. G. (2021b). Fabrication of poly (evans blue) modified graphite paste electrode as an electrochemical sensor for sensitive and instant riboflavin detection. *Moroccan Journal of Chemistry, 9*(1), 007–017. https://doi.org/10.48317/IMIST.PRSM/morjchem-v9i1.18239.

Chen, H. (2011). Electrical and material characterization of tantalum pentoxide (Ta_2O_5) charge trapping layer memory. *Applied Surface Science, 257*(17), 7481–7485. https://doi.org/10.1016/j.apsusc.2011.03.055.

Chen, M., Sun, Y., Liang, J., Zeng, G., Li, Z., Tang, L., et al. (2019). Understanding the influence of carbon nanomaterials on microbial communities. *Environment International, 126*, 690–698. https://doi.org/10.1016/j.envint.2019.02.005.

Chethan, B., Prakash, H. G. R., Ravikiran, Y. T., Vijayakumari, S. C., & Thomas, S. (2019). Polypyyrole based core-shell structured composite based humidity sensor operable at room temperature. *Sensors and Actuators, B: Chemical, 296*. https://doi.org/10.1016/j.snb.2019.126639.

Chethan, B., Raj Prakash, H. G., Ravikiran, Y. T., Vijaya Kumari, S. C., Manjunatha, S., & Thomas, S. (2020). Humidity sensing performance of hybrid nanorods of polyaniline-yttrium oxide composite prepared by mechanical mixing method. *Talanta, 215*. https://doi.org/10.1016/j.talanta.2020.120906, 120906.

Chethan, B., Raj Prakash, H. G., Ravikiran, Y. T., Vijayakumari, S. C., Ramana, C. H. V. V., Thomas, S., et al. (2019). Enhancing humidity sensing performance of polyaniline/water soluble graphene oxide composite. *Talanta, 196*, 337–344. https://doi.org/10.1016/j.talanta.2018.12.072.

Chethan, B., Raj Prakash, H. G., Vijayakumari, S. C., & Ravikiran, Y. T. (2018). Structural and electrical properties of nickel substituted cadmium ferrite. In *AIP Conference Proceedings (Vol. 1953)* American Institute of Physics Inc. https://doi.org/10.1063/1.5032875.

Chng, E. L. K., & Pumera, M. (2013). The toxicity of graphene oxides: Dependence on the oxidative methods used. *Chemistry—A European Journal, 19*(25), 8227–8235. https://doi.org/10.1002/chem.201300824.

Chowdhury, I., Hou, W. C., Goodwin, D., Henderson, M., Zepp, R. G., & Bouchard, D. (2015). Sunlight affects aggregation and deposition of graphene oxide in the aquatic environment. *Water Research, 78*, 37–46. https://doi.org/10.1016/j.watres.2015.04.001.

Chung, H., Son, Y., Yoon, T. K., Kim, S., & Kim, W. (2011). The effect of multi-walled carbon nanotubes on soil microbial activity. *Ecotoxicology and Environmental Safety, 74*(4), 569–575. https://doi.org/10.1016/j.ecoenv.2011.01.004.

Colvin, V. L. (2003). The potential environmental impact of engineered nanomaterials. *Nature Biotechnology, 21*(10), 1166–1170. https://doi.org/10.1038/nbt875.

Curtis, C. E. (1957). Properties of yttrium oxide ceramics. *Journal of the American Ceramic Society, 40*(8), 274–278. https://doi.org/10.1111/j.1151-2916.1957.tb12619.x.

Czimczik, C. I., & Masiello, C. A. (2007). Controls on black carbon storage in soils. *Global Biogeochemical Cycles, 21*(3). https://doi.org/10.1029/2006GB002798.

D'Anna, A. (2009). Combustion-formed nanoparticles. *Proceedings of the Combustion Institute, 32*(1), 593–613. https://doi.org/10.1016/j.proci.2008.09.005.

D'Souza, E. S., Manjunatha, J. G., Raril, C., Tigari, G., Arpitha, H. J., & Shenoy, S. (2021). Electro-polymerized titan yellow modified carbon paste electrode for the analysis of curcumin. *Surfaces, 4*(3), 191–204. https://doi.org/10.3390/surfaces4030017.

da Rocha, A. M., Kist, L. W., Almeida, E. A., Silva, D. G. H., Bonan, C. D., Altenhofen, S., et al. (2019). Neurotoxicity in zebrafish exposed to carbon nanotubes: Effects on neurotransmitters levels and antioxidant system. *Comparative Biochemistry and Physiology Part - C: Toxicology and Pharmacology, 218*, 30–35. https://doi.org/10.1016/j.cbpc.2018.12.008.

De Volder, M. F. L., Tawfick, S. H., Baughman, R. H., & Hart, A. J. (2013). Carbon nanotubes: Present and future commercial applications. *Science, Vol. 339*(6119), 535–539. https://doi.org/10.1126/science.1222453.

Deryabin, D. G., Efremova, L. V., Vasilchenko, A. S., Saidakova, E. V., Sizova, E. A., Troshin, P. A., et al. (2015). Erratum to: A zeta potential value determines the aggregate's size of penta-substituted [60] fullerene derivatives in aqueous suspension whereas positive charge is required for toxicity against bacterial cells [J Nanobiotechnol (2015) 13: 50, doi:10.1186/s12951-015-0112-6]. *Journal of Nanobiotechnology, 13*(1). https://doi.org/10.1186/s12951-015-0130-4.

Dreaden, E. C., Alkilany, A. M., Huang, X., Murphy, C. J., & El-Sayed, M. A. (2012). The golden age: Gold nanoparticles for biomedicine. *Chemical Society Reviews, 41*(7), 2740–2779. https://doi.org/10.1039/c1cs15237h.

Ealias, A. M., & Saravanakumar, M. P. (2017). A review on the classification, characterisation, synthesis of nanoparticles and their application. In *IOP conference series: Materials science and engineering (Vol. 263, Issue 3)* Institute of Physics Publishing. https://doi.org/10.1088/1757-899X/263/3/032019.

Efremova, L. V., Vasilchenko, A. S., Rakov, E. G., & Deryabin, D. G. (2015). Toxicity of graphene shells, graphene oxide, and graphene oxide paper evaluated with escherichia coli biotests. *BioMed Research International, 2015*. https://doi.org/10.1155/2015/869361.

Fawole, O. G., Cai, X. M., & Mackenzie, A. R. (2016). Gas flaring and resultant air pollution: A review focusing on black carbon. *Environmental Pollution, 216*, 182–197. https://doi.org/10.1016/j.envpol.2016.05.075.

Flores-Cervantes, D. X., Maes, H. M., Schäffer, A., Hollender, J., & Kohler, H. P. E. (2014). Slow biotransformation of carbon nanotubes by horseradish peroxidase. *Environmental Science and Technology, 48*(9), 4826–4834. https://doi.org/10.1021/es4053279.

Forbes, M. S., Raison, R. J., & Skjemstad, J. O. (2006). Formation, transformation and transport of black carbon (charcoal) in terrestrial and aquatic ecosystems. *Science of the Total Environment, 370*(1), 190–206. https://doi.org/10.1016/j.scitotenv.2006.06.007.

Giełdoń, A., Witt, M. M., Gajewicz, A., & Puzyn, T. (2017). Rapid insight into C60 influence on biological functions of proteins. *Structural Chemistry, 28*(6), 1775–1788. https://doi.org/10.1007/s11224-017-0957-4.

Guzman, Taylor, M. R., & Banfield, J. F. (2000). Environmental risks of nanotechnology: National nanotechnology intiative funding. *Environmental Science & Technology, 40*, 1401–1407.

Han, L., Sun, K., Jin, J., & Xing, B. (2016). Some concepts of soil organic carbon characteristics and mineral interaction from a review of literature. *Soil Biology and Biochemistry, 94*, 107–121. https://doi.org/10.1016/j.soilbio.2015.11.023.

Hardman, R. (2006). A toxicologic review of quantum dots: Toxicity depends on physicochemical and environmental factors. *Environmental Health Perspectives, 114*(2), 165–172. https://doi.org/10.1289/ehp.8284.

Hareesha, N., Manjunatha, J. G., Amrutha, B. M., Sreeharsha, N., Basheeruddin Asdaq, S. M., & Anwer, M. K. (2021). A fast and selective electrochemical detection of vanillin in food samples on the surface of poly(glutamic acid) functionalized multiwalled carbon nanotubes and graphite composite paste sensor. *Colloids and Surfaces A: Physicochemical and Engineering Aspects, 626*. https://doi.org/10.1016/j.colsurfa.2021.127042, 127042.

Harshiny, M., Iswarya, C. N., & Matheswaran, M. (2015). Biogenic synthesis of iron nanoparticles using Amaranthus dubius leaf extract as a reducing agent. *Powder Technology, 286*, 744–749. https://doi.org/10.1016/j.powtec.2015.09.021.

Helland, A., Wick, P., Koehler, A., Schmid, K., & Som, C. (2007). Reviewing the environmental and human health knowledge base of carbon nanotubes. *Environmental Health Perspectives, 115*(8), 1125–1131. https://doi.org/10.1289/ehp.9652.

Hildebrandt, U., Regvar, M., & Bothe, H. (2007). Arbuscular mycorrhiza and heavy metal tolerance. *Phytochemistry, 68*(1), 139–146. https://doi.org/10.1016/j.phytochem.2006.09.023.

Hilscher, A., Heister, K., Siewert, C., & Knicker, H. (2009). Mineralisation and structural changes during the initial phase of microbial degradation of pyrogenic plant residues in soil. *Organic Geochemistry, 40*(3), 332–342. https://doi.org/10.1016/j.orggeochem.2008.12.004.

Hulteen, J. C., Treichel, D. A., Smith, M. T., Duval, M. L., Jensen, T. R., & Van Duyne, R. P. (1999). Nanosphere lithography: Size-tunable silver nanoparticle and surface cluster arrays. *Journal of Physical Chemistry B, 103*(19), 3854–3863. https://doi.org/10.1021/jp9904771.

Ibrahim, K. S. (2013). Carbon nanotubes-properties and applications: A review. *Carbon Letters*, 131–144. https://doi.org/10.5714/cl.2013.14.3.131.

Imahori, H., Mori, Y., & Matano, Y. (2003). Nanostructured artificial photosynthesis. *Journal of Photochemistry and Photobiology C: Photochemistry Reviews, 4*(1), 51–83. https://doi.org/10.1016/S1389-5567(03)00004-2.

Jang, M. H., & Hwang, Y. S. (2018). Effects of functionalized multi-walled carbon nanotubes on toxicity and bioaccumulation of lead in Daphnia magna. *PLoS One, 13*(3). https://doi.org/10.1371/journal.pone.0194935.

Jin, L., Son, Y., DeForest, J. L., Kang, Y. J., Kim, W., & Chung, H. (2014). Single-walled carbon nanotubes alter soil microbial community composition. *Science of the Total Environment, 466–467*, 533–538. https://doi.org/10.1016/j.scitotenv.2013.07.035.

Kalman, J., Merino, C., Fernández-Cruz, M. L., & Navas, J. M. (2019). Usefulness of fish cell lines for the initial characterization of toxicity and cellular fate of graphene-related materials (carbon nanofibers and graphene oxide). *Chemosphere, 218*, 347–358. https://doi.org/10.1016/j.chemosphere.2018.11.130.

Kayat, J., Gajbhiye, V., Tekade, R. K., & Jain, N. K. (2011). Pulmonary toxicity of carbon nanotubes: A systematic report. *Nanomedicine: Nanotechnology, Biology, and Medicine, 7*(1), 40–49. https://doi.org/10.1016/j.nano.2010.06.008.

Kaynar, Ü. H., Şabikoğlu, I., Kaynar, S.Ç., & Eral, M. (2016). Modeling of thorium (IV) ions adsorption onto a novel adsorbent material silicon dioxide nano-balls using response surface methodology. *Applied Radiation and Isotopes, 115*, 280–288. https://doi.org/10.1016/j.apradiso.2016.06.033.

Keelan, J. A. (2011). Nanotoxicology: Nanoparticles versus the placenta. *Nature Nanotechnology, 6*(5), 263–264. https://doi.org/10.1038/nnano.2011.65.

Kerminen, V. M., Lehtinen, K. E. J., Anttila, T., & Kulmala, M. (2004). Dynamics of atmospheric nucleation mode particles: A timescale analysis. *Tellus Series B: Chemical and Physical Meteorology, 56*(2), 135–146. https://doi.org/10.1111/j.1600-0889.2004.00095.x.

Khan, I., Saeed, K., & Khan, I. (2017). Nanoparticles: Properties, applications and toxicities. *Arabian Journal of Chemistry, 12*(7), 908–931.

Kohler, M., & Fritzsche, W. (2008). *Nanotechnology: An introduction to nanostructuring techniques*. John Wiley & Sons.

Köstner, B. (2001). Evaporation and transpiration from forests in Central Europe—Relevance of patch-level studies for spatial scaling. *Meteorology and Atmospheric Physics, 76*(1–2), 69–82. https://doi.org/10.1007/s007030170040.

Kotchey, G. P., Allen, B. L., Vedala, H., Yanamala, N., Kapralov, A. A., Tyurina, Y. Y., et al. (2011). The enzymatic oxidation of graphene oxide. *ACS Nano, 5*(3), 2098–2108. https://doi.org/10.1021/nn103265h.

Kulmala, M., Vehkamäki, H., Petäjä, T., Dal Maso, M., Lauri, A., Kerminen, V. M., et al. (2004). Formation and growth rates of ultrafine atmospheric particles: A review of observations. *Journal of Aerosol Science, 35*(2), 143–176. https://doi.org/10.1016/j.jaerosci.2003.10.003.

Kumar, A., Billa, S., Chaudhary, S., Kiran Kumar, A. B. V., Ramana, C. V. V., & Kim, D. (2018). Ternary nanocomposite for solar light photocatalyic degradation of methyl orange. *Inorganic Chemistry Communications, 97*, 191–195. https://doi.org/10.1016/j.inoche.2018.09.038.

Laad, M., & Jatti, V. K. S. (2018). Titanium oxide nanoparticles as additives in engine oil. *Journal of King Saud University - Engineering Sciences*, 116–122. https://doi.org/10.1016/j.jksues.2016.01.008.

Li, X., Huang, W., Yao, G., Gao, M., Wei, X., Liu, Z., et al. (2017). Highly sensitive flexible tactile sensors based on microstructured multiwall carbon nanotube arrays. *Scripta Materialia, 129*, 61–64. https://doi.org/10.1016/j.scriptamat.2016.10.037.

Li, W., Sun, J., Xu, L., Shi, Z., Riemer, N., Sun, Y., et al. (2016). A conceptual framework for mixing structures in individual aerosol particles. *Journal of Geophysical Research-Atmospheres, 121*(22), 784–798. https://doi.org/10.1002/2016JD025252.

Lighty, J. A. S., Veranth, J. M., & Sarofim, A. F. (2000). Combustion aerosols: Factors governing their size and composition and implications to human health. *Journal of the Air and Waste Management Association, 50*(9), 1565–1618. https://doi.org/10.1080/10473289.2000.10464197.

Liu, Q., Cui, Q., Li, X. J., & Jin, L. (2014). The applications of buckminsterfullerene C60 and derivatives in orthopaedic research. *Connective Tissue Research, 55*(2), 71–79. https://doi.org/10.3109/03008207.2013.877894.

Liu, G., Zheng, H., Jiang, Z., Zhao, J., Wang, Z., Pan, B., et al. (2018). Formation and physicochemical characteristics of nano biochar: Insight into chemical and colloidal stability. *Environmental Science and Technology, 52*(18), 10369–10379. https://doi.org/10.1021/acs.est.8b01481.

Lowry, G. V., Gregory, K. B., Apte, S. C., & Lead, J. R. (2012). Transformations of nanomaterials in the environment. *Environmental Science and Technology, 46*(13), 6893–6899. https://doi.org/10.1021/es300839e.

Majumdar, S., Nag, P., & Devi, P. S. (2014). Enhanced performance of CNT/SnO$_2$ thick film gas sensors towards hydrogen. *Materials Chemistry and Physics, 147*(1–2), 79–85. https://doi.org/10.1016/j.matchemphys.2014.04.009.

Manjunatha, J. G. (2017). Surfactant modified carbon nanotube paste electrode for the sensitive determination of mitoxantrone anticancer drug. *Journal of Electrochemical Science and Engineering, 7*(1), 39–49. https://doi.org/10.5599/jese.368.

Manjunatha, J. G. (2020a). Fabrication of efficient and selective modified graphene paste sensor for the determination of catechol and hydroquinone. *Surfaces*, 473–483. https://doi.org/10.3390/surfaces3030034.

Manjunatha, J. G. (2020b). Poly (adenine) modified graphene-based voltammetric sensor for the electrochemical determination of catechol, hydroquinone and resorcinol. *The Open Chemical Engineering Journal, 14*, 52–62. https://doi.org/10.2174/1874123102014010052.

Manjunatha, J. G. (2020c). A surfactant enhanced graphene paste electrode as an effective electrochemical sensor for the sensitive and simultaneous determination of catechol and resorcinol. *Chemical Data Collections, 25*. https://doi.org/10.1016/j.cdc.2019.100331.

Manjunatha, S., Chethan, B., Ravikiran, Y. T., & Machappa, T. (2018). Room temperature humidity sensor based on polyaniline-tungsten disulfide composite. In *AIP Conference Proceedings (Vol. 1953)* American Institute of Physics Inc. https://doi.org/10.1063/1.5032431.

Manjunatha, J. G., Deraman, M., Basri, N. H., Nor, N. S. M., Talib, I. A., & Ataollahi, N. (2014). Sodium dodecyl sulfate modified carbon nanotubes paste electrode as a novel sensor for the simultaneous determination of dopamine, ascorbic acid, and uric acid. *Comptes Rendus Chimie, 17*(5), 465–476. https://doi.org/10.1016/j.crci.2013.09.016.

Manjunatha, J. G., Deraman, M., Basri, N. H., & Talib, I. A. (2018). Fabrication of poly (solid red A) modified carbon nano tube paste electrode and its application for simultaneous determination of epinephrine, uric acid and ascorbic acid. *Arabian Journal of Chemistry, 11*(2), 149–158. https://doi.org/10.1016/j.arabjc.2014.10.009.

Manjunatha, J. G., Pushpanjali, P. A., & Hareesha, N. (2021). An overview of recent developments of carbon-based sensors for the analysis of drug molecules. *Journal of Electrochemical Science and Engineering*. https://doi.org/10.5599/jese.999.

Manjunatha, J. G., Raril, C., Prinith, N. S., Pushpanjali, P. A., Charithra, M. M., Tigari, G., et al. (2021). *Fabrication, characterization and application of poly(acriflavine) modified carbon nanotube paste electrode for the electrochemical determination of catechol* (pp. 105–117). Elsevier BV. https://doi.org/10.1016/b978-0-12-820783-3.00022-1.

Mansha, M., Khan, I., Ullah, N., & Qurashi, A. (2017). Synthesis, characterization and visible-light-driven photoelectrochemical hydrogen evolution reaction of carbazole-containing conjugated polymers. *International Journal of Hydrogen Energy, 42*(16), 10952–10961. https://doi.org/10.1016/j.ijhydene.2017.02.053.

Martínez-Paz, P., Negri, V., Esteban-Arranz, A., Martínez-Guitarte, J. L., Ballesteros, P., & Morales, M. (2019). Effects at molecular level of multi-walled carbon nanotubes (MWCNT) in Chironomus riparius (DIPTERA) aquatic larvae. *Aquatic Toxicology, 209*, 42–48. https://doi.org/10.1016/j.aquatox.2019.01.017.

Maurer-Jones, M. A., Gunsolus, I. L., Murphy, C. J., & Haynes, C. L. (2013). Toxicity of engineered nanoparticles in the environment. *Analytical Chemistry, 85*(6), 3036–3049. https://doi.org/10.1021/ac303636s.

Maynard, A. D. (2006). Nanotechnology: Assessing the risks. *Nano Today, 1*(2), 22–33.

Megha, R., Ravikiran, Y. T., Chethan, B., Raj Prakash, H. G., Vijaya Kumari, S. C., & Thomas, S. (2018). Effect of mechanical mixing method of preparation of polyaniline-transition metal oxide composites on DC conductivity and humidity sensing response. *Journal of Materials Science: Materials in Electronics, 29*(9), 7253–7261. https://doi.org/10.1007/s10854-018-8714-z.

Mitrano, D. M., Motellier, S., Clavaguera, S., & Nowack, B. (2015). Review of nanomaterial aging and transformations through the life cycle of nano-enhanced products. *Environment International, 77*, 132–147. https://doi.org/10.1016/j.envint.2015.01.013.

Mondal, K., Balasubramaniam, B., Gupta, A., Lahcen, A. A., & Kwiatkowski, M. (2019). Carbon nanostructures for energy and sensing applications. *Journal of Nanotechnology, 2019*. https://doi.org/10.1155/2019/1454327.

Mondal, A., & Jana, N. R. (2014). Graphene-nanoparticle composites and their applications in energy, environmental and biomedical science. *Reviews in Nanoscience and Nanotechnology*, 177–192. https://doi.org/10.1166/rnn.2014.1051.

Munuswamy, D. B., Madhavan, V. R., & Mohan, M. (2005). Synthesis and surface area determination of alumina nanoparticles by chemical combustion method. *Chemistry of Materials, 172*, 395–401.

Myojo, T., & Ono-Ogasawara, M. (2018). Review; risk assessment of aerosolized SWCNTs, MWCNTs, fullerenes and carbon black. *Kona Powder and Particle Journal, 2018*(35), 80–88. https://doi.org/10.14356/kona.2018013.

Navarro, E., Baun, A., Behra, R., Hartmann, N. B., Filser, J., Miao, A. J., et al. (2008). Environmental behavior and ecotoxicity of engineered nanoparticles to algae, plants, and fungi. *Ecotoxicology, 17*(5), 372–386. https://doi.org/10.1007/s10646-008-0214-0.

Ni, M., Huang, J., Lu, S., Li, X., Yan, J., & Cen, K. (2014). A review on black carbon emissions, worldwide and in China. *Chemosphere, 107*, 83–93. https://doi.org/10.1016/j.chemosphere.2014.02.052.

Nogueira, P. F. M., Nakabayashi, D., & Zucolotto, V. (2015). The effects of graphene oxide on green algae Raphidocelis subcapitata. *Aquatic Toxicology, 166*, 29–35. https://doi.org/10.1016/j.aquatox.2015.07.001.

Pal, S. L., Jana, U., Manna, P. K., Mohanta, G. P., & Manavalan, R. (2011). Nanoparticle: An overview of preparation and characterization. *Journal of Applied Pharmaceutical Science, 1*(6), 228–234. http://www.japsonline.com/admin/php/uploads/159_pdf.pdf.

Panchapakesan, B., Lu, S., Sivakumar, K., Teker, K., Cesarone, G., & Wickstrom, E. (2005). Single-wall carbon nanotube nanobomb agents for killing breast cancer cells. *NanoBiotechnology, 1*(2), 133–139. https://doi.org/10.1385/NBT:1:2:133.

Petersen, E. J., Flores-Cervantes, D. X., Bucheli, T. D., Elliott, L. C. C., Fagan, J. A., Gogos, A., et al. (2016). Quantification of carbon nanotubes in environmental matrices: Current capabilities, case studies, and future prospects. *Environmental Science and Technology, 50*(9), 4587–4605. https://doi.org/10.1021/acs.est.5b05647.

Petersen, E. J., Zhang, L., Mattison, N. T., O'Carroll, D. M., Whelton, A. J., Uddin, N., et al. (2011). Potential release pathways, environmental fate, and ecological risks of carbon nanotubes. *Environmental Science and Technology, 45*(23), 9837–9856. https://doi.org/10.1021/es201579y.

Pratibha, S., Chethan, B., Ravikiran, Y. T., Dhananjaya, N., & Jagadeesh Angadi, V. (2020). Enhanced humidity sensing performance of samarium doped lanthanum aluminate at room temperature. *Sensors and Actuators A: Physical, 304*. https://doi.org/10.1016/j.sna.2020.111903, 111903.

Pratibha, S., Dhananjaya, N., & Lokesh, R. (2020). Investigations of enhanced luminescence properties of Sm^{3+} doped $LaAlO_3$ nanophosphors for field emission displays. *Materials Research Express, 6*.

Pratibha, S., Dhananjaya, N., Manjunatha, C. R., & Narayana, A. (2020). Fast adsorptive removal of direct blue-53 dye on rare-earth doped lanthanum aluminate nanoparticles: equilibrium and kinetic studies. *Materials Research Express, 6*.

Pratibha, S., Dhananjaya, N., Manohara, S. R., & Yadav, L. S. R. (2019). Effect of Sm^{3+}, Bi^{3+} ion doping on the photoluminescence and dielectric properties of phytosynthesized $LaAlO_3$ nanoparticles. *Journal of Materials Science: Materials in Electronics, 30*(7), 6745–6759. https://doi.org/10.1007/s10854-019-00986-x.

Pratibha, S., Dhananjaya, N., Pasha, A., & Khasim, S. (2020). Improved luminescence and LPG sensing properties of Sm^{3+}-doped lanthanum aluminate thin films. *Applied Nanoscience, 10*(6), 1927–1939. https://doi.org/10.1007/s13204-020-01354-6.

Pratibha, S., Dhananjaya, N., Shabaaz Begum, J. P., & Halappa, P. (2020). Modified benign approach for probing the structural, optical and antibacterial activity of Sm^{3+}-doped Bi^{3+}-co-doped $LaAlO_3$ nanoparticles. *The European Physical Journal Plus, 135*(8). https://doi.org/10.1140/epjp/s13360-020-00642-y.

Price, R. L., Haberstroh, K. M., & Webster, T. J. (2004). Improved osteoblast viability in the presence of smaller nanometre dimensioned carbon fibres. *Nanotechnology, 15*(8), 892–900. https://doi.org/10.1088/0957-4484/15/8/004.

Prinith, N. S., Manjunatha, J. G., & Hareesha, N. (2021). Electrochemical validation of L-tyrosine with dopamine using composite surfactant modified carbon nanotube electrode. *Journal of the Iranian Chemical Society*. https://doi.org/10.1007/s13738-021-02283-z.

Pulleman, M. M., Bouma, J., Van Essen, E. A., & Meijles, E. W. (2000). Soil organic matter content as a function of different land use history. *Soil Science Society of America Journal, 64*(2), 689–693. https://doi.org/10.2136/sssaj2000.642689x.

Pushpanjali, P. A., Manjunatha, J. G., Hareesha, N., Souza, E. S. D., Charithra, M. M., & Prinith, N. S. (2021). Voltammetric analysis of antihistamine drug cetirizine and paracetamol at poly(L-Leucine) layered carbon nanotube paste electrode. *Surfaces and Interfaces, 24*. https://doi.org/10.1016/j.surfin.2021.101154, 101154.

Qinglin, H., Bao, J., Xinghua, S. P. V. R., & Guoqing, Y. Z. Y. H. (2015). Effects of graphene oxide nanosheets on the ultrastructure and biophysical properties of the pulmonary surfactant film. *Nanoscale, 7*(43), 18025–18029.

Rahimi-Nasrabadi, M., Khoshroo, A., & Mazloum-Ardakani, M. (2017). Electrochemical determination of diazepam in real samples based on fullerene-functionalized carbon nanotubes/ionic liquid nanocomposite. *Sensors and Actuators, B: Chemical, 240*, 125–131. https://doi.org/10.1016/j.snb.2016.08.144.

Ruiz-Dueñas, F. J., & Martínez, A. T. (2009). Microbial degradation of lignin: How a bulky recalcitrant polymer is efficiently recycled in nature and how we can take advantage of this. *Microbial Biotechnology, 2*(2), 164–177. https://doi.org/10.1111/j.1751-7915.2008.00078.x.

Ryu, C. H., Joo, S. J., & Kim, H. S. (2016). Two-step flash light sintering of copper nanoparticle ink to remove substrate warping. *Applied Surface Science, 384*, 182–191. https://doi.org/10.1016/j.apsusc.2016.05.025.

Salavati-Niasari, M., Davar, F., & Mir, N. (2008). Synthesis and characterization of metallic copper nanoparticles via thermal decomposition. *Polyhedron, 27*(17), 3514–3518. https://doi.org/10.1016/j.poly.2008.08.020.

Saquing, J. M., Yu, Y. H., & Chiu, P. C. (2016). Wood-derived black carbon (biochar) as a microbial electron donor and acceptor. *Environmental Science & Technology Letters, 3*(2), 62–66. https://doi.org/10.1021/acs.estlett.5b00354.

Shvedova, A. A., Kisin, E. R., Murray, A. R., Kommineni, C., Castranova, V., Fadeel, B., et al. (2008). Increased accumulation of neutrophils and decreased fibrosis in the lung of NADPH oxidase-deficient C57BL/6 mice exposed to carbon nanotubes. *Toxicology and Applied Pharmacology, 231*(2), 235–240. https://doi.org/10.1016/j.taap.2008.04.018.

Sigmund, G., Jiang, C., Hofmann, T., & Chen, W. (2018). Environmental transformation of natural and engineered carbon nanoparticles and implications for the fate of organic contaminants. *Environmental Science: Nano, 5*(11), 2500–2518. https://doi.org/10.1039/C8EN00676H.

Sigmund, W., Yuh, J., Park, H., Maneeratana, V., Pyrgiotakis, G., Daga, A., et al. (2006). Processing and structure relationships in electrospinning of ceramic fiber systems. *Journal of the American Ceramic Society, 89*(2), 395–407. https://doi.org/10.1111/j.1551-2916.2005.00807.x.

Song, B., Xu, P., Zeng, G., Gong, J., Wang, X., Yan, J., et al. (2018). Modeling the transport of sodium dodecyl benzene sulfonate in riverine sediment in the presence of multi-walled carbon nanotubes. *Water Research, 129*, 20–28. https://doi.org/10.1016/j.watres.2017.11.003.

Spokas, K. A. (2010). Review of the stability of biochar in soils: Predictability of O:C molar ratios. *Carbon Management, 1*(2), 289–303. https://doi.org/10.4155/cmt.10.32.

Spokas, K. A., Novak, J. M., Masiello, C. A., Johnson, M. G., Colosky, E. C., Ippolito, J. A., et al. (2014). Physical disintegration of biochar: An overlooked process. *Environmental Science & Technology Letters, 1*(8), 326–332. https://doi.org/10.1021/ez500199t.

Su, W. C., Ku, B. K., Kulkarni, P., & Cheng, Y. S. (2016). Deposition of graphene nanomaterial aerosols in human upper airways. *Journal of Occupational and Environmental Hygiene, 13*(1), 48–59. https://doi.org/10.1080/15459624.2015.1076162.

Sun, S., Murray, C. B., Weller, D., Folks, L., & Moser, A. (2000). Monodisperse FePt nanoparticles and ferromagnetic FePt nanocrystal superlattices. *Science, 287*(5460), 1989–1992. https://doi.org/10.1126/science.287.5460.1989.

Sunilkumar, A., Manjunatha, S., Chethan, B., Ravikiran, Y. T., & Machappa, T. (2019). Polypyrrole-tantalum disulfide composite: An efficient material for fabrication of room temperature operable humidity sensor. *Sensors and Actuators, A: Physical, 298*. https://doi.org/10.1016/j.sna.2019.111593.

Syed, B., Prasad, N. M. N., & Satisha, S. (2016). Endogenic mediated synthesis of gold nanoparticles bearing bactericidal activity. *Journal of Microscopy and Ultrastructure*, 162. https://doi.org/10.1016/j.jmau.2016.01.004.

Tabei, Y., Fukui, H., Nishioka, A., Hagiwara, Y., Sato, K., Yoneda, T., et al. (2019). Effect of iron overload from multi walled carbon nanotubes on neutrophil-like differentiated HL-60 cells. *Scientific Reports, 9*(1). https://doi.org/10.1038/s41598-019-38598-4.

Tai, C. Y., Tai, C.-T., Chang, M.-H., & Liu, H.-S. (2007). Synthesis of magnesium hydroxide and oxide nanoparticles using a spinning disk reactor. *Industrial & Engineering Chemistry Research, 46*(17), 5536–5541. https://doi.org/10.1021/ie060869b.

Tenne, R. (2002). Fullerene-like materials and nanotubes from inorganic compounds with a layered (2-D) structure. *Colloids and Surfaces A: Physicochemical and Engineering Aspects, 208*(1–3), 83–92. Elsevier https://doi.org/10.1016/S0927-7757(02)00104-8.

Thomas, S. C., Harshita, Mishra, P. K., & Talegaonkar, S. (2015). Ceramic nanoparticles: Fabrication methods and applications in drug delivery. *Current Pharmaceutical Design, 21*(42), 6165–6188. https://doi.org/10.2174/1381612821666151027153246.

Tigari, G., & Manjunatha, J. G. (2020). Poly(glutamine) film-coated carbon nanotube paste electrode for the determination of curcumin with vanillin: an electroanalytical approach. *Monatshefte Für Chemie - Chemical Monthly, 151*(11), 1681–1688. https://doi.org/10.1007/s00706-020-02700-8.

Tiwari, D. K., Behari, J., & Sen, P. (2008). Application of nanoparticles in waste water treatment 1. *World Applied Sciences Journal, 3*, 417–433.

Tiwari, A. J., & Marr, L. C. (2010). The role of atmospheric transformations in determining environmental impacts of carbonaceous nanoparticles. *Journal of Environmental Quality, 39*(6), 1883–1895. https://doi.org/10.2134/jeq2010.0050.

Uhelski, D., & Miesel, J. R. (2017). Physical location in the tree during forest fire influences element concentrations of bark-derived pyrogenic carbon from charred jack pines (Pinus banksiana Lamb.). *Organic Geochemistry, 110*, 87–91. https://doi.org/10.1016/j.orggeochem.2017.04.014.

Wang, Y., Liu, F., He, C., Bi, L., Cheng, T., Wang, Z., et al. (2017). Fractal dimensions and mixing structures of soot particles during atmospheric processing. *Environmental Science & Technology Letters, 4*(11), 487–493. https://doi.org/10.1021/acs.estlett.7b00418.

Wang, X., Qu, R., Liu, J., Wei, Z., Wang, L., Yang, S., et al. (2016). Effect of different carbon nanotubes on cadmium toxicity to Daphnia magna: The role of catalyst impurities and adsorption capacity. *Environmental Pollution, 208*, 732–738. https://doi.org/10.1016/j.envpol.2015.10.053.

Wei, W., Nong, J., Zhu, Y., Zhang, G., Wang, N., Luo, S., et al. (2018). Graphene/Au-enhanced plastic clad silica fiber optic surface plasmon resonance sensor. *Plasmonics, 13*(2), 483–491. https://doi.org/10.1007/s11468-017-0534-0.

Yang, X., Yang, Q., Zheng, G., Han, S., Zhao, F., Hu, Q., et al. (2019). Developmental neurotoxicity and immunotoxicity induced by graphene oxide in zebrafish embryos. *Environmental Toxicology, 34*(4), 415–423. https://doi.org/10.1002/tox.22695.

Zhang, L., Aboagye, A., Kelkar, A., Lai, C., & Fong, H. (2014). A review: Carbon nanofibers from electrospun polyacrylonitrile and their applications. *Journal of Materials Science, 49*(2), 463–480. https://doi.org/10.1007/s10853-013-7705-y.

Zhang, C., Chen, W., & Alvarez, P. J. J. (2014). Manganese peroxidase degrades pristine but not surface-oxidized (carboxylated) single-walled carbon nanotubes. *Environmental Science and Technology, 48*(14), 7918–7923. https://doi.org/10.1021/es5011175.

Zhang, R., Khalizov, A. F., Pagels, J., Zhang, D., Xue, H., & McMurry, P. H. (2008). Variability in morphology, hygroscopicity, and optical properties of soot aerosols during atmospheric processing. *Proceedings of the National Academy of Sciences of the United States of America, 105*(30), 10291–10296. https://doi.org/10.1073/pnas.0804860105.

Zhang, D., Wang, D., Li, P., Zhou, X., Zong, X., & Dong, G. (2018). Facile fabrication of high-performance QCM humidity sensor based on layer-by-layer self-assembled polyaniline/graphene oxide nanocomposite film. *Sensors and Actuators, B: Chemical, 255*, 1869–1877. https://doi.org/10.1016/j.snb.2017.08.212.

Zhang, L. W., Yu, W. W., Colvin, V. L., & Monteiro-Riviere, N. A. (2008). Biological interactions of quantum dot nanoparticles in skin and in human epidermal keratinocytes. *Toxicology and Applied Pharmacology, 228*(2), 200–211. https://doi.org/10.1016/j.taap.2007.12.022.

Zhao, J., Wang, Z., White, J. C., & Xing, B. (2014). Graphene in the aquatic environment: Adsorption, dispersion, toxicity and transformation. *Environmental Science and Technology, 48*(17), 9995–10009. https://doi.org/10.1021/es5022679.

Part VI

Economics & commercialization of carbon nanomaterials-based sensors

Chapter 20

The use of carbon nanotubes material in sensing applications for H_1-antihistamine drugs

Jessica Scremin, Bruna Coldibeli, Carlos Alberto Rossi Salamanca-Neto, Gabriel Rainer Pontes Manrique, Renan Silva Mariano, and Elen Romão Sartori

Laboratory of Electroanalytical and Sensors, Department of Chemistry, Exact Sciences Center, State University of Londrina, Londrina, Paraná, Brazil

Introduction

Carbon nanotubes (CNTs) are important nanomaterials that have been widely explored in different fields of study due to their attractive physical and chemical properties. In electroanalysis, they are applied in the construction of sensors and biosensors as a base material or surface modifier for the determination of several analytes in different samples.

Over the last decade, a lot of researches have been carried out where carbon nanotube-based sensors were applied in the determination of antihistamine drugs in pharmaceutical and biological samples. Considering the importance of this group of medications in the treatment of different cases of allergy and the side effects associated with their administration, precise and accurate analytical procedures are required for the quality control of the pharmaceutical formulations. Besides, the excretion of unchanged or partially metabolized drugs through the urine makes this matrix an important alternative for the monitoring of patients on prolonged use of this drug and the diagnosis of accidental intoxications.

As well known, electroanalytical techniques have been supported the development of new and improved methods of quantification nowadays. In addition to leading to the development of fast, accurate, simple, and low-cost methods, these techniques can also be allied to countless electrodes, further improving their analytical features. Thus, there is a constant searching for the construction of new sensors that increase the performance of the electroanalytical methods, to guarantee greater reliability in the analyses.

In this context, the CNTs have been proved an excellent material to be employed in the construction of electrochemical sensors, improving the analytical response of the analytes, the accuracy, and the detectability of the methods, such as those applied in the determination of antihistamine drugs as mentioned here.

Considering the aforementioned, the purpose of this review is to gather the papers published in the last decade that presented the application of carbon nanotube-based sensors in the voltammetric determination of drugs belonging to the class of antihistamines, as well as showing the importance and the increasing use of these nanomaterials in electroanalysis.

Carbon nanotube-based sensors

The CNTs are nanostructures in form of sheets of graphene composed of sp^2 carbon units bonded in a hexagonal atomic arrangement in the shape of a honeycomb, rolled up like a hollow and elongated cylinder. They are classified according to the number of these concentric tubes that form the structure; the single-walled carbon nanotubes (SWCNTs) possess a single layer of graphene into a seamless cylinder with a diameter ranging from 1 to 2nm, whereas multiwalled carbon nanotubes (MWCNTs) are formed by more than one layer with a separation of ~0.34nm (Ajayan, 1999; Dresselhaus, Dresselhaus, & Avouris, 2001; Zhao, Gan, & Zhuang, 2002).

The history of CNTs began in 1991 when Sumio Iijima observed it in a carbon soot of graphite electrodes during an arc discharge by applying a current of 100A (Iijima, 1991). Two years later, different researches were simultaneously published reporting the procedure of synthesis of the carbon nanotubes presenting a single layer, the SWCNTs (Bethune et al., 1993; Iijima & Ichihashi, 1993). Since then, the different types of these nanostructured materials have attracted the attention of researchers around the world.

The success in the studies of CNTs and their wide applications was certainly possible due to the advances in production and characterization techniques of this nanomaterial over the years. Different methods for CNTs production have been reported, but the most adopted nowadays are arch discharge, laser ablation, and numerous types of chemical vapor deposition (CVD) (Yan et al., 2015). However, CVD methods stand out in large-scale production due to their advantages over others, such as economically competitive, simple reaction process and reactor design; easy control and manipulation of the process; and abundance of raw material, high yield, and high purity (Mubarak, Abdullah, Jayakumar, & Sahu, 2014).

The great attractiveness of these nanomaterials for different fields of research and industry is related to their fascinating properties, including extremely high tensile strength, high electrical conductivity, high ductility, and relative chemical stability (Jia & Wei, 2017). All these properties together lead to countless possibilities for applying CNTs in electronics, energy storage materials, catalysis, polymer composites, gas storage materials, and sensors (Jia & Wei, 2017; Zaporotskova, Boroznina, Parkhomenko, & Kozhitov, 2016).

In terms of sensing applications, the CNTs demonstrate some interesting attributes that classified them as excellent materials for the construction and modification of electrochemical sensors. Besides the properties mentioned earlier, the CNTs have the ability to enhance electron transfer in the electrochemical process, promote an increase in the electrode active surface area, and consequently improve the analytical response (Agüí, Yáñez-Sedeño, & Pingarrón, 2008).

Although the impressive features and applicability of the CNTs are related to their nanostructures (SWCNTs and MWCNTs), it is reported that the MWCNTs stand out because their production leads to a higher yield and lower cost per unit, besides presenting thermochemical stability with the possibility of improving electrical properties by functionalization processes (Oliveira & Morais, 2018). The aim of functionalizing CNTs to the electrochemical application is to facilitate their dispersion in aqueous media or other solvents to obtain a homogeneous film coated on the surface of the base electrode or to add functional groups to the nanostructures toward specific and selective interactions with a target analyte (Gao, Guo, Liu, & Huang, 2012).

The first report on the use of CNTs as electrode material was published in 1996 when Britto et al. applied this nanomaterial to the study of dopamine oxidation (Britto, Santhanam, & Ajayan, 1996). The results showed that the CNTs were responsible for increasing the reversibility of the oxidation process of dopamine. This study was the beginning of a long journey of application of the CNTs in electroanalysis because from this moment, several other papers were published reporting the employment of CNTs as electrode material or modifiers of conventional electrodes for the determination of numerous analytes (Agüí et al., 2008; Sinha et al., 2018; Wang & Yue, 2017).

Over the last decade, the use of CNTs in the construction of novel sensors for antihistamine determination was particularly made as modifier material of carbon-based electrodes by the drop-casting technique onto glassy carbon electrodes (GCE) or commercially available screen-printed electrode (SPCE) or also its incorporation in the preparation of carbon paste electrodes (CPE). Drop-casting refers to coat the surface of a bare electrode with a stable CNTs suspension until complete drying and formation of a homogeneous thin layer on the electrode surface (Agüí et al., 2008; Trojanowicz, 2006).

Fig. 20.1 presents the example of an unmodified GCE and SPCE widely used to the modification of their surfaces with CNTs and/or metallic nanoparticles by the drop-casting method. The modification with nanostructured materials changes the original characteristics of the base electrode.

Fig. 20.2 displays the schematic representation of the obtention of a CPE as well as its modification with nanostructured materials (CNTs and metallic nanoparticles). Carbon paste is prepared by mixing graphite powder (or some other

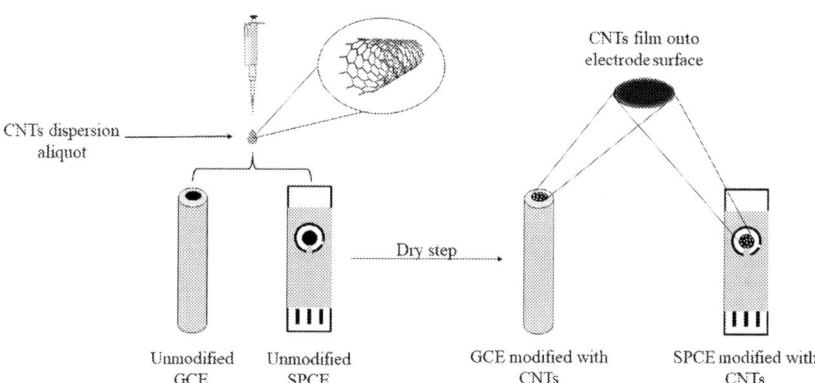

FIG. 20.1 Schematic representation of the modification of the surface of carbon-based electrodes (GCE and SPCE) with CNTs by the drop-casting method. *(No Permission Required.)*

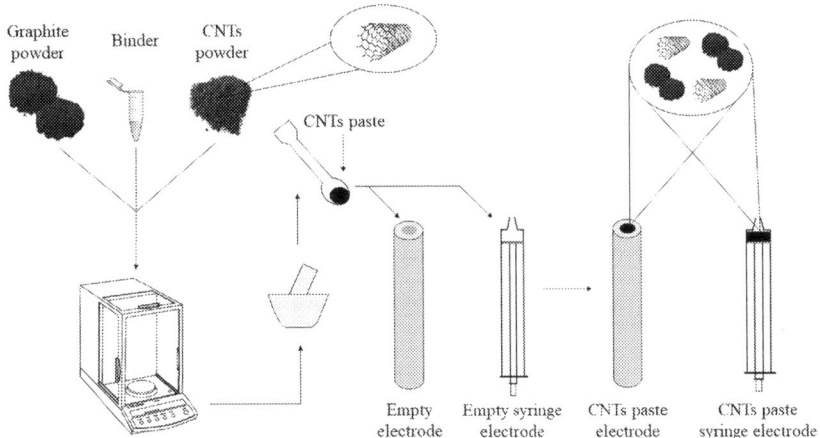

FIG. 20.2 Schematic representation of the obtention of a CPE modified with CNTs. *(No Permission Required.)*

carbonaceous materials) and a water-immiscible binder (ionic liquid, paraffin, or mineral oil). The paste is completely tamped into the electrode well filling air spaces with help with a small spatula and slightly heaped up above the surface of the well. The surface of the electrode is rubbed on a clean area of paper and is polished (smoothed) using circular movements.

H_1-antihistamine drugs: Definition and classes

Allergic diseases affect millions of people around the world. According to the World Allergy Organization (WAO), the prevalence of allergic diseases worldwide is rising dramatically in all countries, it is estimated that between 10% and 40% of the population present some types of allergies, such as asthma; rhinitis; and drug, food, and insect allergy (Pawankar, Canonica, Holgate, Lockey, & Blaiss, 2013). Histamine is released by mast cells through exocytosis and plays a pivotal role in allergic inflammation (Mahdy & Webster, 2011). Furthermore, it has important functions in the regulation of gastric acid secretion and neurotransmission. After the sensitized person comes in contact with the allergen, histamine is found locally in relatively large quantities per 1 million cells (Akdis & Blaser, 2003).

Thereat, antihistamine drugs have been used to alleviate symptoms of allergic reactions. They act in the histamine receptors, where they produce an imbalance between the initiative sites and activated receptors, to rebalance the disorder generated, the activity receptors are dislocated to an initiative condition, and the number of activated receptors decreases, which provide a lower concentration of histamines. Fig. 20.3 shows a representation of this balance (de Lucia, Oliveira-Filho, Planeta, Gallaci, & Avellar, 2007; Leurs, Church, & Taglialatela, 2002). In stage A, it is represented the equilibrium between the inactive state of the histamine H_1-receptor and the active state. In stage B, an agonist, which has an affinity for the active state, stabilizes the receptor and causes a shift in the equilibrium toward the active state. Finally, in stage C, an inverse agonist (antihistamine, in this case) causes a shift in the equilibrium toward the inactive stage.

There are different types of histamine receptors, named H_1-, H_2-, H_3-, and H_4-receptors, and each drug acts in a specific receptor. H_1-antihistamine acts in the H_1-receptors and this class is the most common drug used to relieve allergy

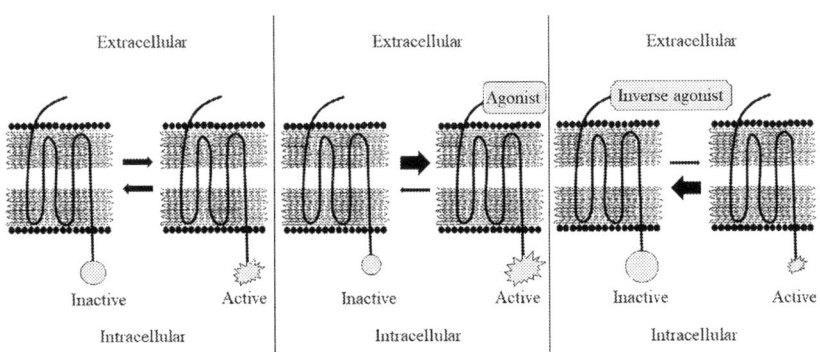

FIG. 20.3 Simplified two-state model of the histamine H_1-receptor. *(No Permission Required. Modified from Leurs, R., Church, M. K., & Taglialatela, M. (2002). H1-antihistamines: Inverse agonism, anti-inflammatory actions and cardiac effects.* Clinical & Experimental Allergy, *32(4), 489–498. https://doi.org/10.1046/j.0954-7894.2002.01314.x.)*

symptoms. This class of antihistamines is indicated for the treatment of certain urticaria, seasonal and perennial rhinitis, allergic conjunctivitis, and solar erythema, in addition to being auxiliary drugs in the treatment of emesis and kinetosis (de Lucia et al., 2007; Junqueira & Carneiro, 2004). H_1-receptors are expressed primarily in vascular endothelial cells and smooth muscle cells, mediating inflammatory and allergic reactions. Specific tissue responses to H_1-receptor stimulation include edema, bronchoconstriction, and sensitization of primary afferent nerve endings (Golan, Tashjian, Armstrong, & Armstrong, 2014; Rang, Dale, Ritter, Flower, & Henderson, 2020).

The first H_1-antihistamine molecule synthesized was thymo-ethyl-diethylamine in 1937. However, due to its weak activity and high toxicity, it was not used in clinical practice. Just in the 1940s, H_1-antihistamines such as phenbenzamine, pyrilamine, and diphenhydramine were used in clinical practice (Mahdy & Webster, 2011). The H_1-antihistamines can be classified in classic or first-generation and second-generation H_1-antihistamine. Although the efficacy of the different H_1-antihistamine drugs in the treatment of allergic patients is similar, even when first- and second-generation H_1-antihistamines are compared, they are very different in terms of their chemical structure, pharmacology, and toxic potential (Del Cuvillo et al., 2006).

The first-generation H_1-antihistamines such as diphenhydramine, chlorpheniramine, pheniramine, meclizine, have an affinity for cerebral H_1-receptors, relatively low molecular weight, and are highly lipophilic; for those reasons, the drug can penetrate the blood-brain barrier and hence have the potential to cause central nervous system depression, which usually manifests as somnolence and impaired cognitive and psychomotor performance. In general, these drugs are rapidly absorbed and metabolized, requiring several administrations during the day (Fukui et al., 2017; Khalilzadeh, Azarpey, & Hazrati, 2017; Mahdy & Webster, 2011).

The second-generation H_1-antihistamine drugs were a big development for the H_1-antihistamine class (Mahdy & Webster, 2011). The second-generation H_1-antihistamines are considerably less likely to cause adverse effects. Due to their high affinity for H_1-receptors, they are less lipophilic than first-generation, causing beneficial characteristics, as no sedation and little or no anticholinergic effect, because they hardly transpose the blood–brain barrier. Another advance is the possibility of a single daily dose, because of a longer half-life (Criado, Criado, Maruta, & Machado Filho, 2010; Simons, 2004).

Determination of H_1-antihistamine drugs

Some types of H_1-antihistamine drugs require a prescription by a licensed healthcare professional, due to higher concentrations of active ingredients or due to the risk of adverse side effects. Overdosing, with antihistamines (first or second generations), whether accidental or intentional, is not rare and can often lead to death. One of the biggest problems is the deficiency of surveillance on these drugs by the authorities. Then, patients often acquire antihistamines without a prescription, considering them completely harmless and exceed the recommended therapeutic dose, ignoring the warnings of potential sedation in the package leaflet.

In case of overdose, symptoms are dependent on age and dose ingested. In adults, it can lead to extreme drowsiness, confusion, delirium, cardiac abnormalities, seizures, and death if the patient is not treated within a few hours after intoxication. Children, on the contrary, may demonstrate some irritability, agitation, insomnia, hallucinations, and seizures that precede a coma (Pastorino, 2010).

Contrarily, the lack of antihistamine dosage in the medication can have serious consequences for the patient, since lack of effectiveness until death. Thus, it is important to correctly determine the quantity of antihistamine in the pharmaceutical formulation so that the patient's health is maintained or restored. In addition, the determination of antihistamines in human urine, blood serum, and plasma helps in the monitoring of patients on prolonged use of the drug, in the control of drug abuse, or the diagnosis of accidental intoxications. Urine is one of the most commonly used biological matrices for drug analysis because of its relative ease of sampling, access to rapid results, and because it is a universal means of excreting the drug, metabolites, or both. Some H_1-antihistamines are excreted unchanged in the human urine and others are excreted in low concentrations (Hardman, Limbird, & Gilman, 1996), but electrochemical systems employing CNT-based materials as working electrodes can detect them.

Working electrodes based on carbon nanotubes for the determination of H_1-antihistamines

Electrochemical techniques, such as voltammetry and amperometry, have become an alternative analytical strategy in the determination of H_1-antihistamines. These techniques achieve some advantages when compared to others, including fast, simple, and low-cost analysis, beyond being classified as environmentally friendly due to the use of minimal volumes of organic solvents just when it is necessary (Skoog, West, Holler, & Crouch, 2007). Voltammetry consists of an

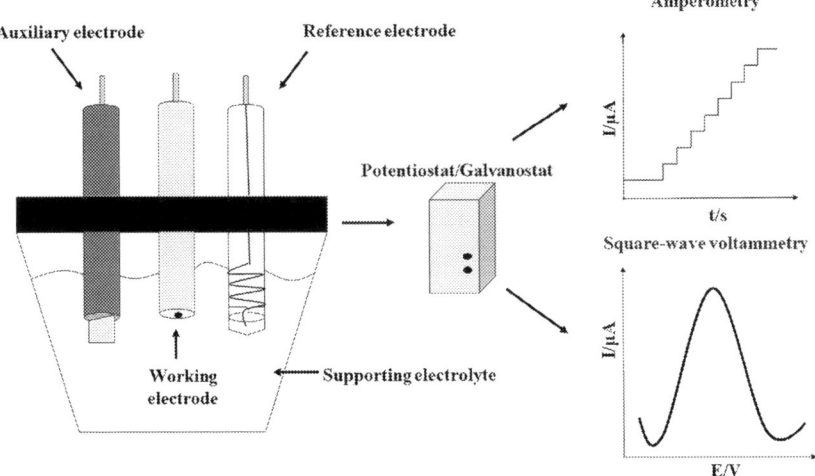

FIG. 20.4 General scheme of the electrochemical determination processes. *(No Permission Required.)*

electroanalytical technique in which a potential interval is swept, whereas the amperometry occurs at a fixed potential. The current generated during a voltammetry or amperometry experiment is recorded and plotted against the potential and time, giving rise to a voltammogram and amperogram, respectively.

The voltammetric methods employing different bare available electrodes for the determination of antihistamine drugs have proven to be precise, accurate, and sensitive (Abdel-razeq, Foaud, Salama, Abdel-atty, & El-Kosy, 2011; Beltagi, Abdallah, & Ghoneim, 2008; Eisele & Sartori, 2015; Golcu, Dogan, & Ozkan, 2005; Khorshed, Khairy, & Banks, 2018; Nagao, Salamanca-Neto, Coldibeli, & Sartori, 2020; Sreedhar, Sreenivasulu, Kumar, & Nagaraju, 2012). But still, there is a continuous search for sensors that present great analytical performance, like those based on the use of the CNTs.

In voltammetric/amperometric methods, it is necessary to use an electrochemical cell, where current measurements are taken. The conventional electrochemical cell incorporates a conventional three-electrode configuration: platinum plate as the auxiliary electrode, an Ag/AgCl (3.0 mol L^{-1} KCl) as the reference electrode, and carbon-based electrode as the working electrode. These electrodes are immersed in a supporting electrolyte, which can be an acid, a base, a buffer, or a salt solution. The electrochemical cell is connected to a potentiostat/galvanostat (Fig. 20.4). When using SPCE, the device configuration itself functions as an electrochemical cell. SPCE incorporates a conventional three-electrode configuration printed on a ceramic substrate, containing a working electrode and auxiliary electrode, both are made of carbon ink, whereas the pseudoreference electrode and electric contacts are made of silver. There are uncovered electrical contacts and a working area, which constitutes the reservoir of the electrochemical cell.

Determination of first-generation H$_1$-antihistamines employing voltammetry

An extensive survey of the literature revealed that several analytical methods have been reported for the determination of several H$_1$-antihistamines using working electrodes based on CNTs. Table 20.1 shows the collection of papers that describe the use of sensors based on CNTs for the determination of first-generation H$_1$-antihistamines. In addition, this table presents some analytical parameters such as linear range and limit of detection (LOD), as well as the types of samples analyzed.

The antihistamine substance diphenhydramine hydrochloride was quantified in pharmaceutical samples (syrup and drug serum) using a CPE modified with the nanocomposite made up of CdO nanoparticles-SWCNTs and the ionic liquid (IL), 1-butyl-3-methylimidazolium hexafluorophosphate ([C$_4$mim]-[PF$_6$]-CdO-SWCNTs/CPE) (Cheraghi & Taher, 2016). According to electrochemical investigations using [Fe(CN)$_6$]$^{-3/-4}$ redox probe, the incorporation of CdO-SWCNTs nanocomposite on the bulk of a CPE in the presence of [C$_4$mim]-[PF$_6$] increased the surface area of the CPE when compared to unmodified CPE. Further, this modification assured the CPE surface had great improvement in the electrochemical oxidation response of diphenhydramine hydrochloride, obtaining a highly sensitive sensor with good selectivity. The system can determine diphenhydramine hydrochloride over the range of 0.05–700 μmol L^{-1}, with a LOD of 9.0 nmol L^{-1}.

MWCNTs were used in three different modifications on the GCE surface for the determination of chlorpheniramine maleate. First, a GCE was modified just with functionalized MWCNTs (MWCNTs/GCE) dispersed in dimethylformamide (DMF) (Pourghobadi & Pourghobadi, 2015). The presence of MWCNTs onto GCE surface remarkably accelerates the electron transfer reactions between chlorpheniramine maleate and the modified electrode. A LOD value of 1.63 μmol L^{-1}

TABLE 20.1 Analytical features of the voltammetric methods developed in the last decade for the determination of first-generation H_1-antihistamines using working electrodes based on carbon nanotubes.

Analytes	Technique	Sensor	Linear range ($\mu mol\,L^{-1}$)	LOD ($\mu mol\,L^{-1}$)	Application	Ref.
Diphenhydramine hydrochloride	SWV[a]	[C_4mim]-[PF_6]-CdO-SWCNTs/CPE	0.05–700.0	0.009	Syrups and drug serum	Cheraghi and Taher (2016)
Chlorpheniramine maleate	DPV[a]	MWCNTs/GCE	5–500	1.63	Human blood serum samples	Pourghobadi and Pourghobadi (2015)
Chlorpheniramine maleate	DPV[a]	[C_4mim]-[PF_6]-MWCNTs/CGE	1.42–99.7	0.4	Pharmaceutical samples (Rinoven tablets)	Pérez-Ortiz, Pizarro, and Álvarez-Lueje (2019)
Chlorpheniramine maleate	LSV[a]	[C_4mim]-[BF_4]-MWCNTs/CGE	1–90	0.7	Pharmaceutical formulations	Khan and Motghare (2019)
Tripelennamine hydrochloride	DPV[a]	CTAB-MWCNTs/GCE	0.01–3.0	0.00238	Pharmaceutical formulations and urine samples	Gowda and Nandibewoor (2016)
Methdilazine	SWV[a]	MWCNTs/GCE	0.01–0.3	0.0018	Pharmaceutical and urine samples	Shetti, Malode, Nayak, Venkata Reddy, and Raghava Reddy (2019)
Pheniramine maleate	SWV[a]	(SLS-MWCNTs/GCE)	561–4209	163.6	Pharmaceutical formulations	Jain and Sharma (2012)
Meclizine hydrochloride	SWV[a]	ZnO-MWCNTs/CPE	0.042–0.22	0.014	Pharmaceutical tablets (Navoproxin) and spiked urine samples	Hendawy, Abdellatef, Hassan, and Shagor (2018)
Promethazine hydrochloride	SWV[a]	DNA-SiAlNb-MWCNTsPE	20–100	1.7	Raw material and pharmaceutical formulations	Marco, Borges, Tarley, Ribeiro, and Pereira (2013)
Promethazine hydrochloride	AdSDPV[a]	dsDNA/bMWCNTs/GCE	0.10–6.0	0.023	Pharmaceutical formulations	Primo, Oviedo, Sánchez, Rubianes, and Rivas (2014)
Promethazine hydrochloride	SWV[a]	PEI-MWCNTs/GCE	0.497–5.03	0.23	Pharmaceutical formulations	De Oliveira, Sousa, Morais, De Lima-Neto, and Correia (2020)
Promethazine hydrochloride	DPV[a]	MWCNTs/CPE	0.02–250.0	0.0019	Pharmaceutical formulations and blood plasma	Honarmand, Motaghedifard and Amani (2015)

[a]AdSDPV, *adsorptive stripping differential pulse voltammetry*; DPV, *differential pulse voltammetry*; LSV, *linear sweep voltammetry*; SWV, *square-wave voltammetry*.

was obtained and the sensor was applied in the determination of the antihistamine in blood serum sample. Second, this same dispersion was deposited onto the GCE and then covered with a black gel film (Pérez-Ortiz et al., 2019). The gel, obtained by the mixture of MWCNTs and the IL [C_4mim]-[PF_6], in the proportion of 1:10 w/w, was mechanically adhered to the surface of the GCE with a spatula ([C_4mim]-[PF_6]-MWCNTs/CGE). CNTs are wrapped up by the IL by van der Waals forces and the insolubility of ionic liquid in water assured stability of the sensor besides facilitating the gel formation and handling. The IL-MWCNTs film onto GCE showed an increased electroactive area, improved the electron transfer, and significantly enhanced the oxidation peak current of chlorpheniramine maleate. Also, a synergistic effect between MWCNTs and IL was detected; it improved the conductivity, resulting in a better response increasing the sensitivity over that of the unmodified electrode (bare GCE) and MWCNTs-modified electrode. LOD value of 0.4 µmol L^{-1} was obtained and applied in the determination of chlorpheniramine maleate in Rinoven tablets. Third, the composite obtained with the ionic liquid, 1-butyl-3-methylimidazolium tetrafluoroborate ([C$_4$mim]-[BF$_4$]), and the MWCNTs were integrated on the surface of GCE ([C$_4$mim]-[BF$_4$]-MWCNTs/CGE) by drop-coating (Khan & Motghare, 2019). This modification promoted sensitivity and selectivity toward chlorpheniramine maleate in the presence of other potential interferents present in pharmaceutical samples. In this case, the linear range and LOD for chlorpheniramine maleate were found to be 1–90 µmol L^{-1} and 0.7 µmol L^{-1}, respectively.

Tripelennamine hydrochloride was determined in the linear range of 0.01–3.0 µmol L^{-1} with LOD value of 0.00238 µmol L^{-1}. For this, the GCE was modified with a dispersion of functionalized MWCNTs in cetyltrimethylammonium bromide (CTAB) by drop-casting (CTAB-MWCNTs/GCE) (Gowda & Nandibewoor, 2016). The cationic surfactant makes the nanotubes positively charged and the combination of MWCNTs with CTAB leads to the formation of a highly porous 3-dimensional nanohybrid structure. The area of CTAB-MWCNTs/GCE was calculated to be 0.117 cm^2, whereas the unmodified one was 0.0448 cm^2.

Methdilazine was determined in pharmaceutical and urine samples employing a GCE modified with a homogeneous suspension of MWCNTs obtained in ethanol (MWCNTs/GCE). Transmission electron microscopy (TEM) technique showed CNTs with graphitic-like structure, well dispersed without aggregation, good crystallinity, and tubular morphology with a diameter of ~20 nm and the length of several microns. With this simple modification, the sensor presented 0.089 cm^2 of area for the modified GCE and 0.042 cm^2 of area for the unmodified GCE; it was able to detect 1.8 nmol L^{-1} of methdilazine (Shetti, Malode, Nayak, Venkata Reddy, et al., 2019).

The quantification of pheniramine maleate was based on its electrochemical response at GCE surface covered with MWCNTs disperse in DMF in the presence of sodium lauryl sulfate (SLS-MWCNTs/GCE). MWCNTs and the surfactant exerted a significant catalytic effect on the electrochemical reduction of pheniramine maleate leading to a decrease in the overpotential by about 300 mV toward the positive values and an enhancement of the peak current when compared to that of the unmodified GCE (Jain & Sharma, 2012). The authors demonstrated that the amount of pheniramine maleate adsorbed on the modified electrode is related to the thickness of the film. Increasing MWCNTs volume is associated with an increasing reduction peak current of pheniramine maleate, but they emphasize that when it is too thick, the film conductivity gets reduced; consequently, the amount of pheniramine maleate adsorbed is small; besides the film becomes unstable, MWCNTs could leave off the electrode surface. This modification promoted good operating characteristics as sensitivity, repeatability, and wide linear range.

The voltammetric behavior of the antihistamine drug meclizine hydrochloride was investigated in a CPE modified with ZnO nanoparticle and MWCNTs (ZnO-MWCNTs/CPE) (Hendawy et al., 2018). The active surface area of ZnO-MWCNTs/CPE (0.23 cm^2) was bigger than MWCNTs/CPE (0.090 cm^2) and MWCNTs/GCE (0.034 cm^2). Besides, the ZnO-MWCNTs/CPE presented the highest peak current and a decrease in peak potential for meclizine hydrochloride compared to these other electrodes. The presence of MWCNTs and metallic nanoparticles in the carbon paste guaranteed satisfactory performance to the determination of meclizine hydrochloride with a linear range of 0.042–0.22 µmol L^{-1} and LOD of 14 µmol L^{-1}. The determination of meclizine hydrochloride was carried out in pharmaceutical tablets and urine samples.

Promethazine hydrochloride was determined in the linear range of 20–100 µmol L^{-1} using a MWCNTs paste electrode (MWCNTsPE) modified with SiO$_2$/Al$_2$O$_3$/Nb$_2$O$_5$ and the biological component DNA (DNA-SiAlNb-MWCNTsPE) (Marco et al., 2013). MWCNTs promoted low charge transfer resistance and silica helped the fixation of DNA molecules, in which basic groups of molecules of DNA are immobilized in acid groups present in the silica. In addition, DNA molecules can promote a preconcentration by adsorption of the promethazine hydrochloride onto its surface. This antihistamine was determined in raw material and coated tablets with a LOD value of 5.9 µmol L^{-1}. Three other sensors have been developed for promethazine hydrochloride determination, all presenting LOD values in the order of µmol L^{-1}. One of them was developed by modifying the GCE with the dispersion of bamboo-like multiwalled carbon nanotubes (bMWCNTs) in calf-thymus double-stranded DNA (dsDNA) (dsDNA/bMWCNTs/GCE) (Primo et al., 2014). The increased density of edge plane defects observed in the bMWCNTs provides excellent electrocatalytic properties to the sensor; at dsDNA/bMWCNTs/GCE, there was a decrease in the overpotential (~0.058 V) for the oxidation of promethazine hydrochloride compared to unmodified GCE as well as a significant increase of 4.1 times in the associated current density. The efficiency of the preconcentration by adsorption of promethazine hydrochloride at dsDNA-bMWCNTs/GCE made possible its sensitive quantification in pharmaceutical products in a linear range of 0.1–6.0 µmol L^{-1}, with a LOD of 0.023 µmol L^{-1}. Another sensor based on carbon nanotubes to promethazine hydrochloride determination was obtained by the modification of the GCE with polyethylenimine (PEI)-functionalized MWCNTs (PEI-MWCNTs/GCE) (De Oliveira et al., 2020). The composite was prepared by mixing MWCNTs within polyethylenimine solution followed by sonication. It was deposited onto the GCE surface by the drop-casting process. In this sensor, the promethazine hydrochloride is oxidized at the modified electrode at more negative potentials and with higher sensitivity than at the unmodified GCE. The linear range was obtained from 0.497 to 5.03 $\mu mol\, L^{-1}$, with a LOD value of 0.23 $\mu mol\, L^{-1}$. The potential applicability of the PEI-MWCNTs/GCE was demonstrated in tablets of Fenergan. Finally, the fourth sensor was prepared by the modification of a CPE with MWCNTs (MWCNTs/CPE) (Honarmand, Motaghedifard, & Amani, 2015). Lower overpotential and an increase in current response were obtained to the MWCNTs/CPE. The anodic peak potential of promethazine hydrochloride was shifted by 28.0 mV to more negative potential in comparison with the unmodified carbon paste electrode. Due to these characteristics, the sensor presented a linear range of 0.02–250.0 $\mu mol\, L^{-1}$, with a LOD of 1.9 $nmol\, L^{-1}$, and was applied in the voltammetric determination of promethazine hydrochloride in pharmaceutical formulations and plasma samples.

Determination of second-generation H_1-antihistamines employing voltammetry or amperometry

The second-generation antihistamine is a newer class of antihistamines, although there are several papers for electrochemical determination for this class. Table 20.2 presents the analytical features, such as linear range and LOD, for the

TABLE 20.2 Analytical features of the voltammetric methods developed in the last decade for the determination of second-generation H_1-antihistamines using working electrodes based on carbon nanotubes.

Analytes	Technique	Sensor	Linear range ($\mu mol\, L^{-1}$)	LOD ($\mu mol\, L^{-1}$)	Application	Ref.
Bilastine	LSV	DHP-MWCNTs/SPCE	0.22–4.31	0.13	Commercial tablets and spiked urine samples	Teixeira and Oliveira (2021)
Cetirizine dihydrochloride	CV	MWCNTs/GCE	0.5–10	0.0707	Urine samples	Patil, Hegde, and Nandibewoor (2011)
Cetirizine dihydrochloride	AdSDPV	PtNPs-MWCNTs/CPE	0.1–200	0.0515	Pharmaceutical, urine, and blood serum samples	Kalambate and Srivastava (2016)
Cetirizine dihydrochloride	DPV	$[C_4mim]$-$[PF_6]$-Chi-MWCNTs/GCE	0.04–480	0.008	Human serum and pharmaceutical tablets	Gholivand, Shamsipur, and Ehzari (2019)
Cetirizine dihydrochloride	SWV	$AgTiO_2NPs$-MWCNTs/CPE	0.3–3.0	0.00876	Pharmaceutical formulations and urine samples	Shetti, Malode, Nayak, Aminabhavi, and Reddy (2019)
Desloratadine	LSV	CMB-MWCNTs/GCE	1.49–32.9	0.88	Commercial pharmaceutical tablets	Salamanca-Neto et al. (2020)
Desloratadine	DPV	ZnO-MWCNTs/GCE	0.02–8.0	0.000769	Pharmaceutical sample (Deloday)	Ozturk, Bakirhan, Ozkan, and Uslu (2020)
Emedastine difumarate	AdSDPV	AgNPs-MWCNTs/GCE	0.1–100	0.00525	Pharmaceutical dosage form (eye drop form)	Imanzadeh, Bakirhan, Habibi, and Ozkan (2020)
Loratadine	Amperometry	AuNPs-Fe_3O_4-MWCNTs/GCE	0.05–5	0.04	Pharmaceutical, human blood serum, and urine samples	Hassanpour, Baj and Abolhasani (2019)

voltammetric and amperometric determination of second-generation H_1-antihistamine using working electrodes based on CNTs. Types of analyzed samples were also displayed.

Commercial SPCE modified with a dispersion of MWCNTs in a solution of dihexadecyl phosphate (DHP) (DHP-MWCNTs/SPCE) was developed to the voltammetric determination of bilastine in commercially existing medicinal tablets and spiked urine samples. This antihistamine drug presented a single oxidation peak, which is much greater (by a factor of 7, approximately) than that obtained in the unmodified sensor. The use of MWCNTs improved the electroanalytical detection of this molecule, as demonstrated in countless works that make use of carbon nanotubes. Besides, the sensor can be reused after a simple and rapid reconditioning procedure. The proposed method was applied to the quantification of bilastine in pharmaceutical formulations and spiked urine samples in the linear range of concentration of 0.22–4.31 $\mu mol\,L^{-1}$ (Teixeira & Oliveira, 2021).

The antihistamine drug, cetirizine dihydrochloride, was determined in several samples using four sensors based on CNTs. The first sensor was obtained by the modification of a bare GCE with functionalized MWCNTs dispersed in acetonitrile (MWCNTs/GCE) (Patil et al., 2011). In the MWCNTs/GCE, the electroactive surface was nearly 3.0 times greater than that of bare GCE, due to the nanometric dimensions of the MWCNTs, the electronic structure, and the topological defects presented on the MWCNTs. The authors highlighted that the modified electrode has no electrochemical activity in the supporting electrolyte (phosphate buffer solution), but the background current increased, indicating that the electrode surface area has increased after the modification. This sensor presented a linear range of 0.5–10 $\mu mol\,L^{-1}$ to cetirizine dihydrochloride. The second sensor was developed by using a modified CPE with platinum nanoparticles (PtNPs) and MWCNTs (PtNPs-MWCNTs/CPE) (Kalambate & Srivastava, 2016). In this work, cetirizine dihydrochloride was determined in the presence of paracetamol and phenylephrine in pharmaceutical formulations, blood serum, and urine samples. The MWCNTs were prepared by the CVD method and the paste was prepared in the composition of 62:8:30 (graphite powder:MWCNTs-PtNPs:mineral oil). SEM and TEM images showed that the PtNPs were uniformly decorated on MWCNTs with little agglomeration. The area of the PtNPs-MWCNTs/CPE increased in a factor of 5.4 times than that of unmodified CPE. It was obtained a linear range of 0.19–193 $\mu mol\,L^{-1}$ and a LOD value of 58.6 $nmol\,L^{-1}$. The third sensor was based on the modification of a surface of GCE with a nanocomposite of chitosan, MWCNTs, and the IL [$C_4 mim$]-[PF_6], [$C_4 mim$]-[PF_6]-Chi-MWCNTs/GCE (Gholivand et al., 2019). Chitosan acts as a dispersant and an adhesion agent of MWCNTs to the electrode surface. The IL presents good ionic conductivity and low instability and toxicity. Uniform distribution of carbon nanotubes on the GCE leads to an increase in the sensor reproducibility and improves the linear range of cetirizine dihydrochloride (0.04–480 $\mu mol\,L^{-1}$) and LOD value of 8 $nmol\,L^{-1}$. The fourth sensor exhibited a linear response in the range of concentration from 0.3 to 3.0 $\mu mol\,L^{-1}$ to cetirizine dihydrochloride with a LOD value of 7.86 $\mu mol\,L^{-1}$. It was prepared by the modification of a CPE using a combination of silver-doped titanium dioxide nanoparticles (Ag-TiO_2NPs) and MWCNTs (AgTiO_2NPs-MWCNTs/CPE) (Shetti, Malode, Nayak, Venkata Reddy, et al., 2019). The modified CPE presented an area of 0.082 cm^2 and 0.042 cm^2 for unmodified CPE. The AgTiO_2NP-MWCNTs nanocomposites demonstrated an exceptional electrocatalytic performance for cetirizine dihydrochloride in contrast to the CPE. This sensor was applied in the voltammetric determination of cetirizine dihydrochloride in pharmaceutical formulations and urine samples.

In the literature, the development of two sensors based on CNTs for the voltammetric determination of desloratadine is described. One of them was obtained by the modification of GCE surface with MWNCTs dispersed in carboxymethyl-botryosphaeran (CMB) solution (CMB-MWCNTs/CGE) (Salamanca-Neto et al., 2020). The carboxymethylated polysaccharide was used to improve the dispersion of MWCNTs in water. The sensor was characterized as homogeneous and presented a higher electroactive area and a lower charge transfer resistance compared to the unmodified GCE. The sensor was employed in the determination of desloratadine in pharmaceuticals and spiked rat serum. The sensor exhibited a linear response in the concentration range from 1.49 to 32.9 $\mu mol\,L^{-1}$, with a LOD of 0.88 $\mu mol\,L^{-1}$. Another sensor for desloratadine determination was obtained by the modification of a GCE surface with the ZnO and MWCNTs composites in the proportion of ZnO:MWCNTs (1:3) (ZnO-MWCNTs/GCE) (Ozturk et al., 2020). Electrochemical impedance study proved that the MWCNTs helped to moor the ZnO films onto the GCE surface, enlarged the surface area of the GCE, and increased the electron transfer rate between the electrode surface and the bulk solution. The sensor presented a linear dependence with a concentration in the linear range of 0.02–8.0 $\mu mol\,L^{-1}$, with a LOD of 0.769 $nmol\,L^{-1}$.

Emedastine difumarate was determined employing a GCE modified by a simple drop-casting method with silver nanoparticles (AgNPs) and functionalized MWCNTs (AgNPs-MWCNTs/GCE) (Imanzadeh et al., 2020). The SEM image of the nanocomposite displayed that the spherical AgNPs are placed uniformly on the surface of the MWCNTs and provide more porosity in the surface area. The AgNPs-MWCNT/GCE has larger surface area (0.113 cm^2) than unmodified GCE (0.037 cm^2). This sensor displayed superior electrocatalytic performance in terms of the anodic peak current of emedastine when compared to unmodified GCE. Linear current response for oxidation of emedastine difumarate was obtained in the range of 0.1–100 μmol L^{-1}, with LOD of 5.25 nmol L^{-1}. It was successfully applied for the determination of emedastine in eye drop samples.

Loratadine was determined with a GCE modified with MWCNTs and Au nanoparticles-Fe$_3$O$_4$ beads/MWCNTs nanocomposites (AuNPs-Fe$_3$O$_4$-MWCNTs/GCE). This modification improved the sluggish electron transfer of the loratadine molecule at the modified sensor. The linear response of the sensor toward loratadine concentration using amperometry was 0.05–5.0 μmol L^{-1}, with a LOD of 0.04 μmol L^{-1}. The sensor was used to quantify loratadine in pharmaceutical, human blood, and urine samples, and satisfactory results were obtained (Hassanpour, Baj, & Abolhasani, 2019).

Conclusion

The research of sensors based on CNTs is continuously growing and its application relies mainly on pharmaceutical formulation and a few other matrices. The application of CNTs in the development of sensors demonstrated their wide satisfactory electroanalytical features with countless works demonstrated the electrocatalytic nature of the working electrodes based on CNTs for the voltammetric/amperometric determination of antihistamine compounds. This behavior is attributed to the higher surface activity of this nanostructured material because of the presence of more surface area in comparison with regular size material. Besides, the electronic structure and topological effects of CNTs surfaces assist better performance to the several electrodes. Although some authors described, the application of CNTs-based sensors in complex biological matrices is still a challenge because of the adsorptive characteristics of this nanomaterial. The overcoming of this barrier is expected to be the next objective of many works to be published.

References

Abdel-razeq, S. A., Foaud, M. M., Salama, N. N., Abdel-atty, S., & El-Kosy, N. (2011). Voltammetric determination of azelastine-HCl and emedastine difumarate in micellar solution at glassy carbon and carbon paste electrodes. *Sensing in Electroanalysis, 6*, 289–306.

Agüí, L., Yáñez-Sedeño, P., & Pingarrón, J. M. (2008). Role of carbon nanotubes in electroanalytical chemistry. A review. *Analytica Chimica Acta, 622*(1–2), 11–47. https://doi.org/10.1016/j.aca.2008.05.070.

Ajayan, P. M. (1999). Nanotubes from carbon. *Chemical Reviews, 99*(7), 1787–1800. https://doi.org/10.1021/cr970102g.

Akdis, C. A., & Blaser, K. (2003). Histamine in the immune regulation of allergic inflammation. *Journal of Allergy and Clinical Immunology, 112*(1), 15–22. https://doi.org/10.1067/mai.2003.1585.

Beltagi, A. M., Abdallah, O. M., & Ghoneim, M. M. (2008). Development of a voltammetric procedure for assay of the antihistamine drug hydroxyzine at a glassy carbon electrode: Quantification and pharmacokinetic studies. *Talanta, 74*(4), 851–859. https://doi.org/10.1016/j.talanta.2007.07.009.

Bethune, D. S., Kiang, C. H., de Vries, M. S., Gorman, G., Savoy, R., Vazquez, J., et al. (1993). Cobalt-catalysed growth of carbon nanotubes with single-atomic-layer walls. *Nature, 363*(6430), 605–607. https://doi.org/10.1038/363605a0.

Britto, P. J., Santhanam, K. S. V., & Ajayan, P. M. (1996). Carbon nanotube electrode for oxidation of dopamine. *Bioelectrochemistry and Bioenergetics, 41*(1), 121–125. https://doi.org/10.1016/0302-4598(96)05078-7.

Cheraghi, S., & Taher, M. A. (2016). Fabrication of CdO/single wall carbon nanotubes modified ionic liquids carbon paste electrode as a high performance sensor in diphenhydramine analysis. *Journal of Molecular Liquids, 219*, 1023–1029. https://doi.org/10.1016/j.molliq.2016.04.035.

Criado, P. R., Criado, R. F. J., Maruta, C. W., & Machado Filho, C. D. A. (2010). Histamina, receptores de histamina e anti-histamínicos: Novos conceitos. *Anais Brasileiros de Dermatologia, 85*(2), 195–210. https://doi.org/10.1590/S0365-05962010000200010.

de Lucia, O, Oliveira-Filho, R. M., Planeta, C. S., Gallaci, M., & Avellar, M. C. W. (2007). *Farmacologia Integrada: uso racional de medicamentos* (1st). São Paulo: Clube de Autores.

De Oliveira, R. C., Sousa, C. P., Morais, S., De Lima-Neto, P., & Correia, A. N. (2020). Polyethylenimine-multi-walled carbon nanotubes/glassy carbon electrode as an efficient sensing platform for promethazine. *Journal of the Electrochemical Society, 167*(10). https://doi.org/10.1149/1945-7111/ab995f.

Del Cuvillo, A., Mullol, J., Bartra, J., Dávila, I., Jáuregui, I., Montoro, J., et al. (2006). Comparative pharmacology of the H1 antihistamines. *Journal of Investigational Allergology and Clinical Immunology, 16*(1), 3–12. http://www.jiaci.org/issues/vol16s1/2.pdf.

Dresselhaus, M. S., Dresselhaus, G., & Avouris, P. (2001). *Carbon nanotubes: Synthesis, structure, properties, and applications*. https://doi.org/10.1007/3-540-39947-X.

Eisele, A. P. P., & Sartori, E. R. (2015). Simple and rapid determination of loratadine in pharmaceuticals using square-wave voltammetry and a cathodically pretreated boron-doped diamond electrode. *Analytical Methods, 7*(20), 8697–8703. https://doi.org/10.1039/c5ay01804h.

Fukui, H., Mizuguchi, H., Nemoto, H., Kitamura, Y., Kashiwada, Y., & Takeda, N. (2017). Histamine H1 receptor gene expression and drug action of antihistamines. In *Vol. 241. Handbook of experimental pharmacology* (pp. 161–169). Springer New York LLC. https://doi.org/10.1007/164_2016_14.

Gao, C., Guo, Z., Liu, J. H., & Huang, X. J. (2012). The new age of carbon nanotubes: An updated review of functionalized carbon nanotubes in electrochemical sensors. *Nanoscale, 4*(6), 1948–1963. https://doi.org/10.1039/c2nr11757f.

Gholivand, M. B., Shamsipur, M., & Ehzari, H. (2019). Cetirizine dihydrochloride sensor based on nano composite chitosan, MWCNTs and ionic liquid. *Microchemical Journal, 146*, 692–700. https://doi.org/10.1016/j.microc.2019.01.068.

Golan, D. E., Tashjian, A. H., Jr., Armstrong, E. J., & Armstrong, A. W. (2014). *Princípios de farmacologia: A base fisiopatológica da farmacoterapia* (3rd). Rio de Janeiro: Guanabara Koogan.

Golcu, A., Dogan, B., & Ozkan, S. A. (2005). Anodic voltammetric behavior and determination of antihistaminic agent: Fexofenadine HCl. *Analytical Letters, 38*(12), 1913–1931. https://doi.org/10.1080/00032710500230871.

Gowda, J. I., & Nandibewoor, S. T. (2016). Electro-oxidation and determination of tripelennamine hydrochloride at MWCNT-CTAB modified glassy carbon electrode. *Electroanalysis, 28*(3), 523–532. https://doi.org/10.1002/elan.201500489.

Hardman, J. G., Limbird, L. E., & Gilman, A. G. (1996). *Goodman & Gilman's—The pharmacological basis of therapeutics* (9th). New York: Mc Graw Hill.

Hassanpour, A., Baj, R. F. B., & Abolhasani, J. (2019). Gold nanoparticles–Fe3O4 beads/multiwalled carbon nanotubes modified glassy carbon electrode as a sensing platform for the electrocatalytic determination of loratadine in biological fluids. *Journal of Analytical Chemistry, 74*(12), 1223–1231. https://doi.org/10.1134/S1061934819120050.

Hendawy, H. A. M., Abdellatef, H. E., Hassan, W. S., & Shagor, A. M. O. K. A. (2018). Voltammetric determination of meclizine HCL and its application in pharmaceuticals and biological fluid using CNTS/ZnO nano-carbon modified electrode. *Journal of the Iranian Chemical Society, 15*(8), 1881–1888. https://doi.org/10.1007/s13738-018-1385-0.

Honarmand, E., Motaghedifard, M. H., & Amani, A. M. (2015). Electrochemical sensor based on multiwall carbon nanotube-paste electrode for determination of promethazine in pharmaceutical formulations and blood plasma. *Journal of Advanced Medical Sciences and Applied Technologies, 61*. https://doi.org/10.18869/nrip.jamsat.1.1.61.

Iijima, S. (1991). Helical microtubules of graphitic carbon. *Nature, 354*(6348), 56–58. https://doi.org/10.1038/354056a0.

Iijima, S., & Ichihashi, T. (1993). Single-shell carbon nanotubes of 1-nm diameter. *Nature, 363*(6430), 603–605. https://doi.org/10.1038/363603a0.

Imanzadeh, H., Bakirhan, N. K., Habibi, B., & Ozkan, S. A. (2020). A sensitive nanocomposite design via carbon nanotube and silver nanoparticles: Selective probing of Emedastine Difumarate. *Journal of Pharmaceutical and Biomedical Analysis, 181*. https://doi.org/10.1016/j.jpba.2020.113096.

Jain, R., & Sharma, S. (2012). Glassy carbon electrode modified with multi-walled carbon nanotubes sensor for the quantification of antihistamine drug pheniramine in solubilized systems. *Journal of Pharmaceutical Analysis, 2*(1), 56–61. https://doi.org/10.1016/j.jpha.2011.09.013.

Jia, X., & Wei, F. (2017). Advances in production and applications of carbon nanotubes. *Topics in Current Chemistry, 375*(1). https://doi.org/10.1007/s41061-017-0102-2.

Junqueira, L. C., & Carneiro, J. (2004). *Histologia básica*. G. Koogan.

Kalambate, P. K., & Srivastava, A. K. (2016). Simultaneous voltammetric determination of paracetamol, cetirizine and phenylephrine using a multiwalled carbon nanotube-platinum nanoparticles nanocomposite modified carbon paste electrode. *Sensors and Actuators, B: Chemical, 233*, 237–248. https://doi.org/10.1016/j.snb.2016.04.063.

Khalilzadeh, E., Azarpey, F., & Hazrati, R. (2017). The effect of histamine h1 receptor antagonists on the morphine-induced antinociception in the acute trigeminal model of nociception in rats. *Asian Journal of Pharmaceutical and Clinical Research, 10*(1), 76–80. https://doi.org/10.22159/ajpcr.2017.v10i1.14092.

Khan, S. I., & Motghare, R. V. (2019). Electrochemical determination of chlorophenaramine based on RTIL/CNT composite modified glassy carbon electrode in pharmaceutical samples. *Journal of the Electrochemical Society, 166*(13), B1202–B1208. https://doi.org/10.1149/2.0911913jes.

Khorshed, A. A., Khairy, M., & Banks, C. E. (2018). Voltammetric determination of meclizine antihistamine drug utilizing graphite screen-printed electrodes in physiological medium. *Journal of Electroanalytical Chemistry, 824*, 39–44. https://doi.org/10.1016/j.jelechem.2018.07.029.

Leurs, R., Church, M. K., & Taglialatela, M. (2002). H1-antihistamines: Inverse agonism, anti-inflammatory actions and cardiac effects. *Clinical & Experimental Allergy, 32*(4), 489–498. https://doi.org/10.1046/j.0954-7894.2002.01314.x.

Mahdy, A. M., & Webster, N. R. (2011). Histamine and antihistamines. *Anaesthesia and Intensive Care Medicine, 12*(7), 324–329. https://doi.org/10.1016/j.mpaic.2011.04.012.

Marco, J. P., Borges, K. B., Tarley, C. R. T., Ribeiro, E. S., & Pereira, A. C. (2013). Development of a simple, rapid and validated square wave voltametric method for determination of promethazine in raw material and pharmaceutical formulation using DNA modified multiwall carbon nanotube paste electrode. *Sensors and Actuators, B: Chemical, 177*, 251–259. https://doi.org/10.1016/j.snb.2012.11.005.

Mubarak, N. M., Abdullah, E. C., Jayakumar, N. S., & Sahu, J. N. (2014). An overview on methods for the production of carbon nanotubes. *Journal of Industrial and Engineering Chemistry, 20*(4), 1186–1197. https://doi.org/10.1016/j.jiec.2013.09.001.

Nagao, K. Y. H., Salamanca-Neto, C. A. R., Coldibeli, B., & Sartori, E. R. (2020). A differential pulse voltammetric method for submicromolar determination of antihistamine drug desloratadine using an unmodified boron-doped diamond electrode. *Analytical Methods, 12*(8), 1115–1121. https://doi.org/10.1039/c9ay02785h.

Oliveira, T., & Morais, S. (2018). New generation of electrochemical sensors based on multi-walled carbon nanotubes. *Applied Sciences, 8*(10), 1925. https://doi.org/10.3390/app8101925.

Ozturk, K., Bakirhan, N. K., Ozkan, S. A., & Uslu, B. (2020). Effect of catalytically active zinc oxide–carbon nanotube composite on sensitive assay of desloratadine metabolite. *Electroanalysis, 32*(1), 50–58. https://doi.org/10.1002/elan.201900193.

Pastorino, A. C. (2010). Revisão sobre a eficácia e segurança dos anti-histamínicos de primeira e segunda geração. *Revista Brasileira de Alergia e Imunopatologia, 33*, 88–92.

Patil, R. H., Hegde, R. N., & Nandibewoor, S. T. (2011). Electro-oxidation and determination of antihistamine drug, cetirizine dihydrochloride at glassy carbon electrode modified with multi-walled carbon nanotubes. *Colloids and Surfaces B: Biointerfaces, 83*(1), 133–138. https://doi.org/10.1016/j.colsurfb.2010.11.008.

Pawankar, R., Canonica, G. W., Holgate, S. T., Lockey, R. F., & Blaiss, M. S. (2013). *World Allergy Organization (WAO). White book on allergy: Update 2013* (p. 115). World Allergy Organization.

Pérez-Ortiz, M., Pizarro, P., & Álvarez-Lueje, A. (2019). Carbon nanotubes–ionic liquid gel. Characterization and application to pseudoephedrine and chlorpheniramine determination in pharmaceuticals. *Journal of the Chilean Chemical Society, 64*(1), 4324–4331. https://doi.org/10.4067/s0717-97072019000104324.

Pourghobadi, Z., & Pourghobadi, R. (2015). Electrochemical behavior and voltammetric determination of chlorpheniramine maleate by means of multiwall carbon nanotubes-modified glassy carbon electrode. *International Journal of Electrochemical Science, 10*(9), 7241–7250. http://www.electrochemsci.org/papers/vol10/100907241.pdf.

Primo, E. N., Oviedo, M. B., Sánchez, C. G., Rubianes, M. D., & Rivas, G. A. (2014). Bioelectrochemical sensing of promethazine with bamboo-type multiwalled carbon nanotubes dispersed in calf-thymus double stranded DNA. *Bioelectrochemistry, 99*, 8–16. https://doi.org/10.1016/j.bioelechem.2014.05.002.

Rang, H. P., Dale, M. M., Ritter, J. M., Flower, R. J., & Henderson, G. (2020). *Rang & Dale's pharmacology* (9th). Rio de Janeiro: GEN Guanabara Koogan.

Salamanca-Neto, C. A. R., Olean-Oliveira, A., Scremin, J., Ceravolo, G. S., Dekker, R. F. H., Barbosa-Dekker, A. M., et al. (2020). Carboxymethyl-botryosphaeran stabilized carbon nanotubes aqueous dispersion: A new platform design for electrochemical sensing of desloratadine. *Talanta, 210*. https://doi.org/10.1016/j.talanta.2019.120642.

Shetti, N. P., Malode, S. J., Nayak, D. S., Aminabhavi, T. M., & Reddy, K. R. (2019). Nanostructured silver doped TiO2/CNTs hybrid as an efficient electrochemical sensor for detection of anti-inflammatory drug, cetirizine. *Microchemical Journal, 150*. https://doi.org/10.1016/j.microc.2019.104124.

Shetti, N. P., Malode, S. J., Nayak, D. S., Venkata Reddy, C., & Raghava Reddy, K. (2019). Novel biosensor for efficient electrochemical detection of methdilazine using carbon nanotubes-modified electrodes. *Materials Research Express, 6*(11). https://doi.org/10.1088/2053-1591/ab4471.

Simons, F. E. R. (2004). Advances in H1-antihistamines. *New England Journal of Medicine, 351*(21), 2203–2217. https://doi.org/10.1056/nejmra033121.

Sinha, A., Dhanjai, Jain, R., Zhao, H., Karolia, P., & Jadon, N. (2018). Voltammetric sensing based on the use of advanced carbonaceous nanomaterials: A review. *Microchimica Acta, 185*(2). https://doi.org/10.1007/s00604-017-2626-0.

Skoog, D. A., West, D. M., Holler, F. J., & Crouch, S. R. (2007). *Fundamentos de Química Analítica* (2nd). São Paulo: Cengage Learning.

Sreedhar, N. Y., Sreenivasulu, A., Kumar, M. S., & Nagaraju, M. (2012). Voltammetric determination of olopatadine hydrochloride in bulk drug form and pharmaceutical formulations. *International Journal of Pharmaceutical Sciences and Research, Vol. 3*, 2517–2521. https://doi.org/10.13040/IJPSR.0975-8232.3.

Teixeira, J. G., & Oliveira, J. (2021). Voltammetric study of the antihistamine drug bilastine: Anodic characterization and quantification using a reusable MWCNTs modified screen printed carbon electrode. *Electroanalysis, 33*(4), 891–899. https://doi.org/10.1002/elan.202060494.

Trojanowicz, M. (2006). Analytical applications of carbon nanotubes: A review. *TrAC - Trends in Analytical Chemistry, 25*(5), 480–489. https://doi.org/10.1016/j.trac.2005.11.008.

Wang, T., & Yue, W. (2017). Carbon nanotubes heavy metal detection with stripping voltammetry: A review paper. *Electroanalysis, 29*(10), 2178–2189. https://doi.org/10.1002/elan.201700276.

Yan, Y., Miao, J., Yang, Z., Xiao, F. X., Yang, H. B., Liu, B., et al. (2015). Carbon nanotube catalysts: Recent advances in synthesis, characterization and applications. *Chemical Society Reviews, 44*(10), 3295–3346. https://doi.org/10.1039/c4cs00492b.

Zaporotskova, I. V., Boroznina, N. P., Parkhomenko, Y. N., & Kozhitov, L. V. (2016). Carbon nanotubes: Sensor properties. A review. *Modern Electronic Materials*, 95–105. https://doi.org/10.1016/j.moem.2017.02.002.

Zhao, Q., Gan, Z., & Zhuang, Q. (2002). Electrochemical sensors based on carbon nanotubes. *Electroanalysis, 14*(23), 1609–1613. https://doi.org/10.1002/elan.200290000.

Chapter 21

Carbon nanomaterial-based sensors: Emerging trends, markets, and concerns

Shalini Menon, Sonia Sam, K. Keerthi, and K. Girish Kumar
Department of Applied Chemistry, Cochin University of Science and Technology, Kochi, Kerala, India

Introduction

Sensors find applications in the monitoring of various fields including health, environment, and food safety (Gupta, Pathak, & Semwal, 2019). For the development of sensors, various spectroscopic, piezoelectric, optical, electrical, and chemical techniques have been published in the literature. Carbon nanomaterials (CNMs) have been used in biosensor platforms for decades, ranging from basic biological detection to complicated signal transduction of the interaction between receptors and analytes. Chemically stable CNMs are an ideal material for biosensors due to their chemical stability, strong electrical conductivity, durable mechanical efficiency, high surface-to-volume ratio, and biocompatibility. Recently, the use of CNMs in biosensor platforms has exploded in popularity (Hwang, Jeong, Kim, & Chang, 2020). CNMs have been used as (i) a sensor recognition factor, providing binding sites for target biomarkers or molecules trapping target biomarkers; (ii) a transducer element, converting the observed molecular activity on the electrode surface to a detectable signal; and (iii) a label for target biomarkers in signal enhancement. CNTs, nanodiamonds (NDs), graphene, and fullerenes are the most often used carbon nanostructures in biosensors. Additionally, some CNMs have been deposited into a bare electrode, such as a screen-printed carbon electrode, which functions as a target analyte recognition and signal transduction component in biosensors (Hwang et al., 2020). Due to the unusual combination of chemical and physical properties of CNMs (i.e., thermal and electrical conductivity, high mechanical strength, and optical properties), their functionalization can allow a broad range of biosensor applications. Additionally, functionalized CNMs exhibit exceptional reproducibility and a high degree of precision and sensitivity for the detection of target analytes at extremely low concentrations.

From 2007 to 2018, there were over 4150 papers on the topic of CNTs in biosensors (Sireesha, Jagadeesh Babu, Kranthi Kiran, & Ramakrishna, 2018). Scientific articles, review articles, book chapters, and conference proceedings are also included in this database. Furthermore, according to the list of publications, China is a leader in the development of CNT-based biosensors (Sireesha et al., 2018). Over the past decade, several publications have appeared demonstrating that functionalized CNTs can solve a variety of biocompatibility and toxicology issues (Sireesha et al., 2018). The number of publications with the title terms sensor, graphene, biosensor, and CNTs has increased dramatically in recent years. These statistical findings demonstrate the significance and utility of CNMs in sensors and biosensors.

In the last few years, a lot of work has gone into designing novel biosensors with high sensitivity and selectivity all over the world. The rapid production of nanomaterials has had a significant impact on biosensor development. The field of nanomaterials is vast, encompassing a wide range of materials with varying properties such as size, nature, form, composition, and chemistry Nanomaterials such as CNTs and graphene are common in biosensors and are at the forefront of research (Zhu, 2017). In the large family of CNMs, these two are the most represented (Zhu, 2017). Because of their special chemical and physical properties, CNTs and graphene have been extensively researched for biosensor applications to date (Zhu, 2017).

One-dimensional (1D), single-walled carbon nanotubes (SWCNTs) and multiwalled carbon nanotubes (MWCNTs) and two-dimensional (2D) graphenes (graphene nanoribbon (GNR), graphene oxide (GO), and reduced graphene oxide (RGO)) have special electrical, mechanical, and optical properties that open up new possibilities for biosensor applications (Tiwari, Vij, Kemp, & Kim, 2015). The use of graphene/CNTs has been greatly expanded by modifying their surface to tune their physicochemical properties (Tiwari et al., 2015). Other CNMs, such as zero-dimensional fullerenes and CDs, have been used as biosensors, but they are not as widely used as 1D and 2D CNMs in biosensing applications (Tiwari et al., 2015). CNTs and graphene are the hardest, lightest, and most conductive fibers known to man, outperforming all other substances in terms of performance-per-weight (ResearchAndMarkets.com, n.d.). In certain markets, they compete directly, and in others, they supplement each other. MWCNTs—once a lucrative nanomaterial, are currently competing (as conducting materials) with graphene and other 2D materials.

Several big MWCNT manufacturers have shut down their operations, but new applications continue to emerge, and LG Chem has built a large-scale production plant (ResearchAndMarkets.com, n.d.). Superaligned CNT arrays, yarns, and films have found applications in consumer electronics, aerospace, polymer composites, sensors, filters, heaters, and biomedicine.

SWCNTs have begun large-scale industrial development, offering new commercial prospects in transparent conductive films, conductive fabrics, transistors, sensors, and memory chips (ResearchAndMarkets.com, n.d.). A lot of manufacturers have shut down, but those who remain are seeing increased demand for their products. SWCNTs are thought to be one of the most promising candidates for use as building blocks in future electronics.

The commercialization of state-of-the-art electronics such as graphene touch screens in smartphones and lightweight radiofrequency (RF) applications on plastics has been aided by new methods of scalable synthesis of high-quality graphene, clean delamination transition, and device integration (ResearchAndMarkets.com, n.d.). NDs are relatively simple and cost-effective to manufacture, and their excellent mechanical, thermal, and chemical properties have pushed them toward large-scale commercialization (ResearchAndMarkets.com, n.d.).

CNMs may be incorporated into biological sensors by developing materials directly on a substrate, drop-casting, inserting CNMs into polymers, codepositing CNMs, and metal nanoparticles, and using CNMs in field-effect transistor (FET)-based structures to increase conductivity (Yang, Hu, et al., 2015; Yang, Denno, Pyakurel, & Venton, 2015). Direct growth of CNMs on electrodes produces a more uniform coating than traditional dip coating or drop-casting processes, and it may make possible batch manufacturing of materials easier. Polymer coatings can alter the physical and chemical properties of CNMs while also assisting in the dispersion of CNMs for deposition (Yang, Hu, et al., 2015). The addition of polymer, on the contrary, has disadvantages such as limiting diffusion, slowing temporal resolution, and lowering conductivity (Yang, Hu, et al., 2015). As new, tunable methodologies for the synthesis and functionalization of CNMs are established, it is expected that a growing range of significant analytical applications in a variety of fields, including rapid and critical medical analyses, drug quality monitoring, food, and environmental protection, will emerge in the immediate future.

Although there have been major advances in the field of sensors, only a few of them have undergone successful miniaturization and commercialization. In the last 20 years, the literature on carbon nanoparticle-based biosensors and nanoscale analytical methods has exploded, with a vast number of publications on the subject of "ultrasensitive," "cost-effective," and "early detection" tools with the potential for "mass production" cited on the Web of research (Bhalla, Pan, Yang, & Payam, 2020). However, none of these methods are currently commercially feasible or practicable for mass development and use of pandemic diseases including coronavirus disease in 2019 (COVID-19) (Bhalla et al., 2020).

Despite their widespread availability and well-known benefits in academia, biosensing, and characterization methods, these CNM-based sensors are yet to reach their full potential in on-site applications, such as during an infectious disease epidemic. Biochemical sensors face scientific obstacles and prospects before they can meet the expectations of reliable, precise, and early detection of diseases as well as several other targets.

This chapter will see the recent trends in carbon nanomaterial-based sensors and subsequently move on to their global market and the challenges faced in the various fields.

Emerging trends in carbon nanomaterial-based optical sensors

An optical sensor is an analytical device, which is based on the changes in the optical property of a probe molecule upon its interaction with the analyte molecule. Simplicity in the signal readout and interpretation offered by the various optical transducers can be said to be the prime factor that makes optical sensors as the widely preferred analytical devices (Law, Marsal, & Santos, 2019). Also, rapidness in analysis, visual detection possibilities, high sensitivity, and broad range of analytes are the other key features of optical transduction. Fluorescence (FL) sensing is a widely accepted optical detection technique, since it is simple, highly accurate, sensitive, and specific (Askim & Suslick, 2017; Sam, Anand, Mathew, & Girish Kumar, 2021). The measurement of variation in FL emission intensity, wavelength, and lifetime of a probe with the analyte concentration is employed in FL sensing and a spectrofluorometer is used for the analysis (Menon, Vikraman, Jesny, & Girish Kumar, 2016; Zhang, Lei, Li, Luo, & Jiang, 2020). The feasibility of blending a biorecognition element with a FL probe makes the technique very common in biosensing applications also (Shin, Gutierrez-Wing, & Choi, 2021). Colorimetry is a promising optical sensing technique as it provides visual detection along with instrumental analysis. A UV-visible spectrophotometer that measures absorbance as a function of wavelength is employed in the quantification of the target (Ebralidze, Laschuk, Poisson, & Zenkina, 2019). Every molecule possessing a change in polarizability under visible light irradiation has a unique Raman spectrum. Surface-enhanced Raman scattering (SERS) is a phenomenon that improves the intensity of Raman scattering when the scatterer is close to a noble metal surface that creates strong electromagnetic fields upon excitation with visible light. This phenomenon which extended the applicability of Raman effect to the field of molecular analysis has proved to study simple molecules to complex proteins (Haynes, McFarland, & Van Duyne, 2005). The spontaneous emission of radiation occurs as a result of a chemical reaction is termed as CL (Ragavan & Neethirajan, 2018). CL spectroscopic technique with simple instrumentation and more sensitivity compared to the photoluminescence-based ones has achieved a great deal of interest in sensor development (Pirsaheb, Mohammadi, & Salimi, 2019).

It is very evident that the emergence of nanotechnology has made revolutionary advancements in optical sensor development. Nanomaterials with intrinsic and unique optical properties are the most explored branch of material science for optical sensors. When metal nanoparticles were at the forefront for this purpose then, CNMs with distinct features are now emerging as the prime choice in various stages of optical sensor development (Fig. 21.1) (Santra, Xu, Wang, & Tan, 2004).

Numerous FL-, absorbance-, SERS-, and CL-based sensors with magnificent sensing features have been reported recently. While CNMs like CDs and GQDs are the prominent probes in the FL sensors, carbon nanohorns (CNHs), graphene, GO, CNTs, etc., act as FL quenchers which have a significant role in such sensors. The successful utilization of CNFs, CNTs, and fullerenes can be seen in some recently developed colorimetric assays. When derivatives of graphene act as supports for various SERS substrates, CDs and GQDs play efficient roles as a reagent and a catalyst in CL. The following is a brief description of these emerging trends in the usage of CNMs in optical sensors.

Carbon nanomaterials as nanoprobes and sensing elements

Simple sensing strategies always remain a cut above in the preferences of sensor research community. In this context, FL sensors in which the probe can also function as the sensing element is very much preferred as it avoids any tedious process of incorporating foreign sensing elements. CDs and GQDs are the emerging zero-dimensional nanomaterials with excellent fluorescent properties in par with metal nanoparticles (Cui, 2018). These materials with the aid of functional groups on their surface can selectively interact with molecules and thus act as a FL turn-on or turn-off sensors for the latter. Hence, functionalization is a key factor in the development of these sensors and it can be achieved during synthesis (Yan et al., 2018; Yan et al., 2018). Most of the recently developed sensors involve a FL turn-off mechanism upon probe-analyte interaction. CDs can function as Forster resonance energy donors in a direct sensing strategy, which offers better selectivity. Hydrogen bonding and π-π interaction between the amino, hydroxyl, or carboxyl groups on the CDs and the analyte molecules make them closer for Forster resonance energy transfer (FRET) to occur in an efficient way (Fu et al., 2021). Furthermore, noncovalent binding/complex formation-induced static quenching of CDs/GQDs can be utilized in the determination of a number of biomolecules and metal ions (Alvand & Shemirani, 2017; Gu, Shang, Yu, & Shen, 2016; Sivasankaran, Jesny, Jose, & Girish Kumar, 2017). The determination of metal ions has been made possible through their selective chelating ability toward the functional moieties on the CDs/GQD surface. The tendency of metal ions to adsorb electrostatically to negatively charged materials like nitrogen-doped CDs has been found to produce a combined static-dynamic FL quenching of the latter, enabling the successful determination of the former (Liu et al., 2019). The stable photoluminescent

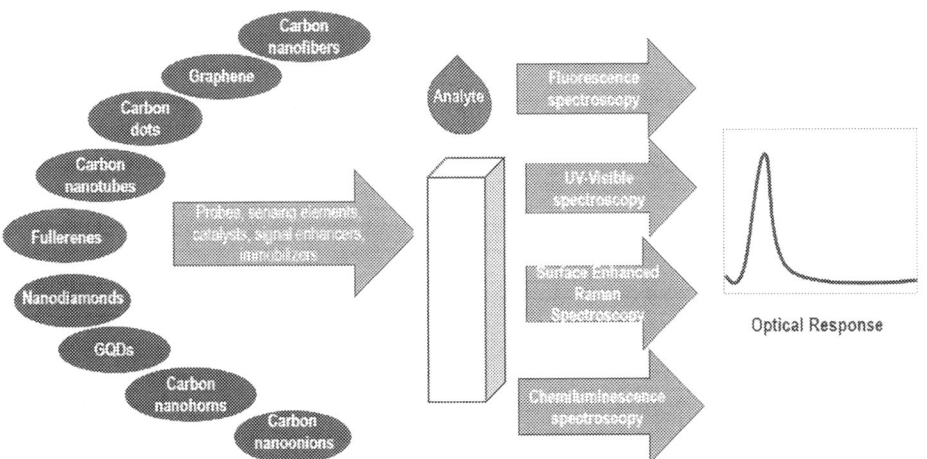

FIG. 21.1 General schematic representation of an optical sensor.

property of these CNMs has also been utilized recently in the development of several sensors, which follows a FL on-off-on strategy, which is an indirect way of sensing. In this case, CDs/GQDs function as the FL probe whose intensity turn-off in the presence of metal ions, which then recovers upon preferential complexation of the metal ion with the target molecule (Ge et al., 2019; Nemati, Zare-Dorabei, Hosseini, & Ganjali, 2018). The fluorescence property of CNOs is also probed in a recently reported FL on-off-on sensors for glutathione and glucose (Revuri et al., 2018; Tripathi et al., 2016).

Nowadays, CNMs like CDs, GQDs, GO, fullerenes, CNFs, and NDs are playing significant roles in the development of the very interesting colorimetric assays. When functionalized CNFs can be utilized as probes, CDs, GO, thiol-modified NDs, etc., can act as efficient sensing elements when successfully incorporated with the colorimetric probes (Beiraghi & Najibi-Gehraz, 2017; Kumar & Talreja, 2019; Shellaiah et al., 2018; Siripongpreda, Siralertmukul, & Rodthongkum, 2020). Surface-functionalized CDs/GO/RGO capped on metal nanoparticles is capable of interacting with analytes selectively to achieve visible colorimetric determination (Amanulla et al., 2017; Zangeneh Kamali et al., 2016; Zheng et al., 2018). CDs can also take part indirectly in the sensing mechanism as aggregation inducers of the colorimetric probe (Wang, Lu, Chen, Sun, & Liu, 2018; Wang et al., 2018). At the same time, GO can function as analyte trap as well as a net to capture the recognition elements (Lee, Lee, & Ahn, 2019). GQDs also have been part of many colorimetric assays as probes and reduction promoter molecules in those that follow in situ formation of colored nanoparticles (Babazadeh, Moghaddam, Keshipour, & Mollazade, 2020).

The fascinating dual-mode sensors in some recent reports that include fluorimetry and colorimetry with a single CD/GQD-based probe are very much interesting. In most of such assays, the direct probe-analyte-induced FL/surfacr plasmon resonance change attained through diverse functionalization is quantified for the sensing (Gao, Zhou, Ma, Wang, & Chu, 2018; Liu et al., 2017; Menon, Vikraman, Jesny, & Girish Kumar, 2016; Zhang, Wang, Deng, & Wang, 2020).

Surface-enhanced Raman scattering (SERS)-based sensors have been attaining much attention for the past few years as they allow rapid analysis of diverse molecules with high sensitivity. Noble metal nanoparticles are the major SERS substrates since their SPR can contribute to signal enhancement. To achieve pronounced ability to adsorb analyte molecules onto the noble metal nanoparticle-based substrates for the chemical enhancement of the Raman signals, the latter is usually coupled with CNMs like graphene. Literature reveals that gold and silver nanoparticles decorated with graphene, GO, RGO, CDs, CNTs are the major substrates of interest in SERS now. This interest is also evident in some recent reports on SERS-based sensors for the determination of metal ions, explosives, amino acids, etc. (Lee, Kim, Lee, Yoon, & Kim, 2019; Ouyang, Hu, Zhu, Cheng, & Irudayaraj, 2017; Poornima Parvathi, Parimaladevi, Sathe, & Umadevi, 2019). The highly delocalized electron system in graphene and its derivatives facilitates π-π interaction to target molecules and thus brings them close to the highly enhanced electromagnetic fields on the substrate. Also, the charge transfer between the adsorbed analytes and graphene-based materials or CNTs can bring about chemical amplification of the Raman signals (Dinh, Huy, Van Vu, Tam, & Le, 2016).

CDs and GQDs are excellent alternatives to the commonly used luminophores found in some recently reported CL sensors. The versatile nature of CDs, which is able to perform a variety of functions, is clear from these reports. Silicon-doped CDs can act as energy acceptors and thus enhance the CL of weak systems like $Fe(II)$-$K_2S_2O_8$ and HCO_3-H_2O_2 employed for the determination of norfloxacin, catecholamines, etc. (Amjadi, Hallaj, Manzoori, & Shahbazsaghir, 2018; Amjadi, Manzoori, Hallaj, & Shahbazsaghir, 2017). The CL can be attributed to the destruction of hole-injected and electron-injected CDs. At the same time, CDs act as an electron donor to an lucigenin system that enhances its CL for the effective determination of the analyte of interest (Wang et al., 2020). CD/GQDs can be also seen as CL emitters when it is coupled with reagents like H_2O_2 which can then be applied as an effective CL probe for sensing purpose (Cao, Zhang, Wang, Ma, & Liu, 2019; Chen et al., 2017; Shah, Dou, Khan, Uchiyama, & Lin, 2019).

In biosensors

A biorecognition element is the salient part of a biosensor, which is able to recognize the target of interest. Since biorecognition elements hardly possess any significant optical property, it has to be either labeled or conjugated with a probe (Wang, Bao, Liu, & Pang, 2011). Initially, organic compounds were used for this purpose. With the emergence of nanotechnology, many nanomaterials with entrancing fluorescent properties like quantum dots, noble metal nanoparticles, and SWCNTs have replaced that position. Evidently, CNMs as probes are more significant in FL biosensors. The use of near-infrared (NIR) emitting SWCNTs as FL probes in biosensor development has initialized some ten years back. The exceptional photostability offered by these semiconducting SWCNTs, which allows long-term measurements, has been still utilizing in FL biosensors. Recognition elements like antibodies are immobilized on the SWCNTs via wrappings like ssDNA on it. The modulations happened in the optical band gap of these SWCNTs upon target binding to the recognition elements, which will then reflect in the FL emission intensity as well as in the wavelength, are monitored for analysis

(Lee, Lee, & Ahn, 2019; Williams et al., 2018; Williams, Lee, & Heller, 2018). It is noteworthy that, among the CNMs, CDs have then come to the fore as a fluorescent probe in the biosensor fabrication in recent years. Numerous FL biosensors developed in these years with different variants of CDs functioning as probes, immobilizers, internal reference in ratiometric sensors, etc., are clear evidence for this (Chen et al., 2019; Xu et al., 2016; You et al., 2019). Also, GQDs as a fluorophore have become part of a number of FL biosensors with similar sensing strategies as that of CDs (Gu et al., 2021; Nana, Ruiyi, Xiulan, Yongqiang, & Zaijun, 2020). Biorecognition elements like aptamers, DNAs, and antibodies can be effectively functionalized onto CDs/GQDs through various interactions like electrostatic, π-π stacking, and covalent conjugations. Since carboxyl-rich CDs can form π-π stacking interactions well with nucleic acids, it is mostly employed in functionalization (Miao, Wang, Zhuo, Zhou, & Yang, 2016; Xu et al., 2016). While in GQDs, 1-ethyl-(3-dimethyaminopropyl) carbodiimide (EDC)—N-hydroxysuccinimide (NHS) interaction is often used (Cao et al., 2017; Nana et al., 2020). Most of the recently developed biosensors have biorecognition elements immobilized onto the CDs and fewer are there without any such direct interaction, where CDs merely act as FL probes (Gao et al., 2020; Wang et al., 2019). Some of the biosensors follow simple sensing strategies, in which the recognition elements that bind to CDs via electrostatic interactions quench their FL and the recovery upon target binding is probed for the determination (Fig. 21.2) (Miao et al., 2016).

This simple strategy applied in biosensors with the aid of cost-effective CDs has been proved for the effective determination of toxic metal ions and cancer biomarkers and even in the detection of cancer cells with a detection threshold of 100 cells/mL (Guo, Zhang, Zhang, & Hu, 2019; Zhang, Li, Wang, Liang, & Jiang, 2019; Zhang, Liu, Han, Luo, & Li, 2019). Most of the other sensors, which have achieved detection limits as low as femtomolar range, are based on a FL on-off-on strategy. In these kinds of sensors, often CDs/GQDs functionalized with the recognition elements act as fluorescent probes and the fluorescence turn-off effect of a quencher (gold nanoparticles (AuNPs), GO, nanoporous carbon (NC)) is restored on target binding to the recognition element (Qaddare & Salimi, 2017; Wang et al., 2017; Wang, Lu, Hu, Wang, & Wu, 2020; Zhu et al., 2016). Sandwich assay format that includes DNA hybridization has been also adopted in one such FL on-off-on mode sensor (Hamd-Ghadareh, Salimi, Fathi, & Bahrami, 2017). Also, some of these assays are designed by assembling CDs onto the quenchers like AuNPs through electrostatic interactions to achieve better response (Wang et al., 2016). Aptamer-functionalized CDs can also be utilized in the protein detection in nanomolar range through protein-induced FL enhancement of CDs (Singh, Chakma, Jain, & Goswami, 2018). Ratiometric FL sensors measure the ratio of intensities of the FL emissions at two wavelengths for the determination of the analyte of interest. Either a dual-emissive fluorophore or a combination of two fluorophores can be used for the purpose in which one spectrum remains constant is referred to as the reference signal. This strategy, which eliminates different background errors, that can occur during the analysis can increase the accuracy of biosensing along with the built-in specificity. The versatile nature of CDs has been employed in a number of recently developed ratiometric FL biosensors as both a reference fluorophore and a responsive fluorescence probe (functionalized with recognition element). The efficient utilization of two differently emitting CDs in a single ratiometric sensor with one as a reference and the other as a responsive probe widens the possibilities of CDs in FL biosensing. Also, CDs can be used as capping over QDs and as nanoquencher in the design of ratiometric bioassays (Bahari, Babamiri, Salimi, & Salimizand, 2021; Ma, Wang, Liu, Shi, & Yang, 2021; You et al., 2019).

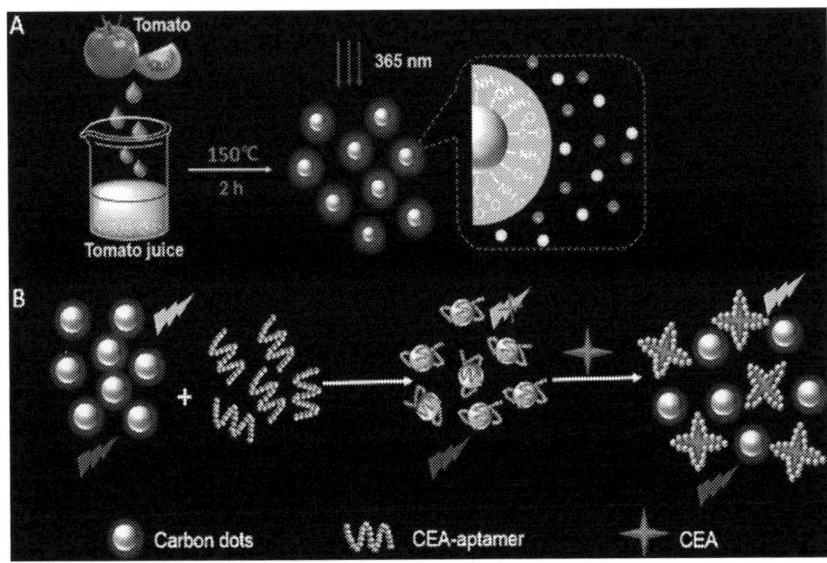

FIG. 21.2 Schematic illustration of assaying tumor biomarker, carcinoembryonic antigen (CEA) using tomato juice-derived green CDs (Miao et al., 2016). *From Miao, H., Wang, L., Zhuo, Y., Zhou, Z. & Yang, X. (2016). Label-free fluorimetric detection of CEA using carbon dots derived from tomato juice. Biosensors and Bioelectronics, 86, 83–89. https://doi.org/10.1016/j.bios.2016.06.043, Copyright (2021), with permission from Elsevier. License Number: 5062601503107.*

In molecularly imprinted polymer-based sensors

Molecularly imprinted polymers (MIPs) are formed by polymerizing functional monomers in the presence of template molecules followed by extraction of the latter. The imprints of the template molecule can act as specific receptors like antibodies; hence, MIPs are also known as "plastic antibodies" (Haupt & Mosbach, 1998). MIPs are an ideal choice to improve the selectivity challenges faced by FL sensors. Many CD/GQD-based FL sensors have the drawback of poor selectivity. Incorporating MIPs of suitable monomers and the template molecule with GQD/CD surface paved the way for them to specifically target molecules. In most of the cases, CDs/GQDs that act as FL probes are embedded on supports like metal-organic framework, covalent organic framework, or a silica matrix, upon which the MIPs are fabricated (Fang et al., 2019; Liu et al., 2019). Determination is made possible through FL quenching of the probes induced by FRET, which further enhances the selective nature of the sensors. Moreover, some MIPs with CD-based sensors that have been reported recently have potential applications in cell imaging also (Fang et al., 2019). Researchers have made possible protein and even cancer cell imaging using these biocompatible and less toxic CD-based MIP sensors (Demir et al., 2018). Also, it is an interesting strategy to incorporate magnetic nanoparticles along with CDs in the sensor fabrication for the rapid separation of MIPs from the sample solution. In addition to CDs and GQDs, GO can be incorporated in MIP-based FL sensors as a support of the probe for signal amplification and to provide a large surface area (Kumar, Karfa, Majhi, & Madhuri, 2020; Li et al., 2017).

Carbon nanomaterials as temperature sensors

Determining temperature or temperature difference in the submicron scale is referred to as nanoscale temperature sensing. It plays a significant role in the applications of life sciences since the cellular compartments are in the submicron scale. In addition to the scale of interest, precise monitoring of temperature is also a challenge in nanoscale temperature sensing. Among the different cellular temperature sensing techniques, FL thermometry is an emerging and well-accepted one owing to its simplicity in working. Thus, fluorescent materials such as quantum dots, polymers, rare earth metal complexes, and dyes that are having the capability to penetrate cellular walls are the best choices to act as nanothermometers. Some recent reports clearly show highly biocompatible NDs as emerging novel nanoscale temperature sensors with high resolution (Wu et al., 2020; Yukawa et al., 2020). Variations in the zero-phonon line of the emission spectra with temperature are the mechanism behind these nanodiamond-based sensors. Now, temperature-dependent optically detected magnetic resonance, exhibited by the nitrogen-vacancy centers of these fluorescent NDs, is preferably used for the same (Wu et al., 2020). Chemical robustness along with excellent resistance to environmental interference on the thermal sensitivity of these NDs is proved to be capable of monitoring the effect of intracellular temperature on different cellular functions. These ND-based nanoscale temperature sensors that follow a simple protocol have been reported to be monitoring the temperature with an accuracy of $\pm 1°C$ (Mitaku & Sawada, 2018).

Carbon nanomaterials as FL quenchers in biosensors

CNMs like CNHs, CNTs, NC, graphene, and GO can act as efficient quenchers of the fluorescent intensity of various probes. This fascinating property exhibited by these materials has been very much utilized in the design strategy of many FL biosensors, which follow signal off-on strategy, so far. Single-walled carbon nanohorns (SWCNHs), which are in horn shape, when aggregate form spherical structures with diameters in the nanometer range. These π-rich SWCNHs are capable of forming strong π-π interactions with nucleobases which results in their well adsorption performance on the nanohorns as well as quenching FL intensity (Wu et al., 2018). A DNA-based sensor is reported recently where SWCNHs act as an efficient nanoplatform for a clamped hybridization chain reaction. The aggregated form of SWCNHs act as a carrier simultaneously for different kinds of DNA strands required for cascaded hybridization chain reaction (c-HCR), a strategy for FL amplification as well as a quencher(Shen et al., 2020). Similarly, SWCNHs are also part of some aptamer-based biosensors as a quencher in the form of a FRET acceptor upon π-π stacking induced proximity with fluorescent probes. Even though graphene and GO are reported as excellent quenchers due to the energy-accepting ability, GOs functionalized with oxygen moieties are found to be more efficient quenchers in the design of many biosensors (Yugender Goud et al., 2017). The remarkable FL quenching capability attained upon coupling multiwalled CNTs and GNR has also been discovered by scientists to obtain sensitivity as low as angstrom units. Even a DNA biosensor based on the nanoquencher, metal-organic framework-derived nanoporous carbon, with excellent real sample applicability, has been developed, when demonstration of application of the former was a challenge to the scientists (Li, Jiang, Liu, Liang, & Jiang, 2019).

Carbon nanomaterials as catalysts

Nanomaterials possessing specific catalytic activity toward chemical and biological reactions can function as cost-effective and recyclable alternatives to biological catalysts. Most of the recently developed chemical sensors that offer remarkable specificity have these catalytic nanomaterials associated with them. CNMs like GO are renowned for this captivating property. Now, CDs, CNFs, and GQDs are emerging as enzyme-mimicking moieties in optical chemical sensors. Literature shows that this fascinating property of these CNMs is more used in association with colorimetric transduction. The basic sensing strategy in these colorimetric sensors is the peroxidase mimicking activity of the GOs/CDs/CNFs/GQDs or their hybrids on the oxidation of 3,3′,5,5′-tetramethylbenzidine (TMB) to form a colored product (Borthakur, Darabdhara, Das, Boukherroub, & Szunerits, 2017; Chandra et al., 2019; Song et al., 2018; Zhou et al., 2018; Zhuo, Fang, Zhu, & Du, 2020). In FL sensors, CDs can be seen as both peroxidase mimics and the FL probes whose emission intensity is quenched by the product of the oxidation (Mathivanan, Tammina, Wang, & Yang, 2020; (Vázquez-González et al., 2017). FL and colorimetric dual-channel sensors by combining the aforementioned strategies are really interesting since they provide more reliable sensing along with specificity (Tang et al., 2019). Different from the early-discussed SERS-based sensors, carbon nanomaterials like CDs are used as a catalyst for the in situ formation of nanoparticles with SPR as SERS-active substrates (Chen et al., 2017; Wang, Lan, et al., 2020; Yao, Li, Wen, Liang, & Jiang, 2020).

Optical biosensors, which include these enzyme-mimicking CNMs, are evidently aimed at improving the specificity in recognizing the target molecules and to improve sensitivity. Thus, aptamer-based as well as immunoassays have been reported recently with a much improved sensing ability. The basic principles seen in these assays are the target-specific aptamer/antibody-induced weakening of the catalytic activity of the enzyme CDs, which are then restored upon target binding (Li, Yang, Wu, Li, & Duan, 2019; Zhang, Lei, et al., 2020). It is much appreciable that the research in this field has successfully resulted in the development of catalytic CD-based aptasensor with SERS, RRS, and FL trivial-mode signal readout followed by a SERS and RRS dual-mode sensor (Li, Li, Liang, Wen, & Jiang, 2021; Wang, Cao, et al., 2019; Wang, Huang, Wen, & Jiang, 2019). These carbonaceous nanomaterials, CDs and GQDs, can also enhance CL through their catalytic effect toward different CL systems (Delnavaz & Amjadi, 2020; Duan, Huang, Chen, Zuo, & Shi, 2019; Li & Han, 2020). Thus, they function as cost-effective substitutes to the commonly used expensive enzyme catalysts. This catalytic activity has evidently enhanced the sensitivity of CL-based sensors. It is noteworthy that this catalytic activity of CDs has been utilizing in aptasensing strategy also, where the target-specific aptamers were immobilized on the CDs (Sun et al., 2018). The strategy of restoring the catalytic activity of the CDs and thereby CL enhancement on aptamer binding to the target is followed in such sensors.

Carbon nanomaterials in bioimaging

The most fascinating application of optical sensing in the biomedical field is the imaging of biological systems, referred to as "bioimaging." It is aimed at extracting biological information by visualizing biological systems such as cells and tissues with the aid of materials with suitable optical properties such as FL and SERS (Eggeling, 2018). The literature clearly reveals that, among these, FL itself is the most utilized optical property in bioimaging. The advent of nanotechnology has helped the research in bioimaging to overcome many of the drawbacks faced by the conventional organic fluorophores such as large size, photobleaching, and interference from cellular environment (Santra et al., 2004). Among these, biocompatibility is the most important property needed for an interference-free bioimaging. It is noteworthy that fluorescent CNMs at various dimensions with excellent biocompatibility are now at the fore in bioimaging compared to metallic nanoparticles. CDs, GQDs, CNTs, CNOs, fullerenes, and NDs are the major carbonaceous nanomaterials employed as FL probes in cellular imaging (Dong et al., 2018; Hamd-Ghadareh et al., 2017; Revuri et al., 2018). Among these, CDs, GQDs, and NDs are the leading FL probes in bioimaging nowadays. These particles having size of around 10 nm, with good FL properties and photostability, have been proven to be suitable candidates in imaging cancer cells, intracellular imaging of molecules, etc. The capability of specifically functionalized CDs in penetrating cancer cells is utilized in cancer cell imaging (Li et al., 2020). It is worth mentioning that the highly biocompatible and less cytotoxic CDs have extended the application of MIP-based FL sensors to cell imaging also (Demir et al., 2018).

Role of carbon nanomaterials in electrochemical sensing

Electrochemical sensors are one of the fastest growing areas of research. It finds a wide range of applications in clinical, environmental, agricultural, and defense fields. Currently, research efforts are focused on the fabrication of sensor devices with improved sensitivity, selectivity, response time, and long-term use (Sengupta, Khanra, Chowdhury, & Datta, 2019).

In this context, electrochemical techniques serve as a powerful analytical tool due to the possibility of miniaturization, their inherent selectivity, and sensitivity to analytes, as well as due to the precision and specificity they provide. They are also less time-consuming, flexible, and simple to set up (Menon, Jesny, & Girish Kumar, 2018; Menon, Mathew, Sam, Keerthi, & Kumar, 2020; Menon, Vikraman, Jesny, & Girish Kumar, 2016; Menon & Kumar, 2017; Shetti, Nayak, Reddy, & Aminabhvi, 2018; Zhang, Wang, et al., 2020). Knowing the fascinating changes that electrochemical sensors can bring about in various fields of science, scientists are in constant effort to guarantee enhanced performance and real-life applicability of electrochemical sensors.

Beyond a doubt, nanomaterials have set the fastest growing field in material science and they have been widely explored in modern smart sensing devices (Arduini, Cinti, Scognamiglio, & Moscone, 2016). Unique electrochemical properties of carbon-based nanomaterials have made great interest among researchers and it is one of the most discussed synthetic materials due to its diverse electronic, optical, and magnetic properties; biocompatibility; chemical versatility, etc. (Baptista, Belhout, Giordani, & Quinn, 2015). Various carbon-based nanomaterials such as graphene and its derivatives, CNTs, fullerenes, CDs, CNHs, and CNOs have been incorporated in electrochemical sensors in different ways to enhance the performance of the sensor by exploiting their advantages such as large surface area, chemical stability, electrical conductivity, and mechanical strength. Compared to the conventional non-nanomaterial-based electrochemical sensors, CNM-incorporated sensors often show enhanced analytical performance (Kour et al., 2020). Other factors that helped to explore CNMs by the electrochemical sensor research community are their properties that are entirely different in bulk materials. Two of the most important properties among them are their large surface area and electrical conductivity at room temperature. Also, apart from contributing to the high selectivity and sensitivity, their biocompatible nature has revolutionized the field of electrochemical biosensors. CNMs, especially graphene, have been widely explored in the biosensor field as an immobilizing platform for biomolecules and also as the sensing element (Fig. 21.3).

A number of strategies have been used for incorporating the nanomaterials onto the electrochemical sensing platform. These include drop-casting, direct development of nanomaterial on the sensing platform and in the form of composites with metal nanoparticles or polymers. Here we aim to discuss the different types of nanomaterials employed in electrochemical sensors.

Graphene and its derivatives

Since its discovery in 2004 by Geim et.al., graphene, the one-atom-thick, two-dimensional allotrope of carbon, has made explosive growth in the field of electrochemical sensors (Zhang & Zhao, 2014). Due to its inherent properties such as large surface area and high conductivity, it has always overshadowed many other carbon materials used in this field. Incorporation of graphene and its derivatives into electrochemical sensors in different ways has enabled the determination of a wide range of analytes owing to its wide electrochemical potential window (~2.50 V) and high electron transfer rate (Raj & John, 2018). Moreover, electronic properties of graphene can be easily tailored by doping and/or functionalization with heteroatoms (Ray, Mishra, Strydom, & Papakonstantinou, 2015).

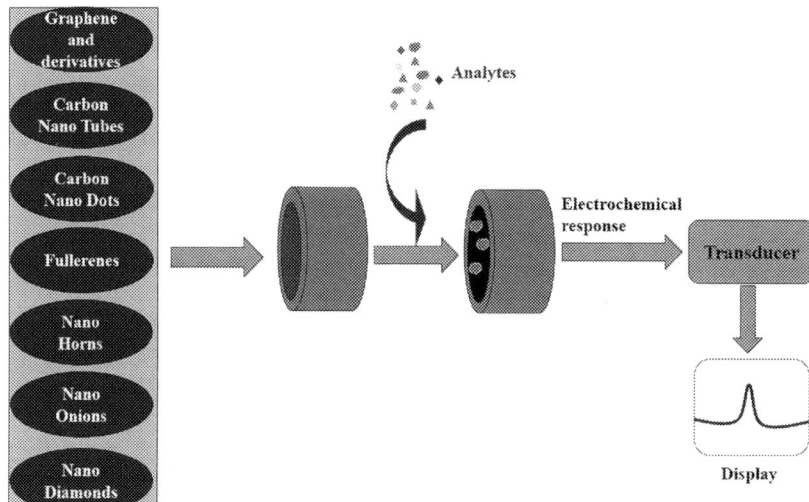

FIG. 21.3 General schematic representation of an electrochemical sensor.

Papakonstantinou and coworkers were the first to explore multilayer graphene nanoflake films as the sensing element for the simultaneous voltammetric determination of dopamine, ascorbic acid, and uric acid (Shang et al., 2008). From there on, plenty of electrochemical sensors have been reported employing graphene and its derivatives such as GO, RGO, functionalized graphene, composites of graphene, GO with metal nanoparticles, and conducting polymers for a wide range of sensing applications.

In most cases, drop-casting method was used for the modification of electrode surface with graphene and its derivatives, though question arises regarding the reproducibility of the method (Mani et al., 2017; Qian, Thiruppathi, Elmahdy, van der Zalm, & Chen, 2020; Xue et al., 2021). The limitations of drop-casting methods can be overcome by chemical vapor deposition (CVD) technique, where graphene is deposited as a thin film onto the conducting substrate (Ruppi & Larsson, 2001). CVD graphene exhibits high purity, lesser defects, and high electrocatalytic activity, and this allows the electrochemical sensors to benefit from these properties (Bai, Xu, & Zhang, 2020; Bai, Zhang, Li, & Luo, 2020; Wang, Sparkman, & Gou, 2017). Apart from these advantages, the fact that CVD graphene can be formed only on certain metallic surfaces and also, the tiresome transfer process, which may lead to contamination and damage to the graphene sheets, limits its applicability in electrocatalytic applications (Awasthi, Madhu, & Singh, 2014).

Derivatives of graphene that finds great applications in electrochemical sensors are GO and RGO. Several oxygen-containing functional groups in GO allow fast electron transfer between the analyte molecules and the electrode surface. Also, better dispersibility and tunable functional properties make it suitable for biosensor fabrication (Bai, Xu, & Zhang, 2020). Thermal, photothermal, and chemical reductions of graphene are used for producing RGO from GO (Bai, Xu, & Zhang, 2020). Another method for the preparation of RGO is via electrochemical technique, where the resulting RGO is more appropriately known as electrochemically reduced GO (ERGO) (Toh, Loh, Kamarudin, & Daud, 2014) . In fact, ERGO exhibits some pristine characteristics of graphene and at the same time possesses functionalities of RGO. To date, a number of electrochemical sensors have been reported using GO, RGO, and ERGO for a wide range of analytes including pollutants, pesticides, and biomolecules (Anand, Mathew, Radecki, Radecka, & Kumar, 2020; Bai, Zhang, Li, & Luo, 2020; Thiruppathi, Sidhureddy, Keeler, & Chen, 2017; Wang et al., 2021).

Current trends in graphene-based electrochemical sensors involve altering the electrochemical properties of graphene and its derivatives by chemical doping (with heteroatoms or by using as composites with metal nanoparticles or polymers (Méndez Romero, 2017). In the case of doping with heteroatoms such as nitrogen, phosphorous, silicon, fluorine, the electrochemical activity is greatly improved by the synergistic effect of high conducting properties of metal atoms as well as the large surface area of graphene (Lee et al., 2016; Thiruppathi et al., 2017). The doped heteroatoms provide new active sites for interaction with the analyte molecules and result in enhanced sensing performance (Kaushal, Kaur, Kaur, Kumari, & Singh, 2020). By cleverly choosing proper dopants for the functionalization of graphene depending on the nature of analyte molecules, researchers can make them suitable for specific sensing applications (Jerome & Sundramoorthy, 2020; Manavalan, Veerakumar, Chen, & Lin, 2020; Zhang, Wang, Zhang, Zhang, & Shen, 2018).

Graphene-polymer nanocomposites are nowadays greatly employed in electrochemical sensors. A composite can be regarded as made up of two or more constituent materials with substantially different physical and chemical properties that, when coupled, generate a material with unique characteristics (Sun et al., 2021). High surface area and unquestionable thermal, electronic, and mechanical properties make graphene and its derivatives a suitable filler material in polymer matrix. Graphene-polymer nanocomposites have emerged to be an excellent material for electrode modification due to their unique properties that are different from the polymer and graphene alone. These include faster electrode kinetics, enhanced selectivity and sensitivity to target analytes, and mechanical stability, and most importantly, it finds greater applications in electrochemical biosensor field. In situ electropolymerization is the widely accepted method for preparing graphene-polymer composites, as it is simple and less time-consuming (Krishnan, Tadiboyina, et al., 2019). To name a few, the most widely used polymers include polythiophene, polyaniline, polypyrrole, and their derivatives (Batool et al., 2019; Chen et al., 2021; Kathiresan et al., 2021; Rajendran, Reshetilov, & Sundramoorthy, 2021). Apart from these, electropolymerizable amino acids are also in trend on account of their low-cost and higher electrochemical response (Anand et al., 2020).

The applications of graphene and its derivatives in the field of electrochemical sensors are far-reaching, and one of the fastest growing areas in this regard is electrochemical biosensors. Integration of graphene or derivatives alone or in the form of composites into the biosensor, along with a high-affinity biorecognition element (aptamer, DNA, etc.), can provide sensitive and specific detection strategy for biomolecules (Liaqat, Riaz, Hasnain Nawaz, Mihaela Badea, & Louis Marty, 2021). Several such biosensors have been reported employing different electrochemical techniques such as amperometry, voltammetry, impedance, and field-effect transistors (Khan & Song, 2021; Park et al., 2021; Villalonga et al., 2021). Apart from acting as a signal amplifier, properly functionalized graphene and its derivatives can act as the immobilizing platform for the biorecognition element. For example, AuNP-RGO composite is widely used in biosensors. The composite acts as an immobilization platform and at the same time helps to anchor target molecules onto it (Khan & Song, 2021; Meng et al., 2021).

The contribution of graphene and its derivatives toward the growth of electrochemical sensors is commendable. Electrochemical sensor research field will be guided in the future by developments in fundamental knowledge on novel graphene-based materials and is going to create a significant breakthrough in this area.

Carbon nanotubes

Because of their unique properties and potential applications in every field of nanotechnology, CNTs have been the subject of research in recent decades. CNTs are allotropes of carbon composed of sp^2 carbon units arranged in the form of single (SWCNT) or multiple (MWCNT) cylindrical rolls of graphene (Oliveira & Morais, 2018). Their peculiar one-dimensional properties make them interesting candidates in analytical, pharmaceutical, and biomedical fields. Large surface area, mechanical stability, biocompatibility, modifiable ends with sidewalls suitable for functionalization, and possibility of ultrasensitive determination and miniaturization are the fascinating features of CNTs that are advantageous for electrochemical sensors (Hu & Hu, 2009). Properties of CNTs can be tailored by proper functionalization. CNTs are basically hydrophobic in nature, proper functionalization (e.g., treatment with a mixture of acids) can make them soluble in aqueous solutions. Introducing defect sites and functionalization by oxygen-containing groups significantly improves their properties and is evident from their fast electrode reaction kinetics (Kour et al., 2020). In such a way, customizing the electrochemical properties of CNTs by incorporating suitable functional groups, metal nanoparticles, or conducting polymers can add positive synergistic effects resulting in an enhanced performance of the sensor. The mechanism behind the electron transfer along the CNT structural networks is still a matter of discussion. Some studies have shown that the open ends and defect sites in CNTs are involved in the mechanism and the sidewalls are inert (Yáñez-Sedeño, Campuzano, & Pingarrón, 2017). Contrary to this, some other studies provide evidence for the charge transfer along the sidewalls (Dumitrescu, Unwin, & MacPherson, 2009).

The unique properties of CNTs are widely used for developing biosensors. CNTs can be easily modified with the biosensing elements such as nucleic acids and peptides. Also, they can be modified with suitable functional groups for selective detection of biomolecules. In these cases, CNTs perform a dual function by acting as an immobilization platform for the biorecognition element and also by contributing to enhanced electrochemical activity (Baptista et al., 2015). Covalent or noncovalent functionalization of CNTs is the most commonly used ones in biosensors. In covalent mode, different functional groups such as carboxylic or amino groups are introduced onto CNTs via chemical reactions, and they can interact with the complementary functional groups of target biomolecules, which leads to its immobilization on the surface of CNTs (Fu et al., 2015; Hasan et al., 2018; Shao, Shurin, Wheeler, He, & Star, 2021; Viswanathan, Rani, Vijay Anand, & Ho, 2009). Though covalent functionalization allows strong interaction between CNTs and biomolecules, it may change the intrinsic properties of CNTs and may affect its mechanical stability.

Interactions such as π-π interactions, electrostatic interactions, and CH-π interactions are made use of in noncovalent functionalization of CNTs. Noncovalent functionalization allows retaining intrinsic properties of CNTs including their structural, electronic, and mechanical properties. Many aromatic compounds such as benzene, pyrene, ferrocene, anthracene, porphyrins, thionine are incorporated into CNTs for immobilization of biomolecules by exploiting π-π stacking interactions between aromatic rings and CNT surface (2016) (Cluff & Blümel, 2016; Kour et al., 2020; McQueen & Goldsmith, 2009). Apart from aromatic compounds, polymers are also extensively used for functionalizing CNTs on account of their conducting nature and possible π-π stacking interaction with CNTs. Polypyrroles, polyethyleneimines, glycolipids, and cellulose-based polymers are greatly employed along with CNTs (Karabiberoğlu, 2016; Lo et al., 2020). Modification of CNTs with polymer can be performed in different ways. There are three most commonly used pathways: (1) electropolymerizing polymer onto CNT-casted electrode surface, (2) casting CNTs onto polymer-coated electrode surface, and (3) modification of CNTs and polymer together on the electrode surface. The synergistic effect offered by the properties of CNTs and polymers is often found to depend on the pathway of modification used (Lo et al., 2020).

Many other different strategies are also employed for modifying CNTs. One among them is by incorporating metal oxides or nanoparticles. Compared with the individual electrocatalytic properties of metal particles and CNTs, the composite shows superior properties due to the synergistic effect of high conductivity of CNTs and electrocatalytic properties of metal particles. To date, several studies have been reported employing a number of metal oxides and nanoparticles Au, Ag, Co, Pt, etc., with CNTs for the determination of a wide range of analytes (Amatatongchai, Sroysee, Chairam, & Napricha, 2017; Erady et al., 2017; Lin, Wang, Zou, Lan, & Ni, 2019).

Fullerenes

Fullerenes, the zero-dimensional allotrope of carbon, has attracted the attention of electrochemical research community owing to their broad range of fascinating properties such as high electronic conductivity, large surface area, biocompatibility, and high adsorption capacity toward molecules (Afreen, Muthoosamy, Manickam, & Hashim, 2015; Yáñez-Sedeño et al., 2017). Buckminsterfullerene (C_{60}) having truncated icosahedron geometry is the most abundant in the fullerene family and is the most widely employed in electrochemical sensors (Kratschmer, 1990). Apart from the advantages listed earlier, what makes fullerenes good candidates in electrochemical sensors are their scope of derivatization and good electron-accepting ability. Though electrocatalyzing properties of fullerenes have been proven without a doubt, poor solubility in aqueous medium limits its applications (Yáñez-Sedeño et al., 2017).

For biosensing applications, the problem of limited solubility in aqueous media has to be overcome and this has created greater interest among researchers for introducing various functionalization methods (Wang & Dai, 2015). As explained earlier in the case of CNTs, covalent and noncovalent functionalization methods are employed for the functionalization of fullerenes. Introduction of polar groups such as hydroxyl, carboxyl, and amine groups into fullerenes significantly improves their solubility and also contributes to enhanced selectivity and sensitivity of the sensor (Han et al., 2013; Pilehvar & Wael, 2015). Noncovalent functionalization of fullerenes involves utilizing supramolecular interaction with the functional groups. Supramolecular interactions are noncovalent interactions such as van der Waals forces, π-π stacking, electrostatic, and host-guest interactions between molecules leading to the formation of complex chemical systems. Studies on fullerene chemistry show that functionalization in fullerenes can be exohedral or endohedral depending on whether the functional groups are attached respectively, outside or inside the cage of fullerenes (Bobrowska & Plonska-Brzezinska, 2021; Filippone, 2020). It is proven that exohedrally functionalized fullerenes contribute more toward high selectivity and sensitivity of electrochemical sensors (Afreen et al., 2015). Currently, there exists a range of methods for modifying electrodes with fullerenes or their derivatives. Drop-casting fullerene onto the surface of the electrode is the most commonly used method (Han, Meng, Zhang, & Zheng, 2018). Apart from this, electrodeposition, electropolymerization, and self-assembled monolayer methods are also used (Bedioui, Devynck, & Claude, 1995; Chang et al., 2014; Imahori et al., 1999).

Fullerenes, on account of their ease of functionalization and ability to act as a signal mediator, find a wide range of applications in biosensors. Different biorecognition elements such as DNAs, aptamers, antibodies, and enzymes can be immobilized onto functionalized fullerenes via adsorption or covalent interaction with the functional groups (Zhang, 2020). Fullerenes incorporated with metal nanoparticles, polymers, and other CNMs are also employed in electrochemical sensors as a modifier and immobilization support (Fig. 21.4) (Anusha, Bhavani, Kumar, & Brahman, 2020; Motoc, Manea, Orha, & Pop, 2019; Rahimi-Nasrabadi, Khoshroo, & Mazloum-Ardakani, 2017; Uygun et al., 2018; You et al., 2020). For example, fullerene-gold nanoparticle composites have been widely encountered in sensors since it offers the synergistic effects of high conductivity of metal nanoparticles and the large surface area of fullerenes (Pilehvar & Wael, 2015; Sutradhar & Patnaik, 2017). Fullerenes and their derivatives exhibit surprising electrical and mechanical properties that attract greater attention and this has contributed to the development of electrochemical sensors employing fullerenes, especially in the field of biosensors. However, their properties are not fully explored until now and need further studies to extend the range of application of fullerenes in the field of electrochemical sensors.

Carbon nanodots

Carbon nanodots are one of the zero-dimensional CNMs containing 10^3 to 10^4 atoms. Carbon nanodots can be categorized into carbon quantum dots (CDs) and graphene quantum dots (GQDs) (Wang & Dai, 2015). Both differ in their origin and show notable differences in structure. CDs can be prepared from any carbon-based materials and contain quasispherical, amorphous, or nanocrystalline particles with sizes lesser than 10 nm. On the contrary, GQDs are derived from graphene-based materials and have lateral sizes less than 100 nm (Hassanvand, Jalali, Nazari, Parnianchi, & Santoro, 2021). The quantum confinement property exhibited by CDs causes them to show luminescent properties, which varies according to the size and functional group attached to the particles. Apart from this, CDs exhibit reducing properties and they are able to act as electron acceptors from other species. This makes CDs an excellent candidate in developing electrochemical sensors (Cesana et al., 2020). Both CDs and GQDs exhibit high surface area, conductivity, low toxicity, biocompatibility, and solubility in aqueous media, and they can be functionalized easily with other molecules due to the presence of abundant edge sites (Hassanvand et al., 2021). So far, CDs have been used in electrochemical sensors in the form of nanocomposites or chemically doped with other materials and this allows tailoring their electrocatalytic, electronic, and thermal properties. Recently, several studies have been reported employing CDs doped with metal nanoparticles. The high selectivity and

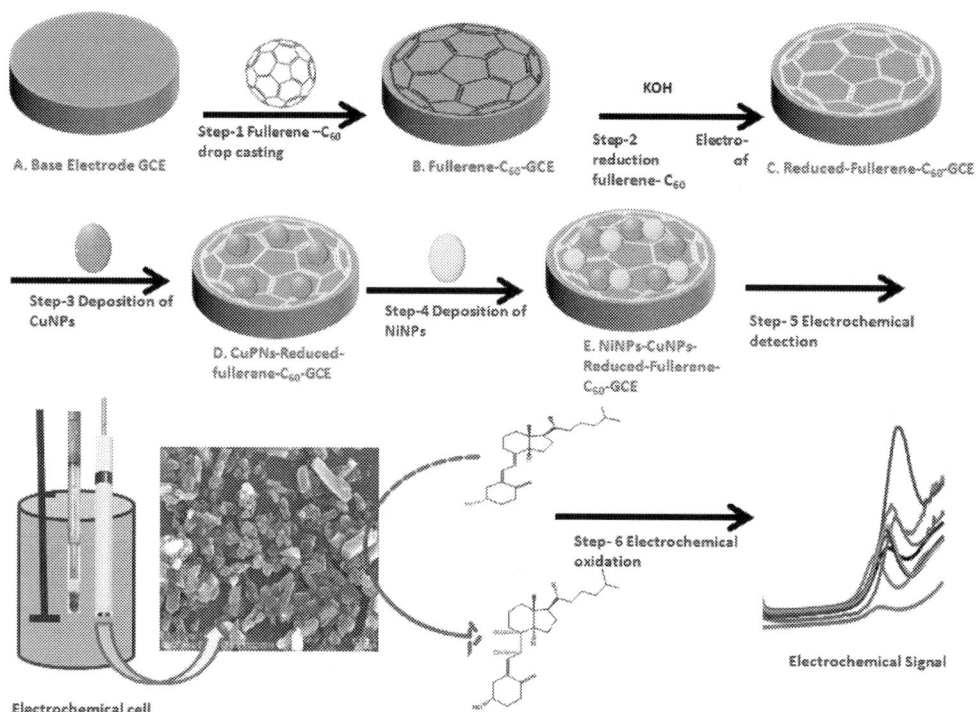

FIG. 21.4 Schematic illustration of the fabrication of CuNP-NiNP@reduced-fullerene-C60/glassy carbon electrode (Anusha et al., 2020). *From Anusha, T., Bhavani, K. S., Kumar, J. V. S. & Brahman, P. K. (2020). Designing and fabrication of electrochemical nanosensor employing fullerene-C60 and bimetallic nanoparticles composite film for the detection of vitamin D3 in blood samples. Diamond and Related Materials, 104. https://doi.org/10.1016/j.diamond.2020.107761, Copyright (2021), with permission from Elsevier. License Number: 5062610387509.*

sensitivity achieved in these sensors suggest the strong synergistic effect of the electrocatalytic activity and conductivity offered by the metal nanoparticles and CDs (Dondo, Shumba, Moyo, & Nyoni, 2020; Gu, Hsieh, et al., 2021).

CDs, the newest member in the family of CNMs, find a wide variety of applications in electrochemical sensors as nanocarriers and nanoprobes. Due to the presence of π-electron conjugation and a large number of functional groups (carboxyl, hydroxyl, epoxy, and carbonyl groups) at their edges, they can act as reaction sites for molecules, ultimately providing high sensitivity to the electrochemical sensor (Benítez-Martínez, López-Lorente, & Valcárcel, 2014). Also, high adsorption capability allows immobilization of various inorganic, organic, and biomolecules on their surface. On account of their fast electron transfer ability and large surface area, CDs and their derivatives modified electrodes have been widely used for the selective and sensitive determination of heavy metal ions such as Cu^{2+}, Cd^{2+}, and Hg^{2+} (Echabaane, Hfaiedh, Smiri, Saidi, & Dridi, 2021; Saisree, Aswathi, Arya Nair, & Sandhya, 2021; Ting, Ee, Ananthanarayanan, Leong, & Chen, 2015). In electrochemical biosensors, CDs and GQDs serve to amplify the electrochemical signal and at the same time act as a platform for anchoring the recognition element (Ganganboina, Dega, Tran, Darmonto, & Doong, 2021; Ganganboina & Doong, 2019; Mansuriya & Altintas, 2021). Apart from these, there are many recent reports involving the use of CDs and their derivatives as nanoprobes for the determination of various biomolecules, pharmaceuticals, and pollutants (Jalali, Hassanvand, & Barati, 2020; Mollarasouli, Majidi, & Asadpour-Zeynali, 2018; Moundzounga et al., 2021; Sanati, Faridbod, & Ganjali, 2017; Wang, Zhang, et al., 2021; Wang, An, Dai, & Luo, 2021; Zhuang, Wang, He, & Chen, 2016).

Another important application of CDs in sensor field is in the development of nonenzymatic sensors for hydrogen peroxide (H_2O_2). H_2O_2 is an oxidant whose higher concentration in blood can be dangerous to humans (Su, Zhou, Long, & Li, 2018). Electrochemical sensing technology is found to be the most promising method for the determination of H_2O_2 and it uses proteins and enzymes such as hemoglobin, myoglobin, and horseradish peroxidase that are highly specific to H_2O_2 (Su et al., 2018). CDs and GQDs can mimic the activity of these enzymes and can replace them in electrochemical sensors since they are costly and difficult to store. Several studies have been reported employing CDs and GQDs along with metal nanoparticles and other CNMs and dyes for the determination of H_2O_2 with high selectivity and sensitivity (Bai, Sun, & Jiang, 2016; Jahanbakhshi & Habibi, 2016; Yang, Hu, et al., 2015).

CDs and GQDs are relatively new members of the carbon nanomaterial family that attracted vast attention among the researchers. Owing to their large surface area, fast electron transfer rate, low toxicity, and high chemical and mechanical stability, CDs find a wide range of applications in the area of electrochemical sensors and biosensors. Many significant properties of CDs and GQDs are still to be explored, and new carbon sources, introducing new functionalization methods, can bring more fascinating applications of CDs and GQDs in electrochemical research field.

Carbon nanohorns

Single-walled CNHs, one of the many allotropes of carbon, are graphitic, horn-shaped tubes, with a diameter in the range of 2 to 3 nm and a tube length of 40 to 50 nm. They are derived from SWCNTs and the horn ends are closed by a five-pentagon conical cap (Ortolani et al., 2019). CNHs are promising nanomaterials for electrochemical sensors due to their high electrocatalytic activity, adsorption capability, microporosity, and high chemical and mechanical stability (Baptista et al., 2015). As in the case of other CNMs, chemical functionalization is necessary to introduce various functional groups onto the surface of CNHs and for manipulating their properties to enhance the performance of the sensor (Karousis, Suarez-Martinez, Ewels, & Tagmatarchis, 2016). Attaching various functional molecules covalently on the surface of CNHs via stable bond formation is one way of functionalizing CNHs. While functionalizing covalently, the functional groups can be directly introduced to the sidewalls of CNHs or the conical ends can be subjected to oxidation in an effort to introduce oxygenated functional groups (Pagona et al., 2007). Moreover, noncovalently, functional groups can be introduced via π-π stacking interactions, electrostatic interactions, or immobilization of inorganic nanoparticles onto the surface of CNHs.

Electrochemical biosensors employing CNHs with immobilized proteins, aptamers, and enzymes have been reported for molecules such as glucose and H_2O_2 (Liu et al., 2010; Liu, Shi, Niu, Li, & Xu, 2008). Also, the enzyme mimicking action of CNHs has been explored and several nonenzymatic CNH-based electrochemical sensors have been reported for H_2O_2 and glucose (Fig. 21.5) (Bracamonte et al., 2017; Zheng et al., 2020). Apart from these, CNHs combined with metal nanoparticles, metal-organic frameworks, conducting polymers, and other CNMs have been widely encountered in electrochemical sensors as a nanoprobe, by exploiting their characteristic properties such as large surface area, high electrical conductivity, microporosity, and biocompatibility, for various biomolecules, heavy metal ions, and environmental pollutants (Tu et al., 2020; Wang, Lan, et al., 2020; Wang, Pan, Liu, Ye, & Yao, 2020; Yao, Wu, & Ping, 2019; Zhu et al., 2018).

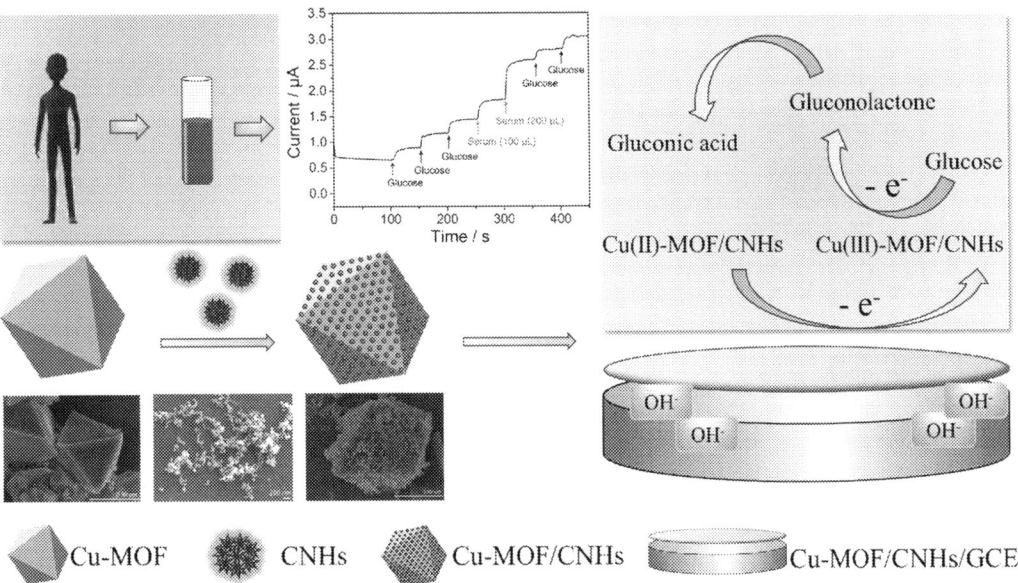

FIG. 21.5 Schematic representation of the fabrication of Cu-MOF/CNH-modified glagssy carbon electrode and the application for glucose detection (Zheng et al., 2020). *From Zheng, W., Liu, Y., Yang, P., Chen, Y., Tao, J., Hu, J. & Zhao, P. (2020). Carbon nanohorns enhanced electrochemical properties of Cu-based metal organic framework for ultrasensitive serum glucose sensing. Journal of Electroanalytical Chemistry, 862, 114018. https://doi.org/10.1016/j.jelechem.2020.114018, Copyright (2021), with permission from Elsevier. License Number: 5062610612360.*

Carbon nanohorns are superior to CNTs and graphene in several properties. They can be synthesized at high yields with high purity without any metallic contaminations. Though several studies are being carried out for tailoring their properties, fundamental knowledge of the unique characteristics of CNHs is necessary to extend their applications in the field of sensors.

Nanodiamonds and carbon nano-onions

NDs and CNOs are the zero-dimensional allotropes of carbon. NDs are nanoparticles of carbon with truncated octahedral geometry. They resemble CDs in size, their diameter being in the range of 2 to 8 nm (Wang & Dai, 2015). NDs contain more sp^3 hybridized carbons and oxygen-containing functional groups. They are characterized by a wide potential window, biocompatibility, high mechanical strength, and chemical inertness (Wang & Dai, 2015). CNOs are one of the emerging CNMs that are nowadays widely encountered in electrochemical sensors. Though CNOs were discovered during the period when CNTs were known by researchers, they are less explored in this field (Zuaznabar-Gardona & Fragoso, 2020). CNOs consist of a hollow spherical core of fullerenes surrounded by concentric graphitic layers. CNOs are considered as an emerging class of CNMs owing to their less toxicity, high surface-to-volume ratio, biocompatibility, and electrocatalytic activity (Sohouli, Shahdost-Fard, Rahimi-Nasrabadi, Plonska-Brzezinska, & Ahmadi, 2020). Of the different methods employed for synthesizing CNOs, those obtained from thermal annealing of nanodiamonds are found to have greater advantages such as larger Brunauer-Emmet-Teller (BET) surface area conductivity and electrocatalytic activity (Breczko, Plonska-Brzezinska, & Echegoyen, 2012; Kuznetsov, Chuvilin, Butenko, Malkov, & Titov, 1994).

Exploiting the characteristic properties exhibited by NDs and CNOs, a number of electrochemical sensors have been reported for a wide variety of molecules. Ease of functionalization of these materials by covalent and noncovalent methods allows tailoring their properties to enhance the performance of the sensor (Baptista et al., 2015). Recently, several sensitive and selective sensors and biosensors based on NDs and CNOs combined with metal nanoparticles, other CNMs, and conducting polymers have been reported for various molecules, wherein the synergistic effect results in enhanced electrochemical response (Sohouli et al., 2020; Mohammadnia et al., 2020; Rizwan, Hazmi, Lim, & Ahmed, 2019; Simioni, Silva, Oliveira, & Fatibello-Filho, 2017; Zuaznabar-Gardona & Fragoso, 2020).

NDs are demonstrated to have low or no toxicity, whereas CNOs exhibit toxicity depending on their size, with smaller CNOs being less toxic (Baptista et al., 2015). Many key features associated with these CNMs are yet to be identified and extended studies on their properties can bring about remarkable changes in the electrochemical sensor field.

Market of carbon nanomaterials: Applications in sensors

Fullerenes, CNTs, graphene and its derivatives, NDs, and CDs are all members of the carbon band. Dimensions are used to categorize nanomaterials in general. None, one, two, and three are the four dimensions that accompany (Georgakilas, 2015). The measurements of a substance that is below the nanoscale (less than 100 nm) spectrum are used to classify it. Nanoparticles are the most basic 0D nanomaterials. One dimension is outside the nanoscale for one-dimensional (1D) nanomaterials. Nanotubes, nanorods, and nanowires are also members of this community. Two dimensions are outside the nanoscale when it comes to two-dimensional (2D) nanomaterials. Graphene, nanofilms, nanolayers, and nanocoatings belong to this category, which have plate-like forms. Materials that are not limited to the nanoscale of any dimension are called three-dimensional (3D) nanomaterials (Georgakilas, 2015). Bulk powders, nanoparticle dispersions, nanowire and nanotube clusters, and multinanolayers will all fall in this category.

These materials have piqued interest in a variety of fields, including biomedical applications, due to their unusual structural measurements and outstanding properties (Patel, Singh, & Kim, 2019).

This section defines, describes, and forecasts the global market of the different types of CNMs for various applications, especially for sensors.

Market overview

CNTs, fullerenes, and mesoporous carbon compounds are all significant CNMs with properties that are significantly different from those of other types of carbon such as graphite and diamond (Georgakilas, 2015). The unique electronic properties of CNT have become a key source of demand for CNMs. Applications of fullerenes in superconductivity, supramolecular assembly, and thin films have resulted in a substantial shift away from CNMs in a wide variety of end-use industries (Transparencymarketresearch, 2021). CNFs, CNTs, fullerenes, graphene, and polyhedral oligomeric silsesquioxane comprise the demand for CNMs (Transparencymarketresearch, 2021). CNMs exhibit semiconductor properties, which distinguish them from traditional graphite. CNMs' semiconductor properties allow catalysis by direct involvement in the charge transfer mechanism. Additionally, CNMs allow for the modification of their charge transfer properties and can be used to design sensors and also hydrogenation catalysts, fuel cells, etc. (Transparencymarketresearch, 2021). CNMs are predicted to expand at a rapid rate over the projected period due to their widespread use in a variety of end-use industries, including health care, automotive, manufacturing, and packaging. CNMs are expected to grow in popularity due to their future applications in electronics and energy storage systems as they possess extraordinary electrical conductivity (Transparencymarketresearch, 2021). The growing need for high-strength and durable structural materials, as well as the widening applications of CNM reach in aircraft, nanomedicines, consumer goods, and water treatment, is expected to drive up market demand for CNMs (Transparencymarketresearch, 2021). Over the next six years, rapid urbanization and increasing investment in the construction and medical industries, especially in emerging economies such as China and India, are likely to drive demand for CNMs (Transparencymarketresearch, 2021). However, the market's growth is expected to be constrained by high processing costs and strict environmental regulations.

Market segmentation

CNTs dominated the CNM industry in 2019, owing to their comprehensive properties such as high thermal conductivity, electrical conductivity, elasticity, tensile strength, and durability, as well as their electron field emitter properties and low thermal expansion coefficient (Transparencymarketresearch, 2021). CNT composites have a large surface area, excellent chemical stability, and a rich electronic polyaromatic composition, which enables them to absorb or conjugate a wide range of biological molecules such as drugs, hormones, antibodies, DNA, and enzymes (Transparencymarketresearch, 2021). These composites are in increasing demand for applications in different fields such as ground emission, thermal conductivity, energy storage, conductive adhesives, molecular electronics, catalyst supports, actuators, sensors, electromagnetic interference (EMI) shielding, and biomedical during the forecast time frame.

CNTs have been segmented into three submarkets: structure, process, and end-user sector. The commercial demand for CNTs has been segmented into single-SWCNTs and MWCNTs. The MWCNT segment dominated the global market of CNTs in terms of production, owing to the comparatively low price of commodity. The sector is expected to contribute more than 22% of total company expansion (Transparencymarketresearch, 2021). The high degree of the molecular structure of graphene results in more robust and solid nanotubes that are used in sensors, polymers, electronics, and energy applications. MWCNT is becoming more prevalent in manufactured polymers due to its ability to repel airborne particulate and increase the resistance of devices to static electricity (Transparencymarketresearch, 2021).

The global industry on CNT has been segmented according to process into chemical vapor deposition (CVD), high-pressure carbon monoxide disproportionation, arc discharge, and laser ablation. In 2019, the CVD segment dominated the CNM industry (Industryarc, 2021). Due to its high throughput, high purity, and low cost of operation, CVD has become the predominant form of film deposition in the semiconductor industry. CVD is often used in optoelectronics, optical coatings, and coatings of wear-resistant components, where it is often used in conjunction with CNMs (Industryarc, 2021). CVD technologies have a number of benefits over physical vapor deposition (PVD), including improved molecular beam evaporation and sputtering. CVD relies on the supply of source materials into the process chamber from external reservoirs that can be refilled without contaminating the growth area (Industryarc, 2021). Additionally, CVD does not need very high vacuum levels and can typically handle larger batches of substrates. Moreover, the CVD approach is more forgiving in terms of its flexibility for precise process conditions during the manufacture of CNMs. Thus, these are the primary drivers of CVD growth over the projected period.

By end-use sector, the global market on CNT is segmented into aeronautics, electrical and electronics, automotive, gasoline, and sports. Electricals and electronics held the highest market share of approximately 42% by volume in 2018, owing to the increasing use of CNTs in the industry due to their superior properties (Market Research Future (MRFR), 2021). In 2019, the sector had a market capitalization of USD 1.12 billion (Market Research Future (MRFR), 2021).

The electronics and semiconductor segment dominated the CNM industry in 2019 and is rising at a 16.4 percent compound annual growth rate, owing to the widespread use of the material in electronic industries as pressure sensor materials (Industryarc, 2021). Thermal conductivity is needed for heat dissipation in electronic devices, and hence, carbon-based nanomaterials have been engineered to enhance thermal, electrical, and mechanical properties. According to the Electronic Industries Association of India (ELCINA), overall production of the electronics sector in India was Rs.3,87,525 crore in 2017–18, up from Rs.3,17,331 crore in 2016–17, a growth rate of approximately 22% (Industryarc, 2021). Electrical and magnetic properties of nanomaterials enhance the electrical conductivity of ceramics and increase the electric resistance of nanometals, making them unique in their ability to be widely used in electronic material fields such as sensor fabrication, EMI shielding, optoelectronics, superconductivity, and memory chips, which is expected to drive the CNM market.

Regional insights

In 2013, North America was the main market, followed by Europe, owing to the extensive use of CNMs in the medical and electronics industries in the region (Grandviewresearch, 2021). Europe had the largest share of the CNM industry in 2019, up to 42%, owing to government funding and investment in the domestic markets (Industryarc, 2021). In 2016, the German government launched The Action Plan Nanotechnology 2020, an interdepartmental initiative for the support of nanotechnology from 2016 to 2020 (Industryarc, 2021). Thus, the goal of sustaining the benefits and promise of nanotechnology in Germany is pursued, thus taking into account any potential threats to humans and the environment. The French Government unveiled plans to spend €600 million (£516 million) in a five-year nanotechnology research and development (R&D) initiative led by French-Italian semiconductor maker STMicroelectronics for the Nano 2017 Program (ST) (Industryarc, 2021). Thus, numerous government investments in the field of nanomaterials in Europe are expected to boost the CNM industry during the forecast era (2021–2026) in the region.

Over the last few years, Japan has been the main supplier of CNMs, followed by China and South Korea. Electronics and automotive are projected to be the primary end-use industries (Transparencymarketresearch, 2021). Europe is expected to follow a similar trajectory, owing to the region's developed automotive industry (Transparencymarketresearch, 2021). Latin America is expected to rise at a steady pace over the next few years as the car and electronics industries expand in countries such as Brazil (Transparencymarketresearch, 2021). Middle East and Africa are projected to be a highly lucrative geographic segment for the CNM market over the forecast period (2016–2024), owing to the GCC countries' strong spending in the defense sector and the expansion of South Africa's automotive industry (Transparencymarketresearch, 2021).

Asia Pacific is expected to expand at the fastest rate due to rapid industrialization and increased investment in the medical and automotive industries in China, India, South Korea, and Singapore, among others (Grandviewresearch, 2021).

The Asia-Pacific CNT industry is expected to expand at an approximately 11.6 percent compound annual growth rate (CAGR) in terms of sales during the forecast period (2020–2027). The presence of a sizable manufacturing base in countries such as China, Japan, Australia, and India is projected to fuel demand for CNTs over this forecast era (Alliedmarketresearch, 2021).

Advances in science and technology, combined with the demand from a diverse range of end-use industries in the field, have prompted companies to ramp up R&D in CNMs, which is expected to benefit the CNM industry.

Competitive market share

Carbon nanomaterial manufacturers; raw material suppliers, traders, distributors, and suppliers; regional manufacturers' associations; general carbon nanomaterial associations; government and regional agencies and research organizations; investment research firms; traders, distributors, and suppliers of technology, chemicals, or components are the primary target audiences of CNMs (MarketsandMarketsTM, 2021). Companies such as Ahlstrom Corporation, E. I. du Pont de Nemours and Company, Hollingsworth & Vose, and Kuraray Co., Ltd. are some of the main players in this business, while others include Arkema, Bayer AG, Showa Denko, Finetex, Elmarco, Nanocyl SA, CNano Technology Ltd., and Hyperion Catalysis International Inc. (Transparencymarketresearch, 2021; Grandviewresearch, 2021). Now when it comes to CNTs, several companies compete in the global industry, including Hyperion Catalysis International Inc, Nanothinx S.A., Continental Carbon Nanotechnologies, Klean Commodities, Nanocyl SA, Arkema SA, and Future Carbon GmbH (Gminsights, 2021). In nature, the demand for CNTs is partly consolidated. The major players are constantly investing in R&D to develop new products and expand their abilities, as well as collaborating with local or regional distributors and manufacturers. Some others among prominent players in the market are Chengdu Organic Chemicals Co. Ltd. (Timesnano), Cabot Corporation, Jiangsu Cnano Technology Co., Ltd., Showa Denko KK, and LG Chem (Mordorintelligence, 2021).

Impact of COVID-19 on the global market of carbon nanomaterials

Since December 2019, the world has been in the grip of a global medical emergency caused by a new strain of coronavirus that is relatively uncommon (SARS-CoV-2) (Market Research Future (MRFR), 2021). Rapidly spreading infectious viruses necessitate the development of novel antiviral agents. However, recognizing the virus alone is insufficient to break the link, and a new wave of antiviral technology is critical for combating this pandemic. CNMs have an excellent record against viruses and can likely overcome a large number of today's healthcare issues (Market Research Future (MRFR), 2021).

Globally, the COVID-19 pandemic epidemic has had an impact on the imports and exports of a variety of commodities. Numerous countries have introduced export controls, greatly impairing supply chains (Alliedmarketresearch, 2021). Additionally, suppliers of CNMs are experiencing delays in obtaining raw materials, which results in postponed shipments to customers. Consequently, the quality of some raw materials has deteriorated significantly, and shipping orders have been severely impacted by vessel shortages and blank sailing, negatively impacting the development of the CNM industry (Alliedmarketresearch, 2021). Likewise, since sectors that rely on CNMs for commodity production, such as automotive and aviation, have been shut down, demand for CNMs has plummeted precipitously, reducing the market development of CNM during the pandemic (Alliedmarketresearch, 2021).

The COVID-19 pandemic has significantly limited the on-going research and development efforts. The low footfall across various sectors is having a negative effect on consumer sales. The rate of consumer growth is predicted to slow significantly during this pandemic. However, this downturn will be temporary, and the economy will quickly reclaim its equilibrium (Futuremarketinsights, 2021; Mordorintelligence, 2021).

Challenges/concerns of carbon nanomaterial-based sensors

In the last five years, research into carbon nanomaterial biosensors has increased. The various new forms of CNMs have fueled this rise in sensor production. There are several successful prototypes in the literature right now, but it is less clear which materials are best and how CNMs and their properties can be designed for potential sensor production. This section explores in detail the various future challenges of CNM-based sensors, which includes combining a deeper theoretical understanding of CNMs into functional nanomaterial applications. Additionally, we will see solutions to various issues.

Toxicity of carbon nanomaterials

CNMs, as one of the major classes of nanomaterials, have a wide range of possible applications across the various forms of nanomaterials. However, the toxicity and side effects of these compounds have lately seen a lot of attention. The findings of the toxicity analysis of carbon nanomaterial show that toxicity is proportional to form, structure, shape, length, diameter, scale, and surface area (Madannejad et al., 2019). The toxicity pathways of CNMs include neurotoxicity, hepatotoxicity, nephrotoxicity, immunotoxicity, cardiotoxicity, genomic and epigenetic toxicity, dermatotoxicity, and carcinogenicity (Madannejad et al., 2019).

CB nanoparticles, CNTs (SWCNTs and MWCNTs), GO, and other forms of CNMs are widely used in the fabrication of various sensors, especially for biomedical applications. However, it is important to know the toxic effects of these before one chooses these materials.

Chemical and biological oxidative effects of CB nanoparticles with mean aerodynamic diameters of 14, 56, and 95 nm have been studied in in vitro analysis, and it was discovered that CB nanoparticles with a 14 nm scale have a greater oxidative potential and stimulate a higher degree of oxidative damage in alveolar epithelial cells than those with larger sizes (56 and 95 nm) (Koike & Kobayashi, 2006). Graphene was the most toxic substance, with overall toxicity of 52.24 percent, based on morphology, concentration, and interaction length, followed by CNTs, fullerene, and carbon nanowire (Koike & Kobayashi, 2006). The surface region of CB nanoparticles and their oxidative potential have a favorable relationship.

Functionalized CNTs (fCNTs), among other CNMs, have been shown to translocate into different subcellular compartments, like brain cells (Bussy et al., 2015). In neuronal cells, MWCNTs triggered DNA damage and overproduction of reactive oxygen species, resulting in a pro-inflammatory response (Bussy et al., 2015).

Nephrotoxicity is characterized as any adverse effect of a drug on renal function. CNTs are compounds that are poorly distributed in various solvents, resulting in immunogenicity in the body and, as a result, limiting their various uses . Surfactants, peptides, polymers, nucleic acids, and oligomers, as well as the incorporation of moieties on the external surface of the nanotube, will significantly increase the solubility of CNTs (Klumpp, Kostarelos, Prato, & Bianco, 2006). The cytotoxicity of polyethylene glycol-modified CNTs on human embryonic kidney cells (HEK-293T) was studied, and it was found that cell viability decreased steadily as the concentration of PEG-CNTs was increased. PEG-CNTs did not cause toxicity in HEK-293T at doses less than 100 μg/mL, but moderate toxicity was observed at concentrations ranging from 100 to 200 μg/mL. Concerns over possible cytotoxicity and adverse effects of GO have grown as human exposure to it has increased. GO-mediated activation of multiple signaling pathways linked to autophagy and necrosis in various types of cells has been identified in various studies (Madannejad et al., 2019). Genotoxicity refers to processes that alter DNA structure, information quality, and segregation but are not always synonymous with mutagenesis (Madannejad et al., 2019). Various studies have shown that MWCNTs and SWCNTs can induce irregular epigenetic regulation in the lung or bronchial cells, such as hyper- and hypomethylation in the promoter region of particular genes (Sierra et al., 2017;

(Ghosh et al., 2018; Tabish et al., 2017). The ability or propensity of a chemical to produce tumors, whether benign or malignant; increase their malignancy; or accelerate tumor incidence is known as carcinogenicity. Several animal experiments have shown that some forms of MWCNTs, including asbestos fibers, are carcinogenic to mesothelial cells and that their pathogenic function is identical to that of asbestos (Madannejad et al., 2019).

Therefore, the inherent toxicity of carbon nanomaterial-based systems in the fabrication of in vivo sensors is a major concern and, like mentioned before, one needs to think twice before incorporating CNMs.

Shortcomings in design and fabrication

While carbon-based materials are excellent candidates for wearable sensors for health tracking, there are still significant challenges in terms of the properties of functional materials and fabrication technologies that allow accurate detection of different human activities as well as physiological signals for practical wearable applications.

Manufacturing technologies should be simple and affordable for the effective development of wearable products. Because of their consistency with a number of solvents, solution-processability, and ease of fabrication, carbon-based nanomaterials have been regarded as suitable suppliers in this regard. Carbon-based nanomaterials can be made in a variety of ways and used in a variety of electronic devices. However, not all of them are appropriate for demonstrating wearable sensors that are lightweight and stretchable.

For the realization of wearable sensor systems, CVD graphene or CNTs grown on Ni or Cu catalyst-based substrates can be patterned using a pattern transferring technique, which is more cost- and time-effective than traditional photolithography (Bae et al., 2013; Marchena et al., 2018; Roy, David-Pur, & Hanein, 2017). The use of adhesive polymers, such as poly(methyl methacrylate) (PMMA) and thermal release tapes to move the target material onto a flexible substrate, still poses the problem of the formation of residual polymers in the pattern transferring process (Marchena et al., 2018). The electrical and mechanical properties of the target material could be damaged by these residual polymers. As a result, additional treatments such as high-temperature thermal annealing and ozone/hydrogen treatment are needed to eliminate the residue while using the pattern transferring technique, which can result in a restriction for a broad array of applications (Majumder, Mondal, & Deen, 2017).

The spray-coating process is one of the most widely used methods for large-area deposition using an airbrush, and it is a simple, versatile, and low-cost method for the deposition of carbon-based films (Majumder et al., 2017). Owing to the use of the airbrush, however, the production precision for supplying the carbon content to the desired location is very poor, which increases the packaging cost and reduces efficiency. To fix the issue, researchers demonstrated CNT-based sensors using spray deposition modeling (SDM) with a digital x–y plotter and a heated substrate (Wang et al., 2017). They were effective in producing CNT patterns and inducing layer-by-layer (LBL) characteristics. Despite the benefits, the adhesion stability of the deposited film is limited since the adhesion to the substrate is still poor. This means that a special functionalized substrate is needed for good adhesion of the carbon-based film deposited by the spray-coating process. It should also be remembered that the methods for functionalizing carbon-based materials could degrade the electrical characteristics of the functionalized film by damaging its conjugated structure.

Screen-printing, gravure printing, and ink-jet printing have also been used to illustrate carbon-based sensor electronics, and there is still a lot of interest in lightweight and stretchable carbon-based sensors. The printing resolution of screen-printing is comparatively poor because it is difficult to produce a desirable carbon-based ink with a controllable viscosity. To achieve high-resolution printing and fine patterning, advanced techniques for carbon-based ink design and printing processes are needed (Zhou & Azumi, 2016). Easy process methods such as drop-casting and vacuum filtration can effectively coat and pattern carbon-based materials onto substrates without any additional processing (Zhou & Azumi, 2016). Drop-casting is an easy and low-cost approach for efficiently shaping a small-area thin film that is more controllable than other coating methods due to the high reliance of droplets used during the drop-casting process (Zhou & Azumi, 2016). However, not only does the method have drawbacks in terms of large-area coverage, but the coffee-ring effect reduces the uniformity of the films on the substrate (Deegan et al., 1997; Zhou & Azumi, 2016).

Different types of CNMs such as CDs, CNFs, graphene, and its forms, which include GNR and graphene flowers, have recently been identified as some of the most commonly used nanomaterials in the fabrication of electrochemical sensors, allowing for strong conductivity and high electron transfer speeds. The fabrication process for CDs, on the contrary, is still in the preliminary stages. There is also work to be done in terms of developing a proper fabrication route that uses sustainable raw materials while delivering high-quality CDs. CDs may be modified for a variety of applications to solve manufacturing problems. Quantum yield and consistency of CDs generated are the key concerns for using CDs (Rani, Ng, Ng, & Mahmoudi, 2020). As a result, future research should focus on streamlining the fabrication process to ensure a consistent product with the desired properties, as well as streamlining the purification system to remove residual reagents.

The main challenges are to comprehend CD functionalities and to rationally build and fabricate CDs using simple methods using green carbon precursors with high quantum yields and a variety of functionalities for a broader range of applications (Rani et al., 2020).

Due to the high mechanical strength and chemical inertness of CNFs, as well as their ability to greatly minimize the oxidation overpotential, CNF-based sensors typically exhibit high stability and selectivity to target molecules (Wang, Cao, et al., 2019; Wang, Wu, Wang, Yu, & Wei, 2019). Although the synthesis and applications of CNFs and CNF-based materials have gained a lot of attention in recent years, we agree that more work should be done in the areas in the following. Firstly, new CNF synthesis methods may be produced. CVD and electrospinning are the two most common approaches for producing CNFs right now (Wang, Cao, et al., 2019). Other approaches, such as template-based synthesis, self-assembly, and chemical hydrothermal methods, may be considered for high-efficiency CNF synthesis (Wang, Cao, et al., 2019).

While nanomaterials have proved to be extremely useful in sensors, their design and fabrication route often necessitate the use of trained technicians to control the sophisticated equipment.

Achieving stability, sensitivity, selectivity, and reproducibility

Stability of the sensor is important, particularly for sensors that rely on antigens, aptamers, or enzymes. As a result, a desirable atmosphere that mimics the cell wall or interior is strongly desired; however, this environment must not decrease electron mobility and rate of sensing, as this will result in either a decline in sensitivity or an increase in fouling (Tiwari et al., 2015). It is possible to solve this problem by incorporating highly conductive CNMs into this biologically favorable environment. Another approach to improve sensitivity in biosensing would be to use labels, i.e., by labeling the analyte, causing an electrochemical reaction to occur, or improving the existing electrochemical reaction used in sensing (Tiwari et al., 2015). By adapting strategies to expand the narrow detection range of biosensors such that delays in disease detection, care, and management can be eliminated.

The following are a few of the major obstacles that have been faced in the production of various sensors: (1) achieving a low detection limit (LOD), (2) suppressing adsorption of nonspecific interfering species, and (3) preserving the reproducibility and stability of the sensor in complex real matrices (Ferrag & Kerman, 2020). The LOD is a critical factor to remember when designing any sensor. LODs as low as picomole and femtomole levels have been reached in certain ultrasensitive sensors due to the development of nanomaterial-modified surfaces (Li et al., 2018; Ponnaiah, Periakaruppan, & Vellaichamy, 2018; Suherman et al., 2017). Nonetheless, these modified surfaces pose a challenge because they are not always as reproducible as one might expect. Since controlling the synthesis and immobilization of nanoparticles with differing populations of size and shape is difficult, the conformation and topology of these nanomaterials can differ between each sensor. Furthermore, as previously said, nanoparticles have a tendency to change their behavior in response to changing environmental conditions. The reproducibility problems that occur as a result of the increased complexity of the modified surfaces are sadly a trade-off. Since it is impractical to test any sensor produced in mass-production facilities, sensor-to-sensor reproducibility is crucial during the manufacturing phase.

Many organisms will quickly adsorb onto the surface of real-world samples. Nonspecific adsorption has been one of the major impediments to using sensors in real-world applications because it reduces the precision, accuracy, and repeatability of the device (Li, Xu, et al., 2017; Lichtenberg, Ling, & Kim, 2019). Researchers are solving this problem at the point-of-care with new materials and methods to increase sensor accuracy. Both passive and active approaches are used for it. Using various organic materials such as polymers, passive approaches seek to build a hydrophilic and noncharged film to obstruct the adsorption of materials like protein on the surface of electrode (Li, Xu, et al., 2017; Lichtenberg et al., 2019). Active techniques, on the contrary, seek to generate surface shear forces that are greater than the adhesion forces of nonspecific biomolecules attached to the surface (Li, Xu, et al., 2017; Lichtenberg et al., 2019). The reliability of sensors, in particular, has been a problem, restricting their use in remote areas with varying temperatures. Sensors are often defined by their shelf life; as a result, it is critical to produce sensors that can operate for an extended period of time. Due to problems with aggregation and flaking of nanomaterial-modified layers, long-term stability can become a major concern when using these nanomaterials. Incorporation of sol-gel materials and ceramics, as well as nanomaterials, has recently been shown to improve the stability of sensors (Ferrag & Kerman, 2020).

Additionally, developing stretchable, self-healing, water-processable, and wearable sensors has recently been a significant development in the sensing field. Rubber-like composites, hydrogels, organogels, and novel polymers are often used to create miniaturized sensors with high stability. As these extremely stretchable materials are combined with the outstanding electrical conductivity of CNM, exceptional sensors with superior analytical performance results (Ferrag & Kerman, 2020).

Concerns over biosensors

CNMs have a wide range of properties that enable them to be used for both therapeutic and diagnostic purposes. They can be lined with anticancer drugs and used as therapeutic platforms in oncology, as well as serving as biosensors for cancer biomarkers (Sireesha et al., 2018). Despite the fact that CNM-based biosensors are promising, they have faced numerous practical challenges in biological applications. For example, biosensor fabrication typically requires CNMs of a certain size and helicity; regulating the size of CNMs during synthesis is difficult (Sireesha et al., 2018). Furthermore, their processing is not cost-effective in large-scale processing of high-purity nanomaterials, which is the primary reason why CNMs are currently too costly in the market, limiting any practical commercial applications (Sireesha et al., 2018). An enzyme must be immobilized onto the surface of carbon nanomaterial-coated electrodes in carbon-based biosensors. As a consequence, biological function, biocompatibility, and structural integrity of biological receptors can be compromised. As a result, designing reversible, reusable, and long-lasting systems to replace existing irreversible and single-use products appears to be challenging.

Since there is an extensive debate in the literature regarding the existence of cytotoxicity of CNMs, the biocompatibility of these nanomaterials remains a challenge. One of the most serious concerns about CNM-based products is their long-term effect on the people who work with them. CNMs are foreign particulate stable compounds that are distinguished by their biopersistence in tissues and the potential for serious health effects on surrounding organs. As a result, calculating or estimating the biocompatibility and toxicity of these nanomaterials in in vivo analysis before commercialization has become critical. CNMs can also change the behavior of proteins by changing their composition, potentially leading to auto-immune effects, which can be prevented when functionalized covalently and noncovalently (Sireesha et al., 2018).

Extracellular vesicles (EV) from tumors have arisen as a new source of cancer biomarkers in liquid biopsies (Martín-Gracia et al., 2020). Traditional approaches for isolation and study have hindered their translation into the field, despite their therapeutic ability. The use of nanomaterial-based biosensors will hasten the development of analytical methods for quantifying extracellular vesicles in a precise, repeatable, stable, fast, and low-cost manner.

One of the major biological problems in designing extracellular vesicle-based blood tests is that the proportion of cancer-derived EVs in the overall pool of EVs present in the blood is poor and highly volatile among patients, resulting in high assay variability (Martín-Gracia et al., 2020). Isolating unique EV subpopulations that may act as biomarkers of their own or be enriched with cancer-derived molecules is one potential alternative. Glypican-1 (GPC1), for example, has been identified as a highly variable marker of pancreatic cancer-derived EVs, with GPC1-positive EV levels providing diagnostic and prognostic significance in pancreatic cancer. Despite advancements in the field, the majority of systems that use NPs to analyze EVs demand that they first be isolated using conventional methods (Melo et al., 2015; Qian, Tan, Zhang, Yang, & Li, 2018). Ultracentrifugation, density-gradient centrifugation, and scale exclusion chromatography are both time-consuming methods that require professional personnel and can result in sample loss. Furthermore, there have been over 1000 different protocols for retrieving EVs from biological fluids published (Van Deun et al., 2017). As a result, establishing precise comparisons to determine the effectiveness of these detection platforms is challenging without normal separation procedures. Many publications neglect even the most basic information about EVs, as well as critical analytical information about the system, such as limit of quantification, precision, accuracy, or LOD. Even when the LOD is published, it is difficult to make a true relation since the units of quantification vary between studies. To solve this issue, follow the guidelines for EV review and reporting (Welsh et al., 2020). Finally, most of the findings, as of now, are proof-of-concept in nature, having been conducted with EVs isolated from cell lines. Despite the fact that some of them used patient surveys, massive patient cohorts should increase statistics. Advances in these areas would hasten the use of EVs in clinics.

There is already a lot of interest in using CNMs like CNTs and graphene in biosensors all over the world. Researchers have shown that CNTs and graphene can provide new sensing protocols while also improving the detection limit of existing protocols. Graphene has high electron mobility in addition to its 2D form, making it a promising substrate for electronic biosensors (Zhu, 2017). Graphene absorbs light from the UV to the NIR ranges and has good FL quenching properties (Zhu, 2017). However, biosensors based on CNTs and graphene also face significant challenges. Since graphene lacks various chiral forms, certain optical properties induced by chiral forms are peculiar to CNTs (Chen et al., 2008). Graphene may require external labels in applications such as selective probing and imaging of living cells, whereas CNTs do not. CNTs and graphene, for instance, need surface functionalization (Zhu, 2017). The subsequent efficiency of sensors is determined by the surface properties. Surface functionalization using proven chemical methods is also far from being widely available and repeatable. In the future, efficient and easy methods for functionalizing CNTs and graphene must be created. Secondly, biosensors are a type of instrument that can convert biological data into other forms of data, such as electric and optical signals. Another aspect that determines the sensor's performance is the integration of the sensing bed with the signal translator and receiver. The translation must be fast, accurate, and understandable. Moreover, in biosensor growth, detecting several target molecules remains a significant challenge. Medical devices are often required to be multifunctional. Also, biosensors with a high level of anti-interference capability are particularly sought after. The actual bioenvironment is much more complex than the one created in the laboratories. Detection of target molecules is often hampered by a variety of biomolecules with similar properties. To address this obstacle, researchers must work to create highly selective targeting agents that are also stable in a variety of bioenvironments. To solve this problem and get CNM-based biosensors closer to a realistic implementation, efforts in materials science and technology, as well as electrical engineering and medical science, are needed.

Lack of point-of-care devices

The sensor device, sampling device, data processing, and readout unit are the key components of a point-of-care testing (POCT) system (Shrivastava, Trung, & Lee, 2020). Mechanical mismatches between these parts, as well as their various shape factors, obstruct smooth integration, posing testing difficulties and signal-to-noise ratio limitations. Some techniques have been designed to bridge the technical gap that exists in the incorporation of several components in a POC biosensor device. The issue remains in designing systems with the desired sensitivity, selectivity, durability, functional automation, and real-time continuous monitoring (Shrivastava et al., 2020). Integrating all of these components into a compact, light-weight package is also a significant challenge.

Carbon nanomaterial-based sensors have a lot of uses in the environmental, medical, and industrial fields. However, instrumentation is a major obstacle for field deployment. For glucose monitoring, portable amperometric meters are available, but for other electrochemical procedures, small devices that can be controlled by nonexperts will be needed. Since certain electrochemical strategies or equilibration times for CNM sensors can be lengthy, an emphasis on fast detection would be beneficial. The range of applications for potential sensors is greater than just clinical testing. Environmental, pharmaceutical, and industrial applications are indeed promising and will allow the use of a number of sensors to detect things easily. Another aim of biosensors is to create implantable sensors that can continuously detect biomolecules for human diseases.

Apart from the tremendous performance of glucose sensors, much more work needs to be performed in order for electrochemical sensors to have a broad effect and use. The ongoing COVID-19 pandemic, for example, has shown the significance and necessity of providing reliable and fast diagnostic instruments. Sensor advances may help to halt the spread of certain infectious diseases and predict the early onset of illnesses including neurodegenerative diseases. Optical and electrochemical sensors have a number of advantages over other diagnostic instruments currently available, including low cost, fast, and real-time identification, and ease of use. They can also be mass-produced and miniaturized for use in portable electronics (Ferrag & Kerman, 2020).

After the sensors are manufactured, random sampling of a subset of the sensors is carried out, and the monitoring and calibration outcomes should be relevant to the whole sample. The "real sample" application is, without a doubt, the most critical test to establish a sensor's validity. It is impossible to verify a sensor as a diagnostic instrument if it is not reliable or usable in actual samples. CNM-based sensors are often used for a number of actual samples, the most important of which are blood serum, urine, sweat, tear, saliva, and interstitial fluid (Ferrag & Kerman, 2020). The matrix effect continues to interact adversely with the identification of a single analyte, lowering recovery values, and sensor sensitivity. Researchers frequently dilute samples to reduce the effect of interferences below a tolerable threshold, which helps to overcome the matrix effect. The more the sample is filtered, the further it is from fact. Sensors should preferably be able to operate in pure real-world samples such as whole blood without any dilution.

In terms of analytical diagnostics, POCT needs more improvement in order to increase diagnostic sensitivity and specificity (reduce false negatives and false positives, respectively) and provide findings equivalent to those obtained from central laboratory-based assays (Shrivastava et al., 2020). For effective commercialization of POCT technology in personal diagnostics and monitoring, high sensitivity and specificity are also needed. To improve the accuracy of POCT, potential sources of false positives, such as nonspecific binding in materials and sensor cartridges, must be eliminated. It is crucial to show extremely precise binding in actual samples, even at very low concentrations, by using newly formed probe molecules. As a consequence, highly selective probes can be useful in the development of new assay methods that avoid false-positive effects (Shrivastava et al., 2020). High sensitivity, in addition to high specificity, can aid with high-precision analyte identification and tracking, even though analytes are found in very low concentrations in body fluids.

Electrochemical and optical signal transduction have been used to create and demonstrate wearable chemical or biochemical sensors for analyte identification in capillary blood, interstitial fluid (ISF), sweat, spit, tears, and wound fluid (Shrivastava et al., 2020). Colorimetry is the most studied optical transduction technique in wearable optical biosensors, due to its low cost, simplicity, and automatic operation (Koh et al., 2016). The devices that result have low identification sensitivity, stability, and the ability to continuously analyze a wide range of analytes in biofluids. The next step in that direction is to incorporate the novel CNMs with excellent signal markers, high sensitivity, and good stability for wearable colorimetric biosensing. These new materials should be able to track proteins, carbohydrates, nucleic acids, small molecules, bacteria, and cells in biofluids, in addition to metabolic analytes. Furthermore, using a color shift to conduct quantitative analysis of low concentration analytes with high precision remains a challenge. Wearable optical biosensors that use FL or SPR as the optical transduction method may be good choices for enhancing the sensitivity and precision of quantitative tests since these methods are up to 1000 times more sensitive than colorimetry, allowing single-bacterium and single-molecule detection (Huang, Liu, Yung, Xiong, & Chen, 2017). Furthermore, they can substantially decrease false-negative signals, resulting in improved accuracy. A spectrometry system must, however, be combined with the sensing unit in a portable form for optical biosensing-based wearable POCT devices. Another technique for designing wearable biochemical sensors is to use electrochemical approaches. However, before they can be used in wearable POCT for personal use, most of these sensors need to be improved in terms of sensor durability, reproducibility, stability, precision, body-friendliness, and cost-effectiveness, as well as human trials.

In addition to integrating novel CNMs in POCT, there is also a need to consider the development of calibration-free sensors. Most existing sensor devices need to be recalibrated because of variations caused by the fabrication of sensors in different batches (Shrivastava et al., 2020).

Researchers must strive to create innovative technologies in order to incorporate more smart sensors into our daily lives if we are to move in the right direction.

Conclusions

The inventions in the field of CNMs have contributed a lot to the advancements in optical sensing. Moreover, CNMs are currently the most studied and discussed in the field of electrochemical sensor research. In the current world scenario, there is a necessity to develop highly specific, sensitive, reliable, quickly responding, and low-cost sensor technologies to be applicable in various disciplines of science including biomedical, environmental, and agricultural fields. The fascinating optical, physicochemical, thermal, and electrochemical properties exhibited by CNMs make them interesting candidates in sensors. The perpetual research for the fruitful utilization of CNMs in optical sensing is very evident from the uncountable assays reported in the last few years. Since health concerns are increasing day by day, the need for biomonitoring is also increasing for effective diagnosis. In this context, the contribution of carbon-based nanomaterials to improve bioimaging is remarkable. These nanomaterials also opened up new possibilities for cost-effective sensing. CNMs such as graphene and its derivatives, CNTs, fullerenes, CDs, CNHs, CNDs, and CNOs are now extensively used in sensors. Though they exhibit properties that are suited for sensors, sometimes, all the key features needed for developing a sensor for a typical target molecule might not be found in the single material and there comes the significance of hybrid materials. When going through the recent trends in carbon nanomaterial-based sensors, most of the CNMs are incorporated in the sensor as composites with metal nanoparticles, conducting polymers, or other CNMs, and the resulting enhanced performance is attributed to the synergistic effect arising from the components of the hybrid used. The sensor research is complete when they are implemented in the real-world systems and CNMs have enough potential to be integrated into miniaturized sensor devices. This chapter also examines the global market patterns for CNMs and their use in sensors across a variety of segments, including form, technology, industry, and region. Sensor market growth is expected to be fuelled by factors such as developments in sensor technology and increased use of smartphones and other electronics products, advancements in the automation industry, increased demand for sensors in the construction of smart cities, and surge in IoT technology. However, the COVID-19 pandemic has severely affected the global market as of now. In conclusion, despite the very promising scenario for the use of carbon nanomaterial-based sensors as analytical instruments in strategic science and technical fields, there are still drawbacks and researchers are expected to focus on finding solutions to these. Thereby in future, development of new CNMs and understanding more about the potential of existing CNMs to find applications in sensors are going to create breakthroughs in the sensing technology. It is to be noted that we have entered the exponential "growth" era, which means further developments are likely to surprise us in the near future.

Conflict of interest

The authors declare that there is no conflict of interests regarding the publication of this paper.

Acknowledgments

Shalini Menon and Sonia Sam hereby acknowledge the Council of Scientific & Industrial Research (CSIR), India, for financial aid in the form of Research Fellowship. Author Keerthi would like to acknowledge the University Grants Commission (UGC), India, for financial aid in the form of Research Fellowship.

References

Afreen, S., Muthoosamy, K., Manickam, S., & Hashim, U. (2015). Functionalized fullerene (C60) as a potential nanomediator in the fabrication of highly sensitive biosensors. *Biosensors and Bioelectronics, 63*, 354–364. https://doi.org/10.1016/j.bios.2014.07.044.

Alliedmarketresearch. (2021). *Carbon nanotubes market size | Industry growth & forecast, 2027*. Retrieved 24, 2021, from: https://www.alliedmarketresearch.com/carbon-nanotube-market.

Alvand, M., & Shemirani, F. (2017). A Fe3O4@SiO2@graphene quantum dot core-shell structured nanomaterial as a fluorescent probe and for magnetic removal of mercury(II) ion. *Microchimica Acta, 184*(6), 1621–1629. https://doi.org/10.1007/s00604-017-2134-2.

Amanulla, B., Palanisamy, S., Chen, S. M., Chiu, T. W., Velusamy, V., Hall, J. M., et al. (2017). Selective colorimetric detection of nitrite in water using chitosan stabilized gold nanoparticles decorated reduced graphene oxide. *Scientific Reports, 7*(1). https://doi.org/10.1038/s41598-017-14584-6.

Amatatongchai, M., Sroysee, W., Chairam, S., & Nacapricha, D. (2017). Amperometric flow injection analysis of glucose using immobilized glucose oxidase on nano-composite carbon nanotubes-platinum nanoparticles carbon paste electrode. *Talanta, 166*, 420–427. https://doi.org/10.1016/j.talanta.2015.11.072.

Amjadi, M., Hallaj, T., Manzoori, J. L., & Shahbazsaghir, T. (2018). An amplified chemiluminescence system based on Si-doped carbon dots for detection of catecholamines. *Spectrochimica Acta-Part A: Molecular and Biomolecular Spectroscopy, 201*, 223–228. https://doi.org/10.1016/j.saa.2018.04.058.

Amjadi, M., Manzoori, J. L., Hallaj, T., & Shahbazsaghir, T. (2017). Application of the chemiluminescence system composed of silicon-doped carbon dots, iron(II) and K2S2O8 to the determination of norfloxacin. *Microchimica Acta, 184*(6), 1587–1593. https://doi.org/10.1007/s00604-017-2139-x.

Anand, S. K., Mathew, M. R., Radecki, J., Radecka, H., & Kumar, K. G. (2020). Individual and simultaneous voltammetric sensing of norepinephrine and tyramine based on poly(L-arginine)/reduced graphene oxide composite film modified glassy carbon electrode. *Journal of Electroanalytical Chemistry, 878*. https://doi.org/10.1016/j.jelechem.2020.114531.

Anusha, T., Bhavani, K. S., Kumar, J. V. S., & Brahman, P. K. (2020). Designing and fabrication of electrochemical nanosensor employing fullerene-C60 and bimetallic nanoparticles composite film for the detection of vitamin D3 in blood samples. *Diamond and Related Materials, 104*. https://doi.org/10.1016/j.diamond.2020.107761.

Arduini, F., Cinti, S., Scognamiglio, V., & Moscone, D. (2016). Nanomaterials in electrochemical biosensors for pesticide detection: Advances and challenges in food analysis. *Microchimica Acta, 183*(7), 2063–2083. https://doi.org/10.1007/s00604-016-1858-8.

Askim, J. R., & Suslick, K. S. (2017). Colorimetric and fluorometric sensor arrays for molecular recognition. In *Vol. 8. Comprehensive supramolecular chemistry II* (pp. 37–88). Elsevier Inc. https://doi.org/10.1016/B978-0-12-409547-2.12616-2.

Awasthi, R., Madhu, & Singh, R. N. (2014). Application of graphene in electrochemical devices. In *Vol. 3. Handbook of functional nanomaterials* (pp. 239–262). Nova Science Publishers Inc. https://www.novapublishers.com/catalog/product_info.php?products_id=46686.

Babazadeh, S., Moghaddam, P. A., Keshipour, S., & Mollazade, K. (2020). Colorimetric sensing of imidacloprid in cucumber fruits using a graphene quantum dot/Au (III) chemosensor. *Scientific Reports, 10*(1). https://doi.org/10.1038/s41598-020-71349-4.

Bae, S. H., Lee, Y., Sharma, B. K., Lee, H. J., Kim, J. H., & Ahn, J. H. (2013). Graphene-based transparent strain sensor. *Carbon, 51*(1), 236–242. https://doi.org/10.1016/j.carbon.2012.08.048.

Bahari, D., Babamiri, B., Salimi, A., & Salimizand, H. (2021). Ratiometric fluorescence resonance energy transfer aptasensor for highly sensitive and selective detection of Acinetobacter baumannii bacteria in urine sample using carbon dots as optical nanoprobes. *Talanta, 221*. https://doi.org/10.1016/j.talanta.2020.121619, 121619.

Bai, J., Sun, C., & Jiang, X. (2016). Carbon dots-decorated multiwalled carbon nanotubes nanocomposites as a high-performance electrochemical sensor for detection of H2O2 in living cells. *Analytical and Bioanalytical Chemistry, 408*(17), 4705–4714. https://doi.org/10.1007/s00216-016-9554-4.

Bai, Y., Xu, T., & Zhang, X. (2020). Graphene-based biosensors for detection of biomarkers. *Micromachines, 11*(1). https://doi.org/10.3390/mi11010060.

Bai, Z. K., Zhang, Y. T., Li, S. H., & Luo, H. X. (2020). A flexible electrochemical sensor based on L-arginine modified chemical vapor deposition graphene platform electrode for selective determination of xanthine. *Chinese Journal of Analytical Chemistry, 2*(1). https://doi.org/10.1016/S1872-2040(20)60042-1.

Baptista, F. R., Belhout, S. A., Giordani, S., & Quinn, S. J. (2015). Recent developments in carbon nanomaterial sensors. *Chemical Society Reviews, 44*(13), 4433–4453. https://doi.org/10.1039/c4cs00379a.

Batool, R., Akhtar, M. A., Hayat, A., Han, D., Niu, L., Ahmad, M. A., et al. (2019). A nanocomposite prepared from magnetite nanoparticles, polyaniline and carboxy-modified graphene oxide for non-enzymatic sensing of glucose. *Microchimica Acta, 186*(5). https://doi.org/10.1007/s00604-019-3364-2.

Bedioui, F., Devynck, J., & Claude, B. C. (1995). Immobilization of metalloporphyrins in electropolymerized films: Design and applications. *Accounts of Chemical Research, 28*(1), 30–36. https://doi.org/10.1021/ar00049a005.

Beiraghi, A., & Najibi-Gehraz, S. A. (2017). Carbon dots-modified silver nanoparticles as a new colorimetric sensor for selective determination of cupric ions. *Sensors and Actuators, B: Chemical, 253*, 342–351. https://doi.org/10.1016/j.snb.2017.06.049.

Benítez-Martínez, S., López-Lorente, Á. I., & Valcárcel, M. (2014). Graphene quantum dots sensor for the determination of graphene oxide in environmental water samples. *Analytical Chemistry, 86*(24), 12279–12284. https://doi.org/10.1021/ac5035083.

Bhalla, N., Pan, Y., Yang, Z., & Payam, A. F. (2020). Opportunities and challenges for biosensors and nanoscale analytical tools for pandemics: COVID-19. *ACS Nano, 14*(7), 7783–7807. https://doi.org/10.1021/acsnano.0c04421.

Bobrowska, D., & Plonska-Brzezinska, M. E. (2021). Endohedral and exohedral single-layered fullerenes. In *Synthesis and applications of nanocarbons* (pp. 25–62).

Borthakur, P., Darabdhara, G., Das, M. R., Boukherroub, R., & Szunerits, S. (2017). Solvothermal synthesis of CoS/reduced porous graphene oxide nanocomposite for selective colorimetric detection of Hg(II) ion in aqueous medium. *Sensors and Actuators, B: Chemical, 244*, 684–692. https://doi.org/10.1016/j.snb.2016.12.148.

Bracamonte, M. V., Melchionna, M., Giuliani, A., Nasi, L., Tavagnacco, C., Prato, M., et al. (2017). H2O2 sensing enhancement by mutual integration of single walled carbon nanohorns with metal oxide catalysts: The CeO2 case. *Sensors and Actuators, B: Chemical, 239*, 923–932. https://doi.org/10.1016/j.snb.2016.08.112.

Breczko, J., Plonska-Brzezinska, M. E., & Echegoyen, L. (2012). Electrochemical oxidation and determination of dopamine in the presence of uric and ascorbic acids using a carbon nano-onion and poly(diallyldimethylammonium chloride) composite. *Electrochimica Acta, 72*, 61–67. https://doi.org/10.1016/j.electacta.2012.03.177.

Bussy, C., Al-Jamal, K. T., Boczkowski, J., Lanone, S., Prato, M., Bianco, A., et al. (2015). Microglia determine brain region-specific neurotoxic responses to chemically functionalized carbon nanotubes. *ACS Nano, 9*(8), 7815–7830. https://doi.org/10.1021/acsnano.5b02358.

Cao, Y., Dong, H., Yang, Z., Zhong, X., Chen, Y., Dai, W., et al. (2017). Aptamer-conjugated graphene quantum dots/porphyrin derivative theranostic agent for intracellular cancer-related microRNA detection and fluorescence-guided photothermal/photodynamic synergetic therapy. *ACS Applied Materials & Interfaces, 9*(1), 159–166. https://doi.org/10.1021/acsami.6b13150.

Cao, J. T., Zhang, W. S., Wang, H., Ma, S. H., & Liu, Y. M. (2019). A novel nitrogen and sulfur co-doped carbon dots-H2O2 chemiluminescence system for carcinoembryonic antigen detection using functional HRP-Au@Ag for signal amplification. *Spectrochimica Acta-Part A: Molecular and Biomolecular Spectroscopy, 219*, 281–287. https://doi.org/10.1016/j.saa.2019.04.063.

Cesana, R., Gonçalves, J. M., Ignácio, R. M., Nakamura, M., Zamarion, V. M., Toma, H. E., et al. (2020). Synthesis and characterization of nanocomposite based on reduced graphene oxide-gold nanoparticles-carbon dots: Electroanalytical determination of dihydroxybenzene isomers simultaneously. *Journal of Nanoparticle Research, 22*(10). https://doi.org/10.1007/s11051-020-05059-3.

Chandra, S., Singh, V. K., Yadav, P. K., Bano, D., Kumar, V., Pandey, V. K., et al. (2019). Mustard seeds derived fluorescent carbon quantum dots and their peroxidase-like activity for colorimetric detection of H2O2 and ascorbic acid in a real sample. *Analytica Chimica Acta, 1054*, 145–156. https://doi.org/10.1016/j.aca.2018.12.024.

Chang, C. L., Hu, C. W., Tseng, C. Y., Chuang, C. N., Ho, K. C., & Leung, M. K. (2014). Ambipolar freestanding triphenylamine/fullerene thin-film by electrochemical deposition and its read-writable properties by electrochemical treatments. *Electrochimica Acta, 116*, 69–77. https://doi.org/10.1016/j.electacta.2013.11.005.

Chen, F., Niu, X., Yang, X., Pei, H., Guo, R., Liu, N., et al. (2021). Self-assembled reduced graphene oxide/polyaniline/sodium carboxymethyl cellulose nanocomposite for voltammetric recognition of tryptophan enantiomers. *Journal of Materials Science: Materials in Electronics*. https://doi.org/10.1007/s10854-021-05809-6.

Chen, J., Ran, F., Chen, Q., Luo, D., Ma, W., Han, T., et al. (2019). A fluorescent biosensor for cardiac biomarker myoglobin detection based on carbon dots and deoxyribonuclease I-aided target recycling signal amplification. *RSC Advances, 9*(8), 4463–4468. https://doi.org/10.1039/C8RA09459D.

Chen, Z., Tabakman, S. M., Goodwin, A. P., Kattah, M. G., Daranciang, D., Wang, X., et al. (2008). Protein microarrays with carbon nanotubes as multicolor Raman labels. *Nature Biotechnology, 26*(11), 1285–1292. https://doi.org/10.1038/nbt.1501.

Chen, H., Wang, Q., Shen, Q., Liu, X., Li, W., Nie, Z., et al. (2017). Nitrogen doped graphene quantum dots based long-persistent chemiluminescence system for ascorbic acid imaging. *Biosensors and Bioelectronics, 91*, 878–884. https://doi.org/10.1016/j.bios.2017.01.061.

Cluff, K. J., & Blümel, J. (2016). Adsorption of ferrocene on carbon nanotubes, graphene, and activated carbon. *Organometallics, 35*(23). https://doi.org/10.1021/acs.organomet.6b00691.

Cui, D. (2018). Nanomaterials for theranostics of gastric cancer. In *Handbook of nanomaterials for cancer theranostics* (pp. 305–349). Elsevier. https://doi.org/10.1016/B978-0-12-813339-2.00011-6.

Deegan, R. D., Bakajin, O., Dupont, T. F., Huber, G., Nagel, S. R., & Witten, T. A. (1997). Capillary flow as the cause of ring stains from dried liquid drops. *Nature, 389*(6653), 827–829. https://doi.org/10.1038/39827.

Delnavaz, E., & Amjadi, M. (2020). An ultrasensitive chemiluminescence assay for 4-nitrophenol by using luminol–NaIO4 reaction catalyzed by copper, nitrogen co-doped carbon dots. *Spectrochimica Acta Part A: Molecular and Biomolecular Spectroscopy, 241*. https://doi.org/10.1016/j.saa.2020.118608, 118608.

Demir, B., Lemberger, M. M., Panagiotopoulou, M., Medina Rangel, P. X., Timur, S., Hirsch, T., et al. (2018). Tracking hyaluronan: Molecularly imprinted polymer coated carbon dots for cancer cell targeting and imaging. *ACS Applied Materials and Interfaces, 10*(4), 3305–3313. https://doi.org/10.1021/acsami.7b16225.

Dinh, N. X., Huy, T. Q., Van Vu, L., Tam, L. T., & Le, A. T. (2016). Multiwalled carbon nanotubes/silver nanocomposite as effective SERS platform for detection of methylene blue dye in water. *Journal of Science: Advanced Materials and Devices, 1*(1), 84–89. https://doi.org/10.1016/j.jsamd.2016.04.007.

Dondo, N., Shumba, M., Moyo, M., & Nyoni, S. (2020). Simultaneous non-steroidal anti-inflammatory drug electrodetection on nitrogen doped carbon nanodots and nanosized cobalt phthallocyanine conjugate modified glassy carbon electrode. *Arabian Journal of Chemistry, 13*(11), 7809–7819. https://doi.org/10.1016/j.arabjc.2020.09.012.

Dong, J., Wang, K., Sun, L., Sun, B., Yang, M., Chen, H., et al. (2018). Application of graphene quantum dots for simultaneous fluorescence imaging and tumor-targeted drug delivery. *Sensors and Actuators, B: Chemical, 256*, 616–623. https://doi.org/10.1016/j.snb.2017.09.200.

Duan, Y., Huang, Y., Chen, S., Zuo, W., & Shi, B. (2019). Cu-doped carbon dots as catalysts for the chemiluminescence detection of glucose. *ACS Omega*, *4*(6), 9911–9917. https://doi.org/10.1021/acsomega.9b00738.

Dumitrescu, I., Unwin, P. R., & MacPherson, J. V. (2009). Electrochemistry at carbon nanotubes: Perspective and issues. *Chemical Communications*, *45*, 6886–6901. https://doi.org/10.1039/b909734a.

Ebralidze, I. I., Laschuk, N. O., Poisson, J., & Zenkina, O. V. (2019). Colorimetric sensors and sensor arrays. In *Nanomaterials design for sensing applications* (pp. 1–39). Elsevier. https://doi.org/10.1016/B978-0-12-814505-0.00001-1.

Echabaane, M., Hfaiedh, S., Smiri, B., Saidi, F., & Dridi, C. (2021). Development of an impedimetric sensor based on carbon dots and chitosan nanocomposite modified electrode for Cu(II) detection in water. *Journal of Solid State Electrochemistry*. https://doi.org/10.1007/s10008-021-04949-3.

Eggeling, C. (2018). Advances in bioimaging—Challenges and potentials. *Journal of Physics D: Applied Physics*, *51*(4). https://doi.org/10.1088/1361-6463/aaa259.

Erady, V., Mascarenhas, R. J., Satpati, A. K., Detriche, S., Mekhalif, Z., Dalhalle, J., et al. (2017). Sensitive detection of Ferulic acid using multi-walled carbon nanotube decorated with silver nano-particles modified carbon paste electrode. *Journal of Electroanalytical Chemistry*, *806*, 22–31. https://doi.org/10.1016/j.jelechem.2017.10.045.

Fang, M., Zhuo, K., Chen, Y., Zhao, Y., Bai, G., & Wang, J. (2019). Fluorescent probe based on carbon dots/silica/molecularly imprinted polymer for lysozyme detection and cell imaging. *Analytical and Bioanalytical Chemistry*, *411*(22), 5799–5807. https://doi.org/10.1007/s00216-019-01960-6.

Ferrag, C., & Kerman, K. (2020). Grand challenges in nanomaterial-based electrochemical sensors. *Frontiers in Sensors*. https://doi.org/10.3389/fsens.2020.583822.

Filippone, S. (2020). Encyclopedia of polymeric nanomaterials. *Encyclopedia of Polymeric Nanomaterials*, *2*, 1–16.

Fu, Y., Huang, L., Zhao, S., Xing, X., Lan, M., & Song, X. (2021). A carbon dot-based fluorometric probe for oxytetracycline detection utilizing a Förster resonance energy transfer mechanism. *Spectrochimica Acta Part A: Molecular and Biomolecular Spectroscopy*, *246*. https://doi.org/10.1016/j.saa.2020.118947, 118947.

Fu, X. C., Zhang, J., Tao, Y. Y., Wu, J., Xie, C. G., & Kong, L. T. (2015). Three-dimensional mono-6-thio-β-cyclodextrin covalently functionalized gold nanoparticle/single-wall carbon nanotube hybrids for highly sensitive and selective electrochemical determination of methyl parathion. *Electrochimica Acta*, *153*, 12–18. https://doi.org/10.1016/j.electacta.2014.11.144.

Futuremarketinsights. (2021). *Carbon nanotubes market: COVID-19 impact assessment and global industry analysis 2020–2030*. https://www.futuremarketinsights.com/reports/global-carbon-nanotubes-market.

Ganganboina, A. B., Dega, N. K., Tran, H. L., Darmonto, W., & Doong, R. A. (2021). Application of sulfur-doped graphene quantum dots@gold-carbon nanosphere for electrical pulse-induced impedimetric detection of glioma cells. *Biosensors and Bioelectronics*, *181*. https://doi.org/10.1016/j.bios.2021.113151.

Ganganboina, A. B., & Doong, R. A. (2019). Graphene quantum dots decorated gold-polyaniline nanowire for impedimetric detection of carcinoembryonic antigen. *Scientific Reports*, *9*(1). https://doi.org/10.1038/s41598-019-43740-3.

Gao, T., Xing, S., Xu, M., Fu, P., Yao, J., Zhang, X., et al. (2020). A peptide nucleic acid–regulated fluorescence resonance energy transfer DNA assay based on the use of carbon dots and gold nanoparticles. *Microchimica Acta*, *187*(7).

Gao, X. X., Zhou, X., Ma, Y. F., Wang, C. P., & Chu, F. X. (2018). A fluorometric and colorimetric dual-mode sensor based on nitrogen and iron co-doped graphene quantum dots for detection of ferric ions in biological fluids and cellular imaging. *New Journal of Chemistry*, *42*(18), 14751–14756. https://doi.org/10.1039/c8nj01805g.

Ge, S., He, J., Ma, C., Liu, J., Xi, F., & Dong, X. (2019). One-step synthesis of boron-doped graphene quantum dots for fluorescent sensors and biosensor. *Talanta*, *199*, 581–589. https://doi.org/10.1016/j.talanta.2019.02.098.

Georgakilas, Vasilios, et al. (2015). Broad Family of Carbon Nanoallotropes: Classification, Chemistry, and Applications of Fullerenes, Carbon Dots, Nanotubes, Graphene, Nanodiamonds, and Combined Superstructures. *Chemical Reviews*, *115*(11), 4744–4822. https://doi.org/10.1021/cr500304f.

Ghosh, M., Öner, D., Duca, R. C., Bekaert, B., Vanoirbeek, J. A. J., Godderis, L., et al. (2018). Single-walled and multi-walled carbon nanotubes induce sequence-specific epigenetic alterations in 16 HBE cells. *Oncotarget*, *9*(29), 20351–20365. https://doi.org/10.18632/oncotarget.24866.

Gminsights. (2021). *Carbon nanotubes market size and share | Statistics-2024*. https://www.gminsights.com/industry-analysis/carbon-nanotubes-market.

Grandviewresearch. (2021). *Carbon nanomaterials market size, share | Industry report, 2019-2025*. Retrieved April 24, 2021, from https://www.grandviewresearch.com/industry-analysis/carbon-nanomaterials-market.

Gu, S., Hsieh, C. T., Kao, C. P., Fu, C. C., Ashraf Gandomi, Y., Juang, R. S., et al. (2021). Electrocatalytic oxidation of glucose on boron and nitrogen codoped graphene quantum dot electrodes in alkali media. *Catalysts*, *11*(1), 1–15. https://doi.org/10.3390/catal11010101.

Gu, D., Shang, S., Yu, Q., & Shen, J. (2016). Green synthesis of nitrogen-doped carbon dots from lotus root for Hg(II) ions detection and cell imaging. *Applied Surface Science*, *390*, 38–42. https://doi.org/10.1016/j.apsusc.2016.08.012.

Guo, Y., Zhang, J., Zhang, W., & Hu, D. (2019). Green fluorescent carbon quantum dots functionalized with polyethyleneimine, and their application to aptamer-based determination of thrombin and ATP. *Microchimica Acta*, *186*(11). https://doi.org/10.1007/s00604-019-3874-y.

Gupta, B. D., Pathak, A., & Semwal, V. (2019). Carbon-based nanomaterials for plasmonic sensors: A review. *Sensors (Switzerland)*, *19*(16). https://doi.org/10.3390/s19163536.

Hamd-Ghadareh, S., Salimi, A., Fathi, F., & Bahrami, S. (2017). An amplified comparative fluorescence resonance energy transfer immunosensing of CA125 tumor marker and ovarian cancer cells using green and economic carbon dots for bio-applications in labeling, imaging and sensing. *Biosensors and Bioelectronics*, *96*, 308–316. https://doi.org/10.1016/j.bios.2017.05.003.

Han, X., Meng, Z., Zhang, H., & Zheng, J. (2018). Fullerene-based anodic stripping voltammetry for simultaneous determination of Hg(II), Cu(II), Pb(II) and Cd(II) in foodstuff. *Microchimica Acta*, *185*(5). https://doi.org/10.1007/s00604-018-2803-9.

Han, J., Zhuo, Y., Chai, Y., Yuan, R., Xiang, Y., Zhu, Q., et al. (2013). Multi-labeled functionalized C60 nanohybrid as tracing tag for ultrasensitive electrochemical aptasensing. *Biosensors and Bioelectronics*, *46*, 74–79. https://doi.org/10.1016/j.bios.2013.02.020.

Hasan, M. R., Pulingam, T., Appaturi, J. N., Zifruddin, A. N., Teh, S. J., Lim, T. W., et al. (2018). Carbon nanotube-based aptasensor for sensitive electrochemical detection of whole-cell Salmonella. *Analytical Biochemistry*, *554*, 34–43. https://doi.org/10.1016/j.ab.2018.06.001.

Hassanvand, Z., Jalali, F., Nazari, M., Parnianchi, F., & Santoro, C. (2021). Carbon nanodots in electrochemical sensors and biosensors: A review. *ChemElectroChem*, *8*(1), 15–35. https://doi.org/10.1002/celc.202001229.

Haupt, K., & Mosbach, K. (1998). Plastic antibodies: Developments and applications. *Trends in Biotechnology*, *16*(11), 468–475. https://doi.org/10.1016/S0167-7799(98)01222-0.

Haynes, C. L., McFarland, A. D., & Van Duyne, R. P. (2005). Surface enhanced Raman spectroscopy. *Analytical Chemistry*, *77*(17). https://doi.org/10.1021/ac053456d.

https://www.businesswire.com/news/home/20201215005871/en/The-Global-Market-for-Carbon-Nanomaterials-2020-2030-Carbon-Nanotubes-Gn-FNs-Gn-Quantum-Dots-2D-Materials-and-NDs- - -ResearchAndMarkets.com. 2020. (Accessed 23 April 2021).

Hu, S., & Hu, C. (2009). Carbon nanotube-based electrochemical sensors: Principles and applications in biomedical systems. *Journal of Sensors*, *2009*. https://doi.org/10.1155/2009/187615.

Huang, X., Liu, Y., Yung, B., Xiong, Y., & Chen, X. (2017). Nanotechnology-enhanced no-wash biosensors for in vitro diagnostics of cancer. *ACS Nano*, *11*(6), 5238–5292. https://doi.org/10.1021/acsnano.7b02618.

Hwang, H. S., Jeong, J. W., Kim, Y. A., & Chang, M. (2020). Carbon nanomaterials as versatile platforms for biosensing applications. *Micromachines*, *11*(9). https://doi.org/10.3390/MI11090814.

Imahori, H., Azuma, T., Ajavakom, A., Norieda, H., Yamada, H., & Sakata, Y. (1999). An investigation of photocurrent generation by gold electrodes modified with self-assembled monolayers of C 60. *The Journal of Physical Chemistry B*, *103*(34), 7233–7237. https://doi.org/10.1021/jp990837k.

Industryarc. (2021). *Carbon nanomaterials market research report: Market share, size and industry growth analysis 2020–2025.* Retrieved April 24, 2021, from: https://www.industryarc.com/Report/15675/carbon-nanomaterials-market.html.

Jahanbakhshi, M., & Habibi, B. (2016). A novel and facile synthesis of carbon quantum dots via salep hydrothermal treatment as the silver nanoparticles support: Application to electroanalytical determination of H2O2 in fetal bovine serum. *Biosensors and Bioelectronics*, *81*, 143–150. https://doi.org/10.1016/j.bios.2016.02.064.

Jalali, F., Hassanvand, Z., & Barati, A. (2020). Electrochemical sensor based on a nanocomposite of carbon dots, hexadecyltrimethylammonium bromide and chitosan for mesalazine determination. *Journal of Analytical Chemistry*, *75*(4), 544–552. https://doi.org/10.1134/S1061934820040061.

Jerome, R., & Sundramoorthy, A. K. (2020). Preparation of hexagonal boron nitride doped graphene film modified sensor for selective electrochemical detection of nicotine in tobacco sample. *Analytica Chimica Acta*, *1132*, 110–120. https://doi.org/10.1016/j.aca.2020.07.060.

Karabiberoğlu, Ş. U., et al. (2016). *Carbon nanotube-conducting polymer composites as electrode material in electroanalytical applications.* Intechopen.

Karousis, N., Suarez-Martinez, I., Ewels, C. P., & Tagmatarchis, N. (2016). Structure, properties, functionalization, and applications of carbon nanohorns. *Chemical Reviews*, *116*(8), 4850–4883. https://doi.org/10.1021/acs.chemrev.5b00611.

Kathiresan, V., Thirumalai, D., Rajarathinam, T., Yeom, M., Lee, J., Kim, S., et al. (2021). A simple one-step electrochemical deposition of bioinspired nanocomposite for the non-enzymatic detection of dopamine. *Journal of Analytical Science and Technology*, *12*(1). https://doi.org/10.1186/s40543-021-00260-y.

Kaushal, S., Kaur, M., Kaur, N., Kumari, V., & Singh, P. P. (2020). Heteroatom-doped graphene as sensing materials: A mini review. *RSC Advances*, *10*(48), 28608–28629. https://doi.org/10.1039/d0ra04432f.

Khan, N. I., & Song, E. (2021). Detection of an il-6 biomarker using a GFET platform developed with a facile organic solvent-free aptamer immobilization approach. *Sensors (Switzerland)*, *21*(4), 1–16. https://doi.org/10.3390/s21041335.

Klumpp, C., Kostarelos, K., Prato, M., & Bianco, A. (2006). Functionalized carbon nanotubes as emerging nanovectors for the delivery of therapeutics. *Biochimica et Biophysica Acta - Biomembranes*, *1758*(3), 404–412. https://doi.org/10.1016/j.bbamem.2005.10.008.

Koh, A., Kang, D., Xue, Y., Lee, S., Pielak, R. M., Kim, J., et al. (2016). A soft, wearable microfluidic device for the capture, storage, and colorimetric sensing of sweat. *Science Translational Medicine*, *8*(366). https://doi.org/10.1126/scitranslmed.aaf2593.

Koike, E., & Kobayashi, T. (2006). Chemical and biological oxidative effects of carbon black nanoparticles. *Chemosphere*, *65*(6), 946–951. https://doi.org/10.1016/j.chemosphere.2006.03.078.

Kour, R., Arya, S., Young, S.-J., Gupta, V., Bandhoria, P., & Khosla, A. (2020). Review—Recent advances in carbon nanomaterials as electrochemical biosensors. *Journal of the Electrochemical Society*, *167*(3).

Kratschmer, W., et al. (1990). Solid C_{60}: a new form of carbon. *Nature*, *347*.

Krishnan, S., Tadiboyina, R., et al. (2019). *Graphene-based polymer nanocomposites for sensor applications.*

Kumar, S., Karfa, P., Majhi, K. C., & Madhuri, R. (2020). Photocatalytic, fluorescent BiPO4@Graphene oxide based magnetic molecularly imprinted polymer for detection, removal and degradation of ciprofloxacin. *Materials Science and Engineering C*, *111*. https://doi.org/10.1016/j.msec.2020.110777.

Kumar, D., & Talreja, N. (2019). Nickel nanoparticles-doped rhodamine grafted carbon nanofibers as colorimetric probe: Naked eye detection and highly sensitive measurement of aqueous Cr3+ and Pb2+. *Korean Journal of Chemical Engineering*, *36*(1), 126–135. https://doi.org/10.1007/s11814-018-0139-0.

Kuznetsov, V. L., Chuvilin, A. L., Butenko, Y. V., Malkov, I. Y., & Titov, V. M. (1994). Onion-like carbon from ultra-disperse diamond. *Chemical Physics Letters*, *222*(4), 343–348. https://doi.org/10.1016/0009-2614(94)87072-1.

Law, C. S., Marsal, L. F., & Santos, A. (2019). Electrochemically engineered nanoporous photonic crystal structures for optical sensing and biosensing. In *Handbook of nanomaterials in analytical chemistry: Modern trends in analysis* (pp. 201–226). Elsevier. https://doi.org/10.1016/B978-0-12-816699-4.00009-8.

Lee, H., Choi, T. K., Lee, Y. B., Cho, H. R., Ghaffari, R., Wang, L., et al. (2016). A graphene-based electrochemical device with thermoresponsive microneedles for diabetes monitoring and therapy. *Nature Nanotechnology, 11*(6), 566–572. https://doi.org/10.1038/nnano.2016.38.

Lee, J., Kim, Y., Lee, S., Yoon, S., & Kim, W. K. (2019). Graphene oxide-based NET strategy for enhanced colorimetric sensing of miRNA. *Sensors and Actuators, B: Chemical, 282*, 861–867. https://doi.org/10.1016/j.snb.2018.11.149.

Lee, K., Lee, J., & Ahn, B. (2019). Design of refolding DNA aptamer on single-walled carbon nanotubes for enhanced optical detection of target proteins. *Analytical Chemistry, 91*(20), 12704–12712. https://doi.org/10.1021/acs.analchem.9b02177.

Li, Y., & Han, S. (2020). Carbon dots-enhanced chemiluminescence method for the sensitive determination of iodide. *Microchemical Journal, 154*. https://doi.org/10.1016/j.microc.2020.104638, 104638.

Li, X., Jiang, X., Liu, Q., Liang, A., & Jiang, Z. (2019). Using n-doped carbon dots prepared rapidly by microwave digestion as nanoprobes and nanocatalysts for fluorescence determination of ultratrace isocarbophos with label-free aptamers. *Nanomaterials, 9*(2). https://doi.org/10.3390/nano9020223.

Li, C., Li, J., Liang, A., Wen, G., & Jiang, Z. (2021). Aptamer turn-on SERS/RRS/fluorescence tri-mode platform for ultra-trace urea determination using Fe/N-doped carbon dots. *Frontiers in Chemistry, 9*. https://doi.org/10.3389/fchem.2021.613083.

Li, X., Peng, G., Cui, F., Qiu, Q., Chen, X., & Huang, H. (2018). Double determination of long noncoding RNAs from lung cancer via multi-amplified electrochemical genosensor at sub-femtomole level. *Biosensors and Bioelectronics, 113*, 116–123. https://doi.org/10.1016/j.bios.2018.04.062.

Li, Y., Xu, Y., Fleischer, C. C., Huang, J., Lin, R., Yang, L., et al. (2017). Impact of anti-biofouling surface coatings on the properties of nanomaterials and their biomedical applications. *Journal of Materials Chemistry B, 6*(1), 9–24. https://doi.org/10.1039/c7tb01695f.

Li, H., Yan, X., Kong, D., Jin, R., Sun, C., Du, D., et al. (2020). Recent advances in carbon dots for bioimaging applications. *Nanoscale Horizons, 5*(2), 218–234. https://doi.org/10.1039/c9nh00476a.

Li, J., Yang, K., Wu, Z., Li, X., & Duan, Q. (2019). Nitrogen-doped porous carbon-based fluorescence sensor for the detection of ZIKV RNA sequences: Fluorescence image analysis. *Talanta, 205*. https://doi.org/10.1016/j.talanta.2019.06.091, 120091.

Liaqat, M., Riaz, S., Hasnain Nawaz, M., Mihaela Badea, A. H., & Louis Marty, J. (2021). Fabrication of electro-active nano-trans surfaces to design label free electrochemical aptasensor for ochratoxin A detection. *Electrochimica Acta, 379*. https://doi.org/10.1016/j.electacta.2021.138172.

Lichtenberg, J. Y., Ling, Y., & Kim, S. (2019). Non-specific adsorption reduction methods in biosensing. *Sensors (Switzerland), 19*(11). https://doi.org/10.3390/s19112488.

Lin, X., Wang, Y., Zou, M., Lan, T., & Ni, Y. (2019). Electrochemical non-enzymatic glucose sensors based on nano-composite of Co3O4 and multiwalled carbon nanotube. *Chinese Chemical Letters, 30*(6), 1157–1160. https://doi.org/10.1016/j.cclet.2019.04.009.

Liu, X., Li, H., Wang, F., Zhu, S., Wang, Y., & Xu, G. (2010). Functionalized single-walled carbon nanohorns for electrochemical biosensing. *Biosensors and Bioelectronics, 25*(10), 2194–2199. https://doi.org/10.1016/j.bios.2010.02.027.

Liu, X., Shi, L., Niu, W., Li, H., & Xu, G. (2008). Amperometric glucose biosensor based on single-walled carbon nanohorns. *Biosensors and Bioelectronics, 23*(12), 1887–1890. https://doi.org/10.1016/j.bios.2008.02.016.

Liu, Y., Duan, W., Song, W., Liu, J., Ren, C., Wu, J., et al. (2017). Red emission B, N, S-co-doped carbon dots for colorimetric and fluorescent dual mode detection of Fe3 + ions in complex biological fluids and living cells. *ACS Applied Materials and Interfaces, 9*(14), 12663–12672. https://doi.org/10.1021/acsami.6b15746.

Liu, Y., Tang, X., Deng, M., Cao, Y., Li, Y., Zheng, H., et al. (2019). Nitrogen doped graphene quantum dots as a fluorescent probe for mercury(II) ions. *Microchimica Acta, 186*(3). https://doi.org/10.1007/s00604-019-3249-4.

Liu, H., Zhang, Y., Zhang, D., Zheng, F., Huang, M., Sun, J., et al. (2019). A fluorescent nanoprobe for 4-ethylguaiacol based on the use of a molecularly imprinted polymer doped with a covalent organic framework grafted onto carbon nanodots. *Microchimica Acta, 186*(3). https://doi.org/10.1007/s00604-019-3306-z.

Lo, M., Seydou, M., Bensghaïer, A., Pires, R., Gningue-Sall, D., Aaron, J. J., et al. (2020). Polypyrrole-wrapped carbon nanotube composite films coated on diazonium-modified flexible ITO sheets for the electroanalysis of heavy metal ions. *Sensors (Switzerland), 20*(3). https://doi.org/10.3390/s20030580.

Ma, Y., Wang, Y., Liu, Y., Shi, L., & Yang, D. (2021). Multi-carbon dots and aptamer based signal amplification ratiometric fluorescence probe for protein tyrosine kinase 7 detection. *Journal of Nanobiotechnology, 19*(1). https://doi.org/10.1186/s12951-021-00787-7.

Madannejad, R., Shoaie, N., Jahanpeyma, F., Darvishi, M. H., Azimzadeh, M., & Javadi, H. (2019). Toxicity of carbon-based nanomaterials: Reviewing recent reports in medical and biological systems. *Chemico-Biological Interactions, 307*, 206–222. https://doi.org/10.1016/j.cbi.2019.04.036.

Majumder, S., Mondal, T., & Deen, M. J. (2017). Challenges in design and fabrication of flexible/stretchable carbon- and textile-based wearable sensors for health monitoring. *Sensors (Switzerland), 17*(1).

Manavalan, S., Veerakumar, P., Chen, S. M., & Lin, K. C. (2020). Three-dimensional zinc oxide nanostars anchored on graphene oxide for voltammetric determination of methyl parathion. *Microchimica Acta, 187*(1). https://doi.org/10.1007/s00604-019-4031-3.

Mani, V., Govindasamy, M., Chen, S. M., Chen, T. W., Kumar, A. S., & Huang, S. T. (2017). Core-shell heterostructured multiwalled carbon nanotubes@reduced graphene oxide nanoribbons/chitosan, a robust nanobiocomposite for enzymatic biosensing of hydrogen peroxide and nitrite. *Scientific Reports, 7*(1). https://doi.org/10.1038/s41598-017-12050-x.

Mansuriya, B. D., & Altintas, Z. (2021). Enzyme-free electrochemical nano-immunosensor based on graphene quantum dots and gold nanoparticles for cardiac biomarker determination. *Nanomaterials, 11*(3), 1–18. https://doi.org/10.3390/nano11030578.

Marchena, M., Wagner, F., Arliguie, T., Zhu, B., Johnson, B., Fernández, M., et al. (2018). Dry transfer of graphene to dielectrics and flexible substrates using polyimide as a transparent and stable intermediate layer. *2D Materials, 5*(3). https://doi.org/10.1088/2053-1583/aac12d.

Market Research Future (MRFR). (2021). *Carbon nanotubes market size to reach USD 6.03 billion by 2027 at 20% CAGR*. Retrieved April 24, 2021, from: http://www.globenewswire.com/news-release/2021/02/16/2176518/0/en/Carbon-Nanotubes-Market-Size-to-Reach-USD-6-03-Billion-by-2027-at-20-CAGR-Market-Research-Future-MRFR.html.

MarketsandMarketsTM. (2021). *Carbon nanotubes (CNT) market analysis | Recent market developments | Industry forecast to 2017–2022*. Retrieved April 24, 2021, from: https://www.marketsandmarkets.com/Market-Reports/carbon-nanotubes-139.html.

Martín-Gracia, B., Martín-Barreiro, A., Cuestas-Ayllón, C., Grazú, V., Line, A., Llorente, A., et al. (2020). Nanoparticle-based biosensors for detection of extracellular vesicles in liquid biopsies. *Journal of Materials Chemistry B, 8*(31), 6710–6738. https://doi.org/10.1039/d0tb00861c.

Mathivanan, D., Tammina, S. K., Wang, X., & Yang, Y. (2020). Dual emission carbon dots as enzyme mimics and fluorescent probes for the determination of o-phenylenediamine and hydrogen peroxide. *Microchimica Acta, 187*(5). https://doi.org/10.1007/s00604-020-04256-0.

McQueen, E. W., & Goldsmith, J. I. (2009). Electrochemical analysis of single-walled carbon nanotubes functionalized with pyrene-pendant transition metal complexes. *Journal of the American Chemical Society, 131*(48), 17554–17556. https://doi.org/10.1021/ja907294q.

Melo, S. A., Luecke, L. B., Kahlert, C., Fernandez, A. F., Gammon, S. T., Kaye, J., et al. (2015). Glypican-1 identifies cancer exosomes and detects early pancreatic cancer. *Nature, 523*(7559), 177–182. https://doi.org/10.1038/nature14581.

Méndez Romero, U. A., et al. (2017). *Graphene derivatives: Controlled properties, nanocomposites, and energy harvesting applications*. Intechopen.

Meng, X., Xu, Y., Zhang, N., Ma, B., Ma, Z., & Han, H. (2021). Ferric hydroxide nanocage triggered Fenton-like reaction to improve amperometric immunosensor. *Sensors and Actuators B: Chemical, 338*, 129840. https://doi.org/10.1016/j.snb.2021.129840.

Menon, S., Jesny, S., & Girish Kumar, K. (2018). A voltammetric sensor for acetaminophen based on electropolymerized-molecularly imprinted poly(o-aminophenol) modified gold electrode. *Talanta, 179*, 668–675. https://doi.org/10.1016/j.talanta.2017.11.074.

Menon, S., & Kumar, K. G. (2017). Simultaneous voltammetric determination of acetaminophen and its fatal counterpart nimesulide by gold nano/L-cysteine modified gold electrode. *Journal of the Electrochemical Society, 164*(9), B482–B487. https://doi.org/10.1149/2.0181712jes.

Menon, S., Mathew, M. R., Sam, S., Keerthi, K., & Kumar, K. G. (2020). Recent advances and challenges in electrochemical biosensors for emerging and re-emerging infectious diseases. *Journal of Electroanalytical Chemistry, 878*, 114596. https://doi.org/10.1016/j.jelechem.2020.114596.

Menon, S., Vikraman, A. E., Jesny, S., & Girish Kumar, K. (2016). Fluorescence determination of nitrite using green synthesized carbon nanoparticles. *Journal of Fluorescence, 26*(1), 129–134. https://doi.org/10.1007/s10895-015-1692-0.

Miao, H., Wang, L., Zhuo, Y., Zhou, Z., & Yang, X. (2016). Label-free fluorimetric detection of CEA using carbon dots derived from tomato juice. *Biosensors and Bioelectronics, 86*, 83–89. https://doi.org/10.1016/j.bios.2016.06.043.

Mitaku, S., & Sawada, R. (2018). Biological meaning of "habitable zone" in nucleotide composition space. *Biophysics and Physicobiology*, 75–85. https://doi.org/10.2142/biophysico.15.0_75.

Mohammadnia, K., Naghian, E., Keihan, E., Sohouli, E., Plonska-Brzezinska, E., et al. (2020). Application of carbon nanoonion-NiMoO4-MnWO4 nanocomposite for modification of glassy carbon electrode: Electrochemical determination of ascorbic acid. *Microchemical Journal, 159*.

Mollarasouli, F., Majidi, M. R., & Asadpour-Zeynali, K. (2018). Amperometric sensor based on carbon dots decorated self-assembled 3D flower-like β-Ni(OH)2 nanosheet arrays for the determination of nitrite. *Electrochimica Acta, 291*, 132–141. https://doi.org/10.1016/j.electacta.2018.08.132.

Mordorintelligence. (2021). Carbon nanotubes market | Growth, trends, COVID-19 impact, and forecasts (2021–2026) https://www.mordorintelligence.com/industry-reports/carbon-nanotubes-market.

Motoc, S., Manea, F., Orha, C., & Pop, A. (2019). Enhanced electrochemical response of diclofenac at a fullerene–carbon nanofiber paste electrode. *Sensors (Switzerland), 19*(6). https://doi.org/10.3390/s19061332.

Moundzounga, T. H. G., Peleyeju, M. G., Sanni, S. O., Klink, M. J., Oseghe, E., Viljoen, E., et al. (2021). A Nanocomposite of graphitic carbon nitride and carbon dots as a platform for sensitive voltammetric determination of 2-chlorophenol in water. *International Journal of Electrochemical Science, 16*, 1–12. https://doi.org/10.20964/2021.05.15.

Nana, L., Ruiyi, L., Xiulan, S., Yongqiang, Y., & Zaijun, L. (2020). Dual amplification in a fluorometric acetamiprid assay by using an aptamer, G-quadruplex/hemin DNAzyme, and graphene quantum dots functionalized with D-penicillamine and histidine. *Microchimica Acta, 187*(3). https://doi.org/10.1007/s00604-020-4127-9.

Transparencymarketresearch. (2021). *Carbon nanomaterials market—Industry analysis 2024*. https://www.transparencymarketresearch.com/carbon-nanomaterials-market.html.

Nemati, F., Zare-Dorabei, R., Hosseini, M., & Ganjali, M. R. (2018). Fluorescence turn-on sensing of thiamine based on Arginine-functionalized graphene quantum dots (Arg-GQDs): Central composite design for process optimization. *Sensors and Actuators, B: Chemical, 255*, 2078–2085. https://doi.org/10.1016/j.snb.2017.09.009.

Oliveira, T. M. B. F., & Morais, S. (2018). New generation of electrochemical sensors based on multi-walled carbon nanotubes. *Applied Sciences (Switzerland), 8*(10). https://doi.org/10.3390/app8101925.

Ortolani, T. S., Pereira, T. S., Assumpção, M. H. M. T., Vicentini, F. C., Gabriel de Oliveira, G., & Janegitz, B. C. (2019). Electrochemical sensing of purines guanine and adenine using single-walled carbon nanohorns and nanocellulose. *Electrochimica Acta, 298*, 893–900. https://doi.org/10.1016/j.electacta.2018.12.114.

Ouyang, L., Hu, Y., Zhu, L., Cheng, G. J., & Irudayaraj, J. (2017). A reusable laser wrapped graphene-Ag array based SERS sensor for trace detection of genomic DNA methylation. *Biosensors and Bioelectronics, 92*, 755–762. https://doi.org/10.1016/j.bios.2016.09.072.

Pagona, G., Sandanayaka, A. S. D., Araki, Y., Fan, J., Tagmatarchis, N., Charalambidis, G., et al. (2007). Covalent functionalization of carbon nanohorns with porphyrins: Nanohybrid formation and photoinduced electron and energy transfer. *Advanced Functional Materials, 17*(10), 1705–1711. https://doi.org/10.1002/adfm.200700039.

Park, S. J., Seo, S. E., Kim, K. H., Lee, S. H., Kim, J., Ha, S., et al. (2021). Real-time monitoring of geosmin based on an aptamer-conjugated graphene field-effect transistor. *Biosensors and Bioelectronics, 174*. https://doi.org/10.1016/j.bios.2020.112804.

Patel, K. D., Singh, R. K., & Kim, H. W. (2019). Carbon-based nanomaterials as an emerging platform for theranostics. *Materials Horizons, 6*(3), 434–469. https://doi.org/10.1039/c8mh00966j.

Pilehvar, S., & Wael, K. D. (2015). Recent advances in electrochemical biosensors based on fullerene-C60 nano-structured platforms. *Biosensors, 5*(4), 712–735. https://doi.org/10.3390/bios5040712.

Pirsaheb, M., Mohammadi, S., & Salimi, A. (2019). Current advances of carbon dots based biosensors for tumor marker detection, cancer cells analysis and bioimaging. *TrAC, Trends in Analytical Chemistry, 115*, 83–99. https://doi.org/10.1016/j.trac.2019.04.003.

Ponnaiah, S. K., Periakaruppan, P., & Vellaichamy, B. (2018). New electrochemical sensor based on a silver-doped iron oxide nanocomposite coupled with polyaniline and its sensing application for picomolar-level detection of uric acid in human blood and urine samples. *Journal of Physical Chemistry B, 122*(12), 3037–3046. https://doi.org/10.1021/acs.jpcb.7b11504.

Poornima Parvathi, V., Parimaladevi, R., Sathe, V., & Umadevi, M. (2019). Application of G-SERS for the efficient detection of toxic dye contaminants in textile effluents using gold/graphene oxide substrates. *Journal of Molecular Liquids, 273*, 203–214. https://doi.org/10.1016/j.molliq.2018.10.027.

Qaddare, S. H., & Salimi, A. (2017). Amplified fluorescent sensing of DNA using luminescent carbon dots and AuNPs/GO as a sensing platform: A novel coupling of FRET and DNA hybridization for homogeneous HIV-1 gene detection at femtomolar level. *Biosensors and Bioelectronics, 89*, 773–780. https://doi.org/10.1016/j.bios.2016.10.033.

Qian, L., Thiruppathi, A. R., Elmahdy, R., van der Zalm, J., & Chen, A. (2020). Graphene-oxide-based electrochemical sensors for the sensitive detection of pharmaceutical drug naproxen. *Sensors (Switzerland), 20*(5). https://doi.org/10.3390/s20051252.

Qian, J. Y., Tan, Y. L., Zhang, Y., Yang, Y. F., & Li, X. Q. (2018). Prognostic value of glypican-1 for patients with advanced pancreatic cancer following regional intra-arterial chemotherapy. *Oncology Letters, 16*(1), 1253–1258. https://doi.org/10.3892/ol.2018.8701.

Ragavan, K. V., & Neethirajan, S. (2018). Nanoparticles as biosensors for food quality and safety assessment. In *Nanomaterials for food applications* (pp. 147–202). Elsevier. https://doi.org/10.1016/B978-0-12-814130-4.00007-5.

Rahimi-Nasrabadi, M., Khoshroo, A., & Mazloum-Ardakani, M. (2017). Electrochemical determination of diazepam in real samples based on fullerene-functionalized carbon nanotubes/ionic liquid nanocomposite. *Sensors and Actuators, B: Chemical, 240*, 125–131. https://doi.org/10.1016/j.snb.2016.08.144.

Raj, M. A., & John, S. A. (2018). Graphene-modified electrochemical sensors. In *Graphene-based electrochemical sensors for biomolecules: A volume in micro and nano technologies* (pp. 1–41). Elsevier. https://doi.org/10.1016/B978-0-12-815394-9.00001-7.

Rajendran, J. R., Reshetilov, A. N., & Sundramoorthy, A. K. (2021). Electrochemically exfoliated graphene/poly(3,4-ethylenedioxythiophene) nanocomposite based electrochemical sensor for detection of nicotine. *Materials Advances, 38*. https://doi.org/10.1039/D0MA00974A.

Rani, U. A., Ng, L. Y., Ng, C. Y., & Mahmoudi, E. (2020). A review of carbon quantum dots and their applications in wastewater treatment. *Advances in Colloid and Interface Science, 278*. https://doi.org/10.1016/j.cis.2020.102124.

Ray, S. C., Mishra, D. K., Strydom, A. M., & Papakonstantinou, P. (2015). Magnetic behavioural change of silane exposed graphene nanoflakes. *Journal of Applied Physics, 118*(11). https://doi.org/10.1063/1.4930932.

Revuri, V., Cherukula, K., Nafiujjaman, M., Jae Cho, K., Park, I. K., & Lee, Y. K. (2018). White-light-emitting carbon nano-onions: A tunable multichannel fluorescent nanoprobe for glutathione-responsive bioimaging. *ACS Applied Nano Materials, 1*(2), 662–674. https://doi.org/10.1021/acsanm.7b00143.

Rizwan, M., Hazmi, M., Lim, S. A., & Ahmed, M. U. (2019). A highly sensitive electrochemical detection of human chorionic gonadotropin on a carbon nano-onions/gold nanoparticles/polyethylene glycol nanocomposite modified glassy carbon electrode. *Journal of Electroanalytical Chemistry, 833*, 462–470. https://doi.org/10.1016/j.jelechem.2018.12.031.

Roy, S., David-Pur, M., & Hanein, Y. (2017). Carbon nanotube-based ion selective sensors for wearable applications. *ACS Applied Materials and Interfaces, 9*(40), 35169–35177. https://doi.org/10.1021/acsami.7b07346.

Ruppi, S., & Larsson, A. (2001). Chemical vapour deposition of κ-Al2O3. *Thin Solid Films, 388*(1–2), 50–61. https://doi.org/10.1016/S0040-6090(01)00814-8.

Saisree, S., Aswathi, R., Arya Nair, J. S., & Sandhya, K. Y. (2021). Radical sensitivity and selectivity in the electrochemical sensing of cadmium ions in water by polyaniline-derived nitrogen-doped graphene quantum dots. *New Journal of Chemistry, 45*(1), 110–122. https://doi.org/10.1039/d0nj03988h.

Sam, S., Anand, S. K., Mathew, M. R., & Girish Kumar, K. (2021). Tannic acid capped copper nanoclusters as a cost-effective fluorescence probe for hemoglobin determination. *Analytical Sciences, 37*(4), 599–603. https://doi.org/10.2116/analsci.20P322.

Sanati, A. L., Faridbod, F., & Ganjali, M. R. (2017). Synergic effect of graphene quantum dots and room temperature ionic liquid for the fabrication of highly sensitive voltammetric sensor for levodopa determination in the presence of serotonin. *Journal of Molecular Liquids, 241*, 316–320. https://doi.org/10.1016/j.molliq.2017.04.123.

Santra, S., Xu, J., Wang, K., & Tan, W. (2004). Luminescent nanoparticle probes for bioimaging. *Journal of Nanoscience and Nanotechnology, 4*(6), 590–599. https://doi.org/10.1166/jnn.2004.017.

Sengupta, P., Khanra, K., Chowdhury, A. R., & Datta, P. (2019). Lab-on-a-chip sensing devices for biomedical applications. In *Bioelectronics and medical devices: From materials to devices—Fabrication, applications and reliability* (pp. 47–95). Elsevier. https://doi.org/10.1016/B978-0-08-102420-1.00004-2.

Shah, S. N. A., Dou, X., Khan, M., Uchiyama, K., & Lin, J. M. (2019). N-doped carbon dots/H2O2 chemiluminescence system for selective detection of Fe2+ ion in environmental samples. *Talanta, 196*, 370–375. https://doi.org/10.1016/j.talanta.2018.12.091.

Shang, N. G., Papakonstantinou, P., McMullan, M., Chu, M., Stamboulis, A., Potenza, A., et al. (2008). Catalyst-free efficient growth, orientation and biosensing properties of multilayer graphene nanoflake films with sharp edge planes. *Advanced Functional Materials, 18*(21), 3506–3514. https://doi.org/10.1002/adfm.200800951.

Shao, W., Shurin, M. R., Wheeler, S. E., He, X., & Star, A. (2021). Rapid detection of SARS-CoV-2 antigens using high-purity semiconducting single-walled carbon nanotube-based field-effect transistors. *ACS Applied Materials and Interfaces, 13*(8), 10321–10327. https://doi.org/10.1021/acsami.0c22589.

Shellaiah, M., Simon, T., Venkatesan, P., Sun, K. W., Ko, F. H., & Wu, S. P. (2018). Nanodiamonds conjugated to gold nanoparticles for colorimetric detection of clenbuterol and chromium(III) in urine. *Microchimica Acta, 185*(1). https://doi.org/10.1007/s00604-017-2611-7.

Shen, F., Zhang, C., Cai, Z., Wang, J., Zhang, X., Machuki, J. O. A., et al. (2020). Carbon nanohorns/Pt nanoparticles/DNA nanoplatform for intracellular Zn^{2+} imaging and enhanced cooperative phototherapy of cancer cells. *Analytical Chemistry, 92*(24), 16158–16169. https://doi.org/10.1021/acs.analchem.0c03880.

Shetti, N. P., Nayak, D. S., Reddy, K. R., & Aminabhvi, T. M. (2018). Graphene-clay-based hybrid nanostructures for electrochemical sensors and biosensors. *Graphene-based electrochemical sensors for biomolecules: A volume in micro and nano technologies* (pp. 235–274). Elsevier. https://doi.org/10.1016/B978-0-12-815394-9.00010-8.

Shin, Y. H., Gutierrez-Wing, M. T., & Choi, J. W. (2021). Review-recent progress in portable fluorescence sensors. *Journal of the Electrochemical Society, 168*(1). https://doi.org/10.1149/1945-7111/abd494.

Shrivastava, S., Trung, T. Q., & Lee, N. E. (2020). Recent progress, challenges, and prospects of fully integrated mobile and wearable point-of-care testing systems for self-testing. *Chemical Society Reviews, 49*(6), 1812–1866. https://doi.org/10.1039/c9cs00319c.

Sierra, M. I., Rubio, L., Bayón, G. F., Cobo, I., Menendez, P., Morales, P., et al. (2017). DNA methylation changes in human lung epithelia cells exposed to multi-walled carbon nanotubes. *Nanotoxicology, 11*(7), 857–870. https://doi.org/10.1080/17435390.2017.1371350.

Simioni, N. B., Silva, T. A., Oliveira, G. G., & Fatibello-Filho, O. (2017). A nanodiamond-based electrochemical sensor for the determination of pyrazinamide antibiotic. *Sensors and Actuators, B: Chemical, 250*, 315–323. https://doi.org/10.1016/j.snb.2017.04.175.

Singh, N. K., Chakma, B., Jain, P., & Goswami, P. (2018). Protein-induced fluorescence enhancement based detection of plasmodium falciparum glutamate dehydrogenase using carbon dot coupled specific aptamer. *ACS Combinatorial Science, 20*(6), 350–357. https://doi.org/10.1021/acscombsci.8b00021.

Sireesha, M., Jagadeesh Babu, V., Kranthi Kiran, A. S., & Ramakrishna, S. (2018). A review on carbon nanotubes in biosensor devices and their applications in medicine. *Nanocomposites, 4*(2), 36–57. https://doi.org/10.1080/20550324.2018.1478765.

Siripongpreda, T., Siralertmukul, K., & Rodthongkum, N. (2020). Colorimetric sensor and LDI-MS detection of biogenic amines in food spoilage based on porous PLA and graphene oxide. *Food Chemistry, 329*. https://doi.org/10.1016/j.foodchem.2020.127165, 127165.

Sivasankaran, U., Jesny, S., Jose, A. R., & Girish Kumar, K. (2017). Fluorescence determination of glutathione using tissue paper-derived carbon dots as fluorophores. *Analytical Sciences, 33*(3), 281–285. https://doi.org/10.2116/analsci.33.281.

Sohouli, E., Shahdost-Fard, F., Rahimi-Nasrabadi, M., Plonska-Brzezinska, M. E., & Ahmadi, F. (2020). Introducing a novel nanocomposite consisting of nitrogen-doped carbon nano-onions and gold nanoparticles for the electrochemical sensor to measure acetaminophen. *Journal of Electroanalytical Chemistry, 871*. https://doi.org/10.1016/j.jelechem.2020.114309, 114309.

Song, N., Ma, F., Zhu, Y., Chen, S., Wang, C., & Lu, X. (2018). Fe_3C/nitrogen-doped carbon nanofibers as highly efficient biocatalyst with oxidase-mimicking activity for colorimetric sensing. *ACS Sustainable Chemistry & Engineering, 6*(12), 16766–16776. https://doi.org/10.1021/acssuschemeng.8b04036.

Su, Y., Zhou, X., Long, Y., & Li, W. (2018). Immobilization of horseradish peroxidase on amino-functionalized carbon dots for the sensitive detection of hydrogen peroxide. *Microchimica Acta, 185*(2). https://doi.org/10.1007/s00604-017-2629-x.

Suherman, A. L., Ngamchuea, K., Tanner, E. E. L., Sokolov, S. V., Holter, J., Young, N. P., et al. (2017). Electrochemical detection of ultratrace (picomolar) levels of Hg^{2+} using a silver nanoparticle-modified glassy carbon electrode. *Analytical Chemistry, 89*(13), 7166–7173. https://doi.org/10.1021/acs.analchem.7b01304.

Sun, Y., Ding, C., Lin, Y., Sun, W., Liu, H., Zhu, X., et al. (2018). Highly selective and sensitive chemiluminescence biosensor for adenosine detection based on carbon quantum dots catalyzing luminescence released from aptamers functionalized graphene@magnetic β-cyclodextrin polymers. *Talanta, 186*, 238–247. https://doi.org/10.1016/j.talanta.2018.04.068.

Sun, X., Huang, C., Wang, L., Liang, L., Cheng, Y., Fei, W., et al. (2021). Recent progress in graphene/polymer nanocomposites. *Advanced Materials, 33*(6). https://doi.org/10.1002/adma.202001105.

Sutradhar, S., & Patnaik, A. (2017). A new fullerene-C60—Nanogold composite for non-enzymatic glucose sensing. *Sensors and Actuators, B: Chemical, 241*, 681–689. https://doi.org/10.1016/j.snb.2016.10.111.

Tabish, A. M., Poels, K., Byun, H. M., Luyts, K., Baccarelli, A. A., Martens, J., et al. (2017). Changes in DNA methylation in mouse lungs after a single intra-Tracheal administration of nanomaterials. *PLoS One, 12*(1). https://doi.org/10.1371/journal.pone.0169886.

Tang, M., Zhu, B., Wang, Y., Wu, H., Chai, F., Qu, F., et al. (2019). Nitrogen- and sulfur-doped carbon dots as peroxidase mimetics: Colorimetric determination of hydrogen peroxide and glutathione, and fluorimetric determination of lead(II). *Microchimica Acta*. https://doi.org/10.1007/s00604-019-3710-4.

Thiruppathi, A. R., Sidhureddy, B., Keeler, W., & Chen, A. (2017). Facile one-pot synthesis of fluorinated graphene oxide for electrochemical sensing of heavy metal ions. *Electrochemistry Communications, 76*, 42–46. https://doi.org/10.1016/j.elecom.2017.01.015.

Ting, S. L., Ee, S. J., Ananthanarayanan, A., Leong, K. C., & Chen, P. (2015). Graphene quantum dots functionalized gold nanoparticles for sensitive electrochemical detection of heavy metal ions. *Electrochimica Acta, 172*, 7–11. https://doi.org/10.1016/j.electacta.2015.01.026.

Tiwari, J. N., Vij, V., Kemp, K. C., & Kim, K. S. (2015). Engineered carbon-nanomaterial-based electrochemical sensors for biomolecules. *ACS Nano*, 46–80. https://doi.org/10.1021/acsnano.5b05690.

Toh, S. Y., Loh, K. S., Kamarudin, S. K., & Daud, W. R. W. (2014). Graphene production via electrochemical reduction of graphene oxide: Synthesis and characterisation. *Chemical Engineering Journal*, *251*, 422–434. https://doi.org/10.1016/j.cej.2014.04.004.

Tripathi, K. M., Bhati, A., Singh, A., Gupta, N. R., Verma, S., Sarkar, S., et al. (2016). From the traditional way of pyrolysis to tunable photoluminescent water soluble carbon nano-onions for cell imaging and selective sensing of glucose. *RSC Advances*, *6*(44), 37319–37329. https://doi.org/10.1039/c6ra04030f.

Tu, X., Gao, F., Ma, X., Zou, J., Yu, Y., Li, M., et al. (2020). Mxene/carbon nanohorn/β-cyclodextrin-metal-organic frameworks as high-performance electrochemical sensing platform for sensitive detection of carbendazim pesticide. *Journal of Hazardous Materials*, *396*. https://doi.org/10.1016/j.jhazmat.2020.122776, 122776.

Uygun, Z. O., Şahin, Ç., Yılmaz, M., Akçay, Y., Akdemir, A., & Sağın, F. (2018). Fullerene-PAMAM(G5) composite modified impedimetric biosensor to detect Fetuin-A in real blood samples. *Analytical Biochemistry*, *542*, 11–15. https://doi.org/10.1016/j.ab.2017.11.007.

Van Deun, J., Mestdagh, P., Agostinis, P., Akay, Ö., Anand, S., Anckaert, J., et al. (2017). EV-TRACK: Transparent reporting and centralizing knowledge in extracellular vesicle research. *Nature Methods*, *14*(3), 228–232. https://doi.org/10.1038/nmeth.4185.

Villalonga, A., Estabiel, I., Pérez-Calabuig, A. M., Mayol, B., Parrado, C., & Villalonga, R. (2021). Amperometric aptasensor with sandwich-type architecture for troponin I based on carboxyethylsilanetriol-modified graphene oxide coated electrodes. *Biosensors and Bioelectronics*, *183*. https://doi.org/10.1016/j.bios.2021.113203.

Viswanathan, S., Rani, C., Vijay Anand, A., & Ho, J. (2009). Disposable electrochemical immunosensor for carcinoembryonic antigen using ferrocene liposomes and MWCNT screen-printed electrode. *Biosensors and Bioelectronics*, *24*(7), 1984–1989. https://doi.org/10.1016/j.bios.2008.10.006.

Vázquez-González, M., Liao, W. C., Cazelles, R., Wang, S., Yu, X., Gutkin, V., et al. (2017). Mimicking horseradish peroxidase functions using Cu^{2+}-modified carbon nitride nanoparticles or Cu^{2+}-modified carbon dots as heterogeneous catalysts. *ACS Nano*, *11*(3), 3247–3253. https://doi.org/10.1021/acsnano.7b00352.

Wang, L., Li, M., Li, W., Han, Y., Liu, Y., Li, Z., et al. (2018). Rationally designed efficient dual-mode colorimetric/fluorescence sensor based on carbon dots for detection of pH and Cu^{2+} ions. *ACS Sustainable Chemistry & Engineering*, *6*(10), 12668–12674. https://doi.org/10.1021/acssuschemeng.8b01625.

Wang, C., Lan, Y., Yuan, F., Fereja, T. H., Lou, B., Han, S., et al. (2020). Chemiluminescent determination of L-cysteine with the lucigenin-carbon dot system. *Microchimica Acta*, *187*(1). https://doi.org/10.1007/s00604-019-3965-9.

Wang, Y., Bao, L., Liu, Z., & Pang, D. W. (2011). Aptamer biosensor based on fluorescence resonance energy transfer from upconverting phosphors to carbon nanoparticles for thrombin detection in human plasma. *Analytical Chemistry*, *83*(21), 8130–8137. https://doi.org/10.1021/ac201631b.

Wang, L., Cao, H. X., Pan, C. G., He, Y. S., Liu, H. F., Zhou, L. H., et al. (2019). A fluorometric aptasensor for bisphenol a based on the inner filter effect of gold nanoparticles on the fluorescence of nitrogen-doped carbon dots. *Microchimica Acta*, *186*(1). https://doi.org/10.1007/s00604-018-3153-3.

Wang, F., Lu, Y., Chen, Y., Sun, J., & Liu, Y. (2018). Colorimetric nanosensor based on the aggregation of AuNP triggered by carbon quantum dots for detection of Ag^+ ions. *ACS Sustainable Chemistry & Engineering*, *6*(3), 3706–3713. https://doi.org/10.1021/acssuschemeng.7b04067.

Wang, H., Huang, X., Wen, G., & Jiang, Z. (2019). A dual-model SERS and RRS analytical platform for Pb(II) based on Ag-doped carbon dot catalytic amplification and aptamer regulation. *Scientific Reports*, *9*(1). https://doi.org/10.1038/s41598-019-46426-y.

Wang, H., Pan, L., Liu, Y., Ye, Y., & Yao, S. (2020). Electrochemical sensing of nitenpyram based on the binary nanohybrid of hydroxylated multiwall carbon nanotubes/single-wall carbon nanohorns. *Journal of Electroanalytical Chemistry*, *862*, 113955. https://doi.org/10.1016/j.jelechem.2020.113955.

Wang, J., Lu, T., Hu, Y., Wang, X., & Wu, Y. (2020). A label-free and carbon dots based fluorescent aptasensor for the detection of kanamycin in milk. *Spectrochimica Acta Part A: Molecular and Biomolecular Spectroscopy*, *226*. https://doi.org/10.1016/j.saa.2019.117651, 117651.

Wang, X., Xu, G., Wei, F., Ma, Y., Ma, Y., Song, Y., et al. (2017). Highly sensitive and selective aptasensor for detection of adenosine based on fluorescence resonance energy transfer from carbon dots to nano-graphite. *Journal of Colloid and Interface Science*, *508*, 455–461. https://doi.org/10.1016/j.jcis.2017.07.028.

Wang, B., Chen, Y., Wu, Y., Weng, B., Liu, Y., Lu, Z., et al. (2016). Aptamer induced assembly of fluorescent nitrogen-doped carbon dots on gold nanoparticles for sensitive detection of AFB1. *Biosensors and Bioelectronics*, *78*, 23–30. https://doi.org/10.1016/j.bios.2015.11.015.

Wang, X., Gao, D., Li, M., Li, H., Li, C., Wu, X., et al. (2017). CVD graphene as an electrochemical sensing platform for simultaneous detection of biomolecules. *Scientific Reports*, *7*(1). https://doi.org/10.1038/s41598-017-07646-2.

Wang, Q., Zhang, J., Xu, Y., Wang, Y., Wu, L., Weng, X., et al. (2021). A one-step electrochemically reduced graphene oxide based sensor for sensitive voltammetric determination of furfural in milk products. *Analytical Methods : Advancing Methods and Applications*, *13*(1), 56–63. https://doi.org/10.1039/d0ay01789b.

Wang, Z., An, R., Dai, Y., & Luo, H. (2021). A simple strategy for the simultaneous determination of dopamine, uric acid, L-tryptophan and theophylline based on a carbon dots modified electrode. *International Journal of Electrochemical Science*, *16*, 1–15. https://doi.org/10.20964/2021.04.39.

Wang, Z., & Dai, Z. (2015). Carbon nanomaterial-based electrochemical biosensors: An overview. *Nanoscale*, *7*(15), 6420–6431. https://doi.org/10.1039/c5nr00585j.

Wang, X., Sparkman, J., & Gou, J. (2017). Strain sensing of printed carbon nanotube sensors on polyurethane substrate with spray deposition modeling. *Composites Communications*, *3*, 1–6. https://doi.org/10.1016/j.coco.2016.10.003.

Wang, Z., Wu, S., Wang, J., Yu, A., & Wei, G. (2019). Carbon nanofiber-based functional nanomaterials for sensor applications. *Nanomaterials*, *9*(7). https://doi.org/10.3390/nano9071045.

Welsh, J. A., Van Der Pol, E., Arkesteijn, G. J. A., Bremer, M., Brisson, A., Coumans, F., et al. (2020). MIFlowCyt-EV: A framework for standardized reporting of extracellular vesicle flow cytometry experiments. *Journal of Extracellular Vesicles*, *9*(1), 1713526. https://doi.org/10.1080/20013078.2020.1713526.

Williams, R. M., Lee, C., & Heller, D. A. (2018). A fluorescent carbon nanotube sensor detects the metastatic prostate cancer biomarker uPA. *ACS Sensors*, *3*(9), 1838–1845. https://doi.org/10.1021/acssensors.8b00631.

Williams, R. M., Lee, C., Galassi, T. V., Harvey, J. D., Leicher, R., Sirenko, M., et al. (2018). Noninvasive ovarian cancer biomarker detection via an optical nanosensor implant. *Science Advances*, *4*(4). https://doi.org/10.1126/sciadv.aaq1090.

Wu, Y., Alam, M. N. A., Balasubramanian, P., Ermakova, A., Fischer, S., Barth, H., et al. (2020). A nanodiamond-based theranostic for light-controlled intracellular heating and nanoscale temperature sensing. *ChemRxiv*. https://doi.org/10.26434/chemrxiv.13334006.v2.

Wu, H., Liu, R., Kang, X., Liang, C., Lv, L., & Guo, Z. (2018). Fluorometric aptamer assay for ochratoxin A based on the use of single walled carbon nanohorns and exonuclease III-aided amplification. *Microchimica Acta*. https://doi.org/10.1007/s00604-017-2592-6.

Xu, M., Gao, Z., Zhou, Q., Lin, Y., Lu, M., & Tang, D. (2016). Terbium ion-coordinated carbon dots for fluorescent aptasensing of adenosine 5′-triphosphate with unmodified gold nanoparticles. *Biosensors and Bioelectronics*, *86*, 978–984. https://doi.org/10.1016/j.bios.2016.07.105.

Xue, J., Yao, C., Li, N., Su, Y., Xu, L., & Hou, S. (2021). Construction of polydopamine-coated three-dimensional graphene-based conductive network platform for amperometric detection of dopamine. *Journal of Electroanalytical Chemistry*, *886*. https://doi.org/10.1016/j.jelechem.2021.115133, 115133.

Yan, F., Jiang, Y., Sun, X., Bai, Z., Zhang, Y., & Zhou, X. (2018). Surface modification and chemical functionalization of carbon dots: A review. *Microchimica Acta*, *185*(9). https://doi.org/10.1007/s00604-018-2953-9.

Yan, F., Bai, Z., Chen, Y., Zu, F., Li, X., Xu, J., et al. (2018). Ratiometric fluorescent detection of copper ions using coumarin-functionalized carbon dots based on FRET. *Sensors and Actuators, B: Chemical*, *275*, 86–94. https://doi.org/10.1016/j.snb.2018.08.034.

Yang, C., Denno, M. E., Pyakurel, P., & Venton, B. J. (2015). Recent trends in carbon nanomaterial-based electrochemical sensors for biomolecules: A review. *Analytica Chimica Acta*, *887*, 17–37. https://doi.org/10.1016/j.aca.2015.05.049.

Yang, C., Hu, L. W., Zhu, H. Y., Ling, Y., Tao, J. H., & Xu, C. X. (2015). RGO quantum dots/ZnO hybrid nanofibers fabricated using electrospun polymer templates and applications in drug screening involving an intracellular H2O2 sensor. *Journal of Materials Chemistry B*, *3*(13), 2651–2659. https://doi.org/10.1039/c4tb02134g.

Yao, D., Li, C., Wen, G., Liang, A., & Jiang, Z. (2020). A highly sensitive and accurate SERS/RRS dual-spectroscopic immunosensor for clenbuterol based on nitrogen/silver-codoped carbon dots catalytic amplification. *Talanta*, *209*. https://doi.org/10.1016/j.talanta.2019.120529, 120529.

Yao, Y., Wu, H., & Ping, J. (2019). Simultaneous determination of Cd(II) and Pb(II) ions in honey and milk samples using a single-walled carbon nanohorns modified screen-printed electrochemical sensor. *Food Chemistry*, *274*, 8–15. https://doi.org/10.1016/j.foodchem.2018.08.110.

You, J., You, Z., Xu, X., Ji, J., Lu, T., Xia, Y., et al. (2019). A split aptamer-labeled ratiometric fluorescent biosensor for specific detection of adenosine in human urine. *Microchimica Acta*, *186*(1). https://doi.org/10.1007/s00604-018-3162-2.

You, H., Mu, Z., Zhao, M., Zhou, J., Yuan, Y., & Bai, L. (2020). Functional fullerene-molybdenum disulfide fabricated electrochemical DNA biosensor for Sul1 detection using enzyme-assisted target recycling and a new signal marker for cascade amplification. *Sensors and Actuators B: Chemical*, *305*. https://doi.org/10.1016/j.snb.2019.127483, 127483.

Yugender Goud, K., Hayat, A., Satyanarayana, M., Sunil Kumar, V., Catanante, G., Vengatajalabathy Gobi, K., et al. (2017). Aptamer-based zearalenone assay based on the use of a fluorescein label and a functional graphene oxide as a quencher. *Microchimica Acta*, *184*(11), 4401–4408. https://doi.org/10.1007/s00604-017-2487-6.

Yukawa, H., Fujiwara, M., Kobayashi, K., Kumon, Y., Miyaji, K., Nishimura, Y., et al. (2020). A quantum thermometric sensing and analysis system using fluorescent nanodiamonds for the evaluation of living stem cell functions according to intracellular temperature. *Nanoscale Advances*, *2*(5), 1859–1868. https://doi.org/10.1039/d0na00146e.

Yáñez-Sedeño, P., Campuzano, S., & Pingarrón, J. (2017). Fullerenes in electrochemical catalytic and affinity biosensing: A review. *C*, *21*. https://doi.org/10.3390/c3030021.

Zangeneh Kamali, K., Pandikumar, A., Jayabal, S., Ramaraj, R., Lim, H. N., Ong, B. H., et al. (2016). Amalgamation based optical and colorimetric sensing of mercury(II) ions with silver@graphene oxide nanocomposite materials. *Microchimica Acta*, *183*(1), 369–377. https://doi.org/10.1007/s00604-015-1658-6.

Zhang, Z., Li, J., Wang, X., Liang, A., & Jiang, Z. (2019). Aptamer-mediated N/Ce-doped carbon dots as a fluorescent and resonance Rayleigh scattering dual mode probe for arsenic(III). *Microchimica Acta*, *186*(9). https://doi.org/10.1007/s00604-019-3764-3.

Zhang, Z. H., Lei, K. N., Li, C. N., Luo, Y. H., & Jiang, Z. L. (2020). A new and facile nanosilver SPR colored method for ultratrace arsenic based on aptamer regulation of Au-doped carbon dot catalytic amplification. *Spectrochimica Acta-Part A: Molecular and Biomolecular Spectroscopy*, *232*. https://doi.org/10.1016/j.saa.2020.118174.

Zhang, J., & Zhao, X. S. (2014). Graphene-based materials for electrochemical energy storage. In *Two-dimensional carbon: Fundamental properties, synthesis, characterization, and applications* (pp. 183–246). Pan Stanford Publishing Pte. Ltd. http://www.panstanford.com/books/9789814411943.

Zhang, X., Wang, K. P., Zhang, L. N., Zhang, Y. C., & Shen, L. (2018). Phosphorus-doped graphene-based electrochemical sensor for sensitive detection of acetaminophen. *Analytica Chimica Acta*, *1036*, 26–32. https://doi.org/10.1016/j.aca.2018.06.079.

Zhang, Y. (2020). An electrochemical benzylpenicillin biosensor based on β-lactamase and fullerene supported by a bilayer lipid membrane. *International Journal of Electrochemical Science*, 12007–12014. https://doi.org/10.20964/2020.12.66.

Zhang, L., Wang, J., Deng, J., & Wang, S. (2020). A novel fluorescent "turn-on" aptasensor based on nitrogen-doped graphene quantum dots and hexagonal cobalt oxyhydroxide nanoflakes to detect tetracycline. *Analytical and Bioanalytical Chemistry*, *412*(6), 1343–1351. https://doi.org/10.1007/s00216-019-02361-5.

Zhang, W. J., Liu, S. G., Han, L., Luo, H. Q., & Li, N. B. (2019). A ratiometric fluorescent and colorimetric dual-signal sensing platform based on N-doped carbon dots for selective and sensitive detection of copper(II) and pyrophosphate ion. *Sensors and Actuators, B: Chemical, 283*, 215–221. https://doi.org/10.1016/j.snb.2018.12.012.

Zheng, W., Liu, Y., Yang, P., Chen, Y., Tao, J., Hu, J., et al. (2020). Carbon nanohorns enhanced electrochemical properties of Cu-based metal organic framework for ultrasensitive serum glucose sensing. *Journal of Electroanalytical Chemistry, 862*. https://doi.org/10.1016/j.jelechem.2020.114018, 114018.

Zheng, M., Wang, Y., Wang, C., Wei, W., Ma, S., Sun, X., et al. (2018). Green synthesis of carbon dots functionalized silver nanoparticles for the colorimetric detection of phoxim. *Talanta, 185*, 309–315. https://doi.org/10.1016/j.talanta.2018.03.066.

Zhou, Z., Zhao, L., Wang, Z., Xue, W., Wang, Y., Huang, Y., et al. (2018). Colorimetric detection of 1,5-anhydroglucitol based on graphene quantum dots and enzyme-catalyzed reaction. *International Journal of Biological Macromolecules, 112*, 1217–1224. https://doi.org/10.1016/j.ijbiomac.2018.02.093.

Zhou, Y., & Azumi, R. (2016). Carbon nanotube based transparent conductive films: Progress, challenges, and perspectives. *Science and Technology of Advanced Materials, 17*(1), 493–516. https://doi.org/10.1080/14686996.2016.1214526.

Zhu, L., Xu, G., Song, Q., Tang, T., Wang, X., Wei, F., et al. (2016). Highly sensitive determination of dopamine by a turn-on fluorescent biosensor based on aptamer labeled carbon dots and nano-graphite. *Sensors and Actuators, B: Chemical, 231*, 506–512. https://doi.org/10.1016/j.snb.2016.03.084.

Zhu, G., Sun, H., Zou, B., Liu, Z., Sun, N., Yi, Y., et al. (2018). Electrochemical sensing of 4-nitrochlorobenzene based on carbon nanohorns/graphene oxide nanohybrids. *Biosensors and Bioelectronics, 106*, 136–141. https://doi.org/10.1016/j.bios.2018.01.058.

Zhu, Z. (2017). An overview of carbon nanotubes and graphene for biosensing applications. *Nano-Micro Letters, 9*(3). https://doi.org/10.1007/s40820-017-0128-6.

Zhuang, X., Wang, H., He, T., & Chen, L. (2016). Enhanced voltammetric determination of dopamine using a glassy carbon electrode modified with ionic liquid-functionalized graphene and carbon dots. *Microchimica Acta, 183*(12), 3177–3182. https://doi.org/10.1007/s00604-016-1971-8.

Zhuo, S., Fang, J., Zhu, C., & Du, J. (2020). Preparation of palladium/carbon dot composites as efficient peroxidase mimics for H2O2 and glucose assay. *Analytical and Bioanalytical Chemistry, 412*(4), 963–972. https://doi.org/10.1007/s00216-019-02320-0.

Zuaznabar-Gardona, J. C., & Fragoso, A. (2020). Electrochemistry of redox probes at thin films of carbon nano-onions produced by thermal annealing of nanodiamonds. *Electrochimica Acta, 353*. https://doi.org/10.1016/j.electacta.2020.136495, 136495.

Chapter 22

Challenges in commercialization of carbon nanomaterial-based sensors

Elif Esra Altuner[a], Merve Akin[a], Ramazan Bayat[a,b], Muhammed Bekmezci[a,b], Hakan Burhan[a], and Fatih Sen[a]

[a]Sen Research Group, Department of Biochemistry, University of Dumlupinar, Kutahya, Turkey, [b]Department of Materials Science & Engineering, Faculty of Engineering, University of Dumlupinar, Kutahya, Turkey

Introduction

The rapid progress of nanotechnology in recent times brings along great advantages and disadvantages in studies. It is important to know the disadvantages as well as advantages since there is a high demand for nanotechnological studies recently (Demirkıran, 2019). Nanoparticles/nanocomposite materials constitute a large part of nanotechnological studies. Studies in the form of nanoparticles (Sen, Demirkan, Şimşek, Savk, & Sen, 2018), carbon-based nanoparticles (Savk et al., 2019), polymeric nanoparticles (Campodoni et al., 2019), metallic nanoparticles (Sen, Demirkan, Şavk, Karahan Gülbay, & Sen, 2018), etc., and their derivatives are available in the literature. Nanoparticle/nanocomposite studies are used in various fields: quantum dot studies (Yoneda et al., 2018), sensors (Ayranci et al., 2019), drugs (Gao, Cui, Levenson, Chung, & Nie, 2004), medicine and health care (Lee, Khan, Park, & Lim, 2012), etc. It has been reported that a wide variety of research on nanoparticles has been performed. It is also used in the commercial field (Demir, Savk, Sen, & Sen, 2017; Ozturk, Sen, Sen, & Gokagac, 2012; Sen et al., 2019; Şen, Demirkan, Levent, Şavk, & Şen, 2018; Sen, Şavk, & Sen, 2018; Şen, Şen, & Gökağaç, 2011; Sert et al., 2017). Therefore, it is important to know the advantages of nanoparticles as well as these disadvantages in the commercial field. In this chapter, explanations are given about the commercial disadvantages and problems of carbon-based nanoparticles, especially carbon-based sensors. Silver nanoparticles from metallic nanoparticles are mostly used in the devices for home jobs (Ahmed et al., 2017); platinum nanoparticles are used in industries (Asztemborska, Steborowski, Kowalska, & Bystrzejewska-Piotrowska, 2015); and nanoparticles of various metals such as palladium (Koskun, Şavk, Şen, & Şen, 2018), gold (Nath & Chilkoti, 2002), ruthenium (Taçyıldız, Demirkan, Karataş, Gulcan, & Sen, 2019), cobalt (Çelik et al., 2016), and iron (Gupta, Jain, Agarwal, & Maheshwari, 2007) are used in sensor studies. Sensor studies of carbon-based nanoparticles have also been reported in the literature (Ayranci et al., 2019; Ozturk et al., 2012; Şen & Gökagaç, 2007; Şen & Gökağaç, 2014; Sert et al., 2016; Sert et al., 2017). Sensors are a very popular subject by researchers in the determination of biological and chemical substances. There are many sensor studies in the literature on this subject (Darder, Colilla, & Ruiz-Hitzky, 2005; Khoo et al., 2009; Nath & Chilkoti, 2002; Sekol et al., 2013). Sensor catalytic studies are available in the literature, especially in studies with nanoparticles/nanocomposites (Frasco & Chaniotakis, 2009; Giraldo et al., 2014; Jaque et al., 2016; Sen et al., 2012). The difficulties encountered in sensor studies of carbon-based nanoparticles have been examined in this work.

Sensors

All living things try to adapt to differences to adapt and continue their lives. This constitutes the in vitro state of the biosensors. Beings exhibit sensitivity that even scientists cannot imagine. For example, dogs' smell is a hundred thousand times ahead of us. Eels detect foreign materials with droplets added to hundreds of thousands of kilograms of water. Butterflies even notice the granular pheromones secreted by their mates. Algae are sensitive to toxins (Bulut, 2011). Biosensors are a system of analytical devices integrated with a physicochemical transducer, that is, a converter, with a biological receptor in its structure. The goal of a biosensor is to produce electrical data that are always numerical and associated with the number of one or a group of analytes (the item to be analyzed) (Augusto Da Costa & Duta, 2001; Bernfeld, 1955; Dönmez & Aksu, 2001; Lin et al., 2009; Otlu, 2011; Uauy, Olivares, & Gonzalez, 1998). There are three main components used for this mechanism. Highly sensitive biomolecules or bioagents interact with the substrate material and give a signal, and these signals are converted into digital data with electronic detectors. These three important components (biomolecule, transducer, and electronic data) are sensitive biological agents that interact with the substance to be determined in a highly selective but reversible manner. Beings have created biosensors with the analysis mechanisms of biological materials that make it possible to detect these data. Biosensor science is developing so rapidly. Therefore the Biosensors Classification and Nomenclature Commission was established by the *International Union of Pure and Applied Chemistry* (Güngör, 2019). Integration of physicochemical analytical systems and biochemical materials is essential for a biosensor. In biosensors, the high specificity of the biological mechanism and the detection sensitivity of the physical analysis mechanism are integrated (Fereidoonnezhad et al., 2018; Haque, 2017; Jaroch, Boyaci, Pawliszyn, & Bojko, 2019; Özdemir, Kilinc, Poli, Nicolaus, & Güven, 2012; Raril et al., 2018; Weigel, Bester, & Hühnerfuss, 2001). Many biosensors have been developed to use in the measurement of generally bioorganic particles and sometimes inorganic particles. Biosensors used in daily life, especially provided that the health field is at the forefront; It is preferred in environmental measurements, military field, food, pharmacology, and chemical industries. A sensor consists of an analyte sample, catalyst, activator, signals, and detectors (Chuma, 2007; Dönmez & Aksu, 2001; Hanaoka et al., 2013; Henn et al., 2010; Hobman & Crossman, 2015; Madrid & Cámara, 1997; Mateos et al., 2016; Raril et al., 2018; Ray, Paul, Bera, & Chattopadhyay, 2006; Satchanska et al., 2005; Srinath, Verma, Ramteke, & Garg, 2002). It is essential to receive continuous signals or data from the sensors. Fig. 22.1 shows the structure of a sensor and an example of a biosensor using tyrosine.

Biosensors are systems obtained by combining the active component that interacts with the sample to be analyzed with a conductive mechanism that sends the signal generated as a result of contact. Reaction and signal data are essential to establish a biosensor mechanism. Biocomponents are called bioreceptors. Nucleic acid and enzyme studies are the most preferred studies. Enzymes, bacteria, microbes, organelles, cell tissues, antibodies, nucleic acids, and receptors on the biological membrane are biocomponents. Enzymes are the most preferred, but bacteria can sometimes be used because they degrade quickly (Erdoğan, 2013).

The history of sensors

Sensing mechanisms such as vision, hearing, smell, taste, and touch, which are important elements of living life, are considered natural and perfect biosensor systems, making them good examples of biosensor studies. The history of biosensors was described in 1950 by L. Clark. It was started when *Clark* followed the oxygen amount in the blood with an electrode during a medical operation at *Cincinnati Hospital (Ohio, USA)*. In 1962, *Clark and Lyons* determined the glucose of the blood by combining the enzyme Glucosidase (GOD) with the O_2 electrode. Thus, a new analytical system was formed. This system, on the one hand, combined the detection sensitivity of the biological system (enzyme), and on the other hand, it combined the detection sensitivity of the physical system (electrode) and displayed a broad-spectrum application

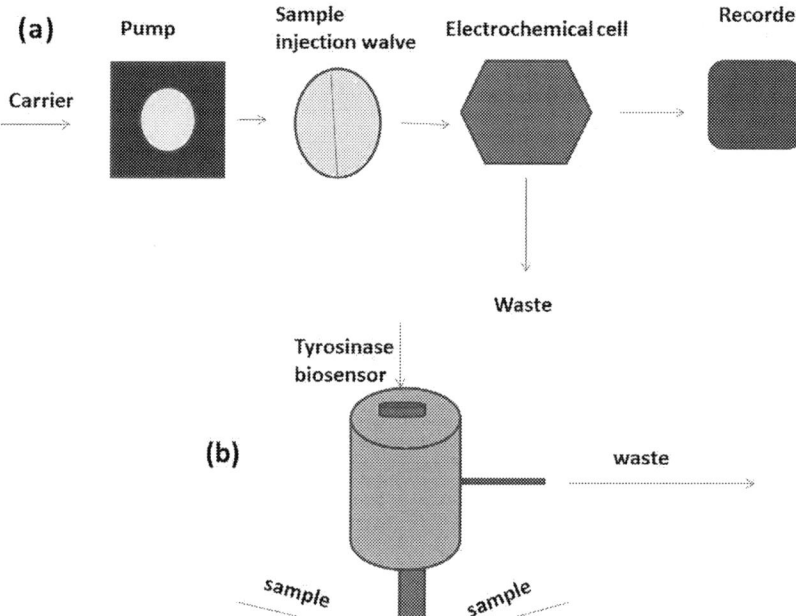

FIG. 22.1 The system of sensors (A), the schematic of one biosensor for tyrosine (Capannesi, Palchetti, Mascini, & Parenti, 2000). *(Reprinted (adapted) with permission from Capannesi, C., Palchetti, I., Mascini, M., Parenti, A. (2000) Electrochemical sensor and biosensor for polyphenols detection in olive oils, Food Chemistry, 71 (4), 553–562. https://doi.org/10.1016/S0308-8146(00)00211-9. Copyright (2021) Food Chemistry—Elsevier.)*

opportunity. In the sample with biological fluid properties, when glucose and dissolved oxygen cross the membrane around the electrode and reach the electrode surface, the sugar oxidizes and turns into gluconic acid. In the meantime, oxygen is consumed. When the glucose in the sample is exhausted, the oxygen dissolved as a result of the initial reaction is measured with the oxygen electrode. When glucose and dissolved oxygen in a biological fluid across the membrane around the electrode reach the electrode surface, the glucose oxidizes into gluconic acid and O_2 is consumed. When the glucose in the environment is exhausted, the dissolved O_2 at the beginning and the end of the reaction is measured with the O_2 electrode. The difference is the O_2 consumed for the oxidation of glucose in the environment, and from here, the amount of glucose in the biological fluid is calculated (Özoğlu, 2019). There are different attempts on this subject. Agriculture, food, medical operations, pharmacological applications are examples (Boz, Paylan, Kızmaz, & Erkan, 2017). First-generation sensors are studied based on enzyme electrodes. The oxidizing-reducing enzyme is in the membrane sandwich in addition to the platinum (Pt) electrode. According to Costa and Dutan's report, a peroxide reaction occurred as a result of an enzyme reaction with the Pt cathode material polarized at 0.7 V. According to the report of Costa and Dutan, a peroxide reaction occurred as a result of an enzyme reaction with the Pt cathode material polarized at 0.7 V. Likewise, electron transfer at the electrode originates from here (Augusto Da Costa & Duta, 2001).

$$Glucose + O_2 + H_2O + glucose\ oxidase \rightarrow gluconic\ acid + H_2O_2.$$

The reaction taking place at the electrode is as follows (Augusto Da Costa & Duta, 2001):

$$O_2 + 2e - +2H^+ \rightarrow H_2O_2.$$

Thus, sensor studies with enzymes have begun. Subsequently, GOD enzymes were used for glucose determination (Ahmed et al., 2017; Briceño, Suarez, & Gonzalez, 2017). These studies are first- and second-generation sensor studies. Electrochemically, the current strength of the platinum wire and the silver electrode reduces the oxygen. Third-generation biosensors are based on the measurement of the sensory property of an analyte by a device and the acquisition of precise numerical data connected to the detector. Surface plasmon resonance, kit studies, and voltammetric electrochemical studies can be given as examples. In surface plasmon resonance biosensors, the agent becomes a component of the sensor. The fourth-generation sensors, on the contrary, mostly cover sensor studies in micro- and nanostructures. They are called "microelectromechanical systems (MEMSs), nanoelectromechanical systems (NEMSs), and bio-nanoelectromechanical systems (BioNEMSs)" (Kokbas, Kayrın, & Tuli, 2013). They are very sensitive and include the latest technological studies.

Types of sensors

Based on the measurement of the sensors, they are divided into five types according to the bioactive layer transmission and measurements (*Handbook of environmental materials management*, 2019; *Handbook of Functionalized Nanomaterials for Industrial Applications*, 2020; *Handbook of nanomaterials for manufacturing applications*, 2020; *Handbook of nanomaterials for sensing applications*, 2021; *Handbook of polymer nanocomposites for industrial applications*, 2021):

(a) Electrochemical-based sensors (amperometry, potentiometry).
(b) Optical-based sensors (photometry, fluorimetry, bioluminescence).
(c) Piezoelectric-based sensors (quartz crystal microbalance, microcantilevers).
(d) Calorimetry-based sensors (thermistors).
(e) Magnetic sensors (Article Derkuş et al., 2018).

Electrochemical-based sensors

Electrochemical biosensors are the most active and preferred sensors compared to other types. The first biosensor studies were based on the measurements made with enzyme electrodes sensitive to enzymes, specially developed for clinical glucose analysis. Heinemann et al. conducted immunoelectrochemical (IEC) studies aiming to improve the sensitivity of electrochemical species related to the enzyme (Aykut & Temiz, 2006). Electrochemical biosensors can be divided into amperometric/voltammetric, potentiometric and conductivity, capacitance, and impedance biosensors. This stage was used in our study. Chitosan-cobalt(II) and palladium-based sensor studies of 3.3', 5.5' tetramethylbenzidine have been performed. In the previous studies, horseradish peroxidase (HRP)-based sensor studies of 3.3', 5.5' tetramethylbenzidine as an enzyme were performed (Lan et al., 2014; Volpe, Compagnone, Draisci, & Palleschi, 1998). 3,3', 5.5' Tetramethylbenzidine (TMB) was investigated as an electrochemical substrate for horseradish peroxidase. HRP activity was determined using flow injection analysis on a glassy carbon working electrode polarized against Ag/AgCl at +100mV in 0.1 mol/L citrate-phosphate buffer (pH 5). In this study, optimum concentrations were found as 2×10^{-4} mol/L TMB and 0.001 mol/L H_2O_2. The detection limit obtained after 15 min of incubation was found to be 8.5×10^{-14} mol/L by amperometric methods. Thus, a sensory study was created with the help of the microplate reader device (Volpe et al., 1998). Sensors of this class are divided into types such as potentiometric, voltammetric, and amperometric. Potentiometric sensors are based on the principle of load voltage, whereas amperometric and voltammetric sensors are based on the current principal method. Photometry, fluorimetry, and bioluminescence methods are in this classification.

Optical-based sensors

Optical methods are among the earliest and most advanced technological studies of biological and chemical reactants among sensor technologies. Different optical techniques have been used in the production of biosensors.

A unique optical biosensor is a group of optical components to create a spectrum of light with certain properties from a light source and a factor that directs or displaces this light, a transformed sensing head (crystals covered with optical fibers or antibodies transformed with dyes and proteins and functional chemical groups), and light consists of the detector (Ragavan, Kumar, Swaraj, & Neethirajan, 2018).

Piezoelectric-based sensors

Piezoelectric sensors are the sensors based on electricity, weight, and details. Thus, commercially available quartz crystal microbalance devices are used. They are superior to other types in terms of sensitivity, versatile application, low cost, and cheapness, and they are ethylated (Aykut & Temiz, 2006).

Magnetic sensors

Magnetic sensors are the sensors based on studies of resistance caused by magnetism in analytical micropipes. Studies with the magnetic resistance of fluid samples, such as magnetic, micro-, or nanoparticles, through micropipes are attempted to be further developed in the future (Khan, Yilmaz, & Soylak, 2016; Lu, He, Wang, Liu, & Hou, 2018; Qiu et al., 2020).

Carbon-based nanomaterials

Carbon-based nanoparticles are divided into derivatives such as carbon nanotubes, carbon nanofibers, graphene and graphene oxide, carbon onions, Vulcan carbons, fullerenes, mesoporous nanocarbons, and macroporous nanocarbons (Gogotsi & Presser, 2014). Carbon-nanostructured materials are also used as a supporting agent in catalytic activators and serve as a very good support material. Therefore, they act as catalysts in catalytic activities in sensor and fuel cells (Basri, Kamarudin, Daud, & Yaakub, 2010).

Carbon nanotube chemical sensors

Carbon nanotubes are of great interest in academic studies (Demir, Sen, & Sen, 2017; Fam, Palaniappan, Tok, Liedberg, & Moochhala, 2011; Nas et al., 2019; Sen, Kuzu, Demir, Akocak, & Sen, 2017). Carbon nanotubes with 1-dimensional volume have a wide research field (Gogotsi & Presser, 2014). They hold great hope in building the structure of electrical and electronic equipment in the future. Carbon nanotubes have advantages such as small volume, high electrical and thermal conductivity, and significant specific area. For this reason, carbon nanotubes are preferred to be used in sensor industries and industrial platforms (Schroeder, Savagatrup, He, Lin, & Swager, 2019). The most used fields of carbon nanotubes are electrical-electronic devices (Schroeder et al., 2019), microelectronic fields (Cao & Rogers, 2009; Park, Vosguerichian, & Bao, 2013; Schroeder et al., 2019), computing (Li et al., 2020; Schroeder et al., 2019), chemical sensors (Kauffman & Star, 2008; Meyyappan, 2016; Schroeder et al., 2019), electrochemical sensors (Schroeder et al., 2019; Wang, 2005), and medicinal therapy fields (Hong, Diao, Antaris, & Dai, 2015; Schroeder et al., 2019). Carbon nanotubes have a hexagonal structure. They have sizes from 1 nm to 1–100 μm. The table regarding the patents of carbon nanotubes in recent years is shown in Fig. 22.2.

Carbon nanotubes are generally synthesized by two synthesis methods, solid and gaseous. Synthesis methods with the solid method are laser etching method, arc discharge method, and solar furnace method (Küçükyıldırım & Akdoğan Eker, 2012). The gaseous synthesis method is chemical vapor deposition, thermal chemical vapor deposition, and plasma-powered chemical vapor deposition. Apart from all these, the hydrothermal synthesis method is also reported in the literature (Küçükyıldırım & Akdoğan Eker, 2012; Qiu et al., 2014). Sensor findings were obtained employing an electrode coated with polypyrrole in sensor studies on carbon nanotubes in the literature, and the signal data given to 18 different

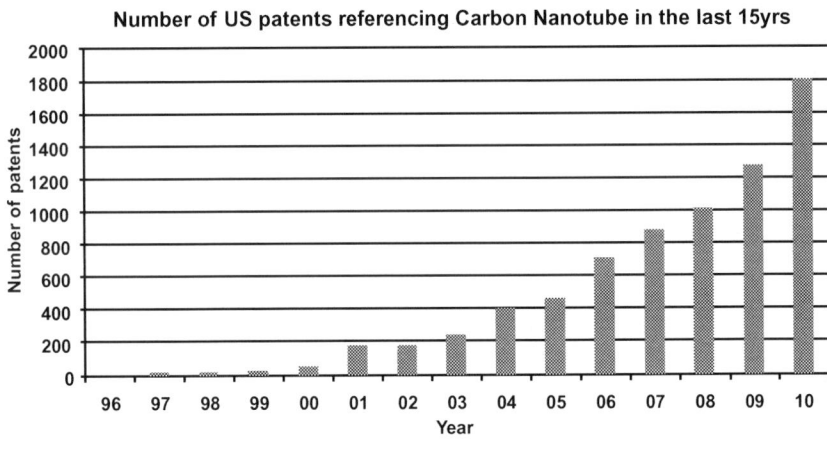

FIG. 22.2 The number of US patents of carbon nanotubes in the last 15 years (Fam et al., 2011). *(Reprinted (adapted) with permission from Fam, D.W.H., Palaniappan, A., Tok, A.I.Y., Liedberg, B., Moochhala, S.M. (2011). A review on technological aspects influencing commercialization of carbon nanotube sensors.* Sensors and Actuators, B: Chemical, 157 *(1), 1–7. https://doi.org/10.1016/j.snb.2011.03.040. Copyright (2021) Sensors and Actuators B: Chemical—Elsevier.)*

phenol types were continuous, and thus, a sensor was developed. Carbon nanotube sensor operation gave signals continuously and more strongly (Kilic & Korkut, 2016).

Ordered mesoporous and macroporous carbon-modified electrodes

The structure surfaces of the electrodes are approximately <2 nm (micro), between 2 and 50 nm (meso), and >50 nm (macro) (Walcarius, 2012). The electrochemical efficiency and sensor efficiency of nanostructured materials, as well as porous materials, also depend on the increase and coherence of the surface of the electrodes in a suitable position in a suitable volume at a suitable performance. Numerous articles on mesoporous carbon-based and carbon-related materials have been reported in the literature (Lee, Han, & Hyeon, 2004; Lee, Kim, & Hyeon, 2006; Liang, Li, & Dai, 2008; Ryoo, Joo, Kruk, & Jaroniec, 2001; Walcarius, 2012; Wan, Shi, & Zhao, 2008). Porous nanoscale carbon materials are prepared by the template carbonization process, featuring nanoscale carbon materials (Lee et al., 2004; Liang et al., 2008; Walcarius, 2012). The template-carbonization process used consists of two subheadings. It is in the form of preparing soft and hard molds with this head. Both of these processes are used for the synthesis and preparation of specific nanoscale carbon materials (Erdoğan, 2013). Macroporous carbons were obtained by carbonization of colloidal crystals in 1998 (Wan et al., 2008). Likewise, mesoporous carbons can be formed by stabilized surfactants and silica colloidal particles by the carbonization and calcination method (Erdoğan, 2013). In Fig. 22.3, the mesoporous carbon synthesis structure is shown in order. In Fig. 22.4, the strategies of mesoporous carbon production are shown in steps. This process scheme is shown in Figs. 22.3 and 22.4.

Carbon quantum dots

Carbon quantum dots are carbon structures that are 10 nm in diameter. Carbon quantum dots show fluorescent properties. They show $\lambda_{ex}/\lambda_{em} = 350/465$ nm oxidation. Carbon quantum dots are produced with two techniques, top-down and bottom-up. When producing materials such as graphite powders or multiwalled carbon nanotubes, they are exposed to harsh chemical or physical techniques. This process is included in the top-down process, for example, techniques such as laser ablation or discharge. The technique of producing small molecules such as carbohydrates by external forces such as ultrasonication, microwave pyrolysis, and heating is an example of the bottom-up technique (Nas et al., 2019). Carbon quantum dots generally show the feature of fluorescent sensors. Algarra et al. have reported studies that carbon quantum dots have florescent sensor properties, and this report has taken its place in the literature (Campos et al., 2016). Wavelength votes of the fluorescent sensor data obtained from the carbon quantum dots obtained from these studies are shown in Figs. 22.5 and 22.6.

Commercialization challenges based on carbon nanomaterials

In sensor studies, one of the most striking nanomaterials among carbon nanomaterials is carbon nanotubes. Sensors, including carbon nanotubes, may have potential toxicity due to used chemicals. Also, interactions with a wide variety of analytes are observed in carbon nanotube-based sensors, and this is versatile. For example, additional materials other than the analytes (other than the interfering substance) used as substrates may also be included in the system. Thus, carbon

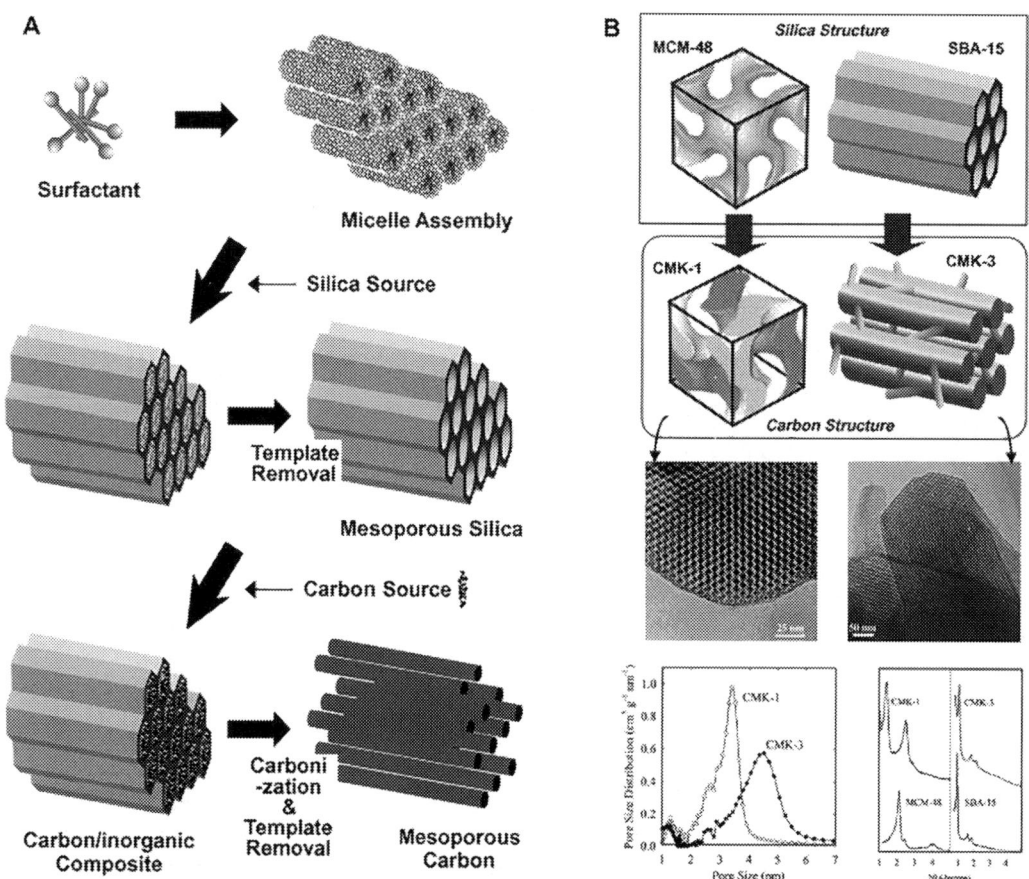

FIG. 22.3 (A) Ordered mesoporous silica hard prototype formed by self-assembly condensation of silica precursors and surfactants as part of a carbon mesostructure by nanotechnological strategy. (B) Mesoporous silica hard templates and ordered mesoporous carbons are the most general (Walcarius, 2012). *(Reprinted (adapted) with permission from Walcarius, A., 2012. Electrocatalysis, sensors and biosensors in analytical chemistry based on ordered mesoporous and macroporous carbon-modified electrodes. TrAC, Trends in Analytical Chemistry, 38, 79–97. https://doi.org/10.1016/j.trac.2012.05.003. Copyright (2021) Trends in Analytical Chemistry—Elsevier.)*

can bind to nanotubes with a nonspecific binding. Thus, this may be a suppressive obstacle to receiving sensor response, and this is a disadvantage for sensor systems and sensor business (Fam et al., 2011). In the industrial trade of graphene and graphene oxide products, new problems arise as we move toward laboratory scale, pilot scale, and industrialization at these stages. Moreover, it is critical to make the improvement parameters of graphene products' quality. The graphene content in the flakes available in the market is less than 50%. Also, the content of sp^2 bonds is lower than 60%. Because there are surface contaminations, graphites obtained from different layers have different structures and partially different properties. In this case, it causes differences in the internal distribution of the products. Small graphene nanosheets are a further advantage, as they are limited to additives and poor performance in commercial applications (Izatt, Christensen, & Rytting, 1971; Khan et al., 2016; Nonaka, Yoon, & Ogo, 2014; Özdemir, Mohamedsaid, Kılınç, & Soylak, 2019; Poole, 2003; Weigel et al., 2001; Yalçinkaya & Erdoan, 2012). Therefore, production, storage, and graphene nanosheets must be applied safely. This situation is important for industrial applications. Also, solvents, surfactants, and graphene products cause pollution and some oxidizing products are also a disadvantage. Graphene products are prohibited to enter the market in commercial terms. High cost and nonrepeatability are the disadvantages. For this reason, the graphene industry in carbon products is not in a good mood. This is a problem for sensor applications (Lin, Peng, & Liu, 2019). Another commercial hurdle is to protect the place of sensor devices in clinical applications. For this reason, carbon materials or additive carbon materials must show a solid performance. To increase the performance in affinity biosensors, nanoscale materials are prepared to increase the number of fixed probes, and support is taken from carbon nanocircuits. The sensitivity of sensors with or without carbon support is particularly important for low abundant cancer biomarkers (Bertok et al., 2019). In the articles published every year in the Web of Science Database, there are studies based on carbon nanofiber-containing smart textiles, clothing fabrics, and fibers. Globally, the smart clothing market was around 674.8

FIG. 22.4 (A) Uniform of mesoporous carbons: (1) resorcinol/formaldehyde (RF) in the presence of cetyl, (2) RF-gel/silica composite at 850 0C to obtain carbon, (3) the silica templates to obtain mesoporous carbons; (B) (a) the layer-by-layer process for preparing stacked colloidal crystal groups and the forming of 3D-ordered macroporous carbon by infiltrating a carbon supply around the spheres and carbonization and (b) typical scanning electron micrograph of macroporous carbon material with arranged pores; and (C) the mesocellular carbon foam (Walcarius, 2012). *(Reprinted (adapted) with permission from Walcarius, A., 2012. Electrocatalysis, sensors and biosensors in analytical chemistry based on ordered mesoporous and macroporous carbon-modified electrodes. TrAC, Trends in Analytical Chemistry, 38, 79–97. https://doi.org/10.1016/j.trac.2012.05.003. Copyright (2021) Trends in Analytical Chemistry—Elsevier.)*

million dollars. At the same time, it is estimated that by 2024, the compound annual rate will reach 8.98 billion dollars with a figure of over 33% (Wang et al., 2020). Versatile sensor applications are available in sports, health, military, and industrial fields. Smartwatches and smartwristbands have less market share than smartwear. This rate can correspond to approximately 1% (Wang et al., 2020). However, when looking at various parameters such as sensor accuracy, heat dissipation, and homogeneity, reliability, and comfort, sensor accuracy is one of the biggest disadvantages of the commercialization of nanofiber carbon materials (Wang et al., 2020). In addition, important studies have been done in the field of nanomaterials (Bilgicli et al., 2020; Eris, Daşdelen, & Sen, 2018; Göksu et al., 2016; Ozturk et al., 2012; Şen et al., 2011; Şen et al., 2019; Şen et al., 2019; Sert et al., 2016; Sert et al., 2017; Taçyıldız et al., 2019; Yildiz et al., 2017).

Conclusion

In 2003, biosensors had a market of $ 7.3 billion. Glucose biosensors are still used in agricultural production, food processing, and environmental monitoring. Enzyme-based electrochemical sensors such as YSI 2700 SELECT are widely used in food analysis. Microfluidic-based portable RAPTOR fiber-optic biosensors from the Research International can detect toxins, chemicals, bacteria, and viruses. Various microbial biological oxygen demand (BOD) biosensor systems are used in wastewater and environmental control processes. Sensors were first discovered by L. Clark in 1950, and the first sensory studies were the studies on enzymes. There are different types of sensors such as electrochemical, optical, magnetic, quartz, and calorimetric sensors. Various sensor studies have been reported in the literature. Along with nanotechnological developments, sensory studies based on nanoparticles have also increased. Therefore, carbon nanoscale materials have also been

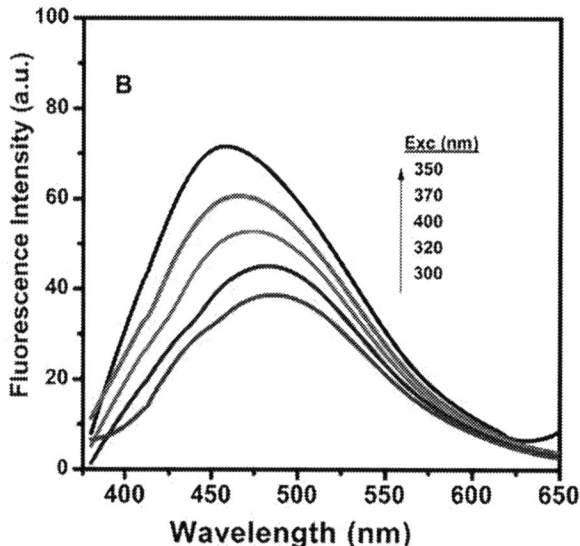

FIG. 22.5 Fluorescence spectra of CQDs@PAMAM-NH2 at different wavelengths (Campos et al., 2016). *(Reprinted (adapted) with permission from Campos, B. B., et al., 2016. Carbon dots as fluorescent sensor for detection of explosive nitrocompounds. Carbon N. Y., 106, 171–178. https://doi.org/10.1016/j.carbon.2016.05.030. Copyright (2021) Carbon—Elsevier.)*

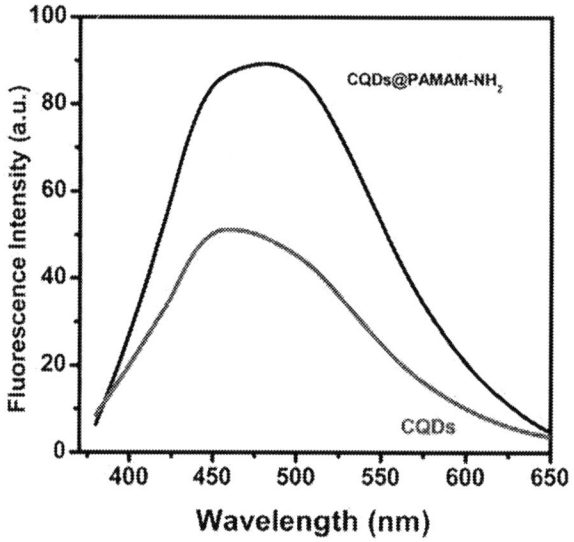

FIG. 22.6 Fluorescence spectra of CQDs and CQDs@PAMAM-NH2 at 350nm (Campos et al., 2016). *(Reprinted (adapted) with permission from Campos, B. B., et al., 2016. Carbon dots as fluorescent sensor for detection of explosive nitrocompounds. Carbon N. Y., 106, 171–178. https://doi.org/10.1016/j.carbon.2016.05.030. Copyright (2021) Carbon—Elsevier.)*

reported in the literature in sensory studies. However, some commercial problems may arise in carbon nanomaterials. Especially in graphene and its oxide types, due to some toxic by-products and due to the possibility of problems in the desired performance, major restrictions and even restrictions have been imposed on the commercialization of graphene and graphene types. In carbon nanotubes, some confusion arising as a result of the bonding of some by-products with nonspecific bonds to the surface of carbon nanotubes together with the analyte used as a substrate is one of the major commercial problems in the industry. On the contrary, there are some disadvantages in carbon nanofibers, especially in smart clothing and in the commercial processing of smart products, because the accuracy of the system sensor is sometimes not clear and its reliability is not complete.

References

Ahmed, S. R., et al. (2017). Size-controlled preparation of peroxidase-like graphene-gold nanoparticle hybrids for the visible detection of norovirus-like particles. *Biosensors & Bioelectronics, 87*, 558–565. https://doi.org/10.1016/j.bios.2016.08.101.

Article Derkuş, R. B., Acar Bozkurt, P., Biol, H. J., Derkus, B., Acar Bozkurt, P., & Derkuş, B. (2018). Multilayer graphene oxide-silver nanoparticle nanostructure as efficient peroxidase mimic Etkili Peroksidaz Taklitçi Olarak Çok Tabakalı Grafen Oksit-Gümüş Nanopartikül Nanoyapısı. *Chem, 46*(2), 159–167. https://doi.org/10.15671/HJBC.2018.225.

Asztemborska, M., Steborowski, R., Kowalska, J., & Bystrzejewska-Piotrowska, G. (2015). Accumulation of platinum nanoparticles by *Sinapis alba* and *Lepidium sativum* plants. *Water, Air, and Soil Pollution, 226*(4), 1–7. https://doi.org/10.1007/s11270-015-2381-y.

Augusto Da Costa, A. C., & Duta, F. P. (2001). Bioaccumulation of copper, zinc, cadmium and lead by *Bacillus* sp., *Bacillus cereus*, *Bacillus sphaericus* and *Bacillus subtilis*. *Brazilian Journal of Microbiology, 32*(1), 1–5. https://doi.org/10.1590/s1517-83822001000100001.

Aykut, U., & Temiz, H. (2006). *Biyosensörler ve Gıdalarda Kullanımı*. Gıda Teknolojileri Elektronik Dergis.

Ayranci, R., Demirkan, B., Sen, B., Şavk, A., Ak, M., & Şen, F. (2019). Use of the monodisperse Pt/Ni@rGO nanocomposite synthesized by ultrasonic hydroxide assisted reduction method in electrochemical nonenzymatic glucose detection. *Materials Science and Engineering: C, 99*, 951–956. https://doi.org/10.1016/j.msec.2019.02.040.

Basri, S., Kamarudin, S. K., Daud, W. R. W., & Yaakub, Z. (2010). Nanocatalyst for direct methanol fuel cell (DMFC). *International Journal of Hydrogen Energy, 35*(15), 7957–7970. https://doi.org/10.1016/J.IJHYDENE.2010.05.111.

Bernfeld, P. (1955). Amylases, α and β. *Methods in Enzymology, 1*(C), 149–158. https://doi.org/10.1016/0076-6879(55)01021-5.

Bertok, T., et al. (2019). Electrochemical impedance spectroscopy based biosensors: Mechanistic principles, analytical examples and challenges towards commercialization for assays of protein cancer biomarkers. *ChemElectroChem, 6*(4), 989–1003. https://doi.org/10.1002/celc.201800848.

Bilgicli, H. G., et al. (2020). Composites of palladium nanoparticles and graphene oxide as a highly active and reusable catalyst for the hydrogenation of nitroarenes. *Microporous and Mesoporous Materials*. 296, https://doi.org/10.1016/j.micromeso.2020.110014.

Boz, B., Paylan, İ. C., Kızmaz, M. Z., & Erkan, S. (2017). Biyosensörler ve Tarım Alanında Kullanımı. *Tarım Makinaları Bilimi Dergisi, 13*, 141–148.

Briceño, S., Suarez, J., & Gonzalez, G. (2017). Solvothermal synthesis of cobalt ferrite hollow spheres with chitosan. *Materials Science and Engineering: C, 78*, 842–846. https://doi.org/10.1016/j.msec.2017.04.034.

Bulut, Y. (2011). *Biyosensörlerin Tanımı ve Biyosensörlere Genel Bakış*. Advanced Technologies Symposium (IATS'11), Elazığ, 2011, 16–18 May. Elazig.

Campodoni, E., et al. (2019). Polymeric 3D scaffolds for tissue regeneration: Evaluation of biopolymer nanocomposite reinforced with cellulose nanofibrils. *Materials Science and Engineering: C, 94*, 867–878. https://doi.org/10.1016/j.msec.2018.10.026.

Campos, B. B., et al. (2016). Carbon dots as fluorescent sensor for detection of explosive nitrocompounds. *Carbon N Y, 106*, 171–178. https://doi.org/10.1016/j.carbon.2016.05.030.

Cao, Q., & Rogers, J. A. (2009). Ultrathin films of single-walled carbon nanotubes for electronics and sensors: A review of fundamental and applied aspects. *Advanced Materials, 21*(1), 29–53. https://doi.org/10.1002/adma.200801995.

Capannesi, C., Palchetti, I., Mascini, M., & Parenti, A. (2000). Electrochemical sensor and biosensor for polyphenols detection in olive oils. *Food Chemistry, 71*(4), 553–562. https://doi.org/10.1016/S0308-8146(00)00211-9.

Çelik, B., Yildiz, Y., Sert, H., Erken, E., Koşkun, Y., & Şen, F. (2016). Monodispersed palladium-cobalt alloy nanoparticles assembled on poly(N-vinyl-pyrrolidone) (PVP) as a highly effective catalyst for dimethylamine borane (DMAB) dehydrocoupling. *RSC Advances, 6*(29), 24097–24102. https://doi.org/10.1039/c6ra00536e.

Chuma, P. A. (2007). Biosorption of Cr, Mn, Fe, Ni, Cu and Pb metals from petroleum refinery effluent by calcium alginate immobilized mycelia of *Polyporus squamosus*. *Scientific Research and Essays, 2*(7), 217–221. https://doi.org/10.5897/SRE.9000242.

Darder, M., Colilla, M., & Ruiz-Hitzky, E. (2005). Chitosan-clay nanocomposites: Application as electrochemical sensors. *Applied Clay Science, 28*(1–4 SPEC. ISS), 199–208. https://doi.org/10.1016/j.clay.2004.02.009.

Demir, E., Savk, A., Sen, B., & Sen, F. (2017). A novel monodisperse metal nanoparticles anchored graphene oxide as counter electrode for dye-sensitized solar cells. *Nano-Structures and Nano-Objects, 12*, 41–45. https://doi.org/10.1016/j.nanoso.2017.08.018.

Demir, E., Sen, B., & Sen, F. (2017). Highly efficient Pt nanoparticles and f-MWCNT nanocomposites based counter electrodes for dye-sensitized solar cells. *Nano-Structures & Nano-Objects, 11*, 39–45. https://doi.org/10.1016/j.nanoso.2017.06.003.

Demirkıran, A. (2019). Nanoteknolojinin İnsan Sağlığına Faydalı ve Zararlı Yönleri. *Ordu Üniversitesi Bilim ve Teknoloji Dergisi, 9*(2), 136–148.

Dönmez, G., & Aksu, Z. (2001). Bioaccumulation of copper(ii) and nickel(ii) by the non-adapted and adapted growing CANDIDA SP. *Water Research, 35*(6), 1425–1434. https://doi.org/10.1016/S0043-1354(00)00394-8.

Erdoğan, Z. Ö. (2013). *Ürik asit tayini için nanopartikül temelli biyosensörler hazırlanması*. Selçuk Üniversitesi Fen Bilimleri Enstitüsü.

Eris, S., Daşdelen, Z., & Sen, F. (2018). Enhanced electrocatalytic activity and stability of monodisperse Pt nanocomposites for direct methanol fuel cells. *Journal of Colloid and Interface Science, 513*(1–2), 767–773. https://doi.org/10.1016/j.jcis.2017.11.085.

Fam, D. W. H., Palaniappan, A., Tok, A. I. Y., Liedberg, B., & Moochhala, S. M. (2011). A review on technological aspects influencing commercialization of carbon nanotube sensors. *Sensors and Actuators, B: Chemical, 157*(1), 1–7. https://doi.org/10.1016/j.snb.2011.03.040.

Fereidoonnezhad, M., et al. (2018). Cover image. *Applied Organometallic Chemistry, 32*(3), e4326. https://doi.org/10.1002/aoc.4326.

Frasco, M. F., & Chaniotakis, N. (2009). Semiconductor quantum dots in chemical sensors and biosensors. *Sensors, 9*(9), 7266–7286. https://doi.org/10.3390/s90907266.

Gao, X., Cui, Y., Levenson, R. M., Chung, L. W. K., & Nie, S. (2004). In vivo cancer targeting and imaging with semiconductor quantum dots. *Nature Biotechnology, 22*(8), 969–976. https://doi.org/10.1038/nbt994.

Giraldo, J. P., et al. (2014). Plant nanobionics approach to augment photosynthesis and biochemical sensing. *Nature Materials*, *13*(4), 400–408. https://doi.org/10.1038/nmat3890.

Gogotsi, Y., & Presser, V. (2014). *Carbon nanomaterials*. Taylor & Francis eBook.

Göksu, H., Yıldız, Y., Çelik, B., Yazıcı, M., Kılbaş, B., & Şen, F. (2016). Highly efficient and monodisperse graphene oxide furnished Ru/Pd nanoparticles for the dehalogenation of aryl halides via ammonia borane. *ChemistrySelect*, *1*(5), 953–958. https://doi.org/10.1002/slct.201600207.

Güngör, S. Y. (2019). *Peroksidaz Aktivitesine Sahip Nanopartikül/Mikroküre Kompozitlerinin Sentezi ve Glutatyon Tayininde Kullanılması*. Fen Bilimleri Enstitüsü.

Gupta, V. K., Jain, A. K., Agarwal, S., & Maheshwari, G. (2007). An iron(III) ion-selective sensor based on a μ-bis(tridentate) ligand. *Talanta*, *71*(5), 1964–1968. https://doi.org/10.1016/j.talanta.2006.08.038.

Hanaoka, Y., Takebe, F., Nodasaka, Y., Hara, I., Matsuyama, H., & Yumoto, I. (2013). Growth-dependent catalase localization in *Exiguobacterium oxidotolerans* T-2-2T reflected by catalase activity of cells. *PLoS ONE*. *8*(10)https://doi.org/10.1371/journal.pone.0076862.

Handbook of environmental materials management. (2019). Springer International Publishing.

Handbook of Functionalized Nanomaterials for Industrial Applications. (2020). Elsevier.

Handbook of nanomaterials for manufacturing applications. (2020). Elsevier.

Handbook of nanomaterials for sensing applications. (2021). Elsevier.

Handbook of polymer nanocomposites for industrial applications. (2021). Elsevier.

Haque, M. (2017). Future medical doctors are ready to prescribe antimicrobials safely and rationally. *Asian Journal of Pharmaceutical and Clinical Research*, *10*(12), 1–2. https://doi.org/10.22159/ajpcr.2017.v10i12.21930.

Henn, B. C., et al. (2010). Early postnatal blood manganese levels and children's neurodevelopment. *Epidemiology*, *21*(4), 433–439. https://doi.org/10.1097/EDE.0b013e3181df8e52.

Hobman, J. L., & Crossman, L. C. (2015). Bacterial antimicrobial metal ion resistance. *Journal of Medical Microbiology*, *64*(5), 471–497. https://doi.org/10.1099/jmm.0.023036-0.

Hong, G., Diao, S., Antaris, A. L., & Dai, H. (2015). Carbon nanomaterials for biological imaging and nanomedicinal therapy. *Chemical Reviews*, *115*(19), 10816–10906. https://doi.org/10.1021/acs.chemrev.5b00008.

Izatt, R. M., Christensen, J. J., & Rytting, J. H. (1971). Sites and thermodynamic quantities associated with proton and metal ion interaction with ribonucleic acid, deoxyribonucleic acid, and their constituent bases, nucleosides, and nucleotides. *Chemical Reviews*, *71*(5), 439–481. https://doi.org/10.1021/cr60273a002.

Jaque, D., Richard, C., Viana, B., Soga, K., Liu, X., & García Solé, J. (2016). Inorganic nanoparticles for optical bioimaging. *Advances in Optics and Photonics*, *8*(1), 1. https://doi.org/10.1364/aop.8.000001.

Jaroch, K., Boyaci, E., Pawliszyn, J., & Bojko, B. (2019). The use of solid phase microextraction for metabolomic analysis of non-small cell lung carcinoma cell line (A549) after administration of combretastatin A4. *Scientific Reports*, *9*(1), 1–9. https://doi.org/10.1038/s41598-018-36481-2.

Kauffman, D. R., & Star, A. (2008). Carbon nanotube gas and vapor sensors. *Angewandte Chemie, International Edition*, *47*(35), 6550–6570. https://doi.org/10.1002/anie.200704488.

Khan, M., Yilmaz, E., & Soylak, M. (2016). Vortex assisted magnetic solid phase extraction of lead(II) and cobalt(II) on silica coated magnetic multiwalled carbon nanotubes impregnated with 1-(2-pyridylazo)-2-naphthol. *Journal of Molecular Liquids*, *224*, 639–647. https://doi.org/10.1016/j.molliq.2016.10.023.

Khoo, E. H., et al. (2009). Plasmonics NanoSensor. *Advanced Materials Research*, *74*, 13–16. https://doi.org/10.4028/www.scientific.net/AMR.74.13.

Kilic, M., & Korkut, S. (2016). Karbon Nanotüp Katkılı Biyosensör Tasarımı ve Biyosensörün Çevresel Kirletici Olan Fenol ve Türevlerinin Tespitinde Kullanılması. *Karaelmas Fen ve Müh Derg*, *6*(2), 399–405.

Kokbas, U., Kayrın, L., & Tuli, A. (2013). Biyosensörler ve Tıpta Kullanım Alanları. *Arşiv Kaynak Tarama Derg*, *22*(4), 499–513.

Koskun, Y., Şavk, A., Şen, B., & Şen, F. (2018). Highly sensitive glucose sensor based on monodisperse palladium nickel/activated carbon nanocomposites. *Analytica Chimica Acta*, *1010*, 37–43. https://doi.org/10.1016/j.aca.2018.01.035.

Küçükyıldırım, B. O., & Akdoğan Eker, A. (2012). Karbon nanotüpler sentezleme ve kullanım alanları. *Makine Mühendisleri Odası*, *53*(630), 34–44.

Lan, J., et al. (2014). Colorimetric determination of sarcosine in urine samples of prostatic carcinoma by mimic enzyme palladium nanoparticles. *Analytica Chimica Acta*, *825*, 63–68. https://doi.org/10.1016/j.aca.2014.03.040.

Lee, J., Han, S., & Hyeon, T. (2004). Synthesis of new nanoporous carbon materials using nanostructured silica materials as templates. *Journal of Materials Chemistry*, *14*(4), 478–486. https://doi.org/10.1039/b311541k.

Lee, E. J., Khan, S. A., Park, J. K., & Lim, K. H. (2012). Studies on the characteristics of drug-loaded gelatin nanoparticles prepared by nanoprecipitation. *Bioprocess and Biosystems Engineering*, *35*(1–2), 297–307. https://doi.org/10.1007/s00449-011-0591-2.

Lee, J., Kim, J., & Hyeon, T. (2006). Recent progress in the synthesis of porous carbon materials. *Advanced Materials*, *18*(16), 2073–2094. https://doi.org/10.1002/adma.200501576.

Li, Q., Jiang, C., Bi, S., Asare-Yeboah, K., He, Z., & Liu, Y. (2020). Photo-triggered logic circuits assembled on integrated illuminants and resonant nanowires. *ACS Applied Materials & Interfaces*, *12*(41), 46501–46508. https://doi.org/10.1021/acsami.0c12256.

Liang, C., Li, Z., & Dai, S. (2008). Mesoporous carbon materials: Synthesis and modification. *Angewandte Chemie, International Edition*, *47*(20), 3696–3717. https://doi.org/10.1002/anie.200702046.

Lin, L., Peng, H., & Liu, Z. (2019). Synthesis challenges for graphene industry. *Nature Materials*, *18*(6), 520–524. https://doi.org/10.1038/s41563-019-0341-4.

Lin, X., Xu, X., Yang, C., Zhao, Y., Feng, Z., & Dong, Y. (2009). Activities of antioxidant enzymes in three bacteria exposed to bensulfuron-methyl. *Ecotoxicology and Environmental Safety*, *72*(7), 1899–1904. https://doi.org/10.1016/j.ecoenv.2009.04.016.

Lu, N., He, X., Wang, T., Liu, S., & Hou, X. (2018). Magnetic solid-phase extraction using MIL-101(Cr)-based composite combined with dispersive liquid-liquid microextraction based on solidification of a floating organic droplet for the determination of pyrethroids in environmental water and tea samples. *Microchemical Journal, 137*, 449–455. https://doi.org/10.1016/j.microc.2017.12.009.

Madrid, Y., & Cámara, C. (1997). Biological substrates for metal preconcentration and speciation. *TrAC, Trends in Analytical Chemistry, 16*(1), 36–44. https://doi.org/10.1016/S0165-9936(96)00075-1.

Mateos, L. M., et al. (2016). Comparative mathematical modelling of a green approach for bioaccumulation of cobalt from wastewater. *Environmental Science and Pollution Research, 23*(23), 24215–24229. https://doi.org/10.1007/s11356-016-7596-y.

Meyyappan, M. (2016). Carbon nanotube-based chemical sensors. *Small, 12*(16), 2118–2129. https://doi.org/10.1002/smll.201502555.

Nas, M. S., Kuyuldar, E., Demirkan, B., Calimli, M. H., Demirbaş, O., & Sen, F. (2019). Magnetic nanocomposites decorated on multiwalled carbon nanotube for removal of Maxilon blue 5G using the sono-Fenton method. *Scientific Reports, 9*(1), 1–11. https://doi.org/10.1038/s41598-019-47393-0.

Nath, N., & Chilkoti, A. (2002). A colorimetric gold nanoparticle sensor to interrogate biomolecular interactions in real time on a surface. *Analytical Chemistry, 74*(3), 504–509. https://doi.org/10.1021/ac015657x.

Nonaka, K., Yoon, K. S., & Ogo, S. (2014). Biochemical characterization of psychrophilic Mn-superoxide dismutase from newly isolated *Exiguobacterium* sp. OS-77. *Extremophiles, 18*(2), 363–373. https://doi.org/10.1007/s00792-013-0621-x.

Otlu, B. (2011). Biyosensörler: Biyoreseptör Moleküller. In *6th international advanced technologies symposium (IATS'11)*.

Özdemir, S., Kilinc, E., Poli, A., Nicolaus, B., & Güven, K. (2012). Cd, Cu, Ni, Mn and Zn resistance and bioaccumulation by thermophilic bacteria, *Geobacillus toebii* subsp. *decanicus* and *Geobacillus thermoleovorans* subsp. *stromboliensis*. *World Journal of Microbiology and Biotechnology*. https://doi.org/10.1007/s11274-011-0804-5.

Özdemir, S., Mohamedsaid, S. A., Kılınç, E., & Soylak, M. (2019). Magnetic solid phase extractions of Co(II) and Hg(II) by using magnetized *C. micaceus* from water and food samples. *Food Chemistry, 271*, 232–238. https://doi.org/10.1016/j.foodchem.2018.07.067.

Özoğlu, Ö. (2019). *Enzim temelli amperometrik laktat biyosensörü üretimi ve tayin sınırının belirlenmesi*. Bursa, Türkiye: Bursa Uludağ Üniversitesi.

Ozturk, Z., Sen, F., Sen, S., & Gokagac, G. (2012). The preparation and characterization of nano-sized Pt-Pd/C catalysts and comparison of their superior catalytic activities for methanol and ethanol oxidation. *Journal of Materials Science, 47*(23), 8134–8144. https://doi.org/10.1007/s10853-012-6709-3.

Park, S., Vosguerichian, M., & Bao, Z. (2013). A review of fabrication and applications of carbon nanotube film-based flexible electronics. *Nanoscale, 5*(5), 1727–1752. https://doi.org/10.1039/c3nr33560g.

Poole, C. F. (2003). New trends in solid-phase extraction. *TrAC, Trends in Analytical Chemistry, 22*(6), 362–373. https://doi.org/10.1016/S0165-9936(03)00605-8.

Qiu, K., et al. (2014). Hierarchical 3D mesoporous conch-like Co_3O_4 nanostructure arrays for high-performance supercapacitors. *Electrochimica Acta, 141*, 248–254. https://doi.org/10.1016/j.electacta.2014.07.074.

Qiu, L., et al. (2020). A review of recent advances in thermophysical properties at the nanoscale: From solid state to colloids. *Physics Reports, 843*, 1–81. https://doi.org/10.1016/j.physrep.2019.12.001.

Ragavan, K. V., Kumar, S., Swaraj, S., & Neethirajan, S. (2018). Advances in biosensors and optical assays for diagnosis and detection of malaria. *Biosensors and Bioelectronics, 105*, 188–210. https://doi.org/10.1016/j.bios.2018.01.037.

Raril, C., et al. (2018). *Surfactant immobilized electrochemical sensor for the detection of indigotine analytical &. [Online]. Available:(2018).* www.abechem.com. Accessed 20 May 2021.

Ray, L., Paul, S., Bera, D., & Chattopadhyay, P. (2006). Bioaccumulation of Pb(II) from aqueous solutions by *Bacillus cereus* M1 16. *Journal of Hazardous Substance Research, 5*(1), 1. https://doi.org/10.4148/1090-7025.1031.

Ryoo, R., Joo, S. H., Kruk, M., & Jaroniec, M. (2001). Ordered mesoporous carbons. *Advanced Materials, 13*(9), 677–681. https://doi.org/10.1002/1521-4095(200105)13:9<677::AID-ADMA677>3.0.CO;2-C.

Satchanska, G., Pentcheva, E. N., Atanasova, R., Groudeva, V., Trifonova, R., & Golovinsky, E. (2005). Microbial diversity in heavy-metal polluted waters. *Biotechnology and Biotechnological Equipment, 19*(3), 61–67. https://doi.org/10.1080/13102818.2005.10817228.

Savk, A., et al. (2019). Multiwalled carbon nanotube-based nanosensor for ultrasensitive detection of uric acid, dopamine, and ascorbic acid. *Materials Science and Engineering: C, 99*, 248–254. https://doi.org/10.1016/j.msec.2019.01.113.

Schroeder, V., Savagatrup, S., He, M., Lin, S., & Swager, T. M. (2019). Carbon nanotube chemical sensors. *Chemical Reviews, 119*(1), 599–663. https://doi.org/10.1021/acs.chemrev.8b00340.

Sekol, R. C., et al. (2013). Pd-Ni-Cu-P metallic glass nanowires for methanol and ethanol oxidation in alkaline media. *International Journal of Hydrogen Energy, 38*(26), 11248–11255. https://doi.org/10.1016/j.ijhydene.2013.06.017.

Şen, B., Akyıldız, B., Aygün, A., Kuyuldar, E., Demirkan, B., & Şen, F. (2019). Nanocarbon-supported catalysts for the efficient dehydrogenation of dimethylamine borane. In *Nanocarbon and its composites* (pp. 615–628). Elsevier.

Şen, B., Aygün, A., Şavk, A., Çalımlı, M. H., Gülbay, S. K., & Şen, F. (2019). Bimetallic palladium-cobalt nanomaterials as highly efficient catalysts for dehydrocoupling of dimethylamine borane. *International Journal of Hydrogen Energy*. https://doi.org/10.1016/J.IJHYDENE.2019.01.215.

Sen, F., Boghossian, A. A., Sen, S., Ulissi, Z. W., Zhang, J., & Strano, M. S. (2012). Observation of oscillatory surface reactions of riboflavin, trolox, and singlet oxygen using single carbon nanotube fluorescence spectroscopy. *ACS Nano, 6*(12), 10632–10645. https://doi.org/10.1021/nn303716n.

Şen, B., Demirkan, B., Levent, M., Şavk, A., & Şen, F. (2018). Silica-based monodisperse PdCo nanohybrids as highly efficient and stable nanocatalyst for hydrogen evolution reaction. *International Journal of Hydrogen Energy, 43*(44), 20234–20242. https://doi.org/10.1016/j.ijhydene.2018.07.080.

Sen, B., Demirkan, B., Şavk, A., Karahan Gülbay, S., & Sen, F. (2018). Trimetallic PdRuNi nanocomposites decorated on graphene oxide: A superior catalyst for the hydrogen evolution reaction. *International Journal of Hydrogen Energy*, *43*(38), 17984–17992. https://doi.org/10.1016/j.ijhydene.2018.07.122.

Sen, B., Demirkan, B., Şimşek, B., Savk, A., & Sen, F. (2018). Monodisperse palladium nanocatalysts for dehydrocoupling of dimethylamineborane. *Nano-Structures and Nano-Objects*, *16*, 209–214. https://doi.org/10.1016/j.nanoso.2018.07.008.

Şen, F., & Gökagaç, G. (2007). Activity of carbon-supported platinum nanoparticles toward methanol oxidation reaction: Role of metal precursor and a new surfactant, tert-octanethiol. *Journal of Physical Chemistry C*, *111*(3), 1467–1473. https://doi.org/10.1021/jp065809y.

Şen, F., & Gökağaç, G. (2014). Pt nanoparticles synthesized with new surfactants: Improvement in C 1-C3 alcohol oxidation catalytic activity. *Journal of Applied Electrochemistry*, *44*(1), 199–207. https://doi.org/10.1007/s10800-013-0631-5.

Sen, B., Kuyuldar, E., Şavk, A., Calimli, H., Duman, S., & Sen, F. (2019). Monodisperse ruthenium–copper alloy nanoparticles decorated on reduced graphene oxide for dehydrogenation of DMAB. *International Journal of Hydrogen Energy*, *44*(21), 10744–10751. https://doi.org/10.1016/j.ijhydene.2019.02.176.

Sen, B., Kuzu, S., Demir, E., Akocak, S., & Sen, F. (2017). Highly monodisperse RuCo nanoparticles decorated on functionalized multiwalled carbon nanotube with the highest observed catalytic activity in the dehydrogenation of dimethylamine – borane. *International Journal of Hydrogen Energy*, *42*(36), 23292–23298. https://doi.org/10.1016/j.ijhydene.2017.06.032.

Sen, B., Şavk, A., & Sen, F. (2018). Highly efficient monodisperse Pt nanoparticles confined in the carbon black hybrid material for hydrogen liberation. *Journal of Colloid and Interface Science*, *520*, 112–118. https://doi.org/10.1016/j.jcis.2018.03.004.

Şen, S., Şen, F., & Gökağaç, G. (2011). Preparation and characterization of nano-sized Pt-Ru/C catalysts and their superior catalytic activities for methanol and ethanol oxidation. *Physical Chemistry Chemical Physics*, *13*(15), 6784–6792. https://doi.org/10.1039/c1cp20064j.

Sert, H., et al. (2016). Monodisperse Mw-Pt NPs@VC as highly efficient and reusable adsorbents for methylene blue removal. *Journal of Cluster Science*, *27*(6), 1953–1962. https://doi.org/10.1007/s10876-016-1054-3.

Sert, H., et al. (2017). Activated carbon furnished monodisperse Pt nanocomposites as a superior adsorbent for methylene blue removal from aqueous solutions. *Journal of Nanoscience and Nanotechnology*, *17*(7), 4799–4804. https://doi.org/10.1166/jnn.2017.13776.

Srinath, T., Verma, T., Ramteke, P. W., & Garg, S. K. (2002). Chromium (VI) biosorption and bioaccumulation by chromate resistant bacteria. *Chemosphere*, *48*(4), 427–435. https://doi.org/10.1016/S0045-6535(02)00089-9.

Taçyıldız, S., Demirkan, B., Karataş, Y., Gulcan, M., & Sen, F. (2019). Monodisperse Ru Rh bimetallic nanocatalyst as highly efficient catalysts for hydrogen generation from hydrolytic dehydrogenation of methylamine-borane. *Journal of Molecular Liquids*, *285*, 1–8. https://doi.org/10.1016/j.molliq.2019.04.019.

Uauy, R., Olivares, M., & Gonzalez, M. (1998). Essentiality of copper in humans. *American Journal of Clinical Nutrition*, *67*(5 Suppl), 952S–959S. https://doi.org/10.1093/ajcn/67.5.952S.

Volpe, G., Compagnone, D., Draisci, R., & Palleschi, G. (1998). 3,3′,5,5′-Tetramethylbenzidine as electrochemical substrate for horseradish peroxidase based enzyme immunoassays. A comparative study. *Analyst*, *123*(6), 1303–1307. https://doi.org/10.1039/a800255j.

Walcarius, A. (2012). Electrocatalysis, sensors and biosensors in analytical chemistry based on ordered mesoporous and macroporous carbon-modified electrodes. *TrAC, Trends in Analytical Chemistry*, *38*, 79–97. https://doi.org/10.1016/j.trac.2012.05.003.

Wan, Y., Shi, Y., & Zhao, D. (2008). Supramolecular aggregates as templates: Ordered mesoporous polymers and carbons. *Chemistry of Materials*, *20*(3), 932–945. https://doi.org/10.1021/cm7024125.

Wang, J. (2005). Carbon-nanotube based electrochemical biosensors: A review. *Electroanalysis*, *17*(1), 7–14. https://doi.org/10.1002/elan.200403113.

Wang, L., et al. (2020). Application challenges in fiber and textile electronics. *Advanced Materials*, *32*(5), 1901971. https://doi.org/10.1002/adma.201901971.

Weigel, S., Bester, K., & Hühnerfuss, H. (2001). New method for rapid solid-phase extraction of large-volume water samples and its application to non-target screening of North Sea water for organic contaminants by gas chromatography-mass spectrometry. *Journal of Chromatography A*, *912*(1), 151–161. https://doi.org/10.1016/S0021-9673(01)00529-5.

Yalçinkaya, Ö., & Erdoan, H. (2012). Preconcentration and determination of manganese and nickel from various water samples by nano zirconium oxide/boron oxide. *Spectroscopy Letters*, *45*(8), 602–608. https://doi.org/10.1080/00387010.2012.656876.

Yildiz, Y., et al. (2017). Highly monodisperse Pt/Rh nanoparticles confined in the graphene oxide for highly efficient and reusable sorbents for methylene blue removal from aqueous solutions. *ChemistrySelect*, *2*(2), 697–701. https://doi.org/10.1002/slct.201601608.

Yoneda, J., et al. (2018). A quantum-dot spin qubit with coherence limited by charge noise and fidelity higher than 99.9%. *Nature Nanotechnology*, *13*(2), 102–106. https://doi.org/10.1038/s41565-017-0014-x.

Part VII

Future of sustainable sensors

Chapter 23

Introduction and overview of carbon nanomaterial-based sensors for sustainable response

Tania Akter[a], Christopher Barile[a], and A.J. Saleh Ahammad[b]
[a]*Department of Chemistry, University of Nevada, Reno, United States,* [b]*Department of Chemistry, Jagannath University, Dhaka, Bangladesh*

Introduction

At present, the applications and necessity of different sensors with specific characteristics have been flourishing in the diverse area of science. The demand for higher sensitivity, selectivity, reliability, cost-effectiveness, and faster response also grabbed enormous attention to develop them (Mehdi Aghaei, Monshi, Torres, Zeidi, & Calizo, 2018). However, extending all the properties in the sensors turned them into more expensive, slower response, and complexity in the operation process. To overcome these difficulties, nanomaterials are extensively taken possession of attention because of their unique properties (Kreyling, Semmler-Behnke, & Chaudhry, 2010).

Among multifarious nanomaterials, carbon nanomaterials (CNs) are the most studied materials. In the last few decades, the applications of CNs in science and technology have increased tremendously (Tiwari, Vij, Kemp, & Kim, 2015). Due to the variety of allotropes and the distinctive physical and chemical properties, each member of the carbon nanomaterials exhibits remarkable features and is extensively practiced in different applications, including biosensing, drug delivery, cancer therapy, tissue engineering, gaseous sensors, electrochemical sensors, and agricultural applications (Chenthattil et al., 2020; Bhattacharya et al., 2016; Mukhopadhyay, Maiti, Saha, & Devi, 2016).

In this chapter, a comprehensive study of sensor applications of various CNs in several areas, including sustainability and the recent development, has been broadly discussed along with the future possibilities.

Carbon-based nanomaterials

Carbon has versatile properties and renowned as the element of living substances. Different allotropes can be formed by following sp^2 and sp^3 hybridization between the individual carbon atoms that accommodate carbon to play an essential role in life also used to produce various structures (Zaytseva & Neumann, 2016). It also possesses a valency of four, allowing it to form single, double, and triple bonds among itself or other elements. These properties are significantly special in nanoscale, where each allotrope possesses some unique characteristics.

Several carbon nanomaterials such as carbon nanotubes, fullerenes, nanographenes, nanodiamonds, and carbon quantum dots were unearthed many decades ago. They have been studied and applied in different fields of science and technologies.

Nanodiamond (ND) was discovered accidentally first in 1963 by a nuclear explosion that was used in carbon-based trigger explosives at the All-Union Research Institute of Technical Physics by Russian scientists (Danilenko, 2004). However, the properties or aspects of NDs were not attracting scientists much at that time. Then in 1982, it was the second time discovered by A.M. Staver and A.I Lyamkin at the Siberian Hydrodynamics institute (Mochalin, Shenderova, Ho, & Gogotsi, 2012). NDs are 5–100-nm nanosized carbons, which are formed by sp^3 hybridization (Bezzon et al., 2019; Nunn, Torelli, McGuire, & Shenderova, 2017). They exhibit excellent mechanical properties, chemical resistance, versatile surface, unique optical properties, and biocompatibility due to the dopant in their structure, which have made the best choice for nanotechnological applications nowadays. Fig. 23.1 shows some of the carbon-based nanomaterials.

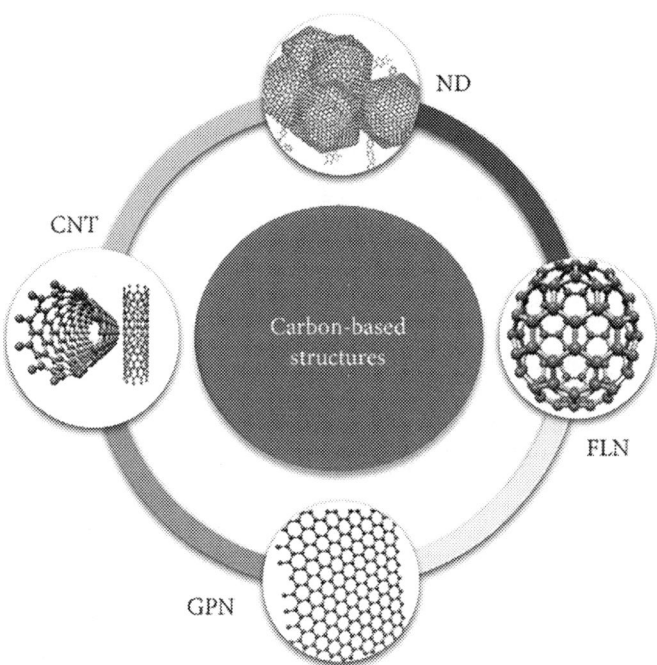

FIG. 23.1 Carbon-based nanomaterials. *(From Bezzon, V. D. N., Montanheiro, T. L. A., De Menezes, B. R. C., Ribas, R. G., Righetti, V. A. N., Rodrigues, K. F. & Thim, G. P. (2019). Carbon Nanostructure-based Sensors: A Brief Review on Recent Advances. Advances in Materials Science and Engineering, 2019. https://doi.org/10.1155/2019/4293073.)*

In 1985, fullerene, another allotrope of carbon, was experimentally discovered by Harry Kroto, Richard Smalley, and Robert Curl of Rice University, and the scientists were awarded the Nobel Prize for this discovery (Kroto, Heath, O'Brien, Curl, & Smalley, 1985). Fullerene molecules exhibit sp^2 and sp^3 hybridized carbon, which is the reason for forming a hollow structure (Hirsch, 2010). Hexagons and 12 pentagons are combined to form this molecule, where the size depends on the number of hexagons (Zaghmarzi et al., 2017). The most abundant fullerene is C_{60}, commonly known as buckminsterfullerene. They have high electron affinity, high charge transportability, and nonlinear optical responses due to having delocalized π electrons (Rao, Seshadri, Govindaraj, & Sen, 1995). They are also well known for their superlative electrical conductivity, which has given them popularity in electrochemical sensors (Fouskaki & Chaniotakis, 2008). In addition, they have brought a revolutionary change to other sensor applications in nanotechnology (Rutherglen, Jain, & Burke, 2009).

Another carbon nanomaterial is a carbon nanotube (CNT), affiliated with the fullerene family, and represents a quasi-one-dimensional structure (Iijima & Ichihashi, 1993). They can be single-walled and multiwalled and possess tubular structures (Iijima, 1991). In CNTs, carbon exhibits sp^2 hybridization on the tube, where carbon–carbon σ bond forms hexagonal honeycomb. Japanese scientist Sumio Iijima discovered multiwalled carbon nanotubes (MWCNTs) in 1991 from Nippon Electric Company (NEC), and in 1993, he and Donald Bethune independently observed single-walled carbon nanotubes (SWCNTs) (Monthioux & Kuznetsov, 2006). They exhibit remarkable electrical properties (Mintmire, Dunlap, & White, 1992), tensile strength (Yu et al., 2000), and thermal conductivity (Kim, Shi, Majumdar, & McEuen, 2001). Their remarkable optical, mechanical, electrical, and chemical properties brought them a center position in nanotechnologies. They are popular materials in sensor applications such as electrochemical sensors, biosensors, and gas sensors (Amrutha, Manjunatha, Bhatt, Raril, & Pushpanjali, 2019; Charithra, Manjunatha, & Raril, 2020; Hareesha, Manjunatha, Raril, & Tigari, 2019; Jg, n.d.; Prinith, Manjunatha, & Raril, 2019; Pushpanjali, Manjunatha, Tigari, & Fattepur, 2020; Tigari & Manjunatha, 2020a, 2020b).

In 2004, a single layer of carbon atoms in a honeycomb-like lattice structure was discovered by Prof. Sir Andre Geim and Prof. Sir Kostya Novoselov and named graphene. It is the first two-dimensional material. The singular graphene unit is the nanographene with a diameter of <100 nm (Bhattacharyya, Seth, Kumar, & Chamkha, n.d.). It is a well-known compound in the gas sensor due to the property of gas molecule absorption (Novoselov et al., 2005).

Zero-dimensional carbon nanomaterials were also found accidentally in 2004 by Xu et al. during the purification of SWCNTs (Xu et al., 2004). This discovery triggered a broad aspect in the fluorescence research area due to their strong photoelectrochemical properties (Zou et al., 2017).

Carbon nanotubes for sensor applications

Carbon nanotubes in gas sensors

Gas sensors are enthralling a massive interest due to extensive environmental analysis, industrial waste management, biomedicine, and atmospheric science. The excellent absorption capability and sensitivity and suitable structure of the CNTs provide the platform for developing gas sensors based on nanotubes (Zaporotskova, Boroznina, Parkhomenko, & Kozhitov, 2016; Zou et al., 2017). Several gas sensors such as sorption, ionization, resonance, and capacitance have been formed on the properties of CNTs. SWCNTs and MWCNTs are both so responsive on the gas sensor (Zhang & Zhang, 2009; Zou et al., 2017).

Among various categories of sensors, sorption gas sensors are most famous for environmental analysis (Boyd, Dube, Fedorov, Paranjape, & Barbara, 2014; Zou et al., 2017). Their main principle is the adsorption or desorption of gaseous molecules by transferring the electrons to or taking from the CNT (Li et al., 2003). The changes in the electrical properties of CNTs can be detected by the detector. It has been reported that SWCNTs are more sensitive to nitrogen-derivative gaseous molecules, e.g., NO_2, NH_3, and some volatile organic molecules (Kong et al., 2000). The gaseous molecules adsorb onto the surface of nanotubes due to the change of conductivity. One project reported field-effect transistor (FET) response to NO_2 and NH_3, where the transistor utilized a conduction channel through the SWCNTs and showed a singular response to the gases (Li et al., 2003). Three models proposed to explain the mechanism of FETs are charge transfer between the nanotube and the adsorbed molecules, molecular strobing of nonpolar molecules, and changing the Schottky barrier between nanotubes and the electrode (Li et al., 2003). By including positive charge the energy barrier can be reduced the CNT adsorption for gaseous molecules by creating electron tunneling. The transistor-based SWCNT sensors successfully utilized NO_2, NH_3, and dimethyl-phosphonate (Zhao, Buldum, Han, & Lu, 2002; Zhao, Gan, & Zhuang, 2002). Fabricated MWCNTs are sensitive to various gaseous molecules such as NH_3, NO_2, NO, H_2, SF_6, and Cl_2. The MWCNTs have been found more effective and sensitive than SWCNTs. In MWCNTs based gas sensors, the p-type conductivity in semiconducting MWCNTs and the Schottky barrier formation of the fabricated MWCNTs attributed the resistivity changes (Fig. 23.2). The CNT-modified electrochemical gas sensors were designed based on monitoring galvanic effects. The sensors of both single-walled and double-walled CNTs have some limitations like slow recovery and lower selectivity. Different functional groups, metal nanoparticles, and oxide-modified CNT change nanotubes' electronic properties and enhance the sensitivity and selectivity to specific gaseous molecules (Chen, Zhu, & Wang, 2006; Varghese et al., 2001; Wei et al., 2004). Recently, polymer-modified CNT gas sensors have attracted more in sensor applications due to good performance at room temperature (Fu et al., 2008). The various converters can be used to detect a broad range of gases at the same time.

In the case of lower adsorptivity, ionization gas sensors are mainly used (Liu, Lin, & Swager, 2016). The main principle in this sensor is to detect the ions that collide with the gas molecules. Cathode and anode are used in this sensor, where CNTs are introduced as anode and aluminum as an anode. Appropriate voltage is required to induce a high electric field between the cathode and anode (Liu et al., 2016). The detection of He, Ar, and CO_2 gases by this method has been reported (Heer, Châtelain, & Ugarte, 1995). The main drawback of this method requires high voltage means high energy consumption.

Capacitance gas sensors have detected susceptible molecules such as vapors of benzene, toluene, alcohol, acetone, and dinitrotoluene. In this sensor, a disoriented array of nanotubes was grown on a silicon plate or layer (De Jonge, Lamy, Schoots, & Oosterkamp, 2002). As a result, a high-magnitude electric field is generated while external voltage is applied between the plates, and nanotubes' termination causes the polarization of adsorbed molecules (Yeow & She, 2006). But in a higher humid environment, the sensor does not perform well; hence, the area of the applications is still restricted.

Carbon nanotubes in biosensors

Biosensors are analytical devices that monitor chemical and biological reactions by measuring the signal in proportion to the analyte concentration. It is a physicochemical device. In 1956, the biosensor was first introduced by Leland Clark Jr., renowned as "Father of Biosensors" (Snow, Perkins, Houser, Badescu, & Reinecke, 2005). He described an electrochemical biosensor, where oxygen reduction occurred in the platinum cathode and monitored the blood oxygen. Clark and Lyons later worked together and developed glucose sensors by incorporating glucose oxidase and oxygen electrodes in a dialysis membrane in 1962 (Snow et al., 2005). In 1967, an enzyme electrode was demonstrated by Updike and Hicks, where glucose was quantified in samples (Clark & Lyons, 1962). Ever since, a broad range of biosensors has been developed for analyzing enzymes, peptides, polypeptides, DNA, antibodies, aptamers, nucleic acids, and so on

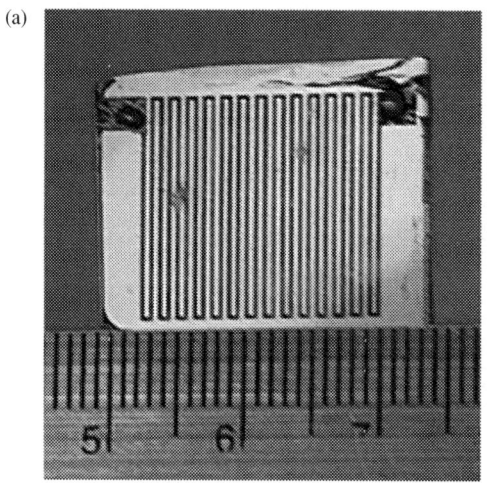
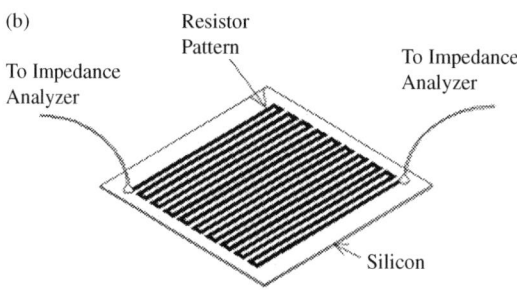

FIG. 23.2 (A) Digital image of the resistive sensor showing the confinement of the MWNTs to the patterned serpentine silicon dioxide pattern. The silicon substrate is the bright portion in the background. (B) Schematic diagram of the serpentine resistor pattern. *(From Ghodrati, M., Mir, A. & Farmani, A. (2020). Carbon nanotube field effect transistors–based gas sensors (pp. 171–183). Elsevier BV. https://doi.org/10.1016/b978-0-12-819,870-4.00036-0.)*

(Updike & Hicks, 1967). With the renaissance of nanotechnology, nanomaterials have been offered a tremendous application in biosensors.

CNTs have been one of the essential and desirable materials in nanotechnology, due to having unique structural, mechanical, optical, and biocompatible properties. Several studies showed the CNTs' capability of enhancing the electrochemical reactivity of the targeted biomolecules, which made them desirable in this field (Trotter, Borst, Thewes, & von Stetten, 2020). Different CNTs have been developed by fabrication or conjugation with other active materials to detect cell surface sugars, DNA biomarkers, enzymes, protein receptors, and lipids molecules (Trotter et al., 2020). Depending on the targeting molecules, the sensors' mechanism developed through several transduction processes subdivided into electrochemical biosensors, electronic transducers, optical-CNT biosensors, and immunosensors (Tîlmaciu & Morris, 2015).

CNTs' incorporation in the biosensing devices enhances the selectivity and sensitivity of electrochemical biosensors for detecting biomarkers of different diseases (Sireesha, Jagadeesh Babu, Kranthi Kiran, & Ramakrishna, 2018). Electrochemical CNT biosensors are more popular due to their cost-effectivity, fast response, and small size (Pasinszki, Krebsz, Tung, & Losic, 2017). SWCNTs and MWCNTs have both been developed to detections, metabolites, and biomarkers. Several CNT-based glucose biosensors have been engineered by conjugating glucose oxidase with nanotubes (Lin, Lu, Tu, & Ren, 2004; Zhao, Buldum, et al., 2002; Zhao, Gan, & Zhuang, 2002). They have further been developed for detecting cholesterol in the blood (Fig. 23.3).

In lung cancer studies, detection of human breath has been designed by introducing SWCNTs in tricosane (Li, Liao, Hu, Ma, & Wu, 2005). Zheng et al. introduced polydopamine-coated carbon nanotubes for electrochemical detection of HeLa and HL60 cancer cells (Zhang, Boghossian, et al., 2011; Zhang, Guo, et al., 2011). CNT immunosensors and optical CNT biosensors are both famous for cancer studies (Zhang, Boghossian, et al., 2011; Zhang, Guo, et al., 2011). Immunosensors recognize antigens by recombining antibodies. Optical biosensors respond to the targeting molecules by monitoring the emission of radiation (UV, UV–visible, and infrared). Several CNT-modified optical biosensors have been developed for studying in vitro dynamic biomolecular processes (Zheng, Zhang, Zou, & Zhu, 2012).

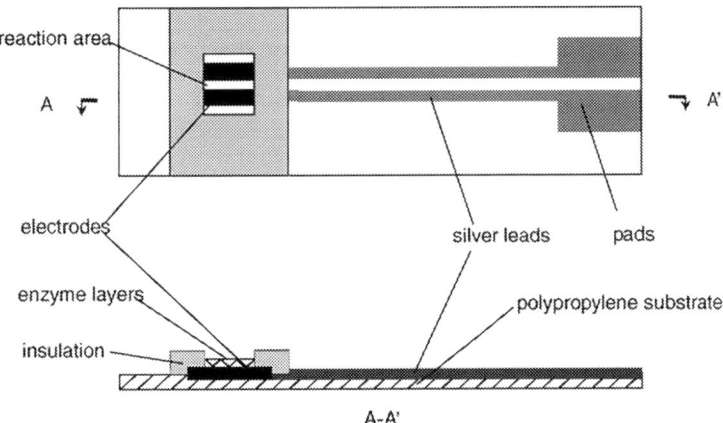

FIG. 23.3 The structure of the cholesterol biosensor. *(From Pasinszki, T., Krebsz, M., Tung, T. T. & Losic, D. (2017). Carbon Nanomaterial Based Biosensors for Non-Invasive Detection of Cancer and Disease Biomarkers for Clinical Diagnosis. Sensors, 17(8), 1919. https://doi.org/10.3390/s17081919)*

Carbon nanotubes in drug delivery applications

Several studies showed that CNTs are suitable for entering the cellular cell due to their morphology and properties. This behavior brought them ideal for various medical applications. Several mechanisms have been developed for taking nanotubes into the cells, including independent and dependent energy pathways (Cooper, 2002). They can conjugate with different functional groups to enter the cellular cells actively or passively.

There are two mechanisms widely used for functionalizing the nanotubes (Chen, Chen, et al., 2008; Chen, Zuo, et al., 2008). Using strong acids can oxidize the CNTs and causes the reduction of length by generating a carboxyl group, which increases their solubility in water. Another approach introduces nanotubes with different active molecules such as a peptide, nucleic acids, and other therapeutic agents. These mechanisms help them to enter the cells during caring for any drug molecules for specific applications (Chen, Chen, et al., 2008; Chen, Zuo, et al., 2008).

Drug particles are conjugated to CNTs through a covalent or noncovalent bond between drug molecules and functionalized nanotubes. On the one hand, covalent interaction makes the drug-CNT more stable that bond breaking is quite hard in the inner cellular microenvironment, which also makes them slow responsive (Chen, Chen, et al., 2008; Chen, Zuo, et al., 2008) and on the other hand, noncovalent interaction makes them fast responsive. However, the bond-breaking possibilities are high (Chen, Chen, et al., 2008; Chen, Zuo, et al., 2008). Some external stimuli have been introduced to functionalize CNTs via temperature, electric field, or light to overcome these problems. For example, Shi et al. introduced chitosan-functionalized CNT with poly-N-Isopropyl acrylamide and 1-butyl-3-21 vinyl imidazolium, which are thermosensitive molecules (Maiti, Tong, Mou, & Yang, 1401). For releasing ibuprofen, an electric field was introduced with functionalized MWCNTs. Khandare et al. reported an anticancer drug, where MWCNT encapsulated with calcium phosphate is considered as a nanocapsule for intercellular drug delivery (Shi et al., 2015).

CNTs are delivered not only by small molecular drugs but also by proteins. MWCNTs have been used as protein carriers (Chen, Chen, et al., 2008; Chen, Zuo, et al., 2008). For tumor cell studies, CNT protein-functionalized molecules are so far famous. Gene studies also grabbed attention to introduce nanotubes for studying DNA and RNA.

Carbon nanotube sensors for food and environmental applications

Several processes have been developed in food management and agricultural production for delivering quality food through sensing functions. Monitoring food storage quality, detecting pesticides and pathogens, preventing food spoilage, and monitoring ripeness time for fresh produces are the primary concerns of these processes (Banerjee et al., 2015). CNT-based sensors offer several assistances in food applications (Banerjee et al., 2015). Their variety of size, low power consumption, and better conjugation ability with other molecules have been widely used in agricultural areas such as fruit ripening (Schroeder, Savagatrup, He, Lin, & Swager, 2019), pesticide detection (Esser, Schnorr, & Swager, 2012), innovative packaging (Chen, Chen, et al., 2008; Chen, Zuo, et al., 2008), and food spoilage time detection (Zhu, Desroches, Yoon, & Swager, 2017). In addition, CNT gas sensors have been developed for characterizing the taste, smell, and quality of food products.

Several studies showed that ethylene is one hormone that is the reason for fruit ripening. Controlling and monitoring ethylene levels are helpful to find the optimal harvesting time and preserving the fresh goods for a longer time (Liu, Petty, Sazama, & Swager, 2015). Many sensors have been developed for tracking the ethylene level. The detection of ethylene concentration by CNT-based sensors has been reported both experimentally and theoretically (Blankenship & Sisler, 1989; Zhang, Huang, Ding, & Li, 2014). Fabricated single-walled and multiwalled nanotubes showed better sensitivity at room temperature than nonfabricated ones. For instance, tin oxide nanoparticle-modified MWCNTs and fluorinated Tris borate copper (I) complex-modified SWCNTs detected ethylene at room temperature (Schroeder et al., 2019).

CNT sensors have also been used to analyze the taste and smell of various foods. For this application, electronic tongues and noses are developed (Chiu & Tang, 2013; Li, Hodak, Lu, & Bernholc, 2017). Usually, metal oxide-, polymer-, bioelectronic-, and electronic-based sensors have been used for testing food. CNTs coordinate with the protein from olfactory receptors of humans (Wasilewski, Gębicki, & Kamysz, 2021), insects (Son et al., 2015), and rodents (Lee et al., 2015) have been studied for developing this sensor. Upon binding with the nanotubes, one research showed that the olfactory receptor gets into an active, negative state from an inactive neutral state to differentiate the butyrate molecules and detect the amyl butyrate (Goldsmith et al., 2011). Another research from the same group developed a nanotube-based taste sensor by mimicking the pathways of the olfactory system (Fig. 23.4; Kim et al., 2009). Some CNT-based sensors have been developed to differentiate similar products based on taste and smell, such as Ni- and Cu-modified CNTs (Jin et al., 2012). Electrochemical sensors can discriminate Chinese rice wine by their age and manufacturing process (Jin et al., 2012). Ensuring food safety, several CNT-based sensors have been reported to find the exact time of food spoilage (Banerjee et al., 2015). For example, chemiresistive sensors containing a series of cobalt meso-aryl-porphyrin complexes have been developed to select ammonia and biogenic amine compounds responsible for meat spoilage (Zhu et al., 2017). Pesticide contamination is also detected and quantified by CNT-based sensors. These sensors rely on the oxidation or reduction of pesticides, and different functionalization processes are also developed for the sensors—for example, AChE-functionalized CNT-based sensors are designed to determine organophosphorus pesticides (Alothman & Wabaidur, 2019). Some CNT nanocomposites have been used to determine the food ingredients (Souza et al., 2020; Tigari & Manjunatha, 2020a, 2020b), for example, carbon nanocomposite-modified anionic surfactant is used for developing a sensor to assess curcumin and riboflavin in natural food supplements (Raril, Manjunatha, & Tigari, 2020; Tigari, Manjunatha, Raril, & Hareesha, 2019).

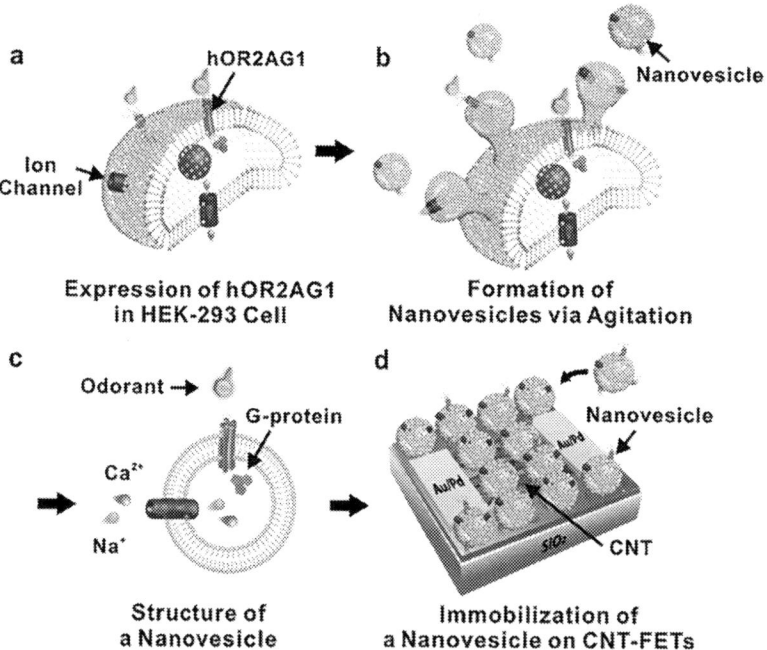

FIG. 23.4 Schematic diagram depicting the preparation of nanovesicles containing hOR2AG1 and the hybridization of the nanovesicles with a carbon nanotube-based FET transducer to build a nanovesicle-based bioelectronic nose (NBN) with high sensitivity and a human-like selectivity. *(From Kim, T. H., Lee, S. H., Lee, J., Song, H. S., Oh, E. H., Park, T. H. & Hong, S. (2009). Single-carbon-atomic-resolution detection of odorant molecules using a human olfactory receptor-based bioelectronic nose. Advanced Materials, 21(1), 91–94. https://doi.org/10.1002/adma.200801435.)*

Nanodiamonds for sensor applications

Nanodiamonds in gas sensors

Nanodiamond (ND) is one of the stable carbon compounds at ambient conditions. It has many unique properties such as superhardness, high thermal conductivity, superrefraction index, high resistivity, permitting it to be an essential compound for various sensor applications. In addition, it has better adaptivity for surface functionalization than other allotropes of carbon (Lisichkin, Korol'kov, Tarasevich, Kulakova, & Karpukhin, 2006), which has allowed it a better material for gas sensing (Zhang, Rhee, Hui, & Park, 2018).

Functionalized NDs have received better attention for this purpose due to the abundance of functional group distribution on its surface. For example, a combined form of pristine ND- and palladium-nanolayered composite has been developed for detecting toxic NO_2 and NH_3 gases (Moncea et al., 2019). This carbon-based sensor allows detecting 50 ppb NO_2 and 25–100 ppm NH_3 gases with a fast response of up to 50% in humid conditions. One research showed that nanocrystalline diamond nanorod-fabricated gas sensors represented high sensitivity to phosgene gas. The process included a deposition of nanodiamond rod on an alumina substrate followed by a dry plasma etching process. A group from the Chinese Academy of Science developed an array of diamond nanoneedles on crystalline diamond film with reactive ions etching to respond to NH_3 and NO_2 gas at room temperature (Lu et al., 2011). The fabrication process for diamond nanoneedle arrays is shown in Fig. 23.5. Detonated nanodiamond (DND) has found a solid response to NH_3 gas (Bevilacqua, Chaudhary, & Jackman, 2009). It has been found that the DND forms two highly conductive pathways through the materials, and during exposure, it forms NH_3 gas with grain boundary and grain interior processes. Monitoring the resistivity of DND has suggested a unique solid-state platform for ammonia sensing applications (Bevilacqua et al., 2009).

Nanodiamonds in energy storage devices

The demands for high energy and power are increasing along with device-friendly daily life. In this situation, NDs have drawn a preferable interest in this area due to having unique physical and chemical properties. Moreover, the development of fabrication methods of ND on conducting substrate has accelerated the growth of ND-based electrode materials in storage applications (Chu, Cui, & Liu, 2017).

The appropriate doping processes, especially acceptor-type doping, can change the nature of NDs (Einaga, Foord, & Swain, 2014). The modification provides them semiconductive, semimetallic, or superconductive dispositions. Acceptor-type doping, such as boron doping, offers them absolute stability with a wide range of potential, including aqueous conditions (Kashiwada, Watanabe, Ootani, Tateyama, & Einaga, 2016; Lee et al., 2017). Boron-doped NDs have several features compared to metal or other carbonaceous electrodes (Kashiwada et al., 2016). One research showed that surface functionalization of NDs can improve the ion storage capability of ND-based electrodes through the additional pseudocapacitive reactions (Wei & Yushin, 2012).

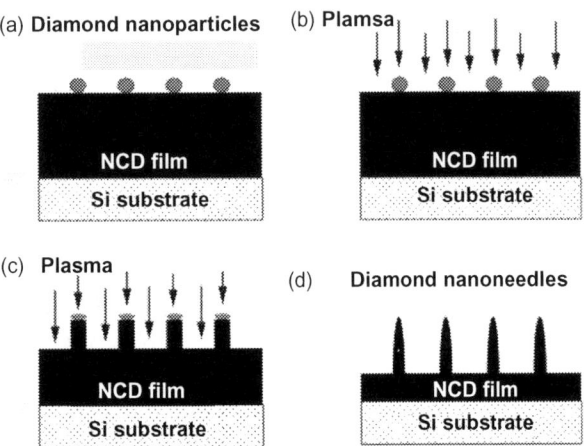

FIG. 23.5 Illustration of the fabrication process for diamond nanoneedle arrays. (A) Seeding of diamond nanoparticles, (B) beginning of RIE process, (C) during RIE process, and (D) formation of nanoneedle structures. *(From Lu, C., Li, Y., Tian, S., Li, W., Li, J. & Gu, C. (2011). Enhanced gas-sensing by diamond nanoneedle arrays formed by reactive ion etching. In Microelectronic Engineering (Vol. 88, Issue 8, pp. 2319–2321). https://doi.org/10.1016/j.mee.2011.02.074.)*

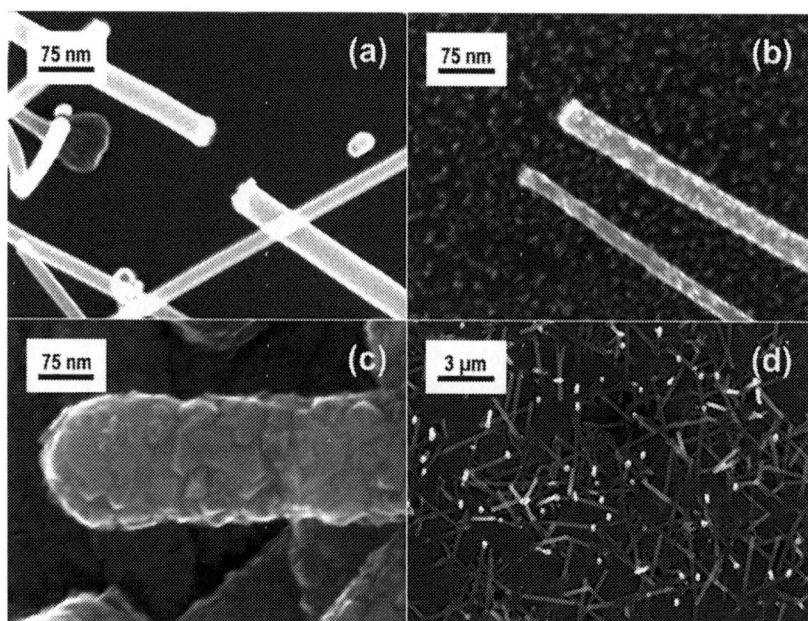

FIG. 23.6 SEM images of (A) bare silicon nanowire, (B) silicon nanowire after the seeding process, (C) diamond-coated silicon wire sample, (D) a large-scale picture of the same sample showing the density and uniformity. *(From Gao, F., Lewes-Malandrakis, G., Wolfer, M. T., Müller-Sebert, W., Gentile, P., Aradilla, D., Schubert, T. & Nebel, C. E. (2015). Diamond-coated silicon wires for supercapacitor applications in ionic liquids. Diamond and Related Materials, 51, 1–6. https://doi.org/10.1016/j.diamond.2014.10.009.)*

In addition, NDs have been widely used in supercapacitor electrodes to increase their physiochemical stability and mechanical strength (Portet, Yushin, & Gogotsi, 2007). A diamond-coated silicon nanowire has been developed by a Franco-German research team (Fig. 23.6). NDs have been used in electrochemically active polymers to increase their capacitance and cyclic stability. For example, one research team reported that PANI (polyaniline)/ND composite electrodes showed improved cyclic strength and capacitance retention at fast sweep rates than pure PANI electrodes (Kovalenko, Bucknall, & Yushin, 2010). As intercalating materials, NDs can be used with graphene oxide (GO) to reduce GO into reduced graphene oxide (rGO), for example, rGO/NDs composite electrodes can be used as electrochemical supercapacitors by the solution-phase process (Kumar et al., 2019). Recently, battery-like supercapacitors are widely using 3D ND networks in a water-soluble redox electrolyte with better specific capacitance and power densities (Yu et al., 2017).

Nanodiamonds in biological applications

Recently, NDs are widely applied in biological applications due to the unique properties of nanoparticles and diamonds. They have been attributed to advanced production, functionalization techniques, and low cytotoxicity (Krueger, 2008). In addition, in recent years, diamond coating has been applied to many medical devices such as heart valves, mandibular joint prostheses, and drug delivery (Puzyr et al., 2007).

The fluorescence properties of NDs made them one of the popular compounds in diagnosis areas. The imaging and therapy are utilizing the fluorescence property for the early diagnosis of the diseases and follow-up treatments (Fu et al., 2007). One research showed that the binding of positive-charged NDs to negative-charged DNA gives red fluorescent NDs (Fu et al., 2007). Using these fluorescence properties, NDs are now widely applied in biosensors. For example, for in vitro diagnostic applications, one research developed spin-enhanced ND biosensors using a microwave field in the biotin-avidin model (Miller et al., 2020). They detected HIV-1RNA by 10-min isothermal amplification steps.

NDs have better conjugation capability with other active functional groups and biological molecules due to a large surface area and better stability, which are appropriate for drug or peptide delivery systems (Chauhan, Jain, & Nagaich, 2020). They are broadly used as a chemotherapeutic agent by loading and releasing mechanics, for example, doxorubicin hydrochloride (DOX) adsorbs on the surface of NDs and later on forms a loose cluster for carrying it to the desired place (Huang, Pierstorff, Osawa, & Ho, 2007). They are also used in hydrophobic drugs as a dispersibility-improving agent. Proteins' immobilization on NDs is also a popular area of application for drug delivery (Chen et al., 2009).

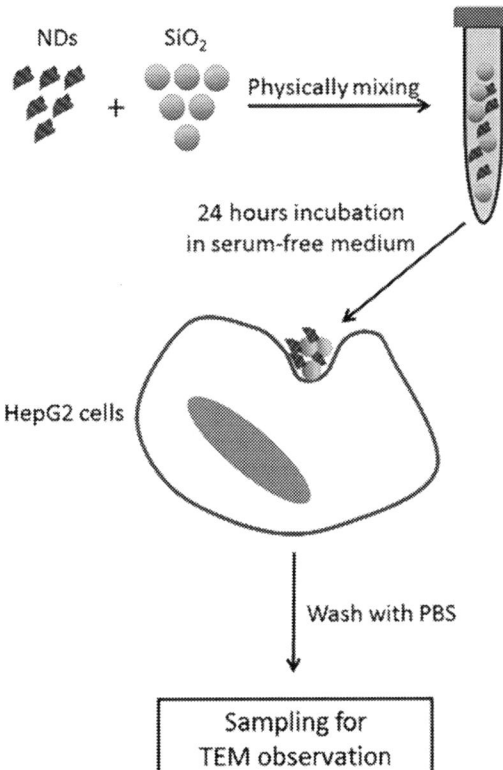

FIG. 23.7 Schematic illustration of the experimental design to identify how nanodiamonds escape the endosomal compartments. *(From Chu, Z., Miu, K., Lung, P., Zhang, S., Zhao, S., Chang, H. C., Lin, G. & Li, Q. (2015). Rapid endosomal escape of prickly nanodiamonds: Implications for gene delivery. Scientific Reports, 5. https://doi.org/10.1038/srep11661.)*

Recently, NDs act as an exciting tool for gene therapy. In this technique, cellular uptake and cytosolic groups require effective release (Fig. 23.7) (Chu et al., 2015). NDs are using as cytosolic carriers. Some studies found that they also potentially deliver small interfering RNAs (Liu & Sun, 2010). For antibacterial studies, NDs have been found effective in killing gram-positive and gram-negative bacteria (Wehling, Dringen, Zare, Maas, & Rezwan, 2014). Some of the studies revealed that their surface composition is appropriate as an antibacterial agent (Wehling et al., 2014). Recently, NDs have been attracted to bone tissue engineering areas (Okamoto, 2013). For example, polymer-functionalized NDs have been used in osteoblast (bone-forming cell) growth and formed bone-like apatite in simulated body fluid (Yang & Webster, 2014).

Fullerenes for sensor applications

Fullerenes in fuel cells

Fuel cells (FCs) have emerged as one of the promising power sources as an alternative to fossil fuels, causing environmental pollution to fulfill the growing energy demands (Yang & Webster, 2014). It is a device that converts chemical energy to electrical energy. Nanomaterials offer some excellent catalytic activities in FCs (Gong, Palmer, Brian, Harvey, & Verstraete, 2016). Due to having unique physical and chemical properties, fullerene is so prevalent in FCs (Coro, Suárez, Silva, Eguiluz, & Salazar-Banda, 2016). Several scientific works have been reported as the most stable and studied fullerene C_{60} used in FCs (Srivastava, Kumar, Singh, Agrawal, & Garg, 2009).

The electro- and thermal conductivities of C_{60} fullerenes are different from those of other carbon allotropes. The pore structure is controllable by changing the fullerene functional groups widely used in FCs as electrocatalysts (Srivastava et al., 2009).

Fullerene derivatives have been used as a catalyst in anode oxidation with methanol in direct methanol fuel cells (DMFCs) (Gong et al., 2016). In DMFCs, fullerene was first used as a catalyst for methanol oxidation, whereas C_{60} was used as a linker with immobilized Pt nanowires with gold nanoparticles (Maroto et al., 2011). Hybrid nanoparticles of Pt/C_{60} and $Pt/Ru-C_{60}$ are also used DMFCs as an oxidation catalyst (Calamba, Ringor, Pascua, & Miyazawa, 2015). The

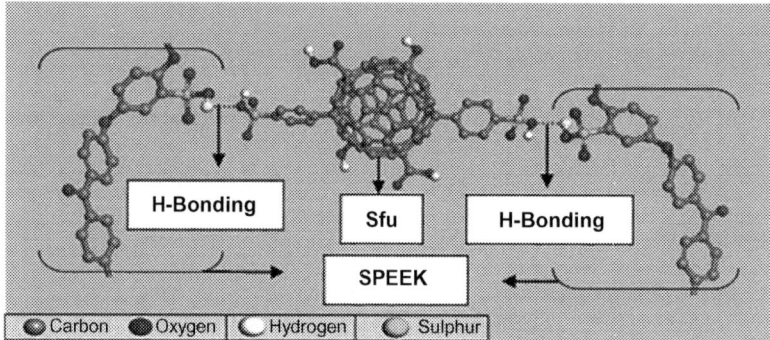

FIG. 23.8 H-bonding interaction of fullerene (Sfu) and SPEEK conductive site. *(From Rambabu, G. & Bhat, S. D. (2015). Sulfonated fullerene in SPEEK matrix and its impact on the membrane electrolyte properties in direct methanol fuel cells. Electrochimica Acta, 176, 657–669. https://doi.org/10.1016/j.electacta.2015.07.045.)*

presence of C_{60} increased the uniformity and dispersity of Pt nanoparticles. In DMFCs, fullerene is also used as coat electrodes in the anode oxidation reaction (Calamba et al., 2015). One study showed that C_{60} coated an optically transparent electrode (OTE) in acetonitrile and toluene mixture (3:1) by electrophoresis (Lee et al., 2009). The electrophoresis shaped the electrode to hollow nanoballs after coating it with C_{60} fullerene on the OTE surface. The coating technique was also applied on indium tin oxide (ITO) by microstructure single-crystal fullerene (Vinodgopal, Haria, Meisel, & Kamat, 2004). Recently, several functionalized fullerenes have been used in the anode oxidation reaction, such as pyridine-functionalized fullerene (PyC_{60}) used with a Pt electrode for methanol oxidation (Guo, Xu, Chen, & Yang, 2015). Carbon nanoonion (one of the essential members of the fullerene family) has been used as an anode electrocatalyst with Pt in FCs (Zhang & Ma, 2015).

Fullerene derivatives are also used in cathode oxygen reduction less frequently (Gong et al., 2016). Fullerene and carbon nanoonion are used as metallic supporters in cathode oxygen reduction reaction (ORR) (Gong et al., 2016). Several studies reported about the fullerene combination with glass-like carbon powder (GCP) and Pt as cathode electrodes, where the catalytic activity toward oxygen reduction was monitored (Xu, 2008).

Fullerene was also reported as proton conductors in FCs due to their high electron affinity that confers a high acidic nature (Shioyama, Ueda, & Kuriyama, 2007). The proton conductivity of fullerene was first reported in 1998 and later in 2001 (Saab, Stucky, Passerini, & Smyrl, 1998; Tasaki et al., 2006). Hence, the composition of C_{60}/Nafion 117 has been used as proton conductor membranes in FCs (Shioyama et al., 2007). The achieved conductivity was improved by loading only 1 wt% fullerene in Nafion (Shioyama et al., 2007). A study showed that the morphology of Nafion changed by the incorporation of fullerene in it, which increased the water uptake (Hinokuma & Ata, 2001). The higher water uptake enhanced the conductivity of fullerene/Nafion composites. Fullerene surface nowadays is functionalized by sulfonation and combines with a sulfonated polyether ether ketone (SPEEK) matrix by forming H-bond in DMFCs as proton-conductive membrane (Fig. 23.8) (Wang et al., 2007). The membrane represented better conductivity and ion-exchange capability (Wang et al., 2007).

Fullerene in biomedical applications

Functionalized fullerenes have been used for versatile purposes in biomedical applications such as drug delivery, reactive oxygen species (ROS) quenching, gene therapy, MRI contrasting, and imaging due to their exceptional properties and stability (Rambabu & Bhat, 2015).

Recently, fullerenes have been attracted to cancer treatment by an appropriate drug delivery system (Rambabu & Bhat, 2015). The water solubility nature and smaller size make them more efficient to carry the drugs through the different environments in our body (Rambabu & Bhat, 2015). For example, water-soluble fullerene derivatives can bind with paclitaxel for treating metastatic breast cancer (Rambabu & Bhat, 2015). Functionalized fullerenes can also carry DNA into the cell for gene transfer analysis, such as water-soluble C_{60} derivative was prepared to transport DNA across the cell for gene therapy (Partha & Conyers, 2009).

Fullerene can be used as an antioxidant agent in biological systems. For example, carboxyfullerenes can be used as neuroprotective agents (Nakamura et al., 2000). Another study showed that they have antioxidant properties and capable of suppressing iron-induced peroxidation in the lipid environment (Murthy, Choi, & Geckeler, 2002). Dendrofullerene (DF-1) is also reported as an antioxidant that acts as a radiation protector (Monti et al., 2000). Functionalized fullerenes also reduce oxidant-induced cytotoxicity, free radical formation, and mitochondrial damage (Daroczi et al., 2006). Gadofullerene derivatives can protect cellular mitochondria from oxidative injury (Daroczi et al., 2006). The extended π conjugation property of fullerenes allows them to absorb visible light and active fluorescence properties, which help them illuminate light during ROS generation (Yin et al., 2009).

Functionalized fullerenes such as endohedral metallofullerenes have been utilized in many medical applications (Rambabu & Bhat, 2015). They can encapsulate the metal atom inside the fullerene cage and protects the metal from any undesired environment in the biological system (Rambabu & Bhat, 2015). Gadolinium-metallofullerenes have been reported as contrast agents to enhance the MRI quality (Mroz et al., 2007). One study proposed that anionic gadofullerene has attractive features in vivo and noninvasive in vivo environments of any mammalian cell through an MRI system (Fatouros et al., 2006).

Fullerenes in biosensors

Fullerene (C_{60}) has made some breakthroughs in biosensing applications for having some unique properties such as absorption of a wide range of visible light, structural angle torsion, doping multiple electrons, electrophilic and nucleophilic combine properties, prolonged life triplet state, and caging properties (Afreen, Muthoosamy, Manickam, & Hashim, 2015). Fullerenes are used as a mediator in the biosensors between the electrode and the bioreceptor (Afreen et al., 2015). The fullerene-modified electrodes are mainly used in biosensors, and the modification occurred by dropping (Pilehvar & Wael, 2015), vapor deposition (Lokesh, Sherigara, Jayadev, Mahesh, & Mascarenhas, 2008), electrochemical deposition (Zhang, Ma, & Zhang, 2015), and electropolymerization processes (Schon, Dicarmine, & Seferos, 2014).

Functionalized fullerenes have been widely used in glucose sensors to detect glucose levels in the blood for diabetes mellitus (Jothi, Jayakumar, Jaganathan, & Nageswaran, 2018). Glucose-detecting biosensors such as piezoelectric and amperometric biosensors used fullerenes by Gavalas and Chaniotakis first (Gavalas & Chaniotakis, 2000). They reported that the incorporation of fullerene increased the sensitivity of the electrodes. Another research developed an electrochemical glucose sensor using cobalt (II) hexacyanoferrate and fullerene C60-enzyme. Another study used palladium nanoparticles incorporated by cysteine-functionalized fullerenes to detect glucose without any enzyme (Zhong, Yuan, & Chai, 2012). Here, the fullerene first covalently bonded with amine groups of cysteine and then formed nanocomposite with Pd nanoparticles.

Fullerene was also reported as a vitamin D_3 detector with bimetallic nanoparticles in electrochemical biosensors (Fig. 23.9). Fullerene-modified glassy carbon electrode (GCE) biosensor was reported by Shetti and colleagues to

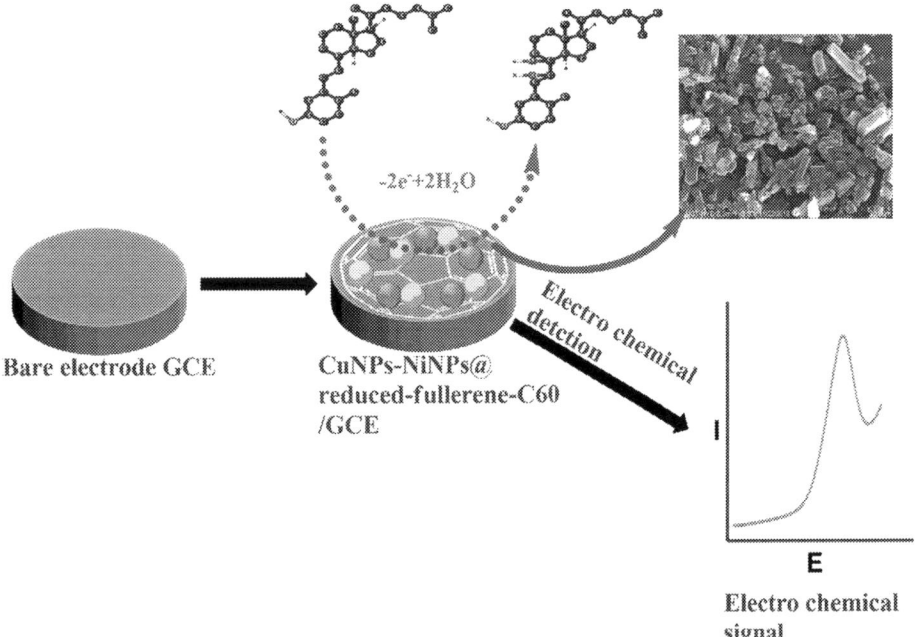

FIG. 23.9 Illustration of electrochemical sensor fabrication procedure for electrochemical oxidation of vitamin D_3. *(From Anusha, T., Bhavani, K. S., Kumar, J. V. S. & Brahman, P. K. (2020). Designing and fabrication of electrochemical nanosensor employing fullerene-C60 and bimetallic nanoparticles composite film for the detection of vitamin D3 in blood samples. Diamond and Related Materials, 104. https://doi.org/10.1016/j.diamond.2020.107761.)*

determine acyclovir (ACV) (Shetti, Malode, & Nandibewoor, 2012). Polymer-modified fullerene is also used in several biosensors such as an impedimetric electrochemical sensor developed by cortisol-imprinted polymer on fullerene used to assess cortisol in saliva (Ertuğrul Uygun, Uygun, Canbay, Girgin Sağın, & Sezer, 2020). Functionalized fullerene was also reported to detect urea in urine. For example, fullerene-cryptand-22-covered PZ quartz crystal developed a piezoelectric sensor to detect urea in urine (Wei & Shih, 2001). Fullerene bioconjugation sensor is also utilized to determine urea in the sample, where acrylic-based hydrogen-ion-sensitive membrane was used (Saeedfar, Heng, Ling, & Rezayi, 2013). C_{60} fullerene was also used in immunosensors to diagnose our body metabolism (Afreen et al., 2015). For example, immunoglobulin and hemoglobin were determined by fullerene-antibody-coated quartz crystal on the silver electrode (Pan & Shih, 2004). Another study reported the determination of mRNA-141 by a fullerene-modified gold electrode (Zhou et al., 2020).

Graphenes for sensor applications

Graphenes in gas sensors

The two-dimensional honeycomb-like structure of graphenes allows them to adsorb gas molecules more effectively than other carbon allotropes (Wang et al., 2016). Compared to carbon nanotubes, it has low electrical noise, thus having better charge fluctuation (Jeong et al., 2010). Its unique properties such as large surface area, excellent thermal and electrical properties, mechanical strength, and optical adsorptivity offer a promising aspect in gas sensors (Tian, Liu, & Yu, 2018).

The first graphene-based gas sensor was developed in 2007 by Schedin, where the stripped graphene was used to detect individual gas molecules such as NH_3, CO, NO, and water vapor (Schedin et al., 2007). After that, the application of graphenes in the gas sensing area increased over the period from 2007. The response toward the specific gas molecules depends on the graphene layers also on the length-to-wide ratio (Hwang et al., 2012). For example, one research showed the response of graphene-based sensors for NH_3 detection intensive for higher graphene layers (Fig. 23.10) (Hwang et al., 2012). The adsorption of gaseous molecules onto the graphene surface changes the signal due to changing morphology, hence the sensor's response. For example, the surface resistivity of monolayer graphenes varies significantly due to the adsorption of O_2 molecules by p-type doping (Chen et al., 2011). Another nanosheet graphene was used in the gas sensor by plasma-enhanced chemical vapor deposition, where the conductivity decreased in NH_3 and increased in NO_2 environments (Yu et al., 2011). In 2017, 3D graphene flower (GF) was prepared by plasma-enhanced chemical vapor deposition for detecting NO_2 gas (Wu et al., 2016). In 2021, one study reported a graphene-based NH_3 sensor, where the aerosol

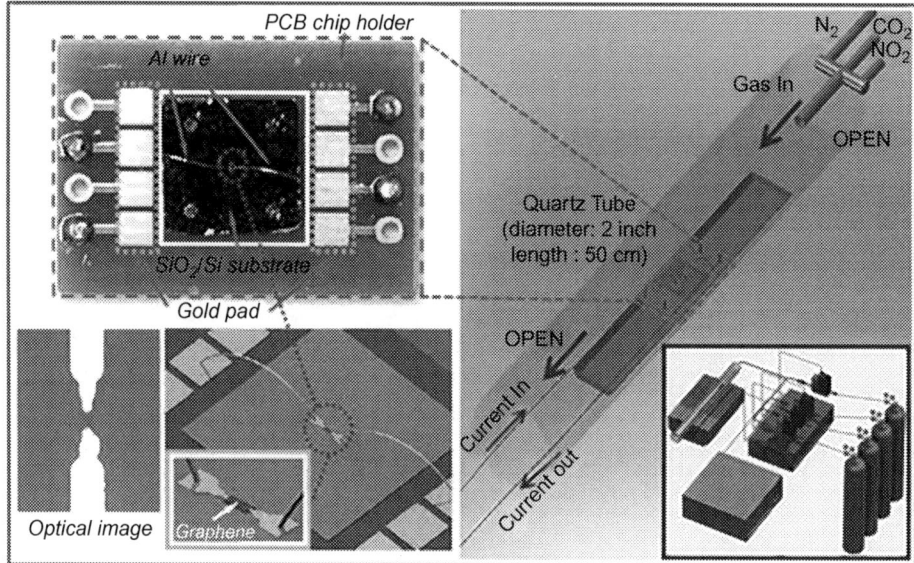

FIG. 23.10 Photographs of a chip-holder with graphenes, a schematic of the SiO_2/Si substrate, an optical image of the graphene device, and a schematic of the measurement system. *(From Hwang, S., Lim, J., Park, H. G., Kim, W. K., Kim, D. H., Song, I. S., Kim, J. H., Lee, S., Woo, D. H. & Chan Jun, S. (2012). Chemical vapor sensing properties of graphene based on geometrical evaluation. Current Applied Physics, 12(4), 1017–1022. https://doi.org/10.1016/j.cap.2011.12.021.)*

jet-printing technique was used (Zhu, Yu, Wu, Lv, & Wang, 2021). The sensor sensitivity was high (4.64% for 4.35 ppm NH_3) and showed good reversibility (Zhu et al., 2021).

Functionalized graphene such as boron-doped graphene, nitrogen-doped graphene, and defective graphene is also widely used in gas sensors (Zhang & Zhang, 2009). Boron doping has enhanced interaction between graphene and NO, NO_2, and NH_3, whereas defective graphene shows strong interactions with CO, NO, NO_2, and weak interaction with NH_3 (Tian et al., 2018). Polycrystalline graphene with graphene line defect was reported as the organic gas detector (Salehi-Khojin et al., 2012). Nowadays, nanosphere-coated graphene reaching popularity in gas sensing, such as NiO-In_2O_3 metal nanosphere-coated reduced graphene, was used for CO_2 gas sensing (Amarnath & Gurunathan, 2021). Research also predicted as Pt-doped armchair graphene nanoribbon could be used for CO and CO_2 gas sensor applications (Salih & Ayesh, 2021).

Graphene in biomedical applications

Graphene and its derivative have shown excellent potentials in many areas. The biomedical application of graphene is a significantly promising new area. The first study of graphene in biomedical applications was reported in 2008 by Dai et al. (Dai, 2002). After that, a lot of exciting research has been explored on graphene for biological applications, including drug delivery, antibacterial studies, biological sensing, and imaging, cell culture, and gene therapy (Shen, Zhang, Liu, & Zhang, 2012).

Graphene has been used as a transducer in biosensors. It has been widely produced for sensing applications to detect the target molecules where the mechanism involved by the interaction of target molecule with graphene honeycomb-like surface. In this case, functionalized graphene responds more actively than nonfunctionalized. The functional group on the graphene can capture molecules more promptly, and their interaction with the specific target is recorded (Fenzl et al., 2017). For example, carboxyl-functionalized graphene can readily interact with any amine groups of proteins (Foo & Gopinath, 2017). This complex can be used in biosensors for target specification (Fig. 23.11).

Graphene oxide (GO) has been used as an ideal carrier for drug and gene delivery. GO has excellent biocompatibility and strong conjugation capability of drugs, genes, and other functional groups. If the system has functional groups, GO first covalently bonds with them, then the drug molecules are incorporated with the π-π conjugation system. For example, first drug delivery research has followed this mechanism, where nanosized GO was first conjugated with amine-terminated six-armed polyethylene glycol, and then, anticancer drug (SN38) was loaded, thereby doing noncovalent π-π stacking process (Liu, Robinson, Sun, & Dai, 2008). Multiple drug delivery systems have been shown excellent response for cancer treatment; however, the mechanism is intricate; only a few materials showed proper answers for this, and GO is one of them (Andersson et al., 1999). At controlled loading, GO acts as a nanocarrier for multiple drugs. One study showed folic acid and SO_3H group-conjugated GO could coloaded doxorubicin (DOX) and camptothecin (CPT) by π-π stacking process (Shen et al., 2012).

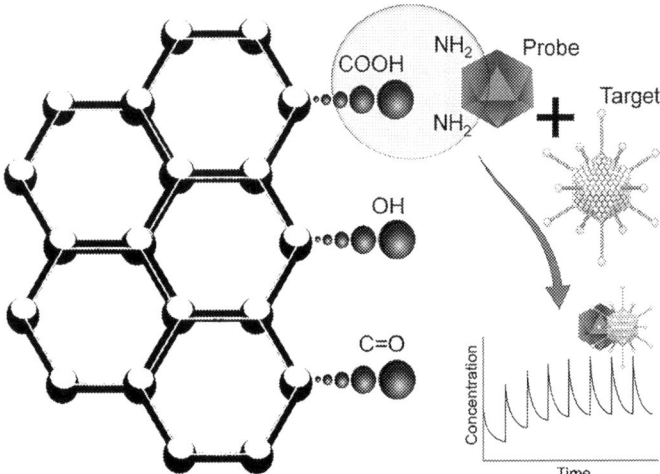

FIG. 23.11 Carboxyl group-functionalized graphene and amines interaction. The surface of this complex is used to authenticate target molecules. *(From Foo, M. E. & Gopinath, S. C. B. (2017). Feasibility of graphene in biomedical applications. Biomedicine and Pharmacotherapy, 94, 354–361. https://doi.org/10.1016/j.biopha.2017.07.122.)*

GO is now broadly applying in gene therapy (Shen et al., 2012). Gene therapy is a novel technique that treats genetic disorders and some chronic diseases. The method requires one gene vector to protect the DNA during cellular uptake of the DNA with good transfection efficiency (Naldini et al., 1996). Functionalized GO can conjugate with DNA plasmid onto the surface of the honeycomb-like structure of GO (Feng, Zhang, & Liu, 2011). For example, polyethyleneimine (PEI)-functionalized GO (GO-PEI) was utilized to conjugate with DNA plasmid (Feng et al., 2011). Here, the DNA plasmid was condensed with the positively charged PEI and then electrostatic interaction (GO-PEI-DNA plasmid) happened with GO nanosheet.

Another study reported chitosan-functionalized GO (GO-CS) for drug and gene delivery applications (Bao et al., 2011). They found that GO-CS has better cancer cell killing ability and condenses the plasma DNA into nanoparticles. GO is also used in antitumor treatment with photothermal and chemotherapies (Zhang, Boghossian, et al., 2011; Zhang, Guo, et al., 2011). PEG-functionalized NGO conjugated with doxorubicin (DOX) was used with photothermal therapy for tumor treatment (Zhang, Boghossian, et al., 2011; Zhang, Guo, et al., 2011).

Tissue engineering has also been exhibited attraction on graphene in this area nowadays. Carbon nanomaterials are combined with bioceramics scaffold for regenerating bone (Wierzbicki et al., 2020). GO and reduced GO (rGO) scaffold were used by one research group to stimulate the propagation of myogenic progenitor cells (Wierzbicki et al., 2020). Antibacterial applications of GO, rGO, and GtO have now been used to control the drug activities based on the ambiance. Recently, researchers are showing interest in preparing graphene quantum dots (GQDs) for bioimaging applications (Pan, Zhang, Li, & Wu, 2010).

Carbon quantum dots in sensor applications

Carbon quantum dots (CQDs) have now extensively experimented in several fields for their excellent physical and chemical properties, such as light absorbance and luminescence properties. They have been offered a promising prospect in sensing, catalysis, bioengineering, optics, and biomedicine (Li et al., 2012).

CQDs have gained plenty of interest in the electrocatalysis field for energy conversion and energy storage applications (Lim, Shen, & Gao, 2015). The surface of functionalized CQD can be formed as an active coordination site with metal ions (Lim et al., 2015). This modified CQD enhances the electrochemical activity by enhancing electron transfer through internal interactions. They may also enhance the oxygen reduction reaction in fuel cells (Li et al., 2017). Enough nitrogen and oxygen-rich functional groups in the zero-dimensional structure make CQD stable in ambiance conditions that may promote the response (Li, Zhou, et al., 2017). They have shown some excellent results in the CO_2 reduction reaction (Zhao et al., 2015). For example, nitrogen-doped functionalized graphite quantum dot was used with gold nanoparticles for CO_2 reduction application (Zhao et al., 2015). The bifunctional catalytic activity of functionalized CQDs has been reported in many electrocatalysis applications (Clancy et al., 2018). Nitrogen-doped graphene quantum dots (NGQDs) combined with Ni_3S_2 on Ni foam was used in water splitting as a catalyst.

The luminescence properties have been made CQDs more popular in biological applications such as drug delivery, bioimaging, and biosensing (Wang & Hu, 2014). Different wavelengths of light can be absorbed by CQDs followed by the fluorescence emission (Cao et al., 2007). This fluorescence image can be used in biological applications. One research reported this experiment to nude mice for collecting images, where they applied several excitation lights from 455 nm to 704 nm and found 595 nm for best fluorescence contrast (Yang et al., 2009). Functionalized CQDs have been used in multi-imaging applications, mainly in MRI techniques (Bourlinos et al., 2012). Ultrafine Gd (III)-doped CQDs showed dual-fluorescence activity and hence were used in MRI techniques (Bourlinos et al., 2012). This fluorescence technique can be used in sensing platforms for nucleic acid and mitochondrial H_2O_2 detection (Du, Min, Zeng, Yu, & Wu, 2014). The mechanism involved π-π interaction between CQDs and any active probes that may occur fluorescence quenching followed by fluorescence imaging (Li, Zhang, Wang, Tian, & Sun, 2011). Doped CQDs are nowadays offered some exciting features in sensing applications. The mechanism involved some doping followed by interaction with target molecules. For example, nitrogen and sulfur codoped carbon quantum dots (N-S-CQDs) were developed as fluorescence probes for detecting silver ions and cysteine (Liao et al., 2018). Here, fluorescence characteristics of N-S-CQDs were induced by target concentration and electron transfer from the surface to silver ion (Fig. 23.12).

CQDs are recently broadly used in drug and gene delivery applications for their unique properties such as, zero-dimensional structure, extra stability, and water solubility (Cheng, Zaki, Hui, Muzykantov, & Tsourkas, 2012). For example, gold nanoparticles combined CQD conjugated with polyethylenimine (PEI) DNA to deliver DNA to cells. First, pDNA quenched CQD-Au nanoparticles and then entered the cells (Cheng, Zaki, Hui, Muzykantov, & Tsourkas, 2012). The mechanism involved quenching followed by recovery of fluorescence signals.

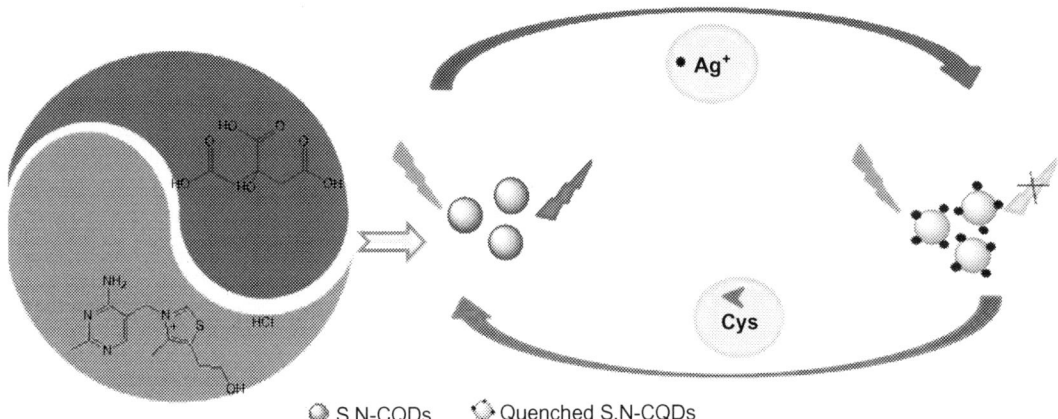

FIG. 23.12 Schematic diagram of N-S-CQDs for the detection of Ag^+ and Cysteine. *(From Liao, S., Zhao, X., Zhu, F., Chen, M., Wu, Z., song, X., Yang, H. & Chen, X. (2018). Novel S, N-doped carbon quantum dot-based "off–on" fluorescent sensor for silver ion and cysteine. Talanta, 180, 300–308. https://doi.org/10.1016/j.talanta.2017.12.040.)*

Conclusion

Over the last two decades, nanomaterials have been contributed a broader prospect in several areas of science and technologies, where the performances are carbon-based nanomaterials which are undisputable. All the carbonaceous materials have the potential to work as sensors in several applications. Their catalytic activities have opened a broader window to be electrochemical, optical, and photochemical sensing applications. They are now offering a more overall prospect in drug delivery, gene therapy, bioengineering, cancer treatment, tumor cell destruction, and many antibacterial applications in biomedical sensing applications. Energy conservation and conversion fields have also shown an enormous interest in them due to their unique characteristics. Solar cells are now one of the best energy sources, where carbon nanomaterial derivatives are widely embedding. Food engineering has a great interest in carbon-based nanoparticles for solving food scarcity. Some environmental issues such as reducing water and air pollution are required a better detection method that can detect the gaseous and any chemicals causing pollution. Finally, carbon-based nanoparticles have the properties, ability, and abundancy to lead the future nanotechnology for building a more comfortable life for the human being.

References

Afreen, S., Muthoosamy, K., Manickam, S., & Hashim, U. (2015). Functionalized fullerene (C60) as a potential nanomediator in the fabrication of highly sensitive biosensors. *Biosensors and Bioelectronics, 63*, 354–364. https://doi.org/10.1016/j.bios.2014.07.044.

Alothman, Z. A., & Wabaidur, S. M. (2019). Application of carbon nanotubes in extraction and chromatographic analysis: A review. *Arabian Journal of Chemistry, 12*(5), 633–651. https://doi.org/10.1016/j.arabjc.2018.05.012.

Amarnath, M., & Gurunathan, K. (2021). Highly selective CO2 gas sensor using stabilized NiO-In2O3 nanospheres coated reduced graphene oxide sensing electrodes at room temperature. *Journal of Alloys and Compounds, 857*, 157584. https://doi.org/10.1016/j.jallcom.2020.157584.

Amrutha, B. M., Manjunatha, J. G., Bhatt, A. S., Raril, C., & Pushpanjali, P. A. (2019). Electrochemical sensor for the determination of alizarin red-S at non-ionic surfactant modified carbon nanotube paste electrode. *Physical Chemistry Research, 7*(3), 523–533. https://doi.org/10.22036/pcr.2019.185875.1636.

Andersson, M., Lindegaard Madsen, E., Overgaard, M., Rose, C., Dombernowsky, P., & Mouridsen, H. T. (1999). Doxorubicin versus methotrexate both combined with cyclophosphamide, 5- fluorouracil and tamoxifen in postmenopausal patients with advanced breast cancer - A randomised study with more than 10 years follow-up from the Danish breast cancer cooperative group. *European Journal of Cancer, 35*(1), 39–46. https://doi.org/10.1016/S0959-8049(98)00354-2.

Banerjee, S. S., Todkar, K. J., Khutale, G. V., Chate, G. P., Biradar, A. V., Gawande, M. B., et al. (2015). Calcium phosphate nanocapsule crowned multiwalled carbon nanotubes for pH triggered intracellular anticancer drug release. *Journal of Materials Chemistry B, 3*(19), 3931–3939. https://doi.org/10.1039/c5tb00534e.

Bao, H., Pan, Y., Ping, Y., Sahoo, N. G., Wu, T., Li, L., et al. (2011). Chitosan-functionalized graphene oxide as a nanocarrier for drug and gene delivery. *Small, 7*(11), 1569–1578. https://doi.org/10.1002/smll.201100191.

Bevilacqua, M., Chaudhary, A., & Jackman, R. B. (2009). The influence of ammonia on the electrical properties of detonation nanodiamond. *Journal of Applied Physics, 106*(12). https://doi.org/10.1063/1.3272912.

Bezzon, V. D. N., Montanheiro, T. L. A., De Menezes, B. R. C., Ribas, R. G., Righetti, V. A. N., Rodrigues, K. F., et al. (2019). Carbon Nanostructure-based Sensors: A Brief Review on Recent Advances. *Advances in Materials Science and Engineering, 2019*. https://doi.org/10.1155/2019/4293073.

Bhattacharya, K., Mukherjee, S. P., Gallud, A., Burkert, S. C., Bistarelli, S., Bellucci, S., et al. (2016). Biological interactions of carbon-based nanomaterials: From coronation to degradation. *Nanomedicine: Nanotechnology, Biology, and Medicine, 12*(2), 333–351. https://doi.org/10.1016/j.nano.2015.11.011.

Bhattacharyya, A., Seth, G.S., Kumar, R., & Chamkha, A.J. (n.d.). Simulation of Cattaneo–Christov heat flux on the flow of single and multi-walled carbon nanotubes between two stretchable coaxial rotating disks. Journal of Thermal Analysis and Calorimetry, 2020(3), 1655–1670.

Blankenship, S. M., & Sisler, E. C. (1989). Ethylene binding changes in apple and morning glory during ripening and senescence. *Journal of Plant Growth Regulation, 8*(1), 37–44. https://doi.org/10.1007/BF02024924.

Bourlinos, A. B., Bakandritsos, A., Kouloumpis, A., Gournis, D., Krysmann, M., Giannelis, E. P., et al. (2012). Gd(III)-doped carbon dots as a dual fluorescent-MRI probe. *Journal of Materials Chemistry, 22*(44), 23327–23330. https://doi.org/10.1039/c2jm35592b.

Boyd, A., Dube, I., Fedorov, G., Paranjape, M., & Barbara, P. (2014). Gas sensing mechanism of carbon nanotubes: From single tubes to high-density networks. *Carbon, 69*, 417–423. https://doi.org/10.1016/j.carbon.2013.12.044.

Calamba, K., Ringor, C., Pascua, C., & Miyazawa, K. (2015). Pleated surface morphology of C60 fullerene nanowhiskers incorporated by polyaniline in N-methyl-2-pyrrolidone. *Fullerenes, Nanotubes, and Carbon Nanostructures, 23*(8), 709–714. https://doi.org/10.1080/1536383X.2014.971118.

Cao, L., Wang, X., Meziani, M. J., Lu, F., Wang, H., Luo, P. G., et al. (2007). Carbon dots for multiphoton bioimaging. *Journal of the American Chemical Society, 129*(37), 11318–11319. https://doi.org/10.1021/ja073527l.

Charithra, M. M., Manjunatha, J. G. G., & Raril, C. (2020). Surfactant modified graphite paste electrode as an electrochemical sensor for the enhanced voltammetric detection of estriol with dopamine and uric acid. *Advanced Pharmaceutical Bulletin, 10*(2), 247–253. https://doi.org/10.34172/apb.2020.029.

Chauhan, S., Jain, N., & Nagaich, U. (2020). Nanodiamonds with powerful ability for drug delivery and biomedical applications: Recent updates on in vivo study and patents. *Journal of Pharmaceutical Analysis, 10*(1), 1–12. https://doi.org/10.1016/j.jpha.2019.09.003.

Chen, J., Chen, S., Zhao, X., Kuznetsova, L. V., Wong, S. S., & Ojima, I. (2008). Functionalized single-walled carbon nanotubes as rationally designed vehicles for tumor-targeted drug delivery. *Journal of the American Chemical Society, 130*(49), 16778–16785. https://doi.org/10.1021/ja805570f.

Chen, C. W., Hung, S. C., Yang, M. D., Yeh, C. W., Wu, C. H., Chi, G. C., et al. (2011). Oxygen sensors made by monolayer graphene under room temperature. *Applied Physics Letters, 99*(24), 243502. https://doi.org/10.1063/1.3668105.

Chen, M., Pierstorff, E. D., Lam, R., Li, S. Y., Huang, H., Osawa, E., et al. (2009). Nanodiamond-mediated delivery of water-insoluble therapeutics. *ACS Nano, 3*(7), 2016–2022. https://doi.org/10.1021/nn900480m.

Chen, Y., Zhu, C., & Wang, T. (2006). The enhanced ethanol sensing properties of multi-walled carbon nanotubes/SnO2 core/shell nanostructures. *Nanotechnology, 17*(12), 3012–3017. https://doi.org/10.1088/0957-4484/17/12/033.

Chen, H., Zuo, X., Su, S., Tang, Z., Wu, A., Song, S., et al. (2008). An electrochemical sensor for pesticide assays based on carbon nanotube-enhanced acetycholinesterase activity. *Analyst, 133*(9), 1182–1186. https://doi.org/10.1039/b805334k.

Cheng, Z., Zaki, A. A., Hui, J. Z., Muzykantov, V. R., & Tsourkas, A. (2012). Multifunctional nanoparticles: Cost versus benefit of adding targeting and imaging capabilities. *Science, 338*(6109), 903–910.

Chenthattil, R., Manjunatha, J. G., Ravishankar, D. K., Fattepur, S., Siddaraju, G., & Nanjundaswamy, L. (2020). Validated electrochemical method for simultaneous resolution of tyrosine, uric acid, and ascorbic acid at polymer modified nano-composite paste electrode. *Surface Engineering and Applied Electrochemistry, 56*(4), 415–426.

Chiu, S. W., & Tang, K. T. (2013). Towards a chemiresistive sensor-integrated electronic nose: A review. *Sensors (Switzerland), 13*(10), 14214–14247. https://doi.org/10.3390/s131014214.

Chu, S., Cui, Y., & Liu, N. (2017). The path towards sustainable energy. *Nature Materials*, 16–22. https://doi.org/10.1038/nmat4834.

Chu, Z., Miu, K., Lung, P., Zhang, S., Zhao, S., Chang, H. C., et al. (2015). Rapid endosomal escape of prickly nanodiamonds: Implications for gene delivery. *Scientific Reports, 5*. https://doi.org/10.1038/srep11661.

Clancy, A. J., Bayazit, M. K., Hodge, S. A., Skipper, N. T., Howard, C. A., & Shaffer, M. S. P. (2018). Charged carbon nanomaterials: Redox chemistries of fullerenes, carbon nanotubes, and Graphenes. *Chemical Reviews, 118*(16), 7363–7408. https://doi.org/10.1021/acs.chemrev.8b00128.

Clark, L. C., & Lyons, C. (1962). Electrode systems for continuous monitoring in cardiovascular surgery. *Annals of the New York Academy of Sciences, 102*(1), 29–45. https://doi.org/10.1111/j.1749-6632.1962.tb13623.x.

Cooper, M. A. (2002). Optical biosensors in drug discovery. *Nature Reviews Drug Discovery, 1*(7), 515–528. https://doi.org/10.1038/nrd838.

Coro, J., Suárez, M., Silva, L. S. R., Eguiluz, K. I. B., & Salazar-Banda, G. R. (2016). Fullerene applications in fuel cells: A review. *International Journal of Hydrogen Energy, 41*(40), 17944–17959. https://doi.org/10.1016/j.ijhydene.2016.08.043.

Dai, H. (2002). Carbon nanotubes: Opportunities and challenges. *Surface Science, 500*(1–3), 218–241. https://doi.org/10.1016/S0039-6028(01)01558-8.

Danilenko, V. V. (2004). On the history of the discovery of nanodiamond synthesis. *Physics of the Solid State, 46*(4), 595–599. https://doi.org/10.1134/1.1711431.

Daroczi, B., Kari, G., McAleer, M. F., Wolf, J. C., Rodeck, U., & Dicker, A. P. (2006). In vivo radioprotection by the fullerene nanoparticle DF-1 as assessed in a zebrafish model. *Clinical Cancer Research, 12*(23), 7086–7091. https://doi.org/10.1158/1078-0432.CCR-06-0514.

De Jonge, N., Lamy, Y., Schoots, K., & Oosterkamp, T. H. (2002). High brightness electron beam from a multi-walled carbon nanotube. *Nature, 420*(6914), 393–395. https://doi.org/10.1038/nature01233.

Du, F., Min, Y., Zeng, F., Yu, C., & Wu, S. (2014). A targeted and FRET-based ratiometric fluorescent nanoprobe for imaging mitochondrial hydrogen peroxide in living cells. *Small, 10*(5), 964–972. https://doi.org/10.1002/smll.201302036.

Einaga, Y., Foord, J. S., & Swain, G. M. (2014). Diamond electrodes: Diversity and maturity. *MRS Bulletin, 39*(6), 525–532. https://doi.org/10.1557/mrs.2014.94.

Ertuğrul Uygun, H. D., Uygun, Z. O., Canbay, E., Girgin Sağın, F., & Sezer, E. (2020). Non-invasive cortisol detection in saliva by using molecularly cortisol imprinted fullerene-acrylamide modified screen printed electrodes. *Talanta, 206*. https://doi.org/10.1016/j.talanta.2019.120225.

Esser, B., Schnorr, J. M., & Swager, T. M. (2012). Selective detection of ethylene gas using carbon nanotube-based devices: Utility in determination of fruit ripeness. *Angewandte Chemie, International Edition, 51*(23), 5752–5756. https://doi.org/10.1002/anie.201201042.

Fatouros, P. P., Corwin, F. D., Chen, Z. J., Broaddus, W. C., Tatum, J. L., Kettenmann, B., et al. (2006). In vitro and in vivo imaging studies of a new endohedral metallofullerene nanoparticle. *Radiology, 240*(3), 756–764. https://doi.org/10.1148/radiol.2403051341.

Feng, L., Zhang, S., & Liu, Z. (2011). Graphene based gene transfection. *Nanoscale, 3*(3), 1252–1257. https://doi.org/10.1039/c0nr00680g.

Fenzl, C., Nayak, P., Hirsch, T., Wolfbeis, O. S., Alshareef, H. N., & Baeumner, A. J. (2017). Laser-scribed graphene electrodes for aptamer-based biosensing. *ACS Sensors, 2*(5), 616–620. https://doi.org/10.1021/acssensors.7b00066.

Foo, M. E., & Gopinath, S. C. B. (2017). Feasibility of graphene in biomedical applications. *Biomedicine and Pharmacotherapy, 94*, 354–361. https://doi.org/10.1016/j.biopha.2017.07.122.

Fouskaki, M., & Chaniotakis, N. (2008). Fullerene-based electrochemical buffer layer for ion-selective electrodes. *Analyst, 133*(8), 1072–1075. https://doi.org/10.1039/b719759d.

Fu, C. C., Lee, H. Y., Chen, K., Lim, T. S., Wu, H. Y., Lin, P. K., et al. (2007). Characterization and application of single fluorescent nanodiamonds as cellular biomarkers. *Proceedings of the National Academy of Sciences of the United States of America, 104*(3), 727–732. https://doi.org/10.1073/pnas.0605409104.

Fu, D., Lim, H., Shi, Y., Dong, X., Mhaisalkar, S. G., Chen, Y., et al. (2008). Differentiation of gas molecules using flexible and all-carbon nanotube devices. *Journal of Physical Chemistry C, 112*(3), 650–653. https://doi.org/10.1021/jp710362r.

Gavalas, V. G., & Chaniotakis, N. A. (2000). [60]Fullerene-mediated amperometric biosensors. *Analytica Chimica Acta, 409*(1–2), 131–135. https://doi.org/10.1016/S0003-2670(99)00887-9.

Goldsmith, B. R., Mitala, J. J., Josue, J., Castro, A., Lerner, M. B., Bayburt, T. H., et al. (2011). Biomimetic chemical sensors using nanoelectronic readout of olfactory receptor proteins. *ACS Nano, 5*(7), 5408–5416. https://doi.org/10.1021/nn200489j.

Gong, A., Palmer, J. L., Brian, G., Harvey, J. R., & Verstraete, D. (2016). Performance of a hybrid, fuel-cell-based power system during simulated small unmanned aircraft missions. *International Journal of Hydrogen Energy*, 11418–11426. https://doi.org/10.1016/j.ijhydene.2016.04.044.

Guo, J., Xu, Y., Chen, X., & Yang, S. (2015). Single-crystalline C60 crossing microplates: Preparation, characterization, and application as catalyst supports for methanol oxidation. *Fullerenes, Nanotubes, and Carbon Nanostructures, 23*(5), 424–430. Taylor and Francis Inc https://doi.org/10.1080/1536383X.2013.843168.

Hareesha, N., Manjunatha, J. G., Raril, C., & Tigari, G. (2019). Sensitive and selective electrochemical resolution of tyrosine with ascorbic acid through the development of Electropolymerized alizarin sodium sulfonate modified carbon nanotube paste electrodes. *ChemistrySelect, 4*(15), 4559–4567. https://doi.org/10.1002/slct.201900794.

Heer, Châtelain, A., & Ugarte, D. (1995). A carbon nanotube field-emission Electron source. *Science, 270*(5239).

Hinokuma, K., & Ata, M. (2001). Fullerene proton conductors. *Chemical Physics Letters, 341*(5–6), 442–446. https://doi.org/10.1016/S0009-2614(01)00549-8.

Hirsch, A. (2010). The era of carbon allotropes. *Nature Materials, 9*(11), 868–871. https://doi.org/10.1038/nmat2885.

Huang, H., Pierstorff, E., Osawa, E., & Ho, D. (2007). Active nanodiamond hydrogels for chemotherapeutic delivery. *Nano Letters, 7*(11), 3305–3314. https://doi.org/10.1021/nl071521o.

Hwang, S., Lim, J., Park, H. G., Kim, W. K., Kim, D. H., Song, I. S., et al. (2012). Chemical vapor sensing properties of graphene based on geometrical evaluation. *Current Applied Physics, 12*(4), 1017–1022. https://doi.org/10.1016/j.cap.2011.12.021.

Iijima, S. (1991). Helical microtubules of graphitic carbon. *Nature, 354*(6348), 56–58. https://doi.org/10.1038/354056a0.

Iijima, S., & Ichihashi, T. (1993). Single-shell carbon nanotubes of 1-nm diameter. *Nature, 363*(6430), 603–605. https://doi.org/10.1038/363603a0.

Jeong, H. Y., Lee, D. S., Choi, H. K., Lee, D. H., Kim, J. E., Lee, J. Y., et al. (2010). Flexible room-temperature NO2 gas sensors based on carbon nanotubes/reduced graphene hybrid films. *Applied Physics Letters, 96*(21). https://doi.org/10.1063/1.3432446.

Jg, M., (n.d.). New, sensor, on, nanotube-graphite, c., mixture, electrode, voltammetric, & resorcinol, O, Asian Journal of Pharmaceutical and Clinical Research 2017(12), 295–300.

Jin, H. J., Lee, S. H., Kim, T. H., Park, J., Song, H. S., Park, T. H., et al. (2012). Nanovesicle-based bioelectronic nose platform mimicking human olfactory signal transduction. *Biosensors and Bioelectronics, 35*(1), 335–341. https://doi.org/10.1016/j.bios.2012.03.012.

Jothi, L., Jayakumar, N., Jaganathan, S. K., & Nageswaran, G. (2018). Ultrasensitive and selective non-enzymatic electrochemical glucose sensor based on hybrid material of graphene nanosheets/graphene nanoribbons/nickel nanoparticle. *Materials Research Bulletin, 98*, 300–307. https://doi.org/10.1016/j.materresbull.2017.10.020.

Kashiwada, T., Watanabe, T., Ootani, Y., Tateyama, Y., & Einaga, Y. (2016). A study on electrolytic corrosion of boron-doped diamond electrodes when decomposing organic compounds. *ACS Applied Materials and Interfaces, 8*(42), 28299–28305. https://doi.org/10.1021/acsami.5b11638.

Kim, T. H., Lee, S. H., Lee, J., Song, H. S., Oh, E. H., Park, T. H., et al. (2009). Single-carbon-atomic-resolution detection of odorant molecules using a human olfactory receptor-based bioelectronic nose. *Advanced Materials, 21*(1), 91–94. https://doi.org/10.1002/adma.200801435.

Kim, P., Shi, L., Majumdar, A., & McEuen, P. L. (2001). Thermal transport measurements of individual multiwalled nanotubes. *Physical Review Letters, 87*(21), 2155021–2155024.

Kong, J., Franklin, N. R., Zhou, C., Chapline, M. G., Peng, S., Cho, K., et al. (2000). Nanotube molecular wires as chemical sensors. *Science, 287*(5453), 622–625. https://doi.org/10.1126/science.287.5453.622.

Kovalenko, I., Bucknall, D. G., & Yushin, G. (2010). Detonation nanodiamond and onion-like-carbon-embedded polyaniline for supercapacitors. *Advanced Functional Materials, 20*(22), 3979–3986. https://doi.org/10.1002/adfm.201000906.

Kreyling, W. G., Semmler-Behnke, M., & Chaudhry, Q. (2010). A complementary definition of nanomaterial. *Nano Today, 5*(3), 165–168. https://doi.org/10.1016/j.nantod.2010.03.004.

Kroto, H. W., Heath, J. R., O'Brien, S. C., Curl, R. F., & Smalley, R. E. (1985). C60: Buckminsterfullerene. *Nature, 318*(6042), 162–163. https://doi.org/10.1038/318162a0.

Krueger, A. (2008). New carbon materials: Biological applications of functionalized nanodiamond materials. *Chemistry - A European Journal, 14*(5), 1382–1390. https://doi.org/10.1002/chem.200700987.

Kumar, S., Nehra, M., Kedia, D., Dilbaghi, N., Tankeshwar, K., & Kim, K. H. (2019). Nanodiamonds: Emerging face of future nanotechnology. *Carbon, 143*, 678–699. https://doi.org/10.1016/j.carbon.2018.11.060.

Lee, M., Jung, J. W., Kim, D., Ahn, Y. J., Hong, S., & Kwon, H. W. (2015). Discrimination of umami Tastants using floating electrode-based bioelectronic tongue mimicking insect taste systems. *ACS Nano, 9*(12), 11728–11736. https://doi.org/10.1021/acsnano.5b03031.

Lee, C. H., Lee, E. S., Lim, Y. K., Park, K. H., Park, H. D., & Lim, D. S. (2017). Enhanced electrochemical oxidation of phenol by boron-doped diamond nanowire electrode. *RSC Advances, 7*(11), 6229–6235. https://doi.org/10.1039/c6ra26287b.

Lee, G., Shim, J. H., Kang, H., Nam, K. M., Song, H., & Park, J. T. (2009). Monodisperse Pt and PtRu/C60 hybrid nanoparticles for fuel cell anode catalysts. *Chemical Communications, 33*, 5036–5038. https://doi.org/10.1039/b911068b.

Li, Y., Hodak, M., Lu, W., & Bernholc, J. (2017). Selective sensing of ethylene and glucose using carbon-nanotube-based sensors: An: Ab initio investigation. *Nanoscale, 9*(4), 1687–1698. https://doi.org/10.1039/c6nr07371a.

Li, G., Liao, J. M., Hu, G. Q., Ma, N. Z., & Wu, P. J. (2005). Study of carbon nanotube modified biosensor for monitoring total cholesterol in blood. *Biosensors and Bioelectronics, 20*(10), 2140–2144. Elsevier Ltd https://doi.org/10.1016/j.bios.2004.09.005.

Li, J., Lu, Y., Ye, Q., Cinke, M., Han, J., & Meyyappan, M. (2003). Carbon nanotube sensors for gas and organic vapor detection. *Nano Letters, 3*(7), 929–933. https://doi.org/10.1021/nl034220x.

Li, H., Zhang, Y., Wang, L., Tian, J., & Sun, X. (2011). Nucleic acid detection using carbon nanoparticles as a fluorescent sensing platform. *Chemical Communications, 47*(3), 961–963. https://doi.org/10.1039/c0cc04326e.

Li, Y., Zhao, Y., Cheng, H., Hu, Y., Shi, G., Dai, L., et al. (2012). Nitrogen-doped graphene quantum dots with oxygen-rich functional groups. *Journal of the American Chemical Society, 134*(1), 15–18. https://doi.org/10.1021/ja206030c.

Li, S., Zhou, S., Li, Y., Li, X., Zhu, J., Fan, L., et al. (2017). Exceptionally high payload of the IR780 iodide on folic acid-functionalized graphene quantum dots for targeted Photothermal therapy. *ACS Applied Materials and Interfaces, 9*(27), 22332–22341. https://doi.org/10.1021/acsami.7b07267.

Liao, S., Zhao, X., Zhu, F., Chen, M., Wu, Z., Song, X., et al. (2018). Novel S, N-doped carbon quantum dot-based "off-on" fluorescent sensor for silver ion and cysteine. *Talanta, 180*, 300–308. https://doi.org/10.1016/j.talanta.2017.12.040.

Lim, S. Y., Shen, W., & Gao, Z. (2015). Carbon quantum dots and their applications. *Chemical Society Reviews, 44*(1), 362–381. https://doi.org/10.1039/c4cs00269e.

Lin, Y., Lu, F., Tu, Y., & Ren, Z. (2004). Glucose biosensors based on carbon nanotube nanoelectrode ensembles. *Nano Letters, 4*(2), 191–195. https://doi.org/10.1021/nl0347233.

Lisichkin, G. V., Korol'kov, V. V., Tarasevich, B. N., Kulakova, I. I., & Karpukhin, A. V. (2006). Photochemical chlorination of nanodiamond and interaction of its modified surface with C-nucleophiles. *Russian Chemical Bulletin, 55*(12), 2212–2219. https://doi.org/10.1007/s11172-006-0574-7.

Liu, S. F., Lin, S., & Swager, T. M. (2016). An Organocobalt-carbon nanotube chemiresistive carbon monoxide detector. *ACS Sensors, 1*(4), 354–357. https://doi.org/10.1021/acssensors.6b00005.

Liu, S. F., Petty, A. R., Sazama, G. T., & Swager, T. M. (2015). Single-walled carbon nanotube/metalloporphyrin composites for the chemiresistive detection of amines and meat spoilage. *Angewandte Chemie, International Edition, 54*(22), 6554–6557. https://doi.org/10.1002/anie.201501434.

Liu, Y. L., & Sun, K. W. (2010). Protein functionalized nanodiamond arrays. *Nanoscale Research Letters, 5*(6), 1045–1050.

Liu, Z., Robinson, J. T., Sun, X., & Dai, H. (2008). PEGylated nanographene oxide for delivery of water-insoluble cancer drugs. *Journal of the American Chemical Society, 130*(33), 10876–10877. https://doi.org/10.1021/ja803688x.

Lokesh, S. V., Sherigara, B. S., Jayadev, Mahesh, H. M., & Mascarenhas, R. J. (2008). Electrochemical reactivity of C60 modified carbon paste electrode by physical vapor deposition method. *International Journal of Electrochemical Science, 3*(5), 578–587. http://www.electrochemsci.org/papers/vol3/3050578.pdf.

Lu, C., Li, Y., Tian, S., Li, W., Li, J., & Gu, C. (2011). Enhanced gas-sensing by diamond nanoneedle arrays formed by reactive ion etching. *Microelectronic Engineering, 88*(8), 2319–2321. https://doi.org/10.1016/j.mee.2011.02.074.

Maiti, D., Tong, X., Mou, X., & Yang, K. (1401). Carbon-Based Nanomaterials for Biomedical Applications: A Recent Study. *Frontiers in Pharmacology, 2019*.

Maroto, E. E., de Cózar, A., Filippone, S., Martín-Domenech, Á., Suarez, M., Cossío, F. P., et al. (2011). Hierarchical selectivity in fullerenes: Site-, Regio-, Diastereo-, and Enantiocontrol of the 1,3-dipolar cycloaddition to C70. *Angewandte Chemie International Edition, 50*(27), 6060–6064. https://doi.org/10.1002/anie.201101246.

Mehdi Aghaei, S., Monshi, M. M., Torres, I., Zeidi, S. M. J., & Calizo, I. (2018). DFT study of adsorption behavior of NO, CO, NO_2, and NH_3 molecules on graphene-like BC 3: A search for highly sensitive molecular sensor. *Applied Surface Science, 427*, 326–333. https://doi.org/10.1016/j.apsusc.2017.08.048.

Miller, B. S., Bezinge, L., Gliddon, H. D., Huang, D., Dold, G., Gray, E. R., et al. (2020). Spin-enhanced nanodiamond biosensing for ultrasensitive diagnostics. *Nature, 587*(7835), 588–593. https://doi.org/10.1038/s41586-020-2917-1.

Mintmire, J. W., Dunlap, B. I., & White, C. T. (1992). Are fullerene tubules metallic? *Physical Review Letters*, *68*(5), 631–634. https://doi.org/10.1103/PhysRevLett.68.631.

Mochalin, V. N., Shenderova, O., Ho, D., & Gogotsi, Y. (2012). The properties and applications of nanodiamonds. *Nature Nanotechnology*, *7*(1), 11–23. https://doi.org/10.1038/nnano.2011.209.

Moncea, O., Casanova-Chafer, J., Poinsot, D., Ochmann, L., Mboyi, C. D., Nasrallah, H. O., et al. (2019). Diamondoid nanostructures as sp(3)-carbon-based gas sensors. *Angewandte Chemie*, *58*(29), 9933–9938.

Monthioux, M., & Kuznetsov, V. L. (2006). Who should be given the credit for the discovery of carbon nanotubes? *Carbon*, *44*(9), 1621–1623. https://doi.org/10.1016/j.carbon.2006.03.019.

Monti, D., Moretti, L., Salvioli, S., Straface, E., Malorni, W., Pellicciari, R., et al. (2000). C60 carboxyfullerene exerts a protective activity against oxidative stress-induced apoptosis in human peripheral blood mononuclear cells. *Biochemical and Biophysical Research Communications*, *277*(3), 711–717. https://doi.org/10.1006/bbrc.2000.3715.

Mroz, P., Pawlak, A., Satti, M., Lee, H., Wharton, T., Gali, H., et al. (2007). Functionalized fullerenes mediate photodynamic killing of cancer cells: Type I versus type II photochemical mechanism. *Free Radical Biology and Medicine*, *43*(5), 711–719. https://doi.org/10.1016/j.freeradbiomed.2007.05.005.

Mukhopadhyay, S., Maiti, D., Saha, A., & Devi, P. S. (2016). Shape transition of TiO_2 Nanocube to Nanospindle embedded on reduced graphene oxide with enhanced photocatalytic activity. *Crystal Growth & Design*, *16*(12), 6922–6932. https://doi.org/10.1021/acs.cgd.6b01096.

Murthy, C. N., Choi, S. J., & Geckeler, K. E. (2002). Nanoencapsulation of [60]fullerene by a novel sugar-based polymer. *Journal of Nanoscience and Nanotechnology*, *2*(2), 129–132. https://doi.org/10.1166/jnn.2002.096.

Nakamura, E., Isobe, H., Tomita, N., Sawamura, M., Jinno, S., & Okayama, H. (2000). Functionalized fullerene as an artificial vector for transfection. *Angewandte Chemie, International Edition*, *39*(23), 4254–4257. https://doi.org/10.1002/1521-3773(20001201)39:23<4254::AID-ANIE4254>3.0.CO;2-O.

Naldini, L., Blömer, U., Gallay, P., Ory, D., Mulligan, R., Gage, F. H., et al. (1996). In vivo gene delivery and stable transduction of nondividing cells by a lentiviral vector. *Science*, *272*(5259), 263–267. https://doi.org/10.1126/science.272.5259.263.

Novoselov, K. S., Geim, A. K., Morozov, S. V., Jiang, D., Katsnelson, M. I., Grigorieva, I. V., et al. (2005). Two-dimensional gas of massless Dirac fermions in graphene. *Nature*, *438*(7065), 197–200. https://doi.org/10.1038/nature04233.

Nunn, N., Torelli, M., McGuire, G., & Shenderova, O. (2017). Nanodiamond: A high impact nanomaterial. *Current Opinion in Solid State and Materials Science*, *21*(1), 1–9. https://doi.org/10.1016/j.cossms.2016.06.008.

Okamoto, M. (2013). Synthetic biopolymer/layered silicate nanocomposites for tissue engineering scaffolds. *Ceramic Nanocomposites*, 548–581. Elsevier Inc https://doi.org/10.1533/9780857093493.4.548.

Pan, N. Y., & Shih, J. S. (2004). Piezoelectric crystal immunosensors based on immobilized fullerene C60-antibodies. *Sensors and Actuators, B: Chemical*, *98*(2–3), 180–187. https://doi.org/10.1016/j.snb.2003.09.034.

Pan, D., Zhang, J., Li, Z., & Wu, M. (2010). Hydrothermal route for cutting graphene sheets into blue-luminescent graphene quantum dots. *Advanced Materials*, *22*(6), 734–738. https://doi.org/10.1002/adma.200902825.

Partha, R., & Conyers, J. L. (2009). Biomedical applications of functionalized fullerene-based nanomaterials. *International Journal of Nanomedicine*, *4*, 261–275.

Pasinszki, T., Krebsz, M., Tung, T. T., & Losic, D. (2017). Carbon nanomaterial based biosensors for non-invasive detection of cancer and disease biomarkers for clinical diagnosis. *Sensors*, *17*(8), 1919. https://doi.org/10.3390/s17081919.

Pilehvar, S., & Wael, K. D. (2015). Recent advances in electrochemical biosensors based on fullerene-C60 nano-structured platforms. *Biosensors*, *5*(4), 712–735. https://doi.org/10.3390/bios5040712.

Portet, C., Yushin, G., & Gogotsi, Y. (2007). Electrochemical performance of carbon onions, nanodiamonds, carbon black and multiwalled nanotubes in electrical double layer capacitors. *Carbon*, *45*(13), 2511–2518. https://doi.org/10.1016/j.carbon.2007.08.024.

Prinith, N. S., Manjunatha, J. G., & Raril, C. (2019). Electrocatalytic analysis of dopamine, uric acid and ascorbic acid at poly(adenine) modified carbon nanotube paste electrode: A cyclic voltammetric study. *Analytical and Bioanalytical Electrochemistry*, *11*(6), 742–756. http://www.abechem.com/No.%206-2019/2019,%2011(6),%20742-756.pdf.

Pushpanjali, P. A., Manjunatha, J. G., Tigari, G., & Fattepur, S. (2020). Poly(niacin) based carbon nanotube sensor for the sensitive and selective voltammetric detection of vanillin with caffeine. *Analytical and Bioanalytical Electrochemistry*, *12*(4), 553–568. http://www.abechem.com/article_39221_5b4cb9d77cdb1bfb6c48fc3dac8c7fc0.pdf.

Puzyr, A. P., Baron, A. V., Purtov, K. V., Bortnikov, E. V., Skobelev, N. N., Mogilnaya, O. A., et al. (2007). Nanodiamonds with novel properties: A biological study. *Diamond and Related Materials*, *16*(12), 2124–2128. https://doi.org/10.1016/j.diamond.2007.07.025.

Rambabu, G., & Bhat, S. D. (2015). Sulfonated fullerene in SPEEK matrix and its impact on the membrane electrolyte properties in direct methanol fuel cells. *Electrochimica Acta*, *176*, 657–669. https://doi.org/10.1016/j.electacta.2015.07.045.

Rao, C. N. R., Seshadri, R., Govindaraj, A., & Sen, R. (1995). Fullerenes, nanotubes, onions and related carbon structures. *Materials Science and Engineering R*, *15*(6), 209–262. https://doi.org/10.1016/S0927-796X(95)00181-6.

Raril, C., Manjunatha, J. G., & Tigari, G. (2020). Low-cost voltammetric sensor based on an anionic surfactant modified carbon nanocomposite material for the rapid determination of curcumin in natural food supplement. *Instrumentation Science and Technology*, *48*(5), 561–582. https://doi.org/10.1080/10739149.2020.1756317.

Rutherglen, C., Jain, D., & Burke, P. (2009). Nanotube electronics for radio frequency applications. *Nature Nanotechnology*, *4*(12), 811–819. https://doi.org/10.1038/nnano.2009.355.

Saab, A. P., Stucky, G. D., Passerini, S., & Smyrl, W. H. (1998). Ionic conductivity of C60-based solid electrolyte. *Fullerene Science and Technology*, *6*(2), 227–242. https://doi.org/10.1080/10641229809350197.

Saeedfar, K., Heng, L. Y., Ling, T. L., & Rezayi, M. (2013). Potentiometric urea biosensor based on an immobilised fullerene-urease bio-conjugate. *Sensors, 13*(12), 16851–16866.

Salehi-Khojin, A., Estrada, D., Lin, K. Y., Bae, M. H., Xiong, F., Pop, E., et al. (2012). Polycrystalline graphene ribbons as chemiresistors. *Advanced Materials, 24*(1), 53–57. https://doi.org/10.1002/adma.201102663.

Salih, E., & Ayesh, A. I. (2021). Pt-doped armchair graphene nanoribbon as a promising gas sensor for CO and CO2: DFT study. *Physica E: Low-dimensional Systems and Nanostructures, 125*. https://doi.org/10.1016/j.physe.2020.114418.

Schedin, F., Geim, A. K., Morozov, S. V., Hill, E. W., Blake, P., Katsnelson, M. I., et al. (2007). Detection of individual gas molecules adsorbed on graphene. *Nature Materials, 6*(9), 652–655. https://doi.org/10.1038/nmat1967.

Schon, T. B., Dicarmine, P. M., & Seferos, D. S. (2014). Polyfullerene electrodes for high power supercapacitors. *Advanced Energy Materials, 4*(7). https://doi.org/10.1002/aenm.201301509.

Schroeder, V., Savagatrup, S., He, M., Lin, S., & Swager, T. M. (2019). Carbon nanotube chemical sensors. *Chemical Reviews, 119*(1), 599–663. https://doi.org/10.1021/acs.chemrev.8b00340.

Shen, H., Zhang, L., Liu, M., & Zhang, Z. (2012). Biomedical applications of graphene. *Theranostics, 2*(3), 283–294. https://doi.org/10.7150/thno.3642.

Shetti, N. P., Malode, S. J., & Nandibewoor, S. T. (2012). Electrochemical behavior of an antiviral drug acyclovir at fullerene-C60-modified glassy carbon electrode. *Bioelectrochemistry, 88*, 76–83. https://doi.org/10.1016/j.bioelechem.2012.06.004.

Shi, X., Zheng, Y., Wang, C., Yue, L., Qiao, K., Wang, G., et al. (2015). Dual stimulus responsive drug release under the interaction of pH value and pulsatile electric field for a bacterial cellulose/sodium alginate/multi-walled carbon nanotube hybrid hydrogel. *RSC Advances, 5*(52), 41820–41829. https://doi.org/10.1039/c5ra04897d.

Shioyama, H., Ueda, A., & Kuriyama, N. (2007). Surface treatment of carbon supports for PEM fuel cell electrocatalyst. *Journal of New Materials for Electrochemical Systems, 10*(4), 201–204.

Sireesha, M., Jagadeesh Babu, V., Kranthi Kiran, A. S., & Ramakrishna, S. (2018). A review on carbon nanotubes in biosensor devices and their applications in medicine. *Nano, 4*(2), 36–57. https://doi.org/10.1080/20550324.2018.1478765.

Snow, E. S., Perkins, F. K., Houser, E. J., Badescu, S. C., & Reinecke, T. L. (2005). Chemical detection with a single-walled carbon nanotube capacitor. *Science, 307*(5717), 1942–1945. https://doi.org/10.1126/science.1109128.

Son, M., Cho, D. G., Lim, J. H., Park, J., Hong, S., Ko, H. J., et al. (2015). Real-time monitoring of geosmin and 2-methylisoborneol, representative odor compounds in water pollution using bioelectronic nose with human-like performance. *Biosensors and Bioelectronics, 74*, 199–206. https://doi.org/10.1016/j.bios.2015.06.053.

Souza, D., Manjunatha, E. S., Raril, J., Tigari, C., Ravishankar, G., & Fattepur, D. K. (2020). Arginine electropolymerized carbon nanotube paste electrode as sensitive and selective sensor for electrochemical determination of vanillin. *Journal of Materials and Environmental Science, 11*(3), 512–521.

Srivastava, M., Kumar, M., Singh, R., Agrawal, U. C., & Garg, M. O. (2009). Energy-related applications of carbon materials-a review. *Journal of Scientific and Industrial Research, 68*(2), 93–96.

Tasaki, K., DeSousa, R., Wang, H., Gasa, J., Venkatesan, A., Pugazhendhi, P., et al. (2006). Fullerene composite proton conducting membranes for polymer electrolyte fuel cells operating under low humidity conditions. *Journal of Membrane Science, 281*(1–2), 570–580. https://doi.org/10.1016/j.memsci.2006.04.052.

Tian, W., Liu, X., & Yu, W. (2018). Research progress of gas sensor based on graphene and its derivatives: A review. *Applied Sciences, 8*(7), 1118. https://doi.org/10.3390/app8071118.

Tigari, G., & Manjunatha, J. G. (2020a). A surfactant enhanced novel pencil graphite and carbon nanotube composite paste material as an effective electrochemical sensor for determination of riboflavin. *Journal of Science: Advanced Materials and Devices, 5*(1), 56–64. https://doi.org/10.1016/j.jsamd.2019.11.001.

Tigari, G., & Manjunatha, J. G. (2020b). Optimized Voltammetric experiment for the determination of phloroglucinol at surfactant modified carbon nanotube paste electrode. *Instruments and Experimental Techniques, 63*(5), 750–757. https://doi.org/10.1134/S0020441220050139.

Tigari, G., Manjunatha, J. G., Raril, C., & Hareesha, N. (2019). Determination of riboflavin at carbon nanotube paste electrodes modified with an anionic surfactant. *ChemistrySelect, 4*(7), 2168–2173. https://doi.org/10.1002/slct.201803191.

Tîlmaciu, C.-M., & Morris, M. C. (2015). Carbon nanotube biosensors. *Frontiers in Chemistry, 3*.

Tiwari, J. N., Vij, V., Kemp, K. C., & Kim, K. S. (2015). Engineered carbon-nanomaterial-based electrochemical sensors for biomolecules. *ACS Nano*, 46–80. https://doi.org/10.1021/acsnano.5b05690.

Trotter, M., Borst, N., Thewes, R., & von Stetten, F. (2020). Review: Electrochemical DNA sensing – Principles, commercial systems, and applications. *Biosensors and Bioelectronics, 154*. https://doi.org/10.1016/j.bios.2020.112069.

Updike, S. J., & Hicks, G. P. (1967). The enzyme electrode. *Nature, 214*(5092), 986–988. https://doi.org/10.1038/214986a0.

Varghese, O. K., Kichambre, P. D., Gong, D., Ong, K. G., Dickey, E. C., & Grimes, C. A. (2001). Gas sensing characteristics of multi-wall carbon nanotubes. *Sensors and Actuators, B: Chemical, 81*(1), 32–41. https://doi.org/10.1016/S0925-4005(01)00923-6.

Vinodgopal, K., Haria, M., Meisel, D., & Kamat, P. (2004). Fullerene-based carbon nanostructures for methanol oxidation. *Nano Letters, 4*(3), 415–418. https://doi.org/10.1021/nl035028y.

Wang, H., DeSousa, R., Gasa, J., Tasaki, K., Stucky, G., Jousselme, B., et al. (2007). Fabrication of new fullerene composite membranes and their application in proton exchange membrane fuel cells. *Journal of Membrane Science, 289*(1–2), 277–283. https://doi.org/10.1016/j.memsci.2006.12.008.

Wang, Y., & Hu, A. (2014). Carbon quantum dots: Synthesis, properties and applications. *Journal of Materials Chemistry C, 2*(34), 6921–6939. https://doi.org/10.1039/c4tc00988f.

Wang, T., Huang, D., Yang, Z., Xu, S., He, G., Li, X., et al. (2016). A review on graphene-based gas/vapor sensors with unique properties and potential applications. *Nano-Micro Letters, 8*(2), 95–119. https://doi.org/10.1007/s40820-015-0073-1.

Wasilewski, T., Gębicki, J., & Kamysz, W. (2021). Bio-inspired approaches for explosives detection. *TrAC Trends in Analytical Chemistry, 142*, 116330. https://doi.org/10.1016/j.trac.2021.116330.

Wehling, J., Dringen, R., Zare, R. N., Maas, M., & Rezwan, K. (2014). Bactericidal activity of partially oxidized nanodiamonds. *ACS Nano, 8*(6), 6475–6483. https://doi.org/10.1021/nn502230m.

Wei, B. Y., Hsu, M. C., Su, P. G., Lin, H. M., Wu, R. J., & Lai, H. J. (2004). A novel SnO2 gas sensor doped with carbon nanotubes operating at room temperature. *Sensors and Actuators, B: Chemical, 101*(1–2), 81–89. https://doi.org/10.1016/j.snb.2004.02.028.

Wei, L. F., & Shih, J. S. (2001). Fullerene-cryptand coated piezoelectric crystal urea sensor based on urease. *Analytica Chimica Acta, 437*(1), 77–85. https://doi.org/10.1016/S0003-2670(01)00941-2.

Wei, L., & Yushin, G. (2012). Nanostructured activated carbons from natural precursors for electrical double layer capacitors. *Nano Energy, 1*(4), 552–565. https://doi.org/10.1016/j.nanoen.2012.05.002.

Wierzbicki, M., Hotowy, A., Kutwin, M., Jaworski, S., Bałaban, J., Sosnowska, M., et al. (2020). Graphene oxide scaffold stimulates differentiation and proangiogenic activities of myogenic progenitor cells. *International Journal of Molecular Sciences, 21*(11), 4173.

Wu, J., Feng, S., Wei, X., Shen, J., Lu, W., Shi, H., et al. (2016). Facile synthesis of 3D graphene flowers for ultrasensitive and highly reversible gas sensing. *Advanced Functional Materials, 26*(41), 7462–7469. https://doi.org/10.1002/adfm.201603598.

Xu, B. (2008). Prospects and research progress in nano onion-like fullerenes. *New Carbon Materials, 23*(4), 289–301. https://doi.org/10.1016/s1872-5805(09)60001-9.

Xu, X., Ray, R., Gu, Y., Ploehn, H. J., Gearheart, L., Raker, K., et al. (2004). Electrophoretic analysis and purification of fluorescent single-walled carbon nanotube fragments. *Journal of the American Chemical Society, 126*(40), 12736–12737. https://doi.org/10.1021/ja040082h.

Yang, S. T., Cao, L., Luo, P. G., Lu, F., Wang, X., Wang, H., et al. (2009). Carbon dots for optical imaging in vivo. *Journal of the American Chemical Society, 131*(32), 11308–11309. https://doi.org/10.1021/ja904843x.

Yang, L., & Webster, T. J. (2014). Small and bright: Nanodiamonds for tissue repair, drug delivery, and biodetection. *IEEE Pulse, 5*(2), 34–39. https://doi.org/10.1109/MPUL.2013.2296800.

Yeow, J. T. W., & She, J. P. M. (2006). Carbon nanotube-enhanced capillary condensation for a capacitive humidity sensor. *Nanotechnology, 17*(21), 5441–5448. https://doi.org/10.1088/0957-4484/17/21/026.

Yin, J. J., Lao, F., Fu, P. P., Wamer, W. G., Zhao, Y., Wang, P. C., et al. (2009). The scavenging of reactive oxygen species and the potential for cell protection by functionalized fullerene materials. *Biomaterials, 30*(4), 611–621. https://doi.org/10.1016/j.biomaterials.2008.09.061.

Yu, M. F., Lourie, O., Dyer, M. J., Moloni, K., Kelly, T. F., & Ruoff, R. S. (2000). Strength and breaking mechanism of multiwalled carbon nanotubes under tensile load. *Science, 287*(5453), 637–640. https://doi.org/10.1126/science.287.5453.637.

Yu, K., Wang, P., Lu, G., Chen, K. H., Bo, Z., & Chen, J. (2011). Patterning vertically oriented graphene sheets for nanodevice applications. *Journal of Physical Chemistry Letters, 2*(6), 537–542. https://doi.org/10.1021/jz200087w.

Yu, S., Yang, N., Zhuang, H., Mandal, S., Williams, O. A., Yang, B., et al. (2017). Battery-like supercapacitors from diamond networks and water-soluble redox electrolytes. *Journal of Materials Chemistry A, 5*(4), 1778–1785. https://doi.org/10.1039/c6ta08607a.

Zaghmarzi, F. A., Zahedi, M., Mola, A., Abedini, S., Arshadi, S., Ahmadzadeh, S., et al. (2017). Fullerene-C60 and crown ether doped on C60 sensors for high sensitive detection of alkali and alkaline earth cations. *Physica E: Low-dimensional Systems and Nanostructures, 87*, 51–58. https://doi.org/10.1016/j.physe.2016.11.031.

Zaporotskova, I. V., Boroznina, N. P., Parkhomenko, Y. N., & Kozhitov, L. V. (2016). Carbon nanotubes: Sensor properties. A review. *Electronic Materials*, 95–105. https://doi.org/10.1016/j.moem.2017.02.002.

Zaytseva, O., & Neumann, G. (2016). Carbon nanomaterials: Production, impact on plant development, agricultural and environmental applications. *Chemical and Biological Technologies in Agriculture, 3*(1). https://doi.org/10.1186/s40538-016-0070-8.

Zhang, J., Boghossian, A. A., Barone, P. W., Rwei, A., Kim, J. H., Lin, D., et al. (2011). Single molecule detection of nitric oxide enabled by d(AT)15 DNA adsorbed to near infrared fluorescent single-walled carbon nanotubes. *Journal of the American Chemical Society, 133*(3), 567–581. https://doi.org/10.1021/ja1084942.

Zhang, W., Guo, Z., Huang, D., Liu, Z., Guo, X., & Zhong, H. (2011). Synergistic effect of chemo-photothermal therapy using PEGylated graphene oxide. *Biomaterials, 32*(33), 8555–8561. https://doi.org/10.1016/j.biomaterials.2011.07.071.

Zhang, Z., Huang, Y., Ding, W., & Li, G. (2014). Multilayer interparticle linking hybrid MOF-199 for noninvasive enrichment and analysis of plant hormone ethylene. *Analytical Chemistry, 86*(7), 3533–3540. https://doi.org/10.1021/ac404240n.

Zhang, X., & Ma, L. X. (2015). Electrochemical fabrication of platinum nanoflakes on fulleropyrrolidine nanosheets and their enhanced electrocatalytic activity and stability for methanol oxidation reaction. *Journal of Power Sources, 286*, 400–405. https://doi.org/10.1016/j.jpowsour.2015.03.175.

Zhang, X., Ma, L. X., & Zhang, Y. C. (2015). Electrodeposition of platinum nanosheets on C60 decorated glassy carbon electrode as a stable electrochemical biosensor for simultaneous detection of ascorbic acid, dopamine and uric acid. *Electrochimica Acta, 177*, 118–127. https://doi.org/10.1016/j.electacta.2015.01.202.

Zhang, Y., Rhee, K. Y., Hui, D., & Park, S. J. (2018). A critical review of nanodiamond based nanocomposites: Synthesis, properties and applications. *Composites Part B: Engineering, 143*, 19–27. https://doi.org/10.1016/j.compositesb.2018.01.028.

Zhang, W. D., & Zhang, W. H. (2009). Carbon nanotubes as active components for gas sensors. *Journal of Sensors, 2009*. https://doi.org/10.1155/2009/160698.

Zhao, J., Buldum, A., Han, J., & Lu, J. P. (2002). Gas molecule adsorption in carbon nanotubes and nanotube bundles. *Nanotechnology, 13*(2), 195–200. https://doi.org/10.1088/0957-4484/13/2/312.

Zhao, Q., Gan, Z., & Zhuang, Q. (2002). Electrochemical sensors based on carbon nanotubes. *Electroanalysis, 14*(23), 1609–1613. https://doi.org/10.1002/elan.200290000.

Zhao, Y., Liu, X., Yang, Y., Kang, L., Yang, Z., Liu, W., et al. (2015). Carbon dots: From intense absorption in visible range to excitation-independent and excitation-dependent photoluminescence. *Fullerenes, Nanotubes, and Carbon Nanostructures*, 23(11), 922–929. https://doi.org/10.1080/1536383X.2015.1018413.

Zheng, T. T., Zhang, R., Zou, L., & Zhu, J. J. (2012). A label-free cytosensor for the enhanced electrochemical detection of cancer cells using polydopamine-coated carbon nanotubes. *Analyst*, 137(6), 1316–1318. https://doi.org/10.1039/c2an16023d.

Zhong, X., Yuan, R., & Chai, Y. (2012). In situ spontaneous reduction synthesis of spherical Pd@Cys-C60 nanoparticles and its application in nonenzymatic glucose biosensors. *Chemical Communications*, 48(4), 597–599. https://doi.org/10.1039/c1cc16081h.

Zhou, L., Wang, T., Bai, Y., Li, Y., Qiu, J., Yu, W., & Zhang, S. (2020). Dual-amplified strategy for ultrasensitive electrochemical biosensor based on click chemistry-mediated enzyme-assisted target recycling and functionalized fullerene nanoparticles in the detection of microRNA-141. *Biosensors*, 150. https://doi.org/10.1016/j.bios.2019.111964.

Zhu, R., Desroches, M., Yoon, B., & Swager, T. M. (2017). Wireless oxygen sensors enabled by Fe(II)-polymer wrapped carbon nanotubes. *ACS Sensors*, 2(7), 1044–1050. https://doi.org/10.1021/acssensors.7b00327.

Zhu, Y., Yu, L., Wu, D., Lv, W., & Wang, L. (2021). A high-sensitivity graphene ammonia sensor via aerosol jet printing. *Sensors and Actuators A: Physical*, 318, 112434. https://doi.org/10.1016/j.sna.2020.112434.

Zou, X., Liu, M., Wu, J., Ajayan, P. M., Li, J., Liu, B., et al. (2017). How nitrogen-doped graphene quantum dots catalyze Electroreduction of CO_2 to hydrocarbons and oxygenates. *ACS Catalysis*, 7(9), 6245–6250. https://doi.org/10.1021/acscatal.7b01839.

Chapter 24

Prospects of carbon nanomaterial-based sensors for sustainable future

P. Karpagavinayagam[a], J. Antory Rajam[b], R. Baby Suneetha[a], and C. Vedhi[a]
[a]*Department of Chemistry, V.O. Chidambaram College, Tuticorin, India,* [b]*St. Mary's College, Tuticorin, India*

Introduction

The energy transition to meet the world's needs for electricity, heating, cooling, and transport sustainably is one of the greatest challenges facing humanity in the 21st century. To address and meet the challenges presented by the global energy demand of an ever-increasing population and associated environmental consequence of a fossil fuel-based society, very innovative, efficient, and increasingly sustainable solutions must be found. In this regard, there are several options by which to generate and store renewable energy from sun, wind, water, geothermal, or biomass—all of which are heavily reliant on materials' technologies to allow for their efficient implementation. Therefore, the development of sustainable materials is extremely important. This is not a trivial task as these new materials should also ideally be low cost, scalable, industrially and economically attractive, and based on renewable and highly abundant resources, while of course achieving application performances in renewable energy conversion or environmental applications that exceed existing technologies.

Even though carbon dioxide is found in small traces in the atmosphere, it plays a vital role in balancing the energy and traps the long-wave radiations from the sun. Therefore, it acts like a blanket over the planet. If the carbon cycle is disturbed, it will result in serious consequences such as climatic changes and global warming.

Carbon-based structures are the most novel materials used in the modern fields of environmental science (e.g., purification/remediation), medical (e.g., diagnostic and treatment), electronics and communication, renewable energy (in both generation and storage), and biology. However, there is a need and indeed a desire to develop increasingly more sustainable variants of classical carbon materials (e.g., activated carbons, carbon nanotubes, carbon aerogels), particularly when the whole life cycle is considered (i.e., from precursor "cradle" to "green" manufacturing and the product end-of-life "grave").

Carbon-based nanomaterials include fullerenes, carbon nanotubes, graphene, and its derivatives, graphene oxide, nanodiamonds, and carbon-based quantum dots.

The natural carbon cycles/production, utilization of natural, abundant, and more renewable precursors, coupled with simpler, lower energy synthetic processes that can contribute in part to the reduction in greenhouse gas emissions or the use of toxic elements, can be considered as crucial parameters in the development of sustainable material manufacturing. Therefore, the synthesis and application of sustainable carbon materials are receiving increasing levels of interest, particularly as application benefits in the context of future energy/chemical industries are becoming recognized. These sustainable carbon materials are used as sensors in environmental, medical, and energy-related fields (Titirici et al., 2015; Xia, Otyepka, Li, Liu, & Zheng, 2015).

Porous carbon materials are applied in many fields including renewable energy storage and generation, due to their superior physical properties and availability. The environmentally friendly production of these materials is crucial for a sustainable future.

Carbon and its transformation

Carbon is an important element of life. It can combine with itself and other elements in three types of hybridizations. This gives the rich diversity of structural forms of solid carbon and is the basis of organic chemistry and life. Carbon can exist in more than one form. Diamond, graphite, carbon nanotube, and fullerene are the elemental forms of carbon. It is also found in combined state as carbonates in minerals and as carbon dioxide gas in the atmosphere (Manjunatha, 2016, 2017; Manjunatha, 2020a, 2020b).

Carbon transformation

To maintain the level of carbon compounds in the earth, it is essential to undergo a regular process of recycling. This process of recycling through various transformations is brought about by the carbon cycle. The dynamics of carbon transformations and transport in soil are complex and can result in sequestration of carbon in the soil as organic matter or in groundwater as dissolved carbonates, increased emissions of CO_2 to the atmosphere, or export of carbon in various forms into aquatic systems.

The carbon cycle is a biogeochemical cycle, in which various carbon compounds are interchanged among the various layers of the earth, namely the biosphere, geosphere, pedosphere, hydrosphere, and atmosphere. The natural carbon cycle is the flow of carbon naturally throughout the globe in various forms, such as carbonates, carbon dioxide, or methane (Climate Change Science Program and the Subcommittee on Global Change Research, 2003; Falkowski et al., 2000; Loh & Baillargeat, 2013).

The natural carbon cycle is kept very nearly in balance. Carbon enters the atmosphere through CO_2 through the respiration of plants and animals, while this CO_2 is absorbed by autotrophs such as green plants or green algae from the atmosphere (or from water) through photosynthesis. Animals incorporate carbon into their system by consuming plants. Dead and decaying organic matter may ferment and release CO_2 or methane (CH_4) or may be incorporated into sedimentary rock, where it is converted to fossil fuels. The burning of coal, hydrocarbon fuels, and fossil fuels returns CO_2 and water (H_2O) to the atmosphere (Smith et al., 1993) (Fig. 24.1).

Types of carbon nanomaterials (CNMs)

Carbon nanomaterials are classified into many types based on their structure. Carbon-based nanomaterials include (Ahmad & Pan, 2015; Cha, Shin, Annabi, Dokmeci, & Khademhosseini, 2013; Sharon & Sharon, 2010):

- Fullerenes
- Carbon nanotubes (CNTs)
- Carbon nanofibers (CNFs)
- Graphene and its derivatives
- Graphene oxide
- Nanodiamonds
- Carbon-based quantum dots
- Nano-onions
- Nanocones

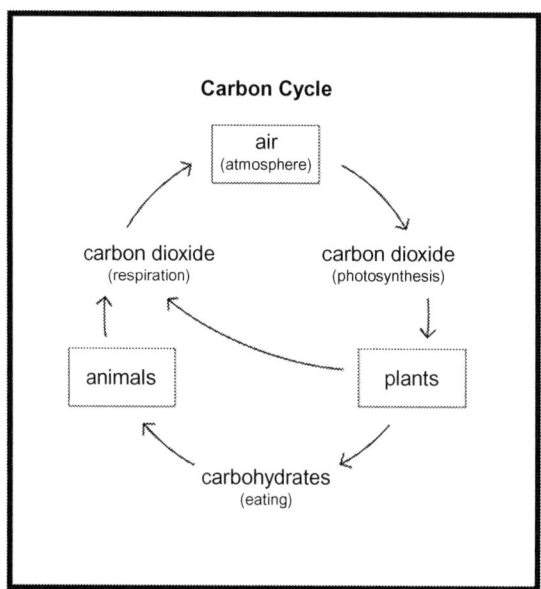

FIG. 24.1 Schematic diagram of the carbon cycle.

- Nanohorns
- Nanobeads
- Nanofibers

Sensors, carbon-based sensors

A sensor is a device, machine, module, or subsystem that detects the events or the changes in the environment responds and gives the output to other electronics, usually a computer processor. A sensorconverts a physical phenomenon into a measurable analog voltage (or a digital signal) converted into a human-readable display or transmitted for reading or further processing. The sensor is also defined as a device used to measure a property, such as pressure, position, temperature, or acceleration, and respond with feedback.

Types of sensors

Different types of sensors are applied in a wide range of industries, including automotive, medical, appliance, aerospace and defense, and industrial and commercial transportation.

The various types of sensors are:

- Chemical sensors
- Electrochemical sensors
- Biosensors
- Biochemical sensors
- Image sensors
- Monitoring sensors

Carbon-based sensors

Sensors designed by basic units of carbon are called carbon-based sensors. Chemically modified glassy carbon electrodes play an important role in fuel cells, oxygen sensors, and biosensors. Glassy carbon electrodes coated with inorganic compounds, polymers, metal nanoparticles, etc., are used to detect the evolution of gases such as oxygen and carbon dioxide; chemicals; diseases; and so on.

Carbon nanotube sensors

Gas sensors based on carbon nanotubes are known as carbon nanotube sensors. As a nanotube is a surface structure, its whole weight is concentrated on the surface of its layers. This feature is the origin of the uniquely large unit surface of turbulence that in turn predetermines their electrochemical and adsorption properties.

Nowadays, increasing medical and biological interest in cheap disposable, analytical, and diagnostic devices has driven research toward the development and adaptation of low-cost electronic sensing devices. Organic semiconductors can be used in flexible devices, including those built on biodegradable and resorbable substrates, while carbon-based nanostructured materials, such as carbon nanotubes and graphene sheets, offer higher performance in terms of field-effect, mobility, and sensitivity. There is an active layer material that allows for the inclusion of biologically active bioreceptors while maintaining electronic performance.

The unique chemical and physical properties of CNT have paved the way to new and improved sensing devices, in general, and electrochemical biosensors, in particular. CNT-based electrochemical transducers offer substantial improvements in the performance of amperometric enzyme electrodes, immune sensors, and nucleic-acid sensing devices. The greatly enhanced electrochemical reactivity of hydrogen peroxide and NADH at CNT-modified electrodes makes these nanomaterials extremely attractive for numerous oxidize- and dehydrogenize-based amperometric biosensors. Aligned CNT "forests" can act as molecular wires to allow efficient electron transfer between the underlying electrode and the redox centers of enzymes. Bioaffinity devices utilizing enzyme tags can greatly benefit from the enhanced response of the biocatalytic-reaction product at the CNT transducer and from CNT amplification platforms carrying multiple tags. The successful realization of CNT-based biosensors requires proper control of their chemical and physical properties, as well as their functionalization and surface immobilization (Manjunatha & Deraman, 2017; Manjunatha, Kumara Swamy, Deraman, & Mamatha, 2012).

Carbon nanofiber sensors

Carbon nanofibers (CNFs) are novel carbon nanomaterials having similar conductivity and stability to carbon nanotubes (CNTs). The important distinguishing characteristic of CNFs from CNTs is the stacking of graphene sheets of varying shapes. This produces more edge sites on the outer wall of CNFs than CNTs, which can promote the electron transfer of electroactive analytes. And the unique physical and chemical properties make CNFs exceptional and promising materials for electrode and as immobilization substrates respectively. Using CNFs as immobilization matrixes, various oxidase-, dehydrogenase-, and other enzyme-based biosensors have been successfully constructed. These biosensors exhibited high sensitivity, and the enzymatic activity was efficiently maintained. The use of CNF molecular wires promoted the direct electron transfer from electrode surfaces to the redox sites of enzymes. We discussed the application of CNFs as electrode material in electroanalysis, as well as their functionalization and surface immobilization. Vertically aligned carbon nanofibers (VACNFs) are used as substrates for the immobilization of biological molecules. VACNF substrates could be effectively functionalized with biomolecules such as DNA and protein either through a photochemical route or by a combined chemical and electrochemical route. These molecular fictionalizations of VACNFs yielded structures with excellent chemical and biological properties, making them very useful for applications such as chemical sensing/biosensing (Titirici et al., 2015) (Fig. 24.2).

Sustainable carbon materials

Sustainability means a limited negative impact on natural resources and people. It avoids degrading or depleting the environment. Materials are called sustainable when they do not harm the environment during their production, usage, or disposal. And also the materials produced by low cost, and resource-saving processes in large-scale production are sustainable materials. High-performing carbon materials are crucial for a sustainable future.

It is desirable for a sustainable society that the production and utilization of renewable materials are net-zero in terms of carbon emissions. Carbon materials with emerging applications in CO_2 utilization, renewable energy storage and conversion, and biomedicine have attracted much attention. However, the preparation process of some new carbon materials suffers from energy consumption and environmental pollution issues. Therefore, the development of low-cost, scalable, industrially and economically attractive, sustainable carbon material preparation methods is required. In this regard, the use of biomass and its derivatives as a precursor of carbon materials is a major feature of sustainability. Recently, a number of sustainable carbon materials have been designed and applied in many fields.

The production of functional nanostructured materials starting from cheap natural precursors using environmentally benign processes is a highly attractive subject in material chemistry today. Recently, much attention has been focused on the use of plant biomass to produce functional carbonaceous materials, encompassing economic, environmental, and social issues. A variety of cheap and sustainable carbonaceous materials with attractive nanostructure and functionalization patterns for a wide range of applications should be synthesized for a sustainable future (Manjunatha et al., 2009; Manjunatha et al., 2010; Manjunatha, Kumara Swamy, Shreenivas, & Mamatha, 2012).

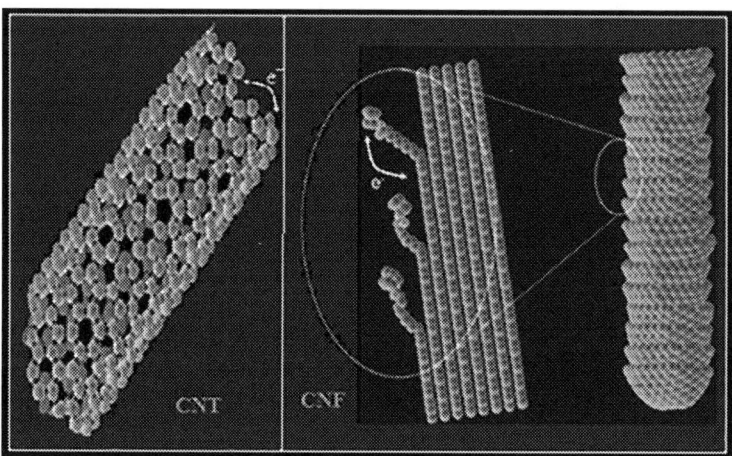

FIG. 24.2 Schematic representation of electron transfer in CNTs and CNFs.

Carbon quantum dots

Quantum dots (QDs) are man-made nanoscale rock crystals that can be carrying electrons. When ultraviolet (UV) light hits these semiconducting nanoparticles, they can emit light of various colors. These synthetic semiconductor NPs have found applications in composites, solar cells, and fluorescent biological labels. Nanoparticles of semiconductors—quantum dots—were hypothesized in the 1970s and firstly formed in the early 1980s. If semiconductor elements are prepared small enough, quantum effects come into play, which limits the energies at which electrons and holes (the absence of an electron) can exist in the particles. As energy is related to wavelength (or color), this means that the photosensitive properties of the particle can be finely tuned depending on its dimensions. Thus, particles can be made to discharge or engage specific wavelengths (colors) of light, merely by controlling their dimensions.

QDs are simulated nanostructures that can possess many varied properties, depending on their material and shape. For instance, due to their particular electronic properties, they can be used as active materials in single-electron transistors (Xie et al., 2016).

The properties of a quantum dot are not only determined by its size but also by its shape, composition, and structure, for instance, if it is solid or hollow. A reliable manufacturing technology that makes use of quantum dots' properties—for a wide-ranging number of applications in such areas as catalysis, electronics, photonics, information storage, imaging, medicine, or sense—needs to be capable of churning out large quantities of nanocrystals where each batch is produced according to the same parameters. Because certain biological molecules are capable of molecular recognition and self-assembly, Nanocrystals could also become an important building block for self-assembled functional nanodevices. The atom-like energy states of QDs furthermore contribute to special optical properties, such as a particle-size-dependent wavelength of fluorescence, an effect that is used in fabricating optical probes for biological and medical imaging.

So far, the use in bioanalytics and biolabeling has found the widest range of applications for colloidal QDs. Though the first generation of quantum dots already pointed out their potential, it took a lot of effort to improve basic properties, in particular colloidal stability in salt-containing solutions. Initially, quantum dots have been used in very artificial environments, and these particles would have simply precipitated in "real" samples, such as blood. These problems have been solved and QDs have found numerous uses in real applications. Quantum dots have found applications in composites, solar cells (Grätzel cells), and fluorescent biological labels (e.g., to trace a biological molecule) that use both the small particle size and tunable energy levels.

Advances in chemistry have resulted in the preparation of monolayer-protected, high-quality, monodispersed, crystalline quantum dots as small as 2 nm in diameter, which can be conveniently treated and processed as a typical chemical reagent (Kumawat, Thakur, Gurung, & Srivastava, 2013) (Fig. 24.3).

Quantum dots in medicine

Quantum dots enable researchers to study cell processes at the level of a single molecule and may significantly improve the diagnosis and treatment of diseases such as cancers. QDs are either used as active sensor elements in high-resolution cellular imaging, where the fluorescence properties of the quantum dots are changed upon reaction with the analyte, or in passive label probes where selective receptor molecules such as antibodies have been conjugated to the surface of the dots.

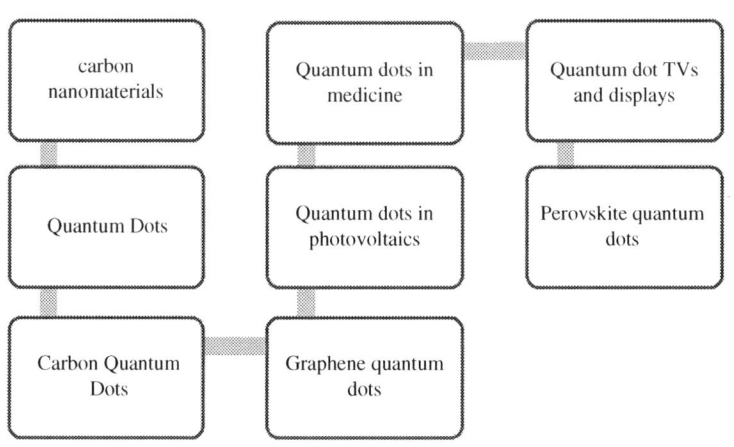

FIG. 24.3 Schematic representation of CNMs and QDs.

Quantum dots could revolutionize medicine. Unfortunately, most of them are toxic. Ironically, the existence of heavy metals in QDs such as cadmium, a well-established human toxicant and carcinogen, poses potential dangers, especially for future medical applications, where Qdots are deliberately injected into the body. As the use of nanomaterials for biomedical applications is increasing, environmental pollution and toxicity have to be addressed, and the development of a nontoxic and biocompatible nanomaterial is becoming an important issue (Wang, 2005; Zhang et al., 2017).

Quantum dots in photovoltaics

The attractiveness of using quantum dots for making solar cells lies in several advantages over other approaches: They can be manufactured in an energy-saving room-temperature process; they can be made from abundant, inexpensive materials that do not require extensive purification, as silicon does; and they can be applied to a variety of inexpensive and even flexible substrate materials, such as lightweight plastics.

Although using quantum dots as the basis for solar cells is not a new idea, attempts to make photovoltaic devices have not yet achieved sufficiently high efficiency in converting sunlight to power. A promising route for quantum dot solar cells is a semiconductor ink with the goal of enabling the coating of large areas of solar cell substrates in a single deposition step and thereby eliminating tens of deposition steps necessary with the previous layer-by-layer method (Manjunatha, Kumara Swamy, Shreenivas, & Mamatha, 2012; Raril & Manjunatha, 2020).

Graphene quantum dots

Graphene, which basically is an unrolled, planar form of a carbon nanotube, therefore, has become a fascinating candidate material for nanoscale electronics. Researchers have shown that it is possible to carve out nanoscale transistors from a single graphene crystal (i.e., graphene quantum dots). Unlike all other known materials, graphene remains highly stable and conductive even when it is cut into devices one nanometer wide. Graphene quantum dots (GQDs) also show great potential in the fields of photoelectronics, photovoltaic, biosensing, and bioimaging owing to their unique photoluminescence (PL) properties, including excellent biocompatibility, low toxicity, and high stability against photobleaching and photoblinking. Scientists still are working on finding efficient and universal methods for the synthesis of GQDs with high stability, controllable surface properties, and tunable PL emission wavelength.

Perovskite quantum dots

Luminescent quantum dots (LQDs), which possess high photoluminescence quantum yields, flexible emission color controlling, and solution processability, are promising for applications in lighting systems (warm white light without UV and infrared irradiation) and high-quality displays.

However, the commercialization of LQDs has been held back by the prohibitively high cost of their production. Currently, LQDs are prepared by the HI method, requiring at high temperature and tedious surface treating in order to improve both optical properties and stability. Although developed only recently, inorganic halide perovskite quantum dot systems have exhibited comparable and even better performances than traditional QDs in many fields. By preparing highly emissive inorganic perovskite quantum dots (IPQDs) at room temperature, IPQDs' superior optical merits could lead to promising applications in lighting and displays.

Quantum dot TVs and displays

The most commonly known use of quantum dots nowadays may be TV screens. Samsung and LG launched their QLED TVs in 2015, and a few other companies followed not long after. Quantum dots, because they are both photoactive (photoluminescent) and electroactive (electroluminescent) and have unique physical properties, will be at the core of the next-generation displays. Compared to organic luminescent materials used in organic light-emitting diodes (OLEDs), QD-based materials have purer colors, longer lifetime, lower manufacturing cost, and lower power consumption. Another key advantage is that, because QDs can be deposited on virtually any substrate, you can expect printable and flexible—even rollable—quantum dot displays of all sizes and quantum dots with gradually stepping emission from violet to deep red.

Applications of sensors

Nanosensors—what they are; what they do. The term nanosensor is not clearly defined. Most definitions refer to a sensing device with at least one of its dimensions being smaller than 100 nm and for the purpose of collecting information on the nanoscale and transferring it into data for analysis. Nanotechnology deals with the physical or chemical properties of matter at the nanoscale, which can be different from their bulk properties. Nanosensors can take advantage of these phenomena. Important characteristics and quality parameters of nanosensors can therefore be improved over the case of classically modeled sensors with merely reduced sensing parts and/or the transducer. Therefore, nanosensors are not necessarily reduced in size to the nanoscale but could be larger devices that make use of the unique properties of nanomaterials to detect and measure events at the nanoscale. For instance, in noble metals such as silver or gold, nanostructures of smaller sizes than the de Broglie wavelength for electrons lead to an intense absorption in the visible/near-UV region that is absent in the spectrum of the bulk material. Nanosensors have been developed for the detection of gases, chemical and biochemical variables, as well as physical variables, and the detection of electromagnetic radiation. The group of Cees Dekker paved the way for the development of CNT-based electrochemical nanosensors by demonstrating the possibilities of SWCNTs as quantum wires and their effectiveness in the development of field-effect transistors. Many studies have shown that although CNTs are robust and inert structures, their electrical properties are extremely sensitive to the effects of charge transfer and chemical doping by various molecules. Most sensors based on CNTs are field-effect transistors (FETs) although CNTs are robust and have inert structures, their electrical properties are extremely sensitive to the effects of charge transfer and chemical doping by various molecules. CNTs-FETs have been widely used to detect gases such as greenhouse gases in environmental applications.

The functionalization of CNTs is important for making them selective to the target analyte. Different types of sensors are based on molecular recognition interactions between functionalism CNTs and target analytes. For instance, researchers have developed flexible hydrogen sensors using single-walled carbon nanotubes (SWCNTs) decorated with palladium nanoparticles.

Nanosensor fabrication

Nanosensors can be prepared by using different methods. Three common methods are top-down lithography, bottom-up fabrication (for instance controlled lateral epitaxial growth and atomic layer deposition), and self-assembled nanostructures (usually done with biomolecules, e.g., liposomes, that combine in such a way that the biochemical detection of an analyte is converted into an electrical signal). Nanosensors are based on nanoparticles and nanoclusters. Nanoparticles, primarily noble metal ones, have outstanding size-dependent optical properties that have been used to build optical nanosensors. The spectrum of a phenomenon called the *localized surface plasmon resonance* (LSPR) depends on the size, shape, and material of the nanoparticles themselves as well as the particle's environment. The high sensitivity of LSPR sensors can approach the single-molecule limit of detection for large biomolecules. Apart from metal nanoparticles, optical nanosensors based on fluorescence measurements have been built with semiconductor quantum dots and other optical sensors have been developed with nanoscale probes that contain dyes whose fluorescence is quenched in the presence of the analyte to be determined; nanoparticle films have been used for gas sensors; magnetic nanoparticles bound to biorecognition molecules (i.e., DNA and enzymes) have been used to enrich the analyte to be detected. For example, researchers have developed an enzyme biomarker test based on gold nanoparticles that can detect enzyme markers of the disease known as proteases in humans, animals, and food products. This nanosensor indicates when proteases are present through a visible color-change reaction (Chong et al., 2015)

Nanosensors based on nanowires, nanofibers, and carbon nanotubes

Most sensors based on carbon nanotubes (CNTs) are field-effect transistors (FETs) because, although CNTs are robust and inert structures, their electrical properties are extremely sensitive to the effects of charge transfer and chemical doping by various molecules. The functionalization of CNTs is important for making them selective to the target analyte—different types of sensors are based on molecular recognition interactions between functionalism CNTs and target analytes. For instance, researchers have developed flexible hydrogen sensors using single-walled carbon nanotubes decorated with palladium nanoparticles. Nanowires and nanofibers have also been used to build chemiresistive sensors for the diagnosis of diseases. They have been used to maximize gas sensor responses in exhaled breath analysis for the detection of volatile organic compounds (which are biomarkers for various diseases; e.g., acetone, hydrogen sulfide, ammonia, and toluene can be used as biomarkers for evaluating diabetes, halitosis, kidney malfunction, and lung cancer, respectively). In one example, porous tin oxide (SnO_2) nanofibers have been demonstrated to detect acetone at levels of around 0.1 ppm, which is eight times lower than the required gas-sensing level for diagnosing diabetes.

Nanosensors based on graphene

Another carbon nanomaterial, functionalized graphene, holds exceptional promise for biological and chemical sensors. Already, researchers have shown that the distinctive 2D structure of graphene oxide (GO), combined with its super permeability to water molecules, leads to sensing devices at an unprecedented speed. Scientists have found that chemical vapors change the noise spectra of graphene transistors, allowing them to perform selective gas sensing for many vapors with a single device made of pristine graphene—no functionalization of the graphene surface required. Researchers also have begun to work with graphene foams—three-dimensional structures of interconnected graphene sheets with extremely high conductivity. These structures are very promising as gas sensors and as biosensors to detect diseases. A graphene nanosensor enables real-time monitoring of insulin at physiologically relevant concentrations—down to a concentration of approximately 35 pM. This device employs an aptamer-based graphene field-effect transistor (GFET) nanosensor that can rapidly respond to external stimuli such as the binding between surface-immobilized aptamer molecules with insulin. This results in significant changes of the electrostatic charge characteristics in the close proximity of the graphene surface (Bandosz et al., 2003; Dhar, Liu, Thomale, Dai, & Lippard, 2008; González-Gaitán, Ruiz-Rosas, Morallón, & Cazorla-Amorós, 2017; Iannazzo et al., 2017; Kim et al., 2016; Landry et al., 2017).

Optoelectronic and photonic applications

While individual nanotubes generate discrete fine peaks in optical absorption and emission, macroscopic structures consisting of many CNTs gathered together also demonstrate interesting optical behavior. For example, a millimeter-long bundle of aligned multiwalled carbon nanotubes (MWCNTs) emits polarized incandescent light by electrical current heating and SWCNT bundles are giving higher brightness emission at lower voltage compared with conventional tungsten filaments (Ahmad & Pan, 2015; Deldicque, Rouzaud, & Velde, 2016; Padovani et al., 2003; Yan et al., 2015).

Nanoinks

Ink formulations based on CNT dispersions are attractive for printed electronics applications such as transparent electrodes, radio frequency identification (RFID) tags, thin-film transistors, light-emitting devices, and solar cells. The physical properties of the wet ink are important in some applications but not in some others, and similarly, the physical properties of the dry ink film are important in some applications but not all. Identifying the function of a dry ink layer that is necessary to enable the desired application requires an understanding of the print process requirements and therefore the wet ink (Calvert, 2001; Harrey, Ramsey, Evans, & Harrison, 2002; Li, Lu, & Wong, 2010; Sangoi et al., 2004; Sirringhaus et al., 2000; Vemula & John, 2006; Wang et al., 2007).

Electrodes

Carbon nanotubes have been widely used as electrodes for chemical and biological sensing applications and many other electrochemical studies. With their unique one-dimensional molecular geometry of a large surface area coupled with their excellent electrical properties, CNTs have become important materials for the molecular engineering of electrode surfaces where the development of electrochemical devices with region-specific electron-transfer capabilities is of paramount importance (Bisker et al., 2016; Eskandari, Hosseini, Adeli, & Pourjavadi, 2014; Landry et al., 2017).

Potential applications of CNMs

The unique physical and chemical properties of carbon-based nanomaterials determine a wide range of options for practical applications, which in turn trigger the increase of their production. The most widespread field of applications has been reported for CNTs. Due to their mechanical properties, namely the high tensile strength, they are incorporated into polymers and other materials in order to create structural and composite materials with advanced properties. CNMs have the numerous potential applications, viz. gene delivery (Uritu et al., 2015), in the automotive industry (Peddini, Bosnyak, Henderson, Ellison, & Paul, 2014), aviation (Luinge, 2011), in cosmetics (Bergeson & Cole, 2012; Kato, Taira, Aoshima, Saitoh, & Miwa, 2010). Graphene has numerous applications as well: it can be used in electronics, various biochemical sensors, in solar cells, and others summarized in a recently published review by Choi, Lahiri, Seelaboyina, and Kang (2010). The most widespread field of applications has been reported for CNTs. In 2013, the industrial production of CNTs already exceeded several thousand tons (Ahmad & Pan, 2015; Ma & Zhang, 2014; Nasibulin et al., 2009)

Environmental applications of carbon-based nanomaterials

The environmental applications of carbonaceous nanomaterials outlined in the review are both proactive (preventing environmental degradation, improving public health, and optimizing energy efficiency) and retroactive (remediation, wastewater reuse, pollutant transformation). We begin with a brief outline of carbonaceous nanomaterials and their corresponding properties. Environmental pollution is one of the major global challenges since pollutants of different nature contaminate urban and agricultural areas. In order to improve pollutant remediation strategies for environmental sustainability, there is a need to increase the efficiency of conventional methods or to introduce innovative approaches. In this context, nanotechnology, especially carbon-based nanomaterials, can greatly contribute because they possess an enormous absorption potential due to their high surface area.

Carbon-based nanomaterials have a promising potential to improve wastewater filtration systems with numerous examples in the available literature (Choi et al., 2010; Das, Ali, Hamid, Ramakrishna, & Chowdhury, 2014; Qu, Alvarez, & Li, 2013). The significant advantages of CNMs comprise their enormous surface area, mechanical and thermal stability, high chemical affinity for aromatic compounds (Yoo, Ozawa, Fujigaya, & Nakashima, 2011). CNMs are used for environmental sensing. Environmental sensing is one of the applications of nanotechnology with potential to bridge this range of scale. Recent endeavors to monitor environmental change through networked sensing systems will inform predictive models and shape future environmental policy. In another application, arrays of aligned MWNTs grown on SiO2 substrate serve as anodes in ionization sensors for gas detection (Modi, Koratkar, Lass, Wei, & Ajayan, 2003). Widely distributed nanobased sensor networks for rapid and sensitive detection of microbes would strengthen domestic security, enhance online response to microbial outbreaks in drinking water systems, extend subsurface monitoring of biodegradation, or capture evolution in microbial communities after environmental perturbation.

Perspectives on sustainable graphitic structures

Graphitic forms of carbon are important in a wide variety of applications, ranging from pollution control to composite materials, yet the structure of these carbons at the molecular level is poorly understood. The discovery of fullerenes and fullerene-related structures such as carbon nanotubes has given a new perspective on the structure of solid carbon. Peter J.F. Harries reviewed and shown how the new knowledge gained as a result of research on fullerene-related carbons can be applied to well-known forms of carbon such as microporous carbon, glassy carbon, carbon fibers, and carbon black (Bandosz et al., 2003; Batson, Dellby, & Krivanek, 2002; Harris, 2004; Tanaka, Yamasaki, Kawai, & Pan, 2004; Thomson & Gubbins, 2000).

Sustainable discovery of CNMs

CNMs can be considered as graphene sheets rolled up to form coaxial tubular structures with several micrometers in length. It was first observed by Iijima while examining carbon samples synthesized by arc discharge of graphite under TEM (Vander Wal, Hall, & Berger, 2002). The precursors used are mainly fossil fuel-derived hydrocarbons such as methane, ethylene, and benzene. This makes the production cost-intensive. Hence, attempts were made to prepare high-quality CNTs and related structures from relatively inexpensive feedstock like waste plastics postconsumer PE and PET, agricultural wastes (sugar cane bagasse and waste corn residue obtained after extracting the starch fraction of corn for bioethanol production) scrap tire chips, etc. An important parameter to be considered while choosing a renewable material as the source for high-quality CNTs is the low H content.

The discovery of CNMs from biomass can be performed sustainably using pyrolysis techniques, but the discovery and preparation costs associated with pyrolysis hinder its wider application. The consumption of renewable creators and waste heat to fabricate high-quality functional CNMs can considerably improve the sustainability and economic viability of this process. This proposes a method to maximize the economic benefits and the sustainability of CNMs by utilizing waste gases and waste heat to prepare high-quality three-dimensional graphene foams. The resulting is exhibit excellent performance in environmental and energy-storage applications. On the basis of a life-cycle assessment, the overall life-cycle impacts of the present synthetic route on human health, ecosystems, and resources are less than those of the conventional chemical vapor deposition process. Overall, incorporating the pyrolysis route for fabricating functional carbonaceous materials into the biomass pyrolysis process improves the sustainability and economic viability of the process and can support wider commercial application of biomass pyrolysis (Liu, Jiang, & Yu, 2015; Yu, Shearer, & Shapter, 2016; Zhang et al., 2020; Zhuo, Alves, Tenorio, & Levendis, 2012)

Conclusion

Recent years have witnessed a fast-growing interest in designing carbon nanomaterials. Over the last two decades, widespread research efforts have been conducted on CNMs as one of the most widely used classes of nanomaterials. Having their inherent mechanical, optical, electrochemical, and electrical properties, CNMs have been extensively used in multiple areas. In addition, owing to their versatile surface properties, sizes, and shapes over the past decade, CNMs have drawn great attention in biomedical engineering. Interestingly, CNMs are becoming promising materials due to the existence of both inorganic semiconducting properties and organic π-π stacking characteristics. Hence, it could effectively interact with biomolecules and respond to light simultaneously. By taking advantage of such aspects in a single entity, CNM-based nanomaterials could be used for developing biomedical applications in the future. Concerning their toxic effect in the biological system, several chemical modification strategies have been developed and successfully used in nanomaterial bioapplications including drug delivery, tissue engineering, detection of biomolecules, and cancer therapy. This review article provides some achievements in the use of CNMs for biomedical applications. Moreover, in this paper, we also focus on some recently found key features of CNMs and their utilizations for superior bioapplications. However, as CNMs still contain toxicity, more systematic studies are needed to determine the toxicity and pharmacokinetics of CNMs. The applicability of various CNMs to solar energy conversion and storage has been demonstrated, and it can be expected that their unique physical and electrical properties may be exploited to further increase the devices' efficiency. Efforts to harness solar energy as one of the major energy sources for humankind in the near future must be accompanied by material development from resources that are likewise renewable.

References

Ahmad, K., & Pan, W. (2015). Microstructure-toughening relation in alumina based multiwall carbon nanotube ceramic composites. *Journal of the European Ceramic Society, 35*(2), 663–671. https://doi.org/10.1016/j.jeurceramsoc.2014.08.044.

Bandosz, T. J., Biggs, M. J., Gubbins, K. E., Hattori, Y., Iiyama, T., Kaneko, K., et al. (2003). Molecular models of porous carbons. *Chemistry and Physics of Carbon, 28*.

Batson, P. E., Dellby, N., & Krivanek, O. L. (2002). Sub-Angstrom resolution using aberration corrected electron optics. *Nature, 418*.

Bergeson, L. L., & Cole, M. F. (2012). Fullerenes used in skin creams. *Nanotechnology Law and Business, 9*(2), 114–120. http://www.nanolabweb.com/index.cfm/action/main.default.download/articleID/391/CFID/5972451/CFTOKEN/35334367/index.html.

Bisker, G., Dong, J., Park, H. D., Iverson, N. M., Ahn, J., Nelson, J. T., et al. (2016). Protein-targeted corona phase molecular recognition. *Nature Communications, 7*. https://doi.org/10.1038/ncomms10241.

Calvert, P. (2001). Inkjet printing for materials and devices. *Chemistry of Materials, 13*(10), 3299–3305. https://doi.org/10.1021/cm0101632.

Cha, C., Shin, S. R., Annabi, N., Dokmeci, M. R., & Khademhosseini, A. (2013). Carbon-based nanomaterials: Multifunctional materials for biomedical engineering. *ACS Nano, 7*(4), 2891–2897. https://doi.org/10.1021/nn401196a.

Choi, W., Lahiri, I., Seelaboyina, R., & Kang, Y. S. (2010). Synthesis of graphene and its applications: A review. *Critical Reviews in Solid State and Materials Sciences, 35*(1), 52–71. https://doi.org/10.1080/10408430903505036.

Chong, Y., Ge, C., Yang, Z., Garate, J. A., Gu, Z., Weber, J. K., et al. (2015). Reduced cytotoxicity of graphene nanosheets mediated by blood-protein coating. *ACS Nano, 9*(6), 5713–5724. https://doi.org/10.1021/nn5066606.

Climate Change Science Program and the Subcommittee on Global Change Research. (2003). *The U.S. climate change science program: Vision for the program and highlights of the scientific strategic plan*.

Das, R., Ali, M. E., Hamid, S. B. A., Ramakrishna, S., & Chowdhury, Z. Z. (2014). Carbon nanotube membranes for water purification: A bright future in water desalination. *Desalination, 336*(1), 97–109. https://doi.org/10.1016/j.desal.2013.12.026.

Deldicque, D., Rouzaud, J. N., & Velde, B. (2016). A Raman-HRTEM study of the carbonization of wood: A new Raman-based paleothermometer dedicated to archaeometry. *Carbon, 102*, 319–329. https://doi.org/10.1016/j.carbon.2016.02.042.

Dhar, S., Liu, Z., Thomale, J., Dai, H., & Lippard, S. J. (2008). Targeted single-wall carbon nanotube-mediated Pt(IV) prodrug delivery using folate as a homing device. *Journal of the American Chemical Society, 130*(34), 11467–11476. https://doi.org/10.1021/ja803036e.

Eskandari, M., Hosseini, S. H., Adeli, M., & Pourjavadi, A. (2014). Polymer-functionalized carbon nanotubes in cancer therapy: A review. *Iranian Polymer Journal (English Edition), 23*(5), 387–403. https://doi.org/10.1007/s13726-014-0228-9.

Falkowski, P., Scholes, R. J., Boyle, E., Canadell, J., Canfield, D., Elser, J., et al. (2000). The global carbon cycle: A test of our knowledge of earth as a system. *Science, 290*(5490), 291–296. https://doi.org/10.1126/science.290.5490.291.

González-Gaitán, C., Ruiz-Rosas, R., Morallón, E., & Cazorla-Amorós, D. (2017). Effects of the surface chemistry and structure of carbon nanotubes on the coating of glucose oxidase and electrochemical biosensors performance. *RSC Advances, 7*(43), 26867–26878. https://doi.org/10.1039/c7ra02380d.

Harrey, P. M., Ramsey, B. J., Evans, P. S. A., & Harrison, D. J. (2002). Capacitive-type humidity sensors fabricated using the offset lithographic printing process. *Sensors and Actuators, B: Chemical, 87*(2), 226–232. https://doi.org/10.1016/S0925-4005(02)00240-X.

Harris, P. J. F. (2004). Fullerene-related structure of commercial glassy carbons. *Philosophical Magazine, 84*(29), 3159–3167. https://doi.org/10.1080/14786430410001720363.

Iannazzo, D., Pistone, A., Salamò, M., Galvagno, S., Romeo, R., Giofré, S. V., et al. (2017). Graphene quantum dots for cancer targeted drug delivery. *International Journal of Pharmaceutics, 518*(1–2), 185–192. https://doi.org/10.1016/j.ijpharm.2016.12.060.

Kato, S., Taira, H., Aoshima, H., Saitoh, Y., & Miwa, N. (2010). Clinical evaluation of fullerene-C 60 dissolved in squalane for anti-wrinkle cosmetics. *Journal of Nanoscience and Nanotechnology, 10*(10), 6769–6774. https://doi.org/10.1166/jnn.2010.3053.

Kim, J., Park, S. Y., Kim, S., Lee, D. H., Kim, J. H., Kim, J. M., et al. (2016). Precise and selective sensing of DNA-DNA hybridization by graphene/Si-nanowires diode-type biosensors. *Scientific Reports, 6*. https://doi.org/10.1038/srep31984.

Kumawat, M. K., Thakur, M., Gurung, R. B., & Srivastava, R. (2013). Graphene quantum dots for cell proliferation, nucleus imaging, and photoluminescent sensing applications. *Analytical Chemistry, 7*, 9148–9155.

Landry, M. P., Ando, H., Chen, A. Y., Cao, J., Kottadiel, V. I., Chio, L., et al. (2017). Single-molecule detection of protein efflux from microorganisms using fluorescent single-walled carbon nanotube sensor arrays. *Nature Nanotechnology, 12*(4), 368–377. https://doi.org/10.1038/nnano.2016.284.

Li, Y., Lu, D., & Wong, C. P. (2010). Conductive nano-inks. In *Electrical conductive adhesives with nanotechnologies*.

Liu, W. J., Jiang, H., & Yu, H. Q. (2015). Development of biochar-based functional materials: Toward a sustainable platform carbon material. *Chemical Reviews, 115*(22), 12251–12285. https://doi.org/10.1021/acs.chemrev.5b00195.

Loh, G. C., & Baillargeat, D. (2013). Graphitization of amorphous carbon and its transformation pathways. *Journal of Applied Physics, 114*(3). https://doi.org/10.1063/1.4816313, 033534.

Luinge, H. (2011). Nanomodifizierte werkstoffe in der luftfahrt: Kohlenstoffnanoröhrchen für leichtere flugzeug-außenhüllen. *Konstruktion, 9*, IW10–IW11.

Ma, P. C., & Zhang, Y. (2014). Perspectives of carbon nanotubes/polymer nanocomposites for wind blade materials. *Renewable and Sustainable Energy Reviews, 30*, 651–660. https://doi.org/10.1016/j.rser.2013.11.008.

Manjunatha, J. G. (2016). Poly (Nigrosine) modified electrochemical sensor for the determination of dopamine and uric acid: A cyclic voltammetric study. *International Journal of ChemTech Research, 9*(2), 136–146. http://sphinxsai.com/2016/ch_vol9_no2/1/(136-146)V9N2CT.pdf.

Manjunatha, J. G. (2017). A new electrochemical sensor based on modified carbon nanotube-graphite mixture paste electrode for voltammetric determination of resorcinol. *Asian Journal of Pharmaceutical and Clinical Research, 10*(12), 295–300. https://doi.org/10.22159/ajpcr.2017.v10i12.21028.

Manjunatha, J. G. (2020a). Fabrication of efficient and selective modified graphene paste sensor for the determination of catechol and hydroquinone. *Surfaces, 3*, 473–483.

Manjunatha, J. G. (2020b). Poly (adenine) modified graphene-based voltammetric sensor for the electrochemical determination of catechol, hydroquinone and resorcinol. *The Open Chemical Engineering Journal, 14*, 52–62. https://doi.org/10.2174/1874123102014010052.

Manjunatha, J. G., & Deraman, M. (2017). Graphene paste electrode modified with sodium dodecyl sulfate surfactant for the determination of dopamine, ascorbic acid and uric acid. *Analytical and Bioanalytical Electrochemistry, 9*(2), 198–213. http://www.abechem.com/No.%202-2017/2017,9(2),198-213.pdf.

Manjunatha, J. G., Kumara Swamy, B. E., Deraman, M., & Mamatha, G. P. (2012). Simultaneous voltammetric measurement of ascorbic acid and dopamine at poly (vanillin) modified carbon paste electrode: A cyclic voltammetric study. *Der Pharma Chemica, 4*(6), 2489–2497. http://derpharmachemica.com/vol4-iss6/DPC-2012-4-6-2489-2497.pdf.

Manjunatha, J. G., Kumara Swamy, B. E., Mamatha, G. P., Gilbert, O., Chandrashekar, B. N., & Sherigara, B. S. (2010). Electrochemical studies of dopamine and epinephrine at a poly (tannic acid) modified carbon paste electrode: A cyclic voltammetric study. *International Journal of Electrochemical Science, 5*(9), 1236–1245. http://www.electrochemsci.org/papers/vol5/5091236.pdf.

Manjunatha, J. G., Kumara Swamy, B. E., Mamatha, G. P., Gilbert, O., Shreenivas, M. T., & Sherigara, B. S. (2009). Electrocatalytic response of dopamine at mannitol and triton x-100 modified carbon paste electrode: A cyclicvoltammetric study. *International Journal of Electrochemical Science, 4*(12), 1706–1718. http://www.electrochemsci.org/papers/vol4/4121706.pdf.

Manjunatha, J. G., Kumara Swamy, B. E., Shreenivas, M. T., & Mamatha, G. P. (2012). Selective determination of dopamine in the presence of ascorbic acid using a poly (nicotinic acid) modified carbon paste electrode. *Analytical and Bioanalytical Electrochemistry, 4*(3), 225–237. http://www.abechem.com/No.%203-2012/2012,4_3_,225-237.pdf.

Modi, A., Koratkar, N., Lass, E., Wei, B., & Ajayan, P. M. (2003). Miniaturized gas ionization sensors using carbon nanotubes. *Nature, 424*(6945), 171–174. https://doi.org/10.1038/nature01777.

Nasibulin, A. G., Shandakov, S. D., Nasibulina, L. I., Cwirzen, A., Mudimela, P. R., Habermehl-Cwirzen, K., et al. (2009). A novel cement-based hybrid material. *New Journal of Physics, 11*. https://doi.org/10.1088/1367-2630/11/2/023013.

Padovani, S., Sada, C., Mazzoldi, P., Brunetti, B., Giulivi, A., D'Acapito, F., et al. (2003). Copper in glazes of renaissance luster pottery: Nanoparticles, ions, and local environment. *Journal of Applied Physics, 93*(12), 10058–10063. https://doi.org/10.1063/1.1571965.

Peddini, S. K., Bosnyak, C. P., Henderson, N. M., Ellison, C. J., & Paul, D. R. (2014). Nanocomposites from styrene-butadiene rubber (SBR) and multiwall carbon nanotubes (MWCNT) part 1: Morphology and rheology. *Polymer, 55*(1), 258–270. https://doi.org/10.1016/j.polymer.2013.11.003.

Qu, X., Alvarez, P. J. J., & Li, Q. (2013). Applications of nanotechnology in water and wastewater treatment. *Water Research, 47*(12), 3931–3946. https://doi.org/10.1016/j.watres.2012.09.058.

Raril, C., & Manjunatha, J. G. (2020). A simple approach for the electrochemical determination of vanillin at ionic surfactant modified graphene paste electrode. *Microchemical Journal, 154*. https://doi.org/10.1016/j.microc.2019.104575.

Sangoi, R., Smith, C. G., Seymour, M. D., Venkataraman, J. N., Clark, D. M., Kleper, M. L., et al. (2004). Printing radio frequency identification (RFID) tag antennas using inks containing silver dispersions. *Journal of Dispersion Science and Technology, 25*(4), 513–521. https://doi.org/10.1081/DIS-200025721.

Sharon, M., & Sharon, M. (2010). *Carbon nanoforms and applications*.

Sirringhaus, H., Kawase, T., Friend, R. H., Shimoda, T., Inbasekaran, M., Wu, W., et al. (2000). High-resolution inkjet printing of all-polymer transistor circuits. *Science, 290*(5499), 2123–2126. https://doi.org/10.1126/science.290.5499.2123.

Smith, T. M., Cramer, W. P., Dixon, R. K., Leemans, R., Neilson, R. P., & Solomon, A. M. (1993). The global terrestrial carbon cycle. *Water, Air, & Soil Pollution, 70*(1–4), 19–37. https://doi.org/10.1007/BF01104986.

Tanaka, N., Yamasaki, J., Kawai, T., & Pan, H. (2004). The first observation of carbon nanotubes by spherical aberration corrected high-resolution transmission electron microscopy. *Nanotechnology, 15*(12), 1779–1784. https://doi.org/10.1088/0957-4484/15/12/015.

Thomson, K. T., & Gubbins, K. E. (2000). Modeling structural morphology of microporous carbons by reverse Monte Carlo. *Langmuir, 16*(13), 5761–5773. https://doi.org/10.1021/la991581c.

Titirici, M.-M., White, R. J., Brun, N., Budarin, V. L., Su, D. S., del Monte, F., et al. (2015). Sustainable carbon materials. *Chemical Society Reviews*, 250–290. https://doi.org/10.1039/C4CS00232F.

Uritu, C. M., Varganici, C. D., Ursu, L., Coroaba, A., Nicolescu, A., Dascalu, A. I., et al. (2015). Hybrid fullerene conjugates as vectors for DNA cell-delivery. *Journal of Materials Chemistry B, 3*(12), 2433–2446. https://doi.org/10.1039/c4tb02040e.

Vander Wal, R. L., Hall, L. J., & Berger, G. M. (2002). Optimization of flame synthesis for carbon nanotubes using supported catalyst. *Journal of Physical Chemistry B, 106*(51), 13122–13132. https://doi.org/10.1021/jp020614l.

Vemula, P. K., & John, G. (2006). Smart amphiphiles: Hydro/organogelators for in situ reduction of gold. *Chemical Communications, 21*, 2218–2220. https://doi.org/10.1039/b518289a.

Wang, J. (2005). Carbon-nanotube based electrochemical biosensors: A review. *Electroanalysis, 17*(1), 7–14. https://doi.org/10.1002/elan.200403113.

Wang, X., Egan, C. E., Zhou, M., Prince, K., Mitchell, D. R. G., & Caruso, R. A. (2007). Effective gel for gold nanoparticle formation, support and metal oxide templating. *Chemical Communications, 29*, 3060–3062. https://doi.org/10.1039/b704825d.

Xia, D., Otyepka, M., Li, X., Liu, W., & Zheng, Q. (2015). Carbon-based materials at nanoscale. *Journal of Nanomaterials, 2015*, 1.

Xie, R., Wang, Z., Zhou, W., Liu, Y., Fan, L., Li, Y., et al. (2016). Graphene quantum dots as smart probes for biosensing. *Analytical Methods, 8*(20), 4001–4016. https://doi.org/10.1039/C6AY00289G.

Yan, Y., Miao, J., Yang, Z., Xiao, F. X., Yang, H. B., Liu, B., et al. (2015). Carbon nanotube catalysts: Recent advances in synthesis, characterization and applications. *Chemical Society Reviews, 44*(10), 3295–3346. https://doi.org/10.1039/c4cs00492b.

Yoo, J., Ozawa, H., Fujigaya, T., & Nakashima, N. (2011). Evaluation of affinity of molecules for carbon nanotubes. *Nanoscale, 3*(6), 2517–2522. https://doi.org/10.1039/c1nr10079c.

Yu, L., Shearer, C., & Shapter, J. (2016). Recent development of carbon nanotube transparent conductive films. *Chemical Reviews, 116*(22), 13413–13453. https://doi.org/10.1021/acs.chemrev.6b00179.

Zhang, S., Jiang, S. F., Huang, B. C., Shen, X. C., Chen, W. J., Zhou, T. P., et al. (2020). Sustainable production of value-added carbon nanomaterials from biomass pyrolysis. *Nature Sustainability, 3*(9), 753–760. https://doi.org/10.1038/s41893-020-0538-1.

Zhang, D. Y., Zheng, Y., Tan, C. P., Sun, J. H., Zhang, W., Ji, L. N., et al. (2017). Graphene oxide decorated with Ru(II)-polyethylene glycol complex for lysosome-targeted imaging and photodynamic/photothermal therapy. *ACS Applied Materials and Interfaces, 9*(8), 6761–6771. https://doi.org/10.1021/acsami.6b13808.

Zhuo, C., Alves, J. O., Tenorio, J. A. S., & Levendis, Y. A. (2012). Synthesis of carbon nanomaterials through up-cycling agricultural and municipal solid wastes. *Industrial and Engineering Chemistry Research, 51*(7), 2922–2930. https://doi.org/10.1021/ie202711h.

Further reading

Benn, T., Cavanagh, B., Hristovski, K., Posner, J. D., & Westerhoff, P. (2010). The release of nanosilver from consumer products used in the home. *Journal of Environmental Quality, 39*(6), 1875–1882. https://doi.org/10.2134/jeq2009.0363.

Bhattacharya, K., Mukherjee, S. P., Gallud, A., Burkert, S. C., Bistarelli, S., Bellucci, S., et al. (2016). Biological interactions of carbon-based nanomaterials: From coronation to degradation. *Nanomedicine: Nanotechnology, Biology, and Medicine, 12*(2), 333–351. https://doi.org/10.1016/j.nano.2015.11.011.

Eatemadi, A., Daraee, H., Karimkhanloo, H., Kouhi, M., Zarghami, N., Akbarzadeh, A., et al. (2014). Carbon nanotubes: Properties, synthesis, purification, and medical applications. *Nanoscale Research Letters, 9*(1), 393. https://doi.org/10.1186/1556-276x-9-393.

Huang, J., Liu, Y., & You, T. (2010). Carbon nanofiber based electrochemical biosensors: A review. *Analytical Methods, 2*(3), 202–211. https://doi.org/10.1039/b9ay00312f.

Manisankar, P., Pushpalatha, A. M., Vasanthkumar, S., Gomathi, A., & Viswanathan, S. (2004). Riboflavin as an electron mediator catalyzing the electrochemical reduction of dioxygen with 1,4-naphthoquinones. *Journal of Electroanalytical Chemistry, 571*(1), 43–50. https://doi.org/10.1016/j.jelechem.2004.04.011.

Rajam, J. A., Gomathi, A., & Vedhi, C. (2016a). Catalytic oxygen reduction on silver nanoparticles modified glassy carbon electrode with 1, 4-naphthoquinone. *Journal of Nanoscience and Technology, 2*(5), 201–203.

Rajam, J. A., Gomathi, A., & Vedhi, C. (2016b). Catalytic oxygen reduction on silver nanoparticle modified glassy carbon electrode with 1, 4-naphthoquinone. *Journal of Nanoscience and Technology, 2*(5), 201–203.

Rajam, A., Gomathi, A., & Vedhi, C. (2018). Electrochemical reduction of oxygen on copper nanoparticle deposited glassy carbon electrode with 1, 4 naphthoquinone and its derivatives. In *Special issue: Conscientious computing technologies* (pp. 972–976).

Sanghavi, B. J., Mobin, S. M., Mathur, P., Lahiri, G. K., & Srivastava, A. K. (2013). Biomimetic sensor for certain catecholamines employing copper(II) complex and silver nanoparticle modified glassy carbon paste electrode. *Biosensors and Bioelectronics, 39*(1), 124–132. https://doi.org/10.1016/j.bios.2012.07.008.

Chapter 25

Sustainable carbon nanomaterial-based sensors: Future vision for the next 20 years

S. Alwin David, R. Rajkumar, P. Karpagavinayagam, Jessica Fernando, and C. Vedhi
Department of Chemistry, V.O. Chidambaram College, Tuticorin, India

Introduction

Nowadays, the necessity to develop new sensors with more specific characteristics has been increasing. Higher sensitivity and reliability, faster response, quicker recovery, low cost, reduced size, in situ analysis, and simple operation are some of the properties required for producing technological sensor devices (Ardila, Oliveira, Medeiros, & Fatibello-Filho, 2014; Mehdi Aghaei, Monshi, Torres, Zeidi, & Calizo, 2018). There is an enormous number of sensors for monitoring gas, heavy metal, humidity, biomolecules, and pressure, among others. However, most of them are expensive, need pretreatment, have difficult operation and slow responses, and do not have the ideal limit of detection, sensibility, and/or selectivity. On the perspective to improve the parameters mentioned above, nanotechnology has sponsored the most promising upgrade on materials' properties, providing significant advances to overcome limitations once experienced by conventional materials. Nanomaterials are produced in the nanoscale range from 1 to 100 nm, and because of their dimension, they significantly change properties from their equivalent counterparts with structure higher than nanoscale, such as the bulk material (Kreyling, Semmler-Behnke, & Chaudhry, 2010).

One of the most studied and currently used materials in the nanotechnology field is the carbon-based one due to its remarkable properties. Carbonaceous structures present numerous advantages compared to other usually employed materials, especially their extraordinary physical–chemical properties. Manufacture processes can be simple, yielding a proper amount of material with low densification defects. Furthermore, carbon-based materials can be considered an alternative to currently expensive electronic compounds, presenting excellent performance and being considered an environment-friendly material (Peng, Zhang, & Wang, 2014). Therefore, carbon nanostructures have been investigated to be used as powerful sensor devices, since they present superior physical and chemical parameters yielding high-quality sensing properties.

Several carbon-based structures discovered many decades ago are studied and applied in technological devices nowadays. Nanodiamond (ND) structure dated from the 1960s (Azevedo, Baldan, & Ferreira, 2012). NDs are nanosized (5–100 nm) carbon structures essentially formed by sp^3 hybridization (Nunn, Torelli, McGuire, & Shenderova, 2017). Their unique optical and electronic characteristics are due to dopants present in the structure, and the excellent surface activity is assigned to structural defects and unsaturated chemical bond arising from carbon atoms (Mochalin, Shenderova, Ho, & Gogotsi, 2012; Osswald, Yushin, Mochalin, Kucheyev, & Gogotsi, 2006; Schrand et al., 2007; Tsai et al., 2016; Zheng et al., 2017). The discovery of fullerene (FLN) occurred in 1985 through Kroto et al.'s (Kroto, Heath, O'Brien, Curl, & Smalley, 1985) work. FLNs are a molecular allotrope of carbon and consist of a three-dimensional closed cage (C_n) made of five- and six-membered rings with 12 pentagons and a different number of hexagons, depending on the FLN size (Hirsch, 2010; Klaus, 2017; Rao, Seshadri, Govindaraj, & Sen, 1995; Troshin & Lyubovskaya, 2008; Zaghmarzi et al., 2017). Another carbon-based structure is carbon nanotube (CNT) that belongs to FLN family showing a quasi-one-dimensional structure (single or multiwalled CNTs) (Iijima & Ichihashi, 1993). Works on CNTs increased after the publication of Iijima in 1991 (Iijima, 1991; Monthioux & Kuznetsov, 2006), which studied and presented their tubular structure. The graphene (GPN) was the first discovered two-dimensional atomic crystal (Novoselov, 2004; Novoselov et al., 2005), formed by a single atom layer of carbon atoms arranged in a honeycomb lattice structure. Fig. 25.1 shows some of the carbon-based structures used in sensor applications.

Developing new devices sometimes requires features not found in a single material. The number of researches that have been dealing with hybrid materials has remarkably grown because of the possibility to improve physical and chemical

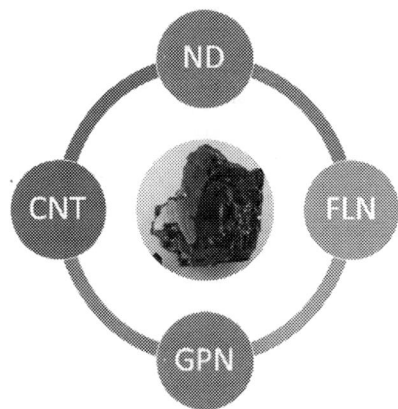

FIG. 25.1 Structures of carbon-based nanomaterials.

properties by composing two or more different phases. Hybrid materials, also known as composites, are systems formed by organic or inorganic components, which constitute a particular multifunctional material class. Compounds forming hybrid materials are usually combined through a synergistic relationship (Chen, Xiao, Qian, Li, & Yu, 2013; Chen, Yuan, Chai, & Hu, 2013), giving rise to unique properties to the new material because of the complementarity of unique features in each compound. Notably, studies on the new hybrid materials based on carbon structures such as CNTs, GPNs, FLNs, graphene oxide (GO), and NDs have been intensively investigated for analytical applications due to their excellent properties, such as high surface area, high mechanical stability, adaptability, and functionality (Yilmaz, Ulusoy, Demir, & Soylak, 2018).

This report presents the latest advances concerning carbon-based sensors in several fields of applications, including mechanical, physicochemical, and biomedicine, regarding different structure arrangements. Most relevant studies will be summarily cited hereafter, and a comparison between the devices previously reported with those in this review will be presented, gathering perspectives for the future of carbon-based sensor devices.

Nanodiamond-based sensors

In recent years, there has been a great development of new materials for electrochemical applications such as carbon nanomaterials (Chen, Xiao, et al., 2013; Chen, Yuan, et al., 2013; Deng et al., 2014; Li et al., 2020; Oliveira, Janegitz, Zucolotto, & Fatibello-Filho, 2013) and metal nanoparticles. The attractiveness of nanomaterials is due, among other characteristics, to the greater surface area and surface energy compared to the same materials on a macroscopic scale (Ferreira & Rangel, 2009), which significantly increase mass transport and the sensitivity of electrochemical measures (Chen, Xiao, et al., 2013; Chen, Yuan, et al., 2013). Among the nanomaterials of carbon, the nanodiamonds (NDs) have optical transparency, chemical inertness, great surface specificity, and high hardness. Due to its surface π bonds and the ability to produce stable dispersions in aqueous media (Kulakova, 2004; Mochalin et al., 2012; Simioni et al., 2017; Simioni, Silva, Oliveira, & Fatibello-Filho, 2017), NDs have become an excellent tool for the development of sensors (Baccarin, Rowley-Neale, Cavalheiro, Smith, & Banks, 2019; Puthongkham & Venton, 2019; Ramos, Carvalho, de Oliveira, & Janegitz, 2020; Simioni, Oliveira, et al., 2017; Simioni, Silva, et al., 2017). This fact is due to the conductivity and the nanometric scale of the material. An example is the work presented by Ramos et al. (2020), in which the presence of NDs increased the conductivity and surface area of the base electrode, allowing a more sensitive determination of serotonin. Another characteristic of NDs is their low toxicity, being used in medicines and treatments for the most varied types of cancer (Ardila et al., 2014; Nazarkovsky, de Mello, Bisaggio, Alves, & Zaitsev, 2020; Terada, Genjo, Segawa, Igarashi, and Shirakawa, 2020). The attractive characteristic of the material can allow its use in the development of modified electrodes for the determination of other substances with environmental concern.

The modification of bare electrodes with nanomaterials can be a challenge because often there are no anchoring sites. Maybe, one of the most common strategies for immobilizing nanoparticles on base electrodes is by the formation of a composite film based on a polymer and a nanoparticulate material, in general polymeric materials (Huang et al., 2020; Silva, Queiroz, & Brett, 2020). The use of films, usually polymeric films, allows a good distribution and incorporation of conductive material or other modifiers onto the base electrode. In the carbon nanotubes, horse radish peroxidase enzyme was made on a carbon-based screen-printed electrode surface, which was only possible using an amphiphilic copolymer. However, there are some important conditions that these materials should present to be used as a modifier of electrodes, for example, low volatility, low solubility in water, and selective miscibility in organic compounds (Svancara, Walcarius, & Vytras, 2012). Also, nowadays, there is a great concern about the decrease in the use of sustainable and environmentally friendly organic compounds (Vishwakarma, 2020). Therefore, researchers have used polymeric films, such as biological origin, for example, biofilm based on manioc starch (Orzari, Santos, & Janegitz, 2018). Manioc is a tuberous plant widely used in Brazilian cuisine. The starch present in cassava (manioc starch (MS)) consists of a mixture of two polysaccharides, amyl pectin and amylose, which are responsible for the gelatinization properties that control the formation of the film (Curá, Jansson, & Krisman, 1995; Hoover, 2001; Jane et al., 1999; Yoo et al., 2009). Some works with manioc starch film have shown promising results for anchoring carbon nanomaterials (Zambianco, Silva, Zanin, Fatibello-Filho, & Janegitz, 2019). The result is due to the easy dispersion of manioc starch in a suitable solvent obtaining a homogeneous mixing of the polymer and the conductive material.

Carbon nanotube-based sensors

wIn description, carbon nanotubes (CNTs) are pseudo-one-dimensional allotropes of carbon, notably recognized as a network of carbon atoms assembled into seamless cylinders of one or more layers with either open or closed ends (Harris, 2009) (De Volder, Tawfick, Baughman, & Hart, 2013). During the past few years, CNTs have drew attention in healthcare system as one of the prospective contenders for medical devices owing to their several unique properties, viz., excellent mechanical strength (Peng et al., 2008), high surface area, excellent electrical conductivity (Wei, Vajtai, & Ajayan, 2001), electrochemically stable in aqueous and nonaqueous solutions, and high thermal conductivity (Pop, Mann, Wang, Goodson, & Dai, 2006). In short, biosensor refers to a diagnostic device that transfigures a biological event into a measurable electrical signal. The remarkable sensitivity of CNTs toward their surface ensures that they serve as an ideal material for highly sensitive nanoscale biosensor devices (Thirumalraj, Kubendhiran, Chen, & Lin, 2017; Wang, 2005; Yun et al., 2007). But the strong intermolecular π-π interactions, and hydrophobic nature of CNTs, is a major obstacle for developing CNT-based biosensors. In order to improve their solubility and stability, different functionalities of the CNTs have been improved by functionalizing through chemical adsorption (Bianco, Kostarelos, Partidos, & Prato, 2005; Kam, Liu, & Dai, 2005; Vardharajula et al., 2012). Many reports have emerged over the last decade showing that functionalized CNTs have an ability to address several concerns related to biocompatibility and toxicology (Colvin, 2003).

CNTs are often chosen as working electrodes for sensing devices owing to their extraordinary low detection limit (Galano, 2010; Kim et al., 2017). The hollow structure of CNTs is good for the adsorption of enzyme. Therefore, in amperometric CNT-based biosensor, CNTs are always used to functionalize with enzyme to generate the enzyme CNT electrodes or to modify the surface of electrodes. For example, Eguílaz et al. reported that covalently functionalized SWCNT with polytyrosine electrode has an efficiency to generate surface quinone groups to facilitate the electrooxidation of nicotinamide adenine dinucleotide-based biosensors. The functionalization of SWCNT with polytyrosine has also shown to be an efficient dispersion medium for these nanostructures (Eguílaz, Gutierrez, González-Domínguez, Martínez, & Rivas, 2016). Burrs et al. demonstrated electrochemical biosensors, which work with functionalized graphene (GR) paper for the detection of small molecules (glucose) or pathogen bacteria (*Escherichia coli* O157:H7) by using fractal platinum nano cauliflower (Burrs et al., 2016). According to different analytes, different enzymes are selected, such as nicotinamide adenine dinucleotide (NADH), glucose oxidase (GOD) (Magner, 1998), aflatoxin oxidase, cholesterol oxidase, urease, lactic acid oxidase, horseradish peroxidase (Koskun, Şavk, Şen, & Şen, 2018; Moyo, Okonkwo, & Agyei, 2014), and so on. The recent and fast growth in CNTs has strongly influenced the medical sector because CNTs very sensitive to be affected by exposure to biomolecules as sensing elements for biosensors. A great deal of effort has been put over in the last decade worldwide in the development of new biosensors with high sensitivity for the rapid detection of targeted substrates.

By virtue of the remarkable structural, optical, electronic, and mechanical properties of CNTs, they offer diverse features of interest to engineer new-generation probes in CNT-based biosensors. As mentioned earlier, CNTs have a large specific surface area that enables the immobilization of a large number of functional units such as receptor moieties for biosensing and CNTs exhibit unique intrinsic optical properties such as photoluminescence in the near infrared (NIR) *Choi* (Lee & Choi, 2019) and strong resonance Raman scattering, making them as an excellent candidate for biological detection. Other interesting property is photothermal response that could reduce the tumor size or even be completely removed by using NIR laser irradiation to generate heat. Different types of CNTs have different properties; for example, SWCNTs have unusual electronic property, based on the one-dimensional quantum effect. CNTs have multiplex functionalization capability through conjugation, thereby potentially enhancing recognition and signal transduction processes. The CNT could be able to conduct electricity $\sim$100 times greater than copper, which supports for transduction of electric signals generated upon the recognition of a target (Holzinger, Le Goff, & Cosnier, 2014). An ideal biosensor should have good physical and chemical as well as biological properties.

CNTs could easily cross the biological membranes, which make them applicable in vivo with minimal invasiveness, and further implemented photoacoustic also. CNTs can be made biocompatible through various dispersion and functionalization methods. The more toxic CNTs the lesser density of surface functional groups, less toxic the more density of surface functional groups. By making the surface of pristine CNTs into hydrophilic f-CNTs, the biocompatibility and biodegradability were increased (Bianco, Kostarelos, & Prato, 2011). Numerous series of CNT biosensors have been established till date to examine an extensive range of cancer biomarkers by means of conjugation of DNA or aptamers, antibodies, peptides, proteins, or enzymes (Choi, Kwak, & Park, 2010). Wang and Dai described an electrochemical biosensor for early-stage discovery of cancer biomarkers (Wang & Dai, 2015), and immune sensor-based CNTs were also reported by Veetil and Ye (2007). However, it is well recognized that still there are quite a few challenges that need to be overcome for successful integration of immune sensor type of biosensors. Optical biosensors, on the other hand, provide a powerful substitute and offer a great advantage over a conventional analytical technique by serving real-time and label-free detection just by analyzing the differences in the radiation of light (UV, visible, or infrared). Sensitivity of the biosensors has been improved by increasing the interaction between the guided light and the sensor surface. Scheme 25.1 shows the different types of CNT-based biosensor, and a clear discussion is given in the following sections.

Buckypaper-based sensors

A high-performance strain sensor based on buckypaper has been fabricated and studied. The sensor with an ultrahigh gauge factor of 20,216 can detect a maximum and a minimum strain range of 75% and 0.1%, respectively. During stretching, the strain sensor achieves a high stability and reproducibility of 10,000 cycles, and a fast response time of less than 87 ms. On the other hand, the sensor shows an excellent sensing performance upon pressure. The pressure range, pressure sensitivity, and loading-unloading cycles are 0–1.68 MPa, 89.7/kPa, and 3000 cycles, respectively. A concept of the optimal value is utilized to evaluate the strain and pressure performances of the sensor. The optimal values of the sensor upon tensile strain and pressure are calculated to be 3.07×10^8 and 1.35×10^7, respectively, which are much higher than those of most strain and pressure sensors reported in the literature. Precise detection of full-range human motions, acoustic vibrations, and even pulse wave at a small scale have been successfully demonstrated by the buckypaper-based sensor. Owing to its advantages, including ultrahigh sensitivity, wide detection range, and good stability, the buckypaper-based sensor suggests a great

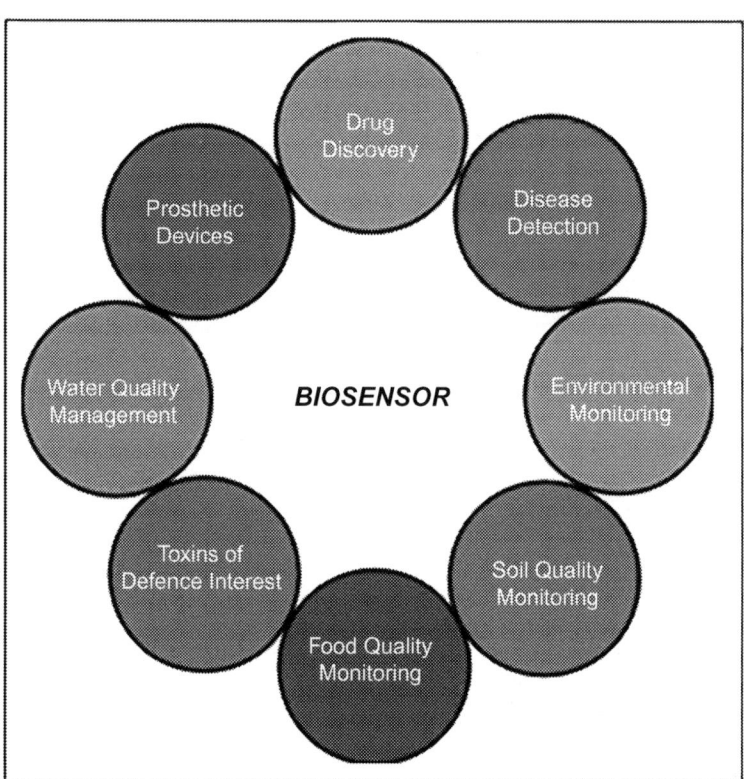

SCHEME 25.1 Different types of CNT-based biosensors.

potential for applications in wearable sensors, electronic skins, micro –/nano-electromechanical systems, vibration-sensing devices, and other strain-sensing devices.

A mixture of the base and curing agent of PDMS (polydimethylsiloxane) with a mass ratio of 10:1 was prepared. After stirring evenly, the mixture was extracted with a syringe and dripped into a square container. The thickness of the mixture was controlled to be 500 μm. Then, the square container was evacuated in a vacuum oven for 30 min to remove the air bubbles inside the mixture. Thereafter, the container was dried in an oven at 60°C for 30 min to form a solid PDMS film. The solid PDMS film was then cut to a size of 4 cm × 2 cm to obtain a solid PDMS substrate. The buckypaper-sensing medium was cut from the as-prepared buckypaper with the size of 8 mm × 8 mm and fixed on the solid PDMS substrate with a pair of copper-conductive tapes.

Similar to the above-mentioned PDMS substrate fabrication, the liquid mixture with a thickness of 500 μm consisting of PDMS and its curing agent was prepared once again. After filled with PDMS mixture, the square-shaped container was placed in a constant temperature oven and dried at 65°C for 17 min to obtain a semi-dry PDMS film, which was viscous but not flowing. After this, the semi-dried PDMS film was combined with the previously prepared solid PDMS substrate, which had the buckypaper on top. The whole stack was then cured in the oven at 60°C for 30 min to produce a buckypaper-based strain sensor. For this sensor containing buckypaper and PDMS, buckypaper with excellent conductivity works as the sensing medium, while the transparent PDMS provides an elastic stretchable supporter to protect the buckypaper from being destroyed upon various mechanical stimuli. The key processes in fabricating the buckypaper-based strain sensor are schematically illustrated in Fig. 25.2.

Graphene-based gas sensors

Graphene is made up of a sp^2 bonded carbon atoms, arranged in a single layer, existing as a 2-D honeycomb structure. It shows properties that are unique, such as quantum Hall effect, very high electron mobility, excellent electrical conductivity, and large surface-to-volume ratio so it can absorb a large amount of aromatic biomolecules by π-π stacking, tunable optical properties, and high mechanical strength, which make it an ideal nanomaterial for potential application in medical and biological fields as a biosensor. It has been used for the detection of different varieties of biomolecules such as deoxyribonucleic acid (DNA), antibodies, enzymes, aptamers, etc.

Desirable and superior properties of graphene and its derivatives are what make it useful as a signaling device for the quantitative detection of biomolecules, antigen, antibody, antibiotics, chemicals, narcotics, toxins, DNA, whole-cell virus/bacteria, etc. In the environmental sector, it is also used for the detection of pesticides, e.g., chlorpyrifos, along with the detection of antibiotics such as chloramphenicol, tetracycline, streptomycin, kanamycin, etc. In the field of diagnostics, graphene-based sensors are being researched into for the diagnosis of various ailments by detecting viruses, bacteria, disease biomarkers, immune bodies, etc. Sensitivity, limit of detection (LOD), and reproducibility of the biosensor can be improved when graphene sheets are used in combination with nanoparticles.

2D structure with extreme high surface area makes graphene compatible for the development of gas sensors. Gas sensing involves absorption and desorption of small gaseous molecules by the thin layer of graphene, resulting in change in the conductance of graphene. Nowadays, breath analysis for noninvasive diagnosis of diseases is attracting widespread attention. Human breath contains biomarkers that can be used for the diagnosis of several diseases like lung cancer, diabetes, etc. (Bajtarevic et al., 2009). According to the research carried out by Clemens et al., compounds like isoprene, acetone, and methanol are expelled in the normal exhalation of breath. However, in the case of certain disorders, these three main components show abnormally high or low concentration as compared to healthy volunteers (Zhou, Xue, et al., 2020). Therefore, high sensitivity of graphene toward gaseous molecules led to the fabrication of a gas-based sensor, which can be utilized for the detection of ethanol, NH_3, NO_2, and O_2 (Kuretake, Kawahara, Motooka, & Uno, 2017). Gas-sensing principle is based on the electrical and ionic conductivities at low and high availability of gas molecules, respectively. The sensor was fixed onto the nonconductive surface to negate the effect of ambient humidity.

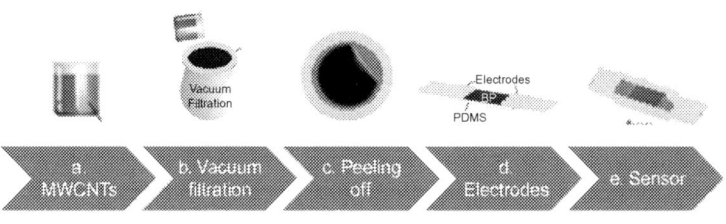

FIG. 25.2 Fabrication process of the strain sensor based on buckypaper.

Graphene oxide-based sensors

Graphene oxide (GO), a nanomaterial also known as graphitic acid, belongs to the graphene superfamily, containing several reactive oxygen functional groups such as epoxy group, hydroxyl group, and carboxyl group, which result in bulking of the plane. Modification of GO with different functional groups set an impact on the properties of the material, which led to its applications in the agnostic field. It has been reported that spacing between the layers is almost twice as that of normal graphene (7 Å), and thus due to this increase in interplanar space, water layers can be seen in between the GO layers and it gets hydrated immediately on immersion into water. Apart from water, polar solvents like alcohol are also easily incorporated between the layers. In terms of electrical conductivity, GO acts as an insulator due to the sp^2 network disruption. However, the presence of oxygen-containing groups makes it thermally unstable. GOs, due to their unique properties, such as dispersion ability in aqueous medium and large surface area, have been used in different bioapplications. They have been utilized in early-stage diagnosis and treatment of various diseases as they can be used for the detection of different target analytes such as proteins, DNA, pesticides, and microbes.

In recent times, there is a huge focus to develop miniaturized diagnostic devices, which are portable and can be used on-site without the need of an elaborate laboratory setup. Graphene enables us to develop such devices for the detection of various parameters as well as health-monitoring devices due to its robustness, added sensitivity, and low cost. Xu et al. (2018) developed a health monitoring device, which can be mounted onto a person's neck (detection of laryngeal prominence motions) for simultaneous in situ detection of various human movements and ion concentration in sweat. The sensor was made by applying an rGO film on top of a porous inverse opal acetyl cellulose (IOAC) film. The rGO film, due to its resistance change, acted as a layer for strain sensing to be able to monitor human motion, while the porous IOAC acted as a flexible microstructured substrate, which enabled the collection and analysis of sweat ion concentration using colorimetric principle/reflection peak shifts besides highly sensitive motion sensing. Similarly, another rGO/single-wall carbon nanotubes (SWCNTs) hybrid fabric-based strain-pressure sensor was developed by Kim et al. In this study, GO was coated on cotton fabric with subsequent chemical reduction to rGO and later coated with SWCNTs to improve durability and electrical conductivity. The fabric showed water-resistant properties (crucial for applicability of wearable electronics) and was used to make a wearable glove, which could detect motion signals when it was bent, gripped, pressed, and wrist-turned (Kim et al., 2018).

Reduced graphene oxide-based sensors

A single layer of carbon atoms arranged in a two-dimensional honeycomb lattice is known as graphene. The oxidized form of graphene is recognized as graphene oxide (GO). Reduced graphene oxide (rGO) is an attractive member of the graphene family. A variety of reductions methods have been used to reduce GO partially to form reduced graphene oxide (rGO). The reduction methods like laser radiation, thermal, chemical, photochemical, photothermal, microbial, microwave, and electrochemical reduction have been applied to obtain rGO from GO by reducing its oxygen content. Based on the reduction method used to produce rGO, the functional groups and conductivity of the material can vary considerably. Reduced graphene oxides have multiple sites for easy surface modification with tunable conductivity, and hence, it is an ideal material for sensors (Kour et al., 2020).

A super-hydrophobic reduced graphene oxide (rGO) for NO_2 detection was synthesized using spark plasma sintering of graphene oxide by Wu et al. The super-hydrophobic reduced graphene oxide sensor exhibited a low limit of detection (9.1 ppb) and high sensitivity (25.5 ppm^{-1}) due to its super-hydrophobicity, high surface area, and large defect sites (Wu, Tao, Miao, & Norford, 2018). Chen et al. utilized rGO combined with benzyl triethyl ammonium chloride (Betec) as a nitrate sensor, which showed a low limit of detection value (1.1 μg/L), quick response (2–7 s), and selectivity against interfering ions (Cl^-, SO_4^{2-}, CO_3^{2-}) (Chen et al., 2018). Sim et al. have reported the synthesis of rGO nanowalls deposited on a graphite rod using electrophoretic deposition method and used it for the detection of four bases of DNA. This sensor exhibited the highly resolved oxidation signals for the four bases of DNA (Sim et al., 2012).

In another study, a Pt NPs/rGO composite was prepared using photochemical reduction of Pt nanoparticles on reduced graphene oxide and utilized for the conductive channel in a solution-gated field effect transistor (FET), by Yin and co-workers. This sensor was used for the real-time detection of hybridization of single-stranded DNA (ssDNA) with high sensitivity (2.4 nM) (Yin et al., 2012). Mao et al. have fabricated field effect transistor biosensor using thermally reduced graphene oxide sheets decorated with gold nanoparticle (Au NP)-antibody conjugates and utilized this biosensor to detect a protein specific to the antibody (Mao, Lu, Yu, Bo, & Chen, 2010). A label-free electrochemical immune sensor was synthesized using rGO/Ag NPs composites as a support material for detecting prostate-specific antigen (PSA) by Han et al. The results have revealed that this electrochemical immune sensor exhibited a low detection limit (0.01 ng m^{-1}), wide linear response range (1.0–1000 ng m^{-1}), outstanding stability, reproducibility, and specificity (Han et al., 2017). Wu et al. constructed gold nanoparticle (AuNP)-dotted reduced graphene oxide (RGO-AuNP) for an aptamer biosensor to selectively detect 3,3′4,4′-polychlorinated biphenyls. The exposure thresholds (uncontaminated water < 0.1 ng L^{-1}) set out by the Environmental Protection Agency (EPA) and International Agency for Research on Cancer (IARC) were satisfied by this aptamer biosensor with a low limit of detection (LOD) of 0.1 pg L^{-1} (Wu, Lu, Fu, Wu, & Liu, 2017).

Chae and co-workers treated reduced grapheme oxide surfaces using oxygen plasma and used it as reactive interfaces for the detection of amyloid-beta (Aβ) peptides, which are the pathological hallmarks of Alzheimer's disease (Chae et al., 2017). An aptamer-immobilized reduced graphene oxide (rGO) field effect transistor (FET) for the detection of immunoglobulin E (IgE) with the detection limit of 8.1 ng m^{-1} was reported by Masaki, Yuki, Yasuhide, Kenzo, and Kazuhiko (2014).

FET biosensor was fabricated using drop casting the reduced graphene oxide suspension onto the sensor surface and used as a biosensor for the detection of DNA through peptide nucleic acid hybridization by Cai and co-workers (Cai et al., 2014). The Pd or Pt nanoparticles were decorated on the TiO_2/RGO hybrid structures using chemical reduction by Esfandiar et al. When comparing to the Pt/TiO_2/RGO sensor, the Pd/TiO_2/RGO sensor exhibited an enhanced sensitivity due to the better rate of spillover effect and dissociation of hydrogen molecules on Pd (Esfandiar, Ghasemi, Irajizad, Akhavan, & Gholami, 2012). Xu and co-workers fabricated rGO and rGO-incorporated composite monolayer ordered porous (MOP) films directly on a ceramic tube by a new template-induced solution-heated method and utilized it as sensors to detect ethanol gas (Xu et al., 2016).

Fullerene-based sensors

Fullerenes are allotropic modification of carbon and consist of a three-dimensional closed cage (C_n) made of 5- and 6-membered rings of sp^2 carbons with the number of pentagons and hexagons. The most abundant and smallest stable fullerene has truncated icosahedron structure and known as buckminsterfullerene C_{60}. Other stable fullerenes are C_{70}, C_{74}, C_{76}, C_{78}, C_{80}, C_{82}, C_{84}, and so on. Fullerenes have excellent properties, such as large specific surface area, high electronic conductivity, inert behavior, good biocompatibility, strong redox activity, remarkable stability, excellent electron-accepting property, good adsorption capacity, and magnificent electrochemical properties. Due to these outstanding properties, fullerenes are utilized in the construction of extremely sensitive sensors and selective detection of various analytes. Wei et al. have reported the synthesis of Au-TiO_2-C_{60} photocatalytic sensor to determine the concentration of organic compounds with high sensibility and precision (Wei, Yang, Mo, & Leng, 2018). Saha and Das have synthesized C_{60}-modified Al_2O_3 and used them as sensor for the detection of humidity in a gaseous environment with high sensitivity and very sharp quasi-linear response (Saha & Das, 2018). Prasad et al. have produced an electrochemical sensor using electroactive C_{60}-monoadduct-micellar molecularly imprinted polymer grafted on ionic liquid-based carbon ceramic electrode for the detection of chlorambucil (an anticancer drug) with high analyte diffusivity and fast electrode kinetics (Prasad, Singh, & Kumar, 2017). Bashiri et al. investigated the interaction of amphetamine (illicit drug) with pristine C_{60} fullerenes and also C_{60} doped with B, Al, Ga, Si, and Ge. Even though pristine C_{60} exhibited good sensitivity to amphetamine, the weak interaction between them limits C_{60} use as an amphetamine drug sensor. The results showed that doping C_{60} with other element improved the sensibility of C_{60} to amphetamine drug (Bashiri, Vessally, Bekhradnia, Hosseinian, & Edjlali, 2017). Moradi et al. investigated the pristine, Al- and Si doped C_{60} fullerenes for the detection of phenyl propanol amine drug. Although pristine C_{60} exhibited good sensitivity to phenyl propanolamine drug, the weak interaction between them limits C_{60} use as a phenyl propanolamine drug sensor. The Al-doped C_{60} exhibited high reactivity with phenyl propanolamine, but less sensitivity. Si-doped C_{60} exhibited excellent reactivity and sensitivity (Moradi, Nouraliei, & Moradi, 2017).

Rahimi-Nasrabadi and co-workers used a sensor-modified electrode with CNT, C_{60}, and 1-butyl-3-methylimidazolium tetrafluoroborate ionic liquid for the determination of diazepam on real samples, like urine, serum, and tablets (Rahimi-Nasrabadi, Khoshroo, & Mazloum-Ardakani, 2017)]. Li et al. synthesized C_{60}-decorated Au nanoparticles @ MoS_2 as a biosensor for adenosine triphosphate (ATP), an essential biomolecule responsible for cellular metabolism with a detection limit of 3.30 fM (Li, Zheng, Liang, Yuan, & Chai, 2017). In another report, functionalized fullerene-C_{60}-thiol-mediated gold nanocomposite has been prepared as a sensor for L-histidine by Sutradhar and co-workers. This L-histidine sensor exhibited superior performance in its linear range, detection limit, stability, sensitivity, and specificity (Sutradhar, Jacob, & Patnaik, 2017). Hao et al. have reported the synthesis of C_{60} fullerene and methylene blue nanoflowers and used it as an electrochemical indicator for the detection of the prostate-specific antigen. The results revealed that the detection limit was 1.7 pg·mL^{-1} (Hao et al., 2017). Zhou and co-workers have developed amino and sulfhydryl multifunctional fullerene nanoparticles and used them as sensor for miRNA-141 (Zhou, Wang, et al., 2020; Zhou, Xue, et al., 2020). Liu et al. have prepared an ECL biosensor based on functionalized fullerenes. In this study, pyrrolidyl fullerenes derivative was used as congruent and unadulterated nano-hub to converge three zinc porphyrins on its monopole. This peculiar assembly was convinced via microimaging and spectrophotometry and used for the detection of miRNA markers (Liu et al., 2020).

In another report, functionalized fullerene nanorod (C_{60}NRs–NH–Ph–GCE)-modified electrodes have been developed by Rather and co-workers and used for the detection of ethylparaben (EP) over a concentration range from 0.01 to 0.52 mM with a detection limit (LOD) of 3.8 nM (Rather et al., 2016). Zhong et al. have reported the synthesis of fullerenes modified with cysteine and palladium nanoparticles (Pd@Cys-C_{60} structure) and utilized it as a sensor for glucose detection without enzyme (Zhong, Yuan, & Chai, 2012). Pan and co-workers have synthesized piezoelectric-based immune sensor system with C_{60} antibodies and used them as sensor for human IgG and hemoglobin.

Mesoporous carbon-based sensors

Ordered mesoporous carbons (OMC) are flexible, extremely open rigid structure, 3D nanostructured porous carbon materials with uniform and tunable pore sizes. Ordered mesoporous carbons have high specific surface areas, good electronic conductivity, intrinsic electrocatalytic properties, chemical inertness, great porosity, large pore volume, large pore sizes with narrow distribution, regular frameworks, good biocompatibility, remarkable functional properties, good chemical stability, excellent thermal stability, significant edge-plane-like sites, and oxygen-rich groups. These attractive features make ordered mesoporous carbons as an outstanding electrode material for electrochemical sensing and biosensing. Balasubramanian and co-workers have synthesized B/N-co-doped mesoporous carbon and used them as sensor for electrochemical detection of isoniazid with better analytical performance in a broad dynamic range 0.02–1783 μM, rapid response time (<4 s), and low detection limit of 1.5 nM. The results revealed that excellency of electrochemical sensor was due to low pore size, large surface area, enhanced electrical conductivity, and abundant active sites (Balasubramanian, Balamurugan, Chen, Chen, & Lin, 2019).

Cao et al. investigated the sensitivity of hemin-ordered mesoporous carbon-based sensor to hydrogen peroxide. This sensor exhibited good reproducibility and stability with the detection limit of 0.3 M (Cao, Sun, Zhang, Hu, & Jia, 2012). Ndamanisha and Guo developed ordered mesoporous carbon functionalized with ferrocenecarboxylic acid-modified glassy carbon electrode for the detection of uric acid in urine sample with good stability, reproducibility, and a detection limit of 1.8 μM (Ndamanisha & Guo, 2008). Lei et al. synthesized ordered mesoporous carbon functionalized with ferrocene-carboxylic acid-modified glassy carbon electrode and used it for the determination of morphine in urine samples with a low detection limit of 50 nM and satisfied recovery of 96.4% samples (Li et al., 2010). Wang and co-workers have prepared electrochemical sensor based on magnetic mesoporous hollow carbon microspheres as an immobilized enzyme material and used them as a sensor for penicillin sodium in milk using a standard addition method. This study revealed that this method has a wide detection range (10^{-8} to 10^{-2} mg/mL), detection limit of 2.655×10^{-7} mg/mL, good precision and stability, enzyme immobilization capacity, convenient cleaning, and recycling of solid materials (Wang et al., 2020).

In another report, hexagonally ordered mesoporous carbon powder gas sensors have been prepared by Lu and co-workers and used it for the detection of carbon monoxide with reproducible sensing behavior (Lu, Liao, Tsai, Liu, & Leu, 2009). Ndamanisha et al. have investigated the electrochemical activity of ordered mesoporous carbon for the determination of glutathione and cysteine. The results revealed that the sensor has acceptable sensitivity and detection limits in a large determination range (Ndamanisha, Bai, Qi, & Guo, 2009). Ju and Guo prepared a voltammetric sensor by modifying glassy carbon electrode with ordered mesoporous carbon for the determination of amaranth with a lower detection limit of 3.2×10^{-8} M (Ju & Guo, 2013).

Bai and co-workers have developed ordered mesoporous carbon-modified glassy carbon electrode and used them as a sensor for the detection of riboflavin. This study showed that ordered mesoporous carbon-modified glassy carbon electrode exhibited high sensitivity of 769 μA mM^{-1}, low detection limit 2×10^{-8} M, good stability, and reproducibility (Bai, Ndamanisha, Liu, Yang, & Guo, 2010). Song et al. prepared mesoporous carbon, mesoporous carbon with carboxylic and amino group-modified electrodes for the determination of dopamine in the presence of ascorbic acid with the lowest detection limits of 4.5×10^{-3}, 4.4×10^{-2}, and 0.33 μM, respectively (Song, Gao, Xia, & Gao, 2008). In another study, a sensitive electrochemical sensor was prepared by Zang et al. to detect nitroaromatic explosives using ordered mesoporous carbon. This study revealed that as low as 0.2 ppb, 1 ppb, and 1 ppb of TNT, 2,4-dinitrotoluene, 1,3-dinitrobenzene can be detected, respectively, on ordered mesoporous carbon-based electrodes (Zang, Guo, Hu, Yu, & Li, 2011). Chrysostome et al. have investigated the ordered mesoporous carbon-modified electrode for the detection of glucose with high sensitivity of 10.81 μA/mM and a low detection limit of 0.02 mM (Ndamanisha & Guo, 2009).

Carbon quantum dot-based sensors

Carbon quantum dots are zero-dimensional (0D) carbon nanomaterials with 1–10 nm diameters. They have attracted significant interest as sensors in recent years due to their properties such as good conductivity, high chemical stability, broadband optical absorption, photostability, low toxicity, excellent biocompatibility, simple synthetic routes, environmental friendliness, good chemical inertness, solubility and unique electronic, chemiluminescent, photoluminescent, fluorescent, and electrochemiluminescent properties. Zhang and co-workers have synthesized carbon dot-based nano-biosensor functionalized with zinc-responsive quinoline derivatives and used them as a sensor for the detection of quinoline with a low detection limit of 6.4 nM (Zhang et al., 2014). Yang et al. have reported the synthesis of reduced graphene oxide-carbon dot composite film for the detection of dopamine with high sensitivity, excellent specificity, good stability, and low detection limit of 1.5 nM (Hu et al., 2014). A DNA sensor was prepared by Wang et al. using N,N'-disuccinimidyl carbonate and carbon dot-dotted nanoporous gold with fruitful performance with a linear range of 10^{-18} to 10^{-14} M and a detection limit of 8.56×10^{-19} M for DNA (Wang et al., 2013)]. Tripathi et al. synthesized a fluorescent carbon dot from environmental pollutant diesel soot and used it as a sensor for cholesterol and in imaging *Escherichia coli* (Tripathi, Sonker, Sonkar, & Sarkar, 2014). Shen et al. fabricated fluorescent boronic acid-modified carbon dot sensor for the determination of nonenzymatic blood glucose with excellent selectivity and effectivity (Shen & Xia, 2014). In another report, polyethylenimine-functionalized carbon quantum dots have been prepared using the one-step hydrothermal method for the detection of Cu^{2+} with a detection limit of 80 nmol/L by Deng et al. (2020).

Zong et al. used carbon dots as fluorescent probes for the detection of copper ion and L-cysteine with a low detection limit of 2.3×10^{-8} M and 3.4×10^{-10} M, respectively (Zong et al., 2014). Wei et al. synthesized nitrogen-doped carbon dots and used it as a biosensor for ovalbumin. Zheng and co-workers have developed a low-cost and highly sensitive fluorescent carbon dot nanosensor for the determination of chromium (VI) (Zheng et al., 2013). Maiti et al. have reported the synthesis of quaternized carbon dot–DNA nano-biohybrid for the detection of this tone with a minimum detection limit of $0.2\,\text{ng mL}^{-1}$ (Maiti, Das, & Das, 2013). Deng et al. prepared carbon dot-functionalized mesoporous Pt/Fe nanoparticles for the detection of human carcinoembryonic antigen in human serum albumin samples with good linear response range from 0.003 to $600\,\text{ng mL}^{-1}$, low detection limit of $0.8\,\text{pg mL}^{-1}$, good specificity, reproducibility, and acceptable stability (Deng et al., 2014).

Costas-Mora et al. developed a nanoprobe based on fluorescent carbon dots using ultrasound-assisted synthesis for the detection of methyl mercury with a detection limit of 5.9 nM (Costas-Mora, Romero, Lavilla, & Bendicho, 2014). Li et al. have reported the detection of organophosphorus pesticide through dual-mode (fluorometric and colorimetric) channels based on acetyl cholinesterase (AChE)-controlled quenching of fluorescence carbon dots with a low detection limit of $0.4\,\text{ng mL}^{-1}$ (Li, Yan, Lu, & Su, 2018). In another research, Dong et al. used fluorescent carbon as a signal probe in conventional ELISA to improve the sensitivity for the detection of residual amantadine in chicken muscle with a detection limit of $0.02\,\text{ng mL}^{-1}$ (Dong et al., 2019). Murru and co-workers have synthesized room-temperature phosphorescent carbon dots and used them as a sensor for hemoglobin detection with good sensitivity and low detection limit of 6.2 nM (Murru et al., 2020). Zhou et al. have developed photoluminescent carbon dots using green hydrothermal method and used it as a sensor for the detection of iron (III) ions (Zhou, Zhou, Gong, Zhang, & Li, 2015). Zhang et al. have investigated the synthesis of terbium-functionalized schizochytrium-derived carbon dots for ratiometric fluorescence determination of 2,6-dipicolinic acid (anthrax biomarker) with the detection limit of 35.9 nM (Zhang et al., 2014).

Sensors based on carbon hybrid structures

Carbon hybrid materials are formed by the integration of one or more carbon allotropes with possible additions of organic or inorganic components. They exhibit superior properties that go extensively beyond those of its building blocks. The building blocks of carbon hybrid materials are typically joined through a synergistic relationship, giving rise to exceptional properties to the new carbon hybrid material due to the complementarity of unique features in each constituent. Research on the new carbon hybrid structures based on nanodiamonds, carbon nanotubes, buckypaper, graphene, graphene oxide, reduced graphene oxide, fullerenes, mesoporous carbon, and carbon quantum dots has been exclusively investigated for sensor applications because of their outstanding properties, such as rapid responsive, high surface area, good selectivity, adaptability, quick response, recovery, unique functionality, immense stability, great repeatability, and advantageous electrochemical properties. Zhang and co-workers have synthesized hybrid nanosensor with nanodiamonds as the core, a shell of hydrogel, and surface-docked magnetic nanoparticles and used it for temperature sensing (Zhang et al., 2018). Lee et al. developed a chemical sensor by integrating carbon nanotubes and nano-layered transition metal dichalcogenides on cellulose paper and utilized it for the detection of NO_2 gas with a sensitivity of $4.57\%\,\text{ppm}^{-1}$ (Lee & Choi, 2019). Danish et al.

have investigated the PI tape-enabled buckypaper hybrid sensor for structural health monitoring applications with excellent stability and reproducibility (Danish & Luo, 2019). In another report, wearable resistive gas sensors based on hybrid graphene/zinc oxide (ZnO) nanocomposites have been prepared by Utari and co-workers and used them as a CO gas sensor, which has high sensitivity, shortest recovery time, rapid response, and selectivity toward CO gas (Utari et al., 2020). Hue et al. developed graphene oxide /graphene hybrid film by a simple drop-casting method and utilized it as a sensor for the detection of ammonia with the detection limit of 1.5 ppb (Hue et al., 2021). Zhang and co-workers have synthesized hybrid composite of tin oxide nanocrystal-decorated reduced graphene oxide by hydrothermal synthesis and used them as a sensor for ethanol gas with a low detection limit of 1 ppm, fast response, rapid recovery time, and good repeatability (Zhang, Liu, Chang, Liu, & Xia, 2015). In another report, a nanohybrid tin oxide quantum dot-C60 fullerene using hydrothermal method has been prepared by Keshtkar and co-workers and used it as a gas sensor for the detection of various gases, including H_2S, methane, and propane. The results revealed that this hybrid sensor exhibited high response and good selectivity (Keshtkar et al., 2018). Barathi et al. have investigated α-Fe_2O_3 nanoleaf-incorporated mesoporous carbon-chitosan hybrid sensor for the detection of nitrite, which exhibited excellent electrocatalytic activity toward nitrite with the detection limit of $31\,nM\,cm^{-2}$ (Barathi, Subramania, & Devaraj, 2020). Mousavi et al. have reported the synthesis of a hybrid fluorescence probe sensor based on Fe–amine framework carbon quantum dots for the detection of 6-mercaptopurine, which exhibited high selectivity, good sensitivity, and low detection limit of $55.70\,ng\,L^{-1}$ (Mousavi, Zare-Dorabei, & Mosavi, 2020).

Vision for next 20 years

(1) *Tuning the properties of carbon materials using surface functionalization*

The important reasons for the use of carbon nanomaterials as sensors are high sensitivity, high selectivity, high surface-to-volume ratio, and excellent electrochemical properties. However, this sensitivity and other properties can be improved even further by surface functionalization of these carbon-based materials. These surface modifications can be done by integrating the carbon materials with nanoparticles, metals, metal oxides, organometallic, ionic liquids, polymers, enzymes, DNA, etc., which will increase the applicability of the carbon materials as sensors. Additionally, by tuning and enhancing the surface architecture, it is possible to attain improvement in the properties such as electrical conductivity, sensor selectivity, surface area, lower limit of detection, detection range, and biocompatibility, which will significantly increase the sensitivity of the carbon-based sensors.

(2) *Determination of more than one analyte*

Another challenging topic is multidetermination. There are few carbon-based sensors that are available for multidetermination. Thus, proper attention should be given to explore new carbon-based sensors for the determination of more than one analyte.

(3) *Detection of disease-causing agents*

It is essential for this world to develop carbon-based biosensors with the characteristic features to detect all the disease-causing agents. Immunoassay detection is a well-known procedure to determine the presence or concentration of various target molecules in biological matrices. The inherent features of carbon materials make them as a strong candidate to detect disease-causing agents using immunoassay detection method with high accuracy. COVID-19 is a major crisis for the health and economy of this world in 21st century. Thus, proper studies are required to use the carbon-based materials along with antibody–antigen–antibody strategy to detect COVID-19.

(4) *Integration of sensors with on-board electronics*

On-board electronics is a subsystem meant for high-end command, control system and data recording applications. Appropriate studies are required to integrate the carbon-based sensors within biochips with on-board electronics. This integration will guide to develop fabricating devices with small size, very low-cost, and simple operation procedure.

(5) *Carbon-based sensors instead of noble metal-based sensors*

Noble metal-based sensors have been used for the detection of organic dyes, DNA/RNA, and protein etc. Since noble metals are very costly, it is necessary to develop low-cost carbon-based sensors with similar sensing features by substituting expensive noble metal substrates with carbon material.

References

Ardila, J. A., Oliveira, G. G., Medeiros, R. A., & Fatibello-Filho, O. (2014). Square-wave adsorptive stripping voltammetric determination of nanomolar levels of bezafibrate using a glassy carbon electrode modified with multi-walled carbon nanotubes within a dihexadecyl hydrogen phosphate film. *Analyst, 139*(7), 1762–1768. https://doi.org/10.1039/c3an02016a.

Azevedo, A. F., Baldan, M. R., & Ferreira, N. G. (2012). Nanodiamond films for applications in electrochemical systems. *International Journal of Electrochemistry*, 1–16. https://doi.org/10.1155/2012/508453.

Baccarin, M., Rowley-Neale, S. J., Cavalheiro, É. T. G., Smith, G. C., & Banks, C. E. (2019). Nanodiamond based surface modified screen-printed electrodes for the simultaneous voltammetric determination of dopamine and uric acid. *Microchimica Acta, 186*(3). https://doi.org/10.1007/s00604-019-3315-y.

Bai, J., Ndamanisha, J. C., Liu, L., Yang, L., & Guo, L. (2010). Voltammetric detection of riboflavin based on ordered mesoporous carbon modified electrode. *Journal of Solid State Electrochemistry, 14*(12), 2251–2256. https://doi.org/10.1007/s10008-010-1065-1.

Bajtarevic, A., Ager, C., Pienz, M., Klieber, M., Schwarz, K., Ligor, M., et al. (2009). Noninvasive detection of lung cancer by analysis of exhaled breath. *BMC Cancer, 9*, 348. https://doi.org/10.1186/1471-2407-9-348.

Balasubramanian, P., Balamurugan, T. S. T., Chen, S. M., Chen, T. W., & Lin, P. H. (2019). A novel, efficient electrochemical sensor for the detection of isoniazid based on the B/N doped mesoporous carbon modified electrode. *Sensors and Actuators, B: Chemical, 283*, 613–620. https://doi.org/10.1016/j.snb.2018.12.020.

Barathi, P., Subramania, A., & Devaraj, A. (2020). Mesoporous carbon/α-Fe2O3 nanoleaf composites for disposable nitrite sensors and energy storage applications. *ACS Omega, 5*(50), 32160–32170. https://doi.org/10.1021/acsomega.0c02594.

Bashiri, S., Vessally, E., Bekhradnia, A., Hosseinian, A., & Edjlali, L. (2017). Utility of extrinsic [60] fullerenes as work function type sensors for amphetamine drug detection: DFT studies. *Vacuum, 136*, 156–162. https://doi.org/10.1016/j.vacuum.2016.12.003.

Bianco, A., Kostarelos, K., Partidos, C. D., & Prato, M. (2005). Biomedical applications of functionalised carbon nanotubes. *Chemical Communications, 5*, 571–577. https://doi.org/10.1039/b410943k.

Bianco, A., Kostarelos, K., & Prato, M. (2011). Making carbon nanotubes biocompatible and biodegradable. *Chemical Communications, 47*(37), 10182–10188. https://doi.org/10.1039/c1cc13011k.

Burrs, S. L., Bhargava, M., Sidhu, R., Kiernan-Lewis, J., Gomes, C., Claussen, J. C., et al. (2016). A paper based graphene-nanocauliflower hybrid composite for point of care biosensing. *Biosensors and Bioelectronics, 85*, 479–487. https://doi.org/10.1016/j.bios.2016.05.037.

Cai, B., Wang, S., Huang, L., Ning, Y., Zhang, Z., & Zhang, G. J. (2014). Ultrasensitive label-free detection of PNA-DNA hybridization by reduced graphene oxide field-effect transistor biosensor. *ACS Nano, 8*(3), 2632–2638. https://doi.org/10.1021/nn4063424.

Cao, H., Sun, X., Zhang, Y., Hu, C., & Jia, N. (2012). Electrochemical sensing based on hemin-ordered mesoporous carbon nanocomposites for hydrogen peroxide. *Analytical Methods, 4*(8), 2412–2416. https://doi.org/10.1039/c2ay25358e.

Chae, M. S., Kim, J., Jeong, D., Kim, Y. S., Roh, J. H., Lee, S. M., et al. (2017). Enhancing surface functionality of reduced graphene oxide biosensors by oxygen plasma treatment for Alzheimer's disease diagnosis. *Biosensors and Bioelectronics, 92*, 610–617. https://doi.org/10.1016/j.bios.2016.10.049.

Chen, X., Pu, H., Fu, Z., Sui, X., Chang, J., Chen, J., et al. (2018). Real-time and selective detection of nitrates in water using graphene-based field-effect transistor sensors. *Environmental Science: Nano, 5*(8), 1990–1999. https://doi.org/10.1039/c8en00588e.

Chen, P., Xiao, T. Y., Qian, Y. H., Li, S. S., & Yu, S. H. (2013). A nitrogen-doped graphene/carbon nanotube nanocomposite with synergistically enhanced electrochemical activity. *Advanced Materials, 25*(23), 3192–3196. https://doi.org/10.1002/adma.201300515.

Chen, S., Yuan, R., Chai, Y., & Hu, F. (2013). Electrochemical sensing of hydrogen peroxide using metal nanoparticles: a review. *Microchimica Acta, 180*(1–2), 15–32. https://doi.org/10.1007/s00604-012-0904-4.

Choi, Y. E., Kwak, J. W., & Park, J. W. (2010). Nanotechnology for early cancer detection. *Sensors, 10*(1), 428–455. https://doi.org/10.3390/s100100428.

Colvin, V. L. (2003). The potential environmental impact of engineered nanomaterials. *Nature Biotechnology, 21*(10), 1166–1170. https://doi.org/10.1038/nbt875.

Costas-Mora, I., Romero, V., Lavilla, I., & Bendicho, C. (2014). In situ building of a nanoprobe based on fluorescent carbon dots for methylmercury detection. *Analytical Chemistry, 86*(9), 4536–4543. https://doi.org/10.1021/ac500517h.

Curá, J. A., Jansson, P., & Krisman, C. R. (1995). Amylose is not strictly linear. *Starch—Stärke, 47*(6), 207–209. https://doi.org/10.1002/star.19950470602.

Danish, M., & Luo, S. (2019). Micro-crack induced buckypaper/PI tape hybrid sensors with enhanced and tunable piezo-resistive properties. *Scientific Reports, 9*(1). https://doi.org/10.1038/s41598-019-53222-1.

De Volder, M. F. L., Tawfick, S. H., Baughman, R. H., & Hart, A. J. (2013). Carbon nanotubes: Present and future commercial applications. *Science, 339*(6119), 535–539. https://doi.org/10.1126/science.1222453.

Deng, X. Y., Feng, Y. L., He, D. S., Zhang, Z. Y., Liu, D. F., & Chi, R. A. (2020). Synthesis of functionalized carbon quantum dots as fluorescent probes for detection of Cu2+. *Chinese Journal of Analytical Chemistry, 48*(10), e20126–e20133. https://doi.org/10.1016/S1872-2040(20)60054-8.

Deng, W., Liu, F., Ge, S., Yu, J., Yan, M., & Xianrang, S. (2014). A dual amplification strategy for ultrasensitive electrochemiluminescence immunoassay based on a Pt nanoparticles dotted graphene–carbon nanotubes composite and carbon dots functionalized mesoporous Pt/Fe. *Analyst, 139*(7), 1713–1720. https://doi.org/10.1039/c3an02084c.

Dong, B., Li, H., Mujtaba Mari, G., Yu, X., Yu, W., Wen, K., et al. (2019). Fluorescence immunoassay based on the inner-filter effect of carbon dots for highly sensitive amantadine detection in foodstuffs. *Food Chemistry, 294*, 347–354. https://doi.org/10.1016/j.foodchem.2019.05.082.

Eguílaz, M., Gutierrez, F., González-Domínguez, J. M., Martínez, M. T., & Rivas, G. (2016). Single-walled carbon nanotubes covalently functionalized with polytyrosine: A new material for the development of NADH-based biosensors. *Biosensors and Bioelectronics, 86*, 308–314. https://doi.org/10.1016/j.bios.2016.06.003.

Esfandiar, A., Ghasemi, S., Irajizad, A., Akhavan, O., & Gholami, M. R. (2012). The decoration of TiO2/reduced graphene oxide by Pd and Pt nanoparticles for hydrogen gas sensing. *International Journal of Hydrogen Energy, 37*(20), 15423–15432. https://doi.org/10.1016/j.ijhydene.2012.08.011.

Ferreira, H. S., & Rangel, M. D. C. (2009). Nanotecnologia: Aspectos gerais e potencial de aplicação em catálise. *Quimica Nova, 32*(7), 1860–1870. https://doi.org/10.1590/S0100-40422009000700033.

Galano, A. (2010). Carbon nanotubes: Promising agents against free radicals. *Nanoscale, 2*(3), 373–380. https://doi.org/10.1039/b9nr00364a.

Han, L., Liu, C. M., Dong, S. L., Du, C. X., Zhang, X. Y., Li, L. H., et al. (2017). Enhanced conductivity of rGO/Ag NPs composites for electrochemical immunoassay of prostate-specific antigen. *Biosensors and Bioelectronics, 87*, 466–472. https://doi.org/10.1016/j.bios.2016.08.004.

Hao, Y., Yan, P., Zhang, X., Shen, H., Gu, C., Zhang, H., et al. (2017). Ultrasensitive amperometric determination of PSA based on a signal amplification strategy using nanoflowers composed of single-strand DNA modified fullerene and methylene blue, and an improved surface-initiated enzymatic polymerization. *Microchimica Acta, 184*(11), 4341–4349. https://doi.org/10.1007/s00604-017-2476-9.

Harris, P. J. F. (2009). *Carbon nanotube science: Synthesis, properties and applications.* Cambridge University Press.

Hirsch, A. (2010). The era of carbon allotropes. *Nature Materials, 9*(11), 868–871. https://doi.org/10.1038/nmat2885.

Holzinger, M., Le Goff, A., & Cosnier, S. (2014). Nanomaterials for biosensing applications: A review. *Frontiers in Chemistry, 2.* https://doi.org/10.3389/fchem.2014.00063.

Hoover, R. (2001). Composition, molecular structure, and physicochemical properties of tuber and root starches: A review. *Carbohydrate Polymers, 45*(3), 253–267. https://doi.org/10.1016/S0144-8617(00)00260-5.

Hu, S., Huang, Q., Lin, Y., Wei, C., Zhang, H., Zhang, W., et al. (2014). Reduced graphene oxide-carbon dots composite as an enhanced material for electrochemical determination of dopamine. *Electrochimica Acta, 130,* 805–809. https://doi.org/10.1016/j.electacta.2014.02.150.

Huang, X., Xu, S., Zhao, W., Xu, M., Wei, W., Luo, J., et al. (2020). Screen-printed carbon electrodes modified with polymeric nanoparticle-carbon nanotube composites for enzymatic biosensing. *ACS Applied Nano Materials, 3*(9), 9158–9166. https://doi.org/10.1021/acsanm.0c01800.

Hue, N. T., Wu, Q., Liu, W., Bu, X., Wu, H., Wang, C., et al. (2021). Graphene oxide/graphene hybrid film with ultrahigh ammonia sensing performance. *Nanotechnology, 32*(11). https://doi.org/10.1088/1361-6528/abd05a, 115501.

Iijima, S. (1991). Helical microtubules of graphitic carbon. *Nature, 354*(6348), 56–58. https://doi.org/10.1038/354056a0.

Iijima, S., & Ichihashi, T. (1993). Single-shell carbon nanotubes of 1-nm diameter. *Nature, 363*(6430), 603–605. https://doi.org/10.1038/363603a0.

Jane, J., Chen, Y. Y., Lee, L. F., McPherson, A. E., Wong, K. S., Radosavljevic, M., et al. (1999). Effects of amylopectin branch chain length and amylose content on the gelatinization and pasting properties of starch. *Cereal Chemistry, 76*(5), 629–637. https://doi.org/10.1094/CCHEM.1999.76.5.629.

Ju, J., & Guo, L.-P. (2013). Sensitive Voltammetric sensor for Amaranth based on ordered mesoporous carbon. *Chinese Journal of Analytical Chemistry, 41*(5), 681–686. https://doi.org/10.1016/s1872-2040(13)60650-7.

Kam, N. W. S., Liu, Z., & Dai, H. (2005). Functionalization of carbon nanotubes via cleavable disulfide bonds for efficient intracellular delivery of siRNA and potent gene silencing. *Journal of the American Chemical Society, 127*(36), 12492–12493. https://doi.org/10.1021/ja053962k.

Keshtkar, S., Rashidi, A., Kooti, M., Askarieh, M., Pourhashem, S., Ghasemy, E., et al. (2018). A novel highly sensitive and selective H2S gas sensor at low temperatures based on SnO2 quantum dots-C60 nanohybrid: Experimental and theory study. *Talanta, 188,* 531–539. https://doi.org/10.1016/j.talanta.2018.05.099.

Kim, J., Park, G., Lee, S., Hwang, S. W., Min, N., & Lee, K. M. (2017). Single wall carbon nanotube electrode system capable of quantitative detection of CD4+ T cells. *Biosensors and Bioelectronics, 90,* 238–244. https://doi.org/10.1016/j.bios.2016.11.055.

Kim, S. J., Song, W., Yi, Y., Min, B. K., Mondal, S., An, K. S., et al. (2018). High durability and waterproofing rGO/SWCNT-fabric-based multifunctional sensors for human-motion detection. *ACS Applied Materials and Interfaces, 10*(4), 3921–3928. https://doi.org/10.1021/acsami.7b15386.

Klaus, D. S. (2017). *Handbook of nanophysics: Clusters and fullerenes.* CRC Press.

Koskun, Y., Şavk, A., Şen, B., & Şen, F. (2018). Highly sensitive glucose sensor based on monodisperse palladium nickel/activated carbon nanocomposites. *Analytica Chimica Acta, 1010,* 37–43. https://doi.org/10.1016/j.aca.2018.01.035.

Kour, R., Arya, S., Young, S. J., Gupta, V., Bandhoria, P., & Khosla, A. (2020). Review—Recent advances in carbon nanomaterials as electrochemical biosensors. *Journal of the Electrochemical Society, 167*(3), 037555.

Kreyling, W. G., Semmler-Behnke, M., & Chaudhry, Q. (2010). A complementary definition of nanomaterial. *Nano Today, 5*(3), 165–168. https://doi.org/10.1016/j.nantod.2010.03.004.

Kroto, H. W., Heath, J. R., O'Brien, S. C., Curl, R. F., & Smalley, R. E. (1985). C60: Buckminsterfullerene. *Nature, 318*(6042), 162–163. https://doi.org/10.1038/318162a0.

Kulakova, I. I. (2004). Surface chemistry of nanodiamonds. *Physics of the Solid State, 46*(4), 636–643. https://doi.org/10.1134/1.1711440.

Kuretake, T., Kawahara, S., Motooka, M., & Uno, S. (2017). An electrochemical gas biosensor based on enzymes immobilized on chromatography paper for ethanol vapor detection. *Sensors (Switzerland), 17*(2), 281. https://doi.org/10.3390/s17020281.

Lee, W. S., & Choi, J. (2019). Hybrid integration of carbon nanotubes and transition metal Dichalcogenides on cellulose paper for highly sensitive and extremely deformable chemical sensors. *ACS Applied Materials and Interfaces, 11*(21), 19363–19371. https://doi.org/10.1021/acsami.9b03296.

Li, F., Song, J., Shan, C., Gao, D., Xu, X., & Niu, L. (2010). Electrochemical determination of morphine at ordered mesoporous carbon modified glassy carbon electrode. *Biosensors and Bioelectronics, 25*(6), 1408–1413. https://doi.org/10.1016/j.bios.2009.10.037.

Li, Q., Xia, Y., Wan, X., Yang, S., Cai, Z., Ye, Y., et al. (2020). Morphology-dependent MnO2/nitrogen-doped graphene nanocomposites for simultaneous detection of trace dopamine and uric acid. *Materials Science and Engineering C, 109.* https://doi.org/10.1016/j.msec.2019.110615.

Li, H., Yan, X., Lu, G., & Su, X. (2018). Carbon dot-based bioplatform for dual colorimetric and fluorometric sensing of organophosphate pesticides. *Sensors and Actuators, B: Chemical, 260,* 563–570. https://doi.org/10.1016/j.snb.2017.12.170.

Li, M. J., Zheng, Y. N., Liang, W. B., Yuan, R., & Chai, Y. Q. (2017). Using p-type PbS quantum dots to quench photocurrent of fullerene-Au NP@MoS2 composite structure for ultrasensitive photoelectrochemical detection of ATP. *ACS Applied Materials and Interfaces, 9*(48), 42111–42120. https://doi.org/10.1021/acsami.7b13894.

Liu, G., Hong, J., Ma, K., Wan, Y., Zhang, X., Huang, Y., et al. (2020). Porphyrin triopendant fullerene guest as an in situ universal probe of high ECL efficiency for sensitive miRNA detection. *Biosensors and Bioelectronics, 150.* https://doi.org/10.1016/j.bios.2019.111963.

Lu, C. C., Liao, K. H., Tsai, J. T. H., Liu, C. Y., & Leu, J. P. (2009). Mesoporous carbon powder gas sensors for detection of carbon monoxide. *Biomedical Engineering - Applications, Basis and Communications, 21*(6), 399–404. https://doi.org/10.4015/S1016237209001684.

Magner, E. (1998). Trends in electrochemical biosensors. *Analyst, 123*(10), 1967–1970. Royal Society of Chemistry https://doi.org/10.1039/a803314e.

Maiti, S., Das, K., & Das, P. K. (2013). Label-free fluorimetric detection of histone using quaternized carbon dot–DNA nanobiohybrid. *Chemical Communications, 49*(78), 8851. https://doi.org/10.1039/c3cc44492a.

Mao, S., Lu, G., Yu, K., Bo, Z., & Chen, J. (2010). Specific protein detection using thermally reduced graphene oxide sheet decorated with gold nanoparticle-antibody conjugates. *Advanced Materials*, 22(32), 3521–3526. https://doi.org/10.1002/adma.201000520.

Masaki, H., Yuki, H., Yasuhide, O., Kenzo, M., & Kazuhiko, M. (2014). Characterization of reduced graphene oxide field-effect transistor and its application to biosensor. *Japanese Journal of Applied Physics*, 53, 05FD05.

Mehdi Aghaei, S., Monshi, M. M., Torres, I., Zeidi, S. M. J., & Calizo, I. (2018). DFT study of adsorption behavior of NO, CO, NO2, and NH3 molecules on graphene-like BC 3: A search for highly sensitive molecular sensor. *Applied Surface Science*, 427, 326–333. https://doi.org/10.1016/j.apsusc.2017.08.048.

Mochalin, V. N., Shenderova, O., Ho, D., & Gogotsi, Y. (2012). The properties and applications of nanodiamonds. *Nature Nanotechnology*, 7(1), 11–23. https://doi.org/10.1038/nnano.2011.209.

Monthioux, M., & Kuznetsov, V. L. (2006). Who should be given the credit for the discovery of carbon nanotubes? *Carbon*, 44(9), 1621–1623. https://doi.org/10.1016/j.carbon.2006.03.019.

Moradi, M., Nouraliei, M., & Moradi, R. (2017). Theoretical study on the phenylpropanolamine drug interaction with the pristine, Si and Al doped [60] fullerenes. *Physica E: Low-dimensional Systems and Nanostructures*, 87, 186–191. https://doi.org/10.1016/j.physe.2016.11.027.

Mousavi, A., Zare-Dorabei, R., & Mosavi, S. H. (2020). A novel hybrid fluorescence probe sensor based on metal-organic framework@carbon quantum dots for the highly selective detection of 6-mercaptopurine. *Analytical Methods*, 12(44), 5397–5406. https://doi.org/10.1039/d0ay01592j.

Moyo, M., Okonkwo, J. O., & Agyei, N. M. (2014). An amperometric biosensor based on horseradish peroxidase immobilized onto maize tassel-multiwalled carbon nanotubes modified glassy carbon electrode for determination of heavy metal ions in aqueous solution. *Enzyme and Microbial Technology*, 56, 28–34. https://doi.org/10.1016/j.enzmictec.2013.12.014.

Murru, F., Romero, F. J., Sánchez-Mudarra, R., García Ruiz, F. J., Morales, D. P., Capitán-Vallvey, L. F., et al. (2020). Portable instrument for hemoglobin determination using room-temperature phosphorescent carbon dots. *Nanomaterials*, 10(5). https://doi.org/10.3390/nano10050825.

Nazarkovsky, M., de Mello, H. L., Bisaggio, R. C., Alves, L. A., & Zaitsev, V. (2020). Hybrid suspension of nanodiamonds-nanosilica/titania in cytotoxicity tests on cancer cell lines. *Inorganic Chemistry Communications*, 111. https://doi.org/10.1016/j.inoche.2019.107673.

Ndamanisha, J. C., Bai, J., Qi, B., & Guo, L. (2009). Application of electrochemical properties of ordered mesoporous carbon to the determination of glutathione and cysteine. *Analytical Biochemistry*, 386(1), 79–84. https://doi.org/10.1016/j.ab.2008.11.041.

Ndamanisha, J. C., & Guo, L. (2008). Electrochemical determination of uric acid at ordered mesoporous carbon functionalized with ferrocenecarboxylic acid-modified electrode. *Biosensors and Bioelectronics*, 23(11), 1680–1685. https://doi.org/10.1016/j.bios.2008.01.026.

Ndamanisha, J. C., & Guo, L. (2009). Nonenzymatic glucose detection at ordered mesoporous carbon modified electrode. *Bioelectrochemistry*, 77(1), 60–63. https://doi.org/10.1016/j.bioelechem.2009.05.003.

Novoselov, K. S. (2004). Electric field effect in atomically thin carbon films. *Science*, 666–669. https://doi.org/10.1126/science.1102896.

Novoselov, K. S., Geim, A. K., Morozov, S. V., Jiang, D., Katsnelson, M. I., Grigorieva, I. V., et al. (2005). Two-dimensional gas of massless Dirac fermions in graphene. *Nature*, 438(7065), 197–200. https://doi.org/10.1038/nature04233.

Nunn, N., Torelli, M., McGuire, G., & Shenderova, O. (2017). Nanodiamond: A high impact nanomaterial. *Current Opinion in Solid State and Materials Science*, 21(1), 1–9. https://doi.org/10.1016/j.cossms.2016.06.008.

Oliveira, G. G., Janegitz, B. C., Zucolotto, V., & Fatibello-Filho, O. (2013). Differential pulse adsorptive stripping voltammetric determination of methotrexate using a functionalized carbon nanotubes-modified glassy carbon electrode. *Central European Journal of Chemistry*, 11(11), 1837–1843. https://doi.org/10.2478/s11532-013-0305-5.

Orzari, L. O., Santos, F. A., & Janegitz, B. C. (2018). Manioc starch thin film as support of reduced graphene oxide: A novel architecture for electrochemical sensors. *Journal of Electroanalytical Chemistry*, 823, 350–358. https://doi.org/10.1016/j.jelechem.2018.06.036.

Osswald, S., Yushin, G., Mochalin, V., Kucheyev, S. O., & Gogotsi, Y. (2006). Control of sp2/sp3 carbon ratio and surface chemistry of nanodiamond powders by selective oxidation in air. *Journal of the American Chemical Society*, 128(35), 11635–11642. https://doi.org/10.1021/ja063303n.

Peng, B., Locascio, M., Zapol, P., Li, S., Mielke, S. L., Schatz, G. C., et al. (2008). Measurements of near-ultimate strength for multiwalled carbon nanotubes and irradiation-induced crosslinking improvements. *Nature Nanotechnology*, 3(10), 626–631. https://doi.org/10.1038/nnano.2008.211.

Peng, L. M., Zhang, Z., & Wang, S. (2014). Carbon nanotube electronics: Recent advances. *Materials Today*, 17(9), 433–442. https://doi.org/10.1016/j.mattod.2014.07.008.

Pop, E., Mann, D., Wang, Q., Goodson, K., & Dai, H. (2006). Thermal conductance of an individual single-wall carbon nanotube above room temperature. *Nano Letters*, 6(1), 96–100. https://doi.org/10.1021/nl052145f.

Prasad, B. B., Singh, R., & Kumar, A. (2017). Synthesis of fullerene (C60-monoadduct)-based water-compatible imprinted micelles for electrochemical determination of chlorambucil. *Biosensors and Bioelectronics*, 94, 115–123. https://doi.org/10.1016/j.bios.2017.02.040.

Puthongkham, P., & Venton, B. J. (2019). Nanodiamond coating improves the sensitivity and antifouling properties of carbon fiber microelectrodes. *ACS Sensors*, 4(9), 2403–2411. https://doi.org/10.1021/acssensors.9b00994.

Rahimi-Nasrabadi, M., Khoshroo, A., & Mazloum-Ardakani, M. (2017). Electrochemical determination of diazepam in real samples based on fullerene-functionalized carbon nanotubes/ionic liquid nanocomposite. *Sensors and Actuators, B: Chemical*, 240, 125–131. https://doi.org/10.1016/j.snb.2016.08.144.

Ramos, M. M. V., Carvalho, J. H. S., de Oliveira, P. R., & Janegitz, B. C. (2020). Determination of serotonin by using a thin film containing graphite, nanodiamonds and gold nanoparticles anchored in casein. *Measurement: Journal of the International Measurement Confederation*, 149. https://doi.org/10.1016/j.measurement.2019.106979.

Rao, C. N. R., Seshadri, R., Govindaraj, A., & Sen, R. (1995). Fullerenes, nanotubes, onions and related carbon structures. *Materials Science & Engineering R: Reports*, 15(6), 209–262. https://doi.org/10.1016/s0927-796x(95)00181-6.

Rather, J. A., Al Harthi, A. J., Khudaish, E. A., Qurashi, A., Munam, A., & Kannan, P. (2016). An electrochemical sensor based on fullerene nanorods for the detection of paraben, an endocrine disruptor. *Analytical Methods*, 8(28), 5690–5700. https://doi.org/10.1039/c6ay01489e.

Saha, D., & Das, S. (2018). Development of fullerene modified metal oxide thick films for moisture sensing application. *Materials Today: Proceedings, 5*(3), 9817–9825. Elsevier Ltd https://doi.org/10.1016/j.matpr.2017.10.172.

Schrand, A. M., Huang, H., Carlson, C., Schlager, J. J., Osawa, E., Hussain, S. M., et al. (2007). Are diamond nanoparticles cytotoxic? *Journal of Physical Chemistry B, 111*(1), 2–7. https://doi.org/10.1021/jp066387v.

Shen, P., & Xia, Y. (2014). Synthesis-modification integration: One-step fabrication of boronic acid functionalized carbon dots for fluorescent blood sugar sensing. *Analytical Chemistry, 86*(11), 5323–5329. https://doi.org/10.1021/ac5001338.

Silva, W., Queiroz, A. C., & Brett, C. M. (2020). Nanostructured poly(phenazine)/Fe_2O_3 nanoparticle film modified electrodesformed by electropolymerization in ethaline-deep eutectic solvent. Microscopic and electrochemical characterization. *Electrochimica Acta*, 136284.

Sim, H., Kim, J. H., Lee, S. K., Song, M. J., Yoon, D. H., Lim, D. S., et al. (2012). High-sensitivity non-enzymatic glucose biosensor based on $Cu(OH)_2$ nanoflower electrode covered with boron-doped nanocrystalline diamond layer. *Thin Solid Films, 520*(24), 7219–7223. https://doi.org/10.1016/j.tsf.2012.08.011.

Simioni, N. B., Oliveira, G. G., Vicentini, F. C., Lanza, M. R. V., Janegitz, B. C., & Fatibello-Filho, O. (2017). Nanodiamonds stabilized in dihexadecyl phosphate film for electrochemical study and quantification of codeine in biological and pharmaceutical samples. *Diamond and Related Materials, 74*, 191–196. https://doi.org/10.1016/j.diamond.2017.03.007.

Simioni, N. B., Silva, T. A., Oliveira, G. G., & Fatibello-Filho, O. (2017). A nanodiamond-based electrochemical sensor for the determination of pyrazinamide antibiotic. *Sensors and Actuators, B: Chemical, 250*, 315–323. https://doi.org/10.1016/j.snb.2017.04.175.

Song, S., Gao, Q., Xia, K., & Gao, L. (2008). Selective determination of dopamine in the presence of ascorbic acid at porous-carbon-modified glassy carbon electrodes. *Electroanalysis, 20*(11), 1159–1166. https://doi.org/10.1002/elan.200804210.

Sutradhar, S., Jacob, G. V., & Patnaik, A. (2017). Structure and dynamics of a dl-homocysteine functionalized fullerene-C60-gold nanocomposite: A femtomolar l-histidine sensor. *Journal of Materials Chemistry B, 5*(29), 5835–5844. https://doi.org/10.1039/c7tb01089c.

Svancara, I. K., Walcarius, K., & Vytras, A. K. (2012). *Electroanalysis with carbon paste electrodes*. CRC Press.

Terada, D., Genjo, T., Segawa, T. F., Igarashi, R., & Shirakawa. (2020). Nanodiamonds for bioapplications-specific targeting strategies. *Biochimica Et Biophysica Acta-GeneralSubjects, 1864*.

Thirumalraj, B., Kubendhiran, S., Chen, S. M., & Lin, K. Y. (2017). Highly sensitive electrochemical detection of palmatine using a biocompatible multiwalled carbon nanotube/poly-L-lysine composite. *Journal of Colloid and Interface Science, 498*, 144–152. https://doi.org/10.1016/j.jcis.2017.03.045.

Tripathi, K. M., Sonker, A. K., Sonkar, S. K., & Sarkar, S. (2014). Pollutant soot of diesel engine exhaust transformed to carbon dots for multicoloured imaging of E. coli and sensing cholesterol. *RSC. Advances, 4*(57), 30100–30107. https://doi.org/10.1039/c4ra03720k.

Troshin, P. A., & Lyubovskaya, R. N. (2008). Organic chemistry of fullerenes: The major reactions, types of fullerene derivatives and prospects for their practical use. *Russian Chemical Reviews, 77*(4), 305–349. https://doi.org/10.1070/RC2008v077n04ABEH003770.

Tsai, L. W., Lin, Y. C., Perevedentseva, E., Lugovtsov, A., Priezzhev, A., & Cheng, C. L. (2016). Nanodiamonds for medical applications: Interaction with blood in vitro and in vivo. *International Journal of Molecular Sciences, 17*(7). https://doi.org/10.3390/ijms17071111.

Utari, L., Septiani, N. L. W., Suyatman, N., Nur, L. O., Wasisto, H. S., & Yuliarto, B. (2020). Wearable carbon monoxide sensors based on hybrid graphene/ZnO nanocomposites. *IEEE Access, 8*, 49169–49179. https://doi.org/10.1109/ACCESS.2020.2976841.

Vardharajula, S., Ali, S. Z., Tiwari, P. M., Eroğlu, E., Vig, K., Dennis, V. A., et al. (2012). Functionalized carbon nanotubes: Biomedical applications. *International Journal of Nanomedicine, 7*, 5361–5374. https://doi.org/10.2147/IJN.S35832.

Veetil, J. V., & Ye, K. (2007). Development of immunosensors using carbon nanotubes. *Biotechnology Progress, 23*(3), 517–531. https://doi.org/10.1021/bp0602395.

Vishwakarma, V. (2020). Impact of environmental biofilms: Industrial components and its remediation. *Journal of Basic Microbiology, 60*(3), 198–206. https://doi.org/10.1002/jobm.201900569.

Wang, J. (2005). Carbon-nanotube based electrochemical biosensors: A review. *Electroanalysis, 17*(1), 7–14. https://doi.org/10.1002/elan.200403113.

Wang, Z., & Dai, Z. (2015). Carbon nanomaterial-based electrochemical biosensors: An overview. *Nanoscale, 7*(15), 6420–6431. https://doi.org/10.1039/c5nr00585j.

Wang, Y., Wang, S., Ge, S., Wang, S., Yan, M., Zang, D., et al. (2013). Facile and sensitive paper-based chemiluminescence DNA biosensor using carbon dots dotted nanoporous gold signal amplification label. *Analytical Methods, 5*(5), 1328–1336. https://doi.org/10.1039/c2ay26485d.

Wang, L., Xiu, Y., Han, B., Liu, L., Niu, X., & Wang, H. (2020). Magnetic mesoporous carbon material based electrochemical sensor for rapid detection of penicillin sodium in milk. *Journal of Food Science, 85*(8), 2435–2442. https://doi.org/10.1111/1750-3841.15328.

Wei, B. Q., Vajtai, R., & Ajayan, P. M. (2001). Reliability and current carrying capacity of carbon nanotubes. *Applied Physics Letters, 79*(8), 1172–1174. https://doi.org/10.1063/1.1396632.

Wei, Z. N., Yang, X. X., Mo, Z. H., & Leng, F. (2018). Photocatalytic sensor of organics in water with signal of plasmonic swing. *Sensors and Actuators, B: Chemical, 255*, 3458–3463. https://doi.org/10.1016/j.snb.2017.09.176.

Wu, L., Lu, X., Fu, X., Wu, L., & Liu, H. (2017). Gold nanoparticles dotted reduction graphene oxide nanocomposite based electrochemical aptasensor for selective, rapid, sensitive and congener-specific PCB77 detection. *Scientific Reports, 7*(1). https://doi.org/10.1038/s41598-017-05352-7.

Wu, J., Tao, K., Miao, J., & Norford, L. K. (2018). Three-dimensional hierarchical and superhydrophobic graphene gas sensor with good immunity to humidity. In *Vol. 2018. Proceedings of the IEEE International Conference on Micro Electro Mechanical Systems (MEMS)* (pp. 901–904). Institute of Electrical and Electronics Engineers Inc. https://doi.org/10.1109/MEMSYS.2018.8346702.

Xu, H., Lu, Y. F., Xiang, J. X., Zhang, M. K., Zhao, Y. J., Xie, Z. Y., et al. (2018). A multifunctional wearable sensor based on a graphene/inverse opal cellulose film for simultaneous,: In situ monitoring of human motion and sweat. *Nanoscale, 10*(4), 2090–2098. https://doi.org/10.1039/c7nr07225b.

Xu, S., Sun, F., Pan, Z., Huang, C., Yang, S., Long, J., et al. (2016). Reduced graphene oxide-based ordered macroporous films on a curved surface: General fabrication and application in gas sensors. *ACS Applied Materials and Interfaces, 8*(5), 3428–3437. https://doi.org/10.1021/acsami.5b11607.

Yilmaz, E., Ulusoy, H.İ., Demir, Ö., & Soylak, M. (2018). A new magnetic nanodiamond/graphene oxide hybrid (Fe3O4@ND@GO) material for preconcentration and sensitive determination of sildenafil in alleged herbal aphrodisiacs by HPLC-DAD system. *Journal of Chromatography B, 1084*, 113–121. https://doi.org/10.1016/j.jchromb.2018.03.030.

Yin, Z., He, Q., Huang, X., Zhang, J., Wu, S., Chen, P., et al. (2012). Real-time DNA detection using Pt nanoparticle-decorated reduced graphene oxide field-effect transistors. *Nanoscale, 4*(1), 293–297. https://doi.org/10.1039/c1nr11149c.

Yoo, S. H., Perera, C., Shen, J., Ye, L., Suh, D. S., & Jane, J. L. (2009). Molecular structure of selected tuber and root starches and effect of amylopectin structure on their physical properties. *Journal of Agricultural and Food Chemistry, 57*(4), 1556–1564. https://doi.org/10.1021/jf802960f.

Yun, Y. H., Dong, Z., Shanov, V., Heineman, W. R., Halsall, H. B., Bhattacharya, A., et al. (2007). Nanotube electrodes and biosensors. *Nano Today, 2*(6), 30–37. https://doi.org/10.1016/S1748-0132(07)70171-8.

Zaghmarzi, F. A., Zahedi, M., Mola, A., Abedini, S., Arshadi, S., Ahmadzadeh, S., et al. (2017). Fullerene-C60 and crown ether doped on C60 sensors for high sensitive detection of alkali and alkaline earth cations. *Physica E: Low-dimensional Systems and Nanostructures, 87*, 51–58. https://doi.org/10.1016/j.physe.2016.11.031.

Zambianco, N. A., Silva, T. A., Zanin, H., Fatibello-Filho, O., & Janegitz, B. C. (2019). Novel electrochemical sensor based on nanodiamonds and manioc starch for detection of diquat in environmental samples. *Diamond and Related Materials, 98*. https://doi.org/10.1016/j.diamond.2019.107512.

Zang, J., Guo, C. X., Hu, F., Yu, L., & Li, C. M. (2011). Electrochemical detection of ultratrace nitroaromatic explosives using ordered mesoporous carbon. *Analytica Chimica Acta, 683*(2), 187–191. https://doi.org/10.1016/j.aca.2010.10.019.

Zhang, D., Liu, J., Chang, H., Liu, A., & Xia, B. (2015). Characterization of a hybrid composite of SnO2 nanocrystal-decorated reduced graphene oxide for ppm-level ethanol gas sensing application. *RSC Advances, 5*(24), 18666–18672. https://doi.org/10.1039/c4ra14611e.

Zhang, T., Liu, G. Q., Leong, W. H., Liu, C. F., Kwok, M. H., Ngai, T., et al. (2018). Hybrid nanodiamond quantum sensors enabled by volume phase transitions of hydrogels. Nature. *Communications, 9*(1). https://doi.org/10.1038/s41467-018-05673-9.

Zhang, Z., Shi, Y., Pan, Y., Cheng, X., Zhang, L., Chen, J., et al. (2014). Quinoline derivative-functionalized carbon dots as a fluorescent nanosensor for sensing and intracellular imaging of Zn2+. *Journal of Materials Chemistry B, 2*(31), 5020–5027. https://doi.org/10.1039/c4tb00677a.

Zheng, T., Perona Martínez, F., Storm, I. M., Rombouts, W., Sprakel, J., Schirhagl, R., et al. (2017). Recombinant protein polymers for colloidal stabilization and improvement of cellular uptake of diamond nanosensors. *Analytical Chemistry, 89*(23), 12812–12820. https://doi.org/10.1021/acs.analchem.7b03236.

Zheng, M., Xie, Z., Qu, D., Li, D., Du, P., Jing, X., et al. (2013). On-off-on fluorescent carbon dot nanosensor for recognition of chromium(VI) and ascorbic acid based on the inner filter effect. *ACS Applied Materials and Interfaces, 5*(24), 13242–13247. https://doi.org/10.1021/am4042355.

Zhong, X., Yuan, R., & Chai, Y. (2012). In situ spontaneous reduction synthesis of spherical Pd@Cys-C60 nanoparticles and its application in nonenzymatic glucose biosensors. *Chemical Communications, 48*(4), 597–599. https://doi.org/10.1039/c1cc16081h.

Zhou, L., Wang, T., Bai, Y., Li, Y., Qiu, J., Yu, W., et al. (2020). Dual-amplified strategy for ultrasensitive electrochemical biosensor based on click chemistry-mediated enzyme-assisted target recycling and functionalized fullerene nanoparticles in the detection of microRNA-141. *Biosensors and Bioelectronics, 150*. https://doi.org/10.1016/j.bios.2019.111964.

Zhou, X., Xue, Z., Chen, X., Huang, C., Bai, W., Lu, Z., et al. (2020). Nanomaterial-based gas sensors used for breath diagnosis. *Journal of Materials Chemistry B, 8*(16), 3231–3248. https://doi.org/10.1039/c9tb02518a.

Zhou, M., Zhou, Z., Gong, A., Zhang, Y., & Li, Q. (2015). Synthesis of highly photoluminescent carbon dots via citric acid and Tris for iron(III) ions sensors and bioimaging. *Talanta, 143*, 107–113. https://doi.org/10.1016/j.talanta.2015.04.015.

Zong, J., Yang, X., Trinchi, A., Hardin, S., Cole, I., Zhu, Y., et al. (2014). Carbon dots as fluorescent probes for \off-on\ detection of Cu2+ and l-cysteine in aqueous solution. *Biosensors and Bioelectronics, 51*, 330–335. https://doi.org/10.1016/j.bios.2013.07.042.

Index

Note: Page numbers followed by *f* indicate figures, *t* indicate tables, and *s* indicate schemes.

A

Acidic oxidation, 69
Activated carbon fibers (ACFs), 266
Adenosine-5-triphosphate (ATP)
　chemosensing of, 66, 435–436
Affinity biosensors, 79
Aflatoxin B1, 99
Ag nanoparticles
　CR on, 171–172, 173*f*
　malachite green (MG) on, 172, 174*f*
　rhodamine 6G (R6G) on, 171, 172*f*
　SAED pattern of, 169, 170*f*
　textile effluent (TE) on, 172–175, 174*f*
　UV-visible spectra, 168, 168*f*
Ag-reduced graphene oxide (Ag-rGO)
　CR on, 171–172, 173*f*
　malachite green (MG) on, 172, 174*f*
　rhodamine 6G (R6G) on, 171, 172*f*
　SAED pattern of, 169, 170*f*
　textile effluent (TE) on, 172–175, 174*f*
　UV-visible spectra, 168, 168*f*
Air pollution monitoring, 255
Air pollution remediation
　carbon nanomaterial-based sensors, 110–118
　　advantages, 111
　　application, 111–118
　　sensing mechanism, 110–111
　CNT-based sensors, 111–115
　　ammonia (NH_3) detection, 114
　　carbon dioxide (CO_2) detection, 113
　　carbon monoxide (CO) detection, 113
　　CH_4 detection, 115
　　hydrogen sulfide (H_2S) detection, 114
　　nitric oxide (NO) detection, 113–114
　　nitrogen dioxide (NO_2) detection, 113–114
　　sensing mechanism, 111–113
　　sulfur dioxide (SO_2) detection, 114
　　volatile organic compound (VOC) detection, 115
　graphene-based sensors
　　ammonia (NH_3) detection, 116
　　carbon dioxide (CO_2) detection, 116
　　carbon monoxide (CO) detection, 116
　　nitrogen dioxide (NO_2) detection, 115–116
　　other air pollutant gases detection, 117
　　sensing mechanism, 115
　nanomaterials in, 105–106, 107*t*
Allergic diseases, 337
Ammonia, 114, 116, 254–255
Amperometric sensors, 32, 32*f*
Antibiotics, 99

Arc discharge method, 36–37, 37*f*, 64
Ascorbic acid, 251
Asthma angle, 309–311

B

Bacteria, 98–99, 324–325
Batch injection analysis (BIA), 131
BET. *See* Brunauer-Emmet-Teller surface analysis (BET)
Bilastine, 343
Biochar, 318
Biochar-based functional materials, 272–273
Biocompatibility, of CNMs, 297–299
　due to surface functionalization, 299
Biodegradation, of CNMs, 297–299
　by enzymes, 298
　by microbes, 298
Biodiesel production, 272
Biodiversity, impact on, 305
Bioelectrochemical (BEC) systems, 265–266
Biofuel cells (BFCs), 272
Bioimaging, 353
Biological applications
　nanodiamonds in, 402–403
Biomarkers, 59
　glucose, 59
　neurotransmitters, 60
　nucleic acids, 59
　proteins, 60
Biomedical applications
　fullerenes in, 404–405
　graphenes in, 407–408
Biosensors, 8, 194–210, 381–382
　amperometric paper-based microfluidic, 282–283
　carbon nanotubes in, 397–398
　concerns over, 366–367
　electrochemical, 383–384
　FL quenchers, 352
　fullerenes in, 405–406
　voltammetric paper-based microfluidic, 276
Bisphenol A, 226
Boron doped diamond (BDD) electrode, 132, 133*f*, 135
Bottom-down technique, 16*f*, 18
Bottom-up approach, 319
Bovine serum albumin (BSA), 306–307
Brunauer-Emmet-Teller surface analysis (BET), 24–26, 25*f*
Buckminsterfullerene, 64–65, 106–108
Buckypaper-based sensors, 432–433

C

Calorimetry-based sensors, 383
Capacitance gas sensors, 397
Carbides, reduction of, 68
Carbon-arc discharge, 152, 267
Carbon-based nanoparticles (CNPs), 315, 317–319
　applications, 320
　characterization, 319
　ecotoxicological characteristics, 324
　engineered, 318–319
　in environment, 322–324
　　biological transformation, 324
　　chemical transformation, 324
　　physical transformation, 323
　influences
　　on human health, 325
　　on terrestrial ecology, 324–325
　toxic, 326*t*
　natural sources, 317–318
　synthesis, 319
Carbon black (CB), 6, 319
　electrode, 220–226
Carbon cycle, 418
Carbon dioxide (CO_2) detection, 113, 116
Carbon dots (CDs), 5
　electrochemical sensors, 220
　in food industry, 98–99
　optical-based sensor, 229
Carbon electrodes
　challenges, 265–266
　development, 265–266
　modification, 265–266
Carbon hybrid structures
　sensors based on, 437–438
Carbon monoxide (CO) detection, 113, 116
Carbon nanodiamonds (CNDs), 5–6
Carbon nanodots, 357–359
Carbon nanofibers (CNFs), 309, 319
　schematic structures, 134*f*
　sensors, 420
Carbon nanofibers-epoxy (CNF-EP) composite electrode, 128, 128*f*
Carbon nanohorns (CNHs), 359–360
Carbon nanomaterial-based sensors, 15, 320–322, 347–348
　biomedical applications
　　biomarkers, 59–60
　　carbon nanotubes (CNT), 62–64, 65*t*
　　carbon quantum dots (CQDs)0, 69–70, 71*t*, 71*f*

445

Carbon nanomaterial-based sensors (Continued)
    fullerene, 64–66
    graphene, 60–62, 63t
    nanodiamonds, 67–68
    overview, 59–60
  challenges, 363–368
    design and fabrication, 364–365
    reproducibility, 365
    selectivity, 365
    sensitivity, 365
    stability, 365
  characterization of, 20–26
  chemical sensors, 320–322
  concerns, 363–368
  on detection ability of toxic substances, molecular modification of, 16f
  future vision, 438
  market, 360–363
    competitive market share, 362
    COVID-19 impact, 362–363
    overview, 360–361
    regional insights, 362
    segmentation, 361–362
  physical sensors, 320
  point-of-care devices (POCT), lack of, 367–368
  safety measures, 326–327
  for sustainable development, 6–9
  toxicity, 363–364
Carbon nanomaterials (CNMs), 191–192
  as carcinogens, 308–309
  chemistry, 192–193
  derivatives, 108
  enzymatic degradation of, 309–310
  ex vitro enzymatic degradation, 310t
  in food industry, 96–97, 100
  history of, 106–108
  on human body, 308
  interaction
    with cells, 306
    with protein, 307
    with tissue, 308
  in nutshell, 311
  overview, 395–396
  properties, 108–110
    carbon nanotubes (CNT), 106f, 108
    fullerenes, 109–110
    graphene, 108–109
  in sensor applications, 150–154
    carbon nanotube (CNTs), 152
    fullerenes, 151–152
    graphene, 151
    graphene oxide (GO), 151
    graphene quantum dots (GQDs), 152–153
    nanodiamonds (NDs), 153–154
    reduced graphene oxide (rGO), 151
  sustainable discovery, 425
  types, 3–6, 6f, 418–419
  uses, 193
Carbon nanomaterial synthesis method, for sensors, 15–20
  bottom-down technique, 16f, 18
  chemical vapor deposition, 19, 19f
  electric arc deposition, 18

laser pyrolysis, 20
lithography, 17–18, 17f
mechanical milling, 16–17, 16–17f
micellization, 19–20, 19f
quantum dots, 18
sol-gel process, 18, 19f
sputtering deposition, 18
top-down technique, 16, 16f
Carbon nano-onions (CNOs)
  electrochemical sensing, 360
  electrochemical sensors, 220–226
Carbon nanotube (CNT), 4, 29, 35–36, 62–64, 65t, 86, 106f, 108, 152, 259–260, 269, 291, 303–304, 319, 396, 429, 431–432, 432s
  in air pollution remediation, 111–115
    ammonia (NH_3) detection, 114
    carbon dioxide (CO_2) detection, 113
    carbon monoxide (CO) detection, 113
    CH_4 detection, 115
    hydrogen sulfide (H_2S) detection, 114
    nitric oxide (NO) detection, 113–114
    nitrogen dioxide (NO_2) detection, 113–114
    sensing mechanism, 111–113
    sulfur dioxide (SO_2) detection, 114
    volatile organic compound (VOC) detection, 115
  biosensor, 78
  chemical sensors, 384–385
  electrochemical biosensors, 398
  electrochemical sensing, 356
    advancement, 262–263
  fluorescence sensors, 158–159
  in food industry, 97–98
  multi-walled, 4
  nanoparticles, 316
  nanosensors based on, 423
  optical-based sensor, 232
  overview, 335–337
  physicochemical properties, 305t
  schematic structures, 134f
  for sensor applications, 397–400
    biosensors, 397–398
    drug delivery, 399
    food and environmental, 399–400
    gas sensors, 397
  sensors, 419
  single-walled, 4
  synthesis, 36–40, 260, 260f
    arc discharge method, 36–37, 37f
    chemical vapor deposition method, 38, 39f
    green synthesis, 39–40, 39f
    laser ablation method, 37–38, 37f
Carbon quantum dots (CQDs), 69–70, 71f, 71t, 385, 437
  electrochemical sensors, 220, 224–225t
  in food industry, 98–99
  in sensor applications, 408
Carbon transformation, 417–418
Carboxyl group-functionalized graphene, 407, 407f
Carboxyl (-COOH) groups, 294–295
Carcinoembryonic antigen (CEA), 280

Carcinogens, 308–309
Catalysts, 353
CB. See Carbon black (CB)
CDs. See Carbon dots (CDs)
Cellulose composite paper, 86
Cellulose-derived nitrocellulose membrane, 78
Ceramic-based nanoparticles, 316
Cetyltrimethylammonium bromide (CTAB), 341
CH_4 detection, 115
Chemical bath deposition (CBD) method, 255
Chemical enhancement mechanism (CM), 172–175
Chemical exfoliation, 61
Chemical sensors, 320–322
Chemical synthesis, 61
Chemical vapor deposition (CVD), 19, 64, 68, 152, 211, 260, 361
  cross-section of, 19f
  method, 38, 39f
  synthesis, 62
  technique, 355
Chemiluminescence (CL), 278
Chemiresistive sensors, 110–111, 400
Chemosensors, 303
Chitosan, 64, 343
Chitosan-functionalized graphene oxide (GO-CS), 408
Chloride ion, 251
Cholesterol biosensor, 398, 399f
Chronoamperometry (CA), 130–132
CNDs. See Carbon nanodiamonds (CNDs)
CNT. See Carbon nanotube (CNT)
Colorimetric detection, 277–278
Colorimetric sensors, 81, 81f, 157
Colorimetry, 367–368
Commercial carbon fibers (CCF), 252
Commercialization challenges, 385–387
Conductometric sensors, 31
Copper, 254
Counter/auxiliary electrode, 155
COVID-19, 362–363
Cutter printer, 86
CVD. See Chemical vapor deposition (CVD)
Cyclic voltammetry (CV), 127–129, 129f, 137, 138f

D
Dendrimer-based nanoparticles, 316–317
Dendrofullerene (DF-1), 404–405
Detonated nanodiamond (DND), 401
Detonation synthesis, 154
D-fructose tracking, 100
Diabetes mellitus, 249
Di-azo dyes, 178
Diclofenac (DCF) detection, 139–144, 143t
Differential pulse voltammetry (DPV), 127–130, 211–220
Digital printing, 83
Dihexadecyl phosphate (DHP), 343
Dipsticks, 276
Direct methanol fuel cells (DMFCs), 403–404
Disease-causing agent detection, 438
Disposable sensor strip fabrication

chemical modification, 85
drawing, 85
  carbon nanotubes/cellulose composite paper, 86
  eyeliner pencil, 85
  pencil, 85–86
  permanent marker pen, 85
fabrication, 80–81
paper cutting methods, 86–87
  cutter printer, 86
  laminated paper-based analytical devices (LPADs), 86–87
  laser cutter, 86
photolithography, 81–82
pollutants, 80, 81t
printing, 82–83
  digital, 83
  flexographic printing, 83, 83f
  inkjet, 84
  screen printing method, 82–83
  wax, 84, 85f
stamping, 84–85
substrate material for, 77–78
DNA damage, 297
Dopamine, 8–9
Doxorubicin (DOX), 138–139
Drawing, 85
  carbon nanotubes/cellulose composite paper, 86
  eyeliner pencil, 85
  pencil, 85–86
  permanent marker pen, 85
Drop-casting method, 355
Drug delivery
  carbon nanotubes in, 399
Dye-related pollution, 77
Dyes
  fluorescence suppression in, 171
  SERS sensing, 171
    CR, 171–172
    malachite green (MG), 172
    rhodamine 6G (R6G), 171
    textile effluent (TE), 172–175
  spectral analysis
    di-azo dyes, 178
    mono azo dyes, 176–177
    textile effluent, 183–184
    thiazine dye, 182
    triarylmethane dye, 178–182
    xanthene dye, 183
Dye-sensitized solar cells (DSSCs)
  using carbon nanomaterials, 271

E

E. coli, 97–98
Edge effect, 129
Electrical properties, from I_V characterization, 26, 26f
Electric arc deposition, 18
Electric arc discharge technique, 105
Electrochemical detection, 127–131
  amperometric technique, 282–283
    clinical analysis, 282–283
    environmental analysis, 283
  in paper-based microfluidic devices, 278–283
  voltammetric technique, 280
    clinical analysis, 280
    environmental analysis, 281–282
Electrochemically reduced graphene oxide (ERGO), 355
Electrochemical sensors, 8, 30, 33–36, 33f, 34t, 80–81, 81f, 193–226, 321, 383–384
  advantages, 264–266
  carbon black electrode, 220–226
  carbon dots, 220
  carbon nanomaterials as, 271
  carbon nano-onions (CNOs), 220–226
  carbon nanotubes, 35–36
  carbon quantum dots, 220, 224–225t
  classification of, 30–32, 31f
  environmental applications, 155–156
  fullerene, 36, 194, 195–197t
  future perspectives, 264–266
  general schematic representation, 354f
  graphene, 35, 211–220, 212–218t
  graphene quantum dots, 211–220, 221–223t
  limitations, 264–266
  multiwalled carbon nanotube electrodes, 194–210, 199–210t
  nanodiamond, 220–226
  optical sensors, 353–360
    carbon nanodots, 357–359
    carbon nanohorns, 359–360
    carbon nano-onions, 360
    carbon nanotubes, 356
    fullerenes, 357
    graphene, 354–356
    nanodiamonds, 360
  overview, 191
  singlewalled carbon nanotube electrodes, 194–210, 199–210t
  working principle of, 30–32, 31f
Electrodeposition, 142–145
Electrogenerated chemiluminescence (ECL), 278
Electrolytes sensing, 251–252
Electromagnetic mechanism (EM) and, 172–175, 183–184
Emedastine difumarate, 343–344
Energy storage devices, 401–402
Engineered carbon nanomaterials
  safety management of, 310
Environmental hazardous sensors, 6–7, 7f
Environmental monitoring
  air pollution monitoring, 255
  heavy metal, 254
  nitrogen monitoring, 254–255
Environmental pollution, 149, 281
Environmental sensor applications, 154–159
  colorimetric sensors, 157
  electrochemical sensor, 155–156
  fluorescence sensors, 157–159
  optical sensors, 157
Enzymatic biosensor, 79
Enzymes, biodegradation by, 298

Eosinophil peroxidase (EPO), 298
Epitaxial method, 61
Ethylene, 400
Extracellular vesicles (EV), 366
Eyeliner pencil, 85

F

Fabrication
  carbon nanomaterial-based sensors, 364–365
  disposable sensor strip (see Disposable sensor strip fabrication)
  nanosensor, 423
  paper-based microfluidics device, 276–277
  of strain sensor based on buckypaper, 433f
Fast lithographic activation of sheets (FLASH), 82
Field effect transistor (FET), 397, 435
Field emission scanning electron microscopy (FESEM), 22–24, 23–24f
Flexographic printing, 83, 83f
Fluorescence (FL)
  quenchers, 352
  sensing, 348–350
  sensors, 157–159
  thermometry, 352
Foodborne illness, 95–96
Food industry
  carbon-based nanomaterials in, 96–97
  carbon dots in, 98–99
  carbon nanomaterials in, 96, 100
  carbon nanosensors in, 99–100
  carbon nanotubes in, 97–98, 399–400
  pesticide residue, 99
Fossil coal, 317
Fuel cells, fullerenes in, 403–404
Fuel combustion, 317
Fullerene–carbon nanofiber–paraffin oil paste (FULL-CNF-P) electrode, 139–141, 140t, 140f
Fullerenes, 3–4, 36, 64–66, 109–110, 151–152, 291, 319, 396, 429, 435–436
  electrochemical sensing, 357
  electrochemical sensors, 194, 195–197t
  gas sensors, 117–118
  nanoparticles, 316
  nanorods, 155
  schematic structures, 134f
  for sensor applications
    biomedical applications, 404–405
    biosensors, 405–406
    fuel cells, 403–404
Functionalization, of CNT
  affinity biosensors, 79
  enzymatic biosensor, 79
  immunosensors, 79–80
Functionalized fullerene, 405–407

G

Gas sensors, 8, 321
  carbon nanotubes in, 397
  graphenes in, 406–407
  nanodiamonds in, 401
  working principle, 110, 111f

GCE. *See* Glassy carbon electrode (GCE)
GCE/GO-DHP sensor
   electrochemical characterization of, 47–48, 48*f*
   preparation of, 45
Genosensors, 198
Giant unilamellar vesicles (GUVs), 306–307
Glassy carbon electrode (GCE), 156, 336, 336*f*
Glucose, 59, 382–383
   monitoring, 249
   sensors, 405
Glucosidase (GOD), 382–383
Glypican-1 (GPC1), 366
GO. *See* Graphene oxide (GO)
GQDs. *See* Graphene quantum dots (GQDs)
Graphene, 4–5, 29, 35, 60–62, 63*t*, 108–109, 151, 291, 303–304, 307, 385–387, 429
   in air pollution remediation
      ammonia (NH_3) detection, 116
      carbon dioxide (CO_2) detection, 116
      carbon monoxide (CO) detection, 116
      nitrogen dioxide (NO_2) detection, 115–116
      other air pollutant gases detection, 117
      sensing mechanism, 115
   electrochemical sensing, 354–356
      advancement, 264
   electrochemical sensors, 211–220, 212–218*t*
   gas sensors, 433
   nanoparticles, 318
   nanoribbons, 61
   nanosensors based on, 423
   nanosheets, 385–387
   optical-based sensor, 226
   physicochemical properties, 305*t*
   schematic structures, 134*f*
   for sensor applications
      biomedical applications, 407–408
      gas sensors, 406–407
      synthesis, 260–262
      temperature sensor, 253
Graphene-based nanomaterials, synthesis of, 40–41
Graphene-based surface enhanced Raman scattering (G-SERS)
   graphene-boosted silver nanocomposites for morphological studies, 169–170, 175–176
   optical studies, 168
   toxic organic dye detection, 168–175
   UV-visible spectral analysis, 175
   vibrational spectral studies, 175
   vibrational spectroscopic studies, 169
Graphene nanowalls (GNWs), 254
Graphene oxide (GO), 5, 43–44, 61, 151, 167–168, 434
   in biomedical applications, 407–408
   redudced, 5
   UV-visible spectra, 168, 168*f*
Graphene oxide-modified glassy carbon electrode, for mephedrone illicit drug determination, 43–44
   materials and methods, 44–46
      apparatus, 44–45
      electrochemical measurements, 46

      GCE/GO-DHP sensor, preparation of, 45
      reagents and solutions, 46–52
      street sample simulation, 46
      synthesis, 45, 45*f*
   results and discussion, 46–52
      4-MMC, electrochemical reduction of, 49, 49*f*
      adulterants in 4-MMC determination, interference of, 52, 53*t*, 53*f*
      DPV technique and analytical features, optimization of, 50–52, 51–52*t*, 51*f*
      experimental parameters, optimization of, 50, 50–51*f*
      GCE/GO-DHP sensor, electrochemical characterization of, 47–48, 48*f*
      GO characterization, 46–47, 46–47*f*
Graphene-polymer nanocomposites, 355
Graphene quantum dots (GQDs), 61–62, 152–153, 422
   doped with nitrogen, 156
   electrochemical sensors, 211–220, 221–223*t*
   fluorescence sensors, 158, 159*t*
   schematic structures, 134*f*
Graphitic carbon
   electrochemical sensing performance advancement, 264
   forms, 425
   nitride synthesis, 262
Gravure printing, 364
Green synthesis, of CNTs, 39–40, 39*f*
G-SERS. *See* Graphene-based surface enhanced Raman scattering (G-SERS)

H

H_1-antihistamine drugs
   classes, 337–338
   definition, 337–338
   determination, 338
      amperometry, 342–344
      carbon nanotubes based electrodes, 338–344
      voltammetry, 339–342
Health monitoring, 248–254
   electrolytes sensing, 251–252
   metabolite sensing, 248–251
      ascorbic acid, 251
      glucose, 249
      lactate monitoring, 250–251
      urea, 250
      uric acid (UA), 249–250
   pressure wearable sensor, 253
   pulse beat monitoring, 254
   respiration monitoring, 252
   strain wearable sensor, 253
   temperature sensor, 253–254
Heavy metal monitoring, 254
Hexamethyldisilane (HMDS), 84
High-energy beam radiation, 68
High-pressure, high-temperature (HPHT) microdiamonds, 154
Histamine, 337
Human health, 325
Humidity sensors, 320
Hummers' method, 5, 151, 211

Hybrid materials, 304
Hydrogel-based sensor, 253
Hydrogen peroxide (H_2O_2), 358
Hydrogen sulfide (H_2S) detection, 114
Hydrothermal treatment, 70
Hydroxyl (-OH) groups, 294–295

I

Ibuprofen (IBP) detection, 139–144, 143*t*
Immunosensors, 79–80
Inflammatory response, 297
Infrared sensors, 321
Inkjet printing, 84, 277*t*, 364
Inorganic nanoparticles, 316–317
Interaction, 110
Ionization gas sensors, 397

L

Lactate monitoring, 250–251
Laminated paper-based analytical devices (LPADs), 86–87
Laser ablation, 37–38, 37*f*, 64, 69, 152, 260
Laser cutter, 86
Laser cutting method, 277*t*
Laser printing, 277*t*
Laser pyrolysis, 20
Lateral flow assays (LFAs), 276
Linear sweep cyclic voltammetry, 78
Liquid–liquid interface method, 155
Lithography, 17–18, 17*f*
Localized surface plasmon resonance (LSPR), 423
Loratadine, 344
Luminescence detection, 278
Luminescent quantum dots (LQDs), 422

M

Magnetic sensors, 384
Malachite green (MG), 172
Manganese peroxidase (MnP), 298
Mechanical exfoliation (ME), 61, 116, 211
Mechanical milling, 16–17, 16–17*f*
Mephedrone. *See* 4-Methyl-methcathinone (4-MMC)
Mephedrone illicit drug determination, graphene oxide-modified glassy carbon electrode for, 43–44
   materials and methods, 44–46
      apparatus, 44–45
      electrochemical measurements, 46
      GCE/GO-DHP sensor, preparation of, 45
      reagents and solutions, 46–52
      street sample simulation, 46
      synthesis, 45, 45*f*
   results and discussion, 46–52
      4-MMC, electrochemical reduction of, 49, 49*f*
      adulterants in 4-MMC determination, interference of, 52, 53*t*, 53*f*
      DPV technique and analytical features, optimization of, 50–52, 51–52*t*, 51*f*
      experimental parameters, optimization of0, 50, 50–51*f*

GCE/GO-DHP sensor, electrochemical characterization of, 47–48, 48f
GO characterization, 46–47, 46–47f
Mesoporous carbon-based sensors, 436
Metabolite sensing, health monitoring, 248–251
  ascorbic acid, 251
  glucose, 249
  lactate monitoring, 250–251
  urea, 250
  uric acid (UA), 249–250
Metal-based nanoparticles, 316, 316t
Metal impurities, 294
Metal oxides based nanoparticles, 316–317, 317t
Methdilazine, 341
4-Methyl-methcathinone (4-MMC), 44
  determination, interference of adulterants in, 52, 53t, 53f
  electrochemical reduction of, 49, 49f
Micellization, 19–20, 19f
Microbes, biodegradation by, 298
Microfluidic paper-based analytical devices (μPADs), 275–276, 279t
Microfluidic paper-based electrochemical device (μPED), 278
Microfluidics, paper-based, 275–276
  analytical devices, formats of, 276
    dipsticks, 276
    lateral flow assays (LFAs), 276
    microfluidic paper-based analytical devices (μPADs), 276
  detection techniques, 277–278
    colorimetric detection, 277–278
    electrochemical detection (see Electrochemical detection, in paper-based microfluidic devices)
    luminescence detection, 278
  device fabrication methods, 276–277
  overview, 275
Microwave methods, 153
Microwave synthesis, 69
Modified Hummers' method, 45, 45f, 175
Molecularly imprinted polymers (MIPs), 352
Mono azo dyes, 176–177
Multi-walled carbon nanotubes (MWCNTs), 4, 106–108, 106f, 111, 114, 259–260, 339–342, 397
  electrodes, 194–210, 199–210t
  glassy carbon electrodes (GCE) with, 343

N

Nanodiamond (ND), 67–68, 69f, 153–154, 395, 429–431
  electrochemical sensing, 360
  electrochemical sensors, 220–226
  pyrazinamide (PZA), 156
  for sensor applications
    biological applications, 402–403
    energy storage devices, 401–402
    gas sensors, 401
Nanoemulsions, 98
Nanofibers, 423
Nanoinks, 424
Nanoprobes, 349–350
Nanostructure classification
  carbon-based nanoparticles (CNPs), 317–319
    applications, 320
    characterization, 319
    engineered, 318–319
    in environment, 322–324
    natural sources, 317–318
    synthesis, 319
  inorganic nanoparticles, 316–317
  one-dimension nanoparticles, 315–316
  organic nanoparticles, 317
  three-dimension nanoparticles, 316
  two-dimension nanoparticles, 316
Nanostructured carbon-epoxy composite (NC-EP) electrodes, 142–145
Nanostructured carbon-modified commercial carbon (NC-MCC) electrodes, 134–137
Nanostructured carbon paste (NC-P) electrodes, 137–141
Nanowires, 423
Nephrotoxicity, 363–364
Neuron-specific enolase (NSE), 280
Neurotransmitters, 60
New psychoactive substances (NPS), 43
Nitric oxide (NO) detection, 113–114
Nitrite, 251
Nitrogen dioxide (NO_2) detection, 113–116
Nitrogen monitoring, 254–255
Nitrophenyl-modified electrode, 155
NPS. See New psychoactive substances (NPS)
Nucleic acids, 59

O

Olaquindox, 155
One-dimension nanoparticles, 315–316
One-pot hot-solution synthesis method, 262–263
Optical sensors, 384
  in bioimaging, 353
  as catalysts, 353
  emerging trends, 348–353
    in biosensors, 350–351
    in molecularly imprinted polymer-based sensors, 352
  environmental applications, 157
  as FL quenchers in biosensors, 352
  general schematic representation, 349f
  as nanoprobes and sensing elements, 349–350
  sensing systems, 226–232, 230–231t
  as temperature sensors, 352
Ordered macroporous carbon-modified electrode, 385
Ordered mesoporous carbon-modified electrode, 385
Ordered mesoporous carbons (OMC), 436
Organic nanoparticles, 317
Ostwald-Freundlich equation, 310
Oxidative stress, 296

P

Paper-based microfluidics, 275–276
  analytical devices, formats of, 276
    dipsticks, 276
    lateral flow assays (LFAs), 276
    microfluidic paper-based analytical devices (μPADs), 276
  detection techniques, 277–278
    colorimetric detection, 277–278
    electrochemical detection (see Electrochemical detection, in paper-based microfluidic devices)
    luminescence detection, 278
  device fabrication methods, 276–277
  overview, 275
Paper-based sensor, 78
Paper cutting methods, 86–87
  cutter printer, 86
  laminated paper-based analytical devices (LPADs), 86–87
  laser cutter, 86
Paste electrodes, 137t
Pencil drawing, 85–86
Permanent marker pen, 85
Perovskite quantum dots, 422
Pesticide
  contamination, 400
  pollution, 77
  residue, 99
Pheniramine maleate, 339–341
Photolithography, 18, 81–82, 277t
Photovoltaics, quantum dots in, 422
Physical sensors, 320
Piezoelectric-based sensors, 384
  biosensors, 77
  calorimetry-based sensors, 383
  magnetic sensors, 384
Plants, 324–325
Plasma treatment, 277t
Plasmonic nanomaterials, 167–168
Platinum nanoparticles (PtNPs), 343
Point-of-care devices, lack of, 367–368
Point-of-care testing (POCT), 78
Pollutants, 80, 81t
Polyaniline-rGO, 255
Polydimethylsiloxane (PDMS), 433
  plotting, 277t
  screen printing, 82
Polyethyleneimine (PEI)-functionalized GO (GO-PEI), 408
Polymer-based nanoparticles, 317
Polymer-modified fullerene, 405–406
Polymorphous materials, 259–260
Porous nanoscale carbon materials, 385
Porphyrin, 115
Potentiometric sensors, 30–31, 31f
Pressure sensors, 253, 320
Printing, 82–83
  digital, 83
  flexographic printing, 83
  inkjet, 84
  screen printing method, 82–83
  wax, 84, 85f
Programmed death-ligand 1 (PD-L1) aptamer, 280
Promethazine hydrochloride, 341–342
Protein-carbon nanomaterial interaction, 307
Proteins, 60
Pulse beat monitoring, 254
Pulsed amperometry (PA), 132
Pyrazinamide (PZA), 156

Q

Quantum dots (QDs), 18, 152–153
   graphene, 422
   in medicine, 421–422
   nanoparticles, 316
   perovskite, 422
   in photovoltaics, 422
   properties, 421
   TVs and displays, 422
Quinoxaline-2-carboxylic acid, 100

R

Radical methods, 153
Raman spectroscopy, 21, 22f
Reactive oxygen species (ROS), 296
Reduced graphene oxide (rGO), 5, 61, 151, 434–435
Reference electrode, 155
Respiration monitoring, 252
rGO. *See* Reduced graphene oxide (rGO)
Rhodamine 6G (R6G), 171, 183

S

Safety management, 310
Safety measures, 326–327
Screen-printed carbon electrode (SPCE), 336, 336f, 339
Screen printing, 82–83, 277t, 364
Semiconductor-based nanoparticles, 317
Sensing elements, 349–350
Sensors, 191, 381–384, 419
   applications, 422–425
      electrodes, 424
      environmental applications, 425
      nanoinks, 424
      optoelectronic application, 424
      photonic application, 424
      potential applications, 424
   carbon-based sensors, 419
   carbon nanofiber, 420
   carbon nanotube, 419
   history, 382–383
   nanosensor, 422–423
      carbon nanotubes, 423
      fabrication, 423
      graphene, 424
      nanofibers, 423
      nanowires, 423
   piezoelectric-based sensors, 384
      calorimetry-based sensors, 383
      magnetic sensors, 384
   structure, 382f
   types, 383–384, 419
      electrochemical-based sensors, 383–384
      optical-based sensors, 384
Shock wave compression, 68
Silane solution method, 64
Silver/copper nanoparticles electrodeposition, 142–145
Single-walled carbon nanotubes (SWCNTs), 4, 106–108, 106f, 114–115, 259–260
   electrodes, 194–210, 199–210t
   transistor-based sensors, 397

Sodium, 252
Solar cells, 271
Sol–gel process, 18, 19f
Soot, 318
SPCE. *See* Screen-printed carbon electrode (SPCE)
Spray-coating process, 364
Sputtering deposition, 18
Square wave anodic stripping voltammetry (SWASV) method, 155
Square-wave voltammetry (SWV), 127–128, 130, 141
Stamping, 84–85
Strain sensor, 253
Sudan I dye, 100
Sulfur dioxide (SO_2) detection, 114
Supercapacitors, 272
Supported lipid bilayers (SLBs), 306–307
Surface-enhanced Raman scattering (SERS), 167–168, 350
Sustainability
   carbon materials, 420
   carbon nanomaterial-based sensors, 270–273
      in biochar-based functional materials, 272–273
      in biodiesel production, 272
      for biofuel cells, 272
      dye-sensitized solar cells, 271
      as electrochemical sensor, 271
      for supercapacitors, 272
      wastewater treatment/purifying, application in, 271–272
   defined, 269
   development, 6–9
   overview, 269
SWCNTs. *See* Single-walled carbon nanotubes (SWCNTs)

T

Temperature coefficient of resistivity (TCR), 254
Temperature sensors, 253–254, 320, 352
Template-carbonization process, 385
TEM. *See* Transmission electron microscopy (TEM)
Terrestrial ecology, 324–325
Tetraethylorthosilicate (TEOS), 84
Textile effluent (TE), 171–175, 171f, 183–184
Thermogravimetric analysis (TGA), 24, 25f
Thiazine dye, 182
Three-dimension nanoparticles, 316
Tissue engineering, 403, 408
Top-down technique, 16, 16f, 319
Toxicity, 324
   administration, mode of, 296
   of carbon nanomaterials, 363–364
   mechanism, 296
      DNA damage, 297
      inflammatory response, 297
      reactive oxygen species (ROS), 296
   metal impurities, 294
   metal ions, 149
   origin of, 292–296
   physical characteristics

      agglomeration, 293–294
      shape, 293
      size, 293
      surface functionalization, 294–296
Transmission electron microscopy (TEM), 22, 23f
Triarylmethane dye, 178–182
2,4,6-trinitrotoluene (TNT), 232
Tripelennamine hydrochloride, 341
Tumor necrosis factor-α (TNF-α), 297
Two-dimension nanoparticles, 316

U

Upconversion nanoparticles (UCNPs), 62
Urea, 250
Uric acid (UA), 8–9
   monitoring, 249–250
Urinary tract infection (UTI), 251
UV-visible spectral analysis, 175

V

Veterinary drug residues, 99
Vitamin C, 251
Volatile organic compound (VOC), 105–106, 115

W

Wastewater treatment, 271–272
Water purification, 271–272
Water treatment
   voltammetric and amperometric sensors, 127–132
   nanostructured carbon-epoxy composite (NC-EP) electrodes, 142–145
   nanostructured carbon-modified commercial carbon (NC-MCC) electrodes, 134–137
   nanostructured carbon paste (NC-P) electrodes, 137–141
   overview, 126
   working electrode, configurations of, 132–145, 134f
Wax dipping, 277t
Wax printing, 84, 85f, 277t
Wearable sensors, 247, 256
   environmental monitoring
      air pollution monitoring, 255
      heavy metal, 254
      nitrogen monitoring, 254–255
   future perspectives, 256
   health monitoring (*see* Health monitoring)
   limited availability, 256
Wildfire charcoal, 317
Wireless pressure sensor, 253

X

Xanthene dye, 171, 183
X-ray diffraction spectroscopy (XRD), 20–21, 20f

Z

Zinc oxide (ZnO), 255

Printed in the United States
by Baker & Taylor Publisher Services